Physical Chemistry

Physical Chemistry

How Chemistry Works

Kurt W. Kolasinski
Department of Chemistry, West Chester University, USA

WILEY

This edition first published 2017

Registered office
John Wiley & Sons Ltd, The Atrium, Southern Gate, Chichester, West Sussex, PO19 8SQ, United Kingdom

For details of our global editorial offices, for customer services and for information about how to apply for permission to reuse the copyright material in this book please see our website at www.wiley.com.

Library of Congress Cataloging-in-Publication Data

Names: Kolasinski, Kurt W.
Title: Physical chemistry : how chemistry works / Kurt W. Kolasinski.
Description: Chichester, West Sussex : John Wiley & Sons, Inc., 2016. | Includes index.
Identifiers: LCCN 2016015389| ISBN 9781118751121 (pbk.) | ISBN 9781118751206 (epub) | ISBN 9781118751213 (adobe PDF)
Subjects: LCSH: Chemistry, Physical and theoretical–Textbooks. | Chemistry–Textbooks.
Classification: LCC QD453.3 .K65 2016 | DDC 541–dc23
LC record available at https://lccn.loc.gov/2016015389

A catalogue record for this book is available from the British Library.

ISBN: 9781118751121

Set in 9.5/13pt Meridien by Aptara Inc., New Delhi, India

1 2017

If you do not communicate your ideas, you are not doing science. You are only engaging in a hobby.

— Kurt W. Kolasinski

Contents

Preface

This book is dedicated to my wife Kirsti, who now knows that the most frightening words in the English language are, "Yes, you're right, I will write another book." And to my daughter Annika, who dutifully kept asking, "How many more pages?"

To all of the readers looking at the daunting task of progressing through this tome, just keep moving one step forward every day, for the more we know, the more we seek. The more we seek, the more we find. The more we find, the more we know what to ask. No matter what new technology is ever invented, the answer to how to perform problem solving in physical chemistry (and anything else) will always be get enough sleep, eat your vegetables, make rational decisions, and do your homework.

I still crave lucidity, though this passage will be anything but lucid, and you may ask the author to explain "aber wann wird man je verstehen"? Potential overdrive, free thought and open minds. As if beams of photons lit the night. Cast adrift. Seas of rage have not yet assailed my concern for the truth. In the dark when I write, I'm sitting here by candle flame. Visions in the night. Show me what is right. Neues Glück liegt auf den Wegen. Die Welt ist groß und sie ist mein. Alte Träume die brach gelegen. Ich lass die Sorgen Sorgen sein. Thus, the "thank- you"s begin with a tip of my hat to the music of Dead Can Dance, Nick Cave, Einstürzende Neubauten, VNV Nation, Judas Priest, In Extremo, and all the Schattenreich artists.

Concerning content I would especially like to thank to Jim Falcone, Bill Gadzuk, Wayne Itano, Eckart Hasselbrink, Erik Månsson, David Pratt, Patrik Schmuki, Greg Sitz, Pat Vaccaro, David Wineland, Dick Zare, and Ahmed Zewail for valuable discussions and comments on the text. My high-school chemistry teacher was John Spina, who proves that dedicated, inspiring teachers exist even in dire school districts. Would it were that society promoted more of the former and fewer of the latter. My high-school physics teacher was the wrestling coach. Thus, I became a physical chemist rather than a chemical physicist; though being one or the other means never having to choose. Jacob Bradley, Abbie Ganas, Paige Kogelman, Shawn Pfeil, and Mark Schuman helped with proofreading. I greatly appreciate the work of Olaf Högermeyer, animator extraordinaire, for creating several of the better-looking illustrations in this book. The sample shown in Fig. 15.2 was prepared and imaged by Abbie Ganas. Data for Fig. 16.1 were taken by William Barclay, Kyle Spivack and Devon Zimmerman. A special thanks to Laurie Butler and Matt Brynteson for providing the data for Fig. 25.16. And to Jin Cheng, Florian Libisch and Emily Carter for calculating the potential surfaces in Fig 17.12. Photos of various gems were taken with the kind assistance of Ivan Kaplan and Teresa Martin of Kaplan Fine Jewelery, West Chester, PA, USA. Dennis Tanjeloff of Astro Gallery of Gems made available the image of pyrite crystals. Many thanks are extended to the myriad students who have weathered my endless tales of burlap and monks, brandy and polar bears; but who nonetheless have asked question after question that helped me clarify what I meant to express and made those explanations more intelligible. Others who have graciously provided me with figures include Flemming Besenbacher, Fleming Crim, Cynthia Friend, Naomi Halas, Sharon Hammes-Schiffer, Marsha Lester, Anne L'Huillier, Erik Månsson, Mike McClain, Hope Michelsen, Jens Nørskov, and Geraldine Richmond.

About the companion website

This book is accompanied by a companion website:

www.wiley.com/go/kolasinski/physicalchemistry

The website hosts a variety of supplementary resources for students and instructors.

For Instructors:

- PowerPoint files of the illustrations presented within this text
- Solutions manual

For Students:

- Worked examples

CHAPTER 1
Introduction

Much of chemistry is motivated by asking 'how'. How do I make a primary alcohol? React a Grignard reagent with formaldehyde. How do I get a solution containing phenolphthalein to change from colorless to pink? Add base. How do I make solid barium sulfate? Add enough barium chloride dissolved in water to a solution of sodium sulfate and the barium sulfate precipitates out. Physical chemistry is motivated by asking 'why'. The Grignard reagent and formaldehyde follow a molecular dance known as a reaction mechanism, in which stronger bonds are made at the expense of weaker bonds. In acidic solutions, the protonated form of phenolphthalein absorbs light in the ultraviolet range but is transparent in the visible range. When base is added, the phenolphthalein is converted to a deprotonated form which absorbs in the green part of the visible spectrum, peaking at 553 nm. The absorption of light occurs when an electron is promoted from its ground state to an excited state. $BaCl_2$ and Na_2SO_4 are both highly soluble in water. $BaSO_4$ is less soluble in water. When enough Ba^{2+} and $SO_4{}^{2-}$ are present in water, it is energetically more favorable for them to form a solid compound, which is denser than water and, therefore, precipitates to the bottom of the beaker. If you are interested in asking why and not just how, then you need to understand physical chemistry.

Introductory chemistry – general chemistry and an introduction to synthesis – has introduced all the major, sweeping, most important aspects of chemistry. In most cases you have been presented these ideas without foundations; that is, you have been told that certain phenomena occur – such as chemical bonding or phase transitions – but you have only had hints given to you about *why* these phenomena occur. You have been introduced to special cases and single-temperature results. However, you have not been given the machinery to move away from ideal cases, or to change parameters so as to predict what will happen next. Let us jog your memory about some key concepts (and perhaps add a few details I wish you would have been taught). This will illuminate where we have to go and motivate the discussion of much of our machinery building. It will also highlight some areas in which the simplifications of introductory chemistry sometimes border on myth building. The role of physical chemistry is to tear down mirages of explanation in order to construct the machinery that results in fundamental understanding. To a large extent, physical chemistry is the study and mastery of k, K, n, and ψ, (rate constant, equilibrium constant, moles, and wavefunction) and how they respond to t, T, and p (time, temperature, and pressure). That is, we want to understand the time evolution and response to experimental conditions of chemical systems. Along the way perhaps the three most important constants that we will have to understand the implications of are k_B, h and N_A (the constants of Boltzmann, Planck, and Avogadro, respectively) and how they relate to the energy, propensity for change and our ability to probe systems at the atomic level.

1.1 Atoms and molecules

Our chemical universe is made up of atoms and molecules. For the purposes of this course we are going to limit ourselves to electrons, protons and neutrons, and not worry about any internal structure of these particles (though we will mention the Higgs boson a couple of times, so don't get the blues!). Atoms are composed of a nucleus, where the protons and neutrons reside, and electrons that occupy the space outside of the nucleus. The space occupied by the electrons is called an orbital. Orbitals have specific shapes; you should recall the spherical s orbitals and dumbbell-shaped p orbitals (as usually represented in chemistry). The d orbitals and f orbitals are more complex yet. Atoms are neutral; that is, the number of negatively charged electrons exactly balances the number of positively charged protons. Neutrons carry no charge. An atom can be excited, for instance by the absorption of light. If the photons in the light are energetic enough, an electron can be removed from the atom to form a positive ion. Negative ions are formed when an atom captures an extra electron from some external source.

Physical Chemistry: How Chemistry Works, First Edition. Kurt W. Kolasinski.

Companion Website: www.wiley.com/go/kolasinski/physicalchemistry

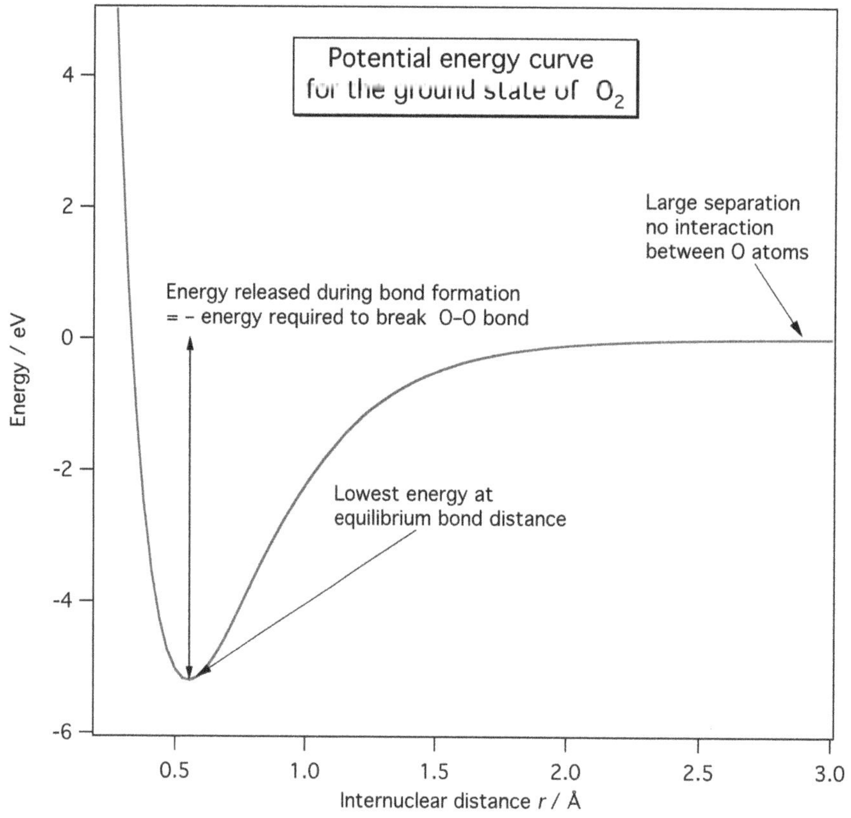

Figure 1.1 A potential energy diagram for the ground electronic state of the O_2 molecule.

Certain atoms will form molecules when they interact with each other. Molecules have well-defined stoichiometries (ratios of how many atoms combine) and well-defined geometries with characteristic bond distances and bond angles. As molecules become bigger and bigger they may be able to take on several different configurations (isomers).

A molecule is held together by chemical bonds. An especially wicked misconception held by many is that energy is released when a chemical bond is broken. As shown in Fig. 1.1, *energy is released when a chemical bond is formed.* Chemical bond formation between two atoms must be an *exothermic* process or else it would never occur. Conversely then, *it takes energy to break a chemical bond,* and this is an *endothermic* process. Chemical bonds in molecules come in two broadly defined flavors. In a covalent bond, electrons are shared. If they are shared equally, a nonpolar bond is formed, but if they are shared unequally a polar bond is formed. Unless the geometry of the molecules has a special symmetry that makes charge distributions cancel out, polar bonds lead to polar molecules, that is, molecules that exhibit dipole moments. In ionic bonds, electrostatic forces hold together the positive and negative charges on ions formed by the transfer of at least one complete electron.

The motions of subatomic particles are governed by quantum mechanics, not classical mechanics. Classical mechanics describes everyday particles: billiard balls and apples falling on the heads of drowsy natural philosophers. Quantum mechanics is different. The energy states of electrons are governed by quantum mechanics. Electrons occupy atomic orbitals in atoms and molecular orbitals in molecules. Because covalent bonding is due to electrons and their sharing by atoms, bonding is inherently quantum mechanical in nature.

1.2 Phases

Moving from a single molecule or atom to collections of particles, the influences of intermolecular forces become important. There are a number of different types of intermolecular interactions: dipole–dipole, van der Waals and hydrogen bonding are some of the most common. The three most commonly encountered phases in chemistry are gases, liquids, and solids. In this course, we will not have to worry about the other two: plasmas and Bose–Einstein condensates, which are most likely to be encountered at very high or very low temperatures, respectively.

You are already familiar with the concept of an ideal gas and its equation of state

$$pV = nRT. \tag{1.1}$$

Intermolecular interactions are absent in an ideal gas. Even in a real gas, intermolecular interactions are weak as long as the pressure and temperature are far from the conditions that induce condensation. Gas-phase molecules are free to translate, rotate, and vibrate. There are also electronic excitations, but usually these are not accessible by thermal excitations. The places where we can put energy (translations, rotations, vibrations, and electronic states) are called *degrees of freedom*.

1.2.1 Directed practice

Throughout this book you will find directed practice, short exercises for the reader to collect their thoughts and reinforce concepts. Calculate the volume in m^3 of one mole of ideal gas at *standard temperature and pressure* (STP is defined as $T = 273.15$ K and 100 kPa) and standard ambient temperature ($T = 298.15$ K) and atmospheric pressure ($p = 101.325$ kPa), denoted SATP.

[Answers: 0.0227 m^3, 0.0245 m^3]

As the temperature is lowered or the pressure increased, the particles in a gas are pushed together, and eventually intermolecular interactions will lead to condensation. Most commonly, a gas will condense into a liquid. Intermolecular interactions are much more important in liquids than gases because the molecules are much closer together. That the molecules are much closer together means that their density is much higher, and that they collide with each other much more frequently than gas-phase molecules. Liquid-phase molecules are still free to translate, rotate and vibrate, but the spectra associated with these degrees of freedom look different than when they are in the gas phase. Liquids are commonly classified as to whether they are polar or nonpolar. Like tends to dissolve like, and polar solvents are capable of dissolving ionic compounds (electrolytes) into their constituent ions. Nonelectrolytes dissolve without the formation of ions. Electrolyte solutions can conduct electrical currents.

At yet lower temperature and higher pressure, liquids condense into solids. Solids are usually denser than liquids, but there is not nearly as big a change during this phase transition as the one from the gas phase to a condensed phase. The atoms and molecules in a solid can no longer translate and rotate freely. There are still vibrations and electronic excitations. Solids are often classified by their electronic structure according to their electrical conductivity properties: metals conduct well and have no band gap between their valence and conduction bands, insulators conduct poorly because they have a large band gap; semiconductors are somewhere in between.

1.3 Energy

When energy is deposited in a system it is distributed among the degrees of freedom available to it. You should recall the basic equation of calorimetry, which states that the amount of heat q required to raise the temperature by ΔT of n moles of a substance is related to the heat capacity C_m (the subscript m denotes a molar quantity)

$$q = nC_m \Delta T. \tag{1.2}$$

You may remember that C_m depends on whether the process occurs at constant volume or constant pressure; thus, we usually denote it as $C_{p,m}$ or $C_{V,m}$. You probably do not appreciate the central importance of the heat capacity to molecular and chemical properties. We will learn much more about this, learn how it is related to molecular and phase structure, and that because of this it is a temperature-dependent property. Because C_m is both temperature- and phase-dependent, we will see that Eq. (1.2) is an approximation and we will have to learn how to treat parameters that are not constant more generally. This is where calculus becomes essential.

1.3.1 Directed practice

Calculate the heat required to raise the temperature of 10.0 g water, initially at 298 K, by 1 °C. The heat capacity of water is 75.3 J K^{-1} mol^{-1}.

[Answer: 41.8 J]

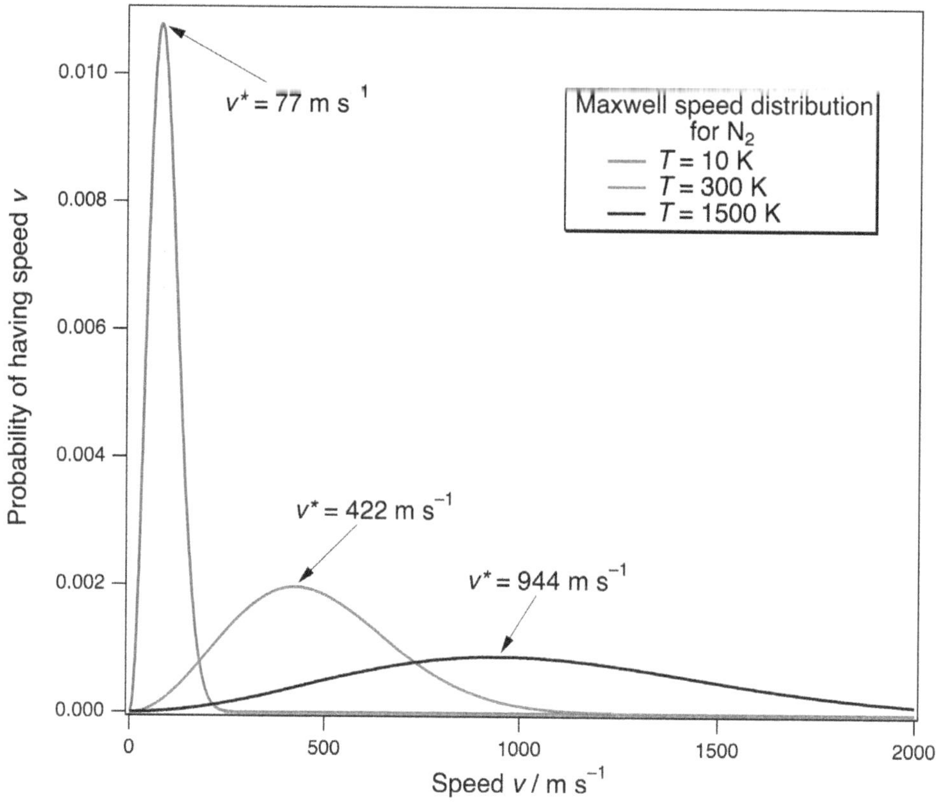

Figure 1.2 Maxwell distributions of molecular speed for gaseous N_2 at three different temperatures.

Equilibrating a sample of gas to a particular temperature leads to a partitioning of energy into the translational, rotational, and vibrational degrees of freedom. Unless we confine our molecules to very small boxes, the partitioning of energy into the translational degree of freedom is described by a continuous distribution known as the Maxwell speed distribution. A Maxwell distribution looks something like Fig. 1.2.

The rotational and vibrational degrees of freedom are inherently quantum mechanical. The energy levels associated with these degrees of freedom are not continuous, but are quantized into discrete energy levels. Energy is partitioned into these degrees of freedom by changing the populations of these levels, that is, by changing the number of molecules in each of the energy levels. This population distribution follows a Boltzmann distribution, about which we need to learn more.

Electronic states are also quantized. The transitions between different energy levels are responsible for lines and bands observed in rotational, vibrational, and electronic spectroscopy. Nuclear magnetic resonance involves spin flips among nuclei.

1.4 Chemical reactions

When atoms and molecules are mixed together, they can react to form products. Some reactions are spontaneous and occur upon mixing, but some reactions only occur if they are given a push. All chemistry involves changes in electron-sharing arrangements. Bonds can both be made and broken during reactions. Chemical reactivity can be classified according to how the reactions proceed. Thermal chemistry occurs when the control of temperature is used to regulate reactivity and the rate of reaction. In other words, the energy that is randomly distributed among the degrees of freedom of the system is sufficient to cause electron-sharing arrangements to change. This is so common that it is usually just called 'chemistry.' Electrochemistry involves the exchange of free charge (electrons and sometimes protons). It responds to temperature but, importantly, it can also respond to voltages placed on electrodes in contact with the reacting species. Stimulated chemistry is induced by excitations caused by photon absorption, electron scattering, or ion scattering. The most common form is photochemistry, in which electronic excitations engendered by photon/matter interactions result in the formation of products.

Reactants are often separated from products by an *activation barrier*. The height of this barrier – the activation energy – must be overcome during the course of the reaction. Higher temperatures lead to more high-energy species being present in the Maxwell–Boltzmann distribution (i.e., the tail to high velocity shown in Fig. 1.2 gets bigger); this we why we commonly see reaction rates increase with increasing temperature. The rates of reactions also depend on the concentrations of reactants, because reactants usually have to collide in order for them to react. The collision rate increases when a higher density of reactants is present. The presence of a *catalyst* can also accelerate a reaction. Catalysts lower activation energies but are not themselves consumed in reactions.

Chemical reactions occur because systems are attempting to reach a minimum in their Gibbs energy. The Gibbs energy, G, contains contributions from enthalpy H and entropy S. Enthalpy is involved with the energy of interactions (such as bonding), while entropy is involved with the randomization of energy among the degrees of freedom and disorder. A system will evolve as long as the Gibbs energy has not reached its minimum; in other words, $\Delta G = 0$ at equilibrium. At equilibrium, the Gibbs energy is no longer changing. The change in Gibbs energy is related to the change in enthalpy ΔH, the absolute temperature T and the change in entropy ΔS by

$$\Delta G = \Delta H - T\Delta S. \tag{1.3}$$

At equilibrium, the rates of forward and reverse reactions are the same and the concentrations of species involved are constant. An equilibrium constant K relates these concentrations to one another. For the reversible reaction

$$\mathrm{A + B \rightleftharpoons C + D} \tag{1.4}$$

we have

$$K = \frac{[\mathrm{C}]\,[\mathrm{D}]}{[\mathrm{A}]\,[\mathrm{B}]}. \tag{1.5}$$

Non-ideal behavior manifests itself in both rate equations and equilibrium calculations. We deal with these deviations from ideal behavior by introducing activity in place of concentrations.

It is important to understand whether a system is in an equilibrium or nonequilibrium state. Nonequilibrium states may be characterized as being in a state of flux, in a steady state, or in a metastable state. Equilibrium is controlled by limits set by thermodynamics. Nonequilibrium states are controlled by reaction kinetics. Another way to phrase this is that thermodynamics makes certain predictions about what is allowed and what is forbidden. While what is forbidden by thermodynamics is forbidden, that which is allowed by thermodynamics is only a limit that, while it cannot be exceeded, this limit might not be approached if the reaction kinetics do not allow the system to evolve to the equilibrium state.

A balanced stoichiometric chemical reaction is written with *stoichiometric coefficients* a, b, and so forth, which are all positive numbers

$$a\mathrm{A} + b\mathrm{B} + \cdots \rightarrow \cdots + y\mathrm{Y} + z\mathrm{Z}.$$

The IUPAC defines *stoichiometric numbers* ν_A, ν_B, and so forth as signed quantities, which are negative for reactants and positive for products. Stoichiometric coefficients are the absolute values of the stoichiometric numbers. In other words, $a = |\nu_\mathrm{A}| = -\nu_\mathrm{A}$ but $z = |\nu_\mathrm{Z}| = \nu_\mathrm{Z}$.

1.4.1 Directed practice

Calculate the molar Gibbs energy of formation $\Delta_\mathrm{f}G^\circ_\mathrm{m}$ of HCl(g) at 298 K if the molar enthalpy of formation $\Delta_\mathrm{f}H^\circ_\mathrm{m}$ is −92.31 kJ mol^{-1} and the molar entropy of formation $\Delta_\mathrm{f}S^\circ_\mathrm{m}$ is 186.90 J K^{-1} mol^{-1}.

[Answer: −148.01 kJ mol^{-1}]

1.5 Problem solving

Observation is the foundation of empirical knowledge and science. There is much to be gained from description. However, if all you want is description than take an art class and don't ask questions, for artists are also looking for meaning. Therefore, we are going to go beyond the descriptive.

Up to this point in your academic career, you have primarily used only the two most basic levels of learning: (1) *remembering*; and (2) *understanding*. To perform well in physical chemistry you will need to become proficient in the next three levels of learning, namely (3) *applying*, (4) *analyzing*, and (5) *evaluating*. Mastering all of these will set you up to perform research, which requires the highest level of learning: (6) *creating*. This course is about *problem solving* – that is, solving the problem of how chemical species react with each other, how they respond to changes in conditions such and temperature and pressure, and what is the basis of a number of their physical properties. I am attempting to teach you methods of chemical problem solving: Applying what you are taught to new situations to analyze variations on problems and evaluate the results. In your future careers you will need to learn a variety of things. It is unlikely that there is a textbook written for most of the things you will encounter. In your professional life you will never find a real problem to solve in which exactly all of the right information is given to you, in the right units. You will always be faced with either too much information or too little information, and have to decide which parts are necessary. Being able to tackle problems under these circumstances is an important skill. It is the foundation of the ability of physical chemists/chemical physicists to tackle problems ranging from atomic physics to thermodynamics, to reaction dynamics in all phases, through nanoscience and beyond biochemistry into astrochemistry. Learning problem solving is one of the most important skills you will acquire in this course.

There is no single method for problem solving. The best way to solve a problem is the one that gives the right answer. If you see the path to an answer, take it. There are often multiple avenues. But if you are stuck, how do you get started? Let me outline the four basic steps for problem solving. These may not always all be applicable, but checking each will help you get past the writer's block of staring at a blank page.

1. Start with a balanced reaction (including phases).
2. Draw a diagram.
3. Draw a straight line.
4. Answer the question asked with appropriate units.

These are obviously oversimplifications for impact and ease of remembrance; but let's consider this distillation. Always start by identifying the system and how it is changing. Organize the information you have available. Evaluate the physical processes that occur. If there is not a diagram to draw, then, for example, chart the amount of substance in moles and how it changes. I call this a *reaction chart*. For the gas-phase reaction of ammonia (initial concentration a_0) with carbon dioxide (initial concentration b_0) to form an equilibrium with urea and water (not initially present), it has the form

	$2NH_3(g)$	+	$CO_2(g)$	$\rightleftharpoons$	$(NH_2)_2CO(g)$	+	$H_2O(g)$
Initial	a_0		b_0		0		0
Eq	a_0-2x		b_0-x		x		x

Note that a properly balanced reaction also includes all phase designations.

Next, determine what equations relate to the data given and answers sought. If you do not communicate your results then you are partaking in a hobby, not performing science. Thus, we are forever presenting graphs or constructing tables. Drawing a straight line (or otherwise presenting results appropriately) for interpretation and clarity is an important part of analysis, evaluating models, and drawing conclusions. In the first instance, you always need to identify what question is being asked and what data you have available. But problem solving does not always start at the beginning and take a linear path to the last step. Sometimes, you need to work backwards: look at the final result you want and find the step that would lead to it. Sometimes, looking at the end is the most important step in showing you where you need to start.

To engage in effective problem solving you need to stop thinking solely of the correct *answer* and start thinking about the correct *method* of finding the answer. A revealing recurring comment on student evaluations of my teaching is "stop doing derivations and show us the answer." Performing derivations is one of the primary means of teaching you the techniques you need to apply to perform effective problem solving. The 'Directed practice' exercises are intended to reinforce this. Directed practices are intended to get you into the habit of interrupting your reading, stopping to think about what you have just read, performing a simple calculation (or filling in missing steps in a derivation), and enhancing comprehension of what you just read. Get into this habit, perform it often – even when a Directed practice is not indicated – and your comprehension and problem-solving prowess will increase. Make your own Directed practices to get you used to and comfortable with manipulating equations and ideas. If you don't understand a Directed practice, ask a question in the lecture.

1.5.1 Directed practice

If $a_0 = b_0 = a$, set up the algebraic equation required to find the value of x for the reaction of ammonia and carbon dioxide to produce urea and water.

Hint: you will have to set up an equation similar to Eq. (1.5) and then collect terms to find a third-order polynomial that you set equal to zero.

[Answer: $4x^3 - \left(8a - \frac{1}{K}\right)x^2 + 5a^2x - a^3 = 0$]

1.5.2 Performing calculations

A calculation a day keeps ignorance away. All calculations were once done by hand. You should look at performing calculations with only a calculator as outdated as using only tables of logarithms and a slide rule. At this point in your scientific education and career it is time to use computers for calculations and the presentation of data. Spreadsheet software is okay for calculating grades. It can be a good introduction to scientific computing but a spreadsheet is hardly sufficient for proper data analysis. You should consider introducing yourself to real data analysis software. There are many options such as Igor Pro, Origin, and SigmaPlot. This type of software allows you to handle large data sets, perform calculations, present data and perform curve fitting. If you are considering further study in physical chemistry then you should also consider learning software that is capable of algebraic and symbolic computing as well as calculus, such as Mathematica, MathCAD, and MATLAB.

1.5.3 Units and reporting data

A number reported without the appropriate units is the scientific equivalent of nihilistic relativism. Forty-two is the answer to everything, we just do not know what the conversion factor is to the units that actually have meaning. This is a dramatic way of saying, no answer is correct without the correct units. Many of your errors will have their basis in not using the right units or the right conversion factors. It is not always easy, but it is always something that you have to check, and a check of the units is one of the surest ways to determine if your calculation has gone astray. Numerical values and units, just like the symbols that represent these physical quantities, can be manipulated with the usual rules of algebra. Let algebra be your friend and use it constantly.

Another basic topic that often causes confusion is the reading of data in tables and figures. Always label column and axes with units. Shorthand is often used to avoid the repetition of orders of magnitude. Best practice is to divide out the units and appropriately and clearly include indications of orders of magnitude. Note the equivalence of the following. Note further that units are in Roman type while variables are italicized, and that a space appears between numbers and units. Manipulation of the magnitudes and units of physical quantities is known as *quantity calculus*. For more on this see the IUPAC Green Book.

T / K	T^{-1} / K^{-1}	K / T	10^3 K / T	T^{-1} / (10^{-3} K^{-1})
300	3.33×10^{-3}	3.33×10^{-3}	3.33	3.33

p / Pa	p / (10^6 Pa)	p / MPa	ln(p/Pa)	ln(p/MPa)
1.45×10^6	1.45	1.45	14.19	0.3716

1.6 Some conventions

This section includes some conventions on notation contained in Tables 1.1 through 1.5 that will be used throughout this book. All recommendations are in line with the IUPAC Green Book. There are several different types of reaction that we will encounter, and we will use symbols to help differentiate these. Reactions that occur on the molecular level in one step are elementary reactions. These reactions are differentiated from one another on the basis of how many species are involved as reactants. *Elementary reactions* may involve one, two or three species, and are termed unimolecular (or monomolecular), bimolecular or trimolecular (termolecular), respectively. An arrow going from left to right is used to show the direction of reaction.

$$\mathrm{A + B \rightarrow Y + Z} \qquad \text{bimolecular elementary reaction}$$

Thus, the number of species on the left determines the *molecularity* of the reaction. If the elementary step is reversible, it can proceed in either the forward or reverse directions and this is indicated with two double-headed arrows.

$$\mathrm{A + B \rightleftarrows Y + Z} \qquad \text{reversible elementary reaction}$$

A *reaction mechanism* is the set of elementary steps that occurs during a net reaction. If the mechanism contains two or more steps, it is a *composite mechanism*. The *net reaction* is the sum of the elementary steps (some of which may have to be counted more than once), and it has the proper stoichiometry for the overall reaction. The net reaction has its reactants and products separated by an equal sign. Thus, the net reaction to form C and D from 2A by a two-step composite mechanism is shown below

$$A \rightarrow B + C \qquad \text{elementary step 1}$$

$$\underline{A + B \rightarrow D} \qquad \text{elementary step 2}$$

$$2A = C + D \qquad \text{net reaction}$$

A is a reactant, C and D are products, B is an intermediate. A net reaction is not described by molecularity, only an elementary reaction is. A net reaction is also indicated by opposing arrows with only a single stroke in their heads; the convention used throughout this book.

$$2A \rightleftharpoons C + D \qquad \text{net reaction}$$

Table 1.1 States of aggregation.

Symbol	State	Symbol	State
a or ads	species adsorbed on a surface	am	amorphous solid
aq	aqueous solution	aq, ∞	aqueous solution, infinite dilution
cd	condensed phase, i.e. liquid or solid	cr	crystalline
f	fluid phase, i.e. liquid or gas	g	gas or vapor
l	liquid	lc	liquid crystal
mon	monomeric form	n	nematic phase
pol	polymeric form	s	solid
sat	saturation or saturated	sat l	saturated liquid
sln	solution	vit	vitreous

Table 1.2 SI base units.

Base quantity	Name	Symbol
length	meter	m
mass	kilogram	kg
time	second	s
electric current	ampere	A
thermodynamic temperature	kelvin	K
amount of substance	mole	mol
luminous intensity	candela	cd

Table 1.3 SI derived units.

Derived quantity	Name	Symbol	Expressed in terms of other SI units
plane angle	radian	rad	$m\ m^{-1} = 1$
solid angle	steradian	sr	$m^2\ m^{-2} = 1$
frequency	hertz	Hz	s^{-1}
force	newton	N	$m\ kg\ s^{-2}$
pressure, stress	pascal	Pa	$N\ m^{-2} = m^{-1}\ kg\ s^{-2}$
energy, work, heat	joule	J	$N\ m = m^2\ kg\ s^{-2}$
power, radiant flux	watt	W	$J\ s^{-1} = m^2\ kg\ s^{-3}$
electric charge	coulomb	C	$A\ s$
electric potential electromotive force	volt	V	$J\ C^{-1} = m^2\ kg\ s^{-3}\ A^{-1}$
electric resistance	ohm	Ω	$V\ A^{-1} = m^2\ kg\ s^{-3}\ A^{-2}$
electric conductance	sieman	S	$\omega^{-1} = m^{-2}\ kg^{-1}\ s^3\ A^2$
electric capacitance	farad	F	$C\ V^{-1} = m^{-2}\ kg^{-1}\ s^4\ A^2$
magnetic flux	weber	Wb	$V\ s = m^2\ kg\ s^{-2}\ A^{-1}$
magnetic flux density	tesla	T	$Wb\ m^{-2} = kg\ s^{-2}\ A^{-1}$
inductance	henry	H	$V\ A^{-1}\ s = m^2\ kg\ s^{-2}\ A^{-2}$
catalytic activity [1]	katal	kat	$mol\ s^{-1}$

[1] Should replace the (enzyme) unit, 1 U = μmol min^{-1} ≈ 16.67 nkat.

Table 1.4 SI prefixes.

Submultiple	Name	Symbol	Multiple	Name	Symbol
10^{-1}	deci	d	10^{1}	deca	da
10^{-2}	centi	c	10^{2}	hector	h
10^{-3}	milli	m	10^{3}	kilo	k
10^{-6}	micro	μ	10^{6}	mega	M
10^{-9}	nano	n	10^{9}	giga	G
10^{-12}	pico	p	10^{12}	tera	T
10^{-15}	femto	f	10^{15}	peta	P
10^{-18}	atto	a	10^{18}	exa	E
10^{-21}	zepto	z	10^{21}	zetta	Z
10^{-24}	yocto	y	10^{24}	yotta	Y

Table 1.5 Symbols, operators, and functions.

Description	Symbol
Signs and Symbols	
left-hand side of equation	LHS
right-hand side of equation	RHS
equal to	$=$
not equal to	$\neq$
identically equal to	$\equiv$
equal to by definition	$\stackrel{\text{def}}{=}$ or $:=$
approximately equal to	$\approx$
asymptotically equal to	$\simeq$
corresponds to	$\hat{=}$
proportional to	$\sim$, $\propto$
tends to, approaches	$\rightarrow$
infinity	∞
less than	$<$
greater than	$>$
less than or equal to	$\leq$
greater than or equal to	$\geq$
much less than	$\ll$
much greater than	$\gg$
Operations	
magnitude of a	$\|a\|$
mean value of a	$\langle a \rangle$, $\bar{a}$
n factorial	$n!$
sum of a_i	$\sum a_i$, $\sum_i a_i$, $\sum_{i=1}^{n} a_i$
product of a_i	$\prod a_i$, $\prod_i a_i$, $\prod_{i=1}^{n} a_i$
Functions	
base of natural logarithms	e
exponential of x	$\exp x$, e^x
logarithm to the base a of x	$\log_a x$
natural logarithm of x	$\ln x$
logarithm to the base 10 of x	$\lg x$
change in x	$\Delta x = x\,(\text{final}) - x\,(\text{initial})$
infinitesimal variation of f	δf
limit of $f(x)$ as x tends to a	$\lim_{x \to a} f(x)$
first derivative of f	$\mathrm{d}f/\mathrm{d}x$, f', $(\mathrm{d}/\mathrm{d}x)f$
second derivative of f	$\mathrm{d}^2 f/\mathrm{d}x^2$, f''
partial derivative of f	$\partial f/\partial x$

(continued)

Table 1.5 (*Continued*)

Description	Symbol
total differential of f	$\mathrm{d}f$
inexact differential of f	$đf$
integral of $f(x)$	$\int f(x)\,\mathrm{d}x,\ \int \mathrm{d}x\, f(x)$
Kronecker delta	$\delta_{ij} = \begin{cases} 1 & \text{if } i = j \\ 0 & \text{if } i \neq j \end{cases}$
Dirac delta function	$\delta(x),\ \int f(x)\,\delta(x)\,\mathrm{d}x = f(0)$
unit step function, Heaviside function	$\varepsilon(x),\ H(x),\ h(x),$ $\varepsilon(x) = 1$ for $x > 0$, $\varepsilon(x) = 0$ for $x < 0$
gamma function	$\Gamma(x) = \int_0^\infty t^{x-1}\,\mathrm{e}^{-t}\,\mathrm{d}t$ $\Gamma(n+1) = n!$ for positive integers n
convolution of functions f and g	$f^*g = \int_{-\infty}^{\infty} f(x-x')\,g(x')\,\mathrm{d}x'$
Complex Numbers	
square root of −1, $\sqrt{-1}$	i
real part of $z = a + \mathrm{i}b$	$\mathrm{Re}\,z = a$
imaginary part of $z = a + \mathrm{i}b$	$\mathrm{Im}\,z = b$
modulus of $z = a + \mathrm{i}b$	$\lvert z\rvert = (a^2 + b^2)^{1/2}$
argument of $z = a + \mathrm{i}b$	$\arg z;\ \tan(\arg z) = b/a$
complex conjugate of $z = a + \mathrm{i}b$	$z^* = a - \mathrm{i}b$
Vectors	
vector ***a***	$\boldsymbol{a},\ \vec{a}$
Cartesian components of ***a***	a_x, a_y, a_z
unit vectors in Cartesian coordinate system	e_x, e_y, e_z or i, j, k
scalar product	$a \cdot b$
vector or cross product	$a \times b$
nabla operator, del operator	$\nabla = e_x \frac{\partial}{\partial x} + e_y \frac{\partial}{\partial y} + e_z \frac{\partial}{\partial z}$
Laplacian operator	$\nabla^2 = \frac{\partial^2}{\partial x^2} + \frac{\partial^2}{\partial y^2} + \frac{\partial^2}{\partial z^2}$
Matrices	
matrix with elements A_{ij}	$\boldsymbol{A}$
product of matrices ***A*** and ***B***	$\boldsymbol{AB}$, where $(\boldsymbol{AB})_{ik} = \sum_j A_{ij} B_{jk}$
unit matrix	$\boldsymbol{E}, \boldsymbol{I}$
inverse of a square matrix ***A***	$\boldsymbol{A}^{-1}$
transpose of matrix ***A***	$\boldsymbol{A}^{\mathrm{T}}, \tilde{\boldsymbol{A}}$
complex conjugate of matrix ***A***	$\boldsymbol{A}^*$
conjugate transpose (adjoint) of ***A*** (hermitian conjugate of ***A***)	$\boldsymbol{A}^{\mathrm{H}}, \boldsymbol{A}^{\dagger}$, where $(\boldsymbol{A}^{\dagger})_{ij} = A_{ji}^*$
trace of a square matrix ***A***	$\sum_i A_{ii},\ \mathrm{tr}\,\boldsymbol{A}$
determinant of a square matrix ***A***	$\det \boldsymbol{A},\ \lvert\boldsymbol{A}\rvert$

SUMMARY OF IMPORTANT EQUATIONS

You should already be familiar with these equations from your previous introduction to chemistry. We will learn much more about them, their limitations and their derivations but you should be prepared to use them now.

$m = M/N_{\mathrm{A}}$	Mass per molecule m is the molar mass M divided by the Avogadro constant N_{A}.
$n = m/M = N/N_{\mathrm{A}}$	Amount of substance (moles) n is equal to mass divided by molar mass as well as number of molecules N divided by the Avogadro constant.

$pV = nRT$	Ideal gas law: pressure p, volume V, amount of substance (moles) n, ideal gas constant R, absolute temperature T.
$R = N_A k_B$	The ideal gas constant per mole R is simply the product of the ideal gas constant per molecule k_B (more commonly called the Boltzmann constant) time Avogadro's constant.
$q = nC_m \Delta T$	Heat evolved q when n moles of substance with heat capacity C_m changes temperature by ΔT.
$\Delta G = \Delta H - T\Delta S$	Gibbs energy change ΔG is related to the changes in enthalpy ΔH and entropy ΔS.
$K = \frac{[C][D]}{[A][B]}$	Equilibrium constant in terms of the concentrations of the participants in the reversible reaction $A + B \rightleftharpoons C + D$.
$\Delta G° = -RT \ln K$	The relationship of standard Gibbs energy change to the equilibrium constant K.
$k = Ae^{-E_a/RT}$	The Arrhenius equation relates the rate constant k to the pre-exponential factor A and the activation energy E_a.
$E = h\nu$	Photon energy is equal to the product of the Planck constant and the speed of light.
$c = \lambda\nu$	The speed of light is equal to the product of wavelength and frequency.
$E_K = \frac{1}{2}mv^2$	Kinetic energy is equal to one half the product of mass and the velocity squared.

The following may not be familiar yet, but as you master the above and those below you will develop an understanding of chemistry.

$E = E° - (RT/zF) \ln Q$	The Nernst equation relates the cell potential E (emf) to the standard potential $E°$, the valence of the electrochemical reaction z, and the reaction quotient Q. F is the Faraday constant.
$K = k_{forward}/k_{reverse}$	The rate constants of a reversible elementary reaction are related to its equilibrium constant.
$R = k[A]^\alpha[B]^\beta$	The rate R of a reaction is given by an empirical rate law composed of a rate constant k and concentration terms, in this case, the concentration of A raised to the α and the concentration of B raised to the β. The form of the rate law depends on the reaction and reaction conditions.
$S = k_B \ln \Omega$	The Boltzmann entropy S is proportional to the Boltzmann constant k_B times the natural logarithm of the number of microstates Ω.
$\mu = \mu° + RT \ln a$	The chemical potential μ is determined by the activity a. $\mu°$ is the standard state chemical potential.

Exercises

1.1 You should know how to write balanced chemical equations. You have been introduced to reaction enthalpy and should be able to calculate the amount of heat released or consumed in a reaction run at constant pressure (as in a beaker on a lab bench). Ethanol has a heat of combustion of 1367 kJ mol^{-1}, a molar mass of 46.07 g mol^{-1} and a density of 0.7893 g cm^{-3}. Octane (a major component of gasoline) has a heat of combustion of 5450 kJ mol^{-1}, a molar mass of 114.23 g mol^{-1} and a density of 0.7025 g cm^{-3}. (a) Write out balanced combustion reactions for both ethanol and octane. (b) Calculate the enthalpy released by burning: (i) 1.00 l of ethanol; and (ii) 1.00 l of octane. (c) How much less efficient (in percent) is burning ethanol compared to octane on a per-volume basis?

1.2 You should know how to write balanced chemical equations. You have already been introduced to reaction enthalpy and should be able to calculate the amount of heat released or consumed in a reaction run at constant pressure (as in a beaker on a lab bench). Methane has a heat of combustion of 890.8 kJ mol^{-1}. Graphite has a heat of combustion of 393.5 kJ mol^{-1}. (a) Write out balanced combustion reactions for both methane and graphite. (b) Calculate the enthalpy released per mole of CO_2 formed by combustion of each. (c) Taking CH_4 as a proxy for natural gas and graphite for coal, what is the percent different in moles of CO_2 produced by coal combustion compared to natural gas in order to produce the same amount of heating?

1.3 Units, conversions and rearranging equations are never-ending topics. We will usually end up with answers in SI units but real-world problems often come in other units and you need to be able to handle different types of units. You also have been exposed to the ideal gas law and should know how to handle it. You should also be able to look at numbers and be able to put them in perspective as to how large or small they are. (a) What are the SI units (and abbreviation for the units) of length, volume, time, quantity of a chemical substance, mass, pressure and concentration? (b) The US consumed 18.8 million barrels (Mbbl) of crude oil per day in 2011. The BP Gulf Oil Spill totaled roughly 200 million gallons of crude oil. (i) There are 42 gal per bbl. The density of crude oil is roughly 0.869 g cm^{-3}. Crude oil is approximately 84.8% carbon by mass. How many moles of C are consumed daily in the US? (ii) Calculate the volume of oil spilled in liters and in barrels, and calculate the percentage of the total of the oil spill compared to daily use in the USA. (c) If all of this carbon were to be converted to CO_2, what volume in m^3 would it occupy at atmospheric pressure and the global mean surface temperature of 14 °C?

1.4 Units, conversions and rearranging equations are never-ending topics. We will usually end up with answers in SI units but real world problems often come in other units and you need to be able to handle different types of units. You also have been exposed to the ideal gas law and should know how to handle it. You should also be able to look at numbers and be able to put them in perspective as to how large or small they are. (a) What are the SI units (and abbreviation for this unit) of length, volume, time, quantity of a chemical substance, mass, pressure and concentration? (b) You should learn how to check answers by unit analysis. Confirm that the molar entropy calculated from the Sackur–Tetrode equation gives the expected answer in SI units of energy per unit of absolute temperature per amount of substance. The Sackur–Tetrode equation is

$$S_m = R \ln \left(\frac{RTe^{5/2}}{\Lambda^3 N_A p} \right)$$

where R is the universal gas constant, Λ is the thermal wavelength in meters, N_A the Avogadro constant, p is pressure, and T the absolute temperature. (c) A regulation football in the National Football League should, according to the rules but not necessarily certain teams, be inflated to '12.5 to 13.5 pounds.' Assuming an initial pressure of 13 'pounds' (that is, 13 psig) at 25 °C and constant volume, what pressure in Pa does the football have at 0 °C?

1.5 You should learn how to check answers by unit analysis. Confirm that the impingement rate (flux) calculated from the Hertz–Knudsen equation gives the expected answer in SI units. The Hertz–Knudsen equation is

$$Z_w = \frac{N_A p}{(2\pi MRT)^{1/2}}$$

1.6 You should know how to handle graphs and fit curves to data points. Get familiar with a data analysis program and it will greatly ease the calculations for some problems. You should know how to handle logs and pH.

a What is the slope and y-intercept of a plot of the natural logarithm of the rate constant (k is the rate constant) versus inverse temperature (that is $1/T$ with T in K) given the following data. Include a plot of the data in your answer.

T / °C	25	50	75	100	150
k / s^{-1}	5.55×10^7	1.41×10^8	3.15×10^8	6.31×10^8	1.98×10^9

b What is the concentration of H^+ and OH^- in water at pH 2.5 and 25°C?

1.7 You should know how to handle graphs and fit curves to data points. Get familiar with a data analysis program and it can greatly ease the calculations for some problems. You should know how to handle logs and pH. (a) What is the slope and y-intercept of a plot of the natural logarithm of the rate constant (k is the rate constant) versus inverse temperature (that is $1/T$ with T in K) given the following data. Include a plot of the data in your answer. (b) What is the concentration of H^+ and OH^- in water at pH 9.72 and 25 °C?

T / °C	250	275	300	325	350
k / s^{-1}	1.07×10^{-7}	8.05×10^{-7}	5.92×10^{-6}	3.42×10^{-5}	1.72×10^{-4}

1.8 You cannot pass this course without being able to do algebra. You also need to perform basic manipulations in integral and differential calculus.

a Solve (find the roots of) the equation $x^2 + x - 20 = 0$.

b Take the derivative of

$$y = c_0 + c_1 x + c_2 x^2, \quad \frac{dy}{dx} =$$

$$y = 22 \exp(-1.9x) = 22e^{-1.9x}, \quad \frac{dy}{dx} =$$

$$y = \ln x, \quad \frac{dy}{dx} =$$

c Integrate

$$\int 2x^3\,dx =$$

$$\int \frac{k}{x}\,dx =$$

$$\int 24\exp(32x)\,dx =$$

1.9 You cannot pass this course without being able to do algebra. You also need to perform basic manipulations in integral and differential calculus. (a) Solve (find the roots of) the equation $2x^2 + 5x - 13 = 0$. (b) Take the derivative of

$$y = c_0 + c_1x^3 + c_2x^5, \quad \frac{dy}{dx} =$$

$$y = a\exp(-2.3x) = a\,e^{-2.3x}, \quad \frac{dy}{dx} =$$

$$y = \ln(a\,x), \quad \frac{dy}{dx} =$$

(c) Integrate

$$\int 2 + x^2\,dx =$$

$$\int \frac{k}{x}\,dx =$$

$$\int 15e^{-1.5x}\,dx =$$

1.10 A vessel of volume 22.4 dm^3 contains 2.0 mol H_2 and 1.0 mol Xe at 273.15 K. Calculate the mole fractions of each component, their partial pressures, and their total pressure.

1.11 Many proteins when well packed have a density of approximately 1400 kg m^{-3}. Approximating the protein as a sphere, use the formula for density ($\rho = m/V$) to calculate the radius of a protein with a molar mass of 100 kDa.

1.12 For the data given in the table below, use Igor Pro (or other program) to plot the data (axes properly labeled with the correct units), change the data points to solid circles, and fit the points to a straight line. Use Add Annotation to write your name and the results of the fit directly on the graph. Hand in a print out of the fully analyzed and annotated graph.

Time / s	Position / m
1.53	1.20
2.62	33.5
3.97	45.2
7.60	157
15.2	280
21.9	440
36.3	890
50.1	1003
62.9	1629

1.13 For the function

$$f(x) = (2x^3 - 6x^2 + 17)e^{-(2x^2+1)}$$

use Mathematica (or other) to integrate the indefinite integral of $f(x)$ and to find the first and second derivatives of $f(x)$. Then integrate the definite integral of $f(x)$ from –2 to 2. Plot $f(x)$ from –2 to 2. Type your name into the workbook and print it out.

1.14 With the aid of balanced chemical reactions, show the difference between combustion of methane performed isothermally at 298 K and isothermally at 400 K.

1.15 The mass of the proton is 1.007 u. The mass of the deuteron is 2.014. Calculate the difference in masses, the ratio of the masses, the percent difference of with respect to the proton, the percent difference compared to the deuteron, and the percent difference with respect to the mean of the masses.

1.16 Calculate the mole fraction and molar concentration when 1.00 g of NaCl dissolved in 100 g of water at 25 °C. The density of water is 0.997 g cm^{-3}.

1.17 Taking the atmosphere to be 78% N_2, 21% O_2 and 1% Ar, calculate the molar mass of the atmosphere. Compare this value to that of Venus, the atmosphere of which is 96.5% CO_2 and 3.5% N_2.

1.18 Calculate the mass in SI units of one atom of Kr from its molar mass.

1.19 Calculate the molar volume of Au(s) at room temperature.

1.20 The mass density of Pt is 21.45 g cm^{-3} at 298 K. Calculate the number density in atoms per cubic meter of solid Pt at 298 K.

Further reading

When I quote IUPAC recommendations, this means an implicit reference to one of the following three works. Complete compendia of terms can be found in the Gold Book and Green Book cited below. If you want to know your eutectoid reaction from your eutonic reaction, these are the references for you.

Gamsjäger, H., Lorimer, J.W., Scharlin, P., and Shaw, D.G. (2008) *Pure Appl. Chem.*, **80** (2), 233–276. This is a reference full of IUPAC recommendations and definitions of everything from absorption coefficient to wet residue method.

International Union of Pure and Applied Chemistry (IUPAC) (1997) Compendium of Chemical Terminology, 2nd edition (the "Gold Book"), compiled by A.D. McNaught and A. Wilkinson. Blackwell Science, Oxford. XML on-line corrected version: <http://goldbook.iupac.org> (2006–) created by M. Nic, J. Jirat, B. Kosata; updates compiled by A. Jenkins.

Quantities, Units and Symbols in Physical Chemistry, IUPAC Green Book (2008) 3rd edition, 2nd printing, prepared for publication by E.R. Cohen, T. Cvitaš, J.G. Frey, B. Holmström, K. Kuchitsu, R. Marquardt, I. Mills, F. Pavese, M. Quack, J. Stohner, H.L. Strauss, M. Takami, and A.J. Thor, IUPAC & RSC Publishing, Cambridge.

If you are going to go beyond being a student who takes physical chemistry and become a student of physical chemistry then after you finish this text you probably want to buy one of these two books next.

Berry, R.S., Rice, S.A., and Ross, J. (2000) *Physical Chemistry*, 2nd edition. Oxford University Press, New York.

Metiu, H. (2006) *Physical Chemistry: Kinetics. Physical Chemistry: Quantum Mechanics. Physical Chemistry: Statistical Mechanics. Physical Chemistry: Thermodynamics* (4 volume set). Taylor & Francis, New York.

For an entertaining and informative historical perspective on the development of physical chemistry, I recommend:

Coffey, P. (2008) *Cathedrals of Science: The Personalities and Rivalries That Made Modern Chemistry*. Oxford University Press, Oxford.

Quinn, S. (1996) *Marie Curie: A Life*. Simon & Schuster, New York.

For a drier, more scholastic history of the field:

Laidler, K.J. (1993) *The World of Physical Chemistry*. Oxford University Press, Oxford.

More focused on quantum theory and its development but quite interesting are:

Jones, S. (2008) *The Quantum Ten: A Story of Passion, Tragedy, Ambition and Science*. Oxford University Press, Oxford.

Baggott, J. (2010) *The Quantum Story: A History in 40 Moments*. Oxford University Press, Oxford.

These websites from NIST are quite useful:

NIST Chemistry Webbook http://webbook.nist.gov/chemistry/

NIST Chemistry Portal http://www.nist.gov/chemistry-portal.cfm

NIST Reference on Constants, Units and Uncertainty http://physics.nist.gov/cuu/index.html

NIST Computational Chemistry Comparison and Benchmark Database http://cccbdb.nist.gov/Summary.asp

From the Royal Society of Chemistry: Chem Spider http://www.chemspider.com/

CHAPTER 2

Ideal gases

PREVIEW OF IMPORTANT CONCEPTS

- Statistical mechanics can relate macroscopic, easily measurable properties such as pressure, temperature, and volume to the microscopic behavior of molecules and assemblies of molecules, but first we establish what pressure, temperature and an ideal gas are.
- An ideal gas is composed of non-interacting point like particles that collide elastically.
- The equation of state of an ideal gas is $pV = nRT = Nk_BT$.
- Ideal gas behavior is a good approximation away from conditions that favor condensation – that is, for low p and high T compared to standard boiling point.
- Molecules can translate, vibrate, and rotate.
- The number of ways a molecule can translate, vibrate, and rotate corresponds to the number of degrees of freedom it has. Electronic and nuclear degrees of freedom can usually be neglected under normal conditions.
- Energy is partitioned among the degrees of freedom.
- The internal energy of a monatomic ideal gas is simply equal to its kinetic energy.
- For ideal gases containing more than one atom, the internal energy also has contributions from rotations and vibrations.
- The heat capacity relates the partitioning of energy into a molecule to changes in temperature. Its value contains contributions from translational energy and rotational and vibrational energy as appropriate.
- Heat capacity has temperature dependence due to the quantum mechanical nature of rotations and vibrations.
- Pressure is defined as the force per unit area exerted by a gas on its container, and is equal to two-thirds the kinetic energy density of the gas.
- Absolute zero can be defined on the basis of ideal gas behavior, and allows us to relate absolute temperature T (in Kelvin) to Celsius temperature t (in degrees Celsius), $T = t + 273.15$. Under almost all circumstance absolute temperature (Kelvin) is used in thermodynamics.

A frequently used approach in physical chemistry is to define an ideal system: a system that is an approximation to a real system, in which the properties can be simply explained. Real systems – those that are actually encountered in the lab or in Nature – are often complex and not fully understood. Our starting point is, therefore, to define a system we can describe exactly. We then determine how good our ideal system represents real systems, over what ranges the ideal system describes real systems accurately, and the limits of its applicability. We need to determine whether, even if it does not deliver quantitatively accurate results, it still describes trends accurately; and whether there are simple corrections that can be added on to enhance the quantitative agreement and range of applicability of the ideal system. Molecular interactions are complex. Therefore, we begin by considering the gas phase, in which molecules are far apart on average. This minimizes the importance of intermolecular interactions compared to condensed phases (liquids and solids). Indeed, our first assignment is to define an ideal gas in which there is a complete absence of intermolecular interactions. Here again we use the term *molecule* to refer to the particles in the gas, regardless of whether they are atoms or molecules.

An ideal gas has the following properties:

- The gas is composed of point-like molecules.
- At any finite temperature the molecules are in constant motion and collide elastically with each other and the walls of their container.
- There are no intermolecular interactions (neither attractions nor repulsions).
- An ideal gas follows the equation of state $pV = nRT$.

Physical Chemistry: How Chemistry Works, First Edition. Kurt W. Kolasinski.

Companion Website: www.wiley.com/go/kolasinski/physicalchemistry

The first point means that even though a sample of gas, which contains n moles of molecules, fills a volume V, each individual particle has zero volume. The gas fills a volume V because each molecule is in motion such that, given enough time, it explores the total extent of the volume available to it in a container. When the molecules collide with each other, or with the walls of the container, they exchange energy and momentum. There is no loss of energy when summed over the collision partners; hence, the collisions are *elastic*. There is conservation not only of energy but also of linear momentum. Because there are no intermolecular interactions, the only way in which ideal gas molecules influence each other is when they collide. Their energy and the energy of the system do not depend on how close or far apart the particles are from one another.

There are different types of gases based on what types of molecule are present. The particles may be atoms or molecules (we consider only neutral species here). A gas of atoms is also called a monatomic gas. The molecules can be classified by the number of atoms they have. We make three distinctions here. *Diatomics* contain two atoms, while *triatomics* and all gases with several atoms are *polyatomic* molecules. There are also 'big' molecules. Although 'big' is not well defined, what is meant here is polymers, proteins, DNA and other such molecules. These very large molecules can sometimes be put in the gas phase by laser desorption or electrospray techniques. Because these are special cases, we need not concern ourselves with big molecules in the gas phase for now. Rather, our discussion will revolve around monatomic, diatomic and polyatomic gases. The reason for dividing gases up along these lines will become apparent as soon as we consider how energy is partitioned in gases.

2.1 Ideal gas equation of state

We begin by discussing the properties of an ideal gas in the *thermodynamic limit*. The thermodynamic limit is defined as letting the volume approach infinity while maintaining a constant temperature (and chemical potential). More meaningfully for this discussion, it is the limit in which the size and shape of the container about the gas have no influence on its properties. There are many more molecules in the bulk of the gas than in contact with any surface, and the walls of the container are far apart compared to the size of the molecule. We shall see later that the influences of breaking such a limit are the hallmarks of nanoscience. An ideal gas obeys the *ideal gas law*

$$pV = nRT \tag{2.1}$$

where p is the pressure, V the volume, n the number of moles, R the gas constant, and T the temperature. Equation (2.1) is also an *equation of state*. It tells us how the state of the gas changes as the conditions of the gas change. It is a mathematical expression of the experimental fact that, for a given amount of any fluid (n moles of gas or liquid, ideal or not), any two of the variables p, V, and T constitute a complete set of independent variables. In other words, all fluids regardless of their microscopic structure have an equation of state $f(p, V, T) = 0$. As we shall see shortly, T is a proxy variable for molecular energy. Therefore, fluids other than ideal ones, which possess intermolecular interactions that influence their energy, will have different equations of state. Deviations from the ideal gas law will be greatest for those species that interact most strongly.

The ideal gas law can be derived from statistical mechanics. However, it can also be derived empirically and independently by developing an equation that is consistent with experimental results. The following experimental observations have been made in the limit of low pressure (the dilute gas limit), that is, they hold for ideal gases:

Boyle's law

$$p_1V_1 = p_2V_2 = \text{constant (for constant } n, T) \tag{2.2}$$

Charles's/Gay-Lussac's law

$$V_1/T_1 = V_2/T_2 \text{ (for constant } n, p) \tag{2.3}$$

$$p_1/T_1 = p_2/T_2 \text{ (for constant } n, V) \tag{2.4}$$

which implies

$$\frac{pV}{T} = \text{constant (for constant } n). \tag{2.5}$$

Avogadro's principle states that

$$V \propto n \text{ (for constant } p, T). \tag{2.6}$$

Therefore, we can construct an equation, the ideal gas law Eq (2.1), which is consistent with all of these conditions. The constant that makes the ideal gas law consistent is R, the *(universal) gas constant*. R is determined by finding the value of the following expression in the limit of low pressure

$$R = \lim_{p \to 0} \frac{pV}{nT} = 8.314 \text{ J mol}^{-1} \text{ K}^{-1} = 0.08206 \text{ L atm mol}^{-1} \text{ K}^{-1}. \tag{2.7}$$

Note that the Boltzmann constant is defined as the gas constant per molecule; thus, division by Avogadro's constant yields its value,

$$k_B = R/N_A = 1.381 \times 10^{-23} \text{ J K}^{-1}, \tag{2.8}$$

and the ideal gas law can be written either on a per mole or per molecule basis

$$pV = nRT = Nk_B T \tag{2.9}$$

Once we have the ideal gas law, the laws of Boyle and Charles and Gay-Lussac are superfluous as they can all be derived from $pV = nRT$. In other words, it is a good rule to learn the most fundamental equation and then manipulate it to derive other useful results. As an example, introducing the mass density ρ defined by

$$\rho = m/V = nM/V, \tag{2.10}$$

where m is the molecular mass, M the molar mass and V the volume, we can rewrite the ideal gas law as

$$\rho = pM/RT. \tag{2.11}$$

2.1.1 Example

At 500 °C and 93.2 kPa, the mass density of sulfur vapor is 3.710 kg m^{-3}. What is the molecular formula of sulfur under these circumstances?

Use the ideal gas law and the definition of density. First, rearrange the definition of density ρ

$$\rho = m/V = nM/V$$

to arrive at an expression for the molar mass M,

$$M = \rho(V/n).$$

Then rearrange the ideal gas law to obtain an expression for V/n

$$V/n = RT/p.$$

Now substitute the expression for V/n into the expression for M

$$M = \rho\frac{RT}{p} = (3710 \text{ g m}^{-3})\frac{8.314 \text{ J K}^{-1} \text{ mol}^{-1}(773 \text{ K})}{93200 \text{ Pa}} = 256 \text{ g mol}^{-1}.$$

The atomic weight of S is 32 g mol^{-1}; 256/32 = 8, so therefore the molecular formula is S_8.

2.1.2 Directed practice

Assuming that the molar mass is a constant, derive an expression for the density of an ideal gas from the ideal gas law as a function of temperature. Why might this formula break down for sulfur at temperatures much above 500 °C?

[Answer: $\rho = pM/RT$]

We should at this point pause and back up. To understand why an ideal gas follows the equation of state given above, we need to define the terms in it and see how these relate to molecular energy. In other words, we need to demonstrate what we really mean by pressure and define a scale for absolute temperature.

2.2 Molecular degrees of freedom

All gas-phase molecules in a macroscopic container, regardless of whether they are monatomic, diatomic or polyatomic, can translate in three directions. In other words, they have three translational *degrees of freedom*. In macroscopic containers at temperatures of interest to any form of usual chemistry, the translational degrees of freedom are completely classical. By that we mean that they are accurately described by the laws of classical mechanics. In principle, they can also possess electronic excitations, although electronic states are generally separated by large differences in energy. Usually, molecules are in their ground electronic state (the *ground state* is the lowest energy state). Only at very high temperatures are a significant number of molecules in electronic excited states (*excited states* are states that are at higher energy than the ground state), so we need not consider them further here. Similarly, we will only consider nuclear degrees of freedom when we deem them necessary, and will not discuss them in this general case. For a monatomic gas that is the end of the story as far as degrees of freedom are concerned.

Diatomic and polyatomic molecules possess bonds between atoms. Molecules are free to rotate in space and the bonds can vibrate. Each atom has three degrees of freedom before it enters the molecule. A molecule with N atoms must therefore have $3N$ degrees of freedom. Three of these degrees are translational, while the remaining $(3N - 3)$ degrees correspond to internal degrees of freedom of rotation and vibration. A diatomic molecule can only vibrate in one way, in that the bond can stretch and contract. The two remaining degrees are rotational degrees of freedom. Think of a pencil. You can rotate it in the plane of the table or perpendicular to the plane of the table. Any other orientation of rotation can be constructed as a combination of in-plane and out-of-plane components. Rotation along the internuclear axis is of no consequence. The same partitioning of three translations and two rotations holds for any linear polyatomic molecule, which leaves $(3N - 5)$ vibrational degrees of freedom. A nonlinear polyatomic molecule has three translational degrees as well as three rotational degrees of freedom (x, y, and z components to both translational and rotational motion). The remaining $(3N - 6)$ degrees are vibrational degrees of freedom. As the molecular size increases, the number of available vibrational states increases rapidly. As a first approximation, we will consider that the degrees of freedom are independent of each other.

Rotations and vibrations are inherently quantum mechanical on a molecular level. We will explore this in detail in later chapters, but for now it will suffice to describe the broad details of the models we use to describe rotating and vibrating molecules.

The first-order approximation to a rotating diatomic molecule is a rigid rotor (Fig. 2.1). A rigid rotor is composed of a massless rod of fixed length corresponding to the equilibrium bond distance R_e. This rod connects two point masses m_1 and

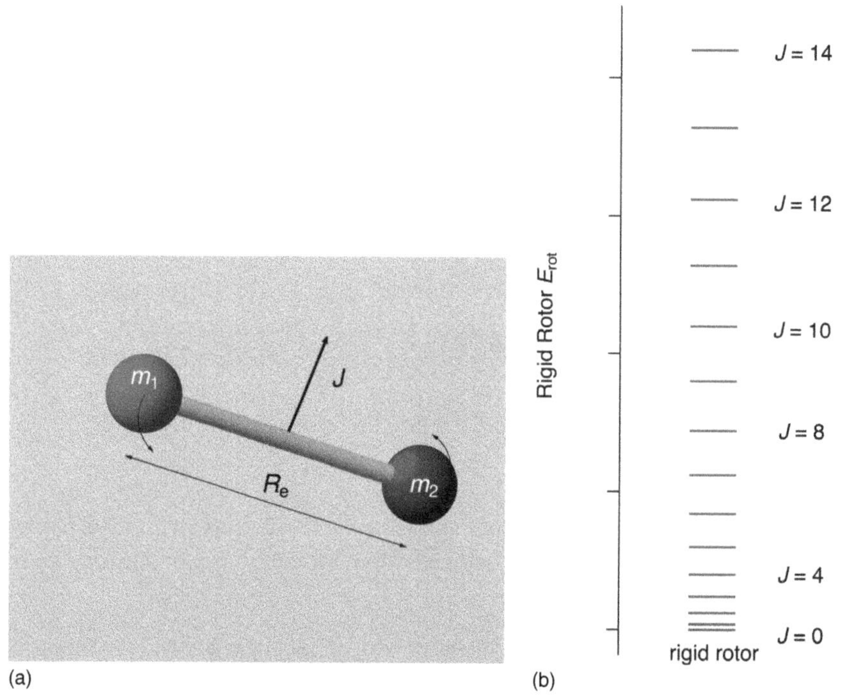

Figure 2.1 (a) Rigid rotor with rotational axis J and equilibrium bond distance R_e between atoms of mass m_1 and m_2. (b) The corresponding quantized energy level structure for states with rotational quantum number $0 \le J \le 14$.

m_2. Quantum mechanically, the rigid rotor is only allowed to attain certain values of energy; that is, the rotational energy level structure is quantized. So, why is rotation quantized but translation is not? Translation occurs on the distance scale of the container in which the molecule resides. As long as this container is large compared to the molecule, translation acts classically and has a continuous distribution of energy levels. As we shall see later, we can model rotation as the motion of a particle on a sphere (or spheroid) of molecular dimensions. Rotation is confined to a distance of molecular dimensions and is, therefore, quantized. Confinement to small dimensions is inherent for bringing out quantized behavior.

The energy level spacing is larger for lighter masses and shorter bonds. Thus, H_2 has much larger energy spacings than I_2. The energy of a rigid rotor is

$$E_{\text{rot}} = J(J+1)hcB \tag{2.12}$$

with a *rotational constant* B defined by

$$B = (\hbar/4c\mu R_{\text{e}}^2) \tag{2.13}$$

where $J = 0, 1, 2, \ldots$ is the rotational quantum number, h is Planck's constant, $\hbar = h/2\pi$ (pronounced h bar and called the reduced Planck constant), and c is the speed of light. (B is conventionally given in cm^{-1} rather than the SI unit of m^{-1}.) By using a value of $c = 3.00 \times 10^{10}$ cm s^{-1}, the unit cm is cancelled in Eq. (2.12) and the energy is given in J. Note that J only takes on discrete values as expected for a quantized variable, rather than a continuous distribution of values as would be expected for a classical variable. The reduced mass is defined by

$$\mu = \frac{m_1 m_2}{m_1 + m_2} \tag{2.14}$$

The energy level spacing between two successive levels is given by

$$\Delta E = E(J+1) - E(J) = 2hcB(J+1). \tag{2.15}$$

Since $B = 0.03737\ \text{cm}^{-1}$ and $60.85\ \text{cm}^{-1}$ for I_2 and H_2, respectively, these spacings are on the order of 0.1–100 cm^{-1} = 0.012–1.2 kJ mol^{-1} = 0.12–12 meV per molecule. This compares to $k_{\text{B}}T = 25$ meV at room temperature. Thus, at sufficiently high temperatures many rotational levels are populated in heavy molecules, and the distribution of molecules that populate these states is approximated by a classical continuous distribution. Diatomic hydrides and lighter molecules have significantly fewer, discretely populated levels (we will discuss this distribution in detail later). Equation (2.15) shows that the energy level spacing increases with increasing J. However, because space is also quantized (or at least the direction into which the rotational angular momentum is quantized), each level contains $(2J+1)$ levels at the same energy (corresponding to the $(2J+1)$ directions into which the rotational angular momentum vector can point). We say that each level has a degeneracy of $(2J+1)$ and that levels at the same energy are *degenerate*. Since both energy level spacing and degeneracy are increasing functions of J, we can ask how the density of states increases – that is, how does the number of states per unit energy change? The density of states $\rho(J)$ is defined (in the limit of large J) by

$$\rho(J) = (2J+1)\frac{\text{d}J}{\text{d}E} = \frac{2\mu R_{\text{e}}}{\hbar^2}. \tag{2.16}$$

Equation (2.16) shows that the rotational density of state is constant.

2.2.1 Directed practice

Perform the conversions of energy units to show that 0.1–100 cm^{-1} = 0.0012–1.2 kJ mol^{-1} = 0.012–12 meV.

The harmonic oscillator (Fig. 2.2) is the first approximation for the vibrational motion of a bond. The harmonic oscillator energy levels are quantized according to

$$E_{\text{vib}} = hc\tilde{\nu}_0\left(v + \tfrac{1}{2}\right) \tag{2.17}$$

where the fundamental wavenumber $\tilde{\nu}_0$ (again in cm^{-1}) depends on the force constant k and reduced mass μ according to

$$\tilde{\nu}_0 = \frac{1}{2\pi c}\left(\frac{k}{\mu}\right)^{1/2}. \tag{2.18}$$

The vibrational quantum number v again takes on integer values $v = 0, 1, 2 \ldots$. Larger values of k and $\tilde{\nu}_0$ correlate with stiffer, stronger bonds. Again, H_2 with the highest fundamental wavenumber $\tilde{\nu}_0 = 4400\ \text{cm}^{-1}$ is anomalously high. I_2

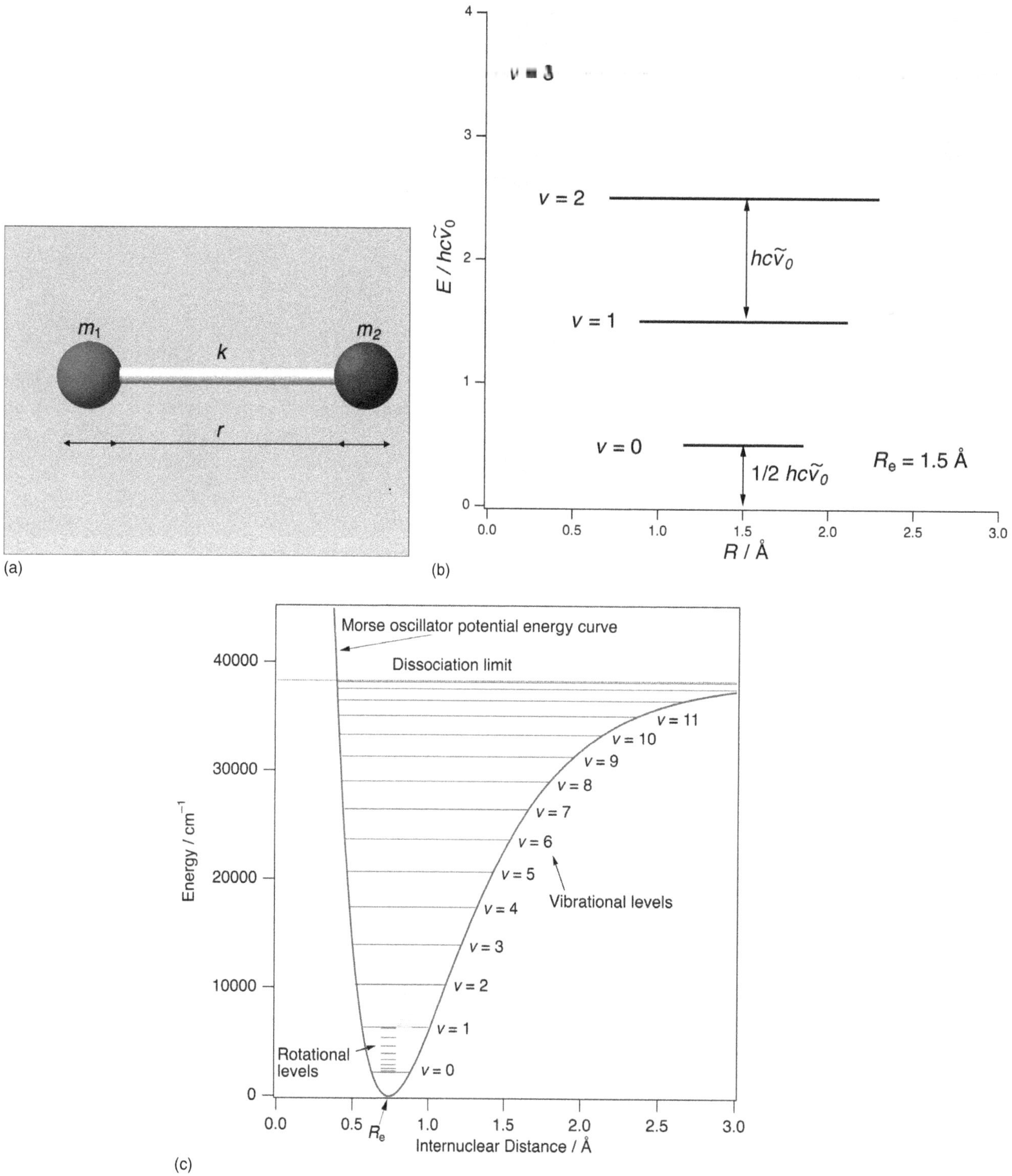

Figure 2.2 (a) An oscillator (i.e., a molecule) with atoms of mass m_1 and m_2 connected by a spring of force constant k at an equilibrium bond distance R_e. The energy level structure of a quantum mechanical (b) harmonic and (c) Morse oscillator. The Morse oscillator shown corresponds to the H_2 molecule. The rotational level spacings are much smaller in energy than the vibrational levels.

exhibits a fundamental wavenumber of only 215 cm^{-1}. Bonds tend to be harmonic in the ground state, but at higher and higher levels of excitation the energy levels are spaced progressively closer together and eventually the bond breaks. In other words, the density of states of a real oscillator increases with increasing vibrational energy. These aspects of anharmonicity are better represented by the Morse oscillator energy level structure shown in Fig. 2.2(c). To a first approximation, we consider each vibrational degree of freedom as corresponding to its own independent oscillator.

Analogous to the rotational energy distribution, there is a high-temperature limit in which vibrational energy will be well approximated by a classical continuous distribution. This is most likely to be reached for polyatomic molecules containing heavy atoms. Since vibrational energy spacings are typically much larger than rotational energy level spacings, quantum mechanical corrections to classical expectations are much more likely when vibrations need to be considered.

2.3 Translational energy: Distribution and relation to pressure

The *kinetic hypothesis* states that the molecules that make up matter are constantly in motion, even when body of matter is at rest. Here, we consider the translational motion of an ideal gas, also known as a *perfect gas*. That molecules are translating means that they have both *kinetic energy* ($=\frac{1}{2}mv^2$) and *linear momentum* (mv), both of which depend on the molecular mass m and its velocity v. Momentum-changing collisions of gas molecules with the container walls cause a force to be exerted on the walls. Pressure is a measure of this force per unit area. Thus, the mass and velocity of the molecules contribute to both the internal energy of the sample and its pressure. To simplify matters further we will assume that our perfect gas is not only colliding elastically (no internal energy changes including electronic excitation) and composed of noninteracting particles, but also that it is dilute (far from condensing) and composed of monatomic particles.

Consider a gas that is isolated from the rest of the universe (no exchange of energy or matter) such that the macroscopic properties are not changing over time. We state that this system is 'at equilibrium.' When averaged over time, its temperature (which we have yet to define) and pressure are constant. Both of these quantities are dependent on the velocity distribution of the gas molecules; therefore, the time average of the velocity distribution must also be constant. This does not mean that the velocity of each individual gas molecule is constant. On the contrary, the molecules are continually colliding with each other and exchanging momentum. However, because there is a huge number of molecules, say, on the order of Avogadro's number, the individual fluctuations experienced by each molecule average out. There will be fluctuations in the velocity distribution, but the magnitude of these fluctuations tends to be small compared to the mean value averaged over the whole distribution. This in turn means that the fluctuations in p and T are small compared to their mean values. At equilibrium, the gas is also uniformly distributed throughout the container and the container is sufficiently large such that any property measured is independent of the shape of the container or the position in the container at which a measurement is performed. Thus, for instance, Pascal's law – that the pressure is equal in all directions – holds at equilibrium.

The distribution of molecular velocities and positions is not only homogeneous (uniform in space) but also isotropic (uniform in all directions) at equilibrium. This means that there are no net flows in the gas, which requires that the average (indicated by angle brackets) of velocity vectors must vanish

$$\langle \mathbf{v} \rangle = \frac{1}{N}\sum_{i}^{N} \mathbf{v}_i = 0, \tag{2.19}$$

which implies

$$\sum_{i}^{N} \mathbf{v}_i = 0 \tag{2.20}$$

We now relate the pressure to the velocity. Take a cubic container with an edge length l, wall area $A = l^2$ and volume $V = l^3$. A Cartesian coordinate system is defined such that the normal of face S of the cube is the z axis, x and y are in the plane of face S. The walls scatter the ideal gas particles elastically such that the initial velocity vector $\mathbf{v}$ with components v_x, v_y, and v_z will at later times have components that are some combination of $\pm v_x$, $\pm v_y$, and $\pm v_z$. Once a molecule strikes S it will return only after it has struck the opposite wall. Thus, it will travel $2l$ along the z axis. The velocity component along z is v_z; thus, the time between collisions is $2l/v_z$ and the frequency of collisions is the inverse of this. During each collision the momentum change along the z direction is from $+mv_z$ to $-mv_z$ and a momentum of $2mv_z$ is exchanged with the wall. The total z component momentum change per unit time at wall S caused by a single molecule is then

$$\left.\frac{\mathrm{d}(mv_z)}{\mathrm{d}t}\right|_{\mathrm{S}} = 2mv_z\frac{v_z}{2l} = \frac{mv_z^2}{l}. \tag{2.21}$$

Assuming that all the molecules have the same velocity distribution (same velocity components $\pm v_x$, $\pm v_y$, and $\pm v$), then summing the contributions of N molecules yields

$$\sum_i^N \left.\frac{\mathrm{d}(mv_z)}{\mathrm{d}t}\right|_{\mathrm{S}} = \frac{Nmv_z^2}{l} = \frac{nVmv_z^2}{l} = nl^2mv_z^2. \quad (2.22)$$

where $n = N/V$ is the number density (note that the IUPAC recommendation for number density is n, which is also the recommendation for amount in moles). The derivative is the rate of change of the momentum, which by definition is the force exerted on wall S. Hence the pressure p, which is the force per unit area, is given by the force divided the area l^2 or

$$p = nmv_z^2. \quad (2.23)$$

In order for the pressure to be equal in all directions

$$v_x^2 = v_y^2 = v_z^2 = \tfrac{1}{3}\left(v_x^2 + v_y^2 + v_z^2\right) = \tfrac{1}{3}v^2. \quad (2.24)$$

Therefore, the pressure in terms of the speed of the molecules v is

$$p = \tfrac{1}{3}nmv^2 = \frac{2N}{3V}\left(\frac{mv^2}{2}\right) = \frac{2E_K}{3V} \quad (2.25)$$

and is equal to two-thirds the total kinetic energy density of the gas.

Pressure is a measure of the force per unit area exerted on the walls of a container. The SI units for pressure, force and area are pascal (Pa), newton (N) and per square meter (m^{-2}), hence,

$$\mathrm{Pa} = \mathrm{N\,m^{-2}} = \mathrm{kg\,m^{-1}\,s^{-2}}.$$

Hydrostatic pressure is the pressure exerted by a column of liquid. Given a liquid of mass density ρ and height h, the pressure is equal to

$$p = \frac{F}{A} = \frac{mg}{A} = \frac{\rho hAg}{A} = \rho hg. \quad (2.26)$$

The gravitational acceleration constant $g = 9.8067\ \mathrm{m\,s^{-2}}$.

2.3.1 Example

A manometer consists of a U-shaped tube containing a liquid. One side is connected to the apparatus and the other is open to the atmosphere. The pressure inside the apparatus is then determined from the difference in heights of the liquid. Suppose the liquid is water, the external pressure is 770 torr, and the open side is 10.0 cm lower than the side connected to the apparatus. What is the pressure in the apparatus in Pa? The density of water at 25 °C is 0.99707 g $\mathrm{cm^{-3}}$.

The external pressure is equal to the internal pressure plus the pressure exerted by the column of water (i.e., the hydrostatic pressure)

$$p_{\mathrm{ext}} = p_{\mathrm{int}} + p_{\mathrm{water}}.$$

Thus

$$p_{\mathrm{int}} = p_{\mathrm{ext}} - p_{\mathrm{water}}.$$

Now convert from torr to pascal and convert the height difference into hydrostatic pressure in Pa from the product of the mass density ρ, gravitational acceleration g and the height h.

$$p_{\mathrm{ext}} = 770\ \mathrm{torr}(101325\ \mathrm{Pa}/760\ \mathrm{torr}) = 102658\ \mathrm{Pa}$$

$$p_{\mathrm{water}} = \rho gh = 0.99707\ \mathrm{g\,cm^{-3}}\left(\frac{1\ \mathrm{kg}}{1000\ \mathrm{g}}\right)\left(\frac{100\ \mathrm{cm}}{1\ \mathrm{m}}\right)^3(9.807\ \mathrm{m\,s^{-2}})(1.10\ \mathrm{m}) = 978\ \mathrm{Pa}.$$

Then, rounding the answer to three significant figures, we obtain

$$p_{\mathrm{int}} = 102658 - 978 = 101680\ \mathrm{Pa} = 102\ \mathrm{kPa}.$$

Note that 1 torr = 1 mmHg of hydrostatic pressure. Since the density of Hg is much higher than the density of water, the hydrostatic pressure exerted by 1 mm of H_2O does not equal the hydrostatic pressure exerted by 1 mm of Hg.

2.4 Maxwell distribution of molecular speeds

The methods of statistical mechanics can be used to derive the exact form of the speed distribution. The fraction of molecules with speeds between v and $v + dv$ is given by

$$f(v)\,\mathrm{d}v = 4\pi v^2 \left(\frac{m}{2\pi k_B T}\right)^{3/2} \exp\left(\frac{-mv^2}{2k_B T}\right) \mathrm{d}v. \tag{2.27}$$

This function is known as the *Maxwell speed distribution*, and is one of the most important equations in physics and chemistry. It is fundamental to the understanding of chemical kinetics and the thermodynamics of matter. The universality and simplicity of the shape of the distribution can be emphasized by expressing it in terms of the reduced variable defined by

$$y = (m/2k_B T)^{1/2} v \tag{2.28}$$

Thus

$$f(y)\,\mathrm{d}y = \frac{4}{\sqrt{\pi}} y^2 e^{-y^2}\,\mathrm{d}y. \tag{2.29}$$

As long as the interaction energy of any two molecules does not depend on the velocity, this distribution is valid for not only ideal gases but also real gases and even condensed phases. You should familiarize yourself with the shape of this curve and how it changes with m and T. This is illustrated in Figs. 2.3 and 2.4.

The Maxwell distribution vanishes at both $v = 0$ and $v = \infty$ with a maximum in between. The curve is much narrower at low temperature. However, it is not symmetrical: it is skewed to high speeds, and therefore the mean value does not correspond to the most probable value. The *most probable speed* is found by setting $\mathrm{d}f(v)/\mathrm{d}v = 0$, which gives

$$v^* = (2k_B T/m)^{1/2} \tag{2.30}$$

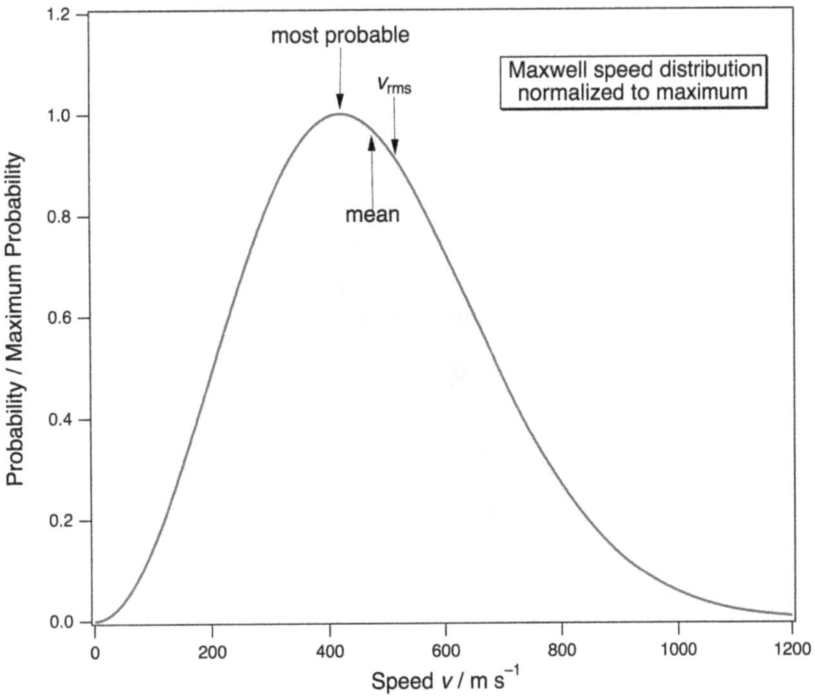

Figure 2.3 The Maxwell distribution and the definition of the most probably speed, the mean speed and the root mean square speed.

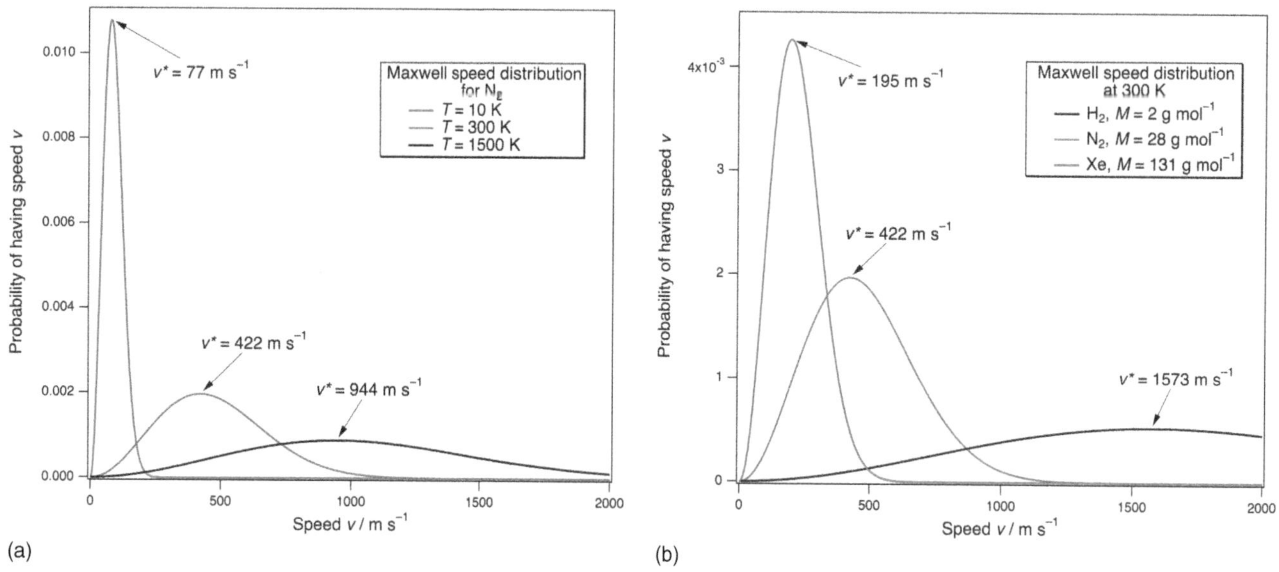

Figure 2.4 The dependence of the Maxwell distribution (a) as a function of temperature for N_2 and (b) at 300 K as the mass changes for H_2, N_2, and Xe.

The mean value is found by using the standard definition of a mean from statistics: the mean of a variable is equal to the integral of the product of the variable and its probability distribution divided by the integral of the product. Accordingly, the *mean speed* is

$$\langle v \rangle \equiv \frac{\int v f(v)\, dv}{\int f(v)\, dv} = \left(\frac{8k_B T}{\pi m}\right)^{1/2} \tag{2.31}$$

The quotient of integrals is known more generally as the first mode of the distribution $f(v)$. The root mean square speed is found by first taking the second mode of the distribution, that is, the mean of v^2

$$\langle v^2 \rangle \equiv \frac{\int v^2 f(v)\, dv}{\int f(v)\, dv} = \left(\frac{3k_B T}{m}\right) \tag{2.32}$$

and then its square root, which as expected is greater than $v*$

$$\langle v^2 \rangle^{1/2} = v_{\text{rms}} = \left(\frac{3k_B T}{m}\right)^{1/2} = \left(\frac{3}{2}\right)^{1/2} v^*. \tag{2.33}$$

2.5 Principle of equipartition of energy

Having derived an expression for the mean square velocity in Eq. (2.32), we can now evaluate the *mean kinetic energy* of a molecule

$$\tfrac{1}{2} m \langle v^2 \rangle = \tfrac{3}{2} k_B T. \tag{2.34}$$

We know that the velocity distribution of an ideal gas must be isotropic. Therefore,

$$\langle v^2 \rangle = \left\langle v_x^2 + v_y^2 + v_z^2 \right\rangle = \langle v_x^2 \rangle + \left\langle v_y^2 \right\rangle + \langle v_z^2 \rangle \tag{2.35}$$

with

$$\langle v_x^2 \rangle = \left\langle v_y^2 \right\rangle = \langle v_z^2 \rangle . \tag{2.36}$$

Thus, for each translational degree of freedom, the mean kinetic energy is

$$\tfrac{m}{2} \langle v_x^2 \rangle = \tfrac{m}{2} \left\langle v_y^2 \right\rangle = \tfrac{m}{2} \langle v_z^2 \rangle = \tfrac{1}{2} k_B T. \tag{2.37}$$

The result in Eq. (2.37) is true for any type of molecule in a fluid phase as long as classical mechanics applies. No need to worry about exceptions to this rule unless, for example, we place the molecule in a box that is comparable in size to molecular dimensions. Equation (2.37) is an expression of the *principle of equipartition of energy*, which states that each degree of freedom of mechanical motion subject to classical mechanics is partitioned an equal amount of the molecule's energy in the amount of $\frac{1}{2}k_B T$. For a monatomic ideal gas in the absence of electronic excitations, there are only translational degrees of freedom. Therefore, the energy available to the gas, the *internal energy* of the gas U – that is, the energy which is to be partitioned among the degrees of freedom available to the gas is equal to its kinetic energy

$$\text{(monatomic ideal gas molecule)} \quad U = E_k = \tfrac{3}{2}k_B T. \tag{2.38}$$

If we want to express this on a molar basis – that is, the internal energy per mole of gas – we need to multiple by Avogadro's number

$$\text{(monatomic ideal gas, mol}^{-1}\text{)} \quad U_m = \tfrac{3}{2}N_A k_B T = \tfrac{3}{2}RT, \tag{2.39}$$

where R is the gas constant. Here, we should pause to consider an important implication of Eq. (2.39). If we had a mole of gas that had no translational energy at all (no internal energy and therefore at an initial temperature $T_i = 0$) and then heated it up to a final temperature T_f, the heat we add to the molecules is partitioned among the available degrees of freedom. The capacity of the gas to absorb that heat is directly related to the number of degrees of freedom available to the molecule. We have already encountered the heat capacity in Chapter 1, and in Chapters 5 and 7 we will define it thermodynamically as the heat capacity at constant volume, which from Eq. (2.39) we can anticipate is

$$\text{(monatomic ideal gas, mol}^{-1}\text{)} \quad C_{V,m} = \tfrac{3}{2}R, \tag{2.40}$$

What if we have a diatomic or polyatomic molecule? In these cases there are $(3N - 3)$ rotational and vibrational degrees of freedom that must be considered. If rotations were behaving classically, $\frac{1}{2}k_B T$ would be partitioned to each rotational degree of freedom. At a sufficiently high temperature that can be defined by the use of statistical mechanics, we expect rotations to act classically and the expected partitioning is found. At a temperature which is low compared to room temperature, we must proceed with an appropriate quantum mechanical description. Therefore, there is a temperature dependence to the partitioning of energy into rotations, although at room temperature and above this partitioning is well approximated by the classical result.

Vibrations act somewhat differently. Vibrational motion is confined by a parabolic potential (or close to parabolic in a Morse oscillator). Therefore, there are both kinetic and potential energy contributions for each vibrational degree of freedom. Each vibrational mode would be partitioned $k_B T$ if it were behaving classically. However, we should anticipate that this high-temperature limit, which can again be defined with statistical mechanics and is significantly above room temperature, is only approached slowly. Each vibrational mode will have to be treated separately. Therefore, the approximation that vibrations act classically is usually a poor one. A quantum mechanical treatment of their contribution to energy partitioning must always be made for accurate results.

Here, we should also anticipate that the heat capacity of a molecule has contributions from all available degrees of freedom. Thus, if translational, rotational, vibrational and electronic degrees of freedom can contribute then

$$C_{V,m} = C_{V,m}^{\text{trans}} + C_{V,m}^{\text{rot}} + C_{V,m}^{\text{vib}} + C_{V,m}^{\text{elec}}, \tag{2.41}$$

and the heat capacity should depend on both temperature and the phase of the material.

2.6 Temperature and the zeroth law of thermodynamics

So what is this thing we call temperature? It is so important that, as seen in Eq. (2.38), it uniquely defines the internal energy of an ideal gas. In Eq. (2.27) along with m, it uniquely defined the velocity distribution. Furthermore, it plays a role in determining almost every chemical property of a system. So fundamental and yet we can now use simple reasoning and the results we have already laid out to establish a basis for thermodynamics. In order to uniquely define temperature, we need to state the *zeroth law of thermodynamics* as follows:

Two systems, each separately in thermal equilibrium with a third system, are in thermal equilibrium with each other.

A restatement of the zeroth law is:

There exists a property called temperature. Any two systems in thermal equilibrium with each other have the same temperature.

As we have seen above, all gases follow an equation of state $f(p, V, T)$ such that, if we specify any two, the third quantity is fixed. Further, we know that at low enough pressure all gases act ideally and, therefore, all gases follow the same equation of state at low pressure. The most easily measured quantities of a gas are its pressure and volume. Thus, we will now explore how these quantities can be used to define temperature because it must be proportional to one of these quantities if the other is held constant.

A constant-volume thermometer can be constructed by building a system that exists at one combination of p, V, and T to act as a reference. The triple point of water is one such system. A fixed volume of water vapor can coexist with ice and liquid water at only one combination of pressure and temperature. The equilibrium pressure at the triple point we call p_3, and the corresponding temperature T_3. At constant volume the temperature is proportional to pressure, $T(p) \propto p$. The pressure of the same volume of gas at different temperatures is used to define the temperature scale. In other words, since we know that the pressure at the triple point is p_3 and the corresponding temperature is T_3, the functional form of $T(p)$ is

$$T(p) = (p/p_3)T_3. \tag{2.42}$$

The temperature at the triple point of water is an arbitrary constant. However, to make the perfect gas temperature scale consistent with the temperature interval of the Celsius scale, and to ensure that the freezing point of pure water corresponds to not only 0 °C but also 273.15 K, T_3 was set to 273.16 K (where K is the abbreviation for the unit kelvin) by the ratification of the International Practical Temperature Scale of 1968 (IPTS-68).

Similarly, the triple point volume V_3 of a given mass of water held at T_3 and a constant pressure p could be measured. The constant pressure temperature scale is then defined by

$$T(V) = (V/V_3)T_3. \tag{2.43}$$

While these two temperature scales depend on the amount of gas as well as the chemical identity of the gas, they approach the same limit when the gas becomes dilute enough. In other words

$$\lim_{\delta p \to 0} T(p) = \lim_{\delta V \to \infty} T(V) = T^*. \tag{2.44}$$

While this type of thermometer is impractical for regular temperature measurements, it does allow for the construction of a primary standard according to which standard temperature points are set (see Table 2.1). In turn, these fixed points allow for the calibration of thermometers based on the resistance of a Pt wire (13.81 K $\leq T \leq$ 903.89 K), a Pt/PtRh thermocouple (903.89 K $\leq T \leq$ 1337.58 K) or black-body radiation ($T >$ 1337.58 K). These secondary standards are then used directly to calibrate conventional thermometers.

By this convention, we can set the conversion factor between the absolute temperature T in kelvin and the Celsius temperature t

$$T = t + 273.15. \tag{2.45}$$

Table 2.1 Fixed point defined by International Practical Temperature Scale of 1968 (IPTS-68). Normal boiling points and melting points measured at 1 atm are referred to.

Reference point	T/K	t/°C
Triple point of equilibrium H_2	13.81	−259.34
Boiling point of equilibrium H_2	20.28	−252.87
Boiling point of Ne	27.102	−246.048
Triple point of O_2	54.361	−218.789
Boiling point of O_2	90.188	−182.962
Triple point of H_2O	273.16	0.01
Boiling point of H_2O	373.15	100.00
Freezing point of Zn	692.73	419.58
Freezing point of Ag	1235.08	961.93
Freezing point of Au	1337.58	1064.43

2.7 Mixtures of gases

Mixtures of ideal gases (or sufficiently dilute real gases) act as if each component is independent. Each component fully occupies the available volume and shares the same temperature at equilibrium. The total pressure is the sum of the *partial pressures*. The partial pressure of component j, p_j, and the total pressure p are then defined as

$$p_j = x_j p \tag{2.46}$$

$$p = \sum_j p_j = p_1 + p_2 + \ldots \tag{2.47}$$

where x_j is the *mole fraction* of component j

$$x_j = n_j/n \tag{2.48}$$

$$n = \sum_j n_j, \tag{2.49}$$

where n is the total number of moles and n_j the moles of component j. Note that completeness requires

$$1 = \sum_j x_j \tag{2.50}$$

and therefore only $j - 1$ of the mole fractions are independent. In other words, in a binary mixture, if we know x_1 then x_2 is set and found from

$$x_2 = 1 - x_1 \tag{2.51}$$

In a mixture of ideal gases, the ideal gas law can be used for each component individually

$$p_j = n_j RT/V. \tag{2.52}$$

2.7.1 Directed practice

Given that the density of air at 0.987 bar and 27 °C is 1.146 kg m^{-3}, calculate the mole fraction and partial pressure of nitrogen and oxygen assuming that air consists only of these two.

[Answers: 0.244 and 0.240 bar for O_2 and 0.756 and 0.747 bar for N_2]

2.8 Molecular collisions

Molecular collisions are required for the initiation of virtually all thermal chemistry and electrochemistry. Thermal chemistry means chemistry initiated by thermal energy (heat) rather than, for instance, photochemistry, which is initiated by the absorption of a photon. Electrochemistry is chemistry involving the exchange of free charge (usually electrons but sometimes also accompanied by proton transfer). Molecular collisions are also often involved in photochemical mechanisms in the steps that follow primary photon absorption. Collisions are the primary mechanism of energy exchange between molecules (radiation emission and absorption again being other paths). Therefore, calculation of the rate of molecular collisions is fundamental to the understanding of thermal equilibration and the rates of chemical reactions.

With the aid of Fig. 2.5, a molecular collision and the collision rate can be defined as follows. Assume that molecules are spherical with diameter d_1 and d_2 for two different types of molecules with masses m_1 and m_2. A collision occurs anytime the two molecules' straight-line distance of closest approach is equal to or less than the sum of the molecular radii. Referring again to Fig. 2.5, think of molecule 2 moving through a static random array of molecules 1. At any time molecule 1 encounters a molecule 2 that is within a disk of area defined by

$$\sigma = \pi d^2, d = \tfrac{1}{2}(d_1 + d_2), \tag{2.53}$$

a collision occurs. Here, d is the *collision diameter* and σ is the collision *cross section*. Let $\langle v_{12} \rangle$ be the mean relative speed of the two molecules. Then in a time t, molecule 2 would on average traverse a distance $\langle v_{12} \rangle t$ and sweep out a volume $\sigma \langle v_{12} \rangle t$. Any other molecule in the cylinder defined by the volume $\sigma \langle v_{12} \rangle t$ collides with molecule 2. The number of collision partners in the cylinder is just the density of collision partners times the volume of the cylinder. We then sum over all molecules of type 2 within the volume of container and divide by the time t to obtain the collision density – that

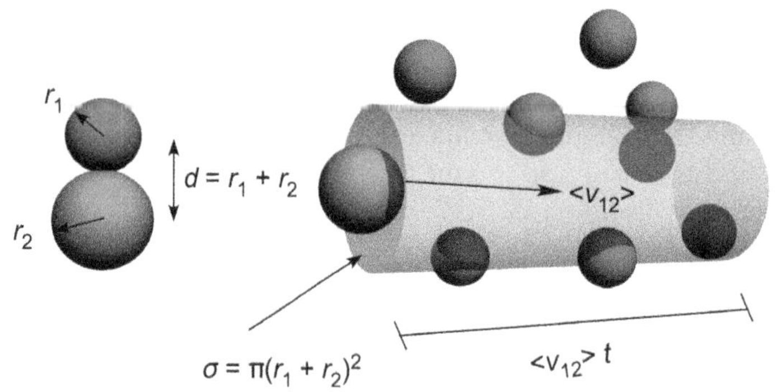

Figure 2.5 Molecules 1 and 2 with radii r_1 and r_2 collide whenever they come into contact. Assuming they are hard spheres and that molecule 2 moves with the mean relative velocity $\langle v_{12}\rangle$ through a gas of stationary molecules 1, a collision occurs whenever molecule 1 lies within a cylinder defined by a circle projected along the relative velocity vector. The cross-sectional area of the circle defines the collision cross-section $\sigma = \pi(r_1 + r_2)$. The length of the cylinder traversed by molecule 2 (assuming it is not deflected) in time t is $\langle v_{12}\rangle t$.

is, the number of collisions per unit time and volume. The *collision density* (or *total collisional frequency*), has two different components (assuming there are just two gases in the mixture). For a mixture of two gases with number densities ρ_1 and ρ_2, Z_{12} is the number of collisions between molecule 1 and molecule 2 per unit time and volume,

$$Z_{12} = \sigma\langle v_{12}\rangle\rho_1\rho_2. \tag{2.54}$$

Z_{11} is the collision density of a gas with molecules of the same type

$$Z_{11} = \sigma\langle v_{11}\rangle\rho_1^2. \tag{2.55}$$

Assuming ideal gas behavior, with a randomized gas phase and that the relative mean speeds $\langle v_{12}\rangle$ and $\langle v_{11}\rangle$ are described by Maxwell distributions, we can substitute from Eq. (2.31) for the mean speed. Instead of the mass in the expression for mean speed, we need the reduced mass μ of the collision partners to obtain the *relative mean speed of collision*. Hence, for two different masses m_1 and m_2

$$\langle v_{12}\rangle = \left(\frac{8k_B T}{\pi\mu}\right)^{1/2} \tag{2.56}$$

whereas for the collision of two molecules of mass m_1, $\mu = m_1^2/2m_1 = m_1/2$

$$\langle v_{11}\rangle = \left(\frac{8k_B T}{\pi\,(m_1/2)}\right)^{1/2} = \sqrt{2}\left(\frac{8k_B T}{\pi m_1}\right)^{1/2} = \sqrt{2}\,\langle v\rangle. \tag{2.57}$$

Use the ideal gas law to convert from density to pressure to obtain

$$Z_{12} = \sigma\left(\frac{8k_B T}{\pi\mu}\right)^{1/2} p_1 p_2 \left(\frac{N_A}{RT}\right)^2, \tag{2.58}$$

where N_A is the Avogadro constant. The self-collision density is

$$Z_{11} = \frac{1}{\sqrt{2}}\sigma\left(\frac{8RT}{\pi M_1}\right)^{1/2}\left(\frac{p_1 N_A}{RT}\right)^2. \tag{2.59}$$

For N_2 (d = 370 pm) at room temperature and pressure, $Z_{11} = 5 \times 10^{34}\ m^{-3}\ s^{-1}$.

The *mean free path* is the distance that a molecule travels on average between collisions. Dividing the distance traveled by one molecule in time t by the number of collisions made by that molecule in that time, we obtain

$$\lambda = \frac{\langle v\rangle t}{\sigma\langle v_{11}\rangle\rho t} = \frac{k_B T}{\sqrt{2}\,\sigma p}. \tag{2.60}$$

If the velocity distributions differ from those described by a randomized Maxwell distribution, then a more rigorous derivation of the collision densities must be used. This can be found, for example, in the book by Berry, Ross, and Rice (see Further Reading).

2.8.1 Directed practice

If the mean free path of the molecules in air at standard ambient conditions is approximately 1.5×10^{-7} m and the mean velocity of a molecule is 450 m s^{-1}, estimate the mean time between collisions.

[Answer: 3.3×10^{-10} s]

SUMMARY OF IMPORTANT EQUATIONS

$pV = nRT = Nk_BT$	Ideal gas law: n = moles, N = number of molecules
$E_{rot} = J(J+1)hcB$	Energy of a rigid rotor with rotational quantum number J and rotational constant B
$\mu = \dfrac{m_1 m_2}{m_1 + m_2}$	Reduced mass of system of two particles of mass m_1 and m_1
$E_{vib} = hc\tilde{\nu}_0\left(v + \frac{1}{2}\right)$	Vibrational energy of a harmonic oscillator
$f(v)\,dv = 4\pi v^2 \left(\dfrac{m}{2\pi k_B T}\right)^{3/2} \exp\left(\dfrac{-mv^2}{2k_B T}\right) dv$	Maxwell speed distribution
$\langle v\rangle = \left(\dfrac{8k_BT}{\pi m}\right)^{1/2}$	Mean speed of a Maxwell distribution
$\frac{1}{2}m\langle v^2\rangle = \frac{3}{2}k_BT$	Mean kinetic energy
$U = E_k = \frac{3}{2}k_BT$	Internal energy of a monatomic ideal gas is equal to its mean kinetic energy
$C_{V,m} = \frac{3}{2}R$	Heat capacity at constant volume of a monatomic ideal gas
$p_j = x_j p$	Partial pressure of gas component j
$x_j = n_j/n$	Mole fraction of component j
$\sigma = \pi d^2, d = \frac{1}{2}(d_1 + d_2)$	Collision cross section in terms of the collision diameter for spherical molecules of diameters d_1 and d_2
$Z_{12} = \sigma\left(\dfrac{8k_BT}{\pi\mu}\right)^{1/2} p_1p_2\left(\dfrac{N_A}{RT}\right)^2$	Collision density between molecules 1 and 2
$Z_{11} = \dfrac{1}{\sqrt{2}}\sigma\left(\dfrac{8RT}{\pi M_1}\right)^{1/2}\left(\dfrac{p_1N_A}{RT}\right)^2$	Self-collision density
$\lambda = \dfrac{k_BT}{\sqrt{2}\sigma p}$	Mean free path λ in terms of the collision cross-section σ

Exercises

2.1 Describe a method to determine the molar mass M of a real gas based on the ideal gas law.

2.2 Show that the spacing between rotational energy levels is given by Eq. (2.15).

2.3 Calculate the density of states for a rigid rotor. That is, derive Eq. (2.16) in the limit of large J.

2.4 Derive Eq. (2.58) from Eq. (2.54).

2.5 Using a hard-sphere potential it can be shown that Helmholtz energy at a given temperature T and volume V is

$$A = A_0 - RT\left(\ln V - 2\pi N_A\sigma^3/3V_m\right)$$

where A_0 is the Helmholtz energy at $T = 0$ K, V_m is the molar volume, and σ is a hard-sphere diameter of the molecule. Use the definition of pressure given by $p = -(\partial A/\partial V)_T$ to derive the ideal gas law.

2.6 Approximately 5 l of CO_2 at standard ambient temperature and pressure (SATP) can escape from a typical 0.75 l champagne bottle. (a) Calculate the corresponding pressure of CO_2 if it were the only component in a bottle of this volume at 7 °C. (b) A typical bubble is 0.5 mm in diameter. Calculate the number of bubbles that can be produced from the number of moles of CO_2 determined in part (a) if the bubbles have an effective pressure of 101 kPa and a temperature of 7 °C.

2.7 The global mean pressure of the atmosphere at the surface of the Earth is $p_S = 98.4$ kPa. The radius of the Earth is 6400 km. Calculate the mass of the atmosphere and the quantity of gas in the atmosphere.

2.8 (a) Which molecular degree of freedom is most likely to behave classically? (b) Which molecular degree of freedom is most likely to behave quantum mechanically?

2.9 Is energy equally partitioned into all molecular degrees of freedom?

2.10 Calculate the mean speed, number density, collision density and mean free path of Ar ($d = 2.86$ Å) at (a) 300 K and 1 bar and (b) 1000 K and 10^{-4} Pa.

2.11 Calculate the mean free path at 300 K of N_2 at pressures of 1 bar (roughly atmospheric pressure), 10 Pa (roughly a vacuum attainable with a mechanical vacuum pump) and 10^{-8} Pa (ultrahigh vacuum requiring for example a turbomolecular pump and baking of the vacuum system).

2.12 A vessel of volume 22.4 dm^3 contains 2.0 mol H_2 and 1.0 mol N_2 at 273.15 K. Calculate: (a) the mole fractions of each component; (b) their partial pressures; and (c) their total pressure.

2.13 At constant temperature, the mean translational energy of the molecules in an ideal gas is constant when averaged over time. At constant temperature, the translational energy of any molecule is not constant. Discuss how these two statements are simultaneously true.

2.14 Given that the density of air at 0.987 bar and 27 °C is 1.146 kg m^{-3}, calculate the mole fraction and partial pressure of nitrogen and oxygen assuming that: (a) air consists only of these two gases; (b) air also contains 1.0 mol percent Ar.

2.15 For ethane at 298 K, what fraction of molecules has a speed between 800 and 1000 m s^{-1}? What is this fraction at $T = 1000$ K? Draw approximately by hand what the limits and distributions must look like.

2.16 At a FIFA World Cup approximately 2500 balls will be used. Their circumference is 69 cm and they are inflated to 0.8 bar above atmospheric pressure. Calculate the mass of N_2 that would be required to fill all of these balls at $T = 25$ °C.

Further reading

Berry, R.S., Rice, S.A., and Ross, J. (2000) *Physical Chemistry*, 2nd edition. Oxford University Press, New York.

CHAPTER 3

Non-ideal gases and intermolecular interactions

PREVIEW OF IMPORTANT CONCEPTS

- The cause of non-ideal behavior is intermolecular interactions.
- There are a number of interactions that change the energy of molecules and ensembles of molecules. Many are electrostatic in nature and therefore depend on the distance between particles.
- Anything that takes a molecule away from the condition 'neutral with a spherical charge distribution' will lead to increased intermolecular interactions.
- A multipole expansion can be used to approximate the charge distribution and interaction energies between molecules as long as their orbitals are not overlapping.
- Some examples of long-range interactions include charge–charge, charge–dipole, charge-induced dipole, dipole-induced dipole, and higher multipole interactions.
- Short-range interactions include chemical bonding, hydrogen bonding and Pauli repulsion.
- A virial expansion represents an approach to represent the equation of state of a real gas.
- Another approximation to the equation of state of a real gas is the van der Waals equation.
- Consequences of non-ideal behavior include not only a change in the equation of state but also that gases can condense into condensed phases with finite volumes.
- A supercritical fluid exists above the critical temperature T_c and cannot be condensed merely by compression at or above T_c.
- If the pressure, temperature and volume are expressed as reduced properties, then the equation of state is the same for all gases. The reduced properties are obtained by dividing the property by its critical value. This is known as the law of corresponding states.

3.1 Non-ideal behavior

Whereas all gases act ideally at low pressure where they are on average far from each other, all gases act non-ideally as they approach conditions that promote condensation. At low enough T and high enough p, real gases should condense into a liquid. Indeed, at 1 atm and low enough T every element except He condenses into a liquid. He condenses into a solid at 1 atm and requires $p = 25$ atm to form a liquid. However, if we examine the ideal gas law in these limits we find

$$\lim_{p\to\infty} \frac{nRT}{p} = 0 \tag{3.1}$$

$$\lim_{T\to 0} \frac{nRT}{p} = 0 \tag{3.2}$$

Therefore an ideal gas would condense into nonexistence. This is an artifact of assuming that the particles are point-like with no intrinsic volume.

If molecules were neutral, nonpolarizable hard spheres, there would be extremely limited deviations from ideal gas behavior. We can add a term to the ideal gas law to account for molecular volume. However, the revised equation would predict that all molecules would condense at $T = 0$ K, rather than having the molecule-specific boiling point T_b that we know exists. The more molecules deviate from neutral nonpolarizable spheres, the more we expect them to act non-ideally. And since the intermolecular interactions increase in strength the closer molecules are to one another, the greater the deviations. Therefore as density increases, we expect deviations from ideal behavior to increase.

Physical Chemistry: How Chemistry Works, First Edition. Kurt W. Kolasinski.

Companion Website: www.wiley.com/go/kolasinski/physicalchemistry

3.2 Interactions of matter with matter

Since interactions are the root cause of non-ideal behavior, now is a good time to review a few qualitative trends from the Periodic Table that will help us to think about interatomic and intermolecular interactions. As we move to the right and down in the periodic chart, the atomic number Z (= positive charge on the nucleus) increases as does the molar mass M (or the mass of a single atom m). The radius of the atom also tends to increase because the number of electrons in the neutral atom must equal Z (there are notable exceptions as in the case of the lanthanide contraction). As a general rule, the heavier the atom, the stronger its interatomic interactions.

The final column of the periodic chart is a bit different from the rest. These are the noble gases. They only partake in chemical bonding in special cases, if at all, because they have filled electronic shells. We will discuss electronic configurations and orbitals in detail later. You are already familiar with these concepts in a descriptive sense; however, we will eventually describe them with a quantum mechanical treatment. Atoms become bigger as more electrons are added because only two electrons can fit in a given orbital. As subsequent orbitals are added to accommodate more electrons, the orbitals have progressively larger radial extents.

The force that holds the electrons in an atom is the attractive electrostatic interaction between the positive nucleus and the negatively charged electron. The form of the potential energy curve $V(r)$ is shown in Fig. 3.1. The negative value indicates an attractive interaction. The interaction energy falls off as $1/r$, in other words, it goes to zero at large electron–nucleus separation r according to

$$V(r) = -Ze^2/4\pi\varepsilon_0 r \tag{3.3}$$

where e is the elementary charge and ε_0 is the vacuum permittivity. The closer the electron gets to the nucleus, the lower the energy. The conundrum of why the electron does not simply spiral into the nucleus can only be answered quantum mechanically. The farther away the electron is from the nucleus, the more weakly it is bound.

All atoms in their ground state are neutral and spherically symmetric. This symmetry means that they are nonpolar. In order to induce a dipole in them, the charge distribution must be distorted from its spherical symmetry: the atom must be polarized. Higher Z atoms are more polarizable than lighter atoms for two reasons. First, since small orbitals close to

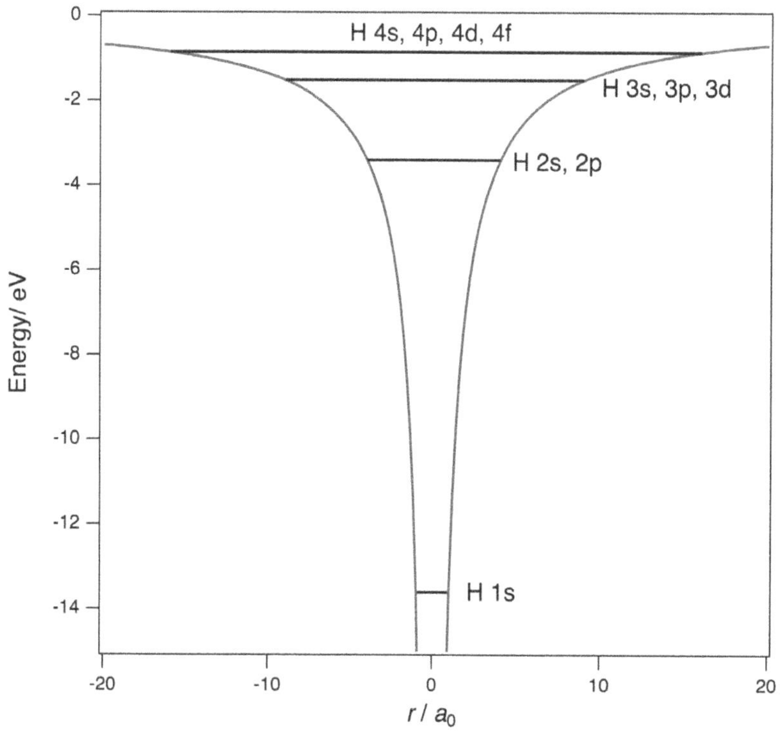

Figure 3.1 A representation of the electrostatic potential that holds an electron in the H atom. The electrostatic interaction that binds the electron to the nucleus behaves classically (obeys classical physics); however, prediction of the energy level structure requires quantum mechanics. In the H atoms, to a first approximation, all electrons with the same principle quantum number are degenerate (at the same energy); whereas this is not true for atoms with more than one electron. a_0 is the Bohr radius, the atomic unit of length.

the nucleus fill first and larger orbitals farther from the nucleus fill last, the outer electrons tend not to experience the full nuclear charge. This effect is called *screening*: the reduction of the effective charge that interacts with a particle of interest as a result of intervening charge. Second, since the outer electrons are further away, they have a lower attraction to the nucleus from the $1/r$ dependence of the electrostatic interaction. Polarizability α is proportional to the volume V occupied by the electronic states of the system (atom or molecule)

$$\alpha = 4\pi\varepsilon_0 V \tag{3.4}$$

where ε_0 is the vacuum permittivity.

The implication for the noble gases is clear. In going from He to Xe and then Rn, the polarizability will increase substantially. Xe is much more susceptible to spontaneous or induced fluctuations in its electron cloud that push it away from spherical symmetry. These fluctuations lead to transient dipoles that can interact with one another and change the energy of the gas. The effects are quite small for He, but reasonably large for Xe. The trend holds for the rest of the periodic chart: as Z increases down a column or across a row, so does polarizability. This is reflected in the physical state of elements at room temperature and pressure in that gases tend to be located in the upper right, and solids are formed as one moves down in the Periodic Table. The only elements that are liquid or close to being liquid at standard ambient conditions are Ga (T_f = 29.77 °C), Hg (T_f = −38.83 °C) and Br_2 (T_f = −7.2 °C). Phosphorus (T_f = 44.1 °C) and the alkali metals, for instance Cs (T_f = 28.5 °C) also have low melting points but are pyrophoric and not encountered in their elemental form (for long) when exposed to air. Recall also that the halogens as well as H_2, N_2, and O_2 exist as diatomic molecules at standard ambient temperature and pressure.

The relationship between polarizability and electronic state volume also holds for molecules. Thus, a small molecule such as H_2 has $\alpha = 0.804 \times 10^{-30}$ m^3 while that of CO is 1.95×10^{-30} m^3 and benzene, with its extended π-bonding system, has $\alpha = 10.3 \times 10^{-30}$ m^3.

There are a number of ways that matter interacts with other matter. The forces of Nature are electromagnetism, the strong force, the weak force and gravity. Electrical and magnetic forces were unified, that is, shown to be fundamentally related by Maxwell's theory of electromagnetism. Most intermolecular interactions result from electromagnetic interactions.

Protons and neutrons, unlike electrons, are composite particles that require a force to hold them together. The strong force holds protons and neutrons together as individual nucleons. The field of quantum chromodynamics (QCD) encompasses the study of the strong force. The distance-dependence of the strong force is unlike that of any other force. Over a range of about 0.8 fm, which is about the radius of a nucleon, the strong force falls off with distance, much like other forces. Beyond this distance, the strong force is constant. However, interactions between the subatomic particles within nucleons screen the strong force, which results in a much smaller and roughly exponentially decaying residual strong force outside of nucleons. It is this residual strong force that binds proton and neutrons together to form nuclei. Furthermore, the screening-induced exponentially decaying distance dependence causes the residual strong force to be effective over no more than 3 fm or so. Thus, even though it is the strongest of all forces and roughly 100 times stronger than electromagnetism, the strong force can be ignored outside of the nucleus and in most of chemistry.

All fermions, such as electrons and protons, interact through the weak force. Weak interactions are involved in radioactive decay, including beta decay, and nuclear fusion. As implied by the name, the weak force is about 1000 times weaker than electromagnetism. In addition, the weak interaction is extremely short-ranged. At distances greater than about 0.1 fm it is negligible. Therefore, the weak force can be ignored outside of radiochemistry. Electromagnetism and the weak interaction were unified by the electro-weak theory of Glashow, Salam and Weinberg in 1968. The electro-weak interaction was unified with the strong interaction by the Standard Model in the 1970s and confirmed by the discovery of the Higgs Boson in 2013.

Gravitation is by far the weakest force with a coupling constant roughly 10^{-37} less than that of electromagnetism. It is also the most ubiquitous force since all matter is attracted by this force according to Newton's law of universal gravitation

$$F = G\frac{m_1 m_2}{r^2}. \tag{3.5}$$

Here, objects with masses m_1 and m_2 are separated by a distance r. The gravitational constant is $G = 6.674 \times 10^{-11}$ N m^2 kg^{-2}. In the (approximately) uniform gravitational field of the Earth, this reduces to

$$F = mg \tag{3.6}$$

for the gravitational force of the Earth that is responsible for weight. The Earth's gravitational acceleration is $g = 9.81\ \mathrm{m\ s^{-2}}$. This value varies with altitude, local geography and latitude. The gravitational potential energy of an object raised a height h is

$$E = mgh. \tag{3.7}$$

Usually, gravity represents a very small correction that has a negligible effect on gases, but this is not so for very cold gases. In fact, the gravitational force is important for the extremely accurate atomic clocks that incorporate a fountain of cesium atoms. The atoms are first laser cooled, then 'thrown' upwards only to fall back down into the fountain under the influence of gravity. The gravitational force can often be neglected for liquids and solids. It is, however, essential for explaining meniscus formation or the operation of a manometer.

3.2.1 Directed practice

The total pressure at a depth z under the surface of water is given by

$$p(z) = p^\circ + \rho g z \tag{3.8}$$

where p° is atmospheric pressure at the surface of the water (taken here to be the standard pressure of 1 bar), ρ is the density of water and g the gravitational acceleration. Calculate the total pressure at a depth of 10 m.

[Answer: 200 kPa, which leads to the rule that for every 10 m dived, the total pressure increases by 1 bar.]

3.3 Intermolecular interactions

We now concentrate on the interactions most important for molecules. As we shall see, these are all the result of electromagnetism and the spatial distribution of electrons relative to the positive nucleus, including how electrons fill atomic and molecular orbitals. We will divide the discussion along the lines of the range of the interactions beginning with long-range interactions.

3.3.1 Long-range interactions

There are three types of long-range interactions: electrostatic, induction, and dispersion. The *electrostatic energy* depends on the magnitude and position of charge. Induction and dispersion are polarization energies that depend on how electrons in a molecule respond to an external electric field.

To calculate the electrostatic potential a priori requires a solution of the Schrödinger equation. However, a *multipole expansion* can be used to approximate its value. This is only valid where there is no orbital overlap. The zeroth moment of the charge distribution can be taken as the charge q itself. The first moment is the dipole moment given by the product of the charge and the distance r by which two charges $-q$ and $+q$ are separated

$$\mu = q\mathbf{R}. \tag{3.9}$$

Note that μ and $\mathbf{R}$ are vectors with the radius vector drawn from the negative to the positive charge. The common unit of dipole moments is the debye, abbreviated D ($1\ \mathrm{D} = 3.336 \times 10^{-30}$ C m). To get some feeling of the magnitude of 1 D, separating charges of $\pm e$ by 1 Å leads to a dipole of 4.80 D. The dipole moment of H_2O is 1.85 D, while that of HCl is 1.11 D. Selected values of dipole moments are listed in Table 3.1. Zwitterionic molecules undergo intramolecular proton transfer in aqueous solution, and this produces a positive and a negative charge within one molecule and correspondingly large dipole moments. The values shown in Table 3.1 loosely conform to the expectation that greater electronegativity differences lead to larger dipoles. However, dipoles are the result of charge distributions in molecular orbitals and the product of charge times distance. Thus, CO has a small dipole (that also points in the 'wrong' direction – that is, with more charge density on the C end of the molecule) and NaF has a smaller dipole moment than NaI.

Next come the quadrupole Θ, octopole and hexadecapole moments. A cigar- or disk-shaped charge distribution gives rise to a quadrupole moment; a nearly spherical molecule will have either an octopole (e.g., CH_4) or hexadecapole (e.g., SF_6) as its first nonzero moment. In a completely spherical charge distribution such as that of a ground-state closed shell atom all moments are zero.

The multipole expansion of the electrostatic energy of two molecules is an expansion in terms of powers of R, where R is the separation of the centers of mass of the two molecules. The interaction energy falls off with distance according

Table 3.1 Dipole moments of selected molecules. Data taken from the 96th edition of the *CRC Handbook of Chemistry and Physics*.

Name	Formula	μ/D
Ammonia	NH_3	1.4718
Hydrogen fluoride	HF	1.826178
Hydrogen chloride	HCl	1.1086
Hydrogen bromide	HBr	0.8272
Hydrogen iodide	HI	0.448
Imidogen	HN	1.39
Hydroxyl	OH	1.655
Mercapto	HS	0.7580
Carbon monoxide	CO	0.10980
Nitric oxide	NO	0.15872
Sodium fluoride	NaF	8.156
Sodium chloride	NaCl	9.00117
Sodium bromide	NaBr	9.1183
Sodium iodide	NaI	9.236
Water	H_2O	1.8546
Methanol	CH_3OH	1.70
Ethanol	CH_3CH_2OH	1.69
Phenol	C_6H_5OH	1.224
Acetic acid	CH_3COOH	1.70
Acetonitrile	CH_3CN	3.92519

to $R^{-(m+n+1)}$, where m and n are the two moments that are interacting. Thus, if the molecules A and B have charges q_A and q_B, the first term in the expansion is the *Coulomb energy*

$$E_1 = q_A q_B / 4\pi\varepsilon_0 R. \tag{3.10}$$

The second term is that between the charge on one molecule and the dipole of the other

$$E_2 = q_B \mu_A \cos\theta_A / 4\pi\varepsilon_0 R^2. \tag{3.11}$$

Next comes the dipole–dipole interaction

$$E_3 = \mu_A \mu_B (2\cos\theta_A \cos\theta_B + \sin\theta_A \sin\theta_B \cos\phi) / 4\pi\varepsilon_0 R^3. \tag{3.12}$$

The coordinate system used to describe multipole interactions is defined in Fig. 3.2. The energy of interaction between dipoles depends on their mutual orientation, as is shown in Fig. 3.3. The most favorable interaction is when the dipoles are aligned collinearly head to tail followed by opposite to one another side-by-side. The least favorable is when they are aligned head-to-head and tail-to-tail. In general, the lowest order nonvanishing term will dominate the interaction energy. However, several terms are required to obtain a truly accurate energy. In other words, the multipole expansion approximates the interaction energy of two charge distributions as a series expansion. Whereas, the first nonzero term in the expansion is generally by far the greatest, the series may require several terms to converge on its asymptotic value for highly accurate calculations.

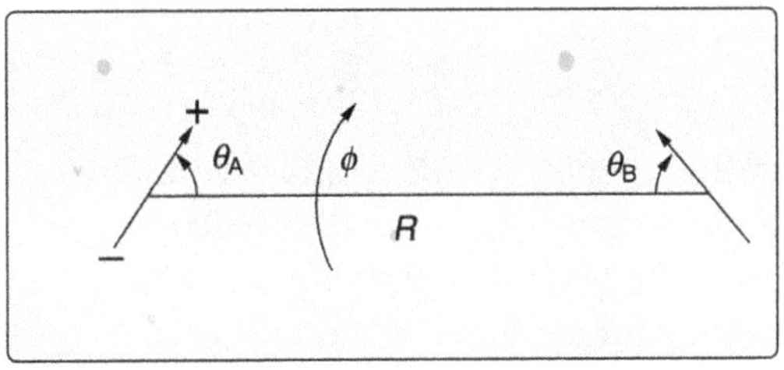

Figure 3.2 The coordinate system for multipole interactions. The centers of mass of the multipoles are separated by a distance R. The multipoles A and B are inclined at angles θ_A and θ_B with respect to this axis. The angle ϕ describes the angle between the planes formed by the multipole axes and the line of centers. A is specified as a dipole to indicate the IUPAC convention.

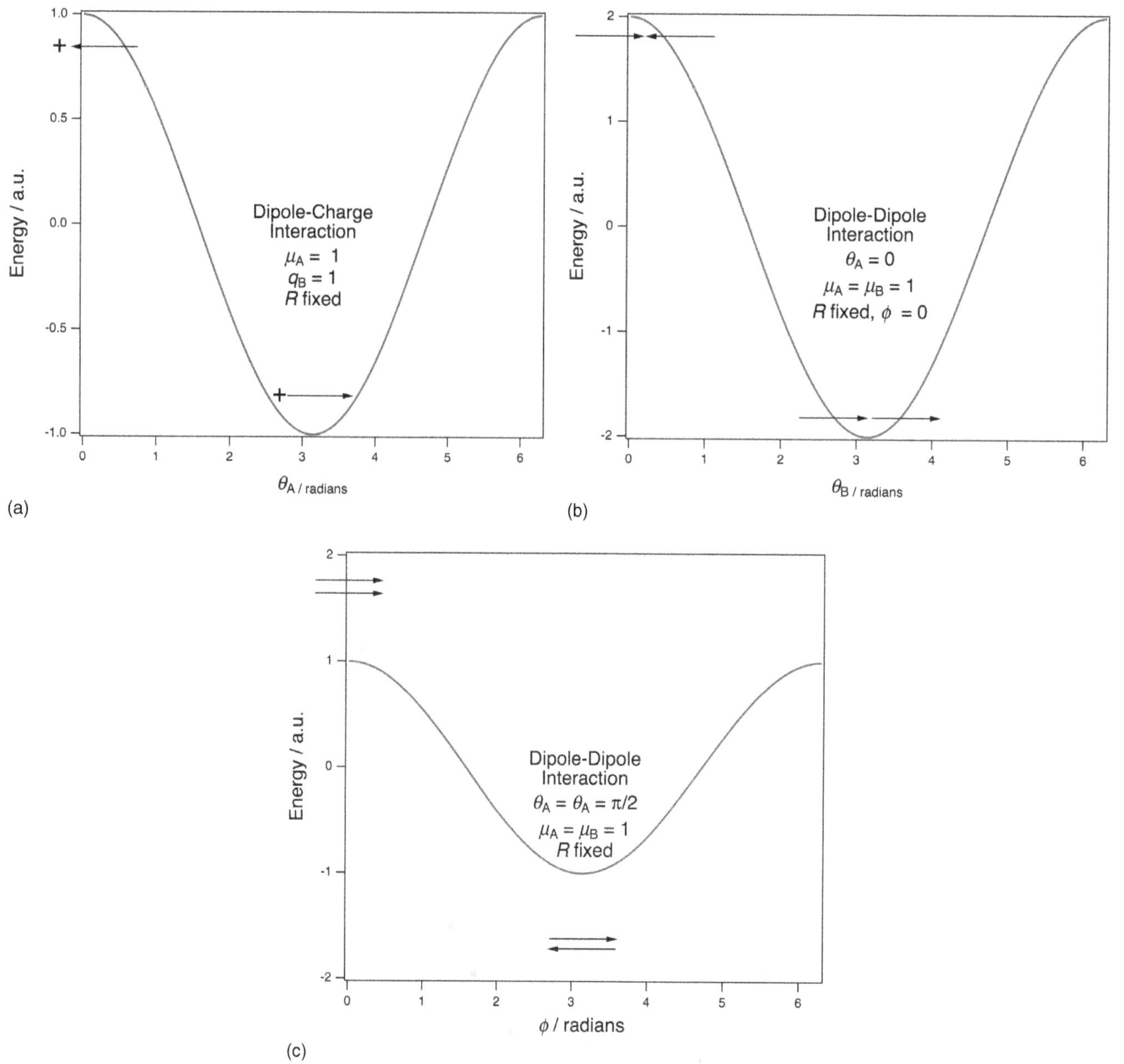

Figure 3.3 (a) Dipole–charge interaction as a function of orientation at fixed separation. (b) Dipole–dipole interaction as a function of orientation in a collinear configuration. (c) Dipole–dipole interaction in a side-by-side orientation. Note that an in-line head-to-tail arrangement in panel (b) is a more favorable configuration than the side-by-side head-to-tail arrangement in panel (c).

The electric field of strength E generated by a charged particle or dipole induces a dipole moment in a neighboring molecule in direct proportion to the polarizability α of the molecule

$$\mu = \alpha E. \tag{3.13}$$

Induction interactions arise when a permanent moment on one molecule induces a moment on the other molecule. This permanent moment-induced moment interaction is attractive. Molecular ions and molecules with large dipoles produce substantial induction energies. Higher-order moments engender progressively weaker induction effects.

The *polarizability* of a molecule is roughly proportional to the molecular volume – that is, the larger a molecule, the greater is its polarizability. The induction energy of a molecule with dipole moment μ_A interacting with a molecule of polarizability α_B when averaged over all orientations is given by

$$E_{ind} = -\mu_A^2 \alpha_B (4\pi\varepsilon_0)^{-2} R^{-6}. \tag{3.14}$$

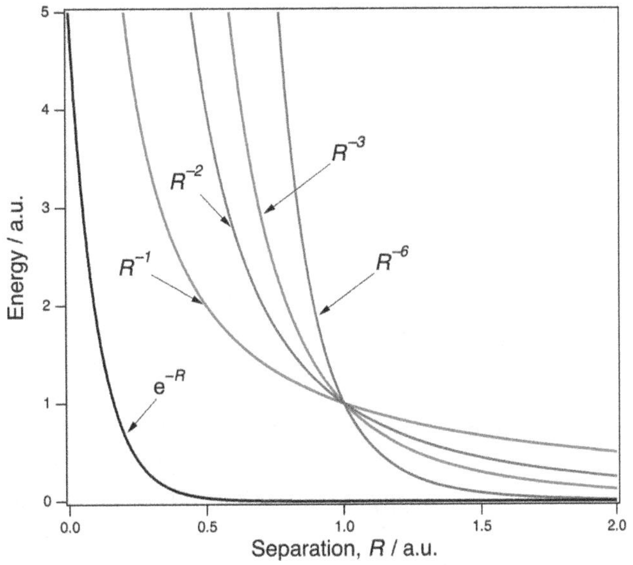

Figure 3.4 The distance-dependence of interactions with different ranges is displayed. Note that all are shown as repulsions (increasing energy with decreasing distance). Dispersion and induction, both of which are always attractive, follow R^{-6} behavior. The absolute value is shown to facilitate comparison on the same scale. Whether charge–charge (R^{-1}), dipole–charge (R^{-2}) and dipole–dipole (R^{-3}) interactions are attractive (negative) or repulsive (positive) depends on the orientation and the sign of the charges. The exchange interaction (Pauli repulsion) is always repulsive.

Instantaneous (transient) moments arise from fluctuations in the electron distribution. These instantaneous moments induce transient moments in their neighbors. These attractive interactions are known as *dispersion interactions*. It is also termed the 'London interaction' as Fritz London was the first to model this interaction accurately with quantum mechanics. Similar to the induction energy, the dispersion energy is given by

$$E_{\text{disp}} = -C_6 R^{-6} \tag{3.15}$$

where C_6 is an empirically determined constant. London showed that the value of C_6 is approximately given by

$$C_6 = \frac{3I_A I_B \alpha_A \alpha_B}{2(4\pi\varepsilon_0)^2(I_A + I_B)} \tag{3.16}$$

where I_A and I_B are the ionization potentials of the molecules. In general, the induction energy is only 2% of the dispersion energy. The constant C_6 is of the order of 10^{-78} J m^6 per atom, and only for small molecules with large dipole moments will the induction energy be important. London dispersion can be particularly important for long hydrocarbon chains interacting with one another. The substitution of F for H in these chains leads to much less polarizable electron distributions and significantly reduces the dispersion interactions. Whereas, the electrostatic energy is precisely pair-additive, the dispersion energy is exactly pair-additive only for the R^{-6} term. Small corrections for higher-order interactions lead to three-body terms. Induction energies are not pair-additive and require higher-body terms.

The catchall term *van der Waals forces* is often used. The IUPAC defines van der Waals forces as the attractive or repulsive forces between molecular entities (or between groups within the same molecular entity) other than those due to bond formation or to the electrostatic interaction of ions or of ionic groups with one another or with neutral molecules. The term includes: dipole–dipole, dipole-induced dipole and London (instantaneous induced dipole-induced dipole) forces. The term is sometimes used loosely for the totality of nonspecific attractive or repulsive intermolecular forces.

3.3.2 Short-range energies

The relative distance behaviors of short-range and long-range energies are shown in Fig. 3.4. Overlapping electron densities lead to short-range interactions. Bonding is, of course, the short-range attraction obtained when unpaired electrons of opposite spin interact. However, if two molecules have no unpaired electrons they have closed shells. Two closed shell species experience *Pauli repulsion*. This is the repulsive interaction that arises because the Pauli exclusion principle forbids two electrons from having the same spatial wavefunction (occupying the same space) and having the same spin. This interaction is also sometimes called an exchange repulsion. The *exchange energy*, which is pair-additive, is approximated by the *Born–Mayer potential*

$$E_{\text{ex}} = A \exp(-bR) \tag{3.17}$$

where A and b are empirically determined constants. We will, of course, spend considerably more time describing bonding interactions after the introduction of quantum mechanics.

3.3.3 The hydrogen bond, halogen bond, and related interactions

The IUPAC recommended definition of a *hydrogen bond* is (see Arunan, *et al.* in Further Reading for more detail):

> *The hydrogen bond is an attractive interaction between a hydrogen atom from a molecule or a molecular fragment X—H in which X is more electronegative than H, and an atom or a group of atoms in the same or a different molecule, in which there is evidence of bond formation.*

The hydrogen bond is an interaction in which a H atom is shared between two electronegative atoms X and Y. The H atom forms a bridge, usually in a linear configuration. Usually, one span of the bridge (X—H) is only slightly elongated compared to its non-hydrogen-bonded bond length, while the second span (H ⋯ Y) is much longer than a covalent Y—H bond. A hydrogen bond contains an acceptor and a donor and may be depicted as X—H ⋯ Y—Z, where the three dots denote the hydrogen bond. X—H is the hydrogen bond donor. The acceptor may be an atom or an anion Y, or a fragment or a molecule Y—Z, where Y is bonded to Z. In some cases, X and Y are the same. In the specific case in which not only X and Y are the same, but also the X—H and H ⋯ Y distances are the same, then the hydrogen bond is symmetric and particularly strong. The acceptor is an electron-rich region such as (but not limited to) a lone pair of Y or π-bonded pair of Y—Z. A hydrogen bond between two water molecules is depicted in Fig. 3.5(a), and its effect on the mean relative orientation of water molecules in the liquid is shown in Fig. 3.5(b).

The forces involved in the formation of a hydrogen bond include: (i) those of an electrostatic origin between permanent multipoles; (ii) inductive forces between permanent and induced multipoles; (iii) forces arising from charge transfer between the donor and acceptor leading to partial covalent bond formation between H and Y; and (iv) those originating from London dispersion forces. The strength of a hydrogen bond is mainly derived from the electrostatic interaction of the dipolar X—H bond with the (partial) negative charge on Y. However, the stronger the interaction, the greater the contribution of covalent interactions arising from electron transfer from Y to X—H. Because of the complex contributions to hydrogen bonding, networks of hydrogen bonds can show the phenomenon of cooperativity, leading to deviations from pair-wise additivity in hydrogen bond properties. If an interaction is primarily due to dispersion forces alone, then it would not be characterized a hydrogen bond. Thus, neither Ar—CH_4 nor CH_4—CH_4 are regarded as hydrogen-bonded systems. The balance of various components of hydrogen bonding varies quite widely from system to system.

The structure of a hydrogen-bonded complex has several typical features. The atoms X and H are covalently bonded to one another and the X—H bond is polarized. The strength of the H ⋯ Y hydrogen bond increases with increasing electronegativity of X. The X—H ⋯ Y angle is often linear (180°) and is usually above 110°. The closer the angle is to 180°, the stronger is the hydrogen bond. The usual correlation of a shorter bond with a stronger bond holds. Thus, the stronger the hydrogen bond, the shorter is the H ⋯ Y distance. The length of the X—H bond usually increases on hydrogen bond formation, and this bond length is anticorrelated with the hydrogen bond strength. The greater the lengthening of the X—H bond in X—H ⋯ Y, the stronger is the H ⋯ Y hydrogen bond.

As hydrogen bonding is an attractive interaction, it stabilizes the system and has several effects on the energetics of the system and its reactions. In order to be detected experimentally, the Gibbs energy of formation for the hydrogen bond should be greater than the thermal energy of the system. Hydrogen-bonding interactions are influenced by the acid—base characteristics of the donor and acceptor. The pK_a of X—H and pK_b of Y—Z in a given solvent correlate strongly with the energy of the hydrogen bond formed between them. Hydrogen bonds often act as a stepping stone in proton-transfer reactions of the type X—H ⋯ Y → X ⋯ H—Y. They can be considered as partially activated precursors to such reactions. This is prevalent in large molecules such as proteins as well as guest–host structures. The hydrogen bond strength correlates well with the extent of charge transfer between the donor and the acceptor.

Hydrogen bonds can strongly influence the energetics of the solid phase, in particular because of their directionality. This means that they can influence packing modes. The crystal packing of a non-hydrogen-bonded solid (say, naphthalene) is often determined by the principle of close-packing, and each molecule is surrounded by a maximum number of other molecules. In hydrogen-bonded solids, there are deviations from this principle to a greater or lesser extent depending upon the strengths of the hydrogen bonds that are involved.

There are other forms of *noncovalent bonds* that are analogous to hydrogen bonds. These include halogen, chalcogen, and pnicogen bonds. These noncovalent bonds are usually of the form H—A ⋯ D, in which D donates electron density into an acceptor level in the HA system. When A belongs to the pnicogens (N, P, As, Sb, Bi), chalcogens (O, S, Se, Te), or

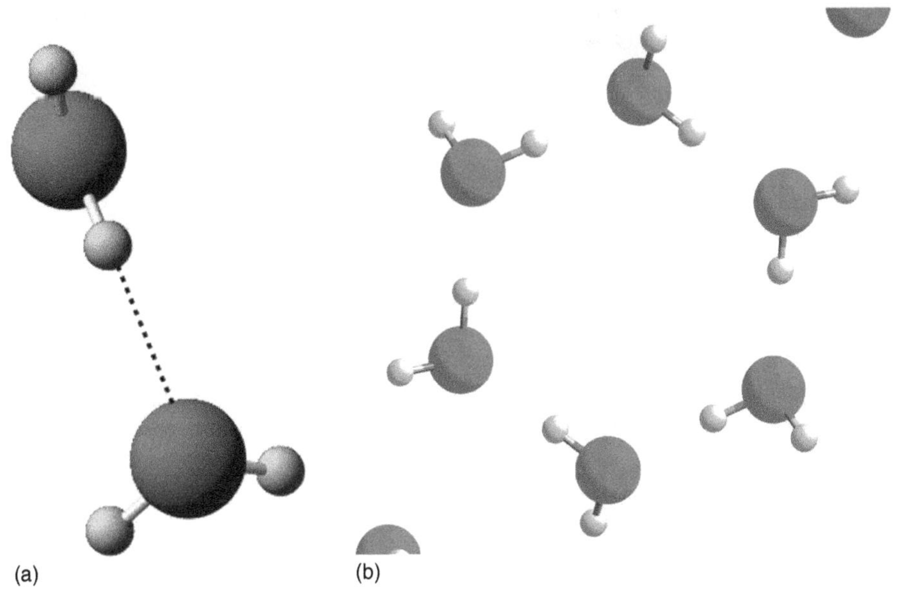

Figure 3.5 Hydrogen bonding in (a) the water dimer and (b) liquid water.

halogens (F, Cl, Br, I), this type of interaction is called a pnicogen, chalcogen, or halogen bond, respectively. Replacing a H atom with the more electron-negative F atom strengthens the noncovalent bond.

All three of these noncovalent bonds are of roughly equal energy. However, whereas the strength of the halogen bond depends sensitively on the halogen represented by the acceptor A, the strength of a pnicogen bond does not depend strongly on the identity of A. The strength of these noncovalent bonds depends much more on geometry than does the hydrogen bond, and they rapidly lose strength when they are distorted away from their preferred geometry. Again, electrostatics, induction and dispersion contribute to the strength of these noncovalent bonds. However, whereas the energetics of hydrogen bonds are generally dominated by electrostatics, in these latter three cases induction tends to play a comparable role to electrostatics with dispersion playing a secondary role.

3.4 Real gases

Real gases experience intermolecular interactions (dipole, hydrogen bonding, van der Waals, etc.) and have a finite volume. This means that they condense into a liquid with a finite volume when the temperature is low enough and the pressure is high enough. Above a certain temperature (the *critical temperature*, T_c) it is impossible to cause condensation merely by compression, even for a real gas.

3.4.1 Virial expansion

Real gases exhibit deviations from the ideal gas law at low temperature and high pressure (conditions that bring the system close to condensing). The ideal gas law is an approximation of real behavior. Under many conditions, it is highly accurate, but under certain conditions it needs to be corrected. A common procedure in physical chemistry is to use a *virial expansion* to create a more generally applicable equation of the form

$$pV_m = RT\left(1 + \frac{B}{V_m} + \frac{C}{V_m^2} + \cdots\right) \tag{3.18}$$

where $V_m = V/n$ is the molar volume. Each term becomes progressively smaller. B is the *second virial coefficient*, C the third, and so on. The virial expansion was introduced by Heike Kamerlingh Onnes, who was awarded the Nobel Prize in Physics 1913, and Willem H. Keesom in 1912. The virial expansion can be view as entirely empirical; that is, the constants are assigned values with no particular physical meaning other than they lead to the best fit of experimental data. However, it emerges that the virial coefficients can be related to parameters of the potential energy surface that describes intermolecular interactions.

Table 3.2 Values of the coefficients used to calculate the second virial coefficient of selected gases over proscribed temperature ranges. Values taken from the 96th edition of the *CRC Handbook of Chemistry and Physics*.

Gas	a_1	a_2	a_3	a_4	a_5	*T* range/ K
Ar	−16	−60	−9.7	−1.5		100–1000
H_2	15.4	−9	−0.21			15–400
HCl	−144	−325	−277	−170		190–470
H_2O	−1158	−5157	−10301	−10597	−4415	300–1200
He	12.44	−1.25				2–700
N_2	−4.3	−55.7	−11.8			75–700
NH_3	−271	−1022	−2717	−4189		290–420
O_2	−16	−62	−8	−3		90–400
Xe	−130	−262	−87			160–650
CO	−9	−58	−18			210–480
CO_2	−127	−288	−118			220–1100
CH_4	−43	−114	−19	−7		110–600
Acetone	−2051	−8903	−18056	−16448		300–480
Ethanol	−4475	−29719	−56716			320–390
Methanol	−1752	−4694				320–400
Isopropanol	−3165	−16092	−24197			380–420

Empirically, it is found that the second virial coefficient B is temperature-dependent and follows the form

$$B = \sum_{k=1}^{n} a_k \left(\frac{T}{T_0} - 1 \right)^{k-1}, \tag{3.19}$$

where the coefficients a_k are given in Table 3.2 for selected gases, $T_0 = 298.15$ K, and the equation is valid for specific temperature ranges.

3.4.1.1 Directed practice

Calculate the second virial coefficients of Ar and ethanol at 375 K.

3.4.2 Empirical equations of state

The van der Waals equation, for which Johannes D. van der Waals won the Nobel Prize in Physics in 1910, is an equation of state that attempts to account for non-ideal behavior by the addition of empirically determined corrections to the ideal gas law. It can be written

$$(p + a/V_m^2)(V_m - b) = RT. \tag{3.20}$$

where $V_m = V/n$ is the molar volume. From Eq. (3.20) we see that the term a/V_m^2, known as the internal pressure, corrects for long-range attractive intermolecular interactions, which scale with the magnitude of the van der Waals constant a. The attractive interactions effectively act like a pressure term that pushes the molecules together. At short range, the molecules repel each other because they cannot occupy the same space. The constant b accounts for this by defining an excluded volume that is four times the total volume of the N molecules in the gas with diameter σ

$$b = \tfrac{2}{3} N \pi \sigma^3. \tag{3.21}$$

The b coefficient is directly related to the critical volume V_c by

$$V_c = 3b. \tag{3.22}$$

The van der Waals constants are also related to the critical temperature T_c and pressure p_c through

$$a = \frac{27 R^2 T_c^2}{64 p_c} \tag{3.23}$$

$$b = \frac{RT_c}{8 p_c}. \tag{3.24}$$

Selected values can be found in Table 3.3.

Table 3.3 Selected values of van der Waals constants a and b, critical temperatures T_c and pressures p_c. The normal boiling point at $p = 101.325$ Pa is also reported. Note that CO_2 sublimes. Values taken from the 96th edition of the *CRC Handbook of Chemistry and Physics*.

Name	a / Pa m^6 mol^{-2}	b / m^3 mol^{-1}	T_b / K	T_c / K	p_c / MPa	V_c / m^3 mol^{-1}
Ar	1.362×10^{-1}	3.220×10^{-5}	87.302	150.687	4.863	9.661×10^{-5}
H_2	2.471×10^{-2}	2.657×10^{-5}	20.388	33.14	1.2964	7.970×10^{-5}
HCl	3.700×10^{-1}	4.061×10^{-5}	188	324.7	8.31	1.218×10^{-4}
H_2O	5.536×10^{-1}	3.049×10^{-5}	373.12	647.10	22.06	9.146×10^{-5}
He	3.461×10^{-3}	2.374×10^{-5}	4.222	5.1953	0.22746	7.122×10^{-5}
N_2	1.370×10^{-1}	3.869×10^{-5}	77.355	126.192	3.39	1.161×10^{-4}
NH_3	4.224×10^{-1}	3.711×10^{-5}	239.82	405.56	11.357	1.113×10^{-4}
O_2	1.382×10^{-1}	3.186×10^{-5}	90.188	154.581	5.043	9.557×10^{-5}
Xe	4.191×10^{-1}	5.154×10^{-5}	165.051	289.733	5.842	1.546×10^{-4}
CO	1.473×10^{-1}	3.952×10^{-5}	81.7	132.86	3.494	1.186×10^{-4}
CO_2	3.658×10^{-1}	4.286×10^{-5}	194.6 s	304.13	7.375	1.286×10^{-4}
CH_4	2.258×10^{-1}	4.223×10^{-5}	111.6	190.56	4.60	1.267×10^{-4}
Acetone	1.602×10^{0}	1.124×10^{-4}	329.23	508.1	4.7	3.371×10^{-4}
Ethanol	1.238×10^{0}	8.564×10^{-5}	351.39	515	6.25	2.569×10^{-4}
Methanol	9.571×10^{-1}	6.652×10^{-5}	337.6	512.7	8.01	1.996×10^{-4}
Isopropanol	1.603×10^{0}	1.124×10^{-4}	355.36	508.3	4.7	3.372×10^{-4}

Alternatively, the van der Waals equation is written

$$p = \frac{nRT}{V - nb} - a\left(\frac{n}{V}\right)^2. \tag{3.25}$$

Note that if both of the van der Waals coefficients a and b are equal to zero, the ideal gas law is recovered, and that n/V is the concentration (for instance, mol m^{-3} in SI units). Thus, the intermolecular interactions increase with increasing concentration as expected since molecules are on average closer together at higher concentration.

3.4.3 Example

Air entering the lungs ends up in tiny sacs called alveoli, from which oxygen diffuses into the blood. The average radius of the alveoli is 0.0050 cm, and the air inside contains a mole fraction of 0.14 O_2. Assuming that the pressure in the alveoli is 1.0 atm and $T = 37$ °C, calculate the number of oxygen molecules in one of the alveoli.

The volume of one alveolus is

$$V = \tfrac{4}{3}\pi r^3 = \tfrac{4}{3}\pi(5 \times 10^{-5}\ \text{m})^3 = 5.2 \times 10^{-13}\ \text{m}^3.$$

The number of moles of air in one alveolus is given by

$$n = \frac{pV}{RT} = \frac{(1.01 \times 10^5\ \text{Pa})(5.2 \times 10^{-13}\ \text{m}^3)}{8.314\ \text{J}\,\text{mol}^{-1}\text{K}^{-1}(37 + 273\ \text{K})} = 2.1 \times 10^{-11}\ \text{mol}$$

The number of oxygen molecules is then

$$N_{O_2} = N_A x_{O_2} n = 6.02 \times 10^{23}\ \text{mol}^{-1}(0.14)(2.0 \times 10^{23}\ \text{mol}) = 1.8 \times 10^{12}\ \text{molecules of } O_2.$$

3.4.4 Example

To synthesize ammonia using the Haber–Bosch process, 2000 mol of N_2 are heated in an 800 l vessel to 625 °C. Calculate the pressure of the gas if N_2 behaves: (a) as an ideal gas; (b) as a van der Waals gas.

a For an ideal gas, use the ideal gas law

$$p = \frac{nRT}{V} = \frac{2000\ \text{mol}(8.314\ \text{J K}^{-1}\ \text{mol}^{-1})(898\ \text{K})}{0.8\ \text{m}^3} = 1.87 \times 10^7\ \text{Pa}$$

b For a van der Waals gas, use the van der Waals equation

$$p = \frac{nRT}{V - nb} - a\left(\frac{n}{V}\right)^2$$

Look up the values of a and b in tables provided by reference works such as the *CRC Handbook of Physics and Chemistry* or *Lange's Handbook of Chemistry*.

$$p = \frac{2000(8.314)898}{0.8 - 2000(3.869 \times 10^{-5})} - 0.137\left(\frac{2000}{0.8}\right)^2 = 1.98 \times 10^7 \text{ Pa}$$

The van der Waals pressure (a much better approximation to the real value) is about 6% higher than that calculated for the ideal gas. Especially in the construction of large high-pressure facilities it is extremely important to have accurate estimates of process pressures.

An improved empirical equation of state is the *Redlich–Kwong equation*

$$p = \frac{nRT}{V - nb} - \frac{an^2}{V(V + nb)T^{1/2}}. \tag{3.26}$$

There is no single function in analytical form that can reproduce the equation of state of a real gas at all densities and pressures. The virial expansion can come closest. An extensive review of equations of state can be found in the book by Metiu.

3.5 Corresponding states

The concept of corresponding states was developed by van der Waals and Pitzer, who showed that it is valid as long as the potential energy function that describes intermolecular interactions requires only two characteristic parameters. A potential of this type is the Lennard–Jones potential

$$U(R) = 4\varepsilon\left[\left(\frac{\sigma}{R}\right)^{12} - \left(\frac{\sigma}{R}\right)^{6}\right], \tag{3.27}$$

which is discussed in more detail in our treatment of liquids. It describes the potential energy U of the interaction of two molecules separated by distance R in terms of the hard-sphere diameter of the molecule σ and an energy parameter ε. The equation of state is different for every gas when it is represented in the usual variables of p, T, and V_m. However, if the pressure, temperature, and molar volume are expressed as reduced properties,

$$p^* = \frac{p}{p_c} \tag{3.28}$$

$$V_m^* = \frac{V_m}{V_c} \tag{3.29}$$

$$T^* = \frac{T}{T_c} \tag{3.30}$$

then the equation of state is approximately the same for all gases. It turns out that any gas for which the intermolecular interactions are described by a potential of the type $U(R) = \varepsilon f(\sigma/R)$ will follow relationships such as those found in Eqs (3.28)–(3.30), which is why the principle of corresponding states is approximately general for all gases. The law of corresponding states demands that different gases at the same reduced volume and reduced temperature exert the same reduced pressure.

The law of corresponding states works particularly well for spherical nonpolar molecules. This means that it is least applicable to polar, nonspherical molecules. The introduction of polarity requires the addition of a third term to the potential energy function. The greater the magnitude of this term, the less accurate the predictions of the law of corresponding states will be.

If we substitute for p, V_m and T in the van der Waals equation from Eqs (3.28)–(3.30) and then express the critical constant p^*, $V_m{}^*$ and T^* in terms of the van der Waals constants a and b, we transform the van der Waals equation into

$$p^* = \frac{8T^*}{3V_m{}^* - 1} - \frac{3}{V_m{}^{*2}}. \tag{3.31}$$

Note that the constants a and b no longer appear in the equation of state. Therefore, isotherms in which p^* is plotted versus $V_m{}^*$ are representative of all gases that follow the van der Waals equation.

3.5.1 Directed practice

Make the substitutions required to derive Eq. (3.31).

3.6 Supercritical fluids

At a combination of temperature and pressure known as the *critical point*, the densities of a gas and liquid are the same. At this point the two phases are indistinguishable. The corresponding temperature T_c is the critical temperature, and the pressure p_c is the critical pressure. Beyond this point, the fluid that exists is called a *supercritical fluid*, which has several peculiar properties. One property is that the constant volume heat capacity tends to infinity at the critical point, as do the isothermal compressibility and the coefficient of thermal expansion. Because the isothermal compressibility becomes infinitely large, the amount of work it takes to compress a volume element within a fluid at the critical point is vanishingly small. This leads to density fluctuations in a supercritical fluid, and density fluctuations, in turn, cause refractive index fluctuations. These can be observed directly by the scattering of visible light; indeed, a fluid at the critical point scatters so much light that it appears milky white. This phenomenon is known as *critical opalescence*.

Another odd property of a supercritical fluid is that its surface tension is zero; thus, it is able to wet any surface. Solubility is a function of the density of the supercritical fluid and can be tuned with changes in p and T. Viscosity is closer to that of a gas than a liquid, which enhances reaction rates by increasing the rate of diffusion as compared to a liquid.

Supercritical CO_2 or sc-CO_2 is particularly important and widely used in chromatography as well as other applications in industry. It acts as a nonpolar solvent. However, it has a large quadrupole moment and polar C=O bonds that can assist is solvating organic species with —OH, —CO, and —F functional groups. Water, salts, and polar molecules such as sugars are not soluble in sc-CO_2, though chelated and organometallic compounds usually are. Ethanol, methanol, isopropanol, acetone, hexane, formic acid and acetic acid are miscible with sc-CO_2 and are often added as co-solvents. They act to keep solutes in solution that otherwise would not dissolve in sc-CO_2. sc-CO_2 is widely used for decaffeinating coffee and the extraction of hops.

SUMMARY OF IMPORTANT EQUATIONS

$E = mgh$	Gravitational potential energy for a body of mass m raised a height h
$\boldsymbol{\mu} = q\mathbf{r}$	Dipole moment = the product of the charge q and the distance r
$E_1 = q_A q_B / 4\pi\varepsilon_0 R$	Electrostatic potential energy (Coulomb energy) of an charges q_A and q_B interacting as a function of distance r
$E_2 = (q_B \mu_A \cos\theta_A)/4\pi\varepsilon_0 R^2$	Charge–dipole interaction energy
$E_3 = \mu_A\mu_B(2\cos\theta_A\cos\theta_B + \sin\theta_A\sin\theta_B\cos\phi)/4\pi\varepsilon_0 R^3$	Dipole–dipole interaction energy
$E_{ind} = -\mu_A^2\alpha_B(4\pi\varepsilon_0)^{-2}R^{-6}$	Induction energy of a molecule with dipole moment μ_A interacting with a molecule of polarizability α_B
$E_{disp} = -C_6 R^{-6}$	London dispersion energy
$C_6 = \dfrac{3I_A I_B \alpha_A \alpha_B}{2(4\pi\varepsilon_0)^2(I_A + I_B)}$	I_A and I_B are the ionization potentials of the molecules
$E_{ex} = A\exp(-bR)$	Born–Mayer potential to describe exchange energy (Pauli repulsion)
$pV_m = RT\left(1 + \dfrac{B_2}{V_m} + \dfrac{B_3}{V_m^2} + \cdots\right)$	Virial expansion of the pressure–molar volume product of a real gas in terms of the virial coefficients B_2, B_3, etc.
$p = \dfrac{nRT}{V - nb} - a\left(\dfrac{n}{V}\right)^2$	van der Waals equation of state for a real gas

Exercises

3.1 Discuss the intermolecular interactions that take place in a mixture I_2 + HI in the gas phase at 300 K and low enough pressure that neither condenses. Will a 1:1 mixture of I_2 + HI behave more or less ideally than pure HI at the same temperature and total pressure?

3.2 Discuss what is meant by the following: (i) electrostatic interactions; (ii) induction interactions; (iii) dispersion interactions; (iv) van der Waals interactions; and (v) noncovalent bonding.

3.3 Describe the limiting behavior of the volume of ideal gas and a van der Waals gas at: (i) low temperature; and (ii) at high pressure. In other words evaluate the volume in the limit $T \rightarrow 0$ or $T \rightarrow \infty$ for the two equations.

3.4 (a) When are the deviations from ideal gas behavior most important? (b) What molecules behave more ideally over larger ranges of parameters?

3.5 The enthalpy of vaporization of methane is 8.19 kJ mol^{-1}. The enthalpy of vaporization of methanol is 35.21 kJ mol^{-1}. Suggest a reason for the difference.

3.6 Calculate the pressure (in Pa) exerted by 1.0 mol C_2H_6 behaving as: (a) an ideal gas; (b) a van der Waals gas when it is confined under the following conditions: (i) at 273.15 K in 22.414 dm^3; (ii) at 1000 K in 0.100 dm^3.

3.7 When evaluated at the critical point, the (p, V_{m}) isotherm has a point of inflexion, which satisfies the criteria

$$\left(\frac{\delta p}{\delta V}\right)_T = 0, \quad \left(\frac{\delta^2 p}{\delta V^2}\right)_T = 0, \quad \left(\frac{\delta^3 p}{\delta V^3}\right)_T < 0.$$

Take the appropriate derivatives of the van der Waals equation to show that the critical pressure, critical temperature and critical molar volume are

$$p_c = a/27b^2, \quad T_c = 8a/27Rb, \quad V_{c,\mathrm{m}} = 3b.$$

3.8 Use the values of the critical pressure, critical temperature and critical molar volume derived from the van der Waals equation to show that the compression factor is the same for all gases when evaluated at the critical point.

$$Z_c = p_c V_{c,\mathrm{m}}/RT_c = 3/8$$

3.9 Using the relations derived in Exercise 3.8, substitute for the values of a and b and express the van der Waals equation in terms of the reduced pressure p_r and the reduced temperature T_r.

$$p_r = \frac{8T_r}{3V_r - 1} - \frac{3}{V_r^2}$$

3.10 The interaction potential between two Ar atoms is represented by a Lennard–Jones potential as given in Eq. (3.27) with parameters $\sigma = 3.542$ Å and $\varepsilon/k_{\mathrm{B}} = 93.3$ K. Calculate the force between two Ar atoms at 2, 5, and 10 Å. Recall that the force is the negative of the derivative of the potential with respect to distance.

Further reading

Arunan, E., Desiraju, G.R., Klein, R.A., Sadlej, J., Scheiner, S., Alkorta, I., Clary, D.C., Crabtree, R.H., Dannenberg, J.J., Hobza, P., Kjaergaard, H.G., Legon, A.C., Mennucci, B., and Nesbitt, D.J. (2011) *Pure Appl. Chem.*, **83** (8), 1637–1641. Definition of the hydrogen bond (IUPAC Recommendations 2011).

Israelachvili, J.N. (2011) *Intermolecular and Surface Forces*, 3rd edition. Academic Press, Burlington, MA.

Metiu, H. (2006) *Physical Chemistry: Thermodynamics*. Taylor & Francis, New York.

Poling, B.E., Prausnitz, J.M., and O'Connell, J.P. (2001) *The Properties of Gases and Liquids*, 5th edition. McGraw-Hill, New York.

Scheiner, S. (2013) The Pnicogen Bond: Its Relation to Hydrogen, Halogen, and Other Noncovalent Bonds. *Acc. Chem. Res.*, **46** (2), 280–288.

CHAPTER 4

Liquids, liquid crystals, and ionic liquids

PREVIEW OF IMPORTANT CONCEPTS

- Real gas molecules have a finite volume and condense into liquids.
- Liquids possess short-range order but long-range disorder, which is described mathematically by a radial distribution function.
- The order and disorder in a liquid is dynamic, with molecules switching in and out of nearest-neighbor shells on the picosecond timescale.
- The importance of intermolecular interactions is much greater because of the closer proximity of molecules to each other.
- The potential governing intermolecular interactions is approximated by a short-range repulsion plus a long-range attraction. This approximation works particularly well for liquids in which the long-range attraction is due mainly to dispersion.
- The coordination number is the mean number of nearest neighbors. It ranges from 8 to 12 for a simple monatomic liquid but is only about four for the open structure of water.
- Liquids are fluids as are gases; thus, they share many transport characteristics.
- Diffusion in a fluid never stops, even at equilibrium.
- A concentration gradient forces diffusion to occur with directionality to make the concentration gradient more uniform.
- The gas/liquid interface is neither sharp nor immobile. A liquid interface at equilibrium continually exchanges material not only between the surface and bulk of the liquid but also between the surface and the gas phase in order to maintain the equilibrium vapor pressure.

4.1 Liquid formation

Decreasing the temperature or increasing the pressure on a gas will lead to condensation. Most commonly, a gas condenses directly to the liquid phase. At one atmosphere of pressure, the temperature of this phase transition is called the *normal boiling point*, while at a pressure of 1 bar (the standard state pressure $p°$) this temperature is called the *standard boiling point*. Decrease the temperature yet further at the standard pressure, and the liquid at equilibrium would crystallize into a solid at the *standard freezing point* T_f (also called the standard melting temperature or *standard melting point*). Because of the very small change in molar volume between the liquid and solid phases, the normal and standard freezing points are essentially equal.

4.2 Properties of liquids

The liquid phase is a fluid, just as a gas is, and will take on the shape of its container. However, it is a much different type of fluid because of intermolecular interactions and the close proximity of the molecules within it. This means that liquids are almost incompressible and that many of their properties can be thought of in terms of a continuous material. That is, as long as we do not look on too fine a scale, many properties of a liquid can be approximated in terms of a continuous distribution of matter rather than discreet molecules.

The density of a liquid is much greater than that of a gas. The molar volume of an ideal gas at *standard ambient temperature and pressure* (SATP, $T = 298.15$ K, $p = 1$ bar $= 100$ kPa) is

$$V_\mathrm{m} = \frac{V}{n} = \frac{RT}{p} = 0.248\ \mathrm{m^3\,mol^{-1}}. \tag{4.1}$$

Physical Chemistry: How Chemistry Works, First Edition. Kurt W. Kolasinski.

Companion Website: www.wiley.com/go/kolasinski/physicalchemistry

Table 4.1 Characteristics of various liquids: molar mass M, melting point T_f, boiling point T_b, mass density ρ, molar volume V_m, number density
C. Values for ρ and V_m refer to 298.15 K except for Ar (90 K), CH_4 (90.68 K), H_2 (20.00 K), HF (293.15 K), He (4.30 K), N_2 (78 K), NH_3 (239.15 K), O_2 (90.00 K), NaCl (1074 K), Ag (1233 K), Na (371 K), and Re (3453 K). Values taken from Poling, Prausnitz and O'Connell (see Further Reading) or the 96th edition of the *CRC Handbook of Chemistry and Physics*.

Species	M / g mol^{-1}	T_f / K	T_b / K	ρ / kg m^{-3}	$V_m \times 10^6$ / m^3 mol^{-1}	$C \times 10^{-28}$ / m^{-3}
Ar	39.948	83.80	87.27	1373	29.10	2.070
Br_2	159.808	265.85	331.90	3102	51.52	1.169
CH_4	16.043	90.69	111.66	451.4	35.54	1.694
C_2H_5OH	46.069	159.05	351.80	785.1	58.68	1.026
C_6H_6	78.114	278.68	353.24	873.7	89.41	0.674
H_2	2.016	13.56	20.38	71.01	28.39	2.121
HF	20.006	189.58	292.68	966.9	20.69	2.911
H_2O	18.015	273.15	373.15	997	18.07	3.333
He	4.003	2.15	4.30	123	32.54	1.850
NH_3	17.031	195.41	239.82	682.3	24.96	2.413
N_2	28.014	63.15	77.35	804.1	34.84	1.729
O_2	31.999	54.36	90.17	1149	27.85	2.162
NaCl	58.44	1074	1738	1556	37.56	1.603
Ag	107.868	1235.08	2485	9320	11.57	5.203
Na	22.9898	370.96	1156.1	927	24.80	2.428
Hg	200.59	234.28	629.73	13534	14.82	4.063
Re	186.2	3453	5900	18900	9.85	6.113

The molar volume of liquid water at SATP is

$$V_m(l) = \frac{M}{\rho} = 1.81 \times 10^{-5}\ \text{m}^3\ \text{mol}^{-1}. \tag{4.2}$$

Therefore, the ratio of the volume of water vapor (acting ideally) to water at SATP is 1370. This is a general observation: the molar volume decreases by a factor of about 1000 upon condensation. Therefore, the mean distance between molecules changes by roughly a factor of 10.

Dividing these values of molar volume by N_A, we find the mean volume occupied by a molecule. The inverse of this quantity is the number density C, which is found in the last column of Table 4.1. Assuming the molecules are spherical with radius r, their volume is $\frac{4}{3}\pi r^3$. Thus, a rough estimate for the radius of the sphere 'occupied' by a water molecule is 21.4 Å in the gas phase and 1.93 Å in the liquid phase. Obviously, a molecule does not shrink by a factor of 10 upon condensation. The molecules are the same size in both cases, but the mean distance between them changes.

We know that, compared to gases, liquids are nearly incompressible. Their volume – and therefore their density – changes slowly as temperature changes in comparison to the large changes observed for a gas. In an ideal gas the relationship is

$$\rho = nM/V = pM/RT. \tag{4.3}$$

An inverse relationship between density and temperature is generally observed for liquids; however, the functional form – known as the Tait equation – is rather complex. For a limited range above the melting point, the density of liquid metals and molten salts follows the equation

$$\rho = \rho_m - k(T - T_m) \tag{4.4}$$

where ρ_m is the density of the liquid at the melting point, k is a constant and T_m is the maximum temperature for which the linear relationship holds. The value of k usually falls in the range of 0.5 to 8 kg m^{-3} K^{-1}.

Water is a very unusual liquid but also the most important. Its maximum density of 999.9749 kg m^{-3} occurs at 277.15 K, that is, 4 K above T_f. Liquid water, like many liquids, can also be supercooled (cooled below T_f) if special precautions (slow cooling, purity, absence of particulates) are taken. An empirical fit to the mass density of water over the entire measured temperature range shown in Fig. 4.1 requires a fourth-order polynomial,

$$\rho_{H_2O} = k_0 + k_1 T + k_2 T^2 + K_3 T^3 + k_4 T^4, \tag{4.5}$$

where $k_0 = -4240$, $k_1 = 63.3$, $k_2 = -0.286$, $k_3 = 5.73 \times 10^{-4}$, and $k_4 = -4.34 \times 10^{-7}$.

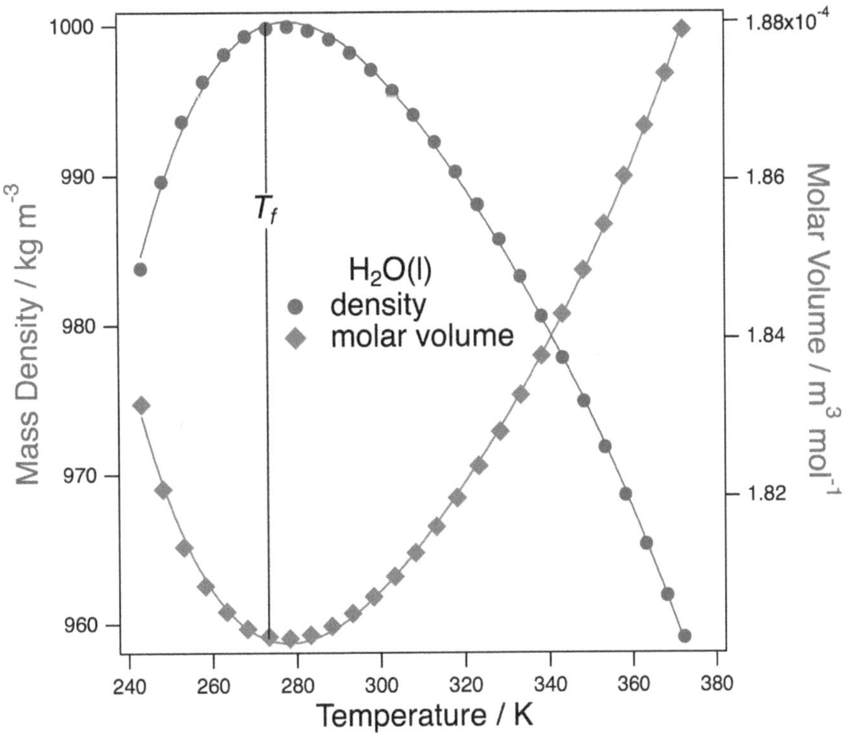

Figure 4.1 The mass density and molar volume of normal water and supercooled water as a function of temperature. Values taken from the 96th edition of the *CRC Handbook of Chemistry and Physics*.

4.3 Intermolecular interaction in liquids

As shown in Table 4.1, which pertains to standard pressure, the liquid phase is stable over much different temperature ranges depending on the nature of the intermolecular forces in the material under consideration. The intermolecular interactions discussed previously that lead to non-ideality in gases are more strongly present in liquids because the molecules are closer together. The interactions depend on the types of molecules present. The interactions in liquids of inert gas atoms or nonpolar molecules are dominated by pairwise-additive terms. Many-body terms are negligible. Polar liquids usually have substantial many-body terms while liquid metals possess very large many-body terms. Constructing a potential energy function to describe many-body interactions is much more complicated than constructing one with just pairwise-additive terms.

The intermolecular interactions of a liquid are perhaps the most difficult to treat of the three common phases. In a gas, the distance between molecules allows us to ignore interactions in the ideal state and then add interactions as a small perturbation to the ideal model. In the solid state, the perfect order of a crystal introduces symmetry, which greatly simplifies the treatment of interactions. The result of these difficulties is that fundamental theories of the properties of liquids have been the most difficult to advance and lag considerably behind those of the gas and solid phases.

Liquid metals and He are exceptional liquids. Helium has two liquid phases, which is also observed for P but still debated for H_2O and S. Helium is, however, the only elemental liquid that cannot be solidified at 1 atm pressure (25 atm are required for solidification), which means it has no liquid/solid phase boundary at atmospheric pressure. Liquid metals are electrically conductive, which indicates that electrons are delocalized because of metallic bonding. Delocalization is a manifestation of the many-body interactions mentioned above. Most molten salts are composed of ions, but there are exceptions such as $HgCl_2$, which has low electrical conductivity and, thus, must be composed of uncharged species.

H_2O is the only naturally occurring inorganic liquid on Earth. It is a truly exceptional liquid. Indeed, if it were not the most important liquid, we might not even study its properties because of the complications of dealing with a heavily hydrogen-bonded liquid with an unusually large relative permittivity that readily autoionizes. However, as it is the most important liquid we will consequently spend considerable effort trying to understand it, and solutions in which it is a solvent.

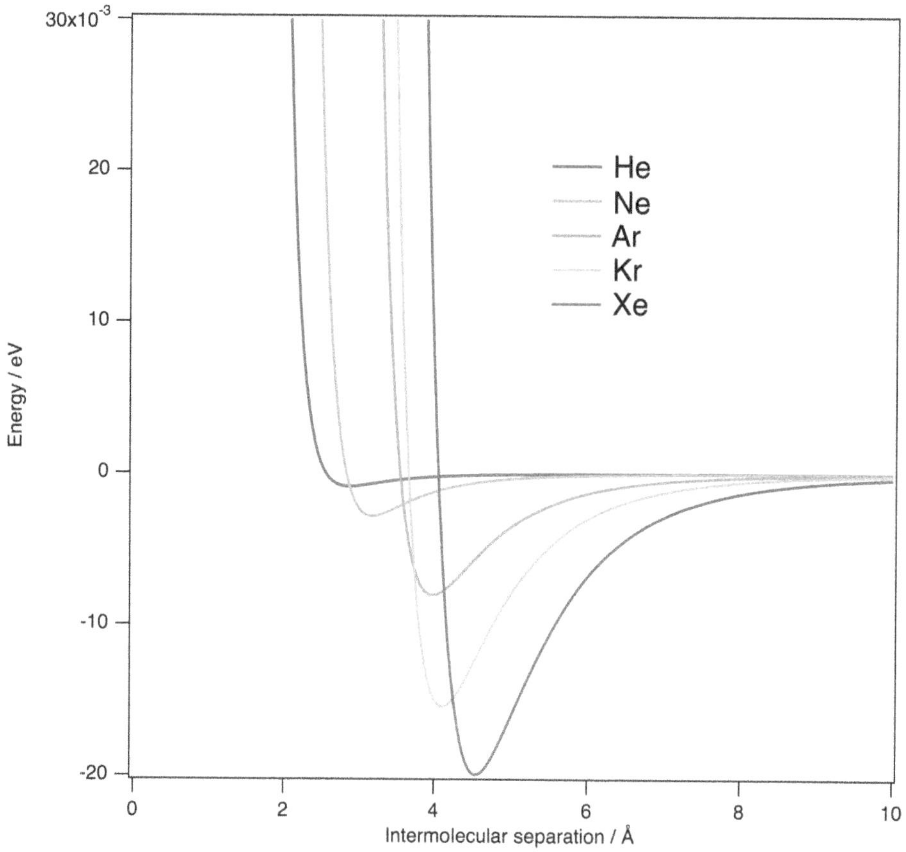

Figure 4.2 The Lennard–Jones potentials for He, Ne, Ar, Kr, and Xe.

Our first-order model of a liquid begins with assuming that the molecules or atoms which comprise the liquid are nondeformable spheres with a fixed diameter. Their interactions with other molecules are independent of orientation. This is known as a hard-sphere model. A hard-sphere potential gives a good approximation to the thermal conductivity and viscosity of a fluid. This is because these properties are mainly determined by long-range repulsions rather than van der Waals attractions. Condensation of a gas into a liquid requires a potential with an attractive well. A simple and widely used semi-empirical potential is the *Lennard–Jones potential* (also known as L-J or 6-12 potential). It is defined in terms of the interparticle separation R and has the form

$$U(R) = 4\varepsilon\left[\left(\frac{\sigma}{R}\right)^{m} - \left(\frac{\sigma}{R}\right)^{n}\right] \tag{4.6}$$

where σ is an empirically determined hard-sphere diameter of the atom/molecule, ε is related to the well depth (minimum of the potential), $m = 12$, and $n = 6$. Examples of Lennard–Jones potentials for the noble gases are shown in Fig. 4.2. The shape should be familiar, as they look much like the attractive potential that is associated with the formation of a chemical bond. The potentials approach zero asymptotically as they approach infinite separation. They pass through a minimum and then again equal zero when the separation is equal to the hard-sphere diameter, that is, at $R = \sigma$. They approach infinity as the intermolecular separation decreases below the effective molecular diameter because two molecules cannot occupy the same space.

The Lennard–Jones parameters for a number of gases are listed in Table 4.2, where some expected trends are immediately obvious. In going from He down the column to Xe, the size of the atom increases, as does its polarizability. Thus, the hard-sphere diameter σ increases as does the well depth ε. Similar trends occur for the hydrogen halides and halogens, except for the well depths of HI and I_2, which are unusually small. The substitution of F for H in either methane or silane leads to a larger molecule but also one with a smaller well depth. Changing from a triple bond to a double bond to a single bond in the C_2H_2—C_2H_4—C_2H_6 series leads to a larger molecule, but also to one that is successively less polarizable and more weakly interacting. Benzene is more compact than cyclohexane, but its extended system of π electrons leads to stronger intermolecular interactions. The difference between benzene and cyclohexane is much larger than the

Table 4.2 Parameters for Lennard–Jones potentials determined from viscosity data. Values from Poling, Prausnitz and O'Connell.

Formula	Molecule	σ / Å	ε/k_B / K
He	helium	2.551	10.22
Ne	neon	2.820	32.8
Ar	argon	3.542	93.3
Kr	krypton	3.655	178.9
Xe	xenon	4.047	231.0
H_2	hydrogen	2.827	59.7
H_2O	water	2.641	809.1
HF	hydrogen fluoride	3.148	330
HCl	hydrogen chloride	3.339	344.7
HBr	hydrogen bromide	3.353	449
HI	hydrogen iodide	4.211	288.7
HCN	hydrogen cyanide	3.630	569.1
F_2	fluorine	3.357	112.6
Cl_2	chlorine	4.217	316.0
Br_2	bromine	4.296	507.9
I_2	iodine	5.160	474.2
NH_3	ammonia	2.900	558.3
N_2	nitrogen	3.798	71.4
NO	nitric oxide	3.492	116.7
NO_2	nitrous oxide	3.828	232.4
O_2	oxygen	3.467	106.7
SF_6	sulfur hexafluoride	5.128	222.1
SiH_4	silane	4.084	207.6
SiF_4	tetrafluorosilane	4.880	171.9
UF_6	uranium hexafluoride	5.967	236.8
CH_4	methane	3.758	148.6
CF_4	tetrafluoromethane	4.662	134.0
CCl_4	tetrachloromethane	5.947	322.7
C_2H_2	ethyne	4.033	231.8
C_2H_4	ethene	4.163	224.7
C_2H_6	ethane	4.443	215.7
C_6H_6	benzene	5.349	412.3
C_6H_{12}	cyclohexane	6.182	297.1
n-C_6H_{14}	n-hexane	5.949	399.3
CH_3OH	methanol	3.626	481.8
C_2H_5OH	ethanol	4.530	362.6
CO	carbon monoxide	3.690	91.7
CO_2	carbon dioxide	3.941	195.2

difference between ethene and ethane. Methanol is smaller and much more polar than ethanol; such greater polarity leads to a much greater well depth for methanol. Substituting a polar OH group for a nonpolar H makes the intermolecular interactions of C_2H_5OH much stronger than those of C_2H_6.

For any combination of m and n for which $m > n$, the depth of the minimum is

$$U_{\min} = 4\varepsilon\left[\left(\frac{n}{m}\right)^{\frac{m}{m-n}} - \left(\frac{n}{m}\right)^{\frac{n}{m-n}}\right] \tag{4.7}$$

and it occurs at

$$R_{\min} = \sigma(m/n)^{1/(m-n)}. \tag{4.8}$$

The R^{-6} term accurately represents the long-range limit of the dispersion energy. The R^{-12} term is a convenient-for-integration term that approximates the short-range repulsion. Because the exchange energy actually is better approximated by an exponential function, a more accurate representation is the *(exp-6) potential*

$$U(R) = A\exp(-kR) - C_6R^{-6}. \tag{4.9}$$

As a first step to understanding the structure of a liquid, let us calculate the potential energy of hard spheres interacting only through a potential that can be described by a Lennard–Jones potential, as shown in Fig. 4.2. Consider N molecules described by the potential $U(R)$. The molecules only interact through pairwise-additive terms, where R is the distance between their centers. The potential energy of their interaction vanishes at $R = \sigma$, which defines the effective diameter of the molecules. The molecules are labeled 1 to N, and we consider two of them, i and j. The number of distinct pairs ij is $\frac{1}{2}N(N-1)$. $U(R_{ij})$ is the potential energy of the interaction between pair ij. Since we have assumed that the molecules in the liquid only interact in a pairwise fashion, in order to calculate the total potential energy of the liquid, Φ, we need to sum the contributions of all the pairs, being careful not to double-count any pairs (indicated by the $i > j$ in the summation),

$$\Phi = \sum_{i>j} \phi(r_{ij}). \tag{4.10}$$

Now we try to build upon this simple starting point to understand the structure and properties of the liquid phase.

4.3.1 Directed practice

Explain the trend in the Lennard–Jones parameters of CH_4, CF_4, and CCl_4.

4.4 Structure of liquids

The structure of a liquid can be approached from two extreme limits. One is to think of it as an extremely dense gas, while the other is to start from the vantage of a disordered solid. To the first model, we must add intermolecular interactions and the potential for many-body interactions. To the second we need to add transport and allowance for short-range order but the lack of long-range order.

To describe the structure of a liquid we introduce two parameters. One is the *coordination number, Z*; this is the mean number of nearest neighbors. The second is the *radial distribution function* $g(R)$, also known as the *pair correlation function*.

In a fluid with no intermolecular forces, every molecule moves through the entire volume of the sample, independent of the presence of any other molecule. The fluid would be homogeneous. For a given volume element ΔV, there is an equal probability of finding a molecule in this volume element anywhere in the total volume V. The probability of finding a molecule in the volume element is $\Delta V/V$. The probability of finding one particular molecule in the volume ΔV at point R_1, and another particular molecule in the volume ΔV at point R_2, is then the product $(\Delta V/V)^2$ independent of R_1 and R_2. If there are N total molecules, then the probability of simultaneously finding any molecule in the volume element at R_1 and any other molecule in ΔV at R_2 is $N(N-1)(\Delta V/V)^2$. Again, this is independent of R_1 and R_2. Thus, the joint probability density of finding any molecule at R_1 and any other molecule at R_2 is

$$n^2 = N(N-1)/V^2, \tag{4.11}$$

where n is the conditional probability density of finding a molecule a distance R from any arbitrarily chosen molecule. For the large number of molecules in a macroscopic sample,$n = N/V = C_0$, where C_0 is the mean number density averaged over the entire liquid. This probability density is independent of R. We conclude that, in the absence of intermolecular interactions, the fluid is structureless. That is, the probability of finding a pair of molecules with any arbitrarily chosen separation R is independent of R. In a structureless fluid the radial distribution function is constant, $g(R) = 1$.

Consider a liquid composed of spherical molecules with a diameter σ. The radial distribution function describes how many molecules are at a given center-to-center distance R from the reference molecule. Because the molecules have a finite size, there is now a strong short-range repulsive interaction between them that excludes one molecule from the volume occupied by the other. As can be seen in Fig. 4.3, this repulsion has a huge influence on the structure of the liquid. The radial distribution function must start at zero, after which at separations greater than σ it increases. Several maxima are observed, with the first occurring near the minimum in $U(R)$ – that is, at the first nearest-neighbor distance. In a perfect lattice of spheres of the same diameter, the nearest neighbors are all equidistant from the reference atom. The next nearest neighbors form a second shell and reside at a larger distance, and so on. The radial distribution function records these positions as peaks in a plot of $g(R)$ versus R. Long-range order in a solid is indicated by an infinite series of such peaks, and thermal motion broadens the peaks. The introduction of disorder turns the series of peaks into a continuous distribution. If the sample is amorphous, as in a liquid, then there is short-range order but a lack of long-range

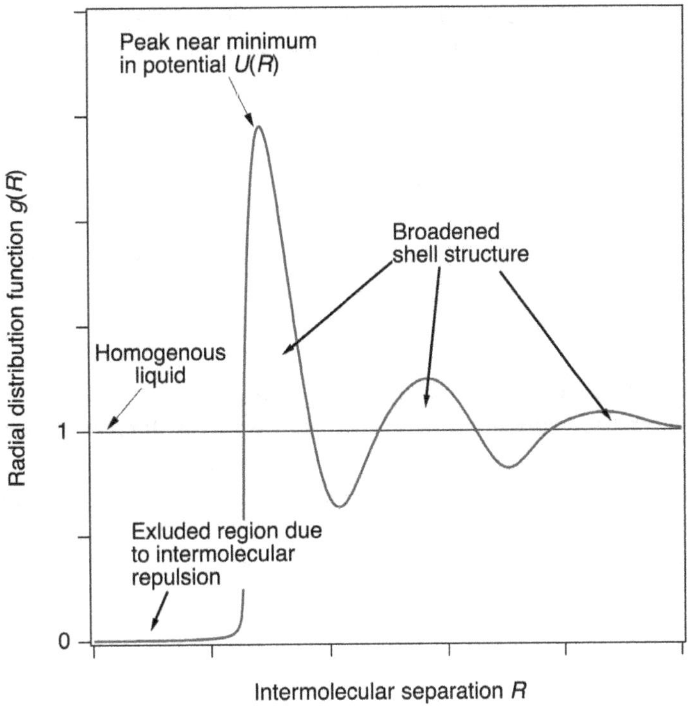

Figure 4.3 A prototypical radial distribution function for a disordered liquid as compared to a homogeneous liquid.

order. This is reflected in $g(R)$ by a decay in the height of the broad peaks, as shown in Fig. 4.3. Usually, at distances greater than roughly three molecular diameters, the oscillations disappear and $g(R)$ converges on 1.

Let us now define the radial distribution function mathematically, which will allow us to develop a rigid sphere model of the liquid. This model was refined by Kirkwood in 1935. We place a molecule at the origin O and measure the radial distance R from this origin. We define the number density $C(R)$ as the number of molecules per unit volume at distance R. The radial distribution function is the ratio of the number density as a function of R to the mean number density C_0,

$$g(R) = C(R)/C_0. \tag{4.12}$$

As can be seen in Fig. 4.3, $g(R)$ exhibits damped oscillations that vary from slightly above the mean density to below it outside of the diameter of the molecule at the origin. The radial distribution function is determined by X-ray or neutron diffraction data. Its value represents a time-averaged quantity. At any given distance over a period of time there will be significant fluctuations in $g(R)$ because of the constant thermal motion of the molecules.

This constant thermal motion means that while the time-averaged distance between one molecule and its nearest neighbor is constant, the identity of the two molecules is not. Recall that each molecule in the liquid is able to sample the entire volume of the liquid. If we were to momentarily stop the molecules and take a picture, any two pictures would look almost identical. But if we were to take a picture, label the molecules, then follow their motions as a function of time, we would see that the configuration of molecules changes on an extremely rapid time scale. The molecule initially at the origin will move and be replaced by another, and the molecules that initially formed the set of nearest neighbors exchange their places with other molecules. The time scale for this decay of the initial structure is several picoseconds (10^{-12} s).

The study of $g(R)$ as a function of temperature and pressure reveals much about the nature of a liquid. The structure of a liquid does not depend strongly on temperature; rather, it is determined mainly by the density. We can now define the *coordination number* Z in terms of the radial distribution function

$$Z = 4\pi C_0 \int_0^{R_m} R^2\, g(R)\, \mathrm{d}R. \tag{4.13}$$

The magnitude of Z varies with the thermodynamic state of a fluid and is mainly determined by the density. The value of Z fluctuates in time because of molecular motion. These fluctuations are on the order of the size of Z. In a simple monatomic liquid the coordination number would range between 8 and 12. It must be remembered that the molecules are continuously exchanging positions and varying their intermolecular separations. Liquid Ar, for example, has a coordination number of roughly 10, which means that the liquid approximates a loosely packed set of spheres.

Water is anything but a simple liquid. It has a coordination number of approximately 4, which indicates that water has a very open structure with tetrahedral coordination. The cause of this is that water is neither spherical in shape, nor are its intermolecular interactions radially symmetric. Instead, its ability to hydrogen bond means that the interactions between water molecules are highly orientation-dependent.

A long thin molecule such as an alkane presents a different sort of geometrical effect and asymmetric interactions. Entropically the molecules prefer to be randomly oriented and there are many randomly ordered structures. However, such a structure has a low density and does not maximize the attractive noncovalent interactions between molecules. The structure that maximizes these intermolecular attractions is highly ordered, with the chains aligned parallel to each other. Such a highly ordered structure, however, has a very poor entropy factor. There is a competition between the entropy and attractive interactions. The attractive interactions also grow in importance as the chain length increases. Therefore, larger alkanes will tend to show more order than shorter ones. Increasing the interaction even further by adding a group capable of hydrogen bonding, such as a terminal alcohol, the energetic factors can become dominant over the entropy term. In such a case, the molecules will tend to align into chains to maximize their attractions.

The structure of a simple fluid varies smoothly and continuously as the density increases from that of a gas to a liquid. Strong intermolecular attractions can lead to deviations from this behavior; for instance, strongly hydrogen-bonding molecules may experience dimer formation. As the density increases, the ordering progressively increases as evidenced by progressively slower decay in $g(R)$.

Two other types of liquids that experience very strong and long-range interactions are molten salts and liquid metals. Metallic bonding is delocalized and highly influenced by many-body interactions. Both, molten salts and liquid metals can conduct charge. Coulomb interactions between charged particles are important in both. To lower the energy of a molten salt, each positive ion is surrounded by negative ions, and vice versa. The pair correlation function can then be decomposed into different combinations of the anions and cations. In NaCl, for example, these are the Na^+—Na^+, Cl^-—Cl^- and Na^+—Cl^- radial distribution functions. In a liquid metal, it is electrons and positive ion cores that comprise the charged species. Electrons move very rapidly compared to the heavy nuclei. Consequently, the ions experience a screened electrostatic potential. The short-range nature of the ion–ion interactions means that the structure of a liquid metal is surprisingly similar to that of a simple liquid.

4.4.1 Directed practice

The radial distribution function vanishes at $R = 0$ for all liquids. Explain.

4.5 Internal energy and equation of state of a rigid sphere liquid

Now that we have an improved understanding of the structure of a liquid, we can improve upon the simplistic calculation of the potential energy derived in Eq. (4.10). Place molecule i at the origin or our coordinate system. The radial density function tells us that, on average, $C(R)\,4\pi R^2\,\mathrm{d}R$ molecules are contained in a shell drawn between R and $R + \mathrm{d}R$ from the origin. The total potential energy is obtained by calculating the potential of the interaction of i with all other molecules in the liquid. Since $C(R)$ is a continuous function in the liquid, the potential energy of this one molecule interacting with all other molecules is calculated by integrating the product of the pairwise potential and the radial density function,

$$\Phi = \int_0^\infty \phi(R)\,C(R)\,4\pi R^2\,\mathrm{d}R. \tag{4.14}$$

The total potential energy of the liquid is obtained by repeating this procedure for each of the N molecules in the liquid, though this would double count each interaction. Thus, the total potential energy is obtained by multiplying Eq. (4.14) by $N/2$.

The internal energy is the kinetic energy plus the potential energy of the intermolecular interactions. The kinetic energy is calculated just as it was for an ideal gas; thus, the sum is

$$U = \tfrac{3}{2}Nk_\mathrm{B}T + \tfrac{1}{2}N\int_0^\infty 4\pi R^2\,\phi(R)\,C(R)\,\mathrm{d}R. \tag{4.15}$$

The equation of state is derived by application of Clausius' virial theorem in a central force,

$$pV = Nk_BT + \frac{1}{3}\overline{\sum_{pairs} F(R_{ij})R_{ij}}. \tag{4.16}$$

The first term is the *kinetic pressure*. Just as for a gas, this is the pressure caused by the collision of molecules with the walls of the container. $F(R_{ij})$ is the central force between the pair ij. We take the product of this force with the distance R_{ij}, sum this over all pairs ij, then take its time average. This may seem an impossible task. However, the sum and averaging is accomplished by an integral of the same type as Eq. (4.15). Further, the force is simply related to the derivative of the potential by

$$F(R_{ij}) = -\mathrm{d}\phi(R)/\mathrm{d}R. \tag{4.17}$$

Thus, we can reformulate Eq. (4.16) as

$$pV = Nk_BT - \frac{2\pi}{3}N\int_0^\infty R^3\,\frac{\mathrm{d}\phi(R)}{\mathrm{d}R}\,C(R)\,\mathrm{d}R. \tag{4.18}$$

In principle, we can calculate p, U and other equilibrium thermodynamic properties but to do so we must determine $\phi(R)$ and $C(r)$ with great accuracy. The form of the potential $\phi(R)$ is not very problematic in this regard. However, the results are highly sensitive to the form of $C(r)$, which is difficult to determine accurately enough from diffraction data. Kirkwood was able to derive an integral equation for the *ab initio* calculation of $C(r)$. This distribution function was found to be very close to what one would expect for a rigid hard sphere, regardless of what was the exact form of the potential. Therefore, even though the molecules in liquids are neither really rigid spheres nor hard spheres, the properties of their liquid phases are approximated fairly well by these assumptions.

The integral in Eq. (4.18) depends on $\mathrm{d}\phi(R)/\mathrm{d}R$ for a potential of the type shown in Fig. 4.2. The derivative is negative at small R, but then turns positive before rapidly falling to zero as R increases. The greatest contribution to the integral is made by the small region near the origin. At high density there are a large number of neighbors in the region where $\mathrm{d}\phi(R)/\mathrm{d}R < 0$. This causes the integral to be negative and the total pressure to be greater than the ideal gas value. At moderate densities, the number of neighbors at larger distance dominates and this changes the sign of the integral's contribution. The pressure of the liquid is then lower than that of an ideal gas.

4.5.1 General equations of state for liquids

There is no generally valid equation of state for liquids with a simple analytical form. The two most widely encountered empirical equations of state are the *Tait equation*

$$\frac{V_{m0} - V_m}{pV_{m0}} = \frac{A}{B+p} \tag{4.19}$$

and the *Huddleston equation*

$$\ln\left(\frac{pV_m^{2/3}}{V_{m0}^{1/3} - V_m^{1/3}}\right) = A + B\left(V_{m0}^{1/3} - V_m^{1/3}\right) \tag{4.20}$$

The constants A and B have different positive values in the two equations. V_m is the molar volume, p the pressure, and V_{m0} is the molar volume at 'zero' pressure.

4.6 Concentration units

Liquids can, of course, form mixtures and solutions. Here, we define the term *mixture* to mean a combination of two or more components, regardless of their relative concentrations. By *solutions*, we mean a mixture in which one component is clearly in excess. In other words, a solution has a clearly defined *solvent* (the majority component) and clearly defined *solutes* (the minority components). We will study the properties of nonelectrolyte (Chapter 12) and electrolyte (Chapter 13) solutions in more detail after we have introduced thermodynamic quantities such as the chemical potential and activity. *Electrolyte solutions* are those solutions that conduct electricity because the molecules that dissolve (i.e., the solute) dissociate into ions. *Nonelectrolyte solutions* do not conduct electricity easily because the solute remains neutral and undissociated. Now is a good time to review some useful notions about composition, as we will use these throughout the remainder of this text.

Consider a mixture with j components. If $j = 2$, we have a binary solution, for $j = 3$ a ternary solution, for $j = 4$ a quaternary solution, and so forth. The total number of moles is simply the sum over all of the individual moles of the components

$$n = \sum_j n_j. \tag{4.21}$$

The mole fractions of individual species (also called 'amount' fractions), for example for species A,

$$x_A = n_A / n, \tag{4.22}$$

must equal unity when summed over all components,

$$x = 1 = \sum_j x_j, \tag{4.23}$$

We will tend to concentrate on binary mixtures, for which from the above we see

$$x_A + x_B = 1 \quad \text{and} \quad x_A = 1 - x_B. \tag{4.24}$$

Ethanol and water are *miscible*: they mix in all proportions. Therefore, we can make binary mixtures of water in which the mole fraction of both components ranges freely from $0 \leq x \leq 1$. The solubility of hexane in water is about 9.5 mg l^{-1}. Therefore, hexane only forms solutions with water in which $x_{\text{hexane}} \ll 1$, that is, in which hexane is clearly the solute and water is clearly the solvent.

Molality b (or *m*; both are suggested by IUPAC but here I use *m* for mass) is defined as moles of solute divided by mass in kg of *solvent*. This unit has the advantage of being independent of temperature. Nowadays, with the wide availability of electronic balances it is also easy to prepare solutions by molality, though historically it was easier to prepare and dispense solutions according to volume. The official SI unit of concentration is moles of solute divided by volume of solution per cubic meter (mol m^{-3}). This is one instance in which SI defers to the convention of usage and allows the use of the more common definition of *molarity* in terms of moles of solute per liter of *solution* (mol l^{-1}, equivalently mol dm^{-3} or M). Molarity is denoted either by c or with square brackets around the component, for example [A]. Note that IUPAC now recommends calling this the *amount concentration*. Because volume is dependent on T, molarity has a slight dependence on T. In addition (as we shall see later), the distinction of volume of solution rather than volume of solvent is important. In non-ideal solutions, volumes are not additive: adding 50 ml of A to 100 ml of B does not lead to a final solution with a volume of 150 ml unless the solution is acting ideally.

Compare in aqueous solutions the molality

$$b = n_{\text{solute}} / m_{\text{solvent}} \tag{4.25}$$

to the molarity

$$c = \frac{n_{\text{solute}}}{V_{\text{solution}}} = \frac{n_{\text{solute}}}{m_{\text{solution}} / \rho_{\text{solution}}} = \frac{\rho_{\text{solution}} n_{\text{solute}}}{m_{\text{solution}}}. \tag{4.26}$$

Therefore, at low concentration in aqueous solutions where $m_{\text{solution}} \approx m_{\text{water}}$ and the density $\rho_{\text{solution}} \approx \rho_{\text{water}} \approx 1$ g cm^{-3}, then $c \approx b$. This does not hold at higher concentrations, as shown in Table 4.3.

Table 4.3 A comparison of concentration units (including weight % (wt%)) for HCl(aq) at $T = 25$ °C.

wt%	c /mol l^{-1}	b /mol kg^{-1}	x_{HCl}	density /g l^{-1}
0	0	0	0	0.9971
0.50	0.137	0.138	0.00249	0.9957
4.00	1.12	1.19	0.0210	0.9772
32.00	10.17	18.98	0.255	0.7884

4.7 Diffusion

Molecules in a fluid are constantly in motion. There is a constant flux of both matter and energy, though no net flux at any given point when the system is at equilibrium. We denote *flux*, the quantity of something passing through a unit surface area per unit time, in general as J. Matter flux has the units of molecules per unit area per unit time ($m^{-2}\ s^{-1}$), while energy flux has units of $J\ m^{-2}\ s^{-1}$. In a pure substance, even though the density is constant, there is *self diffusion* – the matter flux of molecules within a fluid relative to any arbitrarily drawn surface within or bounding the fluid – even at equilibrium. At equilibrium, the composition of a solution (regardless of the phase) must be uniform. Therefore, if a gradient in the concentration of one (or more) species exists, there is a chemical driving force (the chemical potential) that pushes the system to a completely random, uniform concentration. Diffusion is the mechanism that transforms a nonuniform, nonequilibrium concentration profile into a uniform, equilibrium distribution. We have yet to define the chemical potential, but in analogy to the potential energy or the electrical potential, you should have the intuitive feeling that just as a ball rolls to the bottom of a hill to minimize its potential energy, so too do chemical species move in response to differences in chemical potential as they attempt to minimize the energy of the system.

4.7.1 Fick's first law of diffusion

Matter flows from a region of high concentration to low concentration until a uniform concentration profile is achieved at equilibrium. The rate of diffusion is proportional to the concentration gradient, and the proportionality constant is known as the *diffusion coefficient, D*

$$J = \frac{\mathrm{d}n}{\mathrm{d}t} = -DA\frac{\mathrm{d}c}{\mathrm{d}x} \tag{4.27}$$

where J is the matter flux (moles per unit time of the solute passing through the area A) and $\mathrm{d}c/\mathrm{d}x$ is the concentration gradient. The diffusion coefficient is related to the mean distance that a molecule travels between collisions (a distance also know as λ, the *mean free path*) and the mean velocity $\bar{v}$ by

$$D = \frac{1}{2}\lambda\bar{v}. \tag{4.28}$$

At equilibrium, the chemical potential is equal everywhere in the sample. If a concentration gradient is present, the sample is not at equilibrium; rather, the chemical potential depends on the position, and the gradient in chemical potential leads to a thermodynamic force that pushes diffusion in the direction to lower chemical potential.

The flux is related to the concentration c and *drift speed s* by

$$J = sc. \tag{4.29}$$

The *Einstein relation* connects the diffusion coefficient D and the *frictional coefficient f*

$$D = k_\mathrm{B}T/f. \tag{4.30}$$

A typical value of the diffusion coefficient is $D \approx 1 \times 10^{-9}\ m^2\ s^{-1}$ at 298 K for an ion in water. The diffusion coefficient can also be related to the limiting molar conductivity $\lambda°$ as well as to the coefficient of viscosity η of the medium. The *Nernst–Einstein equation* is

$$D = (RT/Fz)\lambda°. \tag{4.31}$$

The *Stokes–Einstein equation* gives

$$D = k_BT/6\pi\eta r \tag{4.32}$$

where F is the Faraday constant, z is the magnitude of the charge on the ion, and r is the radius of the molecule.

4.7.2 Fick's second law of diffusion

This relates the rate of change of concentration at a point to the spatial variation of the concentration at that point

$$\frac{\partial c(x,t)}{\partial t} = D\frac{\partial^2 c(x,t)}{\partial x^2}. \tag{4.33}$$

The exact solution of Eq. (4.33) depends on the boundary conditions. For example, let us say that at time $t = 0$, all of the molecules of interest are at $x = 0$ and packed into an area A. Effectively, all of the molecules are on one wall of the

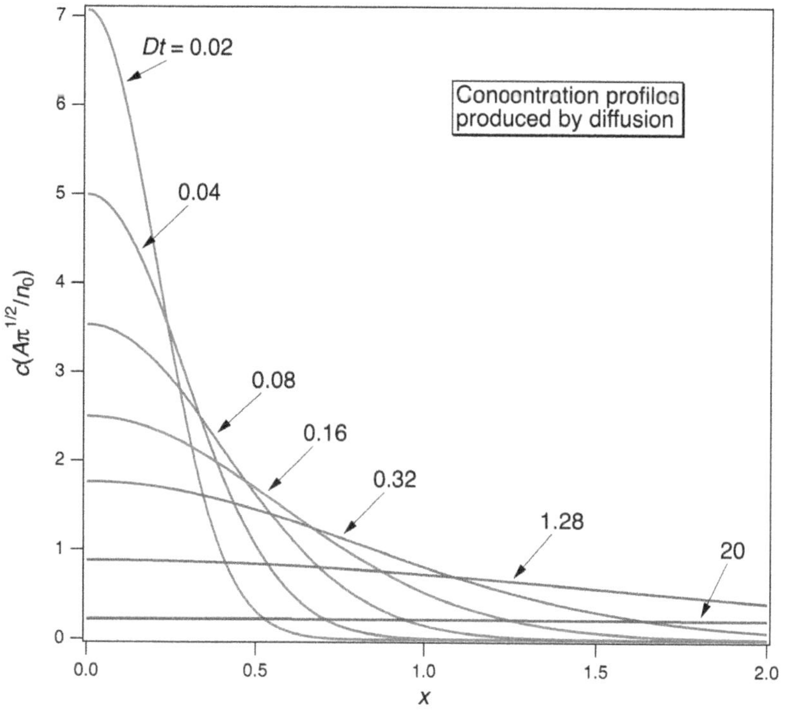

Figure 4.4 Diffusion profiles as a function of distance x and reduced time Dt as indicated.

container, and the area of the wall is A. We then allow the molecules to diffuse as a function of time, with the further conditions that the concentration everywhere is finite and that the total number of moles of molecules is constant at n_0. The solution, the form of which is known as the error function, is given by

$$c(x,t) = \frac{n_0}{A(\pi Dt)^{1/2}} \exp\left(\frac{-x^2}{4Dt}\right) \tag{4.34}$$

The concentration profile given by Eq. (4.34) can be found in Fig. 4.4. The curves are labeled with different values of Dt. Equation (4.34) also allows us to calculate the mean distance traveled by a diffusing particle in time t

$$\langle x \rangle = (4Dt/\pi)^{1/2}, \tag{4.35}$$

which from the Stokes–Einstein equation can be related to the *viscosity* η

$$\langle x \rangle = \left(\frac{2k_B Tt}{3\pi^2 \eta r}\right)^{1/2}. \tag{4.36}$$

The root mean square displacement in one dimension is

$$x_{rms} = \langle x^2 \rangle^{1/2} = (2Dt)^{1/2}. \tag{4.37}$$

In a two-dimensional system the distance traveled r is related to the distance traveled along x and y by

$$r^2 = x^2 + y^2$$

and in three dimensions

$$r^2 = x^2 + y^2 + z^2$$

Therefore

$$r_{rms,2D} = \langle x^2 + y^2 \rangle^{1/2} = (4Dt)^{1/2}, \tag{4.38}$$

and

$$r_{rms,3D} = \langle x^2 + y^2 + z^2 \rangle^{1/2} = (6Dt)^{1/2}. \tag{4.39}$$

Table 4.4 Dependence of the exponent on diffusion type.

Exponent	Diffusion type
$n < \frac{1}{2}$	pseudo-Fickian
$n = \frac{1}{2}$	Fickian
$\frac{1}{2} < n < 1$	Anomalous
$n = 1$	Case II
$n > 1$	supercase II

Diffusion can be thought of as a process in which a series of steps of the same length x_0 are taken in random directions (a random walk or the 'drunken sailor' problem). The diffusion coefficient can then be related to the step length and the rate at which the steps occur. This yields the *Einstein–Smoluchowski equation*

$$D = x_0^2/2\tau = \lambda^2/2\tau \tag{4.40}$$

From this equation we can interpret the diffusion coefficient in terms of steps that are equal in length to the mean free path λ and the time between collisions τ. The *mean free path* for a fluid with hard-sphere diameter σ and number density n is

$$\lambda = 1/\left(\sqrt{2}\pi\sigma^2 n\right). \tag{4.41}$$

4.7.3 Non-Fickian diffusion

Take a membrane and put it in contact with a fluid. The fluid will diffuse into the membrane, causing its mass to change. Take m_t to be the mass at time t and m_∞ the asymptotic limit of the mass for infinite time. The mass uptake follows the general equation

$$m_t/m_\infty = k_n t^n. \tag{4.42}$$

The type of diffusion determines the value of the exponent n. The classifications of diffusion types are listed in Table 4.4. The constant k_n depends on the materials chosen, while Fickian diffusion – the type described in the two sections above – is the most commonly observed. During the swelling of polymers into the gel state in the presence of a solvent, a different type of diffusion – called Case II diffusion – is observed.

An important application of Case II diffusion is in drug delivery. A controlled release drug/polymer blend is made from a low-molecular-weight drug and a *hydrogel*. The latter is a crosslinked hydrophilic network that swells in aqueous solutions. The swelling often responds to pH – a property that can be used to enhance the drug delivery response. For example, a hydrogel that shrinks at low pH will shut down diffusion when exposed to gastric juices, which protects an acid-sensitive drug from being digested in the stomach. Subsequently, the more alkaline environment found in the intestine leads to greater swelling and release of the drug.

4.8 Viscosity

Fluid flow through a tube can be modeled as the flow of a series of layers. The layer in contact with the wall of the tube is stationary, while the velocity at the center of the tube is a maximum. In Newtonian flow, the velocity of the molecules increases linearly moving away from the wall. In Fig. 4.5, we define the x-axis along the direction of flow and the z-axis perpendicular to it. The transport of molecules perpendicular to the flow direction leads to a transfer of molecules with either greater or fewer x-components of velocity between flow layers. The net retarding effect engendered by the transport of slow molecules from the boundary region, as for molecule (ii) in Fig. 4.5, results in the property referred to as *viscosity*. For a molecule that is large compared to the layer thickness, there is an additional contribution to the viscosity. Consider a long rod-like molecule in solution that is oriented perpendicular to the wall, as in case (iii) in Fig. 4.5. The end near the wall collides with slow solvent molecules, while the end near the center collides with faster molecules. This leads to momentum transfer that causes rotation of the molecule. Similarly, a coiled polymer will also experience shear forces than lead to rotational motion and an added frictional drag contribution to the viscosity.

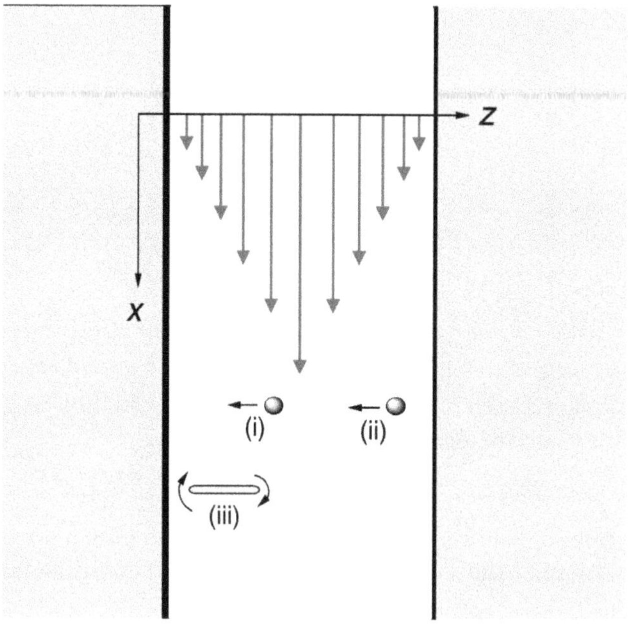

Figure 4.5 Velocity of a fluid increases linearly from the wall. Transport of molecules perpendicular to the flow direction leads to a transfer of molecules with (i) greater or (ii) lesser x-component of velocity between the flow layers. Transport by case (ii) results in the property known as viscosity. A long rod-like molecule in solution that is oriented perpendicular to the wall, as in case (iii), experiences a torque and further increases the viscosity.

The drag arises from transfer of the x-component of linear momentum p_x between layers. If all the flowing layers were moving at the same velocity, there would be no drag. However, the presence of a gradient in z of the flow velocity along x gives rise to the viscosity. The flux of the x-component of linear momentum is related to this gradient by the coefficient of viscosity by

$$J_{p_x} = -\eta \frac{\mathrm{d}v_x}{\mathrm{d}z}. \tag{4.43}$$

The coefficient η is more often simply called the *viscosity*. The units of viscosity are Pa·s or equivalently kg m^{-1} s^{-1}. A common unit is poise (P), where 1 P = 0.1 Pa·s.

A number of different terms are encountered when considering viscosity, particularly the viscosity of solutions. The relative viscosity η_{rel} is the ratio of the solution viscosity η to the solvent viscosity η_0,

$$\eta_{\mathrm{rel}} = \eta/\eta_0. \tag{4.44}$$

The specific viscosity is the relative velocity minus one

$$\eta_{\mathrm{sp}} = \eta_{\mathrm{rel}} - 1. \tag{4.45}$$

The intrinsic viscosity is found by dividing the specific viscosity by concentration and then extrapolating a graph of this ratio to zero concentration, hence,

$$\lim_{c\to 0} \left(\eta_{\mathrm{sp}}/c\right) = [\eta] \tag{4.46}$$

For dilute solutions the relative viscosity is close to unity. Therefore, we can take the logarithm of Eq. (4.45) and then expand using the approximation $\ln(x + 1) = x$, to obtain

$$\ln \eta_{\mathrm{rel}} = \ln\left(\eta_{\mathrm{sp}} + 1\right) = \eta_{\mathrm{sp}}. \tag{4.47}$$

Substituting into Eq. (4.46), we find

$$\lim_{c\to 0} \left(\frac{\ln \eta_{\mathrm{rel}}}{\mathrm{c}}\right) = [\eta]. \tag{4.48}$$

This ratio is called the *inherent viscosity*.

Table 4.5 Values of the Mark–Houwink–Sakurada exponent a.

a	**interpretation**
0	spheres
0.5–0.8	random coils
1	stiff coils
2.0	rods

Assuming that the solution is dilute and the solute is composed of noninteracting spheres, Einstein showed that the solution viscosity is increased compared to the solvent viscosity in proportion to the volume fraction of the spheres according to

$$\eta = (1 + 2.5v_2)\eta_0. \tag{4.49}$$

Note that Eq. (4.49) states that the viscosity of a solution of spheres depends directly on the volume fraction of the spheres, and is independent of their size.

Let us model a coiled polymer macromolecule as a sphere that is impenetrable to solvent to a first approximation. The coil dimensions are approximated by a hydrodynamic sphere of equivalent radius R_e and volume $V_e = \frac{4}{3}\pi R_e^3$. The volume fraction of N_2 molecules in a total volume V is

$$v_2 = N_2 V_e / V. \tag{4.50}$$

The concentration of molecules is $c = N_2/V$ or

$$\frac{N_2}{V} = \frac{cN_A}{M} \tag{4.51}$$

where N_A is the Avogadro constant and M the molar mass. Thus, Eq. (4.49) can be re-arranged to

$$\frac{\eta - \eta_0}{\eta_0} = 2.5\frac{cN_A V_e}{M}. \tag{4.52}$$

Recognizing that $\eta_{sp} = (\eta - \eta_0)/\eta_0$, we see that the intrinsic viscosity is inversely related to the molar mass by

$$\lim_{c\to 0} \frac{\eta_{sp}}{c} = [\eta] = 2.5\frac{N_A V_e}{M}. \tag{4.53}$$

The *Mark–Houwink–Sakurada equation* is an empirical relationship between the viscosity-average molar mass M_V and intrinsic viscosity,

$$[\eta] = KM_v^a. \tag{4.54}$$

The constants K and a depend on the specific combination of polymer and solvent as well as temperature. As shown in Table 4.5, the value of the exponent a varies from 0 to 2 and can be interpreted in terms of the shape of the polymer coil in solution.

4.9 Migration

Under the influence of an electric field, ions in solution will move with directed motion. This motion is in addition to diffusion and, in the absence of convection, is modeled by adding a term to the diffusion equation that is proportional to the electric field strength. By doing so we arrive at the *Nernst–Planck equation*

$$J_k(x) = -D_k\frac{\partial C_k}{\partial x} - \frac{z_k F D_k C_k}{RT}\frac{\partial \phi}{\partial x}. \tag{4.55}$$

If convection is present an additional term is added to the Nernst–Planck equation. In Eq. (4.55), $J_k(x)$ is the flux of species k at distance x from the electrode, D_k is the diffusion coefficient, C_k the concentration, $\partial c_k/\partial x$ the concentration gradient, $\partial\phi/\partial x$ the potential gradient, and z_k the charge on species k. In the absence of a concentration gradient, the diffusion term vanishes. In the absence of an electric field, the migration term vanishes.

The current density flowing as the result of diffusion or mobility is $-z_k F$ times the flux. Thus, the total current density can be decomposed into two components, the diffusion current density and the migration current density

$$z_k F J_k(x) = j_{\mathrm{d},k} + j_{m,k}. \tag{4.56}$$

The migration current density is thus

$$j_{\mathrm{m},k} = \frac{z_k^2 F^2 D_k C_k}{RT}\frac{\partial \phi}{\partial x}. \tag{4.57}$$

This can be rewritten in terms of the *mobility* u_k of species k

$$u_k = \frac{|z_k| F D_k}{RT} \tag{4.58}$$

as

$$j_{\mathrm{m},k} = |z_k| F u_k C_k \frac{\partial \phi}{\partial x}. \tag{4.59}$$

The total current density flowing in a solution as the result of the motion of a number of different ions is the sum of the individual current densities,

$$j = \sum_k j_k. \tag{4.60}$$

Assume now that we have a bulk solution with no concentration gradient so that the total current density is equal to the migration current density, that is, $j_k = j_{\mathrm{m},k}$. The fraction of the total current carried by an individual ion is called the *transference number* t_k, and is given by substitution of Eq. (4.59) into Eq. (4.60) as

$$y_k = \frac{j_k}{j} = \frac{|Z_k| u_k C_k}{\sum_k |z_k| u_k C_k}. \tag{4.61}$$

The direction of the migration current and the diffusion current need not be the same. They depend on the sign of the charge and the electric field gradient for migration and the sign of the concentration gradient for diffusion.

4.10 Interface formation

Having formed a liquid, there now exist two possible interfaces between different phases: the gas/liquid interface, and the liquid/solid interface. Here, we differentiate between a surface and an interface. A *surface* is formed where the last molecules of a liquid or solid exist. As we shall see shortly, the surface of a solid is much better defined than the surface of a liquid. An *interface* is where two phases meet. An interface can be a region of space, whereas a surface is a two-dimensional (though potentially rough) object. The liquid/solid interface is abrupt, as the solid ends in a more or less well-defined atomic or molecular layer. We will discuss the liquid/solid interface further in Chapter 5. The gas/liquid interface, on the other hand, changes more gradually from the composition (density) of the bulk liquid to the composition of the gas. Even in the ideal case, the surface of a liquid is generally not flat due to the presence of capillary waves.

Whereas it is possible for a substance to exist purely in the gas phase, it is impossible to form a substance solely in the liquid phase. Any liquid has a finite vapor pressure. The equilibrium vapor pressure is the partial pressure of a compound that exists at equilibrium above a liquid of that compound. The vapor pressure may have a very small value; for instance, that of Hg is roughly 10 Pa at 298 K and that of liquid Au is 580 Pa at 2400 K. The vapor pressure increases with increasing temperature. The normal boiling point is defined as the temperature at which the vapor pressure is equal to 1 atm (the standard boiling point is the temperature at which the vapor pressure is 100 kPa = 1 bar). In order to determine the temperature dependence of vapor pressure, we need to know something about the equilibrium constant between the liquid and vapor phases. Unfortunately, this requires knowledge of the relative Gibbs energies of the liquid and vapor, and so will have to await a definition of these thermodynamic quantities.

As long as we do not look at a molecular level, we can consider the gas/liquid interface to be smooth. Later, we will encounter the Gibbs thermodynamic model of the gas/solid interface. We will see how it is possible to define a plane – the Gibbs plane – that is called 'the surface', but at the molecular level the transition from gas to liquid is continuous

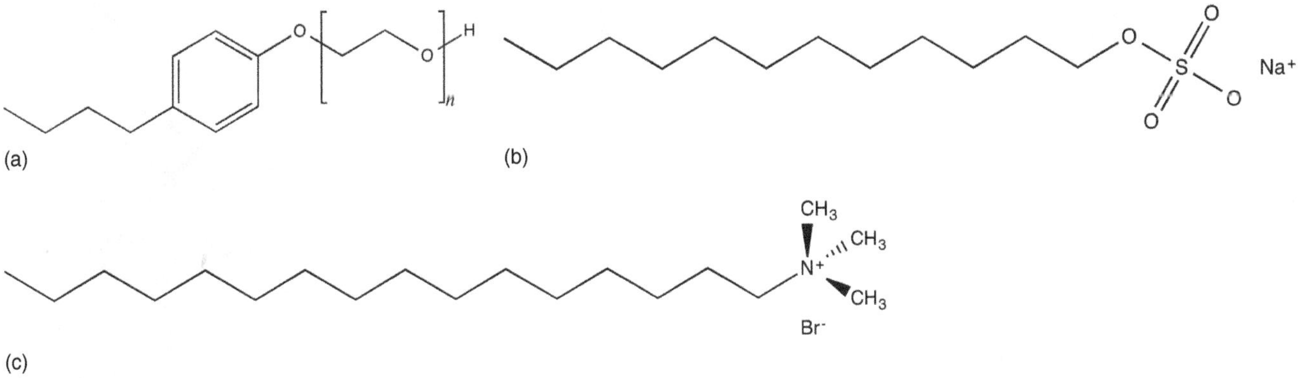

Figure 4.6 Three common surfactants. (a) Triton X-100, a neutral surfactant. (b) Sodium dodecyl sulfate (SDS), an anionic surfactant. (c) Cetyltrimethylammonium bromide (CTAB), a cationic surfactant.

and in general is not flat. Liquids suspended as aerosol particles in a gas assume a spherical shape. They do so because the formation of a surface is an endothermic process. Therefore, to minimize their energy they assume the most compact geometrical form, the sphere.

While liquid interfaces may be curved, we have not yet taken into account dynamic fluctuations in their shape. When considered at a molecular level, the gas/liquid interface is neither sharp nor immobile. A liquid interface at equilibrium continually exchanges material not only between the surface and bulk of the liquid, but also between the surface and the gas phase in order to maintain the equilibrium vapor pressure. At the molecular level, the gas/liquid interface is continually changing. Any snapshot of this interface has a tendency to deviate from a well-defined plane or simple interfacial region, even though any deviation from a smooth profile increases the surface area and energy from their minimum values. These deformations in the liquid surface are known as *capillary waves*.

The properties of capillary waves are changed by the presence of a layer of molecules on the top of the liquid. The calming effect of placing oil on water was studied by Benjamin Franklin. Irving Langmuir noted that monomolecular layers of *amphiphiles* (molecules that are polar on one end and nonpolar on the other, such as an acid with a long hydrocarbon chain) form readily on water. These Langmuir films not only change the properties of the water interface, but they also exhibit interesting properties of their own, such as Newton's rings, the films colored by white light interference effects that form on puddles of water, especially near asphalt-covered streets.

Amphiphiles are commonly found in detergents and many other household products. They may be anionic, cationic, or neutral. Common examples are shown in Fig. 4.6. Triton X-100 is a neutral surfactant with molecular formula $C_{14}H_{22}O(C_2H_4O)_n$, where n is usually 9 or 10. The molecule has a polar —OH head and nonpolar $—CH_3$ tail groups at the other end. Sodium dodecyl sulfate (SDS; $CH_3(CH_2)_{11}OSO_3Na$) is an anionic surfactant, again with a nonpolar $—CH_3$ tail but with a sulfate group as a head. An example of a cationic surfactant is a quaternary ammonium salt such as cetyltrimethylammonium bromide (CTAB; $(C_{16}H_{33})N(CH_3)_3Br$), which is widely used in DNA extraction, nanoparticle synthesis, and as a hair conditioner. Amphiphiles in the form of phospholipids form part of the construction of cell walls. Three structures commonly formed by amphiphiles in aqueous environments are shown in Fig. 4.7. All of these structures aim to minimize the interaction of the less attractive nonpolar end of the molecule with the polar solvent, while simultaneously maximizing the interaction of the polar end of the molecule with water.

At low concentrations, amphiphiles segregate to the surface of the liquid and form Langmuir layers. Some solubility in the solvent (most commonly water) is required for *micelle* formation. Amphiphiles with two alkyl chains generally are insoluble in water and do not form micelles. Micelles form in measureable quantities above a critical concentration known as the *critical micelle concentration* (c_M or cmc). A micelle is a three-dimensional structure ranging from spherical to cylindrical and even dumbbell-like, in which the polar head group points outward and the nonpolar tails face each other on the inside of the sphere. The liposome and bilayer sheet both protect the nonpolar ends by aligning nonpolar tail groups across from one another, while the polar head groups face outwards towards the water. *Liposomes* are spheres that encapsulate water at their center.

Amphiphiles will also segregate to the interface of two immiscible liquids, for example the oil/water interface as shown in Fig. 4.8. The amphiphiles align themselves to maximize the attractive interactions of the polar end with the water and the nonpolar end with the oil.

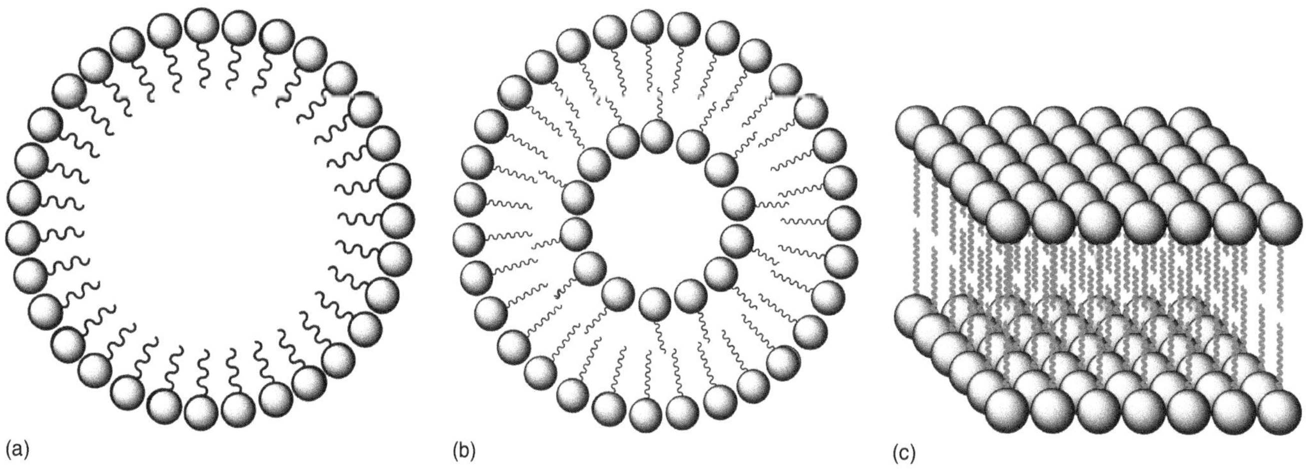

Figure 4.7 Self-assembled structures formed from amphiphiles. (a) Micelle. (b) Liposome. (c) Bilayer sheet. Both micelles and liposomes are three-dimensional structures (spheres, cylinders, even dumbbells). The figures show a cross-section to emphasize the single-layer versus double-layer structure of micelles in comparison to liposomes.

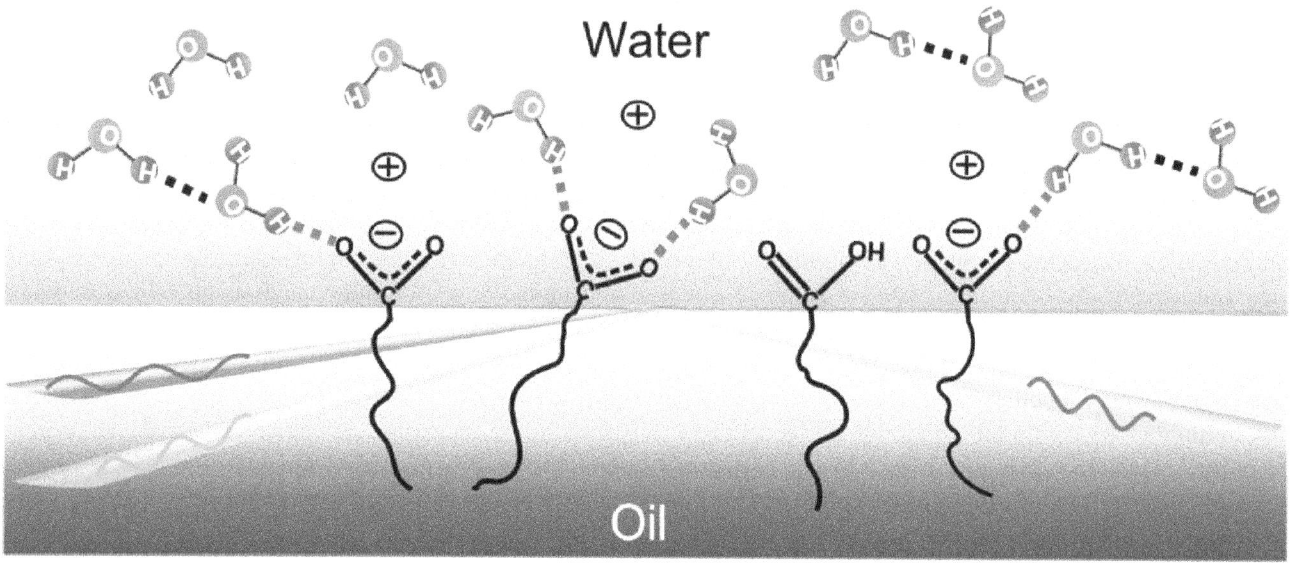

Figure 4.8 Amphiphiles segregate to the oil/water interface and align themselves to maximize the attractive interactions with the two solvents. The structure is similar to that of a Langmuir film formed on water in the absence of the oil. Image provided by Geraldine L. Richmond. Reproduced with permission from Beaman, D.K., Robertson, E.J., and Richmond, G.L. (2011) *J. Phys. Chem. C*, **115**, 12508. © 2011; American Chemical Society.

4.11 Liquid crystals

Liquid crystals are materials that exhibits long-range order in one or two dimensions but not in three. While both small molecules and polymers may exist in the liquid crystalline state, generally special structural features of the molecules are required. Liquid crystal polymers have been known since the 1950s. The formation of liquid crystals is a direct consequence of asymmetrical molecular structure. Primarily driven by entropy, this state occurs because of short-range repulsions that prevent molecules from occupying the same space.

Liquid crystals are composed of rod-like or plate-like molecules. One class of liquid crystalline polymer has rod-shaped side chains. Rods must have an axial ratio of greater than about three to form liquid crystals. Irrespective of the molecular size or shape, the following three requirements must be met for a material to be classified as a liquid crystal.

- There must be a first-order phase transition from a low temperature crystalline state to the liquid crystalline state. This must be followed by another first-order phase transition into the isotropic liquid state. There may be more than one liquid crystalline state, each separated by a first-order phase transition, before the final transition to the isotropic melt.

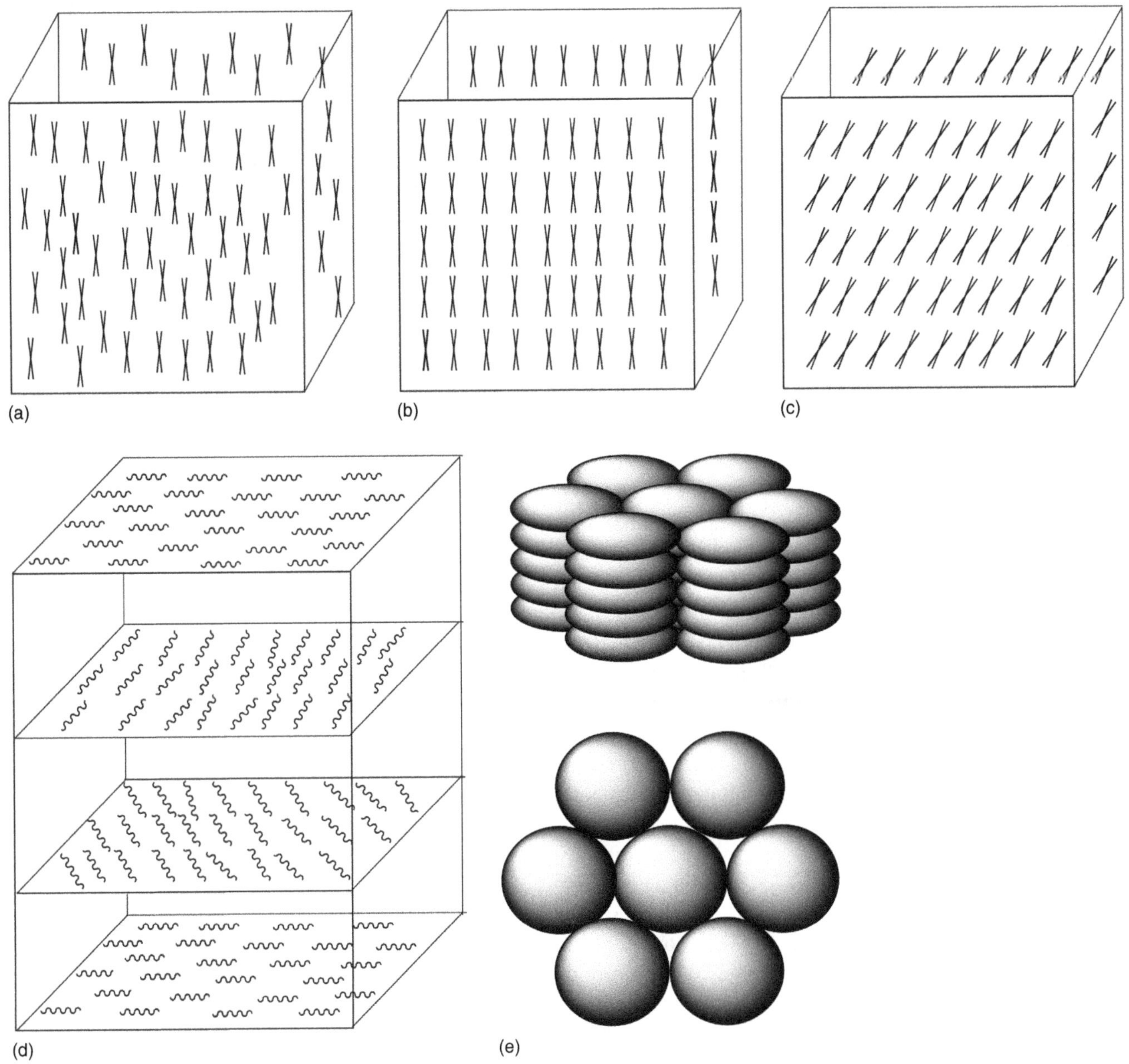

Figure 4.9 Liquid crystal structures. (a) Nematic; (b) smectic A; (c) smectic C; (d) cholesteric; and (e) discotic (side view at top, top view at bottom).

- A liquid crystal exhibits either one- or two-dimensional order. This differentiates it from a true crystal, which has three-dimensional order and a true liquid, which is isotropic.
- While polymer liquid crystals may be viscous, the liquid crystalline phase must exhibit some degree of fluidity.

At low concentration they form a disordered phase. Above a certain coverage they align in a plane or a layer in order to avoid crossing one another. Within the layer, there may be areas (called *domains*) of different azimuthal orientation. An analogy can be drawn to toothpicks placed on the surface of water. At first, the toothpicks are randomly ordered, but as their number density increases they begin to align along their long axis. The preferential alignment in one domain (one region of the water surface) is independent of the alignment of a region. As the domains grow in size, they will eventually meet and a domain boundary or wall is formed separating the region of different orientation.

Liquid crystals are organized into several classes, called *mesophases*, as shown in Fig. 4.9. The toothpick analogy corresponds to parallel alignment of the rod axis in a nematic phase, which is organized in one dimension only. The smectic liquid crystals are organized in two dimensions. Other forms of smectic ordering are known, but types A and C are the most common. The cholesteric mesophase is much like a nematic structure but with a rotation between adjacent layers. The discotic mesophase resembles the stacking of plates.

Polymers that form liquid crystals sometimes form multiple mesophases depending on the temperature and pressure. A first-order phase transition is observed to separate one mesophase from another. The most ordered mesophase is found at low temperature, and increasingly disordered mesophases form at higher temperatures when and if these are observed. The final first-order phase transition from a liquid crystal mesophase to the isotropic melt occurs at what is termed the *clearing temperature*.

When a liquid crystal is sufficiently thermally stable that it is able to pass through the phase transition from liquid crystal to isotropic melt, it is called a *thermotropic* liquid crystal. Liquid crystals that decompose at least partially before melting completely are termed *lyotropic* liquid crystals. Lyotropic liquid crystals are often processed in the solution phase. In concentrated solutions, lyotropic liquid crystals form ordered structures which facilitate the formation of well-ordered liquid crystal phases upon evaporation of the solvent.

4.11.1 Directed practice

Would a spherical molecule form a liquid crystal?

4.12 Ionic liquids

Ionic liquids are salts that are liquid at or near room temperature as opposed to the high temperatures that are required to melt conventional salts; for example, NaCl melts at 1074 K. They are also referred to as room temperature ionic liquid (RTIL), nonaqueous ionic liquid, liquid organic salt, and fused salt. A practical definition is that ionic liquids are salts that melt below 100 °C, whereas those that melt above 100 °C are normal molten salts.

The low melting point and large liquidus range (usually about 200 K) of ionic liquids is due to the structure of the anions and cations of which they are composed. The general form of an ionic liquid is a large asymmetrical organic cation paired with an inorganic anion with various degrees of complexity. Imidazolium, pyridinium, quaternary phosphonium and ammonium salts form the largest classes of ionic liquids. These units are shown in Fig. 4.10, while some of the most common cationic structures are displayed in Fig. 4.11. The alkyl groups associated with the anions and cations range in size from one to 10 carbons, with even-numbered units being more common.

The ability to mix and match anions and cations of various sizes allows for an extraordinary number of potential compounds with tunable properties, such as hydrophobicity, solubility, density, and viscosity. For instance, the density can be increased by decreasing the alkyl chain length in the cation and increasing the molar mass of the anion. Their viscosity is usually comparable to that of an oil. This results in low diffusion rates and concomitantly low reaction rates.

Nonetheless, some general features of the majority of ionic liquids can be stated. Ionic liquids combine high polarity because of the charge separation between anion and cation with bulky nonpolar organic groups. Because of this they dissolve a wide range of organic, organometallic, inorganic compounds, biomolecules, and metal ions. Cellulose is the most widely available biorenewable material. It is insoluble in water and most common organic solvents. However, it can be dissolved in ionic liquids with anions that are strong hydrogen bond acceptors. Nonetheless, many organic solvents as well as water are immiscible in ionic liquids, which makes them suitable as a nonaqueous substitute for phase-transfer processes. The usual rule of solvophilicity (the degree to which one thing dissolves in another) is that like dissolves like. This is difficult to apply to ionic liquids because they are not even 'like' themselves, since they are composed of both nonpolar and ionic regions. Ionic liquids can also be used as biocompatible solvents for microbial transformations.

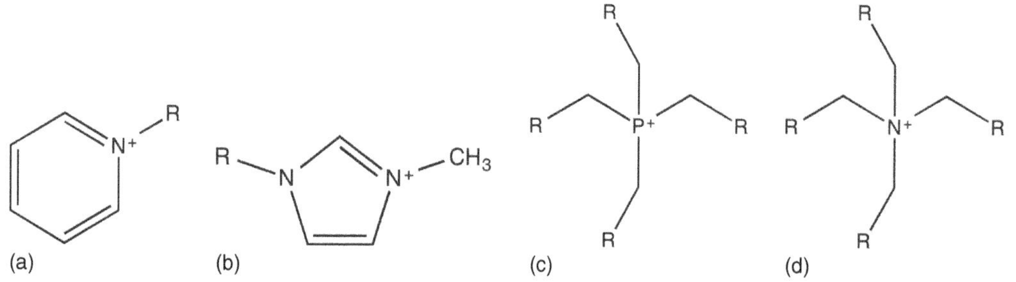

Figure 4.10 Most commonly used cations in ionic liquids. (a) Pyridinium derivative. (b) Imidazolium derivatives. (c) Quaternary phosphonium salt. (d) Quaternary ammonium salt. R is usually a butyl group, though it can range from one to 10 C atoms.

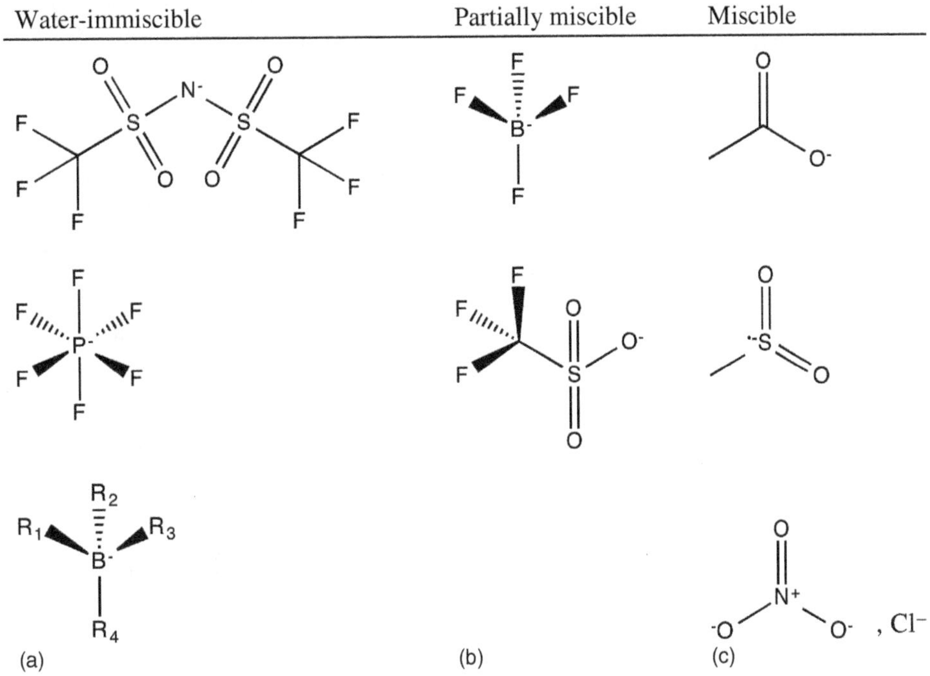

Figure 4.11 Most commonly used anions in ionic liquids. (a) Anions such as bis(trifluoromethylsulfonyl)amide, hexafluorophosphate and tetra-alkylborate lead to water-immiscible ionic liquids. (b) Anions such as tetrafluoroborane and trifluoromethansulfonate have intermediate miscibility in water. (c) Ethanoate, methyl sulfonyl anion, nitrate and chloride lead to fully water-miscible ionic liquids.

Ionic liquids also exhibit very low vapor pressures, on the order of 10^{-11} Pa at room temperature in some cases, and are generally stable up to 600 K. Nonvolatility means that they do not impact air quality nor produce explosive vapors, in contrast to the volatile organic compounds that represent conventional organic solvents. They can also be recycled and repeatedly used. This combination of properties has led to ionic liquids being called 'green' solvents. An extremely low vapor pressure means that they are stable when placed in a vacuum, which means that water can be removed from the liquids by vacuum distillation. Ionic liquids also exhibit large thermal and electrical conductivities, which makes them suitable as solvents for electrochemistry over a wide range of temperatures and operating voltages.

Charge distribution on the ions, H-bonding ability, polarity, and dispersive interactions are factors that determine the solvophilicity of ionic liquids. The structure of the cation usually determines the hydrogen-bonding character; for example, imidazolium-based ionic liquids are highly ordered hydrogen-bonded solvents. The nature of the interaction with water is mainly controlled by the anion, with a lesser but non-negligible contribution from the cation. PF_6-based ionic liquids are immiscible in water, which makes them suitable for biphasic extraction applications. However, other halogen-containing anions, especially BF_4^- are more hydrophilic and can even react with water to form HF. Anions such as $CF_3SO_3^-$ and $(CF_3SO_3)_2N^-$ are much less susceptible to hydrolysis. Imidazolium and ammonium salts are usually hydrophilic, but hydrophobicity can be increased by increasing the length of the alkyl chains.

The combination of nonvolatile, polar ionic liquids with volatile, nonpolar sc-CO_2 represents a two-phase system suitable for separations. Supercritical CO_2 dissolves in ionic liquids, but the reverse is not true. Even at pressures as high as 400 bar, the components remain in two separate phases. Dissolution of CO_2 into an ionic liquid decreases the viscosity of the solution. Organic molecules (which are generally highly soluble in sc-CO_2) when dissolved in an ionic liquid can be extracted when the solution is exposed to sc-CO_2.

Ionic liquids have become more commonly used in biochemical reactions such as the synthesis of pharmaceuticals because of the stability of enzymes in ionic liquids. They also act as optimal solvents in some separation processes, such as the extraction of amino acids. Ionic liquid biphasic systems are used to separate many biologically important molecules such as carbohydrates and organic acids, including lactic acid. Carbohydrates are renewable and inexpensive sources of energy and raw materials for the chemical industry. Underivatized carbohydrates are not soluble in most conventional solvents, although they are soluble in water. Their lack of solubility inhibits the facile chemical transformation of carbohydrates into more useful products. Therefore, the ability of ionic liquids to dissolve carbohydrates increases transformation possibilities.

SUMMARY OF IMPORTANT EQUATIONS

$U(R) = 4\varepsilon\left[\left(\frac{\sigma}{R}\right)^{12} - \left(\frac{\sigma}{R}\right)^{6}\right]$	Lennard–Jones potential. σ = hard-sphere diameter, R = intermolecular separation, ε scales the interaction.
$U(R) = A\exp(-kR) - C_6R^{-6}$	(exp-6) potential.
$g(R) = \rho(R)/\rho_0$	Radial distribution function. $\rho(R)$ = number density as a function of R, ρ_0 = mean number density
$Z = 4\pi\rho_0\int_0^{R_m} R^2 g(R)dR$	Coordination number defined in terms of radial distribution function
$x_A = n_A/n$	Mole fraction of A. n_A = moles of A. $n = \sum_j n_j$ = total moles
$b = n_{solute}/m_{solvent}$	Molality
$c = n_{solute}/V_{solution}$	Molarity
$J = sc$	Matter flux. s = drift speed
$J = \frac{dn}{dt} = -DA\frac{dc}{dx}$	Fick's first law of diffusion. J = matter flux (moles per unit time passing through area A), dc/dx = concentration gradient
$D = \frac{1}{2}\lambda\bar{v}$	Diffusion coefficient. λ = mean free path, $\bar{v}$ = mean speed
$D = k_BT/f$	Einstein relation for diffusion coefficient in terms of the friction coefficient f
$D = (RT/Fz)\lambda°$	Nernst–Einstein relation for diffusion coefficient of an ion of charge z and limiting molar conductivity $\lambda°$, F = Faraday constant
$D = k_BT/6\pi\eta r$	Stokes–Einstein equation. η = coefficient of viscosity, r = radius of molecule
$D = x_0^2/2\tau = \lambda^2/2\tau$	Einstein–Smoluchowski equation. x_0 = step length, λ = mean free path, τ = time between collisions
$\lambda = 1/\left(\sqrt{2}\pi\sigma^2 n\right)$	Mean free path. σ = hard-sphere diameter, n = number density
$\frac{\partial c(x,t)}{\partial t} = D\frac{\partial^2 c(x,t)}{\partial x^2}$	Fick's second law of diffusion
$c(x,t) = \frac{n_0}{A(\pi Dt)^{1/2}}\exp\left(\frac{-x^2}{4Dt}\right)$	Diffusion concentration profile
$r_{rms} = (2\Delta Dt)^{1/2}$	Root mean square displacement due to diffusion in time t. Δ = dimensionality (1 for diffusion on a wire, 2 for a surface, 3 for volume)

Exercises

4.1 Calculate the energetic minimum U_{min} and its position R_{min} for the Lennard–Jones potentials of He, Ar, and Xe.

4.2 The interaction potential between two F_2 molecules is represented by a Lennard–Jones potential. Calculate the force between two F_2 molecules at 3, 6, and 9 Å.

4.3 In a dilute binary solution in which x_1 is the mole fraction of the solvent and x_2 is the mole fraction of the solute, use the definition of mole fraction of x_1 to show that

$$\ln x_1 = -M_1 b_2 \tag{4.55}$$

4.4 For a real fluid described by a hard-sphere potential the following equation of state can be derived

$$p = \frac{RT}{V_m}\left(1 + \frac{2\pi N_A\sigma^3}{3V_m}\right). \tag{4.56}$$

This equation is valid under the assumption that the second term in the parentheses is small compared to 1. Show that for Ar, this equation represents a very good approximation for the equation of state of a gas, but is a poor approximation for the liquid. σ = 3.405 Å, V_m(g, 273 K) = 22.4 dm^3 mol^{-1}, V_m(l, 87 K) = 0.0286 dm^3 mol^{-1}.

4.5 Calculate the mean distance between solute molecules present in a solution at concentration c expressed in molar units.

4.6 The enthalpy of vaporization of water is 40.65 kJ mol^{-1} at 373 K, but only 44.30 kJ mol^{-1} at 298 K Explain on the basis of intermolecular interactions the difference.

4.7 The viscosity of water is 0.890 mPa·s at 298 K. Calculate the diffusion coefficient of a spherical 100 kDa protein with a radius of 3 nm.

4.8 Explain how and why diffusion in a solution in the presence of a concentration gradient is different from diffusion in a solution with a uniform concentration profile.

4.9 If the mean lifetime of an adsorbed CO molecule is 4.80×10^{-5} s, calculate the root mean square distance travelled by a diffusing CO molecule at this temperature.

4.10 Describe the process of diffusion physically and in terms of Fick's first law. Calculate the diffusion constant of Ar at 25 °C and (a) 1.00 Pa, (b) 100 kPa, and (c) 10.0 MPa. If a pressure gradient of 0.10 atm cm^{-1} is established in a pipe, what is the flow of gas due to diffusion?

4.11 The diffusion coefficient of CCl_4 in heptane at 25 °C is 3.17×10^{-9} m^2 s^{-1}. Estimate the time required for a CCl_4 molecule to have a root mean square displacement of 5.0 mm.

4.12 (a) The diffusion coefficient of lysozyme ($M = 14.1$ kg mol^{-1}) is 0.104×10^{-5} cm^2 s^{-1}. How long will it take this protein to diffuse an rms distance of 1 μm? Model the diffusion as a three-dimensional process. (b) You are about to perform a microscopy experiment in which you will monitor the fluorescence from a single lysozyme molecule. The spatial resolution of the microscope is 1 μm. You intend to monitor the diffusion using a camera that is capable of one image every 60 s. Is the imaging rate of the camera sufficient to detect the diffusion of a single lysozyme protein over a length of 1 μm? (c) Assume that in the microscopy experiment of part (b) you use a thin layer of water such that diffusion is constrained to two dimensions. How long will it take a protein to diffuse an rms distance of 1 μm under these conditions?

4.13 Given that the mean lifetime of an adsorbate is

$$\tau = \frac{1}{A} \exp(E_{des}/RT_s)$$

where A is the pre-exponential factor for desorption and E_{des} is the desorption activation energy, show that the mean random walk distance travelled by an adsorbate is

$$\langle x^2 \rangle^{1/2} = \left(\frac{4D_0}{A} \exp\left(\frac{E_{des} - E_{dif}}{RT} \right) \right)^{1/2}.$$

4.14 Estimate the effective radius in nm of a sucrose molecule in water at 25 °C, given that its diffusion coefficient is 5.2×10^{-10} m^2 s^{-1} and the viscosity of water is 1.00 cP (1 cP = 0.001 kg m^{-1} s^{-1}).

4.15 The diffusion coefficient of I_2 in benzene is 2.13×10^{-9} m^2 s^{-1} at 25 °C. The viscosity of benzene is 0.601 cP. How long does it take for I_2 to jump through a length of one molecular diameter? Take one molecular diameter to be the fundamental step size.

4.16 Using the diffusion coefficients from Exercise 4.15, what are the root mean square distances traveled in three dimensions by an I_2 molecule in benzene and by a sucrose molecule in water at 25 °C in 1 s.

4.17 Qualitatively, how would the answers in the previous exercises change if: (a) Br_2 were exchanged for I_2; (b) ethylene glycol were the solvent instead of water; (c) the temperature were decreased to 5 °C; or (d) an electric field were applied.

Further reading

Berry, R.S., Rice, S.A., and Ross, J. (2000) *Physical Chemistry*, 2nd edition. Oxford University Press, New York.

Hayes, R., Warr, G.G., and Atkin, R. (2015) Structure and nanostructure in ionic liquids. *Chem. Rev.*, **115**, 6357.

Israelachvili, J.N. (2011) *Intermolecular and Surface Forces*, 3rd edition. Academic Press, Burlington, MA.

Metiu, H. (2006) *Physical Chemistry: Thermodynamics*. Taylor & Francis, New York.

Murrell, J.N. and Boucher, E.A. (1994) *Properties of Liquids and Solutions*, 2nd edition. John Wiley & Sons, New York.

Poling, B.E., Prausnitz, J.M., and O'Connell, J.P. (2001) *The Properties of Gases and Liquids*, 5th edition. McGraw-Hill, New York.

Wagner, W. and Pruß, A. (2002) The IAPWS Formulation 1995 for the Thermodynamic Properties of Ordinary Water Substance for General and Scientific Use. *J. Phys. Chem. Ref. Data*, **31**, 387.

CHAPTER 5

Solids, nanoparticles, and interfaces

PREVIEW OF IMPORTANT CONCEPTS

- Either gases or liquids can condense to form solids.
- Solids form either crystalline or amorphous (glassy) structures.
- Solid crystalline materials are classified as either ionic, covalent, metallic, or molecular.
- Based on their electronic structure, in particular the magnitude of the band gap, solids are classified as metals, semiconductors, or insulators.
- The unit cell is a building block that can be translated and repeated through space to generate the structure of the entire crystal.
- Depending on the relative lengths of the lattice parameters, the cell angles, and the number of lattice points in a unit cell, solids are categorized within seven lattice systems and fourteen Bravais lattices.
- Each Bravais lattice has a distinct symmetry and number of lattice points.
- A plane of atoms is uniquely defined by the direction that is normal to the plane. A plane is denoted by enclosing the Miller indices of the defining direction in parentheses.
- Within a surface unit cell there are several high-symmetry sites that are geometrically unique and can act as adsorption sites with different coordination numbers.
- The reactivity of solids is usually dominated by their interfaces with gases, liquids, or other solids.
- The bonding at solid surfaces is different than in the bulk, and leads to different properties and chemical reactivity that is useful for catalysis.
- Nanoparticles (particles with radii below 100 nm) can exhibit size-dependent physical and chemical properties.
- Size dependence results from the increasing importance of surface effects and the appearance of quantum confinement effects.
- Carbon forms an unusually wide variety of solid allotropes including diamond, graphite, graphene, fullerenes, and carbon nanotubes.
- Porous solids form in a range of structures that are of interest in heterogeneous catalysis and other applications.
- Macromolecules and polymers form solid phases with properties that vary with the molar masses.

5.1 Solid formation

Decreasing the temperature or increasing the pressure on most liquids will eventually lead to solidification. At one atmosphere of pressure, the temperature of this phase transition is called the *normal melting point* or the *normal temperature of fusion*, T_f. At a pressure of 1 bar (the standard state pressure) this temperature is called the *standard melting point* or the *standard temperature of fusion*. In some cases a gas can be directly condensed into a solid. This is essential to the process of snowflake formation, and is also how solid CO_2 (dry ice) forms at standard pressure. The condensation of a gas directly to a solid phase is called *deposition*. The reverse process – the conversion of a solid to the gas phase – is called *sublimation*.

Solids contract with decreasing temperature, as described by their *expansion coefficient*

$$\alpha_V = \frac{1}{V}\left(\frac{\partial V}{\partial T}\right)_p. \tag{5.1}$$

For a simple cubic solid, the volume expansion coefficient α_V is the cube of the linear expansion coefficient α_l, some typical values of which are listed in Table 5.1. As can be seen from these values, the linear expansion coefficient varies by over one order of magnitude even for simple cubic solids.

Physical Chemistry: How Chemistry Works, First Edition. Kurt W. Kolasinski.

Companion Website: www.wiley.com/go/kolasinski/physicalchemistry

Table 5.1 Selected parameters of elements in the solid state regarding their melting point T_f, mass density ρ_s and linear expansion coefficient α_l. Values taken from the 96th edition of the *CRC Handbook of Chemistry and Physics*.

Symbol	T_f / K	ρ_s(25 °C)/ kg m^{-3}	$\alpha_l \times 10^6$/K^{-1}
Bi	544.56	9790	13.4
Ga	302.92	5910	18
Ge	1211.40	5323	5.8
Si	1687.15	2330	2.6
Ag	1234.93	10500	18.9
Al	933.47	2700	23.1
Au	1337.33	19300	14.2
Fe	1811.05	7870	11.8
Pb	600.61	11300	28.9
Pt	2041.35	21500	8.8
V	2183.15	6000	8.4

Contracting volume leads to an increase in density with decreasing temperature. Equivalently, as shown in Fig. 5.1, the molar volume decreases. Solid water exhibits typical behavior in Fig. 5.1; specifically, the density increases linearly for an extended range below T_f. However, all solids asymptotically achieve a maximum constant density as the temperature approaches absolute zero as a consequence of the third law of thermodynamics.

Solids can be held together by a range of chemical interactions. This leads to four distinct classifications of solid crystalline materials, namely ionic, covalent, metallic, and molecular.

Ionic crystals are held together by the electrostatic interactions that result from full charge transfer between atoms to form cations and anions in the crystal. Large electronegativity differences lead to the formation of ionic crystals. Typical examples are alkali and alkaline earth atoms bound with halides or chalcogenides, such as NaCl and MgO. However, II–VI[1] (e.g., CdS, ZnSe and CdTe) and III–V (e.g., GaN, GaAs, InP and InSb) compound semiconductors also form crystals with significant ionic character, as do a number of other metal oxides, sulfides and halides. Ionic crystals exhibit ionic conductivity at elevated temperatures.

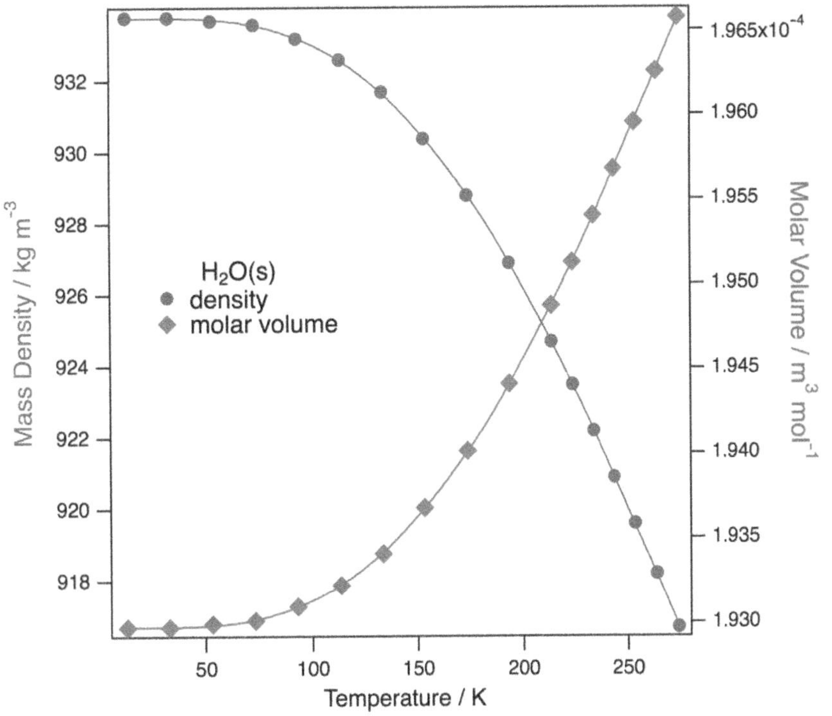

Figure 5.1 The mass density and molar volume of solid water. Values taken from the 96th edition of the *CRC Handbook of Chemistry and Physics*.

Covalent crystals are bound by the sharing of electrons in covalent bonds between atoms. The elemental semiconductors Si and Ge, as well as C-based solids, are examples. Metallic crystals are distinguished from covalent crystals by the delocalized nature of their valence electrons. The kinetic energy of the electrons is the primary binding force of the simple monatomic metals of the first two columns. This also contributes to the binding of transition metals. However, a much greater degree of covalency in their bonding translates into much higher cohesive energies for transition metals compared to the simple metals. The delocalized nature of metal bonding also contributes to the high electrical conductivity of metals. Molecular crystals are bound by weaker intermolecular forces such as van der Waals forces and hydrogen bonding. Thus, they often exhibit low cohesive energies and a lack of electrical and ionic conductivity.

5.2 Electronic structure of solids

An extremely important characteristic of a solid is its electrical resistivity, a parameter that gives us insight into the electronic structure of the solid. Electrical resistivity results from the scattering of electrons as they try to move through the electronic states of a solid. The structure of the electronic states is called its *band structure*. The electrons of an atom are classified as being either valence or core electrons. *Core electrons* in atoms, molecules and solids are much the same. They are tightly bound and localized near the nucleus and are little influenced by chemical changes. The *valence electrons* are strongly influenced by chemical changes and some participate directly in bonding. Valence electrons in solids form delocalized electronic states that are characterized by three-dimensional wavefunctions known as *Bloch waves*. Valence electrons in molecules are localized in molecular orbitals. Some of these may be extended, such as the π-systems in conjugated chains or aromatic rings, while others are highly localized near just one or two atoms. The valence electrons are sometimes localized (as in insulators) but are frequently delocalized (as in semiconductors, superconductors, and metals).

With reference to the band structure, it is instructive to categorize electrons as occupying states either in the *valence band* or the *conduction band*. The valence band is bonding, and states near its maximum energy are in some ways analogous to the highest occupied molecular orbital in a molecule. The conduction band is antibonding, and states near its minimum are analogous to the lowest unoccupied molecular orbital.

The free movement of electrons requires both occupied states (states with an electron in them that act as the source of an electron) and unoccupied states or holes (states into which the electron can move). Either electrons or holes can be thought of as transporting – or 'carrying' – electrical current; thus, they are called *carriers*. The requirement for both results from the Pauli exclusion principle and that electrons are Fermions. As no two electrons can have the same quantum numbers, an empty state is required if an electron is going to hop from one site to the next. Recall that two electrons can occupy one orbital, but when they do so they must have opposite spin quantum numbers.

A gross classification of materials can be made based on their resistivity and its relationship to the band structure shown in Fig. 5.2. A metal conducts electrons readily because its valence band and conduction band overlap. Thus, it has a high density of both occupied and unoccupied states available within k_BT of one another – that is, within the range of thermal excitations available to the electrons. We learn from quantum mechanics that electrons follow Fermi–Dirac statistics, and it is the high-energy tail of this distribution that is important for electrical conductivity. As the temperature increases, the electrons scatter more frequently and the resistivity of a metal increases with increasing temperature as a result. The electrons in a metal fill all the states up to an energy called the *Fermi energy*, E_F. Electrons at E_F are bound in the metal

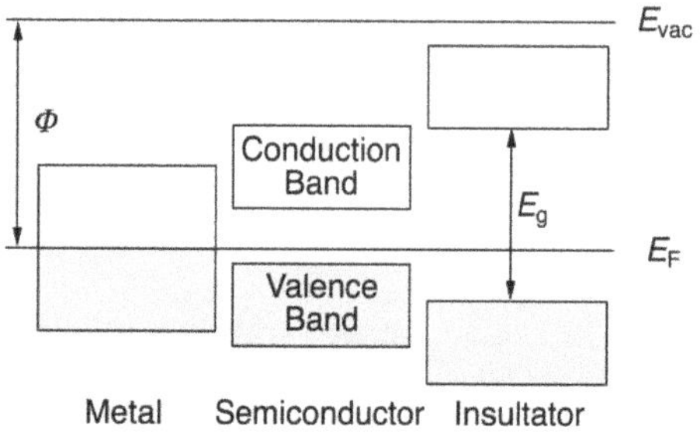

Figure 5.2 The presence and size of a gap between electronic states at the Fermi energy, E_F, determines whether a material is a metal, semiconductor, or insulator. E_g, band gap; Φ, work function, which is equal to the difference between E_F and the vacuum energy, E_{vac}.

by an amount known as the *work function* Φ. The work function is the difference in energy between the Fermi energy and the vacuum energy E_{vac},

$$\Phi = E_{vac} - E_F. \tag{5.2}$$

The *vacuum energy* is the energy of an electron when it is far enough away from the solid that its energy is no longer influenced by it. Measurement of the work function is complicated by the need to remove the electron from the bulk of the material through its surface and into the vacuum. Work functions of real materials vary with the geometry and composition (cleanliness) of their surfaces. Therefore, in order to accurately measure the work function by the photoelectric effect, the surface of the material must be clean and the material must be held in a vacuum chamber so that it remains clean during the measurement.

A semiconductor has a *band gap* with an energetic width E_g that separates the valence band maximum from the conduction band minimum. This gap is small enough that thermal excitations can lead to a small density of carriers. Thermal excitation of an electron to the conduction band leaves behind a valence band hole, both of which can then move to transport electrical current. Because the density of these carriers is low, the resistivity of a semiconductor is high compared to that of a metal. Increasing the temperature increases the carrier density and reduces the resistivity – the opposite trend compared to a metal. Another way to control the resistivity of a semiconductor is to 'dope' electrons or holes into its band structure by introducing impurity atoms. *Doping* – the controlled introduction of electrically active impurities into a semiconductor – is achieved when an atom with a different valence is added to the semiconductor and electrons or holes are made available near the band edges. For example, B is from the group before Si. An atom of B has one fewer valence electrons than Si, but when added to the Si lattice it acts as a p-type dopant. It only adds three electrons, not four, to the valence band, like the Si atom it replaced. Hence, a hole is added to the valence band. A hole is positive. The B atom added a positive carrier so it is a p-type dopant. A P atom has five valence electrons, and when substituted into the Si lattice it adds one electron too many. The valence band is nominally full, and therefore this electron has to go into the conduction band. The P atom added an electron, a negative carrier; hence P is known as an n-type dopant.

Semiconductors are invariably composed of materials that exhibit a high degree of covalency in their bonding. Si and Ge are the most prototypical examples. GaAs and GaN represent mixtures of covalent and ionic bonding. A number of oxides are semiconductors with band gaps below 4 eV such as cerium oxides (ceria, CeO_2 with a fluorite structure but also Ce_2O_3), In_2O_3 (bixbyite-type cubic structure and an indirect band gap of just 0.9–1.1 eV), SnO_2 (rutile, 3.6 eV direct band gap), and ZnO (wurzite, 3.37 eV direct band gap). The ability of ceria to form two different valence states allows its surface and near-surface region to exist in a range of stoichiometries between CeO_2 and Ce_2O_3, producing a so-called 'mixed-valence' oxide. When used as an additive in the support for automotive catalysis, this allows ceria to act as a sponge that absorbs or releases oxygen atoms, based on the reactive conditions. By substituting approximately 10% Sn into indium oxide, the resulting material – indium tin oxide (ITO) – combines two seemingly mutually exclusive properties: a thin film of ITO is both transparent and electrically conductive. This makes ITO very important in display (i.e., touch-screen) and photovoltaic technology.

If the valence band and conduction band are so far apart – that is, if the material has a wide band gap of greater than about 5 eV – then thermal excitations cannot provide carriers for conductivity. The resistance of wide band gap materials is high and they are called *insulators*. The electrons in insulators are not free to move because there are no easily accessible empty states. Ionic crystals such as alkali halide salts form insulators. Many oxides are insulators with band gaps in excess of 6 eV, such as SiO_2 (this is known to form 35 different crystalline polymorphs including silica and quartz, but also glass), Al_2O_3 (also with several polymorphs including alumina, corundum or α-alumina, and γ-alumina), and MgO (magnesia with the same structure as NaCl). These oxides are of particular interest in chemistry because all of them are commonly used as substrates for supported heterogeneous catalysts. α-alumina doped with metal impurities is responsible for sapphire (Cr-, Fe- or Ti-doped) and ruby (Cr-doped).

Most macroscopic-sized, bulk elements are metals, and most macroscopic bulk compounds are insulators. Semiconductors are relatively rare. The responses of these three classes of materials to changes in the size of the structure are different. When materials enter the nanoscale regime (somewhere below 100 nm, with the precise behavior depending on the material and the property under consideration) their electronic structure becomes size-dependent once the wavefunctions of the bands start to extend over distances that are comparable to the size of the material. Semiconductors exhibit this behavior most prominently and over the largest size range. The reason why this is so noticeable with semiconductors is because of the size dependence of their band gap.

5.2.1 Directed practice

Why is the position of the Fermi energy important for determining the properties of a semiconductor?

5.3 Geometrical structure of solids

Crystalline solids are composed of atoms or groups of atoms arranged in an orderly fashion. The three-dimensional periodicity of the patterns distinguishes one crystal type from another. When noninteracting hard spheres are packed into the lowest possible volume – that is, into the highest possible density – they will assume a close-packed, face-centered cubic structure. However, the presence of directional chemical interactions between atoms leads to the wide variety of crystal structures. The unit cell is a small building block (in most cases the smallest) that can be translated and repeated through space to generate the structure of the entire crystal. Once the shape of the unit cell is specified, along with the arrangement of all the atoms within the unit cell, then one has a complete structural description of the crystal.

The concept of a space lattice was introduced by Auguste Bravais. The space lattice, a regular collection of points in three dimensions, is used to generate all possible arrangements of unit cells in a crystal. The periodic translational symmetry of a space lattice is described in terms of three primitive, noncoplanar basis vectors $\boldsymbol{a}_1$, $\boldsymbol{a}_2$, $\boldsymbol{a}_3$, defined such that any lattice point $\boldsymbol{r}(n_1, n_2, n_3)$ can be generated from any other lattice point, for instance the origin at $\boldsymbol{r}(0,0,0)$, in the space translation operation

$$\boldsymbol{r}(n_1, n_2, n_3) = \boldsymbol{r}(0, 0, 0) + \boldsymbol{R} \tag{5.3}$$

where

$$\boldsymbol{R} = n_1\boldsymbol{a}_1 + n_2\boldsymbol{a}_2 + n_3\boldsymbol{a}_3. \tag{5.4}$$

$\boldsymbol{R}$ is the translation vector and n_1, n_2, and n_3 are integers. A lattice generated by a collection of such translation operations is called the simple Bravais lattice. The parallelepiped defined by the three basis vectors is the unit cell of the Bravais lattice. Figure 5.3 displays the three basis vectors and the parallelepiped they define, as well as defining three unit cell angles α, β, and γ. The unit cell with the smallest volume and lattice points located only at the vertices of the parallelepiped is called the primitive cell. The unit cell lengths of basis vectors in this primitive cell – a, b, and c – are called the *lattice parameters*.

The space lattice can be categorized depending on the relative lengths of the lattice parameters, the cell angles, and the number of lattice point in a unit cell. As detailed in Table 5.2 and Fig. 5.4, there are seven lattice systems and 14 Bravais lattices. The 14 Bravais lattices comprise a complete set of all lattices that form in Nature. They are formed by using all possible arrangements of lattice point in each unit cell. Each Bravais lattice is unique. The complete set is generated from the seven simple lattice systems, plus those cases in which a base-centered, a body-centered, or a face-centered lattice can be formed.

A simple Bravais lattice contains lattice points only at the vertices of the parallelepiped. Base-centered means that the Bravais lattice contains a point in the center of the top and bottom faces in addition to points at the vertices. A

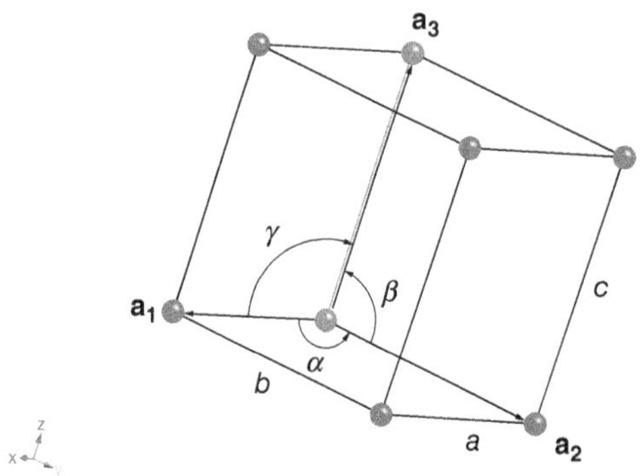

Figure 5.3 The unit cell of a Bravais lattice and the relations of the parameters that define it: basis vectors $\boldsymbol{a}_1$, $\boldsymbol{a}_2$, $\boldsymbol{a}_3$, unit cell angles α, β and γ, lattice parameters a, b, and c.

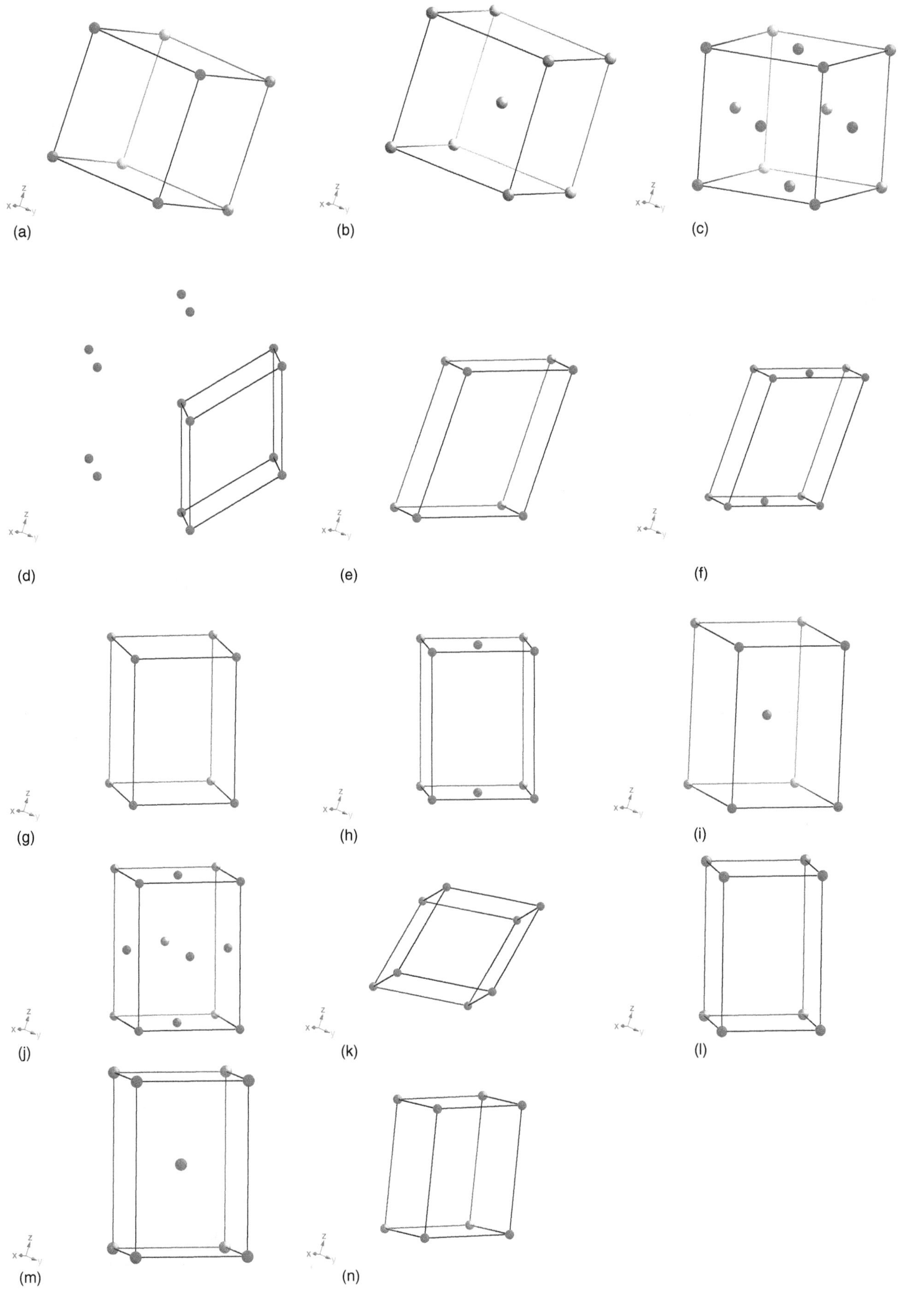

Figure 5.4 The 14 Bravais lattices. (a) Simple cubic; (b) bcc ; (c) fcc; (d) hcp; (e) simple monoclinic; (f) base-centered monoclinic; (g) orthorhombic P; (h) orthorhombic C; (i) orthorhombic I; (j) orthorhombic F; (k) rhombohedral; (l) tetragonal P; (m) tetragonal I; (n) triclinic.

Table 5.2 The seven lattice systems and 14 Bravais lattices. The Bravais lattice is sometimes designated by adding a single-letter abbreviation to the lattice system name: P = simple; C = base-centered; I = body-centered; F = face-centered.

Lattice system	Lattice constants and angles	Bravais lattices
Cubic	$a = b = c$ $\alpha = \beta = \gamma = 90°$	simple, body-centered (bcc), face-centered (fcc)
Hexagonal	$a = b \neq c$ $\alpha = \beta = 90°,\ \gamma = 120°$	simple (hcp)
Monoclinic	$a \neq b \neq c$ $\alpha = \beta = 90° \neq \gamma$	simple, base-centered
Orthorhombic	$a \neq b \neq c$	simple, base-centered, body-centered, face-centered
Tetragonal	$a = b \neq c$ $\alpha = \beta = \gamma = 90°$	simple, body-centered
Triclinic	$a \neq b \neq c$ $\alpha \neq \beta \neq \gamma$	simple
Trigonal	$a = b = c$ $\alpha = \beta = \gamma \neq 90°$	simple

body-centered Bravais lattice has lattice points at the vertices as well as one in the center of the cell. A face-centered Bravais lattice has lattice points in the center of each of the six faces as well as at the eight vertices.

Each Bravais lattice has different symmetry properties. The symmetry properties of the unit cell determine the symmetry properties of the bulk crystal. For example, a cubic crystal expands isotropically (equally along all three basis vectors), whereas this need not be the case for a triclinic crystal in which the basis vectors are, by definition, inequivalent. Each Bravais lattice must have translational symmetry since this is what was used to define it. In addition, it may also exhibit different degrees of rotational, reflection, inversion and improper rotation (rotation followed by inversion) symmetry. These are the same symmetry operation as described in Chapter 26 for molecules. Rotational symmetry means that rotation about one of the crystal axes by $2\pi/n$ radians leads to an equivalent lattice. Such a lattice is said to exhibit n-fold rotational symmetry. A onefold axis is trivial, but the only other allowed values of n are 2, 3, 4, and 6. A fivefold rotational axis is forbidden.

Each Bravais lattice has a distinct number of lattice points and therefore atoms (or constituent units) enclosed in it. To determine the number of atoms in a unit cell, you need to count the number of lattice points in a unit cell, as well as the number of atoms at each lattice point. Corner lattice points are shared by eight adjacent cells. Each only contributes one-eighth of a lattice point to the unit cell. Summing over all eight corners, they contribute just one lattice point in total to the unit cell. Mid-point of edge sites contribute one-quarter of a lattice point. Face-centered sites contribute one-half and body-centered sites contribute one lattice point. Real crystals may have either one or a set of atoms associated with each lattice point. Such a set of atoms is called the *basis*. A basis of a particular crystal is identical in composition, arrangement, and orientation. For example, I_2 forms an orthorhombic crystal with two I atoms at each lattice point. The most stable crystal structures of the elements are detailed in Appendix 8.

As detailed in Table 5.3, the most important crystal structures for semiconductors are the diamond, zinc blende and wurzite structures. Zinc blende is also known as sphalerite. These structures are shown in Fig. 5.5. The diamond lattice is composed of two interpenetrating fcc lattices. The vertex atom of one of the lattices is located at the point (0,0,0). The second lattice is displaced such that its vertex lies at (a/4,a/4,a/4), where a is the lattice constant. The primitive basis of two identical atoms located at (0,0,0) and (a/4,a/4,a/4) is associated with each lattice point of the fcc lattice. For C, the diamond allotrope is metastable compared to graphite. However, for Si and Ge, diamond represents the most stable crystal structure. The zinc blende structure is similar to diamond in being a tetrahedrally coordinated phase. However, the two fcc lattices are occupied by two chemically distinct atoms, for instance, Ga and As in GaAs. Similarly, the wurzite structure is composed of two interpenetrating hexagonal lattices with chemically distinct atoms placed on them. Several compounds, including GaP, ZnS and CdSe, can exhibit either zinc blende or wurzite structures.

The rocksalt (NaCl) structure is common among binary oxides of the general formula MX, such as MgO and NiO. The fluorite structure is common for MX_2 oxides such as CeO_2. An oxide of particular interest is TiO_2, titania. The two most important crystalline forms of titania are rutile and anatase. Rutile has a band gap of ~3.2 eV, which means that it absorbs in the near-ultraviolet by excitation from filled valence band states localized essentially on an O(2p) orbital to the

Table 5.3 Selected properties of various elemental and binary semiconductors. Data taken from the 96th edition of the *CRC Handbook of Chemistry and Physics*.

	Structure	Lattice constant / Å	Band gap / eV
C	diamond	3.56683	5.4
Si		5.43072	1.12
Ge		5.65754	0.67
β-SiC	zinc blende	4.348	2.3
ZnS		5.4093	3.54
ZnSe		5.6676	2.58
CdTe		6.477	1.44
GaP		5.4905	2.24
GaAs		5.65315	1.35
InP		5.86875	1.27
InAs		6.05838	0.36
ZnO	wurtzite	3.24950, 5.2069	3.2
ZnS		3.8140, 6.2576	3.67
CdS		4.1348, 6.7490	2.42
CdSe		4.299, 7.010	1.74
CdTe		4.57, 7.47	1.50
GaN		3.190, 5.189	3.34
InN		3.533, 5.693	2.0

empty conduction band states primarily composed of Ti(3d) states. Such excitations lead to very interesting photocatalytic properties.

5.3.1 Miller indices

A crystal plane can be determined by three integers known as the Miller indices h, k, and l. Take h', k' and l' to be the intercepts (in units of the lattice constant) of a particular crystal plane on the three crystal axes defined by $\boldsymbol{a}_1$, $\boldsymbol{a}_2$ and $\boldsymbol{a}_3$. The three smallest integers h, k and l that satisfy the relation

$$hh' = kk' = ll' \tag{5.5}$$

are the Miller indices. Thus, for the plane that intersects the three crystal axes of a cubic system of lattice constant a at $h' = a$, $k' = a$, $l' = 2a$, the Millar indices are $h = 2$, $k = 2$, and $l = 1$. Such a plane is denoted (221), and in general (hkl) is used to specify a surface termination with an (hkl) bulk plane. Families of symmetry-equivalent planes are designated $\{hkl\}$. Directions are designated $[hkl]$ and families of symmetry-equivalent vector directions are denoted $\langle hkl \rangle$. A plane that is parallel to one of the crystal axes does not intercept it. The corresponding Miller index is 0. When the intersection occurs at a negative number, the negative is indicated by a bar over the number as in $(1\bar{1}1)$, which is read one one bar one. Some examples of the relative orientations of different planes are shown in Fig. 5.6.

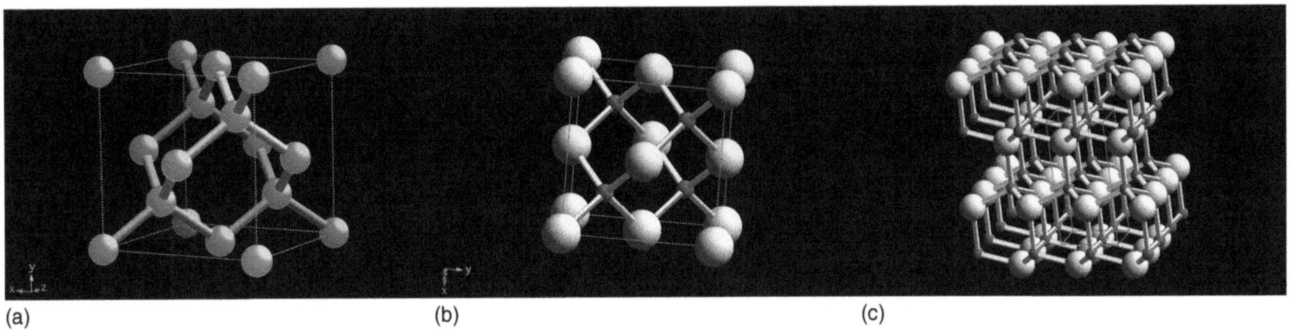

Figure 5.5 The structures most commonly exhibited by semiconductors (a) diamond (b) zinc blende and (c) wurtzite.

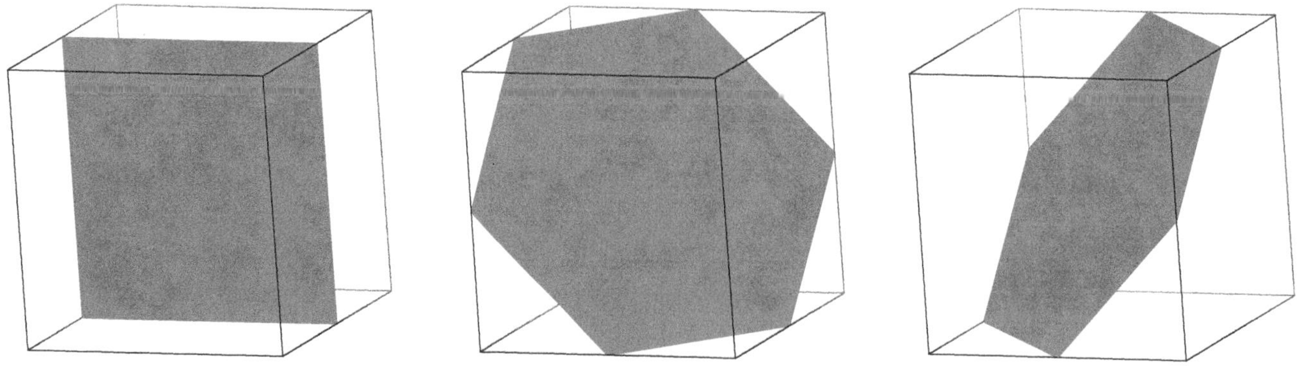

Figure 5.6 Examples of (100), (111) and (1$\bar{1}$1) planes within a cubic unit cell.

5.4 Interface formation

Condensation of a gas or liquid into a solid phase naturally opens up the possibility of forming two types of interfaces: the solid/gas and the solid/liquid interfaces. Unlike a liquid, a solid ends abruptly. Therefore, the solid/gas and solid/liquid interfaces are both abrupt and the geometry of a crystalline solid is well defined.

The formation of a nonplanar interface is influenced by the presence of a solute, which results in what is known as the Mullins–Sekerka instability.[2] A single-component melt solidifying at steady state maintains a planar solid/liquid interface as long as heat is removed through the solid to maintain local interfacial equilibrium. A planar solidifying interface of a two-component solution is unstable above a certain critical concentration of impurity even at equilibrium.

5.4.1 Surface structure and adsorption sites

A large variety of different crystal types exists, and each one of these forms a variety of characteristic surface structures. We will not attempt to be comprehensive here. Instead, we will concentrate on the common features of surface structure and use the face-centered cubic system as a model to convey these common features.

A perfect crystal can be cut along any arbitrary angle, as shown for example in Fig. 5.6. The directions in a lattice are indicated by the Miller indices. A plane of atoms is uniquely defined by the direction that is normal to the plane. To distinguish a plane from a direction, a plane is denoted by enclosing the numbers associated with the defining direction in parentheses, such as (100) The low-index planes displayed in Fig. 5.7 can be thought of as the basic building blocks of surface structure as they represent some of the simplest and flattest of the fundamental planes.

The ideal structures shown in Fig. 5.7 demonstrate several interesting properties. These surfaces are not perfectly isotropic; indeed, we can pick out several high-symmetry sites on any of these surfaces that are geometrically unique. Small-molecule adsorbates tend to chemisorb to these high-symmetry sites, which are then called *adsorption sites*. On the (100) surface we can identify sites of onefold (on top of and at the center of one atom), twofold (bridging two atoms) or fourfold coordination (in the hollow between four atoms). The *coordination number* is equal to the number of surface atoms bound directly to the adsorbate. The (111) surface has onefold, twofold and threefold coordinated sites. Among others, the (110) presents two different types of twofold sites: a long bridge site between two atoms on adjacent rows and a short bridge site between two atoms in the same row. As one might expect based on the results of coordination chemistry,

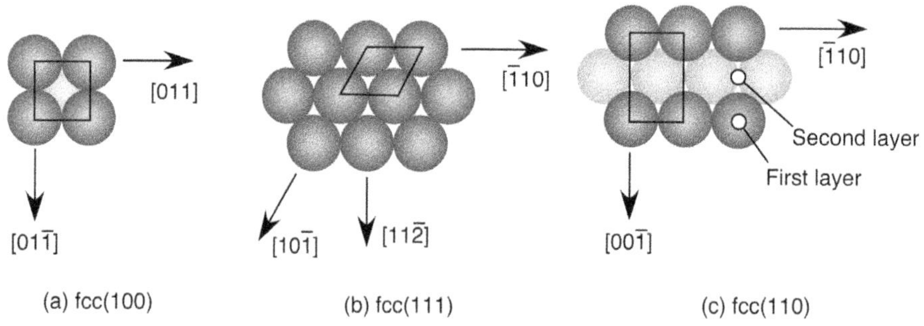

Figure 5.7 Low-index fcc surfaces. Reproduced with permission from Kolasinski, K.W. (2012) *Surface Science: Foundations of Catalysis and Nanoscience*, 3rd edition. John Wiley & Sons, Chichester.

the multitude of sites on these surfaces leads to heterogeneity in the interactions of molecules with the surfaces. This is important in discussions of adsorbate structure and surface chemistry.

The formation of a surface from a bulk crystal is accompanied by great upheaval. Bonds must be broken and the surface atoms no longer have their full complement of coordination partners. Therefore, the surface atoms find themselves in a higher energy configuration compared to being buried in the bulk, and they must relax. Even on flat surfaces, such as the low-index planes, the top layers of a crystal react to the formation of a surface by changes in their bonding geometry. For flat surfaces, the changes in bond lengths and bond angles usually amount to only a few percent. These changes are known as *relaxations*, and can extend several layers into the bulk. The near-surface region, which has a structure different from that of the bulk, is called the *selvage*. This is our first indication that bonding at surfaces is inherently different from that in the bulk, both because of changes in coordination and because of changes in structure. On metal surfaces, the force experienced by surface atoms leads to a contraction of the interatomic distances in the surface layer. In this context it is important to note that *stress is defined as a force*, while *strain is a displacement.*

5.4.2 Surface defects

Surface structure can be made more complex either by cutting a crystal along a higher-index plane or by the introduction of defects. High-index planes (surfaces with h, k, or $l > 1$) often have open structures that can expose second and even third layer atoms. The fcc(110) surface shows how this can occur even for a low-index plane. High-index planes often have large unit cells that encompass many surface atoms. An assortment of defects is shown in Fig. 5.8.

One of the most straightforward types of defects at surfaces is that introduced by cutting the crystal at an angle slightly off of the perfect [*hkl*] direction. A small miscut angle leads to vicinal surfaces, which are close to but not flat low-index planes. The effect of a small miscut angle, as demonstrated in Fig. 5.8, leaves the surface unable to maintain the perfect (*hkl*) structure over long distances. Atoms must come in whole units and in order to stay as close to a low-index structure as possible while maintaining the macroscopic surface orientation, step-like discontinuities are introduced into the surface structure. On the microscopic scale, a vicinal surface is composed of a series of terraces and steps. Therefore, vicinal surfaces are also known as 'stepped' surfaces. The geometric and electronic structure of steps differs from that of terraces, which means that their chemical reactivity is also different.

The natural heterogeneity of solid surfaces has several important ramifications for adsorbates. Simply from a consideration of electron density, we see that low-index planes, let alone vicinal surfaces, are not completely flat. Undulations in the surface electron density exist that reflect the symmetry of the surface atom arrangement, as well as the presence of defects such as steps, vacancies, or impurities. The ability of different regions of the surface to exchange electrons with adsorbates – and thereby form chemical bonds – is strongly influenced by the coordination numbers of the various sites on the surface. More fundamentally, the ability of various surface sites to enter into bonding is related to the symmetry, nature and energy of the electronic states found at these sites. It is a poor approximation to think of the surface atoms of transition metals as having unsaturated valences (dangling bonds) waiting to interact with adsorbates. The electronic states at transition metal surfaces are extended, delocalized states that correlate poorly with unoccupied or partially occupied orbitals centered on a single metal atom. The concept of dangling bonds, however, is highly appropriate for covalent solids such as semiconductors. Non-transition metals, such as aluminum, can also exhibit highly localized surface electronic states.

The heterogeneity of low-index planes presents an adsorbate with a more or less regular array of sites. Similarly, the strength of the interaction varies in a more or less regular fashion that is related to the underlying periodicity of the

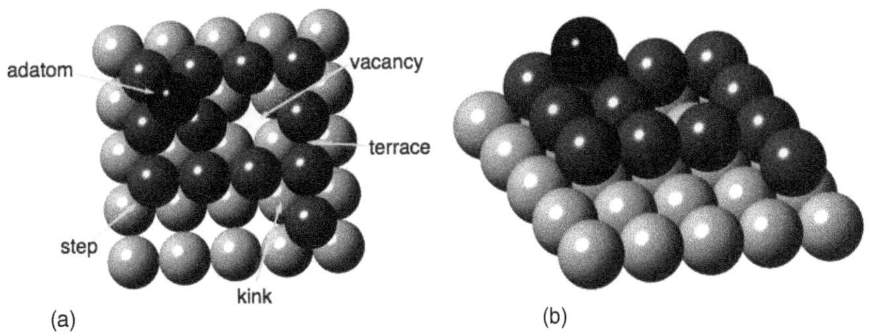

Figure 5.8 Different types of defect sites on a fcc(100) surface observed from (a) above and (b) at an angle.

surface atoms and the electronic states associated with them. These undulations are known as *corrugation*. Corrugation can refer to either geometric or electronic structure. A corrugation of zero corresponds to a completely flat surface, while a high corrugation corresponds to a mountainous topology.

Since the sites at a surface exhibit different strengths of interaction with adsorbates, and since these sites are present in ordered arrays, we expect adsorbates to bind at well-defined sites if they experience chemical interactions with the surface. The symmetry of overlayers of adsorbates may sometimes be related to the symmetry of the underlying surface. We distinguish three structural regimes: random; commensurate; and incommensurate:

- Random adsorption corresponds to the lack of two-dimensional order in the overlayer, even though the adsorbates may occupy (one or more) well-defined adsorption sites.
- Commensurate structures are formed when the overlayer structure corresponds to the structure of the substrate in some rational fashion.
- Incommensurate structures are formed when the overlayer exhibits two-dimensional order; however, the periodicity of the overlayer is not related in a simple fashion to the periodicity of the substrate.

5.5 Glass formation

Some liquids become quite viscous as the temperature nears the melting point. Because of this the molecules in the liquid are not able to move and relax to their most stable configuration easily. With sufficiently rapid cooling, they can freeze into a disordered, randomly aligned solid structure rather than a crystalline structure. This disordered solid structure is known as a *glass*, and the temperature at which the transition to a glass occurs is denoted T_g.

A glass is a nonequilibrium state of a solid. The glass is unstable with respect to crystallization. However, relaxation is subject to internal constraints of the system (e.g., high viscosity) that are not controllable by the experimenter. These internal constraints prevent the formation of the equilibrium state with respect to the applied external constraints. Silica glass devitrifies (in other words, it crystallizes) on a time scale of centuries when held at room temperature. We learn from statistical mechanics that one consequence of this metastability of glasses is that at $T = 0$ K, they possess residual entropy.

Study of the glass transition is particularly important in the fields of polymer and macromolecule research. After introducing more on phase transitions, we will learn much more about this topic in Chapter 13.

5.6 Clusters and nanoparticles

Take an ice cube and cut it in half. Both pieces of this macroscopic sample have the same physical and chemical properties. Their melting point and vapor pressure are both the same, and are also exactly the same as when they were joined together in one piece. If they were grown carefully enough to obtain the lowest energy single-crystal structure, the crystal structure of the whole ice cube would be the same as the crystal structure in each of the two halves. This is true of all materials in the thermodynamic limit: their properties do no depend on the shape and size of their container (in the case of a fluid) or the size of a chunk of solid. However, in the nanoscale regime – that is, for samples with dimensions less than about 100 nm – this is not the case. Properties as fundamental as the melting point and even the equilibrium crystallographic structure depend on the size of nanoparticles. Objects between the size of one molecule and a bulk sample are often called *clusters* or *nanoparticles*.

Why are nanoparticles different and why do their properties depend on size? There are two primary reasons why nano is really different, namely quantum confinement and surface effects. We will learn more about quantum effects later. In atoms or covalent bonds, electrons are localized into orbitals. Similarly in insulators, electrons are strongly localized in the bonds that hold the solid together. However, in metals and semiconductors the valence electrons enter delocalized states (the valence and conduction bands). Once a cluster of atoms in a metal or semiconductor becomes small enough that the electrons can sense the edges of the cluster – that is, once the electrons interact with a potential that confines them in the cluster (rather than being a periodic potential that seems to go on infinitely) – then the properties of those electrons depend on the size of the cluster.

As the size of a cluster changes, so too does the ratio of atoms located at the surface as compared to the number of atoms in the bulk. This is illustrated in Fig. 5.9 and Table 5.4. In the thermodynamic limit $N \gg 10^9$, the number of surface

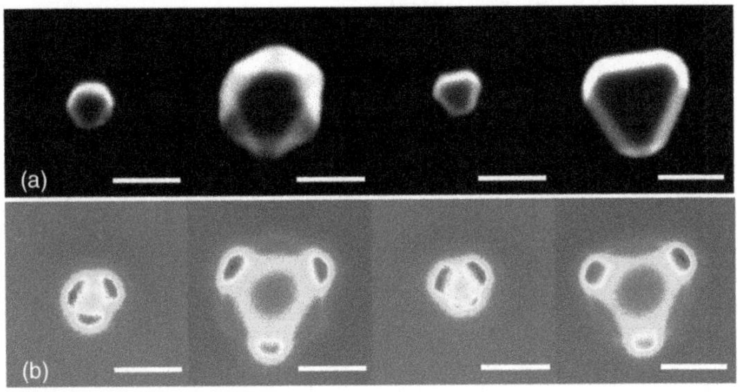

Figure 5.9 (a) The geometrical structure of Al nanoclusters changes with size. (b) The electronic structure, as reflected in cathodoluminescence images, reflects the changes in the geometric structure. Provided by Naomi Halas and Mike McClain. Reprinted with permission from McClain, M.J., Schlather, A.E., Ringe, E., King, N.S., Liu, L., Manjavacas, A., Knight, M.W., Kumar, I., Whitmire, K.H., Everitt, H.O., Nordlander, P., and Halas, N.J. (2015) *Nano Lett.*, **15**, 2751. © 2015, American Chemical Society.

atoms is inconsequential compared to the number of atoms in the bulk. Therefore, the surface does not influence the properties of a macroscopic sample. The properties of a macroscopic sample do not depend on the size and shape of the sample. Even for samples with dimensions on the order of 1 μm or larger, properties do not depend on the size and shape of the container. As the cluster size falls below a diameter of about 5 nm, the surface atoms become the majority. At this point, it will be the surface atoms that determine the properties of the sample, and not the fully coordinated bulk atoms. In the range between 1 nm and 0.5 μm, the influence of surface atoms compared to bulk atoms continuously changes. It should not be surprising now, that the properties of nanoparticles are size-dependent.

Industrial heterogeneous catalytic reactions are performed at the gas/solid or liquid/solid interface. The catalyst is usually composed of metal nanoparticles dispersed over a solid substrate, which is often a porous oxide or zeolite. It is favorable to create a maximum surface area in order to maximize the number of reactive sites. In other words, a general aim is to maximize the *dispersion* of a catalyst. Dispersion D is the fraction of atoms in the catalyst that reside at the surface N_{surf} compared to the total number of atoms in the particle N

$$D = N_{\text{surf}}/N. \tag{5.6}$$

The dispersion depends on the size and shape of the particle, but in general, dispersion increases with decreasing size. For example, if we consider spherical particles of radius r, molar mass M, density ρ, and composed of N spherical atoms, we can calculate the dispersion as follows. We model the particle as a close-packed sphere. Its surface area is then $A = 4\pi r^2$ and volume $V = \frac{4}{3}\pi r^3$. Assume that each surface atom makes a contribution to the surface area of the sphere that is equal to the surface area of a circle that passes through its diameter πr_{atom}^2. Equating these two expressions for the total surface area,

$$A = 4\pi r^2 = N_{\text{surf}}\pi r_{\text{atom}}^2, \tag{5.7}$$

Table 5.4 Clusters can be roughly categorized on the basis of their size. The properties of clusters depend on size. For clusters of different composition the boundaries might occur as slightly different values. The diameter is calculated on the assumption of a close-packed structure with the atoms described by the van der Waals radius (r = 1.86 Å), molar mass (M = 0.02299 kg mol^{-1}) and density (ρ = 970 kg m^{-3}) of the Na atom.

	Small	**Medium**	**Large**
Atoms, N	2–120	120–740	800–10^8
Radius, r / nm	< 0.9	0.9–1.7	1.8–100
Dispersion $D = N_{\text{surf}}/N$	Not separable	0.9–0.5	< 0.5 ~0.45 for N = 1000 0.1 for $N = 9 \times 10^4$ < 0.01 for $N > 8 \times 10^7$
Electronic states	Approaching discrete	Approaching bands	Moving toward bulk behavior
Size dependence	No simple, smooth dependence of properties on size and shape	Size-dependent properties that vary smoothly	Quantum size effects may still be important but properties approaching bulk values

we can solve for the number of surface atoms,

$$N_{surf} = 4r^2/r_{atom}^2. \tag{5.8}$$

From the equation for the volume of a sphere, we can solve for r. The volume of one atom is the total volume divided by the number of atoms. Thus, we can derive an equation for r_{atom} in terms of total volume from

$$V/N = \tfrac{4}{3}\pi r_{atom}^3 \tag{5.9}$$

to obtain

$$r_{atom} = \left(\frac{3V}{4\pi N}\right)^{1/3}. \tag{5.10}$$

Substituting these expressions for r and r_{atom} into Eq. (5.8), we obtain an expression for the number of surface atoms as a function of the total number of atoms

$$N_{surf} = 4N^{2/3}. \tag{5.11}$$

The mass m of the particle is the number of atoms times the mass of one atom; thus, the total number of atoms is given by

$$N = mN_A/M. \tag{5.12}$$

The mass is also equal to the density times the volume

$$m = \rho(4\pi/3)r^3 \tag{5.13}$$

Substituting into the definition of dispersion found in Eq. (5.6) yields our final expression for dispersion as a function of the particle radius.

$$D = \left(\frac{48M}{\pi\rho N_A}\right)^{1/3}\frac{1}{r}. \tag{5.14}$$

As expected, the smaller the particle, the greater the dispersion. For example, $D = 0.84$ for a sphere with $r = 1$ nm, but $D = 0.17$ for $r = 5$ nm when parameters for Na are substituted into Eq. (5.14).

5.6.1 Directed practice

Make the appropriate substitutions and perform the algebra to derive Eq. (5.14) from Eq. (5.6).

Consider the important specific example of water. An isolated water molecule exhibits a bent structure, as shown in Fig. 5.11(a). As water molecules coalesce into clusters, the simplest being the dimer shown in Fig. 5.11(b), they arrange themselves into specific geometries that attempt to maximize the attractive interactions provided by hydrogen bonding. For some sizes, several geometries (isomers) may have energies that are close to each other. However, as shown by Udo Buck and coworkers,[3] it is not until a cluster size of $N = 275 \pm 25$ that water clusters are able to take on the same crystal structure as that exhibited by bulk water and shown in Fig. 5.11(c).

5.7 The carbon family: Diamond, graphite, graphene, fullerenes, and carbon nanotubes

Carbon has unusually flexible bonding. This leads not only to the incredible richness of organic chemistry, but also to a range of allotropes and nanoparticles. The richness of carbon chemistry is related to carbon's ability – in the vernacular of valence bond theory – to participate in three different types of orbital hybridization, namely sp, sp^2, and sp^3.

- sp hybridization is associated with the ability to form triple bonds and is associated with linear structures in molecules. It is the least important for the understanding of carbonaceous solids.
- sp^2 hybridization is associated with the formation of double bonds and planar structures. This is the hybridization that leads to the formation of graphite, graphene, fullerenes and carbon nanotubes.
- sp^3 hybridization is associated with tetrahedrally bound C atoms and forms the basis of C bound in the diamond lattice, which we have discussed above.

Graphite is a layered three-dimensional solid composed of six-membered rings of C atoms with sp^2 hybridization. The honeycomb lattice of the basal plane of graphite is shown in Fig. 5.12(a). This is by far the most stable plane of graphite

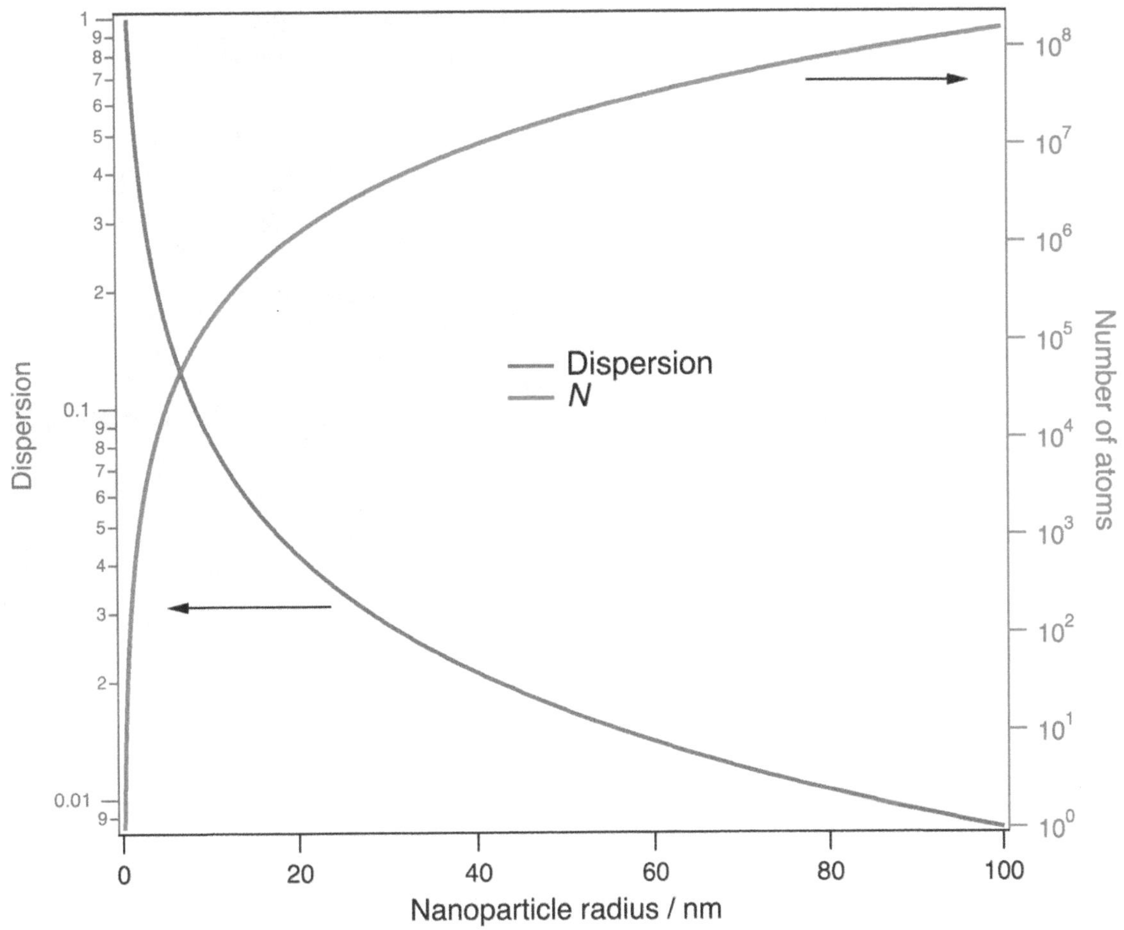

Figure 5.10 Plots of dispersion and number of atoms in a cluster as a function of nanoparticle radius. The nanoparticles are approximated to be composed of atoms with the van der Waals radius of Na atoms.

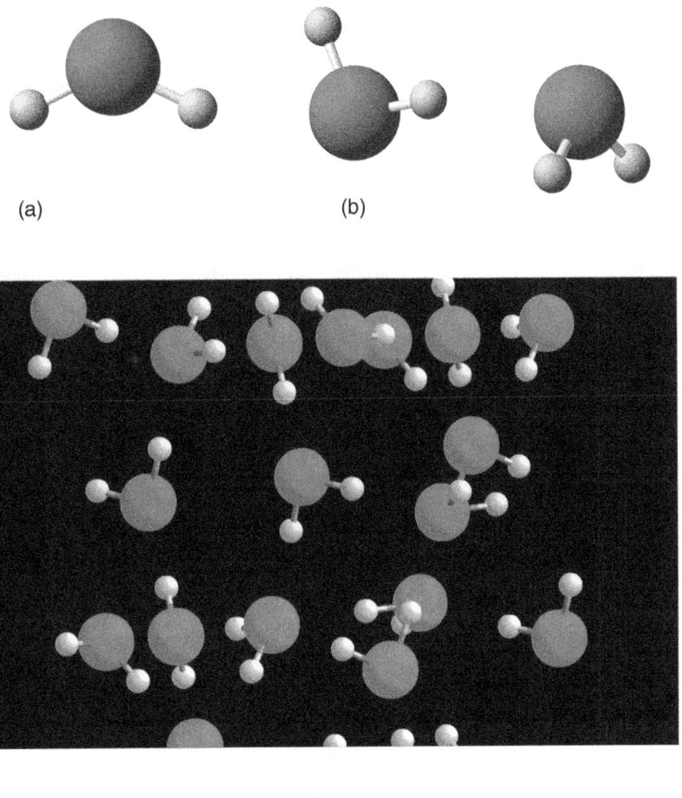

Figure 5.11 (a) An individual water molecule. (b) Water dimer. (c) Crystalline water structure.

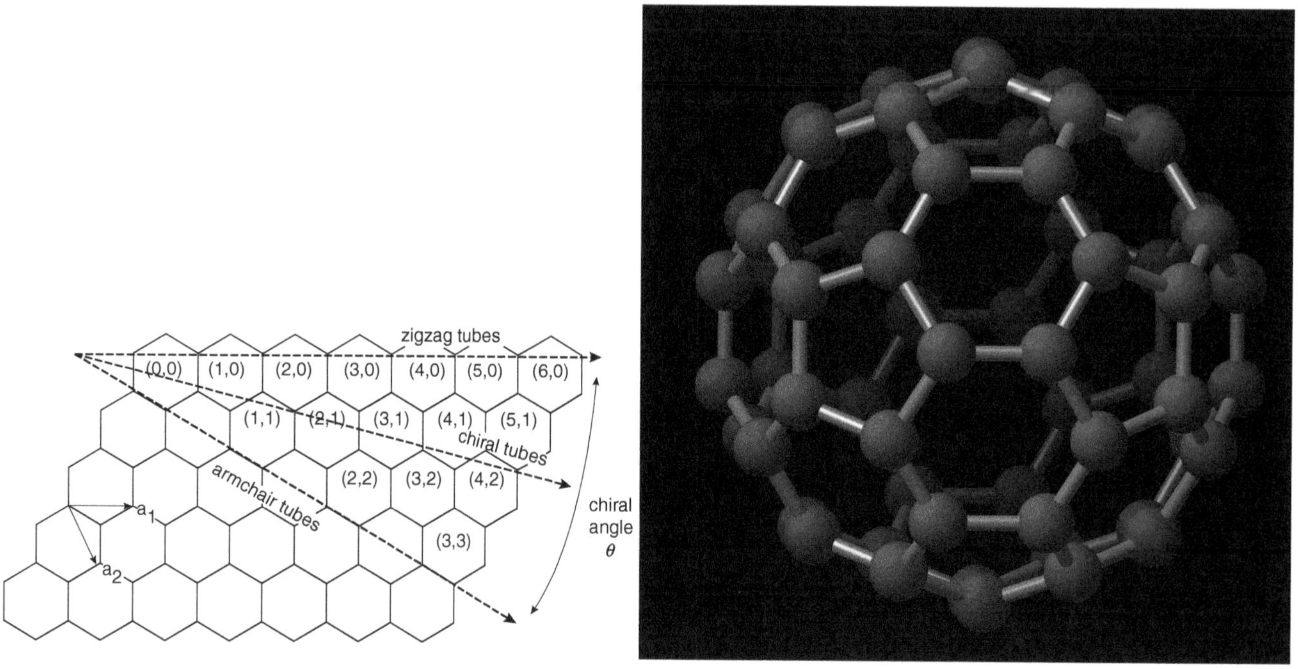

Figure 5.12 (a) The six-membered ring structure that defines the basal plane of graphite or a graphene sheet. $\mathbf{a}_1$ and $\mathbf{a}_2$ are the surface lattice vectors. When rolled to form a seamless structure, the resulting nanotube has either a zigzag, chiral, or armchair configuration depending on the value of the chiral angle. (b) The structure of the fundamental fullerene, that of C_{60}.

because it minimizes the number of reactive sites at unsaturated C atoms. The C atoms in any one layer are strongly covalently bonded to one another, but the layers interact weakly through van der Waals forces. Because of this weak interplane interaction, the surface is easily deformed in an out-of-plane direction, and C atoms in graphite or a graphene sheet can be pulled up by the approach of a molecule or atom as it attempts to adsorb. In their lowest energy form, the layers stack in an ABAB fashion. A single crystal of graphite would, of course, have nearly perfect order. However, the most commonly encountered form of ordered graphite is highly ordered pyrolytic graphite (HOPG), which contains many stacking faults and rotational domains in a macroscopic surface. The rotational domains are platelets of ordered hexagonal rings that lie in the same plane but that are rotated with respect to one another. Defects are formed where the domains meet. The domains can be as large as 1–10 μm across.

Graphene is the name given to a single (or few) layer sample of carbon atoms bound in an hexagonal array. Graphite results from stacking together a large number of these layers. Carbon nanotubes (CNTs) are made by rolling up graphene sheets such that the edge atoms can bind to one another and form a seamless tube with either one (single-wall nanotube, SWNT) or several (multiwall nanotube MWNT) layers. As can be seen from the structure depicted in Fig. 5.12(a), the manner in which the graphene sheet is rolled is very important for determining the structure of the nanotube. A CNT can be characterized by its chiral vector $\boldsymbol{C}$ given by

$$\boldsymbol{C} = n\boldsymbol{a}_1 + m\boldsymbol{a}_2 \tag{5.15}$$

where n and m are integers, and $\boldsymbol{a}_1$ and $\boldsymbol{a}_2$ are the unit vectors of graphene. The hexagonal symmetry of graphene means that the range of m is limited by $0 \le |m| \le n$. The values of n and m determine not only the relative arrangement of the hexagons along the walls of the nanotube, but also the nature of the electronic structure and the diameter. The diameter of a SWNT is given by

$$d = (a/\pi)(n^2 + mn + m^2)^{1/2} \tag{5.16}$$

where the magnitude of the surface lattice vector $a = |\boldsymbol{a}_1| = |\boldsymbol{a}_2| = 0.246$ nm. The chiral angle, which can assume values of $0 \le \theta \le 30°$, is related to n and m by

$$\theta = \sin^{-1}\left(\frac{m}{2}\sqrt{\frac{3}{n^2 + mn + m^2}}\right) \tag{5.17}$$

The translation vector $\boldsymbol{T}$ is the unit vector of the CNT. It is parallel to the tube axis but perpendicular to the chiral vector.

When $m = 0$, $\theta = 0°$ and this leads to the formation of nanotubes in which the hexagonal rings are ordered in a zigzag structure. Most zigzag nanotubes are semiconducting, except for those with an index n divisible by 3, which are conducting (metallic). When $n = m$, $\theta = 30°$ and armchair tubes are formed. All armchair tubes are conducting. Chiral tubes are formed in between these extremes. Most chiral tubes are semiconducting; however, those with $n - m = 3q$ with q an integer are conducting.

Fullerenes (Fig. 5.12(b)) are formed from bending graphene into spherical or near-spherical cages. However, the bending introduces strain into the lattice, which requires the rearrangement of bonds and the introduction of pentagons into the structure to relieve the strain. In 1996, Robert F. Curl Jr, Sir Harold Kroto, and Richard E. Smalley were awarded the Nobel Prize in Chemistry for their discovery of fullerenes. If a nanotube is capped rather than open, then a hemi-fullerene structure closes the end of the tube. The buckyball C_{60} was the first fullerene to be identified, and with a diameter of 0.7 nm it remains a poster child for nanoparticles. Atoms or molecules can be stuffed inside of fullerenes to make a class of compounds known as *endohedral fullerenes*.

Graphene, whether in the form of layers or rolled up into nanotubes, has remarkable materials properties that are much different from those of graphite. Layers are potentially much more easily controllably and reproducibly produced than nanotubes. Graphene's unusual combination of strength, high thermal conductivity and high carrier mobility make it particularly appealing not only for applications to electronic devices but also as a transparent electrode for displays and photovoltaic devices, in sensors and in composite materials.

For these reasons, ready access to graphene layers has ushered in an incredible wave of fundamental and applied research into their creation and properties. The dramatic impact of methods of graphene production and the demonstration of their unusual properties led to Andre Geim and Konstantin Novoselov being awarded the 2010 Nobel Prize in physics for their discoveries with regard to these materials.

Graphene's unusual materials properties can be traced to its combination of covalent bonding with extended band structure. The hexagonal structure is composed of sp^2-hybridized C atoms. Single bonds with σ character hold these atoms together through localized covalent bonds directed in the plane of carbon atoms. In addition, an extended network of orbitals extending above and below the plane is formed with π symmetry. This network results from the highest occupied molecular orbitals (HOMO) overlapping to form the valance band, while the lowest unoccupied molecular orbitals (LUMO) form the conduction band. These π bonds are shared equally by the carbon atoms, effectively giving them a $1^1/_3$ bond order each. More importantly, the electrons are all paired into bonding interactions within one carbon layer. Therefore, interactions within a plane are very strong, whereas out-of-plane interactions are quite weak. As a result, graphite is held together only by van der Waals interactions between the layers, and in-plane values of properties such as the electrical or thermal conductivity are much different than the out-of-plane values.

The 'extra' $^1/_3$ of a bond of each C atom can be removed by saturating it with H atoms; this produces a material known as *graphane*. Once fully saturated, the layer can take up $^1/_3$ of a monolayer of H atoms – that is, one H atom for every three carbon atoms. This forces the layer to pucker and changes the electronic structure from that of a semiconductor/semimetal in which the conduction and valence bands meet at one Dirac point in k-space, to an insulator with a fully developed band gap.

5.7.1 Directed practice

How does a fullerene differ from a single-walled carbon nanotube?

5.8 Porous solids

It is standard usage to refer to a solid substance by adding a suffix to its chemical notation to denote its phase. Hence, crystalline Si is c-Si, amorphous Si is a-Si, alumina in the β structure is β-Al_2O_3 and in the gamma structure γ-Al_2O_3. In the spirit of this convention I denote a nanocrystalline substance as, for example, nc-Si and a porous solid as, for example, por-Si.

A number of parameters are used to characterize *porous solids*. They are classified according to their mean pore size, which for a circular pore is measured by the pore diameter. The International Union of Pure and Applied Chemistry (IUPAC) recommendations define samples with free diameters <2 nm as microporous, between 2 and 50 nm as mesoporous, and >50 nm as macroporous. The term nanoporous is currently in vogue but undefined. The porosity ε is a measure of the relative amount of empty space in the material,

$$\varepsilon = V_p / V \tag{5.18}$$

where V_p is the pore volume and V is the apparent volume occupied by the material. We need to distinguish between the *exterior surface* and the *interior surface* of the material. If we think of a thin porous film on top of a substrate, the exterior surface is the upper surface of the film (taking into account any and all roughness), whereas the interior surface comprises the surface of all the pore walls that extend into the interior of the film. The *specific area* of a porous material, a_p, is the accessible area of the solid (the sum of exterior and interior areas) per unit mass.

Porous materials are extremely interesting for their optical, electronic, and magnetic properties as well as serving as membranes. (A much more in-depth discussion of their use in catalysis can be found in Thomas and Thomas in the Further Reading section.) Porous oxides such as Al_2O_3, SiO_2 and MgO are commonly used as (relatively) inert, high-surface area substrates into which metal clusters are inserted for use as heterogeneous catalysts. These materials feature large pores so that reactants can easily access the catalyst particles and products can easily leave them. Zeolites are microporous (and more recently mesoporous) aluminosilicates, aluminum phosphates (ALPO), metal aluminum phosphates (MeALPO) and silicon aluminum phosphates (SAPO) that are used in catalysis, particularly in the refining and reforming of hydrocarbons and petrochemicals. Other atoms can be substituted into the cages and channels of zeolites, enhancing their chemical versatility. Zeolites can exhibit both acidic and basic sites on their surfaces that can participate in surface chemistry. Microporous zeolites constrain the flow of large molecules, and therefore can be used for shape-selective catalysis. In other words, only certain molecules can pass through them so only certain molecules can act as reactants or leave as products.

Another class of microporous materials with pores smaller than 2 nm is that of metal-organic frameworks (MOFs). These are crystalline solids that are assembled by the connection of metal ions or clusters through molecular bridges. Control over the void space between the metal clusters is easily obtained by controlling the size of the molecular bridges that link them. One implemented strategy, termed 'reticular synthesis,' incorporates the geometric requirements for a target framework and transforms starting materials into such a framework. As synthesized, MOFs generally have their pores filled with solvent molecules because they are synthesized by techniques common to solution-phase organic chemistry. However, the solvent molecules can usually be removed to expose free pores. Closely related are covalent organic frameworks (COF), which are composed entirely of light elements including H, B, C, N, and O.

5.8.1 Directed practice

Why is the exterior surface usually much less than the interior surface of a porous solid?

5.9 Polymers and macromolecules

Polyatomic molecules contain an exactly defined and readily countable number of atoms. Macromolecules are molecules of high molecular weights that are made up of subunits of smaller molecules; the addition of one or a few more of these subunits does not change their properties much. There are, however, some examples of biological macromolecules that are sensitive to the exact composition. Macromolecules include lipids and macrocycles. The terms macromolecule, lipid and macrocycle are all loosely defined by IUPAC. The term macromolecule was coined by Hermann Staudinger, who was awarded the Nobel Prize in Chemistry in 1953 for having recognized that the size of molecules is essentially unlimited. Lipids are substances of biological origin that are soluble in nonpolar solvents. They consist of saponifiable lipids, such as glycerides (fats and oils) and phospholipids, as well as nonsaponifiable lipids, principally steroids. Macrocycles are macromolecules that include cyclical structural elements. Polymers are molecules composed of many repeated subunits. There is no exact definition of when macromolecules become polymers. Usually, when the molar mass exceeds 25 000 g mol^{-1} (25 kilodalton or 25 kD), then it is safe call the material a *polymer*. If there are only 10–100 repeat units, the material is called an *oligomer*. *Telomers* are oligomers formed by chain-transfer reactions. *Biopolymers* are macromolecules (including proteins, nucleic acids and polysaccharides) formed by living organisms

The viscosity and melting points of macromolecules increase with increasing size, consistent with the general remarks made in Table 5.5. For example, linear polyethylenes have a melting point that increases with increasing molar mass, converging on ~145 °C asymptotically. As chain lengths increase, increasing intermolecular interaction leads to

Table 5.5 Properties of the alkane/polyethylene series. Data taken from Sperling, L.H. (2006) *Introduction to Physical Polymer Science*, 4th edition. John Wiley & Sons, Hoboken, NJ.

Number of carbons in chain	State and properties of material	Applications
1–4	Simple gas	Bottled gas for heating and cooking
5–11	Simple liquid	Gasoline
9–16	Medium-viscosity liquid	Kerosene
16–25	High-viscosity liquid	Oil and grease
25–50	Crystalline solid	Paraffin wax candles
50–1000	Semicrystalline solid	Milk carton, adhesives, coatings
1000–5000	Tough plastic solid	Polyethylene bottles and containers
$(3–6) \times 10^5$	Fibers	Surgical gloves, bullet-proof vests

stabilization. Only weak van der Waals forces hold together waxes. These forces also act to form crystalline regions of lamellae in longer-chain polymers. However, longer polymer chains can entangle, which greatly enhances their strength by connecting adjacent crystalline regions with strong covalent bonds.

Polymers generally exhibit a number of solid and highly viscous liquid phases: glassy, glass transition, rubbery plateau, rubbery flow, and viscous flow. At low temperatures, amorphous polymers are glassy, hard, and brittle; this is why latex gloves or plastics cooled with liquid nitrogen can be easily crushed into pieces when dropped or struck with a hammer. Amorphous polymers do not exhibit a first-order melting transition. Crystalline polymers have a first-order melting transition with a temperature above the glass-transition temperature. A first-order phase transition has a discontinuous volume-temperature dependence and a well-defined enthalpy of transition. The glass transition is a second-order transition. The volume-temperature dependence experiences a change in slope at the glass transition, and only the derivative of the thermal expansion coefficient is discontinuous. There is no enthalpy of transition for a second-order phase transition, but there is a change in heat capacity.

Most natural and synthetic polymers (but not proteins) exist in a state characterized not by a single molar mass but by a distribution of masses. The distribution arises from the kinetics of polymerization. The distributions are often characterized by the *number-average molar mass* M_n, defined by

$$M_n = \sum_j N_j M_j \Big/ \sum_j N_j \tag{5.19}$$

and the *weight-average molar mass* M_w

$$M_w = \sum_j N_j M_j^2 \Big/ \sum_j N_j M_j \tag{5.20}$$

where N_j is the number of molecules that have molar mass M_j. The weight-averaged molar mass is always larger. For many applications, the narrower the size distribution, the better the polymer.

SUMMARY OF IMPORTANT EQUATIONS

$\alpha_V = \frac{1}{V}\left(\frac{\partial V}{\partial T}\right)_p$	Volume thermal expansion coefficient at constant pressure
$\Phi = E_{vac} - E_F$	Work function Φ, vacuum energy E_{vac}, Fermi energy E_F
$D = \left(\frac{48M}{\pi \rho N_A}\right)^{1/3} \frac{1}{r}$	Dispersion D, molar mass M, mass density ρ, Avogadro constant N_A, radius r
$d = (a/\pi)(n^2 + mn + m^2)^{1/2}$	SWNT diameter d, surface lattice constant a, chiral vector indices n and m
$\theta = \sin^{-1}\left(\frac{m}{2}\sqrt{\frac{3}{n^2 + mn + m^2}}\right)$	Chiral angle of SWNT θ
$\varepsilon = V_p/V$	Porosity ε, pore volume V_p, volume of solid V
$M_n = \sum_j N_j M_j \Big/ \sum_j N_j$	Number-average molar mass M_n, number of molecules N_j with molar mass M_j
$M_w = \sum_j N_j M_j^2 \Big/ \sum_j N_j M_j$	Weight-average molar mass M_w

Exercises

5.1 Given that the density of V is 6000 kg m^{-3} at 298 K and that α_l is 8.4×10^{-6} K^{-1}, calculate the density and molar volume at 2000 K.

5.2 The density of Si is 2.329 g cm^{-3} at 298 K. The molar mass of Si is 28.09 g mol^{-1} (in units as they are commonly found in tables). Use these values to calculate the volume and radius of a Si atom, assuming it to be spherical.

5.3 Given that the radius of Au is 1.44 Å, calculate the dispersion and number of atoms in nanoparticles with radii of 1, 5, and 10 nm.

5.4 Use crystal structure data to determine the density of Cu (an fcc metal with a = 3.615 Å), Mo (a bcc metal with a = 3.165 Å) and CsCl (a salt with a simple cubic structure with a = 4.129 Å).

5.5 Determine Avogadro's constant from crystallographic data given that the density of Au is 19.3 g cm^{-3} and that it is an fcc metal with a = 4.0786 Å.

5.6 Why do the atoms on the surface of a solid generally have different properties than those in the bulk of the solid?

5.7 Why are the properties of nanoparticles dependent on size?

5.8 What differentiates a semiconductor from an insulator or metal?

5.9 Discuss how a crystal forms from a liquid in contrast to how a glass forms.

5.10 Describe the properties of an amorphous polymer above and below the glass transition.

Endnotes

1. II–VI and III–V are commonly used names. The Roman numerals refer to the old-style group numbering of the periodic table.
2. Mullins, W.W. and Sekerka, R.F. (1964) *J. Appl. Phys.*, **35**, 444.
3. Pradzynski, C.C., Forck, R.M., Zeuch, T., Slavíček, P., and Buck, U. (2012) *Science*, **337**, 1529.

Further reading

Alivisatos, A.P. (1996) Semiconductor clusters, nanocrystals and quantum dots. *Science*, **271**, 933.
Elliott, S. (1998) *The Physics and Chemistry of Solids*. John Wiley & Sons, Chichester.
Haberland, H. (1993) *Clusters of Atoms and Molecules, in: Springer Series in Chemical Physics*, Vol. 52. Springer-Verlag, Berlin.
Haberland, H. (1994) *Clusters of Atoms and Molecules II, in: Springer Series in Chemical Physics*, Vol. 56. Springer-Verlag, Berlin.
Herschbach, D. (1999) Chemical physics: Molecular clouds, clusters, and corrals. *Rev. Mod. Phys.*, **71**, S411.
Johnston, R.L. (2002) *Atomic and Molecular Clusters*. Taylor and Francis, London.
Kolasinski, K.W. (2012) *Surface Science: Foundations of Catalysis and Nanoscience*, 3rd edition. John Wiley & Sons, Chichester.
Ladd, M. (1999) *Crystal Structures – Lattices and Solids in Stereoview*. Woodhead Publishing, Oxford.
Sperling, L.H. (2006) *Introduction to Physical Polymer Science*, 4th edition. John Wiley & Sons, Hoboken, NJ.
Thomas, J.M. and Thomas, W.J. (1996) *Principles and Practice of Heterogeneous Catalysis*. VCH, Weinheim.
Zewail, A.H. (1992) *The Chemical Bond – Structure and Dynamics*. Academic Press, London.

CHAPTER 6

Statistical mechanics

PREVIEW OF IMPORTANT CONCEPTS

- The objective of this chapter is to draw a connection between molecules and macroscopically measureable thermodynamic properties.
- Conventional matter is made up of atoms and molecules. Any macroscopic sample contains an enormous number of atoms approaching the magnitude of the Avogadro constant.
- The thermodynamic macrostate of a system is defined by a set of macroscopic parameters, for instance, the number of molecules, the volume, and the internal energy.
- The energy of a molecule has contributions from kinetic energy plus internal energy (= rotational + vibrational + electronic energy).
- The internal energy of a system (an ensemble of molecules) is equal to the sum of the energy of all the molecules in the system.
- The microstate of a system is a molecular level description of the velocity and internal energy of each molecule at a particular moment in time.
- The equilibrium state of a system corresponds to the most probable set of microstates, which are distributed among energy levels according to a Boltzmann distribution.
- Entropy is related to the number of states over which energy can be distributed.
- Increasing mass leads to closer spacing between translational, rotational, and vibrational energy levels (higher density of states) and higher entropy.
- The partition function tells us how molecules are partitioned into the various levels in accord with their respective Boltzmann factors.
- Introduce yourself to heat capacity, which is one of the most fundamental molecular parameters connecting the distribution of energy to the energy level structure of the molecule. It will play a central role in determining thermodynamic functions.
- Mathematically, the heat capacity is the partial derivative of the internal energy with respect to the temperature.
- Classically, a gaseous diatomic molecule has an internal energy of 3/2 $k_B T$ from translation + $k_B T$ from rotations + $k_B T$ from vibrations = 7/2 $k_B T$ in total. The heat capacity is then simply 7/2 k_B, independent of temperature.
- As long as the temperature is not too low, translations and rotations make classical contributions to internal energy and heat capacity. However, over the temperature range of interest for much of chemistry (200–1000 K), vibrations act quantum mechanically and they are responsible for the temperature-dependence of the heat capacity.
- Fermions have half-integer spin and obey Fermi–Dirac statistics.
- Bosons have integer spin and obey Bose–Einstein statistics.
- The Maxwell–Boltzmann distribution represents the limiting distribution of both Fermi–Dirac and Bose–Einstein distributions in the limit of a dilute gas.

When I took my first course in statistical mechanics from Hans Christian Andersen (yes, they are related) at Stanford, I said to myself, "...why haven't I ever seen this before?" Well here it is – your introduction to stat mech. It is totally ridiculous to start a textbook in physical chemistry with statistical mechanics. It might just scare you off from the rest of the course. But let's get something straight about this chapter. I neither expect nor want to you understand every detail of statistical mechanics – not even everything that is in this chapter. At least, not on your first reading of it. What I want you to learn is the structural basis for a molecular understanding of chemistry. What I really want you to understand is: (i) what is internal energy; (ii) what is a Maxwell–Boltzmann distribution; (iii) how entropy is related to configurations and is nicely presented in the Boltzmann formula and Sackur–Tetrode equation; (iv) what is a partition function; and (v) how does this all relate to the heat capacity? There is a lot of mathematics in between these points, but don't let it get

Physical Chemistry: How Chemistry Works, First Edition. Kurt W. Kolasinski.

Companion Website: www.wiley.com/go/kolasinski/physicalchemistry

in the way. I don't really intend to go into it in detail. However, I do not think it is quite viable in a serious textbook to write

1. There is a thing called the molecule.
2. A miracle happens, that is, statistical mechanics.
3. Here are the equations for entropy, energy distribution, and heat capacity.

In the words of Gary Larson, "I think you should be more explicit here in step two." So I have been. After you have completed your first course in physical chemistry, or before you enter graduate school, come back to this chapter and really try to understand the steps in between. Then go on to learn more. For now, concentrate most on the big picture and you will have a much easier time understanding the rest of physical chemistry.

Here for review and clarity, let us introduce a number of terms and definitions:

- System – the material in the process under study.
- Surroundings – the rest of the universe.
- Open system – system can exchange material with surroundings.
- Closed system – no exchange of material.
- Isolated system – no exchange of material or energy.
- Thermodynamic equilibrium – p, T, composition are constant and the system is in its most stable state. Also called chemical equilibrium.
- Thermal equilibrium – two systems have the same temperature.
- Metastable state – a system in which p, T and composition are constant; however, the system could relax to a lower energy state if it were sufficiently perturbed. A system in a metastable state is in a state of metastable equilibrium, and so can be described consistently by thermodynamic methods.
- Steady state – p, T and composition are constant but are maintained at these values in an open system by a balance of inward and outward fluxes of matter and energy.
- Isothermal – $\Delta T = 0$, isobaric – $\Delta p = 0$, isochoric – $\Delta V = 0$.
- Adiabatic in a thermodynamic sense means that the system exchanges no heat, $q = 0$.
- Diathermal barrier – allows for the transport of heat but not matter between systems.
- Temperature T in kelvin (absolute temperature) is related to temperature t in degrees Celsius by:

$$T(\mathrm{K}) = t(^\circ\mathrm{C}) + 273.15 \tag{6.1}$$

6.1 The initial state of the universe

We start at the beginning – or shortly after the beginning. Immediately after the Big Bang, the universe was in a rather uniform state in which the density of energy exceeded the density of matter. The rather (though not completely) uniform density corresponded to a much more ordered state of the universe than the one we observe now: an extremely inhomogeneous clumping of atoms into molecules and compounds and planets and stars (as well as plants and animals), and stars into galaxies, and galaxies into galactic clusters. From this highly ordered initial state sprang the natural tendency for systems to develop from whatever state they are in to less-ordered states. This is a tendency, not an irrefutable law, and work can be done to overcome this tendency. Experimentalists can use lasers to create initial states that will develop and return to their initial ordered state. But the recurrence time in these so-called 'spin-echo' experiments is extremely short and the number of atoms that will cooperate decays rapidly with time. Thus, even with the most care and best lasers, such an exceptional state of affairs can only be maintained for a fraction of a second.

In order to understand the tendency of isolated systems to equilibrate – to understand how systems evolve chemically – we need to understand the physics that controls and describes the interactions between atomic and subatomic particles. But even this is not enough. We need to understand that the material world around us is composed of microscopic atoms and molecules. Further, there are an enormous number of these molecules. A mere 10 ml of liquid water contains on the order of 10^{23} water molecules. Their microscopic nature (in particular their minute masses and distances between them) means that they do not always play by the rules of the macroscopic world. We will discuss much more about that when we treat quantum mechanics, but for now we concentrate on the effect of the enormous number of molecules.

The sheer number of molecules in a macroscopic sample leads to emergent behavior. The equations that describe the dynamics of individual atoms and molecules are symmetric in time and space. Nonetheless, we know that samples made up of atoms and molecules irreversibly evolve toward equilibrium. In addition, even if we have a complete description of the properties and dynamics of one molecule, there is no way for us to describe the phase transitions that a collection of these molecules undergoes. Both, the irreversible approach to equilibrium and phase transitions are examples of emergent behavior: behavior that cannot be predicted from the properties of one molecule but that emerges from the interactions of collections of molecules.

6.2 Microstates and macrostates of molecules

In order to derive the macroscopic behavior of matter from the microscopic properties of molecules, we need to derive the principles of statistical mechanics. Statistical mechanics starts with the conception of matter as being composed of N particles occupying volume V. This is not how thermodynamics was founded but, as we shall see, it allows us to derive the principles of thermodynamics on a molecular basis. The thermodynamic state of these N particles is called a *macrostate*. The macrostate is a description of the system in terms of collective thermodynamic parameters. We choose our macrostate to be isolated, that is, its internal energy U is held constant, as is N. To specify our system fully, we need only identify its composition in addition to N, V, and U. Should it be more convenient, we can use the methods of thermodynamics to specify our system in terms of other variables, such as N, the pressure p, temperature T, and composition. We will develop these thermodynamic methods in subsequent chapters.

If we want to know details of the individual particles – let us call them molecules though they may be molecules or atoms – then we need to define the *microstate* of the system. The microstate of a system is the description of the system in terms of the characteristics of each individual molecule. Each of the N particles is described quantum mechanically by a wavefunction ψ_k, and the total wavefunction of the system ψ contains a product of all the molecular wavefunctions $\psi_1 \psi_2 \ldots \psi_N$ (it may also contain some other terms to correctly account for the statistics of the system, as we shall see below). At this point we do not need to define what the wavefunction is mathematically. Each individual wavefunction is simply a mathematical function that describes a molecule.

At any finite temperature the molecules are constantly changing their states. The molecules collide, and these collisions change the velocities as well as the rotational and vibrational states of the molecules. Nonetheless, the number of molecules, their internal energy, as well as the volume remain constant. Thus, there is a large number Ω of microstates that correspond to the macrostate defined by N, U, and V. The fundamental postulate of statistical mechanics is the *Gibbs postulate:*

All possible microstates of an isolated assembly are equally probably.

Being a postulate, there is no general proof of this statement, though proofs for specific cases exist. Support of this postulate is derived from the vast number of correct results derived from it.

To describe our system more specifically, we need to know whether the particles are distinguishable or indistinguishable. This makes a difference in the statistics of the system. A one-component gas contains molecules that are identical and indistinguishable. All molecules are able to explore the entirety of the container, and there is no way to label individual molecules. Molecules arranged in a solid lattice are also identical, but because each is fixed to a lattice site, they can be labeled and distinguished.

We allow the N particles to equilibrate in a fixed volume V. In order to equilibrate, the molecules must interact; however, these interactions are not so strong as to interfere with identifying the molecules as individual particles. No reactions occur and for now we ignore any intermolecular interaction energy (essentially we will think in terms of ideal gases rather than real gases). Each of the N particles has an energy ε_j. Anticipating the results of quantum mechanics, we say that there is a level (state) that corresponds to each allowed value of energy and that each molecule occupies one of these levels. The total occupancy of level j is called the *population* n_j. Another term for population is the *occupation number*. The total number of particles N is given by the sum of energy level populations over all allowed states

$$N = \sum_j n_j \tag{6.2}$$

and the total *internal energy* is the sum of the energies of all molecules

$$U = \sum_j n_j \varepsilon_j. \tag{6.3}$$

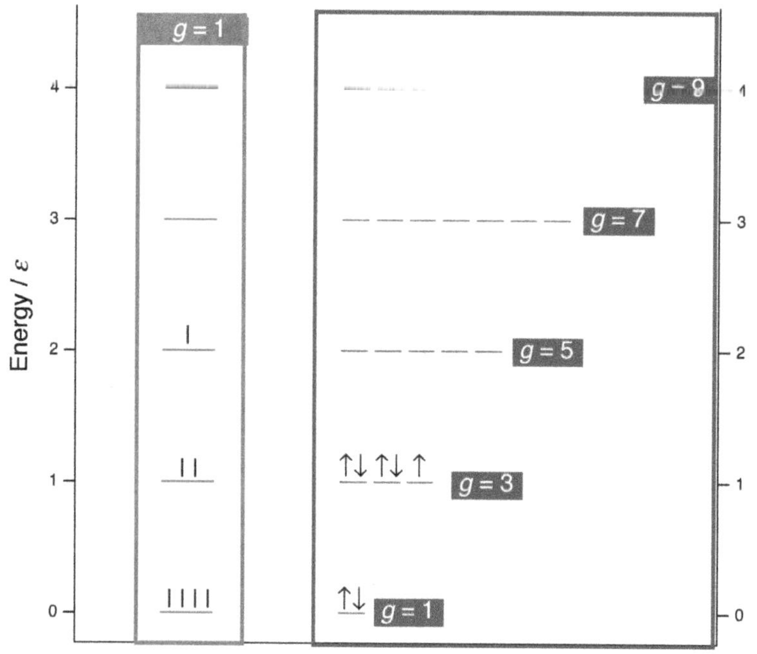

Figure 6.1 A collection of seven objects (i.e., $N = 7$) is distributed over two different sets of states. The red set of states is nondegenerate; that is, there is only one state at each energy (the degeneracy $g = 1$). In the blue set, the degeneracy increases with increasing magnitude of j as $g_j = 2j + 1$. Suggestively in this example, only two objects are allowed in each sublevel within each state. In the red set, the population of the $j = 0$ state is $n_0 = 4$. In the blue set, the population of the $j = 0$ state is $n_0 = 2$.

For any collection of N particles there are different ways for the n_j particles to populate the ε_j energy states. That is, there are numerous microstates described by the different energy arrangements. An example is shown in Fig. 6.1. The number of ways that N distinguishable molecules can be distributed into j energy levels is given by:

$$t(n) = \frac{N!}{n_1!n_2!\dots n_j!} = \frac{N!}{\prod n_j!}. \tag{6.4}$$

The *total number of microstates* Ω is the sum of $t(n)$ for all allowed sets of populations that have a total internal energy U according to

$$\Omega = \sum t(n) = \sum \frac{N!}{n_1!n_2!\dots n_j!}. \tag{6.5}$$

The sum is over all distributions satisfying the constraint on internal energy. The total number of microstates is sometimes referred to as the *weight.*

What types of distribution satisfy the internal energy constraint? Consider a system with $N = 3$ and levels with energies $\varepsilon_0 = 0$, $\varepsilon_1 = \varepsilon, \varepsilon_2 = 2\varepsilon, \varepsilon_3 = 3\varepsilon, \dots$. If the total energy is $U = 3\varepsilon$, then we could put all of the particles in energy level ε_1. Or we could put one particle in ε_0, which requires that the other two go into ε_1 and ε_2, such that the sum of the energies $0 + \varepsilon + 2\varepsilon = 3\varepsilon$. Or we could put two particles in ε_0, which requires that the third occupy ε_3. If the particles were indistinguishable, these would be the three allowed distributions. However, if the three particles are distinguishable – that is, if we have particles that are identifiable as A, B and C – then we also have to keep track of which particle has which energy. In this case, all of the distributions listed in Table 6.1 are possible.

There are a total of $\sum t(n) = 10$ microstates in Table 6.1. That is, there are 10 distinct ways to write the populations of the energy levels. We also see that there are three different classes of distributions (call them classes I, II, and III), but that each class has a certain number of permutations of how it can be written. Since each individual microstate is equally probable, the class with six different permutations is the most likely of the three classes to be present. Let us call this maximum value t_{max}. In this example, there is a 60% chance of finding the macrostate with $U = 3\varepsilon$ in one of the microstates of class II. As N increases, the concentration of probability into the class that corresponds to t_{max} increases. For macroscopic quantities of molecules N is of the order of the Avogadro constant (10^{23} or so). For such astronomically high values of N, to a very good approximation, we can simply take

$$\Omega \simeq t_{max}. \tag{6.6}$$

Table 6.1 Allowed distributions for three distinguishable particles A, B, and C with $U = 3\varepsilon$ and energy level structure $\varepsilon_0 = 0, \varepsilon_1 = \varepsilon, \varepsilon_2 = 2\varepsilon, \varepsilon_3 = 3\varepsilon, \ldots$

A	ε	2ε	2ε	ε	ε	0	0	3ε	0	0
B	ε	ε	0	2ε	0	2ε	ε	0	3ε	0
C	ε	0	ε	0	2ε	ε	2ε	0	0	3ε
n_0	0	1						2		
n_1	3	1						0		
n_2	0	1						0		
n_3	0	0						1		
$t(n)$	1	6						3		

Therefore, we can rewrite Eq. (6.5) as

$$\Omega = t_{\text{max}} = \frac{N!}{n_1!n_2!\ldots n_j!}, \tag{6.7}$$

where the factorial in the denominator must be for the distribution that maximized $t(n)$. Maximization is performed by using the *method of undetermined multipliers* and Stirling's theorem. The method of undetermined multipliers, also called *Lagrange multipliers,* can be found in Appendix 1. From it we obtain the useful expression

$$\frac{\partial \ln \Omega}{\partial n_j} + \alpha + \beta\,\varepsilon_j = 0 \tag{6.8}$$

Stirling's theorem is used over and over again in statistical mechanics. It is derived in Appendix 2, and allows us to approximate the value of the logarithm of a very large number according to

$$\ln N! \simeq N \ln N - N. \tag{6.9}$$

With these methods we find

$$n_j = e^{\alpha}\, e^{\beta\varepsilon_j} = Ae^{\beta\varepsilon_j}. \tag{6.10}$$

We will determine the values of α and β later. The distribution described by Eq. (6.10) is the *Boltzmann distribution.* It describes the most probable constellation of the macrostate, which is the equilibrium state. This distribution is a set of population values – a set of values of n_j – for which the value of Ω is a maximum. Other distributions are possible. These represent fluctuations away from the equilibrium distribution. In the example above, classes I and III represent fluctuations away from the equilibrium distribution of class II. As shown in Fig. 6.2, these fluctuations become increasingly less likely or, at least, the values of the fluctuations become increasingly small compared to the mean values of thermodynamic parameters as the number of particles approaches 10^{23} (a mole of atoms). However, fluctuations are substantial for particles on the nanoscale (<100 nm), and this is one reason why the behavior of nanoparticles can differ from our expectations of macroscopic samples.

Note a subtle difference between the definitions of distribution and microstate. The microstate identifies the state of each molecule. The distribution specifies how many molecules (regardless of their identity) populate each of these microstates.

6.2.1 Directed practice

Chart the allowed distributions for four distinguishable particles A, B, C, and D with $U = 3\varepsilon$ and energy level structure $\varepsilon_0 = 0,\ \varepsilon_1 = 1,\ \varepsilon_2 = 2,\ \varepsilon_3 = 3, \ldots$. Calculate $\sum t(n)$ and t_{max}.

6.3 The connection of entropy to microstates

In 1872, at a time when the very existence of molecules was still very much in doubt, Boltzmann related the entropy S of a gas to the probability of a gas being in a particular microstate. He did so in terms of the distribution function of molecular velocities $f(v_x, v_y, v_z)$. Planck's quantum hypothesis – the idea that the energy states of matter are discreet (quantized) levels rather than continuous distributions – allowed Planck to generalize Boltzmann's results to all thermodynamic systems.

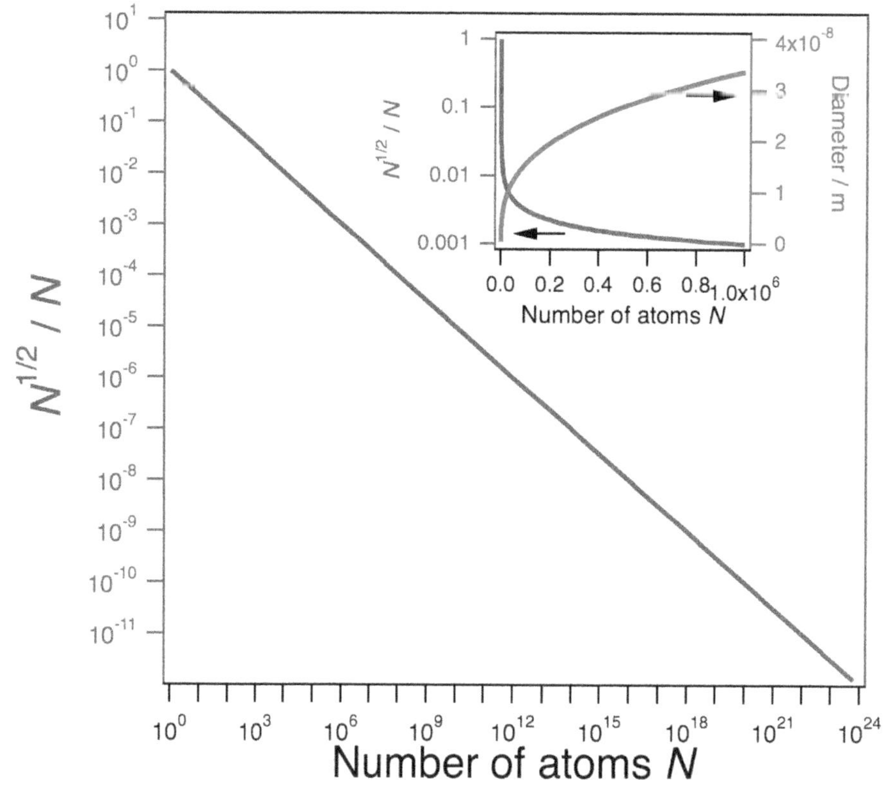

Figure 6.2 The relative probability of a fluctuation (assuming Poisson statistics) is plotted as a function of the number of atoms, N. In the inset, the diameter of a spherical particle with the density of solid Si at 298 K is plotted along the right-hand axis as a function of number of atoms. Fluctuations on the order of 0.1% are to be expected for 40 nm-diameter particles, and this probability increases rapidly as the number of atoms decreases.

From a thermodynamic viewpoint (as we shall define later), equilibrium in an isolated system is obtained when the entropy is a maximum. Statistical mechanics defines the *equilibrium state* as the most probable macrostate, the macrostate that corresponds to the maximum number of microstates. Therefore, the link between thermodynamics and statistical mechanics must lie in an equation that links the entropy S to the number of microstates Ω. The link, as derived by Boltzmann and Planck, is the exceedingly compact and important *Boltzmann formula*:

$$S = k_B \ln \Omega. \tag{6.11}$$

This equation is so fundamental and so groundbreaking that it is literally etched in stone – it appears on Boltzmann's tombstone. The constant k_B is the Boltzmann constant.

An a posteriori derivation of Eq. (6.11) is amazingly simple, but only because we know the answer. The deep insight of the work of Boltzmann and Planck should not be underestimated. We start with the idea that there is a ground energy state, a perfect crystal in internal equilibrium. At $T = 0$, all of the particles must enter this lowest energy state. Thus, each molecule has the same energy and there is only one microstate that describes the system – that is, $\Omega = 1$. This single state is a state of perfect order for which, according to the third law of thermodynamics, $S = 0$. Note that $\ln 1 = 0$, suggesting but not proving a logarithmic relationship

$$\ln \Omega \propto S. \tag{6.12}$$

S is an extensive property. It depends on how much of a system we have. Take two independent assemblies with entropies S_1 and S_2 and corresponding numbers of microstates Ω_1 and Ω_2. The total entropy S of the two systems is given by the sum

$$S = S_1 + S_2. \tag{6.13}$$

The total number of microstates of the combined system Ω is obtained by pairing each of the Ω_1 microstates of system 1 with each of the Ω_2 microstates of system 2. The value of Ω is given by the product

$$\Omega = \Omega_1 \Omega_2. \tag{6.14}$$

Using the *Ansatz* that Eq. (6.11) is satisfied by simply multiplying $\ln \Omega$ by a constant, which we call k_B, we can show that Eqs (6.13) and (6.14) are confirmed. That is, assuming that Eq. (6.11) holds, then on substitution from Eq. (6.14)

$$S = k_B \ln \Omega = k_B \ln \Omega_1 \Omega_2 = k_B \ln \Omega_1 + k_B \ln \Omega_2 = S_1 + S_2, \tag{6.15}$$

we arrive at the result of Eq. (6.13). We will leave it to mathematicians to prove uniqueness. That is, not only are Eqs (6.11), (6.13) and (6.14) consistent, but also the only expression that makes Eqs (6.13) and (6.14) consistent is Eq. (6.11). If the statistical mechanical definition of entropy had been developed first, k_B could have been assigned any arbitrary value. However, to ensure consistency with thermodynamics, its value is set by derivation of the ideal gas law, which we do below, and to satisfy an equation derived from the study of heat, which is shown in Chapter 9 to be a mathematical expression of the second law of thermodynamics,

$$dS = dq_{rev}/T. \tag{6.16}$$

6.3.1 Example

Calculate the entropy change when a mole of CO molecules in a perfect crystal at $T = 0$ K is transformed into an ensemble in which the orientation of the CO dipoles flip randomly between parallel and antiparallel alignments.

There are $N_A = 6.02 \times 10^{23}$ molecules in a mole. A perfect crystal at 0 K has but one microstate, $\Omega = 1$. The randomness introduced by the two orientations of the CO dipole introduces on the order of 2^{N_A}microstates. Thus, the entropy change is

$$\Delta S = S_{final} - S_{initial} = k_B \ln 2^{N_A} - k_B \ln 1 = N_A k_B \ln 2$$

$$\Delta S = 6.02 \times 10^{23} \,(1.38 \times 10^{-23} \text{ J K}^{-1}) \ln 2 = 5.76 \text{ J K}^{-1}$$

Note that this value of ΔS is the value for a sample with 1 mole of molecules, the value of the molar entropy change is $\Delta S_m = 5.76$ J K^{-1} mol^{-1}. This extra entropy that is present due to disorder even at 0 K is known as *residual entropy*.

6.4 The constant α: Introducing the partition function

The population of level j is given by

$$n_j = e^{\alpha} e^{\beta \varepsilon_j}. \tag{6.17}$$

The exponential term involving β is called the *Boltzmann factor*. The total number of molecules is given by the sum of the populations, thus,

$$N = \sum_j n_j = e^{\alpha} \sum_j e^{\beta \varepsilon_j}. \tag{6.18}$$

The exponential term in the summation is sufficiently important that we give it the name *partition function* and the symbol Z,

$$Z = \sum_j e^{\beta \varepsilon_j}. \tag{6.19}$$

The partition function tells us how the population is partitioned into the various levels in accord with their respective Boltzmann factors. Solving for e^{α} and substituting from Eq. (6.19) yields,

$$e^{\alpha} = N/Z. \tag{6.20}$$

Thus, the populations of the individual levels are given by

$$n_j = N e^{\beta \varepsilon_j}/Z. \tag{6.21}$$

The value of α is, therefore,

$$\alpha = \ln N - \ln Z. \tag{6.22}$$

Its meaning is related to the chemical potential, which we will define later.

6.4.1 The value of β

Take a system with constant N and V. Now change the internal energy by adding an infinitesimal amount of heat dq. No pV work is done and the change in internal energy is related to the added heat by

$$dq = dU = d\sum_j n_j \varepsilon_j = \sum_j \varepsilon_j \, dn_j + \sum_j n_j \, d\varepsilon_j. \tag{6.23}$$

The energy level structure does not change at constant volume as shown for example in Eq. (6.57) below. Therefore, $d\varepsilon_j = 0$ and the heat transfer is related to a change in populations according to

$$dq = \sum_j \varepsilon_j \, dn_j. \tag{6.24}$$

The system will equilibrate by maximizing Ω, which will change according to

$$d\ln\Omega = \sum_j \frac{d\ln\Omega}{dn_j} \delta n_j. \tag{6.25}$$

Using an expression found during the derivation of the Boltzmann distribution, Eq. (6.8), we substitute for $d\ln\Omega/dn_j$, which gives

$$d\ln\Omega = -\sum_j (\alpha + \beta\,\varepsilon_j) \, dn_j = -\alpha \sum_j dn_j \; - \beta \sum_j \varepsilon_j \, dn_j \,. \tag{6.26}$$

However at constant N, $\sum_j dn_j = 0$ and then upon substitution from Eq. (6.24)

$$d\ln\Omega = -\beta \, dq. \tag{6.27}$$

Solving for β and substituting from the second law of thermodynamics Eq. (6.16) $dq = T\,dS$,

$$\beta = \frac{-d\ln\Omega}{dq} = \frac{-d\ln\Omega}{T\,dS}. \tag{6.28}$$

Then from Eq. (6.11)

$$\beta = \frac{-d\ln\Omega}{T\,d(k_B \ln\Omega)} = \frac{-d\ln\Omega}{k_B T\,d\ln\Omega} = -\frac{1}{k_B T}. \tag{6.29}$$

Proof that the value of k_B truly is what is known as the Boltzmann constant is actually given unambiguously by the derivation of the equation of state of the ideal gas, which we will perform below.

6.5 Using the partition function to derive thermodynamic functions

For any one arrangement of N identical distinguishable molecules (such as molecules localized at lattice points), the partition function is

$$z = \sum_j e^{\beta\,\varepsilon_j}. \tag{6.30}$$

This assumes that the energy levels ε_j are nondegenerate. We will deal with degeneracy (more than one level at the same energy) in a moment. The localized particles can be arranged in many different ways. For the assembly as a whole, the partition function is

$$Z_W = \sum e^{\beta E_N} \tag{6.31}$$

where the values of E_N include all possible values of the total energy of the assembly. E_N is the sum of all values of ε and the summation is over all possible values of E_N.

Consider two distinguishable particles A and B, in which case

$$z_A = \sum_j e^{\beta\,\varepsilon_{jA}}, \tag{6.32}$$

$$z_B = \sum_j e^{\beta\,\varepsilon_{jB}}, \tag{6.33}$$

and

$$E_N = \varepsilon_A + \varepsilon_B. \tag{6.34}$$

Thus

$$Z_W = \sum_j e^{\beta\,\varepsilon_{jA}} \sum_j e^{\beta\,\varepsilon_{jB}} = Z_A Z_B. \tag{6.35}$$

However, since A and B are identical $Z_A = Z_B = Z$ and $Z_A\ Z_B = Z^2$. This argument is easily generalized to N particles, for which

$$Z_W = Z^N. \tag{6.36}$$

Until now we have considered the energy levels to be nondegenerate; that is, there is one and only one level at each value of energy. However, it is possible for more than one wavefunction – the quantum mechanical description of an energy level – to have the same energy. If more than one level has the same energy, then we say that the levels are *degenerate*. Another term for *degeneracy* is statistical weight, and this gives you an idea of how to account for degeneracy should it occur. If the number of levels at the same energy is g_j, we say the degeneracy of the jth level is g_j and we need to count each degenerate level g_j times.

Including the effects of degeneracy, we can now summarize the results we have derived so far.

$$\beta = -1/k_BT \tag{6.37}$$

$$n_j = e^{\alpha}\, g_j e^{\beta\varepsilon_j} = N g_j e^{\beta\varepsilon_j}/Z \tag{6.38}$$

$$N = \sum_j n_j = e^{\alpha}\, Z \tag{6.39}$$

$$\Omega = \frac{N!}{\prod_j n_j!} \tag{6.40}$$

$$Z = \sum_j g_j e^{\beta\varepsilon_j} \tag{6.41}$$

$$U = \sum_j n_j\, \varepsilon_j \tag{6.42}$$

$$S = k_B \ln \Omega = k_B \left(N \ln N - \sum_j n_j \ln n_j \right) \tag{6.43}$$

To these results derived from statistical mechanics we can now add the following results from thermodynamics, which will allow us to derive all thermodynamic functions directly from statistical (molecular) arguments. These thermodynamic results will be derived in Chapters 7–11.

Helmholtz energy

$$A = U - TS \tag{6.44}$$

Gibbs energy

$$G = H - TS = U + pV - TS = A + pV \tag{6.45}$$

Enthalpy

$$H = U + pV \tag{6.46}$$

Entropy

$$S = -\left(\frac{\partial A}{\partial T}\right)_V \tag{6.47}$$

Pressure

$$p = -\left(\frac{\partial A}{\partial V}\right)_T \tag{6.48}$$

Temperature

$$T = \left(\frac{\partial U}{\partial S}\right)_V \tag{6.49}$$

6.5.1 Example

Take the derivative of the Helmholtz energy at constant volume with respect to temperature and derive an equation that relates it to the entropy.

Taking the derivative of Eq. (6.44), we obtain

$$\left(\frac{\partial A}{\partial T}\right)_V = \left(\frac{\partial U}{\partial T}\right)_V - S - T\left(\frac{\partial S}{\partial T}\right)_V \tag{6.50}$$

The definition of the *heat capacity* at constant volume is

$$(\partial U/T)_V \equiv C_V \tag{6.51}$$

therefore,

$$\left(\frac{\partial A}{\partial T}\right)_V = C_V - S - T\left(\frac{\partial S}{\partial T}\right)_V. \tag{6.52}$$

From the second law of thermodynamics, Eq. (6.16), we know[1] that

$$\mathrm{d}S = \frac{\mathrm{d}q_{\mathrm{rev}}}{T}. \tag{6.53}$$

The heat exchanged along a reversible path at constant volume is $dq_{V,\mathrm{rev}} = C_V\,\mathrm{d}T$. Upon substitution into Eq. (6.53), we obtain

$$\mathrm{d}S = \frac{C_V\,\mathrm{d}T}{T}. \tag{6.54}$$

Therefore

$$T\left(\frac{\partial S}{\partial T}\right)_V = C_V. \tag{6.55}$$

For purists, having specified the path as reversible and isochoric, we can equate $\mathrm{d}S/\mathrm{d}T$ with $(\partial S/\partial T)_V$. Substituting for C_V in Eq. (6.52), yields the desired expression.

$$\left(\frac{\partial A}{\partial T}\right)_V = -S. \tag{6.56}$$

6.6 Distribution functions for gases

The distribution of energy into the various degrees of freedom and among the different microstates of a system is obviously fundamental to determining the values of the thermodynamic functions. To calculate these we need to determine the form of the distribution functions (the partition functions) for each of these degrees of freedom. We begin with the translational energy of various types of ideal gases.

Consider a gas of mass m in a box of volume V with a well-characterized pressure p and temperature T. This corresponds to the *particle in a box* problem, from the quantum mechanical solution of which (see Chapter 20) we know that the energy levels depend on the quantum number n according to

$$\varepsilon_n = \frac{h^2 n^2}{8mV^{2/3}}. \tag{6.57}$$

The spacing between two successive levels n and n+1 is of the order

$$\Delta\varepsilon \sim h^2/mV^{2/3}. \tag{6.58}$$

For 100 kPa of Ar in a 1-liter box at 298 K the spacing is $\Delta\varepsilon = 2 \times 10^{-37}$ J. However, the thermal energy represented by $k_B T$ is 4×10^{-21} J at this temperature. Therefore, multitudinous levels are available at energies much less than $k_B T$. Furthermore, the smearing provided by the thermal energy is so large compared to the energy level spacing that we are justified in considering the distribution to be continuous rather than quantized. Mathematically, this means that we are justified in replacing summations by integrals. This limit is known as the *dilute gas limit*.

The energy level structure is also characterized by a very high *density of states*. Taking the box to be a cube of side length a, the number of levels with an energy less than ε_n is

$$N = \frac{4\pi}{3}\left(\frac{2ma^2\varepsilon_n}{h^2}\right)^{3/2}. \tag{6.59}$$

As we have seen in Fig. 1.2, the velocity spread of molecules in the Boltzmann distribution is very broad and extends far above the mean energy of $3/2\ k_B T$, which equals 6×10^{-21} J in this case. For comparison, up to this value of mean energy there are $N = 3.2 \times 10^{29}$ levels available to the 2.4×10^{22} Ar atoms in the box. In other words, there are 7.5×10^{-8} atoms per energy level. The gas populates these levels very sparsely; the number of levels far exceeds the number of atoms. This result is general for any gas under normal conditions. Because of this, the population of any given level is usually 0, sometimes 1, and rarely – if ever – greater.

6.6.1 Directed practice

Confirm that there are 7.5×10^{-8} atoms per energy level for Ar gas at 100 kPa in a 1 l box at 298 K.

Previously N distinguishable particles were distributed with populations n_j in each of the individual energy levels with energies ε_j. In the present case of N indistinguishable particles most of the n_j values are zero. Therefore, we adopt a method of collecting levels into bundles. Each bundle is characterized by a mean energy ε_k. It contains g_k levels and a total of N_k molecules. Bundles are constructed such that $g_k \gg N_k$, but with N_k sufficiently large that Stirling's theorem can be applied. Finally, the bundles are not too large, which means that each individual level has an energy very close to the mean energy.

Now we calculate how many microstates are available to our bundled system. If there were N_k distinguishable molecules distributed among the g_k levels within one bundle, there would be $g_k^{N_k}$ ways we could distribute the molecules. However, this overestimates by a factor of $N_k!$ the number of microstates available if the particles are labeled. Therefore, for indistinguishable molecules, the number of microstates for the kth bundle is

$$\Omega_k = g_k^{N_k}/N_k!. \tag{6.60}$$

The total number of microstates available for the whole ensemble is obtained by pairing each microstate of one bundle with all of the microstates of every other bundle. This is obtained from the product of the number of microstate available to each bundle

$$\Omega = \prod_k \Omega_k = \prod_k \left(g_k^{N_k}/N_k!\right). \tag{6.61}$$

As before, we restrict the total number of particles and the internal energy with the conditions

$$N = \sum_k N_k \tag{6.62}$$

and

$$U = \sum_k N_k \varepsilon_k, \tag{6.63}$$

respectively. To maximize $\ln \Omega$, we again apply the method of undetermined multipliers to arrive at the condition for the most probable state

$$\frac{\partial \ln \Omega}{\partial N_k} + \alpha + \beta\varepsilon_k = 0. \tag{6.64}$$

Taking the logarithm of Eq. (6.60) and applying Stirling's theorem, we obtain

$$\frac{\partial \ln \Omega}{\partial N_k} = \ln(g_k/N_k). \tag{6.65}$$

Substituting back into Eq. (6.64) and applying the condition of Eq. (6.62), the most probable value of N_k is then

$$N_k = \frac{N g_k\, e^{\beta\varepsilon_k}}{\sum_k g_k\, e^{\beta\varepsilon_k}}. \tag{6.66}$$

The distribution function f_k

$$f_k = N_k/g_k = \exp(\alpha + \beta\varepsilon_k) \tag{6.67}$$

describes the equilibrium number of molecules per level in the kth bundle. The function f_k is a *Maxwell–Boltzmann distribution* function.

6.7 Quantum statistics for particle distributions

A proper quantum mechanical treatment of indistinguishable particles introduces new facets to the energy distributions of particles, which include not only atoms and molecules but also subatomic particles such as electrons. Let us consider two particles described by two wavefunctions $\psi_1(\tau_1)$ and $\psi_2(\tau_2)$ with energies ε_1 and ε_2. The spatial and spin coordinates are described by τ_1 and τ_2. As we will see in the Copenhagen interpretation of Schrödinger's wave equation, the wavefunction represents a probability amplitude, and the absolute square of the wavefunction $|\psi(\tau)|^2$ describes the spatial distribution of the particle, that is, where we can find the particle.

We construct an assembly of these two particles with total energy $\varepsilon_1 + \varepsilon_2$. To construct a proper wavefunction for the assembly, we need to combine the single particle wavefunctions into one composite wavefunction $\psi(\tau_1,\tau_2)$ that properly accounts for particle indistinguishability. The composite wavefunction is formed by taking the product of the two single-particle wavefunctions. However, $\psi_1(\tau_1)\,\psi_2(\tau_2)$, the product of the wavefunction of particle 1 at coordinates τ_1 and particle 2 at coordinates τ_2 is indistinguishable from $\psi_1(\tau_2)\,\psi_2(\tau_1)$, the product of the wavefunction of particle 1 at coordinates τ_2 and particle 2 at coordinates τ_1, because we do not actually know which particle is labeled 1 and which is 2. Thus, the composite wavefunction must be formed from a linear combination of the two products. Furthermore, upon particle exchange (switching τ_1 for τ_2 as done above) the assemblies must be described by the equivalent distributions in the coordinates. That is

$$|\psi(\tau_1\tau_2)|^2 = |\psi(\tau_2\tau_1)|^2. \tag{6.68}$$

Consequently, the wavefunctions themselves must satisfy

$$\psi(\tau_1\tau_2) = \pm\psi(\tau_2\tau_1). \tag{6.69}$$

The + and – signs indicate that there are two different types of behavior when particles are put into assemblies. One set of particles is symmetric with respect to exchange. For these particles, called *bosons*, the (+) sign rules. Bosons exhibit total wavefunctions that are symmetric with respect to particle exchange (they do not change sign). The antisymmetric combination, when the (–) is observed upon exchange, is found for *fermions*. Upon particle exchange, the total wavefunction of a fermion changes sign. Thus, the wavefunction of an assembly of two bosons is constructed from the (+) linear combination

$$\psi_{\text{boson}} = \psi(\tau_1, \tau_2) = \psi_1(\tau_1)\psi_2(\tau_2) + \psi_1(\tau_2)\psi_2(\tau_1) \tag{6.70}$$

and the composite wavefunction of the assembly of fermions is constructed from the (–) combination

$$\psi_{\text{fermion}} = \psi(\tau_1, \tau_2) = \psi_1(\tau_1)\psi_2(\tau_2) - \psi_1(\tau_2)\psi_2(\tau_1). \tag{6.71}$$

Now we generalize for the description of N fermions with N wavefunctions. The composite wavefunction is composed of N terms of the form $\psi_1(\tau_1)\psi_2(\tau_2)\ldots\psi_N(\tau_N)$. The sign connecting two successive terms alternates between + and –. The exchange of coordinates of any two particles changes the sign of the composite wavefunction. Furthermore, since no two particles are permitted to have the same quantum numbers (they cannot be described by the same wavefunctions as this would violate the Pauli exclusion principle), the composite wavefunction must vanish should any two particles have the same wavefunction. Two fermions can occupy the same spatial state; however in this case, their spins must be opposite. These symmetry properties are properly accounted for if we use a determinant to represent the composite wavefunction. Thus the composite wavefunction of an assembly of fermions can be written

$$\psi = \begin{vmatrix} \psi_1(\tau_1) & \psi_1(\tau_2) & \ldots & \psi_1(\tau_N) \\ \psi_2(\tau_1) & \psi_2(\tau_2) & \ldots & \psi_2(\tau_N) \\ \vdots & \vdots & & \vdots \\ \psi_N(\tau_1) & \psi_N(\tau_2) & \ldots & \psi_N(\tau_N) \end{vmatrix}. \tag{6.72}$$

The symmetric wavefunction for an assembly of N bosons is the sum of $N!$ product wavefunctions. However, in this case, all of the terms are added to one another with a + sign. Contrary to fermions, there is no limit to the number of bosons that can occupy the same state. Fermions correspond to particles with half integer spin such as electrons ($s = \frac{1}{2}$) or ^{51}V nuclei ($s = 7/2$). Bosons correspond to particles with integer spin such at the photon ($s = 1$) or the Higgs boson ($s = 0$).

6.7.1 The distribution function for fermions

Using the bundling technique introduced above, we can count the number of microstates for a bundle of fermions. In this case each of the g_k energy levels is allowed to have either 0 or 1 fermion in it until all of the N_k particles in the kth bundle are accounted for. There will be $g_k/N_k!(g_k - N_k)!$ microstates associated with the kth bundle. This implies that $N_k \leq g_k$, which means that the number of fermions must not be greater than the number of states. For all of the bundles, we match each microstate of one bundle with each microstate of the others resulting in a total number of microstates for the assembly of

$$\Omega_{\rm FD} = \prod_k \frac{g_k!}{N_k!(g_k - N_k)!}. \tag{6.73}$$

The subscript FD is added to remind us that fermions follow what is called *Fermi–Dirac statistics*.

The now familiar method is applied to derive the distribution function from the number of microstates. We need to maximize $\ln \Omega$ by applying the condition for the most probable state

$$\frac{\delta \ln \Omega}{\delta N_k} + \alpha + \beta \varepsilon_k = 0 \tag{6.74}$$

subject to the constraints

$$N = \sum_k N_k \tag{6.75}$$

and

$$U = \sum_k N_k \varepsilon_k. \tag{6.76}$$

First, we apply Stirling's theorem to Eq. (6.73) to obtain

$$\ln \Omega_{\rm FD} = \sum_k [g_k \ln g_k - N_k \ln N_k - (g_k - N_k) \ln(g_k - N_k)]. \tag{6.77}$$

Taking the derivative and maximizing $\ln \Omega$ according to Eq. (6.74) yields

$$\ln\left(\frac{g_k - N_k}{N_k}\right) + \alpha + \beta \varepsilon_k = 0. \tag{6.78}$$

Thus

$$N_k = \frac{g_k}{\exp(-\alpha - \beta\varepsilon_k) + 1}, \tag{6.79}$$

which allows us to write the *Fermi–Dirac distribution* function as

$$f_k = \frac{N_k}{g_k} = \frac{1}{\exp(-\alpha - \beta\varepsilon_k) + 1}. \tag{6.80}$$

The distribution function, shown in Fig. 6.3, returns the average number of particles per energy state at equilibrium. Note the satisfying feature that the distribution function does not depend on the arbitrarily chosen value of g_k.

6.7.2 The distribution function for bosons

There is no restriction on how many of the N_k bosons can be distributed into the g_k energy levels of a bundle. The only restrictions are based on how many states there are to occupy and how many bosons there are. Statistical considerations of how many microstates are generated by creating an assembly of bosons yields the expression

$$\Omega_{\rm BE} = \prod_k \frac{(N_k + g_k)!}{N_k! g_k!}. \tag{6.81}$$

Bosons are said to follow *Bose–Einstein statistics*. Again, we need to maximize $\ln \Omega$ subject to the usual constraints on N and U. First, apply Stirling's theorem to $\Omega_{\rm BE}$ to obtain

$$\ln \Omega_{\rm BE} = \sum_k [(N_k + g_k) \ln(N_k + g_k) - N_k \ln N_k - g_k \ln g_k]. \tag{6.82}$$

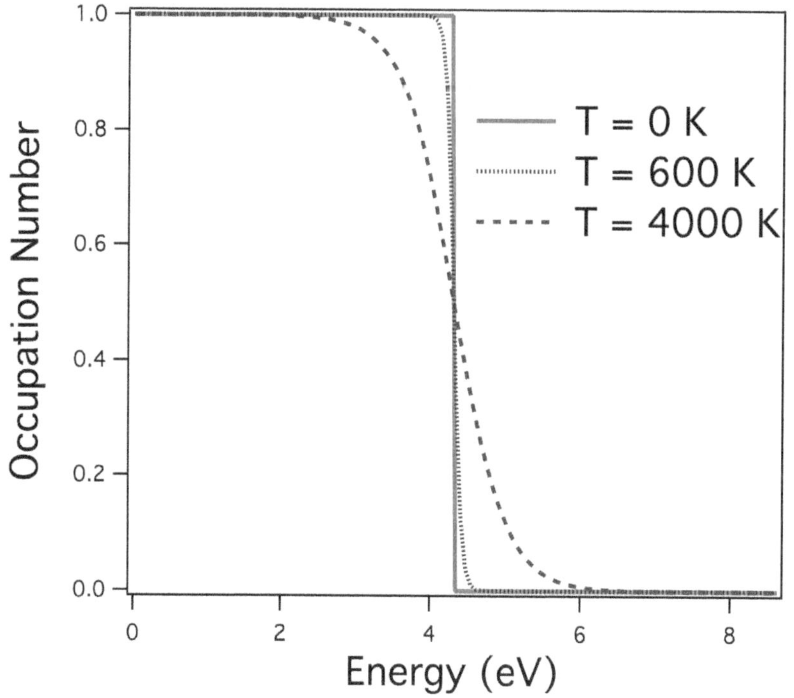

Figure 6.3 Plot of the Fermi–Dirac distribution for a material with a Fermi energy of 4.5 eV.

Thus,

$$\frac{\delta \ln \Omega_{\mathrm{BE}}}{\delta N_k} = \ln\left(\frac{N_k + g_k}{N_k}\right). \tag{6.83}$$

Now substitute this result into Eq. (6.74) to obtain

$$N_k = \frac{g_k}{\exp(-\alpha - \beta\varepsilon_k) - 1}. \tag{6.84}$$

From which we derive the *Bose–Einstein distribution* function

$$f_k = \frac{N_k}{g_k} = \frac{1}{\exp(-\alpha - \beta\varepsilon_k) - 1}. \tag{6.85}$$

6.7.3 Distributions in the dilute gas limit

We have now derived three distribution functions: the Maxwell–Boltzmann distribution, the Fermi–Dirac distribution, and the Bose–Einstein distribution. It is clear why the Fermi–Dirac and Bose–Einstein distributions arise. Particles are described by wavefunctions, and wavefunctions of fermions and bosons behave differently. Fermions follow Fermi–Dirac statistics and bosons follow Bose–Einstein statistics. But if all matter is composed of fermions and bosons, why and how was Boltzmann able to derive a different distribution that accurately describes the distribution of energy in a gas at normal conditions?

Let us look at the Fermi–Dirac and Bose–Einstein distributions in more detail and, specifically, under the conditions of a normal or dilute gas. On substituting $\beta = -1/k_{\mathrm{B}}T$ into the Fermi–Dirac distribution of Eq. (6.80), we obtain

$$g_k/N_k = \exp(-\alpha + \varepsilon_k/k_{\mathrm{B}}T) + 1. \tag{6.86}$$

Under normal conditions T is large (on the order of room temperature or higher to avoid condensation) and the number of states is exceedingly large. Thus, $g_k/N_k \gg 1$, which means that adding 1 on the right-hand side can be neglected or

$$N_k = g_k \exp(\alpha - \varepsilon_k/k_{\mathrm{B}}T). \tag{6.87}$$

This is exactly the Maxwell–Boltzmann distribution.

Now consider the Bose–Einstein distribution, again substituting for β and inverting to obtain

$$g_k/N_k = \exp(-\alpha + \varepsilon_k/k_{\mathrm{B}}T) - 1. \tag{6.88}$$

Again, the number of states is extremely large and subtracting 1 is negligible, giving

$$N_k = g_k \exp(\alpha - \varepsilon_k/k_B T). \tag{6.89}$$

The Maxwell–Boltzmann distribution is again recovered in the limit of sparse population of states that pertains to gases under commonly encountered experimental conditions. In other words, the classical behavior exhibited by inherently quantum mechanical molecules is a byproduct of the conditions under which they are being observed. Classical behavior is observed as the behavior of quantum mechanical particles in the limit of sparse state populations. In this limit the temperatures are high (thus, $k_B T$ is large compared to the translational energy level spacings) and the distances between molecules and the walls are large. Consequently, Boltzmann was able to derive equations to describe their behavior classically even though he was unaware of the need for quantum mechanics. Indeed, quantum behavior would not be postulated for at least three more decades after Boltzmann's work.

To fully specify the distribution functions, we need to determine the value of α. The constant α is determined by the requirement that the total number of particles is fixed. We can rewrite the distribution functions in the form

$$f_k = [\exp((\varepsilon_k - \mu)/k_B T) \pm 1]^{-1}, \tag{6.90}$$

where the (+) is for the Fermi–Dirac distribution and the (−) is for the Bose–Einstein distribution. Here, we have introduced the *chemical potential*

$$\mu = \alpha k_B T. \tag{6.91}$$

At equilibrium, the values of α and β are everywhere the same in the system. That is, for any system at equilibrium, for instance two phases or the reactants and products of a reversible reaction, the chemical potential μ and temperature T are the same everywhere. This will be discussed in greater detail in Chapter 11.

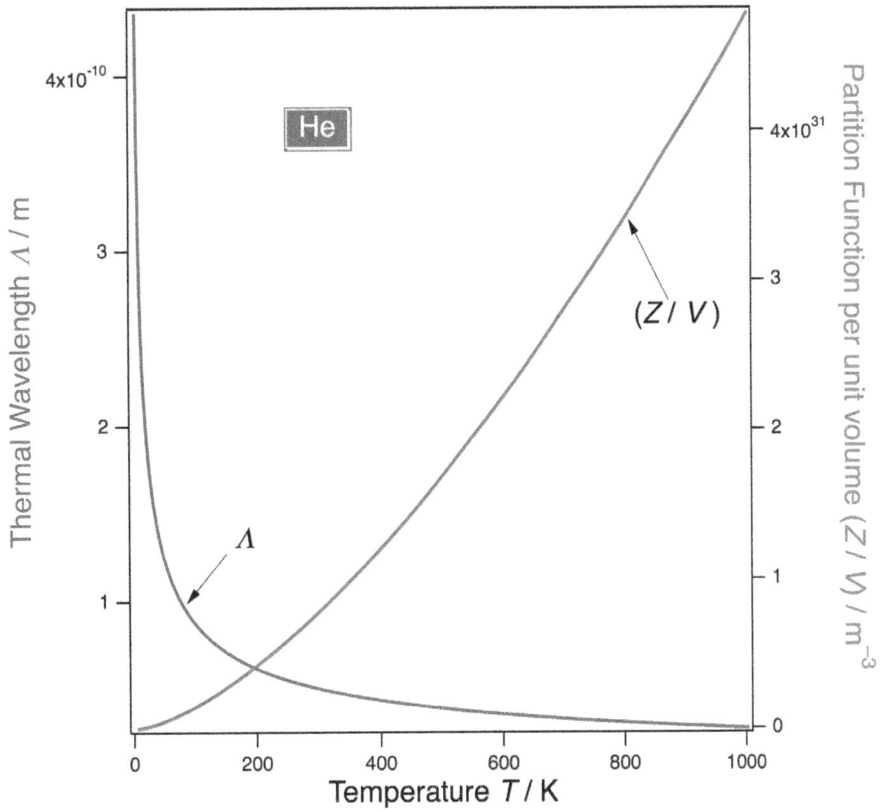

Figure 6.4 The thermal wavelength and translational partition function per unit volume are plotted versus temperature for He. The thermal wavelength is quite small. At temperatures above 77 K, $\Lambda < 1$ Å, which means that Z/V is extremely large.

6.8 The Maxwell–Boltzmann speed distribution

The Maxwell–Boltzmann distribution is derived in the dilute limit where $N_k \ll g_k$. The partition function

$$Z = \sum_k g_k \exp(-\varepsilon_k/k_B T) \tag{6.92}$$

is the partition function per gas molecule. The summation is over the *k* bundles, but the total energy must be the same if instead, the summation is made over all of the individual *j* energy levels, which are packed even more closely together. These states are so closely spaced that the summation may be replaced by an integral. Consider again the particle in a box problem for which the number of levels with energy less than ε is

$$N = \frac{4\pi}{3}\left(\frac{2ma^2\varepsilon}{h^2}\right)^{3/2}. \tag{6.93}$$

Differentiating, we find that the infinitesimal change in the number of states caused by a change $\mathrm{d}\varepsilon$ in the energy is

$$\mathrm{d}N = (2\pi a^3/h^3)(2m)^{3/2}\varepsilon^{1/2}\,\mathrm{d}\varepsilon = \rho(\varepsilon)\,\mathrm{d}\varepsilon. \tag{6.94}$$

The term $\rho(\varepsilon)$ represents the number of states with energies between ε and $\varepsilon + \mathrm{d}\varepsilon$, and is called the *density of states*, which since $V = a^3$ can be written as

$$\rho(\varepsilon)\,\mathrm{d}\varepsilon = (2\pi V/h^3)(2m)^{3/2}\varepsilon^{1/2}\,\mathrm{d}\varepsilon. \tag{6.95}$$

Integrating the partition function, we obtain

$$Z = \int_0^\infty \rho(\varepsilon)\,e^{-\varepsilon/k_B T}\,\mathrm{d}\varepsilon = 2\pi V(2m/h^2)^{3/2}\int_0^\infty \varepsilon^{1/2}\,e^{-\varepsilon/k_B T}\,\mathrm{d}\varepsilon, \tag{6.96}$$

$$Z = V(2\pi m k_B T/h^2)^{3/2}. \tag{6.97}$$

This is the *partition function for the translational degree of freedom* for one molecule of a Maxwell–Boltzmann gas of mass *m* held at temperature *T* in a volume *V*. For a monatomic gas molecule, which only has the translational degree of freedom, this is also the total partition function.

The factor after *V* in Eq. (6.97) is commonly encountered. Because of this, we define the *thermal wavelength* Λ as

$$\Lambda = (h^2/2\pi m k_B T)^{1/2}, \tag{6.98}$$

which has the units of length. Thus, the translational partition function can be written

$$Z = V\Lambda^{-3}. \tag{6.99}$$

To obtain the Maxwell–Boltzmann speed distribution, we need to obtain an expression for the number of molecules at a particular speed $N(v)$ within the velocity increment $\mathrm{d}v$. Since the kinetic energy is $\varepsilon = \frac{1}{2}mv^2$, then $\mathrm{d}\varepsilon = mv\,\mathrm{d}v$ and we can transform the above expression involving the density of states from one dependent on energy to one dependent on speed. We begin by rewriting Eq. (6.89) as

$$N(\varepsilon)\,\mathrm{d}\varepsilon = \rho(\varepsilon)e^{\alpha}e^{-\varepsilon/k_B T}\,\mathrm{d}\varepsilon. \tag{6.100}$$

Substitute for e^α from

$$e^\alpha = \frac{N}{Z} = \frac{N}{V}\left(\frac{h^2}{2\pi m k_B T}\right)^{3/2} \tag{6.101}$$

and for $\rho(\varepsilon)$ from Eq. (6.95) to obtain

$$N(\varepsilon)\,\mathrm{d}\varepsilon = \frac{2N}{\pi^{1/2}(k_B T)^{3/2}}\varepsilon^{1/2}e^{-\varepsilon/k_B T}\,\mathrm{d}\varepsilon. \tag{6.102}$$

Now transform the energy variable into speed to obtain the Maxwell–Boltzmann distribution, that is, the number of molecules with speeds between *v* and $v + \mathrm{d}v$,

$$N(v)\,\mathrm{d}v = 4\pi N\left(\frac{m}{2\pi k_B T}\right)^{3/2} v^2 e^{-mv^2/2k_B T}\,\mathrm{d}v. \tag{6.103}$$

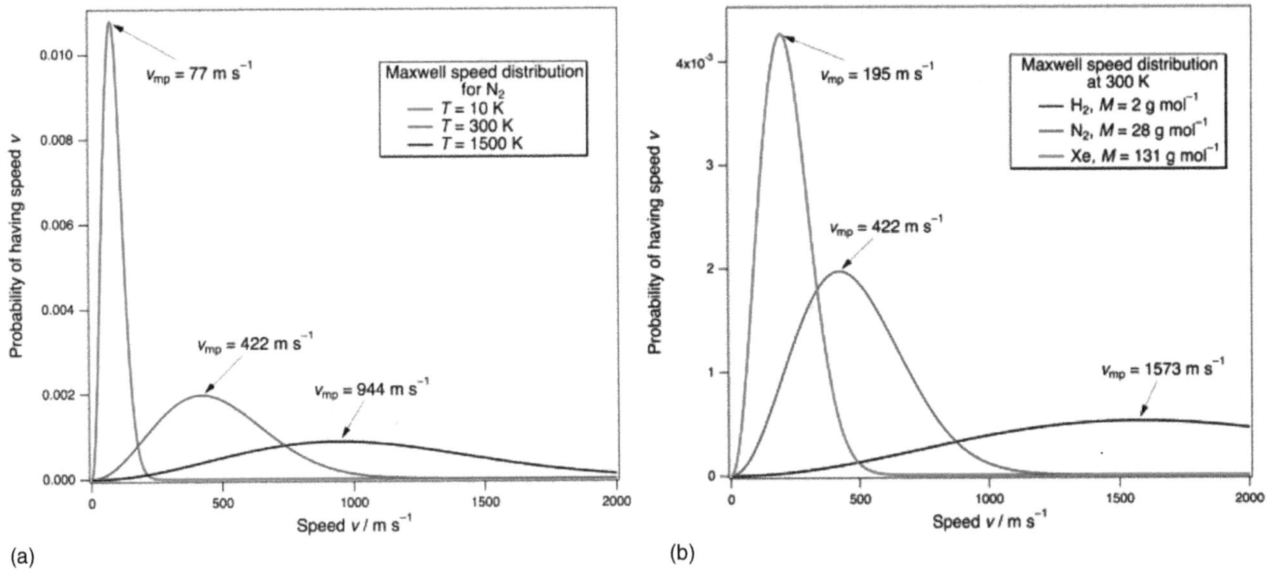

Figure 6.5 Maxwell–Boltzmann speed distributions for: (a) gaseous N_2 at three temperatures; and (b) H_2, N_2 and Xe at 300 K.

The Maxwell–Boltzmann distribution describes the translational degree of freedom of atoms and molecules regardless of whether they have any internal degrees of freedom. An example is shown in Fig. 6.5 for N_2 gas at three different temperatures. Note how the peak of the distribution shifts and the high-energy tail becomes more prominent with increasing T.

6.9 Derivation of the ideal gas law

A particularly important thermodynamic function is the Helmholtz energy. As will be proven in Chapter 11, any system held at constant volume and temperature will continue to change until the Helmholtz energy reaches a minimum and becomes constant. By definition, the Helmholtz energy is

$$A = U - TS. \tag{6.104}$$

Thus, for an infinitesimal change we have

$$\mathrm{d}A = \mathrm{d}U - T\mathrm{d}S - S\mathrm{d}T. \tag{6.105}$$

The *central equation of thermodynamics* is derived from the first and second laws for a reversible process

$$\mathrm{d}U = T\,\mathrm{d}S - p\,\mathrm{d}V. \tag{6.106}$$

Substituting into the expression for $\mathrm{d}A$, we write

$$\mathrm{d}A = -p\,\mathrm{d}V - S\,\mathrm{d}T, \tag{6.107}$$

which allows us to express both p and S in terms of partial differential equation according to

$$p = -\left(\frac{\partial A}{\partial V}\right)_T \tag{6.108}$$

and

$$S = -\left(\frac{\partial A}{\partial T}\right)_V. \tag{6.109}$$

Hence, if we obtain the functional form of A versus T and V from statistical mechanics, we can take the derivatives analytically. Equation (6.104) requires an expression for the entropy, which we found in Section 6.3 to be

$$S = k_B \ln \Omega. \tag{6.110}$$

However, since

$$\ln \Omega = \sum_k [N_k \ln(g_k/N_k)] + N \tag{6.111}$$

$$= -\alpha N - \beta U + N \tag{6.112}$$

$$= N \ln Z - \ln N! \ - \beta U, \tag{6.113}$$

the entropy is

$$S = k_B [N \ln Z - \ln N! + U/k_B T] \tag{6.114}$$

Inserting this into Eq. (6.104) yields

$$A = U - k_B T[N \ln Z - \ln N! + U/k_B T] \tag{6.115}$$

$$= -k_B T[N \ln Z - \ln N!] = -k_B T \ln(Z^N/N!) \tag{6.116}$$

$$A = -k_B T \ln Z_W \tag{6.117}$$

where

$$Z_W = Z^N/N! \tag{6.118}$$

is known as the total *partition function of an assembly of nonlocalized (indistinguishable) molecules.*

To derive the *ideal gas law,* we take the partial derivative of Eq. (6.115) with respect to V and insert this into Eq. (6.108):

$$p = -\left(\frac{\partial A}{\partial V}\right)_T = Nk_B T \left(\frac{\partial Z}{\partial V}\right)_T \tag{6.119}$$

The derivative is found from Eq. (6.97), yielding

$$pV = Nk_B T. \tag{6.120}$$

This is the ideal gas law written for N molecules. On a molar basis, where the number of moles n is found by dividing by the Avogadro constant N_A,

$$n = N/N_A, \tag{6.121}$$

ideal gases are known to follow the equation of state

$$pV = nRT. \tag{6.122}$$

where R is the *ideal gas constant* per mole of gas. These two can only be consistent if

$$k_B = R/N_A. \tag{6.123}$$

In other words, the ideal gas constant per molecule and the *Boltzmann constant* are one and the same. By confirming that k_B is in fact the Boltzmann constant, we also justify our statistical mechanical definitions of temperature, $\beta = -1/k_B T$, and entropy, $S = k_B \ln \Omega$.

6.10 Deriving the Sackur–Tetrode equation for entropy of a monatomic gas

Now that we have both a statistical mechanical definition of entropy

$$S = k_B \ln \Omega \tag{6.124}$$

and the thermodynamic relationship of the entropy to the Helmholtz energy,

$$S = -(\partial A/\partial T)_V \tag{6.125}$$

we can use our microscopic understanding of molecules to derive an expression for the entropy of a monatomic gas in terms of the macroscopic parameters N, T, and V (or p). Starting with Eq. (6.116), we apply Stirling's theorem to obtain

$$A = -Nk_B T[\ln Z - \ln N + 1]. \tag{6.126}$$

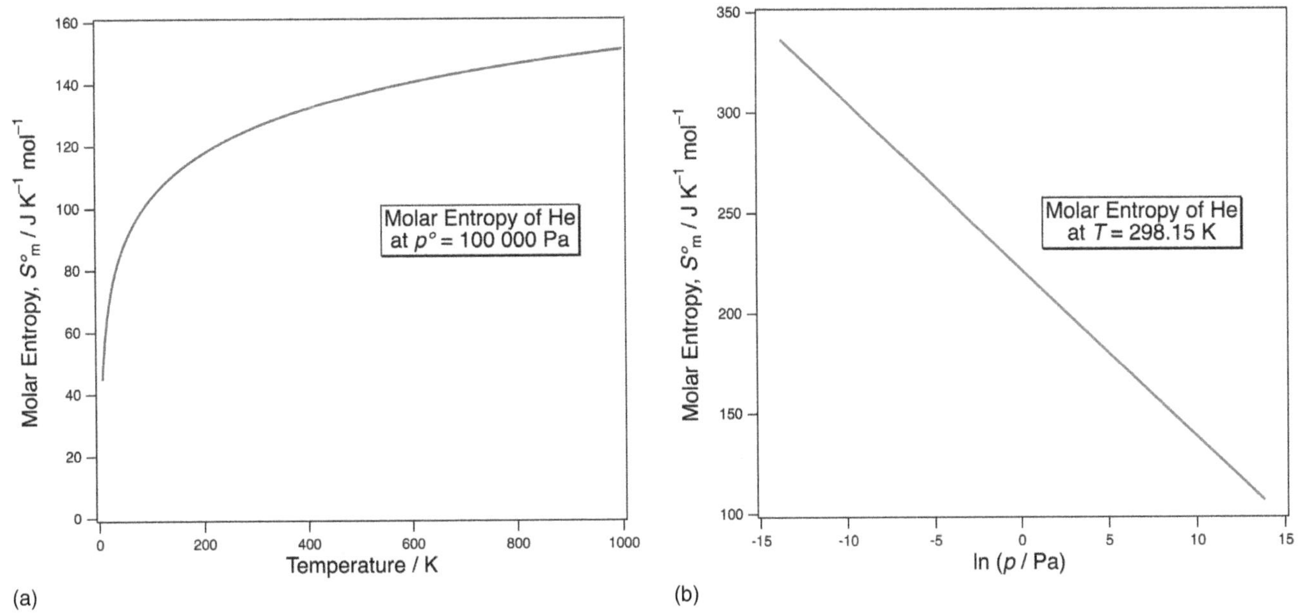

Figure 6.6 The molar entropy of He calculated from the Sackur–Tetrode equation as a function of: (a) temperature at standard pressure; and (b) pressure at 298.15 K.

Again, we use Eq. (6.97) but this time the partial derivative is taken with respect to T to obtain the entropy,

$$S = Nk_B \left[\frac{3}{2} \ln \left(\frac{2\pi m k_B T}{h^2} \right) - \ln \left(\frac{N}{V} \right) + \frac{5}{2} \right]. \tag{6.127}$$

From the ideal gas equation we can substitute $V = Nk_BT/p$ to obtain the dependence of the entropy on pressure,

$$S = Nk_B \left[\frac{3}{2} \ln \left(\frac{2\pi m k_B T}{h^2} \right) - \ln \left(\frac{p}{k_B T} \right) + \frac{5}{2} \right] \tag{6.128}$$

Equation (6.128) is the *Sackur–Tetrode equation* for the entropy of a monatomic gas – that is, a gas that only has three translational degrees of freedom. If the gas has more degrees of freedom, such as rotations and vibrations, the partition function – and therefore also the entropy – will have additional terms. The results for He are shown in Fig. 6.6.

Since $Nk_B = (N/N_A)N_A\, k_B = nR$, we can write the Sackur–Tetrode equation in a compact form by substituting in the thermal wavelength $\Lambda = (h^2/2\pi m k_B T)^{1/2}$,

$$S = nR \ln \left(\frac{e^{5/2} V}{\Lambda^3 N} \right) = nR \ln \left(\frac{RTe^{5/2}}{\Lambda^3 N_A p} \right). \tag{6.129}$$

The Sackur–Tetrode equation is particularly important because it supplies not only an experimental test of statistical mechanics to thermodynamics, but also an independent experimental test of Planck's quantum hypothesis. This is because the equation depends on the value of the Planck constant h as well as the Boltzmann constant k_B. This equation demonstrates that the quantum hypothesis is applicable not only to massless photons but also to massive particles such as molecules. While this is obvious to us since we began this derivation with a quantum mechanical treatment of the particle in a box problem, this was not at all obvious in 1912 when Otto Sackur[2] and Hugo Tetrode[3] independently derived this result from much different arguments, and tested it using data on the vapor pressure of Hg – a monatomic gas that acts fairly ideally under the conditions considered.

6.10.1 Directed practice

Calculate the molar entropy (entropy per mole of gas) of gaseous Hg at 550 K and one atmosphere pressure.

[Answer: 187.7 J K^{-1} mol^{-1}]

6.11 The partition function of a diatomic molecule

The energy of a molecule is distributed between translational energy ε_t and internal energy ε_{int}. There are three components to the internal energy, namely rotational energy ε_r, vibrational energy ε_v, and electronic energy ε_e. To a first approximation we can consider these four contributions independent of each other; hence

$$\varepsilon = \varepsilon_t + \varepsilon_{int} = \varepsilon_t + \varepsilon_r + \varepsilon_v + \varepsilon_e. \tag{6.130}$$

The independence of the four contributions allows us to factor the partition function Z for the molecule. We start by writing the partition function as a sum over all the quantum mechanically allowed states (the eigenstates of the system)

$$Z = \sum_{\text{eigenstates}} \exp(\beta\varepsilon) \tag{6.131}$$

$$= \sum_{\text{eigenstates}} \exp(\beta\varepsilon_t + \beta\varepsilon_r + \beta\varepsilon_v + \beta\varepsilon_e). \tag{6.132}$$

Because the contributions are independent, the exponential contains no cross terms. The sum in the exponential term can simply be written as the product of four exponential terms. The sum of each exponential is independent of the others, which allows us to write

$$Z = \left[\sum \exp(\beta\varepsilon_t)][\sum \exp(\beta\varepsilon_r)][\sum \exp(\beta\varepsilon_v)][\sum \exp(\beta\varepsilon_e)\right] \tag{6.133}$$

$$Z = Z_t Z_r Z_v Z_e. \tag{6.134}$$

6.12 Contributions of each degree of freedom to thermodynamic functions

6.12.1 Translations

The translational partition function of a diatomic molecule is the same as the partition function of a monatomic molecule (in the absence of electronic structure)

$$Z_t = \left(\frac{2\pi m k_B T}{h^2}\right)^{3/2} V. \tag{6.135}$$

We have previously found that the contribution of translations to the internal energy is

$$U_t = \tfrac{3}{2} N k_B T. \tag{6.136}$$

This is found formally from

$$U_t = N k_B T^2 \frac{\mathrm{d}\ln Z}{\mathrm{d}T}. \tag{6.137}$$

The heat capacity at constant volume is defined as the derivative of U with respect to T, hence,

$$C_{V,t} \equiv \left(\frac{\partial U_t}{\partial T}\right)_V = \tfrac{3}{2} N k_B. \tag{6.138}$$

6.12.2 Rotations

A diatomic molecule rotates about an axis extending perpendicular to the internuclear axis with an origin at the center of mass of the molecule. The *moment of inertia* $I = \mu R^2$, where μ is the reduced mass and R is the distance between the atoms, is defined about this axis. The rotational partition function is

$$Z_r = \sum_{J=0}^{\infty} g_J \exp(-\varepsilon_r / k_B T). \tag{6.139}$$

The rotational state is denoted by the rotational quantum number J. Because of space quantization, the rotational axis is not allowed to point arbitrarily in space compared to an imposed z-axis. Instead, the moment of inertia is allowed to make ($2J$+1) distinct projections. Thus, the rotational quantum number defines the energy of a set of $g_J = 2J + 1$ degenerate states.

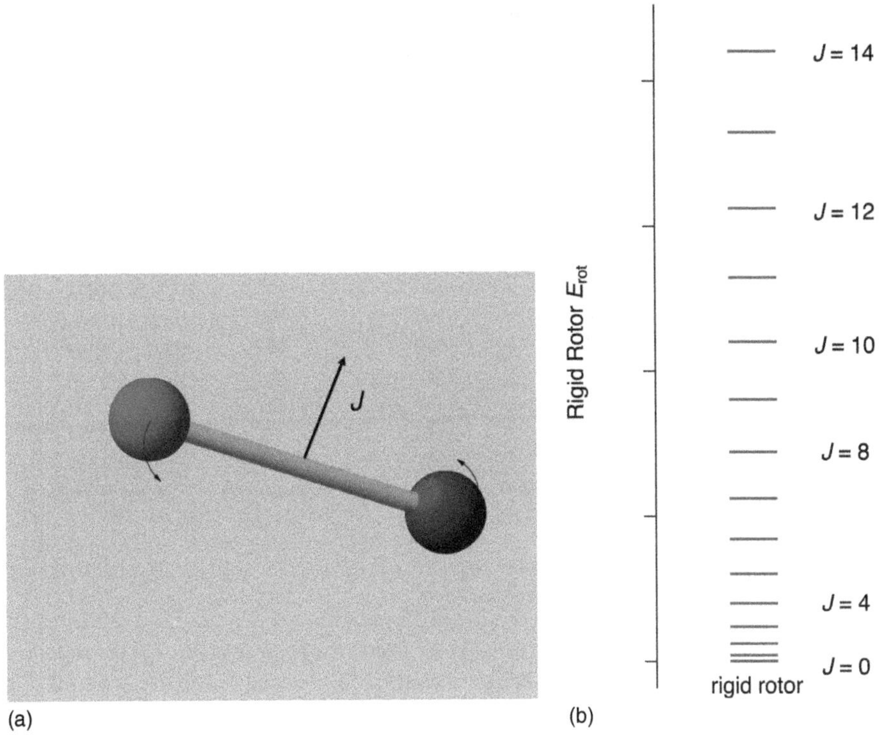

Figure 6.7 (a) Rigid rotor with rotational axis J and (b) the corresponding quantized energy level structure for states with rotational quantum number $0 \le J \le 14$.

The simplest model of a rotating molecule is the rigid rotor (see Fig. 6.7). This is a very good approximation for strong bonds, particularly for low-lying quantum states. The energy of a rigid rotor is given by

$$\varepsilon_r = J(J+1)\frac{h^2}{8\pi^2 I}. \tag{6.140}$$

Because I is directly proportional to the reduced mass of the rotor, *the spacing between rotational levels is inversely proportional to the reduced mass of the atoms in the rotor*. The rotational characteristic temperature is defined as

$$\theta_r = h^2/8\pi^2 I k_B. \tag{6.141}$$

Therefore, the rotational partition function is

$$Z_r = \sum_{J=0}^{\infty} (2J+1)\exp[-J(J+1)\theta_r/T]. \tag{6.142}$$

The spacing between rotational levels is of order $k_B T$. Thus, the introduction of θ_r allows us to split the evaluation of Z_r into two limits. In the limit of high T, defined as $T \gg \theta_r$ for which $\theta_r/T \ll 1$, the summation in Eq. (6.142) can be replaced by an integral. To do so we introduce a change of variables: $x = J(J+1)$ for which $\mathrm{d}x = (2J+1)\mathrm{d}J$; thus,

$$Z_r = \int_0^{\infty} \exp(-x\theta_r/T)\,\mathrm{d}x = T/\theta_r. \tag{6.143}$$

However, we must take account of an element of symmetry in molecules. For a heteronuclear diatomic molecule, rotating the molecule by 2π leaves the system unchanged. For a homonuclear diatomic, rotation by both π and 2π leaves the system unchanged. This factor of 2 is accounted for by introducing the symmetry number $\sigma = 1$ for heteronuclear and $\sigma = 2$ for homonuclear diatomics. The final answer for the high-temperature limit of the rotational partition function is

$$Z_r = T/\sigma\theta_r. \tag{6.144}$$

Thus, in the high-temperature limit, the rotational contributions to the internal energy and constant volume heat capacity are

$$U_r = Nk_B T^2(\partial \ln Z_r/\partial T) \simeq Nk_B T \tag{6.145}$$

Table 6.2 θ_r and θ_v values. These include the boiling point T_b and critical temperature T_c.

Molecule	θ_r / K	θ_v / K	T_b / K	T_c / K
H_2	87.5	6320	20.4	33.18
N_2	2.89	3390	77.4	126
O_2	2.45	2745	90.19	154.8
F_2	1.27	1290	85.04	144.4
Cl_2	0.35	814	239.11	417.0
Br_2	0.12	465	331.93	
I_2	0.05	309	457.5	785
HF	31.5	5990	292.69	461
HCl	15.2	4330	188.3	324.6
HBr	12.2	3820	204.2	363
HI	9.4	3340	237.77	423
CO	2.78	3120	81.7	133
NO	2.45	2745	121.4	180

and

$$C_{V,r} = (\partial U/\partial T)_V \simeq Nk_B \tag{6.146}$$

At very low temperature, $k_B T < k_B \theta_r$ such that molecules cannot be excited from the ground $J = 0$ level to the first excited $J = 1$ level. If only $J = 0$ is populated, then $\lim_{\delta T \to 0} Z_r = 1$. This means that Z_r is roughly constant, its derivative with respect to T vanishes, and therefore both the internal energy U and heat capacity C_V converge to zero as the temperature approaches zero.

Upon inspection of the data in Table 6.2, we note that most gases have θ_r values below their normal boiling point T_b. This means that in the gas phase they are always in the high-temperature limit. The low-temperature behavior of H_2 is often anomalous, regardless of what property is being considered. H_2 has the largest rotational energy level spacing of

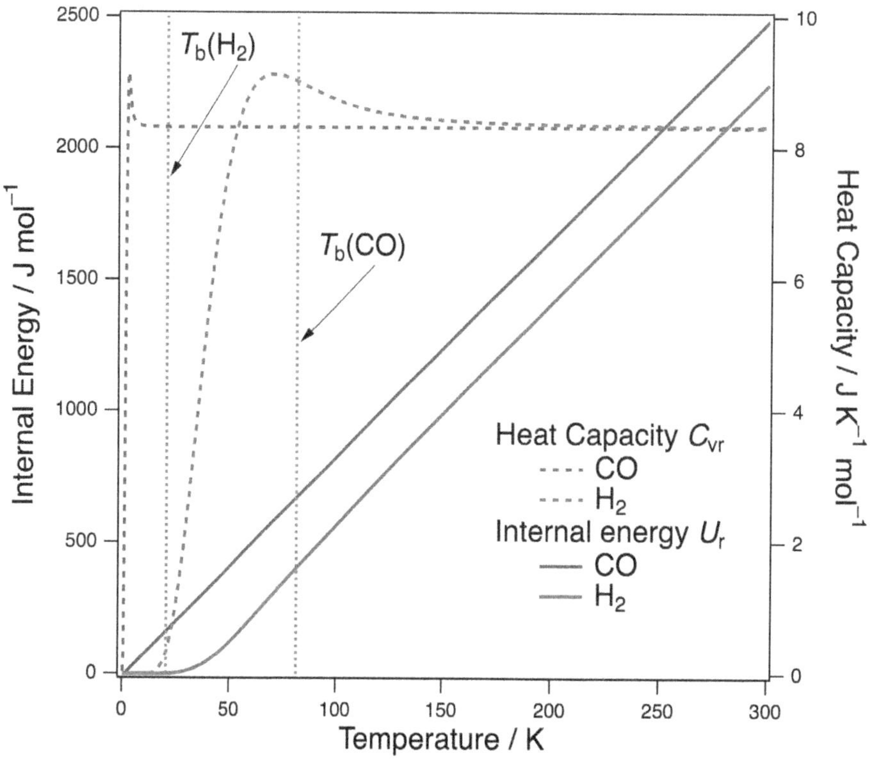

Figure 6.8 A plot of the rotational contributions to the internal energy and heat capacity as a function of temperature for H_2 and CO. The heat capacity rapidly converges on the high-temperature limit of R, and it is at this value above the normal boiling point for all gases except for H_2.

any chemical gas, which leads to much different behavior. Hence, it is not a surprise that low-temperature heat capacity and internal energy of H_2 should deviate from the behavior of other gases.

Another way to consider the high-temperature limit is that it is the classical limit. The low-temperature limit is the regime in which quantum effects predominate.

6.12.2.1 Directed practice

The moment of inertia of CO is $I = 1.45 \times 10^{-46}$ kg m^2. Calculate the difference in the energies of the $J = 5$ and $J = 6$ rotational levels. Compare this to thermal energy as represented by $k_B T$ to show that, at 300 K, the spacing between rotational levels is far smaller than thermal energy.

[Answer: 4.60×10^{-22} J; 4.14×10^{-21} J]

6.12.3 Vibrations

A diatomic molecule has one vibrational mode, corresponding to the stretching and compression of the chemical bond. The partition function for the vibrational degree of freedom is

$$Z_v = \sum_v \exp(-\varepsilon_v / k_B T) \tag{6.147}$$

To a first approximation – and a very good one near the ground state – we treat vibrational motion as that of a harmonic oscillator, depicted in Fig. 6.9. The energy levels of a harmonic oscillator are given by

$$\varepsilon_v = (v + \tfrac{1}{2})\hbar\omega_0. \tag{6.148}$$

Here, we use the reduced Planck constant $\hbar = h/2\pi$ and angular frequency ω_0

$$\omega_0 = (k/\mu)^{1/2} \tag{6.149}$$

where k is the force constant of the oscillator and μ is the reduced mass. The angular frequency ω_0 instead of the frequency ν, where $\omega = 2\pi\nu$, is used so as to avoid confusion between the vibrational quantum number v (italicized v) with ν (Greek

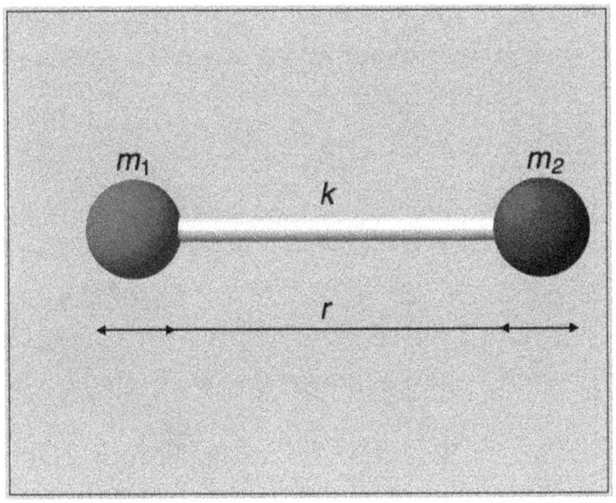

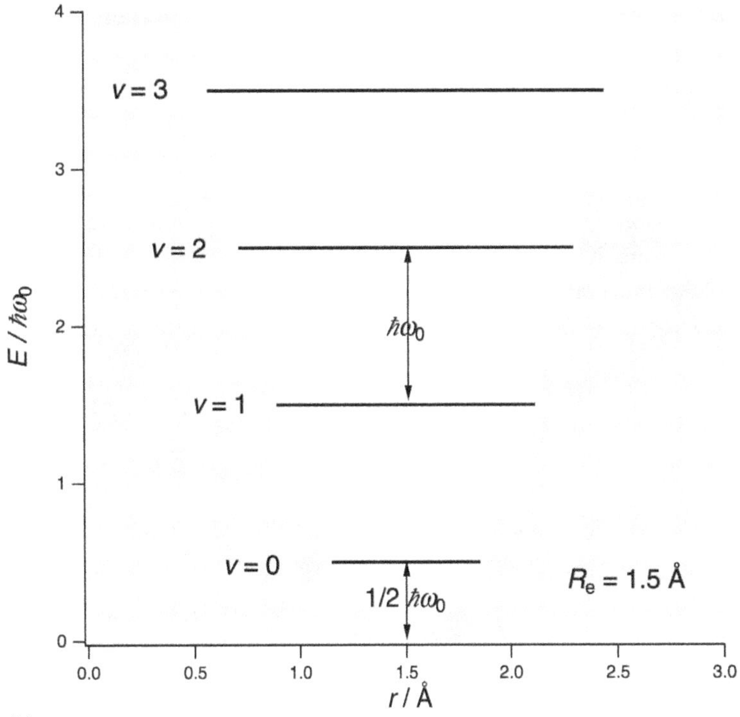

Figure 6.9 (a) A harmonic oscillator with atoms of mass m_1 and m_2 connected by a spring of force constant k at an equilibrium bond distance r_e. (b) The energy levels of a quantum mechanical harmonic oscillator are evenly spaced by $\hbar\omega_0$ and the first level is offset from zero by the zero point energy of $\frac{1}{2}\hbar\omega_0$.

nu). The levels of the harmonic oscillator are nondegenerate. Similar to the rigid rotor, the energy levels of a harmonic oscillator are *spaced more closely together with increasing mass.*

Substituting the energy expression of Eq. (6.148) into Eq. (6.147), we find an expression for the vibrational partition function

$$Z_v = \sum_v \exp\left[-\left(v+\tfrac{1}{2}\right)\hbar\omega_0/k_B T\right] \tag{6.150}$$

$$= e^{-\hbar\omega_0/2k_B T}\sum_v \exp(-v\hbar\omega_0/k_B T). \tag{6.151}$$

The summation in Eq. (6.151) is a geometric progression that simplifies to

$$\sum_v \exp(-v\hbar\omega_0/k_B T) = [1-\exp(-\hbar\omega_0/k_B T)]^{-1}. \tag{6.152}$$

Now introduce the vibrational characteristic temperature

$$\theta_v = \hbar\omega_0/k_B \tag{6.153}$$

so that we can write the vibrational partition function as

$$Z_v = \frac{\exp(-\theta_v/2T)}{1-\exp(-\theta_v/T)}. \tag{6.154}$$

As in Eqs (6.145) and (6.146), we use the appropriate derivatives with respect to T to obtain the internal energy due to vibration and the vibrational heat capacity

$$U_v = Nk_B\theta_v\left(\frac{1}{2}+\frac{1}{\exp(\theta_v/T)-1}\right) \tag{6.155}$$

and

$$C_{V,v} = Nk_B\left(\frac{\theta_v}{T}\right)^2\frac{\exp(\theta_v/T)}{[\exp(\theta_v/T)-1]^2}. \tag{6.156}$$

The first term in Eq. (6.155) represents the zero point energy $\frac{1}{2}\,Nk_B T$, which results from the quantum mechanical nature of the oscillator. As $T \to 0$ this term dominates, which again shows that quantum effects are most obvious at low temperatures. At higher temperatures, Eq. (6.155) converges on the classical result, which is that the internal energy increases linearly with temperature as $Nk_B T$. At low temperatures, the heat capacity again tends toward zero and in the high-temperature classical limit, $C_{V,v} \to Nk_B$. However, as explained below and shown in Fig. 6.8, the heat capacity often varies between these limits in the temperature range of many experiments.

The vibrational energy level spacing is $\hbar\omega_0$. For many diatomics, except those containing two heavy elements, this value is much larger than $k_B T$. As can be seen in Table 6.2, this leads to vibrational characteristic temperatures that are as much as several thousand degrees. Consequently, vibrational population is limited almost exclusively to the ground state near room temperature, and the vibration must be treated within a low-temperature quantum mechanical framework. The behavior of U_v and C_{Vv} are shown in Fig. 6.10.

6.12.3.1 Directed practice

The fundamental vibrational frequency of CO is $\omega_0 = 4.09 \times 10^{14}\ \text{s}^{-1}$. Calculate the difference in the energies of the $v = 0$ and $v = 1$ vibrational levels. Compare this to thermal energy as represented by $k_B T$ to show that, at 300 K, the spacing between vibrational levels is far greater than thermal energy.

[Answer: 4.31×10^{-20} J; 4.14×10^{-21} J]

6.12.4 Electronic excitation

For most molecules – and in particular for most small molecules – electronic excitations occur at significantly higher energies than either rotational or vibrational excitations. The ground electronic state is set to zero, $\varepsilon_0 = 0$, and the excited states are then at energies ε_1, ε_2, and so on, with electronic characteristic temperatures $\theta_{e1} = \varepsilon_1/k_B$, and so forth. For N_2, the first excited state lies at $\varepsilon_1 = 8.59$ eV; hence $\theta_{e1} = 97\,400$ K. Therefore, the electronic partition function,

$$Z_e = g_0 + g_1\exp(-\theta_{e1}/T) + \cdots \tag{6.157}$$

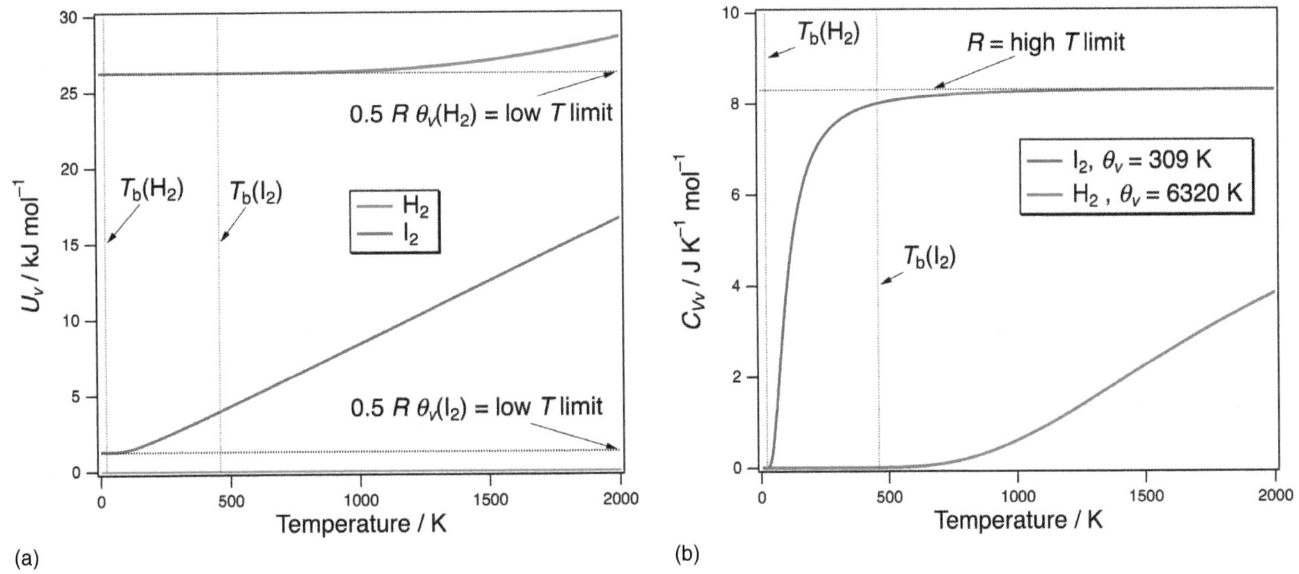

Figure 6.10 A plot of the vibrational contributions to the (a) internal energy and (b) heat capacity as a function of temperature. H_2 has the highest θ_v value and I_2 among the lowest for all diatomic molecule; hence the behavior of all other diatomics falls somewhere in between. While all diatomics have the same high-temperature limiting value of the heat capacity; each has its own characteristic low-temperature limit to the internal energy because of different zero point energies.

almost invariably converges on the first term, which is just the degeneracy of the ground state

$$Z_e = g_0. \tag{6.158}$$

Furthermore, the degeneracies of electronic state are usually 1. Because Z_e is a constant, there is no contribution of electronic excitation to either the internal energy or the heat capacity. There are two important exceptions to the rule that electronic excitation can be ignored. Both O_2 and NO have low-lying excited electronic states that can be excited by thermal energy. Known as the *Schottky anomaly*, the population of the first excited electronic state leads to an unexpected maximum in the C_V versus T curve.

6.13 The total partition function and thermodynamic functions

In the study of thermodynamics, we will see that two of the most important thermodynamic functions are the internal energy and the heat capacity. Therefore, a firm understanding of how their values depend on molecular properties and system parameters such as temperature is extremely important. The total partition function can be used to calculate their values as a function of temperature. The total partition function of a molecule with internal structure is

$$Z = Z_t Z_r Z_v Z_e. \tag{6.159}$$

Taking the logarithm of Eq. (6.159), we see that $\ln Z$ is composed of a number of additive terms for each degree of freedom

$$\ln Z = \ln Z_t + \ln Z_r + \ln Z_v + \ln Z_e. \tag{6.160}$$

As a consequence, a number of thermodynamic functions derive their values from a sum of terms for each degree of freedom. The *Helmholtz energy* is

$$A = -Nk_B T \ln Z + k_B T \ln N!, \tag{6.161}$$

and each degree of freedom supplies an additive term according to

$$A = -Nk_B T \ln Z_t + k_B T \ln N! - Nk_B T \ln Z_r - Nk_B T \ln Z_v - Nk_B T \ln Z_e \tag{6.162}$$

$$A = A_t + A_r + A_v + A_e. \tag{6.163}$$

Note that the translational contribution

$$A_t = -Nk_B T \ln Z_t + k_B T \ln N! \quad (6.164)$$

is the same as the total Helmholtz energy of a monatomic gas with no internal structure.

The *internal energy* is

$$U = Nk_B T \frac{d \ln Z}{dT} = U_t + U_r + U_v + U_e. \quad (6.165)$$

For a general expression valid at all temperatures, we substitute in values for each of the terms individually from above. For most molecules under normal experimental conditions, the translational and rotational terms behave in their classical limits, there is no electronic contribution, and the vibrational term must be treated quantum mechanically. This leads to an approximate quasiclassical result for the total internal energy of a diatomic molecule

$$U = \tfrac{3}{2}Nk_B T + Nk_B T + Nk_B \theta_v \left[\tfrac{1}{2} + \left(\exp\left(\theta_v/T\right) - 1\right)^{-1}\right]. \quad (6.166)$$

Similarly, for the heat capacity

$$C_V = \tfrac{3}{2}Nk_B + Nk_B + Nk_B \left(\frac{\theta_v}{T}\right)^2 \frac{\exp\left(\theta_v/T\right)}{\left[\exp\left(\theta_v/T\right) - 1\right]^2}. \quad (6.167)$$

The dependence of these two distributions on T is displayed in Fig. 6.11.

However, we have seen that both the rotational and vibrational contributions to C_V tend to zero at low temperature. Therefore, there are three limiting values of C_V that should be remembered. For an ideal monatomic gas, $C_V = \frac{3}{2}Nk_B$ at all temperatures. Just above the boiling point, where there is no vibrational excitation and little rotational excitation, the heat capacity of a diatomic molecule also tends to this value, $\lim_{\delta T \to T_b} C_V \to \frac{3}{2}Nk_B$. As the temperature is raised, excited rotational levels begin to contribute, and C_V increases until it reaches it classical limit for fully excited translational and rotational levels, $C_V = \frac{3}{2}Nk_B + Nk_B = \frac{5}{2}Nk_B$. Over a range of higher temperatures not too close to the vibrational characteristic temperature, that is $\theta_r < T < \theta_v$, the heat capacity remains constant until T approaches θ_v. Sufficiently above θ_v, C_V approaches its classical limit $C_V = \frac{3}{2}Nk_B + Nk_B + Nk_B = \frac{7}{2}Nk_B$.

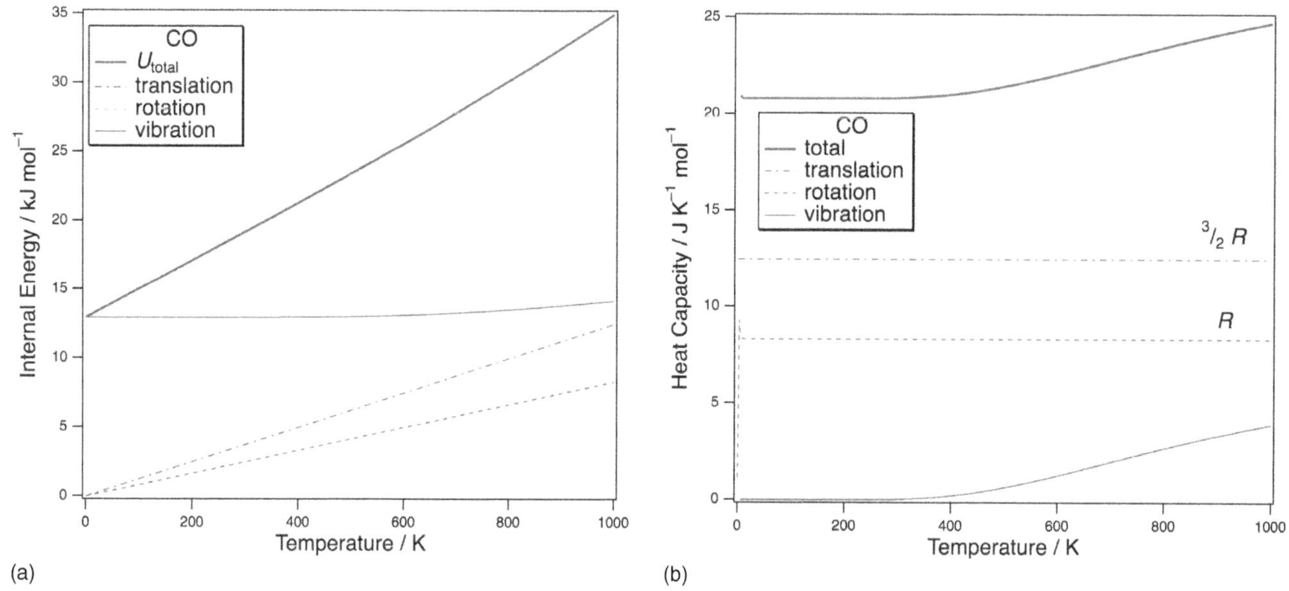

Figure 6.11 A plot of the translational, rotational and vibrational contributions to the (a) internal energy and (b) heat capacity of CO as a function of temperature. CO is taken to be an ideal gas rigid-rotor, harmonic oscillator. Translations and rotations are in the classical limit except at very low temperature. Therefore they make contributions to the internal energy that are linear in T and to the heat capacity that are independent of T. Since it is the only degree that is not yet in the classical limit, the vibrational degree of freedom is completely responsible for the temperature dependence of the heat capacity in the range of interest for most chemistry (200–1000 K).

What about the equation of state of a diatomic molecule? Is the ideal gas law changed as a result of internal structure? The pressure is given by

$$p = Nk_BT\left(\frac{\partial \ln Z}{\partial V}\right)_T \tag{6.168}$$

The total partition function is

$$Z = (2\pi mk_BT/h^2)^{3/2}V\, Z_{\text{int}}. \tag{6.169}$$

Most importantly, the translational portion of the partition function is a function of volume whereas the internal partition function is not. Therefore,

$$\left(\frac{\partial \ln Z}{\partial V}\right)_T = \frac{1}{V} \tag{6.170}$$

and just as for a monatomic gas, the gas law of an ideal diatomic (or an ideal polyatomic) molecule is

$$pV = Nk_BT. \tag{6.171}$$

6.14 Polyatomic molecules

How is the situation changed if we now consider polyatomic molecules? Very large polyatomic molecules such as proteins are not commonly encountered in the gas phase. However small polyatomics such as H_2O, CH_4, SF_6 etc. are. The translational degree of freedom is no different than that of monatomic or diatomic molecules. The partition function is again described by the same formula

$$Z_t = \left(\frac{2\pi mk_BT}{h^2}\right)^{3/2} V. \tag{6.172}$$

The translational contribution to the internal energy is again $\frac{3}{2}Nk_BT$ and that to the heat capacity is $\frac{3}{2}Nk_B$.

The internal degrees of freedom are more complex but, again, their contributions are added to those of translations. Polyatomic molecules have either two (linear polyatomics) or three (nonlinear polyatomics) degrees of rotational freedom. In the classical (high-temperature) limit, these are again are each assigned a contribution of $\frac{1}{2}Nk_BT$ to the internal energy and $\frac{1}{2}Nk_B$ to the heat capacity.

For not too-complex molecules, the rotational and vibrational degrees can be considered independently and each vibration can also be considered independent of each other. If there are x atoms in the molecule, there are a total of $3x$ degrees of freedom. Since 3 of these are translations and either 2 or 3 are rotations, there are $3x - 6$ (nonlinear molecules) or $3x - 5$ (linear) vibrational modes. In the classical limit, each is assigned a contribution of Nk_BT to the internal energy and Nk_B to the heat capacity. Each individual mode j has a corresponding vibrational characteristic temperature $\theta_{v,j} = \hbar\omega_j/k_B$. Each mode being considered independently, they each make a multiplicative contribution to the partition function. Hence

$$Z_v = \prod_{j}^{3x-6 \text{ or } 3x-5} Z_{v,j} \tag{6.173}$$

where each of the individual $Z_{v,j}$ terms is of the same form as Eq. (6.154) with an appropriate value of $\theta_{v,j}$. In a polyatomic molecule, especially as they get bigger, the vibrational frequencies may span from low to high values. This means each vibrational mode has its own high-temperature limit, and therefore it is not uncommon for some vibrations to reach the classical limit while others are still only marginally excited. For sufficiently large molecules, the independent vibration approximation breaks down and a description in terms of phonons (vibrational bands rather than discrete modes) is more appropriate.

The electronic degree of freedom can usually be neglected. However, with increasing size, this approximation again becomes problematic.

SUMMARY OF IMPORTANT EQUATIONS

$$U = \sum_j n_j \varepsilon_j$$
Internal energy of a system U in terms of population n_j and energy ε_j of jth level

$$Z = \sum_j g_j \exp(\beta \varepsilon_j)$$
Partition function of N distinguishable molecules. g_j = degeneracy, $\beta = -1/k_B T$,

$$\Omega = \frac{N!}{\prod_j n_j!}$$
The number of microstates Ω for N distinguishable molecules distributed in j states with populations n_j

$$n_j = N g_j\, e^{\beta \varepsilon_j}/Z$$
Populations n_j for states with degeneracy g_j

$$\ln N! = N \ln N - N$$
Stirling's theorem for large N

$$\frac{\partial \ln \Omega}{\partial n_j} + \alpha + \beta \varepsilon_j = 0$$
Expression used to maximize Ω by the use of Lagrange multipliers

$$S = k_B \ln \Omega = k_B \left(N \ln N - \sum_j n_j \ln n_j \right)$$
Entropy S, number of microstates Ω

$$pV = nRT = Nk_B T$$
Ideal gas law n = moles, N = number of molecules

$$A = U - TS$$
Helmholtz energy

$$G = A + pV$$
Gibbs energy

$$H = U + pV$$
Enthalpy

$$S = -(\partial A/\partial T)_V$$
Entropy

$$p = -\left(\frac{\partial A}{\partial V}\right)_T = Nk_B T \left(\frac{\partial \ln Z}{\partial V}\right)_T$$
Pressure

$$T = (\partial U/\partial S)_V$$
Temperature

$$f_k = N_k/g_k = \left[\exp\left((\varepsilon_k - \mu)/k_B T\right) \pm 1\right]^{-1}$$
The Fermi–Dirac (+) or Bose–Einstein (–) distribution. μ = chemical potential.

$$N(v)\,dv = 4\pi N \left(\frac{m}{2\pi k_B T}\right)^{3/2} v^2 e^{-mv^2/2k_B T}\, dv$$
Maxwell–Boltzmann speed distribution

$$Z = Z_t Z_r Z_v Z_e$$
Total partition function of a molecule Z

$$Z_t = V\left(2\pi m k_B T/h^2\right)^{3/2}$$
Translational partition function

$$Z_r = \sum_{J=0}^{\infty} g(J) \exp(-\varepsilon_r/k_B T)$$
Rotational partition function

$$Z_r = T/\theta_r$$
Rotational partition function in limit of high T, θ_r = rotational characteristic temperature

$$Z_v = \frac{\exp(-\theta_v/2T)}{1 - \exp(-\theta_v/T)}$$
Vibrational partition function, θ_v = vibrational characteristic temperature

$$Z_e = g_0$$
Electronic partition function when first excited state is at a normal energy, g_0 usually equals 1

$$U = Nk_B T^2 \frac{d \ln Z}{dT} = U_t + U_r + U_v + U_e$$
Internal energy of a molecule (U_e is usually zero except at extremely high T)

$$U_t = \tfrac{3}{2} Nk_B T$$
Translational contribution to internal energy (= internal energy of a monatomic gas)

$$U_r = \tfrac{n_r}{2} Nk_B T$$
Rotational contribution to internal energy, $n_r = 2$ for diatomic or linear molecule, $n_r = 3$ for nonlinear polyatomic molecule

$$U_v = Nk_B T\left[\tfrac{1}{2} + \left[\exp(\theta_v/T) - 1\right]^{-1}\right]$$
Vibrational contribution to internal energy (for a polyatomic the contribution is a sum over all vibrations, each with its own value of θ_v)

$$C_V = \left(\frac{\partial U}{\partial T}\right)_V = C_{Vt} + C_{Vr} + C_{Vv} + C_{Ve}$$
Heat capacity at constant volume (C_{Ve} is usually zero except at extremely high T)

$$C_{Vt} = \tfrac{3}{2} Nk_B$$
Translational contribution to the heat capacity (= heat capacity at of a monatomic gas)

$$C_{Vr} = \tfrac{n_r}{2} Nk_B$$
Rotational contribution to heat capacity, $n_r = 2$ for diatomic or linear molecule, $n_r = 3$ for nonlinear polyatomic molecule

$$C_{Vv} = Nk_B \left(\frac{\theta_v}{T}\right)^2 \frac{\exp(\theta_v/T)}{\left[\exp(\theta_v/T) - 1\right]^2}$$
Vibrational contribution to heat capacity (for a polyatomic the contribution is a sum over all vibrations, each with its own value of θ_v)

Exercises

6.1 Chart the allowed distributions for three distinguishable particles A, B and C with $U = 5\varepsilon$ and energy level structure $\varepsilon_0 = 0,\ \varepsilon_1 = 1,\ \varepsilon_2 = 2,\ \varepsilon_3 = 3,\ \varepsilon_4 = 4,\ \varepsilon_5 = 5 \ldots$. Calculate $\sum t(n)$ and t_{max} and the probability in being in each class of distribution.

6.2 The standard molar entropies of four gases at 298.15 K are given below. Explain the trends.

	F_2	Cl_2	HF	HCl
$S_m{}^\circ$ / J K^{-1} mol^{-1}	202.8	223.1	173.8	186.9

6.3 Starting with Eq. (6.43), show that $S = k_B(N \ln Z + U/k_B T)$.

6.4 Starting with Eq. (6.64), apply Stirling's theorem and derive Eq. (6.67).

6.5 Derive the Fermi–Dirac distribution function starting with

$$\frac{\delta \ln \Omega}{\delta N_k} + \alpha + \beta\varepsilon_k = 0$$

and $\Omega_{FD} = \prod_k \dfrac{g_k!}{N_k!(g_k - N_k)!}$.

6.6 Derive the Bose–Einstein distribution function starting with

$$\frac{\partial \ln \Omega}{\partial N_k} + \alpha + \beta\varepsilon_k = 0$$

and $\Omega_{BE} = \prod_k (N_k + g_k)!/(N_k! g_k!)$.

6.7 Integrate $Z = \int_0^\infty \rho(\varepsilon)\, e^{-\varepsilon/k_B T}\, d\varepsilon$ to obtain the translational partition function for a Maxwell–Boltzmann gas.

6.8 Derive the Sackur–Tetrode equation, Eq. (6.127), by taking the derivative with respect to temperature of the Helmholtz energy and substituting into Eq. (6.125).

6.9 Use Eq. (6.129), the Sackur–Tetrode equation, to calculate the molar entropy of He and I_2 at standard conditions of $p = 1$ bar and $T = 298.15$ K. The accepted values are 126.153 J K^{-1} mol^{-1} and 260.687 J K^{-1} mol^{-1} for He and I_2, respectively. Discuss the extent of agreement of your results with the accepted values.

6.10 Use the Sackur–Tetrode equation, to calculate the molar entropy of Ar and CO at standard conditions of $p = 1$ bar and $T = 298.15$ K. The accepted values are 154.8 J K^{-1} mol^{-1} and 197.7 J K^{-1} mol^{-1} for Ar and CO, respectively. Discuss the extent of agreement of your results with the accepted values.

6.11 (a) Calculate the entropy of mixing N_1 molecules of type 1 with N_2 molecules of type 2, given that the number of arrangements of these two molecules in solution is $\Omega = N_0!/N_1!\, N_2!$ where N_0 is the total number of molecules in solution. (b) If the two molecules and their solution act ideally, the enthalpy of mixing is zero. Use this information and the answer from part (a) to calculate the Gibbs energy of mixing $\Delta_{mix}G$.

6.12 Use the Sackur–Tetrode equation to calculate the entropy difference between 1 mol ^{3}He and 1 mol ^{4}He at 298 K and 100 kPa.

6.13 Take the limits of $U_v = Nk_B\theta_v \left[\frac{1}{2} + \left(\exp\left(\theta_v/T\right) - 1\right)^{-1}\right]$ and show that, at low temperature, the vibrational internal energy tends to the zero point energy $\frac{1}{2} Nk_B T$, and that at high temperature the vibrational internal energy increases linearly with temperature.

6.14 Take the appropriate derivative of U and describe the behavior of the heat capacity of an ideal diatomic molecule at low temperature and at high temperature, considering only the rotational, vibrational, and translational contributions.

6.15 We can model a solid with $n = 3N$ harmonic oscillators, where N is the number of atoms. The energy levels are evenly spaced with separation ε. Population is distributed among these states such that the internal energy is $U = u\varepsilon$. Distributing the energy of the system over the n oscillators, the number of microstates of the system is

$$\Omega = \frac{(n+u)!}{u!\,n!}.$$

Show that the entropy of the system is

$$S = k_B \left[n \ln\left(1 + \frac{u}{n}\right) + u \ln\left(1 + \frac{n}{u}\right) \right].$$

Endnotes

1. This will be proved in Chapter 8. Thermodynamics cannot really be understood without statistical mechanics but statistical mechanics is only rendered useful by application of thermodynamics.
2. Sackur, O. (1913) *Ann. Phys.*, **40**, 67.
3. Tetrode, H. (1912) *Ann. Phys.*, **38**, 434; Tetrode, H. (1913) *Ann. Phys.*, **39**, 225.

Further reading

Hill, T.L. (1986) *An Introduction to Statistical Thermodynamics*. Dover Publications, New York.

McQuarrie, D.A. (2000) *Statistical Mechanics*. University Science Books, Mill Valley, CA. This is an excellent next step to a more in-depth study of statistical mechanics.

Metiu, H. (2006) *Physical Chemistry: Statistical Mechanics*. Taylor & Francis, New York. Combines statistical mechanics with a workout in Mathematica.

Renn, J. (2007) *Boltzmann und das Ende des mechanistischen Weltbildes*. Picus Verlag, Wien. Some historical perspective.

Trevena, D.H. (2001) (second printing 2011) *Statistical Mechanics: An introduction*. Woodhead Publishing, Cambridge. This is an excellent text from which I have borrowed heavily.

CHAPTER 7

First law of thermodynamics

PREVIEW OF IMPORTANT CONCEPTS

- The internal energy of a system is the sum of all kinetic and potential energies of the components of the system. It contains contributions from translational, rotational, and vibrational energy as well as intermolecular interactions.
- Changes in variables will always be calculated by taking the final state value minus the initial state value; for example, the internal energy change is $\Delta U = U_f - U_i$.
- Work is the transfer of energy that makes use of uniform motion of atoms.
- Heat is the transfer of energy that makes use of random motion of atoms.
- U is a property of a system known as a state function. Its value depends on the state of the system defined by p, T, V and composition, but not on how that state was achieved.
- q and w are path functions. They are not properties of the system; they are process variables that describe energy flow along a particular path and therefore depend on the nature of that path.
- An adiabatic process involves no exchange of heat, $q = 0$.
- Internal energy is changed by the sum of heat flow and work performed, $\Delta U = q + w$
- The heat exchanged at constant volume is equal to the internal energy change, $\Delta U = q_V$.
- The enthalpy is a state function defined as $H \equiv U + pV$. The enthalpy change is equal to the heat exchanged at constant pressure, $\Delta H = q_p$.
- Thermodynamic state variables such as U, H, p, and T have constant values at equilibrium when averaged over sufficiently long timescales and numbers of molecules.
- Molecules are constantly colliding and exchanging energy. Therefore, over short timescales for the whole distribution or any given molecule, there are fluctuations in thermodynamic variables.
- At each step along a reversible path the system is described by state variables that are described by equilibrium values (or are within a normal fluctuation of these values).
- An irreversible process proceeds in one preferred direction.
- The heat capacity is determined by how energy is partitioned into a system's degrees of freedom. It relates the transfer of heat to a temperature change.
- Partial derivatives are slopes, and these slopes are the coefficients that tell us how large the change du is when we take a step dv in a specific direction with other variables held constant.

Why do we study chemistry? There are three reasons. First, chemistry is fun. That means there are four reasons why we do chemistry. First, chemistry is fun, second, we want to understand why a chemical reaction happens, third we want to harness the energy released in a chemical process, and fourth, we want to exploit the properties of the products formed in a chemical reaction. To understand the second item, we need to understand what the Gibbs energy is and why a system will evolve until its Gibbs energy is a minimum. To understand the third reason we need to understand what makes a reaction exothermic, that is, we need to define enthalpy and what controls it. To understand the fourth reason we need to understand phases, intermolecular interactions and the quantum mechanics that is behind them. Our study of thermodynamics will allow us to address items two and three. As we shall see, we will need to define enthalpy (first law), entropy (second and third laws) and fix the relationship between them (solved in the third law). The next three chapters will define these laws of thermodynamics. The subsequent two will develop these fundamental themes and apply them specifically to the understanding of reactivity, thermochemistry and equilibrium phenomena.

Physical Chemistry: How Chemistry Works, First Edition. Kurt W. Kolasinski.

Companion Website: www.wiley.com/go/kolasinski/physicalchemistry

Why do you need to understand thermodynamics? The good academic reason is that you cannot understand chemistry fundamentally without some knowledge of thermodynamics. The good practical reason is that you might injure or kill yourself or others if you do not. The following is an example that is related to an actual laboratory accident.

Consider the reaction of 15.0 g of MgN with H_2O to produce MgO, N_2 and H_2 in a glass vessel with a volume of 1.00 l. What difference does it make if you use: (a) 10 g, (b) 150 g or (c) 1000 g of H_2O? What about an open versus a closed vessel?

To arrive at approximate answers, let us apply the rules for problem solving that were outlined in Chapter 1.

1 Start with a balanced reaction (include phases):

$$\mathrm{MgN(s) + H_2O(l) \rightarrow MgO(s) + 1/2N_2(g) + H_2(g)}$$

2 Evaluate the physical processes that occur, draw a diagram, or chart the number of moles. Is the reaction exothermic or endothermic? It is exothermic, so the temperature is going to increase. We need to develop methods of calculating the correct answers for both energy release and temperature increase because you cannot look up this reaction in a table.

3 Draw a straight line or otherwise present results appropriately. In this case we are not going to present data in a graph. Instead, a tabular representation of the moles of reactants and products and the reaction conditions (i.e., a reaction chart) is more appropriate. What is the initial composition? What is the final composition? What is the final temperature and pressure? How should the reaction be run?

4 Answer with appropriate units, always use SI units unless specifically asked for other units. Answers: Closed system = grave danger in all three cases. Open system: (a) utter destruction; (b) geyser; (c) toasty.

7.1 Some definitions and fundamental concepts in thermodynamics

- System – the material in the process under study.
- Surroundings – the rest of the universe.
- Open system – system can exchange material with surroundings.
- Closed system – no exchange of material.
- Isolated system – no exchange of material or energy.
- Thermodynamic equilibrium – p, T, composition are constant and the system is in its most stable state. Also called chemical equilibrium.
- Thermal equilibrium – two systems have same temperature.
- Metastable state – a system in which p, T and composition are constant; however, the system could relax to a lower energy state if it were sufficiently perturbed. A system in a metastable state is in a state of metastable equilibrium, and so can be described consistently (but with caution) by thermodynamic methods.
- Steady state – p, T and composition are constant but are maintained at these values in an open system by a balance of inward and outward fluxes of matter and energy.
- Isothermal: $\Delta T = 0$; isobaric $\Delta p = 0$; isochoric $\Delta V = 0$.
- Adiabatic in a thermodynamic sense means that the system exchanges no heat, $q = 0$.
- Diathermal barrier – allows for the transport of heat but not matter between systems.
- Temperature T in kelvin (absolute temperature) is related to temperature t in degrees Celsius by

$$T(\mathrm{K}) = t(^\circ\mathrm{C}) + 273.15 \tag{7.1}$$

7.2 Laws of thermodynamics

To allay the suspense, we will state up front the most important rules in thermodynamics. Combining the laws of thermodynamics with the understanding that all systems evolve to lower their Gibbs energy (or Helmholtz energy, as appropriate), and that the rates of the reactions that lead to equilibration (or a steady state) are governed by concentrations, temperature and activation energies, we can essentially explain all of chemistry. Thus, is it very important that we state the laws of thermodynamics and fully understand their implications and application, as we will over the course of the next few chapters.

- *Zeroth Law of Thermodynamics*

  There exists a property called temperature T. When the temperature of two systems is equal, the two systems are at thermal equilibrium.

- *First Law of Thermodynamics*

  There exists a property called the internal energy U, which is the sum of all kinetic and potential energies as well as all chemical and physical interaction energies. Work and heat change U equivalently.

- *Second Law of Thermodynamics*

  There exists a property called entropy S. The entropy of the universe increases in any spontaneous process, $\Delta S_{tot} > 0$. The entropy of a system is determined by the number of different ways to distribute the available energy over the various states of the system.

- *Third Law of Thermodynamics*

  There exists an absolute zero to the temperature scale. The entropy of a perfect crystal of a pure substance at absolute zero is zero. No finite number of steps in a cyclic process can achieve absolute zero.

We have already encountered the Zeroth Law in Chapter 6. Essentially, we require it and a definition of temperature before we can make any other statements about thermodynamics. With it in hand, we can proceed to a discussion of the First Law.

7.3 Internal energy and the first law

The total energy of a system is its *internal energy, U*. The internal energy is the sum of the kinetic and potential energies of the molecules composing the system as well as rotational and vibrational energy and energy involved in chemical and physical interactions. Changes in internal energy are denoted

$$\Delta U = U_f - U_i. \tag{7.2}$$

Note the sign convention. Whenever we calculate changes – that is, for all delta variables such as ΔU above – we will always take the final state minus the initial state. Equation (7.2) can be used to state the first law of thermodynamics in an alternate fashion:

The internal energy, U, of an isolated system is constant, $\Delta U = 0$.

For a system in contact with its surroundings

$$\Delta U_{\text{total}} = \Delta U_{\text{system}} + \Delta U_{\text{surroundings}} = 0. \tag{7.3}$$

Therefore,

$$\Delta U_{\text{system}} = -\Delta U_{\text{surroundings}}, \tag{7.4}$$

that is, the internal energy change in the system must be equal and opposite to that of the surroundings with which it has thermal contact.

Heat and work, which we shall define precisely shortly, act equivalently with respect to changing a system's internal energy.

$$\Delta U = q + w. \tag{7.5}$$

Equation (7.5) is central to thermodynamics, much of which revolves around determining how to calculate or measure q, the amount of heat transfer to or from a system, and w, the work done on or by a system. The first law is an expression of the conservation of energy. Energy can be converted from one form to another, for example between work and heat, but cannot be created or destroyed. Equation (7.5) means that *the units of work, heat and energy are the same.*

Equation (7.5) also sets the sign convention for q and w. A positive value of q or w leads to an increase in the internal energy (a positive value of ΔU). A negative value of q or w leads to a decrease in internal energy (a negative value of ΔU). In other words, when the system does work on the surroundings the work is done at the expense of the internal energy. This corresponds to a negative value. When work is done by the surroundings on the system, the internal energy increases and a positive value of work is performed.

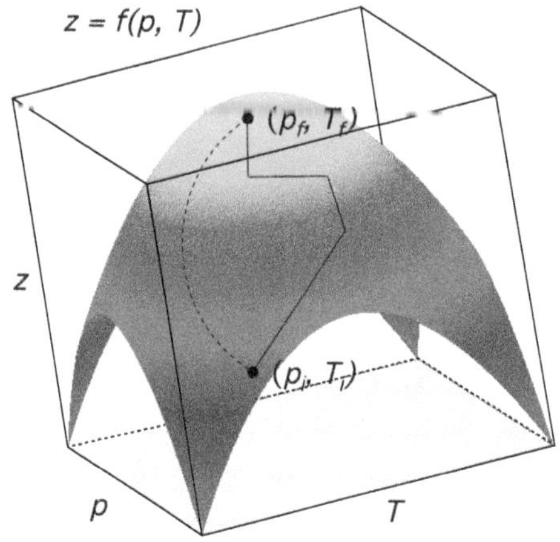

Figure 7.1 Depiction of a state function. Different paths lead from the initial point z_i, characterized by the initial pressure and temperature (p_i, T_i) to the final point z_f characterized by the final pressure and temperature (p_f, T_f). z_i and z_f are not defined by the path that connects them, but by state variables that described them.

7.3.1 State versus path functions and reversibility

The internal energy is a *state function. It depends only on the state of the system and not how it came to that state,* just like the variable z in Fig. 7.1, the value of which depends on p and T but not how the system arrived at this point. Examples of state functions are temperature, volume, pressure, enthalpy, entropy and Gibbs energy. State functions are interrelated and if we know the composition of a system and any two state functions, we can work out the values for all other state functions. Note that since U is a state function, so too is ΔU. The work and heat evolved in going from the initial state to the final state depend on the way the system transforms from initial to final state. Heat and work are *path functions.* For instance, along an isochoric path for a closed system (and no non-pV work), $w = 0$; thus, $\Delta U = q$. On the other hand, *in an adiabatic process there is no exchange of heat.* Thus, $q = 0$ and $\Delta U = w$ for an adiabatic process. When a process is non-adiabatic ($q \neq 0$), the amount of work done is path-dependent and will have to change accordingly if the change in internal energy along this non-adiabatic path is to be the same as along some other path that connects the same initial and final states.

In a *reversible process,* at each step of the process the direction of change can go either forward or backward. In an *irreversible process,* the direction of change is always only in one direction. For example, at equilibrium, room-temperature liquid water and gaseous water are connected by a reversible chemical process. Water molecules are able to evaporate from the surface of the liquid water to enter the gas phase. They do so at exactly the same rate as water molecules condense from the gas phase to enter the liquid phase. However, a piece of iron irreversibly oxidizes to form rust when a polished iron surface is exposed to air. Oxygen and water molecules decompose on the surface of the iron to form various iron oxides. The reverse process (decomposition of the oxides) does not occur spontaneously at room temperature.

How can a system change state along a reversible path? Think of the role of fluctuations. Collisions lead to the establishment of the Maxwell–Boltzmann distribution of populations among energy levels. On average, these distributions and all thermodynamic parameters such as p and T are constant. However, there also exist fluctuations around these mean values. The probability of any given fluctuation in energy decreases exponentially with the magnitude of the fluctuation away from the mean value. Nonetheless, these fluctuations are a part of the equilibrium distribution. As long as a system is evolving slowly enough such that its fluctuations lie within a standard deviation of the mean value, there is no way to tell that the system has departed from equilibrium. Thus, a slowly changing system can maintain a reversible path that is described by equilibrium conditions.

Heat and work are path-dependent. Thus, their values differ for reversible and irreversible paths connecting the same two states. The internal energy change does not depend on the path. Consider infinitesimal changes in heat and work ($đq$ *and* $đw$, respectively) along these two paths that lead to an infinitesimal change in internal energy $\mathrm{d}U$.[1] Denoting the reversible and irreversible paths rev and irrev, respectively, we can write

$$đq_{\mathrm{irrev}} \neq đq_{\mathrm{rev}} \text{ and } đw_{\mathrm{irrev}} \neq đw_{\mathrm{rev}} \text{ but } \mathrm{d}U_{\mathrm{irrev}} = \mathrm{d}U_{\mathrm{rev}}.$$

From the First Law, this implies that

$$đq_{\mathrm{irrev}} + đw_{\mathrm{irrev}} = đq_{\mathrm{rev}} + đw_{\mathrm{rev}}.$$

That is, by changing the path what is taken from heat is given to work to affect the same net change in internal energy. For those of you who are true sticklers for detail, once the path is defined an inexact differential becomes exact. For instance, along an adiabatic path, the change in work is exact. Thus, one is justified in writing dw instead of đw in this case.

Summarizing we state:

Reversible process

- Takes place at an infinitesimal rate (quasi-static).
- Passes through a continuous sequence of equilibrium states that can be described by the appropriate equation of state.
- At every point in the state trajectory, each and every intensive variable is continuous across the boundary at which the type of work corresponding to that variable is being performed.
- $w_{\text{forward}} = -w_{\text{reverse}}$, $q_{\text{forward}} = -q_{\text{reverse}}$ and $\Delta U_{\text{forward}} = -\Delta U_{\text{reverse}}$

Irreversible process

- Need not take place at an infinitesimal rate.
- Passes through nonequilibrium states.
- Intensive variables need not be continuous across boundaries.
- $\Delta U_{\text{forward}} = -\Delta U_{\text{reverse}}$.

7.3.1.1 Example

Calculate the internal energy change and the molar internal energy change when 100 J of heat are transferred into 2 mol of ideal gas while simultaneously 100 J of work are performed by the surroundings on the ideal gas.

In this example $q = +100$ J and $w = +100$ J. Therefore, the internal energy change is

$$\Delta U = q + w = 100\ \text{J} + 100\ \text{J} = 200\ \text{J}$$

and the molar internal energy change is

$$\Delta U_{\text{m}} = \frac{U}{n} = \frac{200\ \text{J}}{2\ \text{mol}} = 100\ \text{J mol}^{-1}.$$

7.4 Work

Work is the transfer of energy that makes use of the uniform motion of atoms. As a reflection of the fact that the work done depends on the path taken, there are numerous expressions for the amount of work done. The correct expression is path-dependent – that is, it depends on how the system is changing and how it is interacting with its surroundings.

First let us define *mechanical work*. The infinitesimal amount of work dw done when an object is displaced by d$\boldsymbol{s}$ under the aegis of a force $\boldsymbol{F}$ is given by

$$\mathrm{d}w = \boldsymbol{F} \cdot \mathrm{d}\boldsymbol{s} = F_s\, \mathrm{d}s \tag{7.6}$$

where F_s is the component of the vector $\boldsymbol{F}$ along the direction of d$\boldsymbol{s}$. Displacing the object from an initial point i to a final point f, the total work performed on it is given by the integral

$$w = \int_i^f \boldsymbol{F} \cdot \mathrm{d}\boldsymbol{s} = \int_i^f F_s\, \mathrm{d}s. \tag{7.7}$$

Work (regardless of whether it is mechanical or thermodynamic) is path-dependent. It depends on the path taken and how the force changes along that path. Work is a *process variable*, it is not a property of the object. The object does not contain a certain amount of work. Instead, work is a quantity that is exchanged between systems. Work is only defined with respect to motion.

Remember that if a parameter is constant it can be pulled out of the integral. Therefore, for a *constant* force moving an object a total distance Δs, the integration is trivial and the work is

Constant force:

$$w = F_s \int_i^f \mathrm{d}s = F_s\,(s_f - s_i) = F_s\, \Delta s. \tag{7.8}$$

Note that for this example of mechanical work, we have taken dw to be an exact differential. As long as the force acting on the object is derivable from a potential, dw is an exact differential. This is the case whenever the system is a *conservative system*. When some or all of the forces acting in the system are not derivable from potentials, the system is *dissipative*. In mechanical systems this arises, for example, when friction is present. When we speak of thermodynamic systems in general, we find that typically the forces are not derivable from potentials. Therefore, we must discuss the general case of thermodynamic work treating it as an inexact differential.

Now we turn to *thermodynamic work*. One of the most frequently encountered forms of work for chemical systems is the expansion of a gas against an external pressure p_{ex}, also known as pV work. In calculating thermodynamic work, we need to take account of the sign convention imposed by the First Law of Thermodynamics: when the system performs work on the surroundings, its internal energy must go down and the sign of work is negative. Let us assume that p_{ex} is applied uniformly over our system (say an ideal gas contained in a box with one movable wall). The infinitesimal work done when the volume changes is

$$đw = -p_{ex}\,dV. \tag{7.9}$$

The negative sign must be inserted to enforce compliance with the First Law: expansion must lead to negative work. We confirm this by performing the integration. If the volume changes from its initial value V_i to a final value V_f, then the total work is

$$w = -\int_{V_i}^{V_f} p_{ex}\,dV. \tag{7.10}$$

For a constant external pressure, we again move the constant out of the integral and evaluate explicitly,

Isobaric work:

$$w = -p_{ex}\int_{V_i}^{V_f} dV = -p_{ex}(V_f - V_i) = -p_{ex}\,\Delta V. \tag{7.11}$$

Consistent with the First Law, expansion corresponding to $\Delta V > 0$ leads to $w < 0$.

Similar to pV work, surface expansion work is given by the product of the *surface tension* γ and the surface area σ

$$w = \int \gamma\,d\sigma. \tag{7.12}$$

Work can be done by moving electrons. This is one of the most common forms of non-pV work. The electrical work of transporting electrical charge Q through an electrical potential ϕ is

$$w = \int_{Q_1}^{Q_2} \phi\,dQ. \tag{7.13}$$

This is equivalent to the work performed to charge a capacitor at the applied potential difference ϕ. If the current I and potential are constant, then the electrical work is the product of electrical power P_{el} (power, electrical or otherwise, has SI units of watts $W = J\,s^{-1}$) and time t

$$w = I\phi t = P_{el}\,t. \tag{7.14}$$

At this point it is good to recall Ohm's law,

$$\phi = I\,R, \tag{7.15}$$

which allows us to relate electrical power to current, resistance R as well as the charge passed,

$$P_{el} = I\phi = I^2R = Q\phi/t. \tag{7.16}$$

7.4.1 Example

Calculate the work done by a gas when expansion is performed isothermally and reversibly.

It is important to define the path to calculate the work appropriately. Defining the path often involves deciphering the vocabulary of thermodynamics and chemical change. If the expansion is *reversible*, then the internal and external pressures must be equal at every step along the path of expansion or compression. This means $p_{int} = p_{ex} = p$. The temperature does not change in an *isothermal process* ($\Delta T = 0$). We know from the ideal gas law that $p = nRT/V$. Substituting this result into Eq. (7.10) we find

$$w = -\int_{V_i}^{V_f} p\, \mathrm{d}V = -\int_{V_i}^{V_f} \frac{nRT}{V}\, \mathrm{d}V = -nRT \int_{V_i}^{V_f} \frac{\mathrm{d}V}{V}.$$

Take a look at the form of this integral. It is a frequently encountered form, which will have a familiar solution $\int_a^b \mathrm{d}x/x = \ln b - \ln a$.

Reversible, isothermal work:

$$w = -nRT(\ln V_f - \ln V_i) = -nRT \ln\left(\frac{V_f}{V_i}\right) = nRT \ln\left(\frac{V_i}{V_f}\right) \tag{7.17}$$

7.4.2 Example

One mole of an ideal gas expands from 5 bar to 1 bar at 298 K. Calculate the work done for (a) a reversible expansion and (b) an expansion against a constant pressure of 1 bar.

a From Eq. (7.17)

$$w_{\text{rev}} = -nRT \ln(V_f/V_i)$$

For an ideal gas, $p_i V_i = p_f V_f$, thus $V_f/V_i = p_i/p_f$ and

$$w_{\text{rev}} = -nRT \ln(p_i/p_f)$$

$$w_{\text{rev}} = -(1\ \text{mol})(8.314\ \text{J K}^{-1}\ \text{mol}^{-1})(298\ \text{K}) \ln \frac{5\ \text{bar}}{1\ \text{bar}} = -3988\ \text{J}.$$

b In this case of irreversible expansion, the external pressure is constant. The internal pressure changes from the initial value and decreases until expansion stops when the internal and external pressures equalize.

$$w_{\text{irrev}} = -p_{\text{ex}}(V_f - V_i).$$

$$w_{\text{irrev}} = -p_f\left(\frac{nRT}{p_f} - \frac{nRT}{p_i}\right) = -nRT\left(1 - \frac{p_f}{p_i}\right) = -1982\ \text{J}.$$

This example indicates what is a general result: *the maximum work is done on the surroundings when the process is carried out reversibly*. Note that the negative sign means that the internal energy is decreasing as a result of the system performing work on the surroundings.

7.4.3 Directed practice

One mole of an ideal gas is compressed from an initial volume of 0.0248 m^3 to 0.00496 m^3 at 298 K. Calculate the work done for (a) reversible compression and (b) irreversible compression with 5 bar external pressure. [(a) 3990 J mol^{-1} (b) 9920 J mol^{-1}]

7.5 Intensive and extensive variables

Some variables in thermodynamics are independent of the mass of the system. These variables, for example pressure and temperature, are known as *intensive variables*. Other variables are dependent upon the mass when the intensive variables are held constant. These so-called *extensive variables* include volume and internal energy. Another way to express the difference is to state that intensive variables can be measured at any point in the system (i.e., the pressure or temperature of a system at thermodynamic equilibrium is everywhere the same), whereas extensive variable are properties of the system as a whole.

Each intensive variable α has a related conjugate variable β. Conjugate variables are related by a particular work process. We can determine what variables are conjugate variables by noting first that the product of a variable and an infinitesimal change in its conjugate variable is equal to work. Work and energy have the same units. Therefore, any combination of variables, the product of which has units of energy, is a candidate pair. Finally, suppose that α has the same value in the system and surroundings. The conjugate extensive variable β is the variable that satisfies the equation

$$\delta w = \alpha\, \mathrm{d}\beta. \tag{7.18}$$

Table 7.1 Examples of extensive quantities and their intensive counterparts expressed as molar quantities.

Quantity	Molar quantity	
Mass, m	Molar mass	$M = m/n$
Volume, V	Molar volume	$V_m = V/n$
Gibbs energy, G	Molar Gibbs energy	$G_m = G/n$
Enthalpy change, ΔH	Molar enthalpy change	$\Delta H_m = \Delta H/n$

Conjugate extensive and intensive variables correspond, respectively, to the generalized coordinates and generalized forces of mechanics. In analogy to expressions for the potential energy of mechanical systems involving generalized coordinates and forces, conjugate variables are related by the negative of the partial derivative of the internal energy, holding all other independent extensive variables constant. That is,

$$\alpha \equiv -\partial U/\partial \beta. \tag{7.19}$$

When an extensive quantity is divided by the amount of substance, it becomes an intensive quantity. Note that 'amount of substance' should be used in place of the colloquial 'number of moles' just as 'mass' is referred to and never 'number of kilograms.' The result is called a *molar quantity*. Some examples are given in Table 7.1. Always check to make sure that you are using consistent units. Molar quantities will often be highlighted with the subscript m. When an extensive quantity is divided by the mass, the result is called a specific quantity. Similarly, if an extensive quantity is divided by volume, area or length, the results are called volumic, areic and lineic, respectively.

7.6 Heat

If the energy of a system changes such that the temperature changes, we say that energy has been exchanged in the form of *heat*. Heat is the exchange of energy that makes use of random motions of atoms. If a process occurs such that it cannot exchange heat with its surroundings, the process is *adiabatic*.

The most fundamental definition of heat is

$$q = \Delta U - w. \tag{7.20}$$

Heat can be generated by rearranging bonds (associated with enthalpies of reaction) or the realignment of intermolecular forces (associated with enthalpies of phase transitions and/or mixing). Again, note the sign convention. If the temperature of the surroundings is lowered, q is positive (heat flows into the system), but if the temperature of the surroundings is raised, then q is negative (heat flows out of the system).

7.6.1 Example

A 73 kg person drinks 500 g of milk, which has a caloric value of approximately 600 cal g^{-1}. If only 17% of the energy in milk is converted to mechanical work, how high (in meters) can the person climb based solely on this energy intake.

The old unit of calories (1 cal = 4.184 J) is still frequently encountered. In the old literature and on packaging in the USA, calories refer to 'big calories' or kcal. The energy intake available for mechanical work is

$$E_{\text{intake}} = 0.17(500\text{ g})(600\text{ cal g}^{-1})(4.184\text{ J cal}^{-1}) = 2.1 \times 10^5\text{ J}$$

The potential energy gained by raising an object of mass m a height h is

$$E_{\text{pot}} = mgh, \tag{7.21}$$

where $g = 9.81$ m s^{-2} is the acceleration due to gravity. Now, equate E_{intake} and E_{pot} and solve for h.

$$h = \frac{E_{\text{intake}}}{mg} = \frac{2.1 \times 10^5\text{ J}}{(73\text{ kg})(9.81\text{ m s}^{-2})} = 290\text{ m}.$$

Recall that J = kg m^2 s^{-2}. If you ever forget the equivalent units of J, just think of the units involved in calculating the kinetic energy

$$E_{\text{kinetic}} = \tfrac{1}{2}mv^2 \Rightarrow \text{kg(m s}^{-1})^2 = \text{kg m}^2\text{ s}^{-2}.$$

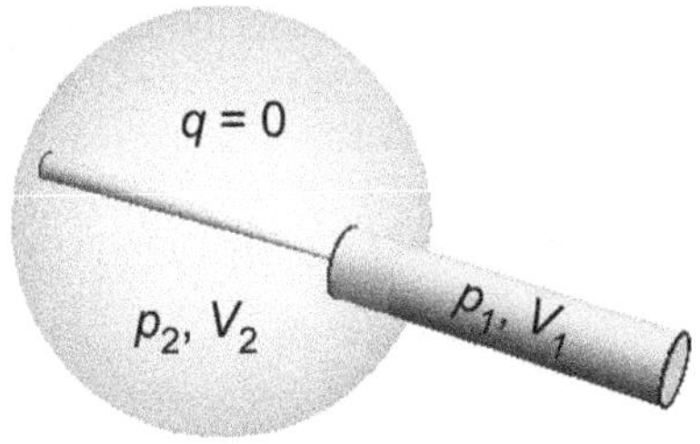

Figure 7.2 Adiabatic expansion of a gas through a nozzle from a region of high pressure at p_1 into a vacuum chamber at p_2 to create a molecular beam.

7.6.2 Example

Consider, as in Fig. 7.2, the adiabatic expansion of gas from a container held at an initial pressure p_1 with volume V_1 through a nozzle into a container of volume V_2 to form a molecular beam.[2] In a *molecular beam* apparatus the pressure p_1 is known as the stagnation pressure. The second container is initially evacuated. (a) Calculate the work, heat and change in internal energy that occurs upon expansion. (b) How could such an apparatus be used to quantify deviation from ideal gas behavior?

a The process is adiabatic, and therefore $q = 0$. The work is given by $w = -\int_{V_1}^{V_2} p_{\text{ext}}\, \mathrm{d}V$. Here, the external pressure is zero because expansion is into a vacuum; thus $w = 0$. From the First Law, $\Delta U = w + q = 0$.

b For an ideal gas the internal energy U is determined by the temperature only. Thus, since $\Delta U = 0$, $\Delta T = 0$ also follows. Any deviation from $\Delta U = 0$ and $\Delta T = 0$ would allow us to quantify the deviation from ideality.

7.7 Non-ideal behavior changes the work

Non-ideal behavior will change the amount of pV-work that is performed during the expansion and compression of a gas. To see how and why, let us consider a van der Waals gas undergoing isothermal, reversible compression and expansion. As before, the work is given by

$$w = -\int_{V_i}^{V_f} p\, \mathrm{d}V. \tag{7.22}$$

The van der Waals equation of state is

$$\left(p + \frac{n^2 a}{V^2}\right)(V - nb) = nRT. \tag{7.23}$$

The constant a is a measure of the attractive forces between the molecules, while b is a measure of the short-range repulsion that keeps one molecule from occupying the space of another molecule. Solving the van der Waals equation for p and substituting into the integral in Eq. (7.22), we obtain

$$w = -\int_{V_i}^{V_f} \frac{nRT}{V - nb} - \frac{n^2 a}{V^2}\, \mathrm{d}V. \tag{7.24}$$

Integration yields the work done on the van der Waals gas,

$$w = -nRT \ln\left(\frac{V_f - nb}{V_i - nb}\right) - n^2 a \left(\frac{1}{V_f} - \frac{1}{V_i}\right). \tag{7.25}$$

For an ideal gas, we previously obtained

$$w_{\text{ideal}} = -nRT \ln(V_f / V_i) \tag{7.26}$$

The difference is

$$\Delta w = w - w_{\text{ideal}} = -nRT \ln\left(\frac{1 - nb/V_f}{1 - nb/V_i}\right) - n^2 a \left(\frac{1}{V_f} - \frac{1}{V_i}\right). \tag{7.27}$$

For compression, $V_f < V_i$. This means the second term is negative. Less work is required to compress the van der Waals gas because the intermolecular attractions aid compression by pulling the molecules together. Conversely, in expansion

Table 7.2 The results of Eq. (7.27) are shown for selected gases when one mole of gas is expanded from $V_i = 0.032823\ \text{m}^3$ to twice this value at 400 K. V_i corresponds to the volume occupied by an ideal gas at 400 K and 101.325 kPa. The work values are given in J mol^{-1}. w_1 corresponds to the short-range repulsive terms (the one involving b). w_2 corresponds to the long-range attraction term (the one involving a).

Gas	w_1	w_2	Δw
Ar	−1.63	−2.07	0.44
H_2	−1.35	−0.38	−0.97
HCl	−2.06	−5.64	3.58
H_2O	−1.55	−8.43	6.89
He	−1.20	−0.05	−1.15
N_2	−1.96	−2.09	0.13
NH_3	−1.88	−6.43	4.55
O_2	−1.62	−2.11	0.49
Xe	−2.61	−6.38	3.77
CO	−2.00	−2.24	0.24
CO_2	−2.17	−5.57	3.40
CH_4	−2.14	−3.44	1.30
acetone	−5.71	−24.40	18.70
ethanol	−4.35	−18.85	14.51
methanol	−3.38	−14.58	11.20
isopropanol	−5.71	−24.42	18.71

less work is extracted from the van der Waals gas because work must be done to pull apart the molecules and overcome their intermolecular attractions.

Now consider the first term. In compression, the first term is positive, which means that additional work must be performed on the van der Waals gas to overcome the short-range repulsions. In expansion, more work is extracted from the van der Waals gas because the short-range repulsions aid expansion. Whether the first or second term dominates depends on the relative values of a and b, as shown in Table 7.2. Under the chosen conditions, all of the gases other than H_2 and He require more work to be done to expand them because the long-range attraction term w_2 (the one that involves the constant a) is greater in magnitude. The differences are most pronounced for polyatomic organic molecules. For N_2, O_2 and CO, the two terms nearly cancel and the work is quite close to the ideal value.

7.7.1 Directed practice

Solve the integral in Eq. (7.24) to obtain Eq. (7.25).

7.8 Heat capacity

When no reactions or phase transitions occur and the number of moles in the system is constant, the heat flow resulting from simple heating of the system is equal to the product of the heat capacity C times the temperature change according to

$$q = \int_{T_i}^{T_f} C\,\mathrm{d}T. \tag{7.28}$$

We know that q is a path function. The temperature is a state function; therefore, the value of the heat capacity must depend on the path. If the heating is done at constant pressure we use C_p (the heat capacity at constant pressure), but for constant volume we must use the heat capacity at constant volume, denoted C_V. As we have already seen, heat capacity depends on the material, its phase, and is a function of temperature because its value depends on the extent to which the degrees of freedom of the material are excited. For an ideal gas, there is a very simple relationship between $C_{p,\mathrm{m}}$ and $C_{V,\mathrm{m}}$:

$$C_{p,\mathrm{m}} - C_{V,\mathrm{m}} = R. \tag{7.29}$$

7.8.1 Example

Calculate the heat exchanged when a monatomic ideal gas is heated at constant pressure from 273 K to 373 K.

In Chapter 6 we learned that $C_{V,\text{m}} = \frac{3}{2}R$ independent of temperature for a monatomic ideal gas. From Eqs (7.28) and (7.29) therefore

$$q_p = \int_{T_1=273\text{ K}}^{T_2=373\text{ K}} C_p\,\text{d}T = C_p \int_{T_1}^{T_2} \text{d}T.$$

$$C_{p,\text{m}} = C_{V,\text{m}} + R = \tfrac{5}{2}R$$

$$q_{p,\text{m}} = \tfrac{5}{2}(8.31\text{ J K}^{-1}\text{ mol}^{-1})(373\text{ K} - 273\text{ K}) = 2080\text{ J mol}^{-1}.$$

7.9 Temperature dependence of C_p

The dependence of the heat capacity on temperature is extremely important. From statistical mechanics, we found that the heat capacity depends on temperature because the population of different quantum states changes with temperature. This leads to a fundamental relationship between the entropy and the heat capacity. Similarly, we shall show below that the temperature dependence of the enthalpy is related to the behavior of the heat capacity. We will demonstrate that, since the temperature dependence of the Gibbs energy is related to the enthalpy, it is an understanding of the temperature dependence of heat capacity that will allow us to understand how equilibrium constants depend on temperature.

The temperature dependence of the heat capacity is related to the structure and phase of the molecule. For a monatomic gas, $C^\circ_{p,\text{m}}$ is strictly independent of temperature with a value of 5/2 R. For a diatomic gas, as long as the temperature is not too low (above the characteristic rotational temperature), there is a weak temperature dependence that converges on 9/2 R. The temperature dependence of polyatomic gases is somewhat stronger and converges on a higher value based on the number of atoms in the molecule. The temperature dependence of the heat capacity of gases can be predicted from the methods of statistical mechanics. As long as we know the rotational and vibrational constants of the molecule and its behavior is close to that of an ideal gas, these predictions will be quite accurate.

For many gases, the heat capacity has been measured accurately over wide temperature ranges. The temperature dependence of $C^\circ_{p,\text{m}}$ is reproduced by the empirical *Shomate equation*. For some substance (see, for example the NIST WebBook) this function is known in its full form

$$C^\circ_{p,\text{m}} = A + Bt + Ct^2 + Dt^3 + Et^{-2}, \tag{7.30}$$

where $t = T/1000$, with T in K. The coefficients are independent of T. Often, it is sufficient to truncate the function as

$$C^\circ_{p,\text{m}} = A + Bt + Et^{-2}. \tag{7.31}$$

For some gases and certain temperature ranges, as for methanol in Fig. 7.3, a simple linear function is sufficient. The advantage of equations such as Eqs (7.30) and (7.31) is that they can be easily and analytically integrated; this is not true of the equations derived from statistical mechanics. Furthermore, these empirical fits incorporate any non-ideal behavior in real gases without the need for further modeling.

Molecular liquids are often only stable over small temperature ranges – just 50 K to 200 K for the solvents most often used in organic synthesis. $C_{p,\text{m}}$ is generally the highest of any phase for liquids, and is not a strong function of temperature until the reduced temperature exceeds $T_r = T/T_c \sim 0.7$. A shallow minimum is often observed just below the boiling point T_b. However, as the temperature approaches the critical point, $C_{p,\text{m}}$ increases sharply and approaches infinity according to

$$C_{p,\text{m}} = c_0 + c_1(T/100) + c_2(T/100)^2. \tag{7.32}$$

The behavior of the heat capacity of water, mercury, graphite and selected gases in the moderate temperature range (far from absolute zero and critical points) is shown in Fig. 7.3. CO and N_2, which have the same mass as well as being isoelectronic, behave similarly. Polyatomics generally have higher heat capacities and the gas-phase curves converge toward asymptotes that increase with increasing number of atoms in the molecule, as expected. The low-temperature behavior is not shown here but can be found in Chapter 6. From the Third Law of Thermodynamics we know that all of the curves converge to zero at low temperature. At phase transitions, the heat capacity changes discontinuously.

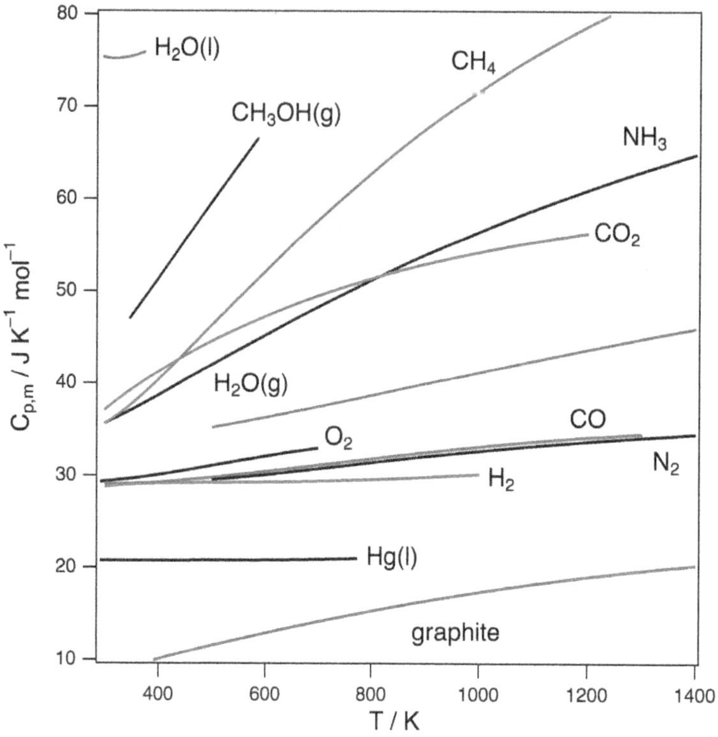

Figure 7.3 The standard heat capacities at constant pressure $C^\circ_{p,\mathrm{m}}$ as a function of temperature for several gases, $H_2O(l)$, Hg(l) and graphite. Curves are calculated from parameters taken from the NIST WebBook.

The heat capacity of solids at low temperatures follows the Debye model. At high temperatures, the heat capacities of atomic solids (simple metals and elemental semiconductors as opposed to solid compounds) converge on a constant value of 3 R (~25 J K^{-1} mol^{-1}). A few atomic solids such as Be, B and C (as both graphite and particularly diamond) have high Debye temperatures and only converge on this value at high temperatures. Compounds and molecular solids have higher values of heat capacity.

7.9.1 Example

Calculate the amount of heat that must be transferred to CH_4 on a molar basis to bring it from 298 K to 1300 K at a constant pressure of 1 bar. (a) Assume a constant heat capacity of $C^\circ_{p,\mathrm{m}} = 35.7$ J K^{-1} mol^{-1}. (b) Assume the behavior of $C^\circ_{p,\mathrm{m}}$ is described by the Shomate equation with $A = -0.703029$, $B = 108.4773$, $C = -42.52157$, $D = 5.862788$, and $E = 0.678565$.

In both cases we need to evaluate the integral

$$q_\mathrm{m} = \int C^\circ_{p,\mathrm{m}}\,\mathrm{d}T.$$

a With constant heat capacity, we simply pull the heat capacity from the integral and evaluate from 298 K to 1300 K.

$$q_\mathrm{m} = \int_{298}^{1300} C^\circ_{p,\mathrm{m}}\,\mathrm{d}T = (35.7\ \mathrm{J\,K^{-1}\,mol^{-1}})T\Big|_{298}^{1300}$$

$$q_\mathrm{m} = 35.8\ \mathrm{kJ\,mol^{-1}}$$

b Now the function

$$C^\circ_{p,\mathrm{m}} = A + Bt + Ct^2 + Dt^3 + Et^{-2}$$

must be substituted explicitly for $C^\circ_{p,\mathrm{m}}$ in the integral, and each of the five terms is evaluated between the limits.

$$q_\mathrm{m} = \int_{298}^{1300} A + Bt + Ct^2 + Dt^3 + Et^{-2}\,\mathrm{d}T$$

$$q_\mathrm{m} = 61.3\ \mathrm{kJ\,mol^{-1}}$$

Assumption of a constant heat capacity leads to a result that is 100%(61.3 − 35.8)/61.3 = 42% below the expected value.

7.10 Internal energy change at constant volume

The change in internal energy with temperature is related to the *heat capacity* at constant volume, C_V. For infinitesimal changes

$$dU = C_V\, dT. \tag{7.33}$$

This is an exact differential equation. Therefore, we can divide both sides by dT to define the heat capacity as a partial differential

$$C_V \overset{\text{def}}{=} (\partial U/\partial T)_V \tag{7.34}$$

For measurable increases in temperature, ΔT, and a heat capacity that is independent of temperature, the change in internal energy is given by

$$\Delta U = C_V\, \Delta T. \tag{7.35}$$

However, if the heat capacity is a function of temperature, we have to integrate Eq. (7.33) explicitly

$$\int dU = \int_{T_1}^{T_2} C_V\, dT. \tag{7.36}$$

$$U(T_2) = U(T_1) + \int_{T_1}^{T_2} C_V\, dT \tag{7.37}$$

In this case, the functional form of the heat capacity C_V has to be determined and substituted into the integral.

At constant volume (for instance in a bomb calorimeter), no work is done ($w = 0$); therefore, the amount of heat generated is equal to the internal energy change. The internal energy change is the difference between the final and initial values, $\Delta U = U(T_2) - U(T_1)$, thus,

$$q_V = \Delta U = \int_{T_1}^{T_2} C_V\, dT. \tag{7.38}$$

The subscript V on q_V indicates the heat exchanged at constant volume. q_V is calculated from the integral over the heat capacity function.

7.10.1 Example

A 0.4089 g sample of benzoic acid (C_6H_5COOH) was burned in a constant volume bomb calorimeter. The temperature measured in the calorimeter rose from $T_i = 20.17$ °C to $T_f = 22.22$ °C. The specific heat capacity of the calorimeter, also called the calorimeter constant, is $C = 5267.8$ J K^{-1}. Calculate the molar internal energy change of combustion $\Delta_c U_m$ for benzoic acid in kJ mol^{-1}.

Start with a balanced reaction including phases

$$C_6H_6COOH(s) + \tfrac{15}{2}O_2(g) \rightarrow 7CO_2(g) + 3H_2O(l).$$

This is a constant volume process, therefore $\Delta U = q_V$. The amount of heat evolved is given by

$$q_V = -C_V\, \Delta T = -5267.8(22.22 - 20.17) = -10.80 \text{ kJ}.$$

Be careful with signs. The heat absorbed by the calorimeter is equal and opposite to the heat evolved from the combustion system, $q_{\text{calorimeter}} = -q_V$. Use the molar mass of benzoic acid ($M = 122.12$ g mol^{-1}) to calculate $\Delta_c U_m$.

$$\Delta_c U_m = \frac{q_V}{n} = \frac{-10.80 \text{ kJ}}{122.12 \text{ g mol}^{-1}/0.4089 \text{ g}} = -3226 \text{ kJ mol}^{-1}$$

Note the sign. *Negative* indicates an *exothermic* process. The final state is lower in energy than the initial state. Always check the sign of the final answer and makes sure it makes sense.

7.11 Enthalpy

Often, we measure energy changes at constant pressure, as in a beaker open to the atmosphere, not constant volume. From the First Law, when the only work that can be done by the system is expansion work, the heat evolved at constant p is

$$q_p = (\Delta U - w)_p = \Delta U + p_{\mathrm{ext}} \int \mathrm{d}V = \Delta U + p\Delta V. \tag{7.39}$$

Thus, we can define a new state function, the *enthalpy*, according to

$$H \stackrel{\mathrm{def}}{=} U + pV, \tag{7.40}$$

such that an infinitesimal change in enthalpy is then given by infinitesimal change in the heat evolved at constant pressure

$$\mathrm{d}H = \partial q_p. \tag{7.41}$$

Enthalpy is a state function. The change in enthalpy with temperature is related to the *heat capacity at constant pressure*, C_p. Analogous to the above for internal energy, we define the heat capacity at constant pressure as the partial derivative of enthalpy with respect to temperature at constant pressure

$$C_p \stackrel{\mathrm{def}}{=} (\partial H/\partial T)_p. \tag{7.42}$$

This is an exact differential and we can rearrange to obtain

$$\mathrm{d}H = C_p\,\mathrm{d}T. \tag{7.43}$$

To find how the enthalpy changes with temperature, we integrate between T_1 and T_2 to obtain

$$\int_{T_1}^{T_2} \mathrm{d}H = \int_{T_1}^{T_2} C_p\,\mathrm{d}T \tag{7.44}$$

$$H(T_2) = H(T_1) + \int_{T_1}^{T_2} C_p\,\mathrm{d}T. \tag{7.45}$$

In words, the enthalpy at an arbitrary temperature T_2, $H(T_2)$, is equal to the enthalpy at some reference temperature $H(T_1)$ plus the integral of the constant pressure heat capacity from T_1 to T_2. Alternatively, the change in enthalpy $\Delta H = H(T_2) - H(T_1)$ is equal to the heat evolved at constant pressure q_p,

$$q_p = \Delta H = \int_{T_1}^{T_2} C_p\,\mathrm{d}T \tag{7.46}$$

If $C_{p,\mathrm{m}}$ has the form of Eq. (7.31) and we substitute this into Eq. (7.45), then we find

$$H_\mathrm{m}(T_2) = H_\mathrm{m}(T_1) + \int_{T_1}^{T_2} (a + bT + eT^{-2})\,\mathrm{d}T. \tag{7.47}$$

Note the change in the constants to reflect expressing the temperature at T instead of t and that now we explicitly indicate molar units for enthalpy. Integration yields

$$H_\mathrm{m}(T_2) = H_\mathrm{m}(T_1) + a(T_2 - T_1) + (b/2)\left(T_2^2 - T_1^2\right) - e\left(T_2^{-1} - T_1^{-1}\right) \tag{7.48}$$

7.11.1 Example

Determine the molar enthalpy change at standard pressure when liquid Hg is heated from 298 K to its boiling point, T_b = 655 K. $C_{p,\mathrm{m}} = 27.4\ \mathrm{J\,K^{-1}\,mol^{-1}}$.

First, write a balanced chemical equation including phases that describes the process.

$$\mathrm{Hg(l)}(T = 298\ \mathrm{K}) \rightarrow \mathrm{Hg(l)}(T = 655\ \mathrm{K}),\ p\ \text{constant},\ C_{p,\mathrm{m}}\ \text{constant}$$

Then evaluate Eq. (7.46) using this information.

$$\Delta H_{\mathrm{m}} = q_{p,\mathrm{m}} = \int_{T_i}^{T_f} C_{p,\mathrm{m}}(T)\,\mathrm{d}T = C_{p,\mathrm{m}} \int_{298\ \mathrm{K}}^{655\ \mathrm{K}} \mathrm{d}T$$

$$\Delta H_{\mathrm{m}} = C_{p,\mathrm{m}}\,\Delta T = 27.4\ \mathrm{J\ K^{-1}\ mol^{-1}}(655\ \mathrm{K} - 298\ \mathrm{K}) = 9.78\ \mathrm{kJ\ mol^{-1}}$$

7.11.2 Example

When 3.0 mol O_2 is heated at a constant pressure of 3.25 atm, its temperature increases from 260 K to 285 K. Given that the molar heat capacity of O_2 at constant pressure is 29.4 J K^{-1} mol^{-1}, calculate q, ΔU, and ΔH.

The change in the system can be represented as

$$O_2(\mathrm{g})(T = 260\ \mathrm{K}) \to O_2(\mathrm{g})(T = 285\ \mathrm{K}), p = 3.25\ \mathrm{atm}, n = 3.0\ \mathrm{mol}, C_p\ \text{constant}$$

Remember that at constant pressure, the heat evolved is equal to the enthalpy change. For an ideal gas at constant moles but changing temperature

$$q = nC_{p,\mathrm{m}}\,\Delta T = 3(29.4)(25) = 2.2\ \mathrm{kJ}$$

This is the heat evolved at constant pressure, thus it is also the enthalpy change, $\Delta H = 2.2$ kJ. The ideal gas law can be used to relate ΔH to ΔU.

$$\Delta U = \Delta H - \Delta(pV) = \Delta H - \Delta(nRT) = \Delta H - nR\,\Delta T.$$

$$\Delta U = \Delta H - nR\,\Delta T = 2.2 - 3(8.314 \times 10^{-3})(25) = 1.6\ \mathrm{kJ}.$$

7.11.3 Directed practice

Calculate the enthalpy change on a molar basis when CH_4 is heated from 298 K to 1300 K at a constant pressure of 1 bar. The behavior of $C^{\circ}_{p,\mathrm{m}}$ is described by the Shomate equation with $A = -0.703029$, $B = 108.4773$, $C = -42.52157$, $D = 5.862788$, and $E = 0.678565$. [61.3 kJ mol^{-1}]

7.12 Relationship between C_V and C_p and partial differentials

In Eqs (7.34) and (7.42) above, we have *defined* the heat capacity at constant volume C_V and the heat capacity at constant pressure C_p in terms of partial differentials,

$$C_V \overset{\mathrm{def}}{=} \left(\frac{\partial U}{\partial T}\right)_V \tag{7.49}$$

$$C_p \overset{\mathrm{def}}{=} \left(\frac{\partial H}{\partial T}\right)_p \tag{7.50}$$

What is the difference between a partial derivative and a derivative? If you can define a function in terms of only one variable, say $y = f(x)$, then you can only take the derivative with respect to one variable, $\mathrm{d}y/\mathrm{d}x$. This derivative corresponds to the slope of the curve. The curve could be a line or some other shape. However, no matter what kind of curve it is, it represents a figure that can be drawn in a plane.

If you have a function of more than one variable, $z = f(x, y)$ as in Fig. 7.4, then the function describes a surface that has to be drawn in three dimensions. If it is a function of three or more variables then it describes a higher dimensional object called a *hypersurface*. On this surface (or hypersurface) you can define the slope of the curve only if you also define the direction in which the variables are changing. Such slopes are known as partial derivatives. Thus, we can conceive of partial derivatives in two equivalent ways:

(i) Partial derivatives are slopes and (ii) these slopes are the coefficients that tell us how large the change $\mathrm{d}z$ is when we take a step $\mathrm{d}x$ in the x-direction or $\mathrm{d}y$ in the y-direction.

The heat capacity at constant volume, C_V, and at constant pressure, C_p, differ by the amount of work required to change the volume of the system. The difference between these two quantities is related to fundamental materials properties.

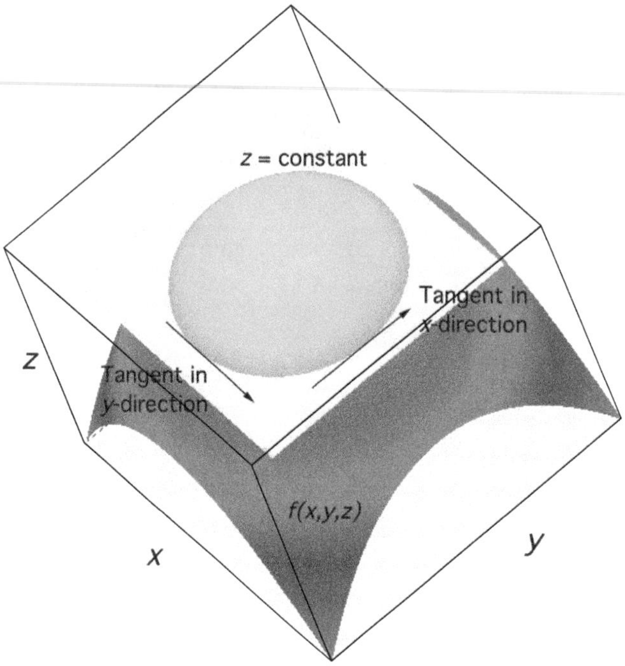

Figure 7.4 First derivatives are related to the slopes of tangents. For simple curves, the slopes are obvious because the tangent can only be drawn one way at a given point. However, for multidimensional figures multiple tangents can be drawn and the slope of the curve depends on the direction in which we look. Partial derivatives are descriptions of the slope that include this directional information.

Both are defined in terms of partial derivatives that relate an energy change to the change in temperature. We use the definitions of C_p and C_V to write

$$C_p - C_V = \left(\frac{\partial H}{\partial T}\right)_p - \left(\frac{\partial U}{\partial T}\right)_V. \tag{7.51}$$

Equation (7.51) could just as easily be written in terms of molar quantities by dividing both sides by the amount of substance n. In this example, we use the heat capacity with units of J K^{-1} instead of molar heat capacity with units J K^{-1} mol^{-1} to simplify the notation.

For an ideal gas the internal energy depends only on temperature, regardless of whether the process is at constant volume or constant pressure. Therefore,

$$C_V = (\partial U/\partial T)_V = (\partial U/\partial T)_p. \tag{7.52}$$

Substituting into Eq. (7.51) we have for an ideal gas

$$C_p - C_V = \left(\frac{\partial H}{\partial T}\right)_p - \left(\frac{\partial U}{\partial T}\right)_p. \tag{7.53}$$

To simplify Eq. (7.53) further, we need to derive an expression for the temperature dependence of enthalpy. Previously, we defined the state function enthalpy $H \stackrel{\text{def}}{=} U + pV$. Consider now a two-step process in which T and p change, and how this affects the enthalpy. Enthalpy is a function of both T and p, that is, it is a function of more than one variable. When we make an infinitesimal change in pressure $\mathrm{d}p$ at constant temperature T, the enthalpy changes an infinitesimal amount $\mathrm{d}H$. The coefficient that relates how large the change $\mathrm{d}H$ is for a given change in pressure $\mathrm{d}p$ is the partial derivative $(\partial H/\partial p)_T$.

Returning to how enthalpy changes with variations in pressure and temperature, we see that we can write the functional form most generally in terms of the partial derivatives

$$\mathrm{d}H = \left(\frac{\partial H}{\partial p}\right)_T \mathrm{d}p + \left(\frac{\partial H}{\partial T}\right)_p \mathrm{d}T. \tag{7.54}$$

From Eq. (7.51), we recognize the partial derivative in the second term as C_p,

$$\mathrm{d}H = \left(\frac{\partial H}{\partial p}\right)_T \mathrm{d}p + C_p\,\mathrm{d}T. \tag{7.55}$$

From the definition of enthalpy Eq. (7.40) and the ideal gas law, we can write the enthalpy as

$$H = U + nRT. \tag{7.56}$$

Taking the partial derivative with respect to T at constant pressure, we find

$$(\partial H/\partial T)_p = (\partial U/\partial T)_p + nR \tag{7.57}$$

Substituting this result into Eq. (7.53)

$$C_p - C_V = (\partial U/\partial T)_p + nR - (\partial U/\partial T)_p = nR. \tag{7.58}$$

Equation (7.58) is true for an ideal gas. More generally, this equation is expressed in terms of two materials properties known as κ_T, the isothermal compressibility and the expansion coefficient α,

$$C_p - C_V = \alpha^2 TV/\kappa_T. \tag{7.59}$$

The *isothermal compressibility* is defined as

$$\kappa_T \overset{\text{def}}{=} \frac{1}{V}\left(\frac{\partial V}{\partial p}\right)_T. \tag{7.60}$$

The *expansion coefficient* is defined by

$$\alpha \overset{\text{def}}{=} \frac{1}{V}\left(\frac{\partial V}{\partial T}\right)_p. \tag{7.61}$$

7.12.1 Example

Find expressions for the isothermal compressibility and expansion coefficient of an ideal gas.

We need to evaluate $(\partial V/\partial p)_T$ and $(\partial V/\partial T)_p$. For an ideal gas, V and p are related by the ideal gas law, which we rewrite in terms of V as a function of p and T as

$$V = nRTp^{-1}.$$

Now substitute this for V and evaluate the partial derivative in the expression for κ_T

$$\kappa_T = \frac{p}{nRT}\left(\frac{\partial}{\partial p}(nRTp^{-1})\right)_T = \frac{p}{nRT}(-nRTp^{-2})$$

$$\kappa_T = p^{-1}.$$

Likewise, substitute for V in the partial derivative of the expression for α and evaluate

$$\alpha = \frac{p}{nRT}\left(\frac{\partial}{\partial T}(nRp^{-1}T)\right)_p = \frac{p}{nRT}(nRp^{-1})$$

$$\alpha = T^{-1}.$$

7.12.2 Directed practice

Show that Eq. (7.59) holds for an ideal gas.

7.13 Reversible adiabatic expansion/compression

Consider adiabatic ($q = 0$) and reversible ($p = p_{\text{external}}$) expansion or compression of an ideal gas ($C_{p,\text{m}} = C_{V,\text{m}} + R$). The process is outlined in Fig. 7.5. What are the values of ΔU, q and w, and how are the initial conditions of T_i, p_i and V_i related to the final conditions T_f, p_f and V_f? If we look at the integrals above for calculating various parameters such as ΔU, q and w, in any integrand only the integration variable is allowed to change. When presented with two changing state variables previously, we have been able to use an equation of state to substitute for one of these variables. In the present case we have three changing state variables. For this, we need a new strategy that takes advantage of the state function nature of U, p, T and V.

Because U, p, T and V are state functions, the values in the initial and final states are not determined by the path connecting them. Since we cannot directly calculate their values when all of these functions are changing simultaneously,

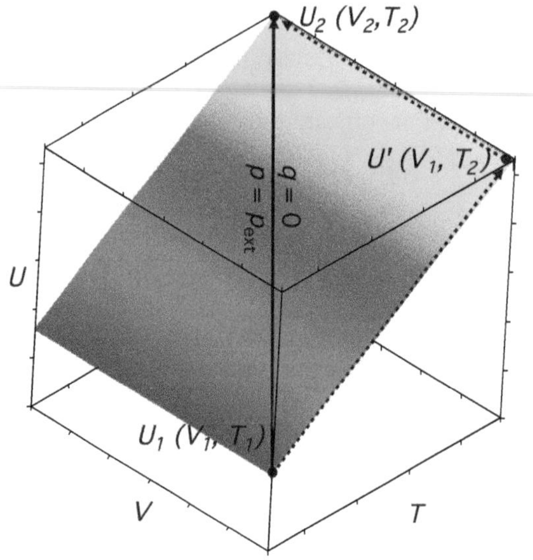

Figure 7.5 Reversible adiabatic expansion can be decomposed into a two-step process. The first step is isochoric ($\Delta V = 0$) while the temperature and pressure change. The second step is isothermal ($\Delta T = 0$) while the volume changes.

we instead decompose the process into two steps for which we can calculate these values. This is illustrated in Fig. 7.5. We first allow the system to evolve along an isochoric path. In a second step, an isothermal path is used to take the system from this intermediate state to the desired final state.

The internal energy is a state function. Therefore, the total internal energy change is equal to the internal energy change of the first step plus the internal energy change of the second step. We use the First Law to write the overall changes in terms of the individual contributions of steps 1 and 2

$$\Delta U = \Delta U_1 + \Delta U_2 = q_1 + w_1 + q_2 + w_2 = q + w. \tag{7.62}$$

The overall process is adiabatic, therefore, $q = 0$. Since the volume does not change in step 1, $w_1 = 0$ and heating leads to $q_1 = n\,C_{v,\mathrm{m}}\,\Delta T$. Since step 2 is isothermal and U of ideal gas depends only on T, the internal energy does not change $\Delta U_2 = 0$, which means that the heat and work on step 2 are equal and opposite $q_2 = -w_2$. Now, we summarize these observations in the corresponding equations.

$$\Delta U = \Delta U_1 + \Delta U_2 = nC_{v,\mathrm{m}}\,\Delta T + 0 = nC_{v,\mathrm{m}}\,\Delta T \tag{7.63}$$

$$q = q_1 + q_2 = nC_{v,\mathrm{m}}\,\Delta T + q_2 = 0 \tag{7.64}$$

$$q_2 = -nC_{v,\mathrm{m}}\,\Delta T$$

$$w = w_1 + w_2 = 0 - q_2 = nC_{v,\mathrm{m}}\,\Delta T \tag{7.65}$$

To calculate the final values of volume and pressure, V_f and p_f, note that

$$nC_{V,\mathrm{m}}\,\mathrm{d}T = -p_{\mathrm{ext}}\,\mathrm{d}V = -nRT\frac{\mathrm{d}V}{V}$$

$$nC_{V,\mathrm{m}} \int \frac{\mathrm{d}T}{T} = -nR \int \frac{\mathrm{d}V}{V}.$$

Thus

$$\ln\left(\frac{T_f}{T_i}\right) = (1-\gamma)\ln\left(\frac{V_f}{V_i}\right) \tag{7.66}$$

where

$$\gamma = C_{p,\mathrm{m}}/C_{V,\mathrm{m}}. \tag{7.67}$$

Using the properties of logs, this is equivalent to

$$T_f/T_i = (V_f/V_i)^{(1-\gamma)} \tag{7.68}$$

or, since

$$T_f/T_i = p_f V_f / p_i V_i,$$
$$p_i V_i^{\gamma} = p_f V_f^{\gamma} \tag{7.69}$$

SUMMARY

$\Delta U = q + w$	First Law
$w = -\int p_{ex}\, dV$	pV work
$q = \int C\, dT$	Simple heating
$H = U + pV$	Definition of enthalpy
$\Delta U = q_V = \int C_V\, dT$ or $\Delta H = q_p = \int C_p\, dT$	

7.13.1 Example

A sample consisting of 1.00 mol Ar is expanded isothermally at 0 °C from 22.4 dm^3 to 44.8 dm^3 (a) reversibly, (b) against a constant external pressure equal to the final pressure of the gas, and (c) freely (against zero external pressure). For the three processes calculate q, w, ΔU, and ΔH.

Remember that in all cases $\Delta U = q + w$ by definition. For this example, $\Delta U = 0$ because we take Ar to be an ideal gas and the internal energy of an ideal gas only depends on temperature. Furthermore $\Delta H = \Delta U + \Delta(pV)$ by definition and $\Delta H = \Delta U + \Delta(nRT)$ because we have an ideal gas. Since there is neither a change in n nor T, $\Delta H = 0$ in all three examples. Any process for which $\Delta H = 0$ is called an *isenthalpic process*. Now, calculate the work and heat appropriately.

a $\Delta U = \Delta H = 0$

$$w = -nRT \ln\left(\frac{V_f}{V_i}\right) = -(1)(8.314)(273)\ln\left(\frac{44.8}{22.4}\right) = -1.57\text{kJ}$$
$$q = \Delta U - w = 0 + 1.57 = +1.57 \text{ kJ}$$

b $\Delta U = \Delta H = 0$

$$w = -p_{ex}\,\Delta V$$
$$p_{ex} = p_f = \frac{nRT}{V_f} = 0.500 \text{ atm}, \quad \Delta V = 44.8 - 22.4 = 22.4 \text{ dm}^3$$
$$w = -0.500 \text{ atm}\left(\frac{1.013 \times 10^5 \text{ Pa}}{1 \text{ atm}}\right)(22.4 \text{ dm}^3)\left(\frac{1 \text{ m}^3}{10^3 \text{ dm}^3}\right) = -1.13 \text{ kJ}$$
$$q = \Delta U - w = 0 + 1.13 = +1.13 \text{ kJ}$$

c $\Delta U = \Delta H = 0$

Free expansion is expansion again zero pressure, therefore, $w = 0$ and $q = \Delta U - w = 0$. Since $q = 0$, free expansion of a gas is also adiabatic expansion.

SUMMARY OF IMPORTANT EQUATIONS

$\Delta U = q + w$	First Law of thermodynamics. Internal energy change is sum of heat plus work.
$w = -\int_{V_i}^{V_f} p_{ex}\, dV$	Work accompanying expansion and contraction is known as pV work and is proportional to the external pressure and the volume change.
$w = -p_{ex}\,\Delta V$	Work performed at constant external pressure for volume change ΔV.
$w = -nRT \ln(V_f/V_i)$	Reversible, isothermal work of n moles of ideal gas, V_i = initial volume, V_f = final volume.
$w = n C_{V,m}\,\Delta T$	Work during reversible adiabatic expansion for n moles of ideal gas and temperature change ΔT.

$w = -\int \phi \, dQ$	Electrical work is proportional to the amount of charge Q that moves through a potential ϕ.
$E_{pot} = mgh$	Potential energy gained by raising an object of mass m a height h. Gravitational acceleration g.
$q = \int_{\text{initial state}}^{\text{final state}} C \, dT$	Heat is given by the integral of the heat capacity integrated over the temperature change. For constant volume use C_V, for constant pressure C_p.
$C_{p,m} - C_{V,m} = R$	Ideal gas, difference of constant pressure and constant volume heat capacities is the gas constant R.
$C_p - C_V = \alpha^2 TV/\kappa_T$	Any material.
$\kappa_T = \frac{1}{V}(\partial V/\partial p)_T$	Isothermal compressibility.
$\alpha = \frac{1}{V}(\partial V/\partial T)_p$	Expansion coefficient.
$\Delta U = \int_{T_i}^{T_f} C_V \, dT$	Internal energy change is equal to the heat evolved at constant volume.
$H = U + pV$	Definition of enthalpy.
$\Delta H = \Delta U + \Delta(pV)$	Relation of enthalpy change to internal energy change.
$\Delta H = \int_{T_i}^{T_f} C_p \, dT$	Enthalpy change is equal to the heat evolved at constant pressure.
$\gamma = C_{p,m}/C_{V,m}$	Heat capacity ratio
$T_f/T_i = (V_f/V_i)^{(1-\gamma)}$	Adiabatic expansion: relation of temperature change and volume change.
$p_i V_i^\gamma = p_f V_f^\gamma$	Adiabatic expansion: relation of pressure change and volume change.

Exercises

7.1 Calculate the work done on a molar basis when a liquid of density ρ_l freezes to form a solid of density ρ_s. Assume the process occurs at the standard melting point.

7.2 Calculate the work done on molar basis when an ideal gas condenses into a liquid of density ρ_l. Assume the process occurs at the standard boiling point.

7.3 Use the First Law in differential form to prove that $C_p = (\partial H/\partial T)_p$ for a reversible constant-pressure process in a system that can only perform pV work.

7.4 Typically 12 g l^{-1} of CO_2 is contained in a 750 ml bottle of champagne. At room temperature and pressure, what volume of CO_2 does this correspond to? Calculate how much work is done by the CO_2 if all were to expand at a cool cellar temperature of 12 °C. If all of this work is converted to kinetic energy of a 5 g cork, calculate the velocity of the cork.

7.5 Calculate the work performed and the internal energy change per mole when Xe is compressed reversibly from 1.50 m^3 to 0.0200 m^3 at 1000 K.

7.6 Take 3.00 mol of an ideal gas at $p_i = 25.0$ bar, $V_i = 4.50$ l and expand it isothermally. Calculate the work done in two different processes: (a) Reversible expansion with $p_f = 4.50$ bar. (b) Irreversible expansion at a constant external pressure of $p_{ext} = 4.50$ bar.

7.7 For Xe, O_2 and benzene, does the short-range repulsion or the attraction lead to a larger change in the work of the real gas compared to the ideal gas?

7.8 The constant-pressure heat capacity of methane in units of J K^{-1} mol^{-1} was found to vary with temperature according to the expression $C_{p,m} = a + b(T/\text{K}) + c(T/\text{K})^2$, with $a = 30.65$, $b = -0.01739$ and $c = 1.3903 \times 10^{-4}$ all with units of J K^{-1} mol^{-1}. (T/K) gets rid of the units of temperature when T is entered in kelvin. Calculate q, w, ΔU, and ΔH on a molar basis (i.e., in kJ mol^{-1}) when the temperature is raised from 150 K to 500 K (a) at constant pressure, (b) at constant volume.

7.9 The constant-pressure heat capacity of CO_2 in units of J K^{-1} mol^{-1} varies with temperature according to the expression $C^\circ_{p,m} = A + Bt + Ct^2 + Dt^3 + Et^{-2}$, with $A = 25.0, B = 55.2, C = -33.7, D = 7.95, E = -0.137$ and $t = T/1000$. Calculate q, w, ΔU, and ΔH on a molar basis (i.e., in kJ mol^{-1}) when the temperature is raised from 298 K to 1200 K (a) at constant pressure, (b) at constant volume.

7.10 (a) Water has $\Delta_{vap}H_m{}^\circ = 44.0$ kJ mol^{-1}. Calculate q, w, ΔU and ΔH when 25.0 g is vaporized at 298 K and 101 kPa. (b) Coffee is placed in an insulated thermos mug in a room with 80% humidity. Explain why the coffee cools much more slowly when a plastic lid is placed on the mug compared to when no lid is in place.

7.11 (a) Ethanol, $M = 46.1$ g mol^{-1}, has $\Delta_{vap}H_m{}^\circ = 38.6$ kJ mol^{-1} at its normal boiling point of 351 K. Calculate q, w, ΔU and ΔH when 25.0 g is vaporized at its normal boiling point against atmospheric pressure. (b) Assuming that ethanol were nonflammable and had the same physiological effects as water, discuss how a sauna in which ethanol is poured over the hot stones would be different than one using water. For water $M = 18.0$ g mol^{-1} and $\Delta_{vap}H_m{}^\circ = 40.65$ kJ mol^{-1} at its boiling point.

7.12 Use Eq. (7.27) and confirm the results in Table 7.2 for N_2 and isopropanol.

7.13 Assuming that the heat capacity of water is constant at 75.3 J K^{-1} mol^{-1} and perfect thermal contact, calculate how long it takes to raise the temperature of 100 g of water by 10 K if a current of 250 mA is passed through a 100 Ω resistor submerged in the water. Recall that electrical power is given by I^2R.

7.14 The parameters for the Shomate equation for SF_6 are given below. Calculate the heat capacity of SF_6 at 298 K and 1000 K. Calculate the heat required to raise the temperature of one mole of SF_6 from 298 K to 1000 K at constant pressure. $A = 58.90$, $B = 255.5$, $C = -252.3$, $D = 88.76$, $E = -1.609$.

7.15 Chinese coal production is in excess of 3 billion metric tons (1 metric ton = 1000 kg) per year. CO_2 sequestration involves creating supercritical CO_2 at 6200 kPa at 45 °C and injecting it into the Earth. Calculate the amount of energy it would take to sequester all of the CO_2 created by burning the annual production of coal in China. Assume that coal is pure graphite and that combustion is complete producing only CO_2. Take the initial conditions to be 101.3 kPa and 25 °C, $C_{p,m}(CO_2) = 37.1$ J K^{-1} mol^{-1}, and the overall process in two steps: reversible isothermal compression followed by isobaric heating.

7.16 Take 350 ml of water at atmospheric pressure. (a) Calculate the work done when the water evaporates at 25 °C. (b) Calculate the work required to reversible compress the vapor from its final volume in part (a) back to 350 ml. Assume the vapor acts as an ideal gas. (c) Why is the answer in (b) wrong, and is it too high or too low?

7.17 (a) Why is the internal energy change in an adiabatic process equal to the work? (b) If a system can only do pV work and the moles are constant, how much work is done in an isochoric process? (c) Can heat be exchanged in an isothermal process?

7.18 Calculate the work done in J for the following circumstances (a) A magnetic stirrer mixes a solution for 4 hours while it draws 0.25 A at 120 V. (b) A water droplet changes from a cube with edge length 2.50 mm to a sphere with the same volume at 298 K. (c) An ideal gas is compressed from an initial volume of 500 l to a final volume of 5.00 l against a constant pressure of 542 bar.

7.19 A sample consisting of 1.00 mol of perfect gas atoms, for which $C_{V,m} = \frac{3}{2}R$ at $P_1 = 100$ kPa and $T_1 = 300$ K, is heated reversibly to 400 K at constant volume. Calculate the final pressure, ΔU, q, and w.

7.20 The difference between the constant pressure and constant volume heat capacities is given by

$$C_p - C_V = \frac{TV\alpha^2}{\beta}$$

where

$$\alpha = \frac{1}{V}\left(\frac{\partial V}{\partial T}\right)_p \quad \text{and} \quad \beta = -\frac{1}{V}\left(\frac{\partial V}{\partial p}\right)_T.$$

Use these equations to show that $C_p - C_V = nR$ for an ideal gas.

7.21 Above room temperature, why is the temperature dependence of the heat capacity of Ar(g) different from that of CH_4(g)?

7.22 The work required to expand methanol vapor at 400 K from 0.1 m^3 to 0.2 m^3 is less than the amount of work required to expand an ideal gas under the same conditions. What leads to the difference and what is the cause?

7.23 A strip of Mg of mass 15 g is dropped into a beaker of dilute hydrochloric acid. Calculate the work done by the system as a result of the reaction. The atmospheric pressure is 1.0 atm and the temperature is 25 °C.

7.24 An average human produces about 10 MJ of heat each day through metabolic activity. If a human body were an isolated system of mass 65 kg with the heat capacity of water, what temperature rise would the body experience? Human bodies are actually open systems, and the main mechanism of heat loss is through the evaporation of water. What mass of water should be evaporated each day to maintain constant temperature?

7.25 A molecular beam apparatus employs supersonic jets that allow gas molecules to expand from a gas reservoir held at a specific temperature and pressure into a vacuum through a small orifice. Expansion of the gas allows one to cool the gas molecules to temperatures of roughly 10 K (or less). The expansion can be treated as adiabatic, with the change in gas enthalpy accompanying expansion $\Delta H_\mathrm{m} = C_{p.\mathrm{m}}\, T_0$ being converted to kinetic energy associated with the flow of the gas. The temperature of the reservoir T_0 is generally much greater than the final temperature of the gas, allowing one to consider the entire enthalpy of the gas to be converted into translational motion. (a) For a diatomic gas, the high-temperature limit is $C_{p,\mathrm{m}} = 9/2\ R$. However, the vibrational degree of freedom is not cooled during the expansion, leading to an effective value of $C_{p,\mathrm{m}} = 7/2\ R$. Using this information demonstrate that the final flow velocity of the molecular beam is related to the initial temperature of the reservoir by $v = \left(\frac{7RT_0}{M}\right)^{1/2}$. (b) Using this expression, what is the flow velocity of a molecular beam of CO when $T_0 = 298$ K?

7.26 Explain why the molecules in a supersonic molecular beam have a high average mean velocity but are at the same time described by a low translational temperature.

7.27 The isothermal compressibility of Cu at 293 K is 7.35×10^{-7} atm^{-1}. Calculate the pressure that must be applied in order to increase its density by 0.08%.

7.28 What is the isothermal compressibility of ice at 273 K, and how much pressure does it take to melt it? What is the density difference between liquid and solid water at 273 K?

7.29 The molar internal energy of an ideal gas only depends on temperature. Label the following expressions as either true or false.

$$\left(\frac{\partial U_\mathrm{m}}{\partial T}\right)_p = 0 \qquad \left(\frac{\partial U_\mathrm{m}}{\partial p}\right)_T = 0 \qquad \left(\frac{\partial U_\mathrm{m}}{\partial V}\right)_T = 0$$

Endnotes

1. Why is an infinitesimal change on a path integral signified with a đ and an infinitesimal change on a state function with a d? The answer is that đq is an inexact differential and dU is an exact differential. More on this in Section 7.4.
2. Molecular beams are one of the great enabling technologies of physical chemistry and they will be discussed in more detail on several occasions. The molecules in a Knudsen beam have properties defined by equilibrium thermal distributions at the temperature of the nozzle or oven from which the gas originates. Knudsen beams rapidly diverge. Supersonic molecular beams are much different. The distribution of energy in the molecules within the beam depends on the conditions of the expansion and is not at all like the distribution in the nozzle. These beams are also highly collimated.

Further reading

Berry, R.S., Rice, S.A., and Ross, J. (2000) *Physical Chemistry*, 2nd edition. Oxford University Press, New York.

Metiu, H. (2006) *Physical Chemistry: Thermodynamics*. Taylor & Francis, New York.

NIST WebBook, http://webbook.nist.gov/chemistry/.

If you would like to see thermodynamics connected to everyday phenomena try R. Müller, Thermodynamik: Von Tautropfen zum Solarkraftwerk. (de Gruyter, Berlin, 2014). From cooking to climate change to power plants, it's all there.

CHAPTER 8
Second law of thermodynamics

PREVIEW OF IMPORTANT CONCEPTS

- The First Law of Thermodynamics is a statement of energy conservation, which does not describe the direction of spontaneous change.
- The Second Law of Thermodynamics states that there exists a property called entropy *S*. The entropy of the universe increases in any spontaneous process, $\Delta S_{tot} > 0$.
- The entropy of a system is determined by the number of different ways to distribute the available energy over the various states of the system.
- For an isolated system, equilibrium is achieved when entropy has a maximum value.
- The statistical mechanical and thermodynamic definitions of entropy are equivalent.
- A reversible Carnot cycle is representative of the most efficient heat engine.
- Partial derivatives and Maxwell's relations can be used to calculate changes in state functions when more than one variable is changing.

When two chemicals are mixed together do they react? This is an existential question for a synthetic chemist. Not much chemistry is performed without making products. From the Zeroth and First Laws of Thermodynamics we have defined a state variable (temperature T) and two state functions (internal energy U and enthalpy H) that we know exist and are rather fundamental. Certainly, temperature changes upon mixing and subsequent reaction are indicative of chemical changes. The temperature increases when an exothermic reaction occurs in a thermos flask and decreases during the course of an endothermic reaction. We quantify the changes in U and H by measuring temperature changes at either constant volume or constant pressure.

We have an instinctive feeling that systems tend to lower their energy. Raise a coffee mug one meter off the ground. Let go. The result is inevitable. The system (ceramic plus finely brewed dark roasted aqueous extract of Arabica beans) spontaneously falls to the ground to lower its energy. Gravitational potential energy is converted into translational energy as the system accelerates toward the floor. Then a fraction of this energy is converted into vibrations and sound waves, while another portion is used to overcome the cohesive energy of the ceramic as the mug strikes the floor. There must be a chemical potential – a concept introduced by J. Willard Gibbs much like the gravitational potential – that indicates which direction is 'downhill' for chemical reactions.

Does U or H act as this chemical potential? This was one of the first arguments used by chemists to explain reactivity. Very often it is the case. Many reactions that proceed spontaneously are exothermic. Add H_2SO_4 to water. It dissolves with great evolution of heat – remember always add acid to water not water to acid when diluting from a concentrated solution. The ΔU and ΔH values for dissolving H_2SO_4 in water are negative. The system moves toward a state of lower energy. Similarly, when KCl is added to water, it dissolves spontaneously. And not just a little bit. Over 35 g of KCl can be dissolved in 100 g of water at room temperature. Nonetheless, the temperature *drops* upon the dissolution of KCl. The reaction is endothermic, ΔH is positive. Take a beaker full of water, it will remain in the liquid phase at atmospheric pressure and room temperature once it has established its equilibrium vapor pressure above the liquid. But if we add heat to the system so that the temperature reaches the boiling point, the water spontaneously evaporates. Again, the process is spontaneous even though it is endothermic.

'Sometimes' is not good enough when establishing a law of Nature. Sometimes, spontaneous reactivity corresponds to minimization of ΔH, but not always. Sometimes, an equilibrium that favors the reactants is established when a process exhibits a positive ΔU, but not always. Therefore, neither ΔU nor ΔH plays the role of the chemical potential. We need to discover one or more new state functions along the way to finding what truly is the chemical potential. The Zeroth and First Laws are insufficient for determining the direction of spontaneous change. They do not determine when a reaction

Physical Chemistry: How Chemistry Works, First Edition. Kurt W. Kolasinski.

Companion Website: www.wiley.com/go/kolasinski/physicalchemistry

Figure 8.1 Satellite image of a hurricane. Study of a closed cycle of compression and expansion of gas leads us to the discovery of a new state function: the entropy. Study of this cycle is also essential to understanding the dynamics and energy flow that lead to the formation of a hurricane. Image courtesy of National Oceanic and Atmospheric Administration (NOAA)/Department of Commerce.

will occur. They are also insufficient to determine whether an equilibrium exists and whether that equilibrium favors the reactants or products.

We need a new state function that describes the evolution of chemical and physical parameters. Some systems react completely, others establish equilibria. Some compounds dissolve completely, others establish reagent-favored equilibria. But even in systems with minimal or no chemical interactions, there are spontaneous changes that occur. We always observe mixing of gases upon opening a valve. Once mixed, ideal gases do not spontaneously unmix even though such a process conserves energy. Once a body or solution is brought into thermal equilibrium at a uniform temperature, spontaneous temperature polarization into hot and cold regions does not occur even, though it conserves energy. There must be another parameter, another state function that remains to be discovered that will unlock the key to describing spontaneity. That state function, as we shall see, is *entropy* and unlocking it will require the Second Law of Thermodynamics.

8.1 The second law of thermodynamics

The First Law is a statement of energy conservation. However, it does not predict the spontaneous direction of change. It does not explain why heat spontaneously flows from a hot to a cold object, or why a gas fills all available space in a container. In our introduction to statistical mechanics, we defined the state function called entropy. Entropy serves as a measure of the number of microstates accessible to a macroscopic system at a given energy. For an isolated system, equilibrium is achieved when entropy has a maximum value. However, if we do not have direct access to quantum states and their relative populations, how can we detect their influence?

Thermochemistry gives us direct access to the energy changes that occur during chemical and physical processes. We will now show that thermochemistry can be used to define and monitor entropy changes. We will establish that the thermodynamic and statistical mechanical definitions of entropy are equivalent. The distribution of population among energy levels is responsible for changes in entropy. Why is all this importance laid at the feet of entropy, energy levels and populations? Because quantifying entropy changes allows us to predict the spontaneous direction of change for a process.

The history of the development of the concept of entropy is long and fraught with the human complications that can accompany the proposal of imaginative ideas and their confirmation into the canon of accepted concepts. Carnot established the fundamental idea of entropy while studying heat engines. His work went largely unrecognized, in part because he described the idea literally not mathematically. Clapeyron developed his conception of entropy in obtaining mathematical relationships to describe the temperature dependence of vapor pressure. His reanimation of the concept allowed Kelvin and Clausius to refine the thermodynamic definition of entropy, and to restate its importance in the form that is now known as the Second Law of Thermodynamics.

Kelvin–Planck statement of the Second Law:

No process is possible in which the sole result is the absorption of heat from a reservoir and its complete conversion into work.

Clausius statement of the Second Law:

It is impossible for an engine to perform work by cooling a portion of matter to a temperature below that of the coldest part of the surroundings. Heat does not flow spontaneously out of a cold reservoir and into a hot reservoir.

In other words, no engine is 100% efficient. Great, but we are interested in the effect of entropy on chemical systems, not heat engines. The Second Law is particularly important since it tells us about the evolution of the universe. Thus, we restate it a third time in a form with more relevance to chemistry.

Second Law of Thermodynamics

There exists a property called entropy S. The entropy of the universe increases in any spontaneous process, $\Delta S_{tot} > 0$. *The entropy of a system is determined by the number of different ways to distribute the available energy over the various states of the system.*

The deep importance of entropy for chemical and physical systems was developed, in particular, in the work of Boltzmann and Gibbs who were able to connect thermodynamics and statistical mechanics. By establishing statistical mechanics as the bedrock upon which the evolution of all chemical and physical systems is based, they set the stage for the understanding of reversible and irreversible chemical changes as well as ushering in the quantum mechanical revolution.

8.2 Thermodynamics of a hurricane

We will now demonstrate the discovery of a state function and proceed to define it thermodynamically. We will then show that this thermodynamic definition is equivalent to the statistical mechanical definition. To do so, we begin by looking up in the sky and pondering a cycle that describes how a tropical storm (hurricane or typhoon) is capable of generating such a large destructive force. In other words, how it is able to do work. The work is done by the expansion and compression of gases.

The processes involved are depicted in Fig. 8.2. In panel (a) the system starts at point A. The xy-plane lies on the surface of the Earth. The z-axis corresponds to altitude. A high-pressure system H is located over a body of water such as the Caribbean Sea. At the same altitude, but removed by some lateral distance, the low-pressure system L is located at point B, which is known as the 'eye' of the storm. A plug of gas moving from a region of high pressure to a region of low pressure expands. Since this gas is in contact with a massive water bath (the sea or ocean), its temperature will remain constant as it expands. Heat flows from the sea into the gas to maintain isothermal conditions in the plug of gas.

You will most likely have seen satellite images of the low-pressure system at B such as that in Fig. 8.1. Circulation of the atmosphere occurs around this point. The Coriolis force associated with the rotation of the Earth forces the atmosphere to rotate counterclockwise about the eye of the storm in the Northern Hemisphere (but clockwise in the Southern Hemisphere). When the plug of gas reaches the eye it is forced upward. The plug is surrounded by a more stagnant layer of air that exchanges little heat with the plug. Essentially, this step from the eye at sea level to some point C in the upper atmosphere happens adiabatically, $q = 0$. The pressure is also lower in the upper atmosphere; therefore this step is an adiabatic expansion.

If the plug of gas that is drawn into the eye is not replaced, a vacuum would appear at A. Therefore, gas must be returned to this point. This occurs in two steps. First, in the upper atmosphere the plug is isothermally compressed at a

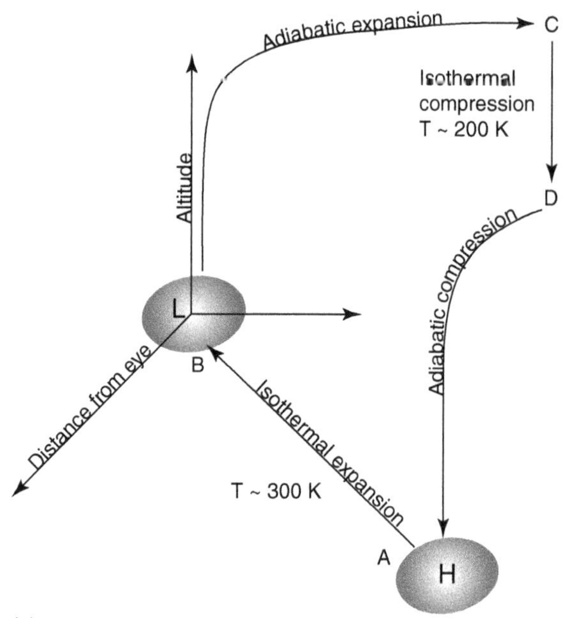

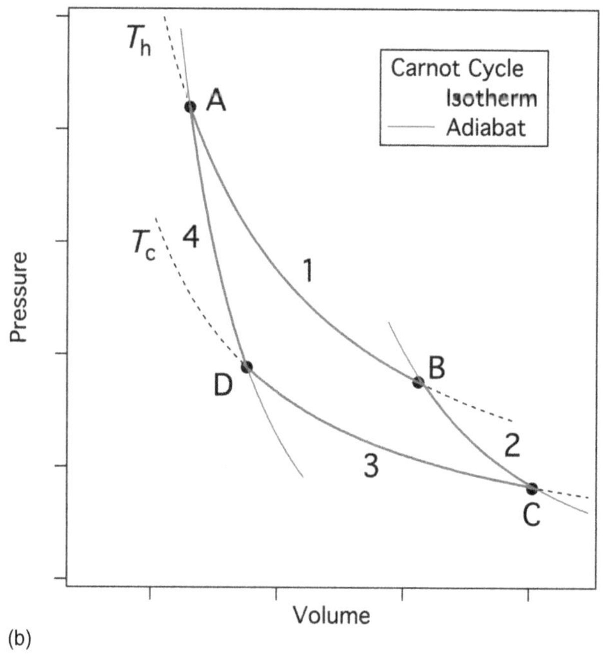

Figure 8.2 (a) Schematic representation of the atmospheric processes that occur during a hurricane. (b) Translation of these phenomena into a plot of pressure versus volume.

temperature of about 200 K. Finally, from this point D the gas is returned to the circulating high-pressure system at A. This final compression step occurs adiabatically.

In Fig. 8.2(b) the changes that occur to the plug of gas are depicted in terms of a plot of pressure versus volume. A succession of two adiabats and two isotherms defines a closed cycle. *Adiabats* are pathways connecting two states of the system for which $q = 0$. *Isotherms* are pathways connecting two states of the system for which $\Delta T = 0$. In a closed cycle, mass and energy are conserved. Indeed, any state function does not change its value in the course of completing one complete cycle. The requirements for a closed system are more or less met by a hurricane. In the laboratory a system can be constructed to meet them more rigorously.

We now define rigorously what is occurring with p, V and energy in this system. What is the energy balance in this cycle? Where does the energy come from to create the winds? How much work is done? How much heat flows? The progression of our system in Fig. 8.2(b) is known as the *Carnot cycle*, which we describe quantitatively in the next section. You will have the opportunity to calculate these parameters for realistic hurricane conditions in the exercises at the end of the chapter.

8.2.1 Carnot cycle

Consider the process depicted in Fig. 8.2 with four reversible steps involving one mole of ideal gas, $n = 1$ mol. We want to calculate the internal energy change, heat and work of each step as well as the cycle in total. Each step can be calculated independently and then summed to obtain the values for the overall process. Since we are dealing with 1 mole of gas, all of the quantities calculated below are equal to the molar values but the subscript m will be dropped for simplification.

Step 1: *Reversible isothermal expansion* at T_h from A to B and V_1 to V_2. T_h is the constant temperature of our hot reservoir (i.e., the sea in our example). The internal energy change is

$$\Delta U = q + w. \tag{8.1}$$

We calculate the two easiest parameters and then use Eq. (8.1) to obtain the third. The internal energy of an ideal gas depends on temperature alone. Therefore, for the isotherm expansion of an ideal gas

$$\Delta U_{A\to B} = 0. \tag{8.2}$$

Per mole of ideal gas the work done in a reversible, isothermal expansion is

$$w_{A\to B} = -RT_h \ln(V_2/V_1). \tag{8.3}$$

Thus, by the First Law, that is, applying Eq. (8.1)

$$q_{A\to B} = RT_h \ln(V_2/V_1). \tag{8.4}$$

The system does work on the surroundings. The energy exchanged as work is exactly balanced by the heat flow into the system from the surroundings; therefore, the internal energy of the system does not change.

Step 2: *Reversible adiabatic expansion* from B to C and V_2 to V_3. No heat leaves the system as the step is adiabatic,

$$q_{B\to C} = 0. \tag{8.5}$$

The temperature falls from T_h to T_c, the temperature of the cold sink. The internal energy change in reversible adiabatic ideal gas expansion is given by

$$\Delta U_{B\to C} = C_V(T_c - T_h). \tag{8.6}$$

Then from the First Law, the internal energy change must equal the work

$$w_{B\to C} = C_V(T_c - T_h). \tag{8.7}$$

The cold temperature is lower than the hot, thus the work is negative. The system does work on the surroundings and the work is done at the expense of the internal energy of the system.

Step 3: *Reversible isothermal compression* from C to D and V_3 to V_4. Analogous to Step 1, we have

$$\Delta U_{C\to D} = 0 \tag{8.8}$$

$$w_{C\to D} = -RT_c \ln(V_4/V_3) \tag{8.9}$$

$$q_{C\to D} = RT_c \ln(V_4/V_3) \tag{8.10}$$

Step 4: *Reversible adiabatic compression* from D to A and V_4 to V_1. Analogous to Step 2, we have

$$q_{D\to A} = 0 \tag{8.11}$$

$$\Delta U_{D\to A} = C_V(T_h - T_c) \tag{8.12}$$

$$w_{D\to A} = C_V(T_h - T_c) \tag{8.13}$$

Now we add up the internal energy changes, the heat flow and the work performed in going around the whole cycle. To evaluate these, we need to find relationships between the various volumes. Steps 2 and 4 are reversible adiabatic expansions, for which the volume change is related to the temperature change and the ratio of the heat capacities

$$\gamma = C_{p,m}/C_{V,m} \tag{8.14}$$

by

$$T_f/T_i = (V_f/V_i)^{(1-\gamma)}. \tag{8.15}$$

8.2.1.1 Example

An ideal gas has a molar heat capacity at constant volume of $C_{V,m} = 3/2\,R$, where R is the gas constant. Given one mole of ideal gas and the pressure and volume conditions in Table 8.1, calculate the work, heat, internal energy change and entropy change associated with each step as well as the whole process.

First, we collect the work and heat expressions for each step in the Carnot cycle, as shown in Table 8.2. Then, we use these expressions to calculate the work, heat, internal energy change and entropy change associated with each step and the whole process. The results are listed in Table 8.3.

Table 8.1 Pressure and volume values at each point in a Carnot cycle for one more of ideal gas for Example 8.2.1.1.

Point	Pressure/bar	Volume/L
A	1.00	24.8
B	0.498	50.0
C	0.398	57.0
D	0.802	28.3

Table 8.2 Expressions for work and heat for each step in a Carnot cycle for one more of ideal gas for Example 8.2.1.1.

Step	Process	Work	Heat
1	isothermal expansion	$w_{A\to B} = -RT_h \ln(V_2/V_1)$	$q = -w$
2	adiabatic expansion	$w_{B\to C} = C_V \Delta T$	$q = 0$
3	isothermal expansion	$w_{C\to D} = -RT_c \ln(V_4/V_3)$	$q = -w$
4	adiabatic expansion	$w_{D\to A} = C_V \Delta T$	$q = 0$

The cycle performs 150 J per mole of gas worth of work on the surroundings. The cycle extracted 150 J mol^{-1} of heat from the surroundings, just as a hurricane would take heat from the water. The internal energy change is zero as expected. A state function must have a constant value for one complete circuit of the cycle. Neither w nor q exhibits this characteristic because they are path functions.

Have we discovered a new state function that corresponds to q/T? The value obtained from the calculation above is close to zero but not exactly zero. It is important to note that this is due to round-off error. We check this by performing the algebra explicitly,

$$\Delta S_{\text{cycle}} = \frac{q_1}{T_h} + 0 + \frac{q_3}{T_c} + 0 = \frac{-w_1}{T_h} + \frac{-w_3}{T_c}$$
$$\Delta S_{\text{cycle}} = R\ln(V_2/V_1) + R\ln(V_4/V_3)$$

This can be evaluated algebraically by applying Eq. (8.15) to determine the ratios of the volumes, from which we find

$$(V_4/V_1)^{1-\gamma} = (V_3/V_2)^{1-\gamma}.$$

If two ratios raised to the same power are equal, the ratios are also equal. Rearranging, we obtain

$$(V_2/V_1) = (V_3/V_4).$$

Now we substitute into the expression for ΔS_{cycle} to find

$$\Delta S_{\text{cycle}} = R\ln(V_2/V_1) - R\ln(V_3/V_4) = 0.$$

Note the change of sign caused by inverting the ratio in the natural logarithm term. Thus, a new state function has been identified. We will define it more explicitly in the next section as not just q/T but q_{rev}/T, where q_{rev} is the heat evolved along a reversible path.

8.2.1.2 Directed practice

Show explicitly by performing the algebra that the net internal energy change, entropy change, heat and work for the Carnot cycle with reversible steps are given by

$$\Delta U_{\text{cycle}} = 0 \quad (8.16)$$
$$\Delta S_{\text{cycle}} = 0 \quad (8.17)$$
$$q_{\text{rev}} = R(T_h - T_c)\ln(V_2/V_1) \quad (8.18)$$
$$w_{\text{rev}} = R(T_h - T_c)\ln(V_1/V_2) \quad (8.19)$$

Table 8.3 Results for work, heat, internal energy change and entropy change for the Carnot cycle defined in Example 8.2.1.1.

Step	w/J mol^{-1}	q/J mol^{-1}	T/K	ΔU/J mol^{-1}	$\Delta S = q/T$/J K^{-1} mol^{-1}
1	−1740	1740	298	0	5.84
2	−312	0		−312	0
3	1590	−1590	273	0	−5.82
4	312	0		312	0
Sum	−150	+150		0	0.02

8.3 Heat engines, refrigeration, and heat pumps

The foundations of thermodynamics were largely set in studies of mechanical work and how it related to heat – either the generation of heat during the performance of work or conversion of thermal energy into work in heat engines. This is an important application that allows us to draw important conclusions regarding processes that affect the energy sector.

8.3.1 Heat engine efficiency

Return now to the Carnot cycle and its relevance to heat engines. The Carnot cycle with reversible steps acts as a limiting case for the perfect heat engine. The efficiency ε of a heat engine is given by the total work performed w divided by the heat transfer with the hot reservoir q_h,

$$\varepsilon = |w|/q_\mathrm{h}. \tag{8.20}$$

But the work is just the difference between the heat supplied by the hot source and returned to the cold sink, thus

$$\varepsilon = \frac{q_\mathrm{h} + q_\mathrm{c}}{q_\mathrm{h}} = 1 + \frac{q_\mathrm{c}}{q_\mathrm{h}}. \tag{8.21}$$

The expressions for q_c and q_h for a reversible Carnot engine are found in Table 8.2, in which the cold step is identified at Step 3 and the hot step is Step 1. Substituting we find

$$\varepsilon_\mathrm{rev} = 1 - \frac{T_\mathrm{c}}{T_\mathrm{h}} = \frac{T_\mathrm{h} - T_\mathrm{c}}{T_\mathrm{h}}. \tag{8.22}$$

8.3.1.1 Directed practice

Derive the expression for the efficiency of a reversible Carnot engine, Eq. (8.22).

This is a statement of *Carnot's theorem*, and it is true for all materials (not just ideal gases) as long as the engine operates reversibly. This is the highest possible efficiency that can be obtained in a heat engine. By equating Eqs (8.21) and (8.22) we find for a reversible engine

$$-T_\mathrm{h}/T_\mathrm{c} = q_\mathrm{h}/q_\mathrm{c} \tag{8.23}$$

8.3.1.2 Example

At a power plant, superheated steam at 560 °C is used to drive a turbine for electricity generation. The steam is discharged to a cooling tower at 38 °C. Calculate the efficiency of this process.

Convert the temperatures to kelvin: T_h = 833 K, T_c = 311 K. Assuming that the generator is reversible,

$$\varepsilon_\mathrm{rev} = 1 - \frac{T_\mathrm{c}}{T_\mathrm{h}} = 1 - \frac{311\ \mathrm{K}}{833\ \mathrm{K}} = 0.63$$

Ideally, the generator is 63% efficient, but in practice friction, heat loss and other inefficiencies reduce the maximum efficiency to about 40% or a little better. These efficiency limitations do not apply for fuel cells, and this is one of the reasons that there is great interest in their development for more efficient and cleaner power production.

8.3.1.3 Example

What is the efficiency of the Carnot engine in Example 8.2.1.1 and to what does it apply?

$$\varepsilon_\mathrm{rev} = 1 - \frac{T_\mathrm{c}}{T_\mathrm{h}} = 1 - \frac{273\ \mathrm{K}}{298\ \mathrm{K}} = 0.084.$$

$$\varepsilon = \frac{|w|}{q_\mathrm{h}} = \frac{|-150\ \mathrm{J}|}{1740\ \mathrm{J}} = 0.086.$$

It applies to the work done on the whole cycle. Note how the efficiency has dropped in comparison to Example 8.3.1.2. The highest efficiency is obtained by the greatest difference between the hot and cold temperatures. In practical systems, materials and environmental considerations limit the values of T_h and T_c. The T_h of an engine must lie below the melting point of the components that make up the engine. T_c can be no lower than the ambient temperature outside of the engine because refrigeration would simply add inefficiency.

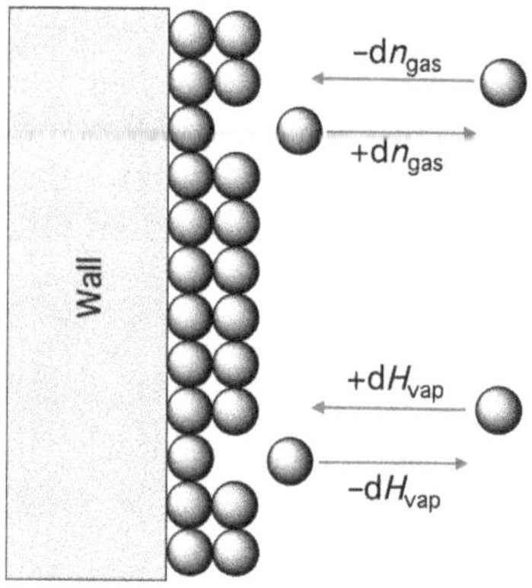

Figure 8.3 The balance between condensation and evaporation leads to mass and energy transfers. Every molecule of H_2O (or any other gas) lost to the gas phase by condensation heats the wall through the enthalpy of vaporization. Every molecule that desorbs from the liquid film condensed on the wall cools the wall by an equal and opposite amount. Precise balancing occurs at equilibrium when the temperatures of the gas and liquid are the same as are the mass and energy balances.

8.3.2 Heat pumps and refrigerators

The physical processes of evaporation and condensation can be harnessed to provide both heating and cooling. Furthermore, the same apparatus is capable not only of doing both, but also performing more efficiently than simple combustion for heating.

The key to these processes is found in Fig. 8.3, and the realization that since condensation and evaporation are the reverse of one another, the enthalpy change accompanying them is equal and opposite (at the same temperature). Evaporation is always an endothermic process. Water has an enthalpy of vaporization of 43.98 kJ mol^{-1} at 298 K, which allows for large heat transfers. We know this instinctively as the evaporation of sweat is used to cool our bodies. Conversely, condensation is always an exothermic process. We do not always make this natural connection but sit in a sauna, throw some water on the rocks, and you rapidly feel the heat released as the vaporized water condenses on your body. Similarly, steam at 100 °C is much more dangerous than water at 100 °C because the transfer of heat to your skin is much more efficient when the enthalpy of condensation is available.

The principle of operation of a heatpump/refrigerator is shown in Fig. 8.4. By performing evaporation on one side of a thermally conductive material, a fluid on the other side of the conductor is cooled. Performing condensation on the same conductor heats the fluid on the other side. Thus, a refrigerator is turned into a heat pump simply by reversal of the processes.

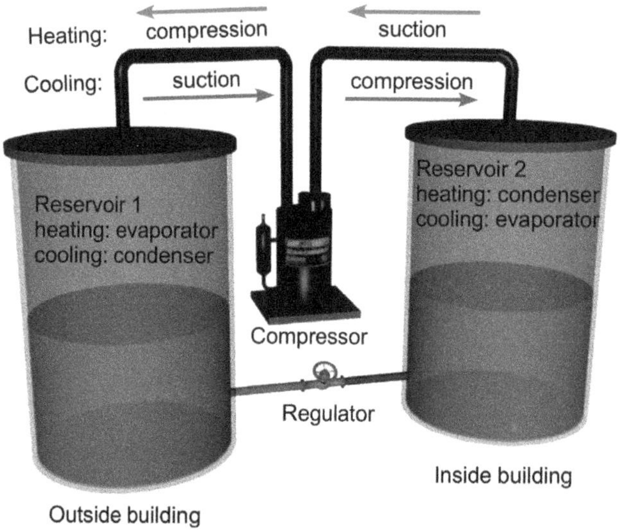

Figure 8.4 Heat pump/refrigerator. A heat-exchange system can be run as either a heat pump or as a refrigerator. The keys are to take advantage of the enthalpy of vaporization and controlling where evaporation and condensation occur.

In order to cool the inside of a building where Reservoir 2 is located, a liquid must be made to evaporate from the reservoir. The walls of the reservoir are thermally conductive and in contact with a fluid that is cooled before it circulates to the rest of the building. The evaporated vapor is then transported to Reservoir 1, where it condenses and deposits heat exterior to the building. The direction of evaporation, vapor transport and condensation is controlled by a compressor. The compressor pumps gas out of Reservoir 2. This lowers the pressure above Reservoir 2 to below the vapor pressure of the liquid. Evaporation ensues. The vapor is then compressed above Reservoir 1. When the pressure exceeds the vapor pressure of the liquid, the vapor condenses.

To heat the building the process is reversed simply my making the compressor work in the opposite direction. The pressure above Reservoir 1 is lowered below the vapor pressure of the liquid, which turns Reservoir 1 into the evaporator. The vapor is compressed in Reservoir 2, turning it into the condenser. The heat liberated by condensation is used to heat the gas in contact with the outside of Reservoir 2, which then flows to the rest of the building to heat it.

Heat pumps and refrigerators require electrical power to run the compressor. Nonetheless, they offer the possibility to deliver heating and cooling extremely efficiently. Say our refrigerator needs to remove 1000 J from its interior at 4 °C and exhaust it into the room at 25 °C. From Eq. (8.4.4), the heat dumped by the condenser into the kitchen is

$$q_{\mathrm{h}} = -q_{\mathrm{c}}\frac{T_{\mathrm{h}}}{T_{\mathrm{c}}} = (1000\ \mathrm{J})\frac{298\ \mathrm{K}}{277\ \mathrm{K}} = 1076\ \mathrm{J}$$

Assuming no thermal loses, 1000 J is transferred by evaporation inside the evaporator to the condenser. The gas then condenses in the condenser to release this 1000 J of heat. The difference is the work done in the Carnot cycle, and is equal to the energy required to run the compressor (again assuming no losses to friction, etc.) The compressor only has to do 76 J of work to achieve 1000 J of cooling!

The coefficient of performance of a refrigerator η_{R} is equal to the absolute value of the ratio of the heat removed from the cold reservoir divided by the work done by the compressor. From the Carnot cycle, we know that the latter is equal to the sum of the heat flows on the two isothermal steps, $w = q_1 + q_3$. A reversible process will give the maximum performance factor, which is given by

$$\eta_{\mathrm{R}} = \frac{T_{\mathrm{c}}}{T_{\mathrm{h}} - T_{\mathrm{c}}} \tag{8.24}$$

A real refrigerator will not achieve this because it is not reversible, and also because the exhaust will generally run much hotter than room temperature (don't grab the back of your refrigerator!) and the evaporator will be much colder than the temperature of the inside of the refrigerator.

For a heat pump, the cycle runs in reverse and the heat flow along the hot Step 1 is used in the ratio to calculate the *coefficient of performance* of a refrigerator η_{HP}

$$\eta_{\mathrm{HP}} = \frac{T_{\mathrm{h}}}{T_{\mathrm{h}} - T_{\mathrm{c}}}. \tag{8.25}$$

8.3.2.1 Directed practice

Derive the equations for η_{R} and η_{HP}.

In a practical system, the important factor is the heat load or the power that actually must be dissipated. Power is energy per unit time. Consider a typical geothermal heat pump system that might be constructed by drilling into the ground in front of a university building. It works for both heating and cooling. Such systems are also known as geoexchange systems to differentiate them from geothermal systems, which rely on extracting energy from a high-temperature source that is heated by volcanism. The ground temperature is a constant 12 °C. One such system at West Chester University has 37 wells, which were drilled 41.8 m deep and lined with 3.81-cm pipes. The pipes are filled with water and a few percent ethanol. These form Reservoir 1. The solution is maintained at 54.4 °C in Reservoir 2. Which reservoir acts as the condenser and which as the evaporator depends on the season and whether the building needs to be heated or cooled. The amount of heating or cooling is determined by the flow through the regulator.

Because the two reservoirs are at a constant temperature, the performance factors are constant

$$\text{heating: } \frac{327}{327-285} = 7.8; \quad \text{cooling: } \frac{285}{327-285} = 6.8.$$

It would seem that obtaining 1076 W of heating or cooling performance from only 76 W of energy expenditure violates the conservation of energy. Indeed it would; thus, something must be missing from this analysis. The cooling example implicitly assumes a constant external temperature. The geoexchange example implicitly assumes a constant ground temperature. When geoexchange is used for heating, the Earth is cooled. If that cooling were not counterbalanced by a heat flow, the ground temperature would sink and the efficiency would also plummet. The core of the Earth is heated by radioactive decay. A remnant of this heat is responsible for effectively constant ground temperature that is observed several meters below the surface. The efficiency of a geoexchange system relies on the Earth to act as an isothermal bath to maintain a constant ground temperature. Similarly, a kitchen refrigerator requires external cooling to maintain the room temperature.

8.3.2.2 Directed practice

Show that the volume of water in the geothermal system is 1.76 m^3. This corresponds (assuming pure water with a density of 1000 kg m^{-3}) to 1760 kg of water. Also, show that the condensation of 443 g of water per second is required to deliver 1 MW of heating, and the compressor must do 128 kW of work to deliver this heat.

8.4 Definition of entropy

8.4.1 Thermodynamic definition

Heat stimulates disorderly motion. In the Carnot cycle, we identified q_{rev}/T as a state function. This ratio thermodynamically defines the *entropy change*. It does not tell us what entropy is, but it defines how we can measure its change. Because this ratio determines the change and not the absolute value, it also does not define the zero of the entropy scale. For this we need the Third Law of Thermodynamics.

A mathematical statement of the Second Law is that for an infinitesimal change,

$$\mathrm{d}S = \mathrm{d}q_{\text{rev}}/T. \tag{8.26}$$

Integration is required to obtain the macroscopic entropy change

$$\Delta S = \int \frac{\mathrm{d}q_{\text{rev}}}{T}. \tag{8.27}$$

Integration of Eq. (8.27), of course, does not resolve the problem of defining the constant of integration. The key to calculating entropy changes is to properly select the functional form of $\mathrm{d}q_{\text{rev}}$ based on the path that the system evolves along. We will return to this question in Section 8.5.

8.4.2 Statistical mechanical definition

Macroscopic systems are composed of an immense number of molecules that can occupy an even more enormous number of energy states. The distribution of molecules over these states exhibits humungous degeneracies. The magnitudes of the numbers, on the order of 10^{23} and much larger, are simply not in line with our everyday experience of counting the number of cans in our beer can collection or even the Chancellor of the Exchequer's experience of counting money in her Majesty's treasury. The statistics of such large numbers have profound implications for chemical systems.

There are an enormous number of energy state distributions that have an energy between E and $E + \mathrm{d}E$ for which $\mathrm{d}E$ is a small fraction of E. An essential tactic of the statistical mechanical approach to describing chemical systems is to find the most probable distribution of molecules over the available energy states for a given set of macroscopic variables describing a system at equilibrium. This most probable distribution can then be used to find the values of other macroscopic properties by taking averages over microscopic quantities. We encounter an example of this when we discuss how the enthalpy of bond dissociation is related to the dissociation energy of a bond in an individual molecule. The former refers to an ensemble property averaged over a thermal distribution at the appropriate temperature. The latter refers to a property of one molecule in a particular rotational, vibrational and electronic state that can be calculated directly from the potential energy surface that describes the atoms involved in the bond.

The total number of microstates associated with a given configuration of energy, or weight of the configuration, is given by

$$\Omega = \sum \frac{N!}{n_1!n_2!\dots n_j!} \tag{8.28}$$

where N is the number of units over which energy is distributed, and the population n_j represents the number of units occupying the jth level of the system. (! indicates a factorial). A system approaches equilibrium by achieving the energy state that maximizes Ω. The equilibrium state also has the maximum entropy S. The proportionality between Ω and S is given by the *Boltzmann formula*

$$S = k_B \ln \Omega \tag{8.29}$$

This equation gives us an easily identifiable definition of entropy. Entropy is a measure of the number of states over which thermal energy is dispersed in the system. But is this physical definition from statistical thermodynamics equivalent to the practical definition obtained from thermodynamics?

8.4.3 Equivalence of thermodynamic and statistical mechanical definitions

We now have two expressions for the entropy. In differential form, the entropy is defined thermodynamically by

$$\mathrm{d}S = \mathrm{d}q_{\text{rev}}/T. \tag{8.30}$$

and statistical mechanically by

$$\mathrm{d}S = k_B \,\mathrm{d}\ln \Omega. \tag{8.31}$$

We now show that these are equivalent by proving that

$$\mathrm{d}q_{\text{rev}} = k_B T \,\mathrm{d}\ln \Omega \tag{8.32}$$

Take a system with constant N and V. The internal energy change engendered by adding an infinitesimal amount of heat along a reversible path is $\mathrm{d}q_{\text{rev}}$. No pV work is done, thus by the First Law of Thermodynamics the heat exchanged is equal to the internal energy change

$$\mathrm{d}q_{\text{rev}} = \mathrm{d}U. \tag{8.33}$$

In Chapter 6, we found that the internal energy change at constant volume is due solely to the change in level populations $\mathrm{d}n_j$ because the energy levels ε_j do not change. Therefore, the internal energy change is

$$\mathrm{d}U = \sum_j \varepsilon_j \,\mathrm{d}n_j \tag{8.34}$$

which when substituted into Eq. (8.33) means that

$$\mathrm{d}q_{\text{rev}} = \sum_j \varepsilon_j \,\mathrm{d}n_j. \tag{8.35}$$

The system equilibrates by maximizing Ω, which changes according to

$$\mathrm{d}\ln \Omega = \sum_j \frac{\partial \ln \Omega}{\partial n_j} \,\mathrm{d}n_j. \tag{8.36}$$

From Eq. (6.8),

$$(\partial \ln \Omega/\partial n_j) = -\alpha + (1/k_B T)\varepsilon_j. \tag{8.37}$$

Therefore,

$$\mathrm{d}\ln \Omega = -\alpha \sum_j \mathrm{d}n_j + (1/k_B T) \sum_j \varepsilon_j \,\mathrm{d}n_j. \tag{8.38}$$

However at constant N, $\sum_j dn_j = 0$. Upon substitution from Eq. (8.35) we obtain

$$\mathrm{d}q_{\text{rev}} = k_B T \,\mathrm{d}\ln \Omega, \tag{8.39}$$

as was to be shown. The thermodynamic and statistical mechanical expressions for entropy are equivalent.

8.5 Calculating changes in entropy

ΔS must be calculated along a reversible path. When an irreversible process is considered, we must construct a reversible path that connects the same initial and final states involved in the irreversible process. Because S is a state function, ΔS is path-independent: as long as the transformation is between the same initial and final states in both processes, we obtain the same answer.

To calculate the change in entropy from thermodynamic data, we start with the definition of entropy

$$dS = dq_{rev}/T \tag{8.40}$$

Then we integrate to obtain the entropy change associated with the change from state A to state B of the system,

$$S_B - S_A = \Delta S_{A\rightarrow B} = \int dq_{rev}/T \tag{8.41}$$

What dq_{rev} is equal to depends on the process. For simple heating at constant pressure $dq_{rev} = C_p\, dT$. For simple heating at constant volume $dq_{rev} = C_V\, dT$. For any process at constant pressure, such as reaction or phase change, the heat evolved is equal to ΔH.

Let us look at how we calculate dq_{rev}. There are two trivial cases:

- Reversible, adiabatic process: $q_{rev} = 0$, therefore, $\Delta S = 0$.
- Cyclical process: $\Delta S = 0$ because the change in any state function is zero around a cycle.

A corollary drawn from this is that for any cycle made up of n steps, only $n - 1$ are independent. If we know the entropy change associated with $n - 1$ steps, the ΔS of the nth step can be calculated since the sum of all steps must be zero.

8.5.1 ΔS for isothermal processes and phase transitions

For any isothermal process, T can be pulled out of the integral in Eq. (8.41) and we obtain

$$\Delta S = q_{rev}/T. \tag{8.42}$$

The internal energy of an ideal gas depends only on temperature. Thus, when expanded isothermally along a reversible path, $\Delta U = 0$ and the heat is equal and opposite to the work

$$q_{rev} = nRT \ln(V_2/V_1). \tag{8.43}$$

Substituting into Eq. (8.42), the entropy change for reversible isothermal expansion of an ideal gas is just

$$\Delta S = nR \ln(V_2/V_1). \tag{8.44}$$

For a phase transition occurring at its standard temperature and standard pressure, the system and surroundings are *at equilibrium* and *any heat transfer is done reversibly*. Therefore, the heat is just the entropy change associated with the transition, and the entropy change is

$$\Delta_{trs}S^\circ = \frac{\Delta_{trs}H^\circ}{T_{trs}}. \tag{8.45}$$

As an example, consider the molar entropy change for fusion and vaporization of water at its standard melting point and boiling point, respectively. The molar enthalpies of fusion and vaporization are 6.01 kJ mol^{-1} and 40.79 kJ mol^{-1}, respectively. Therefore,

$$\Delta_{fus}S^\circ_m = \frac{6010 \text{ J mol}^{-1}}{273 \text{ K}} = 22.0 \text{ J K}^{-1} \text{ mol}^{-1}$$

and

$$\Delta_{vap}S^\circ_m = \frac{40790 \text{ J mol}^{-1}}{373.12 \text{ K}} = 109.3 \text{ J K}^{-1} \text{ mol}^{-1}.$$

Note that the entropy change is positive in both cases but much larger for vaporization than melting. This meets with our expectations. In melting, the system goes from a solid with only vibrational degrees of freedom and well-defined bonding arrangements to a liquid with translational, rotational and vibrational degrees of freedom and constant fluctuations in the nearest neighbors. In vaporization, the gas has an even greater number of microstates associated with it as it expands to fill the entire volume of the container, which for a process at a constant pressure of 1 bar would be approximately 1000-fold greater than the volume of the liquid.

8.5.2 Entropy change during heating at constant pressure or constant volume

The entropy change is defined by

$$\Delta S = \int \frac{dq_{rev}}{T} \tag{8.46}$$

For simple heating at constant pressure $dq_{rev} = nC_{p,m}\, dT$. For simple heating at constant volume $dq_{rev} = nC_{V,m}\, dT$. Thus, in general

$$\Delta S = \int_{T_i}^{T_f} \frac{n\, C_m\, dT}{T}. \tag{8.47}$$

If the temperature range is not too great, or if the heat capacity can otherwise be assumed to be constant, we simply pull it from the integral and integrate to obtain

$$\Delta S = n\, C_m \ln(T_2/T_1) \tag{8.48}$$

The temperature dependence of the heat capacity is discussed in more detail in Chapters 6 and 7. For liquids not too close to phase-transition temperatures nor the critical point, and for solids above their Debye temperature, the heat capacity is a weak function of temperature. In these cases, it can often be taken as a constant and pulled from the integral. For many gases, the heat capacity has been measured accurately over wide temperature ranges. The temperature dependence of $C^\circ_{p,m}$ is reproduced by the empirical *Shomate equation*. For some substances (see, for example the NIST WebBook) this function is known in its full form

$$C^\circ_{p,m} = A + Bt + Ct^2 + Dt^3 + Et^{-2}, \tag{8.49}$$

where $t = T/1000$, with T in K. The coefficients are independent of T. Often, it is sufficient to truncate this to a function of only two or three parameters. The heat capacities at constant pressure and at constant volume are related by

$$C_{V,m} = C_{p,m} - R. \tag{8.50}$$

Therefore, for simple heating at constant pressure or constant volume, in the absence of phase changes or reaction, the entropy change is calculated by integration with the appropriate value of the heat capacity substituted into Eq. (8.47). If phase changes do occur, then the process can be divided into multiple steps, as shown in Example 8.5.2.1.

8.5.2.1 Example

Calculate the change in molar entropy when water ice is heated from 250 K to form liquid water at 300 K. $C_{p,m} = 75.3\ \mathrm{J\ K^{-1}\ mol^{-1}}$ for the liquid and $37.7\ \mathrm{J\ K^{-1}\ mol^{-1}}$ for the solid. $\Delta_{fus}H^\circ_m = 6.02\ \mathrm{kJ\ mol^{-1}}$.

Think of this as a three-step process: heating the solid followed by a phase transition and then heating of the liquid.

$$H_2O(s), T = 250\ \mathrm{K} \xrightarrow[\Delta S_1]{1} H_2O(s), T = 273\ \mathrm{K} \xrightarrow[\Delta S_2]{2} H_2O(l), T = 273\ \mathrm{K} \xrightarrow[\Delta S_3]{3} H_2O(l), T = 300\ \mathrm{K}$$

The total entropy change is the sum of the entropy changes of the three steps.

$$\Delta S_m = \Delta S_1 + \Delta S_2 + \Delta S_3.$$

$$\Delta S_1 = \int_{250}^{273} \frac{C_{p,m}}{T}\, dT = 37.7 \ln\frac{273}{250} = 3.34\ \mathrm{J\ K^{-1}\ mol^{-1}}$$

$$\Delta S_2 = \frac{\Delta_{fus}H^\circ_m}{273\ \mathrm{K}} = 22.04\ \mathrm{J\ K^{-1}\ mol^{-1}}$$

$$\Delta S_3 = \int_{273}^{300} \frac{C_{p,m}}{T}\, dT = 75.3 \ln\frac{300}{273} = 7.06\ \mathrm{J\ K^{-1}\ mol^{-1}}$$

$$\Delta S_m = \Delta S_1 + \Delta S_2 + \Delta S_3 = 32.44\ \mathrm{J\ K^{-1}\ mol^{-1}}$$

Notice how the phase change has a much greater influence on the entropy change than either of the heating steps.

8.5.3 Entropy change when changing *T* and *V*

If both volume and temperature change, then we can go back to more fundamental expressions for dq_{rev} and imagine a two-step process in which temperature is changed in a first step and volume in a second step, as shown in Fig. 8.5. Start

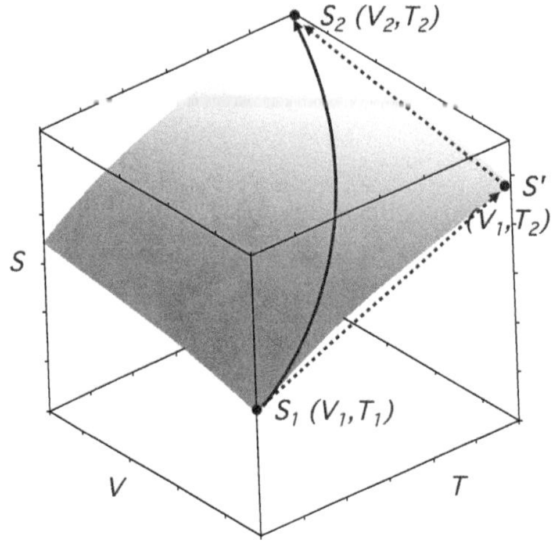

Figure 8.5 A process proceeding from the initial state at S_1 to the final state at S_2 in which both volume and temperature change (curved path) can be decomposed into a two-step process passing through point S'. The first step is isochoric while the temperature changes to its final value. The second step is isothermal while the volume changes to its final value.

with the First Law:

$$dU = đq + đw. \tag{8.51}$$

Along a reversible path with only pV work possible

$$dq_{\text{rev}} = dU + p\,dV. \tag{8.52}$$

Breaking the process into one isochoric step (in which the work is zero and the internal energy changes as the product of heat capacity and dT) and one isothermal step (in which pV work is done but there is no internal energy change), we can write for an ideal gas

$$dq_{\text{rev}} = nC_{V,\text{m}}\,dT + \frac{nRT\,dV}{V} \tag{8.53}$$

Now substitute into Eq. (8.46) and integrate to find ΔS

$$\Delta S = \int_{T_1}^{T_2} \frac{n\,C_{V,\text{m}}\,dT}{T} + \int_{V_1}^{V_2} \frac{nR\,dV}{V} \tag{8.54}$$

If $C_{V,\text{m}}$ is constant we can pull it out of the integral to obtain

$$\Delta S = n\,C_{V,\text{m}} \ln(T_2/T_1) + nR\ln(V_2/V_1). \tag{8.55}$$

Note that the answer is the same as the sum of two steps individually. If the heat capacity depends on temperature, Eq. (8.54) must be integrated with the proper functional form of $C_{V,\text{m}}$.

8.6 Maxwell's relations

To write the two-step process in its most general form we use *partial differentials,*

$$dS = \left(\frac{\partial S}{\partial T}\right)_V dT + \left(\frac{\partial S}{\partial V}\right)_T dV. \tag{8.56}$$

The question now is how do we evaluate the partial differentials. In the discussion of partial differentials in Chapter 7, we discussed two useful definitions

$$C_p \equiv \left(\frac{\partial H}{\partial T}\right)_p \tag{8.57}$$

$$C_V \equiv \left(\frac{\partial U}{\partial T}\right)_V. \tag{8.58}$$

To derive other useful partial differential expressions, we combine the *First Law*

$$dU = đq + đw \tag{8.59}$$

with the *Second Law*

$$q_{rev} = T\,dS \tag{8.60}$$

to obtain the *fundamental equation*

$$dU = T\,dS - p\,dV. \tag{8.61}$$

The fundamental equation is valid for both reversible and irreversible processes because it involves only state functions and the independent variables T and p. The differentials in Eq. (8.61) are exact differentials. We can divide by exact differentials to form partial derivatives. At constant volume, $dV = 0$, thus the partial derivative with respect to T of the fundamental equation is,

$$\left(\frac{\partial U}{\partial T}\right)_V = T\left(\frac{\partial S}{\partial T}\right)_V. \tag{8.62}$$

In Eq. (8.58) we have already seen that $(\partial U/\partial T)_V = C_V$, therefore by transitivity, the heat capacity at constant volume must also equal the right-hand side of Eq. (8.62).

This procedure shows that differentiation of the equation defining changes in the state function U leads to important results that are useful in calculating the changes in state functions when the system experiences changes in variables such at p, T and V. Another important state function that we have encountered is the enthalpy H.

$$dH = T\,dS + V\,dp \tag{8.63}$$

In Chapter 6 we also encountered the *Helmholtz energy A*

$$dA = -S\,dT - p\,dV \tag{8.64}$$

and the *Gibbs energy G*

$$dG = -S\,dT - V\,dp \tag{8.65}$$

Just as for dU, differentiation of these equations in terms of other state functions leads to a series of relationships between partial derivative.

Additional relationships between partial differentials can be derived by applying Legendre transformations to the state functions U, H, A and G. For those of you who like pure math, for any expression of exact differentials of the form

$$dF = X(x,y)\,dx + Y(x,y)\,dy \tag{8.66}$$

then

$$\left(\frac{\partial X}{\partial y}\right)_x = \left(\frac{\partial Y}{\partial x}\right)_y. \tag{8.67}$$

All of the quantities in Eqs (8.61) and (8.63)–(8.65) are exact differentials. Thus, the Legendre transformations can be used to create a set of relationships that are known as *Maxwell's relations*. They are collected here.

$$\frac{\partial^2 U}{\partial S\,\partial V} = \left(\frac{\partial T}{\partial V}\right)_S = -\left(\frac{\partial p}{\partial S}\right)_V \tag{8.68}$$

$$\frac{\partial^2 H}{\partial S\,\partial p} = \left(\frac{\partial T}{\partial p}\right)_S = \left(\frac{\partial V}{\partial S}\right)_p \tag{8.69}$$

$$\frac{\partial^2 A}{\partial V\,\partial T} = \left(\frac{\partial S}{\partial V}\right)_T = \left(\frac{\partial p}{\partial T}\right)_V \tag{8.70}$$

$$\frac{\partial^2 G}{\partial p\,\partial T} = -\left(\frac{\partial S}{\partial p}\right)_T = \left(\frac{\partial V}{\partial T}\right)_p \tag{8.71}$$

To demonstrate the utility of these relationships consider Eq. (8.56) applied to the entropy change of an ideal gas for which both T and V are changing. In Eq. (8.62) we have already seen that

$$(\partial S/\partial T)_V = C_V. \tag{8.72}$$

In Eq. (8.70) we see that $(\partial S/\partial V)_T = (\partial p/\partial T)_V$, and since we have an ideal gas for which $p = nRT/V$, then

$$(\partial p/\partial T)_V = (\partial(nRT/V)/\partial T)_V = nR/V. \tag{8.73}$$

Substituting into Eq. (8.56) yields

$$\mathrm{d}S = nC_{V,\mathrm{m}}\,\mathrm{d}T + (nR/V)\,\mathrm{d}V. \tag{8.74}$$

Integration leads to exactly the same results as in Eqs (8.54) and (8.55). The solutions are equivalent. The partial derivatives allow us to solve for any change in independent variable and for any material for which we have an equation of state. For changes in independent variables S and V, use the relations in Eq. (8.68). For changes in S and p, use Eq. (8.69). For changes in T and V, use Eq. (8.70). For changes in T and p, use Eq. (8.71).

8.6.1 Example

Calculate ΔS (for the system) when the state of 3.00 mol of ideal gas atoms, for which $C_{p,\mathrm{m}} = 5/2\ R$, is changed from 25 °C and 1.00 atm to 125 °C and 5.00 atm. How do you rationalize the sign of ΔS?

Entropy is a state function. Break up the transformation into two steps that are easily calculated. Take these two steps as constant pressure heating (which involves $C_{p,\mathrm{m}}$ and changes T_i to T_f) and constant temperature compression (which involves p_i and p_f as well as V_i and V_f).

Step 1: Constant pressure heating

$$\Delta S_1 = \int_{T_i}^{T_f} \frac{nC_{p,\mathrm{m}}}{T}\,\mathrm{d}t = nC_{p,\mathrm{m}} \int_{T_i}^{T_f} \frac{\mathrm{d}t}{T} = nC_{p,\mathrm{m}} \ln \frac{T_f}{T_i}$$

Step 2: Constant temperature compression

The internal energy of an ideal gas depends only on T. Here T is constant; thus, $\Delta U = 0$ and $q = -w$. Therefore,

$$\Delta S_2 = \int \frac{\mathrm{d}q}{T} = -\int \frac{\mathrm{d}w}{T} = \int_{V_i}^{V_f} \frac{p\,\mathrm{d}V}{T} = \int_{V_i}^{V_f} \frac{nR}{V}\,\mathrm{d}V = nR \int_{V_i}^{V_f} \frac{\mathrm{d}V}{V} = nR \ln \frac{V_f}{V_i}$$

$$\Delta S_2 = nR \ln \frac{V_f}{V_i} = nR \ln \frac{p_i}{p_f} = -nR \ln \frac{p_f}{p_i}$$

Note that in this step we use $\frac{V_f}{V_i} = \frac{p_i}{p_f}$ and $\ln \frac{p_i}{p_f} = -\ln \frac{p_f}{p_i}$ to arrive at the result.

The total entropy change is the sum of the two steps

$$\Delta S = \Delta S_1 + \Delta S_2 = nC_{p,\mathrm{m}} \ln \frac{T_f}{T_i} - nR \ln \frac{p_f}{p_i}$$

$$\begin{aligned}\Delta S &= (3\ \mathrm{mol}) \left(\tfrac{5}{2}\right) (8.314\ \mathrm{J\,K^{-1}\,mol^{-1}}) \ln \frac{398}{298} - (3)(8.314) \ln \frac{5}{1} \\ &= 18.04 - 40.14 = -22.1\ \mathrm{J\,K^{-1}}\end{aligned}$$

The negative ΔS value indicates that the system is changing to a configuration in which energy is dispersed into fewer states. Whereas the temperature change led to a favorable entropy change, the change in pressure causes a larger unfavorable change in entropy that dominates in this case.

8.6.2 Directed practice

Calculate the entropy change for the system in Example 8.6.1 if it is first compressed at constant temperature and then heated at constant pressure from 25 °C and 1.00 atm to 125 °C and 5.00 atm. [$-22.1\ \mathrm{J\,K^{-1}}$]

8.7 Calculating the natural direction of change

Consider an ideal gas in a chamber with thermally conductive walls and initial volume V_i held at temperature T_i, as shown in Fig. 8.6(a). What is the entropy change when the gas collapses to half the initial volume without a force acting upon it? With no force acting upon the system, the process is performed reversibly. Reversibility means that $p = p_{\mathrm{ext}}$.[1] The internal energy U is a function of T but not V for an ideal gas. Therefore, since T does not change U does not change

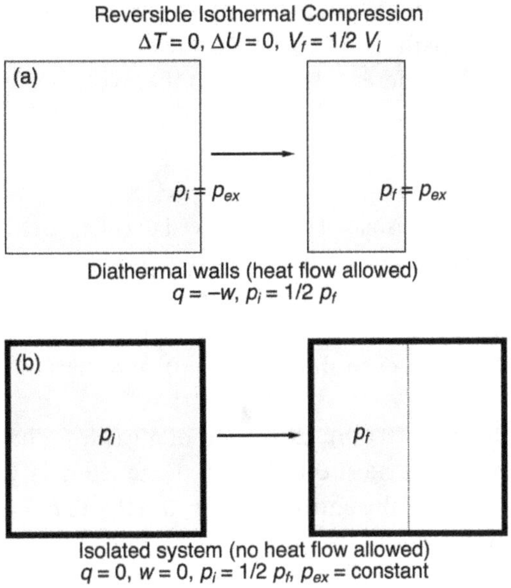

Figure 8.6 Reversible isothermal compression of an ideal gas is envisaged to occur in two different systems. (a) In a system in which heat exchange is allowed. (b) In an isolated system with no exchange of heat. In neither case is matter exchange allowed. The process in (a) can occur but is not spontaneous. It can only occur if we arrange for the external pressure to increase to twice its original value in a manner that allows the internal pressure to remain equal to it at all times. The process in (b) is not only nonspontaneous, it never occurs.

during collapse. From the First Law, $\Delta U = 0$ means $q = -w$. This work is performed and the heat is evolved along a reversible path, which allows us to calculate the change in entropy directly,

$$\Delta S_{\text{collapse}} = \int \frac{dq_{\text{rev}}}{T} = \frac{q_{\text{rev}}}{T} = -\frac{w_{\text{rev}}}{T} = nR\ln\left(\frac{\frac{1}{2}V_i}{V_i}\right) = -nR\ln 2 < 0 \tag{8.75}$$

This is the entropy change in the system, ΔS_{sys}. Since the process occurs reversibly, the heat flow out of the system is equal and opposite to the heat flow into the surroundings, $q_{\text{sys}} = -q_{\text{sur}}$. Therefore, the entropy change in the surroundings, which must be at the same temperature, is equal and opposite to the entropy change in the system, $\Delta S_{\text{sys}} = -\Delta S_{\text{sur}}$, and the total entropy change is zero,

In a reversible process

$$\Delta S_{\text{total}} = \Delta S_{\text{sys}} + \Delta S_{\text{sur}} = 0. \tag{8.76}$$

This process is reversible but not spontaneous. If the surroundings and the system were in true thermodynamic equilibrium, the pressure would be constant. At constant external pressure and with the external pressure in equilibrium with the internal pressure, no change in the volume of the system would occur. Nonetheless, we could engineer changes in the surroundings to affect a reversible change in the system. Such a reversible change is described here. The total entropy change of the universe for the reversible process of compressing the gas is zero.[2]

Now consider the reverse of the process occurring in Fig. 8.6(a), reversible expansion of a gas to twice its initial volume at constant temperature in thermal contact with the surroundings. The entropy change of the system must be equal and opposite to the compression in the first case,

$$\Delta S_{\text{expand}} = nR\ \ln\left(\frac{V_i}{\frac{1}{2}V_i}\right) = nR\ln 2 > 0 \tag{8.77}$$

because we have simply reversed the initial conditions and again allow the system to evolve reversibly. Thus, the change in entropy of the system, $\Delta S_{\text{sys}} = nR\ln 2$, is again equal to and opposite the entropy change in the surrounding, $\Delta S_{\text{sur}} = -nR\ln 2$, because both are at the same temperature and the heat flow is equal and opposite. We again have $\Delta S_{\text{total}} = \Delta S_{\text{sys}} + \Delta S_{\text{sur}} = 0$.

We know that we could very easily engineer the isothermal irreversible expansion or compression of a gas. If we were to place our container from Fig. 8.6(a) in contact with surroundings for which $p_{\text{ex}} = \frac{1}{2}p_i$ holding the wall fixed, then when we let go of the wall the gas will spontaneously expand. This must correspond to a process in which the entropy increases, just as in the reversible process. The work done in this irreversible process would have a smaller magnitude than the work done along the reversible path used to perform the calculation in Eq. (8.77),[3] as would the heat flow. Therefore, the entropy change will also be different. The result is that we can make a statement that is consistent with our intuition but we have not yet proved it.

The total entropy must increase in any spontaneous process.

Now consider the example in Fig. 8.6(b). An isolated gas is allowed to collapse to half of its volume inside of a container. The system is isolated and therefore adiabatic, $q = 0$. A gas never collapses upon itself unless a force acts upon it. It always occupies the complete volume available to it. If the number of moles of gas is constant, then the only way for the collapse described in panel (b) to occur is for the pressure to increase by a factor of two. At constant pressure, the gas would never collapse. This phenomenon is never observed even though $\Delta U = 0$, and there is no reason on energetic grounds for it not to happen.

Consider the reverse of the example in Fig. 8.6(b): expansion of a gas from half the volume of the container to the full volume under isothermal conditions. This most certainly is spontaneous. A gas always expands to occupy the full volume of the container. How do we calculate the entropy change if $\Delta U = 0$, $q = 0$ and $w = 0$ since the gas is essentially expanding into vacuum against zero pressure and the system is adiabatic? Conceptually, the best way is to take our gas as a monatomic ideal gas and return to the Sackur–Tetrode equation,

$$S = nR\left[\frac{3}{2}\ln\left(\frac{2\pi m k_B T}{h^2}\right) - \ln\left(\frac{N}{V}\right) + \frac{5}{2}\right]. \tag{8.78}$$

The entropy change is

$$\Delta S_{sys} = S_f - S_i = nR\left[-\ln\frac{N}{V_f} + \ln\frac{N}{V_i}\right] \tag{8.79}$$

$$\Delta S_{sys} = nR\ln\left(\frac{2V_i}{V_i}\right) = nR\ln 2 > 0 \tag{8.80}$$

As there is no heat flow, the change in entropy in the surroundings is zero, $\Delta S_{sur} = 0$, and the total entropy change is

$$\Delta S_{total} = \Delta S_{sys} + \Delta S_{sur} = nR\ln 2 > 0. \tag{8.81}$$

An equal and opposite entropy change results from the reverse process. Note that the result in Eq. (8.81) is the same as that in Eq. (8.77). As long as the initial and final states are the same in both cases (as they are here because they are at the same temperature and the volumes are the same), the result is the same. This is because entropy is a state function. Furthermore, the result in Eq. (8.77) did not assume the gas to be monatomic. This assumption is not essential because only the translational degree of freedom (and its contribution to entropy) is dependent on volume. Rotations and vibrations are not affected by volume changes. Therefore, we can generalize this result to any ideal gas. Further generalizations are possible, which allow us to state definitively the following principle:

For any irreversible process in an isolated system (no exchange of heat or work with the surroundings) there is a unique direction of spontaneous change: $\Delta S > 0$ for the spontaneous process, $\Delta S < 0$ for the opposite or nonspontaneous direction of change. $\Delta S = 0$ only for a reversible process.

Now take the isolated system to be the universe, for as far as we know, the universe cannot exchange matter or energy with anything outside of it.[4] Therefore, *for any process, a spontaneous change corresponds to one in which the total entropy change of the system plus surroundings is positive.* A system can change spontaneously and have its entropy decrease, but only if the concomitant change in the entropy of the surroundings is sufficiently positive such that the total entropy change is positive. This seems rather complicated and motivates us to find a single parameter that will describe the spontaneous direction of change. We shall see in Chapter 9 that the Gibbs and Helmholtz energies act as these parameters, the former for conditions of constant pressure and temperature, the latter for conditions of constant volume and temperature.

8.7.1 Clausius inequality

Consider the differential form of the first law in which only pV work is possible

$$dU = đq - p_{ex}\,dV \tag{8.82}$$

For a reversible process

$$dU = dq_{rev} - p\,dV \tag{8.83}$$

and since $dS = dq_{rev}/T$

$$dU = T\,dS - p\,dV. \tag{8.84}$$

However, T, S, p and V are state functions, thus Eq. (8.84) does not depend on path and is valid for both reversible and irreversible processes. This allows us to equate Eqs (8.82) and (8.83)

$$dq_{rev} - đq = (p - p_{ex})\,dV. \tag{8.85}$$

If the internal pressure is greater than the external pressure, $p - p_{ex} > 0$ and the system spontaneously expands, $dV > 0$. If the internal pressure is less than the external pressure, $p - p_{ex} < 0$ and the system spontaneously contracts, $dV < 0$. In both cases of a spontaneous change $(p - p_{ex})\,dV > 0$. A reversible process is characterized by $(p - p_{ex})\,dV = 0$. Now equate Eqs (8.82) and (8.84)

$$T\,dS - p\,dV = đq - p_{ex}\,dV \tag{8.86}$$

$$T\,dS - đq = (p - p_{ex})\,dV \ge 0. \tag{8.87}$$

From which we arrive at the *Clausius inequality*,

$$dS \ge \frac{đq}{T}. \tag{8.88}$$

The equal sign holds for reversible processes. The greater than sign hold for irreversible (spontaneous) processes.

SUMMARY OF IMPORTANT EQUATIONS

$\Delta U = q + w$	First Law of Thermodynamics. Internal energy change is sum of heat plus work.
$dS = dq_{rev}/T$	Second Law of Thermodynamics. Entropy change is the heat along a reversible path divided by temperature.
$S = k_B \ln \Omega$	Boltzmann formula: entropy S, number of microstates (weight of configuration) Ω
$\varepsilon = \lvert w \rvert / q_h$	Efficiency of heat engine ε, heat transfer on the hot leg q_h
$\varepsilon_{rev} = 1 - \frac{T_c}{T_h}$	Efficiency of reversible heat engine related to temperature on hot and cold leg
$\Delta S = nR\ln(V_2/V_1)$	Entropy change for reversible isothermal expansion of an ideal gas
$\Delta_{trs}S^\circ = \Delta_{trs}H^\circ/T_{trs}$	Entropy change of a phase transition
$\Delta S = \int_{T_i}^{T_f} \frac{n\,C_m\,dT}{T}$	Entropy change for simple heating. C_m is the molar heat capacity at constant pressure or constant volume as appropriate
$\Delta S = n\,C_m \ln(T_2/T_1)$	Entropy change during heating if heat capacity is independent of temperature
$\Delta S = \int_{T_1}^{T_2} \frac{n\,C_{V,m}\,dT}{T} + \int_{V_1}^{V_2} \frac{nR\,dV}{V}$	Entropy change when both T and V change
$dS \ge \frac{đq}{T}$	Clausius inequality
$C^\circ_{p,m} = A + Bt + Ct^2 + Dt^3 + Et^{-2}$	Shomate equation for the temperature dependence of molar heat capacity at constant pressure
$C_{V,m} = C_{p,m} - R$	Holds for an ideal gas

Exercises

8.1 The enthalpy of vaporization of benzene is 30.72 kJ mol^{-1} at its normal boiling point of 80.09 °C and 33.83 kJ mol^{-1} at 25 °C. (a) Calculate the entropy of vaporization of benzene at these temperatures assuming evaporation is performed reversibly. (b) Explain the trends in $\Delta_{vap}H^\circ_m$ and $\Delta_{vap}S^\circ_m$. How is this consistent with their asymptotic behavior at the critical point?

8.2 The enthalpy of vaporization of chloroform ($CHCl_3$) is 29.4 kJ mol^{-1} at its normal boiling point of 334.88 K. Calculate (a) the entropy of vaporization of chloroform at this temperature and (b) the entropy change of the surroundings.

8.3 Hurricane Katrina had an eye that was up to 75 miles in radius. The pressure at the center of the eye was as low as 920 mbar. The surface temperature at the eye was 36 °C. A high-pressure system that fed the eye had a pressure of 1035 mbar. The temperature at high altitude into which the gas from the eye expanded was 200 K. (a) Assuming that the eye was 500 m high and the number of moles of gas is constant, use a Carnot cycle to complete the following table. The heat capacity of the gas can be taken to be a constant with a value of that of N_2 ($C_{p,\mathrm{m}}$ = 29.2 J K^{-1} mol^{-1}). (b) How much water in moles and grams is evaporated to drive the cycle assuming all the heat is used to evaporate water? Is the assumption of a constant number of moles of gas a good assumption?

Step	***w*/J mol^{-1}**	***q*/J mol^{-1}**	***T*/K**	**Δ*U*/J mol^{-1}**	**Δ*S*/(J K^{-1} mol^{-1})**
A→B					
B→C					
C→D					
D→A					
Sum mol^{-1}					
Sum total					

8.4 Hurricane Irene had an eye that was up to 180 miles in diameter. The pressure at the center of the eye was as low as 942 mbar. The surface temperature at the eye was 35 °C. A high-pressure system that fed the eye had a pressure of 1030 mbar. The temperature at high altitude into which the gas from the eye expanded was 200 K. (a) Assuming that the eye was 500 m high and the number of moles of gas is constant, use a Carnot cycle to complete the following table. The heat capacity of the gas can be taken to be a constant with a value of that of N_2 ($C_{p,\mathrm{m}}$ = 29.21 J K^{-1} mol^{-1}). (b) How much water in moles and grams is evaporated to drive the cycle assuming all the heat is used to evaporate water? Is the assumption of a constant number of moles of gas a good assumption?

Step	***w*/J mol^{-1}**	***q*/J mol^{-1}**	***T*/K**	**Δ*U*/J mol^{-1}**	**Δ*S*/(J K^{-1} mol^{-1})**
A→B					
B→C					
C→D					
D→A					
Sum mol^{-1}					
Sum total					

8.5 State and justify with a thermodynamic argument why in the first step of hurricane formation a higher sea surface temperature leads to a stronger hurricane. How is the energy to create the storm transferred?

8.6 (a) Calculate ΔS for Ne (taken as an ideal gas) when 100 g of it are taken from 300 K and 100 kPa to 1500 K and 300 kPa. (b) Discuss the sign of ΔS and the terms giving rise to it.

8.7 The heat capacity of a monatomic ideal gas is $C_{p,\mathrm{m}} = \frac{5}{2}R$. The heat capacity of Br_2(g) is $C_{p,\mathrm{m}} = A + BT + CT^2$ with $A = 38.527$, $B = -1.977 \times 10^{-3}$, and $C = 1.526 \times 10^{-6}$. The units are such that if temperature T is inserted in kelvin, the equation gives $C_{p,\mathrm{m}}$ in J K^{-1} mol^{-1}. (a) Calculate the entropy change in the ideal gas when it is heated from 350 K to 1000 K. (b) Calculate the entropy change when Br_2(g) is heated from 350 K to 1000 K. (c) Explain in terms of the characteristics of the two gases the sign of ΔS and why the two values of ΔS are different.

8.8 The heat capacity of NO_2(g) is given by $C^\circ_{p,\mathrm{m}} = A + BT + CT^2$ with $A = 16.109$, $B = 7.590 \times 10^{-2}$, and $C = -5.439 \times 10^{-5}$. The units are such that if temperature T is inserted in kelvin, the equation gives $C^\circ_{p,\mathrm{m}}$ in J K^{-1} mol^{-1}. Calculate the standard absolute entropy of NO_2(g) at 1200 K.

8.9 The standard molar entropy of NH_3(g) is 192.45 J K^{-1} mol^{-1} at 298 K, and that of Ag(s) is 42.55 J K^{-1} mol^{-1}. The heat capacities of both species follow an equation of the form $C_{p,\mathrm{m}} = d + eT + fT^{-2}$ with T given in K, $C_{p,\mathrm{m}}$ will have the units of J K^{-1} mol^{-1} if the following coefficients are used: NH_3: $d = 29.75$, $e = 2.51 \times 10^{-2}$, and $f = -1.55 \times 10^5$.

Ag: $d = 21.3$, $e = 8.54 \times 10^{-3}$, and $f = 1.51 \times 10^{5}$. Calculate the standard molar entropy of NH_3 and Ag at 960 °C (note that Ag melts at 962 °C). Compare these two values to each other and their values at 298 K.

8.10 The standard molar entropy of $Cl_2(g)$ is 223.08 J K^{-1} mol^{-1} at 298 K, and that of Fe(s) is 25.10 J K^{-1} mol^{-1}. The heat capacities of both species follow the Shomate equation. The coefficients for both are given below. Calculate the standard molar entropy of Cl_2 and Fe at 600 K. Compare and discuss the trends in these two values to each other and their values at 298 K.

	A	*B*	*C*	*D*	*E*
Cl_2	33.1	12.2	−12.1	4.39	−0.159
Fe	25.7	1.80	6.59	16.5	−0.194

8.11 For a geothermal system with 37 wells that are 41.8 m deep and 3.81 cm in diameter. Calculate the mass of water per second that is required to deliver 1 MW of heating. Calculate the compressor work to deliver this heat.

8.12 Assuming the change in molar volume is negligible, calculate ΔH, ΔU, ΔS, q and w when 14.0 g of Si melts at its normal melting point of 1687 K. $\Delta_{\text{fus}}H^\circ_{\text{m}} = 50.2$ kJ mol^{-1}.

Endnotes

1. The external pressure is increasing but it is increasing in infinitesimal steps. There is no net force because $p = p_{\text{ex}}$.
2. If we have to engineer a pressure rise (rather than waiting for a totally improbably pressure fluctuation that just happens to match these conditions), there must be some work and/or heat flow associated with that, as well as an accompanying entropy change somewhere in the universe. Here, we are just considering the interaction of the system and its immediate surroundings as our universe.
3. It is important to mention the magnitude because work is a signed quantity. Remember that work done by the system on surroundings is negative.
4. Which naturally begs the question, what is outside of the universe? But that we will leave to philosophers and cosmologists.

Further reading

Berry, R.S., Rice, S.A., and Ross, J. (2000) *Physical Chemistry*, 2nd edition. Oxford University Press, New York.
Metiu, H. (2006) *Physical Chemistry: Thermodynamics*. Taylor & Francis, New York.
NIST WebBook, http://webbook.nist.gov/chemistry/.

CHAPTER 9

Third law of thermodynamics and temperature dependence of heat capacity, enthalpy and entropy

PREVIEW OF IMPORTANT CONCEPTS

- At constant T and p, the change in Gibbs energy determines the direction of change of a reaction.
- At constant T and V, the change in Helmholtz energy determines the direction of change of a reaction.
- For any spontaneous isothermal change, the Helmholtz energy (at constant V) or Gibbs energy (at constant p) must decrease.
- The Gibbs energy change during a reaction is related to the enthalpy and entropy changes by $\Delta_r G° = \Delta_r H° - T\Delta_r S°$.
- ΔG is the maximum nonexpansion (e.g., electrical) work that can be done by the system at constant p and T.
- The change in the Helmholtz energy is the maximum amount of work that can be extracted from a system at constant temperature and volume.
- $\Delta_r G°$ and $\Delta_r H°$ can be calculated from values obtained from formation reactions. $\Delta_r S°$ is obtained from absolute entropies.
- The temperature dependence of $\Delta_r H°$ and $\Delta_r S°$ are related to temperature dependence of $C_{p,m}$ (the heat capacity) for the species involved in the reaction.
- The temperature dependence of $\Delta_r G°$ is related to temperature dependence of $\Delta_r H°$.
- For any system in internal equilibrium undergoing an isothermal process, as the temperature approaches absolute zero, both the heat capacity and ΔS approach zero; therefore $\Delta_r G° \rightarrow \Delta_r H°$.
- The entropy of a pure, perfectly crystalline substances is zero at 0 K.
- A state with a temperature of absolute zero is unattainable through any process composed of a finite number of steps.

9.1 When and why does a system change?

Having developed the concepts of entropy and enthalpy, we now seek to define a parameter that will tell us which way a reactive system will go: to the right toward products, or to the left to the reactants. We would also like to know what the equilibrium composition of the system will be. We will find that this parameter is the Gibbs energy for a system held at constant temperature and pressure, or the Helmholtz energy for a system held at constant volume and temperature. These energies were defined in Eqs (8.64) and (8.65) and will be derived in the next section.

We will also develop the related concept of chemical potential. The chemical potential plays the role of any other potential. You already know intuitively that in a gravitational potential a body will fall to the lowest possible gravitational potential energy as long as there is no barrier along the pathway. You can put a vase on a shelf. The shelf provides the barrier that keeps the vase from falling to the minimum of gravitational energy (the floor), at least as long as your cat does not intervene. The chemical potential acts in a similar way for a chemical system. Here, we will forget about any potential barriers. Reaction thermodynamics being path-independent tells us nothing about them anyway. We will discuss the formation and presence of such barriers in the chapters on kinetics and dynamics. Here, we concentrate on the way a chemical system develops either in the absence of such barriers or at sufficiently high temperature that any barriers can be overcome without creating bottlenecks.

Physical Chemistry: How Chemistry Works, First Edition. Kurt W. Kolasinski.

Companion Website: www.wiley.com/go/kolasinski/physicalchemistry

The change in Gibbs energy is the factor that must be calculated to determine the direction of change of a reaction. Another way of phrasing this is that it is a measure of chemical affinity, and we shall prove that in the next chapter. But Gibbs energy changes are difficult to measure if the only two methods at your disposal are equilibrium concentration measurements and the measurement of voltages coupled with application of the Nernst equation (which we discuss in Chapter 15). These were the only two methods known at the end of the 19th century. Henri le Châtelier (as quoted in Coffey, *Cathedrals of Science*) stated the importance of determining the Gibbs energy function as follows.

"The determination of the nature of this function would lead to a complete knowledge of the laws of equilibrium. It would permit us to determine *a priori*, independently of any new experimental data, the full conditions of equilibrium corresponding to a chemical reaction."

Subsequently, G.N. Lewis developed a method for its calculation but his method could not be applied in practice because a constant of integration could neither be measured nor calculated.

Nernst asserted the Third Law of Thermodynamics, which is what we need to determine the constant that eluded Lewis, based on chemical intuition and extrapolations of Gibbs energy and enthalpy changes. But it was left to Einstein to prove that the Nernst heat theorem as it was called was truly the Third Law of Thermodynamics. Nernst's heat theorem requires that the heat capacities of solids in their lowest disorder state, which is a perfect crystal, must go to zero at low temperature. Planck, as we will show in Chapter 19, was able to solve the problem of describing the spectrum of thermal black body radiation by introducing his quantum postulate, which states that vibrational energy is quantized. We have used this idea in the development of statistical mechanics. Einstein applied the quantum postulate to the vibrations of a crystal, and the result delivered a very good approximation to the low-temperature behavior of the heat capacity. Debye would later correct Einstein's model to make it quantitatively accurate over a wider temperature range. Nonetheless, Einstein's model of heat capacities delivered the proof that was required to turn Nernst's heat theorem into the Third Law. It also makes clear that the description of the temperature dependence of heat capacities delivers perhaps the most important aspects of understanding chemical thermodynamics. As we shall see, it is from an accurate description of the temperature dependence of heat capacities that we can measure absolute entropies as well as the temperature dependence of both enthalpy and Gibbs energy. We will show shortly that the change in the Gibbs energy upon reaction determines the equilibrium constant. The description of heat capacities provides the bridge that links the quantum mechanical description of energy level structure to thermodynamics.

9.2 Natural variables of internal energy

By combining the First and Second Laws of Thermodynamics, we can express the change in internal energy in terms of T, S, p and V as

$$dU = T\,dS - p\,dV \tag{9.1}$$

This is known as the *fundamental equation*. This set of variables contains four state variables, as opposed to the path variables δq and δw embodied in the First Law. Therefore, Eq. (9.1) must be applicable to all processes, regardless of whether they are reversible or irreversible.

Equation (9.1) has another interestingly simple property. At constant S, changes in U are related to changes in V by the pressure. That is, a plot of U versus V would have a slope of $-p$, or in mathematical terms,

$$(\partial U/\partial V)_S = -p. \tag{9.2}$$

Similarly, the internal energy at constant volume but changing entropy changes simply as the temperature,

$$(\partial U/\partial S)_V = T. \tag{9.3}$$

In this case the partial derivatives have particularly simple forms: they correspond to a state variable (or the negative of a state variable). From this simplicity of form for relating changes of internal energy by way of Eq. (9.1), we call S and V the *natural independent variables* for the function U. The natural variables are those variables that allow us to determine all other thermodynamic properties by differentiation.

We could, of course, express the internal energy in terms of changes in T and V. Using the appropriate partial derivatives (and results from Eq. (9.1) and another from the Maxwell relations) we find

$$dU = \left(\frac{\partial U}{\partial T}\right)_V dT + \left(\frac{\partial U}{\partial V}\right)_T dV, \tag{9.4}$$

$$= C_V\,dT + \left[T\left(\frac{\partial S}{\partial V}\right)_T - p\right] dV, \tag{9.5}$$

$$= C_V\,dT + \left[T\left(\frac{\partial p}{\partial T}\right)_V - p\right] dV. \tag{9.6}$$

Thus, upon 'simplification' we arrive at a complex function rather than one simply in terms of state variables. Equations (9.1) and (9.6) illustrate that new thermodynamic functions can be defined so as to provide the simplest functional form. Next, we define two new thermodynamic functions of particular importance for determining the natural direction of evolution for a system.

9.3 Helmholtz and Gibbs energies

There are two sets of conditions that are frequently encountered experimentally: constant T and V or constant T and p. Let us now derive new thermodynamic state functions for these two sets of conditions. From the Clausius inequality we have

$$dS - đq/T \geq 0 \tag{9.7}$$

for any reversible (corresponding to the equal sign) or irreversible process (corresponding to the greater than sign). Remember $dU = đq + đw$. Therefore, at constant volume and no non-pV work, $dq = dU$, whereas for constant pressure and no non-pV work, $dq = dH$. Hence, we can develop two parallel derivations, one at constant volume and the other at constant pressure.

Constant volume	*Constant pressure*
$dS - \frac{dU}{T} \geq 0$ (9.8)	$dS - \frac{dH}{T} \geq 0$ (9.9)

We define two new thermodynamic functions

$A = U - TS$ (9.10)	$G = H - TS$ (9.11)

We call the functions the *Helmholtz energy A* and the *Gibbs energy G*. At constant temperature, the two state functions change as

$dA = dU - T\,dS$ (9.12)	$dG = dH - T\,dS$ (9.13)

Introducing these into the inequalities that follow from the Clausius inequality, we obtain

$dA_{T,V} \leq 0$ (9.14)	$dG_{T,p} \leq 0$ (9.15)

The choices for these two new state functions were not arbitrary. Instead we see that Eqs (9.14) and (9.15) can be summarized in words as:

For any spontaneous isothermal change, the Helmholtz energy (at constant V) or Gibbs energy (at constant p) must decrease.

A system will always move to a state of lower A or G if a path is available, as shown in Fig. 9.1. This is because a system is always trying to increase the entropy of the universe, that is, the system + the surroundings. Therefore, the condition that defines *equilibrium* is

$dA_{T,V} = 0$ (9.16)	$dG_{T,p} = 0.$ (9.17)

9.3.1 Example

How much work can be extracted from glucose metabolism performed reversibly at 25 °C given that $\Delta_r U^\circ_m = -2801.3$ kJ mol^{-1} and $\Delta_r S^\circ_m = 260.7$ J K^{-1} mol^{-1}?

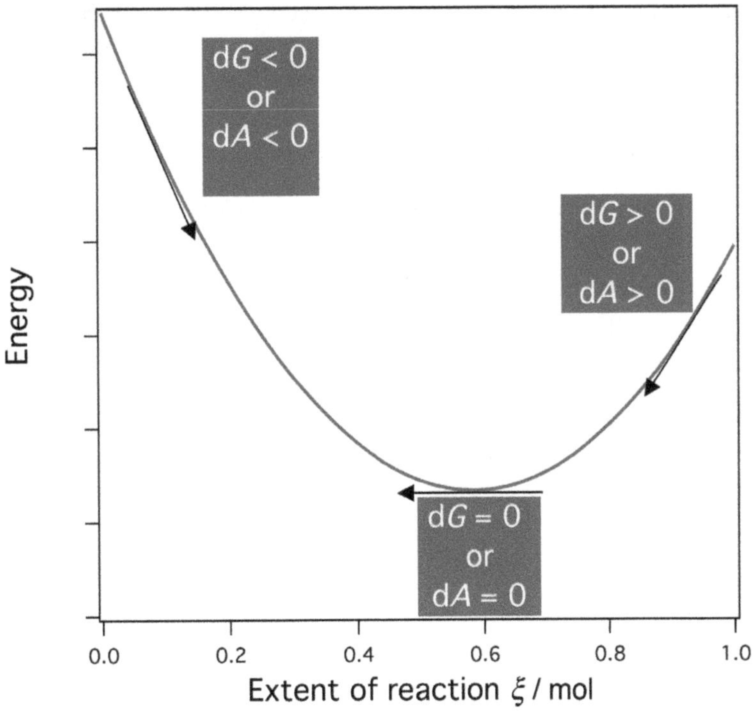

Figure 9.1 A spontaneous process is downhill in either the Gibbs energy, $dG < 0$ at constant pressure and temperature, or Helmholtz energy, $dA < 0$ at constant volume and temperature. Chemical equilibrium is attained when either the Gibbs or Helmholtz energy has reached a minimum, $dG = 0$ or $dA = 0$. The extent of reaction ξ is defined in Chapter 10. In the current example, it is 0 when no reaction has taken place. When the reaction goes to completion, it is 1 mole. When the system settles in to an equilibrium, ξ has some intermediate value.

Treat the body as a constant volume system. The Helmholtz energy is given by

$$dA = dU - T\,dS = \left(dq_{\text{rev}} + dw_{\text{rev}}\right) - dq_{\text{rev}}.$$

Here we have used the first law to substitute for dU. In the second term, we used $dq_{\text{rev}} = T\,dS$. Hence, $dA = dw_{\text{rev}}$ or

$$\Delta A = w_{\text{rev}} = \Delta U - T\Delta S$$

for a finite process. Therefore, the change in the Helmholtz energy is the maximum amount of work that can be extracted from a system at constant temperature and volume. Now calculate the change in Helmholtz energy

$$\begin{aligned}\Delta_r A^\circ_m &= \Delta_r U^\circ_m - T\Delta_r S^\circ_m = -2801.3\ \text{kJ mol}^{-1} - (298\ \text{K})(260.7 \times 10^{-3}\ \text{kJ K}^{-1}\ \text{mol}^{-1})\\ &= -2879.0\ \text{kJ mol}^{-1}\end{aligned}$$

9.4 Standard molar Gibbs energies

A balanced stoichiometric chemical reaction is written with *stoichiometric coefficients a, b,* and so forth, which are all positive numbers

$$A + 2B \rightarrow 3Y + 4Z.$$

The IUPAC defines *stoichiometric numbers* ν_A, ν_B, etc. as signed quantities, which are negative for reactants and positive for products. Thus, for the equation written above, $\nu_A = -1$, $\nu_B = -2$, $\nu_Y = 3$, $\nu_Z = 4$.

Equation (9.11) applies to each substance in a chemical system; therefore changes in the Gibbs energy can be calculated from the changes in enthalpy and entropy that accompany the reaction. Standard entropies and enthalpies of reaction are combined to obtain the *standard Gibbs energy of reaction* according to

$$\Delta_r G^\circ = \Delta_r H^\circ - T\Delta_r S^\circ \tag{9.18}$$

In a *standard formation reaction* there is one and only one product and it has a stoichiometric coefficient of unity. The reactants are elements in their standard state. We can relate the change in Gibbs energy of any reaction to the Gibbs energies of formation of reactants and products. We multiply the Gibbs energy of formation of each species by the corresponding stoichiometric coefficient. We then sum all the values of the products, sum all the values of the reactants, and then subtract the reactant sum from the product sum,

$$\Delta_r G^\circ = \sum_{\text{products}} \nu\, \Delta_f G^\circ - \sum_{\text{reactants}} \nu\, \Delta_f G^\circ. \tag{9.19}$$

More formally correct, we use the fact that the stoichiometric number of the jth species is a signed quantity to write this as

$$\Delta_r G^\circ = \sum_j \nu_j\, \Delta_f G_j^\circ \tag{9.20}$$

By convention, $\Delta_f G^\circ$ for an element in its standard state is zero at all temperatures. Similar relationships are used to relate reaction enthalpy and entropy changes to enthalpies of formation and standard entropies

$$\Delta_r H^\circ = \sum_j \nu_j\, \Delta_f H_j^\circ \tag{9.21}$$

$$\Delta_r S^\circ = \sum_j \nu_j\, S_j^\circ. \tag{9.22}$$

Whereas $\Delta_f H^\circ$ for an element in its standard state is also defined as zero at all temperatures, we shall see below that according to the Third Law of Thermodynamics absolute entropies S° are zero for a perfect crystal at absolute zero.

9.4.1 Example

Determine $\Delta_r G_m^\circ$ (298 K) for the combustion of C_2H_5OH given that $\Delta_f G_m^\circ$ (298 K) for $C_2H_5OH(l)$, $H_2O(l)$ and $CO_2(g)$ are −175, −237 and −394 kJ mol^{-1}, respectively.

First, write a balanced chemical reaction

$$C_2H_5OH(l) + 3O_2(g) \rightarrow 2CO_2(g) + 3H_2O(l).$$

Then apply this stoichiometry to Eq. (9.20)

$$\Delta_r G_m^\circ = 2 \times \Delta_f G_m^\circ(CO_2) + 3 \times \Delta_f G_m^\circ(H_2O) - \Delta_f G_m^\circ(C_2H_5OH)$$
$$\Delta_r G_m^\circ = -1324 \text{ kJ mol}^{-1} (\text{of } C_2H_5\text{ OH at 298 K}).$$

The 'per mole' in the Gibbs energy change refers to 'per mole of the reaction as written,' which in this case happens to be per mole of ethanol combusted but per 3 mol O_2 reacted.

9.4.2 Directed practice

How many products and how much product is made in a standard formation reaction?

9.5 Properties of the Gibbs energy

How does the Gibbs energy change with temperature and pressure? As for any function, we can write this dependence abstractly in terms of partial differential equations

$$dG_m(T,p) = \left(\frac{\partial G_m}{\partial T}\right)_p dT + \left(\frac{\partial G_m}{\partial p}\right)_T dp \tag{9.23}$$

We now evaluate the coefficients represented by the partial derivatives and their functional form. The Gibbs energy is defined by

$$G_m = H_m - TS_m. \tag{9.24}$$

The molar Gibbs energy is a state property (path independent) as are H_m, T, and S_m. For infinitesimal changes, we obtain by differentiation

$$dG_m = dH_m - T\,dS_m - S_m\,dT \tag{9.25}$$

By differentiating the definition of enthalpy, $H_m = U_m + pV_m$, we know that

$$dH_m = dU_m + p\,dV_m + V_m\,dp. \tag{9.26}$$

For a closed system doing no non-expansion work, we can substitute for dU_m from the fundamental equation, Eq. (9.1) above. We then substitute for dH_m in Eq. (9.25) from Eq. (9.26) to find

$$dG_m = V_m\,dp - S_m\,dT \tag{9.27}$$

Therefore, G_m is a function of p and T and the form of this function is given by the partial derivatives

$$(\partial G_m/\partial T)_p = -S_m \tag{9.28}$$

and

$$(\partial G_m/\partial p)_T = V_m. \tag{9.29}$$

Because $S_m > 0$ for all substances, G_m always decreases with increasing T (at constant pressure and composition), as shown in Fig. 9.2. Larger changes in G_m occur with larger values of S_m; therefore, G_m is more sensitive to T changes when gases are involved in the process than when condensed phases are the only ones present.

Similarly $V_m > 0$ for all substances. Therefore, G_m increases when p increases, as shown in Fig. 9.3. The molar Gibbs energy is more sensitive to changes in p for gases than for condensed phases.

9.5.1 Gibbs–Helmholtz equation

From the dependence of G on T in Eq. (9.28) and the definition of G in Eq. (9.11)

$$\left(\frac{\partial G}{\partial T}\right)_p = -S = \frac{G-H}{T} \tag{9.30}$$

We shall see later that the equilibrium constant K is related to $\Delta G_m^\circ/T$ by

$$\Delta_r G_m^\circ = -RT \ln K. \tag{9.31}$$

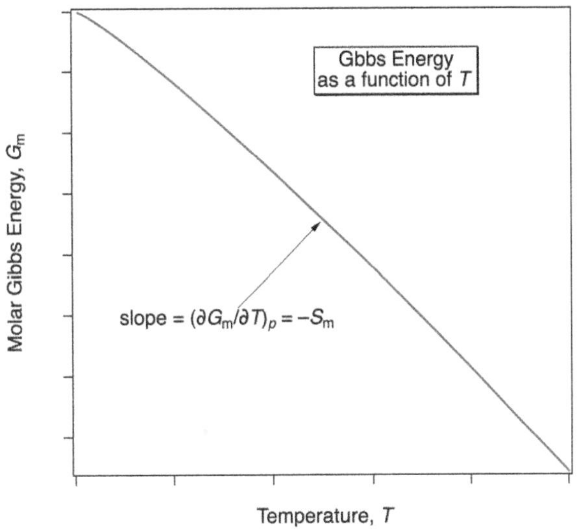

Figure 9.2 The molar Gibbs energy is a decreasing function of temperature at constant pressure. As the molar entropy is of greater magnitude for gases than that of liquids and solids (away from the critical point), the effect of temperature is much greater for gases than condensed phases.

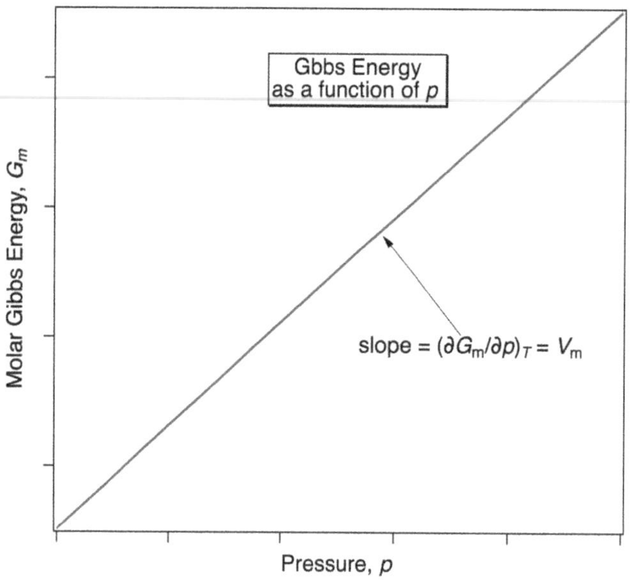

Figure 9.3 The molar Gibbs energy is an increasing function of pressure at constant temperature. As the molar volume of gases is much greater than that of liquids and solids (away from the critical point), the effect of pressure is much greater for gases than condensed phases.

Therefore by determining how G/T changes with temperature, we can also determine how K changes with T. Using the chain rule to differentiate G/T, we obtain

$$\left(\frac{\partial}{\partial T}\frac{G}{T}\right)_p = \frac{1}{T}\left(\frac{\partial G}{\partial T}\right)_p + G\left(\frac{\mathrm{d}}{\mathrm{d}T}\frac{1}{T}\right). \tag{9.32}$$

A useful result to remember is

$$\frac{\mathrm{d}(1/T)}{\mathrm{d}T} = -\frac{1}{T^2}. \tag{9.33}$$

Substituting into Eq. (9.32) from Eq. (9.30) we obtain the differential form of the *Gibbs–Helmholtz equation*

$$\left(\frac{\partial}{\partial T}\frac{G}{T}\right)_p = -\frac{H}{T^2} \tag{9.34}$$

Sometimes written as

$$\left(\frac{\partial(G/T)}{\partial(1/T)}\right)_p = H \tag{9.35}$$

That is, a plot of G/T versus $1/T$ has slope H. The same result holds for ΔG with ΔH in place of H. Thus, by integrating Eq. (9.35) we can find the value of ΔG at any temperature T_2 from a value at a reference temperature T_1 according to

$$\int_{T_1}^{T_2} \mathrm{d}\left(\frac{\Delta G}{T}\right) = \int_{T_1}^{T_2} \Delta H\,\mathrm{d}\left(\frac{1}{T}\right),$$
$$\frac{\Delta G(T_2)}{T_2} = \frac{\Delta G(T_1)}{T_1} + \int_{T_1}^{T_2} \Delta H\,\mathrm{d}\left(\frac{1}{T}\right) \tag{9.36}$$

If ΔH is constant it can be pulled from the integral to yield the *integrated form of the Gibbs–Helmholtz equation*

$$\frac{\Delta G(T_2)}{T_2} = \frac{\Delta G(T_1)}{T_1} + \Delta H\left(\frac{1}{T_2} - \frac{1}{T_1}\right). \tag{9.37}$$

9.5.1.1 Directed practice

Derive the Gibbs–Helmholtz equation in Eq. (9.34) by substituting Eq. (9.30) into Eq. (9.32).

9.5.1.2 Example

Calculate the change in molar Gibbs energy if the pressure is increased from 1.0 bar to 2.0 bar at 298 K for (a) liquid water (incompressible liquid) (b) water vapor (ideal gas).

The molar Gibbs energy depends on changes in pressure and temperature according to

$$dG_m = V_m\,dp - S_m\,dT.$$

The processes are isothermal, therefore $dT = 0$ and

$$dG_m = V_m\,dp.$$

$$\int dG_m = G_m(p_2) - G_m(p_1) = \Delta G_m = \int V_m\,dp$$

incompressible liquid (same for solid)	ideal gas
$V_m = \text{constant} = 1.80 \times 10^{-5}\ \text{m}^3\ \text{mol}^{-1}$	$V_m = RT/p$
$\Delta G_m = \int V_m\,dp = V_m \int dp$	$\Delta G_m = \int V_m\,dp = \int \frac{RT}{p}\,dp$
$\Delta G_m = V_m(p_2 - p_1)$	$\Delta G_m = RT\ln(p_2/p_1)$
$= 1.8\ \text{J mol}^{-1}$	$1.7\ \text{kJ mol}^{-1}$

For the incompressible liquid, the molar volume is a constant and can be pulled from the integral. For the ideal gas, we must substitute explicitly for the molar volume and integrate. The change is positive in both cases but about 1000-fold larger for the gas.

9.5.2 Gibbs energy of an ideal gas at arbitrary pressure

If the initial pressure is taken to be the standard state pressure, $p° = 1$ bar, we may write

$$\Delta G_m = G_m(p_2) - G_m(p_1) = RT\ln(p_2/p_1),$$
$$G_m(p_2) = G_m(p°) + RT\ln(p_2/p°)$$
$$G_m(p) = G_m^{\circ} + RT\ln(p/p°) \tag{9.38}$$

In other words, taking the molar Gibbs energy of an ideal gas at the standard state pressure as the standard molar Gibbs energy G_m°, then the Gibbs energy of an ideal gas at any pressure p is related to the standard state value plus the $RT\ln(p/p°)$ term.

9.5.3 Gibbs energy and electrical work

Equation (9.1) is valid if only pV work is done. In an electrochemical system electrons are generated and electrical work w_{el} is done. Consequently, we have to modify the fundamental equation to include this term

$$dU = T\,dS - p\,dV + dw_{el}. \tag{9.39}$$

Therefore

$$dG = V\,dp - S\,dT + dw_{el}. \tag{9.40}$$

At constant p and T for a finite process,

$$\Delta G = w_{el,rev} = w_{el,max}. \tag{9.41}$$

Hence, ΔG is the maximum nonexpansion (e.g., electrical) work that can be done by the system at constant p and T.

9.5.3.1 Example

In a methane fuel cell, CH_4 undergoes the same redox reaction as in the combustion process to produce CO_2 and H_2O and generates electricity. Calculate the maximum electrical work that can be obtained from 1 mol of CH_4 at 25 °C.

Start with a balanced reaction

$$CH_4(g) + 2O_2(g) \rightarrow CO_2(g) + 2H_2O(l)$$

Using tables to find enthalpies of formation $\Delta_f H_m^{\circ}$ and standard molar entropies S_m°, $\Delta_r H_m^{\circ} = -890.3\ \text{kJ mol}^{-1}$ and $\Delta_r S_m^{\circ} = -242.8\ \text{J K}^{-1}\ \text{mol}^{-1}$ are calculated.

$$\Delta_r G_m^{\circ} = \Delta_r H_m^{\circ} - T\Delta_r S_m^{\circ} = -890.3 - (300)(-242.8 \times 10^{-3}) = -818.0\ \text{kJ mol}^{-1}$$
$$w_{el,max} = n\,\Delta_r G_m^{\circ} = -818.0\ \text{kJ}.$$

The work done is less than the heat generated. This is because of the unfavorable entropy change. If the enthalpy of combustion were used to do work in a heat engine, then the efficiency would be limited, as in a Carnot engine. However, in theory, 100% of the Gibbs energy released in a fuel cell is available for work. Therefore, a fuel cell is in principle more efficient than any internal combustion engine because it is not limited by the Second Law.

9.6 The temperature dependence of $\Delta_r C_p$ and H

Let us assume that in changing temperature from T_1 to T_2, none of the reactants or products in a reaction undergoes a phase change. *Enthalpy* H is a state function defined by

$$H = U + pV. \tag{9.42}$$

Enthalpy is a state function that depends on p and T according to

$$\mathrm{d}H = \left(\frac{\partial H}{\partial p}\right)_T \mathrm{d}p + \left(\frac{\partial H}{\partial T}\right)_p \mathrm{d}T = \left(\frac{\partial H}{\partial p}\right)_T \mathrm{d}p + C_p\,\mathrm{d}T. \tag{9.43}$$

We have used the definition of heat capacity at constant pressure, $C_p \equiv (\partial H/\partial T)_p$. Now consider specifically an isobaric system for which $\mathrm{d}p = 0$ and therefore

$$\mathrm{d}H = C_p\,\mathrm{d}T.$$

As before, we can integrate to determine how the enthalpy at reference temperature T_1 is related to the enthalpy at T_2,

$$H(T_2) = H(T_1) + \int_{T_1}^{T_2} C_p\,\mathrm{d}T \tag{9.44}$$

As discussed in detail in Chapter 7, the heat capacity can be fitted to a polynomial that describes its dependence on temperature. Here, we truncate the Shomate equation to just three terms as

$$C^\circ_{p,\mathrm{m}} = a + bT + eT^{-2}. \tag{9.45}$$

If $C_{p,\mathrm{m}}$ has the form of Eq. (9.45) and we substitute this into Eq. (9.44), then we find

$$H_\mathrm{m}(T_2) = H_\mathrm{m}(T_1) + \int_{T_1}^{T_2} (a + bT + eT^{-2})\,\mathrm{d}T. \tag{9.46}$$

We indicate molar units for enthalpy for clarity. Integration yields

$$H_\mathrm{m}(T_2) = H_\mathrm{m}(T_1) + a(T_2 - T_1) + (b/2)\left(T_2^2 - T_1^2\right) - e\left(T_2^{-1} - T_1^{-1}\right) \tag{9.47}$$

Since Eq. (9.46) is true for each individual substance in the reaction, we can write for the change in enthalpy of reaction with temperature

$$\Delta_r H^\circ_\mathrm{m}(T_2) = \Delta_r H^\circ_\mathrm{m}(T_1) + \int_{T_1}^{T_2} \Delta_r C^\circ_{p,\mathrm{m}}\,\mathrm{d}T \tag{9.48}$$

where

$$\Delta_r C^\circ_{p,\mathrm{m}} = \sum_j \nu_j\, C^\circ_{p,\mathrm{m},j}. \tag{9.49}$$

Equation (9.48) is known as *Kirchhoff's law*. For liquids far from the critical point and solids above the Debye temperature, $C^\circ_{p,\mathrm{m}}$ is a very weak function of temperature. For reactions involving only such materials, it is a good approximation that $\Delta_r C^\circ_{p,\mathrm{m}}$ is independent of temperature. For gases, this is not as good of an approximation unless the gases that appear in the reactants and products are similar or the temperature range is small, as shown in Fig. 9.4.

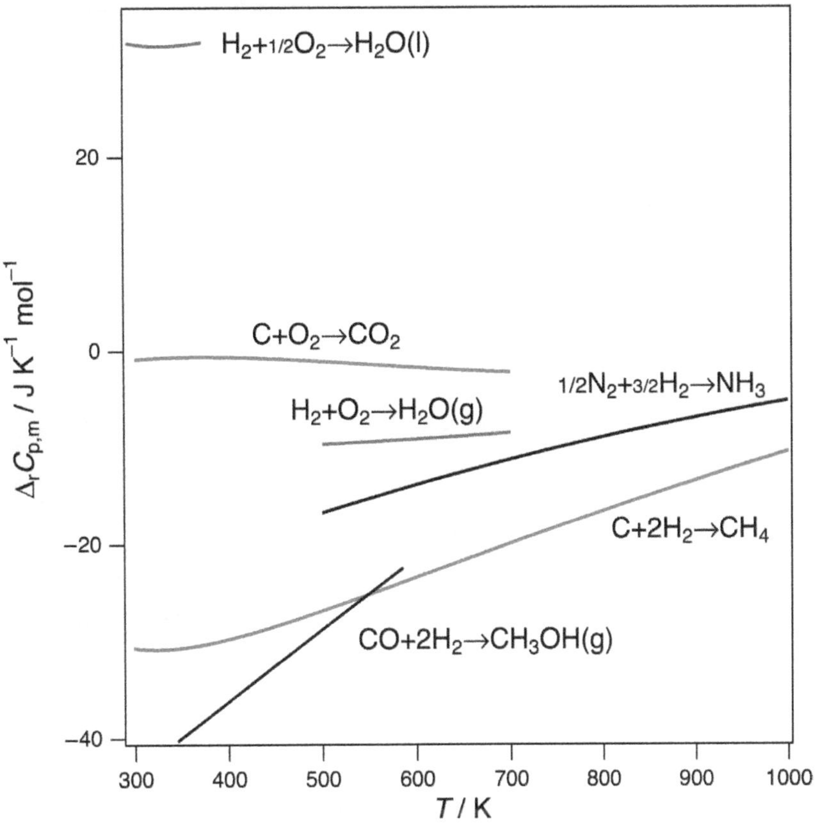

Figure 9.4 The change in reaction heat capacity as a function of temperature. The temperature range over which this quantity can be assumed to be constant depends on the specific reaction under consideration. Calculated with parameters taken from NIST WebBook.

In those instances when it is appropriate to approximate $\Delta_r C^\circ_{p,m}$ as a constant, then we can evaluate the integral in Eq. (9.48) to give

$$\Delta_r H^\circ_m(T_2) = \Delta_r H^\circ_m(T_1) + \Delta T\, \Delta_r C^\circ_{p,m}. \tag{9.50}$$

In those instances when more precision is needed, $\Delta_r C^\circ_{p,m}$ must be evaluated explicitly within the integral. Analogous to the argument used to derive Eq. (9.47), if we use Eq. (9.45) to express the temperature dependence of each individual heat capacity, then the enthalpy will change as

$$\Delta H(T_2) = \Delta H(T_1) + \Delta_r a(T_2 - T_1) + \Delta_r b(1/2)\left(T_2^2 - T_1^2\right) - \Delta_r e\left(T_2^{-1} - T_1^{-1}\right) \tag{9.51}$$

where

$$\Delta_r a = \sum_{\text{products}} \nu a - \sum_{\text{reactants}} \nu a \tag{9.52}$$

$$\Delta_r b = \sum_{\text{products}} \nu b - \sum_{\text{reactants}} \nu b \tag{9.53}$$

$$\Delta_r e = \sum_{\text{products}} \nu e - \sum_{\text{reactants}} \nu e. \tag{9.54}$$

While by hand such a calculation may appear onerous, with a bit of computer programming in an appropriate data analysis software package it is easily implemented.

Figure 9.5 shows the change in enthalpy change as a function of temperature. When the value of $\Delta_r C^\circ_{p,m}$ is positive, $\Delta_r H^\circ_m$ increases with increasing temperature. When the value of $\Delta_r C^\circ_{p,m}$ is negative, $\Delta_r H^\circ_m$ decreases with increasing temperature. Whereas changes of only 2–3 kJ mol^{-1} might be within the uncertainty of many measurements, ignoring larger changes may have a significant impact on experimental interpretations.

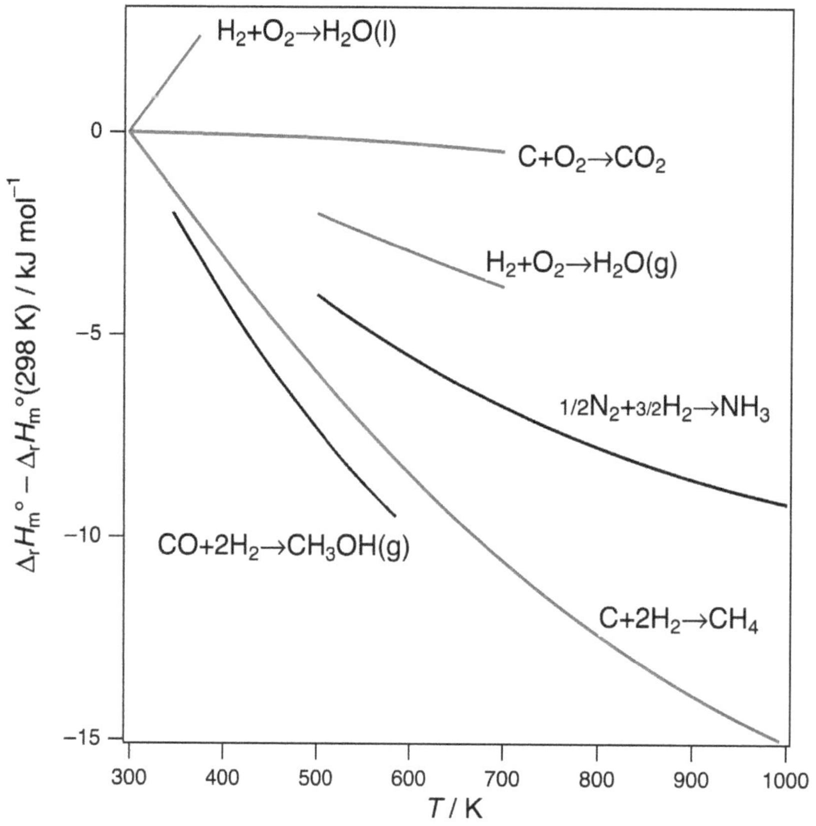

Figure 9.5 The temperature dependence of the enthalpy change during reaction $\Delta_r H_m^\circ$ is shown for several reactions. The value of the enthalpy change at 298 K $\Delta_r H_m^\circ(298\ \mathrm{K})$ has been subtracted from $\Delta_r H_m^\circ$ to emphasize the change that occurs as a function of temperature. Calculated with parameters taken from NIST WebBook.

9.7 Third law of thermodynamics

While the First and Second Laws of Thermodynamics were developed from the study of heat engines, the Third Law was developed from the study of chemistry. The problem can be formulated this way. We have seen that Gibbs energy changes are related to enthalpy and entropy changes by

$$\Delta G = \Delta H - T\,\Delta S. \tag{9.55}$$

In order to calculate ΔG from only thermal data, we would need to know the values of ΔH and ΔS at some reference temperature. Let us take $T = 0$ K as the reference point. Then we can use the temperature dependence of ΔH and ΔS to calculate ΔG at the temperature of interest, call it T_2, using

$$\Delta G = \Delta H_0 - T_2\,\Delta S_0 + \int_0^{T_2} \Delta C_p\,\mathrm{d}T + \int_0^{T_2} \frac{\Delta C_p}{T}\,\mathrm{d}T. \tag{9.56}$$

This equation is valid as long ΔH_0, ΔS_0 and the integrands do not diverge as $T \to 0$.

Nernst noted that the value of ΔG converged on the value of ΔH as the temperature approached 0 K. While low-temperature measurements at the time were limited, they indicated that the value of the heat capacity was very small at low temperature. On the basis of thermodynamic reasoning alone, we can make the following deduction. Assume that the entropy function remains finite, as demanded by Eq. (9.56). Consider a system held at constant pressure. The entropy, S_2, of a system held at T_2 is related to the entropy, S_1, of a system at T_1 by

$$S_2 = S_1 + \int_{T_1}^{T_2} \frac{C_p}{T}\,\mathrm{d}T. \tag{9.57}$$

In the limit $T_1 \to 0$ this becomes

$$\lim_{T_1 \to 0} S_2 = S_0 + \int_0^{T_2} \frac{C_p}{T} \, dT. \tag{9.58}$$

In order for the entropy S_2 to be finite at any temperature T, S_0 must be finite. Furthermore, C_p must not only approach zero as $T \to 0$, but also C_p must approach zero at least as some positive power of T. That is $C_p = k_0 T^m$ with $m > 0$. By this reasoning, then, $C_p = 0$ is the limit of vanishing temperature. By similar reasoning, C_V must also vanish as the temperature vanishes.

Nernst used these observations and reasoning to formulate the *Nernst heat theorem*:

For any system in internal equilibrium undergoing an isothermal process the following are generally valid:

- $\lim_{T \to 0} \Delta G \to \Delta H$
- $\lim_{T \to 0} \left(\frac{\partial \Delta G}{\partial T}\right)_p = \left(\frac{\partial \Delta H}{\partial T}\right)_p$
- $\lim_{T \to 0} \Delta S = 0$
- $\lim_{T \to 0} C_p = 0$
- $\lim_{T \to 0} C_V = 0$

The final three limits are independent of p and any other external variable.

Subsequent theoretical and experimental work has substantiated Nernst's postulate. As formulated by Planck, it is now known as the *Third Law of Thermodynamics*:

The entropy of a pure, perfectly crystalline substance is zero at 0 K.

Mathematically, this was stated by Planck as

The entropy of any system vanishes in the state for which

$$T \equiv \left(\frac{\partial U}{\partial S}\right)_V = 0. \tag{9.59}$$

The advantage of this formulation is that it allows us to calculate absolute molar entropies. It also facilitates the calculation of equilibrium constants solely on the basis of thermochemical measurements. We need to specify that the system is at internal equilibrium because it is possible for a system to get trapped into a long-lived nonequilibrium state. One example of such is a glass. Hence, the Planck formulation specifies a crystalline substance rather than a solid of any form. Nuclear spin interactions in the absence of a magnetic field produce extremely weak interactions. So weak, in fact, that the energy level spacing is still significantly smaller than thermal energy even at 1 K. Randomly oriented nuclear spins can induce significant disorder that does not relax until the temperature is well below 1 K.

9.8 The unattainability of absolute zero

Consideration of the limits of the universe can lead to consideration of our limits to consider. Thus, we are left to ponder a consequence of the Third Law that some consider to be a statement of the Third Law:

A state with a temperature of absolute zero is unattainable through any process composed of a finite number of steps.

Conceptually, we can argue this way. To lower the temperature of a system by an amount dT, we need to extract an amount of heat dq. We know the relationship between these two variables is given by $dq = C\,dT$. However, by the Nernst heat theorem, the heat capacity vanishes as we approach absolute zero. In the limit of zero heat capacity, we have to extract an infinite amount of heat to effect a finite temperature change. Imagine descending a staircase drawn by Duchamp such that for each step you take down, the length of the step is increased by an equal amount. And for every stride you take across the length of a step, an equal length is added to the step below. As the step length grows while the step height remains constant, the slope of the staircase approaches zero. The last step down to absolute zero requires an infinite number of steps be taken across the width of the last step.

The practice of descending a staircase toward absolute zero can be achieved by adiabatic demagnetization. Take a paramagnetic salt with randomly aligned magnetic moments. A strong magnetic field is used to align the moments. This well-ordered state corresponds to a lower entropy than the original state. As long as the magnetization can be carried out isothermally, the reduction in entropy is accompanied by a transfer of heat out of the salt. The magnetic field is then removed adiabatically. Randomization of the moments after demagnetization corresponds to a lowering of the temperature because no heat is allowed to enter the system. This technique was suggested independently by Debye and Giaugue. It is capable of attaining temperature on the order of 1 mK.

Attaining even lower temperatures requires interacting with even lower energy excitations. Nuclear spins provide these interactions. An analogous technique called *adiabatic nuclear demagnetization* orders nuclear spins isothermally and then allows them to relax adiabatically. This process can be used to attain temperatures on the order of 10 μK or so.

Steven Chu, Claude Cohen-Tannoudji, and William D. Phillips were awarded the Nobel Prize in Physics 1997 for the development of laser cooling. Laser cooling does not require the matter being cooled to have any particular nuclear or magnetic spin characteristic. The idea is ingeniously simple, and relies on the Doppler effect and the exchange of momentum that occurs when a photon is absorbed or emitted. Photon-induced momentum transfer is negligible unless the photon energy is high (this is the basis of the Compton effect) or the translation energy of the absorber/emitter is low, as is the case near absolute zero. As we will see later, every transition between quantum states that occurs by the absorption/emission of a photon has associated with it a linewidth. The Doppler effect is one component of this linewidth. Slight shifts in frequency occur based on the direction of the velocity. The red side of the absorption line corresponds to the low-energy side, while the blue side corresponds to the high-energy side.

Imagine now that we take a cold sample of gas, allow it to absorb some laser light of a well-defined frequency (i.e. photon energy), and then observe the gas again after it has relaxed by the re-emission of a photon. If the transition linewidth were infinitely narrow, the energy of the gas would not change. In this case, the energy of the absorbed photons would exactly equal the energy of the emitted photons. By conservation of energy, there would be no change in the energy, that is, the temperature of the gas. Now consider the effect of a finite linewidth – in particular a transition line width that is broad compared to the linewidth of the laser light. We tune the laser to the red side of the line; the photon is absorbed and stays in the excited state for a finite time. During that time the absorber can undergo collisions and exchange energy with its neighbors. When it emits a photon it will do so statistically across the entire transition linewidth. However, since the absorbed photon was on the red side the emitted photon on average has a higher energy than the absorbed photon. After going through the cycle, the collection of absorbers has on average a lower temperature than it did initially.

When laser cooling is combined with magnetic trapping and evaporative cooling, temperatures of just 100–200 nK can be achieved. Eric Cornell, Carl Wieman and coworkers were the first to use these techniques to create a Bose–Einstein condensate from ^{87}Rb atoms. A Bose–Einstein condensate is a macroscopic quantum state in which individual bosons (particles that have zero or integer spin) undergo a phase transition into a common ground state. Cornell, Wieman and Wolfgang Ketterle, who along with coworkers created a Bose–Einstein condensate in ^{23}Na atoms a few months later, won the Nobel Prize in Physics 2001 for their work on Bose–Einstein condensates.

9.9 Absolute entropies

With the Third Law in hand, we are in a position to define a practical scale of absolute entropies. This scale is set by assigning the value zero to a perfect crystal at absolute zero. 'Practical scale' is stated because, as discussed in the section above, we cannot attain absolute zero and therefore we can never measure the entropy at absolute zero. In addition, we must also be assured that in our state as near to absolute zero as we can practically attain, that the system has not been trapped by internal constraints in a long-lived metastable equilibrium state. The use of a practical scale allows us to evade these somewhat philosophical arguments.

In the practical scale of absolute entropies, we define the *calorimetric entropy* of a material at temperature T and pressure p according to

$$S(T,p) = \left(\int_{T_0}^{T_{\text{fus}}} \frac{C_p}{T}\,\mathrm{d}T\right)_{\text{crystal}} + \frac{\Delta_{\text{fus}}H}{T_{\text{fus}}} + \left(\int_{T_{\text{fus}}}^{T_{\text{b}}} \frac{C_p}{T}\,\mathrm{d}T\right)_{\text{liquid}} + \frac{\Delta_{\text{vap}}H}{T_{\text{b}}} + \left(\int_{T_{\text{b}}}^{T} \frac{C_p}{T}\,\mathrm{d}T\right)_{\text{vapor}}. \tag{9.60}$$

If there are additional phase transitions, for instance, from one solid structure to another, then these must be included. Similarly, if the temperature is not yet sufficient for either vaporization or melting to occur, these transitions are excluded. Selected values of absolute entropies are listed in Table 9.1. In general, gases have much higher absolute entropies than solids or liquids. In addition, the more complex a substance, the higher its entropy in a given phase. The structure of a material also plays a role in its absolute entropy as the difference in values for graphite and diamond show.

Table 9.1 Absolute standard molar entropies of selected species at 298.15 K.

Gases	S°_m	Liquids	S°_m	Solids	S°_m
He	126.2	Hg	75.9	Ag	42.6
Ne	146.3	Br_2	152.2	Al	28.3
Ar	154.8	ICl	135.1	Au	47.4
Kr	164.1	H_2O	70.0	B	5.9
Xe	169.7	H_2O_2	109.6	Be	9.5
H_2	130.7	$SiCl_4$	239.7	Cu	33.2
N_2	191.6	$TiCl_4$	252.3	C(graphite)	5.7
O_2	205.2	VCl_4	255.0	C(diamond)	2.4
CO	197.7	CH_3OH	126.8	Fe	27.3
NO	210.8	C_2H_5OH	160.7	Pt	41.6
F_2	202.8	C_6H_6	173.4	Si	18.8
Cl_2	223.1			AgCl	96.3
HF	173.8			GaAs	64.2
HCl	186.9			I_2	260.7
CO_2	213.8			MgO	27.0
NO_2	240.1			ZnO	43.7
NH_3	192.8			MoS_2	62.6
CH_4	186.3			SiO_2(quartz)	41.5
SiH_4	204.6			TiO_2	50.6
SiF_4	282.8			Al_2O_3	50.9
SF_6	291.5			V_2O_5	131.0

Once we are in possession of values such as those in Table 9.1, we need not perform the sum from absolute zero. In other words, using $T = 298.15$ K as our reference state, we can find the absolute entropy at any other temperature T with

$$S(T, p^\circ) = S(T = 298\ \mathrm{K}, p^\circ) + \int_{298}^{T_{\mathrm{fus}}} \frac{C_p}{T}\, \mathrm{d}T + \text{other phases as required.} \tag{9.61}$$

We can also use the methods of statistical mechanics developed to calculate the entropy based on the quantum state energy level structure of the molecule. For this, we need to determine the partition function for each degree of freedom. The partition function of the translation degree of freedom requires only knowledge of the mass. However, for rotations and vibrations we need to know the mass, moment of inertia, vibrational frequencies and models for the rotational and vibrational motions. The electronic ground state is usually the only one that need be considered, but on occasion it may have degeneracy other than one. The energy level structure is determined through spectroscopy. Therefore, the entropy determined via this method is known as the *spectroscopic entropy*. For diatomic molecules these values are easily calculated within the approximations of the rigid rotor and harmonic oscillator, which are good approximations near the ground state. Agreement between calorimetric and spectroscopic absolute entropies is almost always found. Exceptions are easily explained and we have already encountered one. The CO molecule is disordered in the solid state, which leads to residual entropy from the random orientation of the very weak dipoles in the solid.

9.10 Entropy changes in chemical reactions

Now that we have a practical scale of absolute molar entropies, we can use them to calculate the entropy changes involved in reactions. Similar to the process we used for calculating enthalpy changes during reactions, we sum all of the absolute entropies of the products after they have been multiplied by their stoichiometric coefficients. Then we subtract from that the sum of all the absolute entropies of the reactants after they have been multiplied by their stoichiometric coefficients. Formally, we write that in terms of the stoichiometric numbers, which are positive for products and negative for reactants as

$$\Delta_r S^\circ = \sum_j \nu_j S^\circ_j. \tag{9.62}$$

9.10.1 Example

Calculate the standard molar entropy change for the formation of ClF from the elements at 298 K. S°_m = 203, 223, and 218 J K^{-1} mol^{-1} for F_2, Cl_2 and ClF, respectively.

First, write a balanced chemical reaction

$$Cl_2(g) + F_2(g) \rightarrow 2ClF(g)$$
$$\Delta_r S^\circ_m = 2S^\circ_m(ClF) - S^\circ_m(Cl_2) - S^\circ_m(F_2)$$

Thus, $\Delta_r S^\circ_m = 2(218) - 223 - 203 = 10$ J K^{-1} mol^{-1}. Remember that the mol^{-1} refers to per mole of reaction as written, which corresponds to the formation of 2 mol of ClF according to the above reaction.

Consider for a moment what the entropy change upon reaction is telling us. The entropy is proportional to the number of states over which the system can distribute energy. The state density is related to the partition functions for the degrees of freedom available to the system. In other words, the entropy change is related to the phases and quantum state structure of the reactants and products. This reaction involves 2 moles of gas transforming into 2 moles of gas. In fact, the reactants and products are even more similar than that, as 2 moles of diatomic gas are being transformed into 2 moles of diatomic gas. The entropy change upon reaction should be very small, as is confirmed by the calculation. When qualitatively evaluating the entropy change upon reaction, look first at the change in the number of moles of gas. The gas phase has the highest entropy of the three commonly encountered phases; therefore, large reaction entropy changes are associated with changes in the number of moles of gas. Producing a liquid from a solid involves a substantial entropy change, but not as large as producing a gas from either a liquid or solid.

Quantifying the amount of substance in each phase in units of moles is related to accounting for the number of translational degrees of freedom as well as the entropy of mixing. For instance, if a reaction involved 1 mole of liquid forming 2 moles of liquid, the products would be entropically favored by the mixing of those 2 moles. The next step in qualitatively evaluating the change of entropy is to consider the vibrational and rotational degrees of freedom. These will account for smaller effects and they are also not as easy to evaluate by inspection. In general, low-frequency vibrations (bonds between two heavy atoms) are more easily excited (make more states available) than high-frequency bonds (with a light atom such as H). Similarly, rotations with low frequencies associated with heavy–heavy combinations are more readily excited by thermal energy than high-frequency rotations associated with light–light combinations.

9.10.2 Directed practice

Calculate the standard change in molar entropy at 298 K for the formation of IrF_6(s) by direct fluorination of Ir metal. S° (298 K) = 35.5, 203, and 248 J K^{-1} mol^{-1} for Ir(s), F_2(g) and IrF_6(s), respectively. [−396.5 J K^{-1} mol^{-1}]

Let us consider what we can learn from the entropy change during reaction. We have already learned in Chapter 8 that a spontaneous process is one that corresponds to an overall increase in entropy; that is, $\Delta S > 0$ for a spontaneous process. We look in detail at ammonia synthesis performed at 25 °C and a constant pressure of 1 bar. The industrial process is run at 400–500 °C, but this way we can use values directly from standard tables without correcting them for temperature changes.

Start with a balanced chemical reaction

$$N_2(g) + 3H_2(g) \rightarrow 2NH_3(g).$$

Using data tables, we find the S°_m values for the product and reactants, then evaluate

$$\Delta_r S^\circ_m = 2S^\circ_m(NH_3) - S^\circ_m(N_2) - 3S^\circ_m(H_2)$$
$$\Delta_r S^\circ_m = -198 \text{ J K}^{-1}\text{ mol}^{-1}.$$

This is the entropy change of the system, ΔS_{sys}. The entropy change during reaction is clearly unfavorable because 4 moles of gas are transformed into 2 moles of gas.

Now, we evaluate the enthalpy change.

$$\Delta_r H^\circ_m = 2\Delta_f H^\circ_m(NH_3) - \Delta_f H^\circ_m(N_2) - 3\Delta_f H^\circ_m(H_2)$$
$$\Delta_r H^\circ_m = -91.8 \text{ kJ mol}^{-1}.$$

The enthalpy change is clearly favorable for reaction. The strength of the six N–H bonds exceeds the strength of one N–N bond and three H–H bonds. The reaction is exothermic, but is it spontaneous? Neither the First nor the Second Law can answer this – that is, neither the enthalpy change (which is favorable) nor the entropy change (which is unfavorable) determines spontaneity by itself. The control parameter is the Gibbs energy change.

$$\Delta_r G_m^\circ = \Delta_r H_m^\circ - T\,\Delta_r S_m^\circ = -32.7 \text{ kJ mol}^{-1}$$

The standard Gibbs energy of reaction is negative and the reaction favors products. In other words, when N_2 and H_2 are placed in a container at 298 K, their thermodynamic tendency is to react to form NH_3. The reaction is spontaneous under these conditions.

Why is the reaction spontaneous if the entropy change upon reaction is negative? The answer is that the entropy change of the system is –198 J K^{-1} mol^{-1}. However, it is the energy change of the universe that must increase in a spontaneous process. The entropy change of the universe is the entropy change of the system plus surroundings.

To calculate the entropy change in the surroundings, we first need to calculate the enthalpy change in the surroundings. The reaction was run at constant temperature and pressure in contact with the surroundings. The enthalpy released by the system is equal and opposite to the enthalpy change in the surroundings,

$$\Delta H_{\text{sur}} = -\Delta H_{\text{sys}} = -\Delta_r H$$

Thus, the entropy change in the surroundings is

$$\Delta S_{\text{sur}} = \Delta H_{\text{sur}}/T = 308 \text{ J K}^{-1} \text{ mol}^{-1}$$

The change in entropy for the universe is

$$\Delta S_{\text{universe}} = \Delta S_{\text{sur}} + \Delta S_{\text{sys}}$$
$$\Delta S_{\text{universe}} = 308 - 198 = 110 \text{ J K}^{-1} \text{ mol}^{-1}$$

Since $\Delta S_{\text{universe}}$ is positive, thermodynamically the process is spontaneous. The unfavorable entropy change of the system is counterbalanced by the enthalpy flow into the surroundings. The heat from the exothermicity of the reaction is dispersed over so many states in the surroundings that this overcomes the poor entropy factor inherent to the reactive system. This is why the reaction is spontaneous. This example also demonstrates why we do not try to use the Second Law and the Clausius inequality to determine reaction spontaneity directly. The calculation is rather involved and drawn out, and demands that we look not just at the system but also the effect of the system on its surroundings. The change in the Gibbs energy is a single parameter that takes all of this into account. Furthermore, we can relate the Gibbs energy of reaction directly to the equilibrium constant.

However, this reaction does not proceed to equilibrium at room temperature under normal circumstances. Why? Because there is a kinetic constraint. The rate of reaction is very slow due to a high activation barrier, and a catalyst is required to speed up the reaction. Thermodynamics tells us that the reaction should proceed spontaneously but does not tell us how fast the reaction is. The thermodynamics of the reaction – the values of $\Delta_r H$, $\Delta_r S$, and $\Delta_r G$ – are independent of path. The utility of the reaction and how it has to be performed in order to use it in practice, are all about the path. The path is all about the reaction kinetics.

9.11 Calculating $\Delta_r S^\circ$ at any temperature

The methods of adjusting $\Delta_r S$ for changes in temperature are much like those we used to calculate the temperature dependence of $\Delta_r H$. Again, the heat capacity plays a central role and the integrals used look similar to those used before. The entropy change upon constant pressure heating is given by

$$\Delta S_m^\circ = \int_{T_1}^{T_2} C_{p,m}^\circ \frac{dT}{T}. \tag{9.63}$$

This is true for each species in a chemical reaction. Thus, for a chemical reaction, we can find $\Delta_r S_m^\circ$ at any arbitrary temperature as long as we have the reference value at 298 K and $C_{p,m}^\circ$ for each species according to

$$\Delta_r S_m^\circ(T_2) = \Delta_r S_m^\circ(298 \text{ K}) + \int_{298}^{T_2} \Delta_r C_{p,m}^\circ \frac{dT}{T}. \tag{9.64}$$

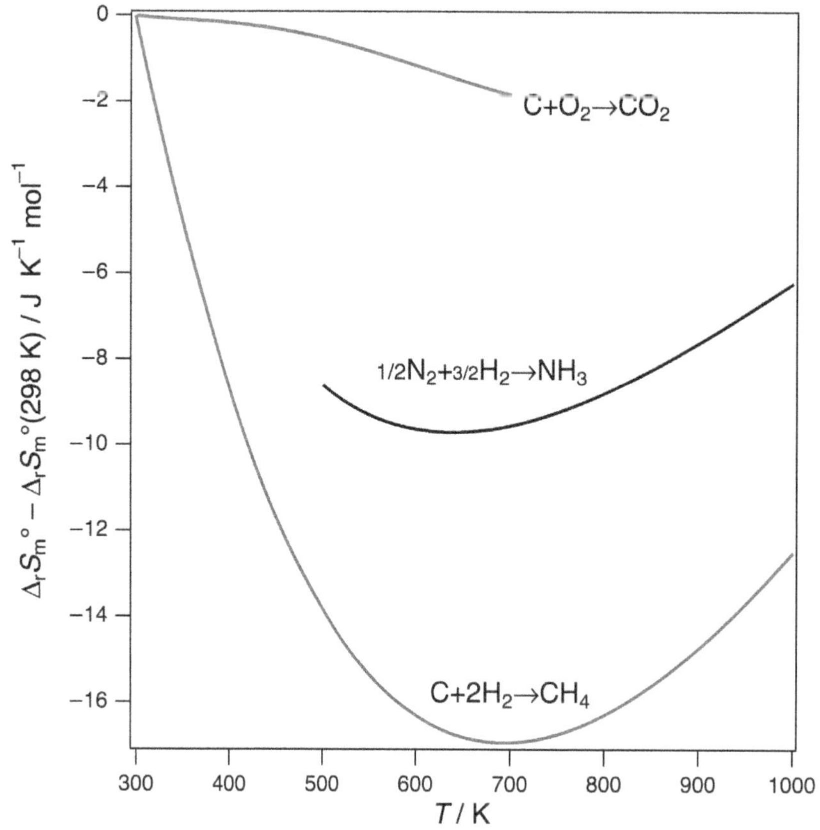

Figure 9.6 The dependence of entropy of reaction on temperature depends on the reactive system. If there is a small change in the reaction heat capacity, as in the case of the CO_2 formation reaction, the dependence is slight. The formation reaction for CH_4 exhibits much stronger dependence. A phase change in a reactant or product would lead to a discontinuous change at that temperature.

The reference value is determined from the absolute entropies as described in the previous section. The change in heat capacities is defined as before

$$\Delta_r C^\circ_{p,\mathrm{m}} = \sum_j \nu_j \, C^\circ_{p,\mathrm{m},j} \tag{9.65}$$

If the change in heat capacities is constant as a function of temperature, then

$$\Delta_r S^\circ_\mathrm{m}(T_2) = \Delta_r S^\circ_\mathrm{m}(298\ \mathrm{K}) + \Delta_r C^\circ_{p,\mathrm{m}} \ln(T_2/298\ \mathrm{K}). \tag{9.66}$$

However, as shown in Fig. 9.4, this approximation is only valid in special cases or for small temperature changes. In general, the functional form of the heat capacity should be substituted explicitly into Eq. (9.65) so that the integration in Eq. (9.64) can be performed.

Alternatively, if the Shomate equation fit is known for S°_m of each species in the reaction, then this can be evaluated for each species at the temperature of interest.

$$S^\circ_\mathrm{m} = A \ln t + Bt + \frac{Ct^2}{2} + \frac{Dt^3}{3} - \frac{E}{2t^2} + G \tag{9.67}$$

Here again, $t = T/1000$. The difference between products and reactants, properly weighted by the stoichiometric coefficients,

$$\Delta_r S^\circ_\mathrm{m} = \sum_j \nu_j \, S^\circ_{\mathrm{m},j} \tag{9.68}$$

can then be evaluated explicitly.

SUMMARY OF IMPORTANT EQUATIONS

$dU = T\,dS - p\,dV$	Fundamental equation: change in internal energy dU equals temperature T times change in entropy dS minus pressure p times change in volume dV.
$A = U - TS$	Definition of the Helmholtz energy A.
$G = H - TS$	Definition of the Gibbs energy. H is the enthalpy.
$\Delta G^\circ = \Delta H^\circ - T\Delta S^\circ$	Gibbs energy change related to enthalpy and entropy changes.
$\Delta_r G^\circ = \sum_j \nu_j \Delta_f G_j^\circ$	Gibbs energy change of reaction. The r refers to reaction; the f to formation. ν_j is the stoichiometric number of species j.
$\Delta_r H^\circ = \sum_j \nu_j \Delta_f H_j^\circ$	Enthalpy change of reaction.
$\Delta_r S^\circ = \sum_j \nu_j S_j^\circ$	Entropy change of reaction.
$\frac{\Delta G(T_2)}{T_2} = \frac{\Delta G(T_1)}{T_1} + \int_{T_1}^{T_2} \Delta H\, d\left(\frac{1}{T}\right)$	Gibbs–Helmholtz equation.
$\Delta_r H_m^\circ(T_2) = \Delta_r H_m^\circ(T_1) + \int_{T_1}^{T_2} \Delta_r C_{p,m}^\circ\, dT$	Dependence of reaction enthalpy on temperature.
$\Delta_r S_m^\circ(T_2) = \Delta_r S_m^\circ(T_1) + \int_{T_1}^{T_2} \Delta_r C_{p,m}^\circ \frac{dT}{T}$	Dependence of reaction entropy on temperature.
$\Delta_r C_{p,m}^\circ = \sum_j \nu_j C_{p,m,j}^\circ$	$C_{p,m,j}^\circ$ is the molar heat capacity at a constant pressure of 1 bar.
$dG = V\,dp - S\,dT$	For a closed system doing no non-expansion work.
$G_m(p) = G_m^\circ + RT\ln(p/p^\circ)$	Molar Gibbs energy of an ideal gas is related to standard state value and pressure p. p° = standard state value of 1 bar, R = gas constant.

Exercises

9.1 Suggest a physical interpretation of the dependence of the Gibbs energy on the pressure. Start with an equation to explain the behavior.

9.2 Starting with an equation, suggest a physical interpretation of the dependence of the Gibbs energy on the temperature.

9.3 As derived from the data of Kopecky (*Pharmazie*, **51**, (1996)), the Gibbs energy change for micelle formation $\Delta_{mic}G_m^\circ$ varies with chain length in quaternary ammonium salts APYCl as given in the table below. Predict the values of $\Delta_{mic}G_m^\circ$ for m = 6, 11, and 16. Which of these values is the most reliable?

m	8	10	12	14
$\Delta_{mic}G_m^\circ$	−3.21	−6.77	−10.26	−14.01

9.4 Ethylene dichloride (1,2-dichloroethane, $C_2H_4Cl_2$) is an important intermediate in the chemical industry and also finds use as a solvent. It is formed along with water from the gas-phase reaction of ethene with HCl and O_2 over a $CuCl_2$ catalyst. Use the information in the table below to answer these questions. (a) Calculate $\Delta_r H_m^\circ$, $\Delta_r S_m^\circ$ and $\Delta_r G_m^\circ$ at 298.15 K. (b) Assuming the difference in heat capacities is not a function of temperature, calculate $\Delta_r H_m^\circ$, $\Delta_r S_m^\circ$ and $\Delta_r G_m^\circ$ at 500 K, where the industrial synthesis is performed.

T = 298 K	C_2H_4(g)	HCl(g)	O_2(g)	$C_2H_4Cl_2$(g)	H_2O(g)
$\Delta_f H_m^\circ$/kJ mol^{-1}	52.4	−92.31		−126.4	−241.8
S_m°/J K^{-1} mol^{-1}	219.32	186.90	205.15	308.4	188.8
$C_{p,m}$/J K^{-1} mol^{-1}	42.9	29.14	29.39	78.7	33.6

9.5 On the basis of the integrated form of the Gibbs–Helmholtz equation, discuss how $\Delta_r G^\circ_m$ changes with increasing temperature based on whether a reaction is exothermic compared to endothermic.

9.6 Consider the reaction $2Br(g) \rightarrow Br_2(g)$ at 350 K. Describe with justification in terms of $\Delta_r H^\circ$, $\Delta_r S^\circ$ and $\Delta_r G^\circ$ whether the reaction favors reactants or products.

9.7 Use the information in the table below to answer these questions for the reaction of SiO_2 with V to form V_2O_5 and Si. (a) Assuming the values above are for the solid phase and that all species remain in the same phase, calculate $\Delta_r H^\circ_m$, $\Delta_r S^\circ_m$ and $\Delta_r G^\circ_m$ at 298 K. (b) Assuming the values above are for the solid phase and that all species remain in the same phase, calculate $\Delta_r H^\circ_m$, $\Delta_r S^\circ_m$ and $\Delta_r G^\circ_m$ at 1000 K.

T = 298 K	SiO_2	V_2O_5	Si	V
$\Delta_f H^\circ_m$/kJ mol^{-1}	−910.7	−1550.6		
$\Delta_f G^\circ_m$/kJ mol^{-1}	−856.3	−1419.5		
S°_m/J K^{-1} mol^{-1}	41.5	131.0	18.8	28.9
$C_{p,m}$/J K^{-1} mol^{-1}	44.4	127.7	20.0	24.9

9.8 Consider the formation of glucose $C_6H_{12}O_6(s)$ and $O_2(g)$ from $CO_2(g)$ and $H_2O(l)$ as in the photosynthesis reaction. Using the enthalpies of formation and absolute entropies at 298 K, calculate the values of $\Delta_r H^\circ_m$ and $\Delta_r S^\circ_m$ at 310 K. The heat capacities $C_{p,m}$ in J K^{-1} mol^{-1} are 37.1, 75.3, 219.2, and 29.4 for $CO_2(g)$, $H_2O(l)$, $C_6H_{12}O_6(s)$, and $O_2(g)$, respectively.

9.9 A student measured a value of $\Delta_{vap} H^\circ_m$ for ethanol at 355 K to be 34.7 ± 0.5 kJ mol^{-1}. Does this agree within error bars of the accepted value of 42.3 kJ mol^{-1} measured at 298.15 K?

9.10 (a) Write down the expression that relates the change in Gibbs energy to the changes in enthalpy and entropy. (b) What change in enthalpy, entropy and Gibbs energy is favorable for a reaction? (c) Two possible metabolic fates of α-D-glucose are given below. Compare the standard Gibbs energy changes at 298 K for these processes. Which process represents the more efficient utilization of α-D-glucose? (i) α-D-glucose (1 M) → 2 C_2H_5OH (1 M) + 2 $CO_2(g)$; (ii) α-D-glucose (1 M) + 6 $O_2(g)$ → 6 $CO_2(g)$ + 6 $H_2O(l)$. $\Delta_f G^\circ$ values in kJ mol^{-1} are $CO_2(g)$ = −394.4, $H_2O(l)$ = −237.2, C_2H_5OH = −181.5, α-D-glucose = −917.2.

9.11 The hydrolysis of ATP $ATP(aq) + H_2O(l) \rightarrow ADP(aq) + P_i^-(aq) + H_3O^+(aq)$ is exergonic at pH = 7.0 and 310 K.

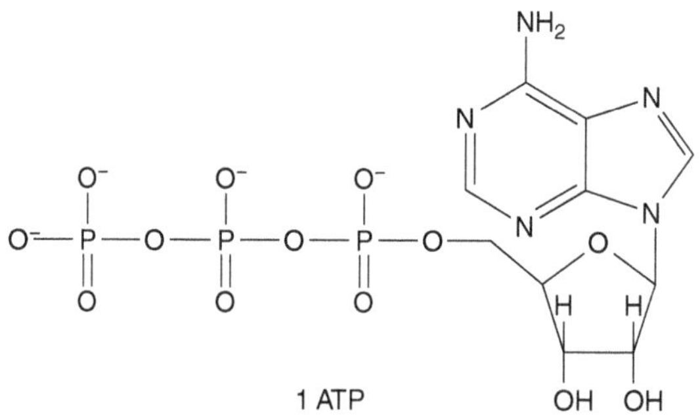

Structure of ATP in the deprotonated ATP^{4-} form.

(a) It is thought that the exergonicity of ATP hydrolysis is due in part to the fact that the standard entropies of hydrolysis of polyphosphates are positive. Why would an increase in entropy accompany the hydrolysis of a triphosphate group into a diphosphate and a phosphate group? (b) Under identical conditions, the Gibbs energies of hydrolysis of H_4ATP and $MgATP^{2-}$, a complex between the Mg^{2+} ion and ATP^{4-}, are less negative than the Gibbs energy of hydrolysis of ATP^{4-}. This observation has been used to support the hypothesis that electrostatic repulsion between adjacent phosphate groups is a factor that controls the exergonicity of ATP hydrolysis. Provide a rationale of the hypothesis and discuss how the experimental evidence supports it. Do these electrostatic effects contribute to the $\Delta_r H$ or $\Delta_r S$ terms that determine the exergonicity of the reaction? Hint: In the $MgATP^{2-}$ complex, the Mg^{2+} ion and ATP^{4-} anion form two bonds: one that involves a negatively charged oxygen belonging to the terminal phosphate group of ATP^{4-} and another that involves a negatively charged oxygen belonging to the phosphate group adjacent to the terminal phosphate group of ATP^{4-}.

9.12 Magnesiothermic reduction of silica may be an economical method for producing nanostructured porous silicon from plants that naturally accumulate silica. Calculate the enthalpy of reaction for this reaction

$$2Mg(g) + SiO_2(s) \rightarrow Si(s) + 2MgO(s)$$

at 1393 K. The enthalpies of formation and heat capacities at 298 K are given below. $\Delta_{sub}H^\circ_m$ (Mg) = 132 kJ mol^{-1} at 1393 K.

	Mg	SiO_2	Si	MgO
$\Delta_f H^\circ_m$/kJ mol^{-1}		−859.4		−601.83
$C_{p,m}$/(J K^{-1} mol^{-1})	24.39	44.98	20.01	37.10

9.13 Urea $CO(NH_2)_2$ is widely used as a fertilizer. It is produced by the Basaroff reaction of NH_3 and CO_2. Using the values in the table below and assuming that the heat capacities are independent of temperature, calculate the enthalpy of reaction for the formation of urea at 298 K and 400 K.

	$\Delta_f H^\circ_m$/kJ mol^{-1}	S°_m/J K^{-1} mol^{-1}	$C_{p,m}$/J K^{-1} mol^{-1}
$H_2O(l)$	−285.8	70.0	75.3
$H_2O(g)$	−241.8	188.8	33.6
$NH_3(g)$	−45.9	192.8	35.1
$CO_2(g)$	−393.5	213.8	37.1
$CO(NH_2)_2$	−333.1	104.26	92.79

9.14 Consider the gas phase reaction (all gases considered ideal)

$$NO + N_2O_5 \rightleftharpoons N_2O_4 + NO_2$$

T = 298 K	NO	N_2O_5	N_2O_4	NO_2
$\Delta_f H^\circ_m$/kJ mol^{-1}	91.3	13.3	11.1	33.2
S°_m/J K^{-1} mol^{-1}	210.8	355.7	304.4	240.1

Calculate $\Delta_r G^\circ_m$ at 298 K.

9.15 $PCl_5(g)$ is mixed with $O_2(g)$ and allowed to equilibrate at 298 K and constant volume. Values of the enthalpies of formation and absolute entropies are given in the table below.

	PCl_5	O_2	$POCl_3$	Cl_2O
$\Delta_f H^\circ_m$/kJ mol^{-1}	−95.35	0	−592.0	76.15
S°_m/J K^{-1} mol^{-1}	352.7	205.0	324.6	266.5

(a) Calculate $\Delta_r G^\circ_m$ at 298 K for the system assuming PCl_5 and O_2 react to form $POCl_3(g)$ and $Cl_2O(g)$ as the only products. (b) Explain whether the enthalpy and entropy changes are favorable or unfavorable for the reaction and what their values mean in molecular terms for the reactants compared to the products.

9.16 Use the information in the table below to answer the remainder of this question for the reaction $N_2(g) + 3H_2(g) \rightarrow 2NH_3(g)$.

T = 298 K	NH_3(g)	N_2(g)	H_2(g)
$\Delta_f H^\circ_m$/kJ mol^{-1}	−45.9		
S°_m/J K^{-1} mol^{-1}	192.8	191.6	130.7
$C^\circ_{p,m}$/J K^{-1} mol^{-1}	35.1	29.1	28.8

(a) Calculate $\Delta_r H^\circ_m$, $\Delta_r S^\circ_m$, and $\Delta_r G^\circ_m$ at 298 K. (b) The industrial synthesis of ammonia is performed at 775 K. Calculate $\Delta_r H^\circ_m$, $\Delta_r S^\circ_m$, $\Delta_r G^\circ_m$, and K at 775 K.

9.17 Use the data in the following table to answer parts (a)–(c) below.

T = 298 K	C_6H_6(l)	O_2(g)	CO_2(g)	H_2O(l)
$\Delta_f H^\circ_m$/kJ mol^{-1}	49.1		−393.5	−285.8
S°_m/J K^{-1} mol^{-1}	173.4	205.2	213.8	70.0
$C^\circ_{p,m}$/J K^{-1} mol^{-1}	136.0	29.4	37.1	75.3

(a) Calculate $\Delta_c H^\circ_m$, $\Delta_c S^\circ_m$ and $\Delta_c U^\circ_m$ at 298 K and $\Delta_r H^\circ_m$ and $\Delta_r S^\circ_m$ at 372 K for the combustion of benzene. (b) Discuss the trend in absolute molar entropies of the four molecules listed. (c) Discuss the meaning of $\Delta_c H^\circ_m$ with respect to the bonding in the reactants and products.

9.18 Use the information in this table for all parts. (a) Calculate ΔH°_m and ΔS°_m at 298 K and 398 K for the gas-phase reaction of CO_2 with H_2 to form carbon monoxide and water. State any assumptions that you make. (b) Calculate $\Delta_{vap} H^\circ_m$ and $\Delta_{vap} S^\circ_m$ at 298 K for water. (c) Does the reaction in part (a) or the vaporization of water have a larger entropy change? Explain the relative magnitudes of these two values. (d) Calculate how much work is done to produce 1 mole of gaseous water at 298 K either: (i) by the vaporization of water; or (ii) from the reaction of CO_2 with H_2.

	$\Delta_f H^\circ_m$ (298 K)/kJ mol^{-1}	S°_m (298 K)/J K^{-1} mol^{-1}	$C_{p,m}$/J K^{-1} mol^{-1}
CO_2(g)	−393.51	213.78	37.11
H_2O(g)	−241.82	188.84	33.58
H_2O(l)	−285.83	69.61	75.29
H_2(g)		130.68	28.82
CO(g)	−110.53	197.67	29.14

9.19 (a) A chemical reaction is characterized by $\Delta_r H^\circ_m = -85.2$ kJ mol^{-1} and $\Delta_r S^\circ_m = -170.2$ J K^{-1} mol^{-1}. Which of the following reactions is more likely to be characterized by the values given in (a)? Justify your answer.

$$\mathrm{A(l) + 2B(s) \rightleftharpoons 3C(s) + D(s)}$$

$$\mathrm{A(g) + 3B(g) \rightleftharpoons C(s) + D(s)}$$

$$\mathrm{A(s) + 4B(g) \rightleftharpoons 3C(g) + D(g)}$$

9.20 Show algebraically that the Gibbs–Helmholtz equation as expressed in Eq. (9.37) is equivalent to assuming that both ΔH°_m and ΔS°_m are constant.

Further reading

Berry, R.S., Rice, S.A., and Ross, J. (2000) *Physical Chemistry*, 2nd edition. Oxford University Press, New York.
Metiu, H. (2006) *Physical Chemistry: Thermodynamics*. Taylor & Francis, New York.
NIST WebBook, http://webbook.nist.gov/chemistry/.

CHAPTER 10

Thermochemistry: The role of heat in chemical and physical changes

PREVIEW OF IMPORTANT CONCEPTS

- The standard state of a substance is its pure form in equilibrium at 1 bar.
- $\Delta_r H^\circ < 0$ defines an exothermic reaction. $\Delta_r H^\circ > 0$ defines an endothermic reaction.
- Energy is not released from bonds when they are broken. Energy is released from a reactive system when weaker bonds are replaced by stronger bonds being formed.
- In an exothermic reaction, weaker bonds are broken on average in the reactants and stronger bonds are made in the products.
- Calorimetry – the science of the measurement of heat exchange associated with a process – is the standard method for determining the enthalpy change during a reaction.
- The heat exchanged at constant pressure is equal to the enthalpy change.
- The heat exchanged at constant volume is equal to the internal energy change.
- Enthalpy is a state function, which leads to Hess's law: The standard enthalpy of an overall reaction is the sum of the standard enthalpies of the individual reactions into which a reaction may be divided.
- The enthalpy change of any reaction can be calculated from the enthalpies of formation of the reactants and products.
- Heat capacity is a temperature-dependent quantity.
- Changes in the heat capacity between reactants and products are responsible for the changes in reaction enthalpy with temperature.

Thermochemistry lies at the heart of one of the areas that makes chemistry so important. Energy can be extracted from a chemical system when bonds are rearranged in the making of products from reactants. This energy can be harvested to perform work, to generate steam and electricity, and to provide warmth in our homes. The energy changes that occur during reactions and phase changes are directly dependent on bonding and intermolecular interactions. Thus, the measurement of temperature changes during these processes is a direct measure of the strength of these interactions. By probing the strength of bonds and intermolecular interactions, thermochemical measurements provide tests to our models of these interactions.

10.1 Stoichiometry and extent of reaction

Extent of reaction is given the symbol ξ (Greek xi). A balanced stoichiometric chemical reaction is written with *stoichiometric coefficients a, b,* and so on, which are all positive numbers

$$a\mathrm{A} + b\mathrm{B} + \cdots \rightarrow \cdots + y\mathrm{Y} + z\mathrm{Z}.$$

The IUPAC defines *stoichiometric numbers* ν_A, ν_B, and so on, as signed quantities, which are negative for reactants and positive for products. Stoichiometric coefficients are the absolute values of the stoichiometric numbers. In other words, $a = |\nu_\mathrm{A}| = -\nu_\mathrm{A}$ but $z = |\nu_z|\nu_z$. The extent of reaction has units of moles and is defined by

$$\xi \equiv \frac{n_i - n_{i0}}{\nu_i} = \frac{\Delta n_i}{\nu_i}. \tag{10.1}$$

Physical Chemistry: How Chemistry Works, First Edition. Kurt W. Kolasinski.

Companion Website: www.wiley.com/go/kolasinski/physicalchemistry

Figure 10.1 Fire is a manifestation of a complex series of radical-driven chain reactions. From cooking, to boiling water to driving turbines, to propulsion from internal combustion engines, the thermochemistry of combustion has a long and intricate relationship with human culture and its development. Photograph provided by Hope Michelsen at the Combustion Research Facility, Sandia National Laboratories.

This is a positive number that is the same for all reactants and products. n_i is the amount of substance (moles) present at a given time, while $n_{i,0}$ is the amount of substance present initially. *The stoichiometric equation must be specified to define the extent of reaction.* From Eq. (10.1) the extent of reaction can be used to calculate the amount of substance present at any given time,

$$n_i = n_{i0} + \nu_i \xi. \tag{10.2}$$

By differentiation, it yields the change in the amount of substance

$$\mathrm{d}n_i = \nu_i \mathrm{d}\xi. \tag{10.3}$$

10.2 Standard enthalpy change

The *standard enthalpy change*, $\Delta H°$, is obtained when the reactants and products of a given reaction are present in their standard states at a specified temperature. The *standard state* is the pure form in equilibrium at 1 bar. Generally, the temperature – if not specified – is taken to be 298.15 K (25 °C). Most elements in the Periodic Table are solids in their standard state at 298 K. The most stable form of the solid is taken as the standard state. For example, graphite is the standard state of carbon, and not diamond or graphene or carbon nanotubes or amorphous carbon. Several other elements have multiple allotropic forms with the standard states as follows: Sb (metallic), As(gray), P(white), Se(gray), S(rhombic) and Sn(white). You should be familiar with the exceptions. The noble gases are all monatomic gases in their standard states. H_2, N_2, O_2, F_2 and Cl_2 are all diatomic gases in the standard state. Br_2 is a diatomic liquid and I_2 is a diatomic solid, but At is an atomic solid. Hg is a monatomic liquid. Ga becomes a liquid just above room temperature at 302.91 K.

Enthalpy changes can be measured for a variety of processes. The process is specified by a subscript after the delta, for example $\Delta_r H$ for the enthalpy change that occurs during a reaction. The subscript m after the H specifies the value as a molar enthalpy. The m is often dropped in the literature, so always check the units to determine whether the enthalpy change is given as an absolute value in kJ or molar value in kJ mol^{-1}. The subscripts used to denote a number of processes are listed in Table 10.1. Recommended superscripts appear in Table 10.2.

Table 10.1 Recommended subscripted abbreviations to denote processes.

Absorption	abs
Adsorption	a or ads
Atomization	at
Combustion reaction	c
Desorption	des
Dilution	dil
Displacement	dpl
Formation reaction	f
Fusion, melting	fus
Hydration	hyd
Immersion	imm
Mixing	mix
Phase transition	trs
Reaction in general	r
Solution	sol
Sublimation	sub
Transition	trs
Triple point	tp
Vaporization	vap

Reactions are classed as either exothermic or endothermic based on the value of the standard enthalpy change for the reaction, $\Delta_r H^\circ$. $\Delta_r H^\circ < 0$ defines an exothermic reaction, $\Delta_r H^\circ > 0$ defines an endothermic reaction, and $\Delta_r H^\circ = 0$ defines a thermoneutral reaction. An increase in temperature tends to favor product formation for endothermic reactions but disfavors products for exothermic reactions.

From the First Law of Thermodynamics and the definition of work at constant pressure for a system in which only pV work is available, the relationship between changes in internal energy and enthalpy is

$$\Delta H^\circ = \Delta U^\circ + \Delta(pV). \tag{10.4}$$

Therefore, at constant pressure where $\Delta p = 0$,

$$\Delta H^\circ = \Delta U^\circ + p\Delta V. \tag{10.5}$$

In other words, internal energy changes and enthalpy changes are virtually identical for any process run at atmospheric pressure that does not involve a change in the volume. If gases are not involved in the process, the change in volume is unlikely to be significant. However, under extremes of pressure – for instance in a geochemical setting or in a diamond anvil cell – the $p\,\Delta V$ term may be significant even if only solids are involved.

10.2.1 Example

By what amount will the change in internal energy differ from the change in enthalpy for the reaction of 0.0500 mol of $MgCO_3$ with excess dilute HCl(aq) at standard temperature and pressure (STP) for which $p = 1$ bar and $T = 273$ K?

Start with a balanced chemical reaction

$$MgCO_3(s) + 2HCl(aq) \rightarrow MgCl_2(aq) + H_2O(l) + CO_2(g) \uparrow$$

Table 10.2 Recommended superscript designations.

Activated complex	‡
Apparent	app
Excess quantity	E
Ideal	id
Infinite dilution	∞
Pure substance	*
Standard	°

Accordingly, 0.05 mol $MgCO_3$ produces 0.05 mol CO_2. The production of a gas leads to a volume change, and the work performed by the system is accomplished by expanding this gas against the pressure of the surroundings. Use the ideal gas law to calculate the volume change and then the work. At constant p and T, the work is given by

$$w = -p\Delta V = \Delta n_{\text{gas}} RT$$

where Δn_{gas} is the change in gas moles = 0.0500 mol.

$$w = -p\Delta V = -0.0500\,\text{mol} \times 8.314\,\text{J K}^{-1}\,\text{mol}^{-1} \times 237\,\text{K} = -113\,\text{J}$$

The work is done by the system on the surroundings and is, therefore, negative. The difference between the internal energy change and the enthalpy change on a molar basis is

$$\Delta U^\circ_{\text{m}} - \Delta H^\circ_{\text{m}} = \frac{-p\Delta V}{n} = \frac{-113\,\text{J}}{0.0500\,\text{mol}} = -2.26\,\text{kJ mol}^{-1}.$$

10.2.2 Directed practice

Show that for the same reaction at 298 K, the work done by the system is -2.48 kJ mol^{-1}.

10.2.3 Temperature dependence of enthalpy of reaction

In Section 9.6 we have shown that the temperature dependence of the enthalpy is given by

$$H(T_2) = H(T_1) + \int_{T_1}^{T_2} C_p \text{d}T \tag{10.6}$$

Since Eq. (10.6) is true for each individual substance in the reaction, we can perform the appropriate sums and differences of the enthalpies of all of the reactants and products to calculate $\Delta_{\text{r}}H$ for the reaction. From this, we can show that the change in enthalpy of reaction with temperature follows *Kirchhoff's law*

$$\Delta_{\text{r}}H^\circ_{\text{m}}(T_2) = \Delta_{\text{r}}H^\circ_{\text{m}}(T_1) + \int_{T_1}^{T_2} \Delta_{\text{r}}C^\circ_{p,\text{m}}\,\text{d}T \tag{10.7}$$

where

$$\Delta_{\text{r}}C^\circ_{p,\text{m}} = \sum_j \nu_j C^\circ_{p,\text{m},j}. \tag{10.8}$$

For all reactions over small ranges of temperature, and for some reactions over extended ranges of temperature change, $\Delta_{\text{r}}C^\circ_{p,\text{m}}$ is constant and can be pulled from the integral. The enthalpy of reaction at T_2 is then given by

$$\Delta_{\text{r}}H^\circ_{\text{m}}(T_2) = \Delta_{\text{r}}H^\circ_{\text{m}}(T_1) + \Delta T\,\Delta_{\text{r}}C^\circ_{p,\text{m}} \tag{10.9}$$

10.3 Calorimetry

The standard method for determining the enthalpy change during a reaction is *calorimetry* – the measurement of temperature rise after a known amount of material has reacted. The temperature rise is related to the heat evolved in the process, q, and the heat capacity, C, by the fundamental equation of calorimetry,

$$q = \int C\,\text{d}T. \tag{10.10}$$

A calorimeter must correspond as close as possible to a closed system. Neither heat nor matter should be exchanged with the surroundings. This is not that difficult to approximate in practice if the reaction can be run in a thermos flask and there is some way to initiate the reaction so that it is clear what the temperature is before the reaction and after the reaction. Initiation of the reaction may occur after mixing of reagents, which requires some sort of mechanical intervention inside of the thermos flask. This may require dropping one container into another, breaking a barrier between containers, or the injection of fluids.

One of the most common types of calorimeter is a constant volume bomb calorimeter. This is shown in Fig. 10.2. The bomb is a stainless steel tube with a removable cap that can be sealed. A solid or liquid sample is placed in a cup near an igniter. The igniter is a wire through which current is passed to heat it sufficiently to initiate combustion. The igniter is consumed in the reaction. The bomb is sealed after the sample is loaded into the cup and a fresh wire has been coiled to

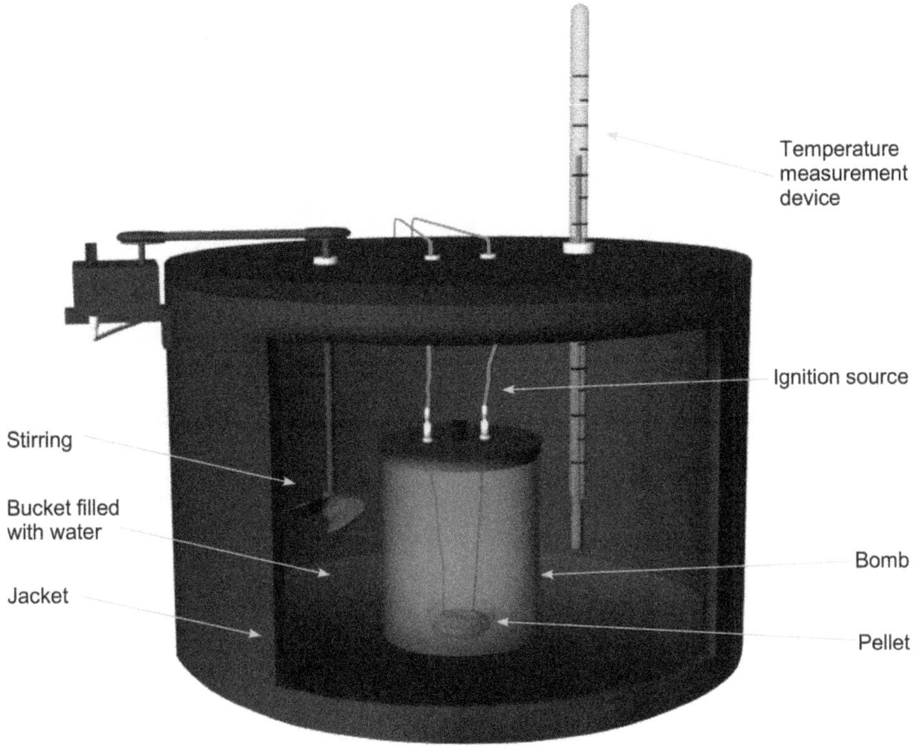

Figure 10.2 Design of a constant volume bomb calorimeter. The sample rests inside the bomb in a cup. Current is passed through a wire to initiate combustion in an excess of O_2. The stainless steel bomb is submerged in water inside of an insulated bucket.

form the igniter. The bomb is pressurized with excess O_2 to a pressure of approximate 3 MPa. The bomb is then placed in a known quantity of water inside of a thermos flask. A thermometer is placed in the water, which is stirred vigorously. The water provides good thermal contact with the stainless steel bomb. The mass of the sample used is matched to the mass of the water and the bomb so that the temperature rise that accompanies the reaction is no more than a few degrees, usually about 2–3 K. In principle, the measurement should be made at close to constant temperature so that the temperature dependence of the heat capacities of materials involved need not be considered. On the other hand, if the temperature rise is too small, uncertainties in the measurement are increased. Hence, the temperature rise of $\Delta T \approx 2$–3 K represents a good compromise.

The bomb, its contents, and the water-filled thermos with temperature measurement are then placed inside of a thermally isolated jacket. Highly accurate work will correct for the energy provided by mixing and the igniter, as well as accounting for thermal drift. The system is allowed to equilibrate thermally, after which the temperature is recorded for several minutes before the ignition and several minutes after ignition, as shown in Fig. 10.3. Extrapolation is then used to determine the temperature change, ΔT. In many practical applications benzoic acid is used as a standard for calibration.

10.3.1 Example

A bomb calorimeter is built with a specific heat capacity of 1.58 kJ K^{-1}. Calculate the molar internal energy change if the combustion of 120 mg of naphthalene, $C_{10}H_8(s)$, leads to a temperature rise of 3.06 K.

Because of the small temperature change, we treat the heat capacity as constant, pull it from the integral in Eq. (10.10) and arrive at Eq. (10.11),

$$q = C\,\Delta T. \tag{10.11}$$

The heat q_V generated by the reaction at constant volume is given by the product of the amount of substance combusted and the internal energy change,

$$q_V = n\,\Delta_c U_m^\circ. \tag{10.12}$$

Note how the subscript c is used to denote this internal energy change specifically corresponds to the change during combustion. With a molar mass of 128.17 g mol^{-1}, 120 mg corresponds to 0.936 mmol of naphthalene. By the First Law

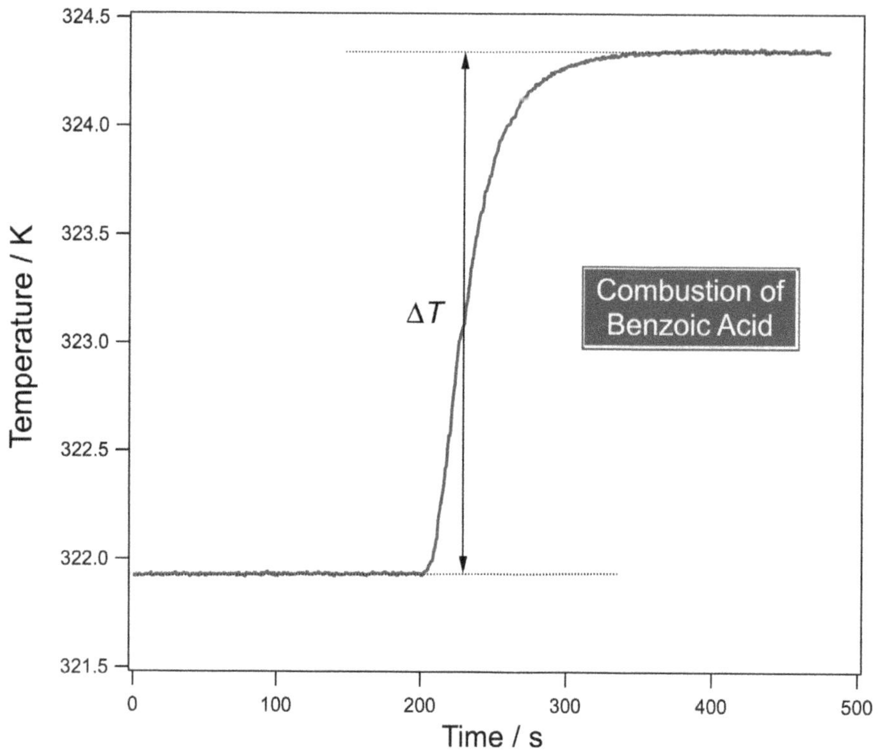

Figure 10.3 Calorimetry data obtained for the combustion of benzoic acid in pure $O_2(g)$ with a bomb calorimeter.

of Thermodynamics, the heat released by combustion $-q_V$ must be equal and opposite to the energy transferred to the calorimeter q,

$$q = -q_V. \tag{10.13}$$

Equating these expressions for heat evolved by the reaction and heat absorbed by the calorimeter – that is, assuming a perfect calorimeter with no contribution from the igniter – we solve the resulting equation for internal energy change

$$\Delta_c U^\circ_m = \frac{C\,\Delta T}{n} = \frac{(1.58\,\text{kJ K}^{-1})(3.06\,\text{K})}{9.36 \times 10^{-3}\,\text{mol}} = 5160\,\text{kJ mol}^{-1}.$$

10.3.2 Example

Calculate $\Delta_c H^\circ_m$ at 298.15 K for the benzoic acid calorimetry experiment in Example 7.9.2.3. Recall that 0.4089 g of C_6H_5COOH were combusted at $T_i = 20.17$ °C with $T_f = 22.22$ °C corresponding to an internal energy change of $\Delta_c U^\circ_m = -3226$ kJ mol^{-1}.

Start with a balanced reaction

$$C_6H_5COOH(s) + 15/2O_2(g) \rightarrow 7CO_2(g) + 3H_2O(l)$$

This is written as a *standard combustion reaction.* A standard combustion reaction corresponds to the reaction of 1 mol of the named reactant with however many moles of gaseous O_2 it takes to oxidize it completely to products. The reaction temperature is 298.15 K unless otherwise stated. If the reactant is a hydrocarbon (or contains only H, C and O), the products are $CO_2(g)$ and $H_2O(l)$. In principle, the same equation applies to a compound containing another element if that element ends in its standard state; for example, N if the product is $N_2(g)$. In general, however, the exact products containing the other elements must be known in order to calculate the enthalpy of combustion.

To calculate the value of $\Delta_c H^\circ_m$, recall that $\Delta_c H = \Delta_c U + \Delta(pV)$. When all reactants and products are in condensed phases at near atmospheric pressure, $\Delta(pV)$ is negligible in comparison with $\Delta_c H$ and $\Delta_c U$. When gases are involved, $\Delta(pV)$ cannot be ignored. Assuming ideal gas behavior, we have

$$\Delta_c H^\circ_m = \Delta_c U^\circ_m + \Delta n_{gas} RT \tag{10.14}$$

$$\Delta n_{gas} = \sum_j \nu_{gas,j}. \tag{10.15}$$

Δn_{gas} is the change in the amount of substance of gas during the course of one full stoichiometric equivalent of reaction. In other words, Δn_{gas} is the difference between the stoichiometric numbers of the gaseous products minus the stoichiometric numbers of the gaseous reactants, $\Delta n_{gas} = \nu_{gas,products} - \nu_{gas,reactants}$. T refers to the final temperature. For the reaction as written above $\Delta n_{gas} = 7 - 7.5 = -0.5$. Hence

$$\Delta_c H^\circ_m = -3226\,\text{kJ mol}^{-1} + (-0.5)(8.314 \times 10^{-3}\,\text{kJ K}^{-1}\,\text{mol}^{-1})(293.32\,\text{K})$$
$$\Delta_c H^\circ_m = -3227\,\text{kJ mol}^{-1}$$

and the enthalpy of combustion is $\Delta_c H^\circ_m = -3227$ kJ mol^{-1} at the final temperature of 293.32 K. To correct this value to the reference temperature of 298.15 K, we apply Eq. (10.9)

$$\Delta_c H^\circ_m(T_2) = \Delta_c H^\circ_m(T_1) + \Delta T\,\Delta_r C^\circ_{p,m}$$

with $T_2 = 298.15\,\text{K}$, $T_1 = 293.32\,\text{K}$, $\Delta T = T_2 - T_1$ and

$$\Delta_r C^\circ_{p,m} = 7C^\circ_{p,m}(CO_2) + 3C^\circ_{p,m}(H_2O) - C^\circ_{p,m}(C_6H_5COOH) - 7.5C^\circ_{p,m}(O_2)$$
$$\Delta_r C^\circ_{p,m} = -6.8\,\text{J K}^{-1}\,\text{mol}^{-1}$$

The temperature correction amounts to only −33 J mol^{-1} and is inconsequential in this case. What does the mol^{-1} refer to? It refers to one full stoichiometric equivalent of reaction as written. Energy in the amount of 3227 kJ is evolved per mole of C_6H_5COOH combusted but per 7.5 mol of O_2 consumed. In other words, per mole for the balanced reaction as written. The enthalpies of combustion of several compounds are listed in Table 10.3.

10.3.3 Directed practice

Why is the determination of the standard molar enthalpy of combustion typically performed with a few grams of material rather than a mole?

10.4 Phase transitions

When heat is added to a substance, its temperature increases in accord with Eq. (10.11). This is shown in Fig. 10.4. As long as no phase transitions or reactions occur, the temperature increases in a simple manner determined by the heat transfer rate and the heat capacity. At a phase transition temperature, further addition of heat does not increase the temperature.

Table 10.3 Standard enthalpy of combustion $\Delta_c H^\circ_m$ and specific heat of combustion at 298.15 K for several compounds. Values taken from the 96th edition of the *CRC Handbook of Chemistry and Physics*.

Compound	Formula	M / g mol^{-1}	$\Delta_c H^\circ_m$ / kJ mol^{-1}	Specific heat of combustion / kJ g^{-1}
Graphite	C	12.011	393.5	32.76
Hydrogen	H_2	2.0159	285.8	141.77
Ammonia	NH_3	17.031	382.8	22.48
Methane	CH_4	16.043	890.8	55.53
Propane	C_3H_8	44.096	2219.3	50.33
Octane	C_8H_{18}	114.229	5470	47.89
Naphthalene	$C_{10}H_8$	128.171	5156.3	40.23
Biphenyl	$C_{12}H_{10}$	154.207	6251	40.54
Methanol	CH_4O	32.042	726	22.66
Ethanol	C_2H_6O	46.068	1366.8	29.67
1-Naphthol	$C_{10}H_8O$	144.17	4957	34.38
Benzophenone	$C_{13}H_{10}O$	182.217	6510	35.73
Benzoic acid	$C_7H_6O_2$	122.122	3228.2	26.43
Glucose	$C_6H_{12}O_6$	180.155	2803	15.56
Lactose	$C_{12}H_{22}O_{11}$	342.296	5630	16.45

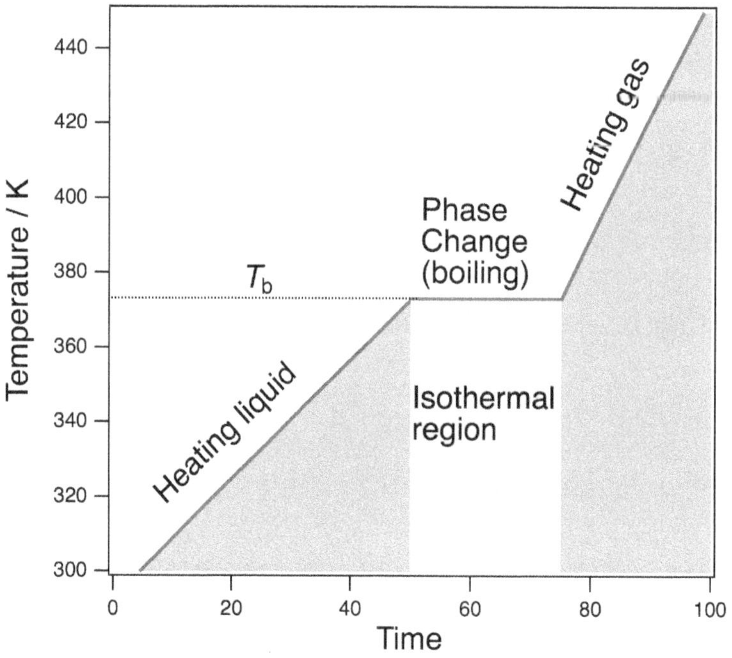

Figure 10.4 Heating of a liquid, water in this example, with heat transfer at a constant rate leads to an increase in the temperature of the liquid to the boiling point. Once a phase transition temperature is reached, the temperature does not increase. The added heat is used to overcome the enthalpy of the transition (vaporization in this case). After all of the liquid has vaporized, added heat increases the temperature of the gas. The gas has a smaller heat capacity than the liquid; therefore, the slope of the heating gas curve is greater than the slope of the heating liquid curve. Slight deviations from linearity in the heating curves result from changes in the heat capacities with temperature.

Instead, the system enters an isothermal region in which the added heat is used to facilitate the phase transition rather than increasing the temperature. Once all of the substance has been converted to the new phase, further heat transfer leads to increases in temperature again. The new phase will have a different heat capacity.

The example of water is shown in Fig. 10.4. starting in the liquid phase. The liquid generally has the highest heat capacity of any phase of a material so the response of temperature to heat transfer is the smallest for this phase – that is, the slope of the temperature versus time curve is the shallowest. However, upon heating, the pattern of increasing temperature, followed by an isothermal region, followed by further increasing temperature is the same for any first-order phase transition. For first-order phase transitions (the most common kind, as discussed and defined further in Chapter 13) there is always a discontinuous change in the heat capacity from one phase to the next. There also will always be an isothermal region because such phase transitions are accompanied by an enthalpy of transition that has to be supplied to make the transition occur (or removed upon cooling).

Phase changes are accompanied by changes in the orientations and distances between molecules (or atoms). The existence of intermolecular interactions means that phase changes will also be accompanied by enthalpy changes. The more drastic the change in molecular orientations and intermolecular distance, the larger the enthalpy changes. The melting of water is endothermic by 6.01 kJ mol^{-1} at 273 K, whereas vaporization of water at 298 K is endothermic by 43.98 kJ mol^{-1}. The vaporization of Au is endothermic by 324 kJ mol^{-1}. Therefore, both the strength of intermolecular interactions and the extent of change of relation of molecules to one another in the two phases determine the magnitude of the enthalpy change during the phase transition.

The enthalpy change for a phase transition is denoted $\Delta_{trs}H$. The heat evolved at constant pressure is equal to the enthalpy change. The *standard temperature of a phase transition* is the temperature at which the phase transition occurs at equilibrium when the pressure is equal to the standard state value of $p° = 1$ bar. In condensed phases it is of little consequence whether the standard state pressure is taken as 1 bar or 1 atm. The difference is more measureable for gases. Therefore, one will encounter the terms *normal transition temperature* (corresponding to $p = 1$ atm) in addition to the standard transition temperature ($p = 1$ bar).

The heat evolved at the standard pressure while a phase transition occurs at its standard transition temperature is equal to the *standard enthalpy of the phase transition*. In molar units this is

$$\Delta_{trs}H^\circ_m = q/n. \tag{10.16}$$

Some common examples are fusion (melting) and vaporization, for instance, for water

$$H_2O(s) \rightarrow H_2O(l) \qquad \Delta_{fus}H^\circ$$

$$H_2O(l) \rightarrow H_2O(g) \qquad \Delta_{vap}H^\circ$$

The standard convention is to write phase transitions with the low-temperature phase on the left and the high-temperature phase on the right. This also sets the convention for the sign of the enthalpy change during a phase transition.

Enthalpy is a state function, therefore, sublimation can be considered the sum of the above two phase changes

$$H_2O(s) \rightarrow H_2O(g) \qquad \Delta_{sub}H^\circ = \Delta_{fus}H^\circ + \Delta_{vap}H^\circ$$

We have already demonstrated that enthalpy is a function of temperature. Thus, to obtain an accurate value for the $\Delta_{sub}H^\circ$ in the sum above, the values of $\Delta_{fus}H^\circ$ and $\Delta_{vap}H^\circ$ have to correspond to the same temperature. Otherwise, the techniques introduced in Chapter 9 would have to be applied to take into account the temperature dependence of the enthalpy change.

10.4.1 Example

A snowball at −15 °C is thrown at a large tree. If we disregard all frictional heat loss, at what speed must the snowball travel so as to melt on impact? Assume that all kinetic energy is transferred into heating only the snowball. The specific heat of fusion of ice is 333 J g^{-1} and the constant pressure heat capacity of water $C_p = 2.09$ J K^{-1} g^{-1}. Hint: The kinetic energy is used to heat the snowball to the melting point and then melt all of it at this temperature.

The kinetic energy is given by

$$E_K = \frac{1}{2}mv^2$$

where m = mass and v = velocity. The total heat energy for this particular problem is

$$q = m\,C_p\Delta T + m\,\Delta_{fus}H^\circ$$

where ΔT = temperature change, C_p = constant pressure heat capacity of water = 2.09 J K^{-1} g^{-1}, and $\Delta_{fus}H$ = 333 J g^{-1}. These two equations are set equal to each other. The kinetic energy must be large enough so that the snowball can increase its temperature from −15 °C to 0 °C (ΔT = 15 K) and then be completely converted to liquid water at 0 °C. Thus,

$$m\,C_p\Delta T + m\Delta_{fus}H = \tfrac{1}{2}mv^2$$

By consistently using SI units, the velocity is certain to be in m s^{-1}.

$$2\left[(15\,\text{K})(2.09\,\text{J}\,\text{K}^{-1}\,\text{g}^{-1})\left(1000\,\tfrac{g}{\text{kg}}\right) + 333\,\text{J}\,\text{g}^{-1}\left(1000\,\tfrac{g}{\text{kg}}\right)\right] = v^2$$

$$v = 854\,\text{m}\,\text{s}^{-1}.$$

A mighty fast snowball, indeed. For reference 100 mph (161 kph) corresponds to 45 m s^{-1}.

10.4.2 Example

Calculate the final temperature of a 5 g sample of Hg after the transfer of 2 kJ of heat at constant pressure from an initial temperature of 298 K. T_b = 655 K, $C_{p,m}(l)$ = 27.4 J K^{-1} mol^{-1}, $C_{p,m}(g)$ = 20.8 J K^{-1} mol^{-1}, $\Delta_{vap}H_m{}^\circ$ = 59.1 kJ mol^{-1}.

We will assume that the 5 g and 2 kJ are exact and that the heat capacities are independent of temperature. The heat will be used to raise the temperature of the sample to its boiling point. If there is heat left over, it will be used to drive the vaporization phase transition – that is, a phase change occurs at constant temperature as the added heat is used to provide the enthalpy change required for the transition. If all the Hg is evaporated, any remaining heat will be used to heat the vapor.

First, calculate the amount of substance.

$$n = m/M = 5\,\text{g}/200.59\,\text{g}\,\text{mol}^{-1} = 2.493 \times 10^{-2}\,\text{mol}$$

The heat required to take the sample to the boiling point is

$$q = nC^{l}_{p,m}\Delta T = 244\,\text{J}.$$

The heat required to vaporize the sample is

$$q_{\text{vap}} = n\,\Delta_{\text{vap}}H^{\circ}_{\text{m}} = (0.02493\,\text{mol})(59.1\,\text{kJ}\,\text{mol}^{-1}) = 1.473\,\text{kJ}.$$

The sum of these two is less than the 2 kJ available so the remaining heat is used to heat the vapor by an amount ΔT_{gas} according to

$$q_{\text{remaining}} = 2\,\text{kJ} - q - q_{\text{vap}} = nC^{\text{g}}_{p,\text{m}}\Delta T_{\text{gas}}.$$

Since $\Delta T_{\text{gas}} = T_f - T_{\text{b}}$, the final temperature is

$$T_f = T_{\text{b}} + \frac{q_{\text{remaining}}}{nC^{\text{g}}_{p,\text{m}}} = T_{\text{b}} + \frac{(2000 - 244 - 1473)\text{J}}{2.493 \times 10^{-2}\,\text{mol}(20.8\,\text{J}\,\text{K}^{-1})} = 1200\,\text{K}.$$

10.5 Bond dissociation and atomization

The *bond dissociation enthalpy* for the species R–X at 298 K, $D^{\circ}_{298}(\text{R}-\text{X})$, is the enthalpy change associated with the dissociation of a specific bond

$$D^{\circ}_{298}(\text{R}-\text{X}) = \Delta_{\text{diss}}H^{\circ}_{\text{m}}(\text{A}-\text{B}) = \Delta_{\text{f}}H^{\circ}_{\text{m}}(\text{R}) + \Delta_{\text{f}}H^{\circ}_{\text{m}}(\text{X}) - \Delta_{\text{f}}H^{\circ}_{\text{m}}(\text{RX}). \quad (10.17)$$

This is the enthalpy change obtained from an ensemble of molecules. Specifically, the enthalpy change when 1 mol of bonds is broken in the standard state. This is related to the bond dissociation energy of one bond in one molecule, but is not exactly the same. The bond dissociation energy of a single molecule is found from the appropriate potential energy curve, similar to the one shown in Fig. 1.1. However, we now know that each molecule has a multitude of rovibrational levels characterized by the vibrational quantum number v and the rotational quantum number J. Each rovibrational level has its own characteristic dissociation energy $D(v,J)$. The largest value is that for the $v = 0$, $J = 0$ state. It is denoted by D_0 and is determined by spectroscopy or from a potential energy surface (PES), as shown in Fig. 10.5, obtained from *ab initio* quantum chemical calculations. An ensemble of molecules at a given finite temperature has a certain amount of thermal energy and a distribution of molecules populating these states. This distribution is temperature-dependent (a Maxwell–Boltzmann distribution at the given temperature), which means that there is a certain probability $P(v,J)$ that each of these states is populated. Even though D_0 is not temperature-dependent, the bond dissociation enthalpy is

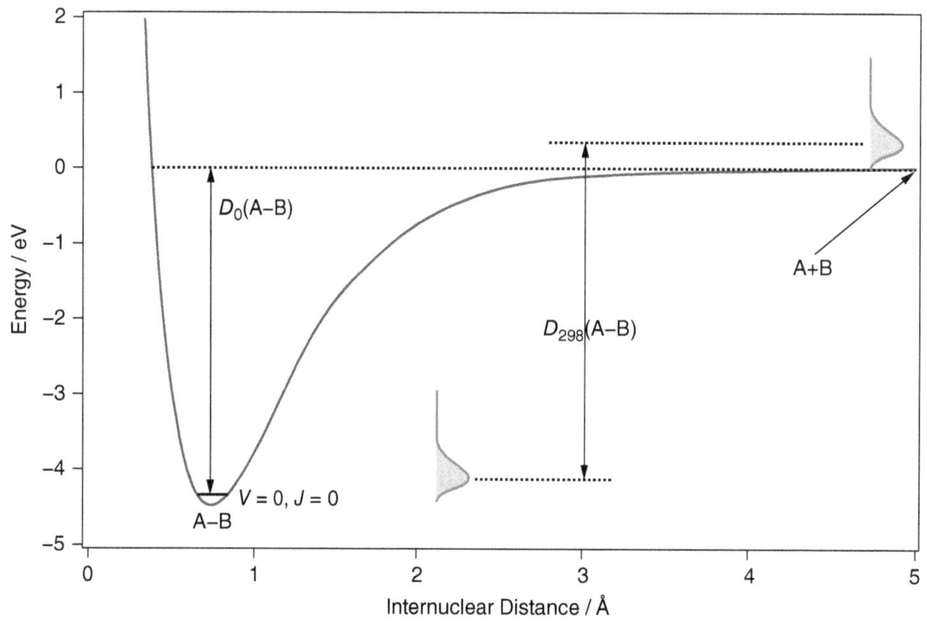

Figure 10.5 Dissociation of an A–B bond from the vibrational ($v = 0$) and rotational ($J = 0$) ground state requires an energy input of D_0 to bring the bond to the dissociated state with zero kinetic energy in the fragments. For an equilibrated sample of AB molecules, their thermal energy is distributed over a Maxwell–Boltzmann distribution of energy, as depicted in the red curve. This leads to a slight different between D_0 and D°_{298}.

because it is obtained by an appropriate thermal average over the dissociation energies of all rovibrational states of the molecule that are populated at that temperature.

At zero kelvin, all molecules are in the lowest energy $v = 0$, $J = 0$ state. Thus, the bond dissociation energy at $T = 0$ K, $D°$(A–B) corresponds to the *bond dissociation energy* of $v = 0$, $J = 0$ state. For a diatomic molecule, the bond dissociation energy at 298 K is related to the absolute zero value by

$$D^{\circ}_{298}(\mathrm{A-B}) = D^{\circ}(\mathrm{A-B}) + \tfrac{3}{2}RT = D^{\circ}(\mathrm{A-B}) + 3.7181\,\mathrm{kJ\,mol^{-1}}. \quad (10.18)$$

Sometimes it is more convenient to speak of an average bond enthalpy, $\overline{D}$. For instance, if we consider the C–C single bond in a number of compounds, we can derive a mean value that is characteristic for the strength of a usual C–C single bond. Thus, D(C–C) = 376 kJ mol^{-1} for ethane, whereas by considering a range of organic compounds we arrive at $\overline{D}$ = 346 kJ mol^{-1}.

The standard *enthalpy of atomization* of an element, $\Delta_{at}H_m°$, is the enthalpy change when 1 mol of gaseous atoms is formed from the element in its standard state. A number of physical processes – including bond breaking, phase transitions and heating and cooling – may make contributions to $\Delta_{at}H°$ (see Example 10.8.2 below). It is also possible for atomization to correspond to other named physical processes, such as dissociation,

$$1/2\,\mathrm{H_2(g)} \rightarrow \mathrm{H(g)} \qquad \Delta_{\mathrm{at}}H^{\circ}_{\mathrm{m}} = 218\,\mathrm{kJ\,mol^{-1}} = 1/2\,D(\mathrm{H{-}H}),$$

vaporization,

$$\mathrm{Hg(l)} \rightarrow \mathrm{Hg(g)} \qquad \Delta_{\mathrm{at}}H^{\circ}_{\mathrm{m}} = 61.5\,\mathrm{kJ\,mol^{-1}} = \Delta_{\mathrm{vap}}H^{\circ}_{\mathrm{m}},$$

or sublimation

$$\mathrm{Ni(s)} \rightarrow \mathrm{Ni(g)} \qquad \Delta_{\mathrm{at}}H^{\circ}_{\mathrm{m}} = 430\,\mathrm{kJ\,mol^{-1}} = \Delta_{\mathrm{sub}}H^{\circ}_{\mathrm{m}}.$$

10.6 Solution

The *molar enthalpy of solution*, $\Delta_{sol}H_m{}^{\infty}$, refers to the complete dissolution of 1 mol of XY in its standard state in an infinite amount of solvent. For a solid electrolyte with water acting as the solvent this corresponds to

$$\mathrm{XY(s)} \rightarrow \mathrm{X^+(aq)} + \mathrm{Y^-(aq)}$$

Dissolution can be either endothermic or exothermic.

The enthalpy of dilution to infinite dilution is also encountered for solutions of common acids, bases, and anything else that dissolves. This is the enthalpy change when a solution of a given molality b at a temperature of 25 °C is diluted with an infinite amount of water. This quantity is almost always exothermic for acids; though, $HClO_4$ in the range of 1–6 mol kg^{-1} is endothermic. We will discuss the energetics of electrolyte dissolution in more detail in Chapter 14.

The molar *enthalpy of hydration* $\Delta_{hyd}H_m{}^{\infty}$ is defined as the enthalpy change when 1 mol of an ideal gas is dissolved in an infinite amount of water. Another term for this quantity is enthalpy of solvation in water at infinite dilution. The enthalpy of hydration influences the distribution of a volatile compound between the aqueous solution phase and air, and is thus important in fields such as environmental science, geochemistry, and chemical engineering. It is related to the enthalpy of solution of the liquid or solid phase, $\Delta_{sol}H_m{}^{\infty}$, by

$$\Delta_{\mathrm{hyd}}H^{\infty}_{\mathrm{m}} = \Delta_{\mathrm{sol}}H^{\infty}_{\mathrm{m}} - \Delta_{\mathrm{vap}}H^{\circ}_{\mathrm{m}}. \quad (10.19)$$

where $\Delta_{vap}H^{\circ}_{m}$ is the molar enthalpy of vaporization. The enthalpy of hydration is always negative.

10.6.1 Directed practice

(a) When 1 g of KCl is dissolved in 500 ml of water at an initial temperature of 298.00 K, the temperature drops by 0.21 K. What is the molar heat of solvation, $\Delta_{sol}H$, for KCl? (b) For HCl, $\Delta_{sol}H$ = −74.8 kJ mol^{-1}. What is the temperature change when 1 g of HCl is dissolved in 500 ml of water? For water C_p = 2.09 J K^{-1} g^{-1}.

[Answers: (a) 16 kJ mol^{-1}; (b) 1.96 K]

10.7 Enthalpy of formation

The standard *enthalpy of formation* $\Delta_f H_m°$ is the heat released or consumed when 1 mol of a product in its standard state is formed from elements in their standard state at a pressure of 1 bar. In a standard formation reaction there is one and only one product and it has a stoichiometric coefficient of unity. Examples include

$$1/2\,N_2(g) + 3/2\,H_2(g) \rightarrow NH_3(g) \qquad \Delta_f H_m = -46\,\text{kJ mol}^{-1}$$
$$U(s) + 3\,F_2(g) \rightarrow UF_6(g) \qquad \Delta_f H_m^\circ = -121\,\text{kJ mol}^{-1}$$

$\Delta_f H_{m°}$ of an element in its standard state is zero at all temperatures.

10.8 Hess's law

The importance of standard enthalpies of formation is fully recognized when we realize that enthalpy is a state function. Because they are path-independent, standard enthalpies of individual reactions, such as formation reactions, can be combined to obtain the enthalpy of another reaction. This is stated as *Hess's law*:

> *The standard enthalpy of an overall reaction is the sum of the standard enthalpies of the individual reactions into which a reaction may be divided.*

It does not matter what combination of reactions we use. As long as the sum of these reactions adds up to the balanced stoichiometric equation of the reaction we are attempting to describe, then the sum of the associated enthalpy changes will also add up to the enthalpy change of the overall reaction. Recall as well that, for a state function, reversing the process corresponds to reversal of sign of the value of the change of that state function. In other words, for a forward reaction with enthalpy change $\Delta_r H$, the reverse reaction has enthalpy change $-\Delta_r H$. Thus, one way to visualize the use of Hess's law is in a tabular form as the sum of a series of reactions,

$$\begin{array}{l} A + B \rightarrow C + D \\ C + E \rightarrow F \\ F + D \rightarrow G \\ \hline A + B + E \rightarrow G \end{array}$$

10.8.1 Example

Calculate the enthalpy of reaction at 298 K and 1 bar for the phase transition of S from the rhombic to the monoclinic structure, given the following, $\Delta_c H_m^\circ(\text{rhombic}) = -296.8$ kJ mol^{-1} and $\Delta_c H_m^\circ(\text{monoclinic}) = -297.1$ kJ mol^{-1}.

The combustion reactions are given by

(1) $$S(\text{rhombic}) + O_2(g) \rightarrow SO_2(g) \qquad \Delta_r H_1$$

(2) $$S(\text{monoclinic}) + O_2(g) \rightarrow SO_2(g) \qquad \Delta_r H_2$$

The required reaction is

(3) $$S(\text{rhombic}) \rightarrow S(\text{monoclinic}) \qquad \Delta_r H_3$$

which is simply Reaction (1) – Reaction (2). Therefore, $\Delta_r H_1 - \Delta_r H_2 = \Delta_r H_3$

$$\Delta_r H_m^\circ(S(\text{rhombic}) \rightarrow S(\text{monoclinic})) = -296.8 - (-297.1) = 0.3\,\text{kJ mol}^{-1}$$

10.8.2 Example

Determine the enthalpy of atomization at $T = 298$ K for Hg using Hess's Law. $T_b = 655$ K, $C_{p,m}(l) = 27.4$ J K^{-1} mol^{-1}, $C_{p,m}(g) = 20.8$ J K^{-1} mol^{-1}, $\Delta_{vap} H_m^\circ = 59.1$ kJ mol^{-1}.

In diagrammatic form, we can draw a Hess's law cycle, as shown in Fig. 10.6. Recall that the change in any state function is precisely zero for one complete circuit of a cycle. For any cycle with n steps, only $n - 1$ steps are independent. That is, if we know $n - 1$ values along the cycle we can use completeness (summation to zero) to calculate the one missing value.

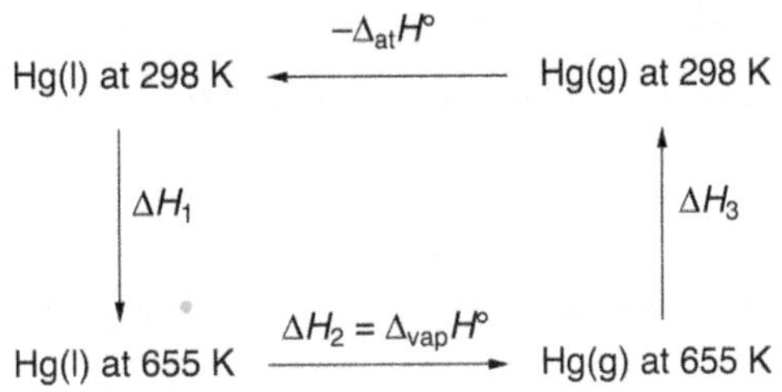

Figure 10.6 A diagrammatic representation of a Hess's law cycle.

By Hess's law

$$\Delta H_1 + \Delta H_2 + \Delta H_3 - \Delta_{at}H_m^\circ = 0$$

The sign on $\Delta_{at}H_m{}^\circ$ is negative since returning from Hg(g) to Hg(l) at 298 K is the reverse of atomization. Thus,

$$\Delta_{at}H_m^\circ = \Delta H_1 + \Delta H_2 + \Delta H_3.$$

Assuming that the heat capacities are constant, we make the identifications

$$\Delta H_1 = C_{p,m}(l)[T_b - T]$$
$$\Delta H_2 = \Delta_{vap}H_m^\circ$$

and

$$\Delta H_3 = C_{p,m}(g)[T - T_b].$$

Substituting in the appropriate values we obtain,

$$\Delta_{at}H_m^\circ = 9.78 + 59.1 - 7.43 = 61.5\,\mathrm{kJ\,mol^{-1}}.$$

The enthalpy change at 298 K differs from that at 655 K because the heat capacity of the gas is less than that of the liquid. This is a general rule: the heat capacity of the liquid phase is usually higher than that of either the gas or the solid phase of the material. Physically, it takes more energy on average to transfer an atom from the liquid to the gas phase at 298 K than at 655 K. The difference in heat capacities means that more energy has to be put into the liquid phase to raise its temperature than the gas phase; thus the enthalpy difference between the liquid and gas phases is lower at higher temperature.

10.9 Reaction enthalpy from enthalpies of formation

Extension of Hess's law allows us to find the enthalpy of any reaction if the enthalpy of formation of each of the products and reactants is know by using

$$\Delta_r H^\circ = \sum_{\text{products}} v\Delta_f H^\circ - \sum_{\text{reactants}} v\Delta_f H^\circ \tag{10.20}$$

where v is the absolute value of the stoichiometric coefficient from the balanced chemical reaction. Equation (10.20) is pedagogically correct: multiply each enthalpy of formation by the stoichiometric coefficient of the corresponding reactant or product. Sum all the products. Sum all the reactants, then subtract the sum of reactants from the sum of products. Note that, according to IUPAC definition, the stoichiometric number is positive for products and negative for reactants. Therefore, it is possible to rewrite Eq. (10.20) in the mathematically and notationally more succinct and accurate form

$$\Delta_r H^\circ = \sum_j v_j \Delta_f H_j^\circ. \tag{10.21}$$

The origin of the enthalpy scale is set by assigning a value of zero to the enthalpy of formation of an element in its standard state.

10.9.1 Example

Calculate the standard enthalpy change for the Boudouard reaction shown below at 298 K.

$$2CO(g) \rightarrow CO_2(g) + C(\text{graphite})$$

$$\Delta_r H_m^\circ = \Delta_f H_m^\circ(CO_2) + \Delta_f H_m^\circ(\text{graphite}) - 2\Delta_f H_m^\circ(CO)$$
$$= -393.5\,\text{kJ}\,\text{mol}^{-1} + 0 - 2(-110.5\,\text{kJ}\,\text{mol}^{-1}) = -172.5\,\text{kJ}\,\text{mol}^{-1}.$$

10.9.2 Example

Solar thermal electricity generation is very efficient in latitudes below about 42°; however, it only works when the sun shines and it requires efficient heat storage for use during dark hours. One method being developed is to use phase-change materials such as molten salts, which are melted during the day and then allowed to fuse and liberate heat at night. Electricity generation requires a high-temperature material such as KNO_3, T_{fus} = 334 °C. For home or pool heating applications, a lower temperature material is required, such as a hydrated salt:

$$Na_2SO_4{\cdot}10H_2O(s) \rightarrow Na_2SO_4(s) + 10H_2O(l).$$

At temperatures above 32.4 °C the reaction goes completely to the right. At night, when the temperature falls below 32.4 °C, the reaction goes completely to the left. If the efficiency of the system is 100%, how much heat could 322 kg of $Na_2SO_4{\cdot}10H_2O$ provide to a home at night? Use the following information:

Compound	$\Delta_f H_m$° (kJ mol^{-1})
$Na_2SO_4{\cdot}10H_2O(s)$	−4324.2
$Na_2SO_4(s)$	−1384.5
$H_2O(l)$	−285.8

This problem can be solved once the amount of heat 1 mol of $Na_2SO_4{\cdot}10H_2O$ absorbs during the day is known. An endothermic reaction is one that absorbs heat from the surroundings. The amount of heat absorbed can be calculated from the standard enthalpies of formation given. The enthalpy of the reaction – that is, the amount of heat absorbed or released – is equal to

$$\Delta_r H_m^\circ = \sum_j \nu_j \Delta_f H_{m,j}^\circ.$$

During the day when the reaction absorbs heat and proceeds to the right,

$$\Delta_r H_m^\circ = \Delta_f H_m^\circ(Na_2SO_4) + 10\Delta_f H_m^\circ(H_2O) - \Delta_f H_m^\circ(Na_2SO_4{\cdot}10H_2O).$$
$$\Delta_r H_m^\circ = (-1384.5) + 10(-285.8) - (-4324.2) = 81.7\,\text{kJ}\,\text{mol}^{-1}.$$

Thus, 1 mol of $Na_2SO_4{\cdot}10H_2O$ absorbs 81.7 kJ of heat from the sun during the day. At night, the reaction is reversed, which means it is exothermic, or releases heat. If the reaction absorbs 81.7 kJ mol^{-1} of heat when it proceeds to the right, it must release 81.7 kJ mol^{-1} of heat to the house at night, when it proceeds to the left. The amount of substance of $Na_2SO_4{\cdot}10H_2O$ in 322 kg can be calculated. The molar mass of $Na_2SO_4{\cdot}10H_2O$ is 322 g mol^{-1}; thus, n = 1000 mol. Therefore, the amount of heat released into the house is

$$q = n\Delta_r H_m^\circ = 8.17 \times 10^4\,\text{kJ}.$$

10.10 Calculating enthalpy of reaction from enthalpies of combustion

Experimentally, especially for hydrocarbons and other organics, the enthalpy of combustion is much more accessible than the enthalpy of formation. The standard enthalpy of combustion, $\Delta_c H_m$°, is the reaction enthalpy for the complete oxidation of 1 mol of an organic compound to CO_2(g) and H_2O(l) if the compound contains C, H and O, and additionally to N_2(g) if N is also present. For instance, for glucose,

$$C_6H_{12}O_6\,(s) + 6\,O_2(g) \rightarrow 6\,CO_2(g) + 6H_2O(l) \qquad \Delta_c H_m^\circ = -2808\,\text{kJ}\,\text{mol}^{-1}.$$

This is the also the same enthalpy change that the body experiences by metabolizing glucose. ΔH is a state function; therefore, regardless of the path taken to oxidize glucose to CO_2 and H_2O the energy gained is the same. By combining enthalpies of formation and phase transitions with enthalpies of combustion in accord with Hess's law, the reaction enthalpy of most any reaction can be obtained.

10.10.1 Directed practice

Given that the standard enthalpies of formation of water and CO_2 are −285.83 and −393.51 kJ mol^{-1}, respectively, calculate the standard enthalpy of formation of glucose from its enthalpy of combustion. [−1273 kJ mol^{-1}]

10.11 The magnitude of reaction enthalpy

We now know how to determine the enthalpy change during a chemical reaction. Experimentally, we would perform a calorimetry experiment. We measure a temperature rise in a vessel with calibrated heat capacity when a known quantity of reactant reacts. Computationally, we can combine standard reaction enthalpies, for instance enthalpies of formation, combustion and phase transitions, with appropriate stoichiometric numbers (positive for products, negative for reactants) to obtain $\Delta_r H°$ for the specific reaction under consideration. We also know how to adjust this value for changes in temperature as long as we know the heat capacities. But what makes a reaction either endothermic or exothermic? Why are some reactions more exothermic than others?

The formal answers to these questions are that when a system moves from a state of high enthalpy to a state of lower enthalpy, the reaction is exothermic. The greater the difference, the greater exothermicity. If a system moves from a state of low enthalpy to a state of higher enthalpy, the reaction is endothermic. But what determines the enthalpy of a chemical system? The answer can be found in the bonds and intermolecular interactions. For the moment let us simplify the discussion by taking an ideal gas (no intermolecular interactions) near zero kelvin (no thermal energy; thus, no rotational, vibrational and electronic excitation and no need to consider the quantum mechanical structure of the molecules). The reactants have certain bonds. The products have a different set of bonds, each described by a potential such as that in Fig. 10.7(a).

Enthalpy change is a state function so we are free to choose any path we want to determine its change. Let us take the reactants and transfer them into an intermediate state in which all of the bonds are broken. It takes a great deal of energy to break bonds. This intermediate state has a much higher enthalpy because we have broken all the bonds that were in the reactants. Energy must be put into the system to break the bonds. Note that chemistry almost never happens this way. The beauty and mystery of chemistry is that the true transition state of a reaction such as A...B...C in Fig. 10.7(b),

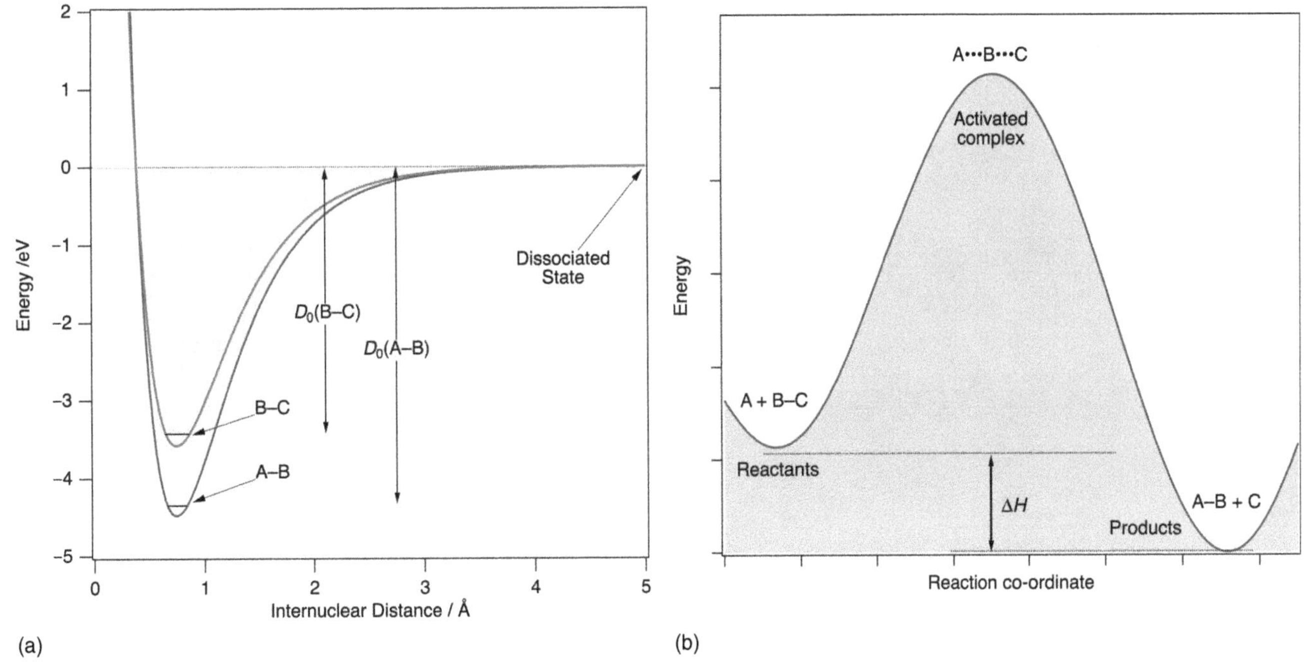

Figure 10.7 The enthalpy of a reaction is determined by the difference in enthalpies of the products and reactants. When stronger bonds replace weaker bonds, the reaction is exothermic.

is never this high in energy. However, this fully dissociated intermediate state represents an upper limit to the energy that we could expect for a transition state.

In the second step of our imagined reaction path, we take all of the atoms that were formed through dissociation of the reactants and we allow them to form all the bonds that are required to form the products. Energy is released when bonds are formed. The formation of bonds lowers the enthalpy of the system. If the bonds that are formed to make the products are stronger on average than the bonds that were broken in the reactants, then the reaction will be exothermic. If, on average, weaker bonds are formed in the products than were broken in the reactants, then the reaction will be endothermic.

I do not like the misleading terminology that energy is stored in chemical bonds. In biology and biochemistry, talk of 'high-energy phosphate bonds' is often heard. This can easily lead to misconceptions.

> *Energy is not released from bonds when they are broken. Energy is released from a reactive system when weaker bonds are replaced by stronger bonds being formed.*

High-energy materials are materials that have weak bonds and that can be reacted to form much more stable compounds. Instead of high-energy materials or high-energy bonds, it is much better to think in terms of high-enthalpy materials. Two hydrogen atoms have much higher enthalpy than a hydrogen molecule that contains two H atoms. How can we possibly talk of the energy stored in the bonds when there is no bond in the reactants of the reaction $H + H \rightarrow H_2$, even though this reaction is exothermic by over 400 kJ mol^{-1}? If you believe that energy is released when bonds are broken you will never understand thermochemistry, transition state theory or reaction dynamics. In other words, you will never understand chemistry and how it works.

10.11.1 Directed practice

How much energy is released by breaking high-energy phosphate bonds?

SUMMARY OF IMPORTANT EQUATIONS

$\xi \equiv \dfrac{n_i - n_{i0}}{\nu_i} = \dfrac{\Delta n_i}{\nu_i}$	Extent of reaction ξ, n_i is moles of species i with initial value n_{i0} and stoichiometric number ν_i.
$q = \int C\,\mathrm{d}T$	Heat q is equal to the integral over temperature T of the heat capacity C.
$q_V = n\Delta U_\mathrm{m}$	The heat evolved at constant volumes is equal to the internal energy change.
$q_p = n\Delta H_\mathrm{m}$	The heat evolved at constant pressure is equal to the enthalpy change.
$\Delta H^\circ = \Delta U^\circ + p\Delta V$	Standard enthalpy change ΔH° at constant pressure p is equal to standard internal energy change ΔU° plus p time volume change. $p\,\Delta V = \Delta n_\mathrm{gas}\,RT$ at constant p and T.
$\Delta H^\circ_\mathrm{m} = \Delta U^\circ_\mathrm{m} + \Delta n_\mathrm{gas}RT$	Enthalpy and internal energy changes during reaction differ by an amount related to the change in the moles of gas.
$\Delta n_\mathrm{gas} = \sum_j \nu_{\mathrm{gas},j}$	
$\Delta_\mathrm{r}H^\circ = \sum_j \nu_j \Delta_\mathrm{f}H^\circ_j$	The enthalpy change of any reaction can be calculated from enthalpies of formation.
$\Delta_\mathrm{r}H^\circ_\mathrm{m}(T_2) = \Delta_\mathrm{r}H^\circ_\mathrm{m}(T_1) + \int_{T_1}^{T_2} \Delta_\mathrm{r}C^\circ_{p,\mathrm{m}}\,\mathrm{d}T$	Temperature dependence of reaction enthalpy is related to the change in heat capacity upon reaction.
$\Delta_\mathrm{r}C_{p,\mathrm{m}} = \sum_j \nu_j C^\circ_{p,\mathrm{m},j}$	$C^\circ_{p,\mathrm{m},j}$ is the molar heat capacity at constant pressure of the jth species, ν_j = stoichiometric number of jth species.

Exercises

10.1 Calculate the final temperature of the system when (a) 10.0 g of $H_2O(l)$ initially at 373 K is allowed to equilibrate with 100.0 g of Ag(s) initially at 300 K and (b) 10.0 g of $H_2O(g)$ initially at 373 K is allowed to equilibrate with 100.0 g of Ag(s) initially at 300 K. $C_{p,\mathrm{m}}(\mathrm{Ag}) = 25.4$ J K^{-1} mol^{-1}. $C_{p,\mathrm{m}}(H_2O(l)) = 76.0$ J K^{-1} mol^{-1}. $C_{p,\mathrm{m}}(H_2O(g)) = 37.5$ J K^{-1} mol^{-1}. $\Delta_\mathrm{vap}H^\circ_\mathrm{m}(H_2O) = 40.65$ kJ mol^{-1}.

10.2 Calculate the final temperature of a 5 g sample of Hg after the transfer of 2 kJ of heat at constant pressure from an initial temperature of 298 K. Take into account the temperature dependence of the heat capacity of the liquid.

10.3 Calculate the amount of heat that must be transferred to CH_4 on a molar basis to bring it from 298 K to 1300 K at constant pressure of 1 bar. (a) Assume a constant heat capacity of $C_{p,\mathrm{m}}° = 35.7\ \mathrm{J\ K^{-1}\ mol^{-1}}$. (b) Assume the behavior of $C_{p,\mathrm{m}}°$ is described by the Shomate equation with $A = -0.703029$, $B = 108.4773$, $C = -42.52157$, $D = 5.862788$, and $E = 0.678565$. Can any of the terms in the integral be neglected to obtain an accurate answer that includes three significant figures?

10.4 (a) Write out a balanced chemical reaction including standard states (i.e., phases) for the combustion of benzene. (b) Write out the balanced chemical reaction which has an enthalpy of reaction given by $\Delta_f H(C_2H_5OH)$. (c) Use enthalpies of formation and/or atomization to calculate a mean bond dissociation energy for the C–H bonds in CH_4(g).

10.5 At 298 K toluene C_7H_8(l) has an enthalpy of formation of 12.0 kJ mol^{-1}, an enthalpy of combustion of −3920 kJ mol^{-1}, an enthalpy of fusion of 6.64 kJ mol^{-1}, and an enthalpy of vaporization of 38.06 kJ mol^{-1}. Write out balanced chemical reactions that correspond to: (i) formation; (ii) combustion; and (iii) condensation directly from the gas phase to the solid phase of toluene. Calculate the enthalpy change when 100 g of toluene is: (i) formed; (ii) combusted; and (iii) condenses directly from the gas phase to the solid phase.

10.6 (a) Write out the balanced chemical reaction which has an enthalpy of reaction given by $\Delta_f H_\mathrm{m}°(K_2SiF_6)$. (b) At 298 K ethane C_2H_6(g) has a molar enthalpy of formation of −84.0 kJ mol^{-1}, a molar enthalpy of combustion of −1560.7 kJ mol^{-1}, and a molar enthalpy of vaporization of 5.16 kJ mol^{-1}. Write out balanced chemical reactions that correspond to: (i) formation; (ii) combustion; and (iii) condensation. Calculate the enthalpy change when 100 g of ethane is: (i) formed; (ii) combusted; and (iii) condensed.

10.7 Use the following information for $SiCl_4$ to answer the questions below. All thermodynamic values are for 298 K apart from $\Delta_\mathrm{vap}H°$, which is given at T_b. Assume that $C°_{p,\mathrm{m}}$ is constant.

$\Delta_f H°$ / kJ mol^{-1}	$\Delta_f G°$ / kJ mol^{-1}	$C°_{p,\mathrm{m}}$ / J mol^{-1} K^{-1}	$\Delta_\mathrm{vap}H°$ / kJ mol^{-1}	T_fus / °C	T_b / °C
−687.0	−619.8	145.3	30.0	−68.76	57.65

(a) Write the standard formation reaction for $SiCl_4$ at 298 K. (b) Calculate the standard molar entropy change of formation for $SiCl_4$(l) at 298 K. State with justification if the entropy change favors reaction. (c) Calculate the standard molar entropy change for vaporization of $SiCl_4$(l) at the normal transition temperature. (d) Calculate the work performed in the standard formation reaction of $SiCl_4$ on a per mole basis at 298 K. (e) Calculate the energy required per mole to take $SiCl_4$ from its standard state at 298 K to the gas phase at 330.8 K.

10.8 For the reaction of MgN(s) + H_2O(l) to form N_2(g), H_2(g) and MgO(s), calculate the final temperature of the mixture if 15.0 g of MgN is reacted to completion with (a) 150.0 g of H_2O or (b) 1000.0 g of H_2O. The heat capacities of water, MgO, N_2 and H_2 are $C_{p,\mathrm{m}} = 75.5$ J K^{-1} mol^{-1}, 66.9 J K^{-1} mol^{-1}, 29.1 J K^{-1} mol^{-1} and 28.8 J K^{-1} mol^{-1}, respectively. The enthalpies of formation of MgN, H_2O and MgO are $\Delta_f H° = 288.7$ kJ mol^{-1}, −241.8 kJ mol^{-1} and −532.6 kJ mol^{-1}, respectively. $\Delta_\mathrm{vap}H°(H_2O) = 44.0$ kJ mol^{-1} and $\Delta_\mathrm{fus}H°(\mathrm{MgO}) = 9.04$ kJ mol^{-1}. $T_i = 300$ K.

10.9 Use the information below for all parts.

	$\Delta_f H°_\mathrm{m}$(298 K) kJ mol^{-1}	$C_{p,\mathrm{m}}$ J K^{-1} mol^{-1}
H_2O(l)	−285.83	see below
H_2O(g)	−241.83	33.58

The mean heat capacity of H_2O(l) in the range 273 to 373 K is 75.52 J K^{-1} mol^{-1}.

The exact heat capacity of H_2O(l) in this range in J K^{-1} mol^{-1} is given by $C°_{p,\mathrm{m}} = a + bT + cT^2 + dT^3$ with $a = 175.5$, $b = -0.842$, $c = 0.00232$ and $d = -2.08 \times 10^{-6}$. If T in K is used, the units of the heat capacity will be correct. The density of liquid water is 999.84 kg m^{-3} at 273 K and 958.63 kg m^{-3} at 373 K.

a Calculate the amount of energy required to convert 1 mol of liquid water at 273 K to steam at 373 K. Assume a constant heat capacity.

b Calculate the work done during the heating step and compare it to the work done during the phase transition by calculating the ratio of the work done during heating to the work.

c Calculate $\Delta S_m°$ for the conversion in part (a). Calculate the % difference between assuming a constant value for the heat capacity compared to using the exact heat capacity in the heating step.

10.10 Given the reactions (1) and (2) below, determine (a) $\Delta_r H°$ and $\Delta_r U°$ for reaction (3), (b) $\Delta_f H°$ for both HCl(g) and $H_2O(g)$ all at 298 K.

(1) $H_2(g) + Cl_2(g) \rightarrow 2HCl(g)$	$\Delta_r H° = -184.62\ \text{kJ mol}^{-1}$
(2) $2H_2(g) + O_2(g) \rightarrow 2H_2O(g)$	$\Delta_r H° = -483.64\ \text{kJ mol}^{-1}$
(3) $4HCl(g) + O_2(g) \rightarrow 2Cl_2(g) + 2H_2O(g)$	$\Delta_r H° = ?\ \text{kJ mol}^{-1}$

10.11 Coal is a heterogeneous substance whose exact composition depends on the geographical location of the mine. The best coal is anthracite, such as that from mines near Shamokin, PA (now nearly empty), which is virtually all C and H with a small amount of O. More conventional bituminous coal (black coal), such as from the strip mines Wyoming or the deep mines of West Virginia, releases less heat when combusted and contains more N and S. Lignite (brown coal) such as that found in Germany and the Latrobe Valley in Australia has the lowest calorific value, contains many impurities especially S and a good deal of water.

Assume that we are working with a sample of anthracite that is 95% by weight C and 5% H. The calorific value (the amount of heat given off upon combustion) is 3.26×10^4 kJ kg^{-1}.

a Calculate the approximate molar mass and chemical formula (C_xH_y) of this coal. Round up to the nearest integer when determining x and y in the chemical formula.

b Write out the combustion reaction for coal and calculate the heat of combustion of coal in kJ mol^{-1}.

c What volume in m^3 at SATP (that is, at standard ambient temperature and pressure = 298 K and 1 atm) of CO_2 is formed when 1 metric tonne of coal is burned? (1 metric ton = 1 Mt = 1000 kg and there are 9891 mol of coal in 1 Mt of coal.)

d The annual production of coal in the USA is about 1.1 billion metric tonnes. How much energy is released by burning this much coal? How much CO_2 (in mass and in volume in m^3 at SATP) is produced by burning this coal?

e To sequester this CO_2 underground it must be compressed to a final pressure of $p_f = 6200$ kPa at $T_f = 45$ °C so that it can be injected into the earth. What is the amount of energy required to compress the CO_2 from 1 tonne of coal from SATP if the expansion is performed reversibly? What percentage of the energy available as electricity made from the burning of coal is required to sequester the CO_2? The efficiency of a modern coal-fired power plant is approximately 45%. $C_p°(CO_2) = 37.11$ J K^{-1} mol^{-1} and assume that it is constant over the required temperature range.

10.12 The 'calories' referred to on the side of food packages are actually kcal. If you drink 1.00 gallon of ice-cold water, how many calories (kcal) does it take to raise the temperature of the water to normal body temperature of 98.6 ° F? If a person weighing 150 lb burns 97.95 kcal per mile, how far would the person need to walk to burn the same amount of calories as drinking the water?

10.13 (a) A strip of Mg of mass 15 g is dropped into a beaker of dilute hydrochloric acid. Calculate the work done by the system as a result of the reaction. The atmospheric pressure is 1.0 atm and the temperature is 25 °C. (b) An average human produces about 10 MJ of heat each day through metabolic activity. If a human body were an isolated system of mass 65 kg with the heat capacity of water, what temperature rise would the body experience? Human bodies are actually open systems, and the main mechanism of heat loss is through the evaporation of water. What mass of water should be evaporated each day to maintain constant temperature?

10.14 (a) $\Delta_r H°$ for the mutarotation of glucose in aqueous solution,

$$\alpha\text{-D-glucose(aq)} \rightarrow \beta\text{-D-glucose(aq)}$$

is −1.16 kJ mol^{-1}. The enthalpy of solution of α-D-glucose(s) is 10.72 kJ mol^{-1} and the enthalpy of solution of β-D-glucose(s) is 4.68 kJ mol^{-1}. Calculate $\Delta_r H°$ for the mutarotation of solid α-D-glucose to solid β-D-glucose.

(b) Using enthalpies of formation, calculate $\Delta_r H°$ for the hydrolysis of aqueous urea into $CO_2(aq)$ and $NH_3(aq)$ at 25 °C.

10.15 When 83.7 mg of biphenyl was burned in a bomb calorimeter the temperature rose by 2.12 K. Calculate the calorimeter constant. By how much will the temperature rise when 50 mg of 1-naphthol is burned in the calorimeter under the same conditions?

10.16 Magnesiothermic reduction of silica may be an economical method for producing nanostructured porous silicon from plants that naturally accumulate silica. Calculate the enthalpy of reaction for this reaction

$$2Mg(g) + SiO_2(s) \rightarrow Si(s) + 2MgO(s)$$

at 1393 K. The enthalpies of formation and heat capacities at 298 K are given below. $\Delta_{sub}H°_m(Mg) = 132$ kJ mol^{-1} at 1393 K.

	Mg	SiO_2	Si	MgO
$\Delta_f H°_m$ / kJ mol^{-1}		−859.4		−601.83
$C_{p,m}$/ (J K^{-1} mol^{-1})	24.39	44.98	20.01	37.10

10.17 The enthalpy of combustion of propane C_3H_8 is 2219.3 kJ mol^{-1}, the heat capacity at constant pressure of water is 75.3 J K^{-1} mol^{-1} in the liquid phase and 33.6 J K^{-1} mol^{-1} in the gas phase. $\Delta_{vap}H^\circ_m = 40.65$ kJ mol^{-1} for water. Take these values to be constant. (a) Calculate the energy released by the combustion of 10.0 g of propane. (b) Assume that all the heat generated in (a) is transferred at constant pressure to 100.0 g of $H_2O(l)$ held initially at 298 K. Calculate the final temperature of the water. (c) Calculate the work performed by water if it expands against a constant pressure of 100 kPa.

10.18 Acetotrophic bacteria can enzymatically catalyze the fermentation of acetic acid according to the reaction $CH_3COOH(aq) \rightarrow CH_4(g) + CO_2(g)$.

T = 298 K	$CH_3COOH(l)$	$CH_4(g)$	$CO_2(g)$
$\Delta_f H°$ / kJ mol^{-1}	−484.3	−74.6	−393.5

(a) Calculate $\Delta_r H_m°$ for the reaction at 298 K if the enthalpy of solvation of $CH_3COOH(l)$ is −1.5 kJ mol^{-1}. (b) A researcher places 100 ml of 1.06 M acetic acid in a 500 ml flask. After adding enough bacteria to completely react all of the acetic acid, the flask was stoppered. Calculate the final temperature and pressure in the flask assuming that the stopper stays in the flask and that all parameters are temperature-independent. Take the heat capacity of the system to be that of water, 75.3 J K^{-1} mol^{-1} (in other words, consider just the heating of water in the temperature change calculation). The solution has a density of 1.01 g cm^{-3} and has an acetic acid mass percent of 6.00%. The initial temperature and pressure are 298.0 K and 101 kPa. (c) How fast will the glass stopper with mass 17.65 g go if all of the work of the gas evolved in the reaction is used to accelerate the stopper? To calculate the work, consider expansion of a gas against the constant lab pressure of 101 kPa.

10.19 Why is the two-phase system of a liquid in contact with its corresponding solid a good candidate for maintaining isothermal conditions?

10.20 Why is the bond dissociation enthalpy a function of temperature but the bond dissociation energy is not?

10.21 Calculate q, w, ΔU and ΔH in kJ mol^{-1} when water ice at 0 °C is heated at standard pressure and is converted to steam at 100 °C. $\Delta_{fus}H^\circ_m(T = 273\ \text{K}) = 6.01$ kJ mol^{-1} and $\Delta_{vap}H^\circ_m(T = 373\ \text{K}) = 40.65$ kJ mol^{-1}. The heat capacity of liquid water in J K^{-1} mol^{-1} is $C^\circ_{p,m} = a + bT + cT^2 + dT^3 + eT^{-2}$, with T in K and $a = -203.6$, $b = 1.523$, $c = -0.003196$, $d = 2.474 \times 10^{-6}$, $e = 3.855 \times 10^{6}$.

10.22 The enthalpy of vaporization of water is 40.65 kJ mol^{-1} at 373 K. The heat capacities of liquid and gaseous water are $C°_{p,m}(l) = 74.57$ J K^{-1} mol^{-1} and $C°_{p,m}(g) = 25.96$ J K^{-1} mol^{-1}. (a) Calculate the energy required to evaporate 145 g of H_2O at 100 °C and 25 °C. (b) Explain on the basis of intermolecular interactions the difference or lack thereof for the two answers in part (a).

10.23 Use the information in the table below to answer these questions for the reaction of B_2O_3 with $AsCl_3$ to form BCl_3 and As_2O_5. Assume that all values given in the table below are independent of temperature over the range 280–425 K. (a) Calculate $\Delta_r H°$, $\Delta_r S°$ and $\Delta_r G°$ at 280 K. (b) Calculate $\Delta_r H°$, $\Delta_r S°$ and $\Delta_r G°$ at 425 K. (c) State with justification whether increasing the temperature increases the production of products.

	BO	$AsCl_3$	O_2	BCl_3	As_2O_5
$\Delta_f H_m°$ / kJ mol^{-1}	−1263.6	−335.6	0	−418.4	−914.6
$\Delta_{vap} H_m°$ / kJ mol^{-1}		35.01	6.82	23.77	
$S_m°$ / J K^{-1} mol^{-1}	54.01	233.5 (l) 327.2 (g)	205.0	209.2 (l) 324.2 (g)	106.2
T_{fus} / K	723	264	55	166	>588
T_b / K	2133	403	90	286	decomposes

10.24 Consider the gas-phase reaction (all gases considered ideal)

$$NO + N_2O_5 \rightleftharpoons N_2O_4 + NO_2$$

$T = 298$ K	NO	N_2O_5	N_2O_4	NO_2
$\Delta_f H_m°$ / kJ mol^{-1}	91.3	13.3	11.1	33.2
$S_m°$ / J K^{-1} mol^{-1}	210.8	355.7	304.4	240.1

(a) Calculate $\Delta_r G_m°$ at 298 K. (b) Calculate the final partial pressures of all the gases when equilibrated at 298 K.

10.25 100 kPa of $PCl_5(g)$ is mixed with 100 kPa of $O_2(g)$ and allowed to equilibrate at 298 K and constant volume. Values of the enthalpies of formation and absolute entropies are given in the table below.

	PCl_5	O_2	$POCl_3$	Cl_2O
$\Delta_f H_m°$ / kJ mol^{-1}	−95.35	0	−592.0	76.15
$S_m°$ / J K^{-1} mol^{-1}	352.7	205.0	324.6	266.5

(a) Calculate the equilibrium partial pressures in the system assuming PCl_5 and O_2 react to form $POCl_3(g)$ and $Cl_2O(g)$ as the only products. (b) Explain whether the enthalpy and entropy changes are favorable or unfavorable for the reaction and what their values mean in molecular terms for the reactants compared to the products.

10.26 The gas-phase stoichiometric reaction between benzoyl bromide (C_7H_5BrO) and water to create HBr and benzoic acid ($C_7H_6O_2$) has an enthalpy of reaction of $\Delta_r H°(g) = -113.1$ kJ mol^{-1}. (a) Write out balanced chemical reactions for the stoichiometric gas-phase reaction and the reaction when carried out in excess water. (b) Calculate using algebraic expressions (use symbols not numbers since you haven't been given all the numbers you need) the value of $\Delta_r H°$ for the reaction as carried out in excess water. Call this value $\Delta_r H°(aq)$. $\Delta_r H°(g)$ will be used in your expression for $\Delta_r H°(aq)$. (c) State with justification whether the aqueous-phase reaction is more, less or equally exothermic than the gas-phase reaction. (d) Calculate q, w, ΔU, and ΔH, when the reaction of 1 mol of benzoyl bromide with 0.5 mol of water is performed in the gas phase at $T = 600$ K and constant pressure.

10.27 The $\Delta H°$ value of an endothermic reaction is positive. What does this tell us about the bonds in the reactive system?

Further reading

Berry, R.S., Rice, S.A., and Ross, J. (2000) *Physical Chemistry*, 2nd edition. Oxford University Press, New York.
Metiu, H. (2006) *Physical Chemistry: Thermodynamics*. Taylor & Francis, New York.
NIST Chemistry WebBook http://webbook.nist.gov/chemistry/.

CHAPTER 11

Chemical equilibrium

PREVIEW OF IMPORTANT CONCEPTS

- Any system held at constant temperature and constant volume will evolve until its Helmholtz energy is a minimum.
- Any system held at constant temperature and constant pressure will evolve until its Gibbs energy is a minimum.
- The Gibbs energy change is directly related to enthalpy and entropy changes by $\Delta_r G = \Delta_r H - T\Delta_r S$.
- The standard Gibbs energy of reaction is related to the equilibrium constant by $\Delta_r G^\circ_m = -RT \ln K$.
- Both the Gibbs energy change and the equilibrium constant are inherently temperature-dependent.
- The chemical potential of a pure substance is its molar Gibbs energy.
- At equilibrium the chemical potential is the same everywhere in a container for all species in the container.
- A system at equilibrium, when subjected to a disturbance, responds in a way that tends to minimize the effect of the disturbance.
- All ionic solids are assumed to be fully dissociated at infinite dilution but precipitate at the point at which the chemical potential of the solid is lower than that of the solvated ions. This point is when the reaction quotient exceeds the solubility product.

As stated by the IUPAC, equilibrium is the state of a system in which the macroscopic properties of each phase of the system become uniform and independent of time. If the temperature is uniform throughout the system, a state of thermal equilibrium has been reached; if the pressure is uniform, a state of hydrostatic equilibrium has been reached; and if the chemical potential of each component is uniform, a state of chemical equilibrium has been reached. If all of these quantities become uniform the system is said to be in a state of complete thermodynamic equilibrium.

11.1 Chemical potential and Gibbs energy of a reaction mixture

The *chemical potential* of a pure substance is defined as

$$\mu = (\partial G/\partial n)_{T,p}. \tag{11.1}$$

In words, the chemical potential shows how the Gibbs energy of a system changes as a substance is added to it. For a pure substance, the Gibbs energy is simply $G = nG_m$, therefore,

$$\mu = \left(\frac{\partial n G_m}{\partial n}\right)_{T,p} = G_m. \tag{11.2}$$

The chemical potential of a pure substance is the same as the molar Gibbs energy.

From Eq. (9.38) we write the chemical potential of an ideal gas

$$G_m(p) = G^\circ_m + RT \ln(p/p^\circ) \tag{11.3}$$

$$\mu = \mu^\circ + RT \ln(p/p^\circ). \tag{11.4}$$

For a system composed of several chemical species in the thermodynamic limit

$$G = \sum_i n_i\,\mu_i. \tag{11.5}$$

Remember what we mean by the thermodynamic limit: for a sample of material that is of such large dimensions that its size and shape no longer affect its properties. For an ideal gas, this simply means that we have a large volume. To write

Physical Chemistry: How Chemistry Works, First Edition. Kurt W. Kolasinski.

Companion Website: www.wiley.com/go/kolasinski/physicalchemistry

Eq. (11.5) more generally we need to explicitly write a term that includes a contribution from the interfaces present in our system. Consider for example a system composed of a solid A in contact with its vapor B. The Gibbs energy of the system is

$$G = n_A \mu_A + n_B \mu_B + \gamma_{AB} A \tag{11.6}$$

where γ_{AB} is the Gibbs energy per unit area of interface and A is the area of the interface. When the number of surface atoms is small, this last term can be neglected. As we have seen in Chapter 4, the ratio of the number of surface atoms to total atoms is a function of the size of the sample, and this ratio only exceeds a non-negligible number, say 10^{-3}, at radii of ≥100 nm for spherical particles. Therefore, as long as our system is larger than ~100 nm, we can neglect the interfacial terms. However, if we are dealing with colloids or many systems of interest in soft matter physics (e.g., micelles, bilayers, etc.), the Gibbs energy per unit surface area is significant and must be considered.

A system evolves until its Gibbs energy has obtained one constant value that is a minimum. By definition then, a system also seeks to minimize its chemical potential. For a given species, transport will occur spontaneously from a region of high chemical potential to one of low chemical potential. The flow of material will continue until the chemical potential has the same value in all regions of the mixture. This property of the chemical potential defines chemical equilibrium.

At equilibrium, the chemical potential of each individual species is the same throughout a mixture.

11.2 The Gibbs energy and equilibrium composition

11.2.1 Gibbs energy and chemical potential determine the direction of change

Consider the equilibrium between two species A and B.

$$A \rightleftharpoons B$$

The reaction Gibbs energy can be expressed as the difference between the chemical potential of the products and the reactants

$$\Delta_r G_m = \sum_{\text{products}} \nu \mu - \sum_{\text{reactants}} \nu \mu = \sum_j \nu_j \mu_j. \tag{11.7}$$

Specifically for the simple reaction in this example,

$$\Delta_r G_m = \mu_B - \mu_A. \tag{11.8}$$

The Gibbs energy change is the control parameter that tells us how the system will evolve. The spontaneous direction of change is the direction that requires no work to be done for the system to proceed in that direction.

$$\begin{aligned} &d_r G < 0 \quad \textit{forward reaction spontaneous, } \mu_A > \mu_B. \\ &d_r G > 0 \quad \textit{reverse reaction spontaneous, } \mu_A < \mu_B. \\ &d_r G = 0 \quad \textit{system at equilibrium, } \mu_A = \mu_B. \end{aligned}$$

The system will evolve from high chemical potential to low chemical potential. When the chemical potential of the reactants and products is the same, equilibrium has been achieved. We have to use $d_r G$ and not $\Delta_r G$ as the criterion for spontaneous direction of change to be absolutely correct. This is illustrated in Fig. 11.1 Even though point C lies at lower Gibbs energy than point A – that is, $\Delta_r G(A \rightarrow C) < 0$ – the system will not evolve from A to C. This is because there is a minimum at B along the way. The system evolves from A to B. Along this path $d_r G < 0$ everywhere until B is reached. At B, $d_r G = 0$. After B, $d_r G > 0$ and the system will not continue to proceed uphill. Note the pathological point D, where the Gibbs energy is equal to that at C. At this point $\Delta_r G(D \rightarrow C) = 0$, but clearly $d_r G < 0$ and the system is not at equilibrium.

Note what spontaneous means in a thermodynamic sense. Spontaneous means that the process occurs irreversibly and it will not go back unless we do work on the system. $d_r G > 0$ means the process is impossible without application of work; that is, the reverse happens. $d_r G \neq 0$ means the extent of reaction has to change, $d\xi \neq 0$. $d_r G = 0$ does not mean that nothing happens; it means that the extent of reaction does not change, $d\xi = 0$, because the rates of the forward and reverse processes are equal. Thus, spontaneous in the thermodynamic sense has a more nuanced meaning than is encountered beyond thermodynamics.

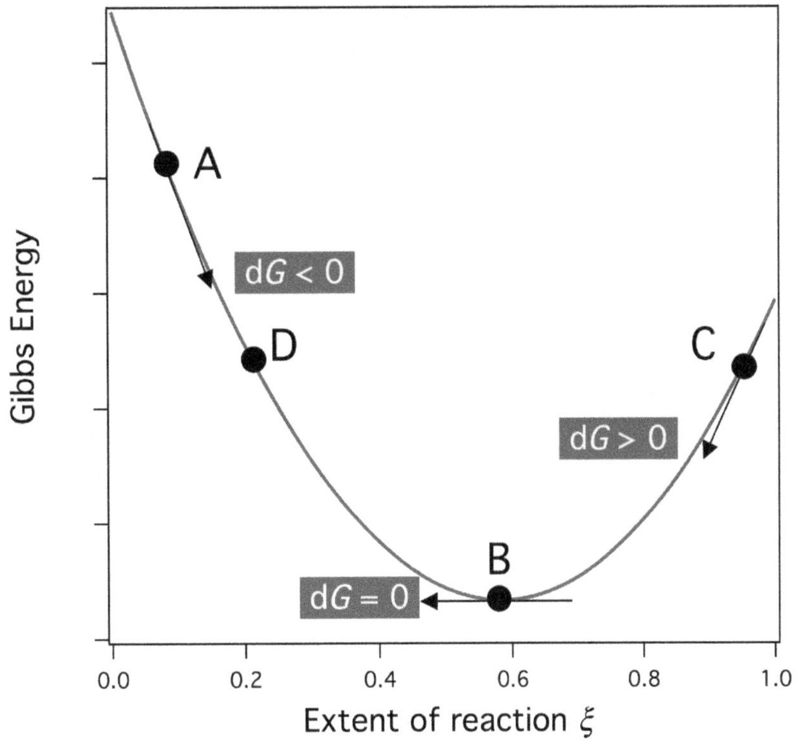

Figure 11.1 The sign of the slope of the Gibbs energy versus extent of reaction determines the direction of spontaneous change. The system will evolve as long as there are no kinetic barriers to the minimum Gibbs energy for the system. Regardless of whether the system starts at point A or point C, the direction of spontaneous change is towards point B.

11.2.2 Ideal gas equilibria

Take A and B to be ideal gases. Substitute from Eq. (9.38) for the chemical potential to obtain,

$$\Delta_r G_m = \mu_B - \mu_A = \mu_B^\circ + RT\ln\left(p_B/p^\circ\right) - \left(\mu_A^\circ + RT\ln\left(p_A/p^\circ\right)\right)$$

$$\Delta_r G_m = \Delta_r G_m^\circ + RT\ln(p_B/p_A). \tag{11.9}$$

The standard *reaction Gibbs energy* is defined as

$$\Delta_r G_m^\circ = G_{m,B}^\circ - G_{m,A}^\circ = \mu_B^\circ - \mu_A^\circ. \tag{11.10}$$

The *reaction quotient* Q is the ratio of partial pressures of products, each raised to the power of its stoichiometric coefficient, divided by the partial pressures of the reactants, each raised to its stoichiometric coefficient. Each partial pressure must also be divided by the standard state pressure to ensure that the reaction quotient is dimensionless. For the reaction of A to B,

$$Q = (p_B/p_A). \tag{11.11}$$

This us allows us to write Eq. (11.9) more generally as

$$\Delta_r G_m = \Delta_r G_m^\circ + RT\ln Q. \tag{11.12}$$

Previously, we have shown that $\Delta_r G_m^\circ$ can be calculated from standard Gibbs energies of formation according to

$$\Delta_r G_m^\circ = \Delta_f G_{m,B}^\circ - \Delta_f G_{m,A}^\circ. \tag{11.13}$$

At equilibrium $\Delta_r G_m = 0$, and the reaction quotient is equal to the equilibrium constant K. Therefore we can write

$$\Delta_r G_m = 0 = \Delta_r G_m^\circ + RT\ln K$$

$$\Delta_r G_m^\circ = -RT\ln K \tag{11.14}$$

This is one of the most important equations in thermodynamics as it links thermochemical data to the value of the equilibrium constant. It is known as the *reaction isotherm*.

It is essential here that the difference between ΔG and $\Delta G°$ is understood. The difference is defined by the difference between the reaction quotient Q and the equilibrium constant K. Q is tied to ΔG. K is linked to $\Delta G°$. Q and ΔG refer to arbitrary composition. The equilibrium constant K refers specifically to the composition at equilibrium, and $\Delta G°$ refers specifically to each species being in its standard state. Large values of the equilibrium constant, $K \gg 1$, are associated with negative values of $\Delta G°$. This favors the products at equilibrium. Small values of K, $K \ll 1$, are associated with positive values of $\Delta G°$. Reactants are favored at equilibrium. In contrast, the meaning of ΔG is to tell us the direction the system evolves spontaneously from the current composition defined by Q.

To emphasize the difference between ΔG and $\Delta G°$, the *affinity* of reaction A, defined by

$$-A = \Delta_r G_m = \Delta_r G_m^\circ + RT \ln Q \tag{11.15}$$

has been introduced. Thus, when A is positive, the reaction proceeds in the forward direction, and when A is negative it proceeds in the reverse direction. Unfortunately, A is also the preferred symbol for the Helmholtz energy.

Equation (11.14) is not limited to ideal gases. Indeed, we can write a generalized reaction quotient in terms of fugacities (real gases) or activities (real mixtures) and derive the same equation. The relation between K and $\Delta_f G_m^\circ$ is general and holds for all real and ideal systems.

11.2.2.1 Directed practice

Calculate K for the ammonia synthesis reaction at 298 K, assuming all gases to be ideal and $\Delta_f G_m^\circ$ for NH_3 to be -16.4 kJ mol^{-1}.

[Answer: $K = 749$ when written as a standard formation reaction but 5.61×10^5 when written $N_2(g) + 3H_2(g) \rightarrow 2NH_3(g)$.]

11.2.2.2 Directed practice

Determine K at 298 K for the reaction $C_2H_4(g) + H_2(g) \rightleftharpoons C_2H_6(g)$, given that $\Delta_f G°$ for $C_2H_4(g)$ and $C_2H_6(g)$ are 68.4 and -32.0 kJ mol^{-1}, respectively.

[Answer: $K = 3.96 \times 10^{17}$]

11.2.2.3 Directed practice

A mixture of $CO(g)$, $H_2(g)$ and $CH_3OH(g)$ at 500 K with $p(CO) = 10$ bar, $p(H_2) = 1$ bar and $p(CH_3OH) = 0.1$ bar is passed over a catalyst. Can more methanol be formed? $\Delta_r G_m^\circ = -21.21$ kJ mol^{-1}.

11.3 The response of equilibria to change

A chemical equilibrium is a system of reactants and products that are coupled to each other by reversible reactions. Any change to the system variables, such as p, T, V and composition, will cause the system to react in response to this change. This idea is stated as *Le Châtelier's principle.*

> *A system at equilibrium, when subjected to a disturbance, responds in a way that tends to minimize the effect of the disturbance.*

Consider methanol synthesis from syn gas (syn gas is short for synthesis gas, a mixture of CO and H_2 that can be made from gasification of coal, biomass or other hydrocarbons),

$$CO(g) + 2H_2(g) \rightleftharpoons CH_3OH(g).$$

Suppose the reaction is run at 298 K but at low enough pressure so that all species are in the gas phase. By inspection of the stoichiometry, we see that 3 mol of gas are converted to 1 mol of gas. This corresponds to an unfavorable entropy factor. Indeed, upon calculation we find $\Delta_r S_m^\circ = -219$ J K^{-1} mol^{-1}. Applying Le Châtelier's principle, increasing the pressure on the system will lead it to relieve this stress by proceeding in the forward direction. Decreasing the amount of substance of gas (at constant T and V) lowers the pressure. If the pressure is increased enough, the methanol will eventually condense, lowering the pressure even further. By inspection, one cannot tell easily what the enthalpy change is. Calculation reveals $\Delta_r H_m^\circ = -90.6$ kJ mol^{-1}. For an exothermic reaction such as this, heat can be thought of as a product. Thus, raising the temperature (at constant p and V) ought to push the system to the left towards reactants. The system reacts to 'use up the heat' added to the system by raising T. Conversely, raising the temperature of an endothermic reaction tends to push it towards the products.

Warm and fuzzy statements like Le Châtelier's principle are helpful to organize our thoughts and spark intuition, but this is physical chemistry and we want to understand the basis of these principles. We can do this by understanding the dependence of the Gibbs energy and the equilibrium constant on changes in variables such as T and p.

11.3.1 Response to temperature

The control parameter that we monitor to tell us how a system will evolve is the Gibbs energy. At constant temperature and pressure a system evolves to minimize the Gibbs energy. The spontaneous direction of change is always the path that leads to a lowering of Gibbs energy. As derived above, the Gibbs energy is related to composition from the equilibrium constant through

$$\Delta_r G^\circ_m = -RT \ln K. \tag{11.16}$$

Now, substitute from the fundamental relationship of Gibbs energy to enthalpy and entropy

$$\Delta_r G^\circ_m = \Delta_r H^\circ_m - T\,\Delta_r S^\circ_m, \tag{11.17}$$

and solve for ln K to obtain

$$\ln K = -\frac{\Delta_r H^\circ_m}{RT} + \frac{\Delta_r S^\circ_m}{R}. \tag{11.18}$$

This is known as the integrated form of the *van't Hoff equation*, also referred to as the *van't Hoff isochore*. R is the gas constant and T the absolute temperature. We see that the Gibbs energy of reaction and the equilibrium constant are inherently temperature-dependent.

Of all the equations in thermodynamics, the two you should memorize and know immediately off the top of your head are Eqs (11.16) and (11.17). They are important in their own right and because so much can be derived from them. In this case, we have used them to derive the van't Hoff equation, which is of particular importance because it relates K – a quantity that is directly related to concentrations and the reaction stoichiometry – to the thermodynamic quantities $\Delta_r H^\circ_m$ and $\Delta_r S^\circ_m$. These quantities are measured directly by thermochemical methods. For some systems, thermochemical methods may be easily performed over a certain temperature range but not over others. For other systems, concentrations might prove much easier to measure than heat flows. Equation (11.18) allows us to pick the experiments that are most accurate and reliable. It also allows for comparisons between methods to enhance accuracy and test assumptions.

Looking at the van't Hoff equation in more depth, we see that measurement of K as a function of temperature can lead to values for $\Delta_r H^\circ_m$ and $\Delta_r S^\circ_m$. As is our want, we can turn this equation into a straight line if we plot ln K versus $1/T$. If $\Delta_r H^\circ_m$ is constant, this so-called *van't Hoff plot* will be linear with slope = $-\Delta_r H^\circ_m/R$ and intercept $\Delta_r S^\circ_m/R$. An example of a van't Hoff plot is shown in Fig. 11.2.

First, a word about slopes and intercepts in general. In many experiments absolute calibration is difficult but relative calibrations are easy. While signals are linear in a certain parameter, they are often subject to an offset (the 'zero' in the machine is not precisely known). The determination of a slope is unaffected by an offset as long as it is constant. In other words, as long as we are adding a constant amount to all points on a line (or any curve for that matter), it is immaterial if the line is a bit higher or a bit lower on our graph, the slope is still the same. On the other hand, the presence of an offset directly changes the value of the intercept. Experimentally, slopes are more reliably determined than intercepts not only because of this insensitivity to offsets, but also because evaluation of the slope does not require an extrapolation. Historically, therefore, preference has been given to methods and equations that rely on slopes rather than intercepts. With increasing experimental precision and accuracy – for instance, knowledge of absolute concentrations rather than merely signals that are proportional to concentration – both slopes and intercepts become more reliably accessible.

Is the integrated form of the van't Hoff equation given in Eq. (11.18) good enough? The answer for that depends on the quality of the experiments that are performed (how big the experimental uncertainties are) and what we are attempting to learn (how accurately we can interpret the data). We know that in general $\Delta_r H^\circ_m$ and $\Delta_r S^\circ_m$ are temperature-dependent. Strictly speaking, a van't Hoff plot should never be linear. Nonetheless, most van't Hoff plots are evaluated with this simple linear form; furthermore, most look linear. Over a small enough range of x values, any curve is linear. Reactions performed in liquid solvents can only be performed over a range in which the solvent is stable. For water, this range is no more than $\Delta T \approx 100$ K, while even a more stable solvent such as toluene can only be used over a range of $\Delta T \approx 200$ K. Practically speaking, the ranges will be much smaller since solvent evaporation over the course of the experiment must be avoided. As shown in Fig. 9.5, temperature ranges of a few tens of kelvin are simply insufficient to observe a significant

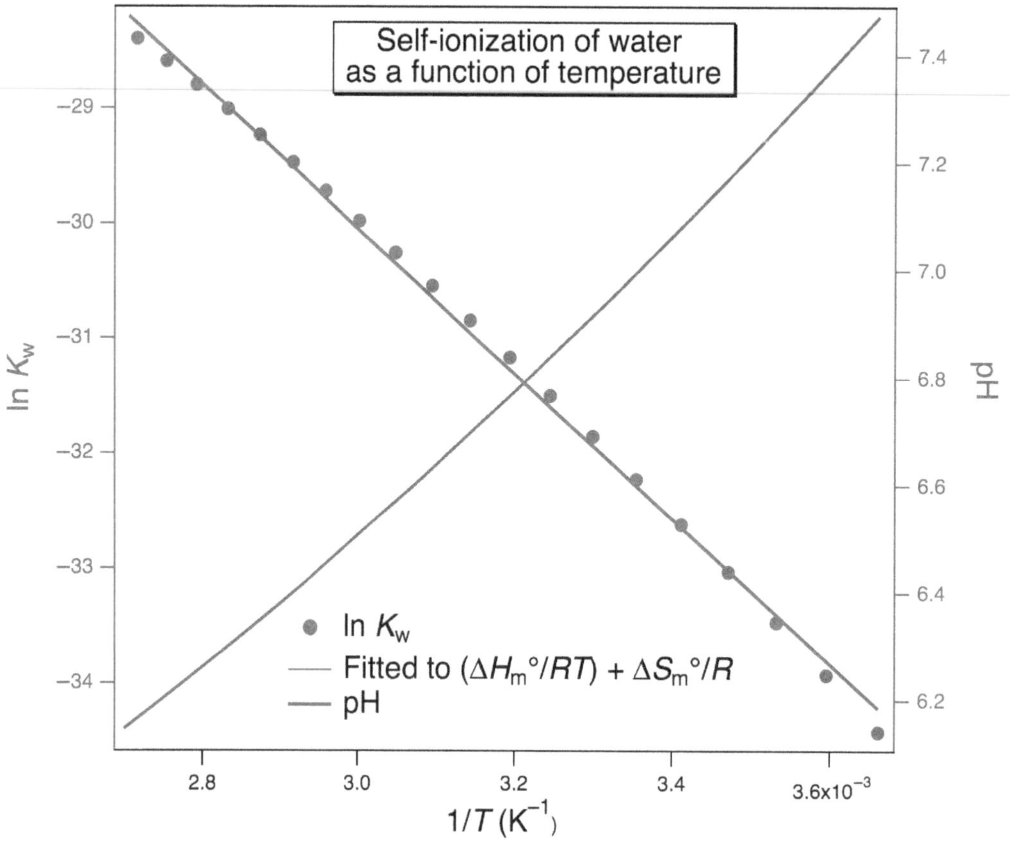

Figure 11.2 The self-ionization of water from 0 °C to 100 °C. Slight curvature in the plot of ln K_w versus $1/T$ indicates that $\Delta_r H_m^\circ$ and $\Delta_r S_m^\circ$ are slightly temperature-dependent over this temperature range. The temperature dependence of K_w means that the pH of neutral water is also T-dependent, as discussed further in Section 11.5.1.

change in $\Delta_r H_m^\circ$. If, on the other hand, a reaction can be followed in the gas phase (or perhaps in a liquid metal or molten salt) over hundreds of kelvin, then the temperature dependence will be significant.

A significant change is one that has a measurable effect on the outcome of experimental results beyond experimental uncertainty. If your experiments represent the first characterization of a new reaction involving novel compounds, then you are probably happy that you can measure the equilibrium constant over any temperature range at all. Your accuracy may only be 5–10 kJ mol^{-1}. Linear analysis of a van't Hoff plot is completely justified. In contrast, if generations of experiments have been performed with progressively better signal-to-noise ratios and lower experimental uncertainty, and you are aiming to test the limits of our understanding of the intricate nature of the system on a microscopic level, then your analysis needs to be as accurate as possible.

The effects of temperature are more generally expressed in terms of the differential form of the *van't Hoff equation*, which is written in two ways

$$\frac{\mathrm{d}\ln K}{\mathrm{d}T} = \frac{\Delta_r H_m^\circ}{RT^2}. \tag{11.19}$$

Upon separation and integration we obtain,

$$\ln K(T_2) = \ln K(T_1) + \int_{T_1}^{T_2} \frac{\Delta_r H_m^\circ}{RT^2}\,\mathrm{d}T. \tag{11.20}$$

We now substitute the functional form of $\Delta_r H_m^\circ$ that we need for our level of theoretical understanding. If we take the enthalpy change to be independent of temperature, we obtain a result that is equivalent to Eq. (11.18),

$$\ln K(T_2) = \ln K(T_1) - \frac{\Delta_r H_m^\circ}{R}\left(\frac{1}{T_2} - \frac{1}{T_1}\right). \tag{11.21}$$

Table 11.1 Thermodynamic parameters related to the self-ionization of water. All values at 298.15 K taken from the 96th edition of the *CRC Handbook of Chemistry and Physics*.

Species	$\Delta_f G^\circ_m$/kJ mol^{-1}	$\Delta_f H^\circ_m$/kJ mol^{-1}	$\Delta_f S^\circ_m$/J K^{-1} mol^{-1}	$C_{p,m}{}^\circ$/J K^{-1} mol^{-1}
$H_2O(l)$	−237.1	−285.830	69.95	75.3
$H^+(aq)$	0.	0.	0.	0.
$OH^-(aq)$	−157.2	−230.015	−10.90	−148.5

If sufficient experimental information is available, we can substitute an appropriate polynomial form for the temperature dependence of $\Delta_r H^\circ_m$, such as that shown in Eq. (9.47). On the other hand, the temperature dependence of $\Delta_r G^\circ_m$, and therefore K, are probably much more easily approximated by using Eq. (11.17), and substituting in directly the temperature-dependent values of the enthalpy and entropy changes, as is done in the following example. Nevertheless, the integrated form of the van't Hoff equation in the form of Eq. (11.21) is quite useful because it allows us to use measurements of K as a function of T to determine $\Delta_r H^\circ_m$. Equation (11.21) predicts that a plot of ln K versus $1/T$ yields a straight line with slope $-\Delta_r H^\circ_m/R$.

11.3.1.1 Example

Using data from the *CRC Handbook of Chemistry and Physics*, plot $\Delta_r G^\circ_m$ versus T over the temperature range 0 to 100 °C for the self-ionization of water. Compare the accuracy of the Gibbs–Helmholtz expression in Eq. (9.37) to a calculation from Eq. (11.17) that includes the temperature dependence of $\Delta_r H^\circ_m$ from Eq. (9.50) and $\Delta_r S^\circ_m$ from Eq. (9.66).

The self-ionization of water can be represented by the reaction

$$H_2O(l) \rightarrow H^+(aq) + OH^-(aq).$$

The *CRC Handbook of Chemistry and Physics* reports values of ln K_w over this temperature range that are plotted in the form of a van't Hoff plot in Fig. 11.2. The self-ionization constant of water is related to the Gibbs energy of reaction, as is any equilibrium constant, by

$$\Delta_r G^\circ_m = -RT \ln K_w.$$

The *CRC Handbook of Chemistry and Physics* also reports the values in Table 11.1, which are used to calculate $\Delta_r G^\circ_m$, $\Delta_r H^\circ_m$, and $\Delta_r S^\circ_m$ for the reaction at $T_1 = 298$ K. These are calculated from $\Delta_r G^\circ_m = \sum_j \nu_j \Delta_f G^\circ_{m,j}$ and similarly for $\Delta_r H^\circ_m$, and $\Delta_r S^\circ_m$.

First, we generate the blue curve in Fig. 11.3 using Eq. (9.36) with $\Delta_r G^\circ_m(T_1) = 79.9$ kJ mol^{-1} and $\Delta_r H^\circ_m(T_1) = 55.815$ kJ mol^{-1} calculated from the appropriate formation values in Table 11.1, and take $T_1 = 298.15$ K. Note that the Gibbs–Helmholtz equation is mathematically equivalent to assuming that $\Delta_r H^\circ_m$ and $\Delta_r S^\circ_m$ are constant and using $\Delta_r G^\circ_m = \Delta_r H^\circ_m - T\,\Delta_r S^\circ_m$ to calculate $\Delta_r G^\circ_m(T_2)$. The blue line is exact at 298 K, but increasingly deviates from the true value (the red square points) as temperature increases. Because the data is of such high quality, clear curvature is recognizable in the plot, which shows that the deviation from the Gibbs–Helmholtz equation is symptomatic of a failure in the assumption that $\Delta_r H^\circ_m$ is constant.

The red curve in Fig. 11.3 is obtained by calculating $\Delta_r G^\circ_m$ from $\Delta_r G^\circ_m = \Delta_r H^\circ_m - T\,\Delta_r S^\circ_m$ substituting $\Delta_r H^\circ_m(T_2) = \Delta_r H^\circ_m(298\,K) + (T_2 - 298.15)\,\Delta_r C^\circ_{p,m}$ for the enthalpy change and $\Delta_r S^\circ_m(T_2) = \Delta_r S^\circ_m(298\,\text{K}) + \Delta_r C^\circ_{p,m} \ln(T_2/298.15)$ for the entropy change, where $\Delta_r S^\circ_m(298\text{ K}) = -80.85$ J K^{-1} mol^{-1}. The fit is far superior to the linear approximation. Nonetheless, a slight systematic deviation as T increases is observed. In order to improve the fit further would require adding a third (or more) term(s) in the expressions to account for the temperature dependence of $\Delta_r C^\circ_{p,m}$.

11.3.2 Response to pressure

Formally, the equilibrium constant does not depend on pressure

$$(\partial K/\partial p)_T = 0. \tag{11.22}$$

However, in reactions involving changes in volume, such as the example of methanol synthesis above or any other reaction that involves a change in the amount of substance of gas in units of moles, this does not mean that the equilibrium concentrations are independent of pressure. As an example, let us treat the gas-phase reaction $N_2O_4 \rightleftharpoons 2NO_2$. We will derive an expression for the total pressure dependence of the equilibrium constant in terms of the *fractional dissociation* α, total pressure p, and the standard state pressure p°. The fractional dissociation is defined as $\alpha = \xi/a_0$. In the reaction

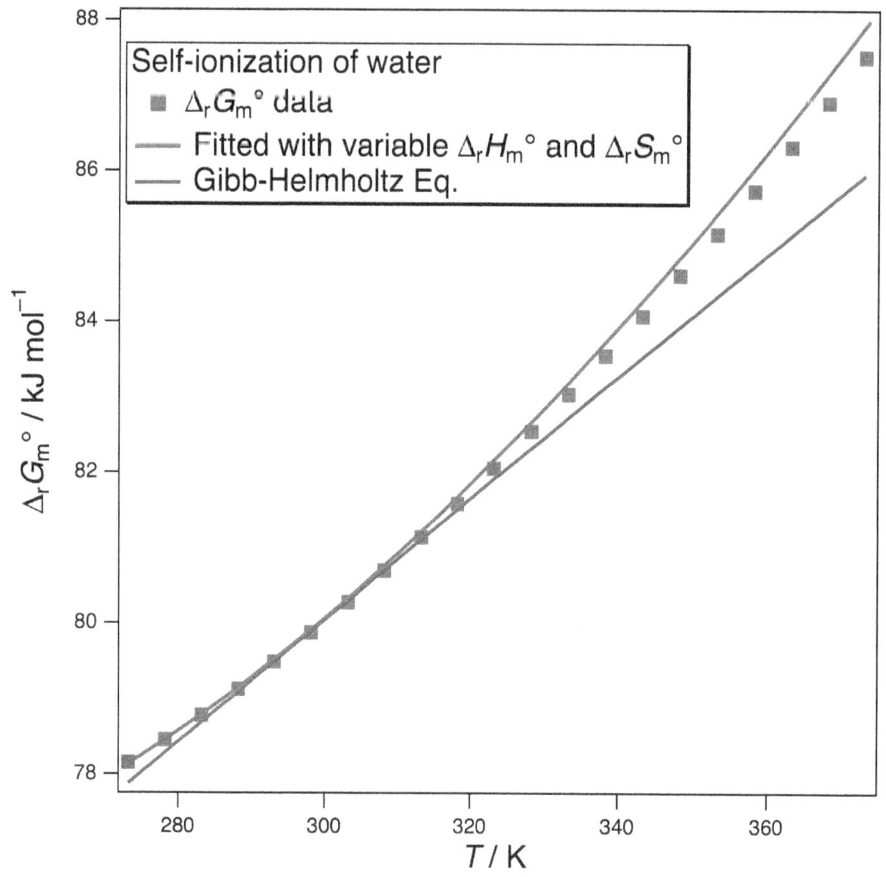

Figure 11.3 The Gibbs energy change for the self-ionization of water exhibits noticeable curvature from the temperature dependence of the enthalpy and entropy of reaction. The blue line is the result of the Gibbs–Helmholtz equation assuming that the enthalpy change is constant at its values from 298 K. Values are taken from the 96th edition of the *CRC Handbook of Chemistry and Physics*.

chart below, ξ quantifies the amount of N_2O_4 that has reacted and a_0 is the initial amount of substance. At constant pressure and temperature we can express the amount of substance in terms of the partial pressure. We will show that the fractional dissociation is a function of the total pressure even though the equilibrium constant itself is independent of p.

The reaction chart is

	$N_2O_4(g)$	$\rightleftharpoons$	$2NO_2(g)$
Initially	a_0		0
Reacts	$-\xi$		$+2\xi$
Eq	$(1-\alpha)a_0$		$2\alpha a_0$

The last line is obtained by substituting $\xi = \alpha a_0$. With a_0 as the initial partial pressure of NO_2, then at equilibrium the partial pressure of NO_2 is $(1-\alpha)a_0$ and $2\alpha a_0$ for N_2O_4. Therefore, the mole fractions at equilibrium are

$$x_{N_2O_4} = \frac{(1-\alpha)a_0}{(1-\alpha)a_0 + 2\alpha a_0} = \frac{(1-\alpha)}{(1+\alpha)}, x_{NO_2} = \frac{2\alpha}{(1+\alpha)}$$

The equilibrium constant is then

$$K = \frac{p_{NO_2}^2}{p_{N_2O_4}p^\circ} = \frac{x_{NO_2}^2 p^2}{x_{N_2O_4} p p^\circ} = \frac{4\alpha^2(p/p^\circ)}{1-\alpha^2}$$

Solving for α yields

$$\alpha = \left(\frac{1}{1+(4p/Kp^\circ)}\right)^{1/2}. \tag{11.23}$$

The equilibrium constant is constant with changing p because it is referenced to p°. As a consequence, when p increases α must decrease according to Eq. (11.23). Increasing p leads to decreasing α because, in accord with Le Châtelier's principle, the reaction shifts to the left because this decreases the amount of substance of gas.

11.4 Equilibrium constants and associated calculations

11.4.1 Use of activities and concentrations

The Gibbs energy of reaction and the equilibrium constant are related by $\Delta_r G_m^\circ = -RT \ln K$. A large negative value of $\Delta_r G_m^\circ$ favors products. Equivalently, for a large value of K, the products are favored. When $\Delta_r G_m^\circ$ is positive or $K \ll 1$, the reactants predominate. For $\Delta_r G_m^\circ$ close to zero or K near one, the equilibrium mixture contains comparable concentrations of all species.

Equilibrium constants are defined in terms of activities/fugacities, not concentrations or partial pressures directly. The equilibrium constant written in terms of activities/fugacities is called the *thermodynamic equilibrium constant*. For the general reaction

$$aA + bB \rightleftharpoons zZ,$$

the thermodynamic equilibrium constant is

$$K = \frac{a_Z^z}{a_A^a a_B^b}. \tag{11.24}$$

Since activities are dimensionless, K is dimensionless. However, for systems behaving ideally such as dilute solutions, ideal gases and real gases at sufficiently low pressure, the approximation is often made of using concentration (or partial pressure) instead of activity. An equilibrium constant written in terms of concentrations K_c is sometimes referred to as the *concentration equilibrium constant*, and one written in terms of partial pressure K_p as the *pressure equilibrium constant*. By IUPAC recommended convention, all equilibrium constants are dimensionless quantities.

To write a thermodynamic equilibrium constant in terms of concentrations, we introduce the activity coefficients (or fugacity and fugacity coefficients for gases). The activity of species J is given the symbol a_J, and its activity coefficient is γ_J. Activity is equal to the activity coefficient times concentration or pressure according to the choice of standard state. We will discuss activities, activity coefficients and their calculation in much greater detail in the chapters on solutions and mixtures. For now we only need to know that the activity coefficient quantifies the deviation from non-ideal behavior. As the system become more dilute and acts more ideally, the activity coefficient approaches one, $\gamma \to 1$, and concentration (or partial pressure) can be used instead of activity. The concentration unit chosen for representing activity depends on the chosen standard state. Since activity is dimensionless, the concentration is divided by the concentration of the standard state to remove the units. Common choices of units and standard states are

$$a_J = \gamma_J \left(b_J/b^\circ\right), \text{for molality } b^\circ = 1 \text{ mol kg}^{-1},$$
$$a_J = \gamma_J([J]/c^\circ), \text{for molarity } c^\circ = 1 \text{ mol l}^{-1},$$
$$a_J = \gamma_J \left(p_J/p^\circ\right), \text{for pressure } p^\circ = 1 \text{ bar (ideal gas)}.$$

While this is usually a good approximation for gases (at not too-high pressure or too-low temperature), this is a poor approximation for liquid solutions, especially for electrolytes at concentrations above ~1 mM.

Volumes are temperature-dependent, masses are not. Therefore, molarity is temperature-dependent whereas molality is not. Historically, volumes were easier to measure than masses; thus, molarity was convenient with volumetric glassware. Now that we have electronic balances, mass is easy to measure and molality is preferred by many experimentalists. In terms of molalities the equilibrium constant is

$$K = \frac{\gamma_Z^z \left(b_Z/b^\circ\right)^z}{\gamma_A^a \left(b_A/b^\circ\right)^a \gamma_B^b \left(b_B/b^\circ\right)^b} = \frac{\gamma_Z^z}{\gamma_A^a \gamma_B^b} \frac{\left(b_Z/b^\circ\right)^z}{\left(b_A/b^\circ\right)^a \left(b_B/b^\circ\right)^b} = K_\gamma K_b. \tag{11.25}$$

The term K_γ accounts for deviations from ideality, while the term K_b is the thermodynamic equilibrium constant for an ideally behaved system. As the solution becomes more dilute $\gamma \to 1$, $K_\gamma \to 1$ and $K \to K_b$. In this chapter, we want to emphasize the structure of concepts and calculations involving the equilibrium constant. Therefore, *we will consider all systems to be behaving ideally unless we explicitly indicate otherwise*. In other words, we assume $\gamma = 1$, $K_\gamma \to 1$ and $K \to K_b$. This will allow us to use concentrations in place of activities.

11.4.1.1 Directed practice

N_2O_4 dissociates according to the following equation

$$N_2O_4(g) \rightleftharpoons 2\,NO_2(g).$$

At 350 K, the equilibrium mixture contains 0.13 mol N_2O_4 and 0.34 mol NO_2. The total pressure is 2 bar. Find the value of K at 350 K.

[Answer: 3.8]

11.4.1.2 Directed practice

$$H_2(g) + I_2(g) \rightleftharpoons 2\,HI(g)$$

At 400 K, $K = 40$. If 2.0 mol of H_2 and I_2 are mixed at a total constant pressure of 1 bar, what is the composition of the equilibrium mixture at 400 K?

[Answer: $H_2 = 0.5$ mol, $I_2 = 0.5$ mol, HI = 3.0 mol]

11.4.2 Heterogeneous/solution phase equilibria

An equilibrium involving two or more phases is a heterogeneous equilibrium, for example

$$AgCl(s) + H_2O(l) \rightleftharpoons Ag^+(aq) + Cl^-(aq)$$
$$\alpha\text{-D-glucose}(s) + H_2O(l) \rightleftharpoons \alpha\text{-D-glucose}(aq)$$
$$CH_3CH_2OH(l) \rightleftharpoons CH_3CH_2OH(g).$$
$$NH_2CO_2NH_4(s) \rightleftharpoons 2\,NH_3(g) + CO_2(g).$$

Any species in its standard state has an activity of unity. Since their activities are one, they can be dropped from the equilibrium constant expression. A solid in its standard state has an activity of unity. A liquid in its standard state has an activity of unity. In a dilute solution, the solvent essentially acts like a substance in its standard state. Thus, the activities of solids and liquids can be dropped from equilibrium constant calculations for heterogeneous equilibria. The activities of solvents can be dropped from equilibrium constant calculations for solution phase equilibria as long as the solution is not too concentrated.

In the above examples, the solubility of AgCl(s) in $H_2O(l)$ at 25 °C is 1.9×10^{-4} g per 100 g of H_2O. As long as AgCl is the only species being dissolved, the ions will never reach a concentration large enough to change the activity of water. Therefore, both the AgCl(s) and the $H_2O(l)$ can be dropped from the equilibrium constant expression. For glucose, on the other hand, the solubility at 80 °C is 81.5 mass% – the 'solution' is no longer a solution but a mixture in which the glucose is the majority component. In such a case the water can no longer be assumed to be acting ideally and the activity of water will differ from its concentration. The equilibrium constant expression for a liquid in contact with its vapor involves only the fugacity of the gas, as does the equilibrium expression for a solid that decomposes into gases.

11.4.2.1 Directed practice

Determine K for the following equilibrium at 298 K if the initial concentration of CH_3NH_2 is 0.10 mol dm^{-3} and the equilibrium concentration of OH^- is 6.6×10^{-3} mol dm^{-3}.

[Answer: $K = 4.7 \times 10^{-4}$]

$$CH_3NH_2(aq) + H_2O(l) \rightleftharpoons CH_3NH_3^+(aq) + OH^-(aq)$$

11.4.3 Biochemical standard state

The *physical chemistry standard state* is defined in terms of molar units $c° = 1$ mol l^{-1} for all ions in solution and $p° = 1$ bar for all gases. In a biological setting, essentially all chemistry occurs in aqueous solution. It is impossible to have both H^+ and OH^- at 1 M concentration simultaneously. Therefore, in the *biochemical standard state*, the standard state concentration is $c' = 1$ mol l^{-1} for all ions other than H^+ and OH^-. The standard pressure is the same in both standard states. In the biochemical standard state, the aqueous phase is taken as neutral with pH 7, that is $[H_3O^+] = [OH^-] = 1.0 \times 10^{-7}$ mol l^{-1}. The difference between these standard states will only be apparent if we need to make comparisons between them or if we try to use values, for example, for standard state Gibbs energies obtained with one standard state in a system using the other standard state. The difference will only be important in reactions that involve either H^+ or OH^-. When looking up values in the literature you must always be aware of how the standard state is defined.

The choice of standard state does not change the relationship of Gibbs energy of reaction to equilibrium, $\Delta G = 0$ still defines a system at equilibrium. Nor does the choice of standard state change the composition at equilibrium. The choice of standard state does change the value of the standard Gibbs energy of reaction. We denote the physical chemistry standard state value of the standard Gibbs energy of reaction in molar units as $\Delta_r G^\circ_m$. We denote the biochemical standard state value of the standard Gibbs energy of reaction in molar units as $\Delta_r G'_m$.

We use the example of a simple solution-phase reaction $A(aq) + B(aq) \rightleftharpoons Z(aq)$. In the physical chemical standard state, we can write the reaction quotient as

$$Q = \frac{(c_Z/c^\circ)}{(c_A/c^\circ)(c_B/c^\circ)} = \frac{Q_c}{Q^\circ} \tag{11.26}$$

where

$$Q_c = \frac{c_Z}{c_A c_B} \tag{11.27}$$

and

$$Q^\circ = \frac{c^\circ}{c^\circ c^\circ}. \tag{11.28}$$

The standard state concentration is c° and the reaction quotient is equal to the ratio of a term involving the concentrations of each component divided by the standard state concentration of each component. The standard state concentration is 1 M for each component. The Gibbs energy of reaction in the physical chemistry standard state is

$$\Delta_r G_m = \Delta_r G^\circ_m + RT\ln Q = \Delta_r G^\circ_m + RT\ln\left(Q_c/Q^\circ\right). \tag{11.29}$$

For the same reaction but now using the biochemical standard state, we write the reaction quotient as

$$Q = \frac{(c_Z/c')}{(c_A/c')(c_B/c')} = \frac{Q_c}{Q'}. \tag{11.30}$$

Q_c is still defined as before but the standard state concentration term is now

$$Q' = \frac{c'}{c'c'}. \tag{11.31}$$

The Gibbs energy of reaction in the biochemical standard state is

$$\Delta_r G_m = \Delta_r G'_m + RT\ln Q = \Delta_r G'_m + RT\ln\left(Q_c/Q'\right) \tag{11.32}$$

At equilibrium $\Delta_r G_m = 0$. Therefore, we can equate the right-hand side of Eq. (11.29) with the right-hand side of Eq. (11.32)

$$\Delta_r G^\circ_m + RT\ln\left(Q_c/Q^\circ\right) = \Delta_r G'_m + RT\ln\left(Q_c/Q'\right). \tag{11.33}$$

By solving Eq. (11.33) we find that the physical chemical standard state value is related to the biochemical standard state by

$$\Delta_r G^\circ_m + RT\ln\left(Q'/Q^\circ\right) = \Delta_r G'_m. \tag{11.34}$$

The standard Gibbs energy of combustion of glucose is $\Delta_c G^\circ_m = -2880\ \text{kJ mol}^{-1}$. The balanced chemical reaction for glucose combustion is

$$C_6H_{12}O_6(aq) + 6\,O_2(g) \rightarrow 6\,CO_2(g) + 6\,H_2O(l)$$

Thus, the reaction quotient is

$$Q = \frac{\left(p_{CO_2}/p^\circ\right)^6}{([\text{glucose}]/c^\circ)\left(p_{O_2}/p^\circ\right)^6}.$$

Thus,

$$Q_c = \frac{(p_{CO_2})^6}{([\text{glucose}])(p_{O_2})^6}$$

and

$$Q^\circ = \frac{(p^\circ)^6}{(c^\circ)(p^\circ)^6}.$$

In the biochemical standard state, Q_c has the same expression (as is always the case) and

$$Q' = \frac{(p^\circ)^6}{(c^\circ)(p^\circ)^6}.$$

Therefore according to Eq. (11.34)

$$\Delta_r G'_m = \Delta_r G^\circ_m + RT\ln(Q'/Q^\circ) = \Delta_r G^\circ_m$$

because $RT\ln(Q'/Q^\circ) = RT\ln(1) = 0$. In other words, if neither H^+ nor OH^- is involved in the reaction, then the standard state value of the Gibbs energy change is the same.

On the other hand, ATP → ADP conversion performed under standard physiological conditions corresponds to the reaction

$$\text{ATP} \rightarrow \text{ADP} + \text{P}_i + \text{H}_3\text{O}^+.$$

The biochemical standard value of the Gibbs energy for ATP hydrolysis is $\Delta_r G'_m = -31$ kJ mol^{-1}. The standard state reaction quotient in the biochemical standard state Q' is

$$Q' = \frac{[ADP]^\circ [P_i]^\circ [\text{H}_3\text{O}^+]^\circ}{[ATP]^\circ} = \frac{1 \times 1 \times 10^{-7}}{1}\ \text{mol}^2\,\text{l}^{-2} = 10^{-7}\ \text{mol}^2\,\text{l}^{-2}.$$

The standard state reaction quotient in the physical chemical standard state Q° is

$$Q^\circ = \frac{1(1)(1)}{1}\ \text{mol}^2\,\text{l}^{-2} = 1\ \text{mol}^2\,\text{l}^{-2}.$$

The value of the standard Gibbs energy of reaction for the physical chemical standard state is

$$\Delta_r G^\circ_m = \Delta_r G'_m - RT\ln(Q'/Q^\circ)$$
$$\Delta_r G^\circ_m = -31\ \text{kJ mol}^{-1} - RT\ln(10^{-7}\ \text{mol}^2\,\text{l}^{-2}/1\ \text{mol}^2\,\text{l}^{-2}) = 9.0\ \text{kJ mol}^{-1}$$

11.5 Acid–base equilibria

11.5.1 Ionization of water

Water dissociates to a small degree into H^+ and OH^-,

$$2\text{H}_2\text{O(l)} \rightleftharpoons \text{H}_3\text{O}^+\text{(aq)} + \text{OH}^-\text{(aq)}$$

The structures of H^+ and OH^- in solution are complex. The short-hand notation H_3O^+ (hydronium) is used to represent the solvated proton in solution. There is no such thing as a free proton in aqueous solution. Instead, the solvated proton interacts unusually strongly with a number of water molecules such that it forms what amounts to a cluster of water molecules bound to the proton. This is a dynamical cluster assembly – a flickering cluster, if you like – in which the water molecules that interact with the H^+ are constantly exchanging and changing their orientations. Nonetheless, in equilibria involving water it is sufficient to consider the solvated proton as a single H^+(aq), which is often denoted H_3O^+(aq) to remind us of the water molecules associated with H^+(aq). The equilibrium is characterized by the *ionization constant*, K_w. Formally,

$$K_w = a_{H^+}\, a_{OH^-}/a^2_{H_2O}. \qquad (11.35)$$

The concentrations of H^+ and OH^- are very small and the activity of water is extremely close to one; therefore, in most instances the activity of water is ignored and the activities of H^+ and OH^- are replaced by their concentrations.

Pure water is always neutral. However, the pH of neutral water is not always 7. It is a fortunate coincidence that $pK_w = 13.995$ at 298 K, which means that $K_w = 1.0116 \times 10^{-14}$ and $[H^+] = \sqrt{K_w} = 1.0058 \times 10^{-7}\,mol\,l^{-1}$. However, as shown in the van't Hoff plot in Fig. 11.2, the ionization constant of water is temperature-dependent. The temperature dependence of K_w means that the pH of neutral water is also temperature-dependent. The slight curvature in the plot of ln K_w versus $1/T$ results from the slight temperature dependence in $\Delta_r H^\circ_m$ and $\Delta_r S^\circ_m$.

It should be mentioned that pH is defined in terms of the activity of H^+ rather than the concentration of H^+,

$$pH = -\lg a_{H^+} = -\lg\left(b_{H^+}\gamma_{H^+}/b^\circ\right). \tag{11.36}$$

Here, it is defined in terms of the molality of H^+ as is preferred by IUPAC. Commonly the activity coefficient is taken to be unity and the pH is taken to be a measure of the molar concentration, which is a good approximation at low concentrations.

11.5.1.1 Example

a The ionic product for the dissociation of water is $pK_w = 14.00$ at 25.0 °C and 13.40 at 40.0 °C. Deduce $\Delta_r H^\circ{}_m$ and $\Delta_r S^\circ{}_m$ for the dissociation (autoionization) of water into H^+ and OH^-. (b) Calculate the value of K_w and pH of neutral water at body temperature (37 °C).

From the van't Hoff isochore as expressed in Eq. (11.21), a plot of ln K_w versus $1/T$ yields a straight line with a gradient $m = -\Delta_r H^\circ_m/R$. The slope of a plot of ln K_w versus $1/T$ is

$$m = \frac{\Delta(\ln K_w)}{\Delta(1/T)} = \frac{\ln K_w(T_2) - \ln K_w(T_1)}{1/T_2 - 1/T_1}$$
$$m = -6552.5\,K$$

Slopes can have units, make sure you get them right. Multiplying by $-R$, we obtain $\Delta H^\circ_m = 54.5\,kJ\,mol^{-1}$. With T = 25 °C = 298.15 K

$$\Delta G^\circ_m = -RT\ln(10^{-14.00}) = 79908\,J\,mol^{-1}$$
$$\Delta S^\circ_m = \frac{\Delta H^\circ_m - \Delta G^\circ_m}{T} = \frac{54477 - 79908}{298.15} = -85.3\,J\,K^{-1}\,mol^{-1}.$$

Note: Using two points close together in T is a very poor way in principle to determine a slope. In this case, we know on the basis of Fig. 11.2 that the data is fairly linear and because of its source it is of high quality. Thus, the values derived here are close to the accepted values of $\Delta_r H^\circ_m(298\,K) = 55.815\,kJ\,mol^{-1}$ and $\Delta_r S^\circ_m(298\,K) = -80.85\,J\,K^{-1}\,mol^{-1}$.

b Go back to the linear equation, the intercept is

$$b = \frac{\Delta S^\circ_m}{R} = \frac{-85.295}{8.314} = -10.259.$$

Now substitute into

$$\ln K_w = m/T + b = -31.386$$
$$K_w(T = 310.15\,K) = e^{\ln K_w} = 2.34 \times 10^{-14}$$
$$pH(37\,°C) = -\lg(\sqrt{K_w}) = 6.82$$

Note: While using just two points to determine $\Delta_r H^\circ_m$ and $\Delta_r S^\circ_m$ is potentially subject to significant error, using these two points and assuming linearity between them is a very reliable method to perform an interpolation. The closer the two points are to the interpolated point, the higher the accuracy.

Finally, a few remarks on values of $\Delta_r H^\circ_m$ and $\Delta_r S^\circ_m$. The enthalpy change is positive as we expect. Water is a stable liquid and self-ionization into H^+ and OH^- is energetically unfavorable. Counter to expectations the entropy change is negative. *It is unfavorable.* The products have fewer microstates available to them than the reactants. The liberation of two particles from one usually leads to a favorable entropy change, just as in the example of generating 2 mol of gas from 1 mol of gas. Water is an unusual liquid. It is highly structured because of hydrogen bonding. Liquid water has the unusual property that it expands upon freezing. The proton and hydroxide generated by self-ionization interact in a unique way with water compared to other ions.

11.5.2 Dissociation of weak acids

For many Arrhenius acids (molecules that produce H^+ in solution), dissociation is not complete. An equilibrium is established in the aqueous solution and the acid dissociation constant, K_a, characterizes the equilibrium. Take the standard state with $c^\circ = 1$ mol dm^{-3} and assume that all activity coefficients equal to unity, $\gamma = 1$. We can replace activities by the magnitude of the molar concentration. For acetic acid

$$CH_3CO_2H(aq) + H_2O(l) \rightleftharpoons H_3O^+(aq) + CH_3CO_2^-(aq)$$

$$K_a = \frac{[H_3O^+]\,[CH_3CO_2^-]}{[CH_3CO_2H]\,c^\circ} \tag{11.37}$$

Water is in vast excess. As a liquid in its standard state, its activity is equal to one and we drop it from the equilibrium constant expression. A common way of reporting K_a values is in terms of pK_a

$$pK_a = -\lg K_a. \tag{11.38}$$

A representative set of values is shown in Table 11.2. The larger the value of pK_a, the weaker the acid. Some acids are multiprotic: they release more than one H^+ into solution with increasing values of pK_a. An unusual example is H_2SO_4, which is a strong acid with respect to the first proton (fully dissociated) but a moderately weak acid for the second proton.

11.5.2.1 Example

Determine $[H_3O^+]$ in a 1.00×10^{-2} mol dm^{-3} solution of CH_3CO_2H at 25 °C.

The reaction chart is

	$CH_3CO_2H(aq)$	$\rightleftharpoons$	$H^+(aq)$	+	$CH_3CO_2^-(aq)$
Initially	a_0		0		0
Eq	$a_0 - \xi$		ξ		ξ
	$(1-\alpha)a_0$		αa_0		αa_0

Thus, $\xi = [H_3O^+] = [CH_3CO_2^-]$ is negligible, therefore

$$K_a = \frac{[H_3O^+]^2}{[CH_3CO_2H]\,c^\circ}$$

Table 11.2 A selection of acid dissociation constants at 25 °C. For multiprotic acids, a number indicates the dissociation step of the associated pK_a value. Values taken from the 96th edition of the *CRC Handbook of Chemistry and Physics*.

Organic acid	Step	pK_a	Inorganic acid	Step	pK_a
CH_2O_2 formic		3.75	NH_3		9.25
CH_4N_2O urea		0.10	$HClO_2$		1.94
CH_5N_3 guanidine		13.6	HCN		9.21
$C_2H_4O_2$ acetic		4.756	HF		3.20
$C_2H_5NO_2$ glycine	1	2.35	H_2O_2		11.62
	2	9.78			
C_2H_7NS cystamine	1	8.27	H_2S	1	7.05
	2	10.53		2	19
$C_3H_4O_3$ pyruvic		2.39	HIO_3		0.78
$C_3H_6O_3$ lactic		3.86	H_3PO_4	1	0.78
				2	2.16
				3	7.21
$C_4H_7NO_4$ L-aspartic	1	1.99	H_2SO_4	2	1.99
	2	3.90			
	3	9.90			
$C_6H_{12}N_2O_4S_2$ L-cystine	1	1	H_2O		13.995
	2	2.1			
	3	8.02			
	4	8.71			

and

$$[H_3O^+] = \sqrt{K_a[CH_3CO_2H]}.$$

Acetic acid is a weak acid ($K_a \ll 1$), therefore very little dissociation occurs and $[CH_3CO_2H] \approx [CH_3CO_2H]_{initial}$. Substitution leads to

$$[H_3O^+] = \sqrt{1.75 \times 10^{-5}(1.0 \times 10^{-2})} = 4.19 \times 10^{-4}\,\text{mol dm}^{-3}.$$

The *degree of dissociation* is

$$\alpha = (4.19 \times 10^{-4}/1.00 \times 10^{-2}) \times 100\% = 4.19\%.$$

The assumption of little dissociation is validated.

11.5.2.2 Directed practice

A chemist mixes 0.5 mol of acetic acid and 0.5 mol of HCN with enough water to make 1 l of solution. Calculate the final concentrations of H_3O^+ and OH^-.

[Answer: $[H_3O^+] = 3.0 \times 10^{-3}$ M, $[OH^-] = 3.3 \times 10^{-12}$ M]

In an aqueous solution of a weak base B, proton transfer is incomplete and the associated equilibrium constant, the base dissociation constant K_b, is defined analogously to that for acid dissociation. Of course, acid and base character are coupled by the self-ionization of water. Whether a species acts as an acid or base depends on the relative strength of the acid or base to which it is added.

Consider the dissociation of acetic acid and the equilibrium established by the acetate ion in aqueous solution.

$$CH_3CO_2H(aq) + H_2O(l) \rightleftharpoons H_3O^+(aq) + CH_3CO_2^-(aq)$$

The reaction of acetic acid with water is the conventional acid dissociation reaction with equilibrium constant

$$K_a = \frac{[H_3O^+]\,[CH_3CO_2^-]}{[CH_3CO_2H]\,c^\circ}.$$

$CH_3CO_2^-$ is the conjugate base of CH_3CO_2H. Its reaction with water is the base dissociation reaction

$$CH_3CO_2^-(aq) + H_2O(l) \rightleftharpoons H_3CO_2H(aq) + OH^-(aq).$$

In this case it is H_2O that dissociates to liberate $OH^-(aq)$ and the base equilibrium constant is

$$K_b = \frac{[CH_3CO_2H][OH^-]}{[CH_3CO_2^-]c^\circ}.$$

But K_a and K_b for conjugate species are not independent because they are couple by the ionization of water,

$$K_aK_b = [H_3O^+][OH^-] = K_w. \qquad (11.39)$$

In terms of the p-values

$$pK_a + pK_b = pK_W = 13.995 \text{ at } 25\,°\text{C} \qquad (11.40)$$

There is no need to tabulate pK_b values because we can calculate pK_b for a conjugate base as long as we know the pK_a of the conjugate acid.

11.5.2.3 Directed practice

Calculate the concentration of OH^- in an aqueous solution of NH_3 of concentration 5.0×10^{-2} mol dm^{-3}.

[Answer: $[OH^-] = 9.5 \times 10^{-3}$ mol dm^{-3}]

11.5.2.4 Example

Calculate the pH of a 5.0×10^{-4} mol dm^{-3} solution of formic acid. $pK_a = 3.75$. Do not assume that dissociation is insignificant compared to the initial concentration of the acid.

The reaction chart is

	$HCO_2H(aq)$	+ $H_2O(l)$	$\rightleftharpoons$ $H_3O^+(aq)$	+ $HCO_2^-(aq)$
Initially	0.00050	xs	0	0
Eq	$0.00050 - \xi$	xs	ξ	ξ

$$K_a = 1.78 \times 10^{-4} = \frac{[H_3O^+]\,[HCO_2^+]}{[HCO_2H]c^\circ} = \frac{[\xi][\xi]}{[0.0005 - \xi]}$$

Substitution leads to

$$\xi^2 + (1.78 \times 10^{-4})\xi - 8.89 \times 10^{-8} = 0.$$

From which

$$\xi = 2.2 \times 10^{-4}$$

and

$$\text{pH} = -\log(2.2 \times 10^{-4}) = 3.65.$$

11.6 Dissolution and precipitation of salts

Ionic compounds are held together by one of the strongest of all intermolecular interactions – the electrostatic interactions of full positive and negative charges. Nevertheless, ionic compounds display a full range of solubility from sparingly soluble to highly soluble. The factor that controls the extent of dissolution in cold thermodynamic terms is the Gibbs energy change. However, to understand dissolution on a molecular level we need to look at the enthalpy and entropy changes and how they relate to molecular processes.

Consider the dissolution of KCl in water at 298 K. The values in Table 11.3 reveal that $\Delta_{sol}H^\circ_m = 17.22$ kJ mol^{-1} and $\Delta_{sol}S^\circ_m = 76.4$ J mol^{-1} K^{-1}. Therefore, $\Delta_{sol}G^\circ_m = -5.56$ kJ mol^{-1}. The negative value of $\Delta_{sol}G^\circ_m$ indicates a spontaneous process, that is, KCl dissolves in water. The process, in this case, is driven by entropy, not enthalpy. The reaction

$$KCl(s) \rightarrow K^+(aq) + Cl^-(aq)$$

is endothermic because the sum of all the intermolecular forces acting on the reactants side is stronger than the sum of the intermolecular forces in the solution. These intermolecular forces include the electrostatic interactions of the ions with each other. These electrostatic interactions are screened in the solvated state by the water molecules that intercede in between the ions to keep them separated. Solvation and these interactions are discussed in greater detail in Chapter 14. Importantly, one other set of intermolecular interactions is being hidden by the way we have written the solvation reaction. We must also consider the effect of the solute on the structure of the solvent and the hydrogen bonding between water molecules. For very dilute solutions when the solution is acting ideally, the electrostatic interactions of ions will be the predominant factor in determining the energetics. At higher concentrations, as non-ideality sets in, changes in the solvent interactions also have to be taken into consideration. Usually, $\Delta_{sol}H^\circ_m$ is small for monovalent ions, with a magnitude of less than 40 kJ mol^{-1}. Depending on the system it can be either endo- or exo-thermic.

Table 11.3 Selected thermodynamic properties of a few ions and ionic compounds in aqueous solutions. $m\%$ represents the solubility in mass percent.

	Na^+	K^+	F^-	Cl^-	Ag^+
S°_m(aq)/J K^{-1} mol^{-1}	59.0	102.5	−13.8	56.5	72.7
	KF	**KCl**	**NaF**	**NaCl**	**AgCl**
S°_m(aq)/J K^{-1} mol^{-1}	66.6	82.6	51.1	72.1	96.3
$m\%$	50.41	26.22	3.97	26.45	0.00019
$\Delta_{sol}H^\circ_m$/kJ mol^{-1}	−17.73	17.22	0.91	3.88	65.4
S°_m(aq)/J K^{-1} mol^{-1}	22.1	76.4	−5.9	43.4	32.9
$\Delta_{sol}G^\circ_m$/kJ mol^{-1}	−24.3	−5.56	2.67	−9.06	55.6

Now a look at the entropy term. It makes sense that dissolution has a favorable entropy change. On the reactants side, 1 mol of solid contains cations and anions in an ordered lattice phase separated from a solvent that is free to take on its ideal structure. On the products side, 2 mol of particles have been liberated into the solvated phase. Much like the liberation of a gas from a solid, the liberation of the ions from a solid is highly entropically favored as far as the ions are concerned. Usually, this is the case in dissolution, as can be read from the values of $\Delta_{sol}S^{\circ}_{m}$ given in Table 11.3. However this is not always the case. The fluoride ion has a particularly strong effect on the structure of the water that is held in its solvation shell. So much so that it leads to a more ordered structure in solution than does even the H^+ ion, which we know interacts so strongly in the solvated state that we call it H_3O^+ instead of H^+. This leads to a negative value of $\Delta_{sol}S^{\circ}_{m}$ for the solution of NaF. In the dissolved state the entropy of the solvent plus the solvated ions is lower than the entropy of pure water in contact with solid NaF. Since the dissolution is also almost thermoneutral, $\Delta_{sol}G^{\circ}_{m}$ is slightly positive and NaF exhibits moderate solubility. We investigate dissolution and precipitation of sparingly soluble salts in the sections below.

11.6.1 Solubility of sparingly soluble salts

The data in Table 11.3 shows that AgCl dissolution is rather endothermic. So much so that it is able to counteract the favorable entropy term and make the Gibbs energy change highly unfavorable. Not surprisingly, AgCl has a low solubility in water. Thus, for the reaction

$$\mathrm{AgCl(s)} \rightleftharpoons \mathrm{Ag^+(aq)} + \mathrm{Cl^-(aq)} \tag{11.41}$$

the equilibrium constant

$$K = \frac{a(\mathrm{Ag^+})a(\mathrm{Cl^-})}{a(\mathrm{AgCl})} \tag{11.42}$$

should be much less than 1.

The activity of AgCl, a solid in its standard state is constant and equal to 1. We rewrite the equilibrium constant in terms of the more commonly encountered *solubility product*

$$K_{sp} = a(\mathrm{Ag^+})a(\mathrm{Cl^-}) = [\mathrm{Ag^+}][\mathrm{Cl^-}]/c^{\circ 2}. \tag{11.43}$$

Assuming ideal behavior, the activity is equal in magnitude to the molar concentration and the standard state concentration c° reminds us of the choice of standard state while keeping K_{sp} dimensionless. For low-solubility salts, the ion concentrations are very low and ideal behavior is to be expected as long as no other electrolytes are present in solution. The solubility of a salt is sensitive to the *common ion effect*, in which a decrease in solubility of the salt occurs when a second non-saturating salt with one ion in common with the salt is added to its saturated solution

The K_{sp} expression must properly take account of the stoichiometry of the dissolution reaction as does any equilibrium constant expression. For example, again assuming ideality,

$$\mathrm{Ca_3(PO_4)_2(s)} \rightleftharpoons 3\,\mathrm{Ca^{2+}(aq)} + 2[\mathrm{PO_4}]^{3-}\mathrm{(aq)} \tag{11.44}$$

$$K_{sp} = [\mathrm{Ca^{2+}}]^3\left[\mathrm{PO_4^{3-}}\right]^2/c^{\circ 5}. \tag{11.45}$$

Sulfides produces S^{2-}(aq), which reacts readily with water,

$$\mathrm{S^{2-}(aq)} + \mathrm{H_2O(l)} \rightleftharpoons \mathrm{HS^-(aq)} + \mathrm{OH^-(aq)}. \tag{11.46}$$

This equilibrium lies far to the right unless the solution is made strongly basic. However, the equilibrium constant is not well known. Therefore, the usual formulation of the solubility product is of little utility. Instead, the dissolution of sulfides is often characterized by the modified solubility parameter K_{spa} that corresponds to the reaction

$$\mathrm{M}_m\mathrm{S}_n\mathrm{(s)} + 2\mathrm{H^+(aq)} \rightleftharpoons m\mathrm{M}^{(2n/m)+} + n\mathrm{H_2S(aq)}, \tag{11.47}$$

$$K_{spa} = \frac{[\mathrm{M}^{(2n/m)+}]^m[\mathrm{H_2S}]^n}{[\mathrm{H^+}]^2}. \tag{11.48}$$

11.6.1.1 Example

Determine K_{sp} and $\Delta_{sol}G^{\circ}_{m}$ for $Fe(OH)_2$ given that the solubility at 298 K is 2.30×10^{-6} mol dm^{-3} at 298 K.

Start with a balanced chemical reaction

$$\mathrm{Fe(OH)_2(s)} \rightleftharpoons \mathrm{Fe^{2+}(aq)} + 2\mathrm{OH^-(aq)}$$

In a saturated solution, the concentration of Fe^{2+} is 2.30×10^{-6} mol dm^{-3}, but the concentration of OH^- is twice this or 4.60×10^{-6} mol dm^{-3}. K_{sp} is, consequently

$$K_{sp} = (2.30 \times 10^{-6})(4.60 \times 10^{-6})^2 = 4.87 \times 10^{-17}.$$

K_{sp} is an equilibrium constant and is related directly to $\Delta_{sol}G^\circ_m$ by

$$\Delta_{sol}G^\circ_m = -RT \ln K_{sp} = 92.0\,\text{kJ mol}^{-1}.$$

11.6.1.2 Directed practice

Determine the solubility of $Ca(OH)_2$ in water given that $K_{sp} = 5.02 \times 10^{-6}$ at 298 K.

[Answer: 1.08×10^{-2} mol dm^{-3}]

11.6.2 Precipitation

Equilibrium is established when the chemical potential of the chemical species is the same in all phases that are in contact with one another. As we add a small amount of solid to a large volume of solvent, the chemical potential in the solution phase is lower than the chemical potential in the solid phase. Therefore the solid dissolves. All ionic solids are fully dissociated in water at infinite dilution. As we continue to add more solid to the solution, the chemical potential in the solid phase will eventually be lower than the chemical potential in the dissolved phase. Ideally, this is the point where *precipitation* occurs. In Chapter 13 we will investigate phase changes in more detail and discover that the nucleation of a new phase is a difficult process that is often delayed from its thermodynamic threshold based on kinetic factors.

Ions in solution are constantly in motion and colliding with one another and the solvent. The point at which precipitation occurs is defined quantitatively by the K_{sp} expression, as shown in Fig. 11.4. Precipitation occurs when the reaction quotient of the species in solution exceeds the solubility product of the solid. Below the threshold shown in Fig. 11.4, whenever ions collide they can only aggregate into a solid particle for a short while. Such a solid particle is not thermodynamically stable compared to the solvated ions. Above the threshold, the equilibrium concentration of the ions in solution is constant at constant temperature. Above this point, when an aggregate forms, it is more stable than the dissolved ions. It will continue to grow in size until it depletes the ions in the solution around it. The particle becomes stable when the concentrations of ions around it reach equilibrium concentrations derived from the K_{sp} expression. If the particle is large enough and has a density greater than that of the solvent, it will precipitate to the bottom of the container. If its density is lower than that of the solvent the particle will float to the top of the liquid.

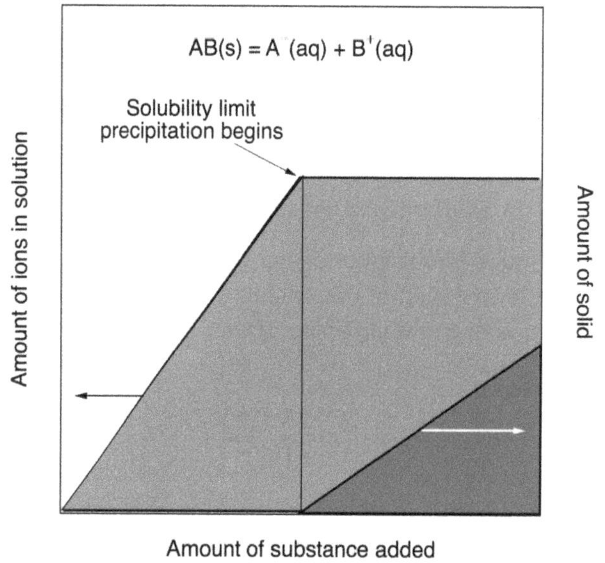

Figure 11.4 The concentration (or amount in moles at constant volume) of ions in a 1:1 electrolyte increases linearly upon addition of a soluble solid until the solubility limit is exceeded. Passing this threshold the onset of precipitation is required to maintain the ion concentrations at their constant equilibrium values.

11.6.2.1 Example

Will precipitation occur when 0.01 mol of Ba^{2+} is added to 1 l of 0.05 M Na_2SO_4 at 298 K? $K_{sp}(BaSO_4) = 1.08 \times 10^{-10}$.

$$BaSO_4 \rightleftharpoons Ba^2 + SO_4^{2-}$$
$$K_{sp} = [Ba^{2+}][SO_4^{2-}]/(c^\circ)^2 = 1 \times 10^{-10}$$

Recall that the standard state concentration is $c^\circ = 1$ M. Sodium sulfate is completely soluble so the initial concentration of sulfate is $[SO_4{}^{2-}]_0 = 0.05$ M, and that of barium is $[Ba^{2+}]_0 = 0.01$ M. Substituting into the reaction quotient

$$Q = [0.01][0.05] = 5 \times 10^{-4}.$$

This value is much greater than 1.08×10^{-10} and precipitation occurs.

11.6.2.2 Example

Calculate the concentrations of calcium ion and fluoride ion in a concentrated solution of calcium fluoride at 298 K, CaF_2. $K_{sp} = 3.45 \times 10^{-11}$.

$$CaF_2(s) \rightleftharpoons Ca^{2+}(aq) + 2F^-(aq)$$

Let $x = [Ca^{2+}]$. Then $[F^-] = 2x$.

$$K_{sp} = [Ca^{2+}][F^-]^2/(c^\circ)^3 = x(2x)^2 = 4x^3 = 3.45 \times 10^{-11}$$
$$x = 2.05 \times 10^{-4}$$

Therefore, $[Ca^{2+}] = 2.05 \times 10^{-4}$ M and $[F^-] = 4.10 \times 10^{-4}$ M.

11.6.2.3 Example

The solubility product of $CsClO_4$ is 3.95×10^{-3} at 298 K. When 500 ml of 0.100 M CsCl is added to 500 ml of 0.200 M $HClO_4$, what are the final concentrations of the components of the solution? Assume that CsCl and $HClO_4$ are fully dissociated and the solution acts ideally.

Recall that Gibbs energy is a state function. Therefore, we can apply Hess's law to coupled chemical reactions and put reactions together in any order as long as they add up stoichiometrically to the overall process. First, allow CsCl and $HClO_4$ to dissolve completely. Then, $CsClO_4$ precipitates until the equilibrium concentrations are achieved. The total volume is 1 l, thus the amount of substance is equal in magnitude to the molar concentration. The reaction chart for the first reaction is

	$Cs^+(aq)$	+	$ClO_4^-(aq)$
Initial/M	[0.1](0.5)		[0.2](0.5)
	0.05		0.1

We now set up the second reaction chart, which corresponds to the precipitation of $CsClO_4$, but written in the form for K_{sp}. In ideal solutions, spectator ions such as H^+ and Cl^- in this case play no role in the equilibria set up by other solution species. We neglect the spectators in writing the second reaction chart.

	$CsClO_4(s)$	$\rightleftharpoons$	$Cs^+(aq)$	+	$ClO_4^-(aq)$
Initial	0		0.05		0.1
Eq	x		$0.05 - x$		$0.1 - x$

$K_{sp} = 3.95 \times 10^{-3} = [Cs^+][ClO_4^-]/c^{\circ 2} = (0.05 - x)(0.1 - x) = 0.005 - 0.15x + x^2$. Using quadratic formula we find $x =$ 0.1426 or 0.00736.

H^+ and Cl^- are spectators. Thus, $[H^+] = 0.1$ M and $[Cl^-] = 0.05$ M. $[Cs^+] = 0.05 - 0.00736 = 0.0426$ and $[ClO_4{}^-] = 0.1 - 0.00736 = 0.0926$. The other root produces unphysical negative concentrations.

11.7 Formation constants of complexes

Consider the formation of a mononuclear binary complex between a central metal ion M (the central group) and a ligand L. The system exists in an aqueous solution. Thus, the metal ion is initially coordinated with water molecules, most likely hexa-coordinated, and the ligands then compete to replace one or more of these water molecules. Many such complexes are known, for example, with $n = 4$ in a square planar arrangement or with $n = 6$ in an octahedral arrangement. More on the solvation of ions in solution can be found in Chapter 14.

An equilibrium between the various species is established by a series of coupled equilibria that can be represented either stepwise

$$\begin{gathered} \mathrm{M + L \rightleftharpoons ML} \\ \mathrm{ML + L \rightleftharpoons ML_2} \\ \vdots \quad \vdots \quad \vdots \\ \mathrm{ML}_{n-1} + \mathrm{L} \rightleftharpoons \mathrm{ML}_n \end{gathered}$$

or in an overall fashion

$$\begin{gathered} \mathrm{M + L \rightleftharpoons ML} \\ \mathrm{M + 2L \rightleftharpoons ML_2} \\ \vdots \quad \vdots \quad \vdots \\ \mathrm{M} + n\,\mathrm{L} \rightleftharpoons \mathrm{ML}_n. \end{gathered}$$

Note that the presence of the water molecules has been omitted.

The associated equilibrium constants for each step are known as *formation constants* (also *stability constants*). We define both a stepwise formation constant K_n according to

$$K_n = \frac{[\mathrm{ML}_n]}{[\mathrm{ML}_{n-1}][\mathrm{L}]} \tag{11.49}$$

and a cumulative formation constant β_n as

$$\beta_n = \frac{[\mathrm{ML}_n]}{[\mathrm{M}][\mathrm{L}]^n}. \tag{11.50}$$

Stepwise formation constants usually decrease as n increases. This can be seen by inspecting the values given in Table 11.4. Deviations from this trend are observed when the coordination number changes (remember that M is initially solvated/coordinated by a number of water molecules) or when the metal ion undergoes a change in magnetic moment by transitioning from a high spin to low spin configuration.

The formation constant is determined by the value of $\Delta G^\circ_\mathrm{m}$, as is any other equilibrium constant. As can be seen by the values in Table 11.5, the trend in $\Delta G^\circ_\mathrm{m}$ is sometimes predominantly determined by the enthalpy change – as in the case of $\mathrm{Cu(NH_3)}_n{}^{2+}$ – or by the entropy change – as found for $\mathrm{AlF_n}^{3-n}$. The strength of the metal–ligand bond is derived mainly from its σ bond character, though some ligands can also participate in π interactions. The bonding in these complexes is thought of in terms of a lone pair donor (the ligand) and an acceptor (the metal). The strength of the $\mathrm{M{-}OH_2}$ interaction is derived from a complex set of parameters. There is no simple explanation for trends; rather, each case must be treated individually. Stabilities are dependent on the differences between M–ligand and $\mathrm{M{-}OH_2}$ interactions; thus, generalizations regarding stability are difficult to make in simple terms.

Table 11.4 Dependence of stepwise formation constants expressed as lg K_n on the number of ligands n at 25 °C. en = $\mathrm{C_2H_4(NH_2)_2}$. Values taken from J. Burgess (1999) *Ions in Solutions: Basic principles of chemical interactions*. Woodhead Publishing, Cambridge, UK.

n	Al^{3+}/F^-	Cr^{3+}/NCS^-	Co^{2+}/NH_3	Cu^{2+}/NH_3	Pb^{2+}/Cl^-	Fe^{2+}/en
1	6.1	3.0	2.1	4.1	1.2	4.3
2	5.0	1.7	1.6	3.5	0.5	3.3
3	3.9	1.0	1.1	2.9	−0.3	2.0
4	2.7	0.3	0.7	2.1	−1.3	
5	1.6	−0.7	0.2			
6	0.5	−1.3	−0.6			

Table 11.5 Enthalpy and entropy contributions to stepwise formation constant at 25 °C for Cu(II)–ammonia and Al(III)–fluoride complexes. en = $C_2H_4(NH_2)_2$. All values are in kJ mol^{-1}. Data from J. Burgess (1999) *Ions in Solutions: Basic principles of chemical interactions*. Woodhead Publishing, Cambridge, UK.

n	1	2	3	4	5	6
$Cu(NH_3)_n^{2+}$						
ΔG°_m	−23	−20	−15	−11		
ΔH°_m	−23	−23	−23	−22		
$-T\,\Delta S^\circ_m$	0	3	7	10		
AlF_n^{3-n}						
ΔG°_m	−35	−29	−21	−15	−7	−2
ΔH°_m	5	3	1	1	−1	−6
$-T\,\Delta S^\circ_m$	−40	−32	−22	−16	−6	4

A number of attempts have been made to rationalize the relative stabilities of complexes in aqueous solutions. One such system is that they follow the hard and soft acids and bases (HSAB) principle. Ligands and metal ions are classified (see Table 11.6) as either hard, borderline or soft acids and bases. Strong metal–ligand interactions occur when like binds with like (hard–hard or soft–soft). Weak interactions occur when a hard–soft combination occurs. Middling interactions occur with borderline cases.

Pearson and coworkers have been able to establish a quantitative relationship between hardness and electronic structure. Indeed, a *principle of maximum hardness* has been developed, as defined by the IUPAC:

> *A chemical system at a given temperature will evolve to a configuration of maximum absolute hardness, η, provided that the potential due to the nuclei, plus any external potential and the electronic chemical potential, remain constant. In terms of molecular orbital theory, the highest value of η reflects the highest possible energy gap between the lowest unoccupied molecular orbital (LUMO) and the highest occupied molecular orbital (HOMO); this value correlates with the stability of the system.*

The *absolute hardness* is defined as the energy difference between the LUMO and the HOMO,

$$\eta = \varepsilon_{\mathrm{LUMO}} - \varepsilon_{\mathrm{HOMO}}. \tag{11.51}$$

In general terms, the following definitions have been put forward:[1]

1. Soft acid – the acceptor atom is of low positive charge, large size and has polarizable outer electrons.
2. Hard acid – the acceptor atom is of high positive charge, small size and has no easily polarized outer electrons.
3. Soft base – the donor atom is of low electronegativity, easily oxidized, highly polarizable and with low-lying empty orbitals.
4. Hard base – the donor atom is of high electronegativity, hard to oxidize, of low polarizability and with only high energy empty orbitals.

Soft–soft combinations depend mainly on covalent bonding and hard–hard combinations, mainly on ionic bonding. A hard molecule has a large HOMO–LUMO gap and a soft molecule has a small HOMO–LUMO gap.

Table 11.6 Classification of chemical species according to the hard and soft acids and bases (HSAB) principle. Data from Huheey, J.E., Keiter, E.A., and Keiter, R.L. (1997) *Inorganic Chemistry: Principles of Structure and Reactivity*, 4th edition. Prentice-Hall, New York.

	Hard	Borderline	Soft
Acids	H^+, Li^+, Na^+, K^+ Be^{2+}, Mg^{2+}, Ca^{2+} Cr^{2+}, Cr^{3+}, Al^{3+}, BF_3 Sn^{4+}, Ti^{4+}, VO^{2+}	Fe^{2+}, Co^{2+}, Ni^{2+} Cu^{2+}, Zn^{2+}, Pb^{2+} SO_2, BBr_3, Sn^{2+} NO_2, NO^+	Cu^+, Au^+, Ag^+ Hg_2^{2+}, Pd^{2+}, Cd^{2+} Pt^{2+}, Hg^{2+}, BH_3
Bases	F^- SO_4^{2-}, OH^- H_2O, NH_3,O_2^- $CH_3CO_2^-$, CO_3^{2-}, NO_3^-	NO_2^-, SO_3^{2-}, Br^- N_3^-, N_2, H_2S C_6H_5N, SCN	H^-, R^-, CN^-, CO I^-, SCN^-, R_3P, C_6H_6 R_2S

11.8 Thermodynamics of self-assembly

Complex formation, as we have seen in the previous section, is the aggregation of a set of ligands with a coordination center. Aggregation of chemical subunits with themselves into larger bodies is related to the formation of micelles, liposomes, bilayers, and membranes. The self-directed process that forms these structures is known as *self-assembly*. Essential to the formation of (close to) perfect self-assembled structures is that the binding energies of the structures formed is just slightly stronger than the thermal energy of roughly $k_B T$. If the structures were bound much more weakly, they would immediately dissociate. If they were bound much more strongly, they would lock themselves into the first strongly bound structure formed, regardless of whether it formed with defects or whether it represented a local minimum in energy rather than a global minimum. With a binding energy close to $k_B T$, error correction can be built into the self-assembly process, allowing for many iterations of reversible steps that allow the system to search for the optimal configuration.

In the state of chemical equilibrium all components of the system have the same chemical potential. Consider an equilibrated solution composed of identical molecules forming aggregates composed of N molecules. The $N = 1$ unit is called a monomer, $N = 2$ dimer, $N = 3$ trimer, and so on. An aggregate of N molecules is an N-mer. The chemical potential of such a solution is given by

$$\mu = \mu_1^\circ + RT\ln x_1 = \tfrac{1}{2}\mu_2^\circ + \tfrac{1}{2}RT\ln x_2 = \tfrac{1}{3}\mu_3^\circ + \tfrac{1}{3}RT\ln x_3 = \cdots, \tag{11.52}$$

where μ_N° is the standard part of the chemical potential, that is, the mean interaction free energy per molecule, for aggregates of number N and with mole fraction x_N, (thus, we are assuming ideal behavior for which concentration measured in mole fraction can be used in place of activity). We can write this more succinctly as

$$\mu = \mu_N = \frac{1}{N}\mu_N^\circ + \frac{RT}{N}\ln x_N, \quad N = 1, 2, 3, \cdots \tag{11.53}$$

At equilibrium for the reaction between N monomers and an aggregate of N molecules

$$N\,\mathrm{X} \underset{k_N}{\overset{k_1}{\rightleftharpoons}} \mathrm{X}_N \tag{11.54}$$

the chemical potential of reactants and products is equal

$$\mu_1^\circ + RT\ln x_1 = \frac{1}{N}\mu_N^\circ + \frac{RT}{N}\ln x_N. \tag{11.55}$$

The standard Gibbs energy change for aggregation into an N-mer is then

$$\Delta_{\mathrm{ag}} G_{\mathrm{m}}^\circ = \mu_N^\circ - N\mu_1^\circ. \tag{11.56}$$

The equilibrium constant for the reaction in Eq. (11.54) is given either by the law of mass action

$$K = \frac{x_N}{x_1^N} \tag{11.57}$$

or by its relationship to the Gibbs energy change

$$-RT\ln K = \Delta_{\mathrm{ag}} G_{\mathrm{m}}^\circ \tag{11.58}$$

Combining these two, we can derive an expression for the mole fraction of the Nth N-mer in terms of the mole fraction of the monomer x_1,

$$x_N = x_1^N \exp\left(\frac{\mu_N^\circ - N\mu_1^\circ}{-RT}\right). \tag{11.59}$$

11.8.1 Directed practice

Derive Eq. (11.59) by making the appropriate substitutions.

The concentration, defined in terms of mole fraction, must satisfy the conservation relation that the total mole fraction sums to unity

$$1 = \sum_{N=1}^{\infty} x_N. \tag{11.60}$$

Note that this derivation is valid as long as the aggregates do not begin to aggregate among themselves.

When do molecules aggregate? If all aggregates experience the same interactions per monomer unit with the solvent, the value of μ_N°/N is constant for all values of N. Substitution of this constant value into Eq. (11.57) means that $\mu_N^\circ - N\mu_1^\circ = 0$ for all N and in this case

$$x_N = x_1^N. \tag{11.61}$$

However, by definition the mole fraction of the monomer (and all other N-mers) must be less than one, $x_1 < 1$. Therefore, all other concentrations must be lower yet, $x_N < x_1$, with the difference becoming increasingly large as N increases. If the standard chemical potential of an N-mer, μ_N°, increases as a function of N, *aggregation* becomes progressively less likely as the size increases.

The condition for aggregation is that μ_N° must be a *decreasing* function of N. If μ_N° is a complicated function of N – for example, one that exhibits a minimum at a certain value of N – then there is a size distribution of aggregates centered about the value of N for which the standard chemical potential is a minimum. The distribution function is controlled by the dependence of chemical potential on the size of the aggregates through Eq. (11.59).

As discussed in Chapter 4, the type of molecule that commonly forms a micelle, liposome or bilayer is an *amphiphile*. If a molecule forms a micelle, for example, then it must exhibit a standard chemical potential of the N-mer that has a minimum at the characteristic size of the micelle. The size distribution of the micelles is affected not only by the nature of the molecule but also the solvent, the presence of other surfactants such as alcohols, as well as the presence of other molecules and, of course, temperature. Below the *critical micelle concentration* (cmc or c_M) amphiphiles predominantly segregate to the surface of the solution to form a Langmuir film. Above the cmc, they spontaneously form micelles in increasing concentration. This is not a phase transition but an equilibrium phenomenon that occurs mainly over a narrow concentration range, as shown in Fig. 11.5. The number of amphiphiles in a micelle is usually of order 100.

Different properties of the system respond to micelle formation, including the surface tension (which saturates at the point where micelles begin to form), light scattering (the turbidity of the mixture increases), electrical conductivity, NMR peak shifts, and diffusion coefficients. Different properties have different sensitivities to the presence of micelles and will return slightly different values of the cmc. This is indicated by the box in Fig. 11.5. The concentration of the monomer rapidly rolls over to approach an asymptotic value. However, since the number of monomers N in a micelle is large the number of micelles formed is much lower than the number of monomers consumed. This translates into a micelle concentration that rises slowly as the monomer concentration saturates. This behavior is also related to why different techniques lead to different values of cmc. Experimental signals related to micelle formation, such as changes in surface tension or light scattering, rise slowly out of the noise rather than appearing like a sharp endpoint in a titration.

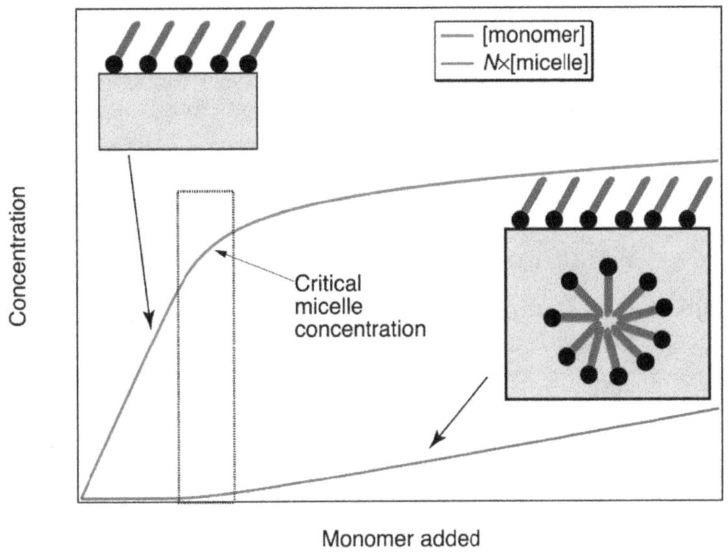

Figure 11.5 Below the critical micelle concentration amphiphiles with a polar head group (black) and nonpolar tail (red) self-assemble into a Langmuir film atop the solvent. Above the critical micelle concentration, material added to the system leads to the formation of micelles composed of N monomers in the bulk of the solvent.

The cmc, expressed in mol m^{-3} or x_M in mole fraction, depends sensitively on the structure of the amphiphile. Values range from 10^{-4} M to 10^{-1} M or $10^{-6} \le x_M \le 10^{-3}$. Its value is equal to the equilibrium concentration of monomer that leads to the formation of the stable micelles. Its value can be related to the standard Gibbs energy change for micelle formation $\Delta_{mic}G^\circ_m$ by solving Eq. (11.59) for x_1. Using the fact that N is large and, therefore, a term in $1/N$ is negligible, we find,

$$x_M = \exp\left(\Delta_{mic}G^\circ_m/RT\right). \tag{11.62}$$

This once again shows that micelle formation must be an exergonic process, that is, $\Delta_{mic}G^\circ_m < 0$ for micelles to form. Inserting a typical value of $x_M = 10^{-4}$ into Eq. (11.62), we find that $\Delta_{mic}G^\circ_m \approx -20$ kJ mol^{-1}. Given that there are ≈ 100 units in a micelle, a binding interaction of just 0.2 kJ mol^{-1} per interaction is required to hold the micelle together. This corresponds to the van der Waals interaction energy of two nonpolar groups. Looking at Fig. 11.5 this makes sense, since in the initial unaggregated state (the Langmuir film), the polar head groups are interacting with water similar to how they interact with water in the aggregated state. The tail groups are in much different configurations in the two states. There are obvious entropic factors that play a role as well as chain–chain interactions, and the value of ln c_M has been found to decrease linearly with increasing alkyl chain length. A difference in the ability of molecules to maximize attractive interactions along the chain and between tail groups is what leads to the wide range of values for the cmc.

Note also the similarity to the idealized behavior for precipitation depicted in Fig. 11.4. There are similarities to the dynamics of formation of a precipitate. Below the cmc, the concentration of monomer increases linearly with increasing addition of monomer to the solution. In this case, saturation of the Langmuir film forces monomers into the bulk of the solution. When the chemical potential of the monomer in the micelle drops below that of the solvated monomer, aggregation into micelles begins.

SUMMARY OF IMPORTANT EQUATIONS

$dU = T\,dS - p\,dV$	Fundamental equation relates changes in internal energy dU to changes in entropy dS and volume dV as well as temperature T and pressure p.
$dG = V\,dp - S\,dT$	Description of the dependence of Gibbs energy on pressure p and temperature T.
$\Delta A = \Delta U - T\,\Delta S$	Changes in the Helmholtz energy ΔA are related to ΔU and ΔS.
$\Delta G = \Delta H - T\,\Delta S$	Changes in the Gibbs energy ΔG are related to changes in enthalpy ΔH and ΔS.
$\dfrac{\Delta G(T_2)}{T_2} = \dfrac{\Delta G(T_1)}{T_1} + \int_{T_1}^{T_2} \Delta H\,d\left(\dfrac{1}{T}\right)$	Integrated form of the Gibbs–Helmholtz equation describes the temperature dependence of the Gibbs energy change.
$\Delta_r G = \sum_{\text{products}} \nu\,\Delta_f G - \sum_{\text{reactants}} \nu\,\Delta_f G$	Gibbs energy of reaction $\Delta_r G$ is obtain from the difference of Gibbs energies of formation of reactants minus products weighted by the stoichiometric coefficients.
$\Delta_r G_m = \sum_{\text{products}} \nu\mu - \sum_{\text{reactants}} \nu\mu$	Gibbs energy of reaction is obtain from the difference of chemical potential of reactants minus products weighted by the stoichiometric coefficients.
$\mu = \mu^\circ + RT\ln(p/p^\circ)$	Chemical potential of an ideal gas μ, standard state potential μ°, gas constant R, pressure p and standard state pressure p°.
$\Delta_r G^\circ_m = -RT\ln K$	Gibbs energy of reaction is related to the thermodynamic equilibrium constant K.
$\ln K = -\dfrac{\Delta_r H^\circ_m}{RT} + \dfrac{\Delta_r S^\circ_m}{R}$	van't Hoff equation. Plot ln K versus $1/T$. slope $= -\Delta_r H^\circ_m/R$, intercept $= \Delta_r S^\circ_m/R$.
$\ln K(T_2) = \ln K(T_1) + \int_{T_1}^{T_2} \dfrac{\Delta_r H^\circ_m}{RT^2}\,dT$	Temperature dependence of equilibrium constant is related to the temperature dependence of the enthalpy change $\Delta_r H^\circ_m$.

Exercises

11.1 Use the relationship between $\Delta_r G$ and chemical potential to show that for the ideal gas reaction $A + B \rightleftharpoons 2C$, the chemical potentials are related to the equilibrium constant by $2\mu^\circ_C - \mu^\circ_A - \mu^\circ_B = -RT\ln K$.

11.2 Explain the difference between $\Delta_r G_m$ and $\Delta_r G^\circ_m$.

11.3 If $\Delta_r H^\circ_m$ and $\Delta_r S^\circ_m$ are independent of temperature, is the equilibrium constant of the reaction independent of temperature?

11.4 For the gas-phase synthesis of methanol from CO and H_2, derive an expression for the total pressure dependence of the equilibrium constant in terms of the fractional dissociation α, total pressure p, and the standard state pressure p°.

11.5 Calculate K for the ammonia synthesis reaction at 298 K, assuming all gases to be ideal.

11.6 Determine K at 298 K for the reaction

$$C_2H_4(g) + H_2(g) \rightleftharpoons C_2H_6(g)$$

given that $\Delta_f G^\circ$ for $C_2H_4(g)$ and $C_2H_6(g)$ are 68.4 and −32.0 kJ mol^{-1}, respectively.

11.7 A mixture of CO(g), H_2(g) and CH_3OH(g) at 500 K with $p(CO) = 10$ bar, $p(H_2) = 1$ bar and $p(CH_3OH) = 0.1$ bar is passed over a catalyst. Can more methanol be formed? $\Delta_r G^\circ = -21.21$ kJ mol^{-1}.

11.8 Integrate the differential form of the van't Hoff equation to obtain and equation that relates how ln K changes with temperature to $\Delta_r H^\circ_m$ assuming (a) $\Delta_r H^\circ_m$ is constant and (b) $\Delta_r H^\circ_m(T_2) = \Delta_r H^\circ_m(T_1) + (T_2 - T_1)\,\Delta_r C^\circ_{p,m}$ with constant difference in heat capacities.

11.9 For the equilibrium $2H_2(g) + O_2(g) \rightleftharpoons 2H_2O(g)$, the equilibrium constant changes with temperature as indicated below. Plot ln K versus $1/T$ to determine an approximate value for enthalpy of combustion of H_2(g). Would it be easy to measure K directly from concentrations for this system?

T/K	400	500	600	700
K	3.00×10^{58}	5.86×10^{45}	1.83×10^{37}	1.46×10^{31}

11.10 N_2O_4 dissociates according to the following equation $N_2O_4(g) \rightleftharpoons 2\,NO_2(g)$. At 350 K, the equilibrium mixture contains 0.13 mol N_2O_4 and 0.34 mol NO_2. The total pressure is 2 bar. Find the value of K at 350 K.

11.11 Determine K for the following equilibrium at 298 K if the initial concentration of CH_3NH_2 is 0.10 mol dm^{-3} and the equilibrium concentration of OH^- is 0.0066 mol dm^{-3}.

$$CH_3NH_2(aq) + H_2O(l) \rightleftharpoons CH_3NH_3^+(aq) + OH^-(aq)$$

11.12 In a study of the equilibrium $H_2(g) + I_2(g)2 \rightleftharpoons HI(g)$, 1 mol of H_2 and 3 mol of I_2 gave rise at equilibrium to x mol of HI. Addition of a further 2 mol of H_2 gave an additional x mol of HI. What is x? What is K at the temperature of the experiment?

11.13 At 400 K, $K = 40$ for the reaction $H_2(g) + I_2(g) \rightleftharpoons 2\,HI(g)$. If 2.0 mol of H_2 and I_2 are mixed at a total constant pressure of 1 bar, what is the composition of the equilibrium mixture at 400 K?

11.14 A chemist mixes 0.5 mol of acetic acid (CH_3COOH) and 0.5 mol of HCN with enough water to make 1 l of solution. Calculate the final concentrations of H_3O^+ and OH^-. Use the following constants: $K_a(CH_3COOH) = 1.8 \times 10^{-5}$, $K_a(HCN) = 4.0 \times 10^{-10}$, $K_w = 1.0 \times 10^{-14}$.

11.15 Lemon juice is very acidic, having a pH of 2.1. (a) If we assume that the only acid in lemon juice is citric acid, HCit, that only the first of citric acid's acid proton dissociates (it is triprotic), and that no citrate salts are present, what is the concentration of citric acid in lemon juice? (b) The acid dissociation constants for citric acid are pK_1 = 3.13, pK_2 = 4.76, and pK_3 = 6.40. Calculate the contribution of second dissociation step.

11.16 The equilibrium constant as a function of temperature for a particular reaction was found to follow the equation $\ln K = (58690\,\mathrm{K}/T) - 12.04$ in the temperature range from 400 to 700 K. Determine $\Delta_r G^\circ$, $\Delta_r H^\circ$ and $\Delta_r S^\circ$ at 600 K.

11.17 (a) On the basis of the integrated form of the Gibbs–Helmholtz equation, discuss how $\Delta_r G^\circ_m$ changes with increasing temperature based on whether a reaction is exothermic compared to endothermic. (b) Use the result in (a) to argue how the equilibrium constant changes with increases in temperature based on whether a reaction is exothermic or endothermic.

11.18 The hydrolysis of adenosine triphosphate to give adenosine diphosphate and phosphate can be represented by ATP $\rightleftharpoons$ ADP + P. The following values have been obtained for the reaction at 37 °C and a standard state concentration of 1 M: $\Delta G° = -31.0$ kJ mol^{-1} and $\Delta H° = -20.1$ kJ mol^{-1}. (a) Calculate $\Delta S°$. (b) Calculate K_c at 37 °C. (c) Assuming that $\Delta H°$ and $\Delta S°$ are temperature-independent, calculate $\Delta G°$ and K_c at 25 °C.

11.19 If the dissolution of a salt is exothermic, does it become more or less soluble as the temperature is increased? Justify your answer.

11.20 The equilibrium constant of the dissolution of ZnF_2 in water is a function of temperature, as indicated by the data below. Determine the values of $\Delta_r G^\circ_m$, $\Delta_r H^\circ_m$ and $\Delta_r S^\circ_m$ for this reaction at 298 K. Discuss what the $\Delta_r H^\circ_m$ and $\Delta_r S^\circ_m$ values mean for this system.

T / °C	10	30	50	70	90
K_{sp}	8.72×10^{-2}	1.90×10^{-2}	4.95×10^{-3}	1.51×10^{-3}	5.26×10^{-4}

11.21 A system reacts according to the equation A + B $\rightleftharpoons$ C + 2D. When equilibrated at 298 K, the concentrations (in mol L^{-1}) are [A] = 2.50×10^{-1}, [B] = 2.50×10^{-1}, [C] = 2.67×10^{-3}, and [D] = 5.34×10^{-3}. (a) Assuming that activity can be replaced by the numerical value of the molar concentration, calculate K and $\Delta_r G^\circ_m$ at 298 K. (b) If $\Delta_r H^\circ_m = 150$ kJ mol^{-1} and both $\Delta_r H^\circ_m$ and $\Delta_r S^\circ_m$ are independent of temperature in the range 298–500 K, calculate the values of K and $\Delta_r G^\circ_m$ and the concentrations of each species when the temperature is changed to 500 K and the system is again equilibrated. (c) Describe with justification what type of reactants and products (with respect to bond strengths) are consistent with the values of $\Delta_r H^\circ_m$ reported here.

11.22 A system reacts according to the equation 2A + B $\rightleftharpoons$ C + D. When equilibrated at 298 K, the concentrations in (mol L^{-1}) are found to be [A] = 2.50×10^{-3}, [B] = 5.00×10^{-3}, [C] = 2.67, and [D] = 2.67. (a) Assuming that activity can be replaced by the numerical value of the molar concentration, calculate K and $\Delta_r G°$ at 298 K. (b) If $\Delta_r H° = -150$ kJ mol^{-1} and both $\Delta_r H°$ and $\Delta_r S°$ are independent of temperature in the range 298–500 K, calculate the values of K and $\Delta_r G°$ and the concentrations of each species when the temperature is changed to 500 K and the system is again equilibrated. (c) Describe with justification what type of reactants and products (with respect to bond strengths) are consistent with the values of $\Delta_r H°$ reported here.

11.23 **a** The bacterium *Nitrobacter* plays an important role in the 'nitrogen cycle' by oxidizing nitrite to nitrate. It obtains the energy it requires for growth from the reaction, $NO_2^-(aq) + \frac{1}{2}O_2(g) \rightarrow NO_3^-(aq)$. Calculate $\Delta_r H^\circ_m$, $\Delta_r G^\circ_m$ and $\Delta_r S^\circ_m$ for this reaction from the following data at 25 °C (all values in kJ mol^{-1}):

Ion	$\Delta_f H^\circ_m$	$\Delta_f G^\circ_m$
NO_2^-	−104.6	−37.2
NO_3^-	−207.4	−111.3

b When the reaction glucose-1-phosphate(aq)$\rightleftharpoons$glucose-6-phosphate(aq) is at equilibrium at 25 °C, the amount of glucose-6-phosphate present is 95% of the total. (i) Calculate $\Delta_r G_m°$ at 25 °C. (ii) Calculate $\Delta_r G_m$ for the reaction in the presence of 0.01 M glucose-1-phosphate and 10^{-4} M glucose-6-phosphate. In which direction does reaction occur under these conditions?

c Suppose there is a biological reactionA + B $\rightleftharpoons$ Z (1) for which $\Delta_r G_m° = 23.8$ kJ mol^{-1} at 37 °C. Suppose that an enzyme couples this reaction with ATP $\rightleftharpoons$ ADP + phosphate (2) for which $\Delta_r G_m° = -31.0$ kJ mol^{-1} at 37 °C. Calculate the equilibrium constant at 37.0 °C for these two reactions and for the coupled reaction A + B + ATP $\rightleftharpoons$ Z + ADP + phosphate (3)

11.24 At 100 °C and $p = 2$ bar, the degree of dissociation of phosgene is 6.30×10^{-5}. Calculate K_p, K_c and K_x for the dissociation reaction

$$COCl_2(g) \rightleftharpoons CO(g) + Cl_2(g).$$

11.25 (a) The ionic product for the dissociation of water is $K_w = 1.00 \times 10^{-14}$ at 25.0 °C and 1.45×10^{-14} at 30.0 °C. Deduce $\Delta_r H°_m$ and $\Delta_r S°_m$ for the dissociation (autoionization) of water into H^+ and OH^-. (b) Calculate the value of the ionic product of water at body temperature (37 °C).

11.26 Use the information in the table below to answer the remainder of this question for the reaction $N_2(g) + 3\,H_2(g) \rightarrow 2\,NH_3(g)$.

T = 298 K	**NH_3(g)**	**N_2(g)**	**H_2(g)**
$\Delta_f H^\circ_m$ / kJ mol^{-1}	−45.9		
S°_m / J K^{-1} mol^{-1}	192.8	191.6	130.7
$C_{p,m}$° / J K^{-1} mol^{-1}	35.1	29.1	28.8

a Calculate $\Delta_r H^\circ_m$, $\Delta_r S^\circ_m$, $\Delta_r G^\circ_m$, and the equilibrium constant K at 298 K. (b) A stoichiometric mixture of $N_2(g)$ and $H_2(g)$ is made with the initial pressure of the N_2 equal to p_0 = 60 bar. Set up but do not solve the algebraic equilibrium constant expression that will allow you to calculate the equilibrium composition of the system. The equation should be written in terms containing only K, p_0 and x, where x is the amount of N_2 that reacts. In other words, the final pressure of N_2 is $p_0 - x$. (c) The industrial synthesis of ammonia is performed at 775 K. Calculate $\Delta_r H^\circ_m$, $\Delta_r S^\circ_m$, $\Delta_r G^\circ_m$, and K at 775 K. (d) Bearing in mind your answer to (c) and its relation to your answer in (a), why must the industrial synthesis of ammonia be performed at 240 bar?

11.27 Consider the gas-phase reaction (all gases considered ideal)

$NO + N_2O_5 \rightleftharpoons N_2O_4 + NO_2$

T = 298 K	NO	N_2O_5	N_2O_4	NO_2
$\Delta_f H^\circ_m$ / kJ mol^{-1}	91.3	13.3	11.1	33.2
S°_m / J K^{-1} mol^{-1}	210.8	355.7	304.4	240.1

The initial partial pressures of NO and N_2O_5 are both 0.5 bar (other gases are zero initially). The system is held at constant temperature and volume. (a) Calculate K and $\Delta_r G^\circ_m$ at 298 K. (b) Calculate the final partial pressures of all the gases when equilibrated at 298 K. (c) The heat capacity of $NO_2(g)$ is given by $C^\circ_{p,m} = A + BT + CT^2$ with $A = 16.109$, $B = 7.590 \times 10^{-2}$, and $C = -5.439 \times 10^{-5}$. The units are such that if temperature T is inserted in kelvin, the equation gives $C°_{p,m}$ in J K^{-1} mol^{-1}. Calculate the standard absolute entropy of $NO_2(g)$ at 1200 K.

11.28 100 kPa of $PCl_5(g)$ is mixed with 100 kPa of $O_2(g)$ and allowed to equilibrate at 298 K and constant volume. Values of the enthalpies of formation and absolute entropies are given in the table below.

	PCl_5	O_2	$POCl_3$	Cl_2O
$\Delta_f H^\circ_m$ / kJ mol^{-1}	−95.35	0	−592.0	76.15
S°_m / J K^{-1} mol^{-1}	352.7	205.0	324.6	266.5

(a) Calculate the equilibrium partial pressures in the system assuming PCl_5 and O_2 react to form $POCl_3(g)$ and $Cl_2O(g)$ as the only products. (b) Explain whether the enthalpy and entropy changes are favorable or unfavorable for the reaction and what their values mean in molecular terms for the reactants compared to the products.

11.29 Why do the lines for the amount of ions and solid in Fig. 11.4 have different slopes?

11.30 The critical micelle concentration of sodium dodecyl sulfate (SDS) $CH_3(CH_2)_{11}OSO_3Na$, an anionic surfactant commonly used in household products (also known as sodium lauryl sulfate) is approximately $c_M \approx 0.0081$ M in water. Calculate the critical mole fraction x_M and the value of $\Delta_{mic} G^\circ_m$.

11.31 As reported previously [Kopecky, F. (1996) *Pharmazie*, **51**, 135–144], lg $(c_M/c°)$ varies with chain length in the quaternary ammonium salt APYCl, as given in the table below. Determine the critical micelle concentration in molar units and the value of the slope of a linear fit to the data. Then determine $\Delta_{mic} G^\circ_m$ and plot it as a function of chain length m. Use this second plot to determine the chain length at which micelles will no longer form.

m	**8**	**10**	**12**	**14**
lg $(c_M/c°)$	−0.563	−1.186	−1.798	−2.456

Endnote

1. Pearson, R.G. (2005) *J. Chem. Sci.*, **117**, 369.

Further reading

Berry, R.S., Rice, S.A., and Ross, J. (2000) *Physical Chemistry*, 2nd edition. Oxford University Press, New York.
Burgess, J. (1999) *Ions in Solutions: Basic principles of chemical interactions.* Woodhead Publishing, Cambridge, UK.
Israelachvili, J.N. (2011) *Intermolecular and Surface Forces*, 3rd edition. Academic Press, Burlington, MA.
Metiu, H. (2006) *Physical Chemistry: Thermodynamics.* Taylor & Francis, New York.
NIST Chemistry WebBook http://webbook.nist.gov/chemistry/.
Pearson, R.G. (2005) Chemical hardness and density functional theory. *J. Chem. Sci.*, **117**, 369–377.

CHAPTER 12

Phase stability and phase transitions

PREVIEW OF IMPORTANT CONCEPTS

- At equilibrium the chemical potential of a substance is the same throughout the sample.
- The chemical potential of a gas responds to changes of temperature and pressure more sensitively than for liquids and solids.
- Two phases coexist at equilibrium if the chemical potential of each component in phase α is equal to the chemical potential of the same component in phase β.
- Ostwald ripening is a phenomenon in which large particles grow at the expense of smaller particles. This effect (i.e., small particles being less stable than larger ones) is important in nucleation phenomena, as varied from the growth of crystals to the formation of raindrops.
- Nucleation of a new phase is a difficult process; homogeneous classical nucleation is difficult to observe in real systems.
- In most real systems some type of defect or impurity is involved in the nucleation process.
- The surface tension of a liquid monotonically drops from its value at the melting point to zero at the critical temperature.
- In the supercritical phase there is no distinction between a liquid and a gas.
- Atomic and small-molecule molecular solids usually do not form glasses (amorphous solids), but glass formation is often observed for polymers.
- The glass transition is accompanied by a change from a brittle to a rubbery character.

12.1 Phase diagrams and the relative stability of solids, liquids, and gases

Did you ever ask yourself why icebergs created from the saltwater of oceans are composed of pure water? What happened to the salt in the saltwater? The process of crystallization, the freezing of water ice from aqueous solutions, naturally expels impurities. We shall see that this occurs because the chemical potentials of solvents and the solutes act differently at the temperature of the phase transition. Once a species is more stable in another phase – that is, once its chemical potential is lower in the other phase – the species makes a transition and passes from one phase to the other. This is the basis of a whole raft of purification techniques: crystallization brought about by lowering temperature; distillation by raising temperature and/or decreasing pressure; and condensation caused by pressure swings.

In the absence of intermolecular forces there would only be ideal gases. The very existence of liquids and solids requires molecules and atoms of finite extent that interact through attractive and repulsive forces. That these forces are not isotropic lends directionality to intermolecular interactions as well as to definite structures in solvation and crystals. Now we look more deeply into the implications of intermolecular interactions with regard to the stability of phases.

A *phase* is a macroscopically homogeneous region of a system, within which all of the intensive variables vary continuously. If the system is composed of two or more phases, the phases are separated by the formation of interfaces between the different phases. At least some of the intensive variables that describe the system vary discontinuously across these interfaces. A *phase transition* (also called phase change) is the process of transferring matter across the interface between phases. A *congruent transition* – including phase transitions between the two-phase equilibria of melting, vaporization, or allotropism of a compound – involves phases of the same composition. *Conjugate phases* are two phases of variable composition in mutual thermodynamic equilibrium.

The most familiar phase transitions are melting and vaporization. The change in volume upon heating is described by the *volume expansion coefficient*, α_V. For cubic materials the volume expansion coefficient is three times the *linear expansion*

Physical Chemistry: How Chemistry Works, First Edition. Kurt W. Kolasinski.

Companion Website: www.wiley.com/go/kolasinski/physicalchemistry

Table 12.1 Melting point T_f, density of solid ρ_s at 25 °C and T_f, density of the liquid at T_f ρ_l, coefficient of linear expansion α_l for various materials. $\%\Delta = (\rho_s(T_f) - \rho_l)/\rho_s(T_f)$. Values taken from the 96th edition of *CRC Handbook of Chemistry and Physics*, except for $\rho_s(T_f)$, which is calculated assuming α_l is constant up to T_f.

Symbol	T_f / K	ρ_s(25 °C)/kg m^{-3}	$\rho_s(T_f)$/kg m^{-3}	ρ_l/kg m^{-3}	$\alpha_l \times 10^6$/K^{-1}	%Δ
H_2O	273.15		916.7	999.85		−9.1
Bi	544.56	979	969.4	1050	13.4	−8.3
Ga	302.92	591	590.8	608	18	−2.9
Ge	1211.40	532.3	524	560	5.8	−6.9
Si	1687.15	232.9	230.4	257	2.6	−11.5
Ag	1234.93	1050	997	932	18.9	6.5
Al	933.47	270	2586	2377	23.1	8.1
Au	1337.33	1930	1848	1731	14.2	6.3
Fe	1811.05	787	747	703.5	11.8	5.8
Pb	600.61	1130	1101	1066	28.9	3.2
Pt	2041.35	2150	2055	1977	8.8	3.8
V	2183.15	600	572.8	550	8.4	4.0

coefficient, α_l. Most materials expand upon melting. For a number of metals shown in Table 12.1, we see that expansion upon melting ranges from about 3% to 8%, but this is not universally so and five such examples are also given in Table 12.1. This, of course, means that the density of the solid is less than the density of the liquid at the melting point. Therefore, silicon 'ice cubes' float in molten silicon just as do water ice cubes in a glass of sparkling water; while solid gold sinks to the bottom of a container filled with molten gold. This is important for fish in lakes because water ice floats to the top of lakes (and rivers) forming an insulating barrier that slows further freezing, rather than dropping to the bottom of the lake where they would promote faster freezing in the winter and increase the probability of never melting in the summer.

The enthalpy of a phase transition is sometimes also called the *latent heat*, though this is not the preferred term any longer. The entropy of a phase transition is the enthalpy divided by the temperature, which in this case is

$$\Delta_{\text{fus}}S^\circ_{\text{m}} = \Delta_{\text{fus}}H^\circ_{\text{m}}/T_{\text{f}}. \tag{12.1}$$

While the enthalpy change upon fusion varies by almost a factor of 1000 for the materials listed in Table 12.1, the entropy change varies by only about a factor of 5. Nonetheless, there is considerable variation in $\Delta_{\text{fus}}S^\circ_{\text{m}}$ that is related to the structure of the material and the nature of the solid-to-liquid transition. The entropy change upon fusion is of the order R ($\approx$ 10 J K^{-1} mol^{-1}) for simple substances, such as metals and solid covalent diatomics, but increases with increasing molecular complexity, for example ionic compounds and polyatomics.

Note that for all of the substances listed in Table 12.2, the phase transition is that of a crystalline solid to a liquid. A crystalline (or polycrystalline) solid is distinct from a liquid. The crystal possesses definite order and symmetry. A liquid has no long-range order and is isotropic. The two are separated by a normal *first-order phase transition* – one in which the enthalpy and density change discontinuously between the two phases. An example of such a discontinuous transition is shown in Fig. 12.1 for the melting of water.

A glass and a liquid are not distinct phases. They are separated by a *second-order phase transition* in which properties change sometimes dramatically within a small temperature range. However, the change is not discontinuous. The difference can be traced back to their structure. Just as a liquid is isotropic with no long-range order, so too is a glass. The glass softens and gradually transitions to the liquid, without discontinuous changes. In particular, there is no sharp change in the entropy or molar volume of a glass and a liquid. In contrast, the change in structural order between a solid and a liquid is distinct and accompanied by a significant change in entropy and molar volume.

The enthalpy change upon vaporization $\Delta_{\text{vap}}H^\circ_{\text{m}}$ varies by about 400 for the materials presented in Table 12.3. This represents somewhat less variation than for melting. However, the $\Delta_{\text{vap}}S^\circ_{\text{m}}$ exhibits much less variation than $\Delta_{\text{fus}}S^\circ_{\text{m}}$, with a factor of just 1.7 covering the spread in values in Table 12.3 (when one excludes the anomalous case of H_2, for which quantum corrections to the rovibrational state distribution must be considered). The entropy change upon vaporization is ~ 85 J K^{-1} mol^{-1}, which is a statement known as *Trouton's rule*. However, *Hildebrand's rule* is more accurate, as it states

Table 12.2 Characteristics of the melting transition for various substances at standard pressure. Values of T_f and $\Delta_{fus}H_m^\circ$ taken from the 96th edition of *CRC Handbook of Chemistry and Physics*. $\Delta_{fus}S_m^\circ$ calculated according to Eq. (12.1).

Symbol	T_f / K	$\Delta_{fus}H_m^\circ$ / kJ mol^{-1}	$\Delta_{fus}S_m^\circ$ / J K^{-1} mol^{-1}
Ar	83.81	1.18	14.08
Xe	161.40	2.27	14.06
C(graphite)	4762.15	117.40	24.65
Si	1687.15	50.21	29.76
Hg	234.32	2.29	9.79
H_2	13.99	0.12	8.58
O_2	54.36	0.44	8.09
NaCl	1073.85	28.16	26.22
$HgCl_2$	550.15	19.41	35.28
$ZnCl_2$	598.15	10.30	17.22
H_2O	273.15	6.01	22.00
CO_2	216.59	9.02	41.65
NH_3	195.42	5.66	28.96
CH_4	90.68	0.94	10.37
C_2H_6	90.36	2.72	30.10
C_3H_8	85.52	3.50	40.93
C_4H_{10}	134.85	4.66	34.56
C_2H_5OH	159.01	4.93	31.01
C_6H_6	278.64	9.87	35.42

that all liquids have the same value of $\Delta_{vap}S_m^\circ$ when vaporization occurs at a temperature for which the vapor density is the same. A value of 93 J K^{-1} mol^{-1} for $\rho_g = 0.0202$ mol l^{-1} is a rough approximation. We can understand the greater regularity of the entropy change for vaporization by recognizing that the entropy change is dominated by the character of the (nearly) ideal gas that is formed in each case

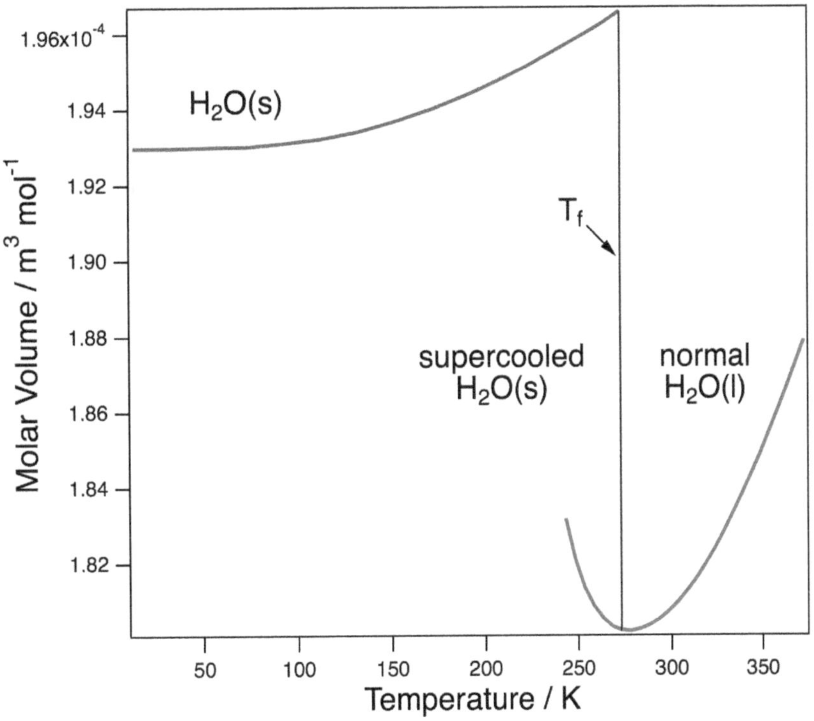

Figure 12.1 The change in molar volume as a function of temperature and phase for water near melting transition. Values taken from the 96th edition of the *CRC Handbook of Chemistry and Physics*.

Table 12.3 Characteristics of the vaporization transition for various substances at standard pressure. * indicates that neither graphite nor CO_2 vaporizes because both sublime, and the transition temperature given for these two is for sublimation. Values of T_b and $\Delta_{vap}H^\circ_m$ taken from the 96th edition of *CRC Handbook of Chemistry and Physics*. $\Delta_{vap}S^\circ_m$ calculated analogously to Eq. (12.1).

Symbol	T_b / K	$\Delta_{vap}H^\circ_m$ / kJ mol^{-1}	$\Delta_{vap}S^\circ_m$ / J K^{-1} mol^{-1}
Ar	87.3	6.43	73.65
Xe	165.1	12.57	76.15
C (graphite)	4098.1	*	*
Si	3538.2	359.00	101.47
Hg	629.8	59.11	93.86
H_2	20.4	0.90	44.14
O_2	90.2	6.82	75.61
NaCl	1738.2	204.50	117.65
$HgCl_2$	577.1	58.90	102.05
$ZnCl_2$	1005.1	126.00	125.35
H_2O	373.1	40.65	108.95
CO_2	194.7	*	*
NH_3	239.8	23.33	97.28
CH_4	111.7	8.19	73.34
C_2H_6	184.5	14.69	79.60
C_3H_8	231.0	19.04	82.41
C_4H_{10}	272.6	22.44	82.30
C_2H_5OH	351.4	38.56	109.72
C_6H_6	353.2	30.72	86.97

12.2 What determines relative phase stability?

Having defined the chemical potential in Chapter 11, we now know what our control parameter is and how to define equilibrium.

At equilibrium the chemical potential of a substance is the same throughout the sample, regardless of how many phases are present.

In other words, the chemical potentials of all phases that are in contact are equal at equilibrium.

Another way to state this criterion for stability is that the Gibbs energy must be minimized. The *molar Gibbs energy* G_m and the *chemical potential* μ are synonymous for a one-component system. We write for a pure substance

$$\mu = (\partial G/\partial n)_{T,p} = G_m. \tag{12.2}$$

We substitute for G_m from the fundamental equation and express this in differential form as

$$d\mu = -S_m\, dT + V_m\, dp. \tag{12.3}$$

Since these are exact differentials, we can express how the chemical potential changes with temperature at constant pressure ($dp = 0$) according to the partial differential

$$(\partial\mu/\partial T)_p = -S_m \tag{12.4}$$

and how the chemical potential changes with pressure at constant temperature ($dT = 0$) according to the partial differential

$$(\partial\mu/\partial p)_T = V_m. \tag{12.5}$$

Since the molar entropy and molar volume of a gas are much greater than for the other two phases, it is clear from Eqs. (12.4) and (12.5) that the chemical potential of a gas responds to changes of temperature and pressure more sensitively than for liquids and gases. The importance of this can be observed directly in a plot of chemical potential versus temperature, as shown in Fig. 12.2. The partial derivative represents the slope of the curve. Each phase has its own characteristic dependence of molar entropy on temperature. Nonetheless, we know that the molar entropy of a gas is greater than that of a liquid at the same temperature. Similarly, the molar entropy of a liquid is always greater than that of a solid of the same material at the same temperature. Stated mathematically, at any given temperature $S_{m,g} > S_{m,l} > S_{m,s}$.

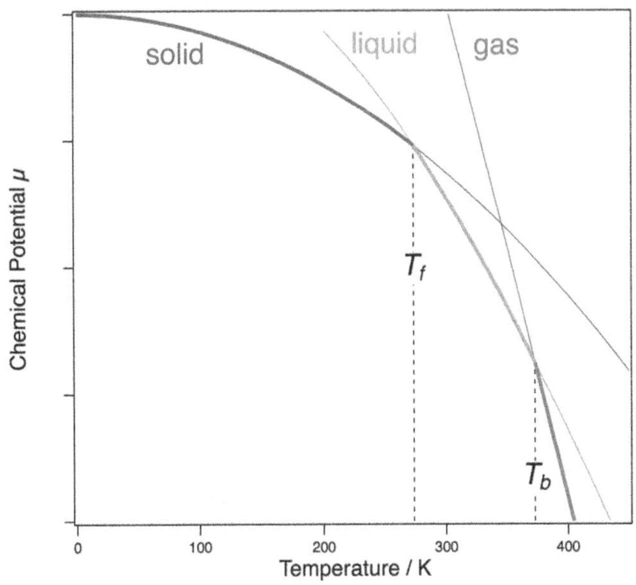

Figure 12.2 Chemical potential–T phase diagram for a material with properties approximately those of water.

The molar entropy of each phase is a function of temperature. In accord with Eq. (12.4) the chemical potential versus temperature of each phase is represented by a curve, as shown in Fig. 12.2.

A system is at equilibrium when its chemical potential is a minimum.

The lowest chemical potential phase is the most stable at any given temperature and pressure. Figure 12.2 represents the chemical potential versus temperature at the standard pressure of $p^\circ = 1$ bar. The curve that represents the chemical potential of the solid starts with a slope of zero at absolute zero. This is the behavior of all perfect crystals according to the Third Law of Thermodynamics. The slope becomes increasingly negative. At the standard melting point, T_f, the chemical potentials of the solid and liquid phases are equal, $\mu(s) = \mu(l)$ at T_f. At this point the curve for the chemical potential of the solid intersects the curve for the chemical potential of the liquid. We do not need to know the absolute value of either slope; however, we do know that the slope of the liquid curve is greater than the slope of the solid curve. Similarly, at the standard boiling point T_b, the chemical potentials of the liquid and gas phases are equal, $\mu(l) = \mu(g)$ at T_b. At this point the curve for the chemical potential of the liquid intersects the curve for the chemical potential of the gas, and the slope of the gas curve is greater than the slope of the liquid curve.

Figure 12.2 demonstrates behavior that in one sense is universal, and in another sense is normal but not always observed. The universal behavior is this:

When the chemical potential of one phase drops below the chemical potential of another, the lower chemical potential phase is more stable.

Figure 12.2 represents a chemical potential–T or μ–T *phase diagram*. The point at which their chemical potentials are the same is the point at which the phase transition occurs at equilibrium. However, we should not conclude that melting always precedes vaporization. Indeed, a direct consequence of Eq. (12.5) is that sublimation can occur rather than the 'normal' ordering of melting followed by vaporization.

According to Eq. (12.5) the chemical potential of a gas is quite sensitive to changes in pressure. Integration of Eq. (12.5) with $dT = 0$ shows us that the chemical potential of an ideal gas increase with increasing pressure according to

$$\mu_g = \mu_g^\circ + RT\ln(p/p^\circ) \tag{12.6}$$

where μ_g° is the chemical potential of an ideal gas at the standard state pressure of $p^\circ = 1$ bar.

Because the molar volumes of the liquid and solid are roughly 1000-fold smaller than the molar volume of the gas, the chemical potentials of liquids and solids are almost independent of pressure when compared on the same scale as the gas. Assuming that the liquid is incompressible – that is, that molar volume is independent of pressure – integration of Eq. (12.5) yields

$$\mu_l = \mu_l^\circ + V_{m,l}\,p. \tag{12.7}$$

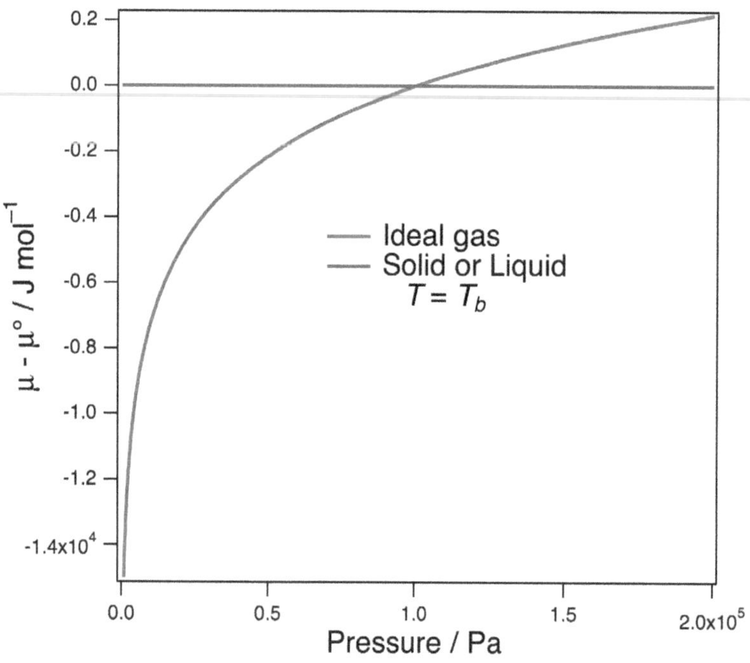

Figure 12.3 Chemical potential versus pressure. The chemical potential of a gas varies strongly with pressure. While not independent of pressure, the changes for liquids and gases are insignificant when compared to those of gases over the same range.

A similar equation holds for the solid phase in which the molar volume of the solid $V_{m,s}$ replaces the molar volume of the liquid $V_{m,l}$, and similarly for the standard potential of the solid μ_s° and that of the liquid μ_l°.

Now look back at Fig. 12.2. When the pressure is decreased, the solid and liquid curves shift only slightly; however, the gas curve shifts significantly lower. At sufficiently low pressure, the gas curve will shift so far that it intersects the solid curve before the liquid curve. In such a case, the material *sublimes* (transitions from solid to gas) rather than melts. At atmospheric pressure, this is true of carbon dioxide, which sublimes at $T_{sp} = 195$ K, some organic compounds, such as caffeine with $T_{sp} = 363$ K, and graphite with $T_{sp} = 3825$ K.

There is another consequence of Fig. 12.3. The chemical potential of a gas can be reduced without limit by reducing its pressure. But this is not the case for solids and liquids, and therefore a means of stabilizing a component in the gas phase exists that does not exist for the other two phases. At any given temperature a gas is able to coexist with either of the condensed phases. It achieves this by adjusting its pressure to the equilibrium *vapor pressure* of the material. In all cases, the vapor is in equilibrium with the liquid or the solid; however, as the temperature is reduced the equilibrium vapor pressure of the gas must decrease to maintain the equality of chemical potential with the condensed phase.

This is a very important result. How do we measure the chemical potential? In the abstract this is a difficult question to answer. However, Eq. (12.6) shows us that it is very simple for an ideal gas: We simply measure the equilibrium partial pressure of the gas. Furthermore, since the chemical potential of all phases must be equal at equilibrium, then an equilibrium measurement of vapor pressure also allows us to quantify the chemical potential of any other phase (e.g., pure liquid, solution or solid) that is in contact with this vapor phase.

12.2.1 Directed practice

Derive Eqs (12.6) and (12.7) by substituting for V_m and using a standard state pressure of $p^\circ = 1$ bar.

12.3 The *p–T* phase diagram

If one plots the equilibrium vapor pressure of a liquid as a function of temperature, a curve such as that displayed in Fig. 12.4 is obtained. We see that the curve is continuous throughout the temperature range, extending from the melting point to the critical point.

The standard boiling point is defined as the temperature at which the vapor pressure is equal to the standard pressure of 100 kPa.

As far as the relationship of vapor pressure to temperature, the boiling point is not a special point.

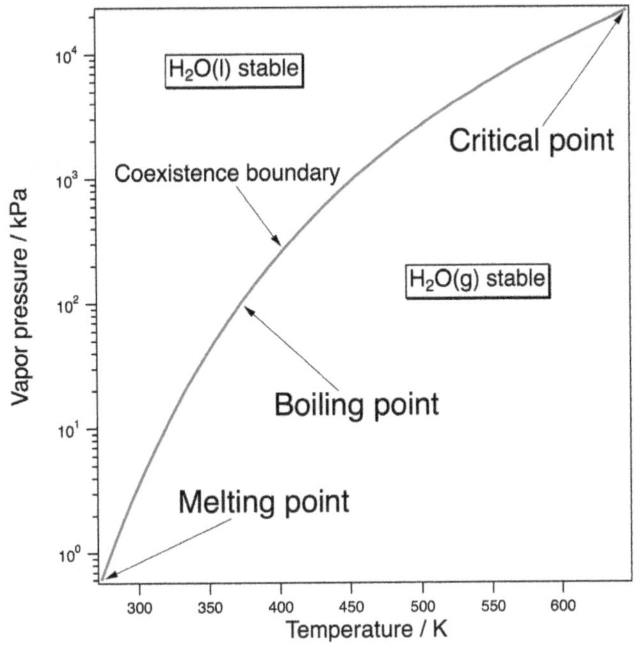

Figure 12.4 The curve that describes the equilibrium vapor pressure of water vapor above liquid water represents the coexistence boundary between these two phases. Data taken from 96th edition of the *CRC Handbook of Chemistry and Physics*.

One, two, or threes phase of a pure substance of fixed composition can coexist. A *pressure–temperature phase diagram* shows us graphically what phases are stable for any combination of p and T. By inspection of such a curve, you are able to identify a *coexistence curve*, standard boiling point, standard melting point, the triple point, and critical point.

The boundary lines in the p–T phase diagram define the *coexistence curves*. These are the combinations of pressure and temperature at which two phases can be in equilibrium with one another. The standard transition temperature corresponds to the temperature at which a particular phase transition occurs when the pressure is equal to the standard pressure of 1 bar. The *triple point* is a unique point in a p–T phase diagram. It is the only combination of pressure and temperature at which the solid, liquid, and gas phases can coexist at equilibrium. Note one other thing about the triple point. If the triple point corresponds to a pressure below standard pressure, then the material exhibits a 'normal' μ versus T phase diagram such as that shown in Fig. 12.2. The solid melts to a liquid and the liquid vaporizes into the gas as T is increased. If, on the other hand, the substance has a triple point that occurs above $p = 1$ bar, then the solid sublimes before it melts.

The water phase diagram also exhibits the presence of different *crystal modifications* or polymorphs – different crystalline forms of the same composition. There are several solid forms of water ice. The presence of two is shown in Fig. 12.6. Only one crystal modification of water is known at atmospheric pressure; all of the others form at higher pressure. The presence of multiple crystal modification means that the phase diagram also exhibits points where two solid phases and the liquid phase intersect. This is a type of triple, however 'the triple point' is the point at which solid, liquid, and gas intersect. If the point in question refers to a different triple intersection of a different set of phases, those phases should be specified. Many other materials exhibit multiple crystal modifications, such as carbon (diamond and graphite), tin (α-Sn and β-Sn, also called 'gray' tin and 'white' tin), phosphorus (red and white phosphorus among others), SiO_2 (quartz, cristobalite, tridymite and stishovite to name a few) and Al_2O_3 (corundum and γ-Al_2O_3).

The *critical point* is another special point in the p–T phase diagram. It occurs at the combination of temperature and pressure when the density of the gas phase is the same as the density of the liquid phase. At this point, the molecules in the two phases are on average the same distance apart and experience the same intermolecular forces; in other words, the two phases are indistinguishable. Thus, other properties such as the heat capacity and molar entropy of the liquid and gas also converge on each other and become indistinguishable. This can be seen in Fig. 12.5 using data for H_2O. Both, the entropy and enthalpy changes go to zero for the transition liquid $\rightleftharpoons$ gas at the critical point, which means that there is no longer a phase transition at and after the critical point. At the critical point and beyond the fluid is called a *supercritical fluid*. It is a fluid that exists beyond the critical point.

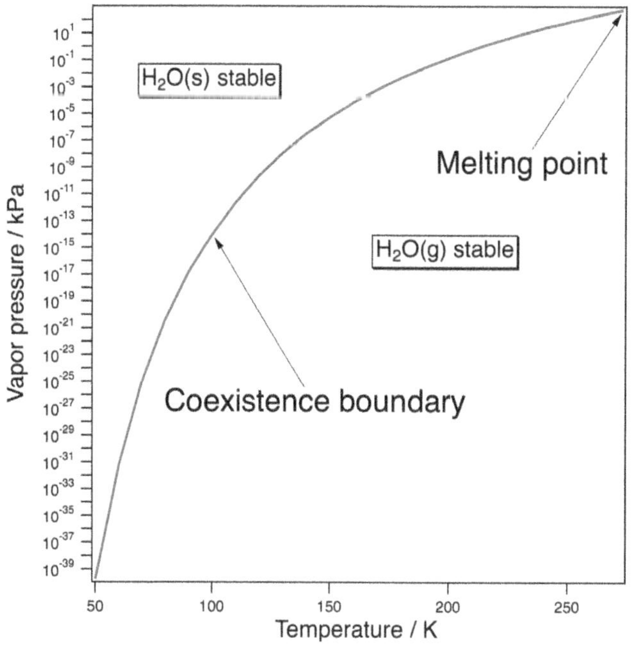

Figure 12.5 The curve that describes the equilibrium vapor pressure of water vapor above solid water represents the coexistence curve for sublimation. Data taken from 96th edition of the *CRC Handbook of Chemistry and Physics*.

Look at the temperature dependence of $\Delta_{\text{vap}}H^\circ_{\text{m}}$. As expected, the enthalpy change depends on temperature, and this dependence is at first linear in T. This is the range for which the correction factor is $\Delta_{\text{r}}C_{p,\text{m}}\ \Delta T$. As the temperature approaches the critical temperature, the difference in heat capacities drops to zero. The change in $\Delta_{\text{r}}C_{p,\text{m}}$ from some finite value to zero is a general phenomenon; thus, we expect curvature of the type observed in Fig. 12.7 to develop for all gases as they approach T_{c}.

A supercritical fluid has a number of remarkable properties. One property is that its existence allows a system to pass from the liquid phase to gas phase along a path that does not include a phase transition. This is extremely important for

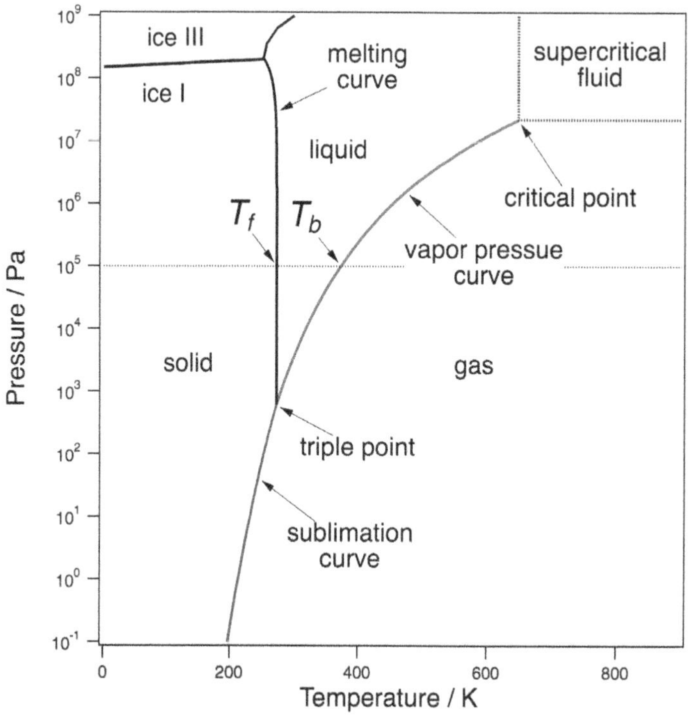

Figure 12.6 p–T phase diagram of water.

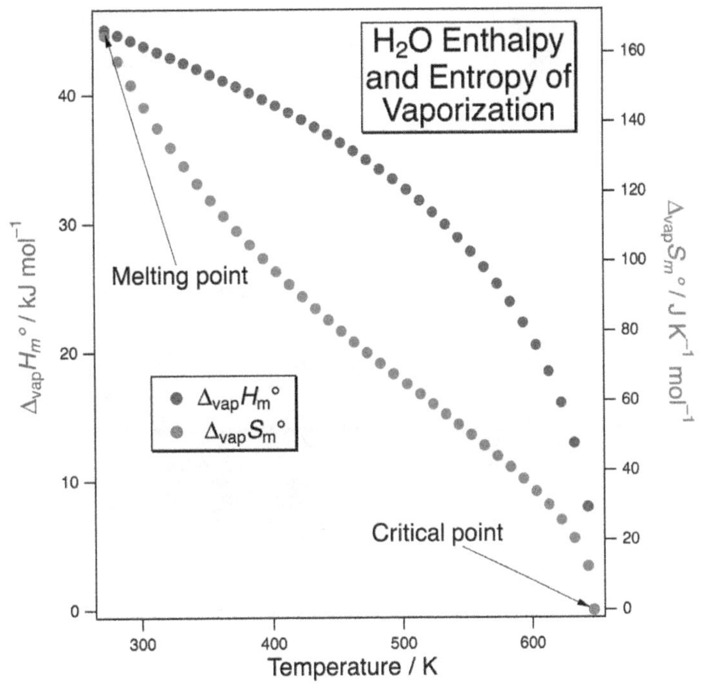

Figure 12.7 Enthalpy and entropy of vaporization of water from 273.16 K (standard melting point) to 647.10 K (critical temperature) at standard pressure.

the field of nanoscience. A technique called *critical point drying* depends on this phenomenon. Ethanol and carbon dioxide are miscible and form a supercritical fluid. Imagine we have a sample – it could be an insect or cell or a nanomaterial – that we want to image to observe its nanoscale details. If the sample is wet because it was made in an aqueous solution, or has fluid inside of it as for the insect or cell, then the small features are subjected to capillary forces that occur when fluids evaporate. The magnitude of these forces increases rapidly as the curvature increases. The equivalent of hundreds or even thousands of atmospheres of pressure can act on the nanoscale features as a result, and this will distort or even destroy the sample.

Instead of allowing the sample to dry in air, we replace the water with ethanol. Then, the sample in ethanol is placed within a stainless steel vessel that can be heated and pressurized with CO_2. The pressure is raised to at least 83 bar at room temperature, after which the temperature is raised to approximately 35 °C. The fluid becomes supercritical. At the elevated temperature, the pressure is reduced such that the gas phase is established, and the temperature is then reduced. Using this procedure, the solution has been removed from the sample without the sample ever experiencing capillary forces. The results of not using such a drying procedure are shown in Fig. 12.8, in which a porous silicon thin film has cracked and exfoliated under the influence of destructive capillary forces.

12.4 The Gibbs phase rule

At equilibrium, all phases that exist must have the same chemical potential. In a single-phase region, T and p can be varied independently, but when two phases α and β coexist the constraint on the chemical potential,

$$\mu_\alpha(p, T) = \mu_\beta(p, T), \tag{12.8}$$

implies that p and T are not independent. Specifying one requires a specific value of the other that can be determined from the phase diagram. These values of p and T are found along a coexistence curve. If three phases coexist,

$$\mu_\alpha(p, T) = \mu_\beta(p, T) = \mu_\gamma(p, T), \tag{12.9}$$

then both p and T are fixed. This corresponds to the triple point. When one phase exists, p and T can be varied freely. The system has two degrees of freedom, F. When two phases coexist, only p or T can be varied. The system has one degree, $F = 1$. If there are three phases, both p and T are fixed, $F = 0$.

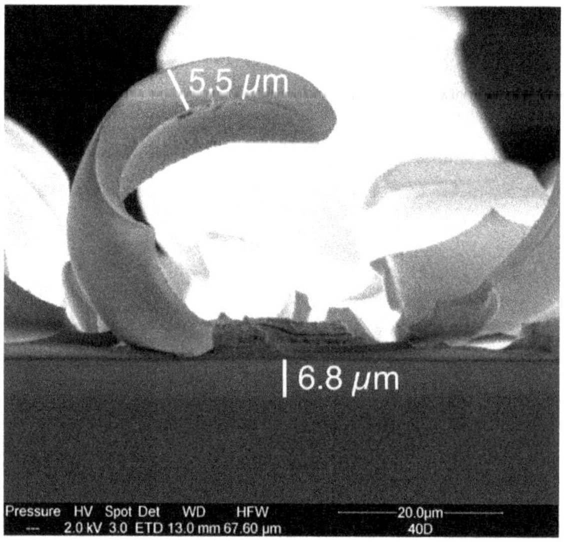

Figure 12.8 Critical point drying is essential for avoiding the deleterious effects of capillary forces on nanomaterials. In this example, exfoliation has occurred, ripping the top of a highly porous nanocrystalline silicon film from a region of the film with lower porosity. The lower portion remains attached to the crystalline Si substrate. For details. see Dudley, M.E. and Kolasinski, K.W. (2009) *Electrochem. Solid State Lett.*, **12**, D22. Reproduced with permission from The Electrochemical Society.

This result can be generalized to a system with an arbitrary number of components.

Two phases coexist at equilibrium if the chemical potential of each component in phase α is equal to the chemical potential of the same component in phase β.

For a binary mixture described by mole fractions for the two components in the two phase, $x_{1\alpha}$, $x_{1\beta}$, $x_{2\alpha}$, $x_{2\beta}$, the equilibrium conditions become

$$\mu_1(\alpha, p, T, x_{1\alpha}) = \mu_1(\beta, p, T, x_{1\beta}) \tag{12.10}$$

$$\mu_2(\alpha, p, T, x_{2\alpha}) = \mu_2(\beta, p, T, x_{2\beta}) \tag{12.11}$$

If we call the number of phases P and the number of components C, then we can express the *Gibbs phase rule* as

$$F = C + 2 - P. \tag{12.12}$$

If there are r reactions in the system that can reach equilibrium, the phase rule becomes

$$F = (C - r) + 2 - P. \tag{12.13}$$

The constant C is now the number of species for which the concentration can be set independently involved in the reactions. Two species in equilibrium have their concentrations tied to one another and count only once.

12.5 Theoretical basis for the p–T phase diagram

The chemical potential depends on p and T (as we have seen before for the Gibbs energy). When the chemical potential of two phases α and β are equal, they are in equilibrium,

$$\mu_\alpha(p, T) = \mu_\beta(p, T). \tag{12.14}$$

Now make an infinitesimal change to the system that keeps it at equilibrium, such that

$$\mathrm{d}\mu_\alpha = \mathrm{d}\mu_\beta. \tag{12.15}$$

The Gibbs energy changes with variations in pressure and temperature according to the fundamental equation, $\mathrm{d}G = V\,\mathrm{d}p - S\,\mathrm{d}T$. Dividing by the amount of substance n, we obtain how chemical potential changes with $\mathrm{d}p$ and $\mathrm{d}T$.

$$\mathrm{d}\mu = V_\mathrm{m}\,\mathrm{d}p - S_\mathrm{m}\,\mathrm{d}T \tag{12.16}$$

Thus, for the two phases α and β we have

$$V_{\alpha,\mathrm{m}}\,\mathrm{d}p - S_{\alpha,\mathrm{m}}\,\mathrm{d}T = V_{\beta,\mathrm{m}}\,\mathrm{d}p - S_{\beta,\mathrm{m}}\,\mathrm{d}T, \tag{12.17}$$

which rearranges to

$$(S_{\beta,\mathrm{m}} - S_{\alpha,\mathrm{m}})\mathrm{d}T = (V_{\beta,\mathrm{m}} - V_{\alpha,\mathrm{m}})\mathrm{d}p. \tag{12.18}$$

Solving this for the derivative of the pressure with respect to temperature, we obtain the *Clapeyron equation*

$$\frac{\mathrm{d}p}{\mathrm{d}T} = \frac{(S_{\beta,\mathrm{m}} - S_{\alpha,\mathrm{m}})}{(V_{\beta,\mathrm{m}} - V_{\alpha,\mathrm{m}})} = \frac{\Delta_{\mathrm{trs}}S_\mathrm{m}}{\Delta_{\mathrm{trs}}V_\mathrm{m}} \tag{12.19}$$

The Clapeyron equation is valid for equilibrium between any two phases, such as vapor/liquid, liquid/solid, vapor/solid, and solid/solid. In other words, it is valid along any coexistence boundary. From this expression we can predict the response of freezing points and melting points to changes in pressure. The utility of the Clapeyron equation is extended by recalling that a phase transition occurs at a constant temperature and pressure. The heat evolved at constant pressure is the enthalpy, thus the change in entropy of a phase transition is given by

$$\Delta_{\mathrm{trs}}S_\mathrm{m} = \Delta_{\mathrm{trs}}H_\mathrm{m}/T_{\mathrm{trs}}. \tag{12.20}$$

Substitution into Eq. (12.19) yields an expression for how the pressure changes with respect to temperature along the coexistence boundary

$$\frac{\mathrm{d}p}{\mathrm{d}T} = \frac{\Delta_{\mathrm{trs}}H_\mathrm{m}}{T\,\Delta_{\mathrm{trs}}V_\mathrm{m}}. \tag{12.21}$$

The implications of Eq. (12.21) can be observed in Fig. 12.9. Phase transitions such as melting and vaporization always have a positive $\Delta_{\mathrm{trs}}H$. Vaporization always corresponds to a positive $\Delta_{\mathrm{trs}}V_\mathrm{m}$, and this is usually the case for melting as well. Therefore, the slope of the p–T boundary for vaporization is always positive, and that for melting is normally also positive. This is illustrated for the melting of Al in Fig. 12.9. Increasing the pressure increases the melting point. However, some materials such as water, Bi, Ga, Ge, and Si expand upon freezing. Thus, the slope is reversed, and the melting point is depressed by the application of pressure. This is also illustrated in Fig. 12.9.

One aspect of the anomalous behavior of water at the solid/liquid phase boundary is very familiar – ice floats on water. This odd behavior is also the reason why beverages can break their container when frozen, and why your ice cubes sometimes are not flat on top when frozen in a conventional ice tray. Ice forms on the top of the freezing liquid

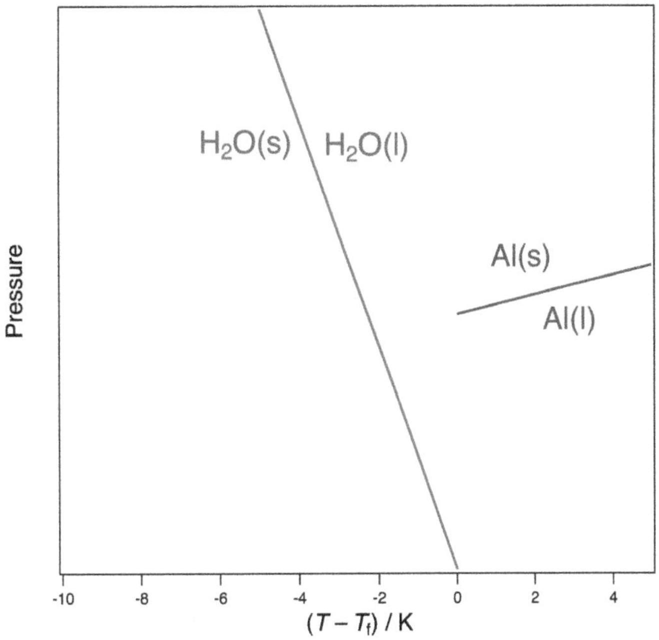

Figure 12.9 Increasing pressure on most materials leads to an increase in the melting point T_f, as is shown for Al. Water behaves anomalously and the melting point is depressed by the application of pressure. The slope of the phase boundary is given by Eq. (12.21).

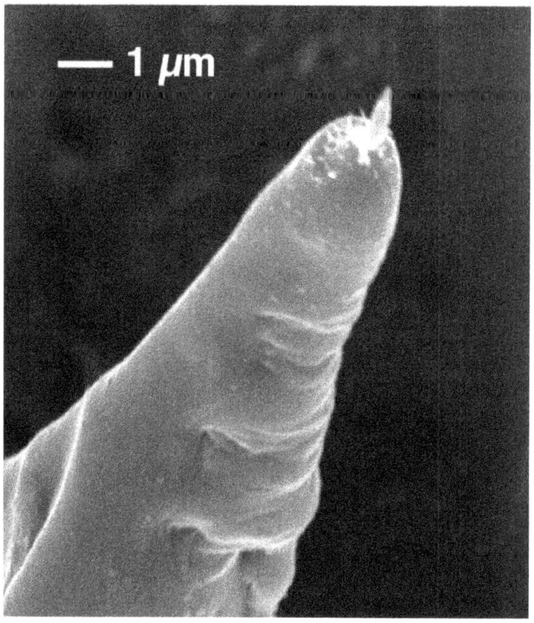

Figure 12.10 A silicon nanospicule atop a silicon pillar formed by laser ablation. The laser is used to melt the Si. A jet of supercooled liquid Si punctures a solid cap on the liquid to form the nanospicule. For details, see Mills, D. and Kolasinski, K.W. (2006) *Nanotechnology*, **17**, 2741. Reproduced with permission from IOP Publishing.

and forms a solid cover above the cold liquid below. The solid encloses the liquid. As the solid grows into the liquid, it expands, placing the liquid under pressure. Eventually, the pressure builds up sufficiently that the liquid bursts through the ice. The resulting fountain can lead to spicule formation – ice spikes that can be several centimeters in length if the conditions are just right. Nanoscale Si and Ge spicules formed by the same mechanism have been observed during laser ablation experiments. An example of such a Si spicule on the end of a laser-heated Si pillar is shown in Fig. 12.10.

12.6 Clausius–Clapeyron equation

Consider the application of the Clapeyron equation to the liquid–vapor boundary, which corresponds to the equilibrium of a pure liquid and its vapor:

$$\mathrm{A}(l) \rightleftharpoons \mathrm{A}(\mathrm{g}).$$

What we would like to know is how the vapor pressure of a liquid changes as a function of temperature. Data for both the equilibrium vapor pressure above liquid water and mercury are shown in Fig. 12.11. We see that both curves have very similar shapes, which hints toward some sort of universal behavior since chemically H_2O and Hg are much different in their intermolecular interactions. What they do have in common is that we can treat the vapor as an ideal gas to a first approximation. Next, we note that the change in molar volume that accompanies vaporization is approximately equal to the molar volume of the vapor $V_\mathrm{m}(\mathrm{g})$ because the molar volume of the liquid $V_\mathrm{m}(l)$ is negligible in comparison. This is true for all vaporization transitions near the boiling point. Therefore, the molar volume change is given by

$$\Delta_\mathrm{vap} V_\mathrm{m} = V_\mathrm{m}(\mathrm{g}) - V_\mathrm{m}(l) \approx V_\mathrm{m}(\mathrm{g}) = RT/p. \tag{12.22}$$

We also know that the entropy change is given by $\Delta_\mathrm{vap} S^\circ_\mathrm{m} = \Delta_\mathrm{vap} H^\circ_\mathrm{m}/T$, which upon substitution into Eq. (12.21) yields

$$\frac{\mathrm{d}p}{\mathrm{d}T} == \frac{\Delta_\mathrm{vap} S^\circ_\mathrm{m}}{\Delta_\mathrm{vap} V^\circ_\mathrm{m}} = \frac{\Delta_\mathrm{vap} H^\circ_\mathrm{m}}{T(RT/p)} = \frac{\Delta_\mathrm{vap} H^\circ_\mathrm{m}}{T}\frac{p}{RT}. \tag{12.23}$$

This rearranges into the *Clausius–Clapeyron equation* for the variation of vapor pressure with temperature

$$\frac{\mathrm{d}p}{p} = \frac{\Delta_\mathrm{vap} H^\circ_\mathrm{m}}{R}\frac{\mathrm{d}T}{T^2}. \tag{12.24}$$

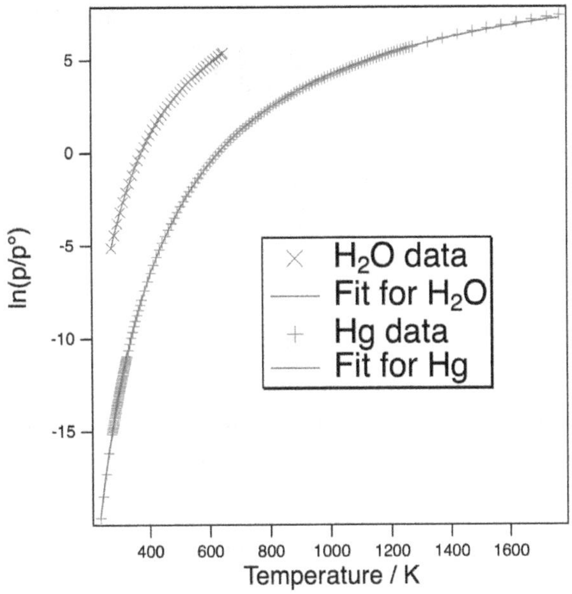

Figure 12.11 The equilibrium vapor pressure above pure $H_2O(l)$ and Hg(l) as a function of temperature. The natural logarithm of the vapor pressure divided by the standard pressure of $p^\circ = 100$ kPa is plotted against absolute temperature. Data taken from the 96th edition of the *CRC Handbook of Chemistry and Physics*.

Integrating this equation we obtain

$$\ln\left(\frac{p_f}{p_i}\right) = \int_{T_i}^{T_f} \frac{\Delta_{\text{vap}} H^\circ_{\text{m}}}{R} \frac{\mathrm{d}T}{T^2}. \tag{12.25}$$

Looking at Fig. 12.7 it is clear that the molar entropy of vaporization is not independent of temperature. Initially, at least, it follows a simple linear relationship

$$\Delta_{\text{vap}} H^\circ_{\text{m}} = (a + bT)R. \tag{12.26}$$

The vapor pressure equals the standard pressure at the standard boiling point, that is $p_i = p^\circ$ at $T_i = T_\text{b}$. Using this relation, substituting the linear dependence of $\Delta_{\text{vap}} H^\circ_{\text{m}}$ on T into Eq. (12.25) and integrating we obtain

$$\ln(p/p^\circ) = c + b \ln T - a/T. \tag{12.27}$$

The parameter c is not free, rather, $c = a/T_\text{b} - b \ln T_\text{b}$.

12.6.1 Directed practice

Perform the integration to derive Eq. (12.27).

The data in Fig. 12.11 have been fitted to Eq. (12.27). The quality of the fit is remarkably good with only two free parameters. The derived parameters a and b can then be substituted back into Eq. (12.26) in order to calculate the enthalpy of vaporization. The fit parameters for both H_2O and Hg are presented in Table 12.4. The predicted value of the enthalpy change is also compared to the accepted literature value. The agreement is quite good based on this very simple set of approximations.

Historically, unfortunately, the relationships in Eqs (12.25) and (12.26) were not recognized. Instead, the formula

$$\log p = A - \frac{B}{T + C} \tag{12.28}$$

Table 12.4 The results of fits of the data presented in Fig. 12.11 to Eq. (12.27).

Liquid	T_b/K	a	b	$\Delta_{\text{vap}} H^\circ_{\text{m}}$/kJ mol^{-1} (fit)	$\Delta_{\text{vap}} H^\circ_{\text{m}}$/kJ mol^{-1} (accepted)
H_2O	373.12	5985.3	-2.6307	41.6	40.7
Hg	629.88	7387.4	-0.26079	60.1	59.1

known as the *Antoine equation* was discovered empirically. The Antoine equation is known to fit vapor pressure data over limited temperature ranges. Tabulations of Antoine equation parameters (often several sets for different temperature ranges) are available in the NIST WebBook, while the *CRC Handbook of Chemistry and Physics* has collected parameters for an equation similar to Eq. (12.27) for metallic elements.

There is one other thing to note about Eq. (12.25). It is equally valid if we were to consider the solid-to-gas transition, that is, sublimation. The only change we need to make is to substitute $\Delta_{sub}H^\circ_m$ for $\Delta_{vap}H^\circ_m$. Thus, if $\Delta_{sub}H^\circ_m$ is assumed to be constant and we know the equilibrium vapor pressure of the solid p_1 at some temperature T_1, the equilibrium vapor pressure p_2 at temperature T_2 would be given by

$$\ln\left(\frac{p_2}{p_1}\right) = \frac{\Delta_{sub}H^\circ_m}{R}\left(\frac{1}{T_1} - \frac{1}{T_2}\right). \tag{12.29}$$

Equation (12.29) is a familiar *integrated form of the Clausius–Clapeyron equation*. It is good for accurate predictions of vapor pressure within a range of about ±20 K of any temperature for which the vapor pressure is known. It is also good for a first approximation to the vapor pressure for temperatures outside of this range.

12.6.2 Directed practice

Does it take as long to cook pasta in boiling water at the top of Mount Vesuvius as it does by a canal in Venice? Use the Clausius–Clapeyron equation to justify your answer.

12.7 Surface tension

It takes energy to create a unit of surface area of liquid or solid. Therefore, a liquid drop will attempt (in the absence of other forces such as gravity or attraction to a solid surface) to minimize its surface area by forming a sphere. The work associated with creating additional surface area at constant p and T is

$$dG = \gamma\, dA \tag{12.30}$$

where G is the Gibbs energy, γ the *surface tension* (SI units J m^{-2} which is equivalent to N m^{-1}) and A is the unit element of surface area. There is a force F generated by surface tension acting inward on the drop that maintains the spherical shape with radius r. This force is given by

$$F = 8\pi\gamma r. \tag{12.31}$$

This force acts normal to the surface of the drop. At equilibrium forces must be equal. Therefore, the pressure inside and outside of the drop must be different since the net of the forces acting on the drop from the outside (from the pressure p_{outer} and F) must balance the force from the inside (from p_{inner}) according to

$$4\pi r^2 p_{outer} + 8\pi\gamma r = 4\pi r^2 p_{inner} \tag{12.32}$$

Hence,

$$p_{inner} = p_{outer} + 2\gamma/r. \tag{12.33}$$

The smaller the radius, the greater the pressure difference. The vapor pressure found in thermodynamic tables refers to the vapor pressure measured above a planar surface. The elevated pressure experienced by molecules in a droplet causes them to evaporate more easily than the molecules contained behind a flat interface. To calculate vapor pressure we start by calculating the molar Gibbs energy of a vapor at its normal vapor pressure p^*,

$$G^*_m = G^\circ_m + RT\ln(p^*/p^\circ). \tag{12.34}$$

G°_m is the standard state molar Gibbs energy and p° is the standard pressure (1 bar). Comparing this to the molar Gibbs energy at an arbitrary pressure p, that is, p is the pressure inside the drop,

$$G_m = G^\circ_m + RT\ln(p/p^\circ) \tag{12.35}$$

the change in the molar Gibbs energy resulting from the curvature of the interface is given by

$$\Delta G_m = G_m - G^*_m = RT\ln(p/p*). \tag{12.36}$$

This change in the Gibbs energy as a function of curvature, and hence also size, is known as the *Gibbs–Thompson effect*. Similarly, we can calculate the change in molar Gibbs energy as a function of the pressure increase brought about by the curved interface at constant temperature according to

$$\Delta G_\mathrm{m} = \int \mathrm{d}G = \int V_\mathrm{m}\,\mathrm{d}p - S_\mathrm{m}\,\mathrm{d}T = \int_0^{\Delta p} V_\mathrm{m}\,\mathrm{d}p = V_\mathrm{m}\,\Delta p. \tag{12.37}$$

Here, we use the usual definition of the change in Gibbs energy at constant composition, isothermal conditions and have assumed that the change in the molar volume V_m is negligible. Substituting from Eq. (12.33) for Δp and equating the result with Eq. (12.36), we arrive at the *Kelvin equation*

$$RT \ln \frac{p}{p^*} = \frac{2\gamma V_\mathrm{m}}{r} \quad \text{or} \quad p = p^* \exp\left(\frac{2\gamma V_\mathrm{m}}{RTr}\right). \tag{12.38}$$

For a droplet (a sphere of liquid surrounded by a fluid), the radius r is positive and for a bubble (a sphere of gas surrounded by a liquid or a membrane of liquid enclosing a gas) r is negative. Therefore, for a droplet in contact with its vapor, the liquid has a higher vapor pressure than normal and the vapor pressure increases with decreasing size. The Kelvin equation is one of the first and most fundamental equations of nanoscience in that it demonstrates that the properties of matter can be size-dependent. Important deviations from size-independent behavior arise prominently in the nanoscale regime of $r < 100$ nm. For example, for water at 300 K the change in vapor pressure is only ~1% for $r =$ 100 nm, but the deviation is ~10% at 10 nm and almost 300% at 1 nm. The Kelvin equation is also fundamental to *Ostwald ripening*, a phenomenon in which large particles, because they are more stable (i.e., they have a lower effective vapor pressure), grow at the expense of high-vapor pressure, less-stable smaller particles. This general effect of small particles being less stable than larger ones is also important in nucleation phenomena, as varied as the growth of crystals to the formation of raindrops and foams. Ostwald ripening should be differentiated from *Smoluchowski ripening*, a process in which small particles aggregate to form larger particles.

Surface tension is a strong function of temperature, as shown in Fig. 12.12. Surface tension decreases with increasing temperature, and goes to zero for supercritical fluids. This means that supercritical fluids wet all surfaces and easily penetrate into porous materials.

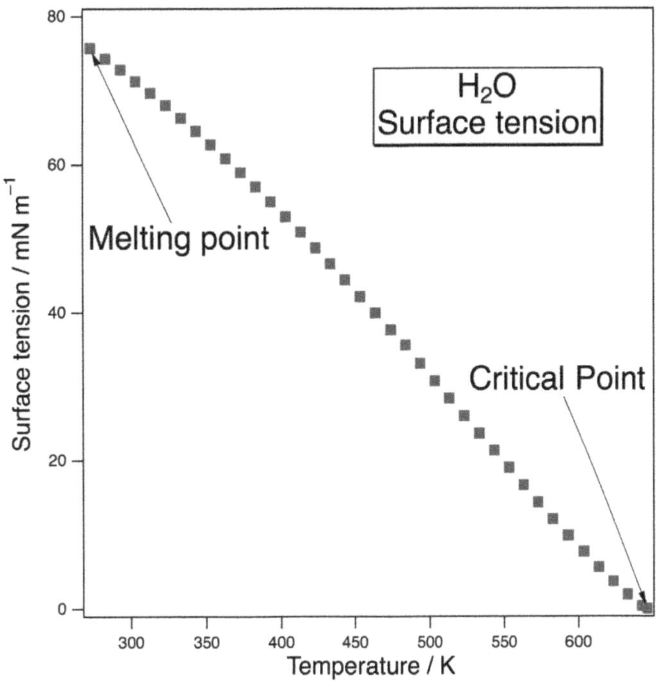

Figure 12.12 The temperature dependence of the surface tension of water over the full stability range of liquid water at standard pressure from T_f to T_c. Data taken from the 96th edition of the *CRC Handbook of Chemistry and Physics*.

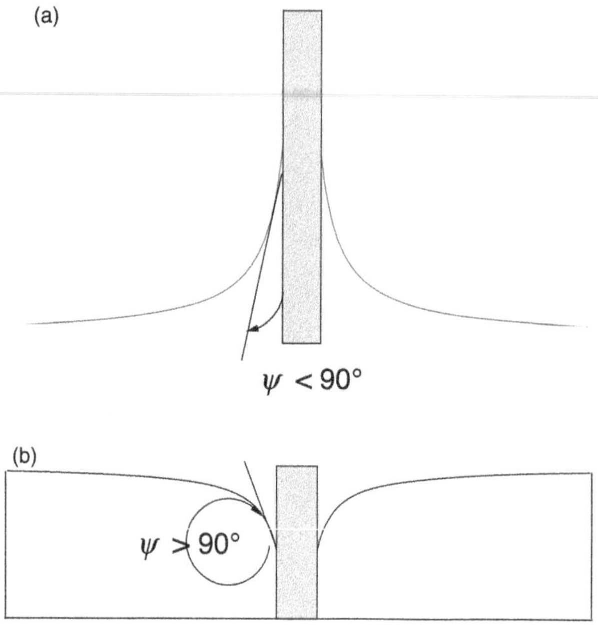

Figure 12.13 Meniscus formation on a planar substrate. (a) A hydrophilic substrate in water. (b) A hydrophobic substrate in water. ψ, contact angle.

12.7.1 Meniscus formation and capillary condensation

A *meniscus* is formed when a planar solid is pushed into a liquid. The balance of surface tension forces at the liquid/solid interface determines the profile of the meniscus. Figure 12.13 illustrates how the relative surface tensions of the liquid and solid lead to either concave or convex meniscuses. The surfaces of oxides tend to be covered with hydroxyl groups. These surfaces are polar and hydrophilic and, therefore, the attractive forces associated with the hydroxyl groups lead to upward-sloping meniscuses in water. Common hydrophilic substrates include Al_2O_3 (alumina, sapphire), Cr_2O_3, SnO_2, SiO_2 (glass, quartz), Au and Ag. A nonpolar surface is not wetted by water and forms a downward-curving meniscus. Si is a particularly versatile substrate in that it can be made hydrophilic when an oxide layer is present, or hydrophobic when the surface is H-terminated or coated with siloxanes. A Si surface normally has a thin (several nanometers) oxide layer on it, known as the native oxide. This surface is readily transformed into a H-terminated surface by dipping in acidic fluoride solutions.

Capillary rise and depression are also consequences of the pressure difference across a curved interface. Here, the radius of curvature of the solid/liquid interface is equal to the capillary radius. The difference in pressure across the curved interface $2\gamma/r$ balances the weight of the liquid column that rises in the gravitational field $\rho g h$, and the capillary rise or depression is given by

$$h = \frac{2\gamma \cos\psi}{\rho g r}. \tag{12.39}$$

where ψ is the *contact angle*, defined in Fig. 12.13. A complete wetting surface (water on a hydrophilic surface) has $\psi = 0°$, while a completely nonwetting surface has $\psi = 180°$.

We move next to the interaction of a vapor with the solid walls of a pore or particles, as shown in Fig. 12.14. Consider first a conical pore with a perfectly wetting surface exposed to a vapor, as in Fig. 12.14(a). The most common case is a pore exposed to humid air and the vapor in question is that of water. Water condenses on a hydrophilic surface, filling the pore up to the point at which a curved interface with radius of curvature r_c is formed. The concave interface can be treated as though a bubble of radius r_c is positioned atop the condensed water. The Kelvin equation allows us to calculate r_c once equilibrium has been established,

$$RT \ln\frac{p}{p^*} = -\frac{2\gamma V_m}{r_c}. \tag{12.40}$$

The formation of a concave liquid/gas interface (the *meniscus*), brought about by the attractive water–surface interaction, reduces the vapor pressure of the liquid and drives condensation of the liquid. The radius r_c corresponds to the capillary radius at the top of the meniscus when the pore walls are wetted by the liquid. The curvature of the interface and the

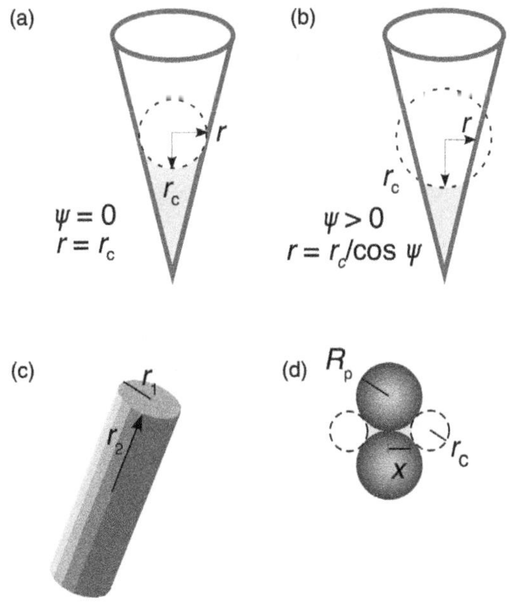

Figure 12.14 (a) A conical pore with a hydrophilic surface induces the condensation of water. (b) Capillary condensation in a pore with a partially wetting surface. (c) A cylindrical pore with a width r_1 significantly smaller than the length. (d) An illustration of meniscus formation in the presence of two spherical particles of radius R_p.

radius of curvature are inversely related. A larger radius of curvature corresponds to a flatter interface. If the surface of the pore is not perfectly wetting but instead exhibits a *contact angle* $\psi > 0$, as shown in Fig. 12.14(b), then the meniscus becomes flatter. The radius of curvature of the interface, r_c, increases and is related to the radius of the conical pore, r, by $r = r_c/\cos\psi$. The level of the liquid is comparatively less than in the fully hydrophilic case because the vapor pressure reduction is less pronounced.

In a cylindrical pore of radius r_1, the sidewalls are straight, which corresponds to an infinite radius of curvature $r_2 = \infty$. Hence, we need to make the substitution $1/r_1 + 1/r_2$ for $2/r_c$ in Eq. (12.40), which allows us to estimate the radius of pores that accommodate *capillary condensation*

$$r_1 = -\frac{\gamma V_m}{RT \ln(p/p^*)}. \tag{12.41}$$

This equation also holds approximately for any other irregular pore shape for which the length is much longer than the width.

The two particles with radius R_p depicted in Fig. 12.14(d) also induce the condensation of vapor and the formation of a meniscus with a radius of curvature r_c. The pressure associated with the presence of this liquid applies a force to the particles, which is known as the *capillary force*. In the case where the particles are not immersed in liquid and the liquid is present due to capillary condensation, $x \gg |r_c|$ and $1/r_1 + 1/r_2 = 1/x - 1/r_c \approx -1/r_c$. The effective pressure acting on the particles due to condensation is

$$\Delta p = \gamma/r_c \tag{12.42}$$

which acts over an area of πx^2 and, therefore, a capillary force of

$$F = \pi x^2 \gamma/r_c, \tag{12.43}$$

which can in turn be related by simple geometry to the particle radius as

$$F = 2\pi\gamma R_p. \tag{12.44}$$

Irregularities in the surface profile of real particles tend to lower the capillary force compared to the value calculated with Eq. (12.44). The capillary force literally forces liquids into powders and porous materials *when the walls are wetted by the liquid*. But if the walls are not wetted by the liquid, the opposite effect is observed. Hydrophilic powders and nanoparticles that are not kept dry will aggregate into larger masses that are held together by the capillary force. Another important consequence of the capillary force is that, as a liquid dries within a porous material, it retreats progressively into smaller and smaller pores. According to Eq. (12.42) the smaller the pore the greater the pressure applied by the liquid to the material. Enormous pressures on the order of hundreds of atmospheres arise in materials with nanoscale pores. As discussed in detail by Bellet and Canham,[1] this can tear apart thin porous films and alter their structure. Similarly, the structure of powders, nanoparticles and nanowires can be changed irreparably as a result of the effects of drying. Cracking of porous films occurs below a certain critical thickness h_c that depends on the surface tension of the liquid γ_L, the surface tension and bulk modulus of the solid γ_S and E_S, respectively, as well as the porosity ε and the pore radius r according to

$$h_c = (r/\gamma_L)^2 E_S (1 - \varepsilon)^3 \gamma_S. \tag{12.45}$$

Drying stress can be avoided if the liquid is replaced by a supercritical fluid. As the supercritical fluid is extracted from the pores, no meniscus forms because there is no vapor/liquid interface and, consequently, no capillary force.

12.8 Nucleation

Nucleation of a new phase is naturally resisted by a pure system. A condensing vapor generally does not form droplets unless the vapor is supersaturated. A liquid does not solidify unless supercooled in the absence of a heterogeneous interface that catalyzes the process. As discussed in relation to the Kelvin equation, Eq. (12.38), the resistance to nucleation is associated with the variation in surface energy with the size of small particles. The surface energy, or change in the Gibbs energy, is unfavorable for the addition of matter to small clusters up to a critical size. After attaining the *critical nucleus size*, further addition of material to the cluster is energetically favorable. The formulation of this size dependence on the change in Gibbs energy is the basis of classical nucleation theory.

In our discussion of nucleation, we will consider the problem of how a bubble of gas forms in a liquid. Here, we seek to answer the question: When do bubbles form? Classical heterogeneous nucleation theory on a flat surface was given a thermodynamic foundation by Max Volmer.[2]

12.8.1 Classical nucleation theory

The usual presentation of classical nucleation theory starts from a consideration of the change in Gibbs energy dG that occurs at constant temperature T and pressure p for an infinitesimal change of moles of the gas phase dn_g and liquid phase dn_l when a spherical droplet is in contact with its vapor. A detailed treatment can be found in Chapter 9 of Adamson and Gast (see Further Reading), and an insightful discussion can be found in Chapter 23 of Metiu (see Further Reading), which we follow here. The droplet of radius R and surface area $A = 4\pi R^2$ is composed of a material with surface tension γ. However, here we wish to consider a bubble, which has a negative radius of curvature, forming inside of a liquid. This slightly complicates the formalism of the mathematics; therefore, below we will take R to be positive.

A transfer of moles from the liquid to the gas changes the radius of the bubble, and the accompanying change in Gibbs energy is given by

$$\mathrm{d}G = \mu_l \,\mathrm{d}n_l + \mu_g \,\mathrm{d}n_g + \gamma \,\mathrm{d}A, \tag{12.46}$$

where μ_l and μ_g are the chemical potentials of the liquid and gas, respectively. Conservation of mass couples the molar changes directly,

$$\mathrm{d}n_l = -\mathrm{d}n_g. \tag{12.47}$$

The amount of gas is given by the product of molar density of gas ρ_g and the volume of the bubble,

$$n_g = \rho_g (4\pi/3) R^3. \tag{12.48}$$

Differentiating, we find d$n_g = 4\pi\rho_g R^2$ dR and d$A = 8\pi R$ dR, which upon substitution into Eq. (12.46) yields,

$$\mathrm{d}G = 4\pi R^2 \left[\rho_g(\mu_g - \mu_l) + \frac{2\gamma}{R} \right] \mathrm{d}R. \tag{12.49}$$

From the Second Law of Thermodynamics, we know that the system is at equilibrium when d$G = 0$ and it will evolve spontaneously in the direction indicated by d$G < 0$. These conditions define the stability of the bubble of radius R according to Eq. (12.49). The chemical potentials are functions of T and p. They are always positive, as is γ. If we choose the temperature and pressure such that $\mu_l < \mu_g$, the behavior of the system is simple: d$G < 0$ can only be achieved by decreasing R, that is, with d$R < 0$. The bubble shrinks and disappears.

However, if we choose T and p such that the gas is more stable than the liquid – that is, such that $\mu_g < \mu_l$ – the behavior is not as clear-cut. Now, the two terms in brackets have opposite signs, and whether the bubble grows (d$G < 0$ with d$R > 0$) or shrinks (d$G < 0$ with d$R < 0$) depends on the relative magnitudes of these two terms. Furthermore, there is a critical radius at which the system is metastable. Metastability is defined by d$G = 0$, noting that the slightest change in R (caused by a fluctuation in n_l and n_g) tips the system toward instability. Setting Eq. (12.49) equal to zero and solving for the critical radius R^* we find,

$$R^* = \frac{2\gamma}{\rho_g(\mu_l - \mu_g)}. \tag{12.50}$$

The behavior of a bubble at a temperature and pressure where the gas is more stable than the liquid thus depends on the radius of the bubble. If $R = R^*$, the bracketed term in Eq. (12.49) vanishes, d$G = 0$, and the droplet is stable. If the bubble is smaller than the critical radius, the bracketed term in Eq. (12.49) is positive, and d$G < 0$ when the bubble shrinks. If the bubble is larger than the critical radius, the bracketed term in Eq. (12.49) is negative, and d$G < 0$ when the bubble grows.

We are now in a position to answer the question: why must a system be supersaturated for homogeneous nucleation to occur? The difference between the chemical potential of an ideal gas dissolved in an ideal solution at a mole fraction x_g and the chemical potential of an ideal gas dissolved at its equilibrium mole fraction x_g^{eq} at a reference pressure (chosen as 1 atm) is given by

$$\Delta\mu_g = RT\ln\left(x_g/x_g^{\text{eq}}\right). \tag{12.51}$$

Note that in Eq. (12.51) R is the ideal gas constant. Thus, two terms are often used to parameterize the degree of supersaturation. The *saturation ratio* is defined as

$$\alpha = x_g/x_g^{\text{eq}} \tag{12.52}$$

and the *supersaturation* is defined as

$$S = x_g/x_g^{\text{eq}} - 1 = \alpha - 1. \tag{12.53}$$

Whereas, the time averaged density of the liquid is uniform, motions of the molecules in the liquid inherent to the Maxwell–Boltzmann distribution of velocities lead to fluctuations in the density at any point in time at any arbitrary location in the liquid. Another way to think of density fluctuations is that clusters of molecular vacancies form in the liquid. A cluster of vacancies leads to a low-density region that defines a bubble. Any bubble with a radius smaller than the critical radius is unstable and decays back to a uniform liquid. Any bubble larger than the critical radius will grow and remain as a bubble. At equilibrium, where $S = 0$, the chemical potential of the gas and the liquid are equal. The critical radius is infinite according to Eq. (12.50). The probability of forming an infinitely large bubble is infinitely small. The phenomena of supersaturation, supercooling and superheating are logical consequences for any system held strictly at equilibrium in the absence of impurities and any other interfaces, which can act as heterogeneous and nonclassical nucleation sites as defined below.

Equation (12.49) shows us that the Gibbs energy is a function of the radius of the bubble. Indeed, dG/d$R = 0$ at $R = R^*$, and evaluation of the second derivative shows that the Gibbs energy has a maximum at the critical radius. The value of the Gibbs energy change at this maximum $\Delta^{\ddagger}G$ is the *Gibbs energy of activation* for the formation of a bubble of the critical radius. According to Boltzmann statistics, the number density of bubbles N_c that attain the critical radius at equilibrium is

$$N_c = N\exp\left(-\Delta^{\ddagger}G/k_BT\right) \tag{12.54}$$

where N is the total number density of molecules and k_B is the Boltzmann constant. The value of $\Delta^{\ddagger}G$ is found by substituting the value of R^* from Eq. (12.50) into Eq. (12.46) after integration, from which we obtain

$$\Delta^{\ddagger}G = \tfrac{4}{3}\gamma R^{*2} = \frac{16\pi\gamma^3}{3\rho^2\left(\mu_l - \mu_g\right)^2}. \tag{12.55}$$

The activation barrier can be lowered from infinity at equilibrium by increasing the chemical potential of the gas. This is accomplished according to Eq. (12.51) by increasing the concentration of the dissolved gas beyond its equilibrium value, that is, by going to large values of the saturation S.

The rate of bubble formation is calculated from Eq. (12.54) by assuming that this rate is equal to by the rate at which one more molecule is added to the critical nucleus. We assume that diffusion-limited transport of molecules to the critical nucleus occurs characterized by a classical prefactor of $k_B T/h$, where h is the Planck constant, and a diffusion barrier of $\Delta_{dif}G$. The rate of bubble formation is then given by the diffusion rate times the number density of critical nuclei

$$J = N(k_B T/h)\exp(-\Delta_{dif}G/k_B T)\exp\left(-\Delta^{\ddagger}G/k_B T\right). \tag{12.56}$$

The rate of formation is, therefore, related to the supersaturation S brought about by a gas pressure p by

$$J = A\exp\left(\frac{-16\pi\gamma^3}{3k_B T(Sp)^2}\right). \tag{12.57}$$

From classical nucleation theory we learn that the nucleation of one phase in another occurs after the formation of a growth nucleus of a critical size R^*, which is opposed by an activation barrier $\Delta^{\ddagger}G$. This is why phenomena such as superheating and supersaturation can occur. Because of the difficulty of nucleating a new phase, nucleation almost never occurs homogeneously in a practical environment. Special circumstances such as large supersaturation, extreme purity and smoothness of interfaces must be constructed to facilitate homogeneous nucleation.

12.8.2 Classical versus nonclassical nucleation

Four types of nucleation have been defined, as shown in Fig. 12.15 specifically for the case of gas bubbles forming in a liquid. Type I nucleation is classical homogeneous nucleation: spontaneous growth of a bubble within a supersaturated liquid following a density fluctuation. Type I nucleation typically requires supersaturation in excess of 100 to occur. Type II nucleation is classical heterogeneous nucleation: spontaneous growth at a liquid/solid interface (either with a random particulate or a container wall) within a supersaturated liquid following a density fluctuation.

However, most studies on nucleation report rates in excess of the predictions of classical nucleation theory. The difficulties that classical nucleation theory has to describe bubble formation rates in real systems are both intrinsic and practical. The conditions required to make a surface conducive to nucleation of bubbles are exactly the conditions required to make detachment of the bubbles difficult. Furthermore, extreme care must be taken to avoid the inclusion of gas-filled cavities within a liquid. These cavities may either be metastable cavities with radii less than the critical nucleus radius (Fig. 12.15(c); Type III nucleation) or they may be stable gas-filled cavities (Fig. 12.15(d); Type IV nucleation) with radii greater than R^*. In either case, the cavities may exist either within the liquid or attached to a solid surface.

The circumstances of practical experiments have led us to the introduction of two additional types of nucleation. Type III nucleation is also called *pseudo-classical nucleation*. It occurs on metastable subcritical nuclei ($R < R^*$) within the bulk of the liquid or attached to a surface. Type IV nucleation is called *nonclassical nucleation*, since no activation energy is required for it to occur. It occurs at stable nuclei that are larger than the critical radius, $R > R^*$. You may have noticed that bubbles stream away from a fixed point on the bottom or side of a glass containing a carbonated beverage. This is the result of Type IV nucleation at a scratch/crevice on the surface of the glass. Type IV nuclei will almost always form when a liquid is poured into a vessel unless precautions are taken.

To illustrate the importance and ubiquity of Types III and IV nucleation, take the case of champagne. Liger-Belair *et al.*[3] have thoroughly studied the properties of bubbles in champagne wines. They instituted a rigorous cleaning strategy in which glasses were thoroughly washed in dilute aqueous formic acid, rinsed using distilled water, and then dried with compressed air. This treatment suppresses the formation of $CaCO_3$ crystals on the glass walls as well as the adsorption of dust particles that might act as nucleation sites. The sparkling wine is still contained within a glass; however, the smooth and clean wine/glass interface is unable to promote bubble formation. The sparkle is lost because at a supersaturation of only $S \sim 4$, as is common with champagne wines, neither classical homogeneous nor classical heterogeneous nucleation occur at an appreciable rate. Laser etching of the glass in a flute can be used to produce sites for Type IV nucleation. These findings are in complete harmony with the conclusion that nucleation observed in most instances, corresponding to supersaturation of ≤ 5, is invariably associated with the existence of metastable gas cavities in the walls of the container or the bulk of the solution, prior to the system being made supersaturated.

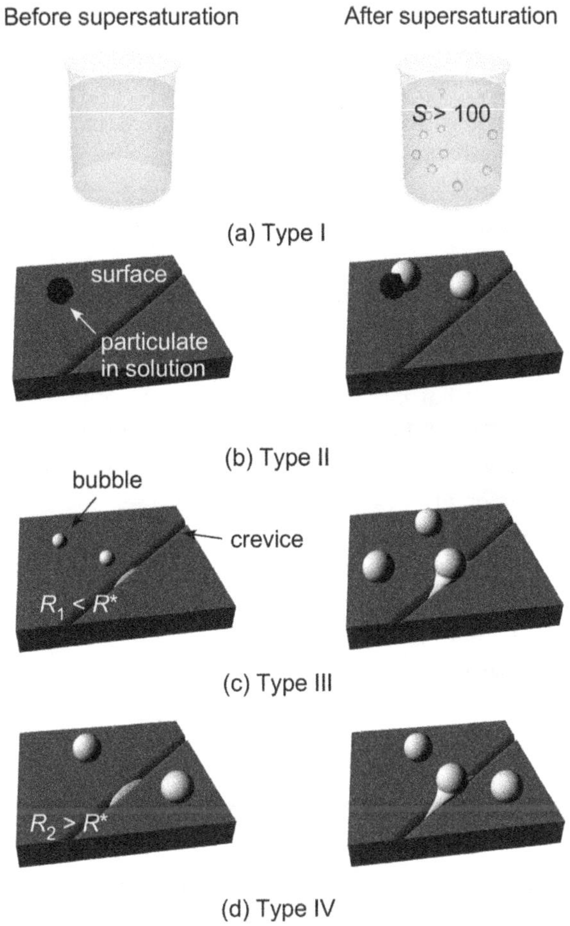

Figure 12.15 (a) Type I, (b) type II, (c) type III, and (d) IV nucleation. S is the value of supersaturation. R_1 and R_2 are radii of curvature. R^* is the critical radius of nucleation. Adapted from Jones, S.F., Evans, G.M, and Galvin, K.P. (1999) *Adv. Colloid Interface Sci.*, **80**, 27. The black hexagonal object represents a random irregular particle suspended in the liquid. Gray regions represent gas-filled cavities. Before supersaturation such cavities only exist in the type III and type IV nucleation scheme. After supersaturation, these cavities are the bubbles that have nucleated and grown. For details, see Kolasinski, K.W. (2014) *Mesoporous Biomater.*, **1**, 49.

12.8.3 Cloud formation: Heterogeneous versus homogeneous nucleation

Nucleation theory makes it clear that the formation of nucleation centers is a difficult process. However, such formation is essential to the creation of a condensed phase from a gaseous source, for instance, the formation of clouds. *Homogeneous nucleation* is the process whereby nucleation of a condensed phase occurs spontaneously in the bulk of a fluid phase. Because of the difficulty of forming nucleation centers homogeneously, it is extremely unlikely for them to form even if the relative humidity slightly exceeds 100% (i.e., the point at which the condensed phase actually forms the more stable phase). High degrees of supersaturation exceeding 150% are required for the homogeneous nucleation of droplets. Homogeneous nucleation can be involved in cloud formation. However, for this to be so, not only must high levels of supersaturation be observed, but also there must be an absence of aerosol particles that can act as catalysts for nucleation.

Heterogeneous nucleation is the process whereby nucleation is initiated at interfaces within or along the boundaries of the fluid phase. Heterogeneous nucleation of water at just below 273 K can commence, in contrast to homogeneous nucleation, as soon as the relative humidity exceeds 100%. This can occur at container walls, but there are no walls in the atmosphere. Nonetheless, a variety of aerosol particles can potentially act as nucleation centers including minerals (aluminosilicates), salts, metals (in elemental, oxide or sulfate forms), elemental carbons, organics, and biological materials. The relative importance of the nucleation mechanism depends on conditions and the type of cloud formed. Cirrus clouds, which are very common and occur at an altitude of 8–17 km, contain condensed water only as ice. In cirrus clouds, heterogeneous nucleation is almost exclusively responsible for cloud formation. Of particular importance are mineral and metal particulates. The efficiency of these aerosols in acting as nucleation centers depends on their surface composition: they work best when they are clean rather than coated with organic contamination.

12.9 Construction of a liquid–vapor phase diagram at constant pressure

Take two miscible liquids and mix these together at a temperature below the normal boiling point of the lowest boiling liquid. For this example we will hold the pressure constant at standard pressure. Since the liquids are miscible, we are free to mix them in any composition. Above the liquids, the equilibrium vapor pressures of the two substances will conform to the fractions determined by Raoult's and Henry's laws (see Chapter 13). The total amount of substance in moles is equal to the sum of the amount of the two components 1 and 2, $n = n_1 + n_2$. The mole fractions in the liquid phase follow $1 = x_1 + x_2$ and in the gas phase $1 = y_1 + y_2$.

Far above the boiling point of the higher boiling point species, both substances must be in the gas phase. Again, we are free to choose any composition of the system, this time in the absence of the liquid phase since it is not stable in this pressure and temperature range.

Now consider either of two scenarios. What if we choose a temperature in between the standard boiling points of the two liquids? What will the composition of the system be? Alternatively, what happens if we start at a temperature below the boiling points of the components and then gradually increase it at constant pressure? We know from studies of colligative properties that the boiling point of the mixture is in between the boiling points of the two components. We also know that in the range where a liquid mixture is stable, the more-volatile component (the one with the lower boiling point) is enriched in the gas phase and the less-volatile component is enriched in the solution phase.

We can answer these questions with the help of a *liquid–vapor phase diagram* such as those shown in Figs 12.16 and 12.17. Start with the phase diagram of methanol–water mixtures in Fig. 12.16. Here, we have plotted versus temperature the mole fractions of methanol in the gas phase y_1 and in the liquid phase x_1. The upper curve is known as the *bubble line* and the lower curve is the *dew line*. Any combination of (p, T, n, n_1) that lies below the dew line results in a stable mixture composed mainly of the liquid phase (plus the equilibrium vapor pressure). Note that we need not specify n_2 if we know n, or vice versa. Any combination of (p, T, n, n_1) that lies above the bubble line results in a mixture that exists solely in the gas phase.

Now make up a mixture that corresponds to point A. In this case it would be (p = 101.3 kPa, T = 354 K, n = 1, n_1 = 0.5 mol). Thus, the initial (nonequilibrium) mole fraction of methanol is $x_{1,0} = 0.5$. Any point situated between the dew line and the bubble line corresponds to an unstable composition. If the system is held at constant temperature and pressure, then the system starting as a liquid at point A has to change the relative compositions of the liquid phase and the gas phase. The compositions change such that the final equilibrated system corresponds to the two points on the tie line that intersects the dew line (which describes the composition of the liquid phase) and the bubble line (which describes the composition of the gas phase). In the case shown, the more-volatile component (methanol) boils off preferentially

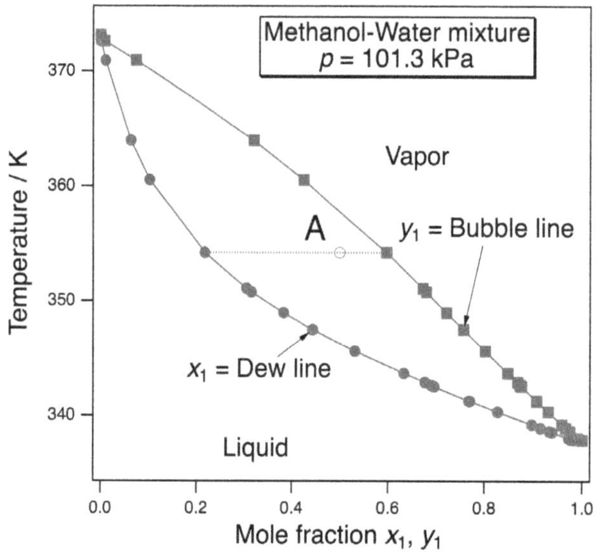

Figure 12.16 The liquid–vapor phase diagram for methanol and water mixtures at a total pressure of 101.3 kPa. x_1 is the mole fraction of methanol in the liquid phase. y_1 is the mole fraction of ethanol in the gas phase. Data from Álvarez, V.H., Mattedi, S., Iglesias, M., Gonzalez-Olmos, R., and Resa, J.M. (2011) *Phys. Chem. Liquids*, **49** (1), 52–71.

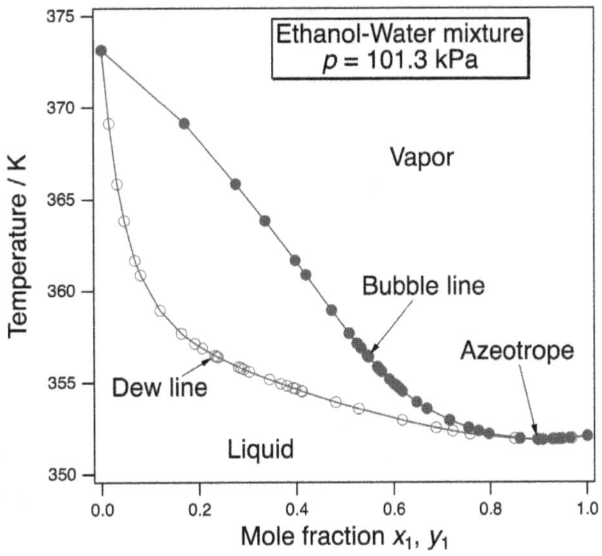

Figure 12.17 The liquid–vapor phase diagram for ethanol and water mixtures at a total pressure of 101.3 kPa. x_1 is the mole fraction of ethanol in the liquid phase. y_1 is the mole fraction of ethanol in the gas phase. Data from Álvarez, V.H., Mattedi, S., Iglesias, M., Gonzalez-Olmos, R., and Resa, J.M. (2011) *Phys. Chem. Liquids*, **49** (1), 52–71.

so that the equilibrated gas phase ends up with a methanol mole fraction $y_1 = 0.596$. The equilibrated liquid phase is enriched in water and has a methanol mole fraction of just $x_1 = 0.215$.

The amount of methanol n_1 is split between the amount in the gas phase $n_{1,g}$ and the amount in the liquid $n_{1,\ell}$. Similarly, the amount of water is divided between the two phases subject to the constrain of constant total amount of substance. The apportionment of the two components into the two phases at equilibrium is described by a set of four coupled equations with four unknowns.

$$(n_{1,g} + n_{2,g})y_1 = n_{1,g} \tag{12.58}$$

$$(n_{1,\ell} + n_{2,1})x_1 = n_{1,\ell} \tag{12.59}$$

$$n_{1,\ell} + n_{1,g} = x_{1,0}n \tag{12.60}$$

$$n_{2,\ell} + n_{2,g} = (1 - x_{1,0})n \tag{12.61}$$

This system of equations can be solved (I would suggest with software for performing algebraic computing unless you really enjoy doing this by hand) to yield the four unknown compositions

$$\frac{n_{1,g}}{n} = \frac{y_1(x_{1,0} - x_1)}{y_1 - x_1} \tag{12.62}$$

$$\frac{n_{1,\ell}}{n} = \frac{x_1(x_{1,0} - y_1)}{x_1 - y_1} \tag{12.63}$$

$$\frac{n_{2,g}}{n} = \frac{(1 - y_1)(x_{1,0} - x_1)}{y_1 - x_1} \tag{12.64}$$

$$\frac{n_{2,\ell}}{n} = \frac{(1 - x_1)(x_{1,0} - y_1)}{x_1 - y_1} \tag{12.65}$$

Substitution of $n = 1$ mol, $x_{1,0} = 0.5$, $x_1 = 0.215$ and $y_1 = 0.596$ yields $n_{1,g} = 0.446$ mol, $n_{1,\ell} = 0.054$ mol, $n_{2g} = 0.302$ and $n_{2,\ell} = 0.198$ mol. The gas phase is enriched with methanol compared to water, as expected.

Now take a look at the phase diagram for ethanol–water mixtures. Chemically, we would expect methanol and ethanol to be rather similar with regards to their interactions with water. Nonetheless, the phase diagrams exhibit a significant difference. Near $x_1 = 0.9$, the dew line and bubble line intersect. This indicates the formation of an *azeotrope*, a mixture that behaves like a pure liquid with regard to its phase change behavior.[4] Specifically, an azeotrope boils to maintain constant composition rather than distilling off one component. If you are trying to purify ethanol by distillation, this is a problem. Once the mixture reaches the azeotropic point, the entire mixture distills resulting in a gas-phase composition that matches the liquid-phase composition without further enrichment of the concentrations in either phase. Note also the effect on Eqs (12.62)–(12.65). If the gas phase and liquid phase mole fractions are the same, $x_1 = y_1$, then the equations all become indeterminate. This is characteristic of the azeotropic point. It can be shown that this point must either correspond to a minimum, as it does for ethanol–water, or a maximum in the phase diagram.

12.10 Polymers: Phase separation and the glass transition

Crystallization from liquid to solid is a first-order phase transition. Small molecules exhibit this routinely as do some polymers. However, many polymers are either partially or exclusively amorphous, which means they vitrify upon cooling and form glasses rather than polycrystalline materials. At the *glass transition temperature* T_g in a polymer, the amorphous regions of the polymer soften. An ideal glass transition is a second-order phase transition. For such an amorphous glassy polymer at low temperature only vibrations are possible, the polymer is hard and brittle. For instance, a glove made from a butadiene–styrene or nitrile rubber is soft and flexible at room temperature; however, when cooled with liquid nitrogen it can be easily crushed to pieces when squeezed by hand. In the glass transition region, the polymer softens to a rubbery state that is accompanied by a reduction of stiffness of the material. This transition is accompanied by a continuous drop in its Young's modulus by three orders of magnitude. Coordinated molecular motion in the polymer chain begins to occur in the glass transition region. The continuous, though nonetheless dramatic, changes that occur upon heating are depicted in Fig. 12.18. The glassy to glass transition is followed by the rubbery plateau and then finally viscous flow.

Polymers may be partly crystalline combined with some admixture of amorphous regions. At any given temperature, the amorphous portion may be above or below its glass transition temperature. Thus, polymers at room temperature can be lumped into four subclasses:

- semicrystalline with amorphous regions above T_g (e.g., polyethylene)
- semicrystalline with amorphous regions below T_g (e.g., cellulose)
- amorphous above T_g (natural rubber)
- amorphous below T_g (poly(methyl methacrylate)).

The glass transition temperature is always lower than the melting point, $T_g < T_f$. The relationship between them depends on the structure of the polymer. The ratio T_g/T_f usually lies in the range of 0.5 to 0.93, with a value of 2/3 being a good first approximation.

Polymer solutions generally exhibit some degree of non-ideality, which is discussed in detail in Chapter 13. Just as for other solutions, consequences of non-ideality include an enthalpy of mixing that is finite rather than zero, and phase separation when the concentration of one component crosses a threshold. Consider a binary solution composed of n_1 and

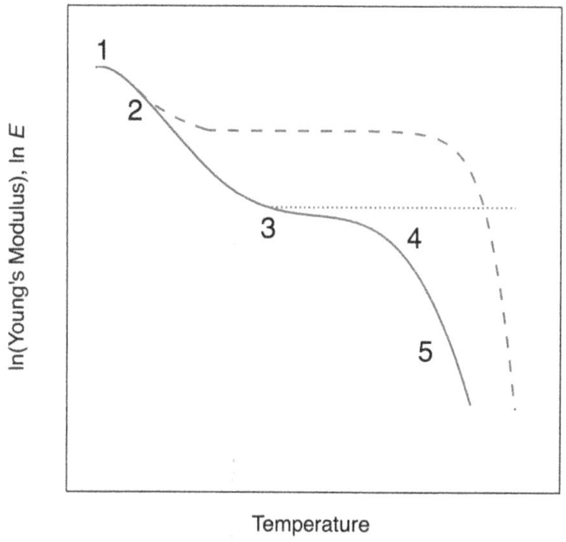

Figure 12.18 Five generic regions of viscoelastic behavior for linear, amorphous polymers. Crystallinity modifies the behavior as shown with the dashed line. Crosslinking introduces the changes represented by the dotted line. 1, Glassy. 2, Glass transition. 3, Rubbery plateau. 4, Transition to viscous flow. 5, Viscous flow.

n_2 moles of the two components with volume fractions v_1 and v_2. In *Flory*[5]*–Huggins theory*, the deviation of the enthalpy of mixing from ideal behavior is quantified by the dimensionless quantity χ_1 (the *polymer–solvent interaction parameter*) defined by

$$\chi_1 = \frac{\Delta_{\text{mix}} H^\circ_{\text{m}}}{RTn_1 v_2}. \tag{12.66}$$

Assuming that the entropy factor behaves ideally, the Gibbs energy of mixing is then

$$\Delta_{\text{mix}} G^\circ_{\text{m}} = RT(n_1 \ln v_1 + n_2 \ln v_2 + \chi_1 n_1 v_2). \tag{12.67}$$

For mixing to be spontaneous, $\Delta_{\text{mix}} G^\circ_{\text{m}} > 0$. Thus, if mixing is endothermic, which corresponds to $\chi_1 > 0$, then there will be a limited range of polymer solubility. The value of χ_1 depends on the specific polymer–solvent combination. Many polymers exhibit endothermic mixing. Polystyrene, for example has $\chi_1 = 0.37$ at 25 °C in toluene and 0.5 at 34 °C in cyclohexane. Exothermic mixing is observed for cellulose nitrate in n-propylacetate at 20 °C for which $\chi_1 = -0.38$. (More on polymer solutions can be found in Chapter 13, Section 9.)

All noncrystalline polymers exhibit a glass transition that meets the Ehrenfest criteria for a second-order phase transition. The temperature and pressure derivatives of both volume and entropy are discontinuous as a function of V or p, respectively, even though the volume and entropy remain a continuous function of V and p. There is a sense in which this phase transition is a kinetic phenomenon. The glass transition temperature is a function of the time scale of the experiment. Slower measurements with slower heating rates result in a lower value of T_g. One key parameter that determines when the glass transition occurs is the free volume available in the solid state. Free volume must be available (a vacancy is required) for the coordinated molecular motion that occurs above T_g. The molecule must have some place to go in order for it to move, and it cannot occupy the same space as another polymer molecule. The free volume as a fraction of the total volume of the solid increases as the temperature is increased. Current estimates are that the free volume must reach at least 2.5% for the glass transition to occur. The exact nature of the glass transition is still an open question. Free-volume, kinetic and thermodynamic theories have been proposed to describe it. All three achieve some level of success for the description of some aspects of the transition and its dependence on variables such as molar mass, crosslinking, and time scale of measurement. However, none of these theories fully explains the nature of the transition.

Fox and Flory used free-volume theory to develop an expression for the dependence of the glass transition temperature T_g on molar mass M. They found

$$T_g = T_{g\infty} - \frac{K}{(\alpha_R - \alpha_G)M} \tag{12.68}$$

where $T_{g\infty}$ is the glass temperature at infinite molar mass, K is a constant determined by the chemical composition of polymer, and α_R and α_G are the volumetric expansion coefficients in the rubbery and glass states, respectively. Equation (12.68) describes accurately the behavior of the particularly well-studied polystyrene system. That the glass temperature drops with decreasing molar mass has important ramifications for the synthesis of polymers.

At the beginning of an isothermal polymerization reaction, the polymer is in the liquid state. However, as polymerization proceeds, vitrification of the polymer occurs. In the glass state, molecular motion is essentially arrested, and therefore vitrification will also halt further polymerization in the regions that have turned glassy.

Such an effect is found in the emulsion polymerization of polystyrene, where the reaction is often run at about 80 °C. The monomer in this chain polymerization acts like a plasticizer mixed between the growing polymer chains. Once the system vitrifies, the reaction is precluded from proceeding to completion.

During a stepwise polymerization, the molar mass increases continuously. An example is epoxy polymerization in which gelation occurs. The polymer is simultaneously polymerizing and crosslinking during the synthesis. The temperature of the reaction is increased in stages to effect complete conversion. The temperature changes the rate of reaction as well as the phase character. The temperature is raised to initiate the reaction and the polymer is allowed to form a gel. As the reaction proceeds the gel vitrifies, bringing the reaction temporarily to a halt. The temperature is then brought above $T_{g\infty}$. This last curing step ensures that the reaction and crosslinking are completed.

12.10.1 Directed practice

Derive Eq. (12.67), assuming that the mole fraction is equal to the volume fraction.

SUMMARY OF IMPORTANT EQUATIONS

$(\partial\mu/\partial T)_p = -S_m$	The chemical potential changes with temperature as the negative of the molar entropy
$(\partial\mu/\partial p)_T = V_m$	The chemical potential changes with pressure as the molar volume
$\mu_g = \mu_g^\circ + RT\ln(p/p^\circ)$	Chemical potential of an ideal gas increase with increasing pressure
$\mu_l = \mu_l^\circ + pV_{m,l}$	Chemical potential of an incompressible liquid depends on pressure
$F = C + 2 - P$	Gibbs phase rule: degrees of freedom, F, number of phases P and the number of components C
$\dfrac{dp}{dT} = \dfrac{\Delta_{trs}S_m}{\Delta_{trs}V_m} = \dfrac{\Delta_{trs}H_m}{T\,\Delta_{trs}V_m}$	Clapeyron equation
$\ln\left(\dfrac{p_f}{p_i}\right) = \int_{T_i}^{T_f}\dfrac{\Delta_{vap}H_m}{R}\dfrac{dT}{T^2}$	Integrated form of Clausius–Clapeyron equation
$\ln\left(\dfrac{p_2}{p_1}\right) = \dfrac{\Delta_{vap}H_m}{R}\left(\dfrac{1}{T_1} - \dfrac{1}{T_2}\right)$	Integrated form of Clausius–Clapeyron equation assuming constant enthalpy change
$\log p = A - \dfrac{B}{T+C}$	Antoine equation
$RT\ln\dfrac{p}{p^*} = \dfrac{2\gamma V_m}{r}$ or $p = p^*\exp\left(\dfrac{2\gamma V_m}{RTr}\right)$	Kelvin equation
$h = \dfrac{2\gamma\cos\psi}{\rho g r}$	Capillary rise h in terms of the contact angle ψ

Exercises

12.1 Suggest a physical interpretation of the dependence of the chemical potential on the pressure. Start with an equation to explain the behavior.

12.2 Again, starting with an equation, suggest a physical interpretation of the dependence of the chemical potential on the temperature.

12.3 With the aid of a chemical potential versus temperature phase diagram, explain why graphite sublimes rather than melts when it is heated from 300 K to 3825 K at standard pressure. Also draw a p–T phase diagram that is compatible with this observation.

12.4 With the aid of a chemical potential versus temperature phase diagram, explain why CO_2 sublimes rather than melts when it is heated reversibly from 100 K to 195 K at standard pressure. Indicate the sublimation temperature T_{sp} on your diagram. (b) Explain how it is possible to get CO_2 to undergo a normal sequence of phase transitions from solid to liquid to gas as the temperature is increased by changing the pressure.

12.5 Phases A and B of a pure compound are in contact and there is no barrier to transport between them. If phase A is at a higher chemical potential than B, how will the system evolve?

12.6 Assuming that $\Delta_{vap}H_m$ is constant over a small temperature range near the normal boiling point T_b, derive an expression for the vapor pressure of a liquid.

12.7 At 1 bar pressure liquid bromine (Br_2) boils at 58.2 °C and at 9.3 °C, its vapor pressure is 0.1334 bar. Assuming ΔH° and ΔS° to be temperature-independent, calculate their values. Calculate the vapor pressure in bar and ΔG° at 25 °C.

12.8 The vapor pressure of naphthalene(s) is 4.92 Pa at 290 K and 182.9 Pa at 330 K. The enthalpy of vaporization is 43.2 kJ mol^{-1} and the normal boiling point is 491 K. The normal melting point is 353 K. Calculate $\Delta_{sub}H_m^\circ$ and $\Delta_{fus}H_m^\circ$ in kJ mol^{-1}, the vapor pressure at 370 K, and comment on their relative values of the enthalpies.

12.9 The vapor pressure of a liquid can be written in the empirical form known as the Antoine equation where a, b and c are constants determined from measurements of pressure in Pa and temperature in K

$$\ln p = a - \frac{b}{T + c}.$$

Starting with this equation, derive an equations that gives the value of $\Delta_{\text{vap}}H^\circ_{\text{m}}$ as a function of temperature. (b) For benzene $a = 20.767$, $b = 2.7738 \times 10^3$ and $c = -53.08$. Calculate the standard boiling point and $\Delta_{\text{vap}}H^\circ_{\text{m}}$ at 298 K.

12.10 For benzene, the Antoine equation parameters are $a = 20.767$, $b = 2.7738 \times 10^3$ and $c = -53.08$. Calculate the standard boiling point and $\Delta_{\text{vap}}H^\circ_{\text{m}}$ at 298 K.

12.11 The vapor pressure of CCl_4(s) is 270 Pa at 232 K and 1092 Pa at 250 K. The entropy of vaporization is 91.8 J K^{-1} mol^{-1} and the normal boiling point is 350 K. Calculate $\Delta_{\text{vap}}H^\circ_{\text{m}}$, $\Delta_{\text{sub}}H^\circ_{\text{m}}$, $\Delta_{\text{fus}}H^\circ_{\text{m}}$ in kJ mol^{-1} and comment on their relative values.

12.12 The enthalpy of vaporization of water is 40.65 kJ mol^{-1} at 373 K. The heat capacities of liquid and gaseous water are $C^\circ_{p,\text{m}}(l) = 74.57$ J K^{-1} mol^{-1} and $C^\circ_{p,\text{m}}(g) = 25.96$ J K^{-1} mol^{-1}. (a) Calculate the energy required to evaporate 145 g of H_2O at 100 °C and 25 °C. (b) Explain on the basis of intermolecular interactions the difference or lack thereof for the two answers in part (a).

12.13 Calculate the vapor pressure in Pa of Hg at 298 K if the enthalpy of vaporization at this temperature is 59.11 kJ mol^{-1} and the normal boiling point is 356.6 °C.

12.14 Discuss why it is difficult for clouds to form when the partial pressure is only slightly above the saturation vapor pressure of water. How might 100 nm sulfate or dust particulates influence this process?

12.15 Gold nanoparticles are placed on a surface of silicon at $T = 1250$ K. The Au atoms do not evaporate into the gas phase. Instead, they can detach from the nanoparticles and enter a mobile adsorbed phase on the silicon, which can be thought of a two-dimensional (2D) gas. The atoms in the 2D gas can also reattach to the nanoparticles. (a) Calculate the effective vapor pressure of the Au atoms for nanoparticles of radius 0.5, 1.0 and 2.0 nm. (b) As the sample is allowed to equilibrate at 1250 K, what happens to the size distribution of the particles, and why? The surface tension of Au is 1.15 J m^{-2}. Its equilibrium vapor pressure at 1250 K is 7.8×10^{-5} Pa. Take the density of Au to be 1.932×10^7 g m^{-3}.

12.16 Calculate the radius of curvature, and discuss capillary condensation in a conical pore that has surfaces with (i) $\psi = 90°$ and (ii) $\psi = 180°$.

12.17 Consider a material with cylindrical pores exposed to air at 25 °C with a humidity of 85%. Into pores of what size will water condense?

12.18 Calculate the effective pressure resulting from capillary forces and the critical film thickness for a porous silicon film with a porosity $\varepsilon = 0.90$ when dried in air after rinsing in water or ethanol. The mean pore diameter is $r_{\text{p}} =$ 5 nm. $\gamma_{\text{EtOH}} = 22.75$ mN m^{-1}, $\gamma_{\text{water}} = 71.99$ mN m^{-1}, $\gamma_{\text{Si}} = 1000$ mN m^{-1}, $E_{\text{Si}} = 1.62 \times 10^{11}$ N m^{-2}.

12.19 Calculate the solubility of polystyrene in cyclohexane at 34 °C.

Endnotes

1. Bellet, D. and Canham, L. (1998) *Adv. Mater.*, **10**, 487.
2. Volmer, M. (1929) *Z. Elektrochem.*, **35**, 555.
3. Liger-Belair, G., Conreux, A., Villaume, S., and Cilindre, C. (2013) *Food Research International*, **54**, 516; Liger-Belair, G. (2012) *Eur. Phys. J. - Special Topics*, **201**, 1.
4. The analogous behavior at the solid/liquid phase transition is represented by the formation of a *eutectic*, a mixture that melts like a pure solid. A eutectic also has the lowest melting point or any mixture of the components.
5. Paul J. Flory was awarded the Nobel Prize in Chemistry 1974 for his fundamental achievements, both theoretical and experimental, in the physical chemistry of the macromolecules.

Further reading

Adamson, A.W. and Gast, A.P. (1997) *Physical Chemistry of Surfaces*, 6th edition. John Wiley & Sons, New York.
Liger-Belair, G. (2012) The Physics Behind the Fizz in Champagne and Sparkling Wines. *Eur. Phys. J. - Special Topics*, **201**, 1.
Kolasinski, K.W. (2007) Solid Structure Formation During the Liquid/Solid Phase Transition. *Curr. Opin. Solid State Mater. Sci.*, **11**, 76.
Kolasinski, K.W. (2014) Bubbles: A review of their relationship to the formation of thin films and porous materials. *Mesoporous Biomater.*, **1**, 49.
Metiu, H. (2006) *Physical Chemistry: Thermodynamics*. Taylor & Francis, New York.
Poling, B.E., Prausnitz, J.M., and O'Connell, J.P. (2001) *The Properties of Gases and Liquids*, 5th edition. McGraw-Hill, New York.
Sperling, L.H. (2006) *Introduction to Physical Polymer Science*, 4th edition. John Wiley & Sons, Hoboken, NJ.

CHAPTER 13

Solutions and mixtures: Nonelectrolytes

PREVIEW OF IMPORTANT CONCEPTS

- A mixture is a combination of two or more substances in any proportion.
- In a solution, the solvent is clearly defined because the solute is dissolved at a concentration that is small compared to the concentration of the solvent.
- Nonelectrolytes dissolve molecularly without producing ions and do not conduct electricity.
- In an ideal solution, all molecular interactions are the same, all component molecules occupy the same volume, mixing is thermoneutral, entropy drives mixing, and all activity coefficients are equal to unity.
- Ideal solutions are miscible (mix in all proportions) at all temperatures.
- Real solutions may exhibit limited solubility or miscibility only over limited ranges of T.
- In real solutions: volumes are not additive, the enthalpy of mixing can be nonzero, entropy may or may not act ideally, and activity coefficients depend on composition.
- The enthalpy of mixing of relatively nonpolar organic molecules is usually positive.
- Activities rather than concentrations are directly related to the chemical potential, and they must be used to understand the properties of real solutions.
- At equilibrium the chemical potential of all species is equal, therefore solution-phase activities of volatile solutes can be determined by measurements in the gas phase.
- The presence of a solute lowers the chemical potential of the solvent compared to the pure liquid.
- Melting point depression, boiling point elevation and osmotic pressure are colligative properties that depend ideally only on the presence of a solute and not on the chemical identity of the solute.
- Osmometry can be used to measure the molar mass of polymers and biomolecules up to values of 500 kDa.

In Chapter 3, we began a discussion of mixtures and solutions of fluids. Many of the points we develop below will be applicable to mixtures in any phase. However, our emphasis here will be on liquid-phase mixtures of nonelectrolytes. In the treatment of equilibria we have seen that activities – not concentrations – are the fundamental quantities that are required to calculate the Gibbs energy (and therefore equilibrium constants, enthalpy, entropy and most other properties). That is especially true in the liquid phase because molecules are in close contact and it is impossible to ignore intermolecular interactions. In this chapter we will study in much more depth the concepts of activity, activity coefficients, and how they are measured. But let us begin with a definition of the ideal case. Note that we will concentrate on the simplest type of mixture, a binary mixture. A binary system has two components, a unary system has one component, a ternary has three components, a quaternary has four components, and so forth.

We will investigate the dependence of mixing parameters for a wide range of systems on a number of variables. Which variables are most important will depend on the system. Temperature and composition are two of the most important and they must always be specified. For liquids mixing with liquids, the pressure dependence of solubilities is negligible up to at least 100 atm or so. For gas/liquid solutions on the other hand, the pressure – or, more accurately, the partial pressure – of the gas is one of the most important parameters for determining the solubility. However, as we approach the critical point where the line between liquid and gas blurs and eventually disappears completely, pressure dependence becomes quite important.

Physical Chemistry: How Chemistry Works, First Edition. Kurt W. Kolasinski.

Companion Website: www.wiley.com/go/kolasinski/physicalchemistry

13.1 Ideal solution and the standard state

Recall the definition of an ideal gas:

- No intermolecular interactions.
- Molecules have no volume.
- All gases obey the same equation of state: $pV = nRT$.

Note the similarities and difference to the properties of an ideal solution (in any phase):

- Intermolecular interactions are the same for all molecules, $(\Delta_{\text{mix}} H^\circ)_{p,T} = 0$.
- Molecules have a finite volume independent of composition, $(\Delta_{\text{mix}} V)_{p,T} = 0$.
- The solvent follows Raoult's law and the solute follows Henry's law.
- The activity coefficient for all species is unity, that is, $\gamma = 1$.
- All mixtures are miscible (mix at all concentrations) for all temperatures.
- Mixing is driven only by a positive entropy change upon mixing, $\Delta_{\text{mix}} S^\circ > 0$.

We will explore the implications of this model and deviations from it throughout this chapter. As we shall see below, all solutions act as ideal dilute solutions when dilute enough (which may be very dilute indeed for solutions with dissimilar intermolecular interactions) but, in general, solutions in the liquid phase are non-ideal. Non-ideal solutions are called *real solutions*.

It is always important to construct an appropriate *standard state*. For a gas, we have chosen a standard state pressure of 1 bar or in SI units, $p^\circ = 100$ kPa. Ideal solutions are miscible; that is, they form mixtures in which the composition may be varied from a mole fraction of 0 to 1 for both components. Therefore, we construct the standard state as the pure liquid. For simplicity, let us concentrate on binary solutions. Neither component can truly be considered a solute or solvent since the roles are reversed on either side of $x = 0.5$, where x is the mole fraction. Thus, in a binary mixture of components 1 and 2, the standard state mole fraction of component 1 is $x_1^\circ = 1$, and similarly for component 2. For solids dissolved in a liquid, it will make more sense to set the standard state equal to a molality of $b^\circ = 1$ mol kg^{-1} or a molar concentration of $c^\circ = 1$ mol l^{-1}. We will discuss the implications of changing the choice of standard state in the next chapter. Below, we will simplify our notation when using mole fraction. That is, there is no need to divide through by the standard state mole fraction of 1 since not only is this value unity, it is also already dimensionless.

13.2 Partial molar volume

In ideal solutions, volumes are additive. It may surprise you that for most real solutions this is not the case. That is, adding 50 ml of one liquid to 50 ml of another does not lead to a mixture of 100 ml final volume. To account for this we introduce the concept of *partial molar volume*. The partial molar volume is the change in volume upon addition of component 2 to component 1, holding all other variables – that is, p, T and moles of 1 – constant. The partial molar volume is not constant but depends on the solution composition. Taking V as the volume of the mixture, we write the partial molar volume of component j as

$$V_j = (\partial V/\partial n_j)_{p,T,n_{i\neq j}}. \tag{13.1}$$

Equation (13.1) reflects the recommended IUPAC notation for partial molar volume. However, this notation is absolutely treacherous, as it implies the use of the same symbol for the partial molar volume of j and the volume of j. Hence, we adopt the notation of Berry, Rice and Ross and follow their derivations closely. In general, a partial molar quantity is defined as

$$\bar{z}_j = (\partial Z/\partial n_j)_{p,T,n_{i\neq j}}. \tag{13.2}$$

Using this notation and definition to denote the partial molar volume of component j as $\bar{v}_j = (\partial V/\partial n_j)_{p,T,n_{i\neq j}}$, we write the total change in volume for infinitesimal changes in molar composition of a binary mixture of 1 and 2 as

$$\mathrm{d}V = \left(\frac{\partial V}{\partial n_1}\right)_{p,T,n_2} \mathrm{d}n_1 + \left(\frac{\partial V}{\partial n_2}\right)_{p,T,n_1} \mathrm{d}n_2 = \bar{v}_1\,\mathrm{d}n_1 + \bar{v}_2\,\mathrm{d}n_2, \tag{13.3}$$

Thus,

$$V = n_1\bar{v}_1 + n_2\bar{v}_2. \tag{13.4}$$

The partial molar volume of component 1 depends on the composition of solution to which it is added because the effective volume of a given number of component 1 molecules depends on the identity of the molecules that surround them. As the concentration changes, the intermolecular forces acting on the molecules change. Therefore, the properties of real solutions and their components change as functions of mole fractions and chemical identities of the components. Molar volume is always positive; however, partial molar quantities need not be – that is, the addition of more liquid can lead to a *contraction* of the total volume. This is often true for small additions of a salt to an aqueous solution because the resulting ions must be strongly solvated. Even nonelectrolyte solutions – especially strongly interacting, hydrogen-bonding components – can exhibit complex partial molar volume changes with composition.

13.3 Partial molar Gibbs energy = chemical potential

13.3.1 Chemical potential of a one-component system

The *molar Gibbs energy* G_m and the *chemical potential* μ are synonymous for a one-component system. Thus, for any pure substance

$$\mu = (\partial G/\partial n)_{T,p} = G_m. \tag{13.5}$$

We substitute for G_m in the fundamental equation and express this in differential form as

$$d\mu = -S_m\, dT + V_m\, dp. \tag{13.6}$$

Since these are exact differentials, we can express how the chemical potential changes with temperature at constant pressure ($dp = 0$) according to the partial derivative

$$(\partial\mu/\partial T)_p = -S_m \tag{13.7}$$

and how the chemical potential changes with pressure at constant temperature ($dT = 0$) according to the partial derivative

$$(\partial\mu/\partial p)_T = V_m. \tag{13.8}$$

Since the molar entropy and molar volume of a gas are much greater than for the other two phases, it is clear from Eqs (13.7) and (13.8) that the chemical potential of a gas responds to changes of temperature and pressure more sensitively than for liquids and solids. The importance of this can be observed directly in a plot of chemical potential versus temperature, as shown in Fig. 13.1. The partial derivative represents the slope of the curve. Each phase has its own characteristic dependence of molar entropy on temperature. Nonetheless, we know that the molar entropy of a gas is greater than that of a liquid at the same temperature. Similarly, the molar entropy of a liquid is always greater than that of a solid of the same material at the same temperature. Stated mathematically, at any given temperature $S_{m,g} > S_{m,l} > S_{m,s}$.

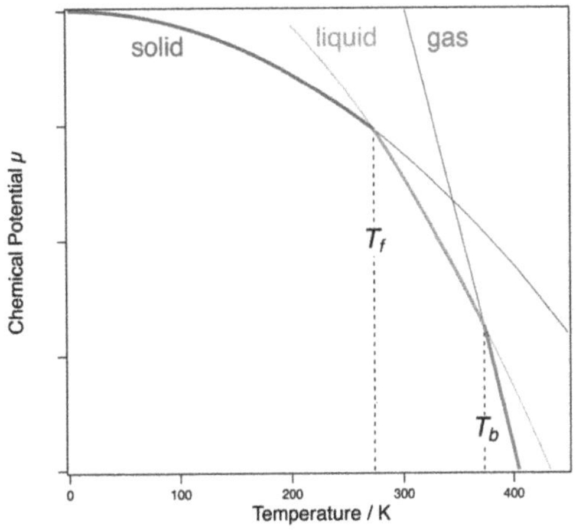

Figure 13.1 μ–T phase diagram for a material with properties approximately those of water. The chemical potential versus temperature of system at equilibrium is described by the curve with the lowest chemical potential.

The molar entropy of each phase is a function of temperature. In accord with Eq. (13.7) the chemical potential versus temperature of each phase is represented by a curve, as shown in Fig. 13.1. A system is at equilibrium when its chemical potential is a minimum. The lowest chemical potential phase is the most stable at any given temperature and pressure. Figure 13.1 represents the chemical potential versus temperature at the standard pressure of $p° = 1$ bar. The curve that represents the chemical potential of the solid starts with a slope of zero at absolute zero. All perfect crystals must behave this way according to the Third Law of Thermodynamics. The slope gradually becomes more negative. At the standard melting point T_f, the chemical potentials of the solid and liquid phases are equal, $\mu(s) = \mu(l)$. At this point the curve for the chemical potential of the solid intersects the curve for the chemical potential of the liquid. We do not need to know the absolute value of either slope; however, we do know that the slope of the liquid curve is greater than the slope of the solid curve. Similarly, at the standard boiling point T_b, the chemical potentials of the liquid and gas phases are equal, $\mu(l) = \mu(g)$. At this point the curve for the chemical potential of the liquid intersects the curve for the chemical potential of the gas, and the slope of the gas curve is greater than the slope of the liquid curve. In the previous chapter we developed the implications of these curves in more detail as we discussed the nature of phase transitions. Here, we focus on mixtures of different chemical components.

13.3.2 Chemical potential of a multicomponent system

Partial molar quantities can be defined using Eq. (13.2) for all thermodynamic properties of a mixture. One of particular importance is the partial *molar Gibbs energy*, which is equivalent to the *chemical potential* of that species. In a mixture with several components, the chemical potential of component j is

$$\mu_j = (\partial G/\partial n_j)_{p,T,n_{i\neq j}}. \tag{13.9}$$

The chemical potential allows us to calculate the total Gibbs energy of an arbitrary mixture with r components

$$G = \sum_{j=1}^{r} n_j \mu_j. \tag{13.10}$$

By definition, the Gibbs energy is related to internal energy, entropy, T, p and V by

$$G = U - TS + pV \tag{13.11}$$

For an infinitesimal change in composition of the mixture, Eq. (13.10) implies

$$\mathrm{d}G = \sum_{j=1}^{r} \mu_j \,\mathrm{d}n_j + \sum_{j=1}^{r} n_j \,\mathrm{d}\mu_j, \tag{13.12}$$

whereas Eq. (13.11) demands

$$\mathrm{d}G = \mathrm{d}U - T\,\mathrm{d}S - S\,\mathrm{d}T + p\,\mathrm{d}V + V\,\mathrm{d}p. \tag{13.13}$$

Equating these last two equations and solving for $\mathrm{d}U$ yields

$$T\,\mathrm{d}S + S\,\mathrm{d}T - p\,\mathrm{d}V - V\,\mathrm{d}p + \sum_{j=1}^{r} \mu_j \,\mathrm{d}n_j + \sum_{j=1}^{r} n_j \,\mathrm{d}\mu_j = \mathrm{d}U. \tag{13.14}$$

However, the internal energy changes as

$$\mathrm{d}U = T\,\mathrm{d}S - p\,\mathrm{d}V + \sum_{j=1}^{r} \mu_j \,\mathrm{d}n_j. \tag{13.15}$$

Substituting into Eq. (13.14) for $\mathrm{d}U$ and cancelling like terms of both sides gives

$$S\,\mathrm{d}T - V\,\mathrm{d}p + \sum_{j=1}^{r} n_j \,\mathrm{d}\mu_j = 0. \tag{13.16}$$

This is known as the *Gibbs–Duhem equation*. Therefore for constant p and T, the Gibbs–Duhem equation simplifies to

$$\sum_{j=1}^{r} n_j \,\mathrm{d}\mu_j = 0 \tag{13.17}$$

or more specifically for a binary solution

$$n_1\,\mathrm{d}\mu_1 + n_2\,\mathrm{d}\mu_2 = 0. \tag{13.18}$$

The Gibbs–Duhem equation shows us that the intensive variables T, p and μ_j are coupled. It places limits on their variability. At constant p and T, the chemical potentials of the components of a binary mixture cannot change independently of one another, only in concert, according to

$$n_1\,\mathrm{d}\mu_1 = -n_2\,\mathrm{d}\mu_2. \tag{13.19}$$

Just as we saw for Eq. (4.23), because the summation in Eq. (13.17) is equal to a constant, there are $r - 1$ independent terms.

It follows from Eq. (13.15) and the usual definitions that

$$\mathrm{d}H = T\,\mathrm{d}S + V\,\mathrm{d}p + \sum_{j=1}^{r} \mu_j\,\mathrm{d}n_j. \tag{13.20}$$

$$\mathrm{d}A = S\,\mathrm{d}T - p\,\mathrm{d}V + \sum_{j=1}^{r} \mu_j\,\mathrm{d}n_j. \tag{13.21}$$

$$\mathrm{d}G = -S\,\mathrm{d}T + V\,\mathrm{d}p + \sum_{j=1}^{r} \mu_j\,\mathrm{d}n_j. \tag{13.22}$$

Let us reflect on these last three equations to obtain a better perspective of the fundamental importance of the chemical potential. Equations (13.20) to (13.22) tell us that the changes in enthalpy H, Helmholtz energy A, and Gibbs energy G as a function of composition are related to the chemical potential according to

$$\mu_j = \left(\frac{\partial H}{\partial n_j}\right)_{p,S,n_{j\neq i}} = \left(\frac{\partial A}{\partial n_j}\right)_{V,T,n_{j\neq i}} = \left(\frac{\partial G}{\partial n_j}\right)_{p,T,n_{j\neq i}}. \tag{13.23}$$

Thus, using the general definition of partial molar quantities in Eq. (13.2), we can also write the chemical potential in terms of the fundamental relationship of H and S

$$\mu_j = \left[\frac{\partial}{\partial n_j}(H - T\,S)\right]_{p,T,n_{j\neq i}} = \bar{h}_j - T\,\bar{s}_j \tag{13.24}$$

where $\bar{h}_j$ and $\bar{s}_j$ are the partial molar enthalpy and partial molar entropy, respectively. We can now investigate the equilibrium properties of multicomponent systems.

13.4 The chemical potential of a mixture and $\Delta_{\mathrm{mix}}G$

Equation (13.19) shows us that the addition of n_1 moles of component 1 to a binary mixture requires that

$$n_1\left(\frac{\partial \mu_1}{\partial n_1}\right)_{p,T,n_2} = -n_2\left(\frac{\partial \mu_2}{\partial n_1}\right)_{p,T,n_2} \tag{13.25}$$

Since both n_1 and n_2 are positive numbers, the derivatives must have opposite signs. The change in Gibbs energy upon addition of n_1 moles is positive $\Delta G > 0$; thus, the derivatives have signs consistent with

$$\left(\frac{\partial \mu_1}{\partial n_1}\right)_{p,T,n_2} > 0 \tag{13.26}$$

$$\left(\frac{\partial \mu_2}{\partial n_1}\right)_{p,T,n_2} < 0. \tag{13.27}$$

Experimentally, it is found that the right-hand side of Eq. (13.25) approaches a constant nonzero value represented by $-B$. Therefore, we substitute this value into Eq. (13.25) to obtain

$$\left(\frac{\partial \mu_1}{\partial n_1}\right)_{p,T,n_2} = \frac{B}{n_1}. \tag{13.28}$$

Integration yields

$$\mu_1 = B\ln n_1 + f(p, T, n_2). \tag{13.29}$$

Here, the integration 'constant' is constant under the chosen conditions of constant p, T and moles of component 2. However, in general, it is allowed to be a function of these variables. To set the value of this constant, recall that the chemical potential is an intensive property that cannot depend on the amount of 1 or 2 but only on their ratio. Furthermore, the chemical potential of the pure solution must correspond to $\mu^\circ(p,T)$, the standard state chemical potential at the chosen values of p and T. This must be so because we have chosen the standard state to be the same as the pure substance. That is, if we introduce the mole fraction $x_1 = n_1/(n_1 + n_2)$, Eq. (13.29) must behave as

$$\mu_1 = B\ln\left(\frac{n_1}{n_1 + n_2}\right) + f(p, T, n_2) = B\ln x_1 + f(p, T, n_2) \tag{13.30}$$

It must also have a limiting behavior of the form

$$\lim_{x_1 \to 1} \mu_1 = \mu^\circ(p, T). \tag{13.31}$$

Therefore, the chemical potential as a function of composition is given by

$$\mu_1 = \mu^\circ(p, T) + B\ln x_1. \tag{13.32}$$

Now it remains to set the value of the constant B. The Gibbs energy of mixing is obtained by the difference of the final mixed state Gibbs energy with the Gibbs energy of the initially unmixed state

$$\Delta G = G_f - G_i \Rightarrow \Delta_{\text{mix}} G = G\,(\text{mixed}) - G\,(\text{not mixed}) = G - G^\circ. \tag{13.33}$$

From Eq. (13.6), we write for the Gibbs energy of mixing

$$\Delta_{\text{mix}} G = \sum_{j=1}^{r} n_j \mu_j - \sum_{j=1}^{r} n_j \mu_j^\circ = \sum_{j=1}^{r} n_j \left(\mu_j - \mu_j^\circ\right). \tag{13.34}$$

Here, we consider the mixing of an ideal solution at constant p and T. Recall that in an ideal solution neither the enthalpy nor volume change upon mixing. Thus, with the aid of the thermodynamic definition of these two variables, we have

$$\Delta_{\text{mix}} H = H - H^\circ = -T^2\left[\frac{\partial}{\partial T}\left(\frac{\Delta_{\text{mix}} G}{T}\right)\right]_{p,n} = 0, \tag{13.35}$$

$$\Delta_{\text{mix}} V = V - V^\circ = \left(\frac{\partial \Delta_{\text{mix}} G}{\partial p}\right)_{T,n} = 0, \tag{13.36}$$

and

$$\Delta_{\text{mix}} G = \Delta_{\text{mix}} H - T\,\Delta_{\text{mix}} S = -T\,\Delta_{\text{mix}} S. \tag{13.37}$$

Equations (13.35)–(13.37) require that the difference of chemical potentials has the form

$$\mu_j - \mu_j^\circ = T\,f_j(n_1, \dots, n_r), \tag{13.38}$$

where the functions $f_j\left(n_1, \dots, n_r\right)$ are functions of composition alone. Looking at Eq. (13.37) and substituting from Eqs (13.34) and (13.18), we obtain

$$\Delta_{\text{mix}} S = -\sum_{j=1}^{r} n_j f_j(n_1, \dots n_r). \tag{13.39}$$

This equation for the entropy of mixing is completely general. We have not assumed any particular phase, only that the solution behaves ideally. Because the functions $f_j(n_1, \dots, n_r)$ do not depend on p and T, the equation must be valid for all ideal solutions, regardless of the values of p and T. At sufficiently high T and low p, the system will behave in the ideal gas limit. As we will confirm below, and have shown before, the entropy of mixing of ideal gases is

$$\Delta_{\text{mix}} S = -R\sum_{j=1}^{r} n_j \ln x_j. \tag{13.40}$$

Therefore, Eqs (13.39) and (13.40) can only be consistent if

$$f_j(n_1, \ldots, n_r) = R \ln x_j, \tag{13.41}$$

which means that the chemical potential of a component in an ideal solution as a function of composition is given by

$$\mu_j = \mu_j^\circ + RT \ln x_j, \tag{13.42}$$

13.5 Activity

In the most general terms, Gibbs energy is related to activity. The *activity a* is related to concentration by the *activity coefficient* γ. Activity can be thought of as the effective concentration of a species. The corresponding activity coefficient is a measure of how close to ideal that species is acting. The type of concentration term that we use depends on the system at hand. It is related to the choice of standard state. In this example, and much of this chapter, we are going to use concentration in terms of mole fraction x. Take A to be the solvent and B the solute:

$$a_A = \gamma_A x_A \tag{13.43}$$

$$a_B = \gamma_B x_B. \tag{13.44}$$

The solute and solvent each have different values for activity coefficients. Since we have a binary solution, $x_A + x_B = 1$. The chemical potential of a liquid A (the solvent in this case) is related to the activity by

$$\mu_A(l) = \mu_A^*(l) + RT \ln a_A. \tag{13.45}$$

As stated above, the standard state is chosen to be the pure substance. Hence, $\mu_A^*(l)$ is the chemical potential of the pure liquid. Substituting from Eq. (13.43) for the activity, we find that the chemical potential of any *real solution* is

$$\mu_A(l) = \mu_A^*(l) + RT \ln \gamma_A x_A \tag{13.46}$$

Real Solution:

$$\mu_A(l) = \underline{\mu_A^*(l) + RT \ln x_A} + \underline{\underline{RT \ln \gamma_A}} \tag{13.47}$$

By making the identification that in an *ideal solution*

$$\gamma_A = 1 \tag{13.48}$$

and therefore

$$a_A = x_A \tag{13.49}$$

then the chemical potential of the solvent in an ideal solution is given by

Ideal Solution:

$$\mu_A(l) = \mu_A^*(l) + RT \ln x_A. \tag{13.50}$$

This, of course, corresponds to what we derived in Eq. (13.42). *The deviation from ideal behavior is accounted for by the RT* ln γ_A *term* in Eq. (13.47). The singly underlined terms represent the ideal contribution to the chemical potential.

All solutions behave more ideally as they become more dilute, that is $a_A \to x_A$ as $x_A \to 1$. Since $x_A < 1$, $\ln x_A < 0$. In other words, according to Eq. (13.50), *the presence of the solute stabilizes the solvent in a solution compared to the pure solvent.*

By analogy with the above for solvents, we perform a similar calculation for the solute

$$\mu_B(l) = \mu_B^\circ(l) + RT \ln a_B \tag{13.51}$$

The value of $\mu_B^\circ(l)$ depends on the standard state we chose for the solute. If instead of mole fraction, we work with molality, then we assume a standard state at molality $b^\circ = 1$ mol kg^{-1}, such that

$$a_B = \gamma_B (b_B / b^\circ). \tag{13.52}$$

Ideal behavior of the solute, as defined by $\gamma_B \to 1$, is observed when the solute is very dilute, or equivalently as $b_B \to 0$. The choice of a *standard state* in terms of molality or molarity is common for electrolytes in solution. It may also be used for immiscible nonelectrolytes. When the standard state is chosen in this manner, we say that the activity coefficient is referenced to Henry's law. As we shall see below, Henry's law describes the behavior we expect of a solute in solution. Specifically, it describes the partial pressure of a volatile solute above a solution.

For miscible nonelectrolytes, it is more common to take the standard state to be the pure substance. In this case the activity coefficient of B approaches 1 when the solution approaches pure B. When the activity coefficient is chosen to be referenced in this manner, we say that the activity coefficient is referenced to Raoult's law. As we shall see below, Raoult's law defines the behavior we expect of the vapor pressure of the majority component above a solution. IUPAC now recommends the use of the symbol f rather than γ when this standard state is chosen.

Officially, while the activity is equal to the product of activity coefficient and concentration, it is *defined* in terms of the chemical potential,

$$a_{\rm B} = e^{(\mu_{\rm B}-\mu_{\rm B}^{\circ})/RT}. \tag{13.53}$$

The term *relative activity* is sometimes also used to differentiate $a_{\rm B}$ from the *absolute activity* $\lambda_{\rm B} = e^{\mu_{\rm B}/RT}$.

13.5.1 Fugacity

Deviations from ideal behavior of real gases, which arise from intermolecular forces and finite molecular volume, are dealt with by introducing the concept of *fugacity*. Fugacity is analogous to activity. It is represented by either f or $\tilde{p}$, and is related to partial pressure according to

$$\tilde{p} = \phi p. \tag{13.54}$$

ϕ is the *fugacity coefficient*. It depends on the identity of the gas, p and T. Unlike activity, which is dimensionless, fugacity has the same units as pressure. All gases behave more ideally as the pressure drops, that is $\phi \to 1$ as $p \to 0$. Formally, we can define the fugacity of component B with partial pressure $p_{\rm B}$ according to

$$\tilde{p}_{\rm B} \equiv p_{\rm B}\ \exp\left[\frac{1}{RT}\int_0^{p_{\rm B}}\left(V_{\rm m} - \frac{RT}{p'}\right)\,{\rm d}p'\right] = \lambda_{\rm B}\lim_{p\to 0}\left(\frac{p_{\rm B}}{\lambda_{\rm B}}\right)_T, \tag{13.55}$$

where $V_{\rm m}$ is the molar volume and $\lambda_{\rm B}$ is the absolute activity. Substituting into the equation for the chemical potential of a gas, we obtain for a real gas

$$\mu({\rm g}) = \mu^{\circ}({\rm g}) + RT\ln\left(\tilde{p}/p^{\circ}\right). \tag{13.56}$$

$$\mu({\rm g}) = \underline{\mu^{\circ}({\rm g}) + RT\ln\left(p/p^{\circ}\right)} + \underline{\underline{RT\ln\phi}}. \tag{13.57}$$

Note the similarity to the expression derived for solvent activity. The chemical potential can be written as the sum of an ideal term (single underline) and a term involving the fugacity coefficient that embodies the deviations from ideal behavior (double underline).

13.6 Measurement of activity

At equilibrium the chemical potential must be the same everywhere in the system, as shown in Fig. 13.2. Therefore, the chemical potentials of two phases in contact must be equal, and for a liquid in contact with its vapor

$$\mu_{\rm A}^{*}(l) = \mu_{\rm A}^{*}(g) \tag{13.58}$$

The * indicates a pure substance. We know how to write the chemical potential for the ideal gas phase

$$\mu_{\rm A}^{*}(g) = \mu_{\rm A}^{\circ} + RT\ln(p_{\rm A}^{*}/p^{\circ}). \tag{13.59}$$

We use this to help us to calculate the chemical potential of the liquid by combining Eqs (13.58) and (13.59). For a pure liquid in contact with its vapor (vapor acts ideally)

pure liquid

$$\mu_{\rm A}^{*}(l) = \mu_{\rm A}^{\circ} + RT\ln(p_{\rm A}^{*}/p^{\circ}) \tag{13.60}$$

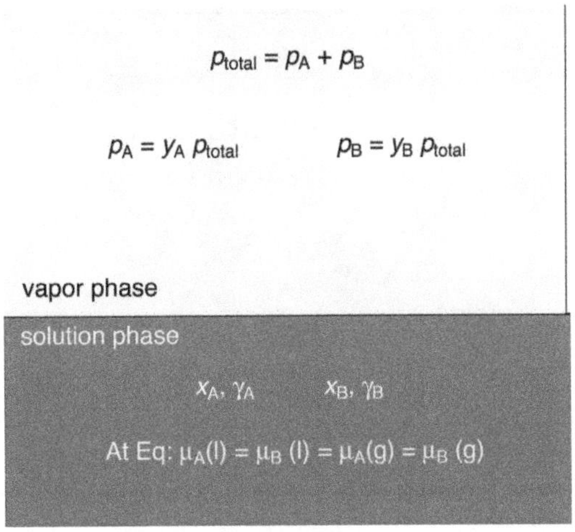

Figure 13.2 When a liquid solution equilibrates with the vapor phase, the concentrations (here in terms of the mole fractions of the two components in solution x_A and x_B as opposed to y_A and y_B in the gas phase) adjust until the chemical potential of every species in every phase is the same. The partial pressures of A and B sum to the total pressure, $p_{total} = p_A + p_B$.

where $\mu_A^*(l)$ is the chemical potential of pure liquid, μ_A° is the chemical potential of standard state, p_A^* is the equilibrium vapor pressure of pure liquid (the saturation vapor pressure), and p° is the pressure of the standard state.

For a solution, we replace $\mu_A^*(l)$ [the value for the pure solvent] with $\mu_A(l)$ [the value for the solvent A in the solution] to obtain

solvent

$$\mu_A(l) = \mu_A^\circ + RT\ln(p_A/p^\circ) \tag{13.61}$$

where p_A is the equilibrium vapor pressure of A above the solution. We use Eq. (13.60) to find an expression for μ_A° and then substitute this into Eq. (13.61) to derive an expression for the chemical potential of the solvent in terms of the solvent's equilibrium vapor pressure above the pure solvent and above the solution

$$\mu_A^\circ = \mu_A^*(l) - RT\ln\left(p_A^*/p^\circ\right) = \mu_A^*(l) + RT\ln\left(p^\circ/p_A^*\right)$$
$$\mu_A(l) = \mu_A^*(l) + RT\ln\left(p^\circ/p_A^*\right) + RT\ln\left(p_A/p^\circ\right)$$
$$\mu_A(l) = \mu_A^*(l) + RT\ln\left(p_A/p_A^*\right) \tag{13.62}$$

If we compare Eq. (13.62) to Eq. (13.45), we see that the ratio $p_A\ /\ p_A^*$ is related to the activity of A. We can *measure the activity* of A in the solution by *measuring the vapor pressure of A above the solution*

Real solution

$$a_A = p_A/p_A^*. \tag{13.63}$$

Furthermore, since we know that $a_A = x_A$ in an ideal solution

Ideal solution

$$x_A = p_A/p_A^*. \tag{13.64}$$

13.6.1 Raoult's law

From Eq. (13.64) we expect an ideal solution to behave such that the vapor pressure of the solvent above the solution is given by

Ideal

$$p_A = x_A\, p_A^* \tag{13.65}$$

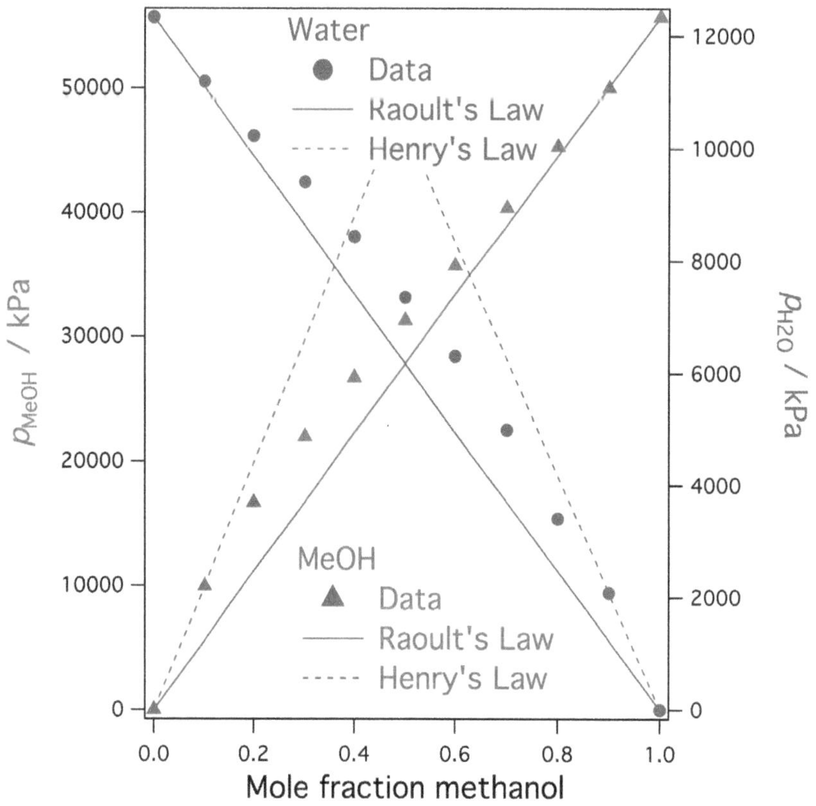

Figure 13.3 Mixtures of methanol and water exhibit positive deviations from Raoult's law. Both components act regularly in that they follow Raoult's law when they are close to pure, and Henry's law as they approach infinite dilution. Data taken from Poling, B.E., Prausnitz, J.M., and O'Connell, J.P. (2001) *The Properties of Gases and Liquids*, 5th edition. McGraw-Hill, New York.

This was first measured by François Raoult and is known as *Raoult's Law*. In words, this law is stated as:

The ratio of the partial pressure of each component in a mixture to its vapor pressure as a pure liquid is a constant and the constant is equal to the mole fraction.

All *solvents* follow Raoult's law in the limit of dilute solutions, that is, as $x_B \to 0$ or equivalently as $x_A \to 1$. All solutions should act ideally at infinite dilution. Therefore, Eq. (13.65) defines the ideal behavior of a solvent in a mixture.

For a real solution and non-ideal gas in the vapor phase, Raoult's law is written more generally as

$$\tilde{p}_A = f_A x_A \tilde{p}_A^* \tag{13.66}$$

where $\tilde{p}$ is fugacity and f_A is the activity coefficient referenced to a standard state of the pure solvent.

13.6.1.1 Directed practice

In an ideal solution the vapor pressure of A and B are both the same. Why is this so and why is it not true for a real solution?

13.6.2 Henry's law

Now look at the behavior of the solute exhibited in Fig. 13.3. With both the vapor and solution acting ideally, the vapor pressure of a volatile *solute* is proportional to the mole fraction of the solute

$$p_B = k_{H,B} x_B \tag{13.67}$$

This was first measured by William Henry and is known as *Henry's law*. It holds for dilute solutions. The Henry's law constant $k_{H,B}$ is not the vapor pressure of pure B but some empirically derived constant. The constant differs from the vapor pressure of pure B because B is surrounded by molecules of A in the solution, while it is surrounded by molecules of B in the pure solute. Since the A–B intermolecular interactions are not the same as the B–B interactions, the Henry's law constant is not the same as the equilibrium vapor pressure of the solute.

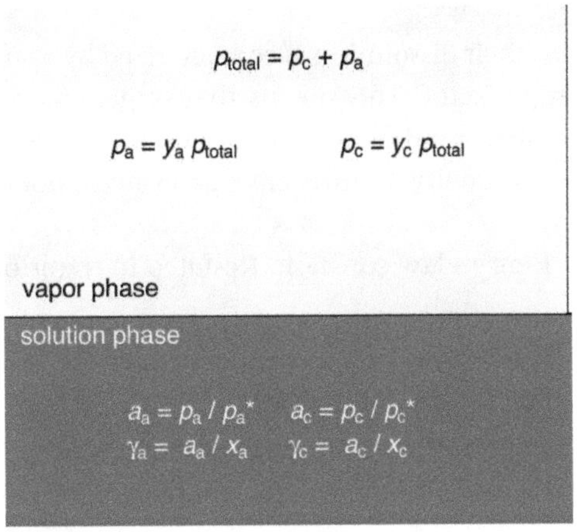

Figure 13.4 The vapor phase total pressure is made up of partial pressure contributions from both volatile components of a binary mixture – chloroform (c) and acetone (a) in this case. The solution-phase mole fractions are denoted with x, and the gas-phase mole fractions with y.

13.6.2.1 Example

A solution of chloroform in acetone is made in which the mole fraction of acetone is 0.713. The total vapor pressure is 29.40 kPa at 28.15 °C. The mole fraction of acetone in the vapor is 0.818. The vapor pressure of pure chloroform at this temperature is 29.57 kPa. Assuming the vapor behaves ideally (but not the solution), calculate the activity and the activity coefficient of chloroform.

The composition of the gas phase is different than the composition of the solution because the components have different vapor pressures, as made clear in Fig. 13.4.

The total pressure p_{total} is made up of the contributions from the partial pressure of chloroform p_c and the partial pressure of acetone p_a. Using Dalton's law of partial pressures

$$p_c = y_c p_{total} = (1 - 0.818)\,29.40\,\text{kPa} = 5.350\,\text{kPa},$$

where y_c is the chloroform mole fraction in the gas phase, as opposed to the solution, which is denoted x_c, as shown in Fig. 13.4. The activity of chloroform is the partial pressure above the solution divided by the vapor pressure above chloroform (the normal equilibrium vapor pressure)

$$a_c = p_c/p_c^* = 5.350/29.57 = 0.181$$

The activity coefficient of chloroform is

$$\gamma_c = a_c/x_c = 0.181/\,(1 - 0.713) = 0.630.$$

13.6.2.2 Example

The mole fraction of a *nonvolatile* solute dissolved in water is 0.010. If the equilibrium vapor pressure of pure water at 293 K is 2.339 kPa and that of the solution is 2.269 kPa, calculate the activity and activity coefficient of water. Is the water in this solution acting more or less ideally than is the chloroform in the solution prepared in Example 13.6.2.1?

The activity of water is

$$a_{water} = p_{water}/p^*_{water} = 2.269/2.339 = 0.9701$$

The activity coefficient of water is

$$\gamma_{water} = a_{water}/x_{water} = 0.9701/\,(1 - 0.010) = 0.980$$

The activity coefficient quantifies the deviation from ideality. The further the value is from 1, the greater the deviation from ideal behavior. Because the activity coefficient of water is closer to 1 than is the activity coefficient of chloroform, the water is acting more ideally.

13.6.3 Solubility of gases in liquids

For gases that do not react with a solvent, their dissolution is characterized by a small enthalpy of mixing. Their solubility is then largely determined by the entropy factor. This means that in general their solubility *decreases* with increasing temperature, just the opposite of the familiar trend of most solids and liquids because entropy is higher in the gas phase rather than the solution phase. Thus, the solubility of unreactive gases near room temperature is close to a mole fraction of order 10^{-5}.

The solubility of a gas is related to its Henry's law constant. Restated in terms of a chemical equation,

$$\mathrm{B}(\mathrm{sat}\ l) \rightleftharpoons \mathrm{B}(\mathrm{g}),$$

we see that the Henry's law constant (when divided by the standard state pressure p°) is the equilibrium constant for the contact of a liquid saturated with B, denoted B(sat l) above, with gaseous B. The opposite reaction describes the solvation of gas B in the liquid solvent

$$\mathrm{B}(\mathrm{g}) \rightleftharpoons \mathrm{B}(\mathrm{sat}\ l),$$

which has an equilibrium constant

$$K_{x,\mathrm{B}} = \frac{x_\mathrm{B}}{p_B/p^\circ}. \tag{13.68}$$

$K_{x,\mathrm{B}}$ is called the *solubility constant* and it is related to the reciprocal of the *Henry's law constant* $k_{\mathrm{H,B}}$ by

$$K_{x,\mathrm{B}} = p^\circ/k_{\mathrm{H,B}} \tag{13.69}$$

For a given gas, $k_{\mathrm{H,B}}$ is temperature-dependent as expected for any equilibrium constant. The temperature dependence of the Henry's law constant is expressed with a van't Hoff-like equation,

$$k_{\mathrm{H,B}} = k_{\mathrm{H,B,298\,K}} \exp\left[\frac{\Delta_{\mathrm{hyd}}H_\mathrm{m}^\infty}{R}\left(\frac{1}{T} - \frac{1}{298}\right)\right], \tag{13.70}$$

where $\Delta_{\mathrm{hyd}}H_\mathrm{m}^\infty$ is the *enthalpy of hydration*. The molar enthalpy of hydration is defined as the enthalpy change when 1 mol of an ideal gas is dissolved in an infinite amount of water.

The solubility of a gas at atmospheric pressure x_s (the mole fraction at saturation) is defined by setting $p_\mathrm{B} = p' = 101325$ Pa in Eq. (13.67). Solving for x_s and taking the logarithm of both sides of the equation, we obtain

$$\ln x_\mathrm{s} = \ln p' - \ln k_{\mathrm{H,B}}. \tag{13.71}$$

Substitute from Eq. (13.70) for $k_{\mathrm{H,B}}$ to arrive at

$$\ln x_\mathrm{s} = a + b/T \tag{13.72}$$

where $a = 12.53 - \ln k_{\mathrm{H,B,298\,K}} - \Delta_{\mathrm{sol}}H_\mathrm{m}^\circ/298R$ and $b = -\Delta_{\mathrm{hyd}}H_\mathrm{m}^\circ/R$. Therefore, a plot of ln x_s versus $1/T$ results in a straight line with a slope of $\Delta_{\mathrm{hyd}}H_\mathrm{m}^\circ/R$ as long as the solvent follows Henry's law and the enthalpy of solution is constant.

13.6.3.1 Directed practice

Make the proper substitutions to derive Eq. (13.72).

The behavior predicted by Eq. (13.72) is followed approximately for most gases. Deviations from Henry's law can be accounted for by replacing the partial pressure of B with the fugacity of B, which we do not do here, and replacing the mole fraction with the activity,

$$p_\mathrm{B} = k_{\mathrm{H,B}}\, a_\mathrm{B} = k_{\mathrm{H,B}}\, \gamma_\mathrm{B}\, x_\mathrm{B}. \tag{13.73}$$

If we allow the activity coefficient to vary according to a power law in temperature,

$$\gamma_\mathrm{B} = \gamma_0 T^{-C}, \tag{13.74}$$

then Eq. (13.71) becomes

$$\ln x_\mathrm{s} = \ln p' - \ln K_\mathrm{B} + \ln \gamma_0 + C \ln T. \tag{13.75}$$

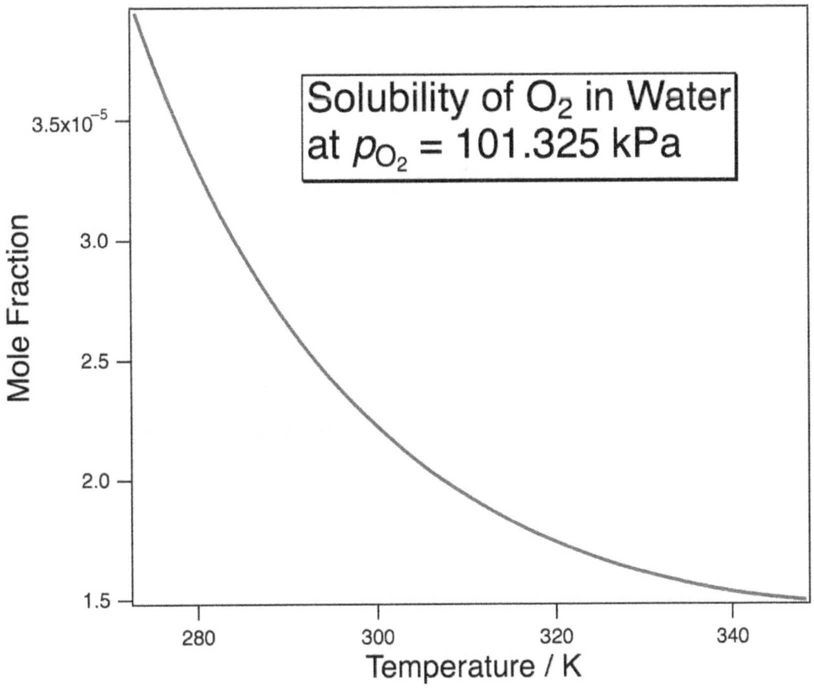

Figure 13.5 The solubility of O_2 in water as a function of temperature for an O_2 partial pressure of 1 atm. Values of A, B, and C for use in Eq. (13.76) were taken from the 96th edition of the *CRC Handbook of Chemistry and Physics*.

Collecting the terms into a new set of temperature-independent constants, we obtain

$$\ln x_s = A + B/T + C \ln T. \tag{13.76}$$

Values for A, B, and C are tabulated in various reference works. The behavior of the saturation mole fraction of O_2 in water is shown in Fig. 13.5. Equation (13.76) is quite successful at representing the solubility of gases in liquids as long as the temperature and pressure do not approach the critical point too closely. However, the coefficients A, B and C obtained from fitting data are highly correlated, which means that it is difficult to interpret them in terms of physically meaningful quantities. In other words, Eq. (13.76) is an empirically reliable equation with a form that is based on physical principles, but it cannot be used unambiguously to determine the values of $\Delta_{sol}H^\circ_m$ and γ_B.

13.6.3.2 Directed practice

At equilibrium, the chemical potential of species A in phases α and β must be equal to a constant μ, the chemical potential of the system. Thus, in an ideal solution,

$$\mu_A^\alpha + RT \ln x_A^\alpha = \mu_A^\beta + RT \ln x_A^\beta = \mu.$$

Show that this condition implies that the equilibrium concentrations of species A and B must follow the *Boltzmann distribution*

$$x_A^\alpha = x_A^\beta \exp\left(-\left[\mu_A^\alpha - \mu_A^\beta\right]/RT\right).$$

The chemical potential μ is the total free energy per mole. It includes contributions from thermal energy as well as all interaction energies.

13.7 Classes of solutions and their properties

13.7.1 Ideal solution/mixture

Recall the characteristics of an ideal gas: (i) no intermolecular interactions; (ii) molecules have no volume; and (iii) all gases follow the same equation of state $pV = nRT$. Now contrast this with the characteristics of an ideal solution: (i) intermolecular interactions are the same for all chemical species; (ii) molecules have a finite volume independent of composition; (iii) both solvent and solute follow Raoult's law; (iv) $\gamma = 1$; (v) miscible (mix at all concentrations); and (vi) $\Delta_{mix}H^{\circ} = 0$, therefore mixing is driven only by entropy, $\Delta_{mix}S^{\circ} > 0$. Now we proceed to derive expressions for the Gibbs energy, entropy and enthalpy of mixing. We use these to define a number of classes of solutions.

13.7.2 Gibbs energy of mixing: Ideal solutions

Following the usual convention, the Gibbs energy of mixing is calculated by subtracting the initial Gibbs energy (the unmixed state) from the final Gibbs energy (the mixed state),

$$\Delta_{mix}G = G_f - G_i.$$

The Gibbs energy is obtained by summing the product of the moles and chemical potential for each component. Here, we take the specific example of a binary solution,

$$G_f = n_A\,\mu_A^f + n_B\,\mu_B^f;\quad G_i = n_A\,\mu_A^i + n_B\,\mu_B^i. \tag{13.77}$$

We use the term solution throughout, though the treatment also applies to mixtures. The standard state is chosen to be that of the pure substance and we use activity in terms of mole fraction. In the initial state we have two pure substances

$$\mu_A^i = \mu_A^{\circ} + RT\ln\left(a_A\right);\qquad \mu_B^i = \mu_B^{\circ} + RT\ln\left(a_B\right)$$

For the pure state $a = 1$, therefore since $\ln(1) = 0$

$$\mu_A^i = \mu_A^{\circ};\qquad \mu_B^i = \mu_B^{\circ} \tag{13.78}$$

In the final state

$$\mu_A^f = \mu_A^{\circ} + RT\ln(a_A);\qquad \mu_B^f = \mu_B^{\circ} + RT\ln(a_B) \tag{13.79}$$

By taking the difference between final and initial state, we determine the change in Gibbs energy

Any solution:

$$\Delta_{mix}G = n_A RT\ln a_A + n_B RT\ln a_B \tag{13.80}$$

The total amount of substance is found from the sum of the components, $n = n_A + n_B$. The amount of any component is equal to the product of total moles and the mole fraction, $n_A = nx_A$ and $n_B = nx_B$. Expressing the activity in terms of the product of activity coefficient and mole fraction, $a_j = \gamma_j x_j$, we obtain,

$$\Delta_{mix}G = nRT[(x_A\ln x_A + x_B\ln x_B) + (x_A\ln\gamma_A + x_B\ln\gamma_B)]. \tag{13.81}$$

However, since we have an ideal solution, $\gamma_A = \gamma_B = 1$, the activity is equal to the mole fraction, $a_A = x_A$ and $a_B = x_B$. Therefore,

Ideal solution

$$\Delta_{mix}G = nRT(x_A\ln x_A + x_B\ln x_B). \tag{13.82}$$

Note that since $x_j \leq 1$ and $\ln x_j < 0$, therefore $\Delta_{mix}G < 0$. This means that mixing is always spontaneous in the ideal case, that is, ideal solutions mix spontaneously in all proportions (which is the definition of miscible).

13.7.2.1 Example

At $T = 25\,°C$, container 1 contains 3.0 mol H_2 at 1.0 atm and container 2 contains 1.0 mol N_2 at 3.0 atm. Assuming ideal gas behavior and calculate $\Delta_{mix}G$.

$$G_i = n_{H_2}\left\{\mu_{H_2}^{\circ} + RT\ln\left(p'_{H_2}/p^{\circ}\right)\right\} + n_{N_2}\left\{\mu_{N_2}^{\circ} + RT\ln\left(p'_{N_2}/p^{\circ}\right)\right\}$$

$$G_f = n_{H_2}\left\{\mu_{H_2}^{\circ} + RT\ln\left(p_{H_2}/p^{\circ}\right)\right\} + n_{N_2}\left\{\mu_{N_2}^{\circ} + RT\ln\left(p_{N_2}/p^{\circ}\right)\right\}$$

$$p_{H_2} = x_{H_2}p, p_{N_2} = x_{N_2}p,\quad p = \frac{nRT}{V}$$

$$V = V_1 + V_2 = \frac{n_{H_2}RT}{p'_{H_2}} + \frac{n_{N_2}RT}{p'_{N_2}} = RT\left(\frac{n_{H_2}}{p'_{H_2}} + \frac{n_{N_2}}{p'_{N_2}}\right)$$

$$\Delta_{mix}G = G_f - G_i = -6.5kJ$$

13.7.3 Entropy of mixing: Ideal solution

By definition, the change in Gibbs energy with respect to temperature is proportional to the entropy according to

$$\left(\frac{\partial G}{\partial T}\right)_{p,n} = -S. \tag{13.83}$$

Therefore

$$\Delta_{\text{mix}}S = -\left(\frac{\partial \Delta_{\text{mix}}G}{T}\right)_{p,n_A,n_B}. \tag{13.84}$$

Substituting from Eq. (13.82), we obtain

Ideal solution

$$\Delta_{\text{mix}}S = -nR(x_A \ln x_A + x_B \ln x_B). \tag{13.85}$$

Again, since ln $x < 0$, $\Delta_{\text{mix}}S > 0$ for all compositions. That is, the entropy factor is always favorable for mixing in ideal solutions.

13.7.4 Enthalpy of mixing: Ideal solution

From the fundamental relationship between Gibbs energy, entropy and enthalpy, we can calculate the enthalpy of mixing

$$\Delta_{\text{mix}}H = \Delta_{\text{mix}}G + T\Delta_{\text{mix}}S$$

Ideal solution

$$\Delta_{\text{mix}}H = 0 \tag{13.86}$$

Thus, the entire driving force for mixing is entropy (the increase thereof) and enthalpy plays no role.

13.7.4.1 Directed practice

Take the derivative of $\Delta_{\text{mix}}G$ with respect to T, substitute directly for $\Delta_{\text{mix}}G$ and $\Delta_{\text{mix}}S$ in Eq. (13.85) to prove that $\Delta_{\text{mix}}H = 0$ for ideal solutions.

13.7.5 Real solutions (non-ideal solutions)

Real solutions deviate from ideal solutions in a number of ways, some of which are systematic and lead to the subclassifications of real solutions defined below. Deviations from ideal behavior include:

- Mixing not always being spontaneous, that is, limited solubility
- $\Delta_{\text{mix}}S$ not equal to maximum value (even < 0 in those cases in which the solution phase separates).
- $\Delta_{\text{mix}}H \neq 0$ resulting from difference in intermolecular interactions between components.
- Deviations from Raoult's Law and Henry's Law.
- Activity coefficients $\neq 1$, which means that we must use activity not mole fraction in the equations defining the changes of thermodynamic variables.

13.7.6 Gibbs energy of mixing: Real solutions

Equation (13.80) and (13.81) hold for real solutions. Now, we must use activities since the activity coefficients differ from unity. They are repeated here to emphasize that we can express the Gibbs energy of mixing for a real solution as the sum of an ideal term (underlined) + the non-ideality (double underlined) term that contains the activity coefficients.

Real solution

$$\Delta_{\text{mix}}G = n_A RT \ln a_A + n_B RT \ln a_B \tag{13.80}$$

$$\Delta_{\text{mix}}G = nRT\left[\underline{(x_A \ln x_A + x_B \ln x_B)} + \underline{\underline{(x_A \ln \gamma_A + x_B \ln \gamma_B)}}\right]. \tag{13.81}$$

13.7.7 Entropy of mixing: Real solution

As before, we take the derivative of the Gibbs energy of mixing as a function of T to find the entropy of mixing.

Real solution

$$\Delta_{\mathrm{mix}}S = -\frac{\partial}{\partial T}[nRT(x_{\mathrm{A}}\ln a_{\mathrm{A}} + x_{\mathrm{B}}\ln a_{\mathrm{B}})] \quad (13.87)$$

In principle, the activity coefficients can depend on temperature so we cannot differentiate this equation further to find a general expression the entropy of mixing. For extremely non-ideal systems, such as oil and water, the entropy term can be unfavorable (negative).

13.7.8 Enthalpy of mixing: Real solution

The fundamental relationship between Gibbs energy, entropy and enthalpy still holds

$$\Delta_{\mathrm{mix}}H = \Delta_{\mathrm{mix}}G + T\Delta_{\mathrm{mix}}S \quad (13.88)$$

While we cannot evaluate this equation analytically, we know what the consequence of non-ideal behavior is

Real solution

$$\Delta_{\mathrm{mix}}H \neq 0. \quad (13.89)$$

Thus, the driving force for mixing is no longer just entropy and enthalpy plays a role in real solutions.

13.7.9 Subclassifications of real solutions

Very commonly we find that for dilute solutions the solute obeys Henry's law while the solvent obeys Raoult's law. Why? A dilute solution is much like the pure liquid since almost every solvent molecule is surrounded by solvent molecules. Therefore, the solvent follows Raoult's law. But the solute is in a completely different environment compared to pure liquid of the solute. This class of solutions is called *ideal-dilute solutions*. A *regular solution* is a solution in which the entropy term acts ideally and the enthalpy acts non-ideally. Ideal-dilute solutions often act like regular solutions.

13.7.9.1 Example

Calculate the molar solubility of CO_2 in H_2O exposed to air at 298 K, which has a CO_2 content of 0.03% (take this as the mole fraction of CO_2 in *air*).

Henry's Law relates the pressure to the mole fraction or concentration in the *solution*

$$p_{\mathrm{CO_2}} = K_{\mathrm{CO_2}} b_{\mathrm{CO_2}}.$$

Whether you use mole fraction or molality depends on the units of the Henry's law constant that you look up. Here, we use the molality version, which is 3010 kPa kg mol^{-1}. Taking the pressure of air to be 1 atm, the partial pressure of 0.03% CO_2 corresponds to

$$p_{\mathrm{CO_2}} = 0.0003\,(101\text{ kPa}) = 0.0303\text{ kPa}$$

From Henry's law

$$b_{\mathrm{CO_2}} = \frac{p_{\mathrm{CO_2}}}{K_{\mathrm{CO_2}}} = \frac{0.0303\,\text{kPa}}{3010\,\text{kPa kg mol}^{-1}} = 1.01\times10^{-5}\text{mol kg}^{-1}$$

If we seek the mole fraction $x_{\mathrm{CO_2}}$ in water, then we can relate molality to mole fraction by

$$b_{\mathrm{CO_2}} = \frac{n_{\mathrm{CO_2}}}{m_{\mathrm{H_2O}}} = \frac{n_{\mathrm{CO_2}}}{n_{\mathrm{H_2O}}M_{\mathrm{H_2O}}}$$

Note that

$$x_{\mathrm{CO_2}} = \frac{n_{\mathrm{CO_2}}}{n_{\mathrm{CO_2}} + n_{\mathrm{H_2O}}} \approx \frac{n_{\mathrm{CO_2}}}{n_{\mathrm{H_2O}}} \quad \text{because } n_{\mathrm{CO_2}} \ll n_{\mathrm{H_2O}}$$

Therefore

$$x_{\mathrm{CO_2}} \approx b_{\mathrm{CO_2}} M_{\mathrm{H_2O}} = 1.01\times10^{-5}\text{ mol kg}^{-1}(0.01802\,\text{kg mol}^{-1}) = 1.82\times10^{-7}.$$

13.7.10 Excess functions

We express the properties of real solutions in terms of the *excess functions* G^E, S^E, H^E, and so on. For instance

$$S^E = \Delta_{mix}S(\text{actual}) - \Delta_{mix}S(\text{ideal}) \tag{13.90}$$

and so forth. Accordingly, the behavior of a *regular solution* is defined by $H^E \neq 0$ but $S^E = 0$. In a regular solution, the solution is random but the intermolecular interactions deviate from ideal (solvent and solute are not identical). The behavior of an ideal solution is defined by vanishing values of all excess functions.

13.7.10.1 Example

Calculate the excess Gibbs energy of a real solution.

Start with the definition of the excess Gibbs energy

$$G^E = \Delta_{mix}G(\text{actual}) - \Delta_{mix}G(\text{ideal}).$$

Then substitute for the actual values, which use activities, and the ideal values, which use mole fractions.

$$\begin{aligned} &= nRT(x_A \ln a_A + x_B \ln a_B) - nRT(x_A \ln x_A + x_B \ln x_B) \\ &= nRT[(x_A \ln \gamma_A x_A + x_B \ln \gamma_B x_B) - (x_A \ln x_A + x_B \ln x_B)] \\ &= nRT(x_A \ln \gamma_A + x_B \ln \gamma_B) \end{aligned}$$

The value of the excess functions represents the deviation of the real value from the ideal value.

13.8 Colligative properties

A *colligative property* is a property that depends on the presence of a solute but not its identity. Here, we will treat three examples: boiling point elevation; melting point depression; and osmotic pressure. A fourth colligative property is vapor pressure lowering, which we have already found to follow Raoult's law. To treat these simply we make two assumptions: (i) the solute is nonvolatile, that is, it makes no contribution to the vapor; and (ii) the solute does not dissolve in solid solvent. Colligative properties have a common origin. They arise from the reduction of the chemical potential of the liquid solvent due to the presence of the solute. This is an entropy effect. Since the solute is not present in the vapor (assumption 1) or solid (assumption 2), it lowers μ(liquid solution) exclusively. This is illustrated in Fig. 13.6.

13.8.1 Boiling point elevation

Consider a solvent A and solute B in equilibrium at standard pressure $p°$. We take the solid and vapor phases to be acting ideally. The chemical potential of the solvent must be equal in the gas phase as well as in the solution at equilibrium, thus, as shown in Section 13.6:

$$\mu_A^*(g) = \mu_A^*(l) + RT \ln x_A. \tag{13.91}$$

A lengthy derivation demonstrates that a linear relationship is expected between the concentration of the solute and boiling point elevation. The linear relationship holds exactly for ideal solutions at low concentration, but is also found for limited concentration ranges in real solutions. The reason for this shift is that adding B with mole fraction x_B lowers the chemical potential of A. The boiling point rises from its normal value T_b to the value found for the solution T_b' according to

$$\Delta T = T_b' - T_b = \left(\frac{RT_b^2}{\Delta_{vap}H}\right) x_B. \tag{13.92}$$

This allows us to rewrite the boiling point shift in terms of molality

$$\Delta T = K_b\, b_B \tag{13.93}$$

where K_b is the *ebullioscopic constant*

$$K_b = \frac{RT_b^2 M}{\Delta_{vap}H} \tag{13.94}$$

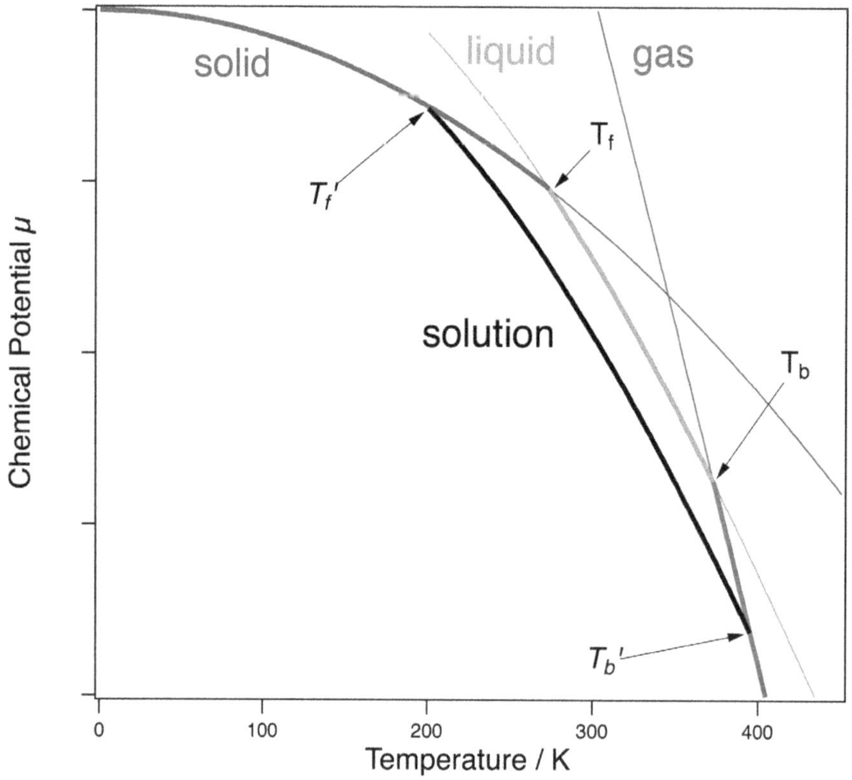

Figure 13.6 Reduction of the chemical potential of a solution compared to the pure solvent is responsible for the appearance of colligative properties. Here it is shown how increased stability of the solution lowers the melting point from the pure liquid value T_f to the solution value T_f' and increases the boiling point of a solution T_b' above the values of the pure solvent T_b.

and M is the molar mass of the solute. Deviations in the actual value of K_b are indications of non-ideal behavior. Note that K_b does not depend on any property of the solute other than its concentration, as would be expected for a colligative property. Historically, the measurement of K_b provided a means of measuring molar masses, for instance, of polymers. However, this is no longer of relevance since it is important to measure the size distribution and not just the mean molar mass of polymers. Modern techniques for molar mass distribution determination are based on light scattering, chromatography, mass spectrometry and, as we show below, osmometry

13.8.2 Freezing point depression

Starting now with liquid and solid phases in equilibrium, we again equate the chemical potential of the solvent in both the solid and solution phases

$$\mu_A^*(s) = \mu_A^*(l) + RT\ln x_A. \tag{13.95}$$

Adding B with mole fraction x_B lowers the chemical potential of A and lowers the melting point from its normal value T_f to the value found for the solution T_f' according to

$$\Delta T = T_f' - T_f = \left(\frac{RT_f^2}{\Delta_{fus}H}\right)x_B \tag{13.96}$$

where T_f is the normal melting point. This is valid for dilute solutions. The melting point shift in terms of molality is

$$\Delta T = -K_f b_B \tag{13.97}$$

where K_f is the *cryoscopic constant.*

13.8.3 Osmosis

Consider the arrangement depicted in Fig. 13.7 in which a semipermeable membrane separates a pure solvent on the left from a solution on the right. Membranes commonly used are made of cellophane-regenerated cellulose, cellulose derivatives as well as porous synthetic plastics and porous glass. Pore diameters of <10 nm are impermeable to molecules

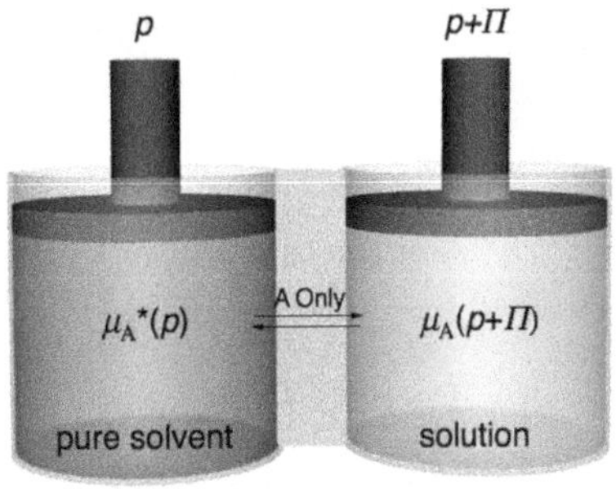

Figure 13.7 Two containers are connected by a semipermeable membrane that only allows for passage of the pure solvent A. The solution has a lower chemical potential μ. To equalize μ in both compartments, the solvent A will flow into the solution. The osmotic pressure Π is the 'extra' component of pressure that must be applied to the solution through the movable piston to stop the flow of A and maintain the two liquids at equal volume.

with M greater than a few thousand daltons. The walls of the roots of plants or cell walls can also act as semipermeable membranes.

Osmosis is the flow of a pure solvent into a solution separated from it by a semipermeable membrane. The osmotic pressure is the pressure Π that must be applied atop the piston in excess of the atmospheric pressure to halt osmotic flow.

At equilibrium, the chemical potential of the pure solvent at pressure p in the left compartment equals the chemical potential of the ideal solvent in the right compartment held at pressure $p + \Pi$ with solvent mole fraction x_A

$$\mu_A^*(p) = \mu_A(x_A, p + \Pi). \tag{13.98}$$

The presence of the solute again lowers the chemical potential according to

$$\mu_A^*(p) = \mu_A^*(x_A, p + \Pi) + RT \ln x_A. \tag{13.99}$$

To calculate the effect of pressure (at constant T and composition) on the pure solvent's chemical potential recall the definition of chemical potential of a pure substance is the molar Gibbs energy then differentiate with respect to p.

$$n\mu = G \tag{13.100}$$

$$\left(\frac{\partial \mu}{\partial p}\right)_{T,x_A} = \frac{1}{n_A}\left(\frac{\partial G}{\partial p}\right)_{T,x_A}. \tag{13.101}$$

We know that the Gibbs energy depends on pressure and temperature according to the fundamental equation $dG = V\,dp - S\,dT$, therefore at constant temperature $dG = V\,dp$. Substituting into Eq. (13.101) yields

$$\left(\frac{\partial \mu}{\partial p}\right)_{T,x_A} = \frac{V}{n_A} = V_{m,A}. \tag{13.102}$$

We use this to evaluate by integration the change in chemical potential brought about by the osmotic pressure,

$$\mu_A^*(p + \Pi) = \mu_A^*(p) + \int_p^{p+\Pi} d\mu_A^* = \mu_A^*(p) + \int_p^{p+\Pi} V_{m,A}\,dp \tag{13.103}$$

Now substitute from Eq. (13.103) into Eq. (13.99) to obtain

$$-RT \ln x_A = \int_p^{p+\Pi} V_{m,A}\,dp \tag{13.104}$$

and if the liquid is incompressible, then V_m is a constant. Being dilute, the molar volume of the solution is equal to the molar volume of the solvent. The molar volume can then be pulled from the integral and the integral can be evaluated to give

$$-RT \ln x_A = \Pi V_m. \tag{13.105}$$

Using $x_B = n_B/n$ and $V_m = V/n$ and noting that not only $x_A = 1 - x_B$, but also for dilute solutions with $x_B \ll 1$, then $\ln x_A = \ln(1 - x_B) \approx -x_B$. We substitute and rearrange to obtain

$$\Pi V = n_B RT. \tag{13.106}$$

This equation resembles the ideal gas equation. However, we certainly expect non-ideality in real solutions. We account for non-ideality by substituting activity $a_A = \gamma_A x_A$ in place of mole fraction in Eq. (13.105),

$$-RT \ln \gamma_A x_A = -g_1 RT \ln x_A = \Pi V_m. \tag{13.107}$$

The parameter g_1 is known as the *osmotic coefficient*. Assume that the dependence of g_1 on the concentration of the solute with molar concentration, $[B] = n_B/V$, is described by a *viral expansion*,

$$g_1 = 1 + B[B] + C[B]^2 + \cdots . \tag{13.108}$$

Then, the *osmotic pressure* depends on the concentration according to

$$\Pi = RT\,[B]\{1 + B[B] + C[B]^2 + \cdots\} \tag{13.109}$$

B is the second virial coefficient, and C is the third virial coefficient. These solution virial coefficients are calculable from the effective interaction potential in the solution in a manner analogous to the treatment of non-ideality by virial expansion in real gases. It can be shown that the second virial coefficient is related to the *excluded volume* of the solute u and its molar mass M by

$$B = N_A\,u/2M^2 \tag{13.110}$$

Here, N_A is the Avogadro constant and the molar mass M is the *number-average molar mass* (also denoted M_n, as discussed in Chapter 5). For small molecules u is negligible. For macromolecules such as polymers and biomolecules, the excluded volume is appreciable and the second virial coefficient B is temperature-dependent. The temperature at which $B = 0$ is known as the Flory θ-temperature.

A common application of *osmometry* is the measurement of molar masses especially of polymers (or other big molecules including biomolecules such as proteins). The apparatus sketched in Fig. 13.8 is of the type that is often used. In this arrangement, the osmotic pressure is equal to the hydrostatic pressure

$$\Pi = \rho g h \tag{13.111}$$

where ρ is the density of the solution, $g = 9.81$ m s^{-2} and h is the height of the solution in the column. Solutions of this type are far from ideal and we use the virial expansion to allow for correction terms.

Historically, measurements of the height of a column of solution were made to determine the osmotic pressure. Modern instrumentation does this step for us and the osmotic pressure is read out directly. Taking c as the mass concentration and the molar mass M, we substitute $[B] = c/M$ and for Π from Eq. (13.109), to obtain

$$\Pi = \frac{cRT}{M}\left\{1 + \frac{cB}{M} + \cdots\right\}. \tag{13.112}$$

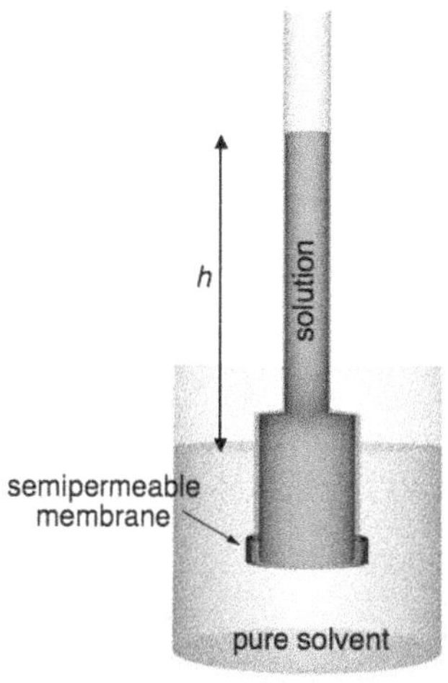

Figure 13.8 Schematic representation of an osmometer used for the determination of polymeric molar masses. In this application the equilibration time often approaches 24 h when the height is relatively large. Automated instruments with very small volumes, requiring essentially zero flow, are available that reduce this time to minutes.

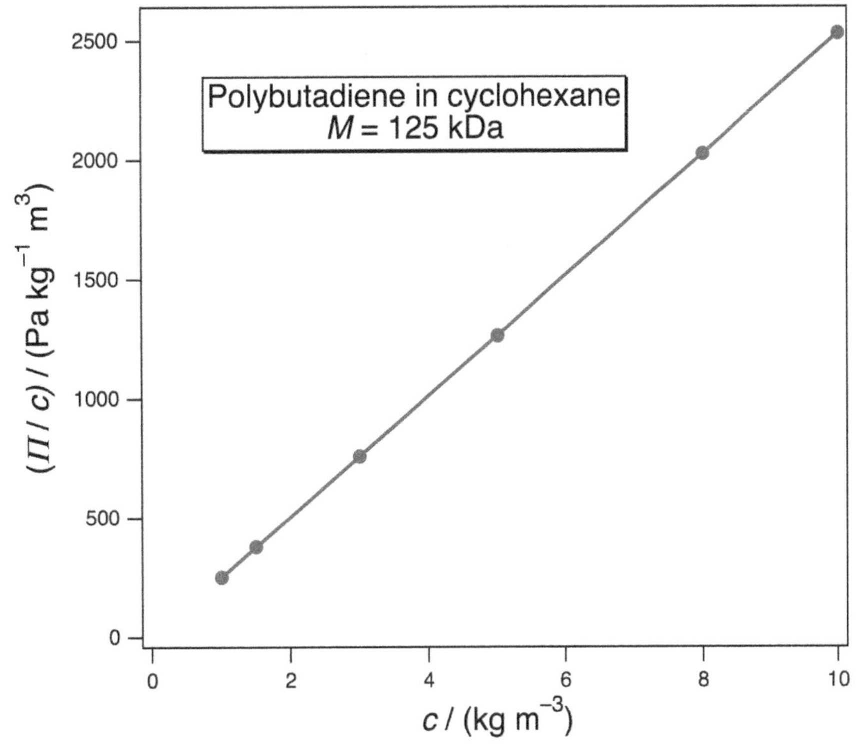

Figure 13.9 Measurement of the osmotic pressure Π as a function of the mass concentration c of polybutadiene in cyclohexane can be used to determine the number-average molar mass and the second virial coefficient of the polymer.

This can be linearized by expressing it in terms of Π/c,

$$\frac{\Pi}{c} = \frac{RT}{M} + \left(\frac{RTB}{M^2}\right)c + \cdots . \tag{13.113}$$

Therefore, a plot of Π/c against c gives a straight line with intercept $a = RT/M$ at $c = 0$ and slope RTB/M^2. As a practical matter, molar masses of no more than 500 kilodaltons (500 kDa) can be measured with osmometry. While osmometry has historically been important for molecular weight determination and continues to be used for this, light scattering is the principle means of molar mass determination for polymers. Small-angle neutron scattering and X-ray scattering are also sometimes used for this purpose.

13.8.3.1 Example

Determine M for polybutadiene dissolved in cyclohexane at 298 K, given the following data:

c / kg m^{-3}	1.00	1.50	3.00	5.00	8.00	10
Π / Pa	253	570	2279	6330	16205	25320

Calculate Π/c and plot versus c as done in Fig. 13.9. The intercept a can then be used to calculate the number-average molar mass from $M = RT/a$. The answer is 125 kDa = 125 000 g mol^{-1}.

13.9 Solubility of polymers

An important example of mixing in a regular solution is the interaction of a solvent S with a polymer P. If a polymer is soluble in a given solvent, then it dissolves in several steps. First, the solvent must wet the polymer. Second, the solvent diffuses into the polymer and makes it swell to form a gel. The *swelling coefficient* Q of a polymer is defined as

$$Q = \frac{m - m_0}{m_0\,\rho_S}. \tag{13.114}$$

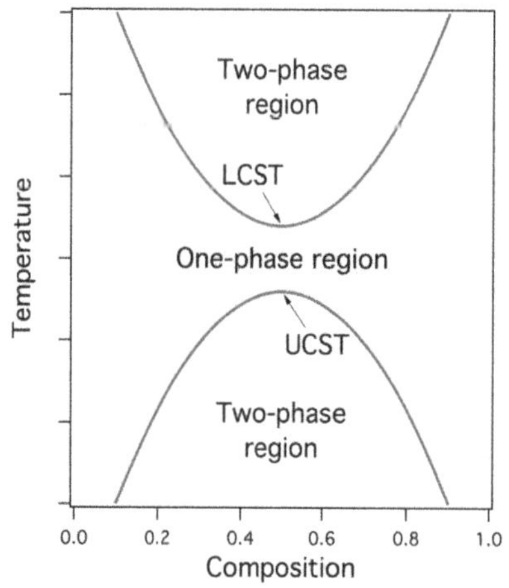

Figure 13.10 A schematic representation of polymer blend phase separation as a function of temperature and composition. An upper critical solution temperature (UCST) is observed in some systems with short chains. A lower critical solution temperature (LCST) is exhibit in some high-molecular-weight blends because of their low entropy of mixing.

where m is the mass of the swollen sample, m_0 is the dry mass of the polymer, and ρ_S is the density of the solvent. Small-molecule nonelectrolytes will typically dissolve by rapidly diffusing into the solvent. For polymers, the process is reversed with the swelling portion occurring over the course of several hours or longer depending on the temperature, sample size and the specific chemical identity of the polymer and solvent.

Solvent diffusion into the polymer responds to whether or not the temperature is above the glass transition temperature T_g of the polymer. In a polymer melt, that is, above T_g, diffusion is Fickian. There is a smooth composition gradient set up as the solvent diffuses into the polymer. Swelling is greatest at the interface between the solvent and polymer and gradually decreases moving into the bulk of the polymer.

Case II swelling is the name of the phenomenon that occurs when a solvent is exposed to a polymer far below T_g. Diffusion into a glassy polymer can be quite slow. In case II swelling, the solvent must first plasticize the polymer. This process is slow but once it occurs, the solvent rapidly diffuses into the plasticized area accompanied by rapid swelling of only the plasticized region. Just beyond this moving boundary of highly swollen material, the polymer is in its normal state. Stresses at the swelling boundary can lead to fracture of the material.

After swelling, the polymer is truly dissolved and diffuses out into the solvent. Crosslinked polymers will only swell, reaching some equilibrium degree of swelling in the gel state but not diffusing into solution as isolated units. The same is also true of other types of high-melting temperature polymers or polymers with particularly strong secondary bonds.

For polymer solutions composed of a polymer and a small-molecule solvent, the entropy factor usually acts approximately ideally. The enthalpy of mixing of relatively nonpolar organic molecules, however, is usually found to be positive. The *Hildebrand–Scatchard equation* describes the behavior of the enthalpy of mixing,

$$\Delta_{\text{mix}}H = V\left[\delta_S^{1/2} - \delta_P^{1/2}\right]^2 \nu_S\, \nu_P. \tag{13.115}$$

The quantity

$$\delta = \left(\frac{\Delta E}{V_m}\right)^{1/2} = \left(\frac{\Delta_{\text{vap}}H_m^\circ - RT}{V_m}\right)^{1/2} \tag{13.116}$$

is called the *Hildebrand solubility parameter*. $\Delta E/V_m$ is the cohesive energy density, ΔE is the energy of vaporization to a gas at zero pressure, V_m the molar volume, V the volume of the mixture and ν_S and ν_P are the volume fractions of the solvent and polymer, respectively. Equation (13.115) breaks down if the enthalpy of mixing is found to be negative. However, it usually provides a good estimate of whether a particular polymer is soluble in a given solvent.

Since Eq. (13.115) only delivers a positive value of $\Delta_{\text{mix}}H$, maximum solubility is obtained when $\delta_S = \delta_P$. Maximum solubility corresponds to maximum swelling. Therefore, if the swelling coefficient Q for a given polymer is obtained for a number of solvents of known solubility parameter, the solubility parameter of the solvent can be obtained by finding the maximum in a plot of Q versus δ_S. An alternative method is to determine the intrinsic viscosity of a polymer solution

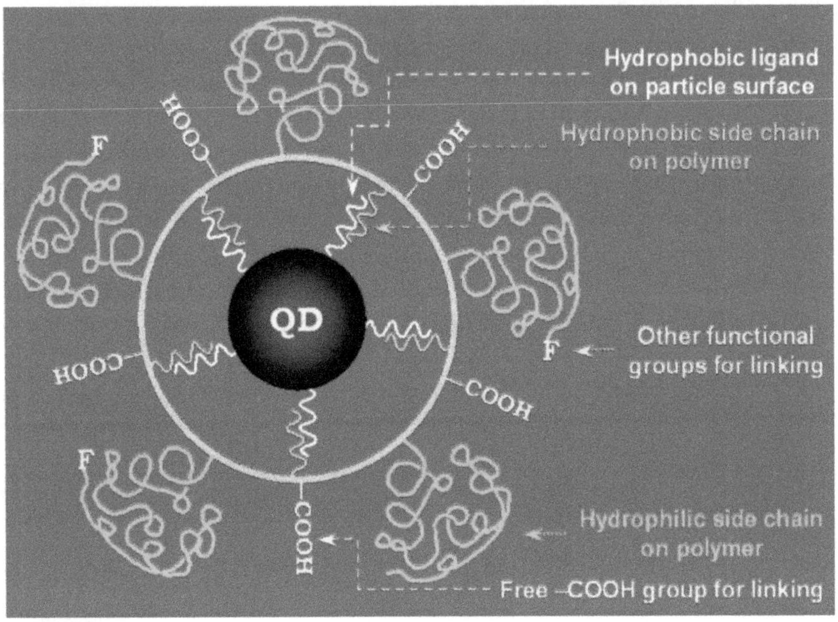

Figure 13.11 Water-solubility of semiconductor quantum dots (QD) can be established by grafting polymer chains to their surfaces. The polymer chains are terminated in functional groups, represented by F, such as —OH, —COOH and —HN_2, among others. Polymer chains are composed of poly(ethylene glycol) (PEG) and poly(maleic anhydride-alt-1-octadecene) (PMAO). Reproduced with permission from Yu, W.W., Chang, E., Falkner, J.C., Zhang, J., Al-Somali, A.M., Sayes, C.M., Johns, J., Drezek, R., and Colvin, V.L. (2007) *J. Am. Chem. Soc.*, **129**, 2871. © 2007, American Chemical Society.

as a function of solvent solubility parameter. The maximum in a plot of intrinsic viscosity versus solubility corresponds to δ_P.

For polymer blends – that is, mixtures of different polymers – the entropy factor *usually does not act ideally*. In most cases, the polymers cannot interchange segments because of covalent bonding. The entropy change is far lower than expected for complete randomization, and total phase separation results. Polymer blends are found to mix in certain cases. Often, a lower critical solution temperature is observed. This behavior is shown in Fig. 13.10. A single-phase region is found at low temperature and the temperature boundary between the single-phase region, and the two-phase region depends on composition. This behavior is characteristic of exothermic mixing combined with a negative excess entropy of mixing. The negative excess entropy is caused by the polymer becoming denser in the blend, which means that in a proper thermodynamic treatment of the Gibbs energy of mixing, the polymers can no longer be considered to be incompressible.

As shown by Vicki Colvin and coworkers,[1] polymers can also be used to modify the solubility of nanoscale semiconductor quantum dots. This is shown in Fig. 13.11. Intermolecular interactions can be tailored such that nanoparticles, which in this case are synthesized in organic solvents, are still soluble in aqueous solutions because of the hydrophilic side chains and other polar groups that are built into the structures.

13.10 Supercritical CO_2

Supercritical CO_2 (sc-CO_2) acts as a nonpolar solvent. However, it has a large quadrupole moment and polar C=O bonds that assist in solvating organic species with —OH, —CO and —F functional groups. Water, salts and polar molecules such as sugars are not soluble in sc-CO_2, though chelated and organometallic compounds usually are. Ethanol, methanol, isopropanol, acetone, hexane, formic acid and acetic acid are miscible with sc-CO_2 and are often added as cosolvents. They act to keep solutes in solution that otherwise would not dissolve in sc-CO_2. sc-CO_2 is widely used for decaffeinating coffee and the extraction of hops and spices because it is nontoxic and easily separated from other fluids simply by changing p and T. Solubility is a function of the density of the supercritical fluid and can be tuned with changes in p and T.

sc-CO_2 is cheap, readily available, nonflammable and recyclable. It has found uses in nanoparticle, thin-film and polymer processing. It is also important in enhanced oil recovery and carbon sequestration.

Solvophilicity is the affinity that a solute has for a given solvent. A solvent that readily dissolves in sc-CO_2 is said to be CO_2-philic, just as a hydrophilic solute is one that dissolves readily in water. The Hildebrand solubility parameter can be used to estimate solvophilicity. The closer the two values of δ, the greater the solvophilicity.

Fluorocarbon and hydrofluorocarbon surfactants can be highly soluble in sc-CO_2. These can be used to create microemulsions, which are thermodynamically stable dispersions of two or more immiscible/partially miscible fluids. The domains created within the emulsion have extents on the nanoscale and are stable in time. A number of polymers also dissolve in sc-CO_2. Increased viscosity is observed in solutions of poly(1-decene) and mixtures of styrene and fluoroacrylate with sc-CO_2. The formation of micelles, lamellar structures and gels is often associated with viscosity increases in solution.

SUMMARY OF IMPORTANT EQUATIONS

$\mu_j = (\partial G/\partial n_j)_{p,T,n_{i\neq j}}$	Chemical potential of component j, μ_j, relates how the Gibbs energy G changes with changes in the number of mole of j, n_j.
$G = \sum_{j=1}^{r} n_j \mu_j$	The Gibbs energy of a mixture is the sum of the product of moles times chemical potential.
$\sum_{j=1}^{r} n_j \, d\mu_j = 0$	Gibbs–Duhem equation at constant p and T.
$dH = T\,dS + V\,dp + \sum_{j=1}^{r} \mu_j \, dn_j$	Dependence of a change in enthalpy on changes in T, p and composition.
$dG = -S\,dT + V\,dp + \sum_{j=1}^{r} \mu_j \, dn_j$	Dependence of a change in Gibbs energy on changes in T, p and composition.
$\mu_j = \mu_j^\circ + RT \ln a_j$	Chemical potential of a component j in a solution at mole fraction x_j.
$a_j = \gamma_j x_j$	Activity of component j in terms of its activity coefficient γ_j and mole fraction x_j.
$\mu_A(g) = \mu_A^\circ + RT \ln (p_A/p^\circ)$	Chemical potential of an ideal gas in terms of its equilibrium partial pressure p_A.
$\mu_A(l) = \mu_A^\circ + RT \ln (p_A/p^\circ)$	Chemical potential of a solvent can be measured by measuring the equilibrium partial pressure above the solution.
$a_A = p_A/p_A^*$	Activity of a solvent or volatile solute can be measured by comparison of the equilibrium vapor pressure above the solution p_A to the equilibrium vapor pressure of the pure substance p_A^*.
$p_A = x_A p_A^*$	Raoult's law for solvents.
$p_B = K_B x_B$	Henry's law for solutes.
$\Delta_{mix} G = n_A RT \ln a_A + n_B RT \ln a_B$	Gibbs energy of mixing for binary solution.
$\Delta_{mix} S = -nR (x_A \ln x_A + x_B \ln x_B)$	Entropy of mixing for binary solution.
$S^E = \Delta_{mix} S(\text{actual}) - \Delta_{mix} S(\text{ideal})$	Excess entropy of mixing.
$\Delta T = K_b b_B$	Boiling point elevation in terms of ebullioscopic constant K_b and molality b_B.
$\Delta T = -K_f b_B$	Melting point depression in terms of cryoscopic constant K_f and molality b_B.
$\Pi = [B]RT$	Osmotic pressure Π in relation to the concentration of solute [B].
$\frac{\Pi}{c} = \frac{RT}{M} + \left(\frac{RTB}{M^2}\right) c$	Osmotic pressure Π divided by the mass concentration c is related to molar mass of solute M, gas constant R, temperature T and the second virial coefficient B.
$Q = \frac{m - m_0}{m_0 \rho_S}$	Polymer swelling coefficient in terms of m the mass of the swollen sample, m_0 the dry mass of the polymer and ρ_S the density of the solvent.
$\Delta_{mix} H = V \left[\delta_S^{1/2} - \delta_P^{1/2}\right]^2 v_S v_P$	Enthalpy of mixing in terms of volume of the mixture V, solubility parameters δ_S and δ_P, and volume fractions v_S and v_P of the solvent and polymer.

Exercises

13.1 Confirm that Eq. (13.38) has the form $\mu_i - \mu_i^\circ = Tf(n_1, \ldots, n_r)$.

13.2 Consider the following four solutions: (i) carbon disulfide and water; (ii) dimethylsulfide and water; (iii) dimethylsulfoxide and water; and (iv) dimethylsulfoxide and dimethylsulfide. (a) Rank the four solutions from most to least ideal. Explain your reasoning. (b) Discuss how the ideal/non-ideal behavior of these four solutions changes their solubilities.

13.3 Consider the following three solutions: (i) *o*-xylene and *p*-xylene; (ii) benzene and water; (iii) phenol and water. (a) Rank the three solutions from most to least ideal. Explain your reasoning. (b) Discuss how the ideal/non-ideal behavior of these three solutions changes their solubilities.

13.4 Calculate the change in the chemical potential of the solvent in a binary mixture $\Delta\mu_A(l)$ when the composition of an ideal solution is changed from $x_A = 1.000$ to $x_A = 0.900$ at 298 K.

13.5 The heat capacity of water is 4.19 J K^{-1} g^{-1}. At 298 K and 1 bar of pressure, where the density of the solution is 1.00 g l^{-1}, when 0.100 mol of KCl is dissolved in 1.00 l of water, the temperature decreases by 0.382 K. Is the solution acting ideally? Justify your answer and why this result is to be expected.

13.6 The vapor pressure of benzene at 25 °C is 12.7 kPa. Calculate the activity and activity coefficient of benzene if the vapor pressure of benzene above a benzene/toluene solution that has $x_{\text{benzene}} = 0.0200$ is 0.246 kPa.

13.7 When 10.0 g of water is mixed with 11.1 g of ethanol at 298 K, the equilibrium vapor pressures of water and ethanol above the mixture are $p_W = 3.61$ kPa and $p_E = 3.85$ kPa, respectively. Calculate the mole fractions of ethanol and water in the solution. Calculate the activities and activity coefficients of ethanol and water in the solution. The vapor pressures of pure water and pure ethanol are 3.17 kPa and 7.87 kPa, respectively.

13.8 (a) Calculate the entropy of mixing N_1 molecules of type 1 with N_2 molecules of type 2 given that the number of arrangements of these two molecules in solution is $\Omega = N_0!/N_1!\,N_2!$, where N_0 is the total number of molecules in solution. (b) If the two molecules and their solution act ideally, the enthalpy of mixing is zero. Use this information and the answer from part (a) to calculate the Gibbs energy of mixing $\Delta_{\text{mix}}G$.

13.9 At 39.9 °C, a solution of ethanol ($x_1 = 0.9006$, $p_1{}^* = 130.4$ Torr) and isooctane ($p_2{}^* = 43.9$ Torr) is equilibrated and the vapor phase has a total pressure of 185.9 Torr. The mole fraction of ethanol in the vapor phase is $y_1 = 0.6667$. (a) Calculate the activity and activity coefficient of each component in the solution. (b) Calculate the total pressure that would have been measured if the solution were ideal.

13.10 The total equilibrium pressure above a toluene/benzene solution is 8.440 kPa. The equilibrium vapor pressures of toluene and benzene are 3.572 kPa and 9.657 kPa, respectively. Calculate the mole fractions of toluene and benzene in the solution and in the vapor.

13.11 Given the following information

Compound	Cryoscopic constant K kg^{-1} mol^{-1}	T_f / K
Water	1.86	273.15
Cyclohexane	20.8	279.35
Ethanol	1.99	158.55

Calculate the melting point of a solution composed of 1 mol of ethanol mixed with: (a) 0.500 kg of water; or (b) 0.500 kg of cyclohexane. Which solution, (a) or (b), will act more ideally? Explain your answer.

13.12 The vapor pressure of pure CCl_4 is 20.7 kPa. The vapor pressures of pure hexane and pure acetic acid are 0.0500 kPa and 3.30 kPa, respectively. All vapor pressure data are taken at 20 °C. Two solutions are made up. One contains 1% by mole fraction of hexane in CCl_4; the other contains 1% by mole fraction acetic acid in CCl_4. In both solutions the activity coefficient of the solvent is $\gamma = 1$. The activity coefficient of one solute is 0.980;

the activity coefficient of the other solute is 0.750. The vapors act ideally. (a) Assign the values of the activity coefficients to the two solutes. Justify your answer based on chemical reasoning. (b) Calculate the activities of the solvent and solute in the solution containing CH_3COOH. (c) Calculate the composition of the vapor phase above the CH_3COOH solution.

13.13 Assume that the entropy acts ideally for the acetic acid in CCl_4 solution in the previous exercise. Calculate the values of $\Delta_{mix}S°$, $\Delta_{mix}G°$ and $\Delta_{mix}H°$ in molar units for the acetic acid solution.

13.14 The normal melting point of pure CCl_4 is 250 K and its enthalpy of fusion is 2.52 kJ mol^{-1}. A solution is made up that contains 1% by mole fraction of hexane in CCl_4. The other contains 1% by mole fraction acetic acid in CCl_4. Calculate the value of the melting point of the hexane solution.

13.15 The partial pressure of CO_2 above champagne being on the order of only 0.0004 bar, and the solubility of CO_2 in champagne being on the order of 1.5 g l^{-1} bar^{-1} at 20 °C. Calculate the equilibrium concentration of CO_2 in champagne and the Henry's law constant for CO_2 in champagne. Note: the equilibrium CO_2 concentration expresses as $c_{eq} = \sim 0.6$ mg l^{-1}.

13.16 The solubility expressed as mole fraction of CO_2 held at 101 kPa in water as a function of temperature is given below. Use this data to calculate the enthalpy of solution $\Delta_{sol}H^\circ_m$ of CO_2.

T / K	288.15	293.15	298.15	303.15	308.15
x_s	8.21×10^{-4}	7.07×10^{-4}	6.15×10^{-4}	5.41×10^{-4}	4.80×10^{-4}

13.17 Consider a solid in contact with a solvent at equilibrium. Calculate the solubility of the solid in the solvent assuming the system acts ideally.

Endnote

1. Yu, W.W., Chang, E., Falkner, J.C., Zhang, J., Al-Somali, A.M., Sayes, C.M., Johns, J., Drezek, R., and Colvin, V.L. (2007) *J. Am. Chem. Soc.*, **129**, 2871.

Further reading

Gamsjäger, H., Lorimer, J.W., Scharlin, P., and Shaw, D.G. (2008) Glossary of terms related to solubility. *Pure Appl. Chem.*, **80** (2), 233–276.

Gamsjäger, H., Lorimer, J.W., Salomon, M., Shaw, D.G., and Tomkins, R.P.T. (2010) The IUPAC-NIST Solubility Data Series: A guide to preparation and use of compilations and evaluations (IUPAC Technical Report). *Pure Appl. Chem.*, **82** (5), 1137–1159.

Poling, B.E., Prausnitz, J.M., and O'Connell, J.P. (2001) *The Properties of Gases and Liquids*, 5th edition. McGraw-Hill, New York.

Pollet, P., Davey, E.A., Urena-Benavides, E.E., Eckert, C.A., and Liotta, C.L. (2014) Solvents for sustainable chemical processes. *Green Chemistry*, **16**, 1034.

Sperling, L.H. (2006) *Introduction to Physical Polymer Science*, 4th edition. John Wiley & Sons, Hoboken, NJ.

CHAPTER 14

Solutions of electrolytes

PREVIEW OF IMPORTANT CONCEPTS

- Electrolytes dissolve in a solvent to produce solvated ions capable of conducting electricity.
- Strong electrolytes dissolve almost completely into ions.
- Weak electrolytes remain mostly in the molecular form.
- The solubility of an electrolyte is determined by the balance between the energy gained by holding the ions in a crystalline lattice as compared to the energy gained by solvating the ions in solution.
- In the limit of infinite dilution, all electrolytes dissociate completely into ions.
- Ions in solution are surrounded by solvation shells. The first shell is fairly well ordered with a characteristic number of solvent molecules.
- The number of molecules in the first solvation shell is called the solvation number. It ranges from about 4 to 12 for simple ions. A value of 6 is quite common for aqueous transition metal cations.
- The most commonly used scale to describe the enthalpy and Gibbs energy of formation and absolute entropy of ions in aqueous solution is relative to the H^+. For H^+ each of these three values is set equal to zero.
- Many polar solvents solvate cations more effectively than water. However, they tend to solvate anions less effectively.
- The activity coefficient is introduced to account for deviations from ideal behavior for ions in solution.
- Deviations from ideal behavior become particularly pronounced when ionic concentrations increase and the solvation shells of individual ions begin to overlap.
- The Debye–Hückel limiting law takes into account the electrostatic interactions of ions in solution to calculate accurate values of the activity coefficient for dilute solutions.
- The activity coefficient is determined by the ionic strength of a solution as well as the charge on the ions.
- Conductivity is a sensitive probe of the number of ions in solution and the ease with which they are transported.

14.1 Why salts dissolve

Salts dissolve to varying degrees in polar solvents, not only water but also others such as alcohols that contain –OH groups. Also, dipolar aprotic solvents such as acetonitrile and dimethylsulfoxide support electrolyte dissolution. Metal cations are not as readily solvated by alcohols as they are in water, but dimethylsulfoxide and pyridine can even exceed the capacity of water to solvate metal ions. Nonetheless, most salts containing simple inorganic anions exhibit very low solubilities in organic solvents because of the difficulty these solvents have in effectively solvating simple inorganic anions.

The solubility of a salt is determined by the Gibbs energy of solvation $\Delta_{sol}G°$, which for the salt MX corresponds to the reaction

$$MX(s) \rightleftharpoons M^+(aq) + X^-(aq). \tag{14.1}$$

As shown in Fig. 14.1, whether or not this Gibbs energy change is negative is determined by the balance between the binding of the ions in the solid, represented by the lattice Gibbs energy $\Delta_{lat}G°$ corresponding to

$$M^+(g) + X^-(g) \rightleftharpoons MX(s) \tag{14.2}$$

and the Gibbs energy of hydration

$$M^+(g) + X^-(g) \rightleftharpoons M^+(aq) + X^-(aq). \tag{14.3}$$

Physical Chemistry: How Chemistry Works, First Edition. Kurt W. Kolasinski.

Companion Website: www.wiley.com/go/kolasinski/physicalchemistry

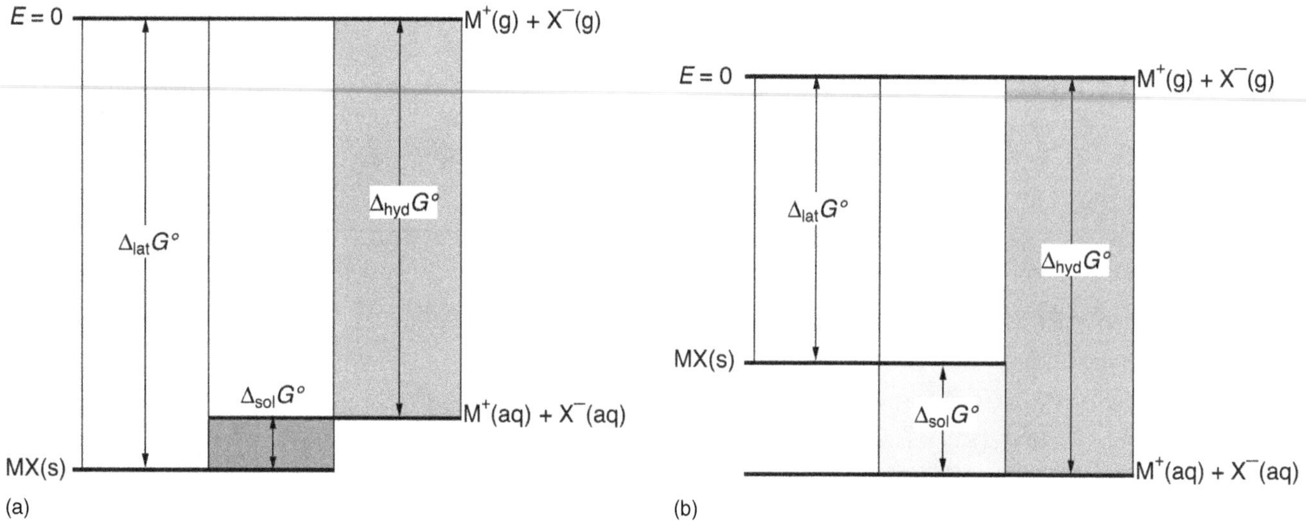

Figure 14.1 The Gibbs energy of solvation $\Delta_{sol}G°$ is the difference between the Gibbs energy of hydration $\Delta_{hyd}G°$ and the lattice Gibbs energy $\Delta_{lat}G°$. (a) When the lattice energy is higher, dissolution is unfavorable. (b) When the hydration energy is greater, dissolution is favorable.

Using Hess's law

$$\Delta_{sol}G^{\circ}_{m} = \Delta_{hyd}G^{\circ}_{m} - \Delta_{lat}G^{\circ}_{m}. \tag{14.4}$$

14.2 Ions in solution

An ion in solution interacts with a solvent to create a range of regions with different structures about it. This is shown in Fig. 14.2, in which four separate regions are depicted: the primary *solvation shell*; the secondary solvation shell; a

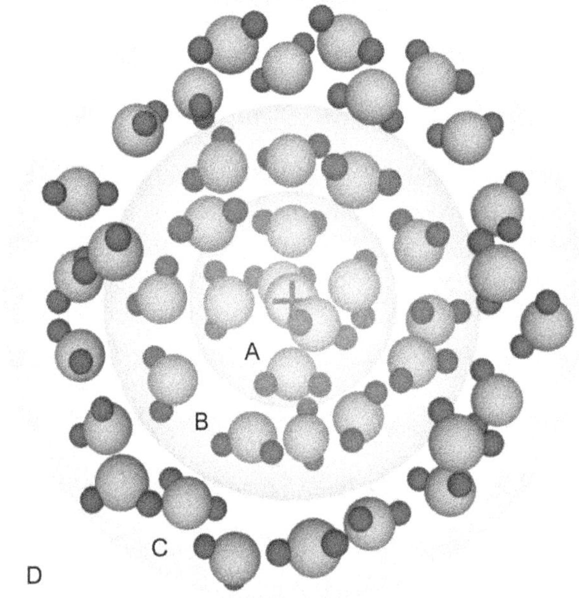

Figure 14.2 The solvation shells of an ion can be segregated into three or four regions, the ordering of which progressively changes from one determined by ion–solvent interactions to solvent–solvent interactions. Region A (primary solvation shell) is highly structured and contains a number of solvent molecules characteristic for the specific ion much like a coordination complex. In region D the ion exhibits no influence on the structure of the bulk liquid. The secondary solvation shell (region B) exhibits some ordering in response to the primary solvation shell. The disordered region C is only observed in highly structured protic solvents.

disordered region; and finally the bulk solvent. In the primary solvation shell, solvent molecules interact directly and strongly with the ion. In the secondary solvation shell, the solvent molecules are not in direct contact with the ion. However, they are influenced by the interaction of dipoles with the electrostatic field generated by the charge on the ion. At some distance away from the ion, the solvent is no longer influenced by the presence of the ion. This is region D, the bulk solvent. Polar protic solvents such as water and alcohols display a good deal of structure. Therefore, the structure of the secondary solvation shell is unlikely to match the structure of the bulk solvent. This leads to a disordered region C buffering the two. In less-structured solvents, this region is missing.

The 'ideal' solvation structure shown in Fig. 14.2 can only be achieved in very dilute solutions, on the order of 1 mM or less. It should also be borne in mind that the solvent molecules in the various regions are not static. They are in constant motion and are continuously being exchanged between regions.

Coordination complexes are well known in inorganic and organometallic chemistry. In these, metal ions take a central position and are coordinated by a number of ligands organized with well-defined symmetries. The first solvation shell, denoted A in Fig. 14.2, can be conceived of in an analogous fashion with water molecules acting as ligands, but in this case they are dynamic ligands. On average there will be a certain number of water molecules about the central ion (frequently six). The water molecules have a preferred orientation but they are free to move and exchange with the second shell and then beyond.

Nuclear magnetic resonance (NMR) can be used to probe the coordination sphere of some, but not all, metal ions in solution. The results of such investigations to determine the *solvation number* – the number of nearest-neighbor solvent molecules in solution – are shown in Table 14.1. It is found that temperature and pressure have very little effect on the solvation number. The clear result from these studies is that a solvation number of 6 is by far the most likely. Very small cations, such as Be^{2+}, or cations that exhibit a peculiar square-planar coordination, such as Pt^{2+} and Pd^{2+}, exhibit a lower solvation number of 4. Rather bulky solvents such as trimethyl phosphate (TMP) can also be forced to assume a solvation number of just 4. Only quite large cations, those of the lanthanides and actinides, are solvated by 8 or 9 molecules.

A number of kinetic methods have been developed to determine values related to the solvation number. These methods depend on the movement of solvated ions by observing diffusion, viscosity, conductivity, or transport numbers. Therefore,

Table 14.1 Cation solvation number determined from NMR spectral analysis. Data taken from Burgess, J. (1999) *Ions in Solutions: Basic principles of chemical interactions*. Woodhead Publishing, Cambridge, UK.

	Solvent						
	Water	**MeOH**	**TMP**	**MeCN**	**DMF**	**DMSO**	**NH_3(l)**
sp-elements							
Be^{2+}	4		4		4	4	
Mg^{2+}	6	6					6
Zn^{2+}	6	6					
Al^{3+}	6	6	4		6	6	6
Ga^{3+}	6	6	4	6	6	6	
In^{3+}, Sc^{3+}			4				
Transition metals							
V^{2+}	6						
Mn^{2+}	6	6		6			
Fe^{2+}, Co^{2+}, Ni^{2+}	6	6		6	6	6	
Ti^{3+}, V^{3+}	6					6	
Cr^{3+}, Fe^{3+}						6	
Pd^{2+}, Pt^{2+}	4						
Lanthanides							
Ce^{3+}, Pr^{3+}, Nd^{3+}					9		
$Tb^{3+} \rightarrow Yb^{3+}$					8		
Actinides							
Th^{4+}	9						

they often deliver numbers much different than 6 since these are measures of the number of solvent molecules that move with the ion. Values ranging from 1 to 12 or more are found.

Conspicuously absent from the NMR results are the important cations of the alkali and alkali earth metals. X-ray diffraction and neutron-scattering studies suggest solvation numbers ranging from 4 to 8.

The number of solvent molecules in the second solvation shell, region B in Fig. 14.2, is less accurately characterized. A value of 12 or more is likely, with larger cations such as the lanthanides and actinides having correspondingly more.

Anion solvation in water exhibits a wider range in solvation numbers in part due to the greater range of structures in the simple anions compared to bare metal cations. Diffraction and scattering show that solvation numbers range from 4 to 12 for most simple anions. A few specific values are: 7 or 8 for SO_4^{2-}; 8 for SeO_4^{2-} and ClO_4^-; 12 for CrO_4^{2-}, MoO_4^{2-} and WO_4^{2-}; and 4 or 5 for acetate. The nitrate ion has proved problematic with estimates ranging from 3 to 18.

In Chapter 4 we previously encountered the *radial distribution function* $g(r)$. This is a measure of the probability of finding a particle at a given distance away from some chosen point. In the present case, we choose the center of the solvated ion as the origin of our coordinates. An example of a radial distribution function is given in Fig. 14.3. The first peak represents the nearest-neighbor distance, in this case for Zn^{2+} in an aqueous solution. The distance from the Zn^{2+} to the H and O atoms of the solvating H_2O molecules are slightly shifted with respect to one another. This is expected because the partially negative O atom in H_2O is attracted to the positive charge on Zn^{2+} while the partially positive H atoms are repelled. The mean distances measured by diffraction and scattering for solvation closely track the distances observed in crystal hydrates. Increasing the charge on the ion shortens the ion–solvent internuclear separation. For example, this distance drops from 0.212 nm in $[Fe(OH_2)_6]^{2+}$ as compared to 0.205 nm in $[Fe(OH_2)_6]^{3+}$. A first approximation to these distances can be made by taking the sum of the ionic radius for the appropriate coordination number – often called the *Shannon radius* – and the mean radius of the water molecule. The radius of a water molecule is approximately 0.138 nm. A number of representative values for the metal ion to oxygen distance are given in Table 14.2.

14.2.1 Directed practice

Why are the peaks for O atoms and H atoms at different positions in Fig. 14.3?

14.2.2 Directed practice

Why is the ionic radius consistently much smaller than the metal ion to oxygen distance in aqueous solution?

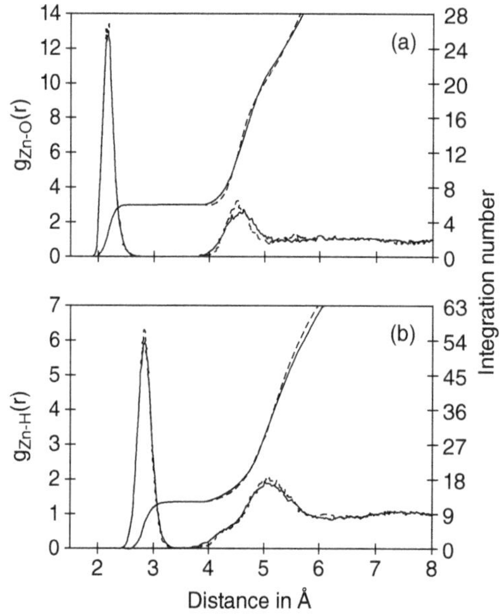

Figure 14.3 Radial distributions functions (RDF) and their running integration numbers obtained from combined quantum mechanical/molecular mechanics simulations at 90 °C (solid line) and 25 °C (dashed line). (a) Zn–O RDF. (b) Zn–H RDF. Reproduced with permission from Fatmi, M.Q., Hofer, T.S., Randolf, B.R., and Rode, B.M. (2006) *J. Phys. Chem. B*, **110**, 616. © 2006, American Chemical Society.

Table 14.2 Metal ion to oxygen distances (in Å) observed by X-ray diffraction in aqueous solutions and crystal hydrates. For comparison the ionic radii for coordination number 6 are given. Metal ion to oxygen distances from Burgess, J. (1999) *Ions in Solutions: Basic principles of chemical interactions*. Woodhead Publishing, Cambridge, UK. Ionic radii from 96th edition of the *CRC Handbook of Chemistry and Physics*.

Ion	Aqueous solution	Crystal hydrate	Ionic radius
Na^+	2.40	2.35–2.52	1.02
K^+	2.87–2.92	2.67–3.22	1.38
Ag^+	2.43		1.15
Mg^{2+}	2.10	2.01–2.14	0.72
Ca^{2+}	2.40	2.30–2.49	1.00
Mn^{2+}	2.20	2.00–2.18	0.83
Fe^{2+}	2.12	1.99–2.08	0.61
Ni^{2+}	2.04	2.02–2.11	0.69
Zn^{2+}	2.08–2.17	2.08–2.14	0.74
Fe^{3+}	2.05	2.09–2.20	0.55
Ce^{3+}	2.55	2.48–2.60	1.01
Nd^{3+}	2.51	2.47–2.51	0.98

14.3 The thermodynamic properties of ions in solution

14.3.1 Enthalpy of hydration

The enthalpy change upon the dissolution of an ionic compound can be readily measured by calorimetry, assuming its solubility is sufficiently high. For a sparingly soluble salt, the temperature dependence of its solubility and application of the integrated form of the van't Hoff equation are used to obtain the enthalpy of solution.

However, an ion cannot be made in solution in the absence of a counterion because a solution must remain neutral. Therefore, the enthalpy/entropy/Gibbs energy of formation of a single ion cannot be measured directly in solution. It is possible to estimate the absolute *enthalpy of hydration* of an ion. If we choose one to estimate accurately, then all other ions could have their values set by transitivity. That is, if we knew the absolute enthalpy change caused by dissolving 1 mol of H^+ (at infinite dilution), then by measuring the enthalpy change caused by dissolving 1 mol of HCl, HBr and HI, we could determine the molar enthalpy of hydration of Cl^-, Br^- and I^-. We could then measure the enthalpy of solution of the metal salts MX_n and so on to determine the molar enthalpies of hydration of the metal ions M^{n+}, and so forth.

This has been done to measure $\Delta_{\rm hyd}H^\circ_{\rm m}(H^+)$ by a series of experiments involving a number of salts HX with increasingly large anions X^-. As the size of the anion increases, the $\Delta_{\rm hyd}H^\circ_{\rm m}(X^-)$ tends to zero. By extrapolating the data to infinitely large and negligibly solvated X^-, the value for $\Delta_{\rm hyd}H^\circ_{\rm m}(H^+)$ is obtained. In this manner, the value $\Delta_{\rm hyd}H^\circ_{\rm m}(H^+) = -1091$ kJ mol^{-1} has been obtained. This value corresponds to the enthalpy change for the reaction

$$H^+(g) \rightleftharpoons H^+(aq).$$

The enthalpy of hydration depends on the size and charge of the ion. The greater the charge and the smaller the ion, the greater the enthalpy of hydration. The value for H^+ is particularly large compared to that of Li^+ (–515 kJ mol^{-1}). As expected, $\Delta_{\rm hyd}H^\circ_{\rm m}(Cs^+) = -263$ kJ mol^{-1} is much less than the value for the smaller Li^+. Doubling the charge greatly increases the enthalpy of hydration; for example, $\Delta_{\rm hyd}H^\circ_{\rm m}(Be^{2+}) = -2487$ kJ mol^{-1} and $\Delta_{\rm hyd}H^\circ_{\rm m}(Ba^{2+}) = -1304$ kJ mol^{-1}. Similar trends also hold for anions with $\Delta_{\rm hyd}H^\circ_{\rm m}(F^-) = -503$ kJ mol^{-1} and $\Delta_{\rm hyd}H^\circ_{\rm m}(I^-) = -298$ kJ mol^{-1}, while $\Delta_{\rm hyd}H^\circ_{\rm m}(SO_4^{2-}) = -1145$ kJ mol^{-1}.

The enthalpy of solvation also depends on the solvent. As mentioned above, many polar solvents actually solvate cations more effectively than water. However, they tend to solvate anions less effectively. This is quantified directly by the *enthalpy of transfer* $\Delta_{\rm trans}H^\circ_{\rm m}$ defined by

$$\Delta_{\rm trans}H^\circ_{\rm m} = \Delta_{\rm sol}H^\circ_{\rm m} - \Delta_{\rm hyd}H^\circ_{\rm m} \tag{14.5}$$

where $\Delta_{\rm sol}H^\circ_{\rm m}$ is the molar enthalpy change of solution in a solvent other than water. Several values for the enthalpy of transfer are shown in Table 14.3.

Table 14.3 Enthalpy of transfer in kJ mol^{-1} for ions transferred from aqueous solution to nonaqueous solution. Values taken from Burgess, J. (1999) *Ions in Solutions: Basic principles of chemical interactions*. Woodhead Publishing, Cambridge, UK.

	NH_3(l)	CH_3OH	CH_3CN	$(CH_3)_2SO$	HMPA
K^+	−30	−30	−26	−47	−64
Ag^+	−102	−35	−61	−69	
Ba^{2+}	−101	−85	−22	−102	−156
Cl^-		+8	+20	+19	

$(CH_3)_2SO$ = DMSO = dimethylsulfoxide; HMPA = hexamethylphosphoramide.

14.3.2 Enthalpy, entropy and gibbs energy of ion formation in solutions

For thermodynamic calculations, we need to establish the values of entropy, enthalpy and Gibbs energy. Rather than using a scale based on absolute entropies and enthalpies, it is easier to construct a scale based on relative values. This is done by selecting one reaction as a standard against which all other values are measured. We chose H^+(aq) as the standard ion. The corresponding the reaction is

$$\mathrm{H(aq)} \rightleftharpoons \mathrm{H^+(aq)} + \mathrm{e^-}$$

for which we set

$$\Delta_f G^\circ(\mathrm{H^+, aq}) = 0 \tag{14.6}$$

for all T. Thus.

$$S^\circ(\mathrm{H^+, aq}) = -\left(\frac{\partial \Delta_f G^\circ(\mathrm{H^+, aq})}{\partial T}\right)_p = 0 \tag{14.7}$$

and

$$\Delta_f H^\circ(\mathrm{H^+, aq}) = \Delta_f G^\circ(\mathrm{H^+, aq}) + T\,S^\circ(\mathrm{H^+, aq}) = 0. \tag{14.8}$$

Now transitivity combined with calorimetry – to measure $\Delta_f H^\circ$ directly – and equilibrium concentration measurements – via $\Delta G^\circ_m = -RT\ln K$ and the fundamental relationship $\Delta G^\circ = \Delta H^\circ - T\Delta S^\circ$ – can be used to determine the formation enthalpy, Gibbs energy and entropy of all species in solution.

For the reaction

$$\tfrac{1}{2}\mathrm{H_2(g)} + \tfrac{1}{2}\mathrm{Cl_2(g)} \rightarrow \mathrm{H^+(aq)} + \mathrm{Cl^-(aq)}$$

$$\Delta_r H^\circ = \Delta_f H^\circ(\mathrm{Cl^-, aq}) + \Delta_f H^\circ(\mathrm{H^+, aq}) - \tfrac{1}{2}\Delta_f H^\circ(\mathrm{H_2, g}) - \tfrac{1}{2}\Delta_f H^\circ(\mathrm{Cl_2, g}) = \Delta_f H^\circ(\mathrm{Cl^-, aq})$$

because the enthalpy of formation of an element in its standard states is equal to zero, just as the enthalpy of formation of H^+ is zero. Similarly, for the Gibbs energy change,

$$\Delta_r G^\circ = \Delta_f G^\circ(\mathrm{Cl^-, aq}).$$

Note, however, that the entropy change is

$$\Delta_r S^\circ = S^\circ(\mathrm{Cl^-, aq}) - \tfrac{1}{2}S^\circ(\mathrm{H_2, g}) - \tfrac{1}{2}S^\circ(\mathrm{Cl_2, g}).$$

Using this convention, all values are measured relative to H^+(aq). If an ion has a negative formation enthalpy, its formation is more exothermic than H^+(aq). If positive, its formation is less exothermic compared to H^+(aq).

The absolute entropy of (crystalline) compounds is set to zero at 0 K by the Third Law of Thermodynamics. However, the entropy of formation of ions in solution is a scale set relative to H^+(aq), not an absolute scale. Therefore, $S^\circ(X^\pm, aq)$ can be either positive or negative because, again, the entropy of formation is measured relative to H^+(aq). If the entropy of formation is negative, the solvation shell is more tightly held than that of H^+(aq) and has fewer microstates over which energy can be distributed. A positive value of $S^\circ(X^\pm, aq)$ means that the solvated ion leads to a system that has more microstates over which energy can be dispersed as compared to the system established by H^+(aq). Note that the choice

Table 14.4 Enthalpy, Gibbs energy and entropy of formation and heat capacity of aqueous ions in kJ mol^{-1}. Data from the 96th edition of the *CRC Handbook of Chemistry and Physics*.

	$\Delta_f H^\circ_m$/ kJ mol^{-1}	$\Delta_f G^\circ_m$/ kJ mol^{-1}	S°_m/ J K^{-1} mol^{-1}	$C_{p,m}$/ J K^{-1} mol^{-1}
Cations				
H^+	0	0	0	0
Ag^+	105.6	77.1	72.7	21.8
Ca^{2+}	–542.8	–553.6	–53.1	
Cs^+	–258.3	–292.0	133.1	–10.5
Cu^+	71.7	50.0	40.6	
Cu^{2+}	64.8	65.5	–99.6	
Fe^{2+}	–89.1	–78.9	–137.7	
Fe^{3+}	–48.5	–4.7	–315.9	
K^+	–252.4	–283.3	102.5	21.8
Li^+	–278.5	–293.3	13.4	68.6
Mg^{2+}	–466.9	–454.8	–138.1	
$NH_4{}^+$	–132.5	–79.3	113.4	79.9
Na^+	–240.1	–261.9	59.0	46.4
Ni^{2+}	–54.0	–45.6	–128.9	
Pb^{2+}	–1.7	–24.4	10.5	
Sn^{2+}	–8.8	–27.2	–17.0	
Zn^{2+}	–153.9	–147.1	–112.1	46.0
Anions				
Br^-	–121.6	–104.0	82.4	–141.8
CH_3COO^-	–486.0	–369.3	86.6	–6.3
Cl^-	–167.2	–131.2	56.5	–136.4
$ClO_4{}^-$	–129.3	–8.5	182.0	
$CO_3{}^{2-}$	–677.1	–527.8	–56.9	
$CrO_4{}^{2-}$	–881.2	–727.8	50.2	
$Cr_2O_7{}^{2-}$	–1490.3	–1301.1	261.9	
F^-	–332.6	–278.8	–13.8	–106.7
$HF_2{}^-$	–649.9	–578.0	92.5	
I^-	–55.2	–51.6	111.3	–142.3
$NO_3{}^-$	–207.4	–111.3	146.4	–86.6
OH^-	–230.0	–157.2	–10.8	–148.5
SO_4^{2-}	–909.3	–744.5	20.1	–293.0

of $S^\circ(H^+, aq) = 0$ demands that the heat capacity has the value $C_p(H^+, aq) = 0$. This means that the heat capacity of ions in solution is also measurable relative to $H^+(aq)$.

A polar protic solvent such as water is highly structured compared to most liquids. Inspecting the values in Table 14.4, we see that large singly charged ions tend to have positive entropies of formation. This is because they disrupt the structure the hydrogen-bonding network. Small ions, especially with multiple charges, exhibit negative entropies of formation since they pull solvating water molecules into a tightly held first solvation shell. Indeed, the attraction of water to such ions is so great that there is actually a reduction in the volume of dilute solutions of these ions compared to pure water.

14.3.3 Directed practice

The value of S°_m for $Zn^{2+}(aq)$ is –112.1 J K^{-1} mol^{-1}, but that of SO_4^{2-} is +20.1 J K^{-1} mol^{-1}. Discuss.

14.4 The activity of ions in solution

The interactions between ions are so strong that the approximation of replacing activities with concentration is valid only in very dilute solutions (<10^{-3} mol kg^{-1} in total ion concentration).

Consider a compound MX that dissolves according to

$$\mathrm{MX(s)} \rightleftharpoons \mathrm{M^+(solvated)} + \mathrm{X^-(solvated)}$$

MX is called a 1–1 electrolyte as both ions have a charge magnitude of 1. The chemical potential of the solute is

$$\mu_{\text{solute}} = \mu_+ + \mu_- \tag{14.9}$$

where μ_+ is the chemical potential of a univalent cation M^+, and μ_- is the chemical potential of a univalent anion X^-. In a real solution we express the chemical potential in terms of activity

$$\mu_+ = \mu_+^\circ + RT\ln a_+ = \mu_+^\circ + RT\ln\frac{b_+}{b^\circ} + RT\ln\gamma_+ \tag{14.10}$$

$$\mu_- = \mu_-^\circ + RT\ln a_- = \mu_-^\circ + RT\ln\frac{b_-}{b^\circ} + RT\ln\gamma_- \tag{14.11}$$

The deviations from ideality are accounted for by the term involving the activity coefficient. There is no way experimentally to separate the individual contributions from the positive and negative ions. Therefore, we introduce the *mean ionic activity coefficient* $\gamma_\pm$ and the *mean ionic activity* $a_\pm$, *mean ionic chemical potential* $\mu_\pm$, and *mean ionic molality* $b_\pm$

$$\gamma_\pm = (\gamma_+\gamma_-)^{1/2}. \tag{14.12}$$

$$a_\pm = (a_+a_-)^{1/2}. \tag{14.13}$$

$$b_\pm = (b_+b_-)^{1/2}. \tag{14.14}$$

$$\mu_\pm = \frac{\mu_{\text{solute}}}{2} = \mu_\pm^\circ + RT\ln a_\pm = \mu_\pm^\circ + RT\ln\frac{b_\pm}{b^\circ} + RT\ln\gamma_\pm \tag{14.15}$$

Now consider of a compound $M_{\nu_+}X_{\nu_-}$ that dissolves according to

$$M_{\nu_+}X_{\nu_-} \rightleftharpoons \nu_+M^{z+} + \nu_-X^{z-}$$

The mean ionic activity coefficient and activity are the geometric means of the individual quantities

$$\gamma_\pm = \left(\gamma_+^{\nu_+}\gamma_-^{\nu_-}\right)^{1/\nu}, \quad \nu = \nu_+ + \nu_- \tag{14.16}$$

$$a_\pm = \left(a_+^{\nu_+}a_-^{\nu_-}\right)^{1/\nu} \tag{14.17}$$

$$b_\pm = \left(b_+^{\nu_+}b_-^{\nu_-}\right)^{1/\nu} \tag{14.18}$$

The chemical potential is still written as before; however, it must reflect the moles of ions released into solution

$$\mu_\pm = \frac{\mu_{\text{solute}}}{\nu} = \mu_\pm^\circ + RT\ln a_\pm = \mu_\pm^\circ + RT\ln\frac{b_\pm}{b^\circ} + RT\ln\gamma_\pm \tag{14.19}$$

14.5 Debye–Hückel theory

When an ion moves in solution, the solvation shells and the counterions around it need to respond. Since they cannot respond immediately, the structure of the solvated ion + the solvation shells + counterions is distorted from the ideal structure. This relaxation effect leads to a decrease in the mobility of the moving ion.

The *ionic atmosphere* (the counterions around the central ion) also has the effect of increasing the viscous drag on the central ion because the ionic atmosphere moves in the opposite direction to the central ion. The enhanced viscous drag is called the *electrophoretic effect* and reduces the mobility of the ions.

Assume that the long-range Coulombic interactions are primarily responsible for deviations from ideality. On a time-averaged basis an ion is surrounded symmetrically by ions of the opposite charge (counterions). The interaction of the ion with the ionic atmosphere (the 'cloud' of counter charge surrounding it) lowers the energy (and chemical potential) of the ion in solution as a result of electrostatic interactions. This lowering of chemical potential is embodied in the $RT\ln\gamma_\pm$ term above. A detailed calculation leads to the simple result contained in the *Debye–Hückel limiting law*

$$\lg\gamma_\pm = -|z_+z_-|BI^{1/2} \tag{14.20}$$

where $B = 0.5092$ for an aqueous solution at 25 °C and I is the dimensionless *ionic strength*

$$I = \frac{1}{2}\sum_i z_i^2\frac{b_i}{b^\circ}. \tag{14.21}$$

If there are only two types of ions in solution

$$I = \frac{1}{2}\left(\frac{b_+ z_+^2 + b_- z_-^2}{b^\circ}\right). \tag{14.22}$$

z_i is the charge number of ion i and b_i is its molality. Recall that $\ln x = (\ln 10)\lg x$. For a 1:1 electrolyte XY, this simplifies to

$$I = b/b^\circ$$

because $b_X = b_Y = b$ and the absolute values of the charges are both 1. The limiting law is accurate up to about 0.01–0.001 mol kg^{-1}, depending on the charges on the ions.

Why the breakdown? Ions are assumed to be point charges. This assumption breaks down as the solvent shells of individual ions start to overlap. This leads to increased repulsive interactions that raise $\gamma_\pm$ compared to prediction, which is based on perfect stabilization of the solution through electrostatic interactions. The solvent is assumed to be a structureless dielectric, a continuous fluid rather than a medium composed of molecules. Ion pairing can occur at high concentrations. If water molecules are bound tightly in a solvation shell, they do not act like bulk water molecules any more.

For higher concentrations, the empirical *Davies equation* extends the estimate of the mean ionic activity coefficient

$$\lg \gamma_\pm = -B|z_+ z_-|\frac{I^{1/2}}{1 + I^{1/2}} - 0.20\,I. \tag{14.23}$$

At higher concentrations yet, empirical formula can be used on a case-by-case basis.

14.5.1 Directed practice

Calculate the ionic strength and mean activity coefficient for 0.05 molal NaCl according to both the Debye–Hückel limiting law and the Davies equation at 25 °C.

[Answers: I=0.05, $\gamma_\pm$=0.769, $\gamma_\pm$=0.789]

14.5.2 Directed practice

The deviations from the values predicted by the Debye–Hückel limiting law in Fig. 14.4 are toward values of $\gamma_\pm$ that are closer to 1 rather than further away from 1. Is this expected based on the limitations of the Debye–Hückel theory?

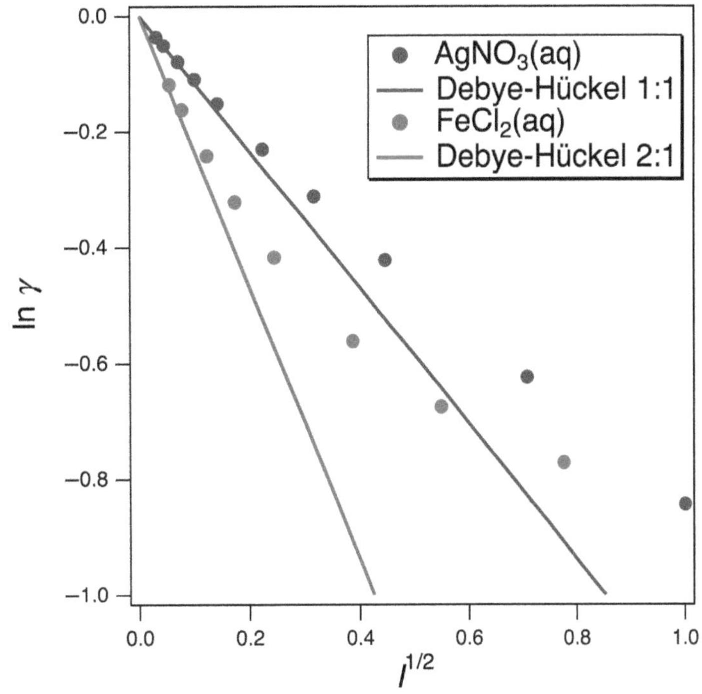

Figure 14.4 The predictions of the Debye–Hückel limiting law are compared to experimental data for a typical 1:1 electrolyte ($AgNO_3$) and a 2:1 electrolyte ($FeCl_2$). Data taken from 96th edition of *CRC Handbook of Chemistry and Physics*.

14.6 Use of activities in equilibrium calculations

Let us apply what we know about equilibria and electrolyte solutions to calculate the composition and activities of electrolytes in solution for three specific examples: an acid–base equilibrium; a buffer solution; and the solubility product. We have treated two of these previously in Chapter 11, assuming that the solutions act ideally. Now, we will add the use of activity and the activity coefficient.

14.6.1 Acid–base equilibrium

Consider 0.1 mol kg^{-1} acetic acid, CH_3COOH, at 25 °C in aqueous solution, for which the acid dissociation constant is $K_a = 1.75 \times 10^{-5}$.

a First, assume that concentration instead of activity can be used and determine the molalities of H_3O^+, CH_3COO^-, and OH^-.

b Use the values found in (a) to approximate the ionic strength. Use this value of the ionic strength to calculate the approximate value of $\gamma_\pm$. Use this value of $\gamma_\pm$ to recalculate the molalities and activities at equilibrium. In all cases assume that the activity coefficient of undissociated CH_3COOH equals 1.

Start with a reaction chart for the balanced chemical reaction:

	HOAc	⇌	AcO$^-$(aq)	+	H$^+$(aq)
Initially	0.1		0		0
Equilibrium	$0.1 - x$		x		x

We need to solve the equilibrium constant expression to determine the molalities at equilibrium

$$K_a = \frac{a_{AcO^-}\, a_{H^+}}{a_{HOAc}} = \frac{\gamma_\pm b_{AcO^-}\, \gamma_\pm b_{H^+}}{\gamma_{HOAc}\, b_{HOAc}} = \frac{\gamma_\pm x\, \gamma_\pm x}{0.1 - x} = \frac{\gamma_\pm^2 x^2}{0.1 - x}$$

First, we assume that $\gamma_\pm = 1$ and solve the resulting quadratic equation for x.

$$x^2 + K_a x - 0.1K_a = 0$$

$$x = \frac{-b \pm \sqrt{b^2 - 4ac}}{2a} = \frac{-K_a \pm \sqrt{K_a^2 - 4\,(-0.1)\,K_a}}{2} = 1.314 \times 10^{-3}.$$

Substituting for x into the expressions we previously derived for the molalities

$$b(H^+) = b(AcO^-) = 1.31 \times 10^{-3}\,\text{mol kg}^{-1},\ b(HOAc) = 1 - x = 0.0987\,\text{mol kg}^{-1}.$$

Since $\gamma_\pm = 1$ and $a = \gamma_\pm\, b$, the activities are numerically equal to the molalities but they are unitless.

$$a(H^+) = a(AcO^-) = 1.31 \times 10^{-3},\ a(HOAc) = 0.0987.$$

The concentration of OH^- is found from

$$b_{OH^-} b_{H^+}/(b^\circ)^2 = 10^{-14},\quad b_{OH^-} = \frac{10^{-14}}{1.31 \times 10^{-3}} = 7.63 \times 10^{-12}\ \text{mol kg}^{-1}$$

(b) First, determine the ionic strength I with $b_+ = b_- = x = 1.314 \times 10^{-3}$ mol kg^{-1}.

$$I = \tfrac{1}{2}(b_+(1)^2 + b_-(1)^2)/b^\circ = \tfrac{1}{2}(2x) = x = 1.314 \times 10^{-3}$$

Use this to determine the mean activity coefficient from the Debye–Hückel limiting law

$$\lg \gamma_\pm = -|z_+ z_-| B I^{1/2} = -|(1)(-1)|(0.509)(1.314 \times 10^{-3})^{1/2} = -0.01845$$

$$\gamma_\pm = 0.9584$$

Now, substitute this value into the K_a expression, and determine a new value of x.

$$K_a = \frac{\gamma_\pm^2 x^2}{0.1 - x}$$

$$\gamma_\pm^2 x^2 + K_a x - 0.1K_a = 0$$

$$x = \frac{-K_a \pm \sqrt{K_a^2 - 4\gamma_\pm^2\,(-0.1)\,K_a}}{2\gamma_\pm^2} = 1.342 \times 10^{-3}.$$

Substituting for x into the expressions we previously derived for the molalities

$$b(H^+) = b(AcO^-) = 1.34 \times 10^{-3}\,\text{mol}\,\text{kg}^{-1},\ b(HOAc) = 1 - x = 0.0987\,\text{mol}\,\text{kg}^{-1}.$$

Since $\gamma_\pm = 0.9584$ and $a = \gamma_\pm\, b$, the activities of the ions are

$$a(H^+) = a(AcO^-) = 0.9584(1.342 \times 10^{-3}) = 1.29 \times 10^{-3}\,\text{mol}\,\text{kg}^{-1}$$

We do not know directly the activity coefficient of undissociated acetic acid. However, since it is uncharged (a non-electrolyte) and not too concentrated, we can assume that that it is one; thus,

$$a(HOAc) = \gamma b(HOAc) \cong (1)(0.0987) = 0.0987.$$

The concentration of OH^-, calculated as before, is 7.46×10^{-12} mol kg^{-1}. Note that in this case the change is not very large because the acid is weak, which produces a very low concentration of ions in the millimolar range.

14.6.2 Buffers

The pK_a of crotonic acid is 4.69. The pH of a *buffer* in terms of the activities of the conjugate base and the undissociated acid is given by the *Henderson–Hasselbalch equation*

$$\text{pH} = \text{p}K_a + \lg \frac{a(A^-)}{a(HA)}. \tag{14.24}$$

We assume the activity coefficient of the undissociated acid can be taken as 1 and the activity is equal to the molarity.

a Assume that the mean activity coefficient is 1 and calculate the hydrogen ion concentration [H^+] in a buffer made from 0.200 M sodium crotonate in 0.200 M crotonic acid.

b Use the values obtained in (a) to estimate the mean ionic activity coefficient using the Debye–Hückel limiting law. Calculate the [H^+] of the buffer made up as in (a). What is the percentage change in the calculated value of [H^+]?

(a) Since there is only one value of pK_a given, crotonic acid must be monoprotic and of the form HA → H^+ + A^-. Since it is a weak acid, the concentration of H^+ is very small, can be neglected, and the concentrations of Na^+ and A^- are set by the sodium crotonate. Therefore [HA] = 0.200 M, [Na^+] = [A^-] = 0.200 M. The pH of the solution is

$$\text{pH} = \text{p}K_a + \lg\left(\frac{[A^-]}{[HA]}\right) = 4.69 + \lg\left(\frac{0.2}{0.2}\right) = 4.69$$

$$[H^+] = 10^{-4.69} = 2.04 \times 10^{-5}\,\text{M}$$

(b) First, calculate the mean activity coefficient from the ionic strength. Because this is a one-to-one electrolyte, the ionic strength is equal to the molarity of the salt. Thus,

$$\lg \gamma_\pm = -\left|z_+z_-\right| BI^{1/2} = -\left|(1)\,(-1)\right|\,(0.509)\,(0.2)^{1/2} = -0.2276$$

$$\gamma_\pm = 0.5921$$

$$a\,(A^-) = \gamma_\pm\,[A^-] = 0.5921\,[0.200] = 0.118$$

$$\text{pH} = \text{p}K_a + \lg\left(\frac{a\,(A^-)}{[HA]}\right) = 4.69 + \lg\left(\frac{0.118}{0.200}\right) = 4.462$$

$$\left[H^+\right] = 10^{-4.462} = 3.45 \times 10^{-5}\,\text{M}$$

$$\%\ \text{change} = 100\%\left(\frac{3.45 \times 10^{-5} - 2.04 \times 10^{-5}}{2.04 \times 10^{-5}}\right) = 69\%$$

The difference is now important; however, in the tenth molar range, the solution may have entered a range in which the Debye–Hückel limiting law is losing its accuracy. The limiting law provides an estimate of the activity coefficients; however if precise values are required, then direct measurement may be required.

14.6.3 Solubility product

The solubility of AgCl in water at 25 °C is 1.34×10^{-5} mol dm^{-3}. Assuming that the Debye–Hückel limiting law applies:

a Calculate $\Delta G°$ for the dissolution of AgCl(s)

b Calculate the solubility of AgCl in water and an 0.05 M solution of K_2SO_4.

Table 14.5 Solubility product and Gibbs energy of solution at 298.15 K for selected sparingly soluble salts. K_{sp} data from 96th edition of the *CRC Handbook of Chemistry and Physics.*

	K_{sp}	$\Delta_{sol}G^\circ_m$/ kJ mol^{-1}
$CaCO_3$	3.36×10^{-9}	48.4
CaF_2	3.45×10^{-11}	59.7
$Ca(OH)_2$	5.02×10^{-6}	30.2
$MgCO_3$	6.82×10^{-6}	29.5
MgF_2	5.16×10^{-11}	58.7
$Mg(OH)_2$	5.61×10^{-12}	64.2
Ag_2CO_3	8.46×10^{-12}	63.2
$AgCl$	1.77×10^{-10}	55.7
Ag_3PO_4	8.89×10^{-17}	91.6
$ZnCO_3$	1.46×10^{-10}	56.1
ZnF_2	3.04×10^{-2}	8.66
$Zn(OH)_2$	3.00×10^{-17}	94.3

The dissolution reaction is

$$\mathrm{AgCl(s)} \rightarrow \mathrm{Ag^+(aq)} + \mathrm{Cl^-(aq)}$$

For each mole of AgCl that dissolves, 1 mol each of the Ag^+ and Cl^- are produced. Hence, if $s = 1.34 \times 10^{-5}$ mol dm^{-3}, then s is also the concentration of Ag^+ and Cl^-. The solubility product expression is thus

$$K_{sp} = a^+_{Ag}a^-_{Cl} = \gamma_\pm[\mathrm{Ag^+}]\gamma_\pm[\mathrm{Cl^-}] = \gamma_\pm^2 s^2.$$

The mean ionic activity coefficient is given by

$$\lg\gamma_\pm = -|z_+z_-|BI^{1/2} = -|(1)(1)|0.509[(1.34\times 10^{-5})^{1/2}]$$
$$\lg\gamma_\pm = -0.001861$$
$$\gamma_\pm = 0.996$$

Now, substitute this into the K_{sp} expression.

$$K_{sp} = \gamma_\pm^2 s^2 = (0.996)^2(1.34\times 10^{-5})^2 = 1.77\times 10^{-10}.$$

Consistent with the values in Table 14.5, the Gibbs energy change is

$$\Delta G^\circ_m = -RT\ln K_{sp} = -8.314(298.15)\ln(1.77\times 10^{-10})$$
$$\Delta G^\circ_m = 55.7\,\mathrm{kJ\,mol^{-1}}.$$

(b) From part (a) we see that the solubility in water is

$$s = \frac{K_{sp}^{1/2}}{\gamma_\pm} = \frac{(1.77\times 10^{-10})^{1/2}}{0.996} = 1.34\times 10^{-5}\,\mathrm{M}$$

Dissolving the sulfate changes the ionic strength I and, therefore, the mean activity coefficient. K_{sp} is a constant but the solubility s changes in response.

$$I = \tfrac{1}{2}(0.1 + 0.05\times 2^2) = 0.15$$
$$\lg\gamma_\pm = -2(0.509)(0.015)^{1/2} = -0.3943$$
$$\gamma_\pm = 0.403$$
$$s = \frac{K_{sp}^{1/2}}{\gamma_\pm} = \frac{(1.77\times 10^{-10})^{1/2}}{0.403} = 3.30\times 10^{-5}\,\mathrm{M}$$

The solubility increases by $\frac{3.30-1.34}{1.34}(100\%) = 146\%$ compared to the initial value. This is an example of a *salt effect* in solubility – a change in the solubility of a solute in aqueous solution on the addition of a salt that does not possess a common ion with the original solute. If the solubility increases on the addition of a salt, the addition is said to cause 'salting-in'; if the opposite, it causes 'salting-out.'

14.6.4 Directed practice

Calculate the solubility of $Zn(OH)_2$ assuming it acts ideally in solution and the standard state is $b° = 1$ molal.

[Answer: 1.96×10^{-6} mol kg^{-1}]

14.7 Charge transport

Under the influence of an electric field, ions migrate in solution. The speed at which the ions migrate is related to their hydrodynamic radius. The size of the hydrodynamic radius depends not only the size of the ion but also on how the ion interacts with its solvent shell. Li, for example, has an anomalously low mobility because it binds water so strongly in its first solvent shell. H^+ and OH^- have extremely high mobilities because they can be transported by rearrangements of water molecules.

We need to understand the conduction of charge in solutions to understand electrochemistry. Solutes that do not produce ions in solutions are *nonelectrolytes* and do not conduct electricity well. *Strong electrolytes* dissociate completely in solutions to create ions. *Weak electrolytes* are partially dissociated species. The electrical *conductance* G of a cell filled with a solution is measured in siemens ($S \equiv \Omega^{-1}$, where Ω is the symbol for ohm the unit of resistance), and depends on the species in solution, its concentration and the geometry of the cell (cross-sectional area A and length l) according to

$$G = \kappa(A/l). \tag{14.25}$$

κ is the *conductivity*. More useful is the *molar conductivity* Λ

$$\Lambda = \kappa/c, \tag{14.26}$$

where c is the molar concentration. The SI units of molar conductivity are S m^2 mol^{-1}.

At infinite dilution, all electrolytes are completely dissociated. The value of molar conductivity at infinite dilution is given the symbol $\Lambda°$. Weak electrolytes dissociate less as concentration increases and the *degree of dissociation* α is directly related to the conductivity

$$\alpha = \Lambda/\Lambda°. \tag{14.27}$$

For a 1:1 weak electrolyte, such as a weak acid

$$AB \rightleftharpoons A^+ + B^-$$

The molar conductivity is described by *Ostwald's dilution law*

$$\frac{1}{\Lambda} = \frac{1}{\Lambda°} + \frac{\kappa}{K_a(\Lambda°)^2}. \tag{14.28}$$

The behavior of strong electrolytes is different because their dissociation character has a different dependence on concentration. The molar conductivity of strong electrolytes is described by the *Debye–Hückel–Onsager equation*

$$\Lambda = \Lambda° - (A + B\Lambda°)\, c^{1/2} \tag{14.29}$$

A and B are constants that can be calculated on the basis of Debye–Hückel theory of strong electrolytes solutions. At 25 °C in aqueous solutions of 1:1 electrolytes their values are $A = 60.20 \times 10^{-4}$ S m^2 mol^{-1} (mol l^{-1})$^{-1/2}$ and $B = 0.229 \times 10^{-4}$ (mol l^{-1})$^{-1/2}$. The Debye–Hückel–Onsager equation is reliable for $c < 0.01$ mol l^{-1} with increasing deviations for higher concentrations that are more exaggerated for higher charged ions, as shown in Fig. 14.5.

Each ion in solution makes a contribution to the conductivity quantified by the molar ionic conductivity of the cations λ_+ and of the anions λ_-. Selected values are found in Table 14.6. The total conductivity is thus made up of contributions from the positive and negative ions, and at infinite dilution, acts according to *Kohlrausch's law* of independent migration of ions

$$\Lambda° = \nu_+\lambda_+^\circ + \nu_-\lambda_-^\circ, \tag{14.30}$$

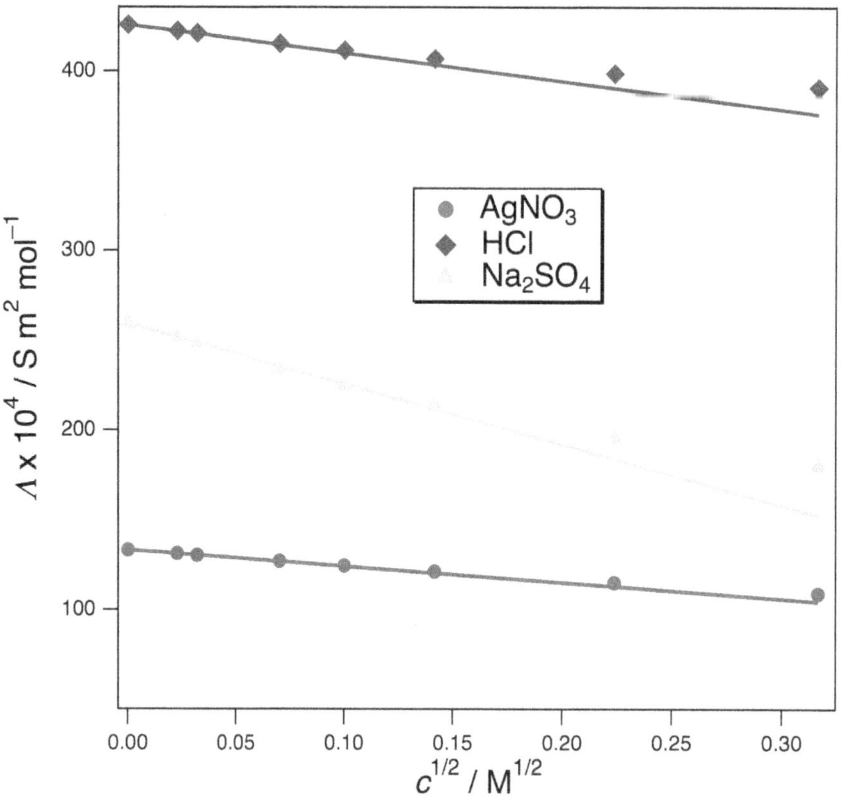

Figure 14.5 Molar conductivity for $AgNO_3$, HCl and Na_2SO_4 are plotted against the square-root of concentration and compared to the results of the Debye–Hückel–Onsager equation (solid lines). Data taken from the 96th edition of the *CRC Handbook of Chemistry and Physics*.

where ν_+ and ν_- are the stoichiometric coefficients in the molecular formula of the salt. For example, the dissolution of $Ca(OH)_2$ corresponds to

$$^1\!/\!_2\,\mathrm{Ca(OH)_2(aq)} \rightleftharpoons {}^1\!/\!_2\,\mathrm{Ca^{2+}(aq)} + \mathrm{OH^-(aq)}$$

$$\frac{1}{2}\Lambda^\circ\left(\mathrm{Ca(OH)_2}\right) = \frac{1}{2}\lambda\left(\mathrm{Ca^{2+}}\right) + \lambda\left(\mathrm{OH^-}\right) = (59.5 + 198) \times 10^{-4}$$

$$\frac{1}{2}\Lambda^\circ(\mathrm{Ca(OH)_2}) = 257.5 \times 10^{-4}\ \mathrm{S\ m^2\ mol^{-1}}$$

Table 14.6 Limiting molar conductivities of selected ions in aqueous solutions at 25 °C. Taken from the 96th edition of the *CRC Handbook of Chemistry and Physics*. $\lambda_\pm$ values given in units of 10^{-4} S m^2 mol^{-1}.

Cation	λ_+	Anion	λ_-
H^+	349.65	OH^-	198
Li^+	38.66	F^-	55.4
Na^+	50.08	Cl^-	76.31
K^+	73.48	Br^-	78.1
Ag^+	61.9	ClO_4^-	67.3
NH_4^+	73.5	CN^-	78
$\frac{1}{2}Mg^{2+}$	53.0	NO_3^-	71.42
$\frac{1}{2}Ca^{2+}$	59.47	$CHOO^-$	54.6
$\frac{1}{2}Fe^{2+}$	54	CH_3COO^-	40.9
$\frac{1}{3}Fe^{3+}$	68	$\frac{1}{2}CO_3^{2-}$	69.3
$\frac{1}{2}Cu^{2+}$	53.6	$\frac{1}{2}SO_4^{2-}$	80.0
$\frac{1}{2}Zn^{2+}$	52.8	$\frac{1}{3}PO_4^{3-}$	92.8

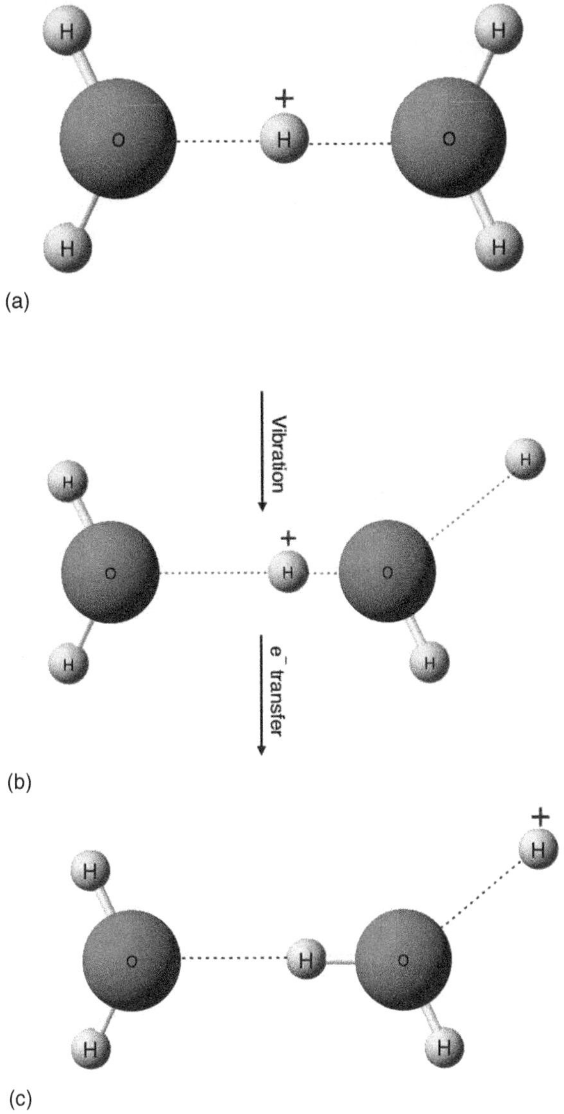

Figure 14.6 The exchange mechanism of proton transport in water. (a) A hydrated proton is held between water molecules (the full solvation shell is not shown). (b) Simple vibrational motion of the proton and a hydrogen atom in a water molecule take the solvation complex to a transition state. Such vibrational motion is significantly faster than physically moving the entire proton from one side of the molecule to the other. (c) After attaining the transition state, an electron hops from H to H^+, which switches the position of which is the H atom bound to water and which is the proton solvated by water molecules. The result is apparent transport of H^+ that is much faster than any other ion in aqueous solution.

The nature of the solvated ion determines the value of the ionic conductivity. The solvated proton has abnormally high equivalent ionic conductivity, $\lambda(H^+) = 349.65 \times 10^{-4}$ S m^2 mol^{-1}. Similarly, the value of $\lambda(OH^-) = 198 \times 10^{-4}$ S m^2 mol^{-1} is much higher than usual for an anion. These anomalously high values are related to these two species being components of water, which means that they have available to them unique exchange mechanisms for the transport of charge in water, explained schematically in Fig. 14.6 for H^+. Most simple inorganic ions of charge z in water have an ionic conductivity of roughly $\lambda_\pm/z \approx 60 \times 10^{-4}$ S m^2 mol^{-1} and falling within the range $(30\text{–}110) \times 10^{-4}$ S m^2 mol^{-1}. Organic ions tend to have lower ionic conductivities in the range of $(20\text{–}60) \times 10^{-4}$ S m^2 mol^{-1}. The anomalously large value of $\lambda(H^+)$ is apparent in Fig. 14.5 as the molar conductivity of HCl is significantly higher than that of either $AgNO_3$ or Na_2SO_4.

14.7.1 Directed practice

Calculate the molar conductivity at infinite dilution of $HClO_4$ and $NaClO_4$. Why are the values so different?

[Answer: 41.7 and 11.7 mS m^2 mol^{-1}]

SUMMARY OF IMPORTANT EQUATIONS

$\Delta_{trans}H_m^o = \Delta_{sol}H_m^o - \Delta_{hyd}H_m^o$	The transfer enthalpy is equal to the difference between the enthalpy of solvation and the enthalpy of hydration
$\mu_\pm = \mu_\pm^\circ + RT\ln a_\pm = \mu_\pm^\circ + RT\ln\frac{b_\pm}{b^\circ} + RT\ln\gamma_\pm$	Mean ionic activity coefficient $\gamma_\pm$, mean ionic activity $a_\pm$, mean ionic chemical potential $\mu_\pm$ and mean ionic molality $b_\pm$
$\lg\gamma_\pm = -\lvert z_+z_-\rvert BI^{1/2}$	Debye–Hückel limiting law: charges on the ions $z_\pm$, $B = 0.5092$ for an aqueous solution at 25 °C
$I = \frac{1}{2}\sum_i z_i^2\frac{b_i}{b^\circ}$	The ionic strength I can be calculated in terms of molality b or molarity c.
$\lg\gamma_\pm = -B\lvert z_+z_-\rvert\frac{I^{1/2}}{1+I^{1/2}} - 0.20I$	Davies equation
$\mathrm{pH} = \mathrm{p}K_a + \lg\frac{a(\mathrm{A^-})}{a(\mathrm{HA})}$	Henderson–Hasselbalch equation
$\Lambda = \kappa/c$	Molar conductivity, Λ, is the conductivity κ divided by the molar concentration c
$\Lambda^\circ = \nu_+\lambda_+^\circ + \nu_-\lambda_-^\circ$	Kohlrausch's law: limiting molar conductivity $\lim_{c\to 0}\Lambda = \Lambda^\circ$, ν_+ and ν_- are the stoichiometric coefficients, molar ionic conductivity of cations λ_+ and anions λ_-
$\alpha = \Lambda/\Lambda^\circ$	The degree of dissociation α of a weak electrolyte
$\frac{1}{\Lambda} = \frac{1}{\Lambda^\circ} + \frac{\kappa}{K_a(\Lambda^\circ)^2}$	Ostwald's dilution law: acid dissociation constant K_a
$\Lambda = \Lambda^\circ - (A + B\Lambda^\circ)c^{1/2}$	Debye–Hückel–Onsager equation: At 25 °C in aqueous solutions of 1:1 electrolytes, $A = 60.20 \times 10^{-4}$ S m^2 mol^{-1} (mol l^{-1})$^{-1/2}$ and $B = 0.229 \times 10^{-4}$ (mol l^{-1})$^{-1/2}$

Exercises

14.1 Calculate $\Delta_r H^\circ$, $\Delta_r S^\circ$, and $\Delta_r G^\circ$ at 298 K for the reaction

$$\mathrm{BaCl_2\,(aq) + Na_2SO_4\,(aq) \rightarrow BaSO_4\,(s) + 2NaCl\,(aq)}.$$

14.2 Calculate $\Delta_r H^\circ$, $\Delta_r S^\circ$ and $\Delta_r G^\circ$ at 298 K for the reaction

$$\mathrm{2AgNO_3\,(aq) + ZnCl_2\,(aq) \rightarrow 2AgCl\,(s) + Zn\left(NO_3\right)_2\,(aq)}.$$

14.3 Describe the Debye–Hückel model and the Debye–Hückel limiting law (write it and identify the parameters). What factors affect the transport of an ion through a solution?

14.4 Calculate the activities of all ions in 2.15 mM solutions of the strong electrolytes (a) NaCl (b) $(NH_4)_2ZrF_6$ and (b) $CeCl_3$ using the Debye–Hückel limiting law. Place your answers in a table such as that below.

<table>
<tr><td>Solution</td><td>NaCl</td><td colspan="2">$(NH_4)_2ZrF_6$</td><td colspan="2">$CeCl_3$</td></tr>
<tr><td>I</td><td></td><td colspan="2"></td><td colspan="2"></td></tr>
<tr><td>$\log\gamma_\pm$</td><td></td><td colspan="2"></td><td colspan="2"></td></tr>
<tr><td>$\gamma_\pm$</td><td></td><td colspan="2"></td><td colspan="2"></td></tr>
<tr><td>a</td><td></td><td></td><td></td><td></td><td></td></tr>
</table>

14.5 Calculate the activities of all ions in 1.25 mM of the strong electrolytes (a) HCl (b) $VOSO_4$ and (b) K_2SO_4 using the Debye–Hückel limiting law. Place your answers in the table below.

<table>
<tr><td>Solution</td><td>NaCl</td><td colspan="2">$VOSO_4$</td><td colspan="2">K_2SO_4</td></tr>
<tr><td>I</td><td></td><td colspan="2"></td><td colspan="2"></td></tr>
<tr><td>$\log \gamma_\pm$</td><td></td><td colspan="2"></td><td colspan="2"></td></tr>
<tr><td>$\gamma_\pm$</td><td></td><td colspan="2"></td><td colspan="2"></td></tr>
<tr><td>α</td><td></td><td></td><td></td><td></td><td></td></tr>
</table>

14.6 A 0.200 mol kg^{-1} aqueous solution of NH_4Cl has a measured mean ionic activity coefficient of 0.718 at $T = 25$ °C. Calculate the percent difference between the measured value and the value obtained from the Debye–Hückel limiting law.

14.7 Chlorous acid, $HClO_2$, has a dissociation constant of $K_a = 1.20 \times 10^{-2}$. Calculate the degree of dissociation for a 0.0755 mol kg^{-1} solution of this acid using the Debye–Hückel limiting law to estimate the mean ionic activity coefficient.

14.8 Lactic acid, HA, has a dissociation constant of pK_a = 3.86. Calculate the degree of dissociation for a 0.100 mol kg^{-1} solution of this acid using the Debye–Hückel limiting law to estimate the mean ionic activity coefficient.

14.9 (a) Silver hydroxide, AgOH, has a solubility product of $K_{sp} = 1.10 \times 10^{-4}$. Calculate the concentrations of Ag^+(aq) and OH^-(aq) in a saturated solution of AgOH assuming all activity coefficients are equal to one. (b) Using your answer from part (a) to calculate the ionic strength, calculate the activities of Ag^+(aq) and OH^-(aq) in a saturated solution of AgOH using the Debye–Hückel limiting law to estimate the mean ionic activity coefficient.

14.10 For $BaSO_4$ $K_{sp} = 1 \times 10^{-10}$. Is a 1 mM solution of $BaSO_4$ stable?

Further reading

Burgess, J. (1999) *Ions in Solutions: Basic principles of chemical interactions.* Woodhead Publishing, Cambridge, UK.

Murrell, J.N. and Boucher, E.A. (1994) *Properties of Liquids and Solutions*, 2nd edition. John Wiley & Sons, New York.

Shannon, R.D. (1976) Revised effective ionic radii and systematic studies of interatomic distances in halides and chalcogenides. Acta Crystallogr., **A32**, 751.

CHAPTER 15

Electrochemistry: The chemistry of free charge exchange

PREVIEW OF IMPORTANT CONCEPTS

- Oxidation is the removal of electrons from a species. Oxidation occurs at the anode.
- Reduction is the addition of electrons to a species. Reduction occurs at the cathode.
- Charge is conserved so the number of electrons in oxidation and reduction must always balance.
- A galvanic cell is an electrochemical cell in which a spontaneous reaction leads to the production of electricity. Chemical energy is converted to electrical energy.
- An electrolytic cell is an electrochemical cell in which a nonspontaneous reaction is driven by an external source of current. Electrical energy is converted to chemical energy.
- An electrochemical cell must have a minimum of two electrodes – an anode (written on the left by convention) and a cathode (written on the right).
- Electrons do not dissolve in aqueous solutions with an appreciable lifetime. Therefore, electron transfer occurs after the formation of a collision complex between donor and acceptor (homogeneous electron transfer), or after the approach of a solution phase species to an electrode surface (heterogeneous electron transfer).
- In a faradic process, the amount of chemical reaction is directly proportional to the amount of current passed.
- At equilibrium no *net* current flows and there is no potential difference between the anode and cathode.
- The electrode potential E_{rev} is related to $\Delta_r G_m$, and the standard potential $E°$ is related to $\Delta_r G^\circ_m$.
- The structure and composition of the solution near the electrode surface (the electric double layer) is different than in the bulk of the solution.

A great deal of chemistry, particularly solution-phase and corrosion reactions, occurs by means of charge exchange. Electrons are swapped back and forth in at least one step of an electrochemical mechanism. This is in contrast to the 'thermal' chemistry of covalent bonds in which electron density is pushed around and rearranged but no free charge is generated or exchanged. Free charge is strongly influenced by electric fields, as is electrochemistry. Because of this the application and measurement of voltages, currents and conductivities are intimately related to electrochemistry. It leads to a whole new set of vocabulary and sign conventions separate from those used in thermal chemistry. The choices of signs are set in accord with the Stockholm convention of 1953.

The commercial applications of electrochemistry are immense. The production of aluminum as well as Cl_2 and NaOH (in the chlor-alkali process) occur via electrolytic syntheses. The electrorefining of Cu and electroplating of Ag, Au or Cr all rely on electrochemistry. That cars do not routinely rust out anymore is because of the successes in understanding the electrochemistry of corrosion that has led to the coating of steel with anticorrosion layers. Batteries, whether rechargeable or single-use, take advantage of spontaneous electrochemical reactions to convert chemical into electrical energy, as do fuel cells. Photovoltaics and solar cells are intertwined with semiconductor electrochemistry. Electrochemistry is also an extremely efficient and versatile tool for producing nanostructures, in particular, nanostructured semiconductors and oxides, as shown in Fig. 15.1.

Physical Chemistry: How Chemistry Works, First Edition. Kurt W. Kolasinski.

Companion Website: www.wiley.com/go/kolasinski/physicalchemistry

Figure 15.1 TiO_2 nanotubes produced by the electrochemical process perfected in the group of Patrik Schmuki.[1] The inset in the lower left shows the open ends of the nanotubes with diameters of roughly 225 nm. A thin film of nanotubes was fractured to reveal the sides of the tubes shown in the main image.

15.1 Introduction to electrochemistry

15.1.1 Defining the terms of electrochemistry

Electrochemical reactions involve the transfer of free charge (electrons). Because of conservation of mass and charge, the transfer of electrons must always balance – the number of donated electrons must equal the number of accepted electrons. Therefore, an electrochemical reaction must always consist of both oxidation and reduction.

- *Oxidation* is the removal of electrons from a species (oxidation state increases). Oxidation occurs at the anode. $D^x \rightarrow D^{x+1} + e^-$.
- *Reduction* is the addition of electrons to a species (oxidation state is reduced). Reduction occurs at the cathode. $A^x + e^- \rightarrow A^{x-1}$.
- A *galvanic cell* is an electrochemical cell in which a spontaneous reaction leads to the production of electricity. Chemical energy is converted to electrical energy.
- An *electrolytic cell* is an electrochemical cell in which a nonspontaneous reaction is driven by an external source of current. Electrical energy is converted to chemical energy.

An electrochemical cell, as depicted in Fig. 15.2, contains an *anode* and a *cathode*. These terms only apply to electrodes through which a net current flows. The anode is the electrode at which the predominating electrochemical reaction is oxidation. Anode $\hat{=}$ oxidation. Electrons are produced spontaneously at the anode of a galvanic cell, and are extracted from the anode of an electrolytic cell. The cathode is the electrode at which the predominating electrochemical reaction is reduction. Cathode $\hat{=}$ reduction. Electrons produced from the anode are consumed at the cathode.

The signs of the anode and the cathode change depending on whether the cell is galvanic or electrolytic, just as the reaction changes from spontaneous to forced. Electrons are negatively charged and flow from low potential to high potential. The spontaneous direction of flow is for electrons to move to the more positive electrode. In a galvanic cell electrons flow from the anode (the negative electrode) to the cathode (the positive electrode). In an electrolytic cell electrons are extracted from the anode (the positive electrode) and flow to the cathode (the negative electrode). The electrons are extracted from the anode by applying an appropriate voltage (bias) with an external power supply.

The electrode at which the reaction we are studying is occurring is called the *working* (or indicator) electrode. This is opposed to a *reference* electrode, which is an electrode that is constructed to have a constant potential and through which no current should flow. The electrodes are connected by an electrolyte that conducts ions. If the electrodes are

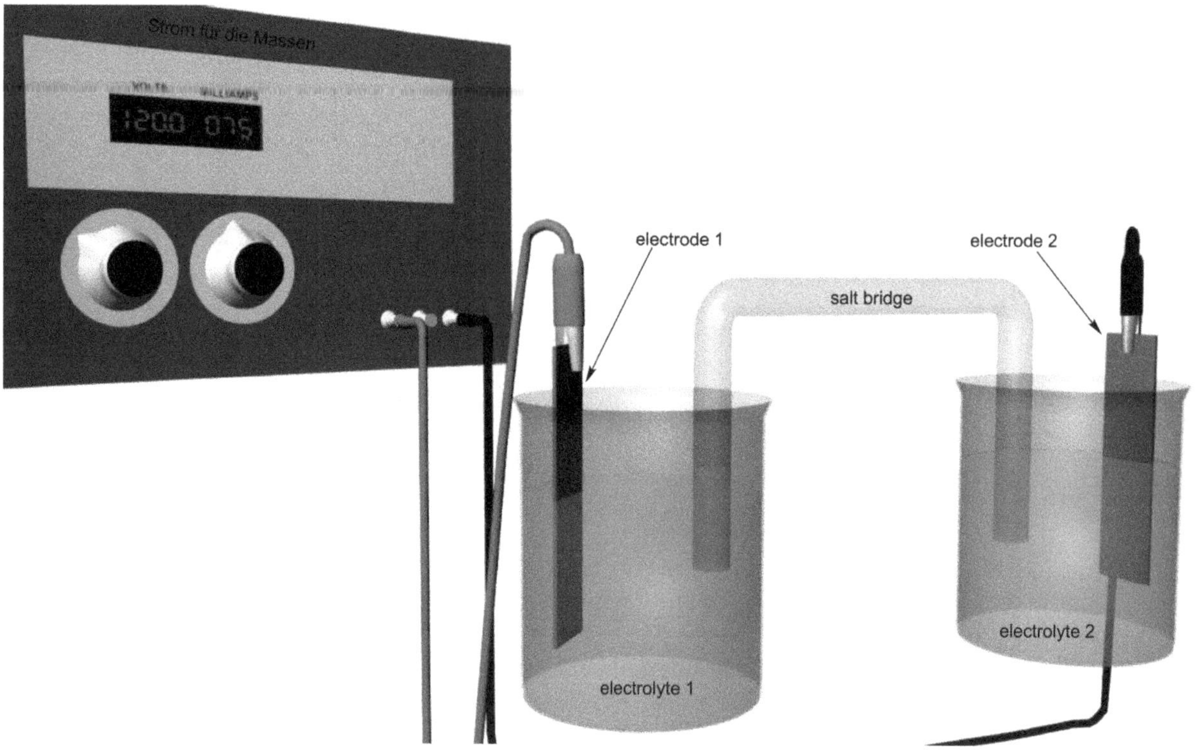

Figure 15.2 An electrochemical cell consists at a minimum of an anode (where oxidation occurs) and a cathode (where reduction occurs). For experimental investigation many cells also contains a reference electrode. These two (or three) electrodes are connected to a potentiostat that measures both current and voltage. A salt bridge is added to provide ion conductivity to maintain charge balance between the two electrolytes. A galvanic cell involves a spontaneous reactions and the potentiostat *measures* the voltage and current. An electrolytic cell is formed when the potentiostat *applies* a bias to the electrodes and forces the electrochemical reaction to run in a particular direction.

in different compartments, they may be connected by a salt bridge (often concentrated KCl in agar jelly), which allows electrical contact but avoids mixing (in other words, ions can flow through the salt bridge but not the solvent).

The redox reaction can be expressed as the difference between two *half-reactions*. The *oxidizing agent* (*oxidant*) is an electron acceptor, while the *reducing agent* (*reductant*) is an electron donor. The reduced and oxidized species in a half-reaction form a *redox couple*. The couple is written Ox/Red and the corresponding reduction half-reaction is

$$\mathrm{Ox} + ze^- \rightleftharpoons \mathrm{Red} \tag{15.1}$$

The constant z is the *electron number* of the electrochemical reaction. It is also called the charge number or valence of the reaction.

By combining a series of half-reactions, we can write a balanced overall equation. Even if the reaction does not occur electrochemically, the electrochemical pathway may be useful for interpreting or calculating state functions because, of course, these are path-independent. As examples, consider the dissolution of AgCl(s), which can be modeled as the reduction of AgCl(s)

$$\mathrm{AgCl(s)} + \mathrm{e}^- \rightleftharpoons \mathrm{Ag(s)} + \mathrm{Cl^-(aq)} \tag{15.2}$$

and the reverse of the reduction of Ag^+(aq) (i.e., the oxidation of Ag(s))

$$-1[\mathrm{Ag^+(aq)} \quad + \mathrm{e}^- \rightleftharpoons \mathrm{Ag(s)}] \tag{15.3}$$

to give the overall reaction

$$\mathrm{AgCl(s)} \rightleftharpoons \mathrm{Ag^+(aq)} + \mathrm{Cl^-(aq)} \tag{15.4}$$

or the reduction of O_2(g) with protons

$$\mathrm{O_2(g)} + 4\mathrm{H^+(aq)} + 4\mathrm{e}^- \rightleftharpoons 2\mathrm{H_2O(l)}. \tag{15.5}$$

with the oxidation of $H_2(g)$

$$-2[2H^+(aq) + 2e^- \rightleftharpoons H_2(g)] \tag{15.6}$$

which, when added stoichiometrically to balance the number of electrons lost with the number of electrons gained, gives

$$2H_2(g) + O_2(g) \rightleftharpoons 2H_2O(l). \tag{15.7}$$

Electrons appear as a reactant or product in a half-reaction. A reduction half-reaction is the addition of one or more electrons to the oxidant. No electrons appear as a reactant or product in a balanced overall electrochemical reaction. The charge on both sides of any reaction must be balanced. For an overall reaction, there can be no net gain or loss of electrons.

When current flows from the *electrode to a solution-phase species*, we say that a reduction current flows; this is also called a *cathodic current*. The flow of electrons from *solution to electrode* is an oxidation current; this is also called an *anodic current*. Which current is taken as positive is a matter of choice, and when reading the literature one should always check which convention the author has chosen. Here, we will take anodic currents as positive and cathodic currents as negative as is preferred by IUPAC.

15.1.1.1 Directed practice

In Eq. (15.7), which species is oxidized and which is reduced?

15.1.2 Relating current flow to amount of substance reacted: Faraday's law

We write a reaction quotient, which we denote Q_r, to distinguish it from the electrical charge Q, for the chemical species in a half-reaction just as we would for any other reaction. For instance, for the following half-reaction

$$O_2(g) + 4\,H^+(aq) + e^- \rightleftharpoons 2\,H_2O(l) \tag{15.8}$$

and assuming O_2 to act as an ideal gas, the reaction quotient is

$$Q_r = \frac{a^2_{H_2O}}{a_{O_2}a^4_{H^+}} = \frac{p^\circ}{p_{O_2}a^4_{H^+}}. \tag{15.9}$$

Michael Faraday's law of electrolysis helps us to understand electrochemistry quantitatively. The mass of species produced at an electrode is proportional to the quantity of electricity Q passed through the electrolyte. Electricity is quantified in coulombs (C), and is defined as the current I in amperes (abbreviated A) multiplied by the time t in seconds (abbreviated s). The mass of a species liberated at an electrode is proportional to the product of the amount of substance (moles) of the species and its charge. The magnitude of the charge on a single electron is $e = 1.602 \times 10^{-19}$ C. The proportionality factor is called the *Faraday constant*, which gives the charge on 1 mol of electrons,

$$F = eN_A = 96\,485 \text{ C mol}^{-1}. \tag{15.10}$$

The charge on a mole of ions with charge z is given by zF. Thus, the quantity of charge required to convert n moles of a species according to a half-reaction that has an electron number z is

$$Q = It = znF. \tag{15.11}$$

15.1.2.1 Example

Electrolysis of an aqueous solution of $Au(NO_3)_3$ deposits 1.200 g of Au at the cathode after passing a current of 25.0 mA. The counter-reaction is the electrolysis of water. Calculate (a) the quantity of electricity passed (the total charge). (b) How long it took to carry out the electrolysis. (c) The volume of O_2 liberated at SATP (also called NSTP, 1 atm, 25 °C) at the anode.

a Start with a balanced reaction. The half-reactions are

$$Au^{3+} + 3e^- \rightleftharpoons Au$$

$$2H_2O \rightleftharpoons O_2 + 4H^+ + 4e^-$$

The overall balanced reaction is found by balancing the number of electrons. If 12 e^- are consumed in the reduction (that is, four times the first half-reaction), then an equal number of 12 e^- will be consumed in the oxidation (that is three times the second half-reaction).

$$4Au^{3+} + 6H_2O \rightleftharpoons 3O_2 + 12H^+ + 4Au$$

Therefore 3 mol electrons lead to the deposition of 1 mol Au. Or 1 mol of electrons leads to the deposition of 1/3 mol of Au. The electron number in the electrolysis of Au^{3+} is $z = 3$. The total charge required to deposit 1.200 g Au is then

$$Q = zFn_{Au} = 3(96485\ \mathrm{C\ mol^{-1}})\frac{1.200\ \mathrm{g}}{197.0\ \mathrm{g\ mol^{-1}}} = 1.76 \times 10^3\ \mathrm{C}$$

b
$$t = \frac{Q}{I} = \frac{1.76 \times 10^3\,\mathrm{A\,s}}{0.025\ \mathrm{A}} = 7.04 \times 10^4\ \mathrm{s}$$

Note that 1 C = 1 A s.

c 4 mol of electrons lead to the liberation of 1 mol of O_2, or 1 mol of electrons lead to the formation of $\frac{1}{4}$ mol O_2. z now equals 4.

$$n_{O_2} = \frac{Q}{zF} = \frac{1.76 \times 10^4\ \mathrm{C}}{4(96485\,\mathrm{C\,mol^{-1}})} = 4.56 \times 10^{-3}\,\mathrm{mol}$$

The volume at NSTP is then

$$4.56 \times 10^{-3}\,\mathrm{mol}(24.8\,\mathrm{dm^3\,mol^{-1}}) = 0.113\,\mathrm{dm^3}.$$

Reactions that involve charge transfer such as the reduction of Au^{3+} and the oxidation of H_2O shown in this example are called *faradaic processes*. Faradaic processes (electrochemical reactions) follow *Faraday's law*.

In a faradic process, the amount of chemical reaction is directly proportional to the amount of current passed, $Q = It = znF$.

Current is the rate of charge flow

$$I = \mathrm{d}Q/\mathrm{d}t \tag{15.12}$$

Therefore, Faraday's law shows us that rate of reaction is directly proportional to the current

$$r(\mathrm{mol\,s^{-1}}) = \frac{\mathrm{d}n}{\mathrm{d}t} = \frac{I}{zF}. \tag{15.13}$$

However, because electrode reactions are heterogeneous reactions, their rate depends on the surface area of the electrode A. Therefore, we introduce the *current density* j

$$j = I/A \tag{15.14}$$

and express the rate in units of mol m^{-2} s^{-1} as

$$r(\mathrm{mol\,m^{-2}\,s^{-1}}) = j/zF. \tag{15.15}$$

There are also *nonfaradaic processes* – that is, processes in which current flows but no electrochemistry happens. These are processes in which the electrode/solution interface acts as a capacitor. The interfacial region can store charge without causing electrochemistry. To understand this we need to understand something about the structure of the electrode/solution interface.

15.1.2.2 Directed practice

What is the rate of Au deposition in Example 15.1.2.1?

[Answer: 8.64×10^{-8} mol s^{-1}]

15.1.3 The electrical double layer

Let us consider specifically the interaction of water and aqueous solutions with a metal surface. Water is a dipolar molecule with a negative charge that is preferentially displaced towards the O atom, and a corresponding amount of positive charge on the opposite side of the molecule. When a dipole approaches a metal surface the electrons in the surface rearrange to lower the energy of the system. Or, looked at from the perspective of the water molecule, if the surface has an excess of negative charge on it the positive H-side of the molecule will preferentially point towards the surface. If the surface has an excess of positive charge, the negative O-end of the molecule will preferentially point towards the surface. The orientation of the molecular water dipoles near the surface responds to the potential placed on the surface.

Anions and cations in an aqueous electrolyte also respond to the surface charge of the metal electrode. In addition, we need to consider the solvation shell of the solvated ions. The ions may interact with the surface with or without loss of their solvation shell. When a solvated ion or molecule approaches a surface and binds to it, but does not lose its solvation shell, it is *nonspecifically adsorbed*. When a solvated species binds and loses all (or part) of its solvation shell it is *specifically adsorbed*. If the surface is biased negative, it will attract cations, but if biased positive it will attract anions. At a certain potential, known as the *potential of zero charge* (pzc), it will attract both equally and the interface is neutral.

The solution must always be neutral overall. Therefore, if the interface has acquired a charge (i.e., a surface charge layer) there must be a layer adjacent to it (i.e., the diffuse charge layer) that contains the opposite charge in order to maintain neutrality. This region (i.e., the combination of the surface charge layer and the diffuse charge layer) is known as the *electrical double layer*. A charged interfacial region is depicted in Fig. 15.3. The surface charge layer is also called the *Stern layer* and lies closest to the surface. The diffuse charge layer is also called the *Gouy layer* and is situated on top of the Stern layer. The *Helmholtz layer* is defined by the plane that contains a layer of fully or partially solvated ions. Whether ions are specifically or nonspecifically adsorbed depends on the electric field strength and the chemical identity of the ions. The outer Helmholtz layer is defined by the plane of fully solvated ions, whereas an inner Helmholtz plane is defined by the plane that passes through the center of specifically adsorbed ions. Depending on the extent to which the charge of the first layer is compensated by excess charge at the solid surface, the diffuse

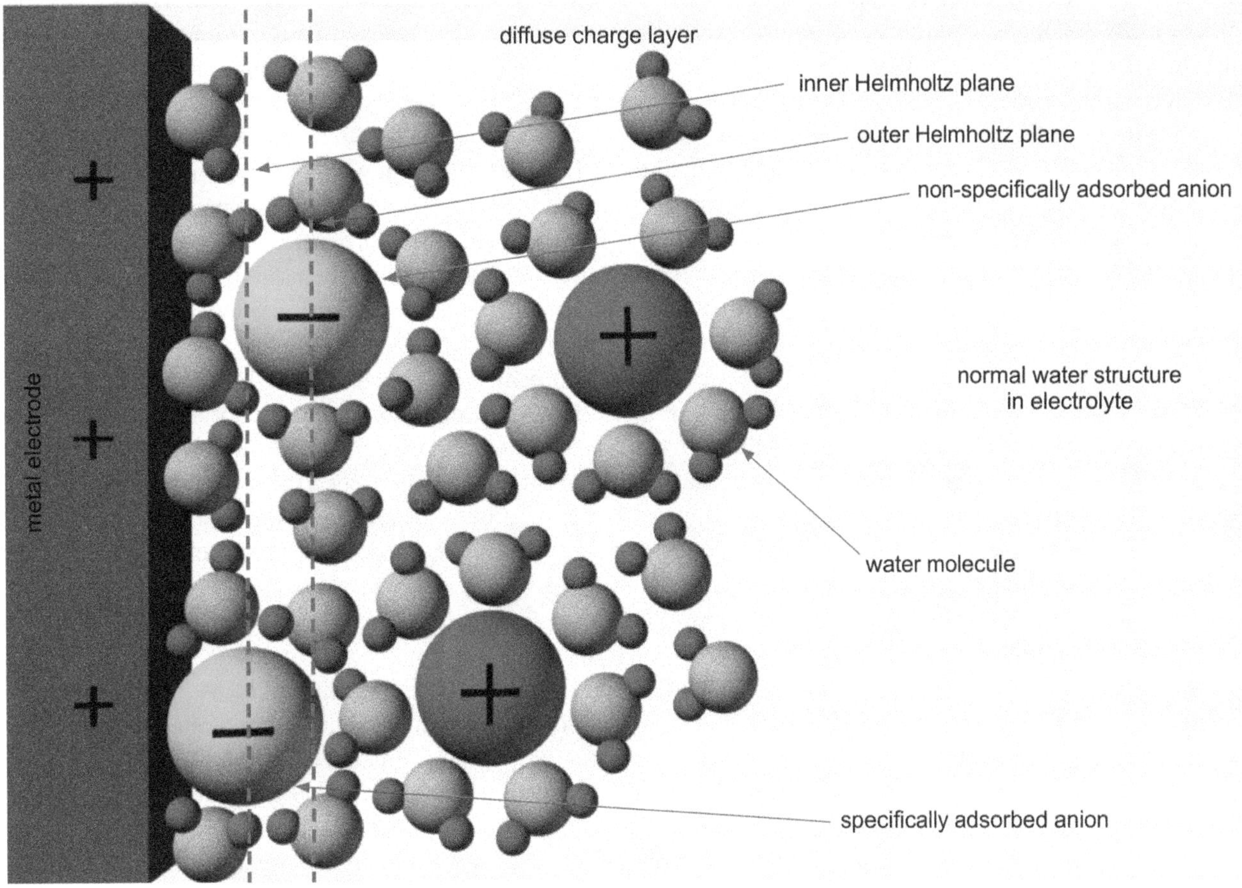

Figure 15.3 The electric double layer. The outer Helmholtz layer is approximately 0.3 nm away from the metal surface (the diameter of a water molecule). The inner Helmholtz layer passes though the center of specifically adsorbed ions (if they are present). The diffuse charge layer provides a transition between the Helmholtz layer and the bulk solution.

charge layer is composed either of like-charged ions (full compensation), counterions (no compensation), or a mixture of the two.

15.2 The electrochemical potential

The *open-circuit potential* (OCP) is the potential difference measured in an electrochemical cell when no bias is placed on the electrodes, and in the absence of current flow. It is also called the *rest potential.* What goes into determining this value and why does an electron move from one species or electrode to the next?

15.2.1 Why does an electron 'hop'?

Any chemical system evolves to lower its Gibbs energy; that is, $dG < 0$ defines the way forward for a reaction as long as there is no kinetic barrier to progression. During electrochemistry, free charge is exchanged. The electron has a negative charge and therefore it is influenced by the presence of an electric field. We need to take this into account in our treatment of the reaction. We do so by introducing the electrochemical potential, which adds a component proportional to the electric field strength to the chemical potential. We will do that in the next section. Now, we want to think about why an electron 'hops' from one species to the next. The answer must be that it lowers the Gibbs energy of the system to move an electron from an occupied orbital on one species to an unoccupied (or partially occupied but not a fully occupied) orbital on another species. We shall learn shortly that electrons are transferred through a transition state in which the electron donor and the electron acceptor are degenerate (have the same energy) in terms of their Gibbs energy. We shall also see that we can change the relative energy of the states in an electrode compared to a species in solution by changing the electrical potential of that electrode. However, because an electron has a negative charge the sign of the electrical potential scale is the opposite to that of the energy scale. That is, electrons move spontaneously from high-energy states to lower-energy states. This corresponds to moving from states of some given electrical potential to states with more positive electrical potential.

This is illustrated in Fig. 15.4 for two different cases. In panel (a) a metal electrode is in contact with a liquid electrolyte solution. The metal has electronic energy levels filled up to the Fermi energy E_F. Below this energy, the electronic energy levels of the metal are full, but above it they are empty. The frontier orbitals of the solution-phase species line up with the Fermi energy of the metal in a manner consistent with the electronic structure of that species, the electronic structure of the metal, and the electrical potential applied to the metal electrode. A little care must be used in interpreting Fig. 15.4 in that we need to consider Gibbs energy and not just electronic energy, but don't let this important subtlety distract us for the moment.

If a positive potential is applied to the metal, the electrochemical potential of the electrons in the metal moves down. If the electrical potential is made positive enough, an electron can be transferred from an occupied orbital in the solution-phase species to the electrode (*anodic current*). If the electrode is biased to a negative potential, then when this potential is negative enough, the electrochemical potential of the electrons in the metal will exceed the electrochemical potential of the unoccupied orbital in the solution species. An electron can then be transferred from the metal to the acceptor in solution (*cathodic current*). Note how the role of donor and acceptor can be changed based on the applied electrical potential.

The situation is somewhat more complicated if we introduce a semiconductor rather than a metal into the solution as our electrode. The presence of a band gap means that there are no electronic states at E_F in the semiconductor. Instead, there is a (nominally) full valence band below E_F that is filled up to the valence band maximum E_V. A (nominally) empty conduction band extends down in electronic energy to the conduction band minimum at E_C. An orbital on the solution species can exchange electrons with either the conduction band or the valence band. However, as always, electron transfer must occur from an occupied state to a less than fully occupied state. The presence of a band gap, as well as a phenomenon called 'band bending' adds considerable complexity to semiconductor electrochemistry. Therefore, we will always assume that the electrodes we are dealing with are metals, unless we explicitly specify otherwise. Nonetheless, semiconductor electrochemistry and photoelectrochemistry are increasingly important areas of research and technology. They represent important methods of nanomaterial production, as well as being relevant to solar cell (photovoltaic) and solar fuels technologies.

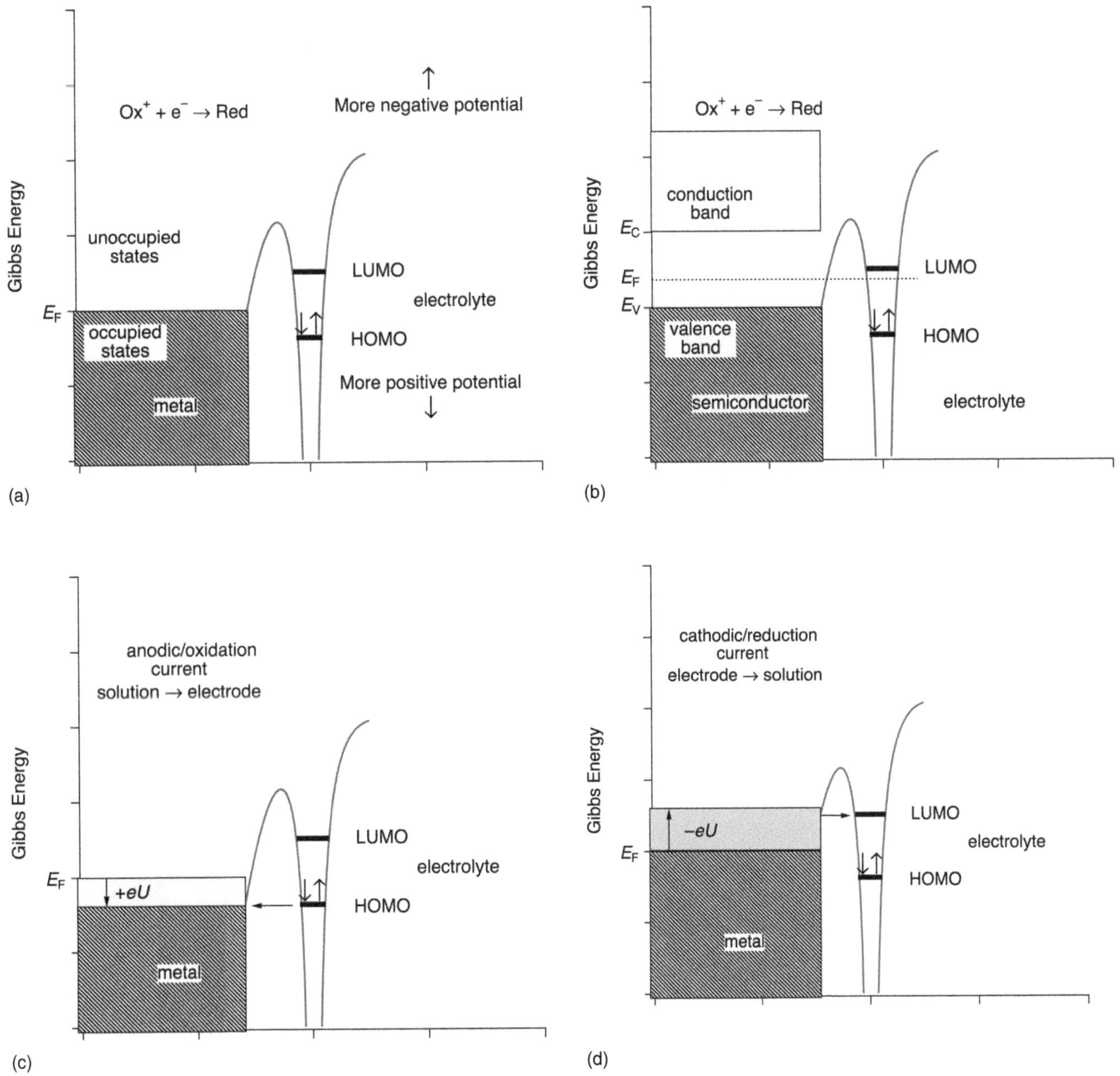

Figure 15.4 The alignment of energy levels of a solution-phase species with an electrode in energy and electrical potential space. (a) Metal electrode. (b) Semiconductor electrode. Whether the highest occupied molecular orbital (HOMO) becomes a donor or the lowest unoccupied molecular orbital (LUMO) becomes an acceptor is determined by the alignment of these frontier orbitals with the bands of the electrode and the applied potential. (c) Applying a positive potential $+U$ to the electrode reduces the Gibbs energy of the electrons in the metal sufficiently to allow electron transfer from the electrolyte species HOMO to the electrode. (d) Applying a negative potential $-U$ to the electrode increases the Gibbs energy of the electrons in the metal sufficiently to allow electron transfer from the electrode to the LUMO of the electrolyte species.

This discussion demonstrates that there are potentials at which we expect a reduction current to flow (working electrode to solution electron transfer), potentials at which an oxidation current flows (solution to electrode electron transfer), and potentials in between at which negligible net current flows. This is shown in the Fig. 15.5, which displays a current–potential curve.

Figure 15.5 makes it clear that current flows in response to changes in the potential placed upon the electrodes. The reactions that occur also depend on the magnitude and sign of the potential. But where does the electron transfer occur? A clue is found when we change the Pt electrode for a Hg electrode in the system shown in Fig. 15.6. H_2 is produced at 0 V in an ideal hydrogen electrode constructed with Pt (assuming, of course, that it is attached to a counter-electrode, not shown). No H_2 production is found near 0 V in the presence of a Hg electrode. Instead, a much greater *overpotential*

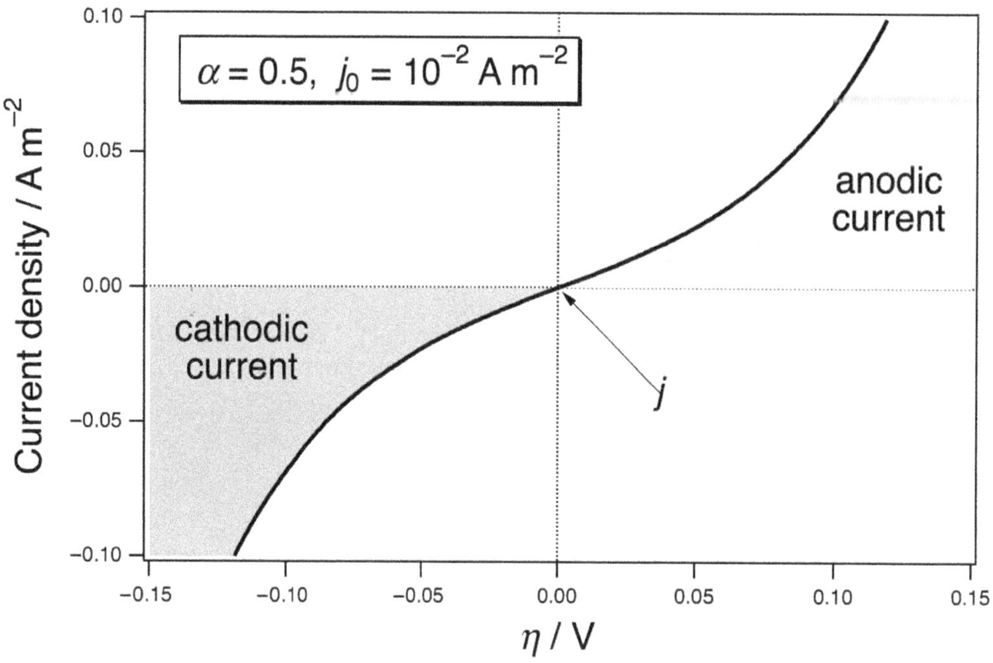

Figure 15.5 A current–potential curve. The direction of current flow in terms of the current density j is determined by the sign and magnitude of the overpotential η placed on an electrode. Cathodic current corresponds to electrode to solution-phase species electron transfer and anodic current corresponds to solution-phase to electrode electron transfer.

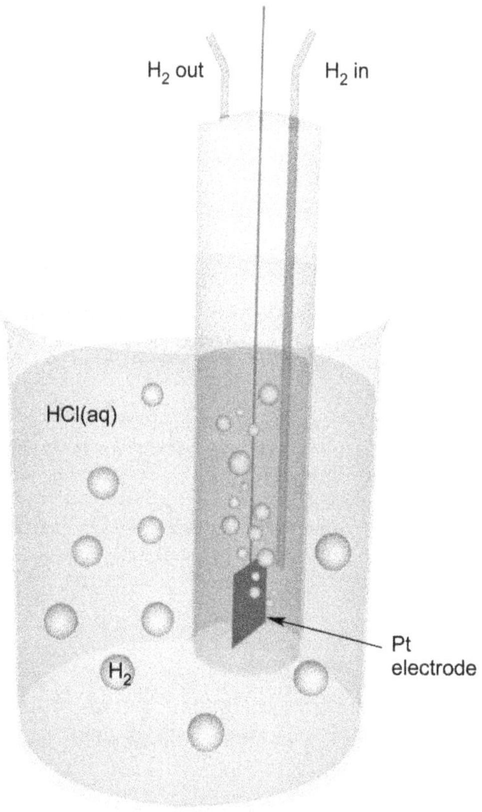

Figure 15.6 A schematic representation of the standard hydrogen electrode (SHE). Hydrogen gas with a fugacity of 1 bar is bubbled through the solution so that there is contact between the H_2, the Pt electrode, and the electrolyte. The electrolyte is an acidic aqueous solution with an H^+ activity of 1.

must be applied in order to overcome a substantial activation barrier for the reduction of H^+. The overpotential is defined as the potential η in excess of the calculated reversible cell potential E_{rev} that would allow for reaction at equilibrium compared to the cell potential E that is required to actually make the reaction occur,

$$\eta = E - E_{rev}. \tag{15.16}$$

This behavior would not be expected if the reaction occurred in solution. In fact, solvated electrons have a very short lifetime in aqueous solutions. Charge exchange is occurring predominantly on the *surface* of the electrode. The general case of heterogeneous electron transfer follows the dynamics that can be described in the following manner. A solvated solution-phase species diffuses to a surface, and then either specifically or nonspecifically adsorbs onto the surface of the electrode. Specific absorption is when the solvated species loses part or all of its solvation shell. Nonspecific adsorption is when the adsorbed species maintains its solvation shell. Electron transfer occurs in the adsorbed phase. Before or after charge transfer there may be additional surface reactions that occur. After charge transfer (and perhaps after subsequent reactions) a reacted species desorbs from the surface; this species may be the final reaction product or further solution-phase chemistry may be required before the final reaction product is obtained.

15.2.2 Relating the electrochemical potential to the chemical potential

Work is done to move charge across a potential difference. The amount of work it takes to reversibly transfer dQ of charge ($dQ = -zF\,dn$) from a phase at potential ϕ_1 to a phase at ϕ_2 is

$$dw_{rev} = (\phi_2 - \phi_1)\,dQ. \tag{15.17}$$

The work performed is reversible, nonexpansion work; therefore, it is equal to the change in Gibbs energy,

$$dw_{rev} = dG. \tag{15.18}$$

The change in Gibbs energy also represents the difference in the *electrochemical potential* $\tilde{\mu}$ of the charged particle in the two phases

$$dG = \tilde{\mu}_2\,dn - \tilde{\mu}_1\,dn. \tag{15.19}$$

The electrochemical potential, introduced by John A.V. Butler[2] and Edward A. Guggenheim,[3] is the sum of the chemical potential and a term that includes the effect of the electric field. The electrochemical potential is defined as

$$\tilde{\mu} = \mu + z\phi F. \tag{15.20}$$

Experimentally, we directly measure the difference in potentials rather than an absolute potential. Thus, if we set ϕ_1 to zero, we find that the electrochemical potential in phase 2 is related to the electrochemical potential in phase 1 by

$$\tilde{\mu}_2 = \tilde{\mu}_1 + z\phi_2 F. \tag{15.21}$$

Charged particles in otherwise identical phases that have different chemical potentials in the two phases are at different electrical potentials. This is the basis for understanding all of electrochemistry and its utility because we can easily change the electrical potential. The equilibrium condition can be written in terms of the electrochemical potential as

$$\Delta_r G_m = \sum_i \nu_i \tilde{\mu}_i = 0. \tag{15.22}$$

The convention is to choose the *standard state* of the solution (and all ions in solution) to be at $\phi = 0$. With this choice of standard state, the electrochemical potential of any ion in solution is equal to its chemical potential,

$$\tilde{\mu}_i = \mu \text{ (ions in solution)}. \tag{15.23}$$

We take the standard state chemical potential of an electron in a metal electrode at zero potential to be zero. Thus, the electron's electrochemical potential at any potential ϕ is given by

$$\tilde{\mu}_{e^-} = -\phi F \text{ (electrons in metal electrodes)} \tag{15.24}$$

The *standard hydrogen electrode* (SHE), depicted in Fig. 15.6, is used as a *reference electrode* against which all standard electrode potentials are measured. It is also called the normal hydrogen electrode (NHE) with a corresponding half-reaction

$$H^+(aq) + e^- \rightleftharpoons \tfrac{1}{2}H_2(g). \tag{15.25}$$

The equilibrium half-cell is described by

$$\mu_{H^+}(aq) + \tilde{\mu}_{e^-} = \frac{1}{2}\mu_{H_2}(g). \tag{15.26}$$

Half-cell reactions are conventionally written as reductions. Now expand Eq. (15.26) in terms of activities and fugacity

$$\mu^\circ_{H^+} + RT\ln a_{H^+} - F\phi(H^+/H_2) = \frac{1}{2}\mu^\circ_{H_2} + \frac{1}{2}RT\ln(\tilde{p}_{H_2}/p^\circ). \tag{15.27}$$

Now we recognize that the potential $\phi(H^+/H_2)$ in Eq. (15.27) is what we call the *cell potential*, which according to convention is denoted $E(H^+/H_2)$. Solving for the cell potential, we obtain

$$E(H^+/H_2) = \frac{\mu^\circ_{H^+} - \frac{1}{2}\mu^\circ_{H_2}}{F} - \frac{RT}{F}\ln\frac{\tilde{p}^{1/2}_{H_2}}{a_{H^+}}. \tag{15.28}$$

If both the activity of H^+ and the fugacity of H_2 are equal to 1, the cell potential has its standard-state potential. We denote this *standard cell potential* $E^\circ(H^+/H_2)$. By definition, the standard chemical potential of H_2 is zero (it is an element in its standard state), $\mu^\circ_{H_2} = 0$, thus,

$$E^\circ(H^+/H_2) = \mu^\circ_{H^+}/F. \tag{15.29}$$

By convention, the Gibbs energy of formation for H^+ set to zero,

$$\Delta_f G^\circ(H^+) = \mu^\circ_{H^+} = 0. \tag{15.30}$$

Therefore, this convention combined with Eq. (15.29) means that the standard potential of the hydrogen electrode must also be zero,

$$E^\circ(H^+/H_2) = 0. \tag{15.31}$$

15.3 Electrochemical cells

An *electrochemical cell* must have a minimum of two electrodes – an anode (written on the left by convention) and a cathode (written on the right). A *reference electrode* is often added for analytical reasons. No current runs through the reference electrode, it is used to measure potential differences. An electrode is composed of an electron conductor in contact with an ion conductor. The electron conductor is usually a metal but can also be a semiconductor. The presence of a semiconductor leads to some very interesting complications arising from its electronic structure. Namely, its valence band is separated from its conduction band by a 'band gap.' For this reason, we will assume that the electron conductor is a metal unless we specify otherwise. The ion conductor is composed of an electrolyte in any of several phases. In most of the examples used here, the ion conductor will be an aqueous solution of electrolytes.

The electrodes can be short-circuited –that is, connected to each other directly – but more usually in physical chemistry they will be connected to each other via meters than can measure both voltage and current. In order to operate an electrolytic cell, a power supply with regulated current and voltage with respect to the reference electrode is also required as a source of electrical energy. This regulated power supply is called a *potentiostat*.

The chemical make-up of a pair of electrodes connected in an electrochemical cell is expressed in terms of *cell notation*. Phase boundaries are indicated by a vertical bar |. A dashed vertical line ⋮ indicates a junction between miscible liquids. A double-dashed vertical line ⋮⋮ indicates a liquid junction in which the liquid junction potential has been eliminated, for example, by a salt bridge. A comma separates two species contained in the same phase. An example of cell notation for the Daniell cell in which an anode containing Zn metal and a soluble Zn salt ($ZnSO_4$ in aqueous sulfuric acid in this

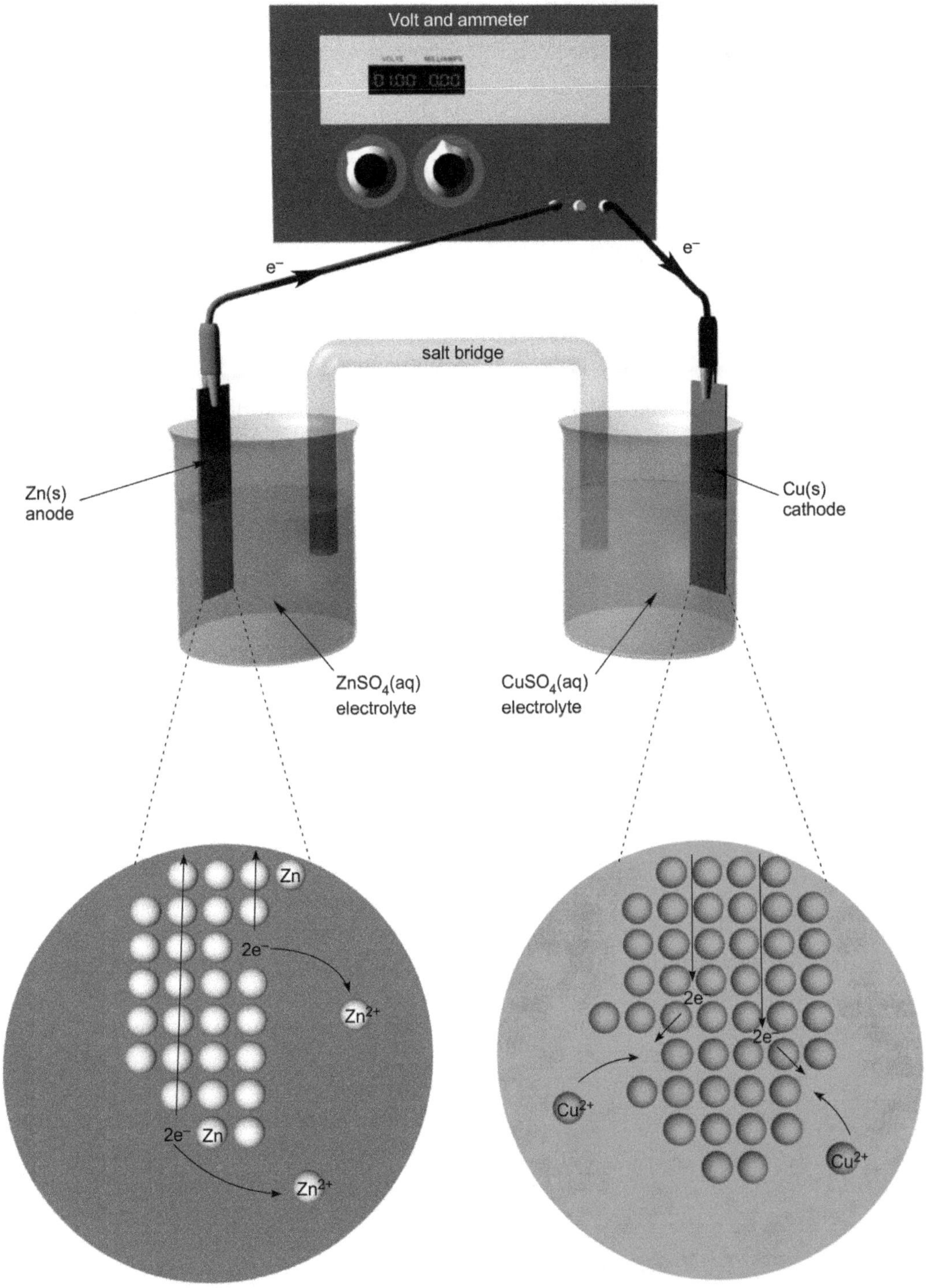

Figure 15.7 A schematic representation of the Daniell cell and the electrochemical processes occurring in it. The Zn electrode corrodes, releasing Zn^{2+} ions into solution. The two electrons lost from a Zn atom are released into the metal and conducted away. The Cu electrode gains mass as Cu^{2+} ions deposit on the surface and are neutralized by the addition of two electrons.

case) is connected to a cathode containing Cu metal and an electrolyte solution containing $CuSO_4$ in $H_2SO_4(aq)$, which is shown in Fig. 15.7, is

$$\mathrm{Zn(s)\,|\,ZnSO_4(aq),\,1\,M\,H_2SO_4(aq)\ \vdots\vdots\ CuSO_4(aq),\ 1\,M\,H_2SO_4(aq)\,|\,Cu(s)}.$$

In this example the $SO_4^{2-}(aq)$ is a *spectator ion*. The aqueous ions are supported in an electrolyte containing $H_2SO_4(aq)$. If the emphasis is on the construction of the electrochemical cell, then a completely specified cell notation as shown above

Table 15.1 Conventions of cell notation.

Electrode type	Designation	Redox couple	Half-reaction
metal/metal ion	$M(s) \mid M^+(aq)$	M^+/M	$M^+(aq) + e^- \rightarrow M(s)$
gas with Pt catalyst	$Pt(s) \mid X_2(g) \mid X^+(aq)$	X^+/X_2	$X^+(aq) + e^- \rightarrow \frac{1}{2}X_2(s)$
	$Pt(s) \mid X_2(g) \mid X^-(aq)$	X_2/X^-	$\frac{1}{2}X_2(g) + e^- \rightarrow X^-(aq)$
metal/insoluble salt	$M(s) \mid MX(s) \mid X^-(aq)$	$MX/M, X^-$	$MX(s) + e^- \rightarrow M(s) + X^-(aq)$
redox with Pt catalyst	$Pt(s) \mid M^+(aq), M^{2+}(aq)$	M^{2+}/M^+	$M^{2+}(aq) + e^- \rightarrow M^+(aq)$

is used. In addition, the temperature must be specified. If the emphasis is on the underlying electrochemistry, the cell notation is stripped of spectator ions and supporting electrolyte, such as in the following example.

$$\mathrm{Zn(s)} \mid \mathrm{Zn^{2+}(aq)} \;\vdots\vdots\; \mathrm{Cu^{2+}(aq)} \mid \mathrm{Cu(s)}$$

To work out the *cell reaction* for this electrochemical cell we subtract (reverse) the left-hand electrode half-reaction from the right-hand electrode half-reaction. Examples of cell notation for different combinations of reactive species are given in Table 15.1.

Right-hand electrode:

$$\mathrm{Cu^{2+}(aq)} + 2\,\mathrm{e^-} \rightleftharpoons \mathrm{Cu(s)}$$

Left-hand electrode:

$$\mathrm{Zn^{2+}(aq)} + 2\,\mathrm{e^-} \rightleftharpoons \mathrm{Zn(s)}$$

Cell reaction:

$$\mathrm{Cu^2{+}(aq)} + \mathrm{Zn(s)} \rightleftharpoons \mathrm{Cu(s)} + \mathrm{Zn^{2+}(aq)}$$

15.4 Potential difference of an electrochemical cell

The potential difference of an electrochemical cell is measured between two identical metal wires. One is attached to the right-hand electrode of the electrochemical cell, while the other is attached to the left-hand electrode. In theory, a true potential difference can only be measured between two metals of the same composition. The junction between two dissimilar metals results in a contact potential caused by the different positions of the Fermi energies in the two metals. In practice, copper conductors are attached to the electrodes and the contact potential that arises is incorporated into the potential to which measurements are referenced.

Liquid *junction potentials* E_j will arise any time two immiscible solutions of different composition contact one another, or when two miscible solutions are separated by a membrane to avoid mixing. These junction potentials are caused by different rates of diffusion between anions and cations. This junction potential will vary in time until equilibrium (or at least a steady state) is established. Salt bridges are commonly employed to stabilize and minimize the magnitude of liquid junction potentials.

Any cell that is not at equilibrium will develop a measureable potential difference between the two electrodes. The potential difference (*cell potential*) is measured in volts ($1\ \mathrm{V} = 1\ \mathrm{J\ C^{-1}}$), and is a measure of the capacity to do work. The maximum amount of work is extracted when the cell works reversibly.

The cell potential measured under reversible conditions (and at constant composition) is the *electromotive force* E_{rev} of the cell (also called *emf*). Constant composition implies that no current is flowing through the cell during the measurement. Under these conditions of local chemical equilibrium for both electrode reactions, the measured potential difference in the absence of junction potentials is related to the Gibbs energy of the overall cell reaction by

$$\Delta_r G_m = -zFE_{\mathrm{rev}} \tag{15.32}$$

F is the Faraday constant and z is the stoichiometric coefficient of the electrons (the electron number) in the half-reaction. Unless otherwise stated, we will assume that junction potentials are zero and local equilibrium is maintained in our cells. In other words, the cell operates reversibly and in accord with Eq. (15.32).

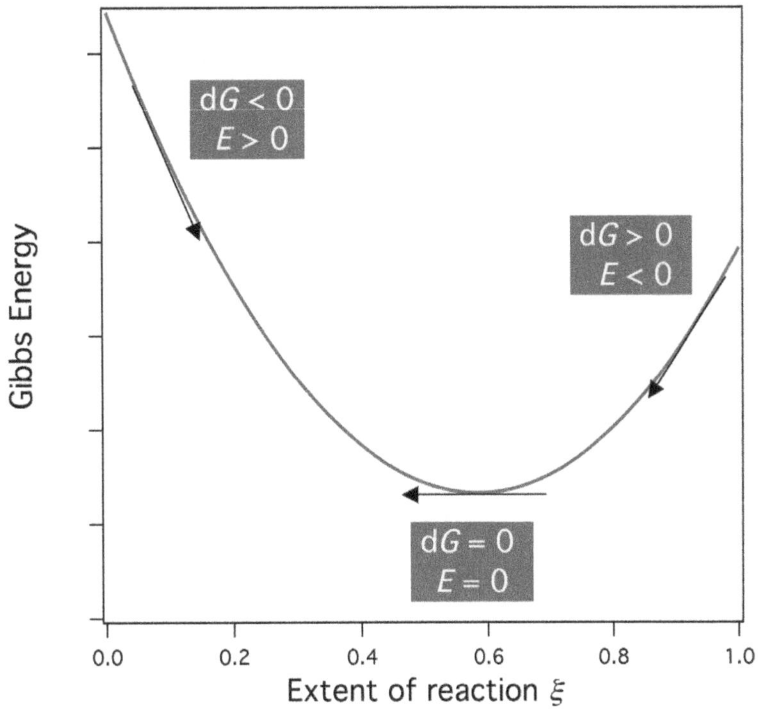

Figure 15.8 The Gibbs energy, its relationship to E, and the direction of spontaneous change. Here, the extent of reaction ξ is normalized to the initial amount of the limiting reagent in moles n_0 so that ξ/n_0 varies from 0 (no reaction) to 1 (reaction to completion).

Thus, by measuring the emf, we obtain a measure of $\Delta_r G_m$ at that specific concentration and temperature of the electrochemical cell. Note the difference in sign between $\Delta_r G_m$ and E_{cell}. A negative $\Delta_r G_m$ indicates a *spontaneous* reaction, which corresponds to a *positive* emf. A positive $\Delta_r G_m$, which indicates a spontaneous *reverse* reaction, corresponds to a *negative* emf. When $\Delta_r G_m = 0$ the system is at equilibrium and the emf = 0. As long as the cell is run reversibly, the sign of the emf will indicate the direction of spontaneous change, as shown in Fig. 15.8. In this manner, emf measurements avoid the possible confusion (discussed in Chapter 11) in results obtained by comparing $\Delta_r G_m$ values instead of $d_r G_m$.

15.4.1 The Nernst equation

We now relate the emf to composition. For the Daniell cell above, the cell reaction is

$$Cu^{2+}(aq) + Zn(s) \rightleftharpoons Cu(s) + Zn^{2+}(aq)$$

and the molar Gibbs energy change is related to the electrochemical potential by

$$\Delta_r G_m = \tilde{\mu}(Zn^{2+}) + \mu(Cu) - \tilde{\mu}(Cu^{2+}) - \mu(Zn). \quad (15.33)$$

Since the two solids are in their standard state, their chemical potentials are equal to zero, and we obtain

$$\Delta_r G_m = \tilde{\mu}^\circ(Zn^{2+}) - \tilde{\mu}^\circ(Cu^{2+}) + RT \ln \frac{a(Zn^{2+})}{a(Cu^{2+})} \quad (15.34)$$

or, in terms of the reaction quotient Q_r

$$\Delta_r G_m = \Delta_r G_m^\circ + RT \ln Q_r. \quad (15.35)$$

The reaction quotient is written more generally in terms of the activity a_j and stoichiometric number ν_j of the jth species as

$$Q_r = \prod_j a_j^{\nu_j}. \quad (15.36)$$

With this result and Eq. (15.32) we obtain the *Nernst equation*

$$E_{rev} = -\frac{\Delta_r G_m^\circ}{zF} - \frac{RT}{zF} \ln Q_r = E^\circ - \frac{RT}{zF} \ln Q_r \quad (15.37)$$

where

$$E^\circ = -\Delta_r G^\circ_m / zF. \tag{15.38}$$

E° is the standard emf (also called standard potential). In terms of activities directly, the Nernst equation is

$$E_{rev} = E^\circ - (RT/zF) \sum_j \nu_j \ln a_j. \tag{15.39}$$

Specifically for the Daniell cell, which has an electron number of two, the Nernst equation is expressed as

$$E_{rev} = E^\circ - \frac{RT}{2F} \ln \frac{a(Zn^{2+})}{a(Cu^{2+})}. \tag{15.40}$$

We reiterate at this point that we are assuming all overpotentials to be zero; thus, the cell potential is given by the reversible cell potential, that is, $E = E_{rev}$. Below, to simplify the notation, we will generally use E instead of E_{rev}.

15.4.1.1 Directed practice

Write the Nernst equation for the reaction system given in Example 15.1.2.1.

15.4.2 Standard potentials

Recall that at equilibrium $Q_r = K$ and $\Delta_r G = 0$. Therefore, the special significance of E° follows from the relationship of the standard Gibbs energy of reaction to equilibrium constant $\Delta_r G^\circ_m = -RT \ln K$. Combining this with Eq. (15.38), we obtain

$$E^\circ = (RT/zF) \ln K \tag{15.41}$$

The standard emf of the Daniell cell (the Zn–Cu example above) is $E^\circ = 1.10$ V, $z = 2$, and therefore at 298 K, $K = 1.5 \times 10^{37}$. This value is impossibly high to measure using concentrations, but straightforward to measure electrochemically.

The scale for E° values is set by assigning the value of zero to the standard hydrogen electrode (SHE)

$$Pt(s)\,|\,H_2(g, \tilde{p} = 1\,bar)\,|\,H^+(aq, a = 1),\ E^\circ = 0.$$

This value of E° is achieved when the H_2 has a fugacity of 1 bar and the $H^+(aq)$ has an activity of 1. All other E° values are measured relative to this. For a metal in equilibrium with solvated ions this corresponds to measuring the voltage generated by the cell represented by

$$Pt\,|\,H_2\,|\,H^+ \,\vdots\vdots\, M^{z-}\,|\,M$$

with the accompanying electrochemical reaction

$$M^{z+} + (z/2)H_2(g) \rightleftharpoons M + zH^+$$

The potential for such a cell for an arbitrary composition is abbreviated $E(M^{z+}/M)$, and when all species in the reaction appear with unit activity this potential corresponds to the standard potential for the M^{z+}/M redox couple $E^\circ(M^{z+}/M)$. The tabulated values of E° correspond to half-reactions performed in water at 25 °C. These values cannot be directly compared to what is to be expected in another solvent.

15.4.3 Measurement of the standard potential $E^\circ(Ag^+/Ag)$

The *electrochemical series*, a few selected entries from which are given in Table 15.2, is composed of standard reduction potentials for all possible electrochemical half-reactions occurring in aqueous solutions. These reduction potentials are measured with respect to the SHE. In principle, these values are arrived at by constructing a cell with the half-reaction of interest on one side and the SHE on the other. The SHE is difficult to construct and work with. In practice, we use transitivity to construct the electrochemical series. As long as we establish the potentials of some small subset of half-reactions with respect to SHE, then we can use the cell potential involving one of the half-reactions in this subset and the reaction of interest to determine the standard reduction potential for that reaction.

Let us consider how we would do this for one such reference reaction, the reduction of $Ag^+(aq)$ to $Ag(s)$. First, construct the cell

$$Pt(s)\,|\,H_2(g)\,|\,HCl(aq) \,\vdots\vdots\, AgNO_3(aq)\,|\,Ag(s).$$

Table 15.2 Selected values from the electrochemical series. Data taken from the 96th edition of the *CRC Handbook of Chemistry and Physics*.

Reaction	E°/V	Reaction	E°/V
$H_2O_2 + 2H^+ + 2e^- \rightleftharpoons 2H_2O$	1.776	$Ag^+ + e^- \rightleftharpoons Ag$	0.7996
$Ce^{4+} + e- \rightleftharpoons Ce^{3+}$	1.72	$Cu^{2+} + 2e^- \rightleftharpoons Cu$	0.3419
$Au^+ + e^- \rightleftharpoons Au$	1.692	$H^+ + 2e^- \rightleftharpoons \frac{1}{2}H_2$	0
$MnO_4^- + 4H^+ + 3e- \rightleftharpoons MnO_2 + 2H_2O$	1.679	$Sn^{2+} + 2e^- \rightleftharpoons Sn$	−0.1375
$Au^{3+} + e^- \rightleftharpoons Au$	1.498	$Ni^{2+} + 2e^- \rightleftharpoons Ni$	−0.257
$Cl_2 + 2e^- \rightleftharpoons 2Cl^-$	1.35827	$Co^{2+} + 2e^- \rightleftharpoons Co$	−0.28
$O_2 + 4H^+ + 4e^- \rightleftharpoons 2H_2O$	1.229	$Fe^{2+} + 2e^- \rightleftharpoons Fe$	−0.447
$2IO_3^- + 12H^+ + 10e- \rightleftharpoons I_2 + 6H_2O$	1.195	$Zn^{2+} + 2e^- \rightleftharpoons Zn$	−0.7618
$IO_3^- + 6H^+ + 6e- \rightleftharpoons I^- + 3H_2O$	1.085	$SiF_6{}^{2-} + 4e^- \rightleftharpoons Si + 6F^-$	−1.24
$AuCl_4^- + 3e- \rightleftharpoons Au + 4Cl^-$	1.002	$Al^{3+} + 3e^- \rightleftharpoons Al$	−1.676
$VO_2^+ + 2H^+ + e^- \rightleftharpoons VO^{2+} + H_2O$	0.991	$Na^+ + e^- \rightleftharpoons Na$	−2.71
$NO_3^- + 4H^+ + 3e- \rightleftharpoons NO + 2H_2O$	0.957	$K^+ + e^- \rightleftharpoons K$	−2.931

The balanced cell reaction is (dropping the spectator ions)

$$\frac{1}{2}H_2(g) + Ag^+(aq) \rightarrow H^+(aq) + Ag(s).$$

Now write the Nernst equation for this reaction,

$$E = E^\circ(Ag^+/Ag) - \frac{RT}{F}\ln\frac{a_{H^+}a_{Ag}}{a_{H_2}^{1/2}a_{Ag^+}}. \tag{15.42}$$

In the SHE $a_{H_2} = 1$ and $a_{H^+} = 1$ and for the solid $a_{Ag} = 1$. Express the activity of Ag^+ in terms of the molality b

$$a_{Ag^+} = \gamma_\pm b/b^\circ, \tag{15.43}$$

the cell potential depends on b according to

$$E = E^\circ(Ag^+/Ag) - \frac{RT}{F}\ln\frac{1}{a_{Ag^+}}, \tag{15.44}$$

$$E = E^\circ(Ag^+/Ag) + \frac{RT}{F}\ln\left(\frac{b}{b^\circ}\right) + \frac{RT}{F}\ln\gamma_\pm \tag{15.45}$$

This rearranges to

$$E - \frac{RT}{F}\ln\left(\frac{b}{b^\circ}\right) = E^\circ\left(Ag^+/Ag\right) + \frac{RT}{F}\ln\gamma_\pm \tag{15.46}$$

From the Debye–Hückel limiting law we know that at 298 K in an aqueous solution of a 1:1 electrolyte the mean ionic activity coefficient is related to b by

$$\ln\gamma_\pm = -1.172\sqrt{b/b^\circ}. \tag{15.47}$$

Therefore, Eq. (15.46) is

$$E - 0.02569\ln(b/b^\circ) = E^\circ(Ag^+/Ag) - 0.03012\sqrt{b/b^\circ}. \tag{15.48}$$

As shown in Fig. 15.9, a plot of $E - (RT/F)\ln(b/b^\circ)$ *vs* $\sqrt{b/b^\circ}$ has an intercept at $\sqrt{b/b^\circ} = 0$ that is equal to E°. This extrapolation to infinite dilution ensures that the experimental result pertains to the concentration range in which the Debye–Hückel limiting law is valid.

Standard electrode potentials vary across the Periodic Table and with chemical species. The factors that contribute to the value of E° can be established by breaking the overall chemical process into a set of elementary steps. We can do this because the value of E° is determined by (and is directly proportional to) the state function ΔG°. Since ΔG° is a state function, so is E°, the magnitude of which must be independent of path. Thus, the half-reaction

$$M^+(aq) + e^- \overset{E^\circ}{\rightleftharpoons} M(s)$$

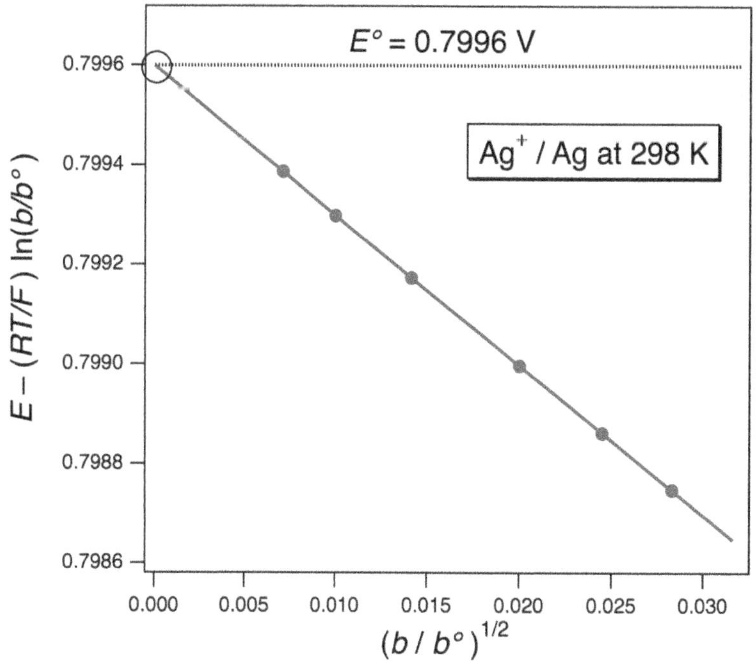

Figure 15.9 An example of the method of extrapolation used to determine $E°$ values. An expression in the form of Eq. (15.46) is formulated and experimental data is acquired for molalities in the range where the Debye–Hückel limiting law can be used reliably.

can be broken up into the elementary steps

$$M^+(aq) \xrightleftharpoons{\Delta_{hyd}H^\circ_m} M^+(g) \xrightleftharpoons{I_1} M(g) \xrightleftharpoons{\Delta_{sub}H^\circ_m} M(s)$$

where $\Delta_{hyd}H^\circ_m$ is the enthalpy of hydration, I_1 is the first ionization energy and $\Delta_{sub}H^\circ_m$ is the enthalpy of sublimation. Trends in $E°$ can then be rationalized in terms of the periodic trends of these energies. However, the fundamental meaning of the magnitude and sign of $E°$ should be kept in mind. As shown in Fig. 15.10, when $E°$ is large and negative, it is thermodynamically more favorable for the species on the right of the reduction potential to give up an electron. In this case, the oxidized form is more stable than the reduced form. When $E°$ is large and positive, the reduced form is

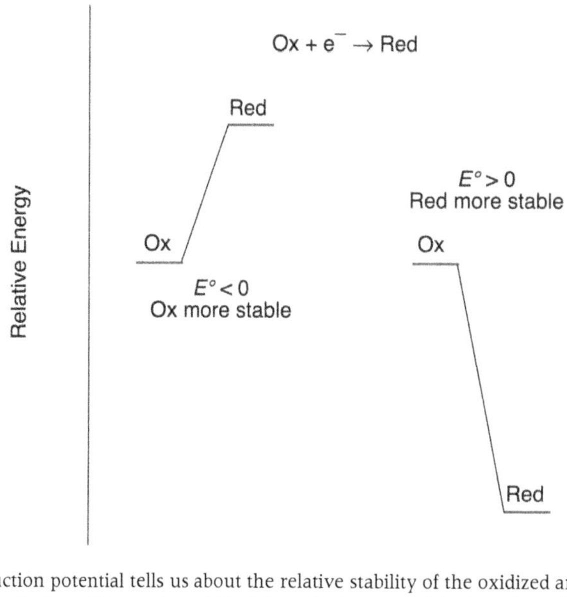

Figure 15.10 The sign of the standard reduction potential tells us about the relative stability of the oxidized and reduced forms of a redox couple. A negative value of $E°$ means that, under standard conditions, the oxidized form Ox is more stable than the reduced form Red. When $E°$ is positive, Red is more stable than Ox. Changes in the chemical environment around Ox and Red can either stabilize or destabilize these two species relative to one another.

Table 15.3 $E°$ values for Fe^{3+}/Fe^{2+} in different chemical environments.

Reaction	$E°$/V
$[Fe(phen)_3]^{3+} + e^- \rightleftharpoons [Fe(phen)_3]^{2+}$	1.147
$[Fe(bipy)_3]^{3+} + e^- \rightleftharpoons [Fe(bipy)_3]^{2+}$	1.03
$[Fe(bipy)_2]^{3+} + e^- \rightleftharpoons [Fe(bipy)_2]^{2+}$	0.78
$Fe^{3+} + e^- \rightleftharpoons Fe^{2+}$	0.771
$[Fe(CN)_6]^{3-} + e^- \rightleftharpoons [Fe(CN)_6]^{4-}$	0.358
In cytochrome c	0.235
In ferredoxin	−0.430
$Fe(OH)_3 + e^- \rightleftharpoons Fe(OH)_2$	−0.56

more stable than the oxidized form. Just as for chemical reactions, the statement that an electrochemical reaction is thermodynamically favored tells us nothing directly about whether this reaction has suitable kinetic parameters for it to proceed at an appreciable rate.

The ionization and sublimation energies are properties of the solid and gas phases. Nothing about the solution properties influences these. However, the enthalpy of hydration is directly related to all those factors that affect the solvation of the ion. Therefore, the ionic strength and the chemical nature of other ions in solution have a direct effect on $E°$. In particular, if anions are introduced that can exchange with the six water molecules that comprise the hexa-aqua complex $[M(OH_2)_6]^{n+}$ that makes up the first solvation shell, then $E°$ will be shifted. Even more strikingly, the M^{n+} ion can be placed within totally different chemical environments by complexing it with ligands, such as CN^-, or by a chemical scaffold as when a crown ether complexes a metal atom or iron is captured within a hemoprotein. This is illustrated by the standard reduction potentials for Fe^{3+}/Fe^{2+} given in Table 15.3.

15.4.4 Practical reference electrodes

A reference electrode is constructed to maintain a constant potential. No current is allowed to pass through the reference electrode, so that concentrations remain constant. Likewise, its temperature must be held fixed. The SHE is not very convenient to work with experimentally, and therefore, in practice a different electrode is normally used as the reference electrode. Two popular alternatives are the *saturated calomel electrode* (SCE) and the *silver–silver chloride electrode*. The SCE corresponds to the half-reaction

$$Hg_2Cl_2(s) + 2e^- \rightleftharpoons 2Hg(l) + 2Cl^-(aq)$$

with cell notation

$$Pt(s) \mid Hg(l) \mid Hg_2Cl_2(s) \mid KCl(sat, aq).$$

The cell potential depends on the Cl^- concentration. When a saturated KCl solution is maintained above a Pt wire immersed in a pool of $Hg(l)$, $Hg_2Cl_2(s)$ forms at the solution/Hg interface and a cell potential of 0.2412 V is maintained at 25 °C. The SCE has the disadvantage of working with Hg.

The silver–silver chloride electrode avoids the use of Hg. A Ag wire encased in AgCl is placed in contact with a KCl(aq) solution. The electrode corresponds to the half-reaction

$$AgCl(s) + e^- \rightleftharpoons Ag(s) + Cl^-(aq)$$

with cell notation

$$Ag(s) \mid AgCl(s) \mid KCl(aq).$$

The standard potential is $E° = 0.22233$ V versus SHE. However, it is commonly run in saturated KCl(aq), for which the potential is then $E° = 0.197$ V.

Since the reference electrode has a fixed potential, the working electrode has its potential measured or controlled with respect to the reference electrode. This lets us know what the energy of the electrons in the working electrode is relative to the energy of the electrons in the reference electrode. Biasing the working electrode negative relative to the reference electrode positions the electrons at a higher energy than in the reference electrode. Biasing the working electrode positive makes the electrons have a lower energy than the reference electrode.

15.5 Surface charge and potential

Biasing changes the effective charge on the surface and changes the coverage of adsorbed ions. The adsorption of anions is enhanced by positive charges at the surface, and is suppressed when the surface is sufficiently biased to surpass some critical value of excess negative charge. Cations behave, of course, in just the opposite manner. Charge transfer and the associated orientation of the dipole moment of molecular adsorbates are also changed with bias.

The approximation of a Helmholtz layer controlling the charge distribution at and near the metal interface pertains to the limit of not too-dilute solutions. At low concentrations, as described in the *Stern–Gouy model*, beyond the Helmholtz layer the bulk electrolyte concentrations are only reached after an extended distance. In the second region, the ion distribution is governed by a Poisson–Boltzmann distribution, which is a long-range gradual sloping of the concentrations from the large excess near the electrode to the limiting bulk values. This model is shown in Fig. 15.2.

Calculation of the charge distribution at and near the interface is an important but difficult task. To determine the value of the electric field as a function of distance, the surface charge and capacitance of the double layer are required. All of these influence the chemical and physical characteristics of the interface, as well as charge transfer and transport to and from the interface. The particulars obviously require a detailed knowledge (or at least a detailed model) of the interfacial structure. The surface charge density is denoted σ^0, while that in the Stern layer is σ^i and that in the Gouy layer is σ^d. When a surface carries no net charge, it is at its point of zero charge (pzc). The pzc depends on the pH and the applied bias.

With a model of the charge density as a function of the distance from the surface, it is possible to calculate the surface potential ψ^0, the potential at the inner Helmholtz plane ψ^{I}, and the potential at the top of the Stern layer ψ^{d}. However, none of these potentials can be measured directly, so what is the potential that enters into equilibrium thermodynamics? The electrical potential energy difference between two points is equal to the electrical work required to move a unit charge from Point A to Point B. The potential is the work divided by the unit charge. The *inner potential* is defined by the work required to move a test charge from infinity to the inside of a phase. This is also known as the *Galvani potential* ϕ and is the potential that enters into equilibrium electrochemistry. It is to be distinguished from the *external potential*, also known as the *Volta potential*, ψ and the *surface potential* χ, which are all related by

$$\phi = \chi + \psi. \tag{15.49}$$

The Volta potential is defined by the work required to move a test charge from infinity (either in vacuum, an inert gas, or a solution) to a point close to the surface (~1 μm away). This distance is sufficiently far away that chemical and image force interactions with the interface can be neglected. Volta potentials and Volta potential differences can be measured directly. However, since we cannot deconvolute the chemical work from electrical work in crossing between two different phases, we cannot unambiguously measure ϕ, χ nor changes in these. We can, nonetheless, measure changes in the *differences* of ϕ and χ.

To relate the Galvani potential to measurable quantities, we use our definition of the electrochemical potential $\tilde{\mu}$ in terms of the chemical potential μ, the charge z_i on the i-th species, and the Faraday constant F,

$$\tilde{\mu}_i = \mu + z_i F\phi. \tag{15.50}$$

The chemical potential is related to the temperature and activity as given in Eq. (13.51). The term $z_i F\phi$ is effectively the work required to move one mole of electrons into a phase with a potential ϕ at its interior. To move charge from phase A to phase B requires the amount of work determined by

$$\Delta\tilde{\mu}_i = \tilde{\mu}_i^{\mathrm{B}} - \tilde{\mu}_i^{\mathrm{A}} = \mu_i^{\mathrm{B}} - \mu_i^{\mathrm{A}} + z_i F(\phi_{\mathrm{B}} - \phi_{\mathrm{A}}). \tag{15.51}$$

At equilibrium the electrochemical potential is everywhere the same, thus $\Delta\tilde{\mu}_i = 0$, which requires that

$$\Delta\phi = \phi_{\mathrm{B}} - \phi_{\mathrm{A}} = \frac{\mu_i^{\mathrm{B}} - \mu_i^{\mathrm{A}}}{z_i F}. \tag{15.52}$$

Whereas the electrochemical potential is everywhere the same at equilibrium, the Galvani potential is not; that is, the Galvani potential difference $\Delta\phi$ does not in general equal zero unless by coincidence the inner potentials of the two phases happen to be the same. A consequence of Eq. (15.52) is that when two dissimilar metals are brought into contact a contact potential $\Delta\varphi$ develops because of the flow of charge to the interface between the two metals. Similarly, the flow of charge and the formation of dipole layers at the electrolyte/electrode interface also lead to a contact potential.

The electrical field in the dipole layer contains contributions not only from the presence of ions, but also due to the polarization of the solvent.

Consider the specific example of a gold electrode in contact with a solution containing Au^{3+} at equilibrium. The appropriate electrochemical reaction is

$$\mathrm{Au(s)} \rightleftharpoons \mathrm{Au^{3+}(aq)} + 3\mathrm{e^-} \tag{15.53}$$

in which the electrons on the right-hand side are in the metal. From the stoichiometry of this reaction, we then write the relationship between chemical potentials of the components of the system as

$$\tilde{\mu}_{\mathrm{Au}} = \tilde{\mu}_{\mathrm{Au^{3+}(aq)}} + 3\tilde{\mu}_{\mathrm{e^-}}. \tag{15.54}$$

We denote the Galvani potential in the solution as ϕ_S and that in the metal as ϕ_M. Assuming that the Au atoms are neutral in the metal such that $\tilde{\mu}_{Au} = \mu_{Au}$, and taking the standard activity $a_0 = 1$, we use Eq. (15.50) to write

$$\mu^\circ_{\mathrm{Au}} + RT\ln a_{\mathrm{Au}} = \mu^\circ_{\mathrm{Au^{3+}}} + RT\ln a_{\mathrm{Au^{3+}}} + 3F\phi_S + \mu^\circ_{\mathrm{e^-}} + 3RT\ln a_{\mathrm{e^-}} - 3F\phi_M. \tag{15.55}$$

By definition, the activity of Au atoms in the bulk is unity, and since the effective concentration of electrons in the metal is constant we can also take their activity to be one; therefore, the Galvani potential difference is

$$\Delta\phi = \phi_M - \phi_S = \Delta\phi^\circ + (RT/3F)\ln a_{\mathrm{Au^{3+}}} \tag{15.56}$$

and the standard Galvani potential difference is

$$\Delta\phi^\circ = \frac{\mu^\circ_{\mathrm{Au^{3+}}} + \mu^\circ_{\mathrm{e^-}} - \mu^\circ_{\mathrm{Au}}}{3F}. \tag{15.57}$$

The potential of one electrode in isolation is not measurable; however, when we introduce a reference electrode whose Galvani potential difference is constant, the potential E of our working electrode with respect to this reference electrode is measurable. Moreover,

$$E - E^\circ = \Delta\phi - \Delta\phi^\circ, \tag{15.58}$$

can be determined experimentally. The value E° is the value of E measured at unit activity under reversible conditions (and similarly for $\Delta\phi^\circ$). Consequently, we are justified in writing the familiar Nernst equation

$$E = E^0 + (RT/zF)\ln a_{\mathrm{M^{z+}}} \tag{15.59}$$

for a metal electrode M in contact with a solution containing its ions M^{z+}. This is, reassuringly, consistent with the result we derived in Section 15.2.

15.6 Relating work functions to the electrochemical series

Charge transfer can only occur between an adsorbate or solution species and a solid substrate if acceptor and donor energy levels exist at the appropriate energy in both systems. How do the energy levels of two disparate systems align? For the gas/solid interface this is a rather straightforward exercise. Its measurement is accomplished using photoelectron spectroscopy, which measures directly the binding energy of any occupied orbital (as discussed in Chapter 23). The presence of a solution complicates the matter greatly. This issue is of particular importance when considering the results of *ab initio* calculations, because without a method of aligning the well-characterized but relative values of the electrochemical series to the absolute scale of work functions and the vacuum level, no quantitative comparison to experimental results can be made. This issue remains controversial and the uncertainty in the absolute scale is as much as ±0.2 eV, which is about 20 kJ mol^{-1}.

The alignment of energy levels is particularly important in semiconductor photochemistry and electrochemistry. Therefore, Fig. 15.11 has been constructed following Reiss and Heller.[4] Prior to forming an electrical connection, the vacuum levels of the semiconductor, E^S_{Vac}, and the standard hydrogen electrode (SHE), E^{SHE}_{vac}, are equal. The difference between the Fermi energy and the vacuum level determines the work function. Noting that the electrochemical potential $\tilde{\mu}$ is equivalent to the Fermi energy and aligning our absolute reference scale by setting $E^{SHE}_{Vac} = 0$, we write

$$\tilde{\mu} = \tilde{\mu}_{\mathrm{redox}} = -\Phi_{\mathrm{redox}}/e, \tag{15.60}$$

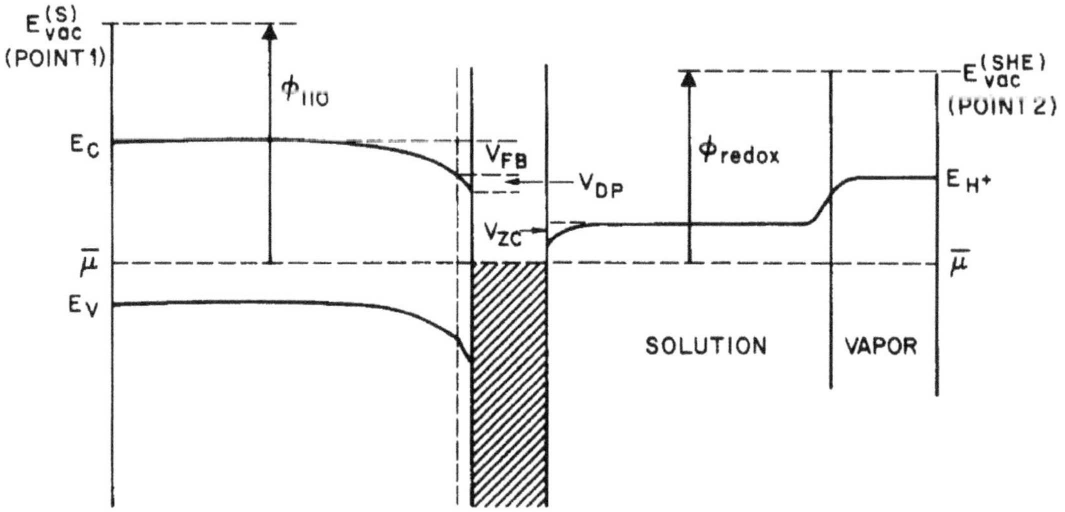

Figure 15.11 An energy level diagram depicting the levels associated with the connection of a semiconductor electrode to a standard hydrogen electrode. Reproduced with permission from Reiss, H. and Heller, A. (1985) *J. Phys. Chem.*, **89**, 4207. © 1985, American Chemical Society.

where Φ_{redox} is the work function of the SHE. Once the two electrodes (the semiconductor and a Pt electrode both immersed in a 1 M solution of $HClO_4$ exposed to 1 atm of H_2 at 298 K) are brought into electrical contact, the vacuum levels shift such that the electrochemical potential is the same everywhere. There is band bending of the conduction band minimum E_c and the valence band maximum E_v in the semiconductor prior to immersion. The band bending is measured in terms of the flat band potential V_{FB}. V_{FB} is measured by finding the slope of a Mott–Schottky plot of capacitance versus applied potential.[5] The situation is slightly less complicated if the semiconductor is replaced by a metal, in that there is no band bending ($V_{FB} = 0$) in the latter case. A further change in potential in front of the semiconductor may occur and is denoted by the dipole layer potential V_{DP}. Similarly, the potential of zero charge V_{zc} represents the response of the Pt electrode to the Stern layer above it. The energy of an electron added to a H^+ ion to produce a H atom is denoted E_{H^+}.

From Fig. 15.11 we can determine that alignment of the Fermi levels leads to

$$E_{Vac}^{S} - E_{Vac}^{SHE} = V_{FB} + V_{DP} - V_{zc} \tag{15.61}$$

From our choice of an absolute reference, substituting for E_{vac}^{S} by using

$$\Phi_S/e = E_{vac}^{S} - \tilde{\mu}, \tag{15.62}$$

and for $\tilde{\mu}$ from Eq. (15.60), we obtain

$$\Phi_{redox}/e = V_{FB} + V_{DP} - V_{zc} - \Phi_S/e. \tag{15.63}$$

Similarly for the chemical potential

$$\tilde{\mu}_{redox} = \tilde{\mu} = V_{FB} + V_{DP} - V_{zc} - \Phi_S/e. \tag{15.64}$$

Equations (15.63) and (15.64) are both referenced to Point 2 in Fig. 15.11. For the specific case of InP(110) in contact with the SHE, Reiss and Heller have found $\Phi_S = 5.78$ eV, $V_{FB} = 1.05$ V, $V_{DP} \approx 0$, and $V_{zc} = -0.3$ V. Therefore,

$$\tilde{\mu}_{redox} = \tilde{\mu}_{H^+/H_2} = -4.43\ \text{eV} \tag{15.65}$$

and the work function of the SHE is $\Phi_{SHE} = 4.43$ eV. This value is in line with the recommendation from the IUPAC of 4.44 ± 0.02 eV.

The absolute determination of Φ_{SHE} is of fundamental importance in physical chemistry because of it relationship not only to the electrochemical series but also because it places limits on the absolute values of the enthalpy and Gibbs energy of hydration. Without an accurate value of the enthalpy and Gibbs energy of hydration of the H^+ ion, we are unable to accurately obtain absolute enthalpies and Gibbs energies of all aqueous ions that are consistent with the electrochemical series. This is one reason why we had to introduce the odd scale of enthalpies and entropies of formation of ions in solution that are all relative to H^+(aq). The uncertainty in the absolute values also greatly complicates the comparison of *ab initio* theoretical calculations to experimental values. It increases the difficulty of accurately predicting equilibrium constants and pK_a values.

Attempts have been made to measure absolute reduction potentials by measuring the energy given off when an ion solvated in a cluster of water molecules in a supersonic molecular beam combines with an electron. The energy is measured by counting the number of water molecules and hydrogen atoms that desorb from the cluster in response to the energy released by ion neutralization. The product of the number of water molecules desorbed times the activation energy for desorption is then equal to the energy released by electron capture of the solvated ion.[6] These direct experimental determinations suggest a value in the range of 4.29–4.45 eV based on two independent methods, or perhaps 4.28 ± 0.09 eV based on an independent set of cluster ion solvation data.[7] Computational methods are becoming comprehensive enough to include solvation accurately and to deliver results with close to chemical accuracy. The value based on cluster ion solvation is far below the accepted value of 4.44 ± 0.02 eV. More recently, Matsui *et al.*[8] achieved much closer value of 4.48 eV.

To place some perspective on the unsettling amount of uncertainty this corresponds to compared to other accurate determination involving hydrogen, consider the following thermodynamic cycle for the calculation of the enthalpy of the standard hydrogen reduction reaction

$$H(g) \rightarrow \frac{1}{2}H_2(g)$$

$$H^+(g) + e^- \rightarrow H(g)$$

$$H^+(aq) \rightarrow H^+(g)$$

Overall

$$H^+(aq) + e^- \rightarrow \frac{1}{2}H_2(g)$$

The first reaction is the reverse of molecular hydrogen dissociation. The best value[9] for the dissociation energy of H_2 is 432.0680589 ± 0.0000044 kJ mol^{-1}. The second reaction is the reverse of the ionization of the H atom. From determination of the Rydberg constant,[10] we know that the energy required for this process is 1311.9156 kJ mol^{-1} and this value is known to 8 parts in 10^{13}. The third reaction is the hydration of the proton[11] with a value of −1091 kJ mol^{-1} with an uncertainty as large as ±15 kJ mol^{-1}. Obviously, chemical accuracy does not yet approach spectroscopic accuracy.

15.7 Applications of standard potentials

15.7.1 Constructing a cell from a redox couple

Take two redox couples Ox_1/Red_1 and Ox_2/Red_2

$$Ox_1 + ze^- \rightarrow Red_1 \rightarrow E_1^\circ$$

$$Ox_2 + ze^- \rightarrow Red_2 \rightarrow E_2^\circ$$

to form the cell

$$Red_1, Ox_1 :: Red_2, Ox_2$$

with the cell reaction

$$Red_1 + Ox_2 \rightarrow Ox_1 + Red_2$$

and cell potential

$$E^\circ = E_2^\circ - E_1^\circ \tag{15.66}$$

The reaction is spontaneous if $E^\circ > 0$ and that, if $E_2^\circ > E_1^\circ$, the cell is spontaneous as written. This leads to the rule LOW REDUCES HIGH. In the specific example of the Daniell cell, since $E^\circ(Zn^{2+}/Zn) = -0.76$ V $< E^\circ(Cu^{2+}/Cu) = +0.34$ V, Zn has a tendency to reduce Cu^{2+}. The electrochemical series can be used to predict when reactions are allowed to occur thermodynamically. By convention, we write a cell so that the spontaneous direction (the one with a positive cell potential) goes left to right according to:

Left electrode | reductant :: oxidant | Right electrode

$$Zn(s) \,|\, ZnSO_4(aq) :: CuSO_4(aq) \,|\, Cu(s).$$

Current flows from the left electrode to the right electrode. The cell potential is determined by

$$E^\circ = E^\circ_{\text{right}} - E^\circ_{\text{left}} \tag{15.67}$$

$$E^\circ = 0.3419\ \text{V} - (-0.7618\ \text{V}) = +1.1037\ \text{V}. \tag{15.68}$$

NOTE: you *do not* multiply E° values by the factor that was used to balance the cell reaction.

15.7.1.1 Example

Write the cell reaction and electrode half-reactions and calculate the standard emf of each of the following cells

a $Zn \mid ZnSO_4(aq) \,\vdots\vdots\, AgNO_3(aq) \mid Ag$
b $Cd \mid CdCl_2(aq) \,\vdots\vdots\, HNO_3(aq) \mid H_2(g) \mid Pt$
c $Pt \mid K_3[Fe(CN)_6](aq), K_4[Fe(CN)_6](aq) \,\vdots\vdots\, CrCl_3(aq) \mid Cr$

The cell notation specifies the right and left electrodes. Note that for balancing of the amount of substance and charge we must equalize the number of electrons in the half-reactions being combined. To calculate the standard emf of the cell we have used $E^\circ = E^\circ_R - E^\circ_L$, with standard electrode potentials from tables.

			E° / V
(a)	R:	$2\ Ag^+(aq) + 2\ e^- \rightarrow 2\ Ag(s)$	+0.80
	L:	$Zn^{2+}(aq) + 2\ e^- \rightarrow Zn(s)$	−0.76
	Overall (R–L):	$2\ Ag^+(aq) + Zn(s) \rightarrow 2\ Ag(s) + Zn^{2+}(aq)$	+1.56
(b)	R:	$2\ H^+(aq) + 2\ e^- \rightarrow H_2(g)$	+0.00
	L:	$Cd^{2+}(aq) + 2\ e^- \rightarrow Cd(s)$	−0.40
	Overall (R–L):	$2\ H^+(aq) + Cd(s) \rightarrow H_2(g) + Cd^{2+}(aq)$	+0.40
(c)	R:	$Cr^{3+}(aq) + 3\ e^- \rightarrow Cr(s)$	−0.74
	L:	$3\ [Fe(CN)_6]^{3-}(aq) + 3\ e^- \rightarrow 3\ [Fe(CN)_6]^{4-}(aq)$	+0.36
	Overall (R–L):	$Cr^{3+}(aq) + 3\ [Fe(CN)_6]^{4-}(aq) \rightarrow Cr(s) + 3\ [Fe(CN)_6]^{3-}(aq)$	−1.10

Those cells with $E^\circ > 0$ will operate as spontaneous cells under standard conditions. Those for which $E^\circ < 0$ may operate in the forward direction as nonspontaneous electrolytic cells if a bias in excess of E° is applied. Otherwise, they run spontaneously in the reverse direction. Recall that E° informs us of the spontaneity of a cell under standard conditions only. For other combinations of concentrations we require E from the Nernst equation.

15.7.1.2 Example

Devise cells in which the following are the reactions and calculate the standard emf in each case.

a $Zn(s) + CuSO_4(aq) \rightarrow ZnSO_4(aq) + Cu(s)$
b $2\ AgCl(s) + H_2(g) \rightarrow 2\ HCl(aq) + 2\ Ag(s)$
c $2\ H_2(g) + O_2(g) \rightarrow 2H_2O(l)$

As conditions (concentrations, temperature, pressure, etc.) were not given, we assume standard conditions. The specification of the right and left electrodes is determined by the direction of the reaction as written. As always, in combining half-reactions to form an overall cell reaction, we must write the half-reactions with equal numbers of electrons to ensure proper balancing. Identify the half-reactions, then set up the corresponding cell.

			E° / V
(a)	R:	$Cu^{2+}(aq) + 2\ e^- \rightarrow Cu(s)$	+0.34
	L:	$Zn^{2+}(aq) + 2\ e^- \rightarrow Zn(s)$	−0.76
	Cell:	$Zn \mid ZnSO_4(aq) \parallel CuSO_4(aq) \mid Cu$	+1.10
(b)	R:	$AgCl(s) + e^- \rightarrow Ag(s) + Cl^-(aq)$	+0.22
	L:	$H^+(aq) + e^- \rightarrow \frac{1}{2}\ H_2(g)$	+0.00
	Cell:	$Pt \mid H_2(g) \mid H^+(aq) \mid AgCl(s) \mid Ag$	+0.22
	or	$Pt \mid H_2(g) \mid HCl(aq) \mid AgCl(s) \mid Ag$	
(c)	R:	$O_2(g) + 4\ H^+(aq) + 4\ e^- \rightarrow 2\ H_2O(l)$	+1.23
	L:	$4\ H^+(aq) + 4\ e^- \rightarrow 2\ H_2(g)$	+0.00
	Cell:	$Pt \mid H_2(g) \mid H^+(aq), H_2O(l) \mid O_2(g) \mid Pt$	+1.23

In all cases $E^\circ > 0$; therefore, all cells are spontaneous as written.

15.7.1.3 Directed practice

Calculate the standard potential if Cd replaces Cu in the Daniell cell.

[Answer: 0.3618 V]

15.7.2 Determining $E°$ and $\Delta_r G°$ for half-reactions

If we required the $E°$ value for a half-reaction that does not appear in a table, it is possible to construct one from other half-reactions. However, some caution must be used, as shown in the examples below.

A reaction in which two atoms of one chemical species form two dissimilar chemical species is called a *disproportionation reaction.* We can calculate the standard electrode potential and $\Delta_r G°$ of the disproportionation half-reaction of Fe^{3+}(aq) into Fe^{2+}(aq) + Fe(s) from two constituent half-reactions. We identify two half-reactions that when added give the balanced the overall reaction.

								$E°$/V
	2 Fe^{3+}(aq)	+	2e⁻	→	2Fe^{2+}(aq)			+0.771
	Fe^{2+}(aq)	+	2e⁻	→	Fe(s)			−0.440
Overall:	2 Fe^{3+}(aq)	+	4e⁻	→	Fe^{2+}(aq)	+	Fe(s)	+0.331

Since we have two reactions with the same number of electrons transferred, we can add $E°$ values to obtain $E°$ for the composite half-reaction. Then we can calculate the Gibbs energy change from this value of $E°$.

$$\Delta_r G^\circ_m = -zFE^\circ = -4(96\,485)(0.331) = -128\,000 \text{ J mol}^{-1}.$$

Now let us calculate $E°$ and $\Delta_r G°$ for a different half-reaction, the reduction of Fe^{3+} to Fe(s). We use the same two half-reactions as above to write the composite half-reaction.

						$E°$ / V	$\Delta_r G° = -zFE°$ / (J mol⁻¹)
	Fe^{3+}(aq)	+	e⁻	→	Fe^{2+}(aq)	+0.771	−(96 485)(0.771) = −74 390
	Fe^{2+}(aq)	+	2e⁻	→	Fe(s)	−0.440	−2(96 485)(−0.44) = 84 907
Overall:	Fe^{3+}(aq)	+	3e⁻	→	Fe(s)	not additive	

The two reactions do not add up to a full cell reaction. They represent a new half-reaction because there is only a reduction step and no accompanying oxidation. Furthermore, the three half-reactions involve *different numbers of electrons.* We cannot add $E°$ from a one-electron half-reaction to the $E°$ of a two-electron half-reaction to obtain the $E°$ of a three-electron transfer. Note the difference in how the sums balance in comparison to the previous example. While we cannot add $E°$ values, we can add the values of $\Delta_r G^\circ_m$ for each step to obtain the composite value. The subtle difference in how we treat additions of $E°$ and $\Delta_r G^\circ_m$ results from the presence of z in the relationship between the two and how we account for stoichiometry. Returning to our example, the sum of molar Gibbs reaction energies

$$\Delta_r G^\circ = 84\,907 - 74\,390 = 10\,517 \text{ J mol}^{-1}$$

allows us to calculate the value of $E°$ for the composite half-reaction,

$$E^\circ = \frac{\Delta_r G^\circ}{-zF} = \frac{10517}{-3(96485)} = -0.036 \text{ V}.$$

This will never happen for a complete *balanced* cell reaction composed of both an oxidation and a reduction because, in a complete *balanced* electrochemical reaction, the number of electrons involved in reduction must equal the number involved in oxidation.

15.7.3 Determination of activity coefficients

The mean ionic activity coefficient can be determined from a solution of known $E°$ and molality b by measurement of E. For instance, for the Ag^+/Ag example above, Eq. (15.45) rearranges to

$$\ln\gamma_\pm = \frac{F}{RT}(E - E^\circ(\text{Ag}^+/\text{Ag})) - \ln\left(\frac{b}{b^\circ}\right). \tag{15.69}$$

Note that this equation is specific for the stoichiometry of the reaction. Substitution of the measured value of E with the known value of the molality b then leads directly to a calculated value of $\gamma_\pm$.

15.7.4 Determination of equilibrium constants

Recall that Eq. (15.41) allows one to convert $E°$ values directly into equilibrium constants.

$$\ln K = \frac{zFE°}{RT}. \tag{15.70}$$

15.7.5 Determination of thermodynamic functions

Just to summarize, the standard reaction Gibbs energy and the standard emf are related by

$$\Delta_r G_m^\circ = -zFE°. \tag{15.71}$$

Therefore, the derivative with respect to T can be evaluated

$$\left(\frac{\partial \Delta_r G_m^\circ}{\partial T}\right)_p = -zF\frac{dE°}{dT} \tag{15.72}$$

and used to evaluate the entropy change because

$$\left(\frac{\partial \Delta_r G_m^\circ}{\partial T}\right)_p = -\Delta_r S_m^\circ. \tag{15.73}$$

We set the right-hand side (RHS) of Eq. (15.72) equal to the RHS of Eq. (15.73) to find

$$\frac{dE°}{dT} = \frac{\Delta_r S_m^\circ}{zF}. \tag{15.74}$$

Thus, a plot of $E°$ versus T has a slope of $\Delta_r S_m^\circ / zF$. The reaction enthalpy is then

$$\Delta_r H_m^\circ = \Delta_r G_m^\circ + T\Delta_r S_m^\circ = -zF\left(E° - T\frac{dE°}{dT}\right). \tag{15.75}$$

15.7.5.1 Example

Use the Debye–Hückel limiting law and the Nernst equation to estimate the potential of the cell Ag | AgBr(s) | KBr(aq, 0.050 mol kg^{-1}) ⋮⋮ $Cd(NO_3)_2$ (aq, 0.010 mol kg^{-1}) | Cd at 25 °C.

Start by determining the cell reaction and $E°$.

			$E°$ / V
(a)	R:	$Cd^{2+}(aq) + 2\,e^- \rightarrow Cd(s)$	−0.40
	L:	$2\,AgBr(s) + 2\,e^- \rightarrow 2\,Ag(s) + 2\,Br^-(aq)$	+0.07
	Overall (R–L):	$Cd^{2+}(aq) + 2\,Ag(s) + 2\,Br^-(aq) \rightarrow Cd(s) + 2\,AgBr(s)$	−0.47

$z = 2$ however, the $E°$ value for the left electrode is not multiplied by 2 when the sum is taken to determine the cell potential. The activity of Cd^{2+} is given by

$$a_{Cd^{2+}} = \gamma_+ b_{Cd^{2+}} = \gamma_\pm(Cd(NO_3)_2) b_{Cd^{2+}},\ b_{Cd^{2+}} = 0.010\,\text{mol kg}^{-1}$$

The activity of Br^- is given by

$$a_{Br^-} = \gamma_- b_{Br^-} = \gamma_\pm(KBr) b_{Br^-},\ b_{Br^-} = 0.050\,\text{mol kg}^{-1}.$$

The mean activity coefficient in the $CdNO_3$ solution is different than $\gamma_\pm(KBr)$ because the two solutions are in separate containers. Likewise the ionic strength is different in the two containers. For the $Cd(NO_3)_2$ solution,

$$I = \frac{1}{2}\left(b_+z_+^2 + b_-z_1^2\right) = \frac{1}{2}(b(2)^2 + 2b(1)^2) = 3(0.010) = 0.030\,\text{mol kg}^{-1}$$

and the mean activity coefficient is

$$\lg \gamma_\pm = -B\,|z_+z_-|\,I^{1/2} = -0.509\,(2)\,(0.030)^{1/2} = -0.176$$

For the KBr solution,

$$I = \frac{1}{2}\left(b_+z_+^2 + b_-z_1^2\right) = \frac{1}{2}\left(b(1)^2 + b(1)^2\right) = b = 0.050\,\text{mol kg}^{-1}$$

and the mean activity coefficient is

$$\lg \gamma_\pm = -B\,|z_+z_-|\,I^{1/2} = -0.509\,(1)\,(0.050)^{1/2} = -0.114$$

To determine E, we use the Nernst equation

$$E = E^\circ - \frac{RT}{zF}\ln Q = E^\circ - \frac{RT}{2F}\ln\frac{1}{a_{Cd^{2+}}(a_{Br^-})2} = E^\circ + \frac{RT}{2F}\ln(a_{Cd^{2+}}(a_{Br^-})^2)$$

$$E = E^\circ + \frac{RT}{2F}\ln\left(b_{Cd^{2+}}\left(b_{Br^-}\right)^2\right) + \frac{2.303RT}{2F}\lg\left(\gamma_{Cd^{2+}}\left(\gamma_{Br^-}\right)^2\right)$$

$$E = -0.62\ \text{V}$$

Note that $RT/F = 25.693$ mV at 25° C.

15.7.5.2 Example

A fuel cell develops an electric potential from the chemical reaction between reagents supplied from an outside source. At 1.0 bar pressure and 298 K, calculate the emf of a cell fuelled by the combustion of butane.

Start with a balanced combustion reaction

$$C_4H_{10} + 13/2\,O_2 \rightarrow \quad 4\,CO_2(g) + 5\,H_2O(l)$$

The Gibbs energy of combustion of C_4H_{10} can either be looked up or calculated from $\Delta_f G^\circ_m$ values as −2746.06 kJ mol^{-1}. We now need to write balanced half-reactions to determine z. Start with the oxygen to water half-reaction used above at the right-hand electrode. Make the left-hand electrode a reaction of $CO_2 + H^+$ to yield the desired product C_4H_{10} $(+H_2O)$. Start by balancing the C atoms, then the O atoms in this half-reaction then add as much H^+ is needed. Then balance the overall reaction by multiplying the O_2 half-reaction to ensure charge neutrality. The result is

R: $13/2\ O_2(g) + 26\ H^+(aq) + 26\ e^- \rightarrow 13\ H_2O(l)$

L: $4\ CO_2(g) + 26\ H^+(aq) + 26\ e^- \rightarrow C_4H_{10}(g) + 8\ H_2O(l)$

Overall: $C_4H_{10}(g) + 13/2\ O_2(g) \rightarrow 4\ CO_2(g) + 5\ H_2O(l)$

Thus, $z = 26$ and the standard emf is

$$E^\circ = \frac{-\Delta_r G^\circ_m}{zF} = \frac{2746.06\times 10^3\ \text{J mol}^{-1}}{26(96485\ \text{C mol}^{-1})} = +1.09\ \text{V}.$$

15.7.5.3 Example

Use these two half-reactions

$$Cr^{3+}(aq) + 3e^- \rightarrow Cr(s) \qquad E^\circ = -0.74\ \text{V}$$

$$Fe(CN)_6^{3-}(aq) + e^- \rightarrow Fe(CN)_6^{4-}(aq) \qquad E^\circ = +0.36\ \text{V}$$

to answer the following questions. The anode and cathode, held at $T = 298$ K, are immersed in 1×10^{-3} M KCl solutions. The concentration of the Cr^{3+}, $Fe(CN)_6^{3-}$ and $Fe(CN)_6^{4-}$ are all 1×10^{-6} M.

a Write out a balanced spontaneous electrochemical reaction involving the two half-reactions.
b Calculate E°.
c Calculate the activities of all species in the reaction.
d Calculate E.
e Write out the cell notation for the reaction.
f Identify the anode and cathode and describe what is happening at the two electrodes.

(a) and (b) $\{Cr^{3+}(aq) + 3e^- \rightarrow Cr(s)\}$ {needs to be reversed} $E^\circ = -0.74$ V
$3Fe(CN)_6{}^{3-}(aq) + 3e^- \rightarrow 3Fe(CN)_6{}^{4-}(aq)$ $E^\circ = +0.36$ V

$Cr(s) + 3Fe(CN)_6{}^{3-}(aq)^- \rightarrow Cr^{3+}(aq) + 3Fe(CN)_6{}^{4-}(aq)$ $E^\circ = +0.36 - (-0.74\ \text{V}) = 1.10$ V

(c) The concentration of KCl, which is fully dissociated, is so much larger than the other ions in solution that we need only consider the concentrations of K^+ and Cl^- in determining the ionic strength and the activity coefficients. Since KCl is a 1:1 electrolyte of concentration 1 mM,

$$I = 1\times10^{-3}.$$

$$\lg\gamma_\pm = 0.509\,|\,(1)(-1)\,|\,(1\times10^{-3})^{1/2} = -0.016096$$

$$\gamma_\pm = 0.9636$$

$$a(Fe(CN)_6^{3-}) = a(Fe(CN)_6^{4-}) = a\left(Cr^{3+}\right) = 0.9636(1\times10^{-6}) = 9.64\times10^{-7}$$

$$a(K^+) = a(Cl^-) = 0.9636(1\times10^{-3}) = 9.64\times10^{-4}$$

$$a(Cr) = 1$$

(d)

$$E - E^\circ - \frac{RT}{zF} \ln \frac{\left[\mathrm{Fe\,(CN)_6^{4-}}\right]^3 \left[\mathrm{Cr^{3+}}\right]}{\left[\mathrm{Fe\,(CN)_6^{3-}}\right]^3}$$

$$E = 1.10\,\mathrm{V} - \frac{0.02568\,\mathrm{V}}{3} \ln \frac{[9.64 \times 10^{-7}]^3 [9.64 \times 10^{-7}]}{[9.64 \times 10^{-7}]^3} = 1.20\,\mathrm{V}$$

(e)

$$\mathrm{Cr(s)\,|\,CrCl_3(aq),\ 1\,M\,KCl}\ \vdots\vdots\ \mathrm{Fe(CN)_6^{3-},\ Fe(CN)_6^{4-},\ 1\,M\,KCl\,|\,Pt(s)}$$

Anode Cathode

Electrons travel from the anode to the cathode through an external circuit. On the surface of the Pt cathode, the $Fe(CN)_6^{3-}$ picks up an electron and becomes $Fe(CN)_6^{4-}$ that then goes back into solution. It is the coordinated Fe^{3+} that is reduced to Fe^{2+}, not one of the CN^- ligands. On the surface of the Cr anode, Cr atoms lose three electrons and desorb from the surface (leave the surface and enter the solution) as Cr^{3+} ions.

15.8 Biological oxidation and proton-coupled electron transfer

Many biological reactions are redox reactions. These can take several forms involving the transfer of not only electrons but also protons. When a hydrogen ion is also transferred, the reaction is sometimes called *proton-coupled electron transfer* (PCET). Biological PCET reactions are fundamental to photosynthesis, respiration, light-driven cell signaling, DNA biosynthesis and nitrogen fixation. Obviously, many enzymatic reactions involve PCET. Three of the most important types of reactions are listed below:

- Transfer of both H^+ and electrons

$$\mathrm{AH_2 + B \rightleftharpoons [A + 2H^+ + 2e^- + B] \rightleftharpoons A + BH_2}$$

- Loss of H^+ and transfer of electrons

$$\mathrm{AH_2 + B \rightleftharpoons [A + 2H^+ + 2e^- + B] \rightleftharpoons A + B^{2-} + 2H^+}$$

- Only transfer of electrons

$$\mathrm{A^{2-} + B \rightleftharpoons [A + 2e^- + B] \rightleftharpoons A + B^{2-}}$$

The mechanisms of such reactions are generally complex and involve several enzymes. Tunneling is often important in these reactions. However, since the mass of the electron is much less than that of the proton, they tunnel over very different length and time scales. In hydrogen atom transfer (HAT) the donor–acceptor pair is the same for both electrons and protons. However, the electron donor–acceptor pair need not be the same as the proton donor–acceptor pair. In some instances electron transfer (ET) and proton transfer (PT) occur in simultaneous, concerted steps, but in other cases ET and PT are sequential in such a manner that the first step promotes the second. The two steps are coupled even though they are temporally separated from one another. Between these two extremes are a range of mixed pathways for which the description is not as clear.

Fluctuations are extremely important in the dynamics. Structural changes can dramatically change the time scale of the motions required for reaction. The importance of fluctuations and conformational changes *during* reaction means that the system may sample a number of transition states and reactive intermediates along the path from reactants to products. The fluctuations involved may include the nuclear dynamics not only of the enzyme and substrate but also of the solvent (water) and any other reactants.

In the terminal respiratory chain, electrons donated by glucose and carried by NADH are transferred from one molecule to another until they are eventually passed on to O_2, which is then converted to H_2O. *Cytochromes* are electron-carrying proteins containing heme groups. Heme groups present several distinct chemical environments that surround a central iron ion, which is involved in the electron transfer,

$$\mathrm{Fe^{3+} + e^- \rightleftharpoons Fe^{2+}}$$

The chemical environment around the iron ion, based on whether it is more or less electron-withdrawing, shifts the standard potential. This stabilizes either the ferric (Fe^{3+}) or ferrous ion (Fe^{2+}) and thereby shifts the more stable species from one side to the other. There are different forms of these cytochrome protein denoted *b*, *c*, *a*, and a_3. Cytochromes *a* and a_3 together are called 'cytochrome oxidase' (respiratory enzyme), and can transfer electrons directly to O_2. Cytochrome *c* oxidase oxidizes cytochrome *c* and reduces and protonates O_2 to water via a PCET reaction pathway.

15.8.1 Biochemical standard electrode potential

Standard electrode potentials $E°$ are referenced to 1 M H^+, that is pH 0. Biochemical standard electrode potentials $E°'$ are references to pH 7. We have seen before the effect of this on the standard Gibbs energies $\Delta_r G°$ as compared to biochemical standard Gibbs energies $\Delta_r G°'$. A different choice of standard state does not change the value of $\Delta_r G$, it only changes the standard state value. Similarly, the new choice of standard state does not change the measured cell potential E. It does, however, change the biochemical standard electrode potential $E°'$ compared to the standard electrode potential $E°$. This is particularly important for reactions involving protons.

Consider the reaction

$$\mathrm{A + B \rightleftharpoons C} + x\mathrm{H^+}.$$

Using the conventional (physical chemistry) standard state with $c° = 1$ M, the Nernst equation for this reaction is

$$\Delta_r G_m = \Delta_r G_m^\circ + RT \ln \frac{[\mathrm{C}]/c°([\mathrm{H^+}]/c°)^x}{[\mathrm{A}]/c°[\mathrm{B}]/c°}. \tag{15.76}$$

However in the biochemical standard state $c° = 1$ M for everything except H^+ and OH^-, which have a standard concentration of 1×10^{-7} M. Thus, the corresponding Nernst equation written according to the convention of the biochemical standard state is

$$\Delta_r G_m = \Delta_r G_m^{\circ\prime} + RT \ln \frac{[\mathrm{C}]/c°([\mathrm{H^+}]/1\times 10^{-7}\,\mathrm{M})^x}{[\mathrm{A}]/c°\,[\mathrm{B}]/c°}. \tag{15.77}$$

$\Delta_r G_m$ is the same in both cases, which means that

$$\Delta_r G_m^\circ = \Delta_r G_m^{\circ\prime} + xRT \ln \frac{1}{1\times 10^{-7}}. \tag{15.78}$$

With $x = 1$ and $T = 298$ K,

$$\Delta_r G_m^\circ = \Delta_r G_m^{\circ\prime} + 39.93\,\mathrm{kJ\,mol^{-1}}. \tag{15.79}$$

If a proton is consumed rather than produced, then $x = -1$ and

$$\Delta_r G_m^\circ = \Delta_r G_m^{\circ\prime} - 39.93\,\mathrm{kJ\,mol^{-1}}. \tag{15.80}$$

If no protons (or hydroxides) are involved in the reaction, $x = 0$ and

$$\Delta_r G_m^\circ = \Delta_r G_m^{\circ\prime}. \tag{15.81}$$

Dividing any of the last three equations by $-zF$ yields the standard potentials for the biological oxidations. For example, for Eq. (15.79)

$$E° = E°' - \frac{39{,}930\,\mathrm{J\,mol^{-1}}}{z(96{,}500\,\mathrm{C\,mol^{-1}})} = E°' - \frac{0.414\,\mathrm{V}}{z}. \tag{15.82}$$

A reaction producing H^+ has a value of $E°$ less than $E°'$ by 0.414 V/z per mole of H^+ formed, and is more spontaneous at pH 7 than at pH 0. Note that corrections of this type only need to be made when trying to compare values of standard electrode potentials from one scale to the other, or when calculations of expected cell potentials must be constructed by using standard potentials from both scales.

15.8.2 Gibbs energy change in terminal respiratory cycle

The conversion of nicotinamide adenine dinucleotide, reduced form (NADH) to its oxidized form (NAD^+) releases two electrons, which are used to reduce O_2 to H_2O. The half-reactions involved are

$$\mathrm{NAD^+(aq) + H^+(aq) + 2e^- \rightleftharpoons NADH(aq)} \qquad E°' = -0.32\,\mathrm{V}$$

and

$$\tfrac{1}{2}O_2(aq) + 2H^+(aq) + 2e^- \rightleftharpoons H_2O(l) \qquad E^{\circ\prime} = +0.816\,V,$$

which corresponds to the overall reaction

$$NADH(aq) + \tfrac{1}{2}O_2(aq) + H^+(aq) \rightleftharpoons NAD^+(aq) + H_2O(l) \qquad E^{\circ\prime} = 1.136\,V.$$

The Gibbs energy change is

$$\Delta_r G_m^{\circ\prime} = -zFE^{\circ\prime} = -2(96500\,C\,mol^{-1})(1.136\,V) = -219\,kJ\,mol^{-1}.$$

Energy can effectively be stored by running an endothermic reaction, such as ATP synthesis ($\Delta_r G^{\circ\prime} = 31.4\ kJ\ mol^{-1}$). Remember that energy is not released by breaking bonds. Rather, it is released when high-enthalpy species are transformed into lower-enthalpy species via chemical reactions, when weak bonds are replaced by stronger bonds. If the above electron transfers were performed in one step, much energy would be lost as heat. Performing it in a number of steps involving NADH, flavin adenine dinucleotide (FAD), cytochrome *b*, cytochrome *c* as well as cytochrome *a* and cytochrome a_3, releases the energy in smaller steps. This allows for the synthesis of several molecules of ATP along the way in a highly efficient manner, with minimal losses to heat generation.

15.8.3 Membrane potentials

Nerve cells and muscle cells are excitable. They can transmit a change of electrical potential along their membranes. The electrical potential established by the difference in ionic concentrations across the membrane is known as the *membrane potential.* Differences in ionic concentrations inside and outside the cell can be established because the cell membrane has different permeabilities for different ions. A semi-permeable membrane allows some species to pass while blocking others.

If two KCl(aq) solutions are separated by a semi-permeable membrane that allows K^+ to diffuse across it but not Cl^-, then a concentration gradient can be established by moving K^+ through the membrane. Moving a species of only one charge type leads to the creation of an electric field that will oppose K^+ migration and eventually stop it when the electrical potential is high enough. The electrical potential generated by a single ion is given by the Nernst equation, which at 25 °C is

$$E = E^\circ - \frac{0.0257\,V}{z}\ln[K^+]. \tag{15.83}$$

The membrane potential ΔE_{K^+} is set by the concentration difference, with a sign determined by the convention internal minus external

$$\Delta E_{K^+} = E_{K^+,in} - E_{K^+,ex} = (0.0257\,V)\ln\frac{[K^+]_{ex}}{[K^+]_{in}}. \tag{15.84}$$

In a real nerve cell, the cell potential is set by the contributions from not only K^+ and Cl^- but also Na^+. The ionic fluxes are regulated not only by the concentration differences and electric fields, but also by a membrane protein called Na^+–K^+ ATPase, which transports Na^+ out of the cell and K^+ into the cell using energy from ATP hydrolysis. This is accounted for in a modified version of the Nernst equation with contributions from these three species along with a factor that quantifies the permeability of the membrane to each ion. This is the *Goldman equation.*

$$E = (0.0257\ V)\ln\frac{[K^+]_{ex}P_{K^+} + [Na^+]_{ex}P_{Na^+} + [Cl^-]_{ex}P_{Cl^-}}{[K^+]_{in}P_{K^+} + [Na^+]_{in}P_{K^+} + [Cl^-]_{in}P_{Cl^-}}. \tag{15.85}$$

Typical concentrations and their distributions inside and outside of a nerve cell are given in Table 15.4. In its resting, unperturbed state, a nerve cell membrane has $P_{K+}/P_{Na^+} \approx 100$ and $P_{Cl^-} \approx 0$. Thus, according to the Goldman equation, the cell potential is

$$E = (0.0257\,V)\ln\frac{[K^+]_{ex}P_K^+/P_{Na^+} + [Na^+]_{ex}}{[K^+]_{in}P_{K^+}/P_{Na^+} + [Na^+]_{in}} = -81\,m\,V, \tag{15.86}$$

which is close to the measured value of about –70 mV.

Table 15.4 Typical concentrations and their distributions inside and outside of a nerve cell. Values taken from Voet, D., Voet, J.G., and Pratt, C.W. (2006) *Fundamentals of Biochemistry*. John Wiley & Sons, Hoboken, NJ.

	Concentration/mM	
Ion	**Intracellular**	**Extracellular**
Na^+	15	150
K^+	150	5
Cl^-	10	110

If the nerve cell is stimulated, the permeability of Na^+ increases dramatically, and the ratio of permeabilities changes to $P_{K^+}/P_{Na^+} \approx 0.17$.

$$E = (0.0257\,\mathrm{V}) \ln \frac{5(0.17) + 150}{150(0.17) + 15} = 34\,\mathrm{mV}. \quad (15.87)$$

This sudden spike in membrane potential is called the *action potential*, and has a rise and fall time of about 1 ms.

SUMMARY OF IMPORTANT EQUATIONS

$Q = It = znF$	Faraday's law relating total current passed Q to product of current I and time t and, equivalently electron number z, moles n and Faraday constant F.
$r(\mathrm{mol\ s^{-1}}) = \dfrac{dn}{dt} = \dfrac{I}{zF}$	Faraday's law relating the rate of a chemical reaction r (change in moles per unit time) to current and electron number.
$\eta = E - E_{rev}$	Overpotential η is equal to the cell potential E minus the thermodynamically predicted reversible potential E_{rev}.
$\tilde{\mu} = \mu + z\phi F$	Electrochemical potential $\tilde{\mu}$ equals sum of chemical potential μ and z times electrical potential ϕ times F.
$\Delta_r G_m = -zFE_{rev}$	The molar Gibbs energy of reaction $\Delta_r G_m$ is directly proportional to the reversible cell potential E_{rev}.
$E_{rev} = E^\circ - (RT/zF)\sum_j \nu_j \ln a_j$	Nernst equation. E° is the standard potential, ν_j stoichiometric number, a_j activity of component j.
$E^\circ = (RT/zF)\ln K$	The standard potential is related to the equilibrium constant K.
$\dfrac{dE^\circ}{dT} = \dfrac{\Delta_r S^\circ_m}{zF}$	The derivative of the standard potential with respect to temperature is equal to the ratio of molar entropy change $\Delta_r S^\circ_m$ divided by zF.
$E = (0.0257\ \mathrm{V}) \ln \dfrac{[K^+]_{ex}P_{K^+} + [Na^+]_{ex}P_{Na^+} + [Cl^-]_{ex}P_{Cl^-}}{[K^+]_{in}P_{K^+} + [Na^+]_{in}P_{K^+} + [Cl^-]_{in}P_{Cl^-}}$	Goldman equation relates the membrane potential to the concentrations of ions exterior and interior to the membrane and their permeability.

Exercises

15.1 Electrolysis of molten KBr generates bromine gas, which can be used in the bromination of an alcohol. How long will it take to convert a 500.0 kg batch of phenol (C_6H_5OH) to monobromophenol using a current of 20 kA?

15.2 Annual world production of NaOH exceeds 66 million metric tonnes (1 metric tonne is 1000 kg). NaOH is produced through electrolysis of NaCl with water to produce H_2, Cl_2 and NaOH. Calculate the volume of Cl_2 that is coproduced annually at 300 K and 100 kPa, as well as the minimum power consumption required.

15.3 Consider the redox couples Cu^{2+}(aq), Cu and Ag^+(aq), Ag. When held at 25 °C and 1 bar pressure $E^\circ[Cu^{2+}, Cu] = 0.339$ V and $E^\circ[Ag^+, Ag] = 0.799$ V. If the initial masses of the electrodes are exactly 10 g each, calculate how much each electrode weighs after a current of 0.250 A flows for 600 s.

15.4 (a) How is it possible to deposit Cu on a Au electrode at a potential lower than that corresponding to the reaction $Cu^{2+}(aq) + 2e^- \rightarrow Cu(s)$? (b) Cu atoms adsorbed from solution onto a well-ordered Au electrode in an electrochemical cell decorate the steps on the surface rather than adsorbing onto the terraces. How can one conclude that Cu atoms can diffuse rapidly over the terraces?

15.5 Consider the reaction $Cu\ (s) + Cu^{2+}\ (aq) \rightleftharpoons 2Cu^+\ (aq)$. If metallic Cu is in equilibrium with a solution of $Cu^{2+}(aq)$ in which $a_{Cu^{2+}} = 1.00 \times 10^{-3}$, what is the activity of $Cu^+(aq)$ at equilibrium and 298.15 K?

15.6 The standard reduction potential of Fe^{3+} is $E° = 0.771$ V. The reduction potential of Fe^{3+} in cytochrome c is 0.254 V, while in ferredoxin it is –0.430 V. Comment on role of the chemical environment in determining the stability of an electroactive species.

15.7 Complex IV of the mitochondrial electron transport chain contains two separate copper-containing complexes. Cu_A contains two Cu ions bridged by S^{2-} ions and each coordinated to two S atoms from cysteine residues. Cu_B contains a Cu ion triply coordinated to N atoms from three different histidine residues. The one-electron reduction potential of Cu^{2+} in Cu_A is 0.254 V, while in Cu_B it is 0.340 V. Comment on role of the chemical environment in determining the stability of the redox species.

15.8 Explain the role of the Pt electrode in the standard hydrogen electrode, and why it is necessary.

15.9 For a cell in which all ionic species have an activity of 0.100 and which follows the reaction $2Mn^{2+}\ (aq) + Zn^{2+}\ (aq) \rightarrow 2Mn^{3+}\ (aq) + Zn\ (s)$, calculate $E°$, E, K and $\Delta_r G°$ at 298 K.

15.10 $La^{3+} + 3e^- \rightarrow La$, $E° = -2.379$ V, and $K^+ + e^- \rightarrow K$, $E° = -2.931$ V. Write a balanced electrochemical reaction for the formation of a stable products when $LaBr_3$ and KBr are dissolved in excess liquid La and K. The reaction is run in the molten state with Br^- as the spectator counter-ion. Calculate $E°$ for this reaction. Calculate the cell potential E at 1075 K if the activities of La^{3+} and K^+ are both 0.500.

15.11 For the reactions in Example 15.7.1.1 calculate the emf E of the cell when all ionic species are present with an activity of 0.1. Comment on whether the reactions are spontaneous.

15.12 Use these two half-reactions

$$VO_2^+(aq) + 2H^+(aq) + e^- \rightarrow VO^{2+}(aq) + H_2O(l) \qquad E° = +0.991\ V$$
$$Pd^{2+}(aq) + 2e^- \rightarrow Pd(s) \qquad E° = +0.951\ V$$

to answer the following questions. The cell is held at $T = 298$ K and is immersed in 5 M HF(aq). The concentrations are $[VO_2^+] = 0.121$ M and $[VO^{2+}] = [Pd^{2+}] = 1 \times 10^{-4}$ M. (a) Write out a balanced spontaneous electrochemical reaction involving the two half-reactions. (b) Calculate $E°$. (c) Assuming that activities can be replaced by the magnitude of molar concentrations and HF is fully dissociated, calculate E. (d) Identify the species and their oxidation states that are being reduced and oxidized. State which is reduced and which is oxidized. (e) Describe what happens on an atomic/molecular level when a Pd wire is dropped into the VO_2^+/HF solution.

15.13 An aqueous mixture of iodic acid (HIO_3) and hydrofluoric acid (HF) causes electroless etching of Si (all in one beaker, no electrodes attached). The Si-containing etch product is H_2SiF_6(aq) (fully dissociated into protons and hexafluorosilicate ions). The corresponding half-reactions are

$$SiF_6^{2-}\ (aq) + 4e^- \rightarrow Si\ (s) + 6F^-\ (aq) \qquad E° = -1.2\ V$$
$$IO_3^-\ (aq) + 6H^+\ (aq) + 5e^- \rightarrow \tfrac{1}{2}I_2\ (s) + 3H_2O\ (l) \qquad E° = +1.195\ V$$

(a) Write a balanced electrochemical reaction for the etching reaction. (b) Calculate $E°$ and $\Delta_r G°$ for the cell reaction. (c) Do you expect the reaction to go to completion based on thermodynamics? (d) Write the cell notation for the reaction. (e) If 5.00 g of Si are added to 40 g of 48 wt% HF(aq), what is the maximum mass of HIO_3 that can react according to this cell reaction?

15.14 Use these two half-reactions

$$NO_3^-(aq) + 4H^+(aq) + 3e^- \rightarrow NO(g) + 2H_2O(l) \qquad E° = +0.957\ V$$
$$Ir_2O_3(s) + 3H_2O(l) + 6e^- \rightarrow 2Ir(s) + 6OH^-(aq) \qquad E° = +0.098\ V$$

to answer the following questions. A Pt electrode is placed in 1.00 mM $HNO_3(aq)$ and an Ir electrode is placed in 1.00 mM KCl(aq) + 1.00×10^{-6} M KOH(aq). Both are held at $T = 298$ K in two separate containers. (a) Write out a balanced spontaneous electrochemical reaction involving the two half-reactions. (b) Calculate $E°$. Write out the correct algebraic expression for E. Calculate the activities of NO_3^-, H^+ and OH^-. (c) Write out the cell notation for the reaction. (d) Identify the anode and cathode and describe what is happening at the two electrodes when an electrical connection is made between them. (e) Calculate the pressure change in Pa above the Pt electrode if the volume initially filled with 101 325 Pa of air is 150 ml and a current of 0.324 A flowed for 145 s?

15.15 Use these two half-reactions

$$Sb_2O_3(s) + 6H^+(aq) + 6e^- \rightarrow 2Sb(s) + 3H_2O(l) \qquad E° = +0.152\ V$$
$$BiO^+(aq) + 2H^+(aq) + 3e^- \rightarrow Bi(s) + H_2O(l) \qquad E° = +0.320\ V$$

to answer the following questions. The Bi and Sb electrodes are held at $T = 298$ K in two separate containers full of aqueous solutions. As the reaction proceeds, no precipitate forms but the Sb electrode changes color. (a) Write out a balanced spontaneous electrochemical reaction involving the two half-reactions. (b) Calculate $E°$. (c) Write out the cell notation for the reaction. (d) Identify the anode and cathode and describe what is happening at the two electrodes. (d) If the Bi electrode changes mass by 0.125 g, how much charge in C has been exchanged? Does the Bi electrode gain or lose mass?

15.16 When a gold electrode is used in an electrochemical cell that evolves $O_2(g)$, the electrochemical reaction does not occur with an appreciable rate unless the potential is raised by 1.02 V above the value predicted on the basis of the standard cell potential, even though all species are present at unit activity. Explain this result.

15.17 The standard potential of the AgCl/Ag, Cl^- couple has been measured very carefully over a range of temperatures, and the results were found to fit the expression

$$E° = a - bT - cT^2 + dT^3$$

where $E°$ is in V, T is expressed in °C and the constants are, $a = 0.23659$ V, $b = 4.8564 \times 10^{-4}$ V °C^{-1}, $c = 3.4205 \times 10^{-6}$ V °C^{-2}, and $d = 5.869 \times 10^{-9}$ V °C^{-3}. Calculate the standard Gibbs energy and enthalpy of formation of $Cl^-(aq)$ and its entropy at 298 K.

15.18 The potential of the cell Pt(s) | $H_2(g)$ | HCl(aq) ⋮⋮ AgCl(s) | Ag(s) is described the equation $E(V) = 0.35510 - 0.3422 \times 10^{-4}t - 3.2347 \times 10^{-6}t^2 + 6.314 \times 10^{-9}t^3$ where t is the temperature in Celsius. Write the cell reaction and calculate E, $\Delta_r G$, $\Delta_r H$ and $\Delta_r S$ for the cell reaction at 62 °C.

15.19 Given the half-reactions GSSG + $2H^+ + 2e^- \rightarrow$ 2GSH with $E°' = -0.240$ V and $NADP^+ + H^+ + 2e^- \rightarrow$ NADPH with $E°' = -0.339$ V, calculate the equilibrium cellular concentration of GSSG. Take the $NADP^+$/NADPH concentration ratio to be a typical value of 0.005, the GSH concentration to be 4.00 mM, pH 7.25 and $T = 37$ °C.

15.20 A certain biological oxidation involved both electron and proton transfer. Why does the choice of standard state not affect the value of the driving force for the oxidation, whereas the value of the pH does?

15.21 The enzyme glycollate oxidase is a catalyst for the reduction of cytochrome c in its oxidized form – denoted cyto-$c(Fe^{3+})$ – by glycollate ions. The relevant standard electrode potentials $E°'$, related to 25 °C and pH 7, are

$$\text{cyto-}c(Fe^{3+}) + e^- \rightarrow \text{cyto-}c(Fe^{2+}) \qquad E°' = 0.250\ V$$
$$\text{glyoxylate}^- + 2H^+ + 2e^- \rightarrow \text{glycollate}^- \qquad E°' = -0.085\ V$$

Write a balanced electrochemical reaction and then calculate the standard emf of the cell, $\Delta G°'$ and the equilibrium constant K' for the reaction at pH 7 and 25 °C.

Endnotes

1. Lee, K., Mazare, A., and Schmuki, P. (2014) *Chem. Rev.*, **114**, 9385.
2. Butler, J.A.V. (1926) *Proc. R. Soc. London, A*, **112**, 129.
3. Guggenheim, E.A. (1929) *J. Phys. Chem.*, **33**, 842.
4. Reiss, H. and Heller, A. (1985) *J. Phys. Chem.*, **89**, 4207.

5. Sze, S.M. (1981) *Physics of Semiconductor Devices*, 2nd edition. John Wiley & Sons, New York.
6. (a) Donald, W.A., Leib, R.D., O'Brien, J.T., and Williams, E.R. (2009) *Chemistry – A European Journal*, **15**, 5926. (b) Leib, R.D., Donald, W.A., O'Brien, J.T., Bush, M.F., and Williams, E.R. (2007) *J. Am. Chem. Soc.*, **129**, 7716.
7. (a) Kelly, C.P., Cramer, C.J., and Truhlar, D.G. (2006) *J. Phys. Chem. B*, **110**, 16066. (b) Tissandier, M.D., Cowen, K.A., Feng, W.Y., Gundlach, E., Cohen, M.H., Earhart, A.D., and Coe, J.V. (1998) *J. Phys. Chem. A*, **102**, 7787.
8. Matsui, T., Kitagawa, Y., Okumura, M., and Shigeta, Y. (2015) *J. Phys. Chem. A*, **119**, 369.
9. Liu, J., Salumbides, E.J., Hollenstein, U., Koelemeij, J.C.J., Eikema, K.S.E., Ubachs, W., and Merkt, F. (2009) *J. Chem. Phys.*, **130**, 174306.
10. (a) Matveev, A., *et al.* (2013) *Phys. Rev. Lett.*, **110**, 230801. (b) Antognini, A., *et al.* (2013) *Science*, **339**, 417.
11. Burgess, J. (1999) *Ions in Solutions: Basic principles of chemical interactions*. Woodhead Publishing, Cambridge, UK.

Further reading

Bard, A.J. and Faulkner, L.R. (2001) *Electrochemical Methods: Fundamentals and applications*, 2nd edition. John Wiley & Sons, New York.

Gerischer, H. (1997), in *The CRC Handbook of Solid State Electrochemistry* (P. Gellings and H. Bouwmeester, eds), CRC Press, Boca Raton, pp. 9–73.

Fletcher, S. (2010) The Theory of Electron Transfer. *J. Solid State Electrochem.*, **14**, 705.

Hamann, C.H., Hamnett, A., and Vielstich, W. (2007) *Electrochemistry*, 2nd edition. Wiley-VCH, Weinheim.

Lee, K., Mazare, A., and Schmuki, P. (2014) One-Dimensional Titanium Dioxide Nanomaterials: Nanotubes. *Chem. Rev.*, **114**, 9385.

Lehmann, V. (2002) *Electrochemistry of Silicon: Instrumentation, Science, Materials and Applications*. Wiley-VCH, Weinheim.

Walter, M.G., Warren, E.L., McKone, J.R., Boettcher, S.W., Mi, Q., Santori, E.A., and Lewis, N.S. (2010) Solar Water Splitting Cells. *Chem. Rev.*, **110**, 6446.

Lewis, N.S. (1998) Progress in understanding electron-transfer reactions at semiconductor/liquid interfaces. *J. Phys. Chem. B*, **102**, 4843.

Migliore, A., Polizzi, N.F., Therien, M.J., and Beratan, D.N. (2014) Biochemistry and Theory of Proton-Coupled Electron Transfer. *Chem. Rev.*, **114**, 3381.

Solis, B.H. and Hammes-Schiffer, S. (2104) Proton-coupled electron transfer in molecular electrocatalysis: theoretical methods and design principles. *Inorg. Chem.*, **53**, 6427.

CHAPTER 16

Empirical chemical kinetics

PREVIEW OF IMPORTANT CONCEPTS

- A rate law relates concentration to time in terms of the concentrations of the species involved in the reaction. The differential form describes the rate of change of concentration with respect to time. The integrated form describes concentration as a function of time.
- We need to differentiate between elementary reactions (ones that occur in a single step) and composite reactions (ones that are composed of a series of elementary steps).
- For an elementary reaction, the kinetic order of reaction is always the same as the molecularity, the partial order with respect to a specific reactant is the same as its stoichiometric coefficient, and products do not appear in the rate equation.
- The kinetic order of a composite reaction is not directly related to the stoichiometry of the balanced chemical reaction.
- A reaction mechanism is a series of elementary reactions that adds up to the stoichiometry of the composite reaction.
- A rate-determining step is the elementary step in the reaction mechanism that has the greatest influence on the reaction rate.
- The rate constant is a positive number, the units of which depend on the reaction order. Many rate constants exhibit Arrhenius behavior, which means they have a temperature dependence given by the Arrhenius equation.
- Thermodynamics determines the concentrations of reactants and products at equilibrium. Kinetics describes the rate of evolution of a system. It determines how long it takes for a system to achieve equilibrium, if at all.
- Collisions almost invariably precede reaction, and the collision rate represents an upper limit to the rate at which a reaction can occur.
- Only collisions that have sufficient energy and which satisfy the proper geometry lead to reaction.

16.1 What is chemical kinetics?

Chemical kinetics is the measurement and description of the rates of chemical reactions. It seeks to develop mathematical equations that describe and predict the course of chemical reactions in terms of the species involved in the reaction. The mathematical equations are called *rate laws*. There are seven types of species that can be involved in chemical reactions: reactants; products; intermediates; activated complexes; spectators (also called third bodies); catalysts; and inhibitors.

- *Reactants* and *products* are stable chemical entities (atoms and molecules) that appear on either the left- or right-hand side of a balanced chemical equation.
- *Intermediates* are also stable species but they only appear after the start of a reaction, and are consumed during the course of a reaction. They do not appear in the balanced stoichiometric equation. By stable, we generally mean something that can be isolated and put in a bottle. However, a better definition might be something with a sufficiently long lifetime. A unit of time of relevance to chemical lifetime is a vibrational period, which as we have seen is on the order of a picosecond. If a species can survive for many vibrational periods, it is essentially a stable chemical species that can be isolated, at least for the sake of spectroscopic investigation, if not for inspection in a bottle.
- An *activated complex* is not a stable chemical species. Its isolation and observation is a 'Holy Grail' in the study of chemical dynamics, and this has only been achieved in a select few examples. An activated complex is a species that may last for no more than the time it takes for a collision to occur, and certainly no more than a few vibrational periods. Nonetheless, it may be the most important species in a description of the dynamics of a reaction since it is the species found in the transition state, which we shall define below.
- A *spectator* is a species that is involved in collisions with reactants, intermediates and products but is not itself transformed chemically during the course of a reaction.
- *Catalysts* change the rate and selectivity of reactions but are not consumed during the reaction.

Physical Chemistry: How Chemistry Works, First Edition. Kurt W. Kolasinski.

Companion Website: www.wiley.com/go/kolasinski/physicalchemistry

- An *inhibitor* is a species that decreases the rate of a reaction. In some cases a species may take on more than one role; for example, a product may also act as an inhibitor.

Chemical kinetics can be completely empirical. That is, we may only want to know the identities and concentrations of the species involved in a reaction at any given time. In this case, the descriptors of chemical kinetics are a series of rate equations in either differential or integrated form. The differential form tells us how the rate is related to concentrations, while the integrated form tells us how the concentrations change with time. The equations are determined by a fit to the experimental data. Since they are empirically derived with little interpretive knowledge of the chemical reaction dynamics, they are generally accurate for interpolation, but their use for extrapolation is limited to conditions only very closely related to the conditions used while acquiring the experimental data. For example, assume that we measured the concentrations of A, B, and C (all behaving as ideal gases) every minute for 10 min after 10 Pa of A and 10 Pa of B are mixed in a constant volume vessel held at 298 K. Curves of the concentrations of A, B, and C versus time are then plotted and fitted to mathematical equations. The derived equations would be highly accurate (assuming the data are good) if we used them to calculate the concentrations at $t = 8.5$ min. This is an *interpolation* – the calculation of concentrations at a time for which we have no data, but which lies within the range of times measured. The equations would probably also be accurate for *extrapolation* to $t = 10.5$ min, a time just outside the range measured. Use of the equations at $t = 100$ min would be dubious. Use of the equations at $T = 500$ K would not be justified without further information.

Empirical rate laws can have any arbitrary mathematical form. Empirical rate laws can be highly accurate and useful, especially if we always run a given reaction under the same conditions – that is, as long as we only need information for interpolation. However, if we want to understand how reactions change when conditions are changed, we either need to increase the range of empirical data, or else we need to understand how the empirical equations will change as we change conditions based on the chemistry that is occurring in our system. This understanding will allow us to make justifiable extrapolations and to judge the certainty with which these extrapolations will be accurate. In order to understand how rate laws change when reaction conditions change, we need to understand the dynamics of the reaction.

Figure 16.1 illustrates what we can learn immediately from kinetics data, and how we can use kinetics data to learn about the way in which chemical reactions occur. Ammonium carbamate was heated to 50 °C and allowed to develop its equilibrium vapor pressure. The pressure was monitored in a small tube held at room temperature, which is shown as the red curve. At a time near 3800 s on the time axis, a valve was opened to allow the warm vapor into the cool region. The pressure shoots up and then decays exponentially. The exponential nature of the decay is confirmed not only by the fitted blue curve to the pressure versus time data, but also by a plot of the natural logarithm of the pressure (normalized to the maximum value of pressure after subtracting the background value), in which a linear relationship is found.

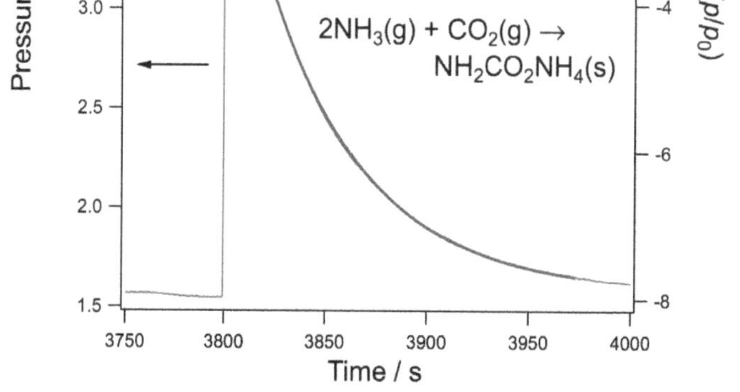

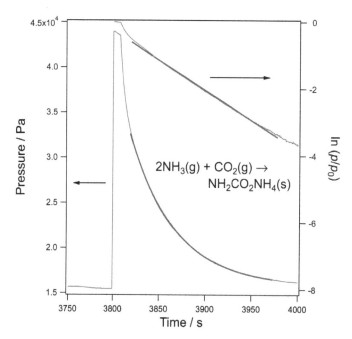

Figure 16.1 Introducing NH_3 and CO_2 into a glass tube at room temperature leads to a reaction to produce ammonium carbamate. The progress of the reaction is followed by measurement of the total pressure.

What does this data mean? Empirically, it means that over the region of the blue curve, the pressure is quite accurately described by $p(t) = p_{bkg} + p_0 e^{-kt}$. However, with further investigation of the relationship of kinetics to dynamics, and with some further experimental input, we will also be able to use this data to construct at least a partial description of the reaction dynamics. In this case, the rate-determining step of the reaction is the adsorption of CO_2. This is a simple first-order process; hence, the exponential decrease in pressure as a function of time as the surface reaction consumes NH_3 and CO_2 and turns them into solid $NH_2CO_2NH_4$. Now, we need to learn how we can relate mathematical descriptions of rates and concentrations versus time to the dynamics of reactions.

16.2 Rates of reaction and rate equations

Chemistry is all about molecules and compounds, so we start with a balanced chemical reaction

$$4A + 3B \rightarrow 2C + D. \tag{16.1}$$

The reaction involves four species: two reactants and two products. There are four individual rates associated with this reaction, each of which we can express in terms of the rate of change of the molar concentration of each species: $d[A]/dt$, $d[B]/dt$, $d[C]/dt$, and $d[D]/dt$. However, these rates are not independent. The stoichiometry of this reaction tells us that when 4 mol of A are consumed, 3 mol of B are consumed while 2 and 1 mol of C and D, respectively, are formed. Therefore, if A were consumed at a rate of $d[A]/dt = -4$ mol l^{-1} s^{-1} (the negative sign indicating consumption), the rate of formation of D must be $d[D]/dt = +1$ mol l^{-1} s^{-1}.

If we know the stoichiometry of a reaction (and there are no side reactions), then knowledge of the rate of change of any one species in the reaction lets us know the rate of change of all the other species. Thus, we can define a single parameter that we call the *rate of reaction R*. If we know the rate of reaction and the stoichiometry of the reaction, we can determine the rate of change of each and every species in the reaction.

The rate of reaction R is the rate of change in the number of moles of a hypothetical product that has a stoichiometric coefficient of 1.

According to this definition, *the reaction rate is always positive*. If we perform a reaction at constant volume, then the amount measured in moles is directly proportional to the concentration. If we perform a gas-phase reaction at constant temperature and constant volume, the partial pressure of a gas is directly proportional to the amount of that gas. Therefore, we can equivalently define the reaction rate in terms of rate of change of amount, concentration or pressure under the appropriate conditions.

For the specific example given above, $R = d[D]/dt$. But not all reactions have a product with a stoichiometric coefficient of 1. We introduce the general mathematical definition of the rate of reaction based on the stoichiometric number of the *i*th species ν_i,

$$R = \frac{1}{\nu_i}\frac{d[i]}{dt} = \frac{1}{V\nu_i}\frac{dn_i}{dt} = \frac{1}{V}\frac{d\xi}{dt} \tag{16.2}$$

where V is the volume and ξ is the extent of reaction. IUPAC recommends the use of R, r or v as symbols for reaction rate, and all three are commonly encountered. Always inspect the rates you determine to ensure that they pass basic tests of sensibility. The reaction rate is always positive by definition. The rate of change of products is also positive, but the rate of change of reactants must be negative. Furthermore, the stoichiometry determines how one rate is related to another. In the above example, the rate of production of C must be twice the rate of formation of D, and so forth.

16.2.1 Example

For the reaction $2NOBr(g) \rightarrow 2NO(g) + Br_2(g)$ under a given set of conditions, the rate of appearance of NO is found to be 1.6×10^{-4} mol dm^{-3} s^{-1}. What are the stoichiometric numbers, the rate of reaction, and the rate of NOBr consumption?

Stoichiometric numbers are signed quantities. Their absolute values appear in the balanced chemical equations. The values for this reaction are $\nu_{NO} = 2$, $\nu_{Br} = 1$, and $\nu_{NOBr} = -2$. The rate of reaction is

$$R = \frac{1}{\nu_{NO}}\frac{d[NO]}{dt} = \frac{1}{2}(1.6 \times 10^{-4}) = 8.0 \times 10^{-5} \text{ mol dm}^{-3}\text{ s}^{-1}.$$

Since Br_2 is a product with stoichiometric coefficient of +1, the rate of reaction is the same as the rate of formation of Br_2. Equation (16.2) is then used to find to the rate of NOBr consumption,

$$R = \frac{1}{\nu_{NOBr}}\frac{d[NOBr]}{dt}.$$

Therefore

$$\frac{d[NOBr]}{dt} = \nu_{NOBr}R = -1.6 \times 10^{-4}\ \text{mol dm}^{-3}\ \text{s}^{-1}.$$

16.2.2 Directed practice

For the reaction $BrO_3^- + 5Br^- + 6H^+ \rightarrow 3Br_2 + 3H_2O$, calculate the rate of reaction and the rates of formation or consumption of all products and reactants if H^+ is being consumed at a rate of 1.0×10^{-6} mol m^{-3} s^{-1}.[1]

Having defined the rate of reaction, we now turn to the rate equations. In general, an empirical rate equation relates the rate to the concentration of each stable species involved in a reaction. In the above example, assuming that only the reactants and products are involved in the reaction then a general form of the empirical rate equation is

$$R = k[A]^{\alpha}[B]^{\beta}[C]^{\chi}[D]^{\delta}. \tag{16.3}$$

The *reaction order* is $n = \alpha + \beta + \chi + \delta$, where α, β, χ, and δ are the *partial orders* or the orders with respect to A, B, C, and D, respectively. The *rate constant* k is defined as a positive number, and its units change depending on the reaction order, the units used for concentration, and the unit of time (though second is the preferred unit). In principle, the partial orders can be positive or negative, integer, or noninteger.

16.3 Elementary versus composite reactions

Having stated that empirical rate laws can have any arbitrary form, we are left with an infinite number of possible mathematical equations. At this point, we need to find some order among this chaos. Fortunately, chemical dynamics is not so random and it allows us to concentrate our mathematical treatment on a few simple cases to start. Here, the distinction between simple and complex reactions helps greatly to clarify the mathematics that we need to know. An *elementary reaction* is a one-step reaction that usually only involves the breaking and making of one or two bonds. An example is

$$NO + O_3 \rightarrow NO_2 + O_2.$$

This reaction occurs when NO strikes O_3. A bond is formed between the N of NO and one of the O atoms in O_3, while simultaneously the bond of that O atom to another O in O_3 is broken. The reaction has a *molecularity* of two. The molecularity of a reaction indicates the number of collision partners involved in the reaction. Common values of molecularity are 1 (a unimolecular reaction) and 2 (bimolecular reaction). A molecularity of 3 (termolecular reaction) or higher is uncommon.

Elementary reactions have a very interesting property regarding their kinetics: their reaction order is the same as their molecularity. The reaction above is a second-order reaction because it has a molecularity of 2. The reaction is second-order because it is first-order in NO and first-order with respect to O_3. As is commonly found for most reactions, the partial orders for the products are both zero and they do not appear in the rate equation. The fundamental principle of chemical kinetics for elementary reactions is this:

For an elementary reaction, the order of reaction is always the same as the molecularity, the partial order with respect to a specific reactant is the same as its stoichiometric coefficient, and products do not appear in the rate equation.

Since we have stated that most elementary reactions have a molecularity of 1 or 2, first-order and second-order rate equations are the most commonly encountered for elementary reactions. Under certain circumstances, when a reaction occurs from a nondepleting reservoir, such as evaporation from a bulk liquid, an elementary reaction may exhibit zeroth-order kinetics.

A composite reaction is a reaction that has a mechanism involving two or more elementary steps. This added complexity means that the fundamental principle of chemical kinetics for composite reactions is:

Neither the reaction order nor the partial order with respect to any reactant or product of a composite reaction can be read from the stoichiometric equation.

In principle, the rate equation of a composite reaction can take any arbitrary form. However, many composite reactions contain a *rate-determining step* (RDS).

A rate-determining step is the elementary step in the reaction mechanism that has the greatest influence on the reaction rate.

The presence of a rate-determining step often simplifies the kinetic behavior of the composite system such that the overall rate equation resembles the rate equation of the elementary step. Therefore, the most commonly encountered reaction orders are first and second, with occasional occurrence of zeroth-order and rare occurrence of third-order. When fractional orders arise, you can be sure that the reaction mechanism is complex and/or that the system does not possess a well-defined rate-determining step. Fractional orders are also commonly obtained for heterogeneous reactions. We will treat the kinetics of heterogeneous reactions in Chapter 18.

16.3.1 Example

Under a certain set of conditions, it was found that the reaction

i $$2N_2O_5 \rightarrow 4NO_2 + O_2$$

followed the first-order rate law

$$-\frac{d[N_2O_5]}{dt} = k_{N_2O_5}[N_2O_5].$$

Under the same conditions, the reaction

ii $$2NO_2 \rightarrow 2NO + O_2$$

followed the second-order rate law

$$-\frac{d[NO_2]}{dt} = k_{NO_2}[NO_2]^2.$$

Can we conclude that either reaction (i) or (ii) is an elementary reaction?

In an elementary reaction, the stoichiometric coefficient of a species is determined by molecularity, and it is the same as the kinetic order with respect to that species. In reaction (i), the stoichiometric coefficient of N_2O_5 is 2 but the kinetic order with respect to N_2O_5 is 1. Therefore, we can conclude unequivocally that reaction (i) is *not* an elementary reaction. It must be a composite reaction.

Both the stoichiometric coefficient and the order with respect to NO_2 are 2 in reaction (ii). This is a necessary condition for reaction (ii) to be an elementary reaction. However, it is not sufficient to prove that reaction (ii) is an elementary reaction. This illustrates a general principle in the scientific study of kinetics. A negative result is both a necessary and sufficient condition to disprove a given hypothesis. However, a single positive result is not sufficient to prove most hypotheses. In this case for instance, we would need to show that both NO and O_2 follow second-order rate laws that are consistent with each other and the rate of disappearance of NO_2. If we were able to detect the presence of an intermediate species, this would be inconsistent with the proposed elementary nature.

There are several features of a reaction system that are characteristic of composite reactions involving a number of steps. These include the following

- multiple bonds (>2) being made and broken;
- the appearance of intermediates;
- measuring instantaneous rates that are different for reactants and products;
- rate laws with fractional orders;
- rates and rate orders that change as a function of time; and
- unusual dependence of the rate on concentration, pH and T or other variables.

16.4 Kinetics and thermodynamics

Kinetics is perhaps the most important topic in all of chemistry. This may seem odd because it is thermodynamics that tells us whether a reaction is possible or not. *Thermodynamics determines the concentrations of reactants and products at equilibrium.*

The laws of thermodynamics are essentially absolute and cannot be violated in an equilibrated system. Even for nonequilibrium systems, conservation of energy and evolution toward higher entropy are difficult to overcome. Therefore, thermodynamics has incredible predictive ability and must always be considered to determine limiting cases that cannot be exceeded. However, the state functions of thermodynamics, such as Gibbs energy change, are path-independent. The statements that all systems evolve to the lowest possible Gibbs energy or that all systems evolve until the change in Gibbs energy is zero have attached to them the caveat "... as long as a path is available."

Kinetics is all about the path. Kinetics describes the rate of evolution of a system. It determines how long it takes for a system to achieve equilibrium, if at all. You may have heard the phrases that a reaction is run under thermodynamic control or under kinetic control. More accurately, the kinetics of some reactions is such that the thermodynamic limit is achieved on the time scale that the reaction was performed. A so-called 'kinetically controlled reaction' simply has not been performed for a sufficiently long period to reach thermodynamic equilibrium. Another way to phrase this is that while what is forbidden by thermodynamics is forbidden, that which is allowed by thermodynamics is only a limit that, while it cannot be exceeded, need not necessarily be approached.

Kinetics also determines in many ways whether a reaction is useful or not, and/or whether it is dangerous or not. Thermodynamics can tell us which reaction are candidates to create explosions, but not if they will explode. Combustion of glucose ignited by a spark in oxygen releases the same amount of energy per mole of glucose as combustion of glucose in the body. The overall thermodynamics are the same, but the utility and effect on the body are not at all the same.

16.5 Kinetics of specific orders

The molecularity of elementary reactions determines their kinetic order. Therefore, we expect first- and second-order kinetics to be commonly observed. Certain experimental conditions lead to zeroth-order kinetics. We now treat the mathematics of these three commonly encountered cases and the related topic of pseudo-order kinetics. In the examples below, we assume our reactants and products are acting ideally and we are free to use concentrations instead of activities.

16.5.1 First-order kinetics

Consider the simple unimolecular elementary reaction of 1 mol of A transforming to 1 mol of P performed at constant volume and temperature. Initially, that is at $t = 0$, A is present at concentration $[A]_0$ and there is no P present. When the concentration of A changes by $-x$, the stoichiometry demands that the concentration of P increases by x. Starting with a stoichiometric reaction, we can represent this in tabular form as

	A	$\rightarrow$	P
$t = 0$	$[A] = [A]_0 = a_0$		$[P] = b_0 = 0$
At time t	$[A] = a_0 - x$		$[P] = x$

Since the reaction is first-order, we write the differential rate law for the consumption of A with effective rate constant k_A as

$$-\frac{d[A]}{dt} = -\frac{d(a_0 - x)}{dt} = \frac{dx}{dt} \tag{16.4}$$

$$\frac{dx}{dt} = k_A(a_0 - x). \tag{16.5}$$

The use and meaning of an effective rate constant will be explained below when we consider reactions of different stoichiometry.

To obtain the integrated form of the rate equation, we use separation of variables to move all terms involving x to one side, and all terms involving t to the other side. Then, the boundary conditions in the table above are used to define the integrals to be evaluated.

$$\int_0^x \frac{dx}{a_0 - x} = \int_0^t k_A \, dt \tag{16.6}$$

$$\ln\left(\frac{a_0}{a_0 - x}\right) = k_A t \tag{16.7}$$

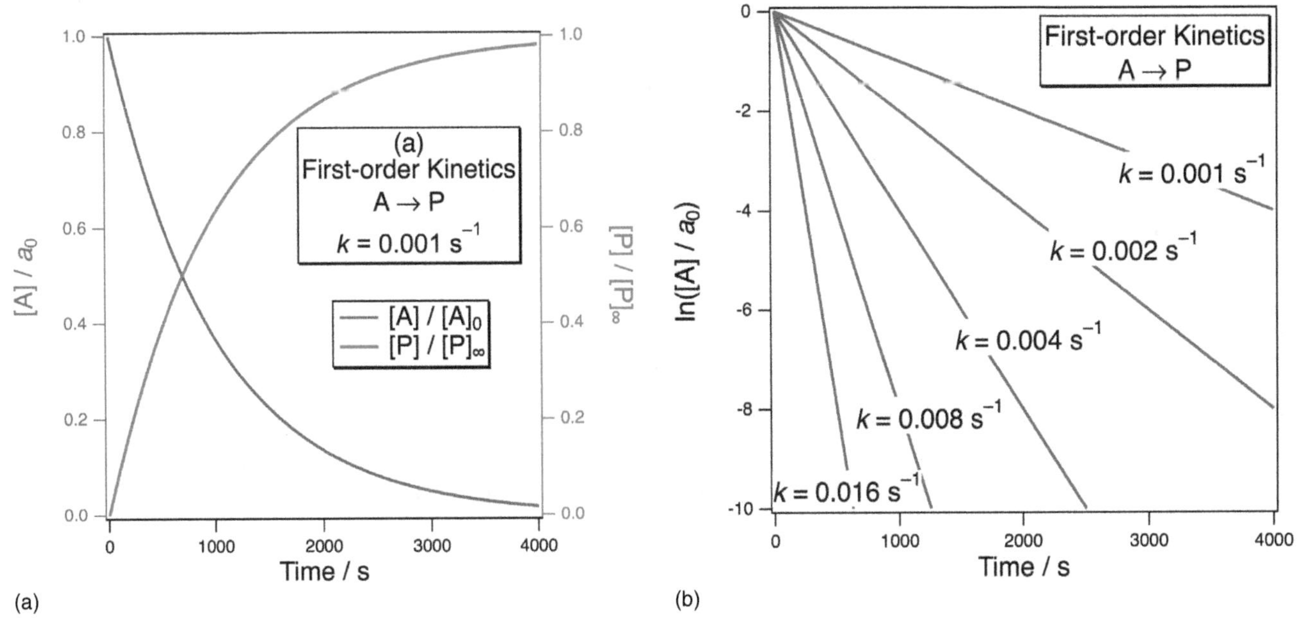

Figure 16.2 (a) The change in the concentration of reactant A and product P for the simple unimolecular first-order reaction A → P. The concentrations are normalized, A to its initial concentration and P to its final concentration, such that they range from 0 to 1. (b) Determination of first-order kinetics by the method of integration is performed by showing that a plot of the natural logarithm of the concentration versus time is linear. A plot of the logarithm of the ratio of concentration to the initial concentration of the reactant A versus time has a slope equal to $-k_A$. Because of the stoichiometry, $k_A = k$. Note how the concentration decays more quickly with increasing magnitude of the rate constant.

The rate of appearance of P is also governed by first-order kinetics. Using the stoichiometry of the reaction (in other words, substituting $[A] = a_0 - x$ or $[P] = x$) we can rewrite Eq. (16.7) for the rates of reagent consumption as well as the rate of product formation,

$$[A] = a_0 \exp(-k_A t) \tag{16.8}$$

$$[P] = a_0(1 - \exp(-k_A t)) \tag{16.9}$$

Equations (16.8) and (16.9) are plotted in Fig. 16.2. The plot shows that the concentration of A decreases as a result of first-order kinetics in a simple exponential fashion. Because of the 1:1 stoichiometry, note that the decay constant is the same as the true rate constant. That is, the *effective rate constants* k_A and k_P are equal to the *true rate constant* k defined by

$$R = k[A]. \tag{16.10}$$

In other words, the true rate constant is the rate constant that determines the rate of reaction. An effective rate constant defines the rate of disappearance of a specific reactant or the rate of appearance of a specific product.

The reason why we need to introduce both effective rate constants with respect to a specific reactant or product and the true rate constant – which is simply called the rate constant if it need not be differentiated from an effective rate constant – is because not all first-order reactions have 1:1 stoichiometry. For example, there are reactions such as 2A → P or A → 2P or A + B → P, and so on. With the definition of x used above, the procedure for obtaining Eqs (16.7) and (16.8) is valid for all stoichiometries. What changes is the relationship of x to [P] (and [B] if a second reagent with kinetic order zero is involved) and the relationship of k_A to k. For instance, the reaction A → 2P has $[A] = a_0 - x$ but $[P] = 2x$. Therefore, $k_A = k = \frac{1}{2}k_P$ and $[P] = 2a_0(1 - \exp(-kt))$.

16.5.1.1 Directed practice

Confirm substitution into Eq. (16.7) leads to Eqs (16.8) and (16.9).

The exponential form of Eq. (16.8) means that if we plot $\ln([A]/a_0)$ versus time, we should obtain a straight line with slope equal to $-k_A$. This is known as the method of integration. The stoichiometry of the reaction then determines the relationship between k_A and k.

16.5.1.2 Example

A first-order reaction A → P is 40% complete after 1 min. (a) What is the value of the rate constant? (b) How long does it take to reach 80% completion?

(a) That the reaction is 40% complete in 1 min means that the concentration at $t = 60$ s, $[A]_{t = 60\ s}$, is (1 – 0.4) of the initial value. In mathematical terms $[A]_{t=60\ s} = (1 - 0.4)[A]_0 = 0.6\ [A]_0$. Rearrange Eq. (16.8) and solve for the rate constant k.

$$\ln([A]/a_0) = -kt \tag{16.11}$$

$$k = -\frac{1}{t}\ln([A]/a_0) = \frac{-1}{60\ s}\ln(0.6) = 8.5 \times 10^{-3}\ s^{-1}$$

(b) Then at 80% completion, $[A] = (1 - 0.8)\ a_0 = 0.2\ a_0$. Solving Eq. (16.11) for t, we find

$$t = -\frac{1}{k}\ln([A]/a_0) = -\frac{1}{8.5 \times 10^{-3}}\ln(0.2) = 190\ s.$$

First-order kinetics is widely observed in any type of phenomenon involving a well-defined lifetime. The *lifetime* is defined by $\tau = 1/k$. Radioactive decay, fluorescence and phosphorescence all follow first-order kinetics with well-defined lifetimes in the simplest cases. The same is often observed for the initial stages of growth of bacterial colonies.

16.5.2 Second-order kinetics

There are four types of second-order reactions to consider. The first type involves the reaction of two different species which both have the same stoichiometric coefficient of one, as in the reaction

	A	+ B	→ P
$t = 0$	$[A]_0 = a_0$	$[B]_0 = b_0$	$[P]_0 = 0$
At time t	$[A] = a_0 - x$	$[B] = b_0 - x$	$[P] = x$

In its most general case, the concentrations of A and B need not be the same. The second-order empirical rate equation takes the form

$$R = k[A][B]. \tag{16.12}$$

The differential rate equation is written most generally by stating it in terms of the effective rate constant k_A as well as x, a_0 and b_0,

$$\frac{dx}{dt} = k_A(a_0 - x)(b_0 - x) \tag{16.13}$$

Integrating the rate equation with the conditions $a_0 \neq b_0$ and $x = 0$ at $t = 0$, we obtain

$$\frac{1}{a_0 - b_0}\ln\left(\frac{b_0(a_0 - x)}{a_0(b_0 - x)}\right) = k_A\, t. \tag{16.14}$$

If we plot the left-hand side of Eq. (16.14) versus t, the slope is equal to k_A. Due to stoichiometry $k = k_A = k_B$.

The second type is a reaction of the same stoichiometry but in the special case of $a_0 = b_0$. Now, however, the boundary conditions of the system are different, which allows us to rewrite Eq. (16.13) as

$$\frac{dx}{dt} = k_A(a_0 - x)^2. \tag{16.15}$$

Integration yields

$$\frac{x}{a_0(a_0 - x)} = k_A\, t. \tag{16.16}$$

A plot of $x/a_0(a_0 - x)$ versus t yields a straight line of slope k_A and again $k = k_A = k_B$.

16.5.2.1 Directed practice

As an alternative to Eqs (16.15) and (16.16), state the rate in terms of $d[A]/dt$ and show that a plot of $1/[A]$ versus t should be linear with slope k_A, as in Eq. (16.17).

If A and B are the same chemical species, then the two reactants must have the same concentration. We now have the reaction

$$2A \rightarrow P.$$

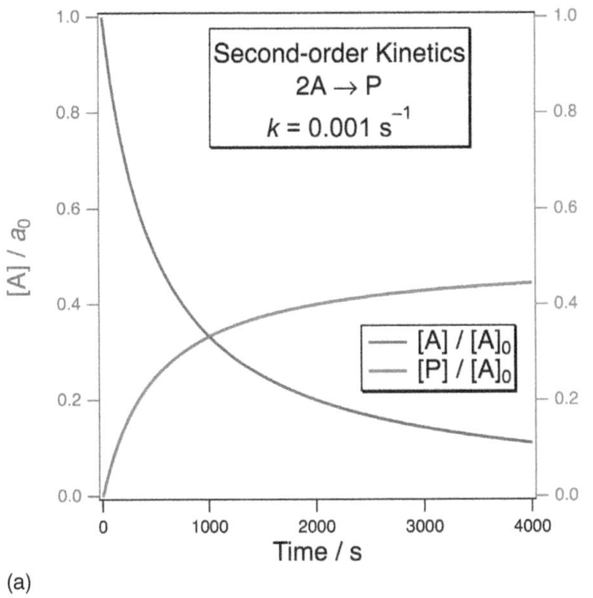

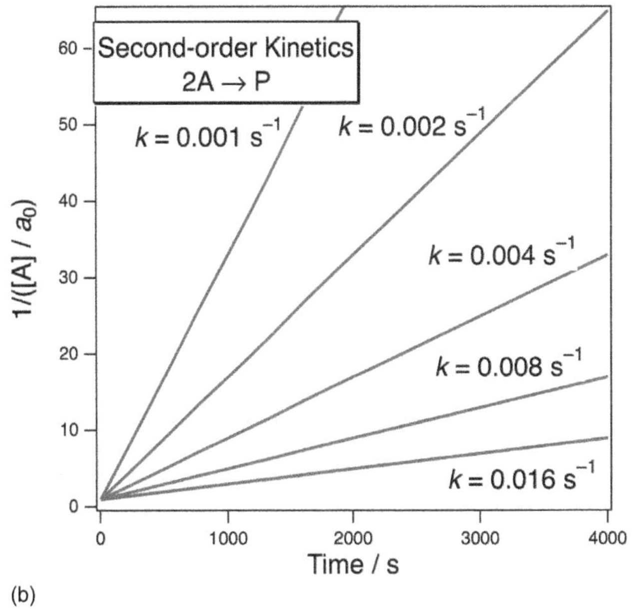

Figure 16.3 (a) Concentration of reactant A in a second-order reaction. (b) Linearized plots of second-order kinetics. The inverse of the ratio of concentration to initial concentration when plotted versus time yields a straight line with a slope equal to the rate constant.

The solution is the same as for A + B with equal concentrations,

$$1/[\mathrm{A}] = 1/a_0 + k_\mathrm{A}\, t. \tag{16.17}$$

but we must take account of the factor of two when evaluating concentrations and the rate constants. A plot of 1/[A] versus t, as shown in Fig. 16.3, yields a straight line with slope equal to k_A, but now $k_\mathrm{A} = 2k$.

Both, Eqs (16.16) and (16.17) describe the change in concentration of a reactant if second-order kinetics are followed. The equations describe the concentration of A versus time for all of the following situations: (i) A + B → P and with partial order of 1 for both A and B; (ii) 2A → P and second-order with respect to A; and (iii) 2A + B + ⋯ → P and second order with respect to A, as long as the concentrations of other reactants are constant. If both Eqs (16.16) and (16.17) hold, which one should be used to determine the effective rate constant k_A? The answer is that either one can be used but one or the other may be easier to use based on the system under investigation. If the concentration of A is directly measureable, then Eq. (16.17) is a sensible choice as [A] is directly related to the measured parameter. If, on the other hand, the concentration of a product is more easily measured, then Eq. (16.16) is a logical choice since x is directly related to the measured parameter. If both concentrations can be measured, then analysis with both forms allows for independent determination of the kinetic parameters.

16.5.2.2 Example

The rate constant at 25 °C and constant volume for the gas-phase reaction

$$\mathrm{NO} + \mathrm{HNO_3} \rightarrow \mathrm{NO_2} + \mathrm{HNO_2}$$

is $k = 4.0 \times 10^{-2}\ \mathrm{kPa^{-1}\ s^{-1}}$. If the initial partial pressures of both NO and HNO_3 are 50 kPa, calculate the total pressure, the partial pressure of NO, and the partial pressure of NO_2 after 1 s, 5 s, 10 s, and 50 s.

As long as the reaction is performed at constant volume and temperature, we may substitute pressure for concentration. Thus,

$$\frac{1}{[\mathrm{A}]} = kt + \frac{1}{a_0} \rightarrow \frac{1}{p_\mathrm{NO}} = kt + \frac{1}{p_0}$$

	NO	+	HNO_3	→	NO_2	+	HNO_2
Initially	p_0		p_0		0		0
at time t	$p_0 - x$		$p_0 - x$		x		x

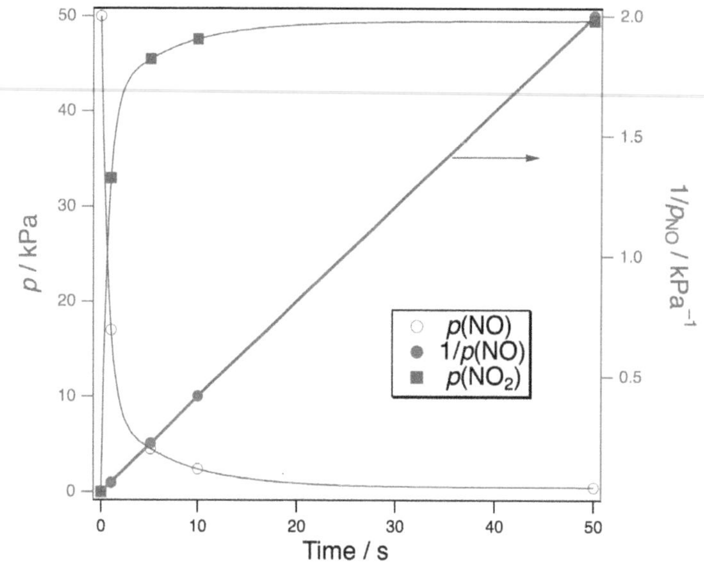

Figure 16.4 An example of the pressure changes in reactants and products for the second-order reaction $NO + HNO_3 \rightarrow NO_2 + HNO_2$.

where $p_0 = 50$ kPa. The results are shown in Fig. 16.4 and the table below.

t/(s)	0	1	5	10	50
p_{total}/(kPa)	100	100	100	100	100
p(NO)/(kPa)	50	17	4.5	2.4	0.50
$p(NO_2)$/(kPa)	0	33	45.5	47.6	49.5

16.5.3 Pseudo-order kinetics

For second- and higher-order reactions in which more than one reactant contributes to the overall order of reaction, the kinetic order is sensitive to the relative concentration of the reactants. If all but one of the reactants is present in great excess, then the concentrations of the species in excess will not change appreciably during the course of the reaction. The reaction kinetics will only depend on the kinetic dependence of the one low-concentration species. This situation is known as *pseudo-order kinetics.*

Consider a reaction of the type

$$A + B \rightarrow P$$

with the second-order empirical rate equation

$$R = k[A][B].$$

If [B] » [A], then [B] does not change significantly (say <5%) during the course of the reaction. Therefore, [B] can be taken as a constant and we can define an effective rate constant according to

$$k^* = k[B] \tag{16.18}$$

The rate equation then becomes

$$R = k^*[A] \tag{16.19}$$

To determine k, we first determine k^* at constant [B] as we would for any first-order reaction. Then, after determining the rate for several different values of [B], a plot of k^* versus [B] should yield a straight line with slope equal to k.

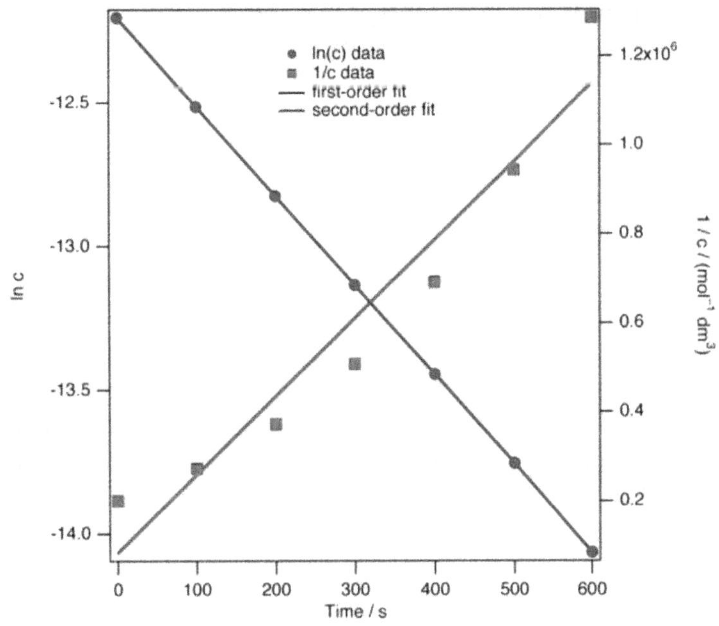

Figure 16.5 First-order and second-order analysis of alkaline bleaching of crystal violet.

This is commonly observed for solvolysis reactions. For example, the reaction of an alkyl halide with water to form an alcohol as in

$$(CH_3)_3CBr(aq) + H_2O(l) \rightarrow (CH_3)_3COH(aq) + HBr(aq).$$

The reaction is run in excess water, thus $[H_2O] \gg [(CH_3)_3CBr]$. The kinetics then becomes pseudo first-order.

16.5.3.1 Example

During the reaction of OH^- with crystal violet, the concentration in mol dm^{-3}, c, of crystal violet in an aqueous solution was measured at 313 K. The data, given in the table below, were recorded. The initial concentration of OH^- is 8.33×10^{-3} mol dm^{-3}.

c/μM	5.00	3.67	2.69	1.97	1.45	1.06	0.778
t/s	0	100	200	300	400	500	600

What is the reaction order? Plot the data and determine the rate constant k.

The concentration of crystal violet is 1000-fold smaller than that of hydroxide. Therefore, we assume that we are operating under pseudo-order conditions – that is, that we are performing an *isolation* experiment in which only the concentration of crystal violet changes. Thus, we try to fit the crystal violet data to first-order ($\ln c$ versus t) and second-order ($1/c$ versus t) expectations. If one of the two fits is linear, we can conclude that that reaction order is observed.

As the fits clearly show in Fig. 16.5, the first-order analysis fits the data extremely well and the second-order analysis provides a poor fit. The rate constant is the negative of the slope $k^* = 0.00310\ s^{-1}$. Note that this is a pseudo-first order rate constant. The true rate constant is give by $k = k^*/[OH^-]^\beta$ and can only be determined after the order with respect to OH^- has been determined.

16.5.4 Zeroth-order kinetics

When a rate law is independent of the concentration of a given species, that species has a kinetic order of zero. In the case of pseudo-order kinetics given above, this behavior can be forced upon a chemical system by having a large excess of one of the reagents. Zeroth-order kinetics is inherent to any system that contains a virtually inexhaustible reservoir of a reagent. For example, evaporation is the process of molecules desorbing from the surface of a liquid. Molecular desorption of this type is a first-order process. The rate of desorption (evaporation) is proportional to the surface area of the liquid/vapor interface. However, if the sample of liquid is a macroscopic liquid being held, for example, in a beaker, then the surface area is constant. The molecules that evaporate from the surface are constantly replaced by other molecules from the bulk of the liquid. Therefore, the rate of evaporation from a macroscopic interface is constant and exhibits zeroth-order kinetics.

Defining x as the change of concentration of A(g) for the evaporation reaction,

$$\mathrm{A}(l) \rightarrow \mathrm{A}(g),$$

we can write the differential rate equation as

$$\frac{dx}{dt} = k \tag{16.20}$$

with $k = x/t$. The rate at which the product concentration increases is constant. There is no change in the concentration of the liquid; however, there is a constant contraction in its volume caused by the constant rate of transformation of liquid into gas.

Note that if the liquid sample evaporates sufficiently to form a droplet, the rate of evaporation still is directly proportional to the surface area. However, evaporation causes the droplet to shrink, and this decreases the radius of the droplet and, thus, also the surface area. The rate then depends on the square of the radius. Such a relationship holds until the nanoscale regime is entered, at which point the equilibrium vapor pressure and binding energy per molecule become size-dependent leading to an acceleration of the evaporation rate as the radius decreases.

16.5.5 Summary of rate equations

The rate equation has both differential and integrated forms. The mathematical form of the rate equation depends on the kinetic order and, in some cases, on the relative concentrations of the species involved. The units of the rate constant also change with kinetic order. The units of an nth order rate constant are (units of concentration)$^{-(n-1)}$ s^{-1}. Other inverse time units are sometimes used, though s^{-1} is preferred. A number of the most frequently encountered relationships are summarized in Table 16.1.

Remember that x is defined by the change in concentration of A. That is, the concentration of A at time zero is a_0 and the concentration of A at time t later is $a_0 - x$. If the stoichiometric coefficients of A and B are different, then this must be accounted for in the rate equations and the relationship of the rate constants. For example, consider the reaction

$$\mathrm{A} + 2\mathrm{B} \rightarrow \mathrm{P}.$$

Table 16.1 Some frequently encountered rate expressions for reactions that exhibit no dependence on the concentrations of products. The rate constant units are based on the assumption that molar concentration is measured.

Reaction	Empirical rate equation	Differential rate equation	Integrated rate equation	Units of k_A	Relationship of rate constants
A → P	$R = k$ zeroth-order	$\frac{dx}{dt} = k$	$k = x/t$	mol dm^{-3} s^{-1}	
A → P	$R = k[A]$ first-order	$\frac{dx}{dt} = k_A(a_0 - x)$	$\ln\left(\frac{a_0}{a_0 - x}\right) = k_A t$	s^{-1}	$k_A = k$
A+B→P	$R = k[A]$ pseudo-first	$\frac{dx}{dt} = k_A(a_0 - x)$	$\ln\left(\frac{a_0}{a_0 - x}\right) = k_A t$	s^{-1}	$k_A = k_B = k$
A+B→P	$R = k[A][B]$ second-order	$\frac{dx}{dt} = k_A(a_0 - x)(b_0 - x),\ a_0 \neq b_0$	$\frac{1}{a_0 - b_0}\ln\left(\frac{b_0(a_0 - x)}{a_0(b_0 - x)}\right) = k_A t$	dm^3 mol^{-1} s^{-1}	$k_A = k_B = k$
A+B→P	$R = k[A][B]$ second-order	$\frac{dx}{dt} = k_A(a_0 - x)^2\ a_0 = b_0$	$\frac{x}{a_0(a_0 - x)} = k_A t$	dm^3 mol^{-1} s^{-1}	$k_A = k_B = k$
2A → P	$R = k[A]^2$ second-order	$\frac{dx}{dt} = k_A(a_0 - x)^2$	$1/[A] = 1/a_0 + k_A t$	dm^3 mol^{-1} s^{-1}	$k_A = 2k$
2A+B→P	$R = k[A]^2$ pseudo-second	$\frac{dx}{dt} = k_A(a_0 - x)^2$	$\frac{x}{a_0(a_0 - x)} = k_A t$	dm^3 mol^{-1} s^{-1}	$k_A = 2k_B = 2k$
A → P	$R = k[A]^n$ nth-order $n \neq 1$	$\frac{dx}{dt} = k_A(a_0 - x)^n$	$\frac{1}{n-1}\left[\frac{1}{(a_0 - x)^{n-1}} - \frac{1}{a_0^{n-1}}\right] = k_A t$	dm^{3n-3} mol^{1-n} s^{-1}	$k_A = k$

If the reaction is first-order in A and B, then the concentration of A at time t is $a_0 - x$, and the concentration of B is $b_0 - 2x$. The rate equation with $a_0 \neq b_0$ is then

$$\frac{dx}{dt} = k_A(a_0 - x)(b_0 - 2x). \tag{16.21}$$

The integrated form must also reflect the stoichiometry,

$$\frac{1}{2a_0 - b_0} \ln\left(\frac{b_0(a_0 - x)}{a_0(b_0 - 2x)}\right) = k_A\, t, \tag{16.22}$$

as does the relationship of the rate constants, $2k_A = k_B = 2k$.

16.6 Reaction rate determination

Reaction rates are determined by calculating how concentrations change as a function of time. *If the stoichiometry of a reaction is known and there are no side reactions,* the disappearance of any one reactant or the appearance of any one product can be measured to determine the kinetics. It will be assumed in this text that no side reactions occur unless this is specifically stated. However, in any real laboratory setting, it must be established whether this is the case. One of the best ways to check for the presence of side reactions is to measure the kinetics of more than one species simultaneously and check for consistency.

The measurement of kinetics must always be performed at a known and controlled temperature. In principle, any property that depends on concentration can be monitored. If only a single gas appears as a reactant or product then total pressure at constant volume or total volume at constant pressure is directly proportional to the rate. If more than one gas is present, partial pressure must be monitored, for example, with a mass spectrometer. Measurement of volume changes (also called *dilatometry*) can also be used in some cases to measure reaction rates in liquids. Such reactions need to exhibit a sufficiently large change in molar volume as is sometimes observed during polymerization and phase changes. Reactions in liquid solutions involving electrolytes can be monitored with conductometric techniques. The concentration of some chemical species can be measured in solution with ion-selective electrodes, for example, pH electrodes.

By far the most accurate measurements can be made by counting either electrons or photons. In electrochemical reactions, the flow of current is directly related to the exchange of electrons. Electrochemical kinetics are discussed in more detail in Chapters 15 and 18.

By far the most versatile, flexible and widely used methods of concentration determination involve spectroscopy. Absorption and fluorescence spectroscopy can both be used to make measurements with high dynamic range (the capacity to measure over a wide range of concentrations), high sensitivity (the capacity to measure low concentrations), and high spatial resolution. Another great advantage of spectroscopic methods (shared by any method that depends on the counting of charged particles) is the ability to measure with high temporal resolution. The measurement of pulses in the microsecond (10^{-6} s) and nanosecond (10^{-9} s) ranges is routine. Progressively higher temporal resolution becomes more difficult and expensive; nonetheless, picosecond (10^{-12} s) and femtosecond (10^{-15} s) regimes are commonly accessible in certain spectral regions with state-of-the-art instrumentation. Using very special laser-based techniques, even the attosecond (10^{-18} s) regime is now accessible in limited circumstances. The femtosecond and attosecond regimes are important timescales for excitations that lead to chemical transformations. The timescale of atomic motions is 100 fs to several picoseconds. Therefore, to follow chemical reactions directly in time on a collision-by-collision basis requires temporal resolution on this scale. More commonly, much longer timescales are studied. Finding conditions that allow a researcher to measure kinetics on an accessible timescale is one of the first experimental necessities faced in any kinetics experiment. The particular property, concentration range, timescale and species probed during a kinetics experiment depend on the equipment available and the information sought.

Below are listed a number of reactions with corresponding rate equations under certain conditions. From this list you should observe several things. One is that chemically similar systems, such as $H_2 + I_2$ and $H_2 + Br_2$, for example, can exhibit extremely different kinetics. There are instances where the kinetic orders are the same as the stoichiometric coefficients, yet at other times they are not related. Sometimes every reactant appears, sometimes not. In summary, we cannot look at a balanced chemical reaction and determine the kinetics from that equation unless we know more about the dynamics of the reaction. But in all cases, the reaction kinetics is a mirror into the reaction dynamics.

(i) $$H_2 + I_2 \rightleftharpoons 2HI$$

(ii) $$-\frac{d[I_2]}{dt} = k[H_2][I_2]$$

(iii) $$H_2 + Br_2 \rightleftharpoons 2HBr$$

(iv) $$-\frac{d[Br_2]}{dt} = \frac{k[H_2][Br_2]^{3/2}}{[Br_2] + k^*[HBr]}$$

(v) $$S_2O_8^{2-} + 2I^- \rightleftharpoons 2SO_4^{2-} + I_2$$

(vi) $$-\frac{d\left[S_2O_8^{2-}\right]}{dt} = k\left[S_2O_8^{2-}\right][I^-]$$

(vii) $$BrO_3^- + 5Br^- + 6H^+ \rightleftharpoons 3Br_2 + 3H_2O$$

(viii) $$-\frac{d\left[BrO_3^-\right]}{dt} = k\left[BrO_3^-\right][Br^-][H^+]^2$$

(ix) $$(CH_3)_3CBr + H_2O \rightleftharpoons (CH_3)_3COH + HBr$$

(x) $$-\frac{d[(CH_3)_3CBr]}{dt} = k[(CH_3)_3CBr]$$

(xi) $$C_2H_4Br_2 + KI \rightleftharpoons C_2H_4 + KBr + KI_3$$

(xii) $$-\frac{d[C_2H_4Br_2]}{dt} = k[C_2H_4Br_2][KI]$$

16.7 Methods of determining reaction order

Given a reaction with the empirical rate equation

$$R = k[A]^\alpha[B]^\beta \tag{16.23}$$

our task is to determine the partial orders α, β, and the true rate constant k. Described below are several methods that allow us to do so. These techniques must often be used together in some combination. In practice they must also be combined with other experimental information to ensure that the system studied is actually reacting according to the assumed stoichiometry. All of the methods described below must be performed at a known and constant temperature.

16.7.1 Integration

The idea is that, for any reaction order and stoichiometry, there is a differential rate law that can be integrated to give us the form of the concentration versus time curve. We have already encountered this method above for first- and second-order kinetics. For kinetics that is first-order in A alone, the concentration of reactant A decays exponentially. Therefore, a plot of $\ln([A]/a_0)$ versus time is linear. For kinetics that is second-order in A alone, a plot of $1/[A]$ versus time is linear. If we plot the data in both manners and find the former is linear but the second does not fit a line, then we can conclude that the reaction is first-order in A alone. If it is clear that one fit is good and the other is bad, this technique works well.

There are several possible complications. The first problem is that if the duration of the experiment is too short, and therefore the percent conversion is too small ($\lesssim$10%), both plots will look linear. We must be able to follow the reaction to a sufficiently high degree of completion to clearly differentiate between different orders.

If neither first- nor second-order fits the data well, we can try another order. There are, however, an infinite number of possible orders if we include all integer and noninteger values. This introduces numerical difficulties and difficulties in interpretation; however, analysis is possible in practice with computer techniques.

If the plots are not well fitted with constant orders, then the kinetic orders are not well defined. This is a strong indication of complicated mechanisms. Complicating factors include catalysis, side reactions, and the involvement of one or more products in the rate law. Two common product-related complications are inhibition and autocatalysis:

- *Inhibition* is when the presence of a product reduces the rate of reaction.
- *Autocatalysis* is a phenomenon in which the presence of a product acts as a catalyst; consequently, the reaction catalyzes itself.

The influence of side reactions, inhibition and autocatalysis all increase with increasing product concentration. If these are present then methods that only rely on a small concentration change (as is possible with the method of initial rates) may be better suited for determining kinetic parameters than integration.

16.7.2 Isolation

The ability to create conditions under which pseudo-order kinetics operates leads to the method known as *isolation*. By adding one reagent in excess, we decouple the kinetics from the concentration dependence of that reagent and isolate the dependence on that of the other reagent alone. For instance, consider a reaction with the empirical rate equation $R = k^*[A]^\alpha$, with $k^* = k[B]^\beta$. First, B is added in excess. Then integration, initial rates or half-lives are used to determine α and k^*. Subsequently, the concentration of B is varied while it still is in excess and changes no more than 5% during the experiment. Taking the natural logarithm of the equation that defines the effective rate constant k^*, we obtain

$$\ln k^* = \ln k + \beta \ln[B]. \tag{16.24}$$

The effective rate constant k^* is obtained by the same method used to obtain α previously. A plot of ln k^* versus ln [B] is constructed. The plot should be linear with a slope equal to β. In theory, the intercept can be used to determine the true rate constant k. However, as the extrapolation may be extensive, in practice, a better estimate is usually obtained by calculating k directly from $k = k^*/[B]^\beta$ for each value of [B] and then taking the mean. Alternatively, A can be placed in excess and methods are then applied to determine β, the effective rate constant and the true rate constant.

Complications associated with isolation are that not all reactions can be run successfully with one reagent sufficiently in excess. In some cases there may be a change of mechanism under such circumstances. A large excess of one reagent might also lead to experimental problems that take the sensitivity, time resolution or dynamic range out of useful limits. A good indication of complications is to compare the results of the two alternative implementations of isolation. If they do not yield consistent results, a mechanism shift or other experimental interference has occurred.

16.7.3 Differential method (method of initial rates)

Taking the logarithm of the empirical rate equation in Eq. (16.23) yields

$$\ln R = \ln k + \alpha \ln[A] + \beta \ln[B]. \tag{16.25}$$

The rate of reaction with respect to a given reactant is obtained by differentiating a plot of the concentration of that reagent versus time. The rate at time $t = 0$ is labeled R_0. We perform two sets of experiments to determine the initial rate of reaction. In the first set of experiments, the concentration of A is varied while that of B is held constant. Just as in isolation experiments, we need to measure a concentration change of less than about 5% to be able to consider the concentration effectively constant. Most accurately, the rate as a function of time is extrapolated back to time zero by determining the initial slope of the concentration versus time graph. A plot of ln R_0 versus ln [A] is constructed. With temperature and [B] held constant, Eq. (16.25) reduces to $\ln R_0 = \alpha \ln[A] +$ constant. The plot should be linear with slope equal to α. The procedure is then repeated holding [A] constant while varying [B]. A plot of ln R versus ln [B] should be linear with slope equal to β. If either of these plots is nonlinear, the order with respect to that reagent is not well-defined.

16.7.4 Half-life

The *half-life* $\tau_{1/2}$ of a reactant is the length of time that it takes for its concentration to drop to one-half of what it was initially – that is, the point in time at which $[A] = \frac{1}{2}a_0$. This is closely related to the *lifetime* τ, which is defined as the time it takes for the concentration to decrease to $1/e$ of its initial value. The half-life is inversely proportional to the rate constant; however, it also has a concentration dependence based on the reaction order. Substituting $[A] = \frac{1}{2}a_0$ into Eq. (16.8) and solving for the time, we obtain the value for the *half-life of a first-order reaction*.

$$\tau_{1/2} = \ln(2)/k \tag{16.26}$$

For a first-order reaction, $\tau_{1/2}$ is independent of concentration; therefore, it is a constant that is characteristic of the kinetics of the reaction. Radioactive decay is a first-order process, and the half-life is used to characterize the radioactivity of different species. Particularly 'hot' radioisotopes have short half-lives, while more stable radioisotopes have long half-lives.

16.7.4.1 Directed practice

Make the appropriate substitutions and show that Eq. (16.26) holds for the half-life of a first-order reaction and that the *lifetime of a first-order reaction* is $\tau = 1/k_A$.

Second-order reactions involving a single reactant or two reactants of equal initial concentration follow the same expression for half-life. The half-life is again obtained by substituting $[A] = a_0/2$ into the integrated rate law to obtain

$$\tau_{1/2} = 1/(k_A\, a_0). \tag{16.27}$$

The *half-life of a second-order reaction* is no longer a constant but depends instead on the initial concentration. Similarly, half-lives can be defined for reactions in which the reactants are mixed stoichiometrically. However, simple formulae for arbitrary mixtures are not available.

Half-lives are of limited use to determine arbitrary reaction orders and rate constants. They are useful for comparing the relative rates of reactions of different orders. They are often quoted to quantify first-order reactions instead of the true rate constant.

16.7.4.2 Example

At 518 °C, the rate of decomposition of a sample of gaseous acetaldehyde, initially at a partial pressure of 48.4 kPa, was 143 Pa s^{-1} when the fractional dissociation was $\phi = 0.05$ and 101 Pa s^{-1} at $\phi = 0.2$. Determine the order of the reaction, assuming that the rate depends only on the partial pressure of acetaldehyde.

The reaction follows A → products. The acetaldehyde partial pressure p_A is

$$p_A = p_{A,0} - \phi p_{A,0} = p_{A,0}(1-\phi).$$

The empirical rate equation with respect to p_A is

$$R = k p_A^\alpha = k\,[p_{A,0}(1-\phi)]^\alpha.$$

Thus, the ratio of the rates for two different conditions is

$$\frac{R_1}{R_2} = \frac{p_{A,1}^\alpha}{p_{A,2}^\alpha} = \frac{[p_{A,0}(1-\phi_1)]^\alpha}{[p_{A,0}(1-\phi_2)]^\alpha} = \left(\frac{1-\phi_1}{1-\phi_2}\right)^\alpha.$$

Taking the logarithm of both sides of this equation

$$\lg\left(\frac{R_1}{R_2}\right) = \alpha \lg\left(\frac{1-\phi_1}{1-\phi_2}\right).$$

Solving for the order α

$$\alpha = \frac{\lg(R_1/R_2)}{\lg((1-\phi_1)/(1-\phi_2))} = \frac{\lg(143/101)}{\lg(0.95/0.80)} = 2$$

The reaction is second order.

16.8 Reversible reactions and the connection of rate constants to equilibrium constants

16.8.1 Reversible reactions

Up to this point all of our reactions have only proceeded in the forward direction. This is not always the case and it changes the nature of the kinetics. Consider a reversible reaction

$$A \underset{k_{-1}}{\overset{k_1}{\rightleftharpoons}} X.$$

Note the convention that k_1 is the rate constant of the forward reaction and k_{-1} that of the reverse reaction. The stoichiometry leads to the following relations for the concentrations:

$$\begin{array}{lll} \text{Initially} & [\mathrm{A}] = a_0 & [\mathrm{X}] = 0 \\ \text{Later} & [\mathrm{A}] = a_0 - x & [\mathrm{X}] = x \end{array}$$

For a reversible reaction, the rate of reaction is given by the net reaction rate of the forward and reverse reactions

$$\frac{\mathrm{d}[\mathrm{A}]}{\mathrm{d}t} = -k_1[\mathrm{A}] + k_{-1}[\mathrm{X}] = -k_1(a_0 - x) + k_{-1}x. \tag{16.28}$$

A net reaction rate is found by adding all terms that create a given species and subtracting all terms that consume that species.

16.8.2 Reversible reaction at equilibrium

At equilibrium, all concentrations have reached the values determined by thermodynamics. Therefore, setting $x = x_{\mathrm{eq}}$, the concentrations at equilibrium are $[\mathrm{A}]_{\mathrm{eq}} = a_0 - x_{\mathrm{eq}}$ and $[\mathrm{X}]_{\mathrm{eq}} = x_{\mathrm{eq}}$. In addition, equilibrium is a steady state in which the concentrations of A and X do not change, implying $\mathrm{d}[\mathrm{A}]/\mathrm{d}t = \mathrm{d}[\mathrm{X}]/\mathrm{d}t = 0$. This allows us to set the right-hand side of Eq. (16.28) equal to zero to obtain

$$k_1(a_0 - x_{\mathrm{eq}}) = k_{-1}x_{\mathrm{eq}}. \tag{16.29}$$

Substitute these relations into the expression for an equilibrium constant, K, to find the relationship between rate constants and the equilibrium constant

$$K = \frac{[X]_{\mathrm{eq}}}{[A]_{\mathrm{eq}}} = \frac{x_{\mathrm{eq}}}{a_0 - x_{\mathrm{eq}}} = \frac{k_1}{k_{-1}} \tag{16.30}$$

This equation is of great importance as it unites the fields of kinetics and thermodynamics:

The equilibrium constant is equal to the ratio of the forward rate constant to the reverse rate constant.[2]

For a reaction composed of a series of reversible steps, the overall reaction is the sum of the steps (as for all mechanisms). An equilibrium constant expression can be written for each step, and the overall equilibrium constant is equal to the product of the constituent equilibrium constants. The relationship between rate constants and the equilibrium constant facilitates the use of thermodynamic data for the calculation of rate constants, as shown in this example.

Consider the follow reaction that has a mechanism composed to two reversible steps.

1. $H_2 + ICl \underset{k_{-1}}{\overset{k_1}{\rightleftharpoons}} HI + HCl$
2. $HI + ICl \underset{k_{-1}}{\overset{k_1}{\rightleftharpoons}} HCl + I_2$

The overall reaction is the sum of the two forward reactions

$$H_2 + 2\ ICl \rightarrow I_2 + 2\ HCl.$$

At equilibrium the steady-state condition applies. This means that the rate of the forward reaction in step 1 is equal to the rate of its reverse, and similarly the forward and reverse rates are the same for the step 2.

$$k_1[H_2][ICl] = k_{-1}[HI][HCl] \tag{16.31}$$

$$k_2[HI][ICl] = k_{-2}[HCl][I_2]. \tag{16.32}$$

Now construct the equilibrium constant expressions for steps 1 and 2 and relate them to the ratios of rate constants.

$$K_1 = \frac{k_1}{k_{-1}} = \frac{[HI][HCl]}{[H_2][ICl]} \tag{16.33}$$

$$K_2 = \frac{k_2}{k_{-2}} = \frac{[HCl][I_2]}{[HI][ICl]} \tag{16.34}$$

The product of Eqs (16.33) and (16.34) gives

$$K_1K_2 = \frac{k_1k_2}{k_{-1}k_{-2}} = \frac{[I_2][HCl]^2}{[H_2][ICl]^2} = K \tag{16.35}$$

The equilibrium constant of the composite reaction is equal to the product of the equilibrium constants of the two coupled reactions. The overall equilibrium constant is also equal to the product of the rate constants of the forward reactions divided by the product of the rate constants of the reverse reactions. For a longer chain of consecutive equilibrated steps, the overall equilibrium constant is the product of all the individual equilibrium constants. It is also equal to the corresponding ratio of the product of forward reactions divided by the product of reverse reactions.

Remember that K can be calculated from the molar Gibbs energy change from

$$\Delta_r G^\circ_m = -RT \ln K. \tag{16.36}$$

Gibbs energy changes for reactions are available from standard tabulated thermodynamic data, both from enthalpy and entropy changes

$$\Delta_r G^\circ_m = \Delta_r H^\circ_m - T\,\Delta_r S^\circ_m \tag{16.37}$$

and from the Gibbs energy of formation

$$\Delta_r G^\circ_m = \sum_{\text{products}} \nu_i\,\Delta_f G^\circ_{m,i} - \sum_{\text{reactants}} \nu_i\,\Delta_f G^\circ_{m,i} \tag{16.38}$$

The relationship between K and rate constants means that thermodynamic data can be used to help calculate the values of the individual rate constants. If we are able to measure either the forward or the reverse rate constant for a given reversible reaction, then the opposite rate constant can be calculated from this value and thermodynamic data. This is important because it can greatly reduce the number of parameters that need to be determined experimentally. In many cases, a reaction in one direction is significantly easier to measure than is its reverse. Therefore, the relationship in Eq. (16.30) allows us to perform just the easy experiment and avoid the difficult one.

16.9 Temperature dependence of rates and the rate constant

The rate of a chemical reaction almost always depends on temperature. Usually, the reaction rate increases with increasing temperature. However, there are special cases in which the reaction rate does not depend strongly on temperature. For some complex reaction mechanisms the rate actually decreases with increasing temperature over certain temperature ranges. To fully understand the temperature dependence of reaction rates we need to develop transition state theory (this is done in Chapter 17). Here, we emphasize a simple description of the empirical observation that most reactions exhibit a more or less exponential increase in reaction rate when the temperature is increased.

16.9.1 The Arrhenius equation

In 1884, van't Hoff noted that the temperature dependence of the concentration equilibrium constant is related to the internal energy change by

$$\left(\frac{\partial \ln K}{\partial T}\right)_p = \frac{\Delta_r U^\circ_m}{RT^2}. \tag{16.39}$$

Writing Eq. (16.30) in logarithmic form

$$\ln K = \ln k_1 - \ln k_{-1}, \tag{16.40}$$

we differentiate with respect to T and then substitute from Eq. (16.39) to obtain

$$\frac{d \ln k_1}{dT} - \frac{d \ln k_{-1}}{dT} = \frac{\Delta_r U^\circ_m}{RT^2}. \tag{16.41}$$

Van't Hoff was not sure what these energies meant, but he recognized that this equation could be split into two equations that define the temperature dependences of k_1 and k_{-1} in terms of two energies E_1 and E_{-1}, such that

$$\frac{d \ln k_1}{dT} = \frac{E_1}{RT^2} \tag{16.42}$$

and

$$\frac{d \ln k_{-1}}{dT} = \frac{E_{-1}}{RT^2} \tag{16.43}$$

with

$$E_1 - E_{-1} = \Delta_r U_m^\circ \tag{16.44}$$

In 1889, Arrhenius integrated Eq. (16.42) to obtain

$$k = A\exp(-E_a/RT). \tag{16.45}$$

Although van't Hoff had also obtained this equation, it is nonetheless known as the *Arrhenius equation* because Arrhenius added essential ideas to the interpretation of the equation. He recognized that the effect of temperature on collision velocities alone is not enough to explain the strong effect of temperature on reaction rates. Instead, he postulated that an equilibrium must exist between activated molecules that go on to react and the batch of reactants that are in the normal state. This postulate forms the basis of transition state theory, as we shall see in Chapter 17. The activated species is now called the *activated complex* , and the equilibrium is formed between normal molecules and activated complexes that occupy a particular region of phase space known as the transition state.

In conventional Arrhenius analysis, the *pre-exponential factor* A and the *activation energy* E_a are taken to be constants. Therefore, taking the logarithm of Eq. (16.45) to obtain

$$\ln k = \ln A - E_a/RT \tag{16.46}$$

we see that a plot called an *Arrhenius plot* of $\ln k$ versus $1/T$ should yield a straight line with an intercept equal to $\ln A$ and slope $-E_a/RT$. A log term should be unitless; hence officially, the value of k is divided by the standard state concentration appropriate to the experimental conditions. The activation energy is then recognized to be equal to the energy required to take a normal molecule to the activated state. The meaning of the pre-exponential factor (it is related to the entropy change for entering the transition state) awaits interpretation in terms of transition state theory. The units of A and k are the same, that is, they depend on the kinetic order. An example of an Arrhenius plot is shown in Fig. 16.6. Examples of first-order and second-order reactions that exhibit Arrhenius behavior, and the temperature ranges over which this behavior is observed, are listed in Tables 16.2 and 16.3, respectively.

16.9.1.1 Directed practice

Use the Arrhenius equation and the values in Table 16.2 to calculate the rate constant for the first-order decomposition of ozone at 400 K.

[Answer: 0.313 s^{-1}]

The assumption of constant values is based on empirical evidence not theory. Most experiments yield linear Arrhenius plots. In point of fact, most experiments do not cover a sufficiently large temperature range to reveal the temperature

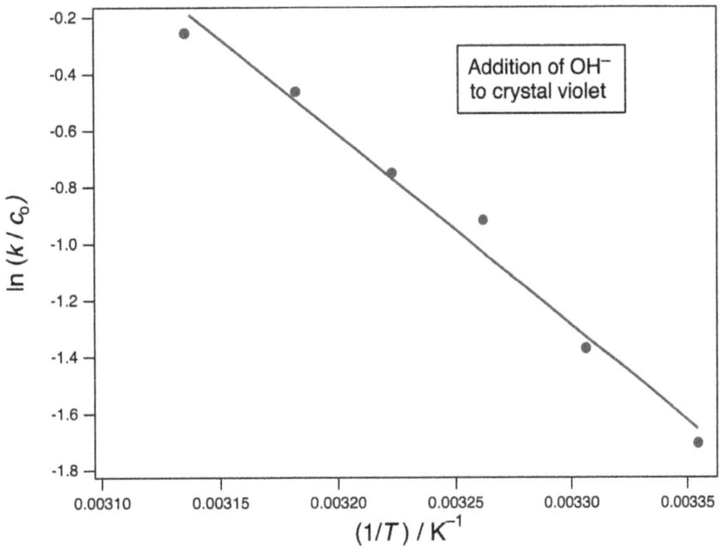

Figure 16.6 An Arrhenius plot of ln (k/c_0) versus $1/T$ for the addition of hydroxide to crystal violet.

Table 16.2 Arrhenius parameters for selected first-order reactions in the gas phase. Values adapted from Houston, P.L. (2001) *Chemical Kinetics and Reaction Dynamics*. Dover Publications, Mineola, NY.

Reaction	A/s^{-1}	E_a/kJ mol^{-1}	Range/K
$CH_3CHO \rightarrow CH_3 + HCO$	3.8×10^{15}	328.4	500–2000
$C_2H_6 \rightarrow 2CH_3$	2.5×10^{16}	367.6	200–2500
$O_3 \rightarrow O_2 + O$	7.6×10^{12}	102.5	383–413
$HNO_3 \rightarrow OH + NO_2$	1.1×10^{15}	199.5	295–1200
$CH_3OH \rightarrow CH_3 + OH$	6.5×10^{16}	386.8	300–2500
$N_2O \rightarrow N_2 + O$	6.1×10^{14}	234.5	1000–3600
$N_2O_5 \rightarrow NO_3 + NO_2$	1.6×10^{15}	234.5	200–384
$C_2H_5 \rightarrow C_2H_4 + H$	2.8×10^{13}	170.3	250–2500
$CH_3NC \rightarrow CH_3CN$	3.8×10^{13}	161.0	393–600
$H_2S \rightarrow SH + H$	6.0×10^{14}	339.4	1965–19700
cyclopropane → propene	2.9×10^{14}	261.1	897–1450
cyclobutane → $2C_2H_4$	2.3×10^{15}	253.5	891–1280

dependence of what thermodynamics tell us must be temperature-dependent quantities. For large temperature ranges, a more accurate expression is

$$k = A\,T^m \exp(-E_0/RT). \tag{16.47}$$

The constant m is a small number. Applying the derivative in Eq. (16.42), we can then show that

$$E_a = E_0 + mRT \tag{16.48}$$

and the E_0 is to be interpreted as the activation energy at absolute zero, whereas E_a is the activation energy at temperature T.

Table 16.3 Arrhenius parameters for selected second-order reactions in the gas phase. Values adapted from Houston, P.L. (2001) *Chemical Kinetics and Reaction Dynamics*. Dover Publications, Mineola, NY.

Reaction	A/m^3 mol^{-1} s^{-1}	E_a/kJ mol^{-1}	Range/K
$CH_3 + H_2 \rightarrow CH_4 + H$	6.7×10^{6}	52	300–2500
$CH_3 + O_2 \rightarrow CH_3O + O$	2.4×10^{7}	121	298–3000
$F + H_2 \rightarrow HF + H$	8.4×10^{7}	4.16	200–300
$H + H_2O \rightarrow OH + H_2$	9.7×10^{7}	86	250–3000
$H + CH_4 \rightarrow CH_3 + H_2$	1.9×10^{8}	50.8	298–2500
$H + O_2 \rightarrow OH + O$	1.4×10^{8}	66.5	250–3370
$H + NO \rightarrow OH + N$	1.3×10^{8}	201	1750–4500
$H + N_2O \rightarrow N_2 + OH$	7.6×10^{7}	63.2	700–2500
$H + CO_2 \rightarrow CO + OH$	1.5×10^{8}	111	300–2500
$H + O_3 \rightarrow O_2 + OH$	8.4×10^{7}	3.99	220–360
$O + H_2 \rightarrow OH + H$	5.2×10^{7}	41.6	293–2800
$O + CH_4 \rightarrow OH + CH_3$	1.0×10^{8}	42.3	298–2575
$O + NO_2 \rightarrow NO + O_2$	3.9×10^{6}	−0.998	230–350
$O + C_2H_2 \rightarrow$ products	3.4×10^{7}	14.6	195–2600
$O + O_3 \rightarrow 2O_2$	7.5×10^{6}	17.8	197–2000
$O + C_2H_4 \rightarrow$ products	9.5×10^{6}	7.87	200–2300
$OH + H_2 \rightarrow H + H_2O$	4.6×10^{6}	17.5	200–450
$OH + CH_4 \rightarrow H_2O + CH_3$	2.2×10^{6}	15.1	240–300
$OH + O \rightarrow O_2 + H$	1.4×10^{7}	−0.915	220–500

16.9.2 Non-Arrhenius behavior

Some complex reactions do not follow the Arrhenius law as a function of temperature, even over moderate temperature ranges. This is often an indication that the reaction mechanism changes as a function of *T*. This is typical of, for example, catalytic reactions including enzyme catalysis, as well as explosions.

16.10 Microscopic reversibility and detailed balance

Neither van't Hoff nor Arrhenius fully recognized it, but they had developed relationships that would lay the basis for the study of chemical dynamics. Equations (16.42) and (16.43) can be understood easily once we understand that a reaction proceeds from reactants to products by traversing a uniquely defined potential energy surface that describes the interactions of all the atoms in the system. A simplified one-dimensional representation of the minimum energy pathway on this surface is shown in Fig. 16.7.

The implications of Fig. 16.7 can be interpreted with the help of two statements of fact that were made in the 1920s by two eminent scientists. As explained in the footnotes, these principles were derived by numerous individuals, but they were named by Tolman and Fowler.

Microscopic reversibility: *In the case of a system in thermodynamic equilibrium not only that the total number of molecules leaving a given state in unit time shall on the average equal the number arriving in that state in unit time, but also that the number leaving by any particular path shall on the average be equal to the number arriving by the reverse of that particular path.*[3]

Principle of detailed balance: *In a system at equilibrium, each collision has its exact counterpart in the reverse direction, and that the rate of every chemical process is exactly balanced by that of the reverse process.*[4]

The consequences of these two statements are extremely important, some of them have already been encountered. All of them will be explored in more depth in following chapters:

- Every molecular species in a system at equilibrium has an exact counterpart, which is moving in the opposite direction and which, on the average, is present at the same concentration.
- The principles apply to single elementary steps as well as overall mechanisms.

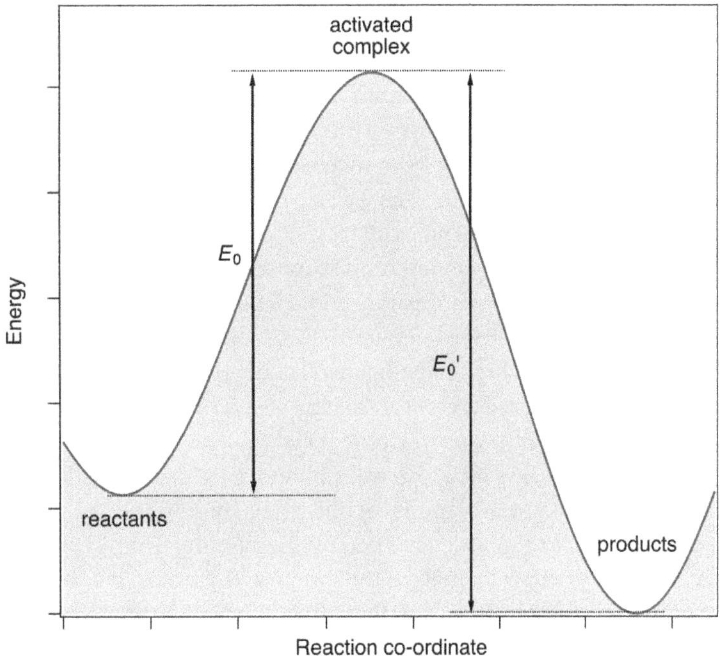

Figure 16.7 A one-dimensional representation of the potential energy change that occurs during the course of a reaction from reactants through a transition state at the maximum of the curve to the products. The reaction coordinate is related in some fashion to the arrangement of atoms in the system.

- The flux of molecules and energy in the forward and reverse directions over a potential energy surface (PES) must be balanced at equilibrium.
- The most probable path in one direction is also the most probable path in the other direction.
- The activated complex and transition state in one direction must be the same in the other direction.
- The forward and reverse activation energies are in general not the same, but the changes in enthalpy or Gibbs energy of the forward reaction much be equal and opposite to the changes in the reverse reaction.
- For an elementary step the difference in Gibbs energies of activation of the forward and reverse reactions is equal to the overall Gibbs energy difference.
- The ratio of forward and reverse rate constants equals the equilibrium constant.

16.11 Rate-determining step (RDS)

A chemical reaction *mechanism* is a complete set of elementary reactions that add up to the overall stoichiometric equation. The individual elementary reactions are also called steps in the mechanism. The vast majority of chemical reactions are composite reactions involving two or more steps. Their kinetics should, in principle, be highly complex and extremely sensitive to experimental conditions. Such a high degree of sensitivity would make the discovery of any inherent structure to chemical kinetics very difficult. Fortunately, it is found that many reactions exhibit understandable regularities, and that many composite reactions follow simple first- or second-order kinetics. This is possible because of the presence of a *rate-determining step*. Sometimes also called a rate-controlling step, *a rate-determining step is the step in a mechanism that has the strongest influence on the rate*. It corresponds to the step for which a change of its rate constant would change the rate of reaction. For a reaction of irreversible steps run at steady state, every step after the RDS has exactly the same rate as the RDS. Therefore, the description 'slowest step' does not define the RDS universally.

Reasons why a step controls the rate include: (i) a low value of the rate constant; or (ii) a highly constrained limiting concentration of one of the reactants in the step. A low value of the rate constant is usually associated with a high value of E_a. Less commonly, a peculiarly low value of the pre-exponential factor can be responsible for the low rate of the RDS. The exponential dependence of reaction rate on activation energy means that small differences in E_a lead to significant differences in rate. This is a factor that enhances the probability that one step will act as a RDS, at least at low and moderate temperatures. At very high temperature (in excess of 1000 K), the exponential factor in the Arrhenius equation is essentially the same for typical values of E_a. At this point the pre-exponential factor determines differences in rates. Of course, most chemistry is run at significantly lower temperatures than this high-temperature range.

Not all reactions have a well-defined RDS, or at least, they do not have a well-defined RDS for all conditions. For example, there may be a change in mechanism as a function of temperature. At low temperatures one mechanism with a well-defined RDS is dominant, but in a higher temperature range a different mechanism with a different RDS will be dominant. In the intermediate temperature range both mechanisms compete to produce products, and the observed kinetics will be a mixture of the two mechanisms. In such a case we would expect an Arrhenius plot to be linear at low and high temperatures (but with different slopes) and exhibit curvature in the intermediate range.

The one step that is controlling the rate must be determined by sensitivity analysis. The RDS is often defined as the slowest step in a reaction, but as alluded to above this is a poor definition. Rather, it is the step that has the greatest influence on the rate that is the RDS. You may have been told in introductory chemistry some silly analogy between the RDS and a train station with only one exit. A better analogy would be to imagine that a Formula 1 racing car, a regular passenger vehicle and a farm tractor are tethered together. Each can travel in its own lane. Asked to go as fast as they can, the tractor driver soon finds him/herself in fifth gear with the pedal floored. The driver shifts to third gear and the racer looks nervously at his or her watch, wondering if he/she will stall in first gear. When the tractor speeds up because they start to go downhill, the other cars are able to respond by accelerating comfortably. All of the cars in the group travel at the same velocity, but the velocity of the group is determined by the rate-determining vehicle, the tractor.

We do not need to identify the RDS in order to analyze the kinetics. If there is one, it can greatly simplify the analysis of kinetics and an understanding of the dynamics. Charles Campbell has defined a method to quantifiably identify the RDS by introducing a parameter he defines as the *degree of rate control* for step i

$$X_{\mathrm{rc},i} = \frac{k_i}{R}\frac{\partial R}{\partial k_i} \tag{16.49}$$

where the overall rate of reaction is R and the partial derivative of R with respect to the rate constant k_i, $\partial R/\partial k_i$, is taken while holding all other rate constants k_j constant. The concept behind $X_{rc,i}$ is simple: Change the forward and reverse rate constant of step i by a small amount that still leads to a measurable change in R, say 1%. Since both forward and reverse rate constants are changed equally, the equilibrium constant of that step is not affected. Subsequent to this change, calculate the change in the overall rate R. The step that makes the greatest difference – that is, the step that changes R the most – is the step with the greatest $X_{rc,i}$ value. This is the RDS.

For any reaction mechanism that is composed of a series of consecutive steps and which has a RDS, $X_{rc,i}$ is large for the RDS and is close to zero for all other steps. Any step with $X_{rc,i} \geq 0.95$ acts as a single RDS. Some complex mechanisms do not have a single RDS; rather, there will then be several RDSs with positive values of $X_{rc,i}$. If a step has a negative value of $X_{rc,i}$ it is an *inhibition* step. This method is extremely general. The only drawback is that, in order to implement it efficiently, the rate equations must be handled using finite difference or other numerical methods. On the other hand, as such kinetic analysis is routine with desktop computers this does not present a serious obstacle to implementation. In other words, if you are routinely performing kinetic analyses on complex multistep reactions you will probably be performing this type of computer analysis; therefore implementation of the above type of analysis should become a routine part of RDS determination.

SUMMARY OF IMPORTANT EQUATIONS

$R = \frac{1}{\nu_i}\frac{d[i]}{dt} = \frac{1}{V\nu_i}\frac{dn_i}{dt} = \frac{1}{V}\frac{d\xi}{dt}$	Rate of reaction, R. ν_i is the stoichiometric coefficient of species i with molar concentration [i], moles n_i. V is the volume and ξ the extent of reaction.
$R = k[A]^\alpha[B]^\beta$	Empirical rate equation for irreversible reaction A + B → P
$\ln R = \ln k + \alpha \ln[A] + \beta \ln[B]$	Logarithmic form of empirical rate equation for irreversible reaction A + B → P
$\tau_{1/2} = \ln(2)/k$	Half-life of a first-order reaction in terms of the rate constant k.
$\tau_{1/2} = 1/(k_A a_0)$	Half-life of a second-order reaction 2A → P or A + B → P with equal initial concentrations a_0 and b_0 in terms of the specific rate constant k_A.
$K_1 = \frac{k_1}{k_{-1}}$	Relationship of equilibrium constant K_1 for a reversible step 1 in terms of the forward, k_1, and reverse, k_{-1}, rate constants of that step.
$K = K_1K_2 = \frac{k_1k_2}{k_{-1}k_{-2}}$	Relationship of equilibrium constant K for a composite reaction composed of two reversible steps to the equilibrium constants and rate constants of those two steps.
$\Delta_r G^\circ_m = -RT \ln K$	Do not forget this.
$\Delta_r G^\circ_m = \Delta_r H^\circ_m - T\Delta_r S^\circ_m$	Or this.
$k = A\exp(-E_a/RT)$	Arrhenius equation. Describes the temperature dependence of the rate constant k to the pre-exponential factor A, the activation energy E_a and the absolute temperature T.
$k = AT^m \exp(-E_0/RT)$	Improved version of Arrhenius equation to include inherent temperature dependence with $E_a = E_0 + mRT$, where E_0 is the activation at $T = 0$ K and m a small number.
$X_{rc,i} = \frac{k_i}{R}\frac{\partial R}{\partial k_i}\partial$	Degree of rate control of step i, $X_{rc,i}$. The overall rate of reaction is R and the partial derivative of R with respect to the rate constant k_i, $\partial R/\partial k_i$,

Exercises

16.1 Write an expression for the net rate in terms of $d[H_2]/dt$ for the elementary reaction $C_2H_6 \underset{k_{-1}}{\overset{k_1}{\rightleftharpoons}} C_2H_4 + H_2$.

16.2 Discuss the difference between an elementary reaction and a composite reaction.

16.3 Discuss the difference between a reversible reaction and an equilibrated reaction.

16.4 Calculate the rate of the elementary reaction $A + B \rightleftharpoons C$ if $k = 0.0253\ M^{-1}\ s^{-1}$ and the concentrations of A and B are both 0.333 M.

16.5 The following are three integrated rate laws

$$[A] = -kt + a$$

$$[A] = b\exp(-kt)$$

$$[A] = \frac{1}{kt + c}$$

for reactions in which a single reagent A is transformed into products. [A] is the concentration of A and *a*, *b*, and *c* are all constants. State which order each equation corresponds to. Manipulate each equation into a linear form and then draw a graph that shows this linear behavior (with properly labeled axes and simulated data points) and how such a plot would be used to determine the rate constant *k*.

16.6 The reaction $S_2O_8^{2-} + 2I^- \rightleftharpoons 2SO_4^{2-} + I_2$ follows the rate law

$$-\frac{d\left[S_2O_8^{2-}\right]}{dt} = k\left[S_2O_8^{2-}\right][I^-].$$

(a) What is the order of the reaction with respect to each reactant. What is the overall order? (b) When $[S_2O_8{}^{2-}]_0 = [I^-]_0 = 2.50 \times 10^{-3}$ mol l^{-1}, the rate of formation of I_2 at $T = 298$ K is 3.67×10^{-5} mol l^{-1} s^{-1}. Calculate the rate of formation of $SO_4{}^{2-}$ and the rate of consumption of $S_2O_8{}^{2-}$ and I^-. (c) Calculate the value of the rate constant at 298 K. (d) Can the reaction as written be described as an elementary reaction? Justify your answer.

16.7 The fluorescence of green fluorescence protein (GFP) decays according to the data given in the table below. Determine the fluorescence lifetime.

t/ns	0.5	0.8	1.5	3	5	10
F	85.7	78.2	63.0	39.7	21.5	4.61

16.8 Flavin adenine dinucleotide (FAD) has a fluorescence lifetime of 2.91 ns in the intercellular medium but <0.01 ns when protein-bound. FAD is protein-bound in normal epithelium but free in some precancerous tissue. Discuss how this difference in lifetimes can be used in imaging to distinguish pre-cancerous cells from non-cancerous cells.

16.9 During the decomposition of C_4H_8 to C_2H_4 at 438 °C, the partial pressure of C_4H_8 drops from 2.53 kPa initially to 1.27 kPa after 2790 s, and to 0.633 kPa after 5580 s. The decomposition occurs in one elementary step. Write out a balanced chemical reaction to describe the decomposition, determine the rate constant, and write out the rate equation for this reaction.

16.10 The gas-phase reaction $OH(g) + CH_4(g) \rightarrow H_2O(g) + CH_3(g)$ is second order overall and first order in both reactants. The initial rate of appearance (measured at constant volume and temperature) of H_2O as a function of temperature is given in the table below. The initial concentration of OH and CH_4 is 0.500 mol m^{-3}. (a) Calculate the Arrhenius parameters for the rate constant. (b) Calculate the initial rate and concentration of OH and H_2O after 1 ms at 250 K for the same initial pressure.

T/K	240	260	280	300
R/mol m^{-3} s^{-1}	284	509	838	1290

16.11 The gas-phase reaction A + 2B → 3C follows the rate equation $-dp_B/dt = 2k\,p_A^{\alpha}\,p_B^{\beta}$.
(a) In an experiment, the following data were obtained:

($-dp_B/dt$)/kPa s^{-1}	**p_A/kPa**	**p_B/kPa**
3.8	1000	1.0
3.8	2000	1.0
15.4	1000	2.0

Calculate *k*, *α*, and *β*. (b) Calculate the pressure of A, B, and C after 10 s and the total pressure if the initial partial pressures are $p_A = 1000$ kPa, $p_B = 1.0$ kPa, and $p_C = 0$.

16.12 The reaction A + B → 2C + D follows the rate equation $d[D]/dt = k[A]^{\alpha}[B]^{\beta}$. In an experiment, the following data were obtained

[A]/mol l^{-1}	[B]/mol l^{-1}	d[D]/dt/l mol^{-1} s^{-1}
0.300	0.300	1.31×10^{-5}
0.600	0.300	3.69×10^{-5}
0.300	0.600	1.85×10^{-5}

Calculate k, α, and β and write the rate equation with the correct orders in terms of reaction rate R. The orders should be rounded off to the nearest integer or half-integer.

16.13 (a) Determine the rate constant and the rate law expression using the data in the table below for the reaction $2NO(g) + 2H_2(g) \rightarrow N_2(g) + 2H_2O(g)$. (b) Is the above reaction an elementary reaction? Justify your answer.

Run	p_0 H$_2$/kPa	p_0 NO/kPa	Rate/kPa s^{-1}
1	53.3	40.0	0.137
2	53.3	20.3	0.033
3	40.0	53.3	0.213
4	23.9	53.3	0.0983

16.14 The second-order rate constants for the reaction of oxygen atoms with aromatic hydrocarbons have been measured. In the reaction with benzene the rate constants are

k/(dm^3 mol^{-1} s^{-1})	1.44×10^7	3.03×10^7	6.90×10^7
T/K	300.3	341.2	392.2

Find the pre-exponential factor and the activation energy of the reaction.

16.15 As shown by the data in the table below, the reversible elementary gas-phase reaction $H + CH_4 \underset{k_{-1}}{\overset{k_1}{\rightleftharpoons}} H_2 + CH_3$ exhibits Arrhenius behavior in the temperature range 350–750 K.

k_1/(dm^3 mol^{-1} s^{-1})	8.78	49.3	162.8	390.8
T/K	450	550	650	750

Find the pre-exponential factor and the activation energy of the reaction.

16.16 What can cause non-Arrhenius behavior and how can one demonstrate it experimentally?

16.17 The rate law for the reaction $BrO_3^-(aq) + 5Br^-(aq) + 6H^+(aq) \rightleftharpoons 3Br_2(aq) + 3H_2O(l)$ is $-d[BrO_3^-]/dt = k[BrO_3^-][Br^-][H^+]^2$. (a) Calculate the rate of change in BrO_3^- and Br_2 concentrations and the reaction rate in a 1.00-liter aqueous solution that is 0.100 M in both $NaBrO_3$ and HBr if $k = 3.75 \times 10^{-4}$ M^{-3} s^{-1}. (b) If the volume of water is doubled, what is the new reaction rate (all other moles held constant)? (c) Is the reaction an elementary or composite reaction? Justify your answer.

16.18 Why is 'the slowest step' a poor definition of the rate-determining step of a reaction, and what is a better definition?

16.19 The forward and reverse rate constants have been measured for the gas-phase reaction

$H_2(g) + I_2(g) \rightleftharpoons 2HI(g)$.

The reaction is bimolecular in both directions with Arrhenius parameters

$A = 4.45 \times 10^5$ dm^3 mol^{-1} s^{-1}, $E_a = 170.3$ kJ mol^{-1} (forward reaction)
$A' = 5.79 \times 10^1$ dm^3 mol^{-1} s^{-1}, $E_a' = 117.6$ kJ mol^{-1} (reverse reaction)

Compute k, k' (rate constant for the reverse reaction), K, $\Delta_r G_m°$, $\Delta_r H_m°$, and $\Delta_r S_m°$ at 298 K.

16.20 The reversible elementary gas-phase reaction $HBr + Br \underset{k_{-1}}{\overset{k_1}{\rightleftharpoons}} H + Br_2$ exhibits Arrhenius behavior in the temperature range 350 750 K. It is characterized in the forward direction by $A = 1.35 \times 10^9$ l mol^{-1} s^{-1} and $E_a = 174.9$ kJ mol^{-1}, and in the reverse direction by $A = 6.76 \times 10^9$ l mol^{-1} s^{-1} and $E_a = 3.8$ kJ mol^{-1}. Use $T = 550$ K and concentrations [H] = 1.25×10^{-6} mol l^{-1}, [Br_2] = 5.25×10^{-2} mol l^{-1}, [HBr] = 6.33×10^6 mol l^{-1}, [Br] = 4.33×10^7 mol l^{-1} to perform the following calculations. (a) Calculate the rate constants for the forward and reverse reactions. (b) Calculate the equilibrium constant of the forward reaction. (c) Calculate the rates of the forward and reverse reactions. (d) Under the given conditions does the reaction proceed in the forward or reverse direction? Justify your answer. (e) Calculate the standard changes in enthalpy, entropy and Gibbs energy for the forward reaction. (f) Draw a diagram (roughly to scale) of energy versus reaction coordinate, which shows the change in potential energy during the course of the reaction. Label the diagram appropriately with the activation energies, reaction enthalpy and species that apply to the diagram. Include a drawing of the activated complex.

16.21 The reversible elementary gas-phase reaction $H + H_2O \underset{k_{-1}}{\overset{k_1}{\rightleftharpoons}} OH + H_2$ exhibits Arrhenius behavior in the temperature range 250–450 K. It is characterized in the forward direction by $A = 9.70 \times 10^7$ m^3 mol^{-1} s^{-1} and E_a = 86.0 kJ mol^{-1}, and in the reverse direction by $A = 4.60 \times 10^6$ m^3 mol^{-1} s^{-1} and $E_a = 17.5$ kJ mol^{-1}. Use $T = 400$ K and concentrations [H] = 5.25×10^{-6} mol m^{-3}, [H_2O] = 5.45×10^{-6} mol m^{-3}, [H_2] = 1.45×10^{-10} mol m^{-3}, [OH] = 5.00×10^{-10} mol m^{-3} to perform the following calculations. (a) Calculate the rate constants for the forward and reverse reactions. (b) Calculate the equilibrium constant of the forward reaction. (c) Calculate the rates of the forward and reverse reactions. (d) Under the given conditions does the reaction proceed in the forward or reverse direction? Justify your answer. (e) Calculate the standard changes in enthalpy, entropy and Gibbs energy for the forward reaction. (f) Draw a diagram (roughly to scale) of energy versus reaction coordinate, which shows the change in potential energy during the course of the reaction. Label the diagram appropriately with the activation energies, reaction enthalpy and species that apply to the diagram. Include a drawing of the activated complex.

16.22 The reversible elementary gas-phase reaction $H_2 + I_2$ $H_2 + I_2 \underset{k_{-1}}{\overset{k_1}{\rightleftharpoons}} 2HI$ exhibits Arrhenius behavior in the temperature range 600–750 K. It is characterized in the forward direction by $A_1 = 4.45 \times 10^5$ l mol^{-1} s^{-1} and E_{a1} = 170.3 kJ mol^{-1}. When equilibrated at $T = 633$ K, the concentrations of H_2 and I_2 are 1.90×10^{-3} mol l^{-1}. $\Delta_r H° =$ 52.7 kJ mol^{-1} and $\Delta_r G° = 3.40$ kJ mol^{-1}. (a) Calculate the rates of the forward and reverse reactions in terms of $d[HI]/dt$ in mol l^{-1} s^{-1} at equilibrium and 633 K. (b) Calculate the equilibrium concentration of HI. (c) Calculate the rate parameters of the reverse reaction: k_{-1}, E_{a-1}, and A_{-1}. (d) Draw a diagram (roughly to scale) of energy versus reaction coordinate, which shows the change in potential energy during the course of the reaction. Label the diagram appropriately with the activation energies, reaction enthalpy and species that apply to the diagram. Include a reasonable drawing of the activated complex and comment upon the implications of the value of $\Delta^‡ S$.

16.23 The reversible elementary gas-phase reaction $NO + O_3$ $NO + O_3 \underset{k_{-1}}{\overset{k_1}{\rightleftharpoons}} NO_2 + O_2$ exhibits Arrhenius behavior at $T =$ 298 K (use this temperature throughout). It is characterized in the forward direction by $A_1 = 1.09 \times 10^9$ l mol^{-1} s^{-1} and E_{a1} = 9.62 kJ mol^{-1}, $\Delta_r H_m° = -200.8$ kJ mol^{-1} and $\Delta_r G_m° = -199.5$ kJ mol^{-1}. (a) Calculate the rate of the forward reaction in terms of d[NO]/dt in mol l^{-1} s^{-1} if the pressures of NO and O_3 are both 10 Pa. (b) Calculate the equilibrium constant. Does the reaction go to completion? (c) Calculate the rate parameters of the reverse reaction: k_{-1}, E_{a-1}, and A_{-1}. (d) Draw a diagram (roughly to scale) of energy versus reaction coordinate, which shows the change in potential energy during the course of the reaction. Label the diagram appropriately with the activation energies, reaction enthalpy and species that apply to the diagram. Include a drawing of the activated complex. (f) Comment on the values of $\Delta^‡ H_m°$ and $\Delta^‡ S_m°$, and suggest reasons why these values are reasonable or unreasonable based on the reaction dynamics.

16.24 The gas-phase reaction $NO + O_3 \rightarrow NO_2 + O_2$ follows Arrhenius behavior with kinetic parameters $A = 323$ Pa^{-1} s^{-1} and E_a = 10.5 kJ mol^{-1}. Calculate the pressure of NO, O_3 and NO_2 after 15 s if the initial pressure of NO is $p_1 = 5.0 \times 10^{-4}$ Pa and the initial pressure of O_3 is $p_2 = 2.0 \times 10^{-4}$ Pa. There are no products initially and the container is isochoric and isothermal at 705 K.

16.25 (a) In the stratosphere, the rate constant for the conversion of ozone to molecular oxygen by atomic chorine follows a one-step elementary reaction with a rate constant given by $k = (1.7 \times 10^{10}\ \mathrm{M}^{-1}\ \mathrm{s}^{-1})e^{-a/T}$, where a = 260 K and T is the absolute temperature. Write out a balanced chemical reaction for this process and determine the rate of this reaction at an altitude of 20 km where [Cl] = 5.00×10^{-17} M, [O_3] = 8.00×10^{-9} M, and T = 220 K.

Endnotes

1. Rates of consumption: -1.7×10^{-7} mol m^{-3} s^{-1}, -8.3×10^{-7} mol m^{-3} s^{-1}, -1.0×10^{-6} mol m^{-3} s^{-1}. Rates of formation: 5.0×10^{-7} mol m^{-3} s^{-1}, 5.0×10^{-7} mol m^{-3} s^{-1}.
2. Remember: Upper-case K for equilibrium constants. Lower-case k for rate constants.
3. Tolman, R.C. (1925) *Proc. Natl Acad. Sci. USA*, **11**, 436. Richard Tolman credits Irving Langmuir (*J. Am. Chem. Soc.*, **38** , 2253 and 2262 (1916)) with the first statement of this principle. After recounting the history of other derivations, he came to the conclusion that the principle was sufficiently important to require a name, which he duly gave it.
4. Fowler, R.H. and Milne, E.A. (1925) *Proc. Natl Acad. Sci. USA*, **11**, 400. These authors recount the history of the development of the principle of detailed balance, name it in the title of their paper (though they do not state the principle literally), and report that it was inspired by the work of Albert Einstein (*Phys. Z.*, **18**, 121 (1917)), in which he used the principle but neither stated it explicitly nor named it. They also explain how the principle was refined and rediscovered in many different fields of physics. Though they recount many of the same references as Tolman, they never mention Tolman nor Langmuir.

Further reading

Henriksen, N.E. and Hansen, F.Y. (2008) *Theories of Molecular Reaction Dynamics: The microscopic foundation of chemical kinetics*. Oxford University Press, Oxford.

Laidler, K.J. (1987) *Chemical Kinetics. HaperCollins*, New York.

Houston, P.L. (2001) *Chemical Kinetics and Reaction Dynamics*. Dover Publications, Mineola, NY.

Campbell, C.T. (2001) Finding the Rate-Determining Step in a Mechanism: Comparing DeDonder Relations with the 'Degree of Rate Control'. *J. Catal.*, **204** (2), 520–524.

CHAPTER 17

Reaction dynamics I: Mechanisms and rates

PREVIEW OF IMPORTANT CONCEPTS

- The collision rate of reactants is an upper limit to the reaction rate in thermal chemistry.
- The reaction rate is less than this because of the need to fulfill a minimum energy requirement and a steric requirement for the collisions that lead to reaction.
- The energy exchanged along the line of centers when the reacting molecules begin to interact chemically determines whether reaction occurs.
- Most reactions exhibit a positive activation energy.
- Some reactions involving radicals and nonactivated adsorption do not have an activation barrier.
- Some complex reactions with stable intermediates exhibit negative activation energies.
- The dynamics of a reaction system is determined by a potential energy surface that describes the interactions between the atoms in the system.
- The transition state is a region of the potential energy surface that lies between the reactants and products. The activated complex is the chemical species in the transition state.
- The activation energy is given by the difference between the mean energy of the reactants and the mean energy of the activated complexes in the transition state.
- Transition state theory describes the rates of many reactions by assuming a quasi-equilibrium between the reactants and the activated complex. The activated complex then creates products by falling into the product valley.
- A reaction mechanism is a complete sequence of elementary reactions (steps), which add together to give the stoichiometric equation of a reaction.
- Within the steady-state approximation at least one intermediate exists at a low concentration, and its concentration is constant.
- A change in reaction kinetics can signal either a change in mechanism or a change in the rate-determining step within the same mechanism.
- Adsorption and desorption are the two most fundamental steps in all heterogeneous catalytic reactions.
- The rate of desorption follows a rate law much like a reaction in any other phase. However, calculation of the rate of adsorption requires a model that appropriately counts the number of active sites.
- The addition of energy in a particular degree of freedom may enhance reactivity. It is the topology of the potential energy surface that determines the efficacy of energy transfer from a degree of freedom towards overcoming the activation barrier.

17.1 Linking empirical kinetics to reaction dynamics

In Chapter 16 we investigated empirical kinetics. We defined and developed the mathematical treatment necessary to handle the description of reaction rates and the dependence of concentrations as a function of time. The purpose of this chapter is to get to the heart of chemistry. We aim to understand how a chemical reaction occurs and how the manner of making and breaking bonds affects reaction rates. To do so, we begin by taking a closer look at how molecular collisions and energy exchange actually lead to the making and breaking of bonds. Beginning with unimolecular and bimolecular elementary reactions, we will develop the foundation of a predictive theory to describe reaction rates, rather than just relying on empirical fits to data that have no predictive power. We will then show how a series of unimolecular and bimolecular processes can be combined to generate an overall reaction mechanism. At that point, we can understand more fully the effects of reaction conditions such as concentrations and temperature on reaction mechanisms. We will

Physical Chemistry: How Chemistry Works, First Edition. Kurt W. Kolasinski.

Companion Website: www.wiley.com/go/kolasinski/physicalchemistry

Figure 17.1 One of the most fundamental tools for the study of chemical dynamics is the supersonic molecular beam apparatus. The gas source is at the far left. The upper inset shows an expanded view of a pulsed nozzle in front of a skimmer. The skimmer has a tiny orifice through which the molecules expand into the next vacuum chamber. The chamber on the left is called the source chamber. The middle chamber (mostly obscured by the gate valve above the diffusion pump) contains a partition and a collimating orifice that allow it to provide two stages of differential pumping. The main chamber on the right is an ultrahigh vacuum chamber suitable for performing laser-induced fluorescence (LIF) and resonance-enhanced multiphoton ionization (REMPI) spectroscopy on either the beam directly or on molecules that are scattered, thermally desorbed or laser-desorbed from the surface of a sample. The lower inset is a schematic diagram of the molecular beam apparatus. Photograph reproduced with permission from Eckart Hasselbrink.

learn to understand when changes in reaction kinetics are produced by changes in mechanism – that is, a change in the steps involved in the reaction – as opposed to a change in which steps are determining the reaction rate. From a fundamental understanding of the relationship of unimolecular and bimolecular processes to reaction mechanisms, we will then be in a position to address complex reaction dynamics of the type described in Chapter 18, which are the reaction dynamics of useful reactions rather than model systems.

17.2 Hard-sphere collision theory

17.2.1 Collision density

Consider a bimolecular elementary reaction between reactants A and B leading to the product P,

$$A + B \rightarrow P, \tag{17.1}$$

which follows the rate law

$$R = k[A][B]. \tag{17.2}$$

For simplicity we take A, B and P as gas-phase species. We know that a reaction can only occur if A and B collide with the right amount of energy in the right orientation. Therefore, the reaction rate is proportional to three factors: the collision rate; the collision energy; and the orientation of the collision, which is known as the *steric factor*. Let us calculate estimates of these factors based on a model in which molecules are represented by hard spheres.

In Chapter 2 we derived that the *collision density* Z_{12} is the number of AB collisions per unit time and volume

$$Z_{12} = N_1 N_2 \sigma \langle v_{12} \rangle \tag{17.3}$$

where $\langle v_{12} \rangle$ is the relative mean speed of A and B

$$\langle v_{12} \rangle = (8 k_B T / \pi \mu)^{1/2} \tag{17.4}$$

where μ is the reduced mass, the number densities N_1 and N_2, and σ is the collision cross-section

$$\sigma = \pi d^2 \tag{17.5}$$

defined in terms of the collision diameter d

$$d = \tfrac{1}{2}(d_1 + d_2) \tag{17.6}$$

for molecules with hard sphere diameters d_1 and d_2. The collision density Z_{12} is an upper limit to the rate of any reaction that requires the collision of A with B.

17.2.2 The energy requirement

The probability that a collision leads to a reaction depends on the energy of the collision. In Chapter 16 we used this premise to derive the Arrhenius equation. By assuming that the cross-section is zero if the collision energy is below the activation energy ε_a (on a per molecule basis, or E_a on a per mole basis), we arrive at an expression for the rate of formation of products by multiplication of the collision density by the fraction of collisions with an energy above the activation energy

$$R_{12} = Z_{12}\exp(-E_a/RT) = N_1 N_2 \sigma\langle v_{12}\rangle \exp(-E_a/RT). \tag{17.7}$$

We are assuming that the products are equilibrated to a Boltzmann distribution described by the absolute temperature T.

Equation (17.7) is in units of molecules per unit volume per unit time. To convert this equation to molar units, we multiply by Avogadro's constant N_A,

$$R_{12} = N_1 N_2 N_A \sigma\langle v_{12}\rangle \exp(-E_a/RT). \tag{17.8}$$

The rate is now expressed in the form

$$R_{12} = N_1 N_2 k. \tag{17.9}$$

in which the rate constant

$$k = N_A \sigma\langle v_{12}\rangle \exp(-E_a/RT) \tag{17.10}$$

is written the form of the Arrhenius equation with the collision frequency factor (pre-exponential factor)

$$A = N_A \sigma\langle v_{12}\rangle. \tag{17.11}$$

This has a weak temperature dependence compared to the exponential term, consistent with our previous discussion of the Arrhenius equation.

17.2.3 The steric requirement

Calculations of the pre-exponential factor A according to Eq. (17.11) are often good for simple reactions, but become increasingly less accurate for reactions involving more complex molecules. This suggests that the geometry of the collision – and not just its energy – is important for reactivity. We introduce the *steric factor, P*, to account for the difference,

$$k = P N_A \sigma\langle v_{12}\rangle \exp(-E_a/RT). \tag{17.12}$$

where P is usually (much) smaller than 1. This is to be expected for hard spheres for which a fraction less than one of all possible orientations of the spheres during collision leads to reaction.

An exception here is a reaction that occurs via what is known as a *harpooning* mechanism. Two classes of reactive system in which this occurs are collisions between an alkali metal and a halogen, such as $K + I_2 \rightarrow KI + I$, or the reaction of an electronically excited rare gas with a halogen to form an *exciplex*, such as $Ar^* + F_2 \rightarrow ArF + F$. An exciplex is a molecule that only exists in an excited electronic state, that is, its ground state is dissociative. In these reactions the cross-section is considerably larger than the collision cross-section, $P > 1$. The reason is that electron transfer between the collision partners with vastly different electronegativities occurs prior to collision at the hard sphere impact parameter. Electron transfer leads to the formation of an ion and a counterion that attract each other at long range. The K atom in the first reaction throws out an electron, much like a harpoon. The electrostatic attraction then acts like the rope that is used to reel in the K^+ ion toward the I_2^- ion, causing the steric factor to be much larger than 1. Exciplexes such as ArF, KrF and

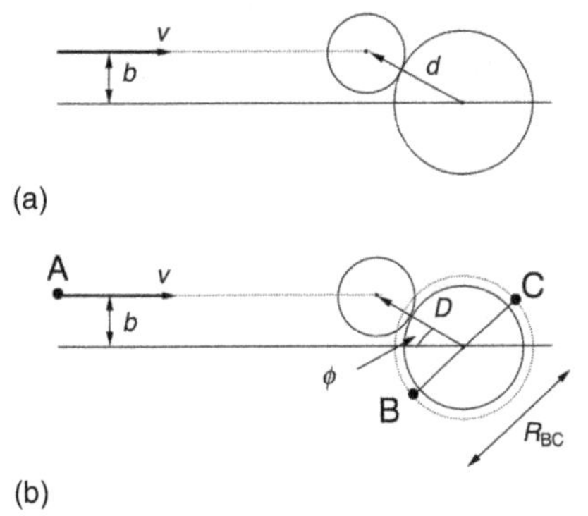

Figure 17.2 The reaction of A + BC. (a) Defining the parameters that characterize a hard-sphere collision. The impact parameter b is the separation between two parallel lines drawn through the centers of mass of the two spheres (the distance of closest approach along the line-of-centers that would result if the particle trajectories were undeflected by the collision), d is the minimum distance attained when one hard sphere meets the other, and v is the relative velocity directed along the line of centers. (b) To calculate an improved value of the steric parameter for reaction the spheres must reach the point at which *chemical interactions are significant*. This distance is D, which is at a distance of closer approach than the hard-sphere collision diameter d; that is, $D < d$. The angle ϕ is the angle between the velocity vector and the internuclear axis of BC, which has length R_{BC}. The circle with a dashed boundary corresponds to the hard-sphere radius of BC.

XeCl are important for the creation of excimer lasers – the dissociative nature of the ground state ensuring a population inversion since the population cannot build up in the ground state.

A simple and intuitive model of how to estimate the steric factor P and improve upon simple collision theory was proposed by Ian W.M. Smith.[1] We introduce two improvements to simple collision theory: (i) taking account of the geometry of the collision; and (ii) allowing the criterion for the minimum energy for reaction to change with this collision geometry. The factors needed to apply the corrections are defined in Fig. 17.2, including the *impact parameter b*. The impact parameter is the shortest distance that would separate the collision partners along the line of the collision if there were no interactions. The relative velocity $\boldsymbol{v}$ of the collision is defined in terms of the velocities of the collision partners $\boldsymbol{v}_1$ and $\boldsymbol{v}_2$ as $\boldsymbol{v} = \boldsymbol{v}_1 - \boldsymbol{v}_2$. Note that $\boldsymbol{v}$ must be positive in order for a collision to occur.

ε_a is identified as the minimum energy for reaction in a collinear geometry. This corresponds to the minimum activation energy from the PES. The B—C bond has length R_{BC}. The angle ϕ between the B—C bond that has to be broken and the line joining the centers of mass of the reactants at impact is assumed to be the geometric property that determines the critical energy for reaction ε_c. The bond that forms is the A—B bond, which has an equilibrium bond length R_{AB}.

Molecular interactions strong enough to initiate chemical reactivity inevitably start at distances below the hard-sphere collision diameter d. Thus, we introduce a *critical separation* $D < d$, which is the distance where chemical interactions are significant. This distance can be found from the position of the transition state on the PES. For the $H + H_2$ reaction, $d =$ 3.5 Å but D is only 1.35 Å. The critical separation is taken as the distance between the centers of mass at the transition state in the collinear geometry

$$D = R^*_{AB} + aR^*_{BC} \tag{17.13}$$

where $a = m_C/(m_B + m_C)$ and has a value of 0.5 when BC is a homonuclear diatomic. The * in the distances indicates that they are the values in the activated complex rather than the equilibrium values.

The most important aspect of the model is that the criterion for reaction to occur is not when the energy of the collision exceeds the minimum barrier height ε_a. Instead, the criterion for reaction is that the component of energy transferred along the line of centers of the collision complex must exceed the minimum barrier height. The critical energy for reaction ε_c is defined in terms of the collision parameters according to

$$\varepsilon_c = \varepsilon(1 - b^2/D^2) \geq \varepsilon_a + 2\varepsilon'(1 - \cos\phi). \tag{17.14}$$

The energy parameter ε' is related to characteristics of the potential energy surface,

$$\varepsilon' = \left(k_{bend} R^*_{AB} R^*_{BC}\right)\left(D/R^*_{AB}\right)^2 \tag{17.15}$$

where k_{bend} is the force constant of bending in the activated complex. The force constant tells us whether the bending motion is floppy (small k_{bend}) or rigid (large k_{bend}). The more rigid the transition state species, the greater the critical energy for reaction. The distances are in the activated complex. The rate constant is then

$$k = N_A(k_B T/2\varepsilon')\, \pi D^2 \langle v_{12}\rangle \exp(-\varepsilon_a/k_B T). \tag{17.16}$$

The exact values of the distances and the force constant will not be known unless we have available a high-quality potential energy surface (PES) from quantum mechanical calculations. Nonetheless, the distances and the force constant can easily be estimated for approximate values. This provides not only a rational estimate of the steric factor, but also a refined conception of the critical energy of reaction.

The critical energy of reaction is not simply the energy of collision; rather, it is the energy transmitted along the line of centers when the molecules begin to interact chemically that is determinative for reaction.

17.3 Activation energy and the transition state

We turn now to a more quantitative definition of the activation energy in terms of the potential energy surface that governs the interaction between atoms in the reactive system. We have seen in Chapter 16 that a single PES describes the reaction, and that microscopic reversibility and detailed balancing can be used to connect the reactants and products through the activated complex in the transition state. The concept of a PES was developed by Henry Eyring and Michael Polanyi.[2] Consider the reversible bimolecular elementary reaction

$$A + B \rightleftarrows Y + Z \tag{17.17}$$

The reaction proceeds from the reactants through a transition state (TS) region of the potential energy surface to the products. The transition state is well defined for a reaction with a positive activation energy. It is at a col in the potential energy surface, which is located at a maximum in the lowest energy path that links the reactants to the products. A look at Figs 17.3 and 17.4 allows one to quickly identify where the transition state is located. The chemical species in the transition state is the activated complex.

When the energy of a molecule is equal to the energy of the potential energy surface of Fig. 17.3, all of the molecule's energy is equal to potential energy. This can only happen in a classical system at $T = 0$ K. At any finite temperature, a molecule in the system has a kinetic energy that corresponds on average to the mean kinetic energy of the ensemble of molecules at that temperature. The mean energy of the system lies slightly above the PES. At $T = 0$ K, the difference between the minimum energy of the reactants and the maximum at the transition state, which corresponds to the minimum energy of the activated complex, is equal to the classical barrier height $E^{c}_{a,0}$. Quantum mechanically, the

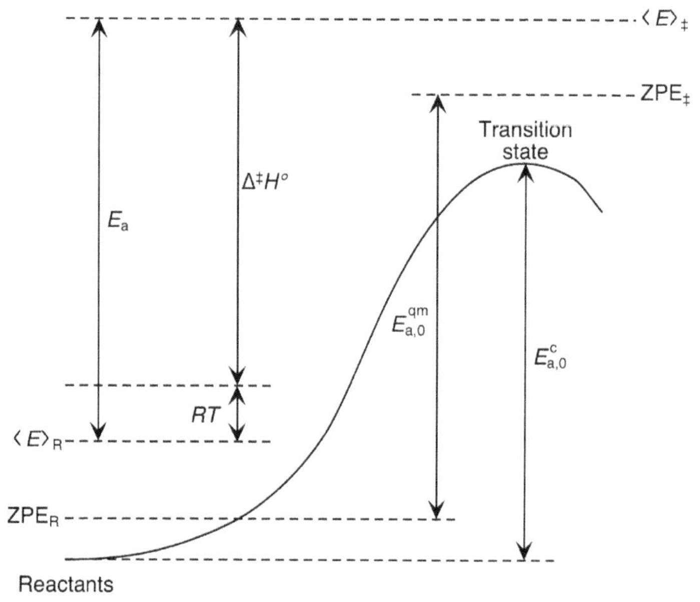

Figure 17.3 The classical barrier height, $E^{c}_{a,0}$, is the energy difference from the bottom of the well to the top of the activation barrier. The quantum mechanical barrier, $E^{qm}_{a,0}$, is a similar difference defined between reactant and transition state zero-point energy levels ZPE_R and $ZPE_{\ddagger}$, respectively. E_a is defined as the difference between the mean energy of the reactants $\langle E\rangle_R$ and the mean energy of the molecules in the transition state $\langle E\rangle_{\ddagger}$. The standard enthalpy of activation $\Delta^{\ddagger}H^{\circ}$ is also shown. Reproduced with permission from Kolasinski, K.W. (2012) *Surface Science: Foundations of Catalysis and Nanoscience*, 3rd edition. John Wiley & Sons, Chichester.

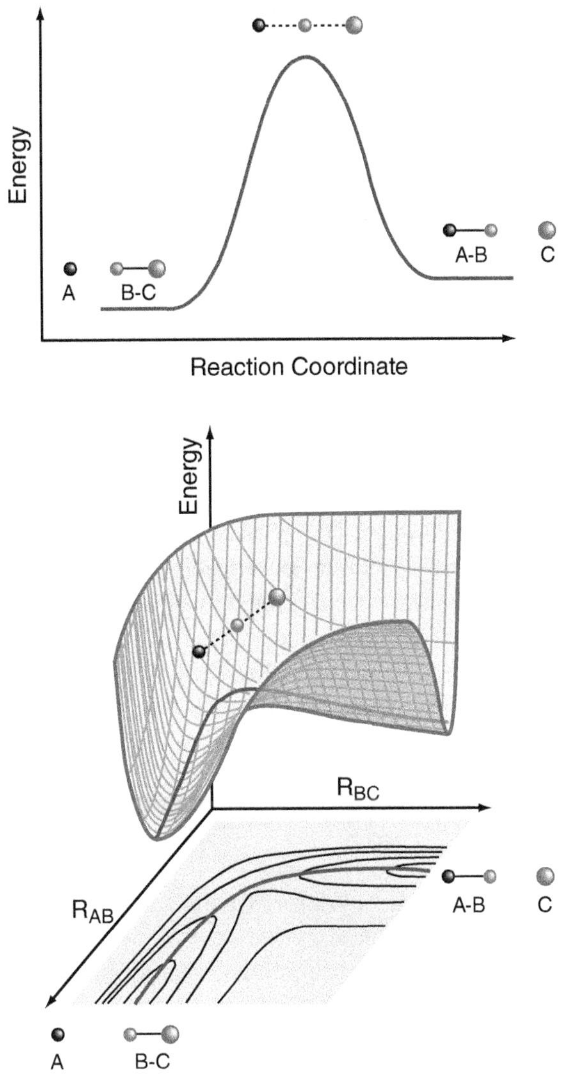

Figure 17.4 A multidimensional potential energy surface (PES, middle panel) is often represented by a simple one-dimensional profile along the path of lowest energy (upper panel). More information is provided by a contour plot (bottom panel), which includes information about the energy away from the lowest energy path along the reaction coordinate. Provided by Fleming Crim and reproduced from Crim, F.F. (2007) *Science*, **317**, 1707, with permission from AAAS.

molecules possess zero-point energy (ZPE) and the minimum energy that they can possess lies slightly above the PES, even at $T = 0$ K. If the reactant molecules and the activated complex had exactly the same degrees of freedom, the thermal behavior of the reactants and activated complex would be exactly the same. The activation energy would be independent of temperature and equal to the classical barrier height. This is the 90% correct answer for the activation energy: it is equal to the minimum energy required for a reaction. This answer is 100% correct at $T = 0$ K, which unfortunately is a temperature at which no reactivity occurs. The 100% correct answer requires a little more subtlety.

As shown by Fowler and Guggenheim,[3] the activation energy is given by the difference between the mean energy of the reactants $\langle E \rangle_R$ and the mean energy of the molecules in the transition state $\langle E \rangle_\ddagger$

$$E_a = \langle E \rangle_R - \langle E \rangle_\ddagger \tag{17.18}$$

Since both $\langle E \rangle_R$ and $\langle E \rangle_\ddagger$ are temperature-dependent, E_a is, in principle, also temperature-dependent. As can be seen in Fig. 17.3, the classical and actual activation energies are identical at $T = 0$ K. At any other temperature, E_a and $E^C_{a,0}$ are slightly different.

To account for the expected temperature dependence, it is useful to introduce a more general mathematical definition of the activation energy

$$E_a = -R\frac{d \ln k}{d(1/T)} = RT^2\frac{d \ln k}{dT}. \tag{17.19}$$

Frequently, it is found experimentally that Eq. (17.19) obeys the form

$$E_a = E_{a,0}^{qm} + mRT, \tag{17.20}$$

which is exactly what we derived in Eq. (16.48). The zero point energy of a molecule depends on its vibrational structure. The slight difference of zero-point energies between the reactants and products causes a slight difference between $E_{a,0}^{qm}$ and $E_{a,0}^{c}$.

Most reactions exhibit a positive activation energy. Some reactions – that is, radical recombination and non-activated adsorption – do not exhibit an activation barrier, so that $E_a = 0$. Some reactions, particularly if they proceed through a stable intermediate, have $E_a < 0$. For a simple reaction such as the dissociative adsorption of H_2 on a metal surface, or the combination of atoms to form a diatomic molecule,[4] the reaction coordinate is simply the distance between the two atoms. For a more complex reaction system, the PES corresponds to a higher dimensional surface (or hypersurface). A diagram of the type shown in Fig. 17.3 is a representation of the minimum energy path through this PES leading from reactants through the activated complex to the products. Such a one-dimensional curve is often called a Lennard–Jones diagram.[5] The relationship of this minimum energy pathway to a contour plot of a PES for a bimolecular reaction is shown in Fig. 17.4.

The trajectory of a reactive collision can be thought of in terms of 'bobsledding' along this PES. If the collision partners could maintain the lowest energy configuration at zero kinetic energy, but still somehow manage to move forward, they would trace out the minimum energy pathway represented in the Lennard–Jones diagram at the top of the figure. Putting enough energy into the system allows the collision partners not only to overcome the activation barrier but also to ride up the walls and round the corner into the product valley. The shape of the PES causes different forms of energy (rotational, vibration and translational) to have different efficacies for promoting reaction. This point is discussed in Section 17.11.

17.4 Transition-state theory (TST)

A general theoretical foundation for the calculation of rate constants was developed by Eyring, Evans and Polanyi[6] and is called *transition-state theory*. Assume that a rapid equilibrium exists between the reactants A and B and the activated complex $AB^{\ddagger}$ before it decays into products P. Rapid equilibrium demands that the first step of the reaction is reversible. The second step is irreversible. This leads to the following set of equations, rate constants and equilibrium constant that describe the system,

$$A + B \underset{k_{-1}}{\overset{k_1}{\rightleftharpoons}} AB^{\ddagger} \xrightarrow{k_2} P \tag{17.21}$$

$$K^{\ddagger} = \frac{[AB^{\ddagger}]c^{\circ}}{[A][B]} \tag{17.22}$$

c^0 is a standard concentration used to keep the equilibrium constant dimensionless. The rate of product formation is found from the unimolecular decay of $AB^{\ddagger}$

$$AB^{\ddagger} \rightarrow P \qquad R = \frac{d[P]}{dt} = k_2[AB^{\ddagger}] \tag{17.23}$$

This expression is not experimentally accessible since we cannot measure the concentration of the activated complex $[AB^{\ddagger}]$. We solve the equilibrium constant expression in Eq. (17.22) for $[AB^{\ddagger}]$

$$[AB^{\ddagger}] = \frac{[A][B]}{c^{\circ}}K^{\ddagger}, \tag{17.24}$$

substitute this into Eq. (17.23), and the rate express becomes

$$R = \frac{d[P]}{dt} = \frac{k_2 K^{\ddagger}}{c^{\circ}}[A][B] \tag{17.25}$$

To derive a value for the rate constant k_2, consider the following. Think of the activated complex as an unstable complex that can dissociate into either the product P or back into the reactants A and B. The dissociation occurs by breaking a bond that has a vibrational frequency ν. Each time the bond extends it has some probability to form a product. The probability

Table 17.1 Dependence of the expressions for activation energy and pre-exponential factor on molecularity and phase for elementary reactions.

Molecularity	Solutions E_a	Solutions A	Gas phase E_a	Gas phase A
Uni	$\Delta^{\ddagger}H + RT$	$e\frac{k_B T}{h}e^{\Delta^{\ddagger}S/R}$	$\Delta^{\ddagger}H + RT$	$e\frac{k_B T}{h}e^{\Delta^{\ddagger}S/R}$
Bi	$\Delta^{\ddagger}H + RT$	$\frac{e}{c^\circ}\frac{k_B T}{h}e^{\Delta^{\ddagger}S/R}$	$\Delta^{\ddagger}H + 2RT$	$\frac{e^2}{c^\circ}\frac{k_B T}{h}e^{\Delta^{\ddagger}S/R}$
Ter			$\Delta^{\ddagger}H + 3RT$	$\frac{e^3}{(c^\circ)^2}\frac{k_B T}{h}e^{\Delta^{\ddagger}S/R}$

of the activated complex forming the product is called the *transmission coefficient* κ. Therefore, the rate constant and rate law are

$$k_2 = \kappa \upsilon \tag{17.26}$$

$$R = \frac{\kappa \nu K^{\ddagger}}{c^\circ}[A][B] = k[A][B] \tag{17.27}$$

The transmission coefficient is often close to unity, and $\kappa = 1$ is generally assumed as a first approximation. There are exceptions. We will see below than an important exception is adsorption – particularly dissociative adsorption – at surfaces.

The change in Gibbs energy for entering the transition state is related to the equilibrium constant by

$$\Delta^{\ddagger}G^\circ_m = -RT \ln K^{\ddagger}. \tag{17.28}$$

We can relate the rate constant to the *Gibbs energy of activation* $\Delta^{\ddagger}G^\circ_m$. With a bit of statistical mechanics (which we do not need to go into here, see Laidler or Houston in Further Reading) we can show that

$$k = \frac{\kappa k_B T}{hc^\circ}\exp\left(-\Delta^{\ddagger}G^\circ_m/RT\right) \tag{17.29}$$

Further analysis using $\Delta^{\ddagger}G = \Delta^{\ddagger}H - T\Delta^{\ddagger}S$ shows us that the pre-exponential factor is related to the *entropy of activation* and the activation energy is related to the *enthalpy of activation*

$$k = \frac{\kappa k_B T}{hc^\circ}\exp(\Delta^{\ddagger}S/R)\exp(-\Delta^{\ddagger}H/RT) \tag{17.30}$$

This is the *Eyring equation,* from which we see that the pre-exponential factor is related to the entropy of activation $A \propto \exp(\Delta^{\ddagger}S^\circ_m)$ and the activation energy is related to the enthalpy of activation $E_a \propto \Delta^{\ddagger}H^\circ_m$. The exact dependence of A and E_a on $\Delta^{\ddagger}S^\circ_m$ and $\Delta^{\ddagger}H^\circ_m$ depends on the molecularity and the phase of the reaction, as stated explicitly in Table 17.1.

The activation parameters $\Delta^{\ddagger}S^\circ_m$, $\Delta^{\ddagger}H^\circ_m$ and $\Delta^{\ddagger}G^\circ_m$ are given in Table 17.2 for selected first-order reactions, and in Table 17.3 for selected second-order reactions. Reviewing these results we see that the behavior of first-order reactions

Table 17.2 Activation parameters derived from the Arrhenius parameters for selected first-order reactions in the gas phase. Values calculated from the values in Table 16.2. No entry is made if the temperature lies outside the valid range of the Arrhenius parameters.

Reaction	$\Delta^{\ddagger}S^\circ_m$ / J K^{-1} mol^{-1}	$\Delta^{\ddagger}H^\circ_m$ / kJ mol^{-1}	$\Delta^{\ddagger}G^\circ_m$ / kJ mol^{-1}	$\Delta^{\ddagger}S^\circ_m$ / J K^{-1} mol^{-1}	$\Delta^{\ddagger}H^\circ_m$ / kJ mol^{-1}	$\Delta^{\ddagger}G^\circ_m$ / kJ mol^{-1}
		$T = 380$ K			$T = 1200$ K	
$CH_3CHO \rightarrow CH_3 + HCO$				33.45	318.4	278.3
$C_2H_6 \rightarrow 2CH_3$	58.67	364.4	342.1	49.11	357.6	298.6
$O_3 \rightarrow O_2 + O$	−8.66	99.3	102.6			
$HNO_3 \rightarrow OH + NO_2$	32.70	196.4	183.9	23.14	189.6	161.8
$CH_3OH \rightarrow CH_3 + OH$	66.62	383.6	358.3	57.06	376.8	308.3
$N_2O \rightarrow N_2 + O$				18.24	224.6	202.7
$N_2O_5 \rightarrow NO_3 + NO_2$	35.82	231.4	217.8			
$C_2H_5 \rightarrow C_2H_4 + H$	2.19	167.1	166.3	−7.38	160.3	169.1
$CH_3NC \rightarrow CH_3CN$	4.72	157.9	156.1			
c-propane → propene				12.06	251.2	236.7
c-butane → $2C_2H_4$				29.28	243.5	208.4

Table 17.3 Activation parameters derived from the Arrhenius parameters for selected second-order reactions in the gas phase at $T = 300$ K. Values for A and E_a adapted from Houston, P.L. (2001) *Chemical Kinetics and Reaction Dynamics*. Dover Publications, Mineola, NY.

Reaction	A / m^3 mol^{-1} s^{-1}	E_a / kJ mol^{-1}	$\Delta^{\ddagger}S^{\circ}_{m}$ / J K^{-1} mol^{-1}	$\Delta^{\ddagger}H^{\circ}_{m}$ / kJ mol^{-1}	$\Delta^{\ddagger}G^{\circ}_{m}$ / kJ mol^{-1}
$CH_3 + H_2 \rightarrow CH_4 + H$	6.7×10^6	52	−100.23	47.0	77.0
$CH_3 + O_2 \rightarrow CH_3O + O$	2.4×10^7	121	−89.62	115.7	142.6
$F + H_2 \rightarrow HF + H$	8.4×10^7	4.16	−79.20	−0.8	22.9
$H + H_2O \rightarrow OH + H_2$	9.7×10^7	86	−78.01	81.0	104.4
$H + CH_4 \rightarrow CH_3 + H_2$	1.9×10^8	50.8	−72.42	45.8	67.5
$H + O_2 \rightarrow OH + O$	1.4×10^8	66.5	−74.96	61.5	84.0
$H + CO_2 \rightarrow CO + OH$	1.5×10^8	111	−74.38	105.6	127.9
$H + O_3 \rightarrow O_2 + OH$	8.4×10^7	3.99	−79.20	−1.0	22.8
$O + H_2 \rightarrow OH + H$	5.2×10^7	41.6	−83.19	36.6	61.5
$O + CH_4 \rightarrow OH + CH_3$	1.0×10^8	42.3	−77.75	37.3	60.7
$O + NO_2 \rightarrow NO + O_2$	3.9×10^6	−0.998	−104.73	−6.0	25.4
$O + C_2H_2 \rightarrow$ products	3.4×10^7	14.6	−86.72	9.6	35.7
$O + O_3 \rightarrow 2O_2$	7.5×10^6	17.8	−99.29	12.8	42.6
$O + C_2H_4 \rightarrow$ products	9.5×10^6	7.87	−97.32	2.9	32.1
$OH + H_2 \rightarrow H + H_2O$	4.6×10^6	17.5	−103.35	12.5	43.5
$OH + CH_4 \rightarrow H_2O + CH_3$	2.2×10^6	15.1	−109.49	10.1	43.0
$OH + O \rightarrow O_2 + H$	1.4×10^7	−0.915	−94.10	−5.9	22.3

is decidedly different than that of second-order reactions. For first-order reactions, the activation enthalpy is always positive and is often the dominant factor in determining $\Delta^{\ddagger}G^{\circ}_{m}$, though the effects of entropy cannot be ignored. Most unimolecular elementary reactions do not exhibit the expected behavior that $\Delta^{\ddagger}S^{\circ}_{m}$ is unfavorable (negative). For second-order reactions, on the other hand, $\Delta^{\ddagger}S^{\circ}_{m}$ is in the range of −110 to −70 J K^{-1} mol^{-1} in the cases given. The entropy factor is unfavorable as expected, and plays a much more important role in determining the value of $\Delta^{\ddagger}G^{\circ}_{m}$. The enthalpy of activation ranges widely from negative to positive values (−6 to 116 kJ mol^{-1}).

Note further an interesting aspect of the data in Table 17.2. Entries in the table are only made when constant values of A and E_a found in Table 16.6 are valid at the chosen temperature. Constant values of A and E_a are able to model the reaction rate of a given reaction over an extended temperature range, even though neither $\Delta^{\ddagger}S^{\circ}_{m}$ nor $\Delta^{\ddagger}H^{\circ}_{m}$ is constant. The reason for this is that $\Delta^{\ddagger}S^{\circ}_{m}$ and $\Delta^{\ddagger}H^{\circ}_{m}$ change sympathetically with temperature. They both decrease with increasing temperature. Whereas, a decrease in $\Delta^{\ddagger}S^{\circ}_{m}$ decreases the rate, a decrease in $\Delta^{\ddagger}H^{\circ}_{m}$ increases the rate. The effects counterbalance one another and allow constant values of A and E_a to approximate the rate quite accurately over extended ranges.

17.4.1 Directed practice

Confirm the values of $\Delta^{\ddagger}S^{\circ}_{m}$, $\Delta^{\ddagger}H^{\circ}_{m}$ and $\Delta^{\ddagger}G^{\circ}_{m}$ in Table 17.2 for the unimolecular dissociation of HNO_3.

17.5 Composite reactions and mechanisms

Many reactions exhibit first-order or second-order kinetics. More often than not, this is not an indication that the reaction is a unimolecular or bimolecular reaction; rather, it is more likely a reflection of a well-defined rate-determining step that is unimolecular or bimolecular. Most reactions follow complicated, multistep mechanisms.

There are a number of clues that a reaction mechanism is composite rather than elementary, and these clues also help us to decipher the mechanism. A clear indication that a reaction is complex is the presence of an *intermediate*. An intermediate does not appear in the balanced stoichiometric equation for the reaction. The intermediate usually forms after a short induction period, with its concentration going through a maximum before disappearing completely as the reaction goes to completion. The presence of an intermediate indicates that there must be at least two steps in the reaction, because one step is required to generate the intermediate and a separate step is required to consume it.

If a reaction has an order other than zero, one, two, or three, it must be complex. In particular, noninteger orders indicate a complex, multistep reaction mechanism. A reaction order that changes as a function of temperature also indicates a multistep reaction. A noninteger order indicates that more than one elementary step is contributing to the reaction rate. If the rate-determining step were first-order at low temperature and second-order at high temperature, then in the intermediate temperature regime the order would be intermediate between first- and second-order. Curvature would also be noted in an Arrhenius plot of ln k versus T. A linear plot would be observed at low T, corresponding to the activation energy of the first rate-determining step, followed by a curved intermediate region, and ending in a linear region with a slope corresponding to the high-T rate-determining activation energy.

Recall the definition of a reaction mechanism.

A reaction mechanism is a complete sequence of elementary reactions (steps) that add together to give the stoichiometric equation of a reaction.

An individual step may have a stoichiometric number that is different from 1. For example, a step may have to occur more than once in a reaction to achieve the proper stoichiometry. The correct mechanism must give the correct mathematic form of the concentration versus time curves for every reactant, intermediate and product in the reaction. Always remember the asymmetry in the burden of proof: If a mechanism makes a prediction about the kinetics of a reaction and an accurate experiment disproves the prediction, the mechanism does not describe the reaction. It is the wrong mechanism. However, if a mechanism makes a prediction that is substantiated by an accurate experiment, this only proves that the mechanism is contained in the set of all possible mechanism that *might* describe the reaction. Proof requires many positive reinforcements. This is the difference between necessary and sufficient conditions.

17.5.1 Types of steps

There are a number of different ways elementary reactions can be combined into a reaction mechanism and we define them here. A mechanism may consist of a series of consecutive steps that occur one after the other. A simple two-step example is as follows.

Step 1: $\quad \mathrm{A} \xrightarrow{k_\mathrm{A}} \mathrm{I}$

Step 2: $\quad \mathrm{I} \xrightarrow{k_\mathrm{I}} \mathrm{P}$

Overall: $\quad \mathrm{A} \longrightarrow \mathrm{P}$

The Arrhenius parameters of the rate constant are the pre-exponential factor and the activation energy. In any reaction composed of a series of consecutive steps, the kinetics of one step may be of singular importance and is known as the rate-determining step (RDS). We defined the RDS in Chapter 16. The concentration versus time curves of A, I and P for a reaction with the simple two-step mechanism defined above is shown in Fig. 17.5. Notice how the concentration of the intermediate progresses through a maximum.

The kinetics of this mechanism lead to the following rate equations. The reactant A is involved in only one step. Therefore, the rate of disappearance of A is related to the rate of the first step alone

$$\frac{\mathrm{d[A]}}{\mathrm{d}t} = -k_\mathrm{A}[\mathrm{A}]. \tag{17.31}$$

The intermediate is produced in the first step and consumed in the second. The net rate of change in the intermediate concentration is thus the difference between these two steps,

$$\frac{\mathrm{d[I]}}{\mathrm{d}t} = k_\mathrm{A}[\mathrm{A}] - k_\mathrm{I}[\mathrm{I}]. \tag{17.32}$$

The product is formed by the second step, which is first-order in the intermediate,

$$\frac{\mathrm{d[P]}}{\mathrm{d}t} = k_\mathrm{I}[\mathrm{I}]. \tag{17.33}$$

Parallel or simultaneous steps may be involved in the mechanism, as in the following two-step mechanism.

Step 1: $\quad \mathrm{A} \xrightarrow{k_\mathrm{B}} \mathrm{B}$

Step 2: $\quad \mathrm{A} \xrightarrow{k_\mathrm{C}} \mathrm{C}$

Overall: $\quad 2\mathrm{A} \longrightarrow \mathrm{B} + \mathrm{C}$

The rate of disappearance of A must reflect the sum of two steps,

$$\frac{\mathrm{d[A]}}{\mathrm{d}t} = -k_\mathrm{B}[\mathrm{A}] - k_\mathrm{C}[\mathrm{A}] = -(k_\mathrm{B} + k_\mathrm{C})[\mathrm{A}], \tag{17.34}$$

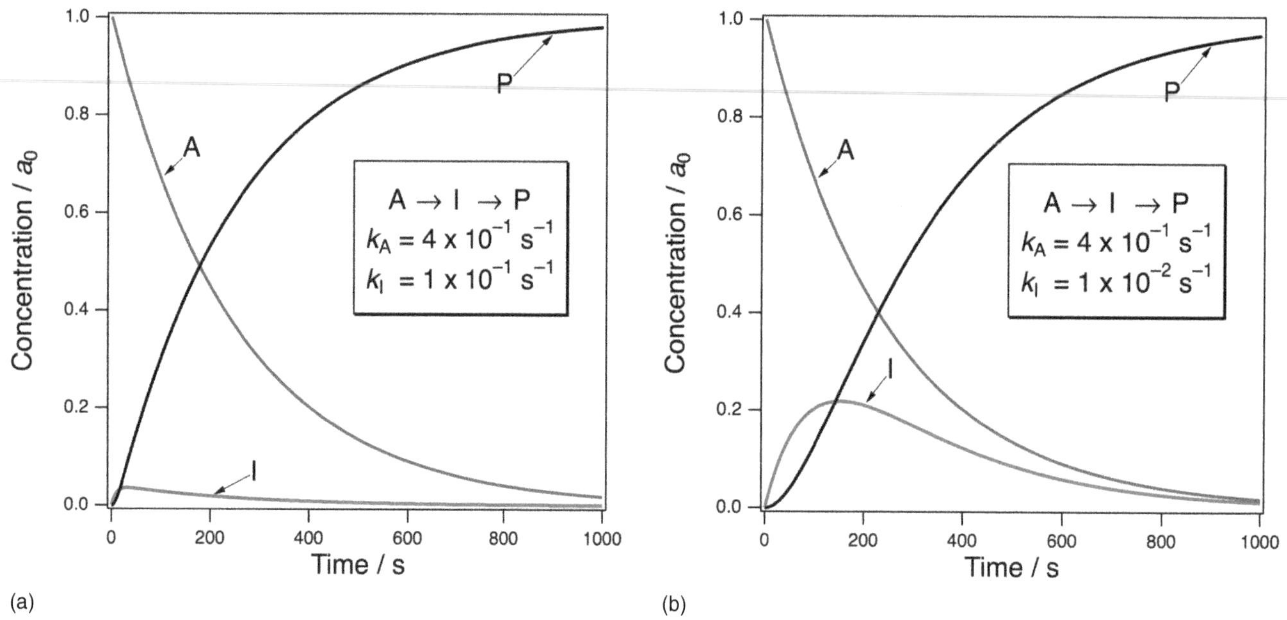

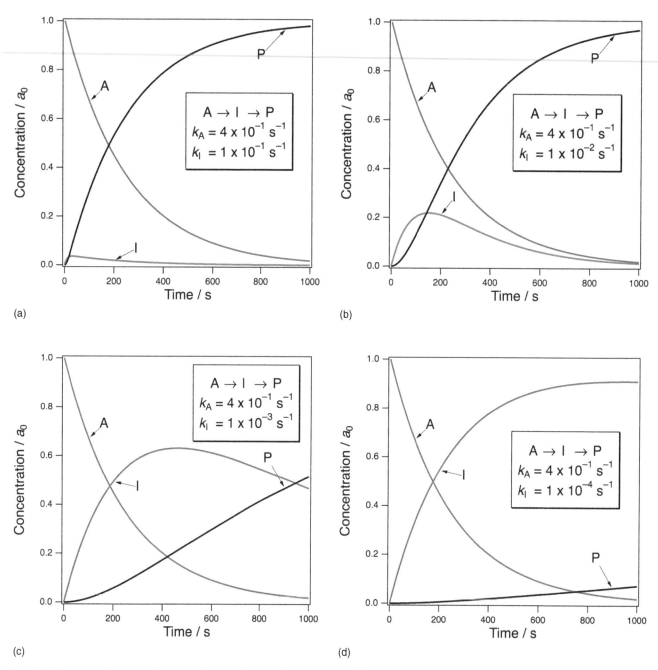

Figure 17.5 Concentration versus time curves for a reaction mechanism composed of two consecutive unimolecular elementary steps in series. In panels (a) through (d) the rate constant of the first step remains constant, but the rate constant of the second step increases by one order of magnitude in each step. The concentration profile of the reactant A is unaffected by this change, but those of the intermediate I and the product P change dramatically in response. All concentrations are normalized to the initial concentration of the reactant a_0.

while the formation of B and C both follow kinetics that is first-order in A,

$$\frac{d[B]}{dt} = k_B[A] \tag{17.35}$$

$$\frac{d[C]}{dt} = k_C[A]. \tag{17.36}$$

Though they exhibit the same partial order with respect to A, it is the competition between these steps that determines the individual rates of formation of B and C. In such a simple example as this, the competition is reflected in the magnitude of the two rate constants, k_B versus k_C.

Some reactions exhibit *feedback*. Feedback is when the presence of a step affects the rate of a previous step. Feedback can be either positive, as in the case of catalysis of a step by a product of a later step, or negative, as is the case for inhibition of a step by the product of a later reaction. Here is an example of autocatalysis in which the intermediate C catalyzes the first step of the reaction.

1.	$A \xrightarrow[C]{} B$
2.	$B \longrightarrow C$
3.	$A + C \longrightarrow D$
Overall	$2A \longrightarrow D$

Many reactions exhibit reversible steps, also called *opposing steps*. For any reaction composed of multiple steps, the rate of change of concentration must take account of the rates of all steps; that is, the net rate must be calculated. For a reversible step, such as the bimolecular reaction,

$$A + B \underset{k_{-1}}{\overset{k_1}{\rightleftharpoons}} X + Y,$$

the net rate of that step is the forward minus the reverse reaction rate

$$\frac{d[X]}{dt} = k_1[A][B] - k_{-1}[X][Y] \tag{17.37}$$

17.5.2 Steady-state approximation (SSA)

The analysis of the kinetics of reactive systems is often simplified by studying them under steady-state conditions. Under steady-state conditions, a system with constant composition is observed. The system may be either closed or open. A closed system at steady state may well correspond to an equilibrium system; however, equilibrium is not a requirement for attaining a steady state.

Within the steady-state approximation (SSA) at least one intermediate exists at a low concentration. The central condition of the SSA is that the concentration of each and every intermediate I is constant,

$$\frac{d[I]}{dt} = 0. \tag{17.38}$$

This allows for significant simplifications in the mathematical treatment of complex reactions.

17.5.2.1 Example

Consider a two-step reaction of consecutive steps involving a single intermediate,

$$A \xrightarrow{k_A} I \xrightarrow{k_I} P.$$

The net reaction rates in terms of the concentrations of A, P, and I are

$$\frac{d[A]}{dt} = -k_A[A], \quad \frac{d[P]}{dt} = k_I[I]$$

$$\frac{d[I]}{dt} = k_A[A] - k_I[I]$$

Now use the SSA to find the intermediate concentration,

$$\frac{d[I]}{dt} = 0 = k_A[A] - k_I[I] \Rightarrow [I] = \frac{k_A}{k_I}[A].$$

This allows us to substitute for [I] and obtain an expression for the rate of formation of the product in terms of the reactant concentration [A], rather than the potentially more difficult to determine concentration of the intermediate [I],

$$\frac{d[P]}{dt} = k_A[A].$$

In this system, the requirements for the SSA to apply (that I is a reactive intermediate with a small and effectively constant concentration) will only be fulfilled if the rate of step $A \rightarrow I$ is much slower than the step $I \rightarrow P$.

17.5.2.2 Example

The following elementary steps comprise a complete reaction mechanism (all species in the gas phase).

$$N_2O_5 \underset{k_{-1}}{\overset{k_1}{\rightleftharpoons}} NO_2 + NO_3$$

$$NO + NO_3 \xrightarrow{k_2} 2NO_2$$

(a) What is the overall reaction? (b) What are the intermediates? (c) Apply the steady-state approximation and find an expression for $[NO_3]$ in terms of the concentrations of N_2O_5, NO_2, and NO.

a The overall stoichiometric equation for the reaction is obtained by addition of the two elementary steps. Note that for a reversible reaction we add only the forward reaction,

$$N_2O_5 + NO \longrightarrow 3NO_2.$$

b NO_3 does not appear in the final stoichiometric equation but it does appear as a product in step 1 and is fully consumed in step 2.

c First apply the SSA to obtain an expression for the net rate of formation of the intermediate,

$$\frac{d[NO_3]}{dt} = k_1[N_2O_5] - k_{-1}[NO_2][NO_3] - k_2[NO][NO_3] = 0.$$

Then solve the resulting equation to obtain an expression for $[NO_3]$,

$$k_1[N_2O_5] = k_{-1}[NO_2][NO_3] + k_2[NO][NO_3]$$

$$[NO_3] = \frac{k_1[N_2O_5]}{k_{-1}[NO_2] + k_2[NO]}.$$

17.6 The rate of unimolecular reactions

The description of the rates of unimolecular reactions in the gas phase possesses several problems. How does a single molecule become sufficiently energized to transform from a relatively stable species to the product? Why is it observed that the order of a unimolecular reaction changes from the expected first-order kinetics at high pressure to second-order kinetics at low pressure? Why is the rate of a unimolecular reaction sensitive to the presence of an unreactive third body? The answers to these questions are found in the Lindeman[7]–Christiansen[8] mechanism. Consider a gas-phase reaction in which a species must be collisionally excited before reaction occurs – that is, a reactive intermediate A* must be formed and reacts according to the two-step mechanism:

$$A + A \underset{k_{-1}}{\overset{k_1}{\rightleftharpoons}} A + A^*. \tag{17.39}$$

$$A^* \xrightarrow{k_2} P. \tag{17.40}$$

The first step represents the reversible activation and deactivation of the reactant A to form the high-energy intermediate A*. The second step is the irreversible decay of the activated species into the product P. The equilibrium constant of the first step is heavily favored toward the reactant, that is,

$$K_1 = k_1/k_{-1} \ll 1. \tag{17.41}$$

One way to think of the first step is that it represents the filling out of the high-energy Boltzmann tail of a thermal distribution.

Now apply the steady-state approximation to A*. We write an expression for the net rate of change of [A*] and set it equal to zero,

$$\frac{d[A^*]}{dt} = k_1[A]^2 - k_{-1}[A][A^*] - k_2[A^*] = 0. \tag{17.42}$$

Rearrange to solve for [A*]

$$[A^*] = \frac{k_1[A]^2}{k_{-1}[A] + k_2}. \tag{17.43}$$

Thus, the rate of product formation, which is given by the rate of the irreversible second step, is

$$\frac{d[P]}{dt} = k_2[A^*] = \frac{k_2 k_1[A]^2}{k_{-1}[A] + k_2} \tag{17.44}$$

There are two limiting cases for this equation. First is the limit of high pressure; that is [A] is sufficiently large that $k_{-1}[A] \gg k_2$, so we can neglect the k_2 in the denominator. This leads to the first-order equation

$$\frac{d[P]}{dt} = \frac{k_2 k_1[A]^2}{k_{-1}[A]} = k'[A]. \tag{17.45}$$

where the effective first-order rate constant k' is

$$k' = \frac{k_2 k_1}{k_{-1}}. \tag{17.46}$$

At high gas pressure, the kinetics of unimolecular decay is predicted to be first order, which is confirmed by experiment.

In the low-pressure limit, that is the limit of small [A] such that $k_{-1}[A] \ll k_2$, we now drop the term involving [A] from the denominator. This yields the second-order equation

$$\frac{d[P]}{dt} = \frac{k_2 k_1[A]^2}{k_2} = k_1[A]^2. \tag{17.47}$$

The behavior of the Lindemann–Christiansen mechanism from second order at low pressure, to noninteger order at intermediate pressure, to first order at high pressure is shown in Fig. 17.6. In the high-pressure limit there exists a fast pre-equilibrium. In other words, the full Boltzmann distribution is maintained all the way to high energy. The concentration of A* is quite low but the conversion of A* into the product is slow. This allows the system to maintain the equilibrium concentration of A* while the rate of product formation is determined by the slow second step. The A* is constantly being formed and relaxed while waiting around to be transformed into the product.

In the low-pressure regime, the rate of forming A* is slow, and the system is unable to maintain the equilibrium concentration of A*. This means that the overall rate of production of the product is constrained by the low concentration

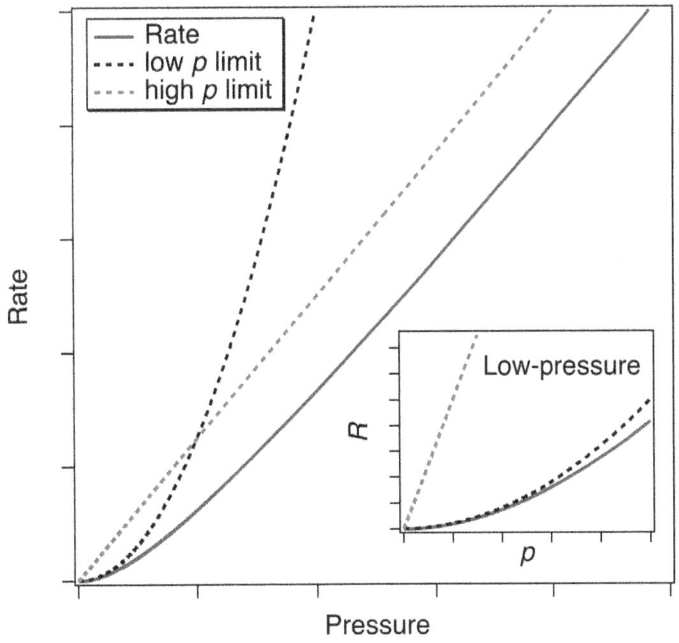

Figure 17.6 A plot of the rate versus pressure predicted for a reaction that follows the Lindemann–Christiansen mechanism. This shows the development of kinetics from the low-pressure, second-order regime to the high-pressure limit in which the rate has the same slope as the high-pressure limit.

of A^*. As soon as it is formed, it transforms rapidly (at least relative to the rate at which it is produced) into the product. The rate-determining step becomes the collision between two A molecules that leads to activation. This is a second-order process.

Note that a change in the kinetics of the reaction is observed as the pressure is changed. However, the mechanism of the reaction itself does not change. The change that is observed to affect the rate is the change of the rate-determining step.

The Lindemann–Christiansen mechanism makes a further prediction that is experimentally verifiable. If an unreactive collision partner M is added to the system, the rate of reaction should respond to the pressure of M. Consider the case where a dilute mixture of A in the majority species M is made. The activation is then caused by collisions of M with A. The kinetics are described by

$$A + M \underset{k_{-1}}{\overset{k_1}{\rightleftharpoons}} A^* + M \tag{17.48}$$

$$A^* \xrightarrow{k_2} P. \tag{17.49}$$

The rate of production of product is

$$\frac{d[P]}{dt} = \frac{k_2 k_1 [A][M]}{k_{-1}[M] + k_2} = k_{uni}[A] \tag{17.50}$$

where the effective unimolecular rate constant k_{uni} is

$$k_{uni} = \frac{k_2 k_1 [M]}{k_{-1}[M] + k_2}. \tag{17.51}$$

Thus

$$\frac{1}{k_{uni}} = \frac{k_{-1}}{k_2 k_1} + \left(\frac{1}{k_1}\right)\frac{1}{[M]} \tag{17.52}$$

and a plot of $1/k_{uni}$ versus $1/[M]$ has a slope $1/k_1$.

The Lindemann–Christiansen mechanism gets the dynamics correct. In particular, it suggests that the dynamics of activation lead to a change from second-order kinetics at low pressure to a high-pressure limit of first-order kinetics. The mechanism is also quantitatively correct in the high-pressure limit. There are, however, shortcomings in the quantitative nature of the model. To alleviate these quantitative failings at low pressure requires a theory that fully treats the statistical randomization of energy into available degrees of freedom in a manner that is fully compatible with transition state theory. This is the province of RRKM theory, named after its inventors Oscar K. Rice, Herman C. Ramsperger,[9] Louis S. Kassel[10] and Rudy A. Marcus.[11] More on RRKM theory can be found in Laidler in Further Reading. Nonstatistical energy partitioning and the related topic of coherent control of chemical reactivity are active areas of research. More on these topics can be found in Further Reading.

17.7 Desorption kinetics

An understanding of the kinetics of surface reactions is essential for an understanding of the kinetics of heterogeneous catalysis, growth and etching of semiconductors (important for integrated circuit and photovoltaic applications), and for the reactions of gases at low pressures for which the walls will dominate reactivity when the mean free path becomes long enough (e.g., comparable to the size of the container). The two most fundamental surface reactions are adsorption and desorption, which must occur in any catalytic or etching reaction. Growth reactions often also include both, though a counter example is molecular beam epitaxy of metals or semiconductors on other metals or semiconductors, a process in which evaporated atoms are deposited and incorporated into the surface of the substrate. Because adsorption and desorption are the two most fundamental surface reactions, we threat them in detail here. We will leave our discussion of multistep surface reactions until the next chapter.

Scrutinizing Fig. 17.7 we see that we need to differentiate several different cases. The kinetics will depend on whether we are considering chemisorption or physisorption, whether the process is nonactivated or activated, and whether we are considering adsorption or desorption. The kinetics of surface reactions has many similarities with kinetics in other areas of chemistry. However, it also has essential differences, the most important of which is that *reactions can only occur on*

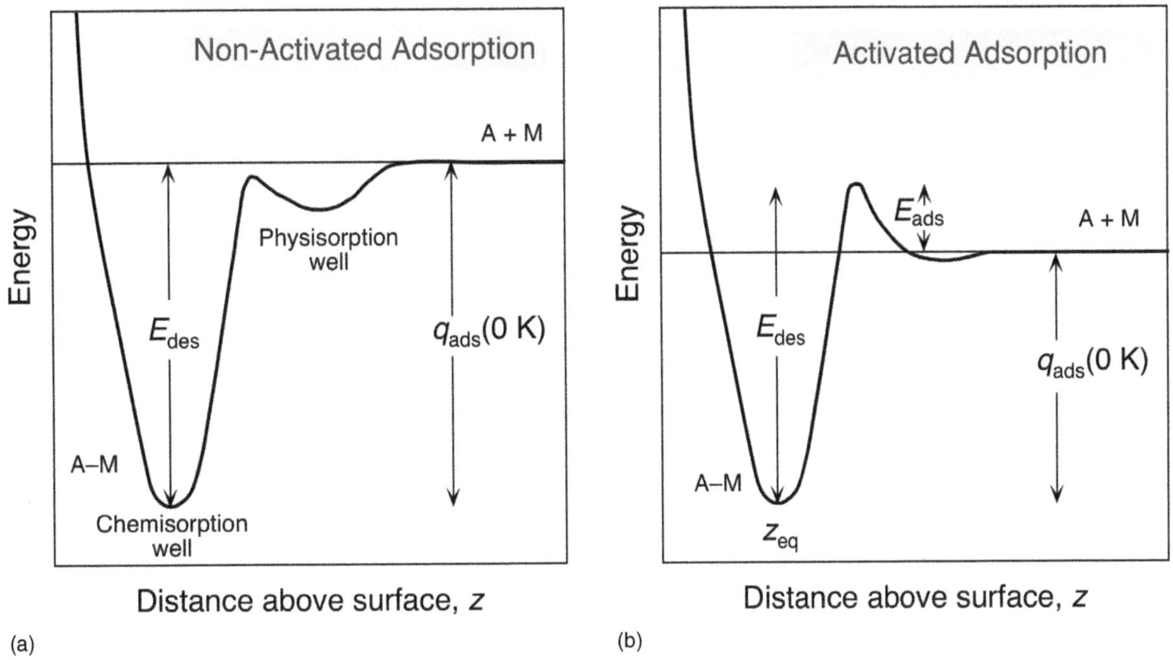

Figure 17.7 (a) Nonactivated adsorption. (b) Activated adsorption. Reproduced with permission from Kolasinski, K.W. (2012) *Surface Science: Foundations of Catalysis and Nanoscience*, 3rd edition. John Wiley & Sons, Chichester.

a limited number of sites. In Chapter 5 we introduced the concept of adsorption sites. The possibility of sites with different reactivities is another complicating factor. The kinetics of adsorption, in particular, is complicated by the mechanism by which molecules find sites. Therefore, we begin by treating the simpler case of desorption.

First, we have to define some new symbols, see Table 17.4, and how we account for sites. Adsorption sites may be either occupied or empty. We need to keep track of both full and empty sites and we do so in terms of *coverage*. Coverage is defined either as the absolute coverage σ, which is the number of sites or adsorbates per unit area, or fractional coverage θ, which is the number of sites or adsorbates divided by σ_0,

$$\theta = \sigma/\sigma_0. \tag{17.53}$$

The fractional coverage is often referred to in 'units' of monolayers (ML). A monolayer is the number of adsorbates per unit area that it takes to fill the surface with one monomolecular layer (a layer that is one adsorbate thick). It is on the order of 10^{19} m^{-2} for small molecules. When working with the liquid/solid interface, you may encounter the absolute coverage given in units of mol m^{-2}, in which case it is denoted by the symbol Γ.

Two different definitions of σ_0 appear in the literature. *The reader must always be attentive as to whether coverage is defined with respect to the number of sites or the number of surface atoms.* These two definitions are equivalent when the number of sites is equal to the number of surface atoms. However, there are instances in which the saturation coverage is, say, one adsorbate for every two surface atoms. A saturated layer of adsorbates then has a fractional coverage of 0.5 ML when defined with respect to the number of surface atoms, but 1.0 ML when defined with respect to sites. The definition in terms of surface atoms is more absolute, but the definition in terms of saturation leads to useful simplifications in kinetics calculations. This definition also has the operational simplification of not requiring any knowledge about the surface structure: a monolayer is simply defined as the number of molecules in the saturated layer. In this textbook I will always try to state explicitly which definition of σ_0 is being used.

17.7.1 Directed practice

The number of surface atoms on Ni(100) is 1.61×10^{19} m^{-2}. If the saturation coverage of CO is 0.5 ML, what is the absolute coverage at saturation?

In the literature, especially when dealing with adsorption on powders and porous substrates, you will also find coverage defined by the number of adsorbates divided by the *mass* of the sorbent. This usage arises because one generally has no specific knowledge of the nature of the surface of the sorbent, or how many surface atoms there are. However,

Table 17.4 Definition of symbols use in surface kinetics.

N_{ads}	Number of adsorbates (molecules or atoms, as appropriate)
N_0	Number of surface sites or atoms (as defined by context)
N_{exp}	Number of atoms/molecules exposed to (incident upon) the surface
A_s	Surface area
σ	Absolute coverage = areal density of adsorbates, $\sigma = N_{ads}/A_s$
σ_0	Absolute coverage of sites or surface atoms (per unit area)
σ^*	Absolute coverage of empty sites (per unit area)
σ_{sat}	Absolute coverage that completes a monolayer (per unit area)
θ	Fractional coverage, fractional number of adsorbates (also called the number of monolayers), $\theta \equiv \sigma/\sigma_0$
θ_{sat}	Saturation fractional coverage, $\theta_{sat} \equiv \sigma_{sat}/\sigma_0$, where σ_0 is the number of surface atoms
ε	Exposure, amount of gas incident on the surface per unit area, units of m^{-2} or Langmuir
Z_ω	Impingement rate in $m^{-2}\ s^{-1}$, $Z_w = p(2\pi m k_B T)^{-1/2}$
E_{ads}	Adsorption activation energy
E_{des}	Desorption activation energy
r_{ads}	Rate of adsorption
r_{des}	Rate of desorption
L	Langmuir, unit of exposure, 1 L = 1.33×10^{-6} mbar × 1 s
ML	Monolayer. A monolayer – one monomolecular layer – is a complete layer that is one adsorbate thick
s	Sticking coefficient, the probability that a molecule that hits the surface remains adsorbed at the surface
s_0	Initial sticking coefficient, sticking coefficient as $\theta \to 0$ ML

the number of surface atoms/sites is proportional to the amount of material; hence, similar concepts of coverage and fractional coverage are used.

The rate of desorption r_{des} is described by the Polanyi–Wigner equation. The rate of desorption r_{des} is the rate of change in the coverage as a function of time. In its most general form, the Polanyi–Wigner equation states that the rate is equal to a concentration term raised to the appropriate reaction order n multiplied by a rate constant k_{des}. The Polanyi–Wigner equation assumes that the rate constant follows Arrhenius behavior, thus we write

$$r_{des} = -d\theta/dt = \theta^n\, k_{des} = \theta^n A \exp(-E_{des}/RT_s). \tag{17.54}$$

Here, we have chosen to write the rate in terms of fractional coverage; however using the conversion factor

$$\sigma = \theta\sigma_0, \tag{17.55}$$

we can just as validly express it in terms of the rate of change of the absolute coverage. A is the pre-exponential or Arrhenius factor, E_{des} is the desorption activation energy, R the gas constant, and T_s the surface temperature.

The equivalent of an ideal gas for an adsorbed layer is a layer that exhibits no interactions between adsorbates. Adsorbate–adsorbate interactions are called *lateral interactions*, which can make both the pre-exponential factor and desorption activation energy dependent on coverage. Unless we state otherwise, we will always assume that the adsorbates are acting ideally, lateral interactions are zero, and there is no coverage dependence to the kinetic parameters.

In the absence of lateral interactions and for well-mixed adlayers, there is no ambiguity in using the Polanyi–Wigner equation to describe desorption. It is trivial to write down the rate law expected for elementary desorption reactions. For evaporation, or desorption from any phase that has a constant coverage because it is being replenished by another state, zero-order kinetics governs desorption

$$A(l) \rightarrow A(g) \quad r_{des} = k_{des}. \tag{17.56}$$

For simple atomic and nonassociative molecular desorption, first-order kinetics is expected

$$A(a) \rightarrow A(g) \quad r_{des} = \theta\, k_{des}. \tag{17.57}$$

In the case of recombinative desorption such as $H + H \rightarrow H_2$, the bimolecular character of the reaction predicts second-order kinetics

$$A(a) + A(a) \rightarrow A_2(g) \quad r_{des} = \theta^2 k_{des}. \tag{17.58}$$

Fractional orders are sometimes observed. These may be an indication of strong lateral interactions, island formation or competing desorption pathways that have different orders.

Transition state theory and statistical mechanics can be used to show (see for example, Kolasinski in Further Reading) that the pre-exponential factor for first-order desorption is

$$A = s_0 \frac{k_B T}{h} \frac{Z_\ddagger}{Z_a}, \tag{17.59}$$

where s_0 is the initial sticking coefficient, which replaces the transmission coefficient κ. The value of s_0 is usually ≈ 1 for nonactivated adsorption. The partition functions in the adsorbed phase and the transition state are Z_a and $Z_\ddagger$, respectively, T is the temperature, k_B is the Boltzmann constant, and h is the Planck constant.

Just as the rate of a chemical reaction can be related to a half-life, the rate of desorption is related to the *surface lifetime*. The expression for first-order desorption is particularly simple as the lifetime for first-order desorption is independent of coverage. The lifetime of the adsorbate on the surface, τ, is given by the *Frenkel equation*:

$$\tau = 1/k_{des} = (1/A) \exp\left(E_{des}/RT_s\right) = \tau_0 \exp\left(E_{des}/RT_s\right). \tag{17.60}$$

The adsorbate lifetime is finite. It depends on the temperature and the desorption activation energy. As the temperature increases, the lifetime drops exponentially, but even at high temperature there is a finite lifetime. Obviously in catalytic chemistry, it is important for the surface lifetime of reactants to be long compared to the time it takes for adsorbates to diffuse across the surface, collide with one another, and react.

Assuming an 'ideal' case in which the sticking coefficient is unity and the partition functions do not change, that is $s_0 = 1$ and $Z_\ddagger/Z_a = 1$, then we obtain $A = k_B T/h = 6.3 \times 10^{12}\ s^{-1}$ at room temperature whence the first-order approximation that a 'normal' value of A is $10^{13}\ s^{-1}$. This is no more than a rough guess. Values of $\sim 10^{13}\ s^{-1}$ can only be expected if no dynamical corrections occur (that is, $s_0 \approx 1$), the adsorbates are noninteracting, and no changes in the vibrational, rotational and translational degrees of freedom occur (i.e., the partition functions do not change). The value of $10^{13}\ s^{-1}$ serves as a useful benchmark. Assuming that $s_0 \approx 1$, values $> 10^{13}\ s^{-1}$ indicate that $Z_\ddagger/Z_a > 1$. This means that the transition state is much 'looser' than the adsorbed phase. A loose transition state in this sense means that it has degrees of freedom that are more easily excited by thermal energy than the adsorbed phase. An example of a loose transition state is a localized adsorbate (i.e., an adsorbate that is bound to a site) that desorbs through an activated complex that is a two-dimensional gas (i.e., can freely diffuse in the plane of the surface but is still bound to it). If $A < 10^{13}\ s^{-1}$, the transition state is constrained. A constrained transition state occurs if the molecule must take on a highly specific configuration in the activated complex. Interestingly – indeed consistent with the principle of microscopic reversibility – the value of the *desorption* pre-exponential factor correlates with whether or not there is a barrier to *adsorption*. As a rule, pre-exponential values in excess of $10^{13}\ s^{-1}$ are found for nonactivated, simple adsorption. Constrained transition states and low values of s_0 are generally associated with activated adsorption, as we shall see shortly.

Alternatively, we can interpret the rate constant in terms of thermodynamic quantities. From the Eyring equation [Eq. (17.30)] we can write the rate constant in terms of the entropy and enthalpy of activation,

$$k_{des} = s_0 \left(k_B T/h\right) \exp\left(\Delta^\ddagger S_m^\circ/R\right) \exp\left(-\Delta^\ddagger H_m^\circ/RT\right). \tag{17.61}$$

The standard enthalpy of activation is related to the desorption activation energy E_{des} by

$$E_{des} = \Delta^\ddagger H_m^\circ + RT, \tag{17.62}$$

and, therefore,

$$k_{des} = e s_0 \frac{k_B T}{h} \exp\left(\frac{\Delta^\ddagger S_m^\circ}{R}\right) \exp\left(\frac{-E_{des}}{RT}\right). \tag{17.63}$$

17.7.2 Directed practice

Substitute Eq. (17.62) into Eq. (17.61) to obtain Eq. (17.63).

Comparing Eq. (17.63) to the Arrhenius form of the rate constant, we see that the pre-exponential factor is determined by $A = e s_0 \left(k_B T/h\right) \exp\left(\Delta^\ddagger S_m^\circ/R\right)$. Pre-exponential factors of $A \approx 10^{13}\ s^{-1}$ are associated with $\Delta^\ddagger S_m^\circ = 0$. Larger pre-exponentials have $\Delta^\ddagger S_m^\circ > 0$ and smaller values correspond to $\Delta^\ddagger S_m^\circ < 0$. A greater entropy is associated with a greater number of accessible states for the system. The direct relationship to the partition functions should now be clear. When a greater number of states is accessible to the transition state, $\Delta^\ddagger S_m^\circ > 0$ and the pre-exponential factor is $A > 10^{13}\ s^{-1}$. When a greater number of states is accessible to the adsorbed state, $\Delta^\ddagger S_m^\circ < 0$ and the pre-exponential factor is $A < 10^{13}\ s^{-1}$.

17.8 Langmuir (direct) adsorption

17.8.1 The Langmuir model of adsorption

To calculate the rate of adsorption r_{ads}, we need a model for the dynamics of adsorption. We must account for the finite number of adsorption sites and how molecules come to occupy them. We begin with an ideal model that was developed by Irving Langmuir,[12] who was awarded the Nobel Prize in Chemistry 1932 for his landmark work in surface chemistry. The fundamental assumptions of this model are the following:

- The surface is homogeneous (only one type of site with a constant binding energy).
- Noninteracting particles strike the surface randomly and adsorb randomly.
- Molecules stick only if they hit an empty site.
- Molecules reflect if they hit a filled site.

The consequences of these assumptions are

1. There is only one site with absolute coverage σ_0.
2. Adsorption at one site does not affect other sites.
3. The fraction of the impinging flux that sticks is proportional to the coverage of unoccupied sites.
4. Adsorption is random (no ordered structures or island formation).
5. Adsorption stops at $\theta = 1$ ML defined in terms of the number of sites.

These assumptions may seem extreme but they provide a good approximation in many systems. Furthermore, the same kinetics is observed in a number of seemingly unrelated systems. For instance, non-cooperative ligand binding to the active site of a receptor protein is described by the Hill equation with $n = 1$. This is the same equation that we shall derive below for the Langmuir isotherm. Langmuirian kinetics also describes O_2 uptake by myoglobin, hormone binding to cell-surface receptors, and small-molecule binding to enzymes.

The rate of adsorption can be no higher than the rate of impingement of molecules onto the surface. The *impingement rate* is given by the Hertz–Knudsen equation

$$Z_w = \frac{N_A p}{\sqrt{2\pi MRT}} = \frac{p}{\sqrt{2\pi m k_B T}} \tag{17.64}$$

where N_A is Avogadro's constant, M is the molar mass (in kg mol^{-1}), m the mass of a particle (kg) and p the pressure (Pa). For $M = 28$ g mol^{-1} at standard ambient temperature and pressure, $Z_w = 2.92 \times 10^{27}$ m^{-2} s^{-1}. Approximately 1 mol per second of molecules of molar mass 28 (e.g., N_2 and CO) hits the surface per square meter at SATP. This value of the impingement rate does not possess an intuitive feel. If, on the other hand, we calculate the impingement rate for $M =$ 28 g mol^{-1}, at 1×10^{-6} torr = 1.33×10^{-6} mbar, we find $Z_w = 3.84 \times 10^{18}$ m^{-2} s^{-1}. Since surfaces have roughly 10^{19} atoms m^{-2}, this is roughly the equivalent of exposing each surface atom to one molecule from the gas phase per second. If all of these molecules had stuck, the coverage would be roughly 1 ML. Hence, a convenient unit of exposure is the Langmuir: 1 L ≡ 1.33×10^{-6} mbar × 1 s = 1.33×10^{-6} mbar s. The Langmuir is a historical and practical unit, which is convenient because of the rule of thumb that 1 L exposure leads to ~1 ML coverage *if the sticking coefficient is unity and independent of coverage.*

17.8.1.1 Directed practice

Confirm that the impingement rate of CO on a surface at SATP is 2.92×10^{27} m^{-2} s^{-1} by using Eq. (17.64).

17.8.2 Nondissociative (molecular) adsorption

The next step is to define the behavior of the sticking coefficient, s, as a function of coverage. In the Langmuir model, we assume $s = 1$ on empty sites and $s = 0$ on filled sites. Therefore, for nondissociative adsorption the sticking coefficient is equal to the probability of striking an empty site

$$s = 1 - \sigma/\sigma_0 = 1 - \theta \tag{17.65}$$

and varies with coverage, as shown in Fig. 17.8. This behavior indicates that the sticking coefficient decreases because of simple *site blocking* rather than because of any chemical or electronic effects. This model does not mean that the adsorbing molecule sticks where it hits. This statement is too restrictive. It merely means that the adsorbing molecule *makes the decision* of whether it sticks or not on the first bounce. The observation of Langmuirian adsorption kinetics does not rule out the possibility of transient mobility after the first collision with the surface. It implies that the molecule

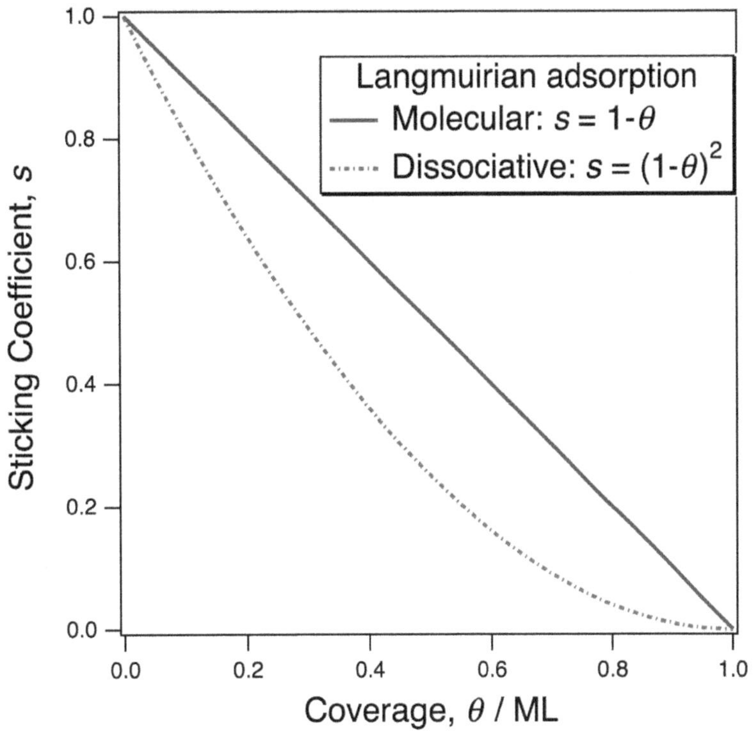

Figure 17.8 The coverage dependence of the sticking coefficient for adsorption that follows Langmuirian kinetics.

loses sufficient momentum normal to the surface that it cannot escape the attractive potential of the molecule–surface interaction.

We now have sufficient information to calculate the rate of adsorption and its dependence on pressure, temperature, and surface coverage. The rate of change of the surface coverage is simply the impingement rate multiplied by the sticking coefficient

$$\frac{d\sigma}{dt} = r_{\text{ads}} = Z_w s = \frac{p}{\sqrt{2\pi m k_B T}}(1-\theta) = c_A k_{\text{ads}}(1-\theta) \tag{17.66}$$

where c_A is the number density of A in the gas phase, and the rate constant for adsorption $k_{\text{ads}} = \left(k_B T/2\pi m\right)^{1/2}$.

Several extensions to Langmuirian adsorption, which do not change its essential features, are quite useful. One is that we might want to define the saturation coverage to be $\theta_{\max}$ rather than strictly 1. This allows for absolute comparison between molecules that have different absolute coverages. Second, the limiting initial sticking coefficient need not be exactly unity; thus, we give it the more general symbol s_0. Finally, we allow for the possibility that adsorption may be activated. The generalized adsorption rate is then

$$r_{\text{ads}} = Z_w s_0(\theta_{\max} - \theta)\exp\left(\frac{-E_{\text{ads}}}{RT}\right) = \frac{p}{\sqrt{2\pi m k_B T}} s_0(\theta_{\max} - \theta)\exp\left(\frac{-E_{\text{ads}}}{RT}\right). \tag{17.67}$$

Note that as defined here, s_0 is the initial sticking coefficient of nonactivated adsorption or the high-temperature limit of the initial sticking coefficient of activated adsorption.[13] The coverage at time t is given by integrating Eq. (17.67)

$$d\sigma = s(\sigma)\, d\varepsilon = s(\sigma)\, Z_w\, dt. \tag{17.68}$$

where ε is the *exposure*. The coverage is linearly proportional to the exposure,

$$\sigma(t) = s\varepsilon = s Z_W t, \tag{17.69}$$

only if the sticking coefficient is constant as a function of coverage. This is not, in general, the case. There are some very important exceptions that do follow Eq. (17.69), including very low coverage, metal-on-metal adsorption, and condensation onto multilayer films (such as the condensation of a vapor onto the surface of the corresponding liquid). Exposure, as we have mentioned above, is often expressed in the experimentally convenient units of pressure times time or Langmuirs. Adsorption within this model leads to growth in coverage versus exposure, as shown in Fig. 17.9.

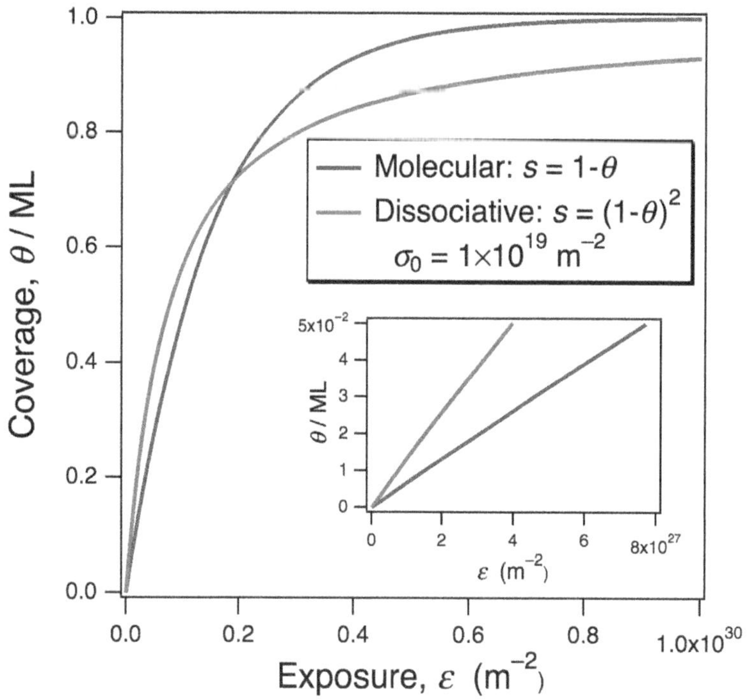

Figure 17.9 Langmuir models (molecular and dissociative) of fractional coverage, θ, as a function of exposure, ε. Note the linearity of both curves at low exposure. Because dissociative adsorption fills two sites, it initially fills the surface at a faster rate. At high coverage, the rate of dissociative is 'poisoned' relative to molecular adsorption by site blocking.

17.8.3 Dissociative adsorption

With the same set of assumptions as defined above, we can also treat dissociative adsorption. Assuming that dissociation requires two adjacent empty sites, the sticking probability varies with coverage, as does the probability of finding two adjacent empty sites. This probability is given by $(1-\theta)^2$ and, therefore,

$$s = s_0\,(1-\theta)^2 \exp\left(-E_{\mathrm{ads}}/RT\right) \tag{17.70}$$

Figures 17.8 and 17.9 show how s changes with θ and how θ increases with ε for Langmuirian dissociative adsorption. Note the difference in behavior between molecular and dissociative adsorption caused by the different site-blocking dependence of the sticking coefficient on coverage.

17.9 Precursor-mediated adsorption

Taylor and Langmuir[14] observed that the sticking coefficient of Cs on W did not follow Eq. (17.8.2) but rather, it followed the form found in Fig. 17.10. They suggested that adsorption is mediated by a precursor state through which the adsorbing atom passes on its way to the chemisorbed state. Precursor-mediated adsorption has subsequently been observed in numerous systems. Precursor states can be classified either as intrinsic or extrinsic. An intrinsic precursor state is one associated with the clean surface, whereas an extrinsic precursor state is due to the presence of adsorbates. The explanation of how a precursor state manages to keep the sticking coefficient higher than the value predicted by the Langmuir model is that the incident molecule enters the precursor state and is mobile. Therefore, the adsorbing molecule can roam around the surface and hunt for an empty site. This greatly enhances the sticking probability because an incident molecule has a certain probability to remain on the surface even if it collides with a filled site. The ability of the adsorbate to search for an unoccupied site not only increases the sticking coefficient compared to the Langmuir model, it also allows adsorbing molecules to form islands rather than adsorb randomly on the surface. In entering the mobile precursor state, the molecule has lost enough energy out of its z-component of translational energy (the direction normal to the surface) so that it cannot escape from the surface. Simultaneously, it loses energy slowly in the x- and y-components (the in-plane components), which allows it to roam on the surface.

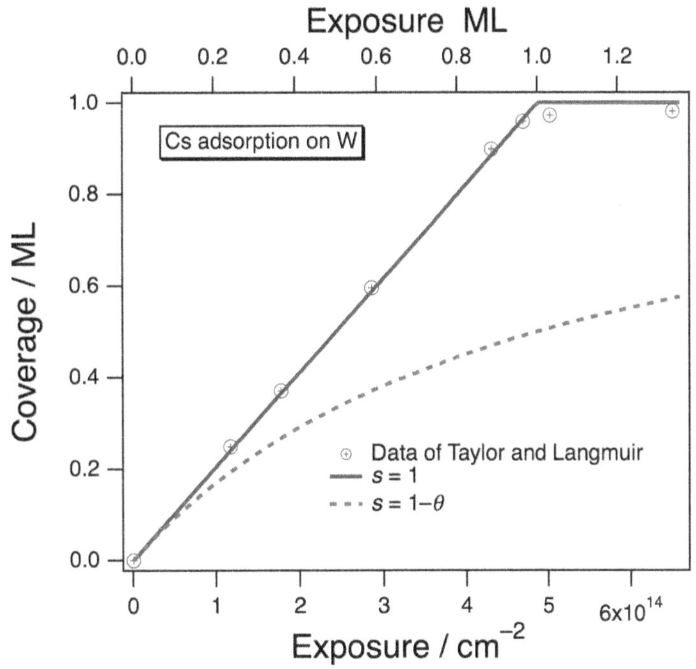

Figure 17.10 Sticking of Cs on W. Replotted from the data of Taylor, J.B. and Langmuir, I. (1933) *Phys. Rev.*, **44**, 423.

17.10 Adsorption isotherms

An adsorption isotherm is the equation of the curve that describes the equilibrium coverage at constant temperature. The basis of isotherms is that we consider a system at equilibrium. Therefore, the surface and the gas are at the same temperature, $T_s = T_g = T$, and the rates of adsorption and desorption balance, $r_{des} = r_{ads}$, such that the coverage is constant, $d\theta/dt = 0$.

Using the set of assumptions posited in the Langmuir model of molecular adsorption, we can derive the Langmuir isotherm. Equating the rates of adsorption, Eq. (17.66), and desorption, Eq. (17.57), yields

$$pk_{ads}(1-\theta) = \theta\, k_{des}. \tag{17.71}$$

Both sides of Eq. (17.71) are written such that the rate is expressed in molecules per unit area per second, thus,

$$k_{ads} = \left(2\pi m k_B T\right)^{-1/2} e^{-E_{ads}/RT} \tag{17.72}$$

and

$$k_{des} = \sigma_0 A\, e^{-E_{des}/RT} \tag{17.73}$$

Note that the adsorption rate constant is written so as to use pressure rather than number density, as was done in Eq. (17.66). The exponential term allows us to account for the possibility of activated adsorption. Rearranging Eq. (17.71) yields

$$\frac{\theta}{1-\theta} = p\frac{k_{ads}}{k_{des}} = pK. \tag{17.74}$$

$K = k_{ads}/k_{des}$is effectively an equilibrium constant. Purists will worry about it having units and how the standard state of the adsorbed layer is defined; thus, it is more properly called the *Langmuir isotherm constant*. Rearranging, we obtain the Langmuir isotherm describing the equilibrium coverage as a function of pressure

$$\theta = \frac{pK}{1+pK}. \tag{17.75}$$

Examples are shown in Fig. 17.11. Alternatively, concentration can be used in place of pressure, such that this expression can also be used to describe adsorption from any fluid (gas, liquid or solution) phase onto a solid substrate as long as the rate constants in Eqs (17.72) and (17.73) are written with the proper units.

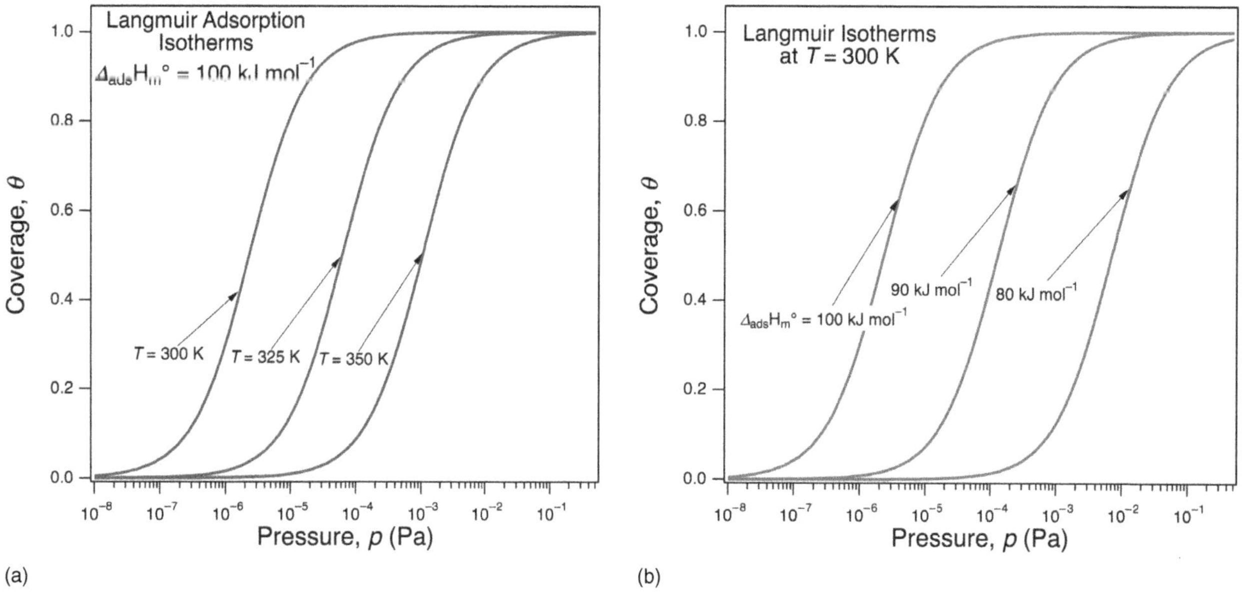

Figure 17.11 Langmuir isotherms as functions of (a) $\Delta_{ads}H_m{}^\circ$ and (b) T. A lower binding energy or a higher temperature require a higher pressure for the system to maintain the same equilibrium coverage, which eventually saturates with high enough pressure.

Multiplying both sides of Eq. (17.71) by σ_0 and recalling that $\sigma = \theta\sigma_0$, we can derive the dependence of the absolute coverage on pressure and temperature,

$$\sigma = \left(1 - \sigma/\sigma_0\right) pb, \tag{17.76}$$

where

$$b = \left(2\pi m k_B T\right)^{1/2} A^{-1} e^{-(E_{ads}-E_{des})/RT}. \tag{17.77}$$

The difference between the desorption and adsorption activation energies is equal in magnitude to the enthalpy of adsorption

$$\Delta_{ads}H_m^\circ = E_{ads} - E_{des}. \tag{17.78}$$

To make Eq. (17.76) useful, we need to rearrange it such that σ only appears on one side. Thus, we arrive at

$$\frac{1}{\sigma} = \frac{1}{bp} + \frac{1}{\sigma_0}. \tag{17.79}$$

A plot of $1/\sigma$ versus $1/p$ can be used to determine σ_0 and b. Since b is determined by the temperature and enthalpy of adsorption, we see that both of these quantities, as well as the pressure, are responsible for determining the equilibrium coverage σ. Note that in the limit of very low coverage, $\sigma \ll \sigma_0$ and $1/\sigma - 1/\sigma_0 \approx 1/\sigma$, hence,

$$\sigma = bp. \tag{17.80}$$

In other words, the coverage is linearly related to the pressure in the limit of low coverage. This relationship is favored at high temperature, low pressure, and for weakly bound systems for which the equilibrium coverage is naturally small.

17.11 Surmounting activation barriers

The energetics and dynamics of a reaction are governed by a multidimensional potential energy surface (PES). Now, we will discuss how the topology of this PES, particularly the region near the transition state, influences the dynamics of a reaction. From the above discussions we understand that molecules must collide to react, and the collision must have sufficient energy and the correct orientation to facilitate reaction. A molecule has a number of degrees of freedom. Whether energy placed in these different degrees of freedom is equally effective at causing reaction depends on the topology of the transition state.

John C. Polanyi was awarded the Nobel Prize in Chemistry 1986 along with Dudley R. Herschbach and Yuan T. Lee for their contributions to the understanding of the dynamics of elementary reactions. Polanyi, whose father Michael was one of the founders of transition state theory, codified the effects of PES topography in what are called the *Polanyi rules*. These rules tell us how the position of the transition state determines the efficacy of energy in any particular degree of freedom

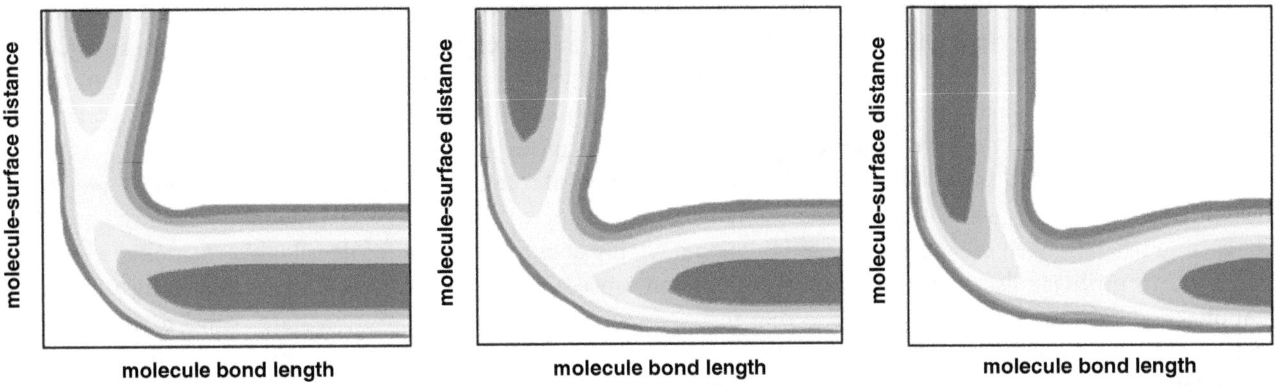

Figure 17.12 Two-dimensional contour plots are used to represent the potential energy surface (PES) for reactions with (a) early, (b) middle) and (c) late barriers. Provided by Jin Cheng, Florian Libisch, and Emily A. Carter. Reproduced with permission.

for promoting reaction. By application of detailed balance, these rules also describe how energy will be distributed into the degrees of freedom of reaction products.

Figure 17.12 displays three different types of PES for the activated dissociative adsorption of a molecule such as H_2. The PESs are distinguished by the position of the activation barrier. The z-direction is perpendicular to the surface and represents the distance of the center of mass of the molecule from the surface. The R-coordinate is the distance between the two H atoms. At the top of the diagram, where z is large, the molecule is far from the surface and motion along R corresponds to vibration of the H—H bond. The bottom of this valley lies at R_e, the equilibrium bond distance of the gas-phase H_2 molecule; this is the reactant valley. The bottom right-hand corner of the PES is the product valley; here, R is large and the H—H bond has broken. The two H atoms are bound to the surface at some equilibrium height above the surface, their adsorption bond distance z_e.

Somewhere between the reactant valley and the product valley is located the transition state. The transition state forms a col or saddle point in the potential energy surface. If we were to draw a minimum energy path from the reactants to the products, the transition state is located at the energy maximum along this path. This path is the familiar one-dimensional potential energy curve, often called a Lennard–Jones curve.[15] If the transition state is located at relatively large z and small R, the transition state resembles the gas-phase molecule; this is corresponds to an *early barrier*. If the transition state is located at small z and large R, the transition state resembles the products more than the reactants; this is a *late barrier*. If the barrier is somewhere between these extremes, it is a *middle barrier*. Note that in the terminology used here, the transition state is a point on the PES and the activated complex is the molecular entity that exists at this point.

Now let's examine the impact on reaction dynamics and energy partitioning. Translational energy is always required for reaction (since we are implicitly discussing normal thermal reactivity, not reactivity induced by radioactivity or photochemistry). We take that as given because the molecules or molecule and surface in the example above must collide. What we want to ask now is: Is putting energy in translation as effective at promoting reaction as putting energy in vibrations or rotations? The answer is that it depends on the location of the barrier.

For an *early barrier*, the activated complex resembles the reactants. We do not need to change the molecule to get it into the transition state, we merely need to push it up the hill to get it there. Therefore, translational energy is the most efficacious at enhancing reactivity. Vibration is a spectator coordinate. Putting energy in vibration does not do much to either help or hinder reactivity. We will delay any discussion of rotational energy until we discuss all three barriers.

For a *late barrier*, the activated complex resembles the products. The H—H bond has to be stretched. True, we still need to push the H_2 up the hill to get to the transition state, but without stretching the H—H bond we cannot turn the corner to access the transition state. For a late barrier vibrational excitation is essential for reactivity. Vibrational efficacy is very high: the addition of more vibrational energy is as efficient – if not more efficient – at promoting reaction as the same amount of energy put into translation.

For a *middle barrier*, the behavior is in between. Both translation and vibration are required. The vibrational efficacy is lower than for a late barrier, but nonetheless vibrational excitation is still required because bond stretching is essential.

Now we turn to rotations. Rotations are unique among the degrees of freedom because they change not only the energy of the molecule but also the orientation of the molecule. Thus, we must also consider the effect of rotation on the steric factor in reactivity. This has been beautifully detailed in the experiments of Charlie Rettner and Dan Auerbach.[16] When the molecule possesses little rotational excitation, the molecule can be reoriented by the forces it experiences as it rides the potential energy surface up to the transition state. A molecule with zero rotational excitation will always end up being perfectly oriented with respect to its rotational axis in the transition state as long as it approaches slow enough to allow time for the forces to reorient the molecule. Addition of rotational energy will make it progressively more difficult for the molecule to reorient. Thus, the addition of rotational energy at low rotational excitation hinders reaction. At very high levels of rotational excitation, the molecule will be rotating so fast that the reaction partner will see an average over all orientations. Thus, the negative effect reaches an asymptotic limit.

On an early barrier PES, the orientational effect is the only effect of rotations. Rotational energy cannot couple into the reaction path and help to access the transition state. On a middle or late barrier PES, rotational energy can couple into the reaction path in much the same way that vibrational energy does. When a molecule is highly rotationally excited, it experiences centrifugal distortion – elongation of the bond caused by a centrifugal force. This bond elongation couples rotational energy into turning the corner so that the molecule can access the transition state. Thus, rotational energy hinders reaction for small amounts of rotational excitation while large amounts of rotational excitation promote reaction.

Now let us use microscopic reversibility to examine the reverse reaction – that of desorption. It is very difficult to generate H_2 specifically in one quantum state for every quantum state that would be populated at a given temperature, put it in a beam, slam it into a surface, and then measure the sticking coefficient for each quantum state. It is much easier (though not easy) to adsorb H atoms onto a surface, heat up the surface and then measure the quantum state distribution of the molecules that desorb from the surface using resonance-enhanced multiphoton ionization (REMPI). Indeed, this is how many of the experiments that led to the above conclusions were performed. The application of detailed balancing and microscopic reversibility allows us to infer information about desorption from adsorption, and vice versa.

For the case of H_2 sticking on Cu surface, a middle-to-late barrier is found. In accord with the Polanyi rules (as discussed above), dissociative adsorption is promoted by increased translational and vibrational energy. Rotational energy hinders adsorption at low excitation but promotes dissociation at high levels of excitation. Reversing the reaction, we examine the energy distribution of the molecules that leave the surface after the recombination of adsorbed H atoms into H_2 molecules. The first observation to note is that, even though the atoms are equilibrated with the surface, the desorbed H_2 molecules do not leave the surface with an equilibrium energy distribution corresponding to the surface temperature. The molecules are much 'hotter' than expected. This is an important but subtle point that is discussed in detail elsewhere.[17] In effect, the transition state is acting as a filter. If only fast molecules with a stretched bond can access it when H_2 approaches the surface, then only fast molecules with a stretched bond can leave it when approached from the recombining atoms side. The molecules that desorb from the surface will be excited translationally and vibrationally. The rotational distribution is slightly 'cold' at low values of the rotational quantum number J, but 'hot' at high values of J.

The partitioning of energy into products is decisive in determining the distribution of products formed in a reaction that has several possible pathways. Consider the gas-phase photodissociation of $BrCH_2CH_2OH$,

$$BrC_2H_4OH + h\nu \rightarrow C_2H_4OH + Br. \tag{17.81}$$

The energy of the C_2H_4OH product is determined by the conservation of energy. The initial internal energy of the BrC_2H_4OH $E_{int}(BrC_2H_4OH)$ plus that of the photon $E_{h\nu}$ must balance the energy required to break the C—Br bond D_0(C—Br), the internal energy of the C_2H_4OH $E_{int}(C_2H_4OH)$ plus the kinetic energies of the two reactants E_K (it is also possible to produce the Br in an excited spin–orbit state, but we can neglect that here)

$$E_{int}\left(BrC_2H_4OH\right) + E_{h\nu} = D_0\left(C - Br\right) + E_{int}\left(C_2H_4OH\right) + E_K \tag{17.82}$$

The internal energy is the sum of the rotational and vibrational energy. Some of the product C_2H_4OH goes on to dissociate further (primarily into C_2H_4 + OH or C_2H_3 + H_2O). Importantly, the vibrational energy of C_2H_4OH couples into further dissociative pathways but the rotational energy does not. Similarly, energy coupled into translational energy is not available to facilitate dissociation of C_2H_4OH.

The results shown in Fig. 17.13 obtained by the group of Laurie Butler detail the energy partitioning for this photochemical reaction. The black curve corresponds to the translational energy distribution function of *nascent* C_2H_4OH. The blue curve corresponds to the translational energy distribution of *stable* C_2H_4OH. Above a translational energy of ~42 kcal mol^{-1} the product must be stable. There simply is not enough energy in the internal modes to facilitate the overcoming of any activation barriers to further dissociation. Below this threshold, the internal energy is sufficient to facilitate dissociation *if the energy is partitioned to vibrational energy*. However, many of the molecules created with translational energy below 42 kcal mol^{-1} have a sufficiently large fraction of the energy partitioned into rotational energy that they remain stable.

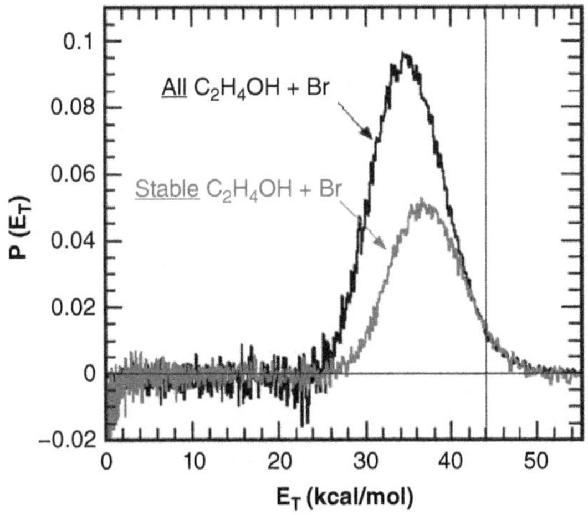

Figure 17.13 Center of mass recoil translational energy distribution for $C_2H_4OH + Br$ resulting from the photodissociation of $BrCH_2CH_2OH$ at 193.3 nm. The black curve is the translational energy distribution of all products. The blue curve is the distribution of stable product that does not go on to react further on the timescale of the experiment. The line at 42 kcal mol^{-1} represents the threshold above which no further dissociation of C_2H_4OH is possible. Provided by Laurie J. Butler. Reprinted with permission from Ratliff, B.J., Womack, C.C., Tang, X.N., Landau, W.M., Butler, L.J., and Szpunar, D.E. (2010) *J. Phys. Chem. A*, **114**, 4934. © 2010, American Chemical Society.

SUMMARY OF IMPORTANT EQUATIONS

$k = P N_A \sigma \langle v_{12} \rangle \exp(-E_a/RT)$	Rate constant k from collision theory, steric factor P, Avogadro's constant N_A, collision cross-section σ, mean relative velocity $\langle v_{12} \rangle$, activation energy E_a.
$E_a = \langle E \rangle_R - \langle E \rangle_\ddagger$	Mean energy of reactants $\langle E \rangle_R$, mean energy of activated complex $\langle E \rangle_\ddagger$
$E_a = E_{a,0}^{qm} + mRT$	Quantum mechanical minimum barrier height inclusive of zero-point energy $E_{a,0}^{qm}$, m an integer
$k = \dfrac{\kappa k_B T}{hc^\circ} \exp(-\Delta^\ddagger G_m^\circ/RT)$	Rate constant from transition state theory k for a bimolecular reaction, transmission coefficient κ, standard state concentration c°, Gibbs energy of activation ,$\Delta^\ddagger G_m^\circ$, see Table 17.1
$\dfrac{d[P]}{dt} = \dfrac{k_2 k_1 [A]^2}{k_{-1}[A] + k_2}$	Unimolecular reaction rate from Lindemann–Christiansen mechanism.
$\dfrac{d[P]}{dt} = k_{uni}[A]$ $\dfrac{1}{k_{uni}} = \dfrac{k_{-1}}{k_2 k_1} + \left(\dfrac{1}{k_1}\right)\dfrac{1}{[M]}$	Unimolecular reaction rate in the presence of a quencher M
$\theta = \sigma/\sigma_0$	Fractional coverage θ, absolute coverage σ, number of sites (or surface atoms) σ_0
$Z_w = \dfrac{N_A p}{\sqrt{2\pi MRT}} = \dfrac{p}{\sqrt{2\pi m k_B T}}$	Impingement rate Z_w, pressure p, molar mass M, molecular mass m
$\tau = 1/k_{des} = (1/A)\exp(E_{des}/RT_s) = \tau_0 \exp(E_{des}/RT_s)$	Frenkel equation: surface lifetime of adsorbate τ, desorption rate constant k_{des}, desorption activation energy E_{des}
$s = 1 - \theta$	Sticking coefficient s within Langmuir model of nonactivated molecular adsorption
$\dfrac{d\sigma}{dt} = r_{ads} = Z_w s$	Rate of adsorption r_{ads} in terms of the rate of change of absolute coverage $d\sigma/dt$
$\sigma(t) = s\varepsilon = sZ_W t$	Absolute coverage as a function of time is the sticking coefficient is constant, exposure $\varepsilon = Z_W t$
$s = s_0(1-\theta)^2 \exp(-E_{ads}/RT)$	Sticking coefficient within Langmuir model of dissociative adsorption s, high-temperature limit of initial sticking coefficient s_0, adsorption activation energy E_{ads}
$\theta = \dfrac{pK}{1+pK}$	Langmuir isotherm in terms of fractional coverage. Langmuir isotherm constant $K = k_{ads}/k_{des}$
$k_{ads} = (2\pi m k_B T)^{-1/2} e^{-E_{ads}/RT}$	Adsorption rate constant
$k_{des} = \sigma_0 A\, e^{-E_{des}/RT}$	Desorption rate constant
$\dfrac{1}{\sigma} = \dfrac{1}{bp} + \dfrac{1}{\sigma_0}$ $b = (2\pi m k_B T)^{1/2} A^{-1} e^{-(E_{ads}-E_{des})/RT}$	Langmuir isotherm in terms of absolute coverage
$\Delta_{ads} H_m^\circ = E_{ads} - E_{des}$	Adsorption enthalpy

Exercises

17.1 For H_2 and N_2 at 500 °C (the temperature at which the ammonia synthesis reaction is run), calculate the fractions of molecules with speeds between 1000 and 1500 m s^{-1}. Draw approximately by hand what the two distributions look like, as well as the relative collision velocity distribution.

17.2 A molecular beam apparatus of the type shown in Fig. 17.1 employs supersonic jets that allow gas molecules to expand from a gas reservoir held at a specific temperature and pressure into a vacuum through a small orifice. Expansion of the gas allows one to cool the gas molecules to temperatures of roughly 10 K (or less). The expansion can be treated as adiabatic, with the change in gas enthalpy accompanying expansion $\Delta H_m = C_{p,m} T_0$ being converted to kinetic energy associated with the flow of the gas. The temperature of the reservoir T_0 is generally much greater than the final temperature of the gas, allowing one to consider the entire enthalpy of the gas to be converted into translational motion. (a) For a diatomic gas, the high-temperature limit is $C_{p,m} = 9/2\ R$. However, the vibrational degree of freedom is not cooled during the expansion leading to an effective value of $C_{p,m} = 7/2$ R. Using this information, demonstrate that the final flow velocity of the molecular beam is related to the initial temperature of the reservoir by $v = \left(\frac{7RT_0}{M}\right)^{1/2}$. (b) Using this expression, what is the flow velocity of a molecular beam of CO when $T_0 = 298$ K?

17.3 Explain why the molecules in a supersonic molecular beam have a high average mean velocity but are at the same time described by a low translational temperature.

17.4 Give two reasons why not all collisions between reactants lead to products.

17.5 Dissociation of ethane into methyl radicals follows first-order kinetics with Arrhenius parameters $A = 2.50 \times 10^{16}\ s^{-1}$ and $E_a = 367.6$ kJ mol^{-1}. Calculate the Gibbs energy of activation, the entropy of activation, and the enthalpy of activation at 500 K.

17.6 On Ru(0001) the pre-exponential factor for H_2 desorption is $A = 4.0 \times 10^{-7}\ m^2\ s^{-1}$. Comparing this value to the value of $A = 9.64 \times 10^{-6}\ m^2\ s^{-1}$ for Ni(110), what can we conclude about the transitions states for H_2 desorption in these two systems?

17.7 The following rate constant versus temperature data have been obtained for the second-order gas-phase elementary reaction $CH_3 + O_2 \rightarrow CH_3O + O$.

Temperature / K	300	400	500	600
k / $m^3\ mol^{-1}\ s^{-1}$	2.29×10^{-14}	4.12×10^{-9}	5.87×10^{-6}	7.42×10^{-4}

Calculate at 450 K, the rate constant, the energy of activation, the enthalpy of activation, the Gibbs energy of activation, and the entropy of activation.

17.8 Are collisions important for determining the rates of unimolecular reactions?

17.9 A unimolecular gas-phase reaction X → Y is activated with collisions by an unreactive collision partner B with the following mechanism:

$$X + B \underset{k_{-1}}{\overset{k_1}{\rightleftharpoons}} X^* + B$$

$$X^* \xrightarrow{k_2} Y.$$

has a rate that follows the effective rate equation $\frac{d[Y]}{dt} = k_{uni}[X]$. (a) Apply the steady-state approximation (SSA) to determine an algebraic expression for k_{uni} in terms of the rate constants k_1, k_{-1} and k_2 as well as [B]. (b) Discuss how the rate law changes in the limit of large and small [B].

17.10 Which of the mechanisms described below is consistent with the reaction

$$3A \rightarrow D + 2E$$

State reasons for making your choice and eliminating the other two mechanisms.

Mech. 1	Mech. 2	Mech. 3
$A \rightarrow B$	$A \rightarrow B$	$2A \rightarrow B + C$
$A \rightarrow 2C$	$B \rightarrow C$	$A + B \rightarrow D$
$2B \rightarrow D$	$A + C \rightarrow 2D$	$C + D \rightarrow 2E$
$2C \rightarrow 2E$	$D \rightarrow 2E$	

17.11 The following elementary steps comprise a complete reaction mechanism (all species in the gas phase).

$$SiF_2 + F_2 \xrightarrow{k_1} SiF_3 + F$$

$$SiF_2 + F \xrightarrow{k_2} SiF + F_2$$

$$SiF + F_2 \xrightarrow{k_3} SiF_3$$

$$SiF_3 \xrightarrow{k_4} SiF_2 + F$$

$$SiF_3 + F \xrightarrow{k_5} SiF_4$$

(a) What is the overall reaction? (b) What are the intermediates? (c) What is the net rate equation for SiF_2? (d) What is the net rate equation for F_2? (e) Apply the steady-state approximation to find an expression for $[SiF_3]$ (f) Apply the steady-state approximation to find an expression for [F].

17.12 The Eigen–Wilkins mechanism of complex formation describes the reaction of a monodentate ligand L and an aqua-metal ion $M(OH_2)_x{}^{n+}$, abbreviated M^{n+}. The first step involves the reversible formation of the complex $M^{n+}{\cdot}L$. The second step is the irreversible formation of the complex $M^{n+}L$ with expulsion of water from the intermediate. (a) Write an expression for the rate of formation of the complex $M^{n+}L$ in terms of the reactant concentrations. (b) Write the rate equations is the limits of $[M^{n+}L] \gg [L]$ and $[L] \gg [M^{n+}L]$.

17.13 The Eigen–Winkler mechanism for complex formation with macrocyclic and cryptand ligands involves three reversible steps: pre-association of the metal ion M^{n+} with the ligand C; interchange with formation of the first metal–ligand bond; and conformational change(s) required to complex or encapsulate M^{n+} fully. The mechanism can be represented as

$$M^{n+} + C \rightleftarrows M^{n+} \cdots C \rightleftarrows MC^{n+} \rightleftarrows \{MC^{n+}\}.$$

(a) Write a general rate equation for the formation of product, assuming all steps are reversible (b) Write a general rate equation for the formation of product, assuming all steps are equilibrated.

17.14 The reversible elementary gas-phase reaction $H_2 + I_2 \underset{k_{-1}}{\overset{k_1}{\rightleftharpoons}} 2HI$ exhibits Arrhenius behavior in the temperature range 600–750 K. It is characterized in the forward direction by $A_1 = 4.45 \times 10^5$ l mol^{-1} s^{-1} and $E_{a1} = 170.3$ kJ mol^{-1}. When equilibrated at $T = 633$ K, the concentrations of H_2 and I_2 are 1.90×10^{-3} mol l^{-1}. $\Delta_r H° = 52.7$ kJ mol^{-1} and $\Delta_r G° = 3.40$ kJ mol^{-1}. (a) Calculate the rates of the forward and reverse reactions in terms of $d[HI]/dt$ in mol l^{-1} s^{-1} at equilibrium and 633 K. (b) Calculate the equilibrium concentration of HI. (c) Calculate the rate parameters of the reverse reaction: k_{-1}, E_{a-1}, and A_{-1}. (d) Calculate $\Delta_r S°$, $\Delta^{\ddagger}H°$, $\Delta^{\ddagger}S°$ and $\Delta^{\ddagger}G°$ for the forward reaction. (e) Draw a diagram (roughly to scale) of energy versus reaction coordinate, which shows the change in potential energy during the course of the reaction. Label the diagram appropriately with the activation energies, reaction enthalpy and species that apply to the diagram. Include a reasonable drawing of the activated complex and comment upon the implications of the value of $\Delta^{\ddagger}S$.

17.15 The reversible elementary gas-phase reaction $H + CH_4 \underset{k_{-1}}{\overset{k_1}{\rightleftharpoons}} H_2 + CH_3$ exhibits Arrhenius behavior in the temperature range 350–750 K. At 450 K the rate constant was measured to be 8.78 l mol^{-1} s^{-1} in the forward direction and 0.443 l mol^{-1} s^{-1} in the reverse direction. At 550 K the rate constant was measured to be 49.3 l mol^{-1} s^{-1}

in the forward direction and 3.38 l mol^{-1} s^{-1} in the reverse direction. Use $T = 550$ K and concentrations [H] = 1.25×10^{-6} mol l^{-1}, [CH_4] = 5.25×10^{4} mol l^{-1}, [H_2] = 6.33×10^{2} mol l^{-1}, [CH_3] = 4.33×10^{-3} mol l^{-1} to perform the following calculations. (a) Calculate Arrhenius parameters (pre-exponential factor and activation energy) for the forward and reverse reactions. (b) Calculate the equilibrium constant of the forward reaction. (c) Calculate the rates of the forward and reverse reactions. (d) Calculate the activation enthalpy, entropy and Gibbs energy of the forward reaction. (e) Calculate the changes in enthalpy, entropy and Gibbs energy for the forward reaction. (f) Draw a diagram (roughly to scale) of energy versus reaction coordinate, which shows the change in potential energy during the course of the reaction. Label the diagram appropriately with the calculated quantities and species that apply to the diagram. Include a drawing of the activated complex. (g) At equilibrium are the reactants or products favored? Under the given conditions are the reactants or products favored? Justify your answers.

17.16 The reversible elementary gas-phase reaction $NO + O_3 \underset{k_{-1}}{\overset{k_1}{\rightleftharpoons}} NO_2 + O_2$ exhibits Arrhenius behavior at $T = 298$ K (use this temperature throughout). It is characterized in the forward direction by $A_1 = 1.09 \times 10^9$ l mol^{-1} s^{-1} and $E_{a1} = 9.62$ kJ mol^{-1}, $\Delta_r H_m{}^\circ = -200.8$ kJ mol^{-1} and $\Delta_r G_m{}^\circ = -199.5$ kJ mol^{-1}. (a) Calculate the rate of the forward reaction in terms of d[NO]/dt in mol l^{-1} s^{-1} if the pressures of NO and O_3 are both 10 Pa. (b) Calculate the equilibrium constant. Does the reaction go to completion? (c) Calculate the rate parameters of the reverse reaction: k_{-1}, E_{a-1}, and A_{-1}. (d) Calculate $\Delta_r S_m{}^\circ$, $\Delta^\ddagger H_m{}^\circ$, $\Delta^\ddagger S_m{}^\circ$ and $\Delta^\ddagger G_m{}^\circ$ for the forward reaction. (e) Draw a diagram (roughly to scale) of energy versus reaction coordinate, which shows the change in potential energy during the course of the reaction. Label the diagram appropriately with the activation energies, reaction enthalpy and species that apply to the diagram. Include a drawing of the activated complex.(f) Comment on the values of $\Delta^\ddagger H_m{}^\circ$ and $\Delta^\ddagger S_m{}^\circ$, and suggest reasons why these values are reasonable or unreasonable based on the reaction dynamics.

17.17 (a) Beginning with the Hertz–Knudsen equation, derive an equation for the rate of impingement of a solute with concentration c. (b) Calculate the initial rate of absorption of dodecylthiol ($C_{12}H_{26}S$) from a 1 mM solution at 20 °C if its initial sticking coefficient is 1×10^{-6}.

17.18 Calculate the collision rate of water with a surface if the surface is exposed to: (a) water vapor with a pressure of 1 atm; and (b) liquid water? The system is held at 350 K.

17.19 Define all fractional coverages with respect to surface atom density. If the sticking coefficient of NO is 0.85 on Pt(111), what is the initial rate of adsorption of NO onto Pt(111) in m^{-2} s^{-1} at 250 K if the NO pressure is 1.00×10^{-6} Pa? $\sigma_0 = 1.50 \times 10^{19}$ m^{-2} on Pt(111)

17.20 Define all fractional coverages with respect to surface atom density. $\sigma_0 = 1.53 \times 10^{19}$ m^{-2} on Pd(111). If the dissociative sticking coefficient of O_2 is constant at a value of 0.25, how many oxygen atoms will be adsorbed onto a Pd(111) surface after it has been exposed to 1.00×10^{-5} Pa of O_2 for 1 s? State your answer in both absolute coverage and fractional coverage. The system is held at 100 K.

17.21 The enthalpy of adsorption of C_2H_6 on Pt(111) is −36.8 kJ mol^{-1}. Assuming that the Langmuir model of adsorption holds and the pre-exponential factor $A = 1 \times 10^{13}$ s^{-1}, calculate the equilibrium coverage in m^{-2} and fractional coverage in ML for $p = 2.00 \times 10^{-3}$ Pa and $T = 140$ K. The surface atom density of Pt(111) is 1.50×10^{19} m^{-2}.

17.22 Define all fractional coverages with respect to surface atom density. All values are for $T = 400$ K on Rh(111). The Rh surface atom density is 1.60×10^{19} m^{-2}. H_2 adsorbs dissociatively following Langmuir kinetics with an initial sticking coefficient of 0.90. CO adsorbs molecularly following precursor-mediated kinetics with $s_0 = 0.85$. For both molecules adsorption is not activated. Calculate the initial rate of adsorption dσ/dt in molecules m^{-2} s^{-1} of CO if $p_{CO} = 1.25 \times 10^{-4}$ Pa. Calculate the rate of formation dθ/dt in ML s^{-1} of H(a) from H_2 adsorption if $p_{H_2} = 1.25 \times 10^{-4}$ Pa and $\theta = 0.25$.

17.23 For Pt(111) the surface atom density is 1.50×10^{19} m^{-2}. The enthalpy of nonactivated molecular adsorption of CO on Pt(111) is −185 kJ mol^{-1}. Calculate the equilibrium coverage in m^{-2} and fractional coverage in ML for $p = 9.50 \times 10^{-7}$ Pa and $T = 575$ K. The pre-exponential factor for desorption is 7.50×10^{13} s^{-1}.

17.24 Define all fractional coverages with respect to surface atom density. For Pt(111) the surface atom density is 1.50 $\times 10^{19}$ m^{-2}. The enthalpy of adsorption of formic acid HCOOH on clean Pt(111) at 100 K is -65 kJ mol^{-1}. This adsorbed layer saturates at 0.5 ML following precursor-mediated adsorption kinetics with a sticking coefficient of 0.95. The adsorption is molecular and nonactivated. (a) A 102 ms pulse corresponding to a pressure of 3.60 $\times$ 10^{-5} Pa is exposed to the surface at 100 K. Calculate the resulting absolute coverage of HCOOH after exposure of Pt(111) to one pulse. (b) HCOOH adsorbs on top of the saturated layer with a constant $\Delta_{ads}H_m = -51.8$ kJ mol^{-1}. It is found to desorb molecularly at 170 K. The pre-exponential factor for desorption is $10^{15.4}$ s^{-1}. At $T = 170$ K and a coverage of $\theta = 1.00$ ML, calculate the rate of HCOOH desorption in ML s^{-1} and m^{-2} s^{-1}. (c) Suggest a credible explanation for the behavior of $\Delta_{ads}H_m$ and how it changes between the first layer and the subsequent layers.

17.25 A gas-phase reaction occurs between the diatomic molecules AB and CD. At low temperature, a linear Arrhenius plot is found and the products of the reaction are ABC + D. As the temperature is raised further, the Arrhenius plots exhibits curvature and a mixture of ABC, D, AD and BC is formed. Discuss a plausible explanation for these observations, which includes a description of the geometry of the transition state(s).

17.26 Consider the reaction between an atom A and a diatomic molecule BC to form AB + C. Discuss the differences between an early barrier and a late barrier, and the roles played by translational, vibrational and rotational energy in changing the reaction probability.

17.27 A gas-phase reaction between atom A and a diatomic molecule BC has a positive activation energy to reach a collinear transition state. Describe how the reaction to form AB occurs. Suggest a plausible explanation for why the product AC is not formed.

17.28 Why does an activation barrier form?

17.29 Discuss why the combination of two H atoms to form H_2 in the gas phase has no appreciable activation barrier.

17.30 H_2 is the most abundant molecular species in the interstellar medium. Discuss why the gas-phase formation of H_2 from two H atoms plays essentially no role in making H_2 in the interstellar medium.

17.31 Calculate the values of $\Delta^{\ddagger}S^{\circ}_m$, $\Delta^{\ddagger}H^{\circ}_m$ and $\Delta^{\ddagger}G^{\circ}_m$ for the reactions $CH_3 + O_2 \rightarrow CH_3O + O$ and $O + NO_2 \rightarrow NO + O_2$. Comment on the values obtained.

Endnotes

1. Smith, I.W.M. (1982) *J. Chem. Educ.*, **59**, 9.
2. Eyring, H. and Polanyi, M. (1931) *Z. Phys. Chem. B*, **12**, 279; Eyring, H. (1931) *J. Am. Chem. Soc.*, **53**, 2537.
3. Fowler, R.H. and Guggenheim, E.A. (1939) *Statistical Thermodynamics*. Cambridge University Press, Cambridge, UK.
4. In order for two atoms to combine and become a stable molecule they must dissipate the energy released by bond formation. If they do not dissipate this energy, they simply dissociate during the next vibrational period. The energy can either be lost radiatively – which almost never happens since photon emission typically takes more than 1 ns while the vibrational period is less than 1 ps – or it must be lost collisionally to a third body. Collisional deactivation can only occur at high pressure or in the presence of a surface. Therefore, the primary means for the formation of neutral molecules such as H_2 in interstellar space is thought to be on the surfaces of dust/ice grains.
5. Lennard-Jones, J.E. (1932) *Trans. Faraday Soc.*, **28**, 333.
6. Eyring, H. (1935) *J. Chem. Phys.*, **3**, 107; Evans, M.G. and Polanyi, M. (1935) *Trans. Faraday Soc.*, **31**, 875.
7. Lindemann, F.A. (1922) *Trans. Faraday Soc.*, **17**, 598.
8. Christiansen, J.A. (1921) *Reaktionskinetiske Studier problemer fra de kemiske Reaktioners Kinetik med en eksperimentel Undersøgelse af Fosgenets termiske Spaltning*. PhD Thesis, University of Copenhagen.
9. Rice, O.K. and Ramsperger, H.C. (1927) *J. Am. Chem. Soc.*, **49**, 1617.
10. Kassel, L.S. (1928) *J. Phys. Chem.*, **32**, 225.
11. Marcus, R.A. (1952) *J. Chem. Phys.*, **20**, 359.
12. Langmuir, I. (1916) *J. Am. Chem. Soc.*, **38**, 2221; Langmuir, I. (1918) *J. Am. Chem. Soc.*, **40**, 1361.
13. Throughout this chapter, unless explicitly stated otherwise, we assume that the temperature of the gas in contact with the surface is the same as the temperature of the surface. It is an interesting dynamical question to consider how the sticking of a molecule depends on its energy and the energy in each degree of freedom rather than just a single temperature. We return to this question in Section 17.11.

14. Taylor, J.B. and Langmuir, I. (1933) *Phys. Rev.*, **44**, 423.
15. Lennard-Jones, J.E. (1932) *Trans. Faraday Soc.*, **28**, 333.
16. Rettner, C.T. and Auerbach, D.J. (1994) *Science*, **263**, 365; Rettner, C.T., Michelsen, H.A., and Auerbach, D.J. (1993) *Faraday Discuss.*, **96**, 17; Michelsen, H.A., Rettner, C.T., Auerbach, D.J., and Zare, R.N. (1993) *J. Chem. Phys.*, **98**, 8294.
17. Kolasinski, K.W. (2012) *Surface Science: Foundations of Catalysis and Nanoscience*, 3rd edition. John Wiley & Sons, Chichester.

Further reading

Carpenter, B.K. (2005) Nonstatistical dynamics in thermal reactions of polyatomic molecules. *Annu. Rev. Phys. Chem.*, **56**, 57.

Gruebele, M. and Wolynes, P.G. (2004) Vibrational energy flow and chemical reactions. *Acc. Chem. Res.*, **37**, 261.

Henriksen, N.E. and Hansen, F.Y. (2008) *Theories of Molecular Reaction Dynamics: The microscopic foundation of chemical kinetics*. Oxford University Press, Oxford.

Houston, P.L. (2001) *Chemical Kinetics and Reaction Dynamics*. Dover Publications, Mineola, NY.

Kolasinski, K.W. (2012) *Surface Science: Foundations of Catalysis and Nanoscience*, 3rd edition. John Wiley & Sons, Chichester.

Laidler, K.J. (1987) *Chemical Kinetics*. HarperCollins, New York.

Levine, R.D. (2005) *Molecular Reaction Dynamics*. Cambridge University Press, Cambridge.

Ohmori, K. (2009) Wave-Packet and Coherent Control Dynamics. *Annu. Rev. Phys. Chem.*, **60**, 487.

Scoles, G. (1988) in *Atomic and Molecular Beam Methods*, Vol. 1. Oxford University Press, New York.

Scoles, G. (1992) in *Atomic and Molecular Beam Methods*, Vol. 2. Oxford University Press, New York.

Shapiro, M. and Brumer, P. (2003) Coherent control of molecular dynamics. *Rep. Prog. Phys.*, **66**, 859.

Utz, A. (2009) Mode selective chemistry at surfaces. *Curr. Opin. Solid State Mater. Sci.*, **13**, 4.

Zewail, A.H. (1992) *The Chemical Bond – Structure and Dynamics*. Academic Press, London.

CHAPTER 18

Reaction dynamics II: Catalysis, photochemistry and charge transfer

PREVIEW OF IMPORTANT CONCEPTS

- Catalysis is used in 90% of all industrially relevant chemical processes.
- Unless the products are high-value-added chemicals such as pharmaceuticals, catalytic processes are almost invariably heterogeneous.
- A good catalyst must be both active and selective.
- Catalysts change the kinetics of reactions, not the thermodynamics.
- In the Langmuir–Hinshelwood mechanism both species involved in reaction are adsorbed onto the surface. This is the most likely mechanism for a heterogeneous catalytic reaction.
- In an Eley–Rideal mechanism a gas-phase molecule (most likely a radical or ion) reacts directly with an adsorbate. This mechanism is sometimes involved in etching reactions and the reactions of energetic environments with surfaces.
- The hot precursor mechanism is intermediate between the extremes of Langmuir–Hinshelwood and Eley–Rideal.
- The Sabatier principle states that the best catalyst is one with intermediate chemical interactions, with binding of reactants that is neither too strong nor too weak.
- Chemisorption rather than physisorption is usually involved in catalysis because it changes the bonding within the adsorbate.
- Chemisorption can be understood in terms that are similar to coordination chemistry, considering the changes in occupation of the frontier orbitals.
- All electronic states that lie below the Fermi energy are filled, those above it are empty, and those state that straddle it are partially filled.
- Promoters enhance catalytic activity by geometric and electronic effects.
- Poisons/inhibitors decrease catalytic activity by geometric and electronic effects.
- It costs energy to break bonds and push occupied states together. Energy is released when bonds are formed. A transition state is formed by pushing molecules together to make and break bonds. When more breaking than making occurs, a large activation barrier will arise.
- Chain reactions are of importance in combustion, some polymerizations and some photochemical reactions. They are carried by the presence of radicals.
- Explosions, whether thermal or chain-branching, are reactions that are self-accelerating such that their rate increases with time.
- Photochemical reactions are induced by the absorption of light and their rate is related to the rate of photon absorption.
- At low concentrations in well-stirred systems where mass transport is not an issue, the electron transfer rate determines the rate of electrochemical reaction.
- At equilibrium the net current flow is zero because the forward and reverse current flows balance. The equilibrium current density in one direction is known as the exchange current density.
- The magnitude and direction of current flow depends on the applied overvoltage.
- An inner-sphere electron transfer occurs via a ligand that bridges the acceptor and donor centers.
- In an outer-sphere electron transfer the collision complex that is formed to facilitate charge transfer does not involve exchange of ligands in the inner solvation shells of the acceptor and donor.
- The solvation shell must first rearrange to form a transition state in which the electron is degenerate in either the acceptor or the donor. Electron transfer then occurs.
- Electron transfer occurs in a Franck–Condon-like transition in which the light, fast electron tunnels from the donor to the acceptor before any of the nuclei of the acceptor, donor and their solvation shells can move.
- The density of states at the Fermi energy of a metal is orders of magnitude higher than the density of states in a semiconductor band; this means that electron transfer kinetics is inherently faster at metal electrodes than at semiconductor electrodes.

Physical Chemistry: How Chemistry Works, First Edition. Kurt W. Kolasinski.

Companion Website: www.wiley.com/go/kolasinski/physicalchemistry

- The voltage drop near a metal electrode occurs completely in the solution because the electrons in the metal are highly polarizable. Changing the voltage on the metal electrode changes the current exponentially because the voltage drop in solution influences the activation energy for charge transfer.
- The voltage drop at a semiconductor electrode occurs completely in the space charge region. The voltage does not change the relative positions of the band edges at the surface compared to the donor/acceptor level in solution. Instead, the current varies exponentially with applied voltage because the surface density of states changes with bias.

18.1 Catalysis

What does the chemical industry actually make? This is a question that I do not believe was ever posed to me during my undergraduate or graduate education. So it came as quite a shock when I actually looked it up. In Table 18.1 you will find the top 13 chemicals, top three gases and top two metals listed roughly in order of the amounts produced. Annually, in excess of 300 million metric tonnes (MMt; i.e., $>10^9$ kg) of gasoline are produced globally, while over 1000 MMt of iron and over 100 MMt of phosphate, sulfuric acid and ammonia are produced. Much of industrial chemistry involves the

Table 18.1 Bulk chemical production ranked roughly in order of mass produced worldwide. Process information taken from *Ullmann's Encyclopedia of Industrial Chemistry*. Wiley-VCH, Weinheim, 2006.

	Material	Process
1.	Gasoline	Fluid catalytic cracking (FCC) over solid acid catalysts (alumina, silica or zeolite) is performed to crack petroleum into smaller molecules. This is followed by naptha reforming at 1–3 MPa, $300 \leq T \leq 450$ °C, performed on Pt catalysts, with other metals (e.g., Re) as promoters.
2.	Phosphate rock (P_2O_5, phosphoric acid)	Digestion of phosphate rock with H_2SO_4, HNO_3 or HCl
3.	H_2SO_4	Contact process. Catalytic oxidation of SO_2 over K_2SO_4 promoted by V_2O_5 on a silica gel or zeolite carrier at 400–450 °C and 1–2 atm, followed by exposure to wet sulfuric acid.
4.	NH_3	Haber–Bosch process, 30 nm Fe crystallites on Al_2O_3 promoted with K and Ca. SiO_2 added as a structural stabilizer, 400–500 °C, 200–300 bar. Recently, a more active but more expensive Ru catalyst has been developed.
5.	Ethylene (C_2H_4)	Thermal steam cracking of naptha or natural gas liquids at 750–950 °C, sometimes performed over zeolite catalysts to lower process temperature.
6.	Propylene	Byproduct from ethylene and gasoline production (FCC). Also produced intentionally from propane over Pt or Cr supported on Al_2O_3.
7.	Ethylene dichloride	$C_2H_4 + HCl + O_2$ over copper chloride catalyst, $T > 200$ °C, with added alkali or alkaline earth metals or $AlCl_3$
8.	Cl_2	Chlor-alkali process, anodic process in the electrolysis of NaCl. Originally at graphite anodes but new plants now exclusively use membrane process involving a Ti cathode (coated with Ru + oxide of Ti, Sn or Zr) or porous Ni-coated steel (or Ni) cathode (with Ru activator).
9.	NaOH	Chlor-alkali process, cathodic process in the electrolysis of NaCl.
10.	Benzene	From catalytic cracking of petroleum. FCC process over solid acid catalysts (silica, alumina, zeolite) sometimes with added molybdena
11.	Urea, $CO(NH_2)_2$	Basaroff reaction of $NH_3 + CO_2$
12.	Ethylbenzene	Ethylene + benzene over zeolite catalyst at $T < 289$ °C and pressures of ~4 MPa
13.	Styrene	Dehydrogenation of ethylbenzene in the vapor phase with steam over Fe_2O_3 catalyst promoted with Cr_2O_3/K_2CO_3 at ~620 °C and as low pressure as practicable
	Gases	
1.	N_2	Liquefaction of air
2.	O_2	Liquefaction of air
3.	H_2	Steam reforming of CH_4 (natural gas) or naptha over Ni catalyst supported on Al_2O_3, aluminosilicates, cement, and MgO, promoted with uranium or chromium oxides at T as high as 1000 °C.
	Metals	
1.	Fe (pig iron)	Carbothermic reduction of iron oxides with coke/air mixture at $T \approx 2000$ °C.
2.	Al	Hall–Heroult process. Electrolysis of bauxite (Al_2O_3) on carbon electrodes at 950–980 °C.

transformation of carbon-containing molecules. There are seven 'building block' molecules – all derived from petroleum – upon which the chemical industry relies, namely benzene, toluene, xylene, ethylene, propylene, 1,3-butadiene, and methanol.

Looking closer at the processes used to produce these materials you will notice a remarkable trend that 14 of the 18 materials listed below (except for ethylene, N_2 and O_2) are produced by heterogeneous reactions – that is, reactions which occur in the presence of multiple phases. Propylene is produced in part as the byproduct of a homogeneous gas-phase reaction, and also intentionally through a heterogeneous catalytic process. Urea is produced by a gas/surface reaction that does not require a catalyst. It is estimated that 90% of all industrial chemical processes require catalysis.

Greater selectivity is generally observed in homogeneous catalysis compared to its heterogeneous counterparts. This is especially true when the control of chirality is important, but for bulk chemicals the catalysts are invariably heterogeneous. The cost of separating a homogeneous catalyst either from the product stream or out of the reaction vessel is simply too great unless the product is a high-value-added molecule, such as a pharmaceutical agent.

Ionic liquids offer the advantages of both homogenous and heterogeneous catalysts with their two main characteristics: selected ionic liquids may be immiscible with the reactants and products, but on the other hand the ionic liquid may dissolve the catalysts. Ionic liquids combine the advantages of a solid for immobilizing the catalyst, and the advantages of a liquid for allowing the catalyst to move freely. While not currently widely used in industrial catalysis, ionic liquids may find a growing role in the future.

The refining of petroleum and the production of petrochemicals are not possible at the scale and required cost without the use of catalysis. The use of zeolites is particularly important in this area as both a catalyst and support. Zeolites are of interest as solid-base catalysts for reactions such as condensations, alkylations, cyclizations and isomerizations. Zeolites can also present acid sites or a combination of both acid and base sites. Their flexibility is further enhanced by the nature of their porosity and active sites in that their micropores can be tuned in a size range comparable to that of molecules. The nature of the active sites within these pores is controlled by the substitution of transition metals and other elements into their walls. A particularly popular type of zeolite known as ZSM-5 has the correct pore structure and active sites to convert biomass-derived molecules into aromatic hydrocarbons and olefins.

Many of the other catalytic systems involve transition or coinage metals dispersed on porous oxide substrates. The production of Cl_2, NaOH and Al are all electrocatalytic processes. How do catalysts work? What are the active sites? Why are some sites active? These are questions that we need to address. Of course, biochemical reactions often involve catalysis by enzymes. We shall see that a great deal can be transferred from an understanding of catalysis by model heterogeneous systems to the understanding of these highly complex homogeneous catalysts.

The prevalence of catalysis in the chemical and pharmaceutical industries can be attributed to two characteristics – activity and selectivity. Both of these characteristics must be present in a good catalyst, as represented schematically in Fig. 18.1. Simply accelerating reactivity by lowering activation barriers is of no use if side reactions are accelerated to a comparable amount. A good catalyst is also long-lived. It should be regenerated in the process once it has helped to produce a product molecule. Ideally, a good catalyst is also inexpensive. This can be an important practical characteristic. Inexpensive has multiple aspects. One is the cost of the raw materials. Iron, aluminum and zinc are cheap, but rhodium, iridium and platinum are expensive. Replacing precious metals with base metals is a constant topic of research. Separations are expensive; hence, a preference for highly selective heterogeneous catalysts. Replacing a catalyst is expensive both because of the material's cost and because the process has to be shut down while the catalyst is replaced or regenerated.

It is important to note that catalysts do not change the thermodynamics of products versus reactants. They do not change the equilibrium constant; rather, they change the kinetics of a reaction. An easier catalytic process to develop is the accelerated production of the thermodynamically most stable species. A more difficult catalytic process to develop is to create a process that selectively creates a thermodynamically less-stable product selectively compared to a more stable product, or to selectively produce just one or a few products out of a selection of molecules that have similar thermodynamic stability. Particularly in the latter case, the best catalyst may not be the most active, but will instead be the one with the most advantageous combination of activity and selectivity.

18.2 Heterogeneous catalysis

A homogeneous catalyst exists in the same phase as the reactants and products, while a heterogeneous catalyst exists in a different (or immiscible) phase. Both types of catalyst enhance activity (accelerate the reaction) and selectivity (make

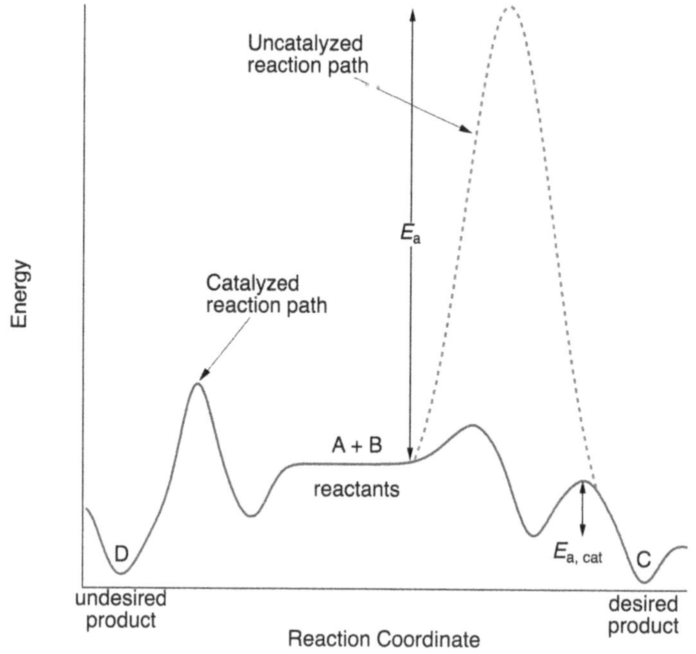

Figure 18.1 The purpose of a catalyst is to produce a desired product not only rapidly but also selectively. This requires the catalyst to lower the activation barrier along the desired reaction path such that it is favored more than any other reaction path.

more of a desired product). To learn how they do this we begin with heterogeneous catalysis and the kinetics of surface reactions. Solid surfaces allow us to create model systems that we can investigate experimentally in great detail. What we learn from these model systems will enable us to better understand reactivity and catalysis in general.

Kinetics on solid surfaces is unlike kinetics in other phases because we need to count the number of sites that can react. We have already encountered this is Chapter 17 in the treatment of adsorption and desorption kinetics. Recall that surface concentrations are often quoted in terms of either *absolute coverage* σ (the number of adsorbates or sites per unit area) or *fractional coverage* θ,

$$\theta = \sigma/\sigma_0. \tag{18.1}$$

The variable σ_0 is defined as either the total number of adsorption sites or number of surface atoms in a single layer. Let us now discuss three general classes of surface reaction mechanisms and their effects on kinetics.

18.2.1 Surface reaction mechanisms

Reactions occurring on solid surfaces are generally classified as occurring according to one of three classes of mechanisms: Langmuir–Hinshelwood, Eley–Rideal, and hot precursor. Here, we will discuss these three mechanisms in terms of bimolecular reactions. A complete catalytic surface reaction mechanism will include adsorption, reaction, and desorption. Adsorption can be either molecular (nondissociative) or dissociative.

In the *Langmuir–Hinshelwood mechanism* both species involved in reaction are adsorbed onto the surface,

$$\mathrm{A(a)} + \mathrm{B(a)} \rightarrow \mathrm{AB(a)} \rightarrow \mathrm{AB(g)}. \tag{18.2}$$

The mechanism, depicted with an example from Cynthia Friend's group in Fig. 18.2(a), is named after Irving Langmuir, who received the Nobel Prize in Chemistry 1932, and Cyril N. Hinshelwood, who received the Nobel Prize in Chemistry 1956. The two adsorbates are fully equilibrated with the surface. At least one of these species must be mobile on the surface so that surface diffusion brings about a collision between A and B on the surface. The product AB is then formed at the surface. The product may also be adsorbed onto the surface. However, in some cases the product may have such a short lifetime on the surface that it does not fully equilibrate before it desorbs into the gas phase.

The defining characteristic of the Langmuir–Hinshelwood mechanism is that both reactants are adsorbed. As we shall explain below, at least one of the reactants will be chemisorbed. Physisorption is not excluded as a possibility for the reactants; however, chemisorption is usually required to sufficiently perturb a molecule such that it will be more reactive than it would be in the absence of a catalyst. Diffusion of adsorbates is also always involved in Langmuir–Hinshelwood

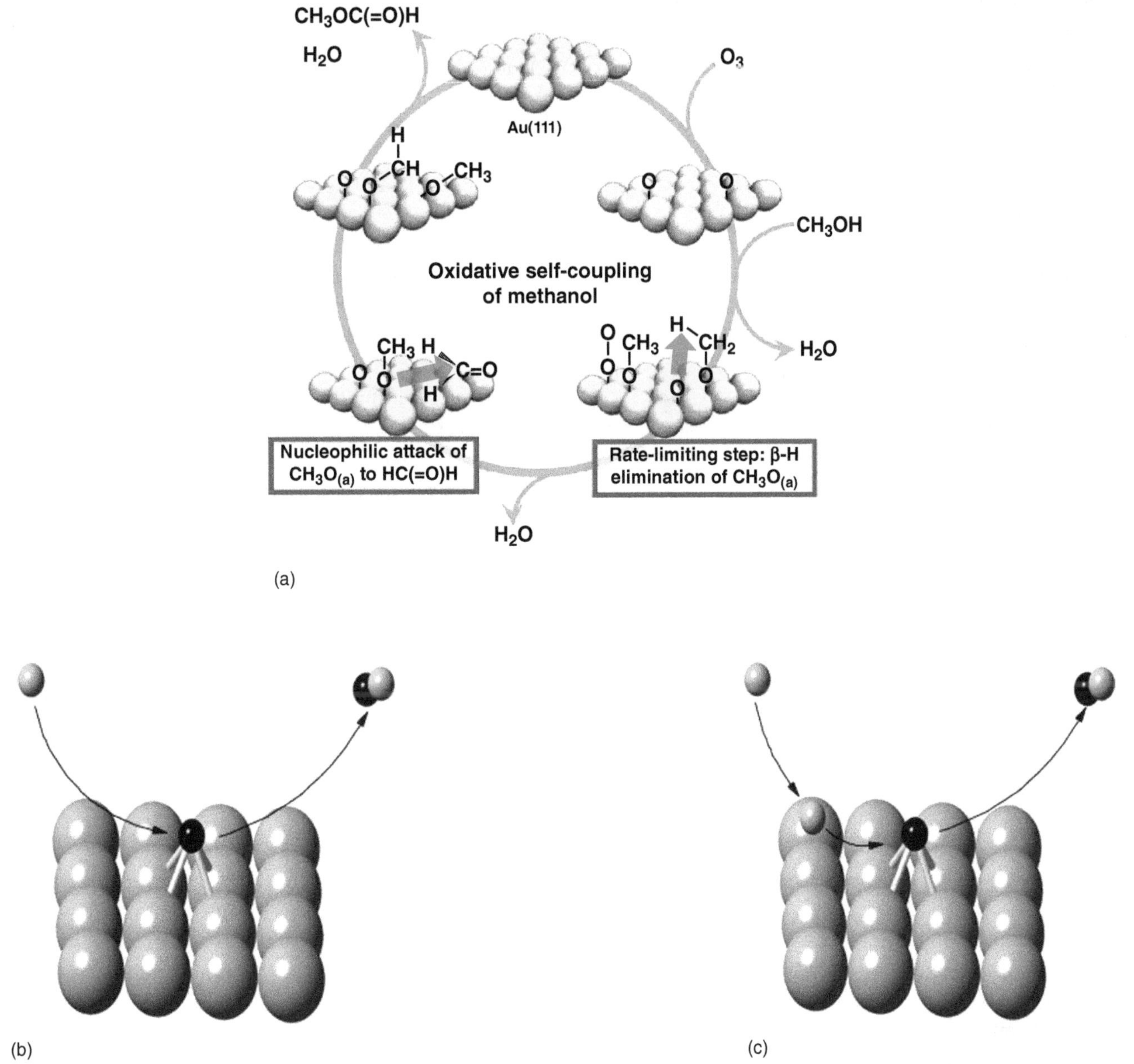

Figure 18.2 (a) Langmuir–Hinshelwood reaction between adsorbates. Provided by Cynthia M. Friend. Reproduced with permission from Xu, B.J., Madix, R.J., and Friend, C.M. (2014) *Acc. Chem. Res.*, **47**, 761. © 2014, American Chemical Society. (b) Eley–Rideal reaction between one adsorbed molecule and one molecule from the gas phase. (c) Hot precursor mechanism between one adsorbed molecule and one molecule that has not yet fully equilibrated (the hot precursor).

reactions. Because both reactants must be adsorbed onto the surface, our first approximation is that the kinetics of a reaction that follows a Langmuir–Hinshelwood mechanism will depend on the coverages of the reactants

$$r = k\theta_A^{\alpha}\,\theta_B^{\beta}. \tag{18.3}$$

Just as for gas-phase elementary reactions, a surface elementary reaction has an order of 1 if it is unimolecular, and an order of 2 if it is bimolecular. For a reaction involving a fluid in contact with a surface (either liquid or gas) the coverage at equilibrium is related to the concentration in the fluid phase by the adsorption isotherm. Thus, the rate will also usually depend on partial pressures or concentrations. However, this dependence may be quite complex and change dramatically with temperature and the magnitude of the concentration because it is the surface coverage and the mixing of adsorbates by diffusion and adsorption that control the rate directly. *This mechanism is by far the most commonly observed in catalytic chemistry*. Reactions that exhibit Langmuir–Hinshelwood kinetics are expected to depend strongly on the surface temperature.

Dan Eley and Eric Rideal proposed[1] an alternative mechanism, as shown in Fig. 18.2(b), in which an adsorbed species reacts directly with a gas-phase species. The rate of reaction should then be directly related to the pressure of the gas-phase species and the coverage of the adsorbed species. Little surface temperature dependence is expected in such a reaction because the incident species must already have sufficient energy to facilitate reaction. It can be shown[2] that any reaction with a substantial activation barrier will preferentially react by a Langmuir–Hinshelwood mechanism rather than an *Eley–Rideal mechanism*.

Ironically, Eley–Rideal dynamics probably was not involved in the experiments for which it was proposed. Clear evidence of the occurrence of a reaction via the Eley–Rideal mechanism did not occur until decades later, the best examples being provided by Charles T. Rettner and Daniel J. Auerbach[3] in which H atoms react with adsorb H, D, or Cl atoms. Their work also shows us where an Eley–Rideal reaction might be expected to occur: in the reaction of a radical with an adsorbed species. Reactions involving radicals such as H atoms often exhibit a small or even negligible activation barrier. These direct reactions are known as *abstraction reactions*. It is usually observed that the abstraction of an atom, as in $H + RH \rightarrow H_2 + R$, occurs with a much higher rate and lower activation barrier than the abstraction of a radical.

The most convincing manner to prove the occurrence of Eley–Rideal dynamics is to measure the product energy distribution. Because one of the species is not adsorbed, the reaction product must have a 'memory' of the energy distribution of the incident molecule. This is exactly what Rettner and Auerbach were able to prove by measuring translational and rovibrational energy state distributions of highly energetic products.

A third mechanism lies intermediate between these two, as shown in Fig. 18.2(c). In the *hot precursor mechanism* proposed by Bengt Kasemo and John Harris,[4] one species is fully equilibrated and adsorbed. The second reactant lands on the surface, enters a mobile precursor state, and before it can fully equilibrate it reacts with the adsorbed reactant. Any time there is the occurrence of Eley–Rideal dynamics there will almost certainly be hot precursor dynamics happening in parallel. At very high temperatures, the surface residence time of adsorbed molecules is extremely short – so short, in fact, that they may not be able to completely equilibrate during the few collisions they have with the surface. Therefore, hot precursor dynamics may be important for understanding the kinetics of high-temperature surface chemistry. For both Eley–Rideal and hot precursor mechanisms, surface diffusion is not required.

18.2.2 Examples of Langmuir–hinshelwood dynamics

The reaction of CO and O_2 to produce CO_2 is important as a model reaction. It is involved in combustion and is an extremely important reaction that occurs within the catalytic converter in the emissions control system of motor vehicles. Its study has also been the basis of a huge variety of experiments that have advanced our knowledge of catalytic chemistry.

Mix CO and O_2 in a glass container at room temperature and 1 bar pressure and the reaction does not proceed at an appreciable rate. Heating the mixture will eventually promote reaction. However, if a film of a platinum group metal is evaporated onto the walls of the container, the reaction occurs rapidly at a much lower temperature. The presence of the metal catalyzes the reaction, making it occur more rapidly at a lower temperature than for the homogeneous gas-phase system.

Now consider the temperature-programmed desorption data in Fig. 18.3. We detect molecules that desorb from a surface as a function of surface temperature and identify them using quadrupole mass spectrometry (QMS). Isotopically labeled $^{13}C^{16}O$ and $^{18}O_2$ are exposed to a palladium sample that has been polished to expose the (111) plane. Vibrational spectroscopy shows that both molecules chemisorb to the surface in molecular form when the surface is held at 100 K. When a pure O_2 layer is heated, most of the O_2 desorbs molecularly below 200 K. A fraction of the molecularly adsorbed O_2 dissociates on the surface above 180 K and remains on the surface as chemisorbed O atoms, which recombine to desorb as O_2 at high temperature. The CO is more strongly bound than the O_2. If only CO were placed on the surface, heating would lead to desorption of molecular CO at around 500 K. When CO is exposed to an O_2-covered surface it can actually displace the O_2(a) from the surface. When the temperature of a coadsorbed CO + O_2-layer is increased, we observe that O_2 desorbs molecularly at low temperature just as it did from the pure layer. For the conditions used in this experiment there is some desorption of CO in the 500 K peak expected of a pure CO-layer. We also observe the presence of a new peak, which corresponds to the production of CO_2 above 350 K. Importantly, the CO_2 desorbs at $^{18}O^{13}C^{16}O$. No CO_2 is detected while the surface is held at 100 K and exposed to CO and O_2 simultaneously, when CO is exposed to an O_2 layer, or when O_2 is exposed to chemisorbed CO. Vibrational spectroscopy reveals that CO_2(a) is not stable on the

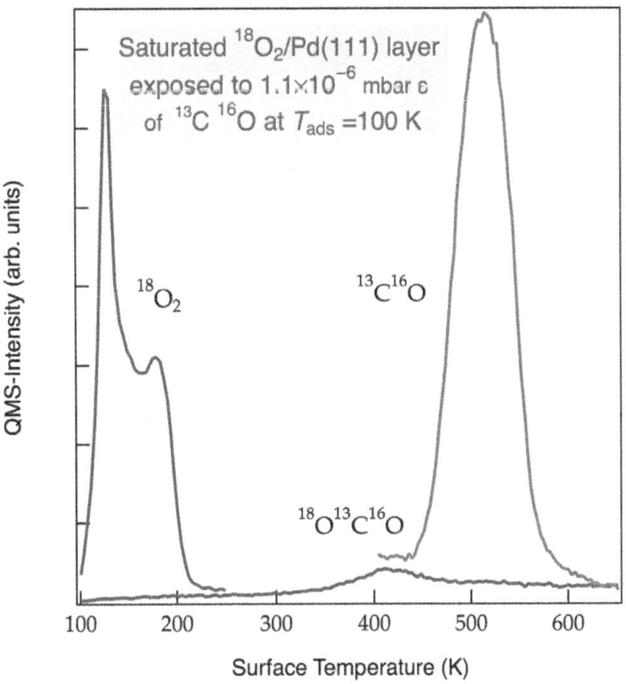

Figure 18.3 CO oxidation on Pd(111) clearly follows a Langmuir–Hinshelwood mechanism because the product CO_2 is only detected in a mass spectrometer after O_2 and CO have adsorbed onto the surface, and then the surface is heated sufficiently to overcome the barrier to reaction. The reaction only commences after some O_2 has desorbed molecularly and some of the adsorbed O_2 had dissociated to adsorbed atoms. The O(a) then reacts with CO(a) within the peak that has a maximum just above 400 K. Not all of the CO(a) reacts; hence, some is left to desorb molecularly in the peak centered at 500 K. Reproduced with permission from Kolasinski, K.W., Cemič, F., de Meijere, A., and Hasselbrink, E. (1995) *Surf. Sci.*, **334**, 19. © 1995, Elsevier.

surface at 350 K (its desorption peak is below 100 K). Therefore, the desorption of CO_2 is reaction-limited – as soon as the CO_2 is formed it immediately desorbs.

These observations allow us to conclude that this reaction occurs via a Langmuir–Hinshelwood mechanism, and the reaction product is only formed after both reactants are adsorbed onto the surface. Furthermore, the reaction occurs between O(a) that results exclusively from the dissociation of O_2 and neither from any dissociation of CO nor the reaction of CO with CO. This is proven by the observation of isotopic mixing in the CO_2 and the lack of mixing in the desorbed CO. The CO oxidation reaction, therefore, exhibits the following mechanism

$$\tfrac{1}{2}O_2(g) \rightarrow \tfrac{1}{2}O_2(a)$$
$$\tfrac{1}{2}O_2(a) \rightarrow O(a)$$
$$CO(g) \rightarrow CO(a)$$
$$CO(a) + O(a) \rightarrow CO_2(g)$$

Overall:

$$CO(a) + \tfrac{1}{2}O2(g) \rightarrow CO_2\,(g).$$

Recall that a mechanism must always sum to produce the stoichiometry of the overall reaction. Hence the need for the stoichiometric number ½ in the first two steps.

As long as this reaction is performed at low coverage conditions (a combination of a high enough temperature and low enough partial pressures of CO and O_2), the rate-determining step is the surface reaction in the final step. As alluded to above, the chemisorption of CO and O_2 is competitive. CO can displace O_2 from the surface. CO oxidation exhibits asymmetric inhibition. CO can adsorb onto a surface that is covered with O_2 or O. Their saturation coverage is low and provides empty sites. However, CO forms a dense layer upon which O_2 cannot dissociate (dissociation requires adjacent empty sites). This asymmetry means that, under the correct conditions of pressure and temperature, this system exhibits oscillatory kinetics and spatiotemporal pattern formation[5] such as target patterns, spiral waves and soliton waves.

The ammonia synthesis reaction

$$N_2(g) + 3H_2(g) \rightarrow 2\,NH_3(g)$$

in the Haber–Bosch process is an extremely important industrial reaction since it underpins modern agriculture and without it, the world's population could not be fed. It is estimated that roughly 3% of world energy consumption is used to run this reaction. Its importance is reflected in history: Fritz Haber was awarded the Nobel Prize in Chemistry 1918 for the synthesis of ammonia from its elements. Carl Bosch was awarded the Nobel Prize in Chemistry 1931 for developing a practical high-pressure processes to carry out the reaction industrially. Gerhard Ertl was awarded the Nobel Prize in Chemistry 2007 for his studies of chemical processes on solid surfaces, in particular elucidating the mechanism of the ammonia synthesis reaction. Under industrially relevant conditions, the rate-determining step is the dissociation of N_2. A large activation barrier stands between $N_2(g)$ and the adsorbed atoms. After the dissociation of N_2 and H_2, the chemisorbed atoms react to form NH(a), $NH_2(a)$, then NH_3 that immediately desorbs. But why does adsorption represent the rate-determining step in one Langmuir–Hinshelwood reaction, while a surface reaction between adsorbates determines the rate in another? These details will always be specific to the reaction system of interest, but to address them more generally we need to understand more about the chemisorption bond and how a catalyst works.

Before we move on to a discussion of chemisorption, let us consider the reaction of $CO_2 + NH_3$. Kinetic data for this reaction is depicted in Fig. 16.1. CO_2 is created in large quantities during ammonia synthesis because H_2 is produced by the steam reforming of hydrocarbons or coal. As an example, H_2 is produced from natural-gas-sourced methane,

$$CH_4(g) + H_2O(g) \rightleftharpoons CO(g) + 3H_2(g),$$

after which CO_2 is formed by CO oxidation. The reaction of $CO_2 + NH_3$ to produce ammonium carbamate is the first step in the Basaroff reaction to form urea. The second step is decomposition at high temperature and pressure to form $CO(NH_2)_2 + H_2O$. The data in Fig. 16.1 show that ammonium carbamate formation exhibits simple first-order kinetics when formed from a stoichiometric mixture of $CO_2 + NH_3$ at 45 kPa and room temperature. A plausible explanation of the simple kinetics for this complex surface reaction can be found in the discussion above. The first two steps of the mechanism are the adsorption of NH_3 and CO_2

$$NH_3(g) \rightarrow NH_3(a)$$
$$CO_2(g) \rightarrow CO_2(a)$$
$$NH_3(a) + CO_2(a) \rightarrow \cdots \rightarrow \cdots \rightarrow NH_2CO_2NH_4(a).$$

We do not know what the surface reaction steps are without further investigation. NH_3 is bound strongly enough at room temperature that is rapidly achieves its saturation coverage under these conditions, and this coverage is little affected by changes in NH_3 pressure over a wide range. CO_2 is extremely weakly bound. Its saturation coverage is much smaller than that of NH_3, and because of this its coverage is limiting the reaction rate. The adsorption of CO_2, a process that is first order in CO_2 partial pressure, is the rate-limiting step of the reaction.

18.2.2.1 Directed practice

Why is the CO_2 peak so small and the CO peak so large in Fig. 18.3?

18.2.3 Molecular chemisorption: The Blyholder model of CO chemisorption

Molecules and atoms can either physisorb or chemisorb to surfaces. *Physisorption* is controlled by weak intermolecular interactions. The properties of the molecule are little changed by physisorption. The electronic levels do not shift or distort much compared to a gas-phase molecule. The vibrational frequencies do not change much either. In other words, the bonding in the molecule does not change much and, therefore, the reactivity of the molecule does not change much. Physisorbed molecules also tend to have a short lifetime on the surface because of the weak binding. The only time that physisorption will be important for catalysis is if the molecule is reactive enough already but we need to bring it into contact with a reactive species at the surface of the catalyst.

Chemisorption is much more important for activating molecules. Chemisorption involves the establishment of new chemical interactions with a surface. Much like there is conservation of mass and charge and energy, there is in a certain sense also a conservation of bonding. If an adsorbate makes new chemical bonds with a surface and it was fully saturated before interacting with the surface, then the chemisorbed species makes its chemisorption bond at the expense of the

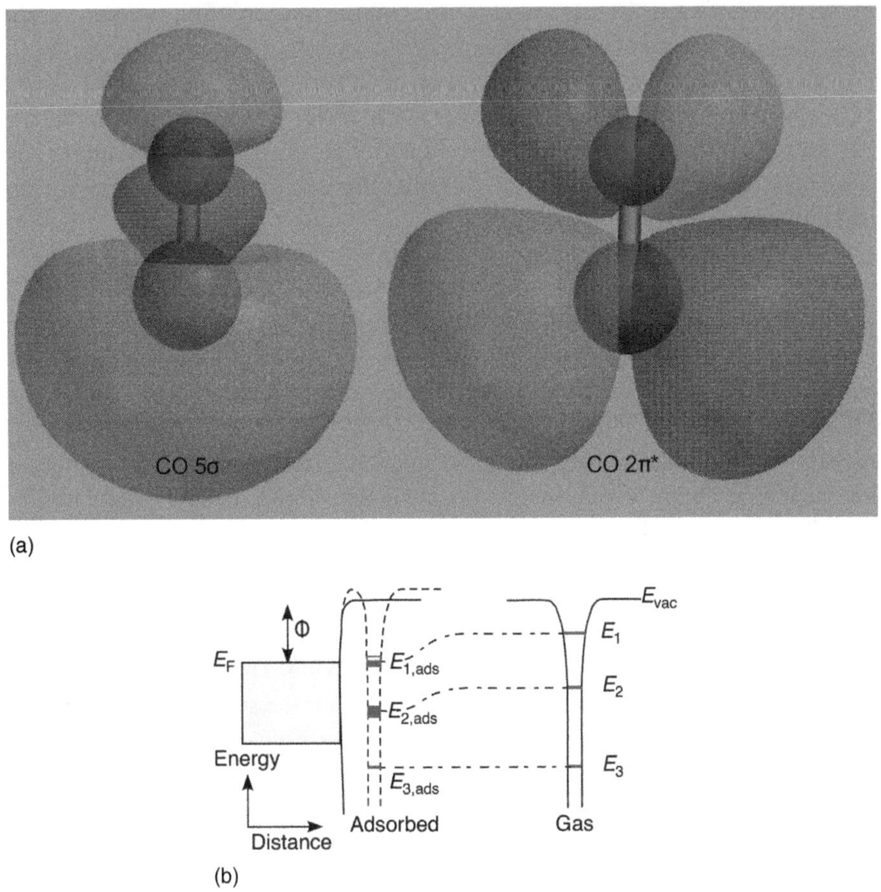

Figure 18.4 (a) Quantum chemical calculations reveal the shapes of the 5σ (HOMO) and $2\pi^*$ (LUMO) molecular orbitals of CO. (b) Diagram of broadening and shifting of adsorbate levels as they approach a surface. E_F, Fermi energy; E_{vac}, vacuum energy; Φ work function of the surface material; E_1, E_2, E_3, energies of molecular orbitals 1, 2 and 3, respectively, of the molecule far from the surface; $E_{1,ads}$, $E_{2,ads}$, $E_{3,ads}$, energies of molecular orbitals 1, 2 and 3, respectively, of the adsorbed molecule; shaded area, occupied band (e.g., valence band). Reproduced with permission from Kolasinski, K.W. (2012) *Surface Science: Foundations of Catalysis and Nanoscience*, 3rd edition. John Wiley & Sons, Chichester.

strength of other bonds in the molecule. This and the *Sabatier principle* are the essence of catalysis. Enunciated by Paul Sabatier,[6] this principle states:

> *The best catalyst is one with intermediate chemical interactions. One that binds too strongly may never allow the reactants to react or may not allow release of products. One that binds too weakly may not activate the reactants or may not have a long enough residence time to allow them to meet and react.*

How does the chemisorption bond form? We answer this with the help of Fig. 18.4 and take CO as a specific example. George Blyholder[7] was the first to describe chemisorption along these lines. A molecule in the gas phase has energy levels with well-defined (sharp) energies. Each one of these levels contains two electrons, or one electron if the molecule is also a radical. There are also normally unoccupied orbitals, which contain no electrons when the molecule is in the ground state. These orbitals are occupied in excited states. Each molecular orbital is described as being bonding, antibonding, or nonbonding. A bonding orbital stabilizes the molecule. The occupation of such an orbital lowers the energy of the molecule by prompting bond formation. Bonds form when electron–electron and nucleus–nucleus repulsions are minimized and electron–nucleus interactions are maximized. An antibonding orbital destabilizes the bonding in the molecule. One half of the sum of all the electrons in bonding orbitals minus the number of electrons in antibonding orbitals is equal to the bond order. The higher the bond order, the stronger the bonding in a molecule. The occupation of nonbonding orbitals does little or nothing to change the bond order. An example is a lone pair of electrons.

We will take our surface to be that of a transition metal. All states are filled up to the Fermi energy E_F. The valence electron states form bands with a width of several electron volts rather than sharp levels. As a molecule approaches the surface, its sharp levels begin to interact with the surface electronic states, and this causes the molecular levels to shift and broaden in energy. The broadening and shifting are caused by mixing the molecular wavefunctions with the metallic wavefunctions. The molecular states remain primarily localized on the molecule and we can think of the two systems as largely separate.

Once the molecule is attached to the surface, the occupation of electronic states follows the same rule: all states below E_F must be occupied. However, now that the electronic states of the molecule are broadened, there are three possibilities: (i) A state that lies fully below E_F will be completely occupied with two electrons; (ii) A state that lies fully above E_F will be fully empty with zero electrons; and (iii) A state that straddles the Fermi energy will be partially occupied with more than zero but fewer than two electrons.

As in coordination chemistry, the first orbitals to consider are the *frontier orbitals* – the highest occupied molecular orbital (HOMO) and the lowest unoccupied molecular orbital (LUMO). Core levels are strongly localized about their nuclei and do not participate in bonding directly. We will discuss bonding and molecular orbitals in detail in Chapter 25. For CO the HOMO is a 5σ orbital; the LUMO is a $2\pi^*$ orbital. The 5σ is nonbonding with respect to the C—O bond. The $2\pi^*$ orbital is antibonding with respect to the C—O bond. The 5σ orbital is localized on the C end of the molecule. The $2\pi^*$ is symmetrically distributed along the molecular axis.

In the chemisorbed state, the 5σ orbital is completely occupied as it lies below E_F. The $2\pi^*$ is partially occupied. The degree to which it is occupied depends on which metal the CO is binding to. The 5σ orbital interacts strongly with the metal electrons. Effectively, the electron density of the 5σ orbital is donated to the metal and new hybrid electronic states are formed. The state created by donation from the 5σ orbital is bonding with respect to the chemisorption bond. This state is predominately localized about the C end of the molecule. The $2\pi^*$ orbital accepts electron density from the metal through a process called *backdonation*. New hybrid electronic states are constructed which are primarily localized about the CO molecule and are bonding with respect to the chemisorption bond. The overlap of the 5σ orbital with the metal states is most favorable if the molecule is oriented with the C end toward the surface. The overlap of the $2\pi^*$ orbital with the metal states is most favored by a linear geometry. Consequently, we predict that the CO molecule should chemisorb C-end down, with its axis along the normal to the surface. This expectation is confirmed by experiment. CO is nearly always bound in an upright geometry.

Donation from the nonbonding 5σ orbital has little effect on the strength of the C—O bond. In contrast, backdonation into the antibonding $2\pi^*$ orbital decreases the bond order of the C—O bond. Decreasing bond order means that the CO molecule is more weakly bound and is reflected in a decrease in the vibrational frequency of the C—O stretch. Decreasing the bond order also makes CO in the chemisorbed state more reactive than CO in the gas phase. The metal catalyzes reactions involving CO – not only CO oxidation but the rich chemistry known as Fischer–Tropsch chemistry that starts with mixtures of CO + H_2 – by producing a more active chemical form of CO with a weaker C—O bond.

18.2.4 Dissociative chemisorption of H_2: The Nørskov model

Some catalytic reactions require not just the weakening of bonds, but the breaking of them. The simplest case is that of H_2, which has just two electrons in one bonding molecular orbital, the 1σ orbital. To break the H–H bond we could imagine sucking the electrons out of the bonding orbital to reduce the bond order. Alternatively, in analogy with what happened with backbonding to CO, we might add electron density to the next higher molecular orbital, the antibonding $1\sigma^*$ orbital. A series of calculations by Jens K. Nørskov and coworkers[8] demonstrated that it is the latter that occurs. They also developed the following description of the dissociation dynamics that we trace in Fig. 18.5.

As a closed-shell species such as H_2 approaches a surface, a repulsive interaction develops due to the Pauli repulsion of the electrons in the solid and the fully occupied 1σ molecular orbital in the molecule. For a rare gas atom this is the end of the story. The interaction of the electrons simply becomes more repulsive as we push the atom closer to the surface. There is a weak physisorption minimum relatively far from the surface caused by van der Waals forces; however, no chemical bond is formed. For H_2, on the other hand, electron density can be donated from the metal into the $1\sigma^*$ antibonding orbital if its energy falls below the Fermi level. Electron transfer leads to a weakening of the bond, and eventually dissociation will occur if enough electron density can be transferred.

We see that there is a competition between the Pauli repulsion, which raises the energy of the system, and the donation of electron density into the $1\sigma^*$, which lowers the energy of the system. Donation occurs because the $1\sigma^*$ drops in energy and broadens as H_2 approaches the surface. When it drops below E_F, electron density is transferred from the metal to the H_2 molecule. This interaction is bonding with respect to the surface but antibonding with respect to the H–H bond. If the energetic stabilization achieved by donation into the $1\sigma^*$ exceeds the destabilization engendered by the Pauli repulsion, the energy landscape is downhill all the way. Adsorption is nonactivated, analogous to the potential energy curve in Fig. 17.7(a). In contrast, if the destabilization caused by Pauli repulsion exceeds the stabilization accompanying the

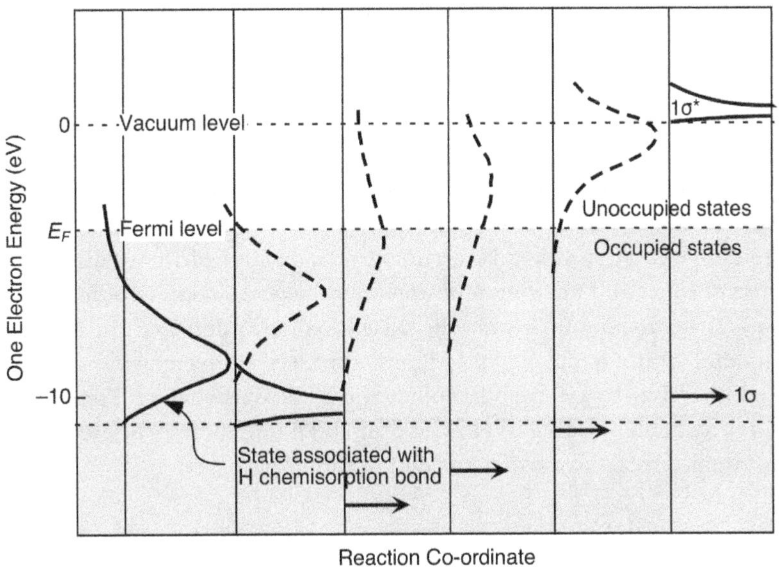

Figure 18.5 Progressive filling of the antibonding molecular orbital of H_2 as it approaches the surface weakens the H–H bond and eventually leads to dissociative hydrogen adsorption. The presence of an activation barrier to dissociative adsorption depends on the interplay between the energetic cost of Pauli repulsion and the energetic gain of forming new bonds to the surface. Redrawn after the work of Nørskov, J.K., Houmøller, A., Johansson, P.K., and Lundqvist, B.I. (1981) *Phys. Rev. Lett.*, **46**, 257. Original figure © American Physical Society.

formation of the chemisorption bond, adsorption will be activated and follow a potential energy curve of the type shown in Fig. 17.7(b). The presence or absence of a barrier depends on which metal is chosen. A high barrier is found on the coinage metals Cu, Ag and Au, while a small or no barrier is found on the platinum group metals, Pt, Pd, Ir, and Rh. The height of the barrier is quite sensitive to electronic structure. Therefore, some sites on a surface, such as steps, may exhibit a lower barrier than others, such as terraces.

From this dynamics we can state a general principle regarding activation barriers in chemical reactions.

It costs energy to break bonds and push occupied states together. Energy is released when bonds are formed. A transition state is formed by pushing molecules together to make and break bonds. When more breaking than making occurs, a large activation barrier will arise.

18.2.5 Promoters and poisons

The geometrical and electronic structure of catalysts can be manipulated to tailor the reactivity of catalysts. Agents that enhance reaction rates are called *promoters*, while those that decrease reactivity are called *poisons*. Promoters are usually beneficial. Poisons must often be avoided, though they can also be employed to shut down unwanted side reactions.

The reactivity of a catalyst – its ability to partake in donation and backdonation – is strongly correlated with its geometric and electronic structure. Promoters change (or stabilize) some aspect of geometrical or electronic structure that enhances reaction rates. For example, in the ammonia synthesis reaction N_2 dissociation is rate-limiting. Just as for H_2 dissociation, N_2 dissociation is facilitated by shifting electron density from the surface to antibonding orbitals on the molecule. Increasing the electron density on the metal that is available for donation should help activate the N_2 molecule by lowering the barrier to dissociation. The addition of an electropositive metal such as K to the catalyst, which is made of Fe, has precisely this beneficial effect. The adsorbed K atom, which resembles a positive ion when it is chemisorbed may also be able to stabilize the transition state of the dissociating molecule by electrostatic interactions. The addition of K to the ammonia synthesis catalyst is an example of electronic promotion – the addition of promoters to enhance the reactivity of the catalyst.

The Fe catalyst only remains highly active if the Fe is dispersed in nanoscale particles on the surface of a porous Al_2O_3 substrate. At the high temperatures associated with the Haber–Bosch process, the Fe has a tendency to ball up into larger particles as expected from Ostwald ripening; this process is called *sintering*. The addition of CaO and MgO to the catalyst is found to stabilize the structure of the catalyst and reduce the rate of sintering. This is an example of geometric promotion – the addition of promoters that keep the catalyst in its active form.

Poisons also act through both geometrical and electronic mechanisms. Geometrical poisoning is most commonly a short-range direct interaction called *site blocking*. We have already encountered the effects of site blocking in the Langmuir model of adsorption and in the asymmetrical inhibition of CO oxidation. If sites are occupied by atoms or molecules that do not participate in the reaction, then those sites are lost as far as enhancing the rate of reaction. Particularly sensitive to the effects of site blocking poisons are catalysts that only exhibit reactivity at minority active sites. If these active sites are blocked the reactivity can be shut down entirely.

Poisons can also act through long-range electronic effects. Imagine that we add an electronegative species such as As, P or S rather than an electropositive species to our ammonia synthesis catalyst. The electron-withdrawing nature of these atoms deactivates a number of atoms around them, not just the one site they are bound to. Thankfully, both S and Pb poison the threeway automotive catalyst. Immediately upon introduction of the catalytic converter, which has as its active component a mixture of Pd, Pt and Rh, it became apparent that the S content of fuel has to be reduced and the Pb content had to be removed. The reduction of S is nice because it reduces odors and the formation of particulate matter. Humans will be forever grateful for the removal of Pd from motor fuels, as Pb exposure – particularly at an early age – is correlated with a decreased IQ and increases in both violent and addictive behavior. The removal of Pb from motor fuel has greatly reduced the environmental loading of Pb. The drop in Pd pollution associated with automotive exhaust has been measured directly in Alpine glacial lake sediments and Antarctic ice cores.

18.2.6 Directed practice

What is meant by a short-range or a long-range effect of a promoter or poison?

18.3 Acid–base catalysis

Reactions in aqueous solutions can often be homogeneously catalyzed by acid, base or both, as is the case for ester hydrolysis. A hydrolysis reaction will exhibit pseudo-first-order kinetics in the concentration of the substrate [S]. In specific acid–base catalysis, the rate is equal to the sum of the uncatalyzed reaction, with rate constant k_0, plus the rates of the reactions catalyzed by acid and base, with rate constants k_H and k_{OH}, respectively,

$$R = k_0[S] + k_H[S][H^+] + k_{OH}[S][OH^-]. \tag{18.5}$$

The concentrations of H^+ and OH^- are linked by the water autoionization equilibrium

$$K_w = [H^+][OH^-]. \tag{18.6}$$

Therefore, the rate of reaction can be written in terms of an effective first-order rate equation

$$R = k[S] \tag{18.7}$$

where

$$k = k_0 + k_H[H^+] + \frac{k_{OH}\,K_w}{[H^+]}. \tag{18.8}$$

Obviously, at pH values far from neutral, only acid or base catalysis – but not both – will contribute significantly to the reaction rate. If at a low pH the reaction rate is dominated by acid catalysis, then we can neglect the base-related term. Taking the logarithm of Eq. (18.8) yields

$$\lg k = \ln k_0 + \lg K_H + \lg[H^+] = \ln k_0 + \lg K_H - \text{pH}. \tag{18.9}$$

A plot of lg k versus pH should then have a slope of −1 at sufficiently low pH. If the reaction rate is dominated by the uncatalyzed reaction near neutral pH, the line will become a curve and then a horizontal line when the acid-catalyzed pathway is no longer important due to the low concentration of H^+. If increasing the pH further leads to reaction dominated by the base-catalyzed mechanism, the rate constant will then be described by

$$\lg k = \ln k_0 + \lg K_{OH} + \lg K_w - \lg[H^+] = \ln k_0 + \lg K_{OH} + \lg K_w + \text{pH}. \tag{18.10}$$

The plot of lg k versus pH will curve upward and approach a slope of +1 at sufficiently high pH. This ideal behavior is depicted in Fig. 18.6. Obviously, if either the acid-catalyzed or base-catalyzed pathway is negligible, one branch of the curve will be missing. Should the uncatalyzed reaction rate be negligible, the middle portion of the plot exhibits curvature rather than a horizontal line.

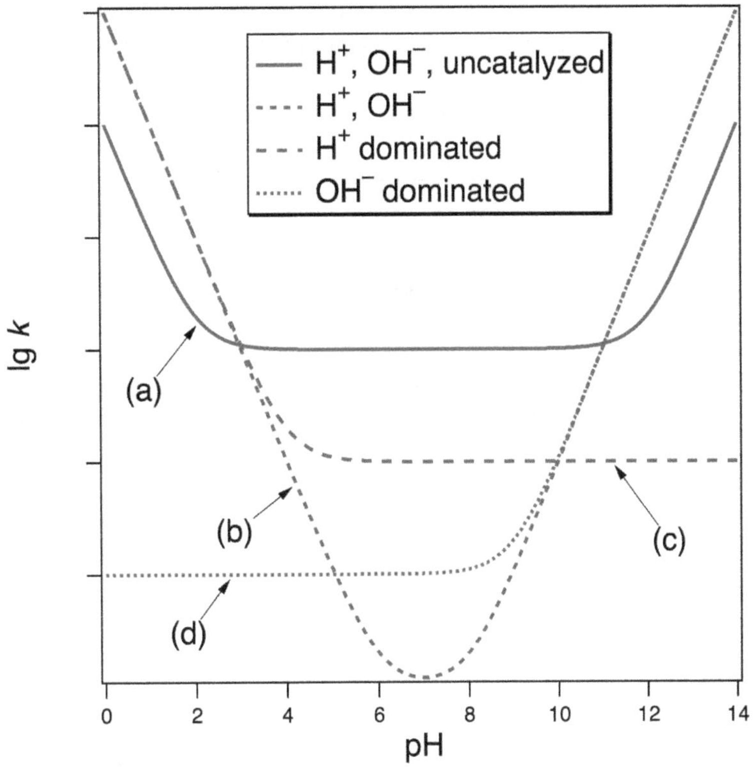

Figure 18.6 The effective rate constant is plotted versus pH for cases in which the balance between acid-catalyzed, base-catalyzed and uncatalyzed pathways changes. (a) All three contribute, $k_H = k_{OH} = 10^2 k_0$. (b) The uncatalyzed pathway is negligible, $k_H = k_{OH} = 10^6 k_0$. (c) The acid-catalyzed pathway dominates, $k_H = 10^{10} k_{OH} = 10^4 k_0$. (d) The base-catalyzed pathway dominates, $10^{11} k_H = k_{OH} = 10^5 k_0$.

An important example of acid catalysis occurs in atmospheric chemistry. Sulfuric acid is produced in a clear sky by the reaction mechanism (all species in the gas phase)

$$SO_2 + OH + M \rightarrow HSO_3 + M \tag{18.11}$$

$$HSO_3 + O_2 \rightarrow SO_3 + HO_2 \tag{18.12}$$

$$SO_3 + H_2O + M \rightarrow H_2SO_4 + M \tag{18.13}$$

M is an unreactive third-body that affects only energy transfer. The rate-determining step is the first reaction, which is sufficiently slow that the atmospheric lifetime of SO_2 should be one to two weeks. This contradicts the experimental observation that $SO_4{}^{2-}$ concentrations in acid rain are at a maximum over the SO_2 source (usually fossil fuel combustion or metal ore smelting). The answer to this riddle is that SO_2 dissolves in water droplets found in clouds,

$$SO_2(g) + H_2O(l) \rightleftarrows HSO_3^-(aq) + H^+(aq) \tag{18.14}$$

where it can react with hydrogen peroxide that also reversibly dissolves in the droplets

$$H_2O_2(g) \rightleftarrows H_2O_2(aq) \tag{18.15}$$

$$HSO_3^-(aq) + H_2O_2(aq) + H^+(aq) \rightarrow SO_4^{2-}(aq) + 2H^+(aq) + H_2O(l) \tag{18.16}$$

The acid-catalyzed reaction in Eq. (18.16) is extremely rapid, which results in the localization of $SO_4{}^{2-}$ production to the atmosphere close to the source rather than allowing for more widespread dispersion.

18.4 Enzyme catalysis

Certain proteins act as biological catalysts and are known as *enzymes*. Evolution has optimized enzymes to selectively accelerate biochemical reactions with amazing efficiency. For example, the hydrolysis of peptide bonds occurs uncatalyzed in aqueous solutions with a half-life of about 300–600 years. But in the presence of the appropriate enzyme catalyst, this half-life is reduced to roughly 1 ms if the reaction is not limited by diffusion. Evolution has incorporated promoters and structural sensitivity to optimize the performance of enzymes by constructing the active site and its surroundings in response to performance constraints. It has optimized them to work under quite specific conditions. Enzymatic activity is strongly dependent on temperature and pH. The conditions of highest activity are referred to as the *optimum temperature* and *optimum pH*.

Proteins that are folded into the conformation that corresponds to their functional form are called *native proteins*. Many native proteins are stable to temperatures above 100 °C. Above this temperature they undergo a process akin to a melting transition in which the protein unfolds into a nonfunctional form; this process is called *denaturation*. Enzymes are sensitive to denaturation and some lose their activity only a few degrees above their optimum temperature. Reversible changes in enzymatic activity occur over a range of pH values near the optimum pH. Many enzymes experience irreversible denaturation at extremes of high and low pH.

The first-order approximation to how enzymes work is based on the proposal from Emil Fischer in 1890 of the 'lock-and-key' mechanism. A certain substrate S fits into the active site of enzyme E and this facilitates the production of the product P. Whilst there is some truth to this, such as mechanism is literally too rigid. The chemical functionality of enzymes is better understood in terms of five important points:

1. Proximity and configuration are critical. One of the most important things that an enzyme does is to bring the reactants into a favorable geometry.
2. The enzyme is flexible such that only a specific substrate induces the proper interactions that lead to catalysis. These changes were proposed by David Koshland[9] in a stark break with the unforgiving 'lock-and-key' mechanism. The nature of the conformational relaxation into the configuration that characterizes the enzyme–substrate complex falls somewhere between two extremes. One corresponds to the *induced fit mechanism* as originally proposed by Koshland in which the initial recognition step *causes* the subsequent conformational relaxation. The opposite extreme is called the *conformational selection mechanism*, in which the proteins explore a number of different conformations. Binding only occurs *after* the correct configuration is obtained.
3. The protein structure surrounding the active site creates a microenvironment that optimizes reactivity of the catalytic groups. This microenvironment shields the active site from the bulk solvent (water). Furthermore, the dynamics of the protein may actually facilitate procession of the activated complex from the transition state to the products by favoring movement along the reaction coordinate. It does so while disfavoring vibrational motions that do not advance the activated complex along the catalytic path.
4. Enzymes can cause the substrate to adopt an activated configuration that is much more reactive than its equilibrium configuration.
5. Enzymes can stabilize the transition state, often by electrostatic interactions. Stabilization can also occur through hydrogen-bonding interactions that take place because of precise matching of the pK_a values of donor and acceptor moieties.

The role of structural dynamics – topology and conformational changes – in enzyme catalysis is central to understanding their reactivity. Thus, as shown in Fig. 18.7 it is better to represent the potential energy diagram leading from reactants to products not in terms of a single one-dimensional Lennard–Jones type curve, but rather in terms of a two-dimensional landscape of such curves. This picture emphasizes that the catalytic energy profile (up and over the barrier) is influenced by the conformational landscape. Changes in conformation change the catalytic energy profile. The existence of multiple conformations allows the reactive system to sample multiple transition states (and reactive intermediates) as it searches the catalytic–conformational energy landscape. The path from reactants to products is anything but a simple up and over the barrier trajectory. If the reaction corresponds to *proton-coupled electron transfer* (as is often the case in enzyme-catalyzed reactions), the generalized reaction coordinate needs to be further resolved into a collective solvent coordinate along the electron transfer path and another along the proton transfer path, as done in the theoretical work of Sharon Hammes-Schiffer and coworkers.[10] Structural changes can also dramatically change the time scales of the motions involved in the reaction path.

Enzymes display different degrees of specificity. Some are exquisitely selective and can transform only one substrate. An example of this type is urease, which only catalyzes the hydrolysis of urea,

$$CO(NH_2)_2 + H_2O \rightarrow CO_2 + 2\,NH_3.$$

Cleavage of a peptide bond by enzyme-catalyzed hydrolysis proceeds by a sequence of several steps. The proteolytic enzymes that perform this exhibit *group specificity*. They catalyze hydrolysis only if the substrate possesses certain structural

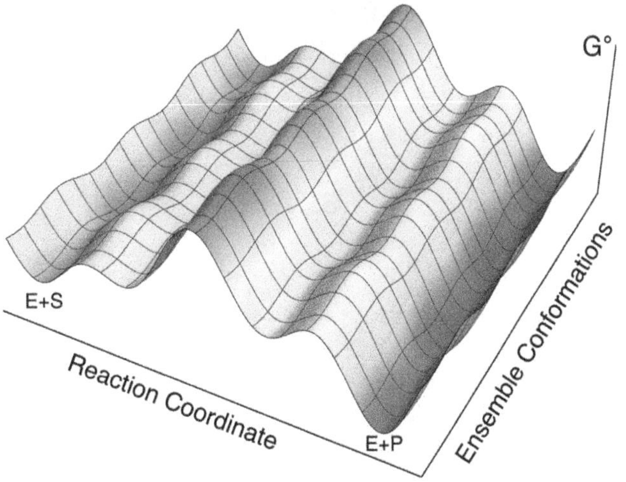

Figure 18.7 A schematic representation of the Gibbs energy landscape of an enzyme-catalyzed reaction. Conformational changes occur along both axes. The conformational changes occurring along the reaction coordinate axis correspond to the environmental reorganization that facilitates chemical reaction. The conformation changes directed along the ensemble conformations axis represent the ensembles of configurations existing at all states along the reaction coordinate. Provided by Sharon Hammes-Schiffer. Reproduced with permission from Benkovic, S.J., Hammes, G.G., and Hammes-Schiffer, S. (2008) *Biochemistry*, **47**, 3317. © 2008, American Chemical Society.

characteristics in the neighborhood of the linkage to be hydrolyzed. They also exhibit *stereochemical specificity*, and only catalyze the hydrolysis of peptides containing amino acids in the L configuration. The enzyme first binds to the site on the protein that is targeted for cleavage. A water molecule must also achieve the proper position to participate in the reaction. This configuration of enzyme, protein and water corresponds to a local minimum on the potential energy surface for the reaction. A fluctuation in the configuration of the protein then drives the system into the transition state, where electron transfer occurs. The appropriate bonds are broken and/or made and the system relaxes into the product valley. The products are released to the solution and the enzyme returns to its initial configuration.

Many catalytic biological reactions that involve a single substrate follow a similar type of kinetics. The rate varies linearly at low concentration (first-order kinetics) and becomes independent of substrate concentration at higher concentration (zero-order kinetics). Leonor Michaelis and Maud Menten[11] were the first to explain kinetics of this type in terms of what is now called *Michaelis–Menten kinetics*. The enzyme E and the substrate S form an addition complex ES in a reversible reaction, followed by reaction of this complex with formation of the product P.

$$\mathrm{E} + \mathrm{S} \underset{k_{-1}}{\overset{k_1}{\rightleftharpoons}} \mathrm{ES} \tag{18.17}$$

$$\mathrm{ES} \xrightarrow{k_2} \mathrm{E} + \mathrm{P} \tag{18.18}$$

Usually, the substrate is present in great excess and the steady state approximation can be applied,

$$\frac{\mathrm{d[ES]}}{\mathrm{d}t} = k_1[\mathrm{E}][\mathrm{S}] - k_{-1}[\mathrm{ES}] - k_2[\mathrm{ES}] = 0. \tag{18.19}$$

$[\mathrm{E}]_0$ is the total concentration of enzyme. [E] is the concentration of free enzyme, that is, the concentration of the enzyme not bound in the complex,

$$[\mathrm{E}]_0 = [\mathrm{E}] + [\mathrm{ES}]. \tag{18.20}$$

Substitution of Eq. (18.20) into Eq. (18.19) leads to

$$[\mathrm{ES}] = \frac{k_1[\mathrm{E}]_0\,[\mathrm{S}]}{k_{-1} + k_2 + k_1[\mathrm{S}]}. \tag{18.21}$$

The rate is

$$v = k_2[\mathrm{ES}] = \frac{k_1 k_2[\mathrm{E}]_0\,[\mathrm{S}]}{k_{-1} + k_2 + k_1[\mathrm{S}]} = \frac{k_2[\mathrm{E}]_0\,[\mathrm{S}]}{(k_{-1} + k_2)/k_1 + [\mathrm{S}]}. \tag{18.22}$$

This is usually rewritten in the form of the *Michaelis–Menten equation*

$$v = \frac{V[S]}{K_M + [S]} \tag{18.23}$$

where V, the *limiting rate*, is

$$V = k_2[E]_0 \tag{18.24}$$

and K_M, the *Michaelis constant*, is

$$K_M = \frac{k_{-1} + k_2}{k_1} \tag{18.25}$$

The Michaelis–Menten equation can be transformed to

$$v/[S] = (-1/K_M)R + v/K_M. \tag{18.26}$$

Hence, as shown in Fig. 18.8, a plot of $v/[S]$ versus v results in a straight line with slope = $-1/K_M$ and intercept/slope = $-V$.

It is instructive to consider the Michaelis–Menten equation in two limiting cases. When $[S] \gg K_M$

$$v = k_2[E]_0 = V, \tag{18.27}$$

which is zeroth-order in the substrate concentration. This is why V is known as the *limiting rate*. The rate constant k_2 can be determined from Eq. (18.24) and is known as the *turnover number*, the maximum number of substrate molecules per unit time that can be converted into product. It is also referred to as the *catalytic constant*.

The second limiting case occurs when $[S] \ll K_M$. Equation (18.23) then reduces to

$$v = \frac{V[S]}{K_M} = \frac{k_2}{K_M}[E]_0\,[S] = \frac{k_1 k_2}{k_{-1} + k_2}[E]_0\,[S], \tag{18.28}$$

which is first-order in the substrate concentration. Since both k_1 and k_2 appear in the rate equation, there is no rate-controlling step. Instead, both steps 1 and 2 contribute to the rate. Only in the limit of high substrate concentration is step 2 truly the rate-limiting step. The simple form of Michaelis–Menten kinetics may actually conceal more complex kinetics in which several intermediate complexes contribute.

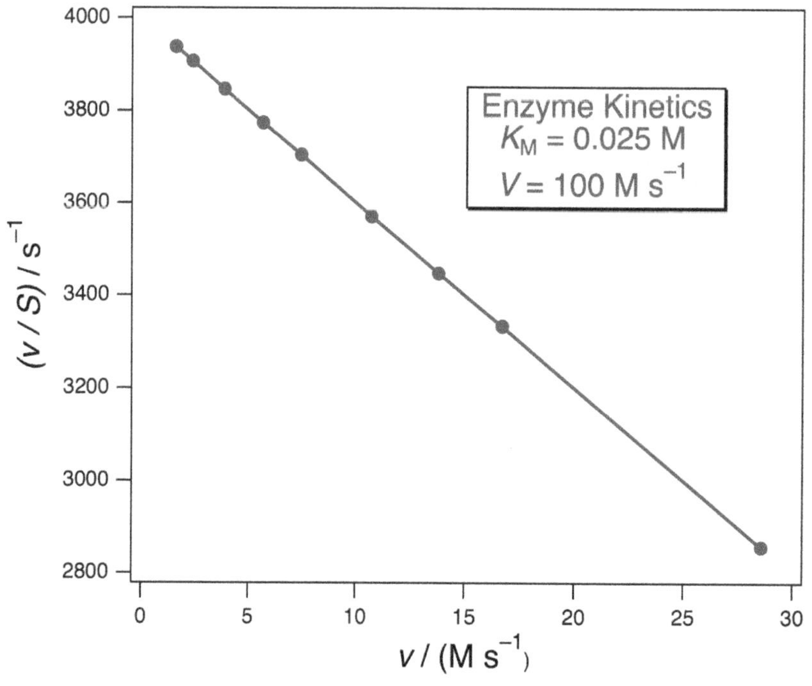

Figure 18.8 A plot of the Michaelis–Menten equation for enzyme kinetics.

A reciprocal plot of the reaction rate in Eq. (18.23) in terms of the initial rate v_0 and the initial concentration of substrate $[S]_0$ yields the *Lineweaver–Burk equation,*

$$\frac{1}{v_0} = \frac{1}{V} + \frac{K_M}{V}\frac{1}{[S]_0}. \tag{18.29}$$

Poisons can be just as important as in the case of heterogeneous catalysis. The term *inhibitor* is more commonly used to denote a species that can bind to the enzyme and block the substrate from reacting. The kinetics of the reaction will be slowed down in proportion to how strongly the inhibitor binds to the enzyme. In a working biochemical environment there will be any number of molecules that can potentially competitively bind at the enzyme's active site. Part of the engineering that has occurred to optimize the performance of catalysts is construction of the protein structure around the active site that gives the appropriate substrate an advantage in competitive binding.

18.4.1 Directed practice

Derive Eq. (18.29) from Eq. (18.23).

18.5 Chain reactions

Many reactions involving *free radicals* (species that contain unpaired electrons) and atoms follow a chain reaction mechanism. *Chain reactions* are also of great importance in combustion reactions. Much of the initial understanding of chain reactions resulted from the work of Cyril Hinshelwood and Nikolai Semenov, who shared the Nobel Prize in Chemistry 1956 for these studies. One example is the gas-phase reaction of H_2 + Br_2, which follows a five-step mechanism to produce HBr. The reaction is initiated by collisional activation of Br_2.

1. $Br_2 + M \xrightarrow{k_1} 2Br + M$ initiation

2. $Br + H_2 \xrightarrow{k_2} HBr + H$ } chain propagation

3. $H + Br_2 \xrightarrow{k_3} HBr + Br$ } chain propagation

4(−2). $H + HBr \xrightarrow{k_4} H_2 + Br$ inhibition

5(−1). $2Br + M \xrightarrow{k_{-1}} Br_2 + M$ termination or chain ending

The initiation reaction starts the sequence by creating two radicals. Initiation must increase the number of radicals. Chain propagation steps conserve the number of radicals while producing a product of the overall reaction. The termination step reduces the number of radicals. Step 4 inhibits the reaction by consuming a product (HBr) and reproducing one of the original reactants. After one initiation reaction, the chain propagation steps can occur many times before an inhibition or termination step occurs.

The kinetics of a chain reaction is in general very complex and leads to a complex rate equation, though orders are usually ½, 1, 3/2, or 2. In the case of the H_2 + Br_2 reaction, we can find rate expressions by applying the steady-state approximation. Assuming for the moment that inhibition can be neglected, the net rate of HBr production is obtained by summing the rates of the two elementary reactions that produce it,

$$\frac{d[HBr]}{dt} = k_2[Br][H_2] + k_3[H][Br_2]. \tag{18.30}$$

We now proceed to eliminate the concentrations of the intermediates by applying the steady state approximation, first for the H atom

$$\frac{d[H]}{dt} = k_2[Br][H_2] - k_3[H][Br_2] = 0 \tag{18.31}$$

then the Br atom

$$\frac{d[Br]}{dt} = 2k_1[Br_2][M] - 2k_{-1}[Br]^2[M] - k_2[Br][H_2] + k_3[H][Br_2] = 0 \tag{18.32}$$

Adding these two expressions, we find that the Br concentration at steady state is given by

$$[Br] = (k_1/k_{-1})^{1/2}[Br_2]^{1/2} \tag{18.33}$$

Equation (18.31) allows us to eliminate the term involving [H] from Eq. (18.30)

$$\frac{d[HBr]}{dt} = 2k_2[Br][H_2], \quad (18.34)$$

Which, upon substitution for [Br] from Eq.(18.33), yields a final rate expression in terms of only of elementary rate constants and reactant concentrations

$$\frac{d[HBr]}{dt} = 2k_2(k_1/k_{-1})^{1/2}[Br_2]^{1/2}[H_2] = k[Br_2]^{1/2}[H_2]. \quad (18.35)$$

If inhibition cannot be neglected, the rate equation is modified such that

$$\frac{d[HBr]}{dt} = \frac{2k_2\left(k_1/k_{-1}\right)^{1/2}[H_2][Br_2]^{1/2}}{1 + (k_4/k_3)([HBr]/[Br_2])}. \quad (18.36)$$

With its concentration appearing in the denominator, HBr clearly plays the role of an inhibitor, as increasing its concentration decreases the overall rate of production of HBr.

18.5.1 Directed practice

Perform the missing algebra steps above to confirm Eq. (18.35).

Chain reactions are strongly affected by additives. Chain reactions that rely on the production of H atoms such as the $H_2 + O_2$ reaction are catalyzed by the presence of finely divided platinum group metals. A number of substances can act as initiators – species that when added increase the rate of reaction but that are not true catalysts because they are not regenerated. For example, the rate of organic decomposition reactions can be enhanced by the presence of CH_3 radicals. These enhance the chain propagation steps. One source of methyl radicals is the highly reactive azomethane molecule that decomposes upon heating to produce methyl radicals and N_2 according to

$$CH_3N_2CH_3 \rightarrow N_2 + 2\,CH_3.$$

Peroxides, whether hydrogen peroxide or organic peroxides such as benzoyl peroxide, are often used as radical initiators as the labile peroxide bond easily generates reactive RO• radicals. On the other hand, molecules that produce stable radicals such as molecular oxygen or various organic radicals containing phenyl groups that are stabilized by resonance act as inhibitors. Inhibition by impurities occurs when reactive radicals that would propagate the chain are consumed by less-reactive radicals that reduce the rate.

18.5.2 Chain polymerization

An important variant of chain reaction is the chain reaction that leads to polymer formation. *Chain polymerization* begins with an initiation step to form a radical

$$I \xrightarrow{k_i} R\bullet. \quad (18.37)$$

The radical goes on to react with the monomer M

$$R\bullet + M \xrightarrow{k_2} RM\bullet. \quad (18.38)$$

Assuming that the rate constant for the propagation step is independent of the degree of polymerization, we then write the propagation step as

$$RM_n\bullet + M \xrightarrow{k_p} RM_{n+1}\bullet. \quad (18.39)$$

Termination occurs by combination of polymerized radicals which, in principle, have different lengths n and m,

$$RM_n\bullet + RM_m\bullet \xrightarrow{k_t} RM_{n+m}R. \quad (18.40)$$

where I is the initiator, R• is the initiator radical, and M is the monomer.

The kinetics described above lead to a rate of polymerization R_p that is first-order in monomer concentration but half-order in initiator according to

$$R_p = k_p(k_i/k_t)^{1/2}[M][I]^{1/2}, \quad (18.41)$$

which is consistent with the result we derived for HBr above.

Table 18.2 Values for rate constants of chain polymerization at 60 °C. Taken from Sperling, L.H. (2006) *Introduction to Physical Polymer Science*, 4th edition. John Wiley & Sons, Hoboken, NJ.

Monomer	k_p / L mol^{-1} s^{-1}	k_t / 10^{-6} L mol^{-1} s^{-1}
Methyl methacrylate	573	2.0
Styrene	187.1	29.4
Acrylonitrile	1960	782

Representative values of the rate constants are given in Table 18.2. A typical initiator is benzoyl peroxide, which has $k_i = 2.76 \times 10^{-6}$ s^{-1} at 60 °C. The average number of monomers consumed for each radical initiating a chain is known as the *kinetic chain length* ν. At steady state this is given by the ratio of the rate of propagation R_p to either the rate of initiation R_i or termination R_t,

$$\nu = \frac{R_p}{R_i} = \frac{R_p}{R_t} = \frac{k_p[M]}{2(\phi k_i k_t [I])^{1/2}}. \quad (18.42)$$

The initiator efficiency factor ϕ is the fraction of initiator molecules that successfully initiate a polymerization after their decomposition into the radical initiator R•. In a well-run polymerization reaction, ϕ is typically about 0.8.

18.5.3 Self-catalyzed step polymerization

Polyesterification reactions are often run at equilibrium. Carboxyl and hydroxyl groups react to form a polyester with the removal of water

$$\text{—COOH} + \text{—OH} \overset{k}{\rightleftharpoons} \text{—COO—} + H_2O$$

The equilibrium constant expression is

$$K = \frac{[COO][H_2O]}{[COOH][OH]} \quad (18.43)$$

The fractional conversion (or degree of reaction) is

$$[COO] = \alpha[COOH]_0 = \alpha M_0, \quad (18.44)$$

where M_0 is the initial concentration of carboxyl groups.

Self-catalyzed polyesterification follows a third-order rate law

$$\frac{d[COOH]}{dt} = -k[COOH]^2[OH] \quad (18.45)$$

If the initial concentrations of carboxyl and hydroxide groups are equal, the degree of reaction α is

$$\frac{1}{(1-\alpha)^2} = 2M_0 kt + 1 \quad (18.46)$$

The degree of reaction is related to the *number-average degree of polymerization* DP_n by

$$DP_n = \frac{1}{1-\alpha}. \quad (18.47)$$

The kinetics of polymerization determines the attributes of the molar mass distribution. One parameter used to characterize this is the *polydispersity index PDI*,

$$PDI = M_w/M_n, \quad (18.48)$$

where M_w is the weight-average molar mass and M_n is the number-average molar mass (defined in Chapter 4). If disproportionation is the termination step in a chain polymerization, $PDI = 2$ in the ideal case. Termination by combination leads to $PDI = 1.5$. Stepwise polyesterification has $PDI = 2$. Anionic polymerizations are characterized by much narrower distributions corresponding to $PDI \leq 1.05$. However, cationic polymerizations are characterized by very broad distributions, as is often the case for Ziegler–Natta polymerization as well. The only truly monodisperse polymers are proteins. In contrast, natural cellulose exhibits a very broad distribution. Branching is one source of non-ideality that leads to broader molar mass distributions.

18.5.3.1 Directed practice

Why are the properties of polymers sensitive to the value of the polydispersity index?

18.6 Explosions

Explosions are reactions that are self-accelerating such that their rate increases with time. There are two general types of explosion: thermal and chain-branching. Some explosive reactions deliver their explosive force by the production of large quantities of gas from solid-state sources. This is true of potassium nitrate (NH_4NO_3) and TNT (2,4,6-trinitrotoluene). Thermal explosions can occur in highly exothermic reactions. If heat generated by the reaction is not dissipated sufficiently rapidly, the temperature increases and, in accord with Arrhenius behavior, so too does the rate. The increasing rate and exothermicity provide feedback, which accelerates the reaction further until an explosion occurs. This type of explosion can be avoided by providing sufficiently high, well-cooled surface area in the reaction vessel and good mixing to ensure even dissipation of the heat generated by the exothermic reaction.

A chain-branching explosion occurs if the number of radical intermediates increases with time in a chain reaction. Consider a chain reaction with rate constants of initiation, branching and termination k_i, k_b and k_t according to:

$$\mathrm{A + B} \xrightarrow{k_i} \mathrm{R\bullet} \tag{18.49}$$

$$\mathrm{R\bullet} \xrightarrow{k_b} \phi\mathrm{R\bullet} + \mathrm{P_1} \tag{18.50}$$

$$\mathrm{R\bullet} \xrightarrow{k_t} \mathrm{P_2} \tag{18.51}$$

A and B are reactants, R• is a radical, and P_1 and P_2 are nonreactive products. ϕ is the efficiency of producing other radicals. If R• is converted into two radicals in the branching step, then $\phi = 2$. The differential rate equation for R• is

$$\frac{\mathrm{d[R\bullet]}}{\mathrm{d}t} = k_i[\mathrm{A}][\mathrm{B}] - k_b[\mathrm{R\bullet}] + \phi k_b[\mathrm{R\bullet}] - k_t[\mathrm{R\bullet}] \tag{18.52}$$

$$= \Gamma + k_{\mathrm{eff}}[\mathrm{R\bullet}] \tag{18.53}$$

where

$$\Gamma = k_i[\mathrm{A}][\mathrm{B}] \quad \text{and} \quad k_{\mathrm{eff}} = k_b(\phi - 1) - k_t. \tag{18.54}$$

Equation (18.53) can be integrated and solved for [R•]

$$[\mathrm{R\bullet}] = \frac{\Gamma}{k_{\mathrm{eff}}}(e^{k_{\mathrm{eff}}t} - 1). \tag{18.55}$$

There are two limiting cases. First when $k_t \ll k_b(\phi - 1)$, we have$k_{\mathrm{eff}} = k_b(\phi - 1)$ and

$$[\mathrm{R\bullet}] = \frac{\Gamma}{k_b(\phi - 1)}\left(e^{k_b(\phi-1)t} - 1\right). \tag{18.56}$$

In this limit, branching dominates, [R•] grows exponentially with time and there is an explosion. However if $k_t \gg k_b(\phi - 1)$, then $k_{\mathrm{eff}} = -k_t$ and the concentration of radical intermediates is given by

$$[\mathrm{R\bullet}] = \frac{\Gamma}{k_t}(1 - e^{k_t t}). \tag{18.57}$$

Now termination dominates, [R•] grows slowly to an asymptotic value, and there is no explosion.

These results demonstrate that whether or not a system reacts explosively depends not only on the chemical nature of the reaction but also on the reaction conditions. Temperature, pressure, and the presence of initiators and inhibitors (as well as the conductance of heat) all contribute to the presence or absence of explosions. A good example is that of mixture of $H_2 + O_2$, which goes in and out of being explosive as the temperature and pressure are increased. The occurrence of these explosion limits can be explained as follows.

18.6.1 Explosive limits

- At low p the mean free path λ is large. The reaction is homogeneous and no explosion occurs. Recombination of chain carriers at the walls avoids the explosion.
- As p is increased at constant T, the mixture will eventually become explosive. This combination of p and T is known as the *first explosion limit*.

- Between the first and second explosion limits the reaction proceeds explosively. λ is not sufficient to allow wall quenching of radical chain carriers, and branching proceeds explosively.
- As p is increased at constant T, the mixture will reach a point at which it is no longer explosive. This combination of p and T is known as the *second explosion limit.*
- Above the second explosion limit, third-body collisions lead to less-reactive species being formed. They can survive several collisions without reaction. This allows them to travel to the walls where they are quenched. This leads again to homogeneous reaction.
- As p is increased further at constant T, the mixture will eventually become explosive. This combination of p and T is known as the *third explosion limit.*
- Above the third explosion limit, collisions are now so frequent that even less-reactive species can produce more chain carriers before reaching the walls. Explosive reactivity is again observed.

18.7 Photochemical reactions

Photochemical reactions include an array of important processes such as photosynthesis, photodissociation, bleaching from exposure to sunlight, analog photography, and vision. We will discuss the dynamics of photoexcitation and photochemistry in more depth after we introduce quantum mechanics. Here, we focus on the kinetics of simple photochemical reactions. First let's remind ourselves of a few basic characteristics of light. The speed of light c is related to the wavelength λ and frequency ν by

$$c = \lambda\nu. \tag{18.58}$$

The energy E of a photon is related to its wavelength or frequency by

$$E = \frac{hc}{\lambda} = h\nu. \tag{18.59}$$

The intensity (photons per second also called *photon flow*) times the energy of a photon (joules per photon) is equal to the power (J s^{-1} = W, the unit of power) of the photon source

$$P = IE = \frac{N_{\text{photon}} h\nu}{t}, \tag{18.60}$$

where N_{photon} is the number of photons (sometimes denoted as N_γ) and t the time. The IUPAC name and symbol for this form of intensity is photon flow Φ_{p}. However, so as not to confuse this with the quantum yield, I will use I for photon flow (intensity).

In order to cause photochemistry, photons must be absorbed. This is known as the *Grotthuss–Draper law.* Thus, the rate of a photochemical process is proportional to the absorbed power, which is known as the *law of photochemical equivalence.* Photochemistry in the linear regime (in other words, photochemistry initiated by the absorption of a single photon) is most commonly caused by visible or ultraviolet photons. Photons in this energy range cause electronic excitations among the valence electrons. By causing transitions of electrons involved in bonding from one orbital to another, photon absorption can lead to bond dissociation or the formation of reactive excited states, as illustrated in Fig. 18.9.

18.7.1 Kinetics of photochemical reactions

Let us consider the intensity of the light source measured in terms of the flux of photons. The IUPAC recommended name for *photon flux* in m^{-2} s^{-1} is the *photon irradiance,* with symbol E_{p}. Here, so as not to confuse this symbol with the photon energy, I will use J for photon flux since J is the general symbol for flux. Consider the flux of absorbed photons J_{abs} in units of moles per unit area per unit time. Photons have to be absorbed to cause photochemistry, so it is the flux of adsorbed light J_{ads} rather than the flux of the incident light J_0 that determines the kinetics.

$$J_{\text{abs}} = J_0 - J_{\text{trans}} = J_0(1 - 10^{-\varepsilon l[\text{A}]}) \tag{18.61}$$

where ε = the molar absorptivity, l is the path length and [A] is the concentration of species that absorbs the light. This should be familiar from the *Beer–Lambert Law.* When the reactant concentration is not too large, Eq. (18.61) simplifies to

$$J_{\text{abs}} = 2.303\, J_0\, \varepsilon l[\text{A}]. \tag{18.62}$$

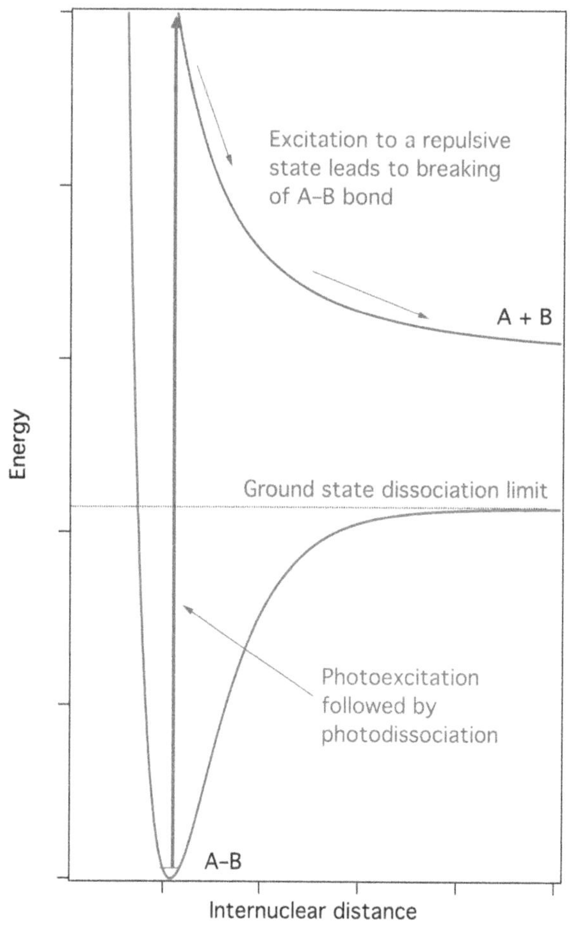

Figure 18.9 Absorption of visible and ultraviolet (UV) photons leads to electronic excitations. Photoexcitation can lead to the formation a bound electronically excited state. However, if the excited state is repulsive – that is, an excited state with no well, as shown in blue – then photodissociation can occur.

For a simple photochemical reaction induced by the absorption of one photon,

$$A \xrightarrow{h\nu} P,$$

this yields a first-order rate law (in light intensity) for photochemical processes that are initiated by the absorption of one photon

$$R = -\frac{d[A]}{dt} = \frac{J_{abs}}{l}. \tag{18.63}$$

Integrating over the length l yields the dependence of the concentration of A versus time

$$[A] = [A]_0 e^{-2.303 J_0 \varepsilon t} = [A]_0 e^{-kt}. \tag{18.64}$$

Be careful with units. If J_{abs} is in mol cm^{-2} s^{-1} and l is in cm, then [A] is given in mol l^{-1} and the rate is in mol l^{-1} s^{-1} if Eq. (18.63) is multiplied by 1000 to convert cm^3 to liters. ε is often given in cm^{-1} mol^{-1} L. Note that SI units are different and sometimes other units (such as pressure or molecules rather than moles) will be used.

Now for a specific example. In the absence of light, hydrogen iodide decomposes by an elementary thermal reaction. The thermal reaction rate is described by the second-order rate law

$$HI + HI \rightarrow H_2 + I_2 \qquad \frac{-d[HI]}{dt} = 2k[HI]^2. \tag{18.65}$$

A photochemical reaction occurs when the system is irradiated at wavelengths below about 400 nm according to the following three-step mechanism

1 $HI + h\nu \rightarrow H + I$ $\quad R_1 = I_{abs}$ (18.66)

2 $H + HI \rightarrow H_2 + I$ $\quad R_2 = k_2[H][HI]$ (18.67)

3 $I + I \rightarrow I_2$ $\quad R_3 = k_3[I]^2$ (18.68)

The net rate of change of HI concentration is

$$\frac{-d[HI]}{dt} = I_{abs} + k_2[H][HI]. \tag{18.69}$$

From the steady-state approximation

$$\frac{d[H]}{dt} = 0 = I_{abs} - k_2[H][HI]. \tag{18.70}$$

Combining these two equations, the net rate of consumption of HI in photochemical reaction is given by

$$-\frac{d[HI]}{dt} = 2I_{abs}. \tag{18.71}$$

The absorption of one photon leads to the decomposition of two HI molecules. To quantify this photon efficiency, we introduce the *quantum yield* Φ

$$\Phi = \frac{\text{rate of process}}{\text{rate of absorption of radiation}} = \frac{R}{I_{abs}}. \tag{18.72}$$

Specifically for HI photodecomposition,

$$\Phi = \frac{-d[HI]}{dt}/I_{abs} = \frac{2I_{abs}}{I_{abs}} = 2. \tag{18.73}$$

The interesting point in the photochemical reaction above is that the rate is twice the adsorbed intensity and is not dependent on the concentration of HI. A quantum yield greater than one is possible if secondary reactions occur (complex mechanisms). Quantum yields less than one are commonly found when *quenching* of the excited state occurs, or for reactions that branch into the production of several products. Quenching is nonreactive de-excitation of the excited state. *Collisional quenching* occurs when the electronic excitation is lost by collision of the excited-state species with another molecule or a surface such as the walls of the container. *Radiative quenching* occurs if the excited state fluoresces; thus, losing its electronic energy by photon emission.

A reaction such as that in Eq. (18.66) is known as *photodissociation*; this is a photochemical reaction in which photoexcitation directly leads to bond scission. An alternative is a photoinitiated reaction, which occurs when the electronically excited state is a bound state with a greater reactivity than the ground state. After excitation, the electronically excited state species collides with another reactant, which leads to the formation of products.

18.7.1.1 Example

The quantum efficiency for the formation of ethane from heptane with 313 nm light is 0.21. (a) Calculate the moles per second of ethane that are formed when the sample is irradiated with a 50 W, 313 nm source under conditions of total absorption of the light. (b) How many moles per second of heptane are decomposed under these conditions?

a The energy of the 313 nm photon is

$$E = h\nu = hc/\lambda = 6.35 \times 10^{-19}\ \text{J}.$$

A 50 W source generates photons at a rate of

$$I_{abs} = \frac{P_{abs}}{E} = \frac{50\ \text{J s}^{-1}}{6.35 \times 10^{-19}\ \text{J}} = 7.9 \times 10^{19}\ \text{s}^{-1}.$$

The number of ethane molecules formed per second is therefore 0.21 times this quantity or

$$R(\text{ethane}) = \Phi I_{abs} = (0.21)(7.9 \times 10^{19}\ \text{s}^{-1}) = 1.7 \times 10^{19}\ \text{s}^{-1},\ \text{which is } 2.8 \times 10^{-5}\ \text{mol s}^{-1}.$$

b The quantum yield is related to a specific process. We were given the quantum yield for the production of ethane. We were not given the quantum yield for the decomposition of heptane. These two numbers need not be the same. At a minimum, the rate of heptane decomposition is equal to the rate of production of ethane. However, we do not know the fate of the other excited heptane molecules, the fraction of which is (1 – 0.21) = 0.79. These molecules may have reacted to form other products. Alternatively, they may have relaxed by emitting a photon (fluorescence) or by deactivation by collision with another molecule or with a wall.

18.7.1.2 Example

Calculate the quantum yield of the photochemical decomposition of HI irradiated with light of wavelength 253.7 nm in the gas phase. 307 J of incident energy were absorbed and 1.3×10^{-3} mol of HI were decomposed.

The energy of the photon is

$$E = \frac{hc}{\lambda} = \frac{6.63 \times 10^{-34}\,\mathrm{J\,s} \times 3 \times 10^{8}\,\mathrm{m\,s^{-1}}}{253.7 \times 10^{-9}\,\mathrm{m}} = 7.84 \times 10^{-19}\,\mathrm{J}$$

The number of photons adsorbed is

$$N_{\mathrm{abs}} = \frac{307\,\mathrm{J}}{7.84 \times 10^{-19}\,\mathrm{J}} = 3.916 \times 10^{20}\,\text{photons}.$$

The number of molecules destroyed is

$$N_{\mathrm{HI}} = (1.30 \times 10^{-3}\,\mathrm{mol} \times 6.02 \times 10^{23}\,\mathrm{mol^{-1}}) = 7.826 \times 10^{20}\,\text{molecules}$$

$$\Phi = \frac{N_{\mathrm{HI}}}{N_{\mathrm{abs}}} = \frac{7.826 \times 10^{20}}{3.916 \times 10^{20}} = 2.$$

18.7.1.3 Example

A photochemical reaction obeys the mechanism written below. (a) If the quantum yield for the formation of B is $\Phi = 0.24$ at $\lambda = 313$ nm, and 1.30×10^{-5} mol of B are produced in 36.5 min, what is the intensity of the absorbed light in photons per second? (b) If the light is tuned to 254 nm where the absorption coefficient of A is three times smaller than at 313 nm, what would have been the rate of formation of B given the same conditions otherwise? (c) What is the role of M in the reaction, and how does it affect the rate? (d) Given that the quantum yield is the ratio of the rate of B molecules formed to I, where I is the intensity of absorbed light, derive a steady-state expression for the quantum yield. (e) What are the values of the quantum yield in the limits of very large and very small concentrations of M?

$$\mathrm{A} + h\nu \xrightarrow{\Phi_1} \mathrm{A}^*$$
$$\mathrm{A}^* + \mathrm{M} \xrightarrow{k_2} \mathrm{A} + \mathrm{M}$$
$$\mathrm{A}^* \xrightarrow{k_3} \mathrm{B} + \mathrm{C}$$

a The quantum yield times the intensity of absorbed light is equal to the rate (number of molecules produced per second)

$$\Phi I_{\mathrm{abs}} = \frac{N_{\mathrm{A}}\,n}{t}.$$

Solving for I_{abs},

$$I_{\mathrm{abs}} = \frac{N_{\mathrm{A}}\,n}{\Phi t} = \frac{6.02 \times 10^{23}(1.30 \times 10^{-5})}{0.24\,(36.5)\,(60)} = 1.49 \times 10^{14}\,\text{photons}\,\mathrm{s^{-1}}.$$

b $R \propto I_{\mathrm{abs}}$ and the intensity of absorbed light is directly proportional to the absorption coefficient. Therefore, the rate drops in proportion to the intensity of the absorbed light,

$$\frac{\mathrm{d[B]}}{\mathrm{d}t} = \left(\frac{1}{3}\right)\frac{1.30 \times 10^{-5}\,\mathrm{mol}}{36.5\,(60)\,\mathrm{s}} = 1.98 \times 10^{-9}\,\mathrm{mol\,s^{-1}}.$$

c The rate of production of B is given by

$$R = \frac{\mathrm{d[B]}}{\mathrm{d}t} = k_3[\mathrm{A}^*].$$

M is a quencher. It reduces the concentration of A* by collisional relaxation. By lowering [A*] it will reduce the rate of production of B since the rate is proportional to [A*].

d The quantum yield of the overall process is given by

$$\Phi = \frac{R}{I_{abs}}.$$

Applying the steady-state approximation, we can find an expression for $[A^*]$.

$$\frac{d[A^*]}{dt} = \Phi_1 I_{abs} - k_2[A^*][M] - k_3[A^*] = 0$$

$$[A^*] = \frac{\Phi_1 I_{abs}}{k_3 + k_2[M]}$$

Therefore

$$\Phi = \frac{\Phi_1 k_3}{k_3 + k_2[M]}.$$

e In the limit of vanishing concentration of M, $[M] = 0$

$$\Phi = \frac{\Phi_1 k_3}{k_3} = \Phi_1.$$

Thus, the quantum yield converges on the quantum yield of formation of the proper excited state in the limit of no quencher.

In the limit of large [M]

$$\Phi = \frac{\Phi_1 k_3}{k_3 + k_2[M]} = \frac{\Phi_1 k_3}{k_2[M]} \to 0 \quad \text{when} \quad k_2[M] \gg k_3.$$

The quantum yield is inversely proportional to the concentration of quencher and will tend to zero if the concentration becomes sufficiently large.

18.8 Charge transfer and electrochemical dynamics

Ions are the primary carrier of current in solutions, whereas *electrons* carry current in metals and semiconductors. The ions are surrounded by solvation shells, as discussed in detail in Chapter 14. When ions come into contact with an electrode, they may (i.e., specific adsorption) or may not (i.e., nonspecific adsorption) release their solvation shell. The anions and cations will line up near the surface (generally with concentrations very different than the bulk solution) to form the electric double layer. The structure of the electric double layer is discussed in Chapter 15.

An exchange of charge between ions and the electrode occurs at the interface between the electrode and the solution. Surface reactions such as $H + H \rightarrow H_2$ may also occur. Whether electron transfer occurs depends on how the electronic states of the ion line up with the electronic structure of the electrode. This alignment is determined by the electrochemical potential of the ion and the band structure (work function and positions of the valence and conduction bands) of the electrode, as well as the potential placed on the electrode. The interplay between a number of chemical and physical phenomena acts to control the rate of current flow. These phenomena include:

- Mass transfer – the exchange of material between the interface and the bulk solution by diffusion.
- Electron transfer at the electrode surface.
- Chemical reactions both before and after the electron transfer event. These reactions may be catalytic heterogeneous reactions on the electrode surface, or homogeneous solution-phase reactions.
- Other surface reactions include adsorption, desorption and surface restructuring, especially in the case of deposition or etching.
- The electrochemical process may result in the formation of a porous layer on the electrode – as is the case under some circumstances when etching and/or deposition occur. Then, distinctions between the solution in the pores and in the bulk solution, as well as the internal surface on the pore walls and the external surface directly interfacing the bulk solution, must be made and these can lead to further complications. Similarly, if gas bubbles are formed and linger on the surface these can exclude regions of the surface from interacting with the solution, impeding transport. It should be

noted, however, that rapid bubble formation can actually be a very efficient method of mixing the solution to ensure mass transport between the interfacial and bulk regions.

To simplify our discussion somewhat, we will concentrate on reactions that do not change the electrode structure by deposition or etching of more than one monolayer (a layer that is only one atom or molecule thick).

The *overvoltage* η is defined as

$$\eta = E - E_{\mathrm{rev}} \tag{18.74}$$

where E is the potential at which electrolysis begins for a specific electrode material and E_{rev} is the potential at which the process should occur reversibly. For H_2 evolution, η is essentially 0 for a high-surface area clean Pt electrode, but is 0.15 V for a Ag electrode and 0.70 V for a Zn electrode.

All of the processes that contribute to the rate can also make a contribution to the overvoltage: (i) diffusion of ions from the solution to the surface; (ii) charge transfer on the surface of the electrode; (iii) surface reactions; and (iv) bubble formation if the product is a gas, such as H_2. Each process that contributes to the overvoltage can be thought of as having a resistance associated with it. These resistances, as well as the resistances associated with any junctions in the electrochemical cell and the electrolyte, then add in series to produce an overall cell resistance. The solution resistance R_s is a true resistance that, by Ohm's law, has associated with it a potential drop that is given by the product $I\,R_s$. Therefore, in general, we can write the applied (measured) cell potential E_{appl} as the sum of the reversible (also called the equilibrium or Nernstian) cell potential E_{rev}, the overpotential, and the cell resistance term,

$$E_{\mathrm{appl}} = E + I\,R_s = E_{\mathrm{rev}} + \eta + I\,R_s. \tag{18.75}$$

Note that the $I\,R_s$ term depends on the construction of the cell; thus, it can also be minimized by clever construction of the cell. These tricks include using three electrodes (anode, cathode and a reference through which no current passes), placing the reference as close to the working electrode as possible, and the use of ultramicroelectrodes through which currents of only a few nanoamps can be measured.

The evolution of H_2 is a very common electrochemical reaction, and its rate is usually controlled by the rate of surface charge transfer. However, for many electrochemical reactions at metal electrodes the electron transfer step is rapid and mass transport limits the rate of reaction. There are three types of transport that need to be considered:

- *Migration*: movement of charged species under the influence of an electric field. This responds to gradients in the electric field.
- *Diffusion*: movement of any chemical species under the influence of the chemical potential. This responds to gradients in concentration.
- *Convection*: movement of chemical species under the influence of hydrodynamics. This responds to gradients in density (natural convection) and mechanical agitation (stirring, forced convection).

If diffusion controls the rate and the overvoltage, the kinetics is diffusion controlled by what is called *concentration polarization*. One-dimensional mass transport along an axis perpendicular to the electrode surface (taken to be the x-axis in this example) is described by the *Nernst–Planck equation*,

$$J_k(x) = -D_k\frac{\partial C_k(x)}{\partial x} - \frac{z_k F D_k C_k}{RT}\frac{\partial \phi(x)}{\partial x} + C_k v(x) \tag{18.76}$$

Here, $J_k(x)$ is the flux of species k at distance x from the electrode, D_k is the diffusion coefficient, C_k the concentration, $\partial C_k(x)/\partial x$ the concentration gradient, $\partial \phi(x)/\partial x$ the potential gradient, z_k the charge, and $v(x)$ is the velocity of the solution along the x-axis.

At low concentrations in well-stirred systems, the electron transfer rate determines the rate of reaction. In this region, where mass transport is not an issue, Tafel observed the following relationship between the overvoltage and the *current density* j (current per unit area of the electrode):

$$\eta = a + b\ln j \tag{18.77}$$

where a and b are constants depending on the system, and the current density is the current divided by the electrode area, $j = I/A$. This is known as the *Tafel equation*. At the reversible overvoltage, $\eta = 0$ because the net current (sum of forward and reverse charge transfer reactions) is zero. The equal and opposite current flowing at this point is known as the *exchange current density* j_0.

The rate of an electrode reaction is usually measured in terms of the current or current density. The rate of reaction per unit area of electrode r (in mol m^{-2} s^{-1}) multiplied by the number of electrons transferred by electron exchange in the half-reaction z, the Faraday constant F (in C mol^{-1}), and the area of the electrode A (in m^2) will equal the current I (in A; recall that 1 ampere is 1 C s^{-1}). Conventionally, a rate is expressed in terms of a concentration term multiplied by a rate constant.

To make this discussion more specific, consider the one-electron half-reaction of a species Ox accepting an electron to become Red,

$$\mathrm{Ox} + \mathrm{e}^- \rightleftharpoons \mathrm{Red}. \tag{18.78}$$

For this reaction $z = 1$ and the rate equations for the forward and back reactions are written

$$r_\mathrm{f} = k_\mathrm{f} C_\mathrm{Ox} = \frac{I_\mathrm{f}}{zFA} = \frac{j_\mathrm{f}}{F} \tag{18.79}$$

$$r_\mathrm{b} = k_\mathrm{b} C_\mathrm{Red} = j_\mathrm{b}/F \tag{18.80}$$

The net rate is just $r_\mathrm{net} = r_\mathrm{f} - r_\mathrm{b}$; thus, the measured current density is

$$j = j_\mathrm{f} - j_\mathrm{b} = F(k_\mathrm{f} C_\mathrm{Ox} - k_\mathrm{b} C_\mathrm{Red}). \tag{18.81}$$

The application of a voltage to the electrode changes the relative electrical energy of the reactants and products. Butler and Volmer developed a model that showed that this shift is in the amount $F\eta$. The height of the activation barrier that opposes charge transfer also changes. The barrier height for the forward reaction is changed by an amount $\alpha F\eta$, where α is a fraction, called the *transfer coefficient* with a value less than 1 (usually ≈ ½) and F is the Faraday constant.

The current density for the forward reaction j_f follows Arrhenius-type behavior such that

$$j_\mathrm{f} = F C_\mathrm{Ox} k_0 \exp(\alpha F\eta/RT) \tag{18.82}$$

$$j_\mathrm{f} = j_0 \exp(\alpha F\eta/RT) \tag{18.83}$$

The remaining fraction $(1 - \alpha)$ of the electrical energy change causes the activation energy for the reverse reaction to increase by $(1 - \alpha)F\eta$, and the reverse current density is

$$j_\mathrm{b} = j_0 \exp(-(1 - \alpha)F\eta/RT). \tag{18.84}$$

Substituting into Eq. (18.81), the net current density is

$$j = j_0[\exp(\alpha F\eta/RT) - \exp(-(1 - \alpha)F\eta/RT)]. \tag{18.85}$$

As can be seen in Fig. 18.10, the net current is precisely 0 at $\eta = 0$ because the rates of forward and reverse reactions are balanced. The forward reaction dominates as the overpotential is pushed to positive values. The reverse reaction is virtually unobservable at large positive values of η, but dominates when the overpotential is negative.

Expanding on this in Fig. 18.11, we see that the characteristic shapes of the current density versus overpotential curves depend on the value of j_0, the *exchange current density*. The exchange current density is the value of j_f and j_b at $\eta = 0$. In other words, it is the current density that flows in the forward and reverse directions at equilibrium.

This is known as the *Butler–Volmer equation*. When $\eta = 0$, the net current is zero (as expected from the Tafel equation). When the overvoltage is large, the second exponential is negligible and the current density is

$$j = j_0 \exp(\alpha F\eta/RT). \tag{18.86}$$

This can be rearranged to obtain

$$\eta = \frac{RT}{\alpha F} \ln \frac{j}{j_0} = \frac{RT}{\alpha F} \ln j_0 + \frac{RT}{\alpha F} \ln j. \tag{18.87}$$

Note that Eq. (18.87) has the same form as the Tafel equation with

$$a = \frac{RT}{\alpha F} \ln j_0; \quad b = \frac{RT}{\alpha F}. \tag{18.88}$$

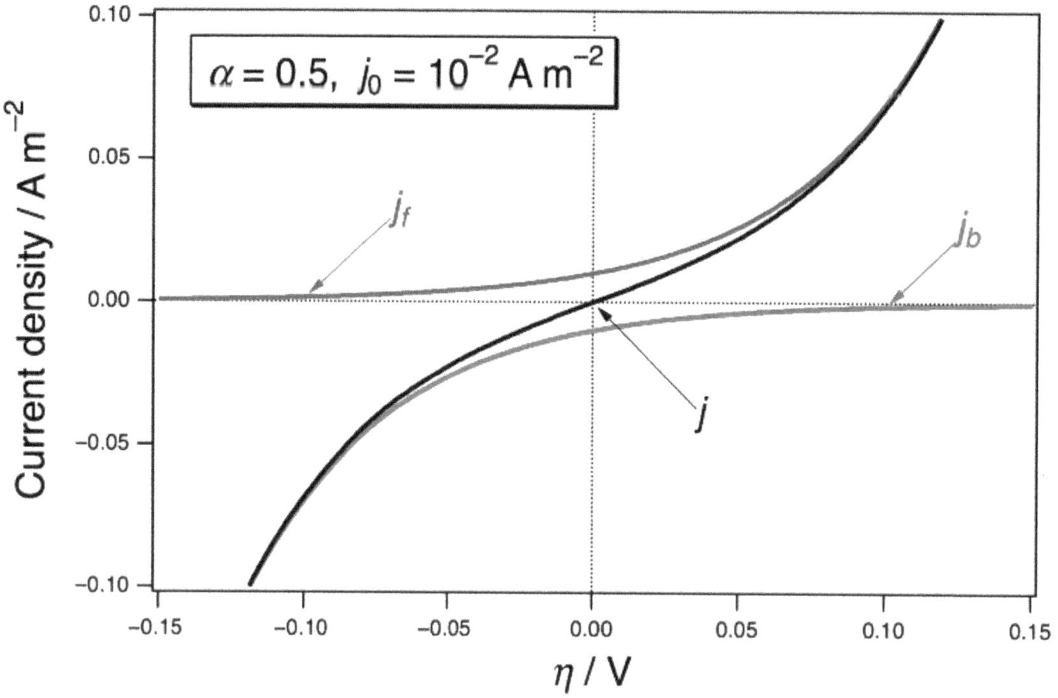

Figure 18.10 The behavior of the current density (forward j_f, back j_b and total j) as a function of the overpotential η. Calculated according to the Butler–Volmer equation with a transfer coefficient of $\alpha = 0.5$ and an exchange current density of $j_0 = 0.01$ A m^{-2}.

18.8.1 Rates of electrochemical reactions

Electrochemical reactions involve the exchange of free charge in at least one step of their reaction mechanism. However, they may also involve any number of chemical steps that result in the making and breaking of bonds in the usual manner. Any one of these steps can, in principle, be the rate-determining step. Simple electron exchange between ions coordinated only by water molecules is generally fast, approaching the diffusion-controlled limit. However, some ions exhibit notoriously slow electron exchange, such as $S_2O_8^{2-}$ and ClO_4^-.

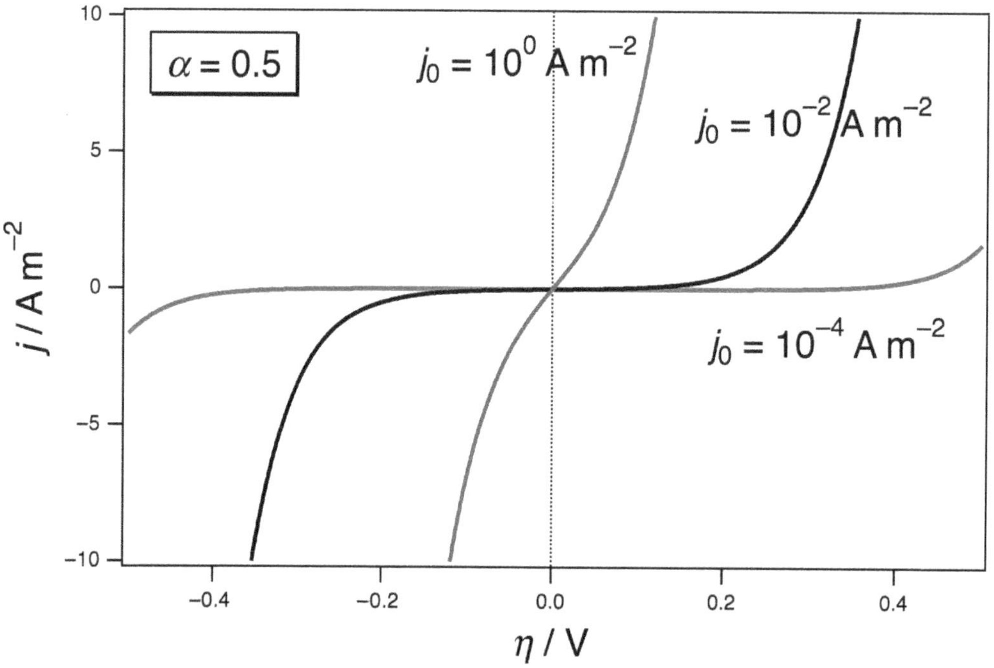

Figure 18.11 Plots of the net current density versus overvoltage as predicted by the Butler–Volmer equation.

Table 18.3 Rate constants for aqua-metal cation reductions of Co(III) complexes at 25 °C. All values in $M^{-1}\ s^{-1}$ taken from Burgess, J. (1999) *Ions in Solutions: Basic principles of chemical interactions*. Woodhead Publishing, Cambridge, UK.

	Cr^{2+}	V^{2+}	Eu^{2+}	Cu^{+}	U^{3+}
Inner-sphere					
$[CoCl(NH_3)_5]^{2+}$	6×10^5	7.6	400	4.9×10^4	4.9×10^4
$[CoBr(NH_3)_5]^{2+}$	1.4×10^6	25	250	4.5×10^5	1.4×10^4
$[CoI(NH_3)_5]^{2+}$	3×10^6	120	120		
$[CoOH(NH_3)_5]^{2+}$	1.5×10^6			3.8×10^2	
Outer-sphere					
$[Co(NH_3)_6]^{3+}$	9×10^{-5}	3.7×10^{-3}	1.7×10^{-3}		1.3
$[Co(en)_5]^{3+}$	2×10^{-5}	2×10^{-4}	8×10^{-4}	$\leq 4 \times 10^{-4}$	0.13

Two general limiting mechanisms of electron transfer are encountered. These are called inner- and outer-sphere electron transfer. The *outer-sphere* mechanism is simple electron transfer with no change in the primary solvation shell. *Inner-sphere* electron transfer involves substitution in the coordination sphere and is often, but not always, accompanied by atom or group transfer.

18.8.2 Inner-sphere mechanism

This mechanism was described by Henry Taube in 1954,[12] who received the Nobel Prize in Chemistry 1983 for his work on the mechanisms of electron transfer reactions, especially in metal complexes. A specific example is shown below in which both Co and Cr ions are bonded to Cl^- in the transition state during electron transfer. Cl^- is said to act as a *bridging ligand*. The reaction proceeds as follows. The two reactants collide to form a precursor complex in which Cl^- displaces a water molecule from the solvation shell about Cr^{2+}.

$$(NH_3)_5CoCl^{2+} + Cr(H_2O)_6^{2+} \rightleftharpoons \underset{\text{(precursor complex)}}{(NH_3)_5CoCl^{2+} \ldots Cr(H_2O)_5^{2+}} + H_2O$$

The coordination spheres of the bridged precursor complex jostle around until they attain a transition-state configuration in which the electron to be transferred is degenerate on either the Co^{3+} or the Cr^{2+} ions

$$\underset{\text{(precursor complex)}}{(NH_3)_5Co^{III}Cl \ldots Cr^{II}(H_2O)_5} \rightarrow \underset{\text{(activated complex)}}{[(NH_3)_5Co \ldots Cl \ldots Cr(H_2O)_5]^{4+}}$$

Electron transfer occurs in the activated complex from Cr^{2+} to Co^{3+}, followed by irreversible transfer of Cl^- and the formation of a successor complex

$$[(NH_3)_5Co \ldots Cl \ldots Cr(H_2O)_5]^{4+} \rightarrow \underset{\text{(successor complex)}}{(NH_3)_5Co^{II} \ldots ClCr^{III}(H_2O)_5}$$

After separation, loss of the 5 NH_3 rapidly ensues at which time they are protonated in solution. The overall reaction is then

$$(NH_3)_5CoCl^{2+} + Cr(H_2O)_6^{2+} + 5H^+(aq) \xrightarrow{HClO_4} Co(H_2O)_6^{2+} + 5NH_4^+ + (H_2O)_5CrCl^{2+}.$$

Many ligands can act as bridging ligands. An essential prerequisite is that they have a lone pair of electrons free on the opposite side to bond to the approaching metal center. This is true for halides or for many ligands containing O^{2-} or S^{2-}. When bound as ligands, NH_3, pyridine, amines or phosphines do not have an available lone pair and do not form bridges. Pyrazine and azide N_3^- can bridge. Inner-sphere processes are orders of magnitude faster than outer-sphere processes, as seen by the values reported in Table 18.3. In a few cases, the bridge-bound super complex is stable enough to be considered a reaction intermediate rather than as a transition state.

The inner-shell mechanism is summarized more generally in the reaction steps below and in Fig. 18.12. The two reactants are an electron donor with charge *m*, D^m, and an acceptor with charge *n*, A^n. The reactants and products are assumed to be hexacoordinated for this example. Note that if the reaction were occurring in aqueous solution, all of the

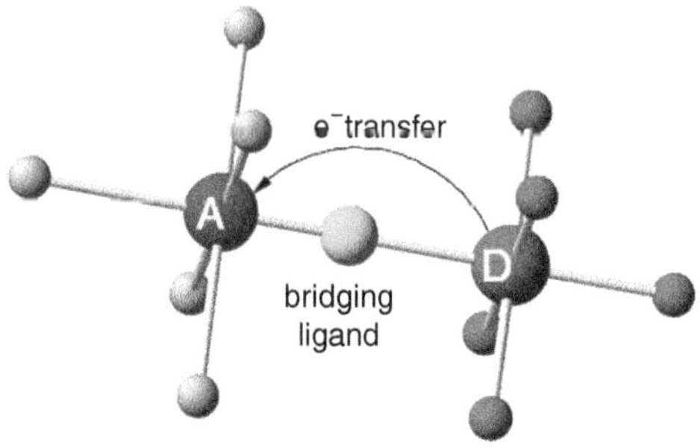

Figure 18.12 In the inner-sphere mechanism of electron transfer, one ligand from the first coordination shell is replaced by coordination to a bridging ligand. Electron transfer then occurs, followed by transfer of the bridging ligand to the donor. Separation of the successor complex is accompanied by replacement of the bridging ligand by another ligand on the acceptor.

ligands apart from the bridging ligand L_B could potentially be water molecules. Thus, L, L′ and the leaving ligand L_L, may or may not be chemically distinct based on the specific reaction under consideration. The reaction proceeds via complex formation, followed by replacement of the leaving ligand with formation of the bridge bond. The species so formed may be stable enough to be considered a reaction intermediate rather than a transient transition state species, depending on the specific reaction system.

Electron transfer between the redox active centers within the activated complex occurs at the transition state. Note the change in charge on the redox centers consistent with the negative charge on the electron. This is followed by complex separation. The site formerly coordinated by the bridge-bound ligand needs to be replaced by a new ligand, which again, may be a water molecule.

$$L_5D^mL_B + L_LA^nL'_5 \underset{k_{-d}}{\overset{k_d}{\rightleftharpoons}} L_5D^mL_BL_LA^nL'_5 \qquad \text{Complex formation}$$

$$L_5D^mL_BL_LA^nL'_5 \underset{k_{-b}}{\overset{k_b}{\rightleftharpoons}} L_5D^m \cdots L_B \cdots A^nL'_5 + L_L \qquad \text{Bridge formation}$$

$$L_5D^m \cdots L_B \cdots A^nL'_5 \underset{k_{-et}}{\overset{k_{et}}{\rightleftharpoons}} L_5D^{m+1} \cdots L_B \cdots A^{n-1}L'_5 \qquad \text{Electron transfer}$$

$$L_5D^{m+1} \cdots L_B \cdots N^{n-1}L' + L'' \xrightarrow{k_{sep}} L_5M^{m+1}L'' + L_BN^{n-1}L' \qquad \text{Complex separation}$$

$$L_5D^mL_B + L_LA^nL'_5 + L'' \rightleftharpoons L_5D^{m+1}L'' + L_BA^{n-1}L' + L_L \qquad \text{Overall}$$

18.8.3 Outer-sphere mechanism: Marcus theory of electron transfer

There are cases in which the formation of bridge linking between metal centers is impossible, yet electron transfer still occurs. In these cases, outer-sphere electron transfer must be occurring. Consider electron exchange between a donor molecule D and an acceptor molecule A. The first step of complex formation is the same in both inner-sphere and outer-sphere mechanisms. However, in an outer-sphere process, there is no bridge bond formed. The outer-sphere mechanism is described by the following steps,

$$D + A \underset{k_{d'}}{\overset{k_d}{\rightleftharpoons}} DA \qquad \text{Complex formation}$$

$$DA \underset{k_{b-et}}{\overset{k_{et}}{\rightleftharpoons}} D^+A^- \qquad \text{Electron transfer}$$

$$D^+A^- \xrightarrow{k_{sep}} D^+ + A^- \qquad \text{Complex separation}$$

$$D + A \rightleftharpoons D^+ + A^- \qquad \text{Overall}$$

Applying the steady-state approximation to both intermediates (DA and D^+A^-), the rate, which must be equal to the rate of the irreversible complex separation step,

$$R = k_{sep}[D^+A^-], \tag{18.89}$$

is described by

$$R = \frac{k_{sep}\, k_{et}\, k_d}{k_{b-et}\, k_{d'} + k_{sep}\, k_{d'} + k_{sep}\, k_{et}}[D][A] = k_{exp}\,[D][A]. \tag{18.90}$$

The second-order rate law is first-order in both the donor and acceptor concentrations. The exact form of the experimentally determined rate constant k_{exp} will depend on the relative magnitudes of the individual rate constants.

The qualitative outline of the outer-sphere mechanism was constructed with important contributions from Noel S. Hush. A quantitative predictive theory, known as the *Marcus theory* of electron transfer, awaited a conceptual breakthrough, which was provided by Rudolph A. Marcus. He was awarded the Nobel Prize in Chemistry 1992 for his contributions to the theory of electron transfer reactions in chemical systems. The remarkable thing about Marcus theory is that it allows a simple classical picture to be drawn to explain an inherently quantum mechanical phenomenon. Furthermore, this simple picture can explain broad trends – predicting the majority of rate constants for electron transfer to an acceptable degree of accuracy – while also predicting previously unknown phenomena such as the existence of a rate constant maximum and the 'inverted region,' which is in part responsible for the incredible efficiency of a number of biological electron transfer processes. Such a predictive capacity is generally accepted as one of the quintessential aspects of a ground-breaking theory. It should be kept in mind that the simplest form of Marcus theory does not predict rate constants with what is considered chemical accuracy by modern standards. Nonetheless, the theory clearly and cogently describes outer-sphere electron transfer, and lays bare those assumptions that can be optimized by quantum mechanical corrections.

Consider a reversible one-electron transfer from a reduced donor species D to an oxidized acceptor species A in an aqueous solution. Both D and A exist in solution with a tightly bound and highly organized inner solvation shell of water molecules, as shown in Fig. 18.13. Looser outer shells surround these. The inner shells have much different configurations depending on whether they surround D or A. It takes a significant amount of energy to reorganize the equilibrium solvent shell surrounding D into the configuration that would surround A at equilibrium. Although there is a high degree of order in the inner shell, it is subject to the normal fluctuations expected from a thermal Boltzmann distribution at equilibrium. Counterions are constantly in flux in the outer solvation shell. Therefore, *while on average a system at equilibrium finds itself at the bottom of its Gibbs energy well, instantaneously a system can have virtually any energy from zero to infinity*. The probability that the system is found at any particular Gibbs energy G is given by the Boltzmann distribution. It depends, among other things, on the energy difference from the ground state and the excited state, and the temperature in an exponential fashion.

Because the electron has a mass that is at least a thousand times less than the nuclei involved in electron transfer, its velocity is significantly greater than that of the nuclei, and it is transferred in a Franck–Condon-like transition, much like a transition between two electronic states accompanied by the absorption or emission of a photon. That is, the electron can transfer so quickly that *the nuclei do not have time to adjust their positions during electron transfer*. This means that if an electron were to hop from D to A, the solvation shells of D and A would be fixed in their initial configurations while D

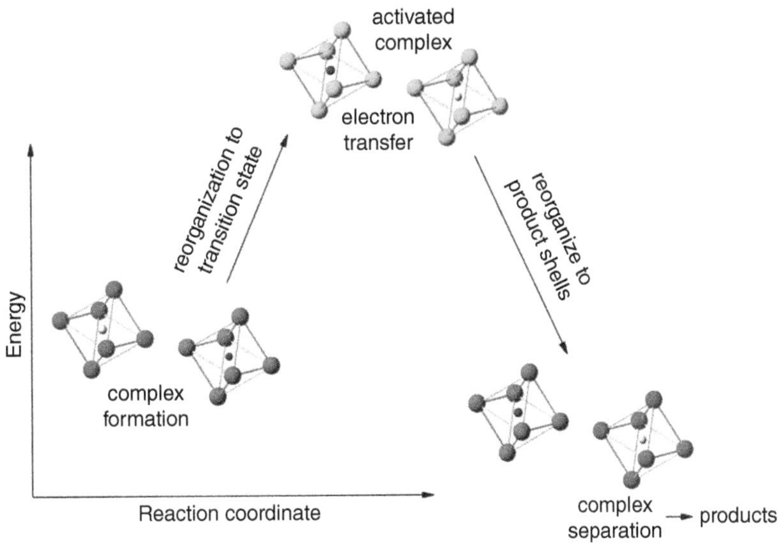

Figure 18.13 Complex formation without the formation of bridge bond or the exchange of ligands characterizes an outer-sphere electron transfer process. The first solvation shell is represented by the octahedron – blue for the equilibrium configuration about the acceptor, red for the equilibrium configuration about the donor, and green for the transition state configuration that makes the electron to be transferred degenerate on either species. The solvation shell first reorganizes to the transition state configuration, after which electron transfer occurs.

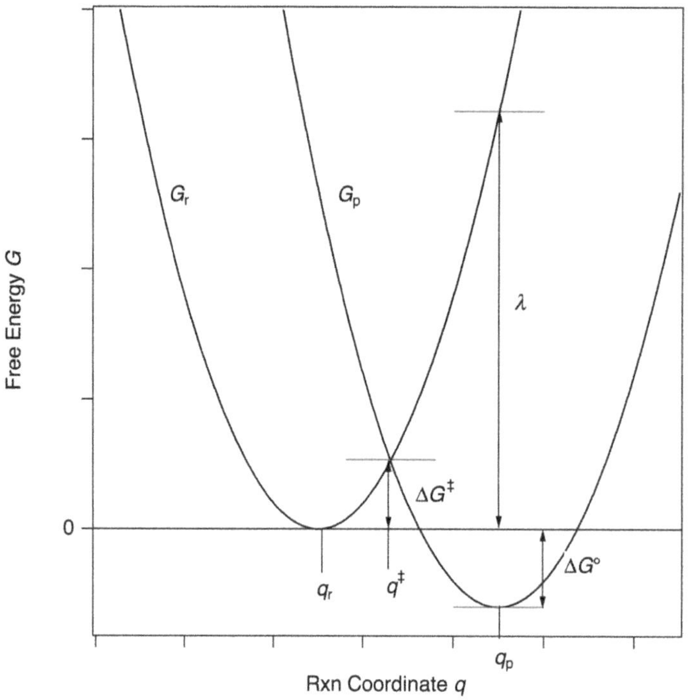

Figure 18.14 The displaced oscillators used to define the parameters required to calculate electron transfer rate constants within Marcus theory. The Gibbs energy curves of the reactants and products (G_r and G_p respectively) intersect at $q^\ddagger$ to form the transition state, which lies $\Delta G^\ddagger$ above the ground state of the reactants. The reorganization energy λ is also shown. Reproduced with permission from Kolasinski, K.W. (2012) *Surface Science: Foundations of Catalysis and Nanoscience*, 3rd edition. John Wiley & Sons, Chichester.

is transformed into D^+ and A is transformed into A^-. These solvation atmospheres do not correspond to the equilibrium configurations about D^+ and A^-. Therefore, the system finds itself in an excited state, which violates the conservation of energy during the electron transition.

Marcus realized that the Franck–Condon nature of the transition could only be maintained if the electron transfer were made between degenerate states. He proposed[13] the following dynamics for the electron transfer even:

> *The solvation shell first reconfigures itself. Once the solvent shell of the donor D reconfigures itself so that the system has the same energy, regardless of whether the electron is on D or A, electron transfer can occur degenerately. The transition state for electron transfer is, therefore, this configuration of the solvent shell, and the activation energy for electron transfer is the Gibbs energy change required to bring about this reorganization.*

This simple assertion would have been greeted with approval, and likely relegated to the closet of good but not great ideas if Marcus has not followed up with a series of theoretical papers that justified the simple picture shown in Fig. 18.14. Let G_r represent the Gibbs energy of the reactants and G_p the Gibbs energy of the products. These are connected by a reaction coordinate q. The nature of the reaction coordinate is ill-defined. At equilibrium the reactants would find themselves at $q = q_r$ and the products at $q = q_p$. We are free to place the zero of Gibbs energy wherever we choose, and for this exercise we set $G_r(q_r) = 0$. In other words, the Gibbs energy origin is placed at the equilibrium position of the reactants. The products are then located ΔG° below this value at equilibrium, $\Delta G^\circ = G_p(q_p) - G_r(q_r) = G_p(q_p)$. The transition state occurs at $q = q^\ddagger$, and the activation Gibbs energy $\Delta G^\ddagger$ is the energy at this point referenced to the Gibbs energy origin.

If we assume a linear response – which one might term a harmonic approximation – between the Gibbs energy and the reaction coordinate, the Gibbs energy exhibits a parabolic dependence on the reaction coordinate. Furthermore, with our choice of origin and by defining the *reorganization energy* λ as the energy difference between D when it is surrounded by its equilibrium solvent shell configuration and D when it is surrounded by the solvent shell associated with A at equilibrium, we find that the model reduces to only two parameters, namely ΔG° and λ according to

$$G_r = \lambda q^2 \tag{18.91}$$

$$G_p = \lambda (q - 1)^2 + \Delta G^\circ \tag{18.92}$$

$\Delta G°$ is readily available from tables and is directly related to the difference in $E°$ values by $\Delta G^\circ_m = -zFE°$, with z the electron number and F the Faraday constant. The reorganization energy λ can be determined experimentally. The transition state occurs at

$$q^\ddagger = 0.5 + \Delta G°/2\lambda. \tag{18.93}$$

The activation energy is found by solving for the intersection point defined by $\Delta^\ddagger G = G_r(q^\ddagger) = G_p(q^\ddagger)$, and has the value

$$\Delta^\ddagger G° = (\lambda + \Delta G°)^2/4\lambda. \tag{18.94}$$

Within transition state theory, the rate constant for electron transfer is then

$$k = \kappa A \exp(-\Delta^\ddagger G^\circ_m/RT), \tag{18.95}$$

where κ is the transmission coefficient, A is a pre-exponential factor with appropriate units, and R and T are as usual the gas constant and absolute temperature, respectively. In many instances, $\kappa \approx 1$ for outer shell electron transfer in solution.

The TST expression for the rate constant combined with Eq. (18.94) makes for an extraordinary prediction, which is displayed graphically in Fig. 18.15. As the driving force for the reaction, represented by $-\Delta G°$, increases, the rate constant increases up until the point where $\Delta G° = -\lambda$. *At this point, k reaches a maximum and decreases as the reaction becomes increasingly exergonic* – that is, more thermodynamically favored. This so-called *inverted region* is intimately related to the efficiency of natural and artificial photosynthetic systems. Photosynthesis is shut down if the electron initially excited by photon absorption immediately relaxes from the acceptor state back into the donor state from which it was excited. However, for sufficiently large donor–acceptor energy differences, the greater the energy stored by electron transfer from the donor to the acceptor, the slower back electron transfer will be. Molecular chemiluminescence also relies on the inverted region. Chemiluminescence occurs when an electronic excited state intervenes between the transition state and the product ground state. If the excited state couples more strongly to the transition state than the ground state, the excited state can be populated as a result of the charge transfer reaction. Radiative relaxation then leads to the conversion of chemical energy (effectively the exothermicity of the reaction) into light as the system drops down to the ground state.

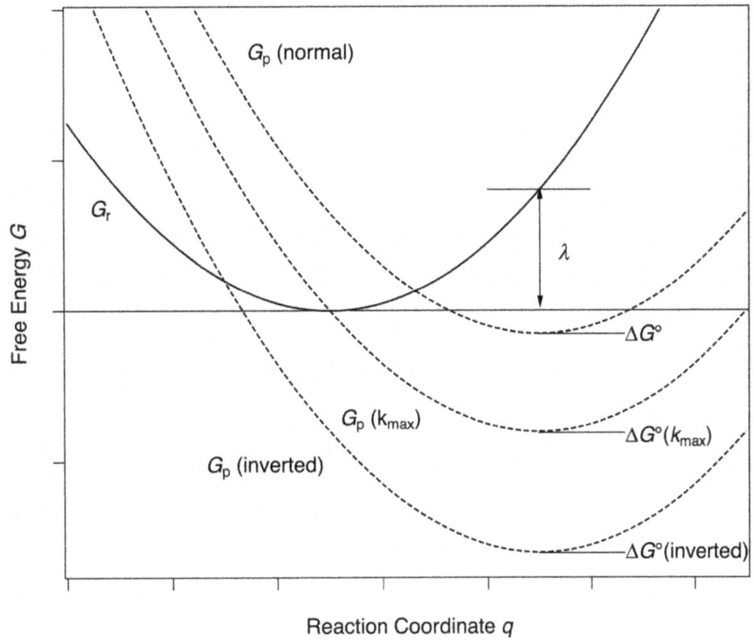

Figure 18.15 The rate constant for electron transfer passes through a maximum at $\Delta G° = -\lambda$ as the relative values of λ and $\Delta G°$ change, and the system passes from the normal region ($\Delta G° < |\lambda|$) to the inverted region ($\Delta G° > |\lambda|$). Reproduced with permission from Kolasinski, K.W. (2012) *Surface Science: Foundations of Catalysis and Nanoscience*, 3rd edition. John Wiley & Sons, Chichester.

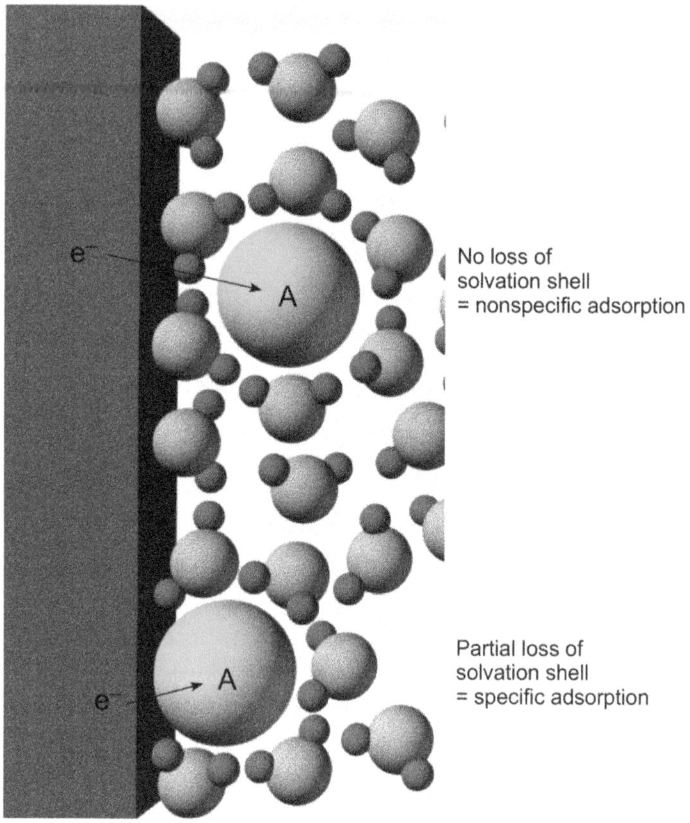

Figure 18.16 Schematic representations of inner-sphere and outer-sphere heterogeneous electron transfer from solid electrode to an acceptor species A. The acceptor is fully solvated in solution, but whether it is fully solvated when adsorbed at the surface (known as nonspecific adsorption) or partially solvated when adsorbed (known as specific adsorption) depends on the specific acceptor molecule.

18.8.4 Heterogeneous electron transfer

We now turn to charge transfer between a solute and a solid electrode. We can imagine two limiting cases for electron transfer with an electrode occurring either with specific or nonspecific adsorption. Here, I denote the specific adsorption case 'inner-sphere' and the nonspecific adsorption case 'outer-sphere' electron transfer, in analogy with homogeneous phase reactions. These are illustrated in Fig. 18.16. The inner sphere electron transfer case involves direct specific adsorption of A on the surface of the electrode. Specific adsorption proceeds with A losing part or all of its solvation shell, and leads to the formation of a strong chemical interaction (chemisorption) of A with the substrate. Effectively, the chemisorption bond and a surface atom act as the bridge in the heterogeneous inner sphere process. Our discussion is framed in terms of electron transfer from a donor level $|D\rangle$ in the electrode to an acceptor level $|A\rangle$ on the solute.

Electron transfer decreases exponentially with the distance between A and D. The range parameter is on the order of $\beta \approx 1\ \text{Å}^{-1}$. Thus, essentially all (> 90%) electron transfer occurs within ~3 Å of the surface. Since the width of a water molecule is roughly this size (diameter = 2.76 Å), this means that essentially all electron transfer occurs when the acceptor is adsorbed on the surface. Or, put more precisely, electron transfer occurs in the distance range between the closest approach of A with the surface (which may include its inner solvation shell) and this distance, plus the width of one water molecule.

Outer-sphere electron transfer involves the close approach to the surface of the solute with its full inner shell of solvation, which may, for example, contain six or so water molecules. The electron transfer occurs by tunneling from an occupied state at or below E_F into the not-fully-occupied acceptor level $|A\rangle$. In the case of a metal, we expect the electron to come from a state close to E_F. From a semiconductor with conventional doping and temperatures compatible with aqueous solutions, the conduction band has a very low density of occupied states, and the band gap is devoid of states (with the possible exception of surface states). Thus, the electron is most likely to come from the top of the valence band at energy E_V. Because this is a tunneling event, its probability falls off exponentially with distance. It also requires favorable orbital overlap, which means that the symmetry of the orbitals $|D\rangle$ and $|A\rangle$ can influence the tunneling probability. The case of weak coupling between these states has been treated by Marcus, Heinz Gerischer[14] and Nathan S. Lewis.[15]

There are major differences between electron transfer at metal and semiconductor electrodes. These are due to the different band structures, which influence the mobility, concentration and energy of electrons at their surfaces. In general, the rate of electron transfer can be written as

$$r_{et} = c_{red}\, c_{ox}\, k_{et} \tag{18.96}$$

where c_{red} is the concentration of the reductant (the donor D), c_{ox} is the concentration of the oxidant (the acceptor A). Because the density of electron states at and below E_F is so large and constant, electron transfer kinetics is effectively pseudo first order for metals, which yields

$$r_{metal} = c_{ox}\, k_{metal} \tag{18.97}$$

However, at semiconductors, the electron density in the conduction band or the hole density in the valence band is very small and, therefore, second-order kinetics is expected

$$r_{sc} = n_s\, c_{ox}\, k_{sc} \tag{18.98}$$

where n_s is the density of occupied states at the band edge (E_V for a semiconductor). This means that the units of the rate constants k_{metal} and k_{sc} differ.

A second major difference between electron transfer kinetics at metal electrodes and semiconductors is that *the voltage drop near a metal electrode occurs completely in the solution* because the electrons in the metal are so highly polarizable. Therefore, changing the voltage on the metal electrode changes the current exponentially (as shown explicitly below), because the relative position of the Fermi level with respect to the donor/acceptor level in the solution varies. Since this energy difference influences the activation energy for charge transfer, the current changes exponentially. On the other hand, *the voltage drop at a semiconductor electrode occurs completely in the space charge region inside the semiconductor.* Therefore, changing the voltage does not change the relative positions of the band edges at the surface compared to the donor/acceptor level in solution. Instead, the current varies exponentially with applied voltage because the surface density of states changes with bias.

These kinetic differences between metal/solution and semiconductor/solution electron transfer kinetics can thus be stated as follows:

- *Bias voltage changes the rate of electron flow through a metal electrode because it changes the rate constant for electron transfer.*
- *In contrast, bias voltage changes the rate of electron flow through a semiconductor electrode because it changes one of the concentrations in the rate equation.*

To obtain a feeling for the qualitative implications of these differences, consider, for example the work of Lewis and coworkers[16] on the ferrocenium/ferrocene ($Fe^{+/0}$) redox system. The difference in dielectric constant between a metal electrode such as Pt and a semiconductor such as Si is quite large. Consequently, the image charge effects in the two systems are much different with efficient screening at a metal surface. This reduces λ_0, the inner shell contribution to λ, at a Pt electrode to about half the value found for a homogeneous electron transfer, but leaves λ_0 little changed at a Si electrode. Thus, $\lambda_0 = 0.5$ eV for a Pt electrode versus 1.0 eV for a Si electrode. Using $n_{s0} = N_c \exp[(-0.9\text{ eV})/k_B T]$, with $N_c = 10^{19}$ cm^{-3} for the effective density of states in the Si conduction band, n_{s0} is 10^4 cm^{-3}, which yields $k^\circ_{sc} = 10^{-12} - 10^{-13}$ cm s^{-1}, and exchange current densities of 10^{-11} A cm^{-2} at [Fe^+] = 1.0 M. In contrast $k^\circ_{metal} = 400$ cm s^{-1} for $Fe^{+/o}$ at Pt, which would lead to an exchange current density of $j_0 = 4 \times 10^4$ A cm^{-2} at [Fe^+] = 1.0 M. The latter prediction cannot be measured directly because the current density is restricted by diffusion before it reaches this ideal limiting value. The 15 orders of magnitude difference is related primarily to the reduction in the electron density in the semiconductor electrode relative to that at the metal, not to a change in mechanism.

Within a transition state theory framework, the rate constant of electron transfer can be written in terms of a pre-exponential factor and a Gibbs energy of activation as usual

$$k_{et} = A \exp\left(-\Delta^{\ddagger}G^\circ / k_B T\right). \tag{18.99}$$

The units of A are different for a metal (s^{-1}) and for a semiconductor (m^4 s^{-1}) because of the change in kinetic order, as discussed above. Due to the differences in band structure, the expressions for the activation energy are also different. For a metal electrode held at a potential E

$$\Delta^{\ddagger}G^\circ = (E - E_{ox} + \lambda)^2 / 4\lambda. \tag{18.100}$$

For a semiconductor transferring an electron from its valence band

$$\Delta^{\ddagger}G^{\circ} = (E_V - E_{ox} + \lambda)^2/4\lambda. \quad (18.101)$$

E_{ox} is the Nernst potential of the oxidant,

$$E_{ox} = E^{\circ} - (RT/zF)\ln Q \quad (18.102)$$

λ is the reorganization energy, and E_F or E_V is referenced to a common vacuum level along with E_{ox}.

After electron transfer has occurred, the acceptor molecule with an additional electron A^- can undergo further reactions that may be either heterogeneous or homogeneous. The rate-determining step for any given reaction may be either the electron transfer step or one of these chemical reactions.

As long as the acceptor level $|A\rangle$ approaches the donor level $|D\rangle$ from above – that is, from a higher energy – it can be assumed that electron transfer occurs either at the Fermi level for a metal or else at the appropriate band edge for a semiconductor after the acceptor level has fluctuated to an appropriate energy to facilitate degenerate electron transfer. In this case, all holes are injected into the electrode in a narrow energy range at E_F, the valence band maximum or conduction band minimum, regardless of the ground-state energy of the acceptor level on the oxidant. However, once the acceptor level lies below the energy of the occupied band, there are always states that are degenerate with respect to electronic energy available for electron transfer from the electrode. The solvation shell still needs to reorganize; therefore, there is still an activation energy.

18.8.4.1 Example

The VO_2^+(aq) species has a standard reduction potential of 0.91 V, which places it below the Si valence band maximum under standard conditions. The electron transfer rate from the Si valence band to VO_2^+(aq) exhibits Arrhenius behavior, as demonstrated in Fig. 18.17. Use this data to determine the reorganization energy for VO_2^+(aq).

The Gibbs energy of activation is related to the reorganization energy by

$$\Delta^{\ddagger}G^{\circ} = (E_V - E_{ox} + \lambda)^2/4\lambda.$$

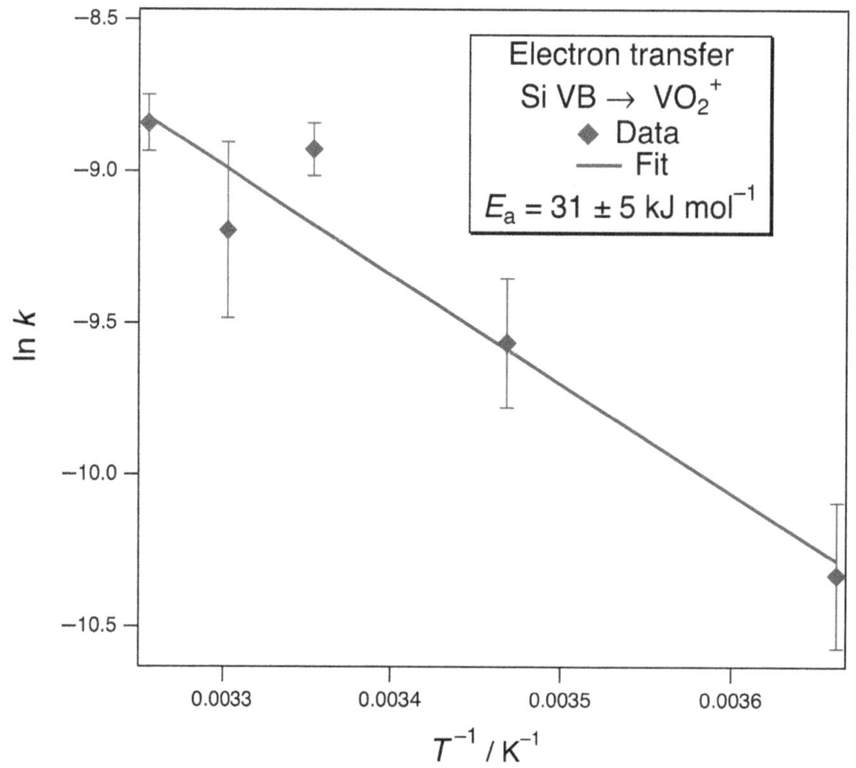

Figure 18.17 An Arrhenius plot of ln k versus $1/T$ for electron transfer from the Si valence band to VO_2^+(aq) (formed in an aqueous solution of HF + V_2O_5). The activation energy results from the barrier to charge transfer that initiates an etching reaction. Reproduced with permission from Kolasinski, K.W. and Barclay, W.B. (2013) *Angew. Chem., Int. Ed. Engl.*, **52**, 6731.

In this case, E_{ox} lies below the Si valence band maximum, E_V. Therefore, the acceptor is degenerate with states in the Si valence band. This allows us to take $E_V - E_{ox} = 0$, which means

$$\Delta^{\ddagger}G^{\circ} = \frac{\lambda^2}{4\lambda} = \frac{\lambda}{4}$$

$$\lambda = 4\Delta^{\ddagger}G^{\circ}$$

This corresponds to $\lambda = 124$ kJ mol^{-1} or 1.29 eV per particle.

SUMMARY OF IMPORTANT EQUATIONS

$R = k_0[S] + k_H[S][H^+] + k_{OH}[S][OH^-]$	Rate of reaction in the presence of specific acid–base catalysis, rate constant of uncatalyzed, acid-catalyzed and base-catalyzed reactions k_0, k_H and k_{OH}, respectively
$R = k[S]$	Effective first-order rate for acid–base catalysis
$k = k_0 + k_H[H^+] + \frac{k_{OH}\,K_w}{[H^+]}$	Effective first-order rate constant for acid–base catalysis k, water self-ionization constant K_w
$\lg k = \ln k_0 + \lg k_H - \text{pH}$	Low pH limit of effective rate constant for acid catalysis
$\lg k = \ln k_0 + \lg K_{OH} + \lg K_w + \text{pH}$	High pH limit of effective rate constant for base catalysis
$R = \frac{V[S]}{K_m + [S]}$	Michaelis–Menten equation: rate of enzyme-catalyzed reaction R, $V = k_2[E]_0$, substrate concentration [S]
$K_m = \frac{k_{-1} + k_2}{k_1}$	Michaelis constant
$R_{max} = k_2[E]_0$	Limiting rate R_{max}, turnover number k_2, initial enzyme concentration $[E]_0$
$\frac{1}{R_0} = \frac{1}{R_{max}} + \frac{K_m}{R_{max}}\frac{1}{[S]_0}$	Lineweaver–Burk equation: initial rate R_0, initial substrate concentration $[S]_0$
$R_p = k_p(k_i/k_t)^{1/2}[M][I]^{1/2}$	Rate of chain polymerization R_p, monomer concentration [M], initiator concentration [I], rate constants for propagation, initiation and termination k_p, k_i, k_t, respectively
$PDI = M_w/M_n$	Polydispersity index PDI, weight-average molar mass M_w, number-average molar mass M_n
$E_{photon} = h\nu = hc/\lambda$	Planck equation: photon energy E_{photon}, Planck constant h, frequency ν, wavelength λ
$P_{abs} = I_{abs}E_{photon}$	Absorbed power P_{abs}, absorbed intensity I_{abs}
$Pt = N_{photon}h\nu$	Power of light source P, time of exposure t, number of photons N_{photon}
$\Phi = \frac{\text{rate of process}}{\text{rate of photon absorption}} = \frac{R}{I_{abs}}$	Quantum yield Φ, rate in molecules per second R, photon flow (intensity) in photons per second I
$\eta = E - E_{rev}$	Overvoltage η, potential at which electrolysis is performed E, potential developed by cell running reversibly E_{rev}
$\eta = a + b\ln j$	Tafel equation: current density $j = I/A$
$a = \frac{RT}{\alpha F}\ln j_0;\ b = \frac{RT}{\alpha F}$	transfer coefficient α, exchange current density j_0, Faraday constant F, gas constant R, absolute temperature T
$j = j_0[e^{(\alpha F\eta/RT)} - e^{(-(1-\alpha)F\eta/RT)}]$	Volmer–Butler equation
$k = \kappa A\exp(-\Delta^{\ddagger}G^{\circ}/k_BT)$	Rate constant from transition state theory k, transmission coefficient κ, pre-exponential factor A, Gibbs energy of activation $\Delta^{\ddagger}G^{\circ}$
$\Delta^{\ddagger}G^{\circ} = (\lambda + \Delta G^{\circ})^2/4\lambda$	Gibbs energy of activation for homogeneous outer-sphere electron transfer, reorganization energy λ
$\Delta^{\ddagger}G^{\circ} = (E - E_{ox} + \lambda)^2/4\lambda$	Gibbs energy of activation for outer-sphere electron transfer at a metal electrode held at potential E
$\Delta^{\ddagger}G^{\circ} = (E_V - E_{ox} + \lambda)^2/4\lambda$	Gibbs energy of activation for outer-sphere electron transfer from a semiconductor electrode with valence band maximum at E_V
$E_{ox} = E^{\circ} - (RT/zF)\ln Q$	Nernst potential of an oxidant E_{ox}, standard reduction potential E°, electron number z, reaction quotient Q

Exercises

18.1 Why does an activation barrier form?

18.2 Describe how a catalyst works in terms of its effects on bonding and why, therefore, chemisorption is usually more important in heterogeneous catalysis than is physisorption.

18.3 When 0.5 Langmuir of H_2 is dosed onto Ir(110)–(2 × 1) at 100 K, it adsorbs dissociatively. Its TPD spectrum exhibits one peak at 400 K. When propane is dosed to saturation at 100 K on initially clean Ir(110)–(2 × 1), the H_2 TPD spectrum exhibits two peaks, one at 400 K and one at 550 K. Interpret the origin of these two H_2 temperature programmed desorption (TPD) peaks from C_3H_8 adsorption.

18.4 Considering the HCOOH/Pt(111) system, O atoms are adsorbed onto the surface before exposure to a pulse of HCOOH. It is found that the sticking coefficient of HCOOH does not change. The enthalpy change is again found to be −65 kJ mol^{-1}, but a second exothermic signal is found shortly after the pulse is over. When the surface is heated, no HCOOH desorbs at 170 K. Instead, only H_2O is observed to desorb at 205 K. Suggest a plausible explanation for these observations, including chemical reactions to support your answer.

18.5 K acts as a promoter in the Fischer–Tropsch synthesis of liquid hydrocarbon fuels from synthesis gas (a mixture of CO + H_2). Describe the two mechanisms by which an electropositive element such as K can act as a promoter.

18.6 Describe the role of poisons and promoters in catalysis.

18.7 Ammonia synthesis on an Fe catalyst follows a Langmuir–Hinshelwood mechanism after dissociative adsorption of the reactants. Write out the complete mechanism, the overall reaction, and identify the rate-determining step.

18.8 S acts as a poison in the ammonia synthesis reaction. Describe the two ways in which S acts as a poison in this reaction. Does S change the kinetics or thermodynamics of the reaction?

18.9 Oxygen and carbon monoxide chemisorb on Rh and physisorb on graphite. The reaction of $^{18}O_2$ with C ^{16}O leads to the formation of C ^{18}O ^{16}O when these gases are exposed at 400 K to a graphite surface covered partially with Rh nanoparticles. Write out a mechanism for the reaction. Where does the reaction occur?

18.10 In the presence of a transition metal catalyst, the reaction A(g) + B(g) → C(g) is found to follow the rate equation $R = k\vartheta_A\vartheta_B$. The coverages of A, ϑ_A, and B, ϑ_B, are found to follow the relations:

$$\vartheta_A = \frac{K_A p_A}{1 + K_A p_A + K_B p_B} \quad \text{and} \quad \vartheta_B = \frac{K_B p_B}{1 + K_A p_A + K_B p_B}.$$

K_A and K_B are constants that reflect how strongly adsorbed A and B are. A large value of K_A or K_B indicates strong adsorption. C does not bind to the surface under the reaction conditions. (a) State with justification what type of reaction mechanism is being followed. Write out a set of elementary steps that constitute the reaction mechanism. (b) Write out an expression for the rate R in terms of the pressures p_A and p_B. (c) Determine the rate law in the limit of low pressure. What are the reaction orders with respect to p_A, p_B and overall? (d) Determine the rate law in the limit of weakly bound A and strongly bound B. What are the reaction orders with respect to p_A and p_B? What is the kinetic role of B in the reaction?

18.11 An enzyme-catalyzed reaction between an enzyme E and substrate S is run at steady state with a large excess of substrate compared to the enzyme. After the establishment of steady state for the reaction in which the enzyme binds reversibly to the substrate to form a complex ES and a product P is produced irreversibly from the complex, an inhibitor I that irreversibly binds to the enzyme is introduced into the mixture. The concentration of the inhibitor is roughly equal to the initial concentration of the enzyme. Write out a set of elementary steps that describes the mechanism. Discuss with justification from rate equations the rate of production of P as a function of time after the introduction of I for: (i) the case in which the rate constant of formation of ES, k_1, is large compared to the rate at which ES returns to uncomplexed enzyme k_{-1}; or (ii) $k_1 \ll k_{-1}$. In both cases, the rate constant for binding to the inhibitor k_2 is comparable to k_{-1}.

18.12 The following rates have been obtained for the reaction of urea catalyzed by 2 μM urease at various substrate concentrations:

[S] / mM	1.00	2.00	5.00	10.0	20.0
v / mM s^{-1}	0.764	1.48	3.33	5.71	8.89

Use the Lineweaver–Burk equation to determine the Michaelis constant K_M, the values of the limiting rate V, and the turnover number k_2.

18.13 The following rates have been obtained for the reaction of H_2O_2 catalyzed by 10 μM catalase at various substrate concentrations:

$10^3 \times$ [S] / mol dm^{-3}	0.4	0.6	1.0	1.5	2.0	3.0	4.0	5.0	10.0
Rate, v / M s^{-1}	1.57	2.34	3.85	5.66	7.41	10.71	13.79	16.67	28.57

(a) Use the Michaelis–Menten equation to show algebraically that a plot of v/[S] against v should be linear with a slope of $-(1/K_M)$. (b) Determine the Michaelis constant K_M, the values of the limiting rate V, and the turnover number k_2.

18.14 The base hydrolysis of Co(III)–amine and amine–halide complexes is believed to follow the conjugate base mechanism. The first step involves proton transfer to OH^- from the complex. This step is equilibrated with equilibrium constant K_1. The second rate-limiting step is the unimolecular decay of the ammonium-containing complex to form products with rate constant k_2. For the specific case of $[CoCl(NH_3)_5]^{2+}$ reacting with OH^-(aq), derive an expression for the rate of formation of products d[P]/dt in terms of the reactant concentrations.

18.15 The Eigen–Wilkins mechanism of complex formation describes the reaction of a monodentate ligand L and an aqua-metal ion $M(OH_2)_x^{n+}$, abbreviated M^{n+}. The first step involves the reversible formation of the complex $M^{n+}{\bullet}L$. The second step is the irreversible formation of the complex $M^{n+}L$ with expulsion of water from the intermediate. (a) Write an expression for the rate of formation of the complex $M^{n+}L$ in terms of the reactant concentrations. (b) Write the rate equations is the limits of $[M^{n+}L] \gg [L]$ and $[L] \gg [M^{n+}L]$.

18.16 The Eigen–Winkler mechanism for complex formation with macrocyclic and cryptand ligands involves three reversible steps: pre-association of the metal ion M^{n+} with the ligand C; interchange with formation of the first metal–ligand bond; and conformational change(s) required to complex or encapsulate M^{n+} fully. The mechanism can be represented as

$$M^{n+} + C \rightleftarrows M^{n+} \cdots C \rightleftarrows MC^{n+} \rightleftarrows \{MC^{n+}\}.$$

(a) Write a general rate equation for the formation of product, assuming that all steps are reversible. (b) Write a general rate equation for the formation of product, assuming that all steps are equilibrated.

18.17 Describe the mechanism of two different types of explosions. How can these mechanisms be quenched/inhibited?

18.18 A photochemical reaction obeys the mechanism written below. (a) If the quantum yield for the formation of B is $\Phi = 1.50$ at $\lambda = 354$ nm and 2.36×10^{-6} mol of B are produced in 900 s, what is the intensity of the absorbed light in photons s^{-1}? (b) Apply the steady-state approximation and derive an expression for the quantum yield B formed. (c) What are the values of the quantum yield in the limits of large and small fluorescence lifetimes of A?

$$A + h\nu_1 \xrightarrow{I_{ads}} A^*$$

$$A^* \xrightarrow{k_2} A + h\nu_2$$

$$A^* \xrightarrow{k_3} 2B$$

$$\Phi = \frac{R_B}{I_{abs}}.$$

18.19 A photochemical reaction obeys the following mechanism.

$$A + h\nu \xrightarrow{\Phi_1=0.45} 2B$$
$$2B \xrightarrow{k_2=1.45\times10^3\ \mathrm{Pa^{-1}\ s^{-1}}} C + D$$

(a) Write out the overall reaction and identify the intermediate(s). (b) Apply the steady-state approximation and determine an expression for d[C]/dt that does not include the concentration of any intermediates. (c) When the light intensity is low, the rate is found to be linearly proportional to the intensity of the light source. As the intensity is increased, the rate approaches saturation, and the rate increases less than linearly with light intensity. Discuss this behavior in terms of the rate-determining step and the reaction mechanism. (d) The molar absorptivity of A is 50 000 $\mathrm{dm^3\ mol^{-1}\ cm^{-1}}$ at 355 nm. If p_A = 145 kPa and P_{abs} = 333 μW, calculate the rate of formation of C and the rate of disappearance of A in molecules per second. Calculate the partial pressure of D in Pa after 5000 s irradiation with 355 nm light. The kinetics are operating in the low intensity limit, the volume is 0.250 $\mathrm{m^3}$, and T = 300 K.

18.20 A photochemical reaction obeys the following mechanism:

$$A + h\nu \xrightarrow{\Phi_1=0.75} A^*$$
$$A^* + B \xrightarrow{k_2=6.0\times10^5\ \mathrm{Pa^{-1}\ s^{-1}}} C$$
$$B + C \xrightarrow{k_3=5.0\times10^3\ \mathrm{Pa^{-1}\ s^{-1}}} D + E$$

(a) Write out the overall reaction and identify the intermediate(s). (b) Apply the steady-state approximation and determine an expression for d[D]/dt that does not include the concentration of any intermediates. (c) Identify with justification the rate-determining step under these conditions. (d) Given that $p_A = p_B$ = 25.0 kPa, calculate the rate of formation of D in molecules per second when 145 mW of 632 nm light is absorbed by A.

18.21 A photochemical reaction obeys the following mechanism:

$$A + h\nu \xrightarrow{\Phi_1=0.0500} A^*$$
$$A^* + B \xrightarrow{k_2=4.5\times10^6\ \mathrm{Pa^{-1}\ s^{-1}}} C$$
$$B + C \xrightarrow{k_3=2.7\times10^2\ \mathrm{Pa^{-1}\ s^{-1}}} D + E$$

(a) Write out the overall reaction and identify the intermediate(s). (b) Apply the steady-state approximation and determine an expression for d[E]/dt that does not include the concentration of any intermediates. (c) Identify with justification the rate-determining step under steady-state conditions. (d) Calculate the rate of formation of E in molecules per second when 250 mW of 193 nm light is absorbed by A. (e) Given that the initial pressure in a 1.00-liter vessel is 101 kPa at 298 K, calculate the partial pressure of E after 1 min irradiation during which the adsorbed power at 193 nm is constant at 250 mW.

18.22 A photochemical reaction run at 300 K obeys the following mechanism:

$$A + B \xrightarrow{k_1=7500\ \mathrm{Pa^{-1}\ s^{-1}}} C + D$$
$$C + h\nu \xrightarrow{\Phi=0.75} C^*$$
$$B + C^* \xrightarrow{k_3=2500\ \mathrm{Pa^{-1}\ s^{-1}}} E$$

(a) Write out the overall reaction and identify the intermediate(s). (b) Apply the steady-state approximation and determine an expression for d[E]/dt that does not include the concentration of any intermediates. (c) Irradiation at 357 nm leads to a constant adsorbed power of 3.49 mW. If the volume of the reaction vessel is 0.500 $\mathrm{m^3}$ and the initial partial pressures of the A and B are both 1.00 kPa, calculate the rate of formation of E (in molecules $\mathrm{s^{-1}}$) and the partial pressures of A, B and E after irradiation for 6000 s.

18.23 Write the complete inner-sphere mechanism for the reaction

$$\mathrm{CoBr(NH_3)_5^{2+}(aq) + CuBr(s) \overset{acid}{\rightleftharpoons} CuBr_2(aq) + Co^{2+}(aq) + 5\ NH_4^+(aq)}.$$

18.24 Describe how outer-sphere electron transfer occurs between two solution-phase species (donor D and acceptor A) in the framework of Marcus theory.

18.25 Write the complete outer-sphere mechanism for the reaction

$$Co(NH_3)_6^{3+}(aq) + V^{2+}(aq) \xrightleftharpoons{\text{acid}} V^{3+}(aq) + Co^{2+}(aq) + 6\,NH_4^+(aq).$$

18.26 Use the complete reaction mechanism including inhibition given in Section 18.5 to derive the rate expression given in Eq. (18.36).

18.27 For electron transfer in which the solvent has a linear harmonic response described by the force constant k_0, the reactant Gibbs energy curve is $G_r(q) = k_0(q - q_r)^2$ and the product Gibbs energy curve is $G_p(q) = k_0(q - q_p)^2 + \Delta G$. (a) Show that the transition state occurs at the reaction coordinate

$$q^{\ddagger} = \frac{q_p^2 - q_r^2 + \Delta G/k_0}{2q_p - 2q_r}.$$

(b) Show that for such a system the activation Gibbs energy is given by Eq. (18.94).

Endnotes

1. Eley, D.D. and Rideal, E.K. (1940) *Nature (London)*, **146**, 401.
2. Weinberg, W.H. (1991) in *Dynamics of Gas-Surface Interactions* (eds C.T. Rettner and M.N.R. Ashfold), The Royal Society of Chemistry, Cambridge, p. 171.
3. Rettner, C.T. (1992) *Phys. Rev. Lett.*, **69**, 383; Rettner, C.T. (1994) *J. Chem. Phys.*, **101**, 1529; Rettner, C.T. and Auerbach, D.J. (1995) *Phys. Rev. Lett.*, **74**, 4551.
4. Harris, J. and Kasemo, B. (1981) *Surf. Sci.*, **105**, L281; Harris, J., Kasemo, B., and Törnqvist, E. (1981) *Surf. Sci.*, **105**, L288.
5. Ertl, G. (2009) *Reactions at Solid Surfaces*. John Wiley & Sons, Hoboken, NJ; Imbihl, R. and Ertl, G. (1995) *Chem. Rev.*, **95**, 697.
6. Sabatier, P. (1920) *La catalyse en chimie organique*. Bérange, Paris; Sabatier, P. (1911) *Ber. Deutsch. Chem. Gesellschaft*, **44**, 1984.
7. Blyholder, G. (1964) *J. Phys. Chem.*, **68**, 2772.
8. Lundqvist, B.I., Nørskov, J.K., and Hjelmberg, H. (1979) *Surf. Sci.*, **80**, 441; Nørskov, J.K., Houmøller, A., Johansson, P.K., and Lundqvist, B.I. (1981) *Phys. Rev. Lett.*, **46**, 257.
9. Koshland, D.E., Jr (1960) *Adv. Enzymol.*, **22**, 45.
10. Hanoian, P., Liu, C.T., Hammes-Schiffer, S., and Benkovic, S. (2015) *Acc. Chem. Res.*, **48**, 482; Soudackov, A.V. and Hammes-Schiffer, S. (2014) *J. Phys. Chem. Lett.*, **5**, 3274; Solis, B.H. and Hammes-Schiffer, S. (2014) *Inorg. Chem.*, **53**, 6427.
11. Michaelis, L. and Menten, M.L. (1913) *Biochem. Z.*, **49**, 333.
12. Taube, H. and King, E.L. (1954) *J. Am. Chem. Soc.*, **76**, 4053; Taube, H. (1955) *J. Am. Chem. Soc.*, **77**, 4481.
13. Marcus, R.A. (1956) *J. Chem. Phys.*, **24**, 966; Marcus, R.A. (1957) *J. Chem. Phys.*, **26**, 867; Marcus, R.A. (1960) *Faraday Discuss.*, **29**, 21; Marcus, R.A. (1963) *J. Phys. Chem.*, **67**, 853; Marcus, R.A. (1963) *J. Phys. Chem.*, **67**, 2889; Marcus, R.A. (1993) *Rev. Mod. Phys.*, **65**, 599.
14. Gerischer, H. (1997), in *The CRC Handbook of Solid State Electrochemistry* (eds P.I.J. Gellings and H.J.M. Bouwmeester), CRC Press, Boca Raton, p. 9.
15. Lewis, N.S. (1998) *J. Phys. Chem. B*, **102**, 4843; Royea, W.J., Fajardo, A.M., and Lewis, N.S. (1997) *J. Phys. Chem. B*, **101**, 11152; Lewis, N.S. (1991) *Annu. Rev. Phys. Chem.*, **42**, 543.
16. Lewis, N.S. (1991) *Annu. Rev. Phys. Chem.*, **42**, 543.

Further reading

Adams, D.M., Brus, L., Chidsey, C.E.D., Creager, S., Creutz, C., Kagan, C.R., Kamat, P.V., Lieberman, M., Lindsay, S., Marcus, R.A., Metzger, R.M., Michel-Beyerle, M.E., Miller, J.R., Newton, M.D., Rolison, D.R., Sankey, O., Schanze, K.S., Yardley, J., and Zhu, X.Y. (2003) Charge transfer on the nanoscale: Current status. *J. Phys. Chem. B*, **107** (28), 6668–6697.

Ertl, G. (2009) *Reactions at Solid Surfaces*. John Wiley & Sons, Hoboken, NJ.

Fletcher, S. (2010) The theory of electron transfer. *J Solid State Electrochem.*, **14**, 705.

Hammer, B. and Nørskov, J.K. (2000) Theoretical surface science and catalysis – Calculations and concepts. *Adv. Catal.*, **45**, 71.

Henriksen, N.E. and Hansen, F.Y. (2008) *Theories of Molecular Reaction Dynamics: The microscopic foundation of chemical kinetics*. Oxford University Press, Oxford.

Houston, P.L. (2001) *Chemical Kinetics and Reaction Dynamics*. Dover Publications, Mineola, NY.

Hush, N.S. (1971) *Reactions of Molecules at Electrodes*. Wiley-Interscience, London.

Kolasinski, K.W. (2012) *Surface Science: Foundations of Catalysis and Nanoscience*, 3rd edition. John Wiley & Sons, Chichester.

Laidler, K.J. (1987) *Chemical Kinetics*. HarperCollins, New York.
Layfield, J.P. and Hammes-Schiffer, S. (2014) Hydrogen tunneling in enzymes and biomimetic models. *Chem. Rev.*, **114**, 3466.
Levine, R.D. (2005) *Molecular Reaction Dynamics*. Cambridge University Press, Cambridge.
Levine, R.D. and Bernstein, R.B. (1987) *Molecular Reaction Dynamics and Chemical Reactivity*. Oxford University Press, New York.
Marcus, R.A. and Sutin, N. (1985) Electron transfers in chemistry and biology. *Biochim. Biophys. Acta*, **811**, 265.
Rettner, C.T. and Ashfold, M.N.R. (1991), in *Advances in Gas-Phase Photochemistry and Kinetics*. The Royal Society of Chemistry, Cambridge.
Ringe, D. and Petsko, G.A. (2008) How Enzymes Work. *Science*, **320**, 1428.
Zare, R.N. (1998) Laser Control of Chemical Reactions. *Science*, **279**, 1875.
Zewail, A.H. (1992) *The Chemical Bond – Structure and Dynamics*. Academic Press, London.

CHAPTER 19

Developing quantum mechanical intuition

PREVIEW OF IMPORTANT CONCEPTS

- Light is thought of classically as an electromagnetic wave.
- The quantum of the electromagnetic field is called the photon. It is neither particle nor wave, but is has properties that in certain limits behave like classical waves or classical particles.
- The energy of a photon is related to its frequency by the Planck equation $E = h\nu$.
- The number of photons present determines the intensity of the light.
- Interaction (reflection or transmission) of light with apertures/objects with spacings comparable to the wavelength of light leads to interference phenomena.
- Correspondence principle – in the limit of high quantum numbers, systems behave classically. Equivalently, in the limit of large energies and large orbits, quantum calculations deliver the same results as classical calculations.
- Planck explained black-body radiation by proposing the quantum hypothesis. The excitation of oscillations that emit photons are quantized – that is, they have discrete rather than continuous energies.
- Einstein explained the photoelectric effect by combining the idea of quantized photons with the conservation of energy. Each photon absorbed causes one excitation and results in the photoemission of one electron when the photon energy exceeds the work function.
- Energy loss spectroscopy can deliver information similar to absorption and emission spectroscopy.
- de Broglie demonstrated that wave-particle duality is not limited to photons. Microscopic particles can also exhibit a wavelength that leads to significant effects.
- Bohr demonstrated that requiring quantized angular momentum leads to an accurate description of the quantized energy level structure of the H atom.
- The line spectra of atoms and molecules can be understood as the result of transitions made between quantized energy levels. By conservation of energy, the energy of the photon associated with the transition is equal to the energy difference between the levels.

The world of subatomic particles simply does not behave how everyday macroscopic objects do. Because of this you will need to be introduced to numerous new phenomena and a new vocabulary. One of the best ways to be ready for the new ideas of quantum mechanics is to be comfortable with the physics of classical electromagnetic waves and classical mechanics of particles. Hence, here we first review a few items of terminology and then proceed to review wave behavior and classical mechanics.

The use of quantum mechanics is most necessary for atomic scale distances and masses. A convenient length scale is the Ångström (1 Å = 0.1 nm = 1×10^{-10} m), in particular because bonds are on the order of 1 Å in length. Of course, this unit is not recommended within the SI system of units; however, it is simply so convenient and widespread, it still finds use. Another non-SI unit that is extremely useful is the electron volt, eV. Covalent bond energies are on the order of several eV. The electron volt is equal to the kinetic energy of an electron (or any other particle with a charge of one elementary unit $e = 1.602 \times 10^{-19}$ C) that is accelerated through a potential difference of 1 V. Thus, the conversion factor between the clumsy (on the atomic scale) joule and the convenient electron volt is 1.602×10^{-19} J eV^{-1}. The electron volt is convenient when speaking of the energy of a single subatomic particle, while kJ mol^{-1} is convenient when dealing with on the order of Avagadro's number worth of particles, 1 eV = 96.485 kJ mol^{-1}.

Spectroscopy is the primary means by which we investigate the quantum world. The interpretation of spectroscopic data is key to our understanding. Remember that spectrum is singular and its plural is spectra.

Physical Chemistry: How Chemistry Works, First Edition. Kurt W. Kolasinski.

Companion Website: www.wiley.com/go/kolasinski/physicalchemistry

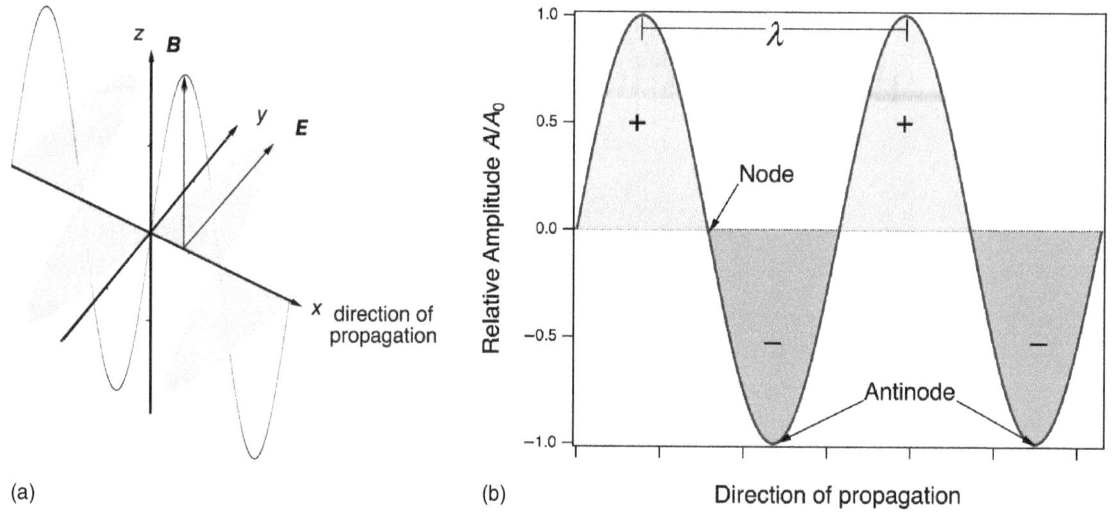

Figure 19.1 (a) A classical electromagnetic wave is composed of an oscillating electric field $\boldsymbol{E}$ and a magnetic field $\boldsymbol{B}$ that oscillates perpendicular to the electric field. The z-axis is chosen to lie perpendicular to the direction of the electric field, and the electric field oscillates in the xy-plane. (b) A wave of amplitude A_0 (absolute value of the maximum or minimum field strength) at the antinodes. The wavelength λ is the distance that it takes for the wave to complete a full cycle. A node is a point at which the field strength goes to zero. The sign of the electric or magnetic field changes sign on either side of a node.

19.1 Classical electromagnetic waves

19.1.1 Characteristics of light

Light is described as an oscillation of 'something' at right-angles to the direction of propagation – a *transverse* wave. The 'something' that is oscillating is the amplitude of electric and magnetic fields, as shown in Fig. 19.1. They oscillate with a frequency ν and a wavelength λ. Sitting at one point in space and counting how many complete oscillations the wave makes in a given length of time is a measure of the *frequency*, the units of which are per unit time, s^{-1} or hertz (Hz). The distance between two successive maxima (or two successive minima) is equal to the *wavelength*, which is measured in units of distance. Maxima and minima collectively are known as extrema and also as antinodes. The sign of the oscillating field changes from (+) to (−) and passes through zero at a node. Another commonly encountered descriptor of electromagnetic waves is the *wavenumber* $\tilde{\nu}$, which is measured in inverse length, m^{-1} in SI. However, because it is so ingrained in the literature, the more commonly encountered unit is cm^{-1} and this will be used here.

The speed of light was the first physical constant to be assigned an exact value. This is because it is used as the basis for defining many other fundamental constants. The *speed of light in vacuum* is represented by c, and is

$$c = 2.99792458 \times 10^8 \text{ m s}^{-1} (\text{exact}). \tag{19.1}$$

If we need to differentiate it from the speed of light in a medium, we will denote the speed of light in vacuum as c_0 and the speed of light in a medium as c. The *refractive index* is the ratio between the speed of light in vacuum and the speed of light in a medium

$$n = c_0/c. \tag{19.2}$$

The product of *wavelength* λ and *frequency* ν is equal to the speed of light

$$c = \lambda\nu. \tag{19.3}$$

The *wavenumber* $\tilde{\nu}$ is the inverse of the wavelength thus,

$$\tilde{\nu} = 1/\lambda = \nu/c. \tag{19.4}$$

Implicit in our discussion below, and throughout this book, is that we are considering propagation of light *in vacuo* – that is, in a medium with refractive index $n = 1$. This is a simplifying approximation so that we do not have to keep track of this factor of n and label everything with a subscript 0. When light enters a dense medium, its speed decreases in proportion to the refractive index, as shown in Eq. (19.2). The electromagnetic field of light establishes a polarization in the medium that oscillates at the same frequency as the incident light. Therefore, the frequency of light in the medium remains constant at ν. However, in accord with Eq. (19.3) the wavelength of light in the medium λ must be reduced from its value *in vacuo* λ_0 such that $\lambda = \lambda_0/n$. The refractive index is a slowly varying function of frequency in a transparent material, but changes rapidly near absorption features. The refractive index of air is $n_{air} = 1.0003$ at 589 nm (a standard wavelength associated with the familiar yellow D line emission of Na). When performing high-resolution spectroscopy it is important to correct for this factor that would change the D_1 line from $\lambda_0 = 589.76$ nm to $\lambda = 589.58$ nm for light propagating in air at atmospheric pressure. The refractive index is proportional to density. Therefore, it depends not only on the composition of the medium but also on temperature and pressure. For example, water at atmospheric pressure and 20 °C has $n = 1.3330$.

James Clerk Maxwell developed the mathematical theory of magnetic fields and showed that electricity and magnetism cannot exist in isolation. He suggested that light was composed of oscillating electric $\boldsymbol{E}$ and magnetic $\boldsymbol{B}$ fields, which are perpendicular to each other, and that these field can be represented by sine or cosine waves,

$$E(x,t) = E_0 \cos\left[\frac{2\pi}{\lambda}x - 2\pi\nu t\right] = E_0 \cos[kx - \omega t] \tag{19.5}$$

$$B(x,t) = B_0 \cos\left[\frac{2\pi}{\lambda}x - 2\pi\nu t\right] = B_0 \cos[kx - \omega t] \tag{19.6}$$

where $\omega = 2\pi\nu$ is the angular frequency (units of radians per second represented by s^{-1} and never Hz), and $k = 2\pi/\lambda$ is the wavevector. The maximum values of the amplitudes of the electric and magnetic fields are E_0 and B_0, respectively. The argument of the cosine function is known as the *phase* of the wave ϕ_0

$$\phi_0 = 2\pi\left(\frac{x}{\lambda} - \nu t\right) = kx - \omega t. \tag{19.7}$$

Light propagating in a uniform medium has a constant phase. Thus, differentiating this equation with respect to time and setting it equal to zero, we obtain the *phase velocity* of the wave

$$c = \frac{dx}{dt} = \frac{\omega}{k}. \tag{19.8}$$

What is commonly called the 'speed of light' is the speed at which a point of constant phase (an antinode, for example) propagates past a fixed observer. An electromagnetic wave is two-dimensional and c is the propagation velocity of a plane of constant phase along the x-axis.

The intensity of a classical wave is proportional to $|A|^2$, where A is the amplitude of the wave. Thus, the intensity of an electric field (what is commonly measured in any light detector since the interaction of matter with the magnetic field is generally much weaker) is proportional to E_0^2. ϕ describes the *relative phase* (*phase difference*) of the waves. The term 'relative phase' describes the displacement of one wave with respect to another. Waves that have constant displacement are coherent waves – that is, they are in phase with each other. Waves that are in phase add constructively, while waves that are out of phase add destructively. A shift in phase by 90° (= $\pi/2$ radians) changes a sine wave into a cosine wave

$$E_0 \sin[kx - \omega t] = E_0 \cos[kx - \omega t - \pi/2] \tag{19.9}$$

Recall that a cosine function can also be represented by a complex exponential function

$$\cos x = \frac{1}{2}(e^{ix} + e^{-ix}),\ e^{\pm ix} = \cos x \pm i \sin x \tag{19.10}$$

and that $i^2 = -1$. Therefore, an equivalent way to write the wave equation of the electric field is

$$E(x,t) = E_0 \exp[-i(kx - \omega t + \phi)] = E_0\, e^{-i(kx-\omega t+\phi)} \tag{19.11}$$

where the relative phase ϕ has been included explicitly. The real part of E represents the physical wave. Using this complex notation, the intensity is proportional to the absolute square of the field, that is by the product of the complex conjugate of the field E^* with the field E

$$I(x,t) = E^*(x,t)\, E(x,t).^1 \tag{19.12}$$

Electromagnetic radiation can possess a near infinite range of wavelengths. The span of wavelengths is termed the *electromagnetic spectrum*; this is shown in Fig. 19.2. The electromagnetic spectrum ranges from radio waves at low energy up to cosmic rays at the high end. The length unit of choice depends on the spectral range. Different parts of the spectrum

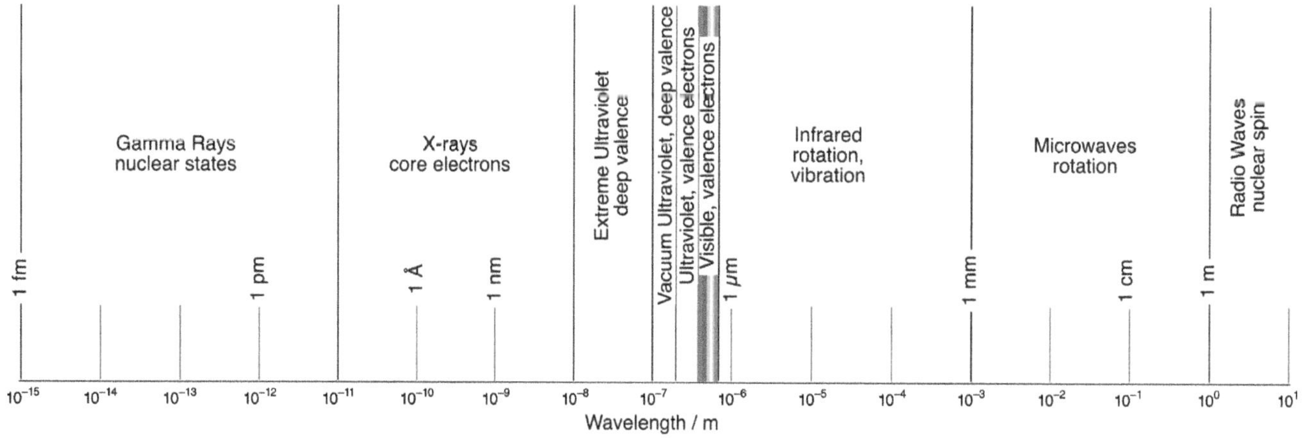

Figure 19.2 The electromagnetic spectrum.

are associated not only with different names, but also with different characteristic molecular/atomic excitations as noted in Table 19.1. This means that different regions are also associated with different spectroscopic techniques, which also tend to have different preferred units. Hence, pure rotational spectroscopy often uses wavelength in μm, frequency in MHz, or wavenumber in cm^{-1}; vibrational spectroscopy (also called infrared spectroscopy) is most commonly conducted in cm^{-1} but sometimes also in μm. Visible light is commonly referenced by wavelength in nm, though wavenumber in cm^{-1} is also used. This is true for electronic spectroscopy throughout the UV/Visible range. X-ray spectroscopy and X-ray diffraction often reference the wavelength in Å (now sometimes in pm) or photon energy in eV. Gamma spectroscopy is often performed in energy units of keV or MeV.

Several other terms are commonly associated with light and electromagnetic waves. White light is comprised of waves of many frequencies. White light can be characterized by a central frequency and a broad bandwidth of frequencies of random phase on either side of this central value. This is in contrast to monochromatic light (light of a single frequency) or narrow bandwidth light (light of a small frequency range). Plane-polarized light has its E field in a well-defined plane. The plane of the E field is termed the *plane of polarization*. In circularly polarized light, the E field rotates in a plane orthogonal to the propagation direction.

19.1.2 Superposition of waves

In addition to oscillating with a definite frequency and wavelength, one of the essential defining characteristics of a wave is its ability to exhibit interference phenomena as a result of the relative phase behavior of combined waves. In Figs 19.3–19.5 we see the results of the superposition of two waves of equal amplitude propagating collinearly. Three different cases are shown: Fig. 19.3 shows equal frequency and in-phase; Fig. 19.4 shows equal frequency and out-of-phase; and Fig. 19.5 shows initially in-phase but unequal wavelength.

All electromagnetic waves of the same wavelength travel at the same speed in the same medium. Therefore, waves of equal wavelength that are in-phase at the origin remain in phase everywhere. Note that waves with a well-defined phase relationship need to be added coherently. That is, the electric field of the combined wave is obtained by adding the

Table 19.1 The electromagnetic spectrum divided into different wavelength ranges.

Spectral range	Wavelength	Energy / eV	Excitation
Radio	100 km–100 cm	10^{-11}–10^{-6}	nuclear spin
Microwave	100 cm–1 mm	10^{-6}–10^{-3}	rotation
Infrared	1 mm–700 nm	10^{-3}–1.8	rotation, vibration
Visible	700–400 nm	1.8–3.1	valence electrons
Ultraviolet	400–200 nm	3.1–6.2	valence electrons
Vacuum ultraviolet	200–100 nm	6.2–12.4	deep valence
Extreme ultraviolet	100–10 nm	12.4–124	deep valence
X-ray	10–0.01 nm	125–125 k	core electrons
Gamma ray	<10 pm	>125 k	nuclear states

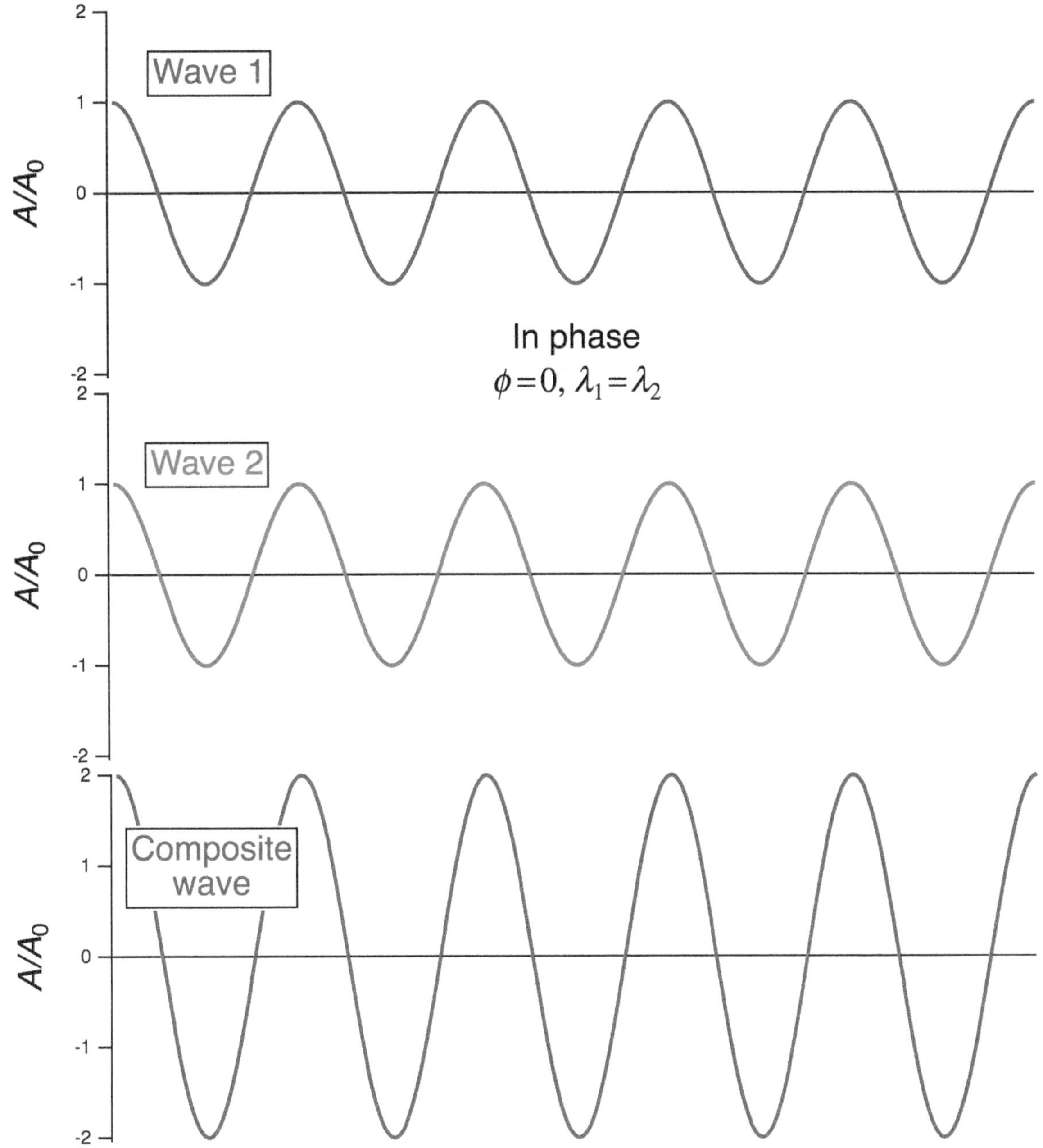

Figure 19.3 The constructive interference of in-phase waves of the same wavelength.

electric fields of the component waves. The intensity of the combined wave is then obtained by squaring the summed wave. *In a coherent sum, add waves then square to obtain the intensity.*

Consider two waves that they have the same frequency and amplitude but a phase difference ϕ. They may be written

$$E_1 = E_0\, e^{-i(kx-\omega t)} \tag{19.13}$$

$$E_2 = E_0\, e^{-i(kx-\omega t+\phi)} \tag{19.14}$$

The electric field of the superposed wave is obtained by addition of the two waves

$$E = E_0\, e^{-i(kx-\omega t)}\,(1 + e^{-i\phi}) \tag{19.15}$$

Factoring out a term of $e^{-i\phi/2}$ from the term in parenthesis and expanding the result using Eq. (19.10), we rewrite the electric field of the superposed waved as

$$E = 2E_0\, e^{-i(kx-\omega t+\phi/2)}\cos(\phi/2) \tag{19.16}$$

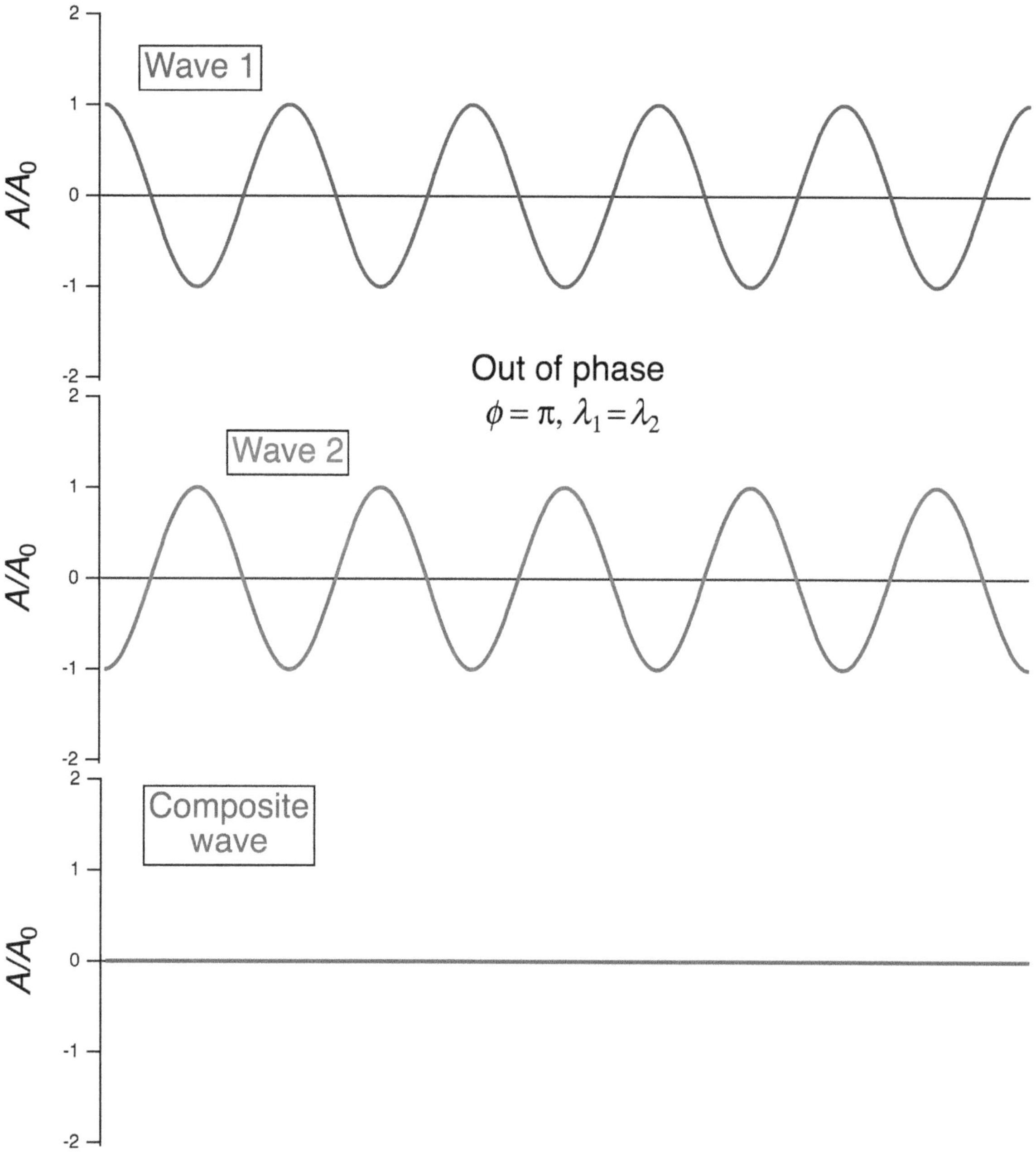

Figure 19.4 The destructive interference of out-of-phase waves of the same wavelength.

This allows us to write E as a product of real functions and complex exponential functions only. This simplifies the result because in taking the product involving the *complex conjugate*,

$$(e^{i\theta})^* e^{i\theta} = e^{-i\theta} e^{i\theta} = 1. \tag{19.17}$$

Therefore, the intensity (irradiance) is

$$I = |E|^2 = E^* E = 4E_0^2 \cos^2[\phi/2]. \tag{19.18}$$

19.1.2.1 Directed practice

Complete the algebraic steps necessary to derive Eq. (19.18).

If the waves did not have a well-defined phase relationship, an *incoherent sum* would be performed. In an incoherent sum, the intensity of the each wave is first obtained by squaring the field strength, then the component intensities are added to obtain the resultant intensity of the combined wave. In a coherent sum of two waves with equal amplitude

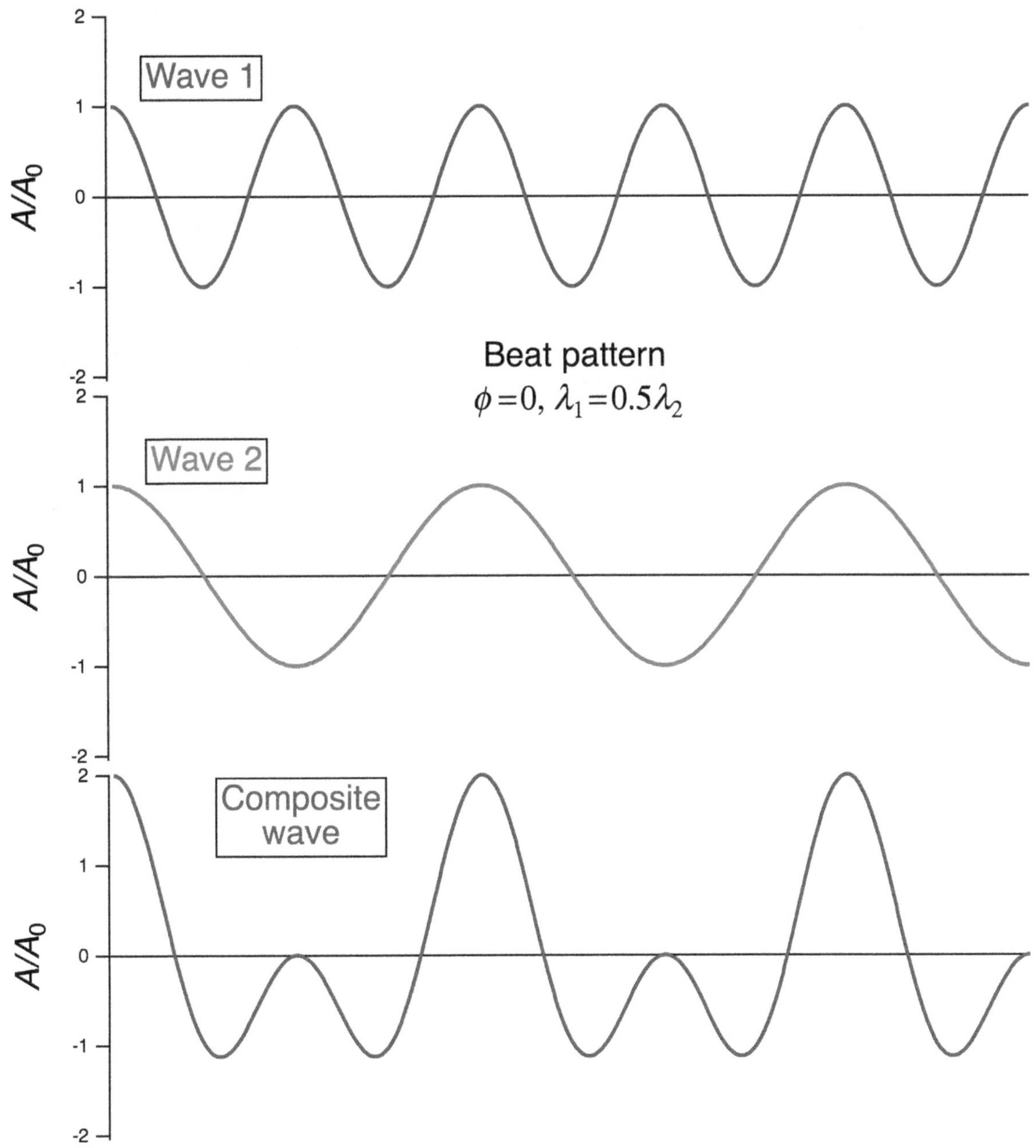

Figure 19.5 The variable interference (beat pattern) of incoherent waves – waves with different wavelengths and therefore variable phase.

E_0, the maximum intensity is $4E_0^2$. In an incoherent sum of two waves with equal amplitude E_0, the maximum intensity is $2E_0^2$. The output of two conventional light bulbs adds incoherently: two 100 W bulbs have the same intensity as one 200 W bulb (avoiding for the moment any consideration of distance dependences and shadowing from the construction of the bulbs).

However, if we have coherent sources – most easily constructed from lasers but also possible with classical sources under special conditions – then we need to consider the relative phases of the constituent waves. Constructive interference is perfect for waves of the same wavelength and amplitude that are in phase $\phi = 0$ or, more generally, differ in phase by $\phi = 2n\pi$, where n is an integer. Constructive interference becomes progressively weaker as the phase difference increases. Time invariant interference between two sources of light can only take place if they are coherent sources, that is, they must have the same frequency and have a constant phase difference.

When the phase difference increases to π radians, the interference becomes perfectly destructive for waves of the same wavelength and amplitude. This is shown in Fig. 19.4. Interference is a periodic phenomenon and destructive interference becomes increasingly less effective as the phase difference increases beyond π radians. Destructive interference is a maximum at $\phi = (2n+1)\pi$, with n again an integer.

Waves with different frequencies and a well-defined initial phase relationship display variable interference because the relative phase difference changes with time. This is shown in Fig. 19.5 and is known as a *beat pattern*. Note that energy is conserved. Interference leads to changes in the distribution of radiant energy. Destructive interference at the node redistributes energy into the antinodes. Interference phenomena can only occur where beams overlap and the occurrence of destructive interference at one point in space and time must be accompanied by constructive interference elsewhere to balance it. For complete destructive interference to occur at all points in time and space it would mean that the two beams would have to interfere at their sources such that no energy is radiated at all.[2] We will see in Chapter 27 that fitting cycles of light inside an optical cavity such that they interfere constructively is important in the design of lasers.

If the two waves have slightly different values of wavelength described by $k + \Delta k$ and $\omega + \Delta\omega$, the total electric field that results from the two waves propagating collinearly is

$$E = E_0\, e^{-\mathrm{i}(kx-\omega t)}\,(1 + e^{-\mathrm{i}(\Delta kx - \Delta\omega t)}). \tag{19.19}$$

This equation has the same form as Eq. (19.15) but with the relative phase difference replaced by

$$\phi = \Delta kx - \Delta\omega t. \tag{19.20}$$

The superposed intensity is then described by

$$I = 4E_0^2 \cos^2[(\Delta kx - \Delta\omega t)/2]. \tag{19.21}$$

The cosine factor describes an envelope function that modulates the traveling waves. This traveling wave propagates with a *group velocity* of

$$c_g = \Delta\omega/\Delta k. \tag{19.22}$$

This represents the velocity at which a plane of constant phase of the envelope travels. It need not be equal to the phase velocity of the individual waves.

19.1.3 Double slit interference

Pick a plane of constant phase, say a maximum, and follow it as the wave propagates. We refer to this plane as the *wavefront* (we could pick any plane of constant phase but the maximum is the most commonly chosen). The wavefront is normal to the direction of propagation. A plane wave is a collimated beam with plane wavefronts.

Consider the situation depicted in Fig. 19.6 in which a monochromatic plane wave is incident on two infinitesimal slits a distance d apart. Each slit acts as a point source and we observe the light transmitted from these slits on a screen at a distance L away. The electric field at point P is the sum of the fields emanating from our two sources,

$$E = A(e^{-\mathrm{i}kr_1} + e^{-\mathrm{i}kr_2})e^{\mathrm{i}\omega t}, \tag{19.23}$$

where A is the amplitude and r_1 and r_2 are the distances of the two slits from the point P. When L is large compared to d, the paths from the two slits are effectively parallel and r_1 and r_2 differ by only $d = \sin\theta$, a quantity that is called the *optical path difference* (OPD). Therefore,

$$E = A\, e^{-\mathrm{i}kr_1}\,(1 + e^{-\mathrm{i}kd\sin\theta})e^{\mathrm{i}\omega t}, \tag{19.24}$$

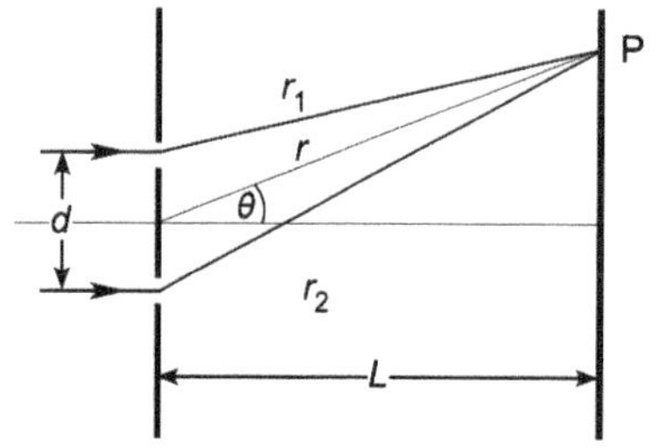

Figure 19.6 The geometry of the two-slit interference phenomenon. Slits that are small compared to their separation d are situated a distance L from a screen. A dark fringe is observed at the point P on a screen when the pathlength difference $r_2 - r_1$ meets the requirement for destructive interference.

which corresponds to a phase difference between the two waves of

$$\phi = kd \sin \theta. \tag{19.25}$$

The intensity is obtained from the absolute square of the electric field,

$$I \propto 4|A|^2 \cos^2\left(\frac{\pi d \sin \theta}{\lambda}\right). \tag{19.26}$$

For large L, the angle θ is small and

$$\sin \theta = x/L. \tag{19.27}$$

The light incident on the screen exhibits an interference pattern that has a $\cos^2$ variation with the distance x from the origin along the screen. Maxima (instances of constructive interference) occur whenever the OPD is equal to an integral multiple of π,

$$OPD = m\lambda \tag{19.28}$$

and minima (instances of destructive interference) occur when

$$OPD = \left(m + \frac{1}{2}\right)\lambda. \tag{19.29}$$

The resulting intensity pattern is shown in Fig. 19.7. The fringes do not extend to infinity in any real case because the source cannot be perfectly monochromatic and collimated (though lasers can get extremely close to this) and the slits cannot be infinitesimally narrow, lest no light pass. This second effect causes the light passing through the slits to be emitted in cones rather than hemispheres. Because of this, the intensity is damped as the distance from the origin increases.

19.1.4 Multiple-slit interference

If we generalize the geometry of Fig. 19.6 from one slit to N slits, we find that the OPD between rays emanating from adjacent slits is again $d \sin \theta$. The OPD between the first and the jth slit is then $(j-1)d \sin \theta$. Summing the electric fields of the individual slits and squaring, we find the intensity of the resulting interference pattern to be given by

$$I(\theta) = A^2 \frac{\sin^2[(\pi/\lambda)Nd \sin \theta]}{\sin^2[(\pi/\lambda)d \sin \theta]}. \tag{19.30}$$

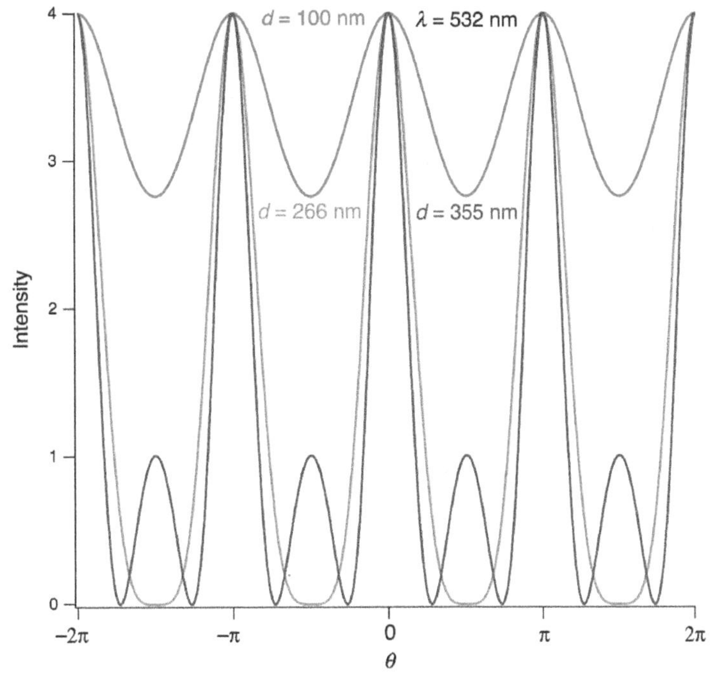

Figure 19.7 The modulation depth of $\cos^2$ interference fringes observed in two-slit diffraction increases as d increases when the spacing is small compared to the wavelength. It reaches a limiting value at $d = \lambda/2$ at which point subsidiary maximum begin to grow in as d increases further.

The large principal maximum occurs at

$$\frac{\pi d}{\lambda}\sin\theta = m\pi \tag{19.31}$$

where $m = 0, \pm 1, \pm 2, \ldots$. This is more commonly presented as the *grating equation*,

$$m\lambda = d\sin\theta \tag{19.32}$$

where m is known as the *order*. The grating equation predicts the angle at which different wavelengths will leave a diffraction grating. As is obvious from this derivation, a diffraction grating does not actually diffract the light. The light is subject to multiple slit interference in reflection. Interference is the result of several beams of light interacting with one another. Diffraction is more accurately defined as the interaction of light with a single aperture.

19.1.4.1 Example

Determine the angular location of the first three orders for a grating with 6000 lines cm^{-1} and incident yellow light with $\lambda = 589$ nm.

The *first-order* reflection is found at an angle given by the solution of the grating equation for $m = 1$, $\sin\theta = \lambda/d$. The distance between slits is the inverse of the ruling density, $d = 1/6000$ cm $= 1.667 \times 10^{-6}$ m. Solving for the angle, we find, $\sin\theta = 589 \times 10^{-9}$ m$/1.667 \times 10^{-6}$ m $= 0.3534$, which corresponds to $\theta = 20.7°$.

Second order, $m = 2$, $\sin\theta = 0.3534 \times 2 = 0.7068 \Rightarrow \theta = 45.0°$

Third order, $m = 3$, $\sin\theta = 0.3534 \times 3 = 1.0602$, an unphysical result for θ – that is, it cannot be observed with this grating.

19.1.5 'Diffraction' in crystals

The lattice structure of a crystal resembles a diffraction grating where the 'slits' are the spaces between the ordered atoms of the crystal, as shown in Fig. 19.8. If $d = 2$ Å (a distance on the order of a bond length or the spacing of planes in a crystal) and $\sin\theta = 0.3534$ (which from the above example seems like a reasonable sort of angle), then the wavelength of light that would diffract into this angle is

$$\lambda = 2 \times 10^{-10}\ \text{m} \times 0.3534 = 7.068 \times 10^{-11}\ \text{m} = 0.7068\ \text{Å}.$$

This is in the X-ray region.

Because of this, X-ray diffraction is used to determine the structures of crystals. Diffraction patterns are collected from crystals irradiated with X-rays of a known wavelength and spacings are found by solving the *Bragg equation*,

$$n\lambda = 2d_{hkl}\sin\theta \tag{19.33}$$

We recognize this as the grating equation, with the order now represented by n and d_{hkl} is the spacing between planes designated by the Miller indices *hkl*. In Fig. 19.8, the sum of the red line segments PB and BQ represents the pathlength difference between the ray that reflects from the lattice plane at point A and the ray that reflects from the lattice plane below it at point B (similarly and successively for the planes below it). When the pathlength difference corresponds to integer values, constructive interference is observed. When it corresponds to half-integer values, the waves associated with the two rays are out-of-phase and destructively interfere with one another.

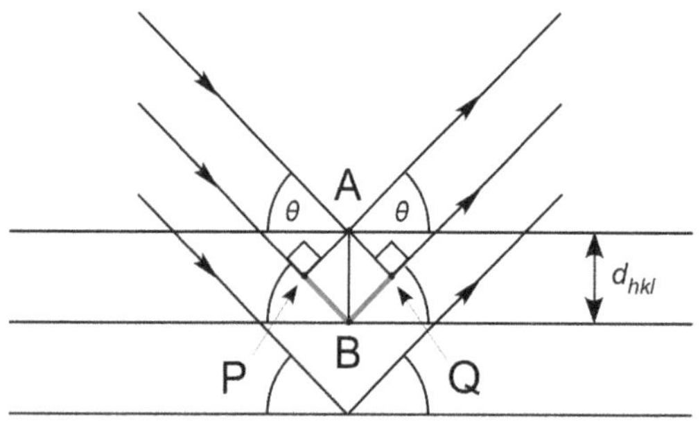

Figure 19.8 Diffraction (actually multiple-source interference) from a series of parallel planes as in a crystal.

19.2 Classical mechanics to quantum mechanics

The last decade of the 19th century and first decade of the 20th century were an incredible period in science that could have benefited from a greater sense of humility. Albert Michelson, quoting Lord Kelvin, stated in 1894 that science had "... exhausted new discoveries and that advances were only to be made in more precision of measurement." At about the same time, Ernst Mach and Wilhelm Ostwald were fighting vociferously against the idea that molecules existed. Against this backdrop one has to interpret how revolutionary Ludwig Boltzmann's ideas of statistical mechanics and the necessity of atoms and molecules really were at the time. By 1900, Max Planck would apply statistical mechanics to demonstrate the need for the quantization of energy, specifically the quantization of oscillators to explain black-body radiation. By 1905, Albert Einstein would use Planck's hypothesis to demonstrate that light behaves not only like a wave, but also like a particle with quantized energy. In 1907, Einstein combined statistical mechanics with quantum theory to explain the low-temperature behavior of solids. And by 1913 Niels Bohr declared his theory of quantized orbits for electrons in atoms. Our perception of the world would never be the same.

Now of course, we know that the materials around us are composed of atoms and molecules. Chemically, the world is quantized: there is a smallest unit (the atom/molecule) that is the smallest stable clump of matter with a given set of chemical properties. From Einstein, we know that energy and mass are equivalent

$$E = mc^2.$$

Therefore, since matter is 'quantized' in molecules why should we be surprised that energy is quantized as photons or that molecules can have quantized energy states? This is easy to say in hindsight, but it is not our experience from the classical world and macroscopic objects. We can kick a football at any speed (up to some limit fortunately for goalkeepers). We can spin in circles at any rotation rate. It is not our experience that we can only spin at angular frequencies with specified values or with our rotational axis pointed only in certain directions. Again, some perspective is in order. That is how *macroscopic* objects behave, and while many precepts of classical physics are still valid, subatomic and atomic particles simply behave in a manner that is different from what our senses experience.

Because we live in a world of classical physics, none of us comes to the study of quantum mechanics with quantum intuition. We possess classical intuition. In order to develop quantum intuition, we need to be ready to jettison treasured ideas from classical physics and everyday intuition. To develop quantum intuition we need to use something that looks like what we will later call the 'variational principle.' We need to introduce a set of terms (call them a basis set) that approximates the answers to our questions. If the approximation works well and describes our solutions to the level of accuracy that is available, then our basis set was appropriate. If the answer is not quite accurate enough, we can add a few more terms that modify our description. If the answer is way off we might try to add a few more terms. However, eventually we may need to throw out the original basis set and start with new terms and a better approximation to begin with. In other words, we will use classical analogies and push them as far as we can. But when they fail, we must be ready to accept new quantum mechanical ideas – new terms – that we add to our classical language.

With a perspective that was not available to the intrepid souls who pioneered quantum mechanics, we can aid the development of quantum intuition by stating a few principles that will remain inviolable. The laws of thermodynamics apply to quantum systems. Thus, conservation of energy still holds. Indeed, Noether's first theorem, proven by A. Emmy Noether in 1915 but published later,[3] demands that time invariance in physical laws (that is, when you perform your experiments neither changes the laws of physics nor the spectrum of hydrogen) requires the conservation of energy and linear momentum, just as rotationally symmetric physical laws require conservation of angular momentum. Causality still holds. Combined with relativity, this may lead to some very interesting paradoxes, but the application of unbalanced forces still accelerates particles. Quantum mechanics converges on classical mechanics for macroscopic objects. With this in mind, we can review a few concepts derived from classical mechanics.

Classical mechanics is based on an *assumption* of exact trajectories. We know where a particle is, its velocity, its mass, and therefore its momentum. The kinetic energy of a particle of mass m is

$$E_k = \tfrac{1}{2}mu^2 = p^2/2m \tag{19.34}$$

where the linear momentum p is given by

$$p = mu = (2mE_k)^{1/2}. \tag{19.35}$$

The velocity is denoted u rather than v so as to avoid confusion with the frequency ν. Newton's laws of motion describe the kinematic of particles, for example, relating force F to the acceleration a,

$$F = \frac{dp}{dt} = ma. \tag{19.36}$$

The potential energy of a system depends on the interactions in that system. The electrostatic potential between two particles of charge q separated by a distance r is

$$V = q^2/4\pi\varepsilon_0 r. \tag{19.37}$$

Opposite charges attract, like charges repel, interaction increases with decreasing distance.

The potential of a harmonic oscillator in terms of the displacement from equilibrium x bound by a spring with force constant k is

$$V = \tfrac{1}{2}kx^2. \tag{19.38}$$

All components of energy (translational, rotational, vibrational) are continuous in classical mechanics.

That classical mechanics is deterministic is a widely held assumption. Yet chaos (chaotic trajectories, nonlinear dynamics) is possible in a system that follows classical mechanics. The simplest example is three charged particles. Born objected to the assumption of determinism in his Nobel Lecture.[4] The determinism of classical mechanics lapses completely into indeterminism as soon as the slightest inaccuracy in the data on velocity is permitted. The question becomes, whence this uncertainty? Is it simply measurement error, as discussed by Born, which could presumably be removed by better instruments, or is it fundamental? As we shall see below, this lack of determinism is fundamental. The scattering of blackbody radiation photons provides a source of fundamental uncertainty and inherently stochastic long-term dynamics for microscopic systems. Quoting Born, "According to the heuristic principle used by Einstein in the theory of relativity, and by Heisenberg in the quantum theory, concepts which correspond to no conceivable observation should be eliminated from physics. … I should like only to say this: the determinism of classical physics turns out to be an illusion, created by overrating mathematico-logical concepts. It is an idol, not an ideal in scientific research and cannot, therefore, be used as an objection to the essentially indeterministic statistical interpretation of quantum mechanics."

19.3 Necessity for an understanding of quantum mechanics

To free yourself from the restrictions of purely classical ideas (including a renunciation of determinism), you need to develop a quantum vocabulary based on the properties and behavior of four model systems: the particle in a box; the harmonic oscillator; the rigid rotor; and the H atom (which is equivalent to a particle with an intrinsic magnetic moment of $\frac{1}{2}$ in a spherical 'box' with soft walls). Using these models as your basis set (rather than billiard balls and springs and a rigid, continuous interpretation of Newtonian laws of motion) you can construct a path to understanding once you learn their behavior and how to model deviations from their ideal behavior.

Whereas $F = ma$ is the guiding equation of classical mechanics, $\hat{H}\psi = E\psi$ is the guiding equation of quantum mechanics. It describes the motion of everything but converges on the results of classical mechanics for macroscopic objects. This is an expression of the *correspondence principle*:

> *In the limit of high quantum numbers, systems behave classically. Equivalently, in the limit of large energies, large masses and large orbits, quantum calculations deliver the same results as classical calculations.*

The correspondence principle was first used by Bohr in 1913 and formally proposed in 1920.[5]

Paul Ehrenfest proved in 1927 that quantum theory converges on classical mechanics if the limit of $h \to 0$ is applied correctly. Thus, if we are only concerned with macroscopic objects, we can use classical mechanics to treat their kinematics. However, microscopic objects must be described by quantum mechanics, and classical mechanics is misleading and inaccurate for microscopic systems.

Even though the motions of macroscopic objects are described by classical mechanics, we observe the consequences of quantum mechanical interactions every day. This is because quantum mechanics governs the interactions of atoms and molecules and the electrons in chemical bonds. In additions, quantum mechanics governs the interactions of electromagnetic radiation (light) with atoms and molecules. Thus, all the colors you see and even your ability to see at all are completely inadequately described unless we use quantum mechanics.

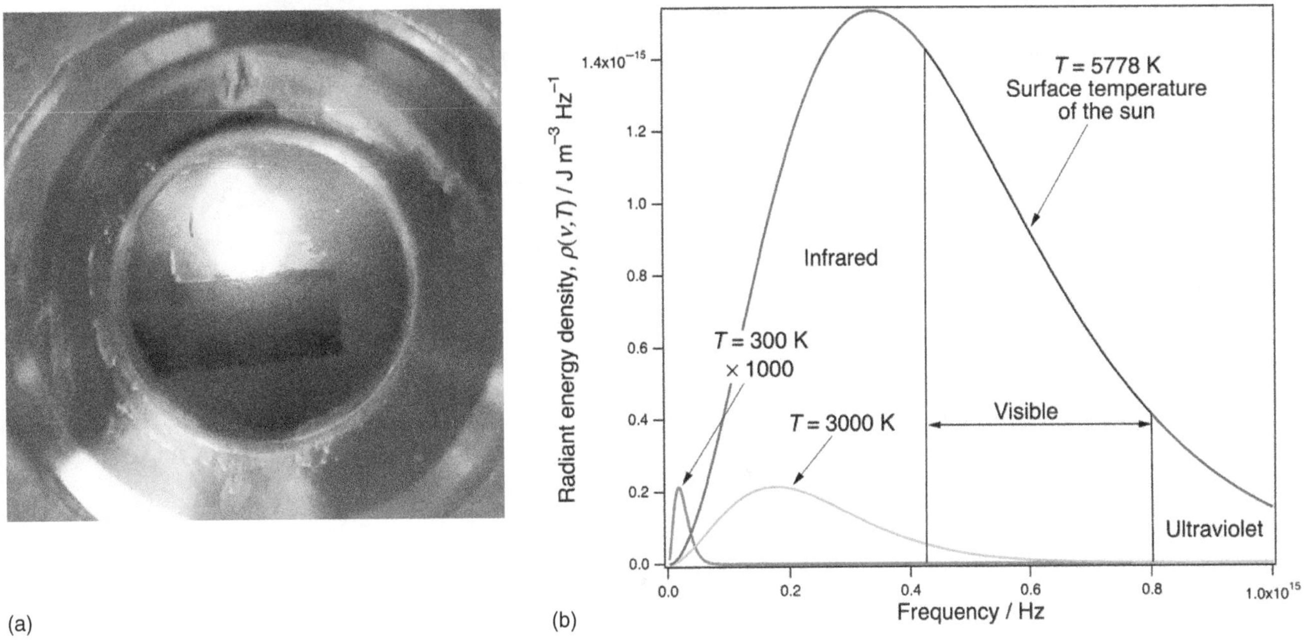

Figure 19.9 (a) Black-body radiation emitted by plasma created with laser ablation of Ti with a Nd:YAG laser. The photograph is taken through an orange filter that removes the scattered laser light at 532 nm. (b) The spectrum of black-body radiation and it dependence on temperature.

Until about 1900, as reflected in the words of Michelsen and Kelvin above, the ideas of Newton and Maxwell were considered to explain all the phenomena of science. However, these theories failed to explain several experimental results related to very small particles. We shall now review several of these phenomena to understand what quantum theory must describe, and what the nature of the failings of classical mechanics are when we observe microscopic systems. The specific examples we will investigate are black-body radiation; photoelectric effect; nature of light and wave-particle duality; and atomic energy level structure as revealed by spectroscopy. The variation of heat capacity with temperature is also incompatible with classical mechanics. However, we have dealt with this phenomenon in the context of thermodynamics.

19.3.1 Black-body radiation

A black body is a perfect emitter and absorber of light of all wavelengths. One way to create a black body is to take a hollow sphere of metal that has pinhole orifice from which light can be emitted. In order to study its radiation, the black body must also be held at a constant temperature. As shown in Fig. 19.9, the spectral density of light emitted from a black body has a characteristic shape that looks something like a Maxwell–Boltzmann distribution. The spectral density passes through a maximum then declines to zero at high frequency/short wavelength. The position of the maximum shifts to higher photon energy with increasing temperature.

The classical prediction of the spectrum of black-body radiation fits the distribution at low frequency but is completely wrong at high frequency. This result, known as the *ultraviolet catastrophe*, is a true failure of classical physics. In fact, a direct prediction of the classical model for black-body radiation is that there should be no darkness: high-frequency photons should saturate all space.

In 1900, Planck[6] succeeded in properly describing the behavior of a black-body radiator. A refined version of his derivation is presented in Appendix 5. In his explanation, Planck hypothesized that the matter oscillations in the walls of the black body are quantized. In other words, their energy comes in packets rather than being continuous. A necessary – though somewhat unbelievable – consequence was that the light energy E that they emit is also not continuous. Instead, it comes in packets

$$E = h\nu \tag{19.39}$$

where ν is the frequency. The constant h came to be known as the Planck constant. This is the *Planck quantization hypothesis*, and the equation is known as the *Planck equation*. Planck, somewhat disbelieving of his own invention, was not sure how literally this quantization condition had to be taken. It was, however, an unavoidable *Ansatz* that had to be made in order to get the correct distribution function.

Planck's reasoning was something along the following lines. The origin of black-body radiation is the oscillation of electric dipoles in the material that makes up the walls of the black-body emitter. The frequency of light emitted is equal to the frequency of the vibration. Planck used statistical mechanics to describe the energy distribution in a collection of oscillators that gives rise to the radiation. The energy in the oscillators is assumed to follow a Boltzmann distribution. For a simple two-level system, the population in the first excited state N_1 is related to the population in the ground state N_0 by

$$N_1 = N_0 \exp(-E_1/k_B T) = N_0 \exp(-h\nu/k_B T). \tag{19.40}$$

The oscillators are assumed to be harmonic and have a fundamental frequency ν. N_1 is the number of oscillators in the first excited state, N_0 the number in the ground (unexcited) state, E_1 is the energy of first excited state (the ground state energy is set to zero), k_B is the Boltzmann constant, and T is the absolute temperature.

Using the results derived in Appendix 5, the mean energy of an oscillator is given by the *Planck distribution*

$$\bar{E}_{\text{osc}} = \frac{h\nu}{\exp(h\nu/k_B T) - 1}. \tag{19.41}$$

In the limit of high temperature when $h\nu \ll k_B T$, this reduces to the classical limit of $k_B T$. In the limit of low temperature, the average energy goes to zero. The resulting radiant energy density (plotted in Fig. 19.9) is

$$\rho(\nu, T)\mathrm{d}\nu = \frac{8\pi h\nu^3}{c^3} \frac{1}{\exp(h\nu/k_B T) - 1} \mathrm{d}\nu. \tag{19.42}$$

The success of Planck's distribution law was the birth of the quantum revolution in science. It instigated serious thought about whether noncontinuous distributions should be taken literally. It also made scientists more confident in using argumentation based on statistical mechanics, even if they did not appear to be directly related to thermodynamics.

19.3.1.1 Directed practice

The black-body radiant energy distributions comparable to solar irradiation are given in Appendix 5. Calculate the photon energy at the maximum of the radiant energy density per unit wavelength bandwidth distribution, and the number density of photons per unit frequency bandwidth in both J and eV.

[Answers: 3.80×10^{-19} J, 2.37 eV; 1.22×10^{-19} J, 0.760 eV]

19.3.2 Photoelectric effect

In the 1880s, Heinrich Hertz and Wilhelm Hallwachs[7] discovered that irradiating a metal plate with ultraviolet (UV) light caused the plates to lose negative charge. This was a surprising and, at the time, inexplicable phenomenon. The electron was still waiting to be discovered by Sir Joseph John 'J.J.' Thomson in 1897. Even with the discovery of the electron, the photoelectric effect eluded explanation because the number of electrons released correlated with the intensity of the light, while whether or not electrons were emitted depended on the frequency of the light. The threshold of electron emission also appeared to be determined by which metal was being irradiated.

Einstein[8] applied Planck's quantum hypothesis to explain the *photoelectric effect* in 1905. This was the work that actually garnered the Nobel Prize in Physics for Einstein. He proposed that all photons, not just those emitted from quantized oscillators, are quantized with an energy given by the *Planck equation*

$$E = h\nu. \tag{19.43}$$

Furthermore, when a photon is absorbed, it delivers this full energy to the body that absorbs it.

As shown in Fig. 19.10(a), electrons are bound in a metal by a minimum amount known as the *work function*. The work function is the difference between the highest energy that an electron can have in the metal (an energy known as the *Fermi energy*) and the vacuum energy. The vacuum energy is where we set our origin. An electron at rest far removed from the metal so that it does not interact with the metal is at the vacuum energy, a value that we take as the zero of our energy scale.

When photons from a low-intensity source (a lamp, sychrotron or a low-intensity laser) are absorbed by a metal, no electrons are ejected, regardless of the intensity of the light unless the frequency of the radiation exceeds a characteristic value.[9] This characteristic value is set by the identity of the metal (and the chemical state of its surface). When photons with an energy above the threshold energy are absorbed, photoelectrons are ejected from the metal. The process is called

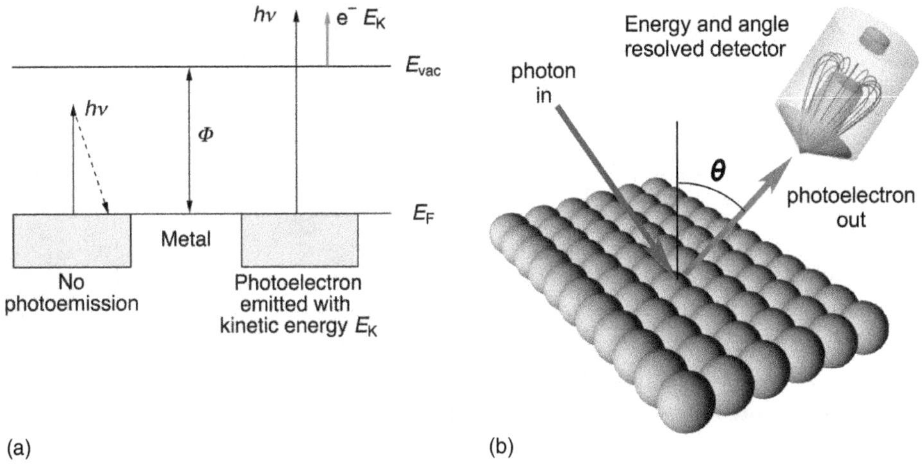

Figure 19.10 (a) The relationship of photon energy to work function and the kinetic energy of photoelectrons. Below threshold (left), $h\nu < \Phi$, no photoelectrons are created. The excited electron relaxes with the release of heat in the metal. When the photon energy exceeds the work function (right), photoelectrons escape the material and leave with sufficient kinetic energy to conserve energy. (b) The electron kinetic energy is measured in an electron spectrometer that can often resolve not only the energy but also the angular distribution of the photoelectrons. Implicit in this figure is that we are discussing the linear regime in which the absorption of one photon leads to the ejection of one electron. Such single-photon events are associated with conventional light sources (i.e., lamps). Lasers can attain much higher intensities that make nonlinear effects possible. Nonlinear (multiphoton) events are discussed in detail in Chapter 27.

photoemission or the *photoelectric effect*. For above threshold radiation, the kinetic energy of the photoelectrons depends on the frequency of the light. The number of photoelectrons depends on the intensity of light.

We introduce the minimum binding energy of electrons, Φ, which is called the *work function* defined above as the difference between the vacuum energy E_{vac} and the *Fermi energy* E_{F},

$$\Phi = E_{\text{vac}} - E_{\text{F}}. \tag{19.44}$$

Photons have an energy $h\nu$ and give up all of their energy upon absorption, never a fraction. If this energy is lower than the binding energy of the electron – that is, lower than the work function – no electrons are emitted. If the photon energy is larger than Φ, the electron is ejected with kinetic energy

$$E_{\text{k}} = \frac{1}{2}m_{\text{e}}u^2 = h\nu - \Phi. \tag{19.45}$$

This is the *Einstein equation*. For each absorbed photon with energy greater than the work function, one electron is emitted. No electrons are emitted by absorption of photons with energy less than the work function. In this case, the electrons are excited but simply relax back into the metal and the absorbed energy is registered as heat, as is shown on the left side of Fig. 19.10(a).

19.3.2.1 Example

The work function for metallic cesium is 2.14 eV. Calculate the kinetic energy and the speed of the electrons ejected by light of wavelength (a) 700 nm, (b) 300 nm.

a Eq. (19.45) can also be expressed as

$$E_{\text{k}} = E_{\text{photon}} - \Phi.$$

If $E_{\text{k}} < 0$, then no photoelectrons are created; in other words, the photon does not have enough energy to eject electrons from the metal. At 700 nm the photon energy is

$$E_{\text{photon}} = \frac{hc}{1.602 \times 10^{-19}\ \text{J eV}^{-1}\ \lambda} = \frac{1.986 \times 10^{-25}\ \text{J m}}{1.602 \times 10^{-19}\ \text{J eV}^{-1}\ (700 \times 10^{-9}\ \text{m})} = 1.77\ \text{eV}$$

$$E_{\text{k}} = E_{\text{photon}} - \Phi = 1.77 - 2.14 < 0.$$

Therefore, no electrons are created by 700-nm photons.

b

$$E_{\text{photon}}(\text{at } 300\text{ nm}) = 4.13\text{ eV}.$$

$$E_k = E_{\text{photon}} - \phi = 4.13 - 2.14 = 1.99\text{ eV}.$$

$$u = \left(\frac{2E_k}{m_e}\right)^{1/2} = \left(\frac{2(1.99\text{ eV})(1.602\times10^{-19}\text{ J eV}^{-1})}{9.11\times10^{-31}\text{ J}}\right)^{1/2} = 8.37\times10^5\text{ m s}^{-1}$$

Note the advantage of expressing the photon energy in eV when approaching problems in photoelectron spectroscopy.

19.3.3 Electron energy loss spectroscopy

In 1914, James Franck and Gustav Hertz[10] reported that when Hg atoms were irradiated with electrons with a controlled velocity, the emission of light occurred only after the electron kinetic energy passed a certain threshold value. This is the inverse of the photoelectric effect. In this case, irradiation with electrons leads to the emission of light. Franck and Hertz found that, coincident with the emission of light, the incident electrons lost kinetic energy and that the kinetic energy loss was equal to the energy of the photon that was emitted from the Hg atoms,

$$E_{\text{initial}} - E_{\text{final}} = E_{\text{loss}} = h\nu. \qquad (19.46)$$

The threshold for Hg was equal to an energy loss of 4.89 eV, which corresponds to a photon at 254 nm. At higher kinetic energy other thresholds were found and, correspondingly, the electrons were found to have an energy loss equal to the emitted photon energy. Thus, we observe that discrete energy transfer can occur in particle–particle collisions.

19.3.3.1 Directed practice

Electrons with an initial kinetic energy of 5 eV are incident on Hg atoms. Calculate the final energy and velocity of electrons that have excited Hg atoms that emit at 254 nm.

[Answers: 0.11 eV, 1.76×10^{-20} J, 1.97×10^5 m s^{-1}]

19.4 Quantum nature of light

Planck's quantum hypothesis – required to explain the behavior of black-body radiation – and Einstein's explanation of the photoelectric effect make it clear that the following three principles must be obeyed to understand light–matter interactions:

- *The quantum of the electromagnetic field is called the photon.*[11] *It is neither particle nor wave (get over it, it is quantum mechanical and does not correspond to either one of these classical absolutes) but is has properties that in certain limits behave like classical waves or classical particles.*
- *The energy of a photon is related to its frequency, angular frequency, wavenumber or wavelength, respectively by the Planck equation*[12]

$$E = h\nu = \hbar\omega = hc\tilde{\nu} = hc/\lambda \qquad (19.47)$$

- *The number of photons present determines the intensity of the light. The power P of a light source is then given by the rate at which photons emanate from a source (number of photons N per unit time t) multiplied by the photon energy,*

$$P = (N/t)E. \qquad (19.48)$$

19.4.1 Example

(a) Calculate the frequency in Hz, the energy per photon in eV, and the energy per mole of photons in kJ mol^{-1} for radiation of $\lambda = 550$ nm (yellow/green).(b) Calculate the energy and wavelength of an infrared photon of wavenumber 2400 cm^{-1}.(c) How many photons per second radiate from a 1 mW HeNe laser operating at 632.8 nm?

a In these calculations the product $hc = (6.62608\times10^{-34}\text{ J s})(2.997926\times10^8\text{ m s}^{-1}) = 1.986\times10^{-25}$ J m, will be of use. The frequency is

$$\nu = \frac{c}{\lambda} = \frac{3.00\times10^8\text{ m s}^{-1}}{550\times10^{-9}\text{ m}} = 5.46\times10^{14}\text{ s}^{-1}.$$

The photon energy is

$$E = \frac{hc}{\lambda} = 3.61\times10^{-19}\text{ J}(6.02\times10^{23}\text{ mol}^{-1}) = 217\text{ kJ mol}^{-1}.$$

$$E = \frac{3.61\times10^{-19}\text{ J}}{1.602\times10^{-19}\text{ J eV}^{-1}} = 2.26\text{ eV}.$$

b The energy and wavelength are

$$E = hc\tilde{\nu} = 1.986 \times 10^{-13}\ \text{J cm}(2400\ \text{cm}^{-1}) = 4.77 \times 10^{-20}\ \text{J} = 0.298\ \text{eV}$$

and

$$\lambda = \frac{1}{\tilde{\nu}} = \frac{1}{2400} = 4.17 \times 10^{-4}\ \text{cm} = 4.17\ \mu\text{m}$$

c The rate of photon emission from a HeNe laser with constant power of 1 mW is

$$\lambda = 632.8\ \text{nm},\ P = 1\ \text{mW} = 0.001\ \text{J s}^{-1}$$

$$E = h\nu = \frac{hc}{\lambda} = \frac{(6.626 \times 10^{-34}\ \text{J s})(2.998 \times 10^{8}\ \text{m s}^{-1})}{632.8 \times 10^{-9}\ \text{m}} = 3.14 \times 10^{-19}\ \text{J}$$

$$P = \frac{NE}{t}$$

$$\frac{N}{t} = \frac{P}{E} = \frac{P}{hc/\lambda} = \frac{P\lambda}{hc} = \frac{(1 \times 10^{-3}\ \text{J s}^{-1})(632.8 \times 10^{-9}\ \text{m})}{(6.626 \times 10^{-34}\ \text{J s})(3 \times 10^{8}\ \text{m s}^{-1})} = 3.19 \times 10^{15}\ \text{photons s}^{-1}$$

19.5 Wave–particle duality

The exhibition of both particle-like and wave-like behavior is not a unique property of photons. Diffraction is a characteristic property of waves. It is commonly observed for light when it strikes a slit with a size comparable to the wavelength of light. Any interference phenomenon requires the interaction of waves with shifted phase relations to occur. Thus, the observation of diffraction is taken as proof of the wave-like behavior of an object.

In 1927, Davisson and Germer observed the angular distributions of electrons scattered from Ni and explained their data in terms of the diffraction of the electrons from crystallites in their sample.[13] Davisson, who had begun his work on low-energy electron diffraction in 1922, and G.P. Thomson, who studied high-energy electron diffraction from thin films concurrently,[14] were awarded the Nobel Prize in Physics 1937 "...for their discovery of the interference phenomena arising when crystals are exposed to electronic beams." Thus, G.P. Thomson, son of J.J. Thomson who was awarded the Nobel Prize in Physics 1906 for discovering that *the electron was a particle*, was awarded the Nobel Prize in Physics for discovering that *the electron was a wave*. In 1929, after the observation of electron diffraction, Stern observed[15] the diffraction of He atoms from single-crystal salt surfaces. This dispelled any lingering doubt that diffraction could only be observed for photons or subatomic particles. These results confirmed de Broglie's outlandish prediction of the wave-like character of matter, which we will derive below.

19.5.1 Photon linear momentum and the de broglie relation

To begin our understanding of the wave-like nature of matter, let us first look more deeply at the particle-like behavior of the photon, and its implications. By symmetry, we will see that the inverse of these ideas leads to the natural conclusion that matter must have wave-like character in the limit of small particles.

Einstein's theory of relativity states that there is an energy equivalent to matter

$$E = mc^2, \tag{19.49}$$

where m is the rest mass and c is the speed of light. It implies that light and mass can be interconverted, as indeed is observed when matter and antimatter annihilate with the formation of a photon. But the consequences of this equation do not end there. The linear momentum p of a particle is given by

$$p = mu. \tag{19.50}$$

Here, quite intentionally I have used u for the velocity so that we do not confuse it later with the frequency ν. Perhaps then a photon, which seems to have particle-like characteristics, has a linear momentum given by mc. This cannot be strictly correct. The mass m is the rest mass and there is no frame of reference in which the photon is at rest. However, we can circumvent this by invoking the equivalence of energy and mass. A molecule that absorbs a photon has its mass increased not because it has swallowed a massive particle but because its energy has increased by exactly the photon

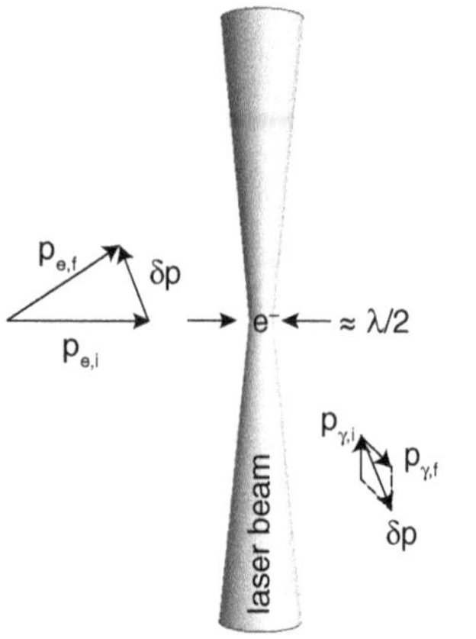

Figure 19.11 A laser beam is focused optimally to a beam waist with a diameter no less than roughly $\lambda/2$, where λ is the wavelength of the light. The photon with initial linear momentum $\mathbf{p}_{\gamma,\mathrm{i}}$ scatters off an electron with initial momentum $\mathbf{p}_{\mathrm{e,i}}$. The momentum transfer δp must be equal and opposite for the two particles, resulting in final momenta $\mathbf{p}_{\gamma,\mathrm{f}}$ and $\mathbf{p}_{\mathrm{e,f}}$ for the photon and electron, respectively.

energy. This change in mass has the net effect of giving the molecule a kick, as if linear momentum had been transferred to it by a collision with a massive particle.

This proposition is proven by the Compton effect in which high-energy photons (X-rays and gamma rays) participate in linear momentum transfer with a massive particle (i.e., an electron) exactly in the proportion demanded by relativity. Therefore, relativity gives the energy of a photon in terms of the magnitude of its linear momentum,

$$E = pc. \tag{19.51}$$

The ramifications of this are quite profound. Any time a photon is absorbed or scattered from a massive object – that is, an electron, an atom, or a molecule – the conservation of linear momentum must be observed. *Photons are not only deflected by their interactions with matter, photons deflect matter when they scatter from it.*

Suppose we want to detect where an object is by using light scattering, as we would in a microscope. To detect the position of the object, we use a laser in the geometry drawn in Fig. 19.11. We focus the laser down to its minimum spot size, which is roughly $\lambda/2$, where λ is the wavelength of the laser light. Light scatters off the particle when it traverses the focal spot and scatters light back into our detector. In order to collect all of the backscattered light, our detector has to be close to hemispherical. When scattered light is detected, we know that the particle was in the focal spot – that is, we know the position of the particle to within the width of the focal spot. Thus, the precision of our measurement is to a good approximation given by $\delta x = \lambda/2$. In principle, we can make our precision arbitrarily high by choosing the wavelength of our light arbitrarily small. We shall see later that it is exceedingly difficult to make an X-ray laser, and nigh-on impossible to make a gamma ray laser, but let that not trouble us for the moment.

We assume that our photons are only elastically scattered and not absorbed by the object. Their wavelength is the same before and after the collision. Nonetheless, momentum is transferred to the object because of the change in propagation direction of the photon. The vector diagram shows that the momentum change δp in the backscattered photon must lie somewhere between $\sqrt{2}\, p_\gamma \le \delta p \le 2p_\gamma$, where p_γ is the linear momentum of the photon.

Just how large is the linear momentum of the photon? Equating the energy of a photon in terms of its wavelength from Eq. (19.47) with its energy in terms of its momentum in Eq. (19.51), we can derive an equation for the linear momentum of a photon in terms of its wavelength,

$$h/\lambda = p. \tag{19.52}$$

This leads to a bit of a conundrum. In order to accurately determine the position of the particle, we need to use a wavelength that is significantly smaller than the radius of the particle. However, the shorter the wavelength of light, the greater the momentum transfer. In other words, the more accurately we want to know the position of the object, the greater the disturbance from making the measurement. Note that the kinematics of this argument is completely classical. The one and only quantum mechanical idea that we have used is that the photon has an energy given by hc/λ. We are using Newtonian formulations of conservation of linear momentum and particle trajectories. We also use a relativistic formula to derive the linear momentum of a photon, but relativity is a classical not a quantum mechanical theory (formulation of relativistic quantum kinematics brought Dirac the Noble prize as we shall see later; a quantum mechanical description of gravity remains at the frontier of theoretical physics).

But the Planck constant is small; hence, the momentum of a photon is small. Is this of any consequence, you might ask. The results in Table 19.2 allow us to answer this question. Let us consider three objects: a snooker ball; a C_{60} molecule (Buckminsterfullerene or buckyball); and an electron. A snooker ball is a decidedly classical object. Its trajectory can be controlled with seemingly infinite precision by the likes of Stephen Hendry (though the weekend billiards player may disagree). The snooker ball has a mass of about 100 g and a radius of just over 26 mm. Taking the velocity to be $u = 5\ \mathrm{m\ s^{-1}}$, the linear momentum of the ball is $p = mu = 0.5\ \mathrm{kg\ m\ s^{-1}}$. To detect the ball, we use green light from a Nd:YAG laser at 532 nm. This wavelength is far below the radius of the ball, and it also imparts precious little momentum to the ball. The relative momentum change $\delta p/p$ is so small as to be completely negligible for this classical particle. The accuracy of the position determination relative to the radius of the ball is quite good $\delta x/r = 1 \times 10^{-5}$, and could easily be improved by decreasing the wavelength into the UV region with no significant change in the momentum transfer. It would *seem* that for classical objects, we can make measurements to an arbitrary degree of accuracy for any property we choose to measure. This, of course, is our intuition from the classical world. To make a better measurement of length, we just need a finer ruler. But this is not the world of the atom.

Now consider the buckyball, C_{60} again moving at $5\ \mathrm{m\ s^{-1}}$. This object is considerably smaller, though big compared to an electron. The buckyball has a radius of 0.35 nm and a mass of 1.9×10^{-24} kg. Because of its small size we need to choose a photon in the X-ray region. For point of argument we take a photon with a wavelength $\lambda = r/2 = 0.17$ nm to detect the buckyball. This limits the precision of our position determination to be just 25% of the size of the buckyball. Just as importantly, the relative momentum transfer is now 19% of the original value. Perhaps we could construct our apparatus to improve these numbers. Nonetheless, we see that the observation now represents a significant perturbation to our system.

Does it get worse than this if we look at a really small particle? Consider an electron with a classical radius of 93 fm (quantum mechanically it is much smaller) and mass of just 9.1×10^{-31} kg. Now, our gamma-ray photon has a momentum that is a billion times greater than the momentum of the electron moving at $5\ \mathrm{m\ s^{-1}}$. In order to detect its position to a precision of only 25% compared to the size of the electron, we must completely obliterate any knowledge of where the electron might be after the detection event.

Heisenberg had been grilled during his doctoral exam by Wien, who presented him with a trick question regarding the resolving power of a microscope that he was unable to answer. Wien did not believe that a physicist could be 'just' a theoretician without practical aptitude, and Heisenberg had revealed himself to be less than accomplished in the laboratory. To prove a point, Wien made sure that Heisenberg was awarded his doctorate with the lowest possible passing mark. A lesser character would have been broken; however, Heisenberg would later use an argument based on the resolving power of a microscope (similar to the one presented above) to begin the derivation of the principle that bears his name. We will return to that derivation shortly.

We turn now to another troubled dissertation thesis, this one of de Broglie.[16] We have seen above that the marriage of the quantum hypothesis with relativity means that the massless photon must carry linear momentum. This proposition was confirmed by the observation of the Compton effect[17] – the recoil of electrons caused by the inelastic scattering of photons. It occurred to de Broglie in 1923 that the converse must be true. Mass is concentrated in bits we call *particles*. Electromagnetic energy is concentrated in bits we call *photons*. Just as photons are characterized by linear momentum as if they had mass, so too must particles be characterized by a wavelength as if they were waves. This wavelength is expressed in the *de Broglie relation*,

$$\lambda = \frac{h}{p}, \tag{19.53}$$

where $p = mu$, the linear momentum. Yes, this equation is so important that it has been repeated after rearranging Eq. (19.52).

What this wavelength represented, de Broglie had no idea. The particle must be delocalized in some sense over this length and the wave must possess a certain phase. Then, drawing on his knowledge of music he suggested that these phase waves would have to fit into the orbits that Bohr had proposed to describe the H atom. We will use this notion to make a compact derivation of the Bohr atom in the next section.

Looking at the final column in Table 19.2, the de Broglie wavelength of a snooker ball is imperceptibly small compared to the radius of the particle. A snooker ball acts classically for all intents and purposes. The same holds for a rocket. That is

Table 19.2 Comparison of light-scattering-induced momentum transfer and position uncertainty measurements for particles with a velocity of 5 m s^{-1}. Here, λ is the wavelength of the light used to detect the position of the particle of mass m and radius r. λ_{dB} is the de Broglie wavelength.

Particle	m/kg	r/m	λ/m	$\delta p/p$	$\delta x/r$	λ_{dB}/m
Snooker ball	1.0×10^{-1}	2.6×10^{-2}	5.3×10^{-7}	2.5×10^{-27}	1.0×10^{-5}	1.3×10^{-33}
C_{60}	1.9×10^{-24}	3.5×10^{-10}	1.7×10^{-10}	1.9×10^{-1}	2.5×10^{-1}	6.8×10^{-11}
e^-	9.1×10^{-31}	9.3×10^{-14}	4.6×10^{-14}	3.1×10^{9}	2.5×10^{-1}	1.5×10^{-4}

why classical mechanics is completely sufficient when modeling how to shoot a rocket to Mars or put a satellite in orbit. Classical mechanics is not wrong in this case. Classical mechanics correctly describes the kinematics of objects whose de Broglie wavelength is small compared to the size of the particle. Because classical mechanics is a correct limit of quantum mechanics, there must be correspondences between the two. There will be elements that are transferable between the two, such as the conservation laws of energy, linear momentum and angular momentum. We will need to determine what elements transfer and how to develop the principle of correspondence.

The de Broglie wavelength of a slow electron, as shown in Table 19.2, is large compared to the nominal size of the particle. Under these circumstances the behavior of the particle is inherently quantum mechanical. From these two limiting cases, we can develop a principle to determine when we expect to observe quantum mechanical effects:

> *We can expect to observe quantum mechanical effects and the insufficiency of a classical mechanical description when at least one dimension of the system under consideration approaches the magnitude of its de Broglie wavelength.*

As a coda, we return to de Broglie's dissertation. Upon receipt by his board of examiners, there was consternation. The idea of matter waves is so, well, nonclassical as to be absurd. Fortunately, the skeptical panel was chaired by the eminent physicist Paul Langevin. Not knowing what to make of this absurd idea and not wanting to deny outright a PhD to the son of the fifth duc de Broglie, he turned to Einstein, who responded, "...[de Broglie] has lifted a corner of the great veil."[18] The dissertation was promptly accepted and published in its entirety.

19.5.1.1 Directed practice

Calculate the de Broglie wavelength of the three particles listed in Table 19.2.

19.5.2 Elementary quantum particles and composite quantum particles

The photon and electron both exhibit wave-like and particle-like behaviors. They are neither waves nor particles in a classical sense. Thus, if we insist upon a name, they are quantum particles. They are also *elementary quantum particles*. They are devoid of internal structure and not composed of further particles. Nuclei, protons and neutrons are all composite quantum particles. A *composite quantum particle* is made up of even smaller elementary quantum particles. The nuclei are made up of protons and neutrons, which in turn are made up of three quarks each. Composite quantum particles act as individual quantum particles as long as they are not probed (or interacted with) on an energy scale that is comparable to the forces that hold them together. The strong interaction holds the nucleus together. It also binds together the quarks that make up the protons and neutrons. In chemistry we generally are not slamming into the nuclei and nucleons with energies approaching their binding energy; therefore, we can consider these composite quantum particles to behave as unitary quantum particles.

As we have seen in Chapter 6, and shall see again in Chapter 20, all quantum particles can be classified as either fermions or bosons. This relates to whether they have half-integer (fermions) or integer (bosons) spin. Spin, in a quantum mechanical context, is an intrinsic angular momentum associated with a magnetic moment that has nothing to do with a body turning about an axis. It has no classical analog. However, spin, as a name, is as close as we can get so we use the name as an abbreviation for the more accurate (but awkward) intrinsic magnetic moment.

Other types of composite quantum particles also exist and they are bound by different interaction mechanisms. One such composite particle is the Cooper pair. This is a composite particle made up of two electrons interacting with a phonon (a lattice vibration). These Cooper pairs are responsible for conventional superconductivity observed, for instance, in Nb. However, the binding energy of Cooper pairs is quite weak. Nb goes through the superconducting transition at 9.2 K, at which point thermal energy is sufficient to break up the Cooper pairs. Cooper pairs and a number of other composite quantum particles (always ending in -on as in exciton, polaron, polariton, and plasmon) are examples of quasiparticles (if they are composed of fermions) or collective excitations (if they are composed of bosons). These composite quantum particles are encountered frequently in solid-state chemistry and condensed-matter physics. They are mathematical constructs, not real particles, that allow for simplifications in the mathematics of handling the problem of many mutually interacting particles (the many body problem). One other example of a composite quantum particle is the *hole* h^+ – the absence of an electron in the band structure of a solid or a vacancy in an atomic or molecular orbital. A hole acts in many ways like an inverse electron or an electron with a positive charge. For example, conduction of a negative electron from position 1 on the left to position 2 on the right can also be thought of as the motion of a positive hole from position 2 on the right to position 1 on the left.

19.6 The Bohr atom

The emission and absorption of light from atoms had been known since the work of Robert Bunsen and Gustav Kirchhoff to be composed in large part of very narrow lines, so-called because of the impression they left on photographic plates that were used to record their spectra in early spectrometers. Similarly, these lines can be observed if a simple prism is used to disperse the light emitted from a discharge sparked in an atomic lamp. Lines are observed because the light is emitted at well-defined frequencies instead of continuously over a wide range of frequencies. Crucially, Bunsen and Kirchhoff demonstrated that emission and absorption occurred at the same wavelengths in characteristic patterns for different atoms. Classically, lines should never be observed. The simplest of these spectra is that of the H atom, though similarities are found in the spectra of alkali metals. The basis for these similarities will become obvious when we understand atomic structure more deeply. It was noted by several spectroscopists (Balmer, Rydberg, Paschen, Beck) that there was a certain regularity to these lines; namely, that the frequencies of the lines fitted a pattern involving the square of integers.

Ritz[19] collected all of the various formulae and proposed his *combination principle*: spectral lines arise from the difference of spectral terms that are associated with the atom. Thus, the spectrum of the H atom shown in Fig. 19.12 can be expressed in modern notation in terms of the *Balmer–Rydberg–Ritz formula*

$$\tilde{\nu} = \frac{1}{\lambda} = R_{\mathrm{H}}\left(\frac{1}{n_1^2} - \frac{1}{n_2^2}\right). \tag{19.54}$$

Here, the wavenumber of each transition is equal to the Rydberg constant for the H atom $R_{\mathrm{H}} = 109\ 677.581\ \mathrm{cm}^{-1}$ multiplied by the difference of the inverse of two integers n_1 and n_2. The wavenumber is always positive, so $n_1 < n_2$.

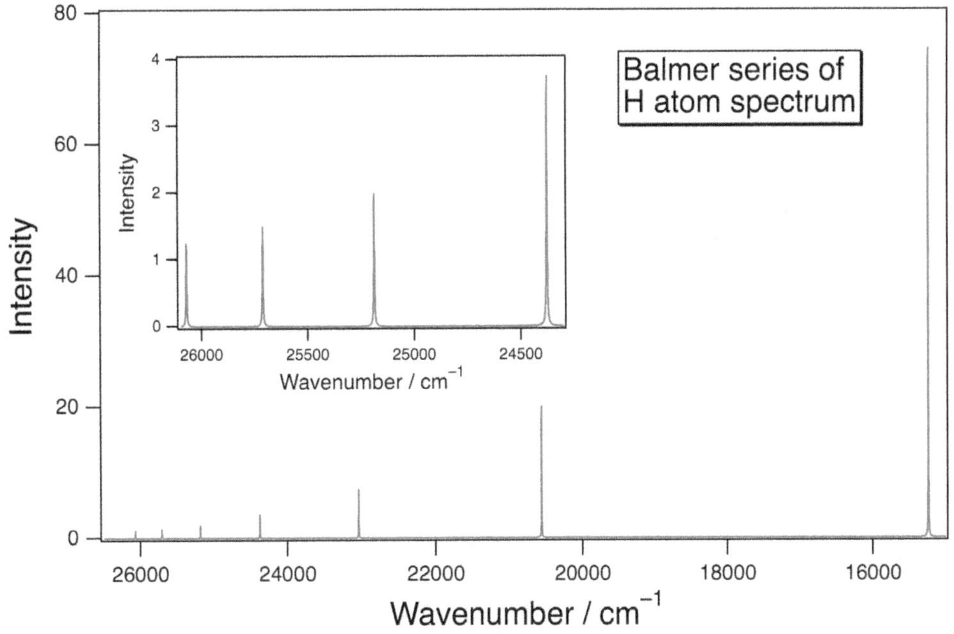

Figure 19.12 A portion of the H atom emission spectrum from the ultraviolet through the visible and into the infrared regions is reproduced. The spectral lines are not infinitely narrow but correspond to peaks with a 4 cm^{-1} linewidth determined in this case by instrumental resolution.

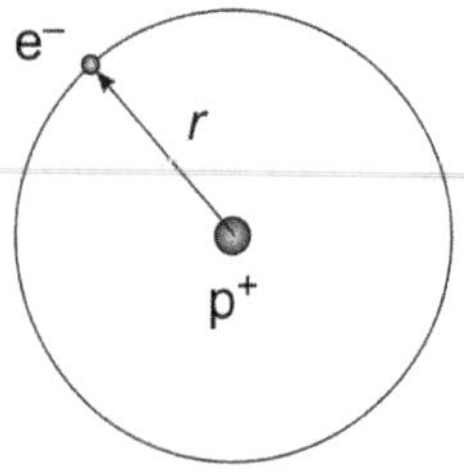

Ceci n'est pas un atome

Figure 19.13 The Bohr model of the hydrogen atom. An electron with a negative charge orbits the nucleus (a proton with a single positive charge) in circular orbits of radius r. This is not an atom, not even a correct model of an atom. Nonetheless, it is the first word in a new language to represent atoms.

Uniting the Ritz combination principle with the Planck relation for the quantized energy of the photon suggests strongly that the energy levels of atoms must be quantized, and that the energy of the photon that is emitted when atoms relax must correspond to the energy difference between these levels,

$$E_{\text{photon}} = \Delta E = E_2 - E_1 = hc\tilde{\nu}. \tag{19.55}$$

The problem is, why are the energy levels quantized into terms and why do atoms make jumps between levels rather than continuous transitions?

In 1913, Bohr[20] produced a refined model of the hydrogen atom in order to explain its emission spectrum. This is an *ad hoc* empirical model, as depicted in Fig. 19.13, rather than an a priori model. An *ad hoc* model is one that fits the experimental data by using a set of assumptions that are only justified because they produce a correct set of results. Bohr's model is not based on a theory that is justifiable on its own from which the results are derived. An explanation of why Bohr's model works so well will have to wait until we introduce the more fundamental theories of Dirac, Heisenberg, and Schrödinger.

We start with what is expected classically. Electrons carry a unit of negative charge ($-e$) and a mass (m_e). In the hydrogen atom they are bound by a much more massive nucleus that contains one proton of mass m_p and has a positive charge of $+e$. We label the nuclear charge Z. Since $m_p/m_e = 1836$ the electron travels at much higher velocity than the nucleus when both have the same kinetic energy. Thus, we can consider the nucleus to be stationary as a first approximation. In a stable atom the electrostatic attraction between the positive nucleus and negative electron must balance the centrifugal force experienced by the electron. If the electron orbits the atom at a distance r and with a velocity u, then balance of forces leads to the equation

$$\frac{e^2}{4\pi\varepsilon_0 r^2} = \frac{m_e u^2}{r}, \tag{19.56}$$

where ε_0 is the vacuum permittivity.

The total energy of the atom is the sum of the kinetic and potential energies. With a stationary nucleus, the kinetic energy is that of the electron and the potential energy is that of the electrostatic attraction,

$$E = T + V = \frac{m_e u^2}{2} - \frac{e^2}{4\pi\varepsilon_0 r} = -\frac{e^2}{8\pi\varepsilon_0 r}. \tag{19.57}$$

The energy of this system is depicted in Figure 19.14(a). Thermodynamics tells us that a system should evolve to the lowest energy state. A look at Eq. (19.57) tells us that the minimum energy is at $r = 0$. The electron should spiral into the nucleus and land on top of the proton. *Classical mechanics predicts that the atom is not stable.*

The H atom thus presents two fundamental challenges that classical physics is incapable of answering: line spectra rather than continuous spectra and stable atoms. Bohr proposed[21] that the angular momentum L of the electron about the proton is quantized, that is, that it can only take certain values given by

$$L = m_e u r = n\hbar \tag{19.58}$$

where $n = 1, 2, 3, \ldots$.

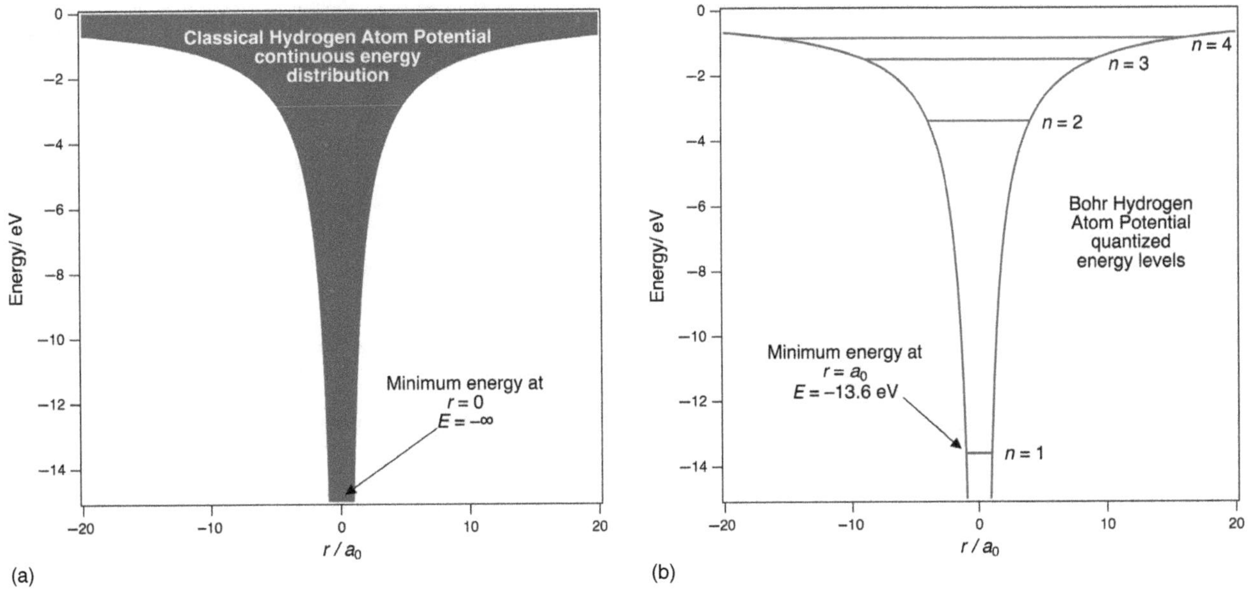

Figure 19.14 (a) The classical $1/r$ potential of the hydrogen atom is unstable because it predicts that the electron should fall into the nucleus. (b) The Bohr atom rectifies this instability with the introduction of quantized angular momentum, which leads to quantized levels with well-defined values of energy and radius.

Now, solve for u^2 in Eqs (19.56) and (19.58) and set these equal to one another

$$u^2 = \frac{n^2\hbar^2}{m_e^2 r^2} = \frac{e^2}{m_e(4\pi\varepsilon_0)r}. \tag{19.59}$$

Next, solve for r

$$r = \frac{4\pi\varepsilon_0\hbar^2 n^2}{m_e e^2} = n^2 a_0 \tag{19.60}$$

Here, we introduce the *Bohr radius* a_0, which is defined as

$$a_0 = \frac{4\pi\varepsilon_0\hbar^2}{m_e e^2} \tag{19.61}$$

and has a value of 52.92 pm. The Bohr radius is the mean radius of an orbit for the electron in the $n = 1$ state of the H atom. It represents a natural yardstick for the measurement of atomic scale distances. Since n is an integer, we see the first striking result of the Bohr model. The electron is only allowed to revolve about the nucleus at certain distances. In a more accurate model of the atom, it turns out that this is the mean distance from the nucleus rather than the radius of a circular orbit.

The second decidedly nonclassical result of the Bohr model is that the energies of the allowed electronic states are also only certain discrete values. These energies are found by substituting for r from Eq. (19.60) into Eq. (19.57) to obtain

$$E_n = -\frac{m_e e^4}{32(\pi\varepsilon_0\hbar)^2}\frac{1}{n^2} = -hcR_\infty\frac{1}{n^2}. \tag{19.62}$$

with $n = 1, 2, 3, \ldots \infty$. The energy of this system is depicted in Figure 19.14(b). At $n = \infty$ the energy goes to zero and the electron is no longer bound. Above this point the energy of the electron becomes positive and continuous. The positive energy is equal to the kinetic energy of the electron, which is continuous rather than quantized because there is no restriction on the kinetic energy of the electron once there is no longer a potential to confine the electron to a certain region of space.

This equation can be generalized for any nucleus with charge Z and just one electron surrounding it such that

$$E_n = -\frac{Z^2 m_e e^4}{32(\pi\varepsilon_0\hbar)^2}\frac{1}{n^2} = -hcR_\infty\frac{Z^2}{n^2}. \tag{19.63}$$

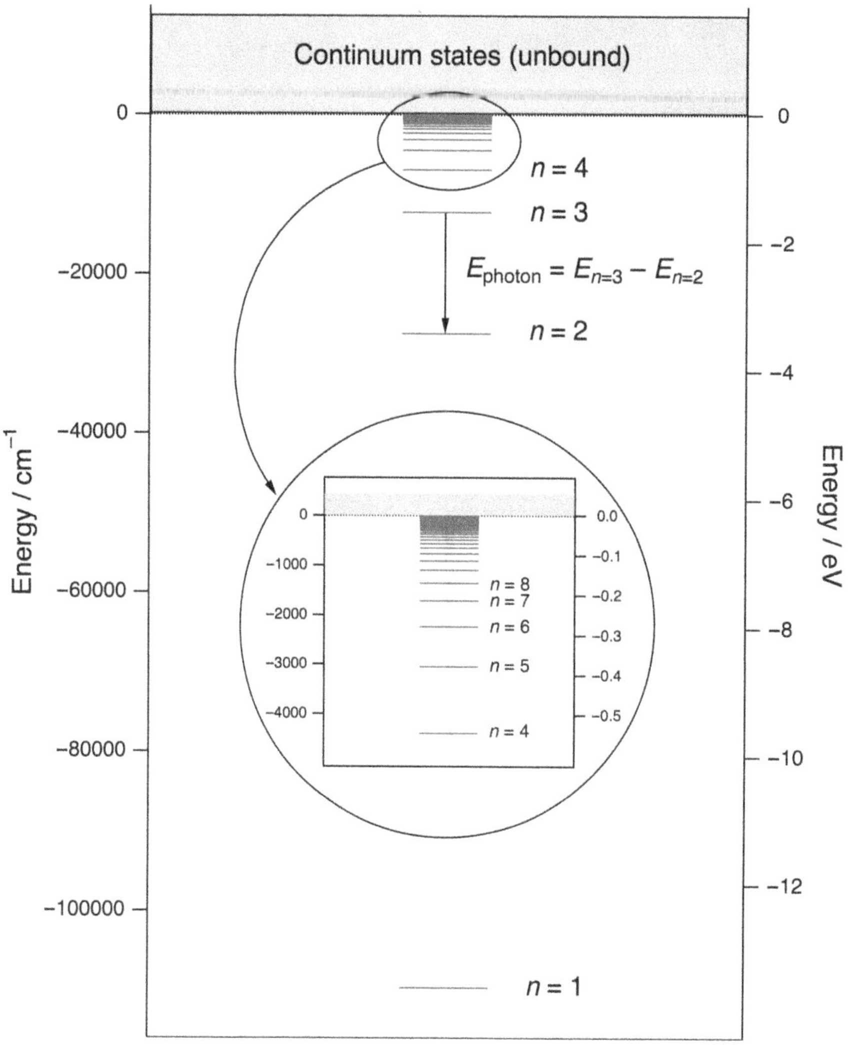

Figure 19.15 The energy levels of the hydrogen atom according to the Bohr model. Photon emission or absorption accompanies a transition between levels. The energy of the photon must equal the energy difference between the atomic levels.

R_∞ is the Rydberg constant for a nucleus of infinite mass

$$R_\infty = \frac{e^2}{8\pi\varepsilon_0 a_0 hc} = 109\,737\ \text{cm}^{-1} \tag{19.64}$$

This corresponds to a value of 2.180×10^{-18} J = 13.61 eV.

The energy level structure described by Eq. (19.62) now allows us to interpret why atomic spectra are made up of discrete lines. A line corresponds to the photon energy that connects (is resonant with) two states, as shown in Fig. 19.15. The states both have well-defined energies (their term values), and therefore the energy difference is well defined and equal to the photon energy. If the energies of the states are infinitely well-defined, the lines will be infinitely sharp, that is, their difference would be an exact number. The various mechanisms involved in determining the finite widths of these lines will be discussed in Chapter 23.

There was an aspect to the Bohr model that most physicists found somewhere between distasteful and disturbing. If there really are discrete energy levels and nothing in between, do atoms really make discontinuous jumps between these levels? It really was stretching the imagination of classically trained scientists to accept this. One might say, it would take a quantum leap of faith to accept the quantum jumps implied by Bohr's model. How does an electron get from one orbit to the next if there is an excluded region in between? Obviously, a much deeper and more profound theory is required to explain how these transitions are made and, indeed, to explain the changes that occur from the initial to final states.

It would take until 1986 before direct observation of quantum jumps, as displayed in Fig. 19.16, was possible.[22] This was performed by David J. Wineland and his group, who accomplished the feat while observing transitions of the Hg^+

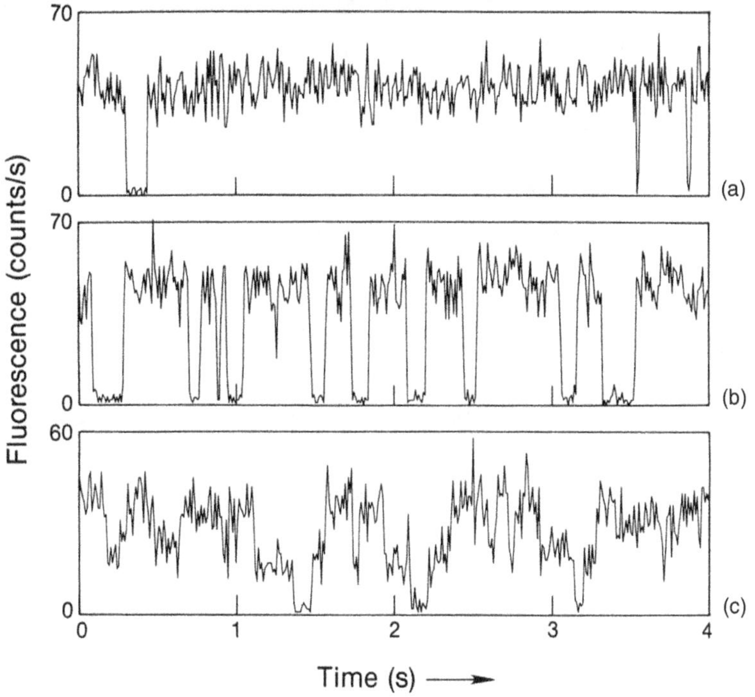

Figure 19.16 The observation of quantum jumps. Bergquist, J.C., Hulet, R.G., Itano, W.M., and Wineland, D.J. (1986) *Phys. Rev. Lett.*, **57**, 1699. Figure provided by Dave Wineland and Wayne Itano. Reproduced with permission from The American Physical Society.

ion. Wineland received the Nobel Prize in Physics 2012 in part for this work. Essentially simultaneously, the group of Dehmelt and that of Toschek, both working with Ba^+, and Finn and Greenlees, who worked with neutral Ba atoms, also observed quantum jumps.

19.6.1 Example

Calculate the ionization energy in eV of the hydrogen atom on the basis of the Bohr theory.

The *ionization energy* of the H atom in its ground state is equal to the minimum amount of energy required to remove an electron from the $n = 1$ state. This is the negative of the energy of the $n = 1$ level

$$I(\mathrm{H}, n = 1) = -E_{n=1} = -\left(-\frac{R}{n^2}\right) = \left(\frac{109678\ \mathrm{cm}^{-1}}{1^2}\right)\left(\frac{1\ \mathrm{eV}}{8065\ \mathrm{cm}^{-1}}\right) = 13.61\ \mathrm{eV}.$$

This corresponds to a transition from the $n = 1$ state of the Bohr atom to the $n = \infty$ state.

The Bohr model of the H atom puts the proton at the center and has an electron orbiting it in a circular orbit. Just as Planck's oscillators had quantized vibrations, Bohr proposed that each of these orbits should have quantized angular momentum. We clearly need a better model because it is ridiculous that electrons jump from here to there but are never in between. Furthermore, the circular orbits are supposed to represent stationary states of the system. So let's meld this *ad hoc* model (one that seems to have no basis in fact other than it gives some correct answers) with another preposterous idea – the de Broglie wavelength. Imagine a circular orbit of radius r and circumference $2\pi r$. If we try to fit a wave into a circular orbit, it will be a traveling wave if an integral number of wavelengths does not fit into the circumference. Such a travelling wave would be subject to destructive interference and could not possibly be a stable state of the system. However, if an integral number of waves fit exactly into the circumference, the wave will rejoin itself and form a standing wave, just like a guitar string does. The wavelength of the electron would then have to satisfy the condition

$$\lambda = \frac{h}{mu} = \frac{hr}{n\hbar} = \frac{2\pi r}{n}. \tag{19.65}$$

Could a standing wave correspond to the stationary state? And wouldn't this wave start to fill in the space between the orbits? Obviously, a new approach to the mechanics is required. A wave solution would be one – a matrix solution would be another. We shall soon see that they are, in fact, equivalent.

SUMMARY OF IMPORTANT EQUATIONS

$c = \lambda\nu$	Speed of light c, photon wavelength λ and frequency ν
$E = E_0\, e^{-i(kx-\omega t+\phi)}$	Electric field E, amplitude E_0, wavevector k, angular frequency ω, time t, relative phase ϕ
$I(x) \propto 4A^2 \cos^2(\pi dx/\lambda L)$	Intensity of light I in two-slit interference, A amplitude, d slit spacing, x distance on screen, L distance to screen
$m\lambda = d\sin\theta$	Grating equation: order m, spacing d, angle θ
$n\lambda = 2d_{hkl}\sin\theta$	Bragg equation: order n, plane spacing d_{hkl}
$E_k = \frac{1}{2}mu^2 = p^2/2m$	Kinetic energy E_K, mass m, velocity u, linear momentum p
$p = mu = (2mE_k)^{1/2}$	
$V = q^2/4\pi\varepsilon_0 r$	Electrostatic potential energy V, charge q, vacuum permittivity ε_0, distance r
$V = \frac{1}{2}kx^2$	Harmonic potential, spring constant k, displacement x
$E = h\nu = hc\tilde{\nu} = hc/\lambda = \hbar\omega$	Planck–Einstein relation, photon energy E, Planck constant h, wavenumber $\tilde{\nu}$
$\bar{E}_{osc} = \dfrac{h\nu}{e^{h\nu/k_BT}-1}$	Planck distribution: mean oscillator energy $\bar{E}_{osc}$, Boltzmann constant k_B, absolute temperature T
$\rho(\nu,T)d\nu = \dfrac{8\pi h\nu^3}{c^3}\dfrac{1}{\exp(h\nu/k_BT)-1}d\nu$	Black-body radiant energy distribution
$\Phi = E_{vac} - E_F$	Work function Φ, vacuum energy E_{vac}, Fermi energy E_F
$E_k = h\nu - \Phi$	Photoelectron kinetic energy E_K
$Pt = NE_{photon}$	Power of light source P, time t, number of photons N, photon energy E_{photon}
$\lambda = h/p$	de Broglie relation: wavelength of particle λ
$\tilde{\nu} = \dfrac{1}{\lambda} = R_H\left(\dfrac{1}{n_1^2} - \dfrac{1}{n_2^2}\right)$	Balmer–Rydberg–Ritz formula: Rydberg constant for the hydrogen atom R_H, principal quantum number of levels 1 and 2, n_1 and n_2
$E_n = -hcR_\infty Z(1/n^2)$	Energy of nth level in the Bohr model E_n, Rydberg constant for infinitely heavy nucleus R_∞, nuclear charge Z
$r = a_0 n^2$	Radius of nth Bohr orbit, Bohr radius $a_0 = 52.92$ pm

Exercises

19.1 (a) State one property of light that is wave-like and explain why the property is wave-like. (b) (b) State one property of light that is particle-like and explain why the property is particle-like.

19.2 A Nd:YAG laser operates at a wavelength of 532 nm with a pulse width of 6 ns and a pulse energy of 160 mJ. (a) Calculate the frequency in Hz as well as the energy per photon in eV and J and the energy of a mole of photons in kJ mol^{-1}. (b) What is the power associated with one pulse? (c) How many photons are in one pulse?

19.3 The work functions of Pt, Nb and U are 6.35, 4.3 and 3.6 eV, respectively. Describe an experiment using light from a fluorine laser ($\lambda = 157$ nm) that would identify the three metals.

19.4 The distribution in wavelength of the light emitted from a radiating black-body is a sensitive function of the temperature. This dependence is used to measure the temperature of hot objects, without making physical contact with those objects in a technique called *optical pyrometry*. In the limit $(hc/\lambda k_BT) \gg 1$, the maximum in a plot of $\rho(\lambda,T)$ versus λ is given by

$$\lambda_{max} = \frac{hc}{5k_BT}.$$

The irradiance (power per unit area) of a black body is given by

$I = \sigma T^4$ with $\sigma = 5.67 \times 10^{-8}$ Wm^{-2} K^{-4}.

a If the wavelength of light at the peak of the Sun's black-body emission is 500 nm and its radius is $r_S = 6.95 \times 10^5$ km, calculate the mean surface temperature of the Sun and the power radiating from the Sun, assuming it is a perfect black-body emitter.

b The Sun's irradiance at the Earth is given by the power emitted divided by the surface area over which that power is distributed. Effectively, only one-quarter of the Earth's surface is illuminated at any one time by the Sun, and the Earth's reflectivity (called the albedo) is $\alpha = 0.306$. Assuming that both the Earth and Sun are perfect black-body emitters, estimate the mean steady-state surface temperature of the Earth T_E by equating the power absorbed by the Earth to the power it emits.

19.5 Assume that water absorbs light of wavelength 4.00 μm with 100% efficiency. How many photons are required to heat 2.50 g of water by 1.00 K? The heat capacity of water is 75.3 J mol^{-1} K^{-1}.

19.6 If an electron passes through an electrical potential difference of 1 V, it has an energy of 1 electron-volt = 1 eV. What potential difference must it pass through in order to have a wavelength of 0.225 nm?

19.7 The spacing between atoms on the surface of a crystal is on the order of 1 Å. In order for diffraction to occur, the wavelength of the scatterer must roughly match the order of magnitude of the spacing of the diffracting elements. Calculate the velocity (in m s^{-1}) and kinetic energy (in eV) that an electron must have to exhibit diffraction from the surface of a crystal.

19.8 The Cr K_α X-ray emission line has a wavelength of 2.29 Å. When X-ray photons of this wavelength are diffracted from a β-brass sample, photons are detected in first-order diffraction peaks with 2θ values of 45.8°, 66.7°, and 84.7°. These three peaks correspond to diffraction from different planes in the crystal structure. These planes are labeled (100), (110) and (111), respectively. Calculate the interplane distance (the distance between atomic layers) for the (100), (110) and (111) planes.

19.9 (a) Neutron diffraction is one of the most powerful modern techniques for studying the structure of materials. This technique involves generating a collimated beam of neutrons at a particular temperature from a high-energy source (i.e., an accelerator). If the speed of a neutron is given as $u = (3k_BT/m)^{1/2}$, where m is the mass of the neutron and u is the root mean squared speed of a Maxwell distribution, what temperature must the neutron have in order for their wavelength to be 1 Å? (b) Calculate the value of the de Broglie wavelength associated with an electron with a speed of 6.0×10^7 m s^{-1} and an oxygen molecule with the root mean squared speed of a 300 K Boltzmann distribution.

19.10 In an X-ray photoelectron experiment, the absorption of an X-ray ejects an electron from the inner shell of an atom. An oxygen atom core electron in the K shell has a binding energy of 543.1 eV. Use conservation of energy to calculate the kinetic energy in eV and the speed in m s^{-1} of the electron that emerges from an O atom after the absorption of a photon of wavelength $\lambda = 1.50$ nm.

19.11 A He lamp operating with a wavelength of 58.43 nm was used to measure the photoelectron spectrum of a single-crystal sample of Cu with the (100) plane exposed to the vacuum. From a clean surface electrons were emitted with a kinetic energy of 16.1 eV. When a small amount of Na, which is an electropositive metal, is evaporated onto the Cu(100) surface, the emitted electrons display a higher kinetic energy. Calculate the work function of the Cu(100) surface and explain the effect of Na adsorption on it.

19.12 A Nd^{3+}:YAG laser when operating can be made to produce light at 532 nm and 266 nm. It produces pulses with a duration of 6 ns. The work function of clean cadmium is 4.08 eV. Calculate the number of photoelectrons produced when a 1 nJ pulse of either 532 nm or 266 nm light is incident upon a clean Nb surface. What is the kinetic energy in J of these electrons? Calculate the power in W of the laser pulse.

19.13 A He lamp mounted in a vacuum chamber emits photons with an energy of 40.82 eV. It delivers 10^{13} photons s^{-1} onto the sample. Calculate the photon energy in J and the radiant power on the sample in W. Calculate the wavelength and frequency of the photons. Nb(111) has a workfunction of 4.36 eV and that of Ta(111) is 4.00 eV. Describe quantitatively an experiment in which the He lamp could be used to identify these two metals.

19.14 Millikan in 1916 measured the threshold frequency for the emission of photoelectrons from sodium to be $\nu_0 = 43.9 \times 10^{13}$ s^{-1}. Calculate the work function of sodium in eV. Calculate the wavelength of this light and state what type of light this corresponds to. In a modern experiment in which the sodium is carefully outgassed in a vacuum chamber held at a better pressure than attainable by Millikan, the threshold value is found to be $\nu_0 = 5.5 \times 10^{13}$ s^{-1}. Explain the difference.

19.15 The visible emission spectrum of the H atom displays lines at 656.2852, 486.133, 434.047 and 410.174 nm. Use the Rydberg formula to identify the lines and determine the value of the Rydberg constant. Hint: the transitions all end in the same level n_1 and the series of excited states begins at $n_2 = n_1 + 1$. Start with $n_1 = 1$ and work out the possibilities.

19.16 The energy levels for hydrogenic ions (one-electron systems) are given by

$$E_n = -Z^2e^2/8\pi\varepsilon_0 a_0 n^2$$

with $n = 1, 2, 3, 4, \ldots$. Calculate the ionization energies of H^+, He^+, Li^{2+}, Ca^{19+} and Rn^{85+} in their ground states in units of electron-volts (eV).

19.17 What is the ionization energy of the $n = 7$ level of the H atom in eV? What is the wavelength in nm of the photon required to ionize this level? What region of the electromagnetic spectrum does this belong to?

19.18 Explain the difference between a bound-bound and a bound-free transition. How can you tell the difference by looking at a spectrum?

19.19 The wavenumber of the photon emitted or absorbed when an H atom makes a transition between electronic states with principal quantum numbers n_1 and n_2 is given by $\tilde{\nu} = 109865\ \text{cm}^{-1}\left(\frac{1}{n_1^2} - \frac{1}{n_2^2}\right)$. Calculate the wavelength in m and energy in J corresponding to the $n = 3$ to $n = 6$ transition in the hydrogen atom.

19.20 Calculate the time it would take for an electron to orbit the $n = 1$ Bohr orbit, assuming a circular orbit.

19.21 Calculate the Rydberg constant for the H atom and the D atom in which the mass of the electron in R_∞ is replaced by the reduced mass of the electron–nucleus system. Calculate the wavenumber of the $n = 2$ to $n = 3$ transition frequencies of the H atom, D atom and a hydrogenic atom with an infinitely massive nucleus.

Endnotes

1. Recall that we are assuming $n = 1$, whereas more generally $I(x, t) = nE^*(x, t)\ E(x, t)$. We are also glossing over issue of units and associated constants that relate the irradiance I to field strength and amplitude.
2. Never try this by sending a laser beam back into the laser cavity that generated it. You will probably blow up the cavity. However, it has been shown that spontaneous emission from a single atom [Kleppner, D., *Phys. Rev. Lett.*, **47**, 233 (1981)] or electron [Gabrielse, G. and Dehmelt, H., *Phys. Rev. Lett.*, **55**, 67 (1985)] trapped in a cavity can be suppressed. Such results, as well as enhancement of spontaneous emission [Haroche, S., *Rev. Mod. Phys.*, **85**, 1083 (2013)], are in the realm of cavity quantum electrodynamics (cavity QED) and relate to how the electromagnetic waves of the emitter interact constructively or destructively with the modes of the vacuum field contained in the cavity.
3. Noether, A.E. (1918) Nachr. König. Gesellsch. Wiss. Göttingen: Math.-phys. Klasse, pp. 235–257.
4. Born, M. (1955) *Science*, **122**, 675.
5. Bohr, N. (1913) *Philos. Mag.*, **26**, 1; Bohr, N. (1920) *Z. Phys.*, **2**, 423.
6. Planck, M. (1900) *Verhand. Deutschen Phys. Gesellschaft*, **2**, 202, 237.
7. Hertz, H.R. (1887) *Ann. Phys.*, **31**, 983.
8. Einstein, A. (1905) *Ann. Phys.*, **17**, 132.
9. With lasers of sufficient intensity, multiphoton events in which more than one photon is absorbed either coherently or sequentially are possible. Such events are discussed in detail in Chapter 27 and in Kolasinski, K.W. (2012) *Surface Science: Foundations of Catalysis and Nanoscience*, 3rd edition. John Wiley & Sons, Chichester.
10. Franck, J. and Hertz, G. (1914) *Verhand. Deutschen Phys. Gesellschaft*, **16**, 457.
11. Curiously, the photon was named neither by Planck nor Einstein but by Gilbert N. Lewis.
12. The last two equalities are strictly true only *in vacuo*. In other media and at high resolution, the effects of refractive index must be accounted for.
13. Davisson, C.J. and Germer, L.H. (1927) *Phys. Rev.*, **30**, 705.
14. Thomson, G.P. and Reid, A. (1927) *Nature (London)*, **119**, 890; Thomson, G.P. (1928) *Proc. R. Soc. London, A*, **117**, 600.
15. Stern, O. (1929) *Die Naturwissenschaften*, **21**, 391.
16. de Broglie, L. (1923) *Compt. Rend. Acad. Sci. Paris*, **177**, 506, 548, 640; de Broglie, L. (1927) *J. Phys. (Paris)*, **VIII** (5), 225.
17. Compton, A.H. (1923) *Phys. Rev.*, **21**, 483.
18. Baggott, J. (2010) *The Quantum Story: A History in 40 Moments*. Oxford University Press, Oxford, p. 41.
19. Ritz, W. (1908) *Phys. Z.*, **9**, 521.
20. Bohr, N. (1913) *Philos. Mag.*, **26**, 1.

21. Bohr, N. (1913) *Philos. Mag.*, **26**, 1.
22. Itano, W.M., Bergquist, J.C., and Wineland, D.J. (2015) Early observations of macroscopic quantum jumps in single atoms. *Int. J. Mass Spectrom.*, **377**, 403–409.

Further reading

The history of the development of quantum mechanics is worthy of study in its own right. Below you will find three interesting and informative accounts.

Baggott, J. (2010) *The Quantum Story: A History in 40 Moments*. Oxford University Press, Oxford.
Heisenberg, W. (1979) *Quantentheorie und Philosophie*. Reclam, Stuttgart.
Jones, S. (2008) *The Quantum Ten: A Story of Passion, Tragedy, Ambition and Science*. Oxford University Press, Oxford.

For an introduction to the strange world of entanglement and quantum measurement, try these:

Zeilinger, A. (2007) *Einsteins Spuk*. Goldmann Verlag, München.
Zeilinger, A. (2010) *Dance of the Photons: From Einstein to Quantum Teleportation*. Farrar, Straus and Giroux, New York (in English).

For more on the wave nature of light:

Born, M. and Wolf, E. (1999) *Principles of Optics: Electromagnetic Theory of Propagation, Interference and Diffraction of Light*, 7th edition. Cambridge University Press, Cambridge.
Young, M. (2000) *Optics and Lasers: Including Fibers and Optical Waveguides*, 5th edition. Springer-Verlag, Berlin.

CHAPTER 20

The quantum mechanical description of nature

PREVIEW OF IMPORTANT CONCEPTS

- We can expect to observe quantum mechanical effects and the insufficiency of a classical mechanical description when at least one dimension of the system under consideration approaches the magnitude of its de Broglie wavelength.
- A quantum mechanical description should be used when the thermal energy $k_B T$ is small compared to the energy level spacing $\varepsilon_i - \varepsilon_j$.
- In the limit of high quantum numbers, systems behave classically. Equivalently, in the limit of large energies, large masses and large orbits, quantum calculations deliver the same results as classical calculations.
- The postulates of quantum mechanics cannot be derived from more fundamental principles and form the basis of the theory from which its framework is derived.
- The wavefunction describes everything there is to know about a system.
- The absolute square of the wavefunction describes a probability distribution.
- Operators act on wavefunctions to determine the values of observables.
- If a system is in an eigenstate of an operator, the corresponding observable has an exact value. If the system is not in an eigenstate of the operator, the observable is described by an expectation value that is equal to the mean of many measurements.
- The eigenstates of a system form an orthonormal basis set.
- Eigenstates can be superposed in linear combinations to create states of the system. Wavefunctions have relative phases, which can lead to interference effects when they are combined.
- The variational principle states that the correct wavefunction for the system yields the correct expectation value of the operator. Any other wavefunction gives a value for the expectation value that is greater than the correct value.
- Solution of the Schrödinger equation yields the complete set of eigenfunctions for the corresponding Hamiltonian.
- The Hamiltonian is the total energy operator and contains operators that account for the kinetic and potential energy of the system.
- Quantization of energy levels is linked to confinement of the system by a potential.
- *Some operators do not commute.* Such noncommuting operators have corresponding observables that cannot be determined simultaneously to arbitrary precision. They are subject to an uncertainty relation.
- *Some operators do commute.* The corresponding observables can be determined simultaneously to arbitrary precision. They are not subject to an uncertainty relation.

The purpose of this chapter is to establish the mathematical foundations of quantum theory. In the previous chapter we became acquainted with 'old' quantum theory. This theory resembles the structure of quantum theory but it does so with *ad hoc* assumptions. Now, we will attempt to introduce a set of postulates – fundamental statements of fact that cannot be derived from more basic tenets – from which all other aspects of quantum theory are derived. In doing so we hope to establish reasons for the assumptions made in 'old' quantum theory. These postulates should also allow us to derive new results and make predictions about previously unknown phenomena.

This chapter may appear overly mathematical and formal. And it is. That is the nature of the foundations of quantum mechanics. This is essential hard work. However, the purpose of this chapter is not just to make you familiar with the mathematics of quantum mechanics. It is also meant to make you familiar with the vocabulary and concepts of quantum mechanics. Read it out loud to yourself – preferably alone, as muttering about quantum mechanics in public spaces is not necessarily the best way to win friends and influence people!

Physical Chemistry: How Chemistry Works, First Edition. Kurt W. Kolasinski.

Companion Website: www.wiley.com/go/kolasinski/physicalchemistry

20.1 What determines if a quantum description is necessary?

We have already encountered one principle to guide us in determining the necessity for a quantum description:

We can expect to observe quantum mechanical effects and the insufficiency of a classical mechanical description when at least one dimension of the system under consideration approaches the magnitude of its de Broglie wavelength.

A second condition for the necessity for the use of quantum mechanics rather than classical mechanics can be stated as so:

A quantum mechanical description should be used when the thermal energy k_BT is small compared to the energy level spacing $\varepsilon_i - \varepsilon_j$.

Classical mechanics can be derived from quantum mechanics in the limit in which allowed energy values are continuous rather than discrete. As shown by Ehrenfest, this occurs mathematically in the limit of $\hbar$ approaching zero. In a practical sense, this occurs when the thermal energy k_BT is large compared to the energy level spacing $\varepsilon_i - \varepsilon_j$ between successive levels i and j. You will recall from our discussion of statistical mechanics in Chapter 6 that a thermal distribution is characterized by a Boltzmann distribution,

$$\frac{n_i}{n_j} = \frac{g_i}{g_j} \exp(-(\varepsilon_i - \varepsilon_j)/k_BT). \tag{20.1}$$

Classically, this is a continuous distribution, but quantum mechanically this is a discrete distribution. For H_2 gas filling a macroscopic container at room temperature, the translational degree of freedom acts decidedly classically because the spacing between translational levels is much smaller than the thermal energy. The rotational and vibrational degrees of freedom, on the other hand, display characteristics of their quantum state distributions that are readily detected by spectroscopy, as we will explore in more detail in the chapters to follow.

The above statements are, of course, expressions of Bohr's *correspondence principle*:

In the limit of high quantum numbers, systems behave classically. Equivalently, in the limits of large energies, large masses and large orbits, quantum calculations deliver the same results as classical calculations.

Thus, it is in the opposite limits (low temperature, small mass and small length scale) in which quantum calculations are unavoidable.

20.2 The postulates of quantum mechanics

The laws of thermodynamics cannot be derived from more fundamental principles. They simply are the basis of thermodynamics from which the remainder of the theoretical framework of thermodynamics is derived. Similarly, the postulates of quantum mechanics cannot be derived from more fundamental principles. They form the basis of the theory from which its framework is derived. As with the laws of thermodynamics, I present the postulates first and then we will delve into their meaning and implications in subsequent chapters. The content of the postulates is more or less resolved. However, unlike the laws of thermodynamics, the numbering (and even the number) of the postulates of quantum mechanics is not set by convention. Reading the postulates, you will see that we need to define several terms including eigenfunction, eigenvalue, Hermitian and operator. This we will do in the remainder of the chapter.

Postulate 1: Completeness and Repeatability

Every physical system of N particles is completely described by a mathematical function of the coordinates $\tau_1, \tau_2, \ldots, \tau_N$ of all the particles in the system and of time t. This finite, continuous and single-valued function is called the system's wavefunction, $\Psi(\tau_1, \tau_2 \ldots, \tau_N, t)$. The state described by this wavefunction is a predictive tool. An immediately repeated measurement yields the same result as long as the state described by this wavefunction is an eigenfunction.

Postulate II: Born's Rule

The wavefunction is a probability amplitude that potentially contains a phase factor (an imaginary part). Therefore, the probability distribution described by the wavefunction is obtained from its absolute square and $\Psi(\tau, t)^*\Psi(\tau, t)\mathrm{d}\tau$ is proportional to the probability that the system is found in the range $\mathrm{d}\tau$ about τ at time t.

Postulate III: Collapse Axiom A

Each physically observable quantity is represented mathematically by a unique operator that acts on the wavefunction in a prescribed manner. Operators that correspond to physical properties must be linear and Hermitian and, therefore, the eigenstates of the operator (and wavefunctions that describe the system) must be orthogonal.

Postulate IV: Collapse Axiom B

A wavefunction ψ that describes a system forms an eigenvalue equation with the operator $\hat{A}$ such that $\hat{A}\psi = \alpha\psi$, where α is the value of the physical quantity represented by $\hat{A}$ that would be measured experimentally for the system. Thus, any system whose wavefunction is an eigenfunction of $\hat{A}$ has a definite value of α.

Postulate V: Unitarity

To every physical system there corresponds a unique operator representing the total energy of the system. This operator, the Hamiltonian $\hat{H}$, and the observable physical states of the system are described by those wavefunctions that satisfy the time-dependent Schrödinger equation $\hat{H}\Psi = i\hbar(\partial\Psi/\partial t)$.

Postulate VI: Expectation Values

Consistent with the correspondence principle and the interpretation of the wavefunction as a probability amplitude, the average measurement of a physical observable is represented by the quantum analog of the classical mean of the operator $\hat{A}$ operating on the wavefunction ψ according to $\langle A\rangle = \int \psi^* \hat{A} \psi \, d\tau / \int \psi^* \psi \, d\tau$. $\langle A\rangle$ is known as the expectation value of the operator $\hat{A}$.

Postulate VII: Superposition Principle

Since the eigenfunctions of a given Hamilitonian form a complete set of orthogonal, normalizable functions, they correspond to vectors in a Hilbert space such that, for any two good wavefunctions ψ_1 and ψ_2, the superposition state represented by $\psi = c_1\psi_1 + c_2\psi_2$ is also a good wavefunction.

Postulate VIII: Composition Postulate

Composite states, such as those formed between the system S and the environment E can be expressed as superpositions of the form $\sum_{k,l} \gamma_{k,l}\psi_{s,k}\psi_{e,l}$ where $\psi_{s,k}$ and $\psi_{s,l}$ are the basis functions that describe the system and the environment.

In the following chapters we will deal with all of these postulates directly, even briefly in the case of Postulate VIII, which is the basis for entanglement. The phenomenon of entanglement brings to light some rather fundamental paradoxes of quantum mechanics in which quantum behavior simply does not comport with our classical experiences.[1] Schrödinger identified it as "...the characteristic trait of quantum mechanics, the one that enforces its entire departure from classical lines of thought."[2] These issues are embodied in the so-called Einstein–Podolsky–Rosen paradox[3] and tests of Bell's inequalities.[4] This postulate has deep implications for the meaning of and our perception of reality. An excellent discussion of quantum paradoxes is given by Griffiths.[5] While we will not deal with the mathematics of entanglement in detail, we will return to the topic when we discuss photosynthesis as entanglement and long-lived coherent excitation transfer appear to play a role in photosynthesis in the process of transferring the energy of a photon absorbed in a light-harvesting center to the reaction center.[6]

20.3 Wavefunctions

20.3.1 Mathematical requirements

Postulate I requires the existence of a wavefunction, and physical reality constrains the mathematical nature of these functions. As we shall see below, Postulate V allows us to write the complete wavefunction, which is a function of space and time, as the product of a function of the spatial coordinates alone and a time-dependent function,

$$\Psi(\tau, t) = \psi(\tau)\phi(t), \tag{20.2}$$

(as long as the potential is independent of time). The composite variable τ represents the spatial coordinates of each particle in the system. The physical constraints on the wavefunction lead to the following set of mathematical boundary conditions:

- Ψ must be continuous, finite (contain no singularities) and single-valued.

- Ψ is a function that may contain complex numbers.
- Ψ must be quadratically integrable and normalizable.
- All partial first derivatives of Ψ must be continuous (except for when an infinitely high potential barrier is encountered), contain no singularities, and be single-valued.

We shall see shortly that the time-dependent function is the simple exponential function $\phi(t) = \exp(Et/i\hbar)$, hence, all of the constraints on the *spacetime wavefunction* Ψ also apply to the *spatial wavefunction* ψ.

A continuous function is smooth, containing no jumps. A jump corresponds to an infinite first derivative. Since the linear momentum of a particle is proportional to the first derivative of the wavefunction, an infinite first derivative implies an infinite momentum. This is impossible in any realistic physical system.

The value of the wavefunction must be finite even at infinity. In fact, the wavefunction must tend to zero as it approaches infinity, otherwise its integral would not be finite. A single-valued wavefunction can have one, and only one, value at any set of coordinate values. A complex number contains both a real and an imaginary part. Postulate II requires the product of a wavefunction with itself to result in a real number. This can only occur in general for complex numbers if we form the product between a function and its complex conjugate in which i is replaced by –i in all imaginary terms. This product is denoted the absolute square of the wavefunction $|\Psi|^2 = \psi^*\psi$. Furthermore, since we need to integrate this product to determine the probability of the existence of the state, the wavefunction must be quadratically integrable to a finite real number. If the wavefunction is quadratically integrable to a finite real number, then we can always scale it such that the wavefunction is normalized, that is, $\int_{-\infty}^{\infty} \psi^*\psi \, d\tau = 1$.

Finally, Ψ will be continuous and smooth except for at a finite number of points that will be clearly identifiable within the model we are using (such as when the potential goes to infinity, as it does at the infinitely high walls of the particle in a box problem below); hence, its partial first derivatives are also continuous except at certain boundaries imposed by the model. A discontinuity in the first derivative implies an infinite second derivative. Since the energy operator is proportional to the second derivative of the wavefunction, an infinite second derivative implies an infinite energy, which is not physically realistic.

20.3.2 Born interpretation of the wavefunction

Schrödinger proposed the wave equation that we examine in detail below without really understanding what the wavefunction was. Initially, it was unclear whether this function was merely a mathematical construct that delivered the correct answers, or whether there was a deeper physical meaning to the wavefunction. It was Born who realized how the wavefunction could be interpreted physically.[7] His proposal was that ψ is a probability amplitude and its absolute square

$$|\psi(x)|^2 = \psi^*(x)\psi(x) \tag{20.3}$$

is proportional to the probability of finding the particle at position x. $|\psi(x)|^2$ is a probability density, as illustrated in Fig. 20.1.

The implications of the *Born interpretation* are rather unsettling for someone steeped in the seeming determinism of classical mechanics. If a particle is described by $\psi(x)$ and its position by a probability distribution, then the particle must have wave-like character and the concept of a single well-defined trajectory has to be abandoned. This will become all the more disruptive to classical intuition when we encounter further consequences that are known as the Heisenberg uncertainty principle.

Linking the square of the wavefunction to a probability density means that the product of the square of the wavefunction times a volume is equal to the probability of finding the particle in that volume. Thus, the probability of finding a particle in a certain region of space between τ_1 and τ_2 is

$$P = \int_{\tau_1}^{\tau_2} |\psi(\tau)|^2 d\tau. \tag{20.4}$$

If there is one particle described by the wavefunction, then there is unit probability of finding that particle when considering all of the space in which that particle can be found. In other words, a good wavefunction is normalized to

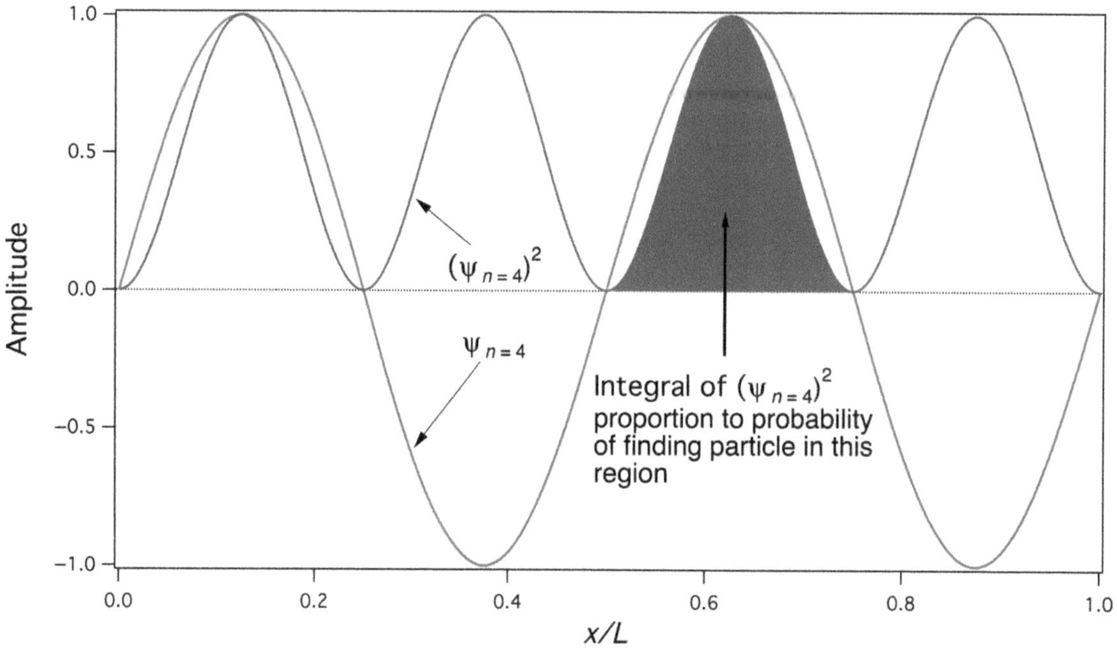

Figure 20.1 Illustrations of ψ and $|\psi|^2$ for the $n = 4$ state of a particle in a box. A probability amplitude can be either positive or negative, which is related to the potential for interference effects upon combination with another wavefunction. The probability density is always positive, and its integral is proportional to the probability of finding the particle between the limits of integration. By symmetry, it is easy to see that the integral of the blue region is 0.25 of the normalized area of the complete integral of $|\psi|^2$.

1 when integrated over all space.[8] The *normalization* condition is that we should be able to scale our wavefunctions by a factor N such that when the square of the wavefunction is integrated over all space, the result is unit probability,

$$\int_{-\infty}^{\infty} N\psi^* N\psi \mathrm{d}\tau = N^2 \int_{-\infty}^{\infty} \psi^* \psi \mathrm{d}\tau = 1 \tag{20.5}$$

The volume element $\mathrm{d}\tau$ depends on the coordinates and dimensionality. Assuming that we have properly normalized the wavefunction and incorporated the constant N into ψ, in one-dimensional linear coordinates, the normalization condition is

1D, linear coordinates
$$\int_{-\infty}^{\infty} \psi^* \psi \mathrm{d}x = 1, \tag{20.6}$$

while in three dimensions it reads

3D, linear coordinates
$$\int_{-\infty}^{\infty}\int_{-\infty}^{\infty}\int_{-\infty}^{\infty} \psi^* \psi \mathrm{d}x\,\mathrm{d}y\,\mathrm{d}z = 1. \tag{20.7}$$

For central field problems, such as an atom, spherical polar coordinates are more appropriate,

3D, spherical polar
$$\int_{0}^{\infty}\int_{0}^{\pi}\int_{0}^{2\pi} \psi^* \psi r^2 \sin\theta \,\mathrm{d}r\,\mathrm{d}\theta\,\mathrm{d}\phi = 1. \tag{20.8}$$

Recall that if a function is variable separable and can be factored such as,

$$\psi(r,\theta,\phi) = f(r)f(\theta)f(\phi) \tag{20.9}$$

then the integrals can also be factored, for example,

$$\int_{0}^{\infty}\int_{0}^{\pi}\int_{0}^{2\pi} Nf(r)f(\theta)f(\phi)r^2 \sin\theta \,\mathrm{d}r\,\mathrm{d}\theta\mathrm{d}\phi = N\int_{0}^{\infty} r^2 f(r)\mathrm{d}r \int_{0}^{\pi} f(\theta)\sin\theta\,\mathrm{d}\theta \int_{0}^{2\pi} f(\phi)\mathrm{d}\phi. \tag{20.10}$$

Both 1 and N are dimensionless quantities. Since the normalization integral depends on the dimensionality of the system, the dimensions of ψ must also change with the dimensionality of the system.

20.3.2.1 Example: particle in a one-dimensional box

The wavefunction for a particle confined in a one-dimensional box of length L, as we shall prove in the next chapter, is (in the range $0 \leq x \leq L$)

$$\psi_n(x) = \left(\frac{2}{L}\right)^{1/2} \sin\left(\frac{n\pi x}{L}\right).$$

Take $L = 10.0$ nm. Now let's calculate the probability that a ground-state electron (that is, an electron in the $n = 1$ state) is (a) between $x = 4.95$ nm and 5.05 nm, and (b) in the left half of the box. (c) What is the probability of finding the particle at $x = 0$ or $x = 5$ nm?

The $n = 1$ wavefunction is $\psi = (2/L)^{1/2} \sin(\pi x/L)$. The probability is found by integration of the square of the wavefunction over the appropriate limits.

$$P = \int_a^b \psi^* \psi \, dx = \left(\frac{2}{L}\right) \int_a^b \sin^2\left(\frac{\pi x}{L}\right) dx = \left[\frac{x}{L} - \frac{1}{2\pi} \sin\left(\frac{2\pi x}{L}\right)\right]_a^b.$$

a In this case $a = 4.95$ nm and $b = 5.05$ nm.

$$P = \left[\frac{x}{L} - \frac{1}{2\pi} \sin\left(\frac{2\pi x}{L}\right)\right]_a^b = \frac{b-a}{L} - \frac{1}{2\pi}\left[\sin\left(\frac{2\pi b}{L}\right) - \sin\left(\frac{2\pi a}{L}\right)\right]$$

Substituting for a and b we find $P = 0.010$.

b In this case $a = 0$ and $b = L/2 = 5$. However, since the wavefunction is symmetric about the center of the box and the integral over the whole of the box is 1 (normalized wavefunction), the integral over half of the box is 0.5. Look back at Fig. 20.1. The $n = 4$ state is shown, which has four extrema and two full cycles of the sine function. The $n = 1$ state corresponds to just one half-cycle of the sine function.

c The wavefunction is zero at $x = 0$. The probability is also zero at $x = 0$. The wavefunction is a maximum at $x = 5$ nm; nonetheless, the probability of finding a particle at a point – any point – is zero. The integral of any *finite* function over a single point is zero.[9] Ponder that for a moment. The particle does not exist at any point in the box. Yet the particle is in the box. Ergo, the particle is not point-like. The probability of finding the particle in any region of space is nonvanishing. In the vicinity of a node, this probability is exceedingly small. The smaller we make the region, the smaller the probability, but it only goes uniquely to zero at a point. Thus, there is no difficulty for this non-point-like particle to move from one side of the box to the other, even though the wavefunction contains nodes in it.

20.4 The Schrödinger equation

The wave-like nature of small particles means that they do not follow the well-defined trajectories inherent to classical mechanics. Instead, quantum mechanics uses wavefunctions to describe particles and their motions. A wavefunction describes a probability amplitude and the square of a wavefunction tells us something about the probability of a particle being at a certain point in space and time. We must abandon a description in terms of particle trajectories, or – as conceived of in the path integral formulation of quantum mechanics – all trajectories are potentially possible for the particle, but they are not equally probable. We need to learn how to calculate the probabilities and to understand the implications of particle dynamics being described by a probability amplitude that contains both magnitude and phase information.

The *Schrödinger equation* was proposed by Schrödinger[10] in 1926. His argument for what was at the time a very preposterous idea – that we really do have to take this wave-particle duality of de Broglie seriously when discussing the dynamics of subatomic particles – was to state that geometrical optics (treating light as rays) and physical optics (treating light as waves) emerge from one another in different limits. Furthermore, William Hamilton used as an analogy the theory of the propagation of light in a nonhomogeneous medium to derive theories in pure mechanics, most notably the Hamilton function of action and how it is used to describe the evolution of a mechanical system. It was left to Schrödinger to reverse engineer from Hamilton's work the implications for atomic-scale systems. In doing so he would derive the Planck relation of energy $E = \hbar\omega$, Bohr's energy level structure of the H atom $E_n = -2\pi^2 me^4/h^2n^2$, and both the intensities and polarizations of spectral lines observed in the Stark effect as natural consequences of these dynamics. Something this good just could not be wrong, so he forged ahead without a clear conception about the nature of the wavefunction.

For a particle of mass m moving in one dimension x with a possibly time-dependent potential described by $V(x,t)$, the *time-dependent Schrödinger equation* has the form

$$-\frac{\hbar}{2m}\frac{\partial^2\Psi(x,t)}{\partial x^2} + V(x,t)\Psi(x,t) = \mathrm{i}\hbar\frac{\partial\Psi(x,t)}{\partial t}. \tag{20.11}$$

The time-dependent wavefunction $\Psi(x,t)$ is the solution to this equation and $\hbar = h/2\pi$ is the reduced Planck constant. In many of our applications, the potential is independent of time. This allows us to factor the wavefunction into a spatial component and a time-dependent component,

$$\Psi(x,t) = \psi(x)\phi(t) \tag{20.12}$$

The function $\psi(x)$ is the spatial part of the wavefunction. These functions are also called the *stationary states* of the system.

Upon substitution of the factored wavefunction back into Eq. (20.11), the time-dependent Schrödinger equation becomes

$$\frac{1}{\psi(x)}\left[-\frac{\hbar}{2m}\frac{\mathrm{d}^2\psi(x)}{\mathrm{d}x^2} + V(x)\psi(x)\right] = \frac{1}{\phi(t)}\mathrm{i}\hbar\frac{\mathrm{d}\phi(t)}{\mathrm{d}t} \tag{20.13}$$

Each side of Eq. (20.13) is a function of one variable only: x for the LHS and t for the RHS. They can only be equal to each other if they are both equal to the same constant. We call this constant E, and we will discover that it is equal to the total energy of the state described by the wavefunction. Setting the RHS equal to E, we obtain,

$$i\hbar\frac{\mathrm{d}\phi(t)}{\mathrm{d}t} = E\phi(t). \tag{20.14}$$

The solution to this differential equation is the exponential function,

$$\phi(t) = \exp(Et/\mathrm{i}\hbar), \tag{20.15}$$

which describes a sinusoidally oscillating wave with angular frequency $\omega = E/\hbar$. For a potential that does not vary in time, the total energy $E = \hbar\omega$ is constant and the time-dependent wavefunction has the form

$$\Psi(x,t) = \psi(x)\exp(Et/\mathrm{i}\hbar). \tag{20.16}$$

From the Born interpretation, we learn that the wavefunction is a probability amplitude, the absolute square of which gives the probability of finding the particle. However, to take the absolute square, we must take into consideration that the wavefunction is a complex function; thus, the square of the function is taken by multiplying the complex conjugate of the wavefunction by the wavefunction,

$$|\Psi(x,t)|^2 = \Psi^*(x,t)\Psi(x,t). \tag{20.17}$$

The complex conjugate is taken by switching the sign of each term involving an imaginary number $\mathrm{i} \to -\mathrm{i}$; hence,

$$|\Psi(x,t)|^2 = \psi^*(x)\exp(Et/(-\mathrm{i}\hbar))\psi(x)\exp(Et/\mathrm{i}\hbar) = |\psi(x)|^2. \tag{20.18}$$

This explains why the $\psi(x)$ are called the stationary states – the probability distribution described by the square of the wavefunction is not time-dependent.

The LHS of Eq. (20.13) is also equal to the total energy E. This allows us to write the *time-independent Schrödinger equation,*

$$-\frac{\hbar}{2m}\frac{\mathrm{d}^2\psi(x)}{\mathrm{d}x^2} + V(x)\psi(x) = E\psi(x). \tag{20.19}$$

We introduce the *Hamiltonian operator* $\hat{H}$

$$\hat{H} = \frac{\hbar}{2m}\frac{\mathrm{d}^2}{\mathrm{d}x^2} + V(x) \tag{20.20}$$

which allows us to write the time-independent Schrödinger equation in the compact form

$$\hat{H}\psi(x) = E\psi(x). \tag{20.21}$$

Most of the questions we will try to answer will have their solution in the time-independent Schrödinger equation, which is why the short form in Eq. (20.21) has become so familiar. Essentially all of chemistry and physics is contained within the Schrödinger equation. This statement comes with the two very significant 'howevers.' The first is that we need to write down a complete Hamiltonian to describe *all* the interactions in our system. The kinetic energy term is easy, but the potential energy term requires detailed knowledge and a mathematical expression for every kind of interaction (electrostatic, magnetic, gravitational, chemical, van der Waals, etc.) experienced in the system. Once we can formulate all of the interactions, 'all' we have to do is find the wavefunctions. However – and this one is even tougher – the Schrödinger equation cannot be solved exactly for any atom or molecule containing more than one electron. Obviously, a major component of quantum mechanics is the pursuit of ever better approximations to the exact solution of the Schrödinger equation.

20.4.1 Directed practice

The time-dependent Schrödinger equation is sometimes written

$$\Psi(x,t) = \psi(x)e^{-\mathrm{i}Et/\hbar}.$$

Show that this is equivalent to Eq. (20.16).

20.5 Operators and eigenvalues

Postulates III and IV intentionally mention operators and eigenvalues. To mathematicians, these are terms that are loaded with meaning invoking a Pavlovian response. They are built into the language of the postulates because they bring with them immediate clues as to how to formulate and solve the Schrödinger equation.

An *operator* is something that carries out mathematical manipulations on the function ψ. For example, to find the momentum, apply the momentum operator $\hat{p}$ to the wavefunction that describes the system. The hat (or circumflex) on top of $\hat{p}$ (read 'p-hat') is there to differentiate the linear momentum operator $\hat{p}$ from the magnitude of the linear momentum p, just as using bold-italics differentiates the linear momentum vector $\boldsymbol{p}$ from its magnitude. To find the energy, we apply the energy operator, given the special name of the Hamiltonian $\hat{H}$ because of its singular importance. $\hat{H}$ is a second-order differential operator. It contains terms like $\mathrm{d}^2/\mathrm{d}x^2$ but not higher derivatives. The form of $\hat{H}$ depends on the system. We have to consider all factors that contribute to energy – both potential and kinetic – to construct the Hamiltonian.

Whenever an operator is involved in an equation, *the order of operation is important*. For instance, in Eq. (20.21) we cannot first factor out ψ and then operate. We must first operate, and then solve the resulting equation. The order of operation is also indicated in the operator. The operator $x(\mathrm{d}/\mathrm{d}x)$ means first take the derivative with respect to x, then multiply by x. The operator $(\mathrm{d}/\mathrm{d}x)x$ means first multiply by x and then take the derivative with respect to x. Note that the order of operation is not important for multiplication or division by constants. In addition, *operators operate on functions, not on other operators*. Hence, $\hat{A}\hat{B}\psi$ means operate first with $\hat{B}$ on ψ, then with $\hat{A}$ on the function that results from $\hat{B}\psi$.

The order of operation is important because operators may or may not commute. Operators are said to commute if, for all ψ,

$$\hat{A}\hat{B}\psi = \hat{B}\hat{A}\psi. \tag{20.22}$$

Operators do not commute if, for all ψ,

$$\hat{A}\hat{B}\psi \neq \hat{B}\hat{A}\psi. \tag{20.23}$$

The difference has profound implications. Because of this, we define the *commutator* of $\hat{A}$ and $\hat{B}$, which is itself an operator, as

$$[\hat{A},\ \hat{B}] = \hat{A}\hat{B} - \hat{B}\hat{A}. \tag{20.24}$$

When applied to a function, the commutator is evaluated by operating on ψ with $\hat{A}\hat{B}$ then subtracting from this the result of operating on ψ with $\hat{B}\hat{A}$,

$$[\hat{A},\ \hat{B}]\psi = \hat{A}\hat{B}\psi - \hat{B}\hat{A}\psi. \tag{20.25}$$

Operators act on functions to produce new functions. Postulate III stipulates that the operators are linear, which means that they obey the following rules

$$\hat{A}(\psi_1 + \psi_2 + \ldots) = \hat{A}(\psi_1) + \hat{A}(\psi_2) + \ldots \tag{20.26}$$

and

$$\hat{A}(c\psi_1) = c\hat{A}\psi_1, \tag{20.27}$$

where c is a constant, again reaffirming that the order of operation for multiplication (or division) by a constant is unimportant.

Postulate III also stipulates that the operators are *Hermitian*. An operator is Hermitian if for two functions ψ_1 and ψ_2

$$\int \psi_1^* \ \hat{A}\psi_2 \ \mathrm{d}\tau = \int \psi_2 \ (\hat{A}\psi_1)^* \ \mathrm{d}\tau \tag{20.28}$$

For operators that correspond to physically observable quantities, the above two integrals are equal. The LHS corresponds to operating first on ψ_2 then multiplying by the complex conjugate of ψ_1. The RHS corresponds to operating on ψ_1, taking the complex conjugate then multiplying by ψ_2. It is important that the operators are Hermitian because Hermitian operators have real eigenvalues.

20.5.1 Example

Show that the eigenvalue α of a Hermitian operator $\hat{A}$ operating on the normalized wavefunction ψ is real.

According to the given set of assumptions, our operator and wavefunction must satisfy the eigenvalue equation

$$\hat{A}\psi = \alpha\psi.$$

Now, multiply both sides of this eigenvalue equation from the left with ψ^* and integrate over all space,

$$\int \psi^*\hat{A}\,\psi \ \mathrm{d}\tau = \int \psi^*\alpha\,\psi \ \mathrm{d}\tau.$$

α is a constant and can be moved outside of the integral,

$$\int \psi^*\hat{A}\,\psi \ \mathrm{d}\tau = \alpha \int \psi^* \ \psi \ \mathrm{d}\tau.$$

The wavefunction is normalized, which means that the integral on the right is equal to 1,

$$\int \psi^*\hat{A}\,\psi \ \mathrm{d}\tau = \alpha.$$

Now, take the complex conjugate of the eigenvalue equations, multiply both sides from the right by ψ and integrate over all space to obtain,

$$\int \ (\hat{A}\psi)^*\psi \ \mathrm{d}\tau = \int \alpha^*\psi^*\psi \ \mathrm{d}\tau$$

$$\int \ (\hat{A}\psi)^*\psi \ \mathrm{d}\tau = \alpha^* \int \psi^*\psi \ \mathrm{d}\tau$$

$$\int \ (\hat{A}\psi)^*\psi \ \mathrm{d}\tau = \alpha^*.$$

The definition of a Hermitian operator allows us to link the first result with the second result by transitivity,

$$\int \psi^*\hat{A}\psi \ \mathrm{d}\tau = \int \psi(\hat{A}\psi)^* \ \mathrm{d}\tau = \int \ (\hat{A}\psi)^*\psi \ \mathrm{d}\tau$$

Having performed the integration already to evaluate the left and right integrals we can then set

$$\alpha = \alpha^*,$$

which can only be true if α is a real number.

The RHS of the Schrödinger equation is particularly simple. It contains the product of a constant, the eigenvalue E, and a function, the eigenfunction ψ. A priori, we do not know either the eigenfunctions or the eigenvalues. The eigenfunction is the function that properly satisfies the Schrödinger equation. Once we know it, then according to Postulate I we know everything there is to know about the system (assuming we have already figured out how to write all possible operators). The trick here is how do we find the eigenfunctions that satisfy this equation?

Let's repeat that statement because it is so important. We only need to solve the Schrödinger equation once to determine the wavefunctions. Once we have the wavefunctions of the system, we can apply known operators to determine all experimentally observable properties of the system. For example, having solved the Schrödinger equation for the energy levels of the H atom, we can apply operators to those wavefunctions to determine the angular momentum of a particular level or to find the transition probability between any two levels, and so forth.

The solution to all quantum mechanical problems is now at hand in principle. Notice, however, that our solution contains a substantial shift in our conception of understanding, and it is the essential feature that Heisenberg introduced, which would so fundamentally shape the development of quantum mechanics. As stated by Born in his Nobel Prize lecture, "According to the heuristic principle used by Einstein in the theory of relativity, and by Heisenberg in the quantum theory, concepts which correspond to no conceivable observation should be eliminated from physics."[11] If we can no longer simultaneously determine the position and velocity of an electron, we can no longer define its trajectory. Bohr orbits are not just incorrect, they do not exist because they cannot be defined, and cannot be observed. Classical analogies die hard, but many of these simply do not exist on subatomic scales. Similarly, decoherence and the collapse of wavefunctions mean that many quantum phenomena are not (in the case of superposition or wavelength of objects) or not easily (in the case of entangled particles) observed on the macroscopic scale.

As an aside it should be mentioned that Heisenberg's matrix mechanics[12] and Dirac's symbolic formulation[13] both predate Schrödinger's wave mechanics by about a year. It was quickly realized that Heisenberg and Dirac were describing the same operator mathematics. Heisenberg's colleagues from Göttingen – Born and Jordan – worked with Heisenberg to formulate the concepts of noncommuting operators and observables into a matrix representation. Dirac did not know matrix mathematics so he derived it all on his own. After a couple of stormy and sometimes acrimonious years, Schrödinger demonstrated that wave mechanics and matrix mechanics were equivalent, with each approach having advantages and disadvantages in implementation depending on the problem at hand. Born would then coin the term *quantum mechanics* as the name of this unified description of microscopic dynamics.

20.6 Solving the Schrödinger equation

Postulate V tells us that solving the Schrödinger equation will provide us a complete description of the system and how it develops in time. Therefore, the following procedure should become very familiar to you.

- First, write out a complete Hamiltonian that accounts for all interactions in the system.
- Second, substitute the resulting Hamiltonian into the Schrödinger equation.
- Third, solve the Schrödinger equation for the wavefunctions.
- Fourth, brew coffee, interpret and publish results.[14]

You should note that practice can be much different than principle, and decidedly more difficult to execute than to plan. Nonetheless, let's see how this works by solving a few simple examples.

20.6.1 Free particle in one dimension

A free particle is a particle of mass m that is unconfined by a potential. This means that the potential energy is constant, so we are free to chose any value. In this example we choose $V(x) = 0$ everywhere. The Hamiltonian is the sum of kinetic and potential energy operators and in this case is given by

$$\hat{H} = \hat{T} + \hat{V} = \frac{\hbar^2}{2m}\frac{d^2}{dx^2} \quad (20.29)$$

By inspection, a good candidate for a solution to this differential equation is

$$\psi = Ne^{ikx} \quad (20.30)$$

where both N and k are constants (k can be either positive or negative[15]). Now prove that this is a solution to the Schrödinger equation,

$$\hat{H}\psi = \hat{E}\psi, \tag{20.31}$$

by showing that operating on this function with the Hamiltonian returns a constant times the original function as required. Begin by taking the second derivative of ψ.

$$\frac{d^2\psi}{dx^2} = N\frac{d}{dx}(ike^{ikx} = N(ik)(ike^{ikx}) = -k^2Ne^{ikx} \tag{20.32}$$

Thus, the LHS of the Schrödinger equation is

$$\hat{H}\psi = \frac{-\hbar^2}{2m}\frac{d^2\psi}{dx^2} = \frac{-\hbar^2}{2m}(-k^2Ne^{ikx}) = \frac{\hbar^2k^2}{2m}Ne^{ikx} = \frac{\hbar^2k^2}{2m}\psi. \tag{20.33}$$

Substituting this into Eq. (20.31) gives

$$\frac{\hbar^2k^2}{2m}\psi = E\psi. \tag{20.34}$$

The result in Eq. (20.34) proves that our guess – or, in the parlance of the field, our *trial wavefunction* – was indeed a solution of the differential equation because application of the Hamiltonian operator led to a constant ($\hbar^2k^2/2m$) times the wavefunction. The constant is equal to the energy,

$$E = (\hbar k)^2/2m. \tag{20.35}$$

Note that the normalization constant is irrelevant for the calculation of the energy.

Take a look at the resulting energy in Eq. (20.35). Is it quantized? No, as shown in Fig. 20.2. We have applied quantum mechanics but the result is that the energy can take on any value because there is no restriction on the value of k. This highlights two essential features of quantum mechanics. First, the methods of quantum mechanics are also valid in the region where classical results are obtained. Second, quantization of energy levels is *not* observed in the absence of confinement. The potential in the example was zero (actually it need only be constant). The particle is free. Only if the particle is confined by the potential to a region of space comparable to its de Broglie wavelength in at least one dimension will quantization be important.

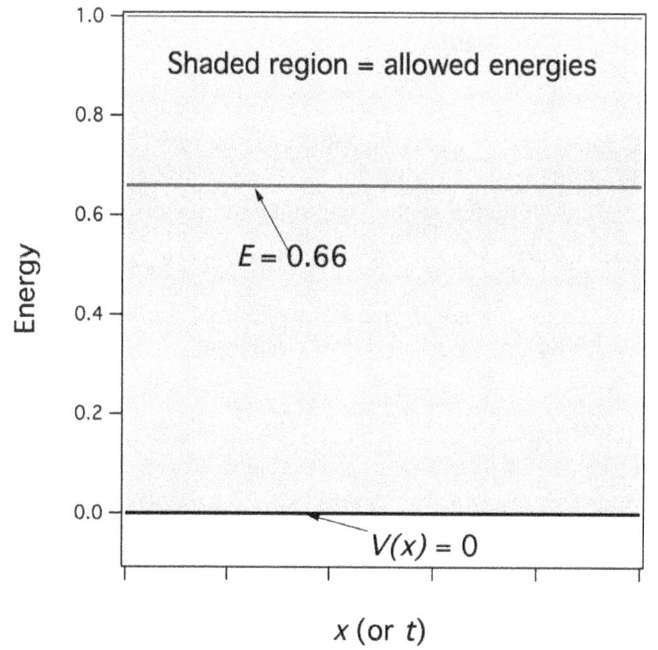

Figure 20.2 A particle moving in a potential that is constant both in space and time. Since the potential is constant, we are free to choose the value of the potential as zero. With no other interactions in the system, the energy of the particle is constant. The energy of one particular particle that has an energy of 0.66 (in units where 1 is the maximum allowed) is depicted. The shaded region, which is continuous, represents the energies that the particle may have.

20.6.2 Linear momentum of a free particle

The extreme usefulness of the wavefunction is that once we have found the correct wavefunction to describe our system by solving the Schrödinger equation, we can use this solution to calculate all other observable properties of the system. The same wavefunction that solves the Schrödinger equation also solves all other eigenvalue problems for the system. We follow the same procedure as before. Apply the appropriate operator to the wavefunction and solve the resulting eigenvalue equation for the eigenvalue.

Now let's calculate the linear momentum p_x of a free particle. We have determined that a free particle is described by the wavefunction

$$\psi = Ne^{ikx}. \tag{20.36}$$

Apply the linear momentum operator $\hat{p}_x = (\hbar/\mathrm{i})(\mathrm{d}/\mathrm{d}x)$ to the wavefunction ψ and evaluate the corresponding eigenvalue equation for the eigenvalue of linear momentum p_x

$$\hat{p}_x\psi = p_x\psi \tag{20.37}$$

$$\frac{\hbar}{\mathrm{i}}\frac{\mathrm{d}\psi}{\mathrm{d}x} = \frac{\hbar}{\mathrm{i}}\frac{\mathrm{d}}{\mathrm{d}x}Ne^{ikx} = \frac{\hbar}{\mathrm{i}}N(\mathrm{i}k)e^{ikx} = \hbar kNe^{ikx} = k\hbar\psi. \tag{20.38}$$

We confirm that the eigenvalue equation yields a constant times the original function and that the constant is equal to the linear momentum along the x axis

$$p_x = k\hbar. \tag{20.39}$$

Classically,

$$E_\mathrm{k} = p^2/2m, \tag{20.40}$$

thus,

$$E_\mathrm{k} = (k\hbar)^2/2m \tag{20.41}$$

which is precisely the answer we found in Eq. (20.35). The quantum mechanical approach leads to the same answer as classical mechanics. An unconfined particle exhibits continuous not quantized energy and linear momentum. As shown in Fig. 20.3, the energy depends quadratically on the wavevector. Classical mechanics and quantum mechanics converge on the same answers in the limit of classical behavior.

What is k? Substituting the spatial part of the wavefunction into the time-dependent expression for the wavefunction, we obtain

$$\Psi(x,t) = N\exp(\mathrm{i}[kx - \omega t]), \tag{20.42}$$

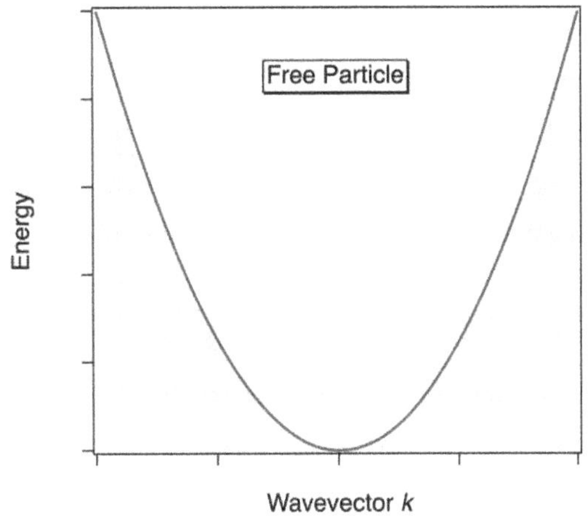

Figure 20.3 Energy versus wavevector dispersion relation. The parabolic form is characteristic of the free particle (a particle moving in a constant potential). The form of the potential changes the dispersion relation. For example, an electron moving in the periodic potential of a solid will exhibit a much different dispersion relation characteristic of the specific electronic state that it occupies. The relationship between E and k for electrons in a material can be determined by angle-resolved photoemission.

whereas before $\omega = E/\hbar$. This is a classical expression for a plane wave with wavelength λ and wavevector k (also called the circular frequency). The two are related by

$$|k| = 2\pi/\lambda. \tag{20.43}$$

A positive value of k corresponds to a particle translating to the right and a negative value of k to translation to the left. From Eq. (20.39) we know how k is related to the linear momentum p, which upon substitution leads to

$$\lambda = h/p. \tag{20.44}$$

This is, of course, the *de Broglie relation*. The wave behavior of a classical unconfined particle had been hiding in plain sight within the mathematics of classical mechanics. However, there had never been any reason to convey physical meaning to this wavelength since: (i) its magnitude is far smaller than classical objects; and (ii) they had no reason to discover the Planck constant while formulating classical mechanics.

20.6.2.1 Directed practice

Make the appropriate substitutions and derive Eq. (20.44).

20.6.3 Wavepackets

Equation (20.42) represents a 'problem' wavefunction. By defining k, we define the energy of the particle through Eq. (20.41). Now, we try to normalize the wavefunction in Eq. (20.42),

$$\int_{-\infty}^{\infty} \Psi^*\Psi \, dx = N^2 \int_{-\infty}^{\infty} dx = N^2(\infty). \tag{20.45}$$

The wavefunction is not normalizable. A free particle cannot exist in a stationary state. There is no such thing as a free particle with a definite energy. This is, of course, is a consequence of the Heisenberg uncertainty principle.

Sinusoidal waves of the type described by the wavefunctions in Eq. (20.42) or equivalently Eq. (20.36) extend to infinity and cannot be normalized. However, superpositions of such wave lead to interference and can be normalized. This is well known from Fourier analysis. We can use the techniques of Fourier analysis to define a wavepacket. A *wavepacket* is a group of wavefunctions with a range of k values – and therefore a range of energies – that represent a localized particle. Because a wavepacket has a range of component speeds, it spreads in time. We cannot precisely define its speed and position at the same time, as required by the uncertainty relationship.

The wavefunction of a wavepacket is described by

$$\Psi(x,t) = \frac{1}{\sqrt{2\pi}} \int_{-\infty}^{\infty} \phi(k) \exp\left(\mathrm{i}\left[kx - \frac{\hbar k^2}{2m}t\right]\right) dk. \tag{20.46}$$

This wavefunction can be normalized. In general, if we know the wavefunction at time $t = 0$,

$$\Psi(x,0) = \frac{1}{\sqrt{2\pi}} \int_{-\infty}^{\infty} \phi(k) e^{ikx} \, dx, \tag{20.47}$$

then by application of Plancherel's theorem in Fourier analysis, we find

$$\phi(k) = \frac{1}{\sqrt{2\pi}} \int_{-\infty}^{\infty} \Psi(x,0) \, e^{ikx} \, dx. \tag{20.48}$$

While this equation cannot be solved in general, it can be solved numerically. Alternatively, if we make a wavepacket out of a set of wavefunctions, we can use this set of equations to observe the development of the system as a function of time. The integration has to be performed numerically; however, it does provide a method to perform simulations of the dynamical development of a system. This method can be extended to realistic potentials and is not limited to free particles.

A particularly interesting form of the initial wavepacket in this respect is the Gaussian wavepacket described by

$$\Psi(x,0) = Ne^{-ax^2} \tag{20.49}$$

in which N and a are both constants and a is both real and positive. Having calculated, for instance, an excited state potential energy surface (PES) by quantum chemical calculations, it is possible to deposit a Gaussian wavepacket on this PES. Integration of the development of the wavepacket as a function of time allows the quantum chemist to track the excited state dynamics of the system. These dynamics may include photochemistry, for example.

20.7 Expectation values

Postulate VI tells us that the *expectation value* of an operator is derived from the classical value of the mean according to

$$\langle \Omega \rangle = \frac{\int_{-\infty}^{\infty} \psi^* \hat{\Omega} \psi \mathrm{d}\tau}{\int_{-\infty}^{\infty} \psi^* \psi \mathrm{d}\tau} \tag{20.50}$$

The expectation value is the weighted average value we expect to measure in the laboratory when we evaluate the observable Ω.

Nuances between the meanings of eigenvalues, expectation values, means as well as discrete and continuous distributions are shown in Fig. 20.4. An expectation value is the only value measured from a system that is prepared in an eigenstate of the operator. The probability of measuring this value is 1. If the system is prepared in a non-eigenstate, then measurement will lead to either a continuous or discrete probability distribution for the measurements. In either case, Eq. (20.50) will deliver the mean of the distribution, which is equal to the expectation value of the measurement.

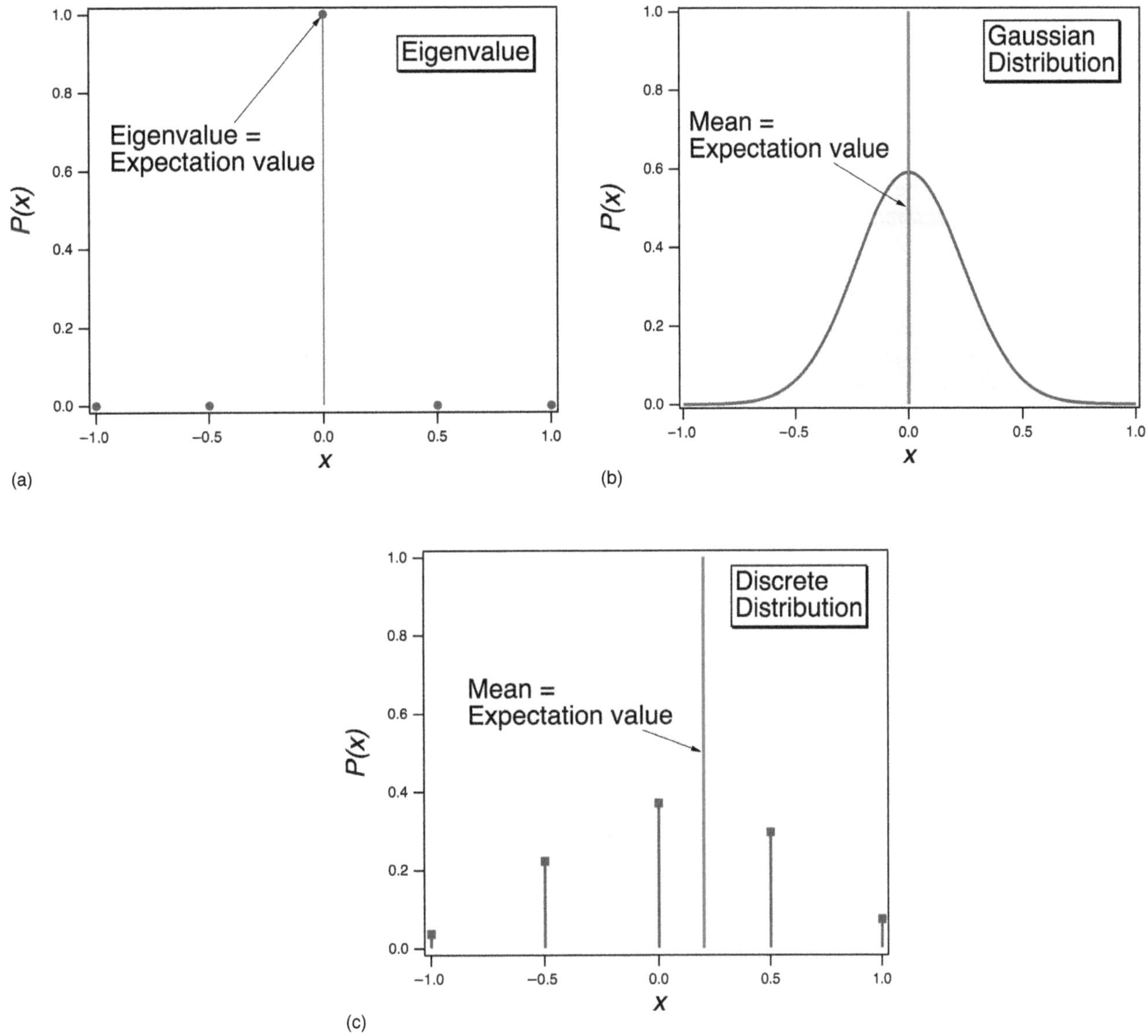

Figure 20.4 (a) When a system is in an eigenstate of the operator, the only value returned upon measurement is the eigenvalue. (b) A continuous Gaussian distribution. The mean of a symmetrical distribution is equal to the most probable value of the distribution. (c) A discrete distribution skewed toward increasing values. The mean of a skewed distribution is different than the most probable value.

20.7.1 Example

A system is in a pure state that is described by the normalized wavefunction ψ_1, with eigenvalue ω_1 for the operator $\hat{\Omega}$. What value of ω (the expectation value of $\hat{\Omega}$) will be obtained for a correctly performed experiment?

For a normalized wavefunction, the denominator in Eq. (20.50) is equal to one. The expectation value is given by

$$\omega = \langle \Omega \rangle = \int_{-\infty}^{\infty} \psi_1{}^* \hat{\Omega} \psi_1 \mathrm{d}\tau = \int_{-\infty}^{\infty} \psi_1{}^* \omega_1 \psi_1 \mathrm{d}\tau = \omega_1 \int_{-\infty}^{\infty} \psi_1{}^* \psi_1 \mathrm{d}\tau = \omega_1.$$

This is a trivial result that is also predicted by the repeatability postulate (Postulate I). When the system is in a pure state that is an eigenstate of the operator, the value of the observable is equal to the eigenvalue of the operator that corresponds to that observable. Always – for each and every measurement made on that system as long as the system is described by that same wavefunction.

20.7.2 Example

The normalized wavefunction of the ground state of a *harmonic oscillator* with reduced mass μ is

$$\psi_0 = N_0 \, e^{-\alpha^2 x^2}.$$

This wavefunction is an eigenstate of the Hamiltonian operator with eigenvalue $E_0 = \frac{1}{2}\hbar\omega$, where the angular frequency is related to the force constant k and reduced mass by $\omega = (k/\mu)^{1/2}$. The kinetic energy and potential energy operators are

$$\hat{T} = -\frac{\hbar^2}{2\mu}\frac{\mathrm{d}^2}{\mathrm{d}x^2}, \quad \hat{V} = \frac{1}{2}kx^2$$

where x is the displacement coordinate. Calculate the expectation values of the total energy, kinetic energy and potential energy of the harmonic oscillator

The Hamiltonian is the sum of the kinetic and potential energy operators,

$$\hat{H} = \hat{T} + \hat{V}.$$

We are also told that the wavefunction is an eigenstate of the Hamiltonian with eigenvalue $E_0 = \frac{1}{2}\hbar\omega$. Therefore, it satisfies the Schrödinger equation,

$$\hat{H}\psi_0 = E_0\psi_0 = \left(\tfrac{1}{2}\hbar\omega\right)\psi_0.$$

The expectation value of the total energy is obtained by integrating the product of the complex conjugate of the wavefunction with the Hamiltonian operating on wavefunction as in Eq. (20.50),

$$\langle E \rangle = \int_{-\infty}^{\infty} \psi_0^* \, \hat{H} \, \psi_0 \, \mathrm{d}x \Big/ \int_{-\infty}^{\infty} \psi_0^* \, \psi_0 \, \mathrm{d}x = \int_{-\infty}^{\infty} \psi_0^* \, \hat{H} \, \psi_0 \, \mathrm{d}x.$$

The integral in the denominator is equal to one by normalization. Since ψ_0 is an eigenfunction of the Hamiltonian, we can substitute from the line above and then solve the integral,

$$\langle E \rangle = \int_{-\infty}^{\infty} \psi_0^* \, E_0 \, \psi_0 \, \mathrm{d}x = E_0 \int_{-\infty}^{\infty} \psi_0^* \, \psi_0 \, \mathrm{d}x = E_0 = \tfrac{1}{2}\hbar\omega.$$

The expectation value is equal to the eigenvalue if the wavefunction is an eigenstate of the operator.

Now, we calculate the expectation values of the kinetic and potential energies,

$$\langle E_\mathrm{k} \rangle = \int_{-\infty}^{\infty} \psi_0^* \, \hat{T} \, \psi_0 \, \mathrm{d}x \Big/ \int_{-\infty}^{\infty} \psi_0^* \, \psi_0 \, \mathrm{d}x$$
$$\langle V \rangle = \int_{-\infty}^{\infty} \psi_0^* \, \hat{V} \, \psi_0 \, \mathrm{d}x \Big/ \int_{-\infty}^{\infty} \psi_0^* \, \psi_0 \, \mathrm{d}x. \tag{20.51}$$

The kinetic energy operator involves taking the second derivative, whereas the potential energy operator involves only multiplication. The second integral will be easier to calculate. Furthermore, we need not perform both integrations because, just as in classical mechanics, the sum of kinetic and potential energies is equal to the total energy. Since we know the total energy already, once we have the potential energy we can calculate the kinetic energy from

$$\langle E_\mathrm{k} \rangle = \langle E \rangle - \langle V \rangle.$$

Now, however, ψ_0 is not an eigenfunction of the potential energy operator, only of the total energy operator. Therefore, we must explicitly evaluate the integrals in Eq. (20.51),

$$\langle V \rangle = \frac{\int_{-\infty}^{\infty} N_0\, e^{-\alpha^2 x^2} \left(\frac{1}{2}kx^2\right) N_0\, e^{-\alpha^2 x^2}\, \mathrm{d}x}{\int_{-\infty}^{\infty} N_0\, e^{-\alpha^2 x^2} N_0\, e^{-\alpha^2 x^2}\, \mathrm{d}x}$$

Evaluating the integral in the expression for the potential energy we find

$$\langle V \rangle = \frac{k}{8\alpha^2}.$$

Substituting in for α, we find

$$\langle V \rangle = \frac{\hbar\omega}{4} = \frac{E_0}{2}$$

and, therefore, the kinetic energy is

$$\langle E_\mathrm{k} \rangle = E_0 - E_0/2 = E_0/2.$$

For the harmonic oscillator, the mean value of the potential and kinetic energies are equal. Any one measurement of the kinetic or potential energy will yield a value for each. These values will not always be the same. The kinetic and potential energies of the harmonic oscillator are continually changing. They are not eigenvalues of the wavefunction. However, their expectation value – the value obtained by averaging many measurements – is constant.

20.8 Orthonormality and superposition

Postulate VII brings with it another heavy dose of vocabulary. Eigenfunctions of a given Hamilitonian form a complete set of orthogonal, normalizable functions. They correspond to vectors in a Hilbert space. Any two good wavefunctions ψ_1 and ψ_2 can form a superposition state represented by $\psi = c_1\psi_1 + c_2\psi_2$ that is also a good wavefunction. Let's explore this in more detail.

Good operators[16] are *Hermitian*: their eigenvalues are real and their eigenvectors[17] are orthogonal. For orthogonal wavefunctions, the following integral vanishes.

Orthogonality

$$\int_{-\infty}^{\infty} \psi_k^* \, \psi_j \, \mathrm{d}\tau = 0 \quad \text{if } j \neq k. \tag{20.52}$$

A physical interpretation of this is that there is no net overlap of these wavefunctions. Good wavefunctions are also usually written to be normalized

Normalization

$$\int_{-\infty}^{\infty} \psi_k^* \, \psi_j \, \mathrm{d}\tau = 1 \quad \text{if } j = k. \tag{20.53}$$

A physical interpretation for this is that one particle in a state described by a wavefunction must have unit probability of being found in that state when all space is considered. These two statements are often combined into one expression for *orthonormality* (the state of being both orthogonal and normalized) by introducing the *Kronecker delta function* δ_{kj},

Orthonormality

$$\int_{-\infty}^{\infty} \psi_k^* \, \psi_j \, \mathrm{d}\tau = \delta_{kj} \tag{20.54}$$

where

$$\delta_{kj} = \begin{cases} 1 & \text{if } k = j \\ 0 & \text{if } k \neq j \end{cases}. \tag{20.55}$$

What does it mean that a good wavefunction can also form superposition states? A superposition is a mixture of the two eigenstates, in this case ψ_1 and ψ_2. We can construct a normalized wavefunction from a linear combination of these two functions,

$$\psi = c_1\psi_1 + c_2\psi_2 \tag{20.56}$$

where c_1 and c_2 are real numbers that describe how much of each of the component wavefunctions is in the superposition wavefunction. The wavefunction ψ is said to be a linear combination of the basis functions ψ_1 and ψ_2 as we do not allow terms in which the basis functions are raised to a power. A *basis set* is a set of function that is used as a basis for forming superpositions. This basis set forms a complete set if any and all states of the system can be represented by linear combinations of the eigenfunctions in the set. Very often, the basis functions are chosen to be an orthornormal set of functions. To ensure that the superposition is also normalized, the coefficients are subject to the constraint,

$$c_1^2 + c_2^2 = 1. \tag{20.57}$$

20.8.1 Example

Two wavefunctions ψ_1 and ψ_2 are eigenfunctions of the operator $\hat{\Omega}$ with eigenvalues ω_1 and ω_2, respectively. They form an orthonormal basis set. Calculate the expectation value associated with operator $\hat{\Omega}$ that will be obtained for an experiment performed on the superposition state described in Eq. (20.56).

The answer is obtained by substituting for the wavefunction directly into the expectation value expression.

$$\begin{aligned}
\langle\Omega\rangle &= \int_{-\infty}^{\infty} \psi^* \,\hat{\Omega}\, \psi \,\mathrm{d}\tau = \int_{-\infty}^{\infty} \left(c_1\psi_1 + c_2\psi_2\right)^* \hat{\Omega} \left(c_1\psi_1 + c_2\psi_2\right) \mathrm{d}\tau \\
&= \int_{-\infty}^{\infty} (c_1\psi_1 + c_2\psi_2)^* \left(c_1\hat{\Omega}\psi_1 + c_2\hat{\Omega}\psi_2\right) \mathrm{d}\tau \\
&= \int_{-\infty}^{\infty} (c_1\psi_1 + c_2\psi_2)^* \left(c_1\omega_1\psi_1 + c_2\omega_2\psi_2\right) \mathrm{d}\tau \\
&= c_1^2\omega_1 \int_{-\infty}^{\infty} \psi_1^*\psi_1 \,\mathrm{d}\tau + c_2^2\omega_2 \int_{-\infty}^{\infty} \psi_2^*\psi_2 \,\mathrm{d}\tau + c_2c_1\omega_1 \int_{-\infty}^{\infty} \psi_2^*\psi_1 \,\mathrm{d}\tau + c_1c_2\omega_2 \int_{-\infty}^{\infty} \psi_1^*\psi_2 \,\mathrm{d}\tau
\end{aligned}$$

Now use the orthonormality condition of Eq. (20.54) to evaluate the integrals,

$$\langle\Omega\rangle = c_1^2\omega_1(1) + c_2^2\omega_2(1) + c_2c_1\omega_1(0) + c_1c_2\omega_2(0) = c_1^2\omega_1 + c_2^2\omega_2$$

20.8.2 Example

Two wavefunctions ψ_1 and ψ_2 are orthonormal. Normalize the superposition function defined by

$$\psi = N(\psi_1 + \psi_2)$$

The normalization conditions means that

$$N^2 \int (\psi_1 + \psi_2)^*(\psi_1 + \psi_2)\,\mathrm{d}\tau = 1$$

$$\int |\psi_1|^2\,\mathrm{d}\tau + \int |\psi_2|^2\,\mathrm{d}\tau + \int \psi_1^*\psi_2\,\mathrm{d}\tau + \int \psi_2^*\psi_1\,\mathrm{d}\tau = 1/N^2$$

The first two integrals are equal to one because ψ_1 and ψ_2 are normalized. The third and fourth integrals, the cross-terms, vanishes because ψ_1 and ψ_2 are orthogonal. Therefore,

$$N = 1/\sqrt{2}$$

and the normalized wavefunction is $\frac{1}{\sqrt{2}}(\psi_1 + \psi_2)$. That is, a wavefunction formed as a linear combination of equal amounts of N basis functions has a normalization constant equal to $1/\sqrt{N}$.

In general, the basis set can contain an arbitrary number of functions and the coefficients can be complex numbers that need not be equal. This means that a general state of the system ψ can be written in terms of the linear combination of eigenfunctions ψ_n with $n = 1, 2, 3, \ldots$, as

$$\psi = \sum_n c_n \psi_n. \tag{20.58}$$

In such a general case, ψ will be normalized if the coefficients obey

$$\sum_n c_n^* c_n = 1. \tag{20.59}$$

Let's examine the result from Example 20.8.1 for this more general case. If the system is in a superposition state described by Eqs (20.58) and (20.59), and the expectation values for the operator $\hat{\Omega}$ operating on the basis functions ψ_n are ω_n, then the expectation value for the superposition state is

$$\langle \Omega \rangle = \int_{-\infty}^{\infty} \psi^* \hat{\Omega} \psi \, d\tau = \int_{-\infty}^{\infty} \left(\sum_i c_i^* \psi_i^* \right) \hat{\Omega} \left(\sum_j c_j \psi_j \right) d\tau \tag{20.60}$$

$$= \sum_i \sum_j c_i^* c_j \int_{-\infty}^{\infty} \psi_i^* \, \hat{\Omega} \, \psi_j \, d\tau \tag{20.61}$$

$$= \sum_i \sum_j c_i^* c_j \int_{-\infty}^{\infty} \psi_i^* \, \omega_j \, \psi_j \, d\tau \tag{20.62}$$

$$= \sum_i \sum_j c_i^* c_j \omega_j \int_{-\infty}^{\infty} \psi_i^* \, \psi_j \, d\tau \tag{20.63}$$

$$= \sum_i \sum_j c_i^* c_j \omega_j \delta_{ij} \tag{20.64}$$

because all of the basis functions are orthonormal. Therefore, the double sum reduces to

$$\langle \Omega \rangle = \sum_i c_i^* c_i \omega_i = \sum_i |c_i|^2 \omega_i. \tag{20.65}$$

If the system were in a pure state, say for instance ψ_1, then $c_1 = 1$ and all other $c_{i,\, i \neq 1} = 0$. Measurement always returns ω_1 and the expectation value is also equal to this eigenvalue. If the system is in a 50:50 mixture of the first two states, as in Example 20.8.2, then $c_1 = c_2 = 1/\sqrt{2}$, all other coefficients equal zero and the expectation value is $\langle \Omega \rangle = \frac{1}{2}(\omega_1 + \omega_2)$. This is the mean of the two values. The factor $c_i^* c_i$ is equal to the contribution the eigenstate ψ_i to the superposition state, which is also the weighting factor that goes into determining the mean required to obtain the expectation value.

Each measurement made on the system must return a value equal to one of the eigenvalues. We never know which of these eigenvalues will be measured. The mean of many measurements is the expectation value because the probability distribution of the measurements is determined by the admixture of eigenfunctions represented by $c_i^* c_i$.

20.8.3 Variational principle

An interesting property of the expectation value, eigenvalue equation, orthonormality and linear combinations of wavefunctions is embodied in the *variational principle*. The variational principle is a theorem that states

> *The correct wavefunction for the system yields the correct expectation value of the operator. Any other wavefunction gives a value for the expectation value that is greater than the correct value.*

Let us prove this principle using the Hamiltonian and the total energy as a concrete example. Given a Hamiltonian $\hat{H}$ that has eigenvalue (the correct value of the energy) E_j in the state described by the wavefunction ψ_j, we know that the expectation value equation yields

$$\langle E \rangle_j = \int_{-\infty}^{\infty} \psi_j^* \, \hat{H} \, \psi_j \, d\tau \Bigg/ \int_{-\infty}^{\infty} \psi_j^* \, \psi_j \, d\tau = E_j. \tag{20.66}$$

All allowed states of the system are described by orthonormal wavefunctions ψ_j, where j is an integer starting at $j = 0$, which represents the ground state (lowest energy). Take a random wavefunction ϕ that is arbitrarily defined as a linear combination of these orthonormal functions,

$$\phi = \sum_j c_j\, \psi_j. \tag{20.67}$$

Substitute this wavefunction back into the expectation value equation. Since the ψ_j are normalized, the denominator is precisely one. Combining this with the orthogonality of the wavefunctions allows us to simplify Eq. (20.66) as

$$\langle E\rangle = \int_{-\infty}^{\infty} \phi^*\, \hat{H}\, \phi\, \mathrm{d}\tau = \sum_j \sum_k c_j^*\, c_k \int_{-\infty}^{\infty} \psi_j^*\, \hat{H}\, \psi_k\, \mathrm{d}x = \sum_j c_j^*\, c_j\, E_j. \tag{20.68}$$

Subtracting the ground-state energy E_0 from both sides of this equation yields

$$\langle E\rangle - E_0 = \sum_j c_j^*\, c_j\, E_j - E_0 = \sum_j c_j^*\, c_j(E_j - E_0). \tag{20.69}$$

The last step follows from the identity $\sum_j c_j^*\, c_j = 1$. We know that E_0 is the lowest energy and must be greater than or equal to E_j. Further, the product $c_j^*\, c_j \geq 0$; therefore, the far RHS of Eq. (20.69) must also be greater than or equal to zero. This leads to

$$\langle E\rangle - E_0 \geq 0 \tag{20.70}$$

or

$$\langle E\rangle \geq E_0. \tag{20.71}$$

As was to be proved, the expectation value of an arbitrarily defined wavefunction must be greater than or equal to the lowest value for the system. This is a powerful principle and provides the basis of the variational method of calculating wavefunctions and determining eigenvalues. The variational principle tells us that, for any eigenvalue problem defined in terms of the expectation value equation, we will always approach the correct answer from above. If we arbitrarily define a wavefunction in terms of a linear combination, then the combination of coefficients that minimizes the energy gives the best approximation to the true wavefunction of that system (for the given basis functions and given Hamiltonian).

20.9 Dirac notation

The last three sections have shown us that we are going to solve a lot of integrals with a similar form. It would be nice to develop some shorthand notation to make working with these integrals easier, especially because often we do not need to actually evaluate the integral analytically to determine its value. Dirac had a particular fascination with developing his own notation – if not his own mathematics – and he created a succinct form of notation that we will use on occasion. In *Dirac notation*, we represent the normalization condition by

$$\int \psi^* \psi\, \mathrm{d}\tau = \langle\psi|\psi\rangle = 1 \tag{20.72}$$

Here, $\langle\psi|$ is called a bra and it represents the complex conjugate of the wavefunction ψ. $|\psi\rangle$ is a ket and it represents the wavefunction ψ. When the two are placed together as in $\langle\psi|\psi\rangle$, this is called a *bracket* and it represents the integral defined in Eq. (20.72). The wavefunction ψ_{nlm}, where n, l, and m are quantum numbers that define the wavefunction, is abbreviated $|nlm\rangle$.

When an operator is added to the integral it is place between the bra and the ket. For the operator $\hat{\Omega}$, the corresponding expectation value is given by

$$\langle\Omega\rangle = \frac{\langle\psi|\hat{\Omega}|\psi\rangle}{\langle\psi|\psi\rangle} \tag{20.73}$$

If the wavefunction is normalized as in Eq. (20.72), then

$$\langle\Omega\rangle = \langle\psi|\hat{\Omega}|\psi\rangle. \tag{20.74}$$

The orthonormality condition can then be represented by

$$\langle \psi_j | \psi_k \rangle = \delta_{jk} \tag{20.75}$$

20.9.1 Example

Two kets $|\psi_1\rangle$ and $|\psi_2\rangle$ are eigenfunctions of the operator $\hat{\Omega}$ with eigenvalues ω_1 and ω_2, respectively. They form an orthonormal basis set. Use Dirac notation to calculate the expectation value associated with operator $\hat{\Omega}$ that will be obtained for an experiment performed on the superposition state described in Eq. (20.56).

Translating Example 20.8.1, we have

$$\langle \Omega \rangle = \langle c_1\psi_1 + c_2\psi_2 | \hat{\Omega} | (c_1\psi_1 + c_2\psi_2) \rangle$$

$$\langle \Omega \rangle = \langle c_1\psi_1 + c_2\psi_2 | c_1\omega_1\psi_1 + c_2\omega_2\psi_2 \rangle$$

$$= c_1^2\omega_1\langle \psi_1|\psi_1\rangle + c_2^2\omega_2\langle \psi_2|\psi_2\rangle + c_2c_1\omega_1\langle \psi_2|\psi_1\rangle + c_1c_2\omega_2\langle \psi_1|\psi_2\rangle$$

$$\langle \Omega \rangle = c_1^2\omega_1 + c_2^2\omega_2$$

20.9.2 Directed practice

Rewrite the evaluation of the expectation value in Eq. (20.60) in Dirac notation.

Summarizing Dirac notation: A normalized wavefunction can be represented either as ψ or, in Dirac notation, $|\psi\rangle$. $|\psi\rangle$ is called a ket. A normalized wavefunction integrates to unity when integrated over all space. In integral notation this is represented as $\int \psi^*\psi \, d\tau = 1$. In Dirac notation this is represented $\langle \psi|\psi\rangle = 1$. An expectation value is a number. This number is the quantum mechanical analog of the classical mean. We denote an expectation value by angled brackets (i.e., $\langle \Omega \rangle$). For a normalized wavefunction the expectation value is calculated from the integral of an operator acting on a wavefunction, which is then multiplied by the complex conjugate of the wavefunction. In integral notation this is $\langle \Omega \rangle = \int \psi^* \, \hat{\Omega} \, \psi \, d\tau$. In Dirac notation this is $\langle \Omega \rangle = \langle \psi|\hat{\Omega}|\psi\rangle$.

20.10 Developing quantum intuition

Now let's turn to a few more topics that will help us to approach and understand model quantum mechanical systems and eventually atoms and molecules. These are topics that aid the development of a quantum sense of intuition. The uncertainty principle is a natural consequence of the postulates of quantum mechanics and provides a fertile basis to explore quantum logic.

20.10.1 The uncertainty principle

In classical mechanics, it is assumed that all observables can be measured precisely in all combinations. Classical operators are said to commute – that is, is does not matter in what order we apply them or in what combination, we will always get the following answer:

Classically $$\hat{\Omega}_1\hat{\Omega}_2\psi - \hat{\Omega}_2\hat{\Omega}_1\psi. \tag{20.76}$$

The commutator is defined as before by

$$[\hat{\Omega}_1, \hat{\Omega}_2] = \hat{\Omega}_1\hat{\Omega}_2 - \hat{\Omega}_2\hat{\Omega}_1. \tag{20.77}$$

The commutator of classical operators is zero,

Classically $$[\hat{\Omega}_1, \hat{\Omega}_2] = 0 \tag{20.78}$$

In quantum mechanics, the commutator is not always zero. The order of operation is important! Some operators do not commute (they are called *complementary observables* or incompatible observables), while others do commute,

Quantum Mechanically $$\hat{\Omega}_1\hat{\Omega}_2\psi \stackrel{?}{=} \hat{\Omega}_2\hat{\Omega}_1\psi. \tag{20.79}$$

Complementary observables obey an uncertainty relationship as demonstrated by Heisenberg in 1927.[18] This means that some properties of a system can be measured exactly but certain combinations of observables can only be known within limits. For example, it is impossible to specify simultaneously, with arbitrary precision, both the momentum and the position of a particle

$$\Delta p\ \Delta q \geq \hbar/2 \tag{20.80}$$

where Δp is the uncertainty in linear momentum as defined by its root mean squared (rms) deviation

$$(\Delta p)^2 = \langle(\hat{p} - \langle p\rangle)^2\rangle = \int \psi^*(\hat{p} - \langle p\rangle)^2\psi\ \mathrm{d}\tau \tag{20.81}$$

and similarly Δq is the rms deviation in position parallel to $\boldsymbol{p}$,

$$(\Delta q)^2 = \langle(\hat{q} - \langle q\rangle)^2\rangle = \langle\psi|(\hat{q} - \langle q\rangle)^2|\psi\rangle. \tag{20.82}$$

It should be clear now that the integral on the RHS of Eq. (20.81) is expressing analogous mathematics as the RHS of Eq. (20.82), but with p and q interchanged.

20.10.1.1 Example

Show that $(\Delta p)^2 = \langle(\hat{p} - \langle p\rangle)^2\rangle = \langle p^2\rangle - \langle p\rangle^2$.

Let us set up the problem in Dirac notation

$$\langle(\hat{p} - \langle p\rangle)^2\rangle = \langle\psi|(\hat{p} - \langle p\rangle)^2|\psi\rangle.$$

Now expand the squared term and simplify recalling that $\langle p\rangle$ is just a number (and therefore the order of operation is not important when multiplying by it) and that the wavefunction is taken to be normalized.

$$\begin{aligned}\langle\psi|(\hat{p} - \langle p\rangle)^2|\psi\rangle &= \langle\psi|\hat{p}^2 - \hat{p}\langle p\rangle - \langle p\rangle\hat{p} + \langle p\rangle^2|\psi\rangle \\ &= \langle\psi|\hat{p}^2|\psi\rangle - \langle\psi|\hat{p}\langle p\rangle|\psi\rangle - \langle\psi|\langle p\rangle\hat{p}|\psi\rangle + \langle\psi|\langle p\rangle^2|\psi\rangle \\ &= \langle p^2\rangle - \langle p\rangle\langle\psi|\hat{p}|\psi\rangle - \langle p\rangle\langle\psi|\hat{p}|\psi\rangle + \langle p\rangle^2\langle\psi|\psi\rangle \\ &= \langle p^2\rangle - 2\langle p\rangle\langle p\rangle + \langle p\rangle^2 \\ \langle(\hat{p} - \langle p\rangle)^2\rangle &= \langle p^2\rangle - \langle p\rangle^2 \text{ QED}\end{aligned} \tag{20.83}$$

20.10.2 Directed practice

Perform the proof in Example 20.10.1.1, but this time use integral notation.

The uncertainty principle does *not* mean that everything is uncertain. In fact, some things can be known exactly. For example, the magnitude of orbital angular momentum and its z-component can be determined exactly simultaneously because

$$\left[\hat{L}^2, \hat{L}_z\right] = 0. \tag{20.84}$$

Indeed, both of these operators also commute with the Hamiltonian. Thus, total energy, magnitude and projection of angular momentum can all be simultaneously determined in, for example, the H atom. However, there are certain questions regarding incompatible observables about which our knowledge is limited. A much better name would be the 'indeterminacy principle.' The Heisenberg relations and the Schrödinger equation define which questions can be determined exactly and which have approximate answers.

The uncertainty principle is not an assumption. Rather, it is a direct consequence of the postulates of quantum mechanics, in particular the statistical interpretation inherent in the Born rule and the definition of expectation values. The most general form of the uncertainty principle is written in terms of the expectation value of the commutator

$$(\Delta A)^2(\Delta B)^2 \geq \left(\frac{1}{2\mathrm{i}}\left\langle[\hat{A},\ \hat{B}]\right\rangle\right)^2 \tag{20.85}$$

for any pair of operators whose commutator comprises an operator of the form

$$[\hat{A},\ \hat{B}] = -\mathrm{i}\hbar. \tag{20.86}$$

Is the strange behavior observed for an electron (that the measurement perturbs the object of measurement) but not for a snooker ball (or other macroscopic object) a property of measurement or a property of the particle itself? We have already encountered this question in Section 19.5.1. The answer in quantum mechanics is both. Measurement interactions affect the particle and the accuracy of the measurement – even for macroscopic particles such as the roughly 10 kg objects that are to be used to measure gravitation waves.[19] However, the collapse of the wavefunction by the interaction of the particle with the environment is also a fundamental property of the particle and the basis of one of the postulates of quantum mechanics.[20]

20.10.3 Position–momentum uncertainty relationship

Assume that we have a particle in a state described by the wavefunction

$$\psi(x) = a^{1/2}e^{-ax}, \tag{20.87}$$

where a is a normalization constant and $0 \leq x \leq \infty$. Let us determine the expectation value of the commutator of the position and momentum operators so as to derive their uncertainty relation. The position and momentum operators are $\hat{x} = x$ and $\hat{p} = (\hbar/\mathrm{i})\mathrm{d}/\mathrm{d}x$. The commutator is evaluated more easily if we let it act on a dummy wavefunction ψ,

$$[\hat{x},\hat{p}]\psi = \left[x, \frac{\hbar}{\mathrm{i}}\frac{\mathrm{d}}{\mathrm{d}x}\right]\psi = x\frac{\hbar}{\mathrm{i}}\frac{\mathrm{d}\psi}{\mathrm{d}x} - \frac{\hbar}{\mathrm{i}}\frac{\mathrm{d}}{\mathrm{d}x}(x\psi) = \frac{\hbar}{\mathrm{i}}x\frac{\mathrm{d}\psi}{\mathrm{d}x} - \frac{\hbar}{\mathrm{i}}\psi - x\frac{\hbar}{\mathrm{i}}\frac{\mathrm{d}\psi}{\mathrm{d}x} = -\frac{\hbar}{\mathrm{i}}\psi = \mathrm{i}\hbar\psi.$$

Thus, the commutator is

$$[\hat{x},\hat{p}] = -\hbar/\mathrm{i} = \mathrm{i}\hbar \tag{20.88}$$

The expectation value of the commutator is

$$\langle [\hat{x},\hat{p}] \rangle = \int_0^\infty \psi^*\mathrm{i}\hbar\psi\,\mathrm{d}x = \int_0^\infty a^{1/2}e^{-ax}\mathrm{i}\hbar\, a^{1/2}e^{-ax}\,\mathrm{d}x = \mathrm{i}\hbar a\int_0^\infty e^{-2ax}\,\mathrm{d}x$$

$$\langle [\hat{x},\hat{p}] \rangle = \mathrm{i}\hbar a\left(\frac{e^{-2ax}}{-2a}\right)_0^\infty = \frac{1}{2}\mathrm{i}\hbar \tag{20.89}$$

20.10.3.1 Directed practice

Show that $[\hat{x},\hat{p}] = -[\hat{p},\hat{x}]$.

20.10.4 Energy–time uncertainty relationship

A similar relationship holds between energy and time.

$$\Delta E\,\Delta t \geq \hbar/2 \tag{20.90}$$

As there is no time operator it is important to realize that what we mean by time is actually lifetime, or the time it takes for the system to undergo substantial change. A more precise definition of Δt is the amount of time it takes a given expectation value of the system to change by one standard deviation.

Equation (20.90) has two immensely important consequences. The first is what is known as the *intrinsic width* of a spectral feature. A peak in spectroscopically measured data cannot be infinitely narrow. Instead, the width of the peak – which is related to how well defined the transition is in terms of energy – is ultimately determined by the lifetime of the states involved in the transition. The second consequence is that if the energy of a state is very broad, then energy conservation need only apply to an extent ΔE over short timescales Δt. In this context it is useful to convert the reduced Planck constant to units of eV s, $\hbar = 0.658$ eV fs. That is, a system can 'borrow' an eV of energy to enter an excited state, but can only do so for less than a femtosecond. The interpretation of this is somewhat controversial. Energy conservation is exact for long time scales, and quantum mechanics does not license the violation of energy conservation, particularly for long-lived states with well-defined energy. However for short-lived states with ill-defined energies, the restrictions on what processes can occur may be relaxed, but they must always fall within the constraints of the uncertainty principle.

If Δt is the lifetime of an excited state and ΔE is the uncertainty in the energy of that state, then the smaller Δt the larger ΔE. The width of a spectral line is affected by any process that alters the lifetime of an excited state, for example, collisions between molecules and chemical interaction. The study of spectral linewidths yields information about these basic processes.

20.10.5 Coordinate systems

In three dimensions, we introduce the *Laplacian operator* for shorthand

$$\hat{H} = (-\hbar^2/2m)\nabla^2 + V(x,y,z) \tag{20.91}$$

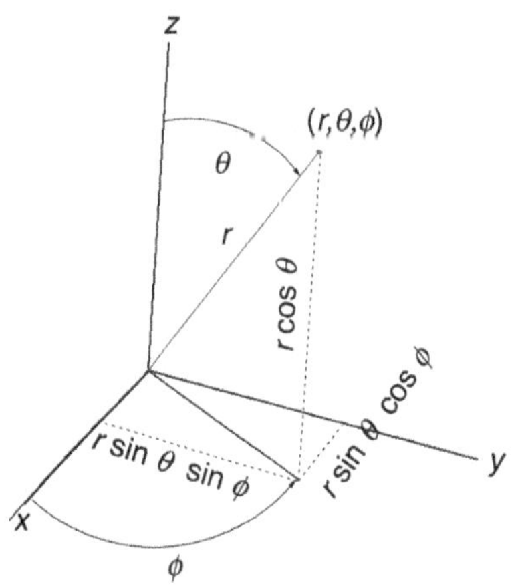

Figure 20.5 The relation of spherical polar coordinates to linear coordinates in three dimensions.

where

$$\nabla^2 = \frac{\partial^2}{\partial x^2} + \frac{\partial^2}{\partial y^2} + \frac{\partial^2}{\partial z^2}. \tag{20.92}$$

When working with a symmetry similar to that of a three-dimensional box, it makes sense to use linear coordinates x, y and z. However, the atoms are spherically symmetric so it is much easier to use the spherical coordinates r, θ, ϕ rather than the linear coordinates x, y, z to describe them. This requires a transformation to spherical polar coordinates, as shown in Fig. 20.5.

Mathematically this transformation is carried out by the following equations,

$$x = r\sin\theta\cos\phi \tag{20.93}$$
$$y = r\sin\theta\sin\phi \tag{20.94}$$
$$z = r\cos\theta. \tag{20.95}$$

The inverse transformation is

$$r = (x^2 + y^2 + z^2)^{1/2} \tag{20.96}$$
$$\theta = \cos^{-1}(z/r) \tag{20.97}$$
$$\phi = \tan^{-1}(y/x) \tag{20.98}$$

with ranges $0 \le r \le \infty$, $0 \le \theta \le \pi$, $0 \le \phi \le 2\pi$. The differential volume element is

$$\mathrm{d}x\,\mathrm{d}y\,\mathrm{d}z = r^2\sin\theta\,\mathrm{d}r\,\mathrm{d}\theta\,\mathrm{d}\phi. \tag{20.99}$$

The Laplacian then becomes

$$\nabla^2 = \frac{1}{r^2}\frac{\partial}{\partial r}\left(r^2\frac{\partial}{\partial r}\right) + \frac{1}{r^2}\left[\frac{1}{\sin\theta}\frac{\partial}{\partial\theta}\left(\sin\theta\frac{\partial}{\partial\theta}\right) + \frac{1}{\sin^2\theta}\frac{\partial^2}{\partial\phi^2}\right] \tag{20.100}$$
$$\nabla^2 = \frac{1}{r^2}\frac{\partial}{\partial r}\left(r^2\frac{\partial}{\partial r}\right) + \frac{1}{r^2}\Lambda^2 \tag{20.101}$$
$$\Lambda^2 = \frac{1}{\sin\theta}\frac{\partial}{\partial\theta}\left(\sin\theta\frac{\partial}{\partial\theta}\right) + \frac{1}{\sin^2\theta}\frac{\partial^2}{\partial\phi^2} \tag{20.102}$$

Note that the Λ^2 term, known as the *Legendrian operator*, only operates on angular variables.

Other operators include the position operator, the linear momentum operator and the angular momentum operator for the projection of angular momentum along the z axis,

$$\hat{x} = x, \quad \hat{p}_x = \frac{\hbar}{\mathrm{i}}\frac{\mathrm{d}}{\mathrm{d}x}, \quad \hat{l}_z = \frac{\hbar}{\mathrm{i}}\frac{\mathrm{d}}{\mathrm{d}\phi}.$$

More examples of operators and the conventions on notation in quantum mechanics can be found in Appendix 6.

20.10.6 Evaluation of brackets by inspection

Quantum mechanical calculations often involve turning words into calculus, and vice versa. We need to learn to read equations. This is part of the reason for the introduction of Dirac notation. Often, we will be able to evaluate the required integrals by inspection. We should always look at the integrals we need to evaluate and determine whether we can give them a numerical value without having to integrate them.

We have already encountered two examples of integrals that can be evaluated by inspection. Those are integrals of the type that are determined by the orthonormality condition,

$$\langle \psi_j | \psi_k \rangle = \delta_{jk}. \tag{20.103}$$

This integral is 1 if $j = k$ and 0 if $j \neq k$. We do not even have to write out the functions.

It is a good habit to look at integrals and always ask these two questions: (i) Are the limits symmetrical?; and (ii) is the function in the integrand even or odd? If the limits of the integral are symmetrical and the integrand is an odd function, the integral vanishes. No need to evaluate further. If a triple integral is separable into a product of three integrals, the product vanishes if any one of the three integrals is zero.

A function is even if $f(x) = f(-x)$ and odd if $f(x) = -f(-x)$. Examples of even functions are

$$\cos x,\ x^n \text{ with } n \text{ even},\ e^{x^2}.$$

Examples of odd functions are

$$\sin x,\ x^n \text{ with } n \text{ odd}.$$

Multiplication of even and odd follows the rules of multiplication of sign. Assigning even to + and odd to –, the product of even and even is even $+ \times + = +$, odd and odd is even $- \times - = +$ and cross-products are odd $+ \times - = -$. e^x is an example of a function that is neither even nor odd.

SUMMARY OF IMPORTANT EQUATIONS

$\dfrac{n_i}{n_j} = \dfrac{g_i}{g_j}\exp(-(\varepsilon_i - \varepsilon_j)/k_B T)$	Boltzmann distribution: populations n_i and n_j in state i and j, degeneracies g_i and g_j, energies ε_i and ε_j Boltzmann constant k_B, absolute temperature T
$\Psi(x,t) = \psi(x)\exp(Et/i\hbar)$	Time-dependent wavefunction $\Psi(x,t)$, wavefunction ψ, energy E
$P = \int_{\tau_1}^{\tau_2} \lvert\psi(\tau)\rvert^2 d\tau$	Probability P of finding system between τ_1 and τ_2, coordinates τ
$-\dfrac{\hbar}{2m}\dfrac{\partial\Psi(x,t)}{\partial x^2} + V(x,t)\Psi(x,t) = i\hbar\dfrac{\partial\Psi(x,t)}{\partial t}$	Time-dependent Schrödinger equation: reduced Planck constant $\hbar$, mass m, potential $V(x,t)$
$\hat{H}\psi = E\psi$	Time-independent Schrödinger equation: Hamiltonian operator $\hat{H}$
$\hat{H} = -(\hbar/2m)\nabla^2 + V(\tau)$	
$\nabla^2 = \dfrac{\partial^2}{\partial x^2} + \dfrac{\partial^2}{\partial y^2} + \dfrac{\partial^2}{\partial z^2}$	Laplacian operator in linear coordinates
$\nabla^2 = \dfrac{1}{r}\dfrac{\partial^2}{\partial r^2}r + \dfrac{1}{r^2}\Lambda^2$	Laplacian operator in spherical polar coordinates
$\Lambda^2 = \dfrac{1}{\sin^2\theta}\dfrac{\partial^2}{\partial\phi^2} + \dfrac{1}{\sin\theta}\dfrac{\partial}{\partial\theta}\sin\theta\dfrac{\partial}{\partial\theta}$	Angular portion of Laplacian, the Legendrian operator
$[\hat{A}, \hat{B}] = \hat{A}\hat{B} - \hat{B}\hat{A}$	Commutator of operators $\hat{A}$ and $\hat{B}$.
$E = h\nu = \dfrac{hc}{\lambda}$	Energy of a *photon*. *Do not use this to calculate the energy of a massive particle.*
$E_k = \dfrac{p^2}{2m} = \dfrac{(k\hbar)^2}{2m}$	Kinetic energy of a free particle E_k, linear momentum p, wavevector k
$\lambda = \dfrac{h}{p}$	de Broglie wavelength λ of a *massive* particle. *Do not use this to calculate λ of a photon.*
$\langle A\rangle = \dfrac{\int_{-\infty}^{\infty}\psi_0^*\hat{A}\psi_0\,d\tau}{\int_{-\infty}^{\infty}\psi_0^*\psi_0\,d\tau} = \dfrac{\langle\psi\lvert\hat{A}\rvert\psi\rangle}{\langle\psi\lvert\psi\rangle}$	Expectation value of operator $\hat{A}$

$\int_{-\infty}^{\infty} \psi_k^* \psi_j \, d\tau = \delta_{kj}$	Orthonormality condition
$\delta_{kj} = \begin{cases} 1 & \text{if } k = j \\ 0 & \text{if } k \neq j \end{cases}$	Kronecker delta function
$\psi = \sum_n c_n \psi_n$	Linear combination of basis functions ψ_n with coefficients c_n to form the superposition wavefunction ψ
$\Delta p \, \Delta q \geq \hbar/2$	Uncertainty relationships for the rms deviation in linear momentum and position
$\Delta E \, \Delta t \geq \hbar/2$	Uncertainty relationships for the rms deviation in energy and lifetime
$(\Delta A)^2 (\Delta B)^2 \geq \left(\frac{1}{2i} \langle [\hat{A}, \hat{B}] \rangle \right)^2$	Generalized form of Heisenberg uncertainty relations
$x = r \sin\theta \cos\phi$ $y = r \sin\theta \sin\phi$ $z = r \cos\theta$	Transformation of linear coordinates to spherical polar coordinates

Exercises

20.1 According to the Copenhagen interpretation, what is the meaning of the wavefunction and the wavefunction squared?

20.2 State the Heisenberg uncertainty principle and its implications.

20.3 Describe two conditions in which quantum behavior is likely to be observed.

20.4 Write out the Schrödinger equation, defining each term in general.

20.5 Why does zero point energy arise?

20.6 Why does a quantum mechanical system with discrete energy levels behave as if it has a continuous energy spectrum if the energy difference between energy levels ΔE satisfies the relationship $\Delta E \ll k_B T$?

20.7 Calculate the normalization constant for the free particle wavefunction.

20.8 Show that $\psi = N(e^{ikx} + e^{-ikx})$ is an acceptable wavefunction for the free particle.

20.9 Consider a particle moving unconfined in a one-dimensional region of space with constant potential energy V. Solve the Schrödinger equation and use the result to derive the de Broglie relationship for the wavelength of a particle.

20.10 Show that the function $A \exp(-2ax)$ is an eigenfunction of the operator d^2/dx^2 and determine its eigenvalue.

20.11 Two (unnormalized) excited state wavefunctions of the H atom are

$$\psi = \left(2 - \frac{r}{a_0}\right) e^{-r/2a_0} \quad \text{(i)}$$

and $$\psi = r \sin\theta \cos\phi e^{-r/2a_0}. \quad \text{(ii)}$$

(a) Normalize both wavefunctions to 1. (b) Confirm that these two functions are mutually orthogonal.

20.12 Assume that the three real function ψ_1, ψ_2, and ψ_3 are normalized and orthogonal. Normalize the following functions.

(a) $\psi = N(\psi_1 + \psi_2)$ (b) $\psi = N(\psi_1 - \psi_2)$ (c) $\psi = N(\psi_1 + \psi_2 + \psi_3)$

20.13 Determine the commutators of: (i) the operators $(\hbar/i)d/dx$ and $1/x$, (ii) the operators $(\hbar/i)d/dx$ and x^2.

20.14 Operator $\hat{\Omega}$ returns the value $\omega_1 = 6\hbar$ when it operates on wavefunction ψ_1 and $\omega_2 = 15\hbar$ when it operates on ψ_2. If the system is in a superposition state described by

$$\psi = c_1 \psi_1 + c_2 \psi_2$$

$\hat{\Omega}$ has an expectation value $\omega = 11\hbar$. Assuming that ψ_1 and ψ_2 form an orthonormal basis set and with the condition

$$c_1^2 + c_2^2 = 1,$$

(this condition ensures that the superposition wavefunction is also normalized), determine the values of c_1 and c_2.

20.15 ψ_1 and ψ_2 are orthonormal functions with eigenvalues ω_1 and ω_2, respectively, when operated on by $\hat{l}_z$. They form a complete set and two normalized superposition states are constructed from these,

$$\psi_A = c_1\psi_1 + c_2\psi_2, \quad \psi_B = c_3\psi_1 + c_4\psi_2.$$

ψ_A contains 2/3 of ψ_1. The operator for the projection of the angular momentum is $\hat{l}_z = \frac{\hbar}{\mathrm{i}}\frac{\mathrm{d}}{\mathrm{d}\varphi}$. Its expectation values for the superposition states are $\langle\omega_A\rangle = 25\hbar$ and $\langle\omega_B\rangle = 35\hbar$. Calculate values for c_1, c_2, c_3, c_4, ω_1 and ω_2.

20.16 Use the eigenfunction $\psi(x) = Ae^{\mathrm{i}kx} + Be^{-\mathrm{i}kx}$ rather than $\psi(x) = C\sin kx + D\cos kx$ to apply the boundary conditions for the particle in the box. (a) Use the boundary conditions determine the acceptable values of A, B and k. (b) Do these two functions give different probability densities if each is normalized?

20.17 The operator for the linear momentum is $\hat{p} = \frac{\hbar}{\mathrm{i}}\frac{\mathrm{d}}{\mathrm{d}x}$. The normalized wavefunction for the particle in a box can be written

$$\psi(x) = A(e^{\mathrm{i}kx} - e^{-\mathrm{i}kx}).$$

Calculate the expectation value of the linear momentum of a particle in a box.

20.18 Calculate the energy separations in J, kJ mol^{-1}, eV and cm^{-1} between the levels (a) $n = 6$ and $n = 7$ of an electron in a box of length $L = 1.5$ nm. (b) What wavelength of photon is absorbed when the electron is excited from the $n = 6$ to the $n = 7$ level? (c) An electron can scatter off of a particle, transfer its kinetic energy and cause a transition much like a photon can. Calculate the maximum wavelength of an electron that can cause a transition from the $n = 6$ to the $n = 7$ state.

20.19 What are the most likely locations of a particle in a box of length L in the $n = 4$ state?

20.20 The wavefunction for a particle confined in a one-dimensional box of length L is

$$\psi_n(x) = \left(\frac{2}{L}\right)^{1/2} \sin\left(\frac{n\pi x}{L}\right).$$

Take $L = 3$ nm. Calculate the probability that a ground state electron (that is, an electron in the $n = 1$ state) is between $x = 1.95$ nm and 2.05 nm.

20.21 Calculate (a) the mean potential energy and (b) the mean kinetic energy of an electron in the ground state of a hydrogenic atom. Recall that the mean value corresponds to an expectation value. The ground-state wavefunction is

$$\psi = \left(\frac{1}{\pi a_0^3}\right)^{1/2} e^{-r/a_0}.$$

20.22 The operator for the projection of the angular momentum is $\hat{l}_z = \frac{\hbar}{\mathrm{i}}\frac{\mathrm{d}}{\mathrm{d}\varphi}$ and its eigenvalue is denoted m_l. The wavefunctions of one of the $2d$ orbitals ψ_{21-1} and a $1s$ orbital ψ_{100} are given by

$$\psi_{21-1} = (8\sqrt{\pi})^{-1}\left(\frac{1}{a_0}\right)^{3/2}\left(\frac{r}{a_0}\right)e^{-r/2a_0}\sin\theta\, e^{-\mathrm{i}\phi} \quad \text{and} \quad \psi_{100} = (\sqrt{\pi})^{-1}(1/a_0)^{3/2}e^{-r/a_0}$$

Show that these wavefunctions are eigenfunctions of the operator $\hat{l}_z$. Calculate the eigenvalues. Set up the definite integral in spherical polar coordinates that shows that these wavefunctions are orthogonal. Simplify them sufficiently and explain why the integral vanishes. You do not need to evaluate the integral analytically, use reasoning.

Endnotes

1. Zeilinger, A. (2010) *Dance of the Photons: From Einstein to Quantum Teleportation*. Farrar, Straus and Giroux, New York. Zeilinger, A. (2007) *Einsteins Spuk*. Goldmann Verlag, München.
2. Schrödinger, E. (1935) *Proc. Cambridge Philos. Soc.*, **31**, 555.
3. Einstein, A., Podolsky, B., and Rosen, N. (1935) *Phys. Rev.*, **47**, 777.
4. Bell, J.S. (1987) *Speakable and Unspeakable in Quantum Mechanics*. Cambridge University Press, Cambridge.
5. Griffiths, D.J. (2005) *Introduction to Quantum Mechanics*. Pearson Education, Upper Saddle River, NJ.
6. Sarovar, M., Ishizaki, A., Fleming, G.R., and Whaley, K.B. (2010) *Nat. Phys.*, **6**, 462.
7. Born, M. (1926) *Z. Phys.*, **37**, 863; Born, M. (1926) *Z. Phys.*, **38**, 803.
8. We see in Section 20.6.3 that a free particle challenges this definition of a good wavefunction as far as normalization goes.
9. A good wavefunction must be finite and continuous everywhere the particle is allowed to be, as must be its first partial derivatives. This is a mathematical way to say the question is moot. More fundamentally, the uncertainty principle also tells us the question is moot. We cannot know where the particle is and what state it is in exactly, other than to say it is in the box. The question 'where is the particle exactly' is classical; however, the set of questions that can be asked is not the same as the set of questions that have physically meaningful and quantum mechanically justifiable answers.
10. Schrödinger, E. (1926) *Phys. Rev.*, **28**, 1049.
11. Born, M. (1955) *Science*, **122**, 675. www.nobelprize.org/nobel_prizes/physics/laureates/1954/born-lecture.html
12. Heisenberg, W. (1925) *Z. Phys.*, **33**, 879.
13. Dirac, P.A.M. (1925) *Proc. R. Soc. London, A*, **109**, 642.
14. Brewing coffee may be required for each step. Substitution of other beverages is possible, results may vary.
15. Alternatively, we could choose k to be only positive in which case $\psi = Ae^{ikx} + Be^{-ikx}$ would have to be chosen for the trial wavefunction to account for motion to the right ($+k$) and motion to the left ($-k$). When passage above or through barriers is possible, there are advantages to this more general formulation.
16. 'Good' in this context means operators that correspond to physical observables.
17. Eigenvector and eigenfunction are synonymous in quantum mechanics. As we saw in the previous section, eigenfunctions are a particular type of wavefunction.
18. Heisenberg, W. (1927) *Z. Phys.*, **43**, 172.
19. Purdy, T.P., Peterson, R.W., and Regal, C.A. (2013) *Science*, **339**, 801.
20. Bassi, A., Lochan, K., Satin, S., Singh, T.P., and Ulbricht, H. (2013) *Rev. Mod. Phys.*, **85**, 471.

Further reading

Atkins, P.W. and Friedman, R.S. (2010) *Molecular Quantum Mechanics*, 5th edition. Oxford University Press, Oxford.
Griffiths, D.J. (2005) *Introduction to Quantum Mechanics*. Pearson Education, Upper Saddle River, NJ.
Heitler, W. (2010) *The Quantum Theory of Radiation*, 3rd edition. Dover Publications, New York.
Metiu, H. (2006) *Physical Chemistry: Quantum Mechanics*. Taylor & Francis, New York.
Schatz, G.C. and Ratner, M.A. (2002) *Quantum Mechanics in Chemistry*. Dover Publications, New York.
Winn, J.S. (1995) *Physical Chemistry*. HarperCollins College Publishers, New York.

CHAPTER 21

Model quantum systems

PREVIEW OF IMPORTANT CONCEPTS

- Confinement in one dimension requires one quantum number to determine the energy. Confinement in three dimensions requires three quantum numbers.
- Confinement leads to zero point energy in the lowest energy level.
- More nodes in a wavefunction are correlated with higher energy.
- As the de Broglie wavelength of a particle approaches the size of the container formed by the potential that confines it, the more important quantum confinement effects become.
- When the Hamiltonian can be written as the sum of terms, each of which depends only on a set of coordinates unique to that term, the corresponding wavefunction can be written as the product of component wavefunctions, each of which depends only on the corresponding set of coordinates.
- When barriers are not infinitely high, particles can penetrate into classically forbidden regions.
- Tunneling occurs when particles with an energy less than the barrier height are able to pass through the barrier.
- Tunneling is favored by low-mass, thin, smooth barriers and energy close to the top of the barrier. Tunneling exhibits an exponential distance dependence.
- Most potentials are roughly harmonic near a local minimum.
- The energy levels of a harmonic oscillator are evenly spaced with alternating even and odd symmetry wavefunctions.
- The ground state of the harmonic oscillator has a zero-point energy of half the level spacing.
- All angular momenta share a common algebra. The form to their eigenvalue equations and pattern of degeneracy is also the same.
- The Hamiltonian commutes with the magnitude and z-component of angular momentum, but only one component of angular momentum can be specified at a time.
- The solution of the particle on a sphere problem allows us to understand the energy level structure of rotational motion as well as the angular dependence of electron orbitals.
- Spin is a relativistic effect that arises naturally when quantization is performed with time treated equivalently to the three spatial dimensions.
- In a classical body, both orbital and spin angular momenta are functions of r, θ, and ϕ. In quantum mechanics, orbital angular momentum is a function of r, θ, and ϕ; however, spin is not a function of r, θ, and ϕ.
- Spin is an intrinsic property of elementary particles. While its projection can be flipped, it magnitude cannot change without changing the particle. Composite particles, such as nuclei, can have excited states with different values of spin.
- Spin interacts with a magnetic field, which is the basis of electron spin resonance and nuclear magnetic resonance spectroscopies.

There are two different aspects of this chapter to emphasize. One is that it provides practice in the application of quantum mechanical methods to model physical systems. We learn how to apply the mathematics of quantum mechanics to the solution of physical problems. We learn how to interpret quantum mechanical results and behavior. But more than that, the systems presented here also represent the first-order approximations to the behavior of a wide variety of actual physical systems. They are our first attempts at developing the language that will allow us to describe quantum mechanical phenomena. Applying perturbations to these model systems quite frequently allows us to model and understand the behavior of real systems with terms familiar to us from the classical world.

We have already treated one model system – the free particle – in detail in Chapter 20. We also encountered the particle in a box and the harmonic oscillator in Chapter 20. The model atomic system, the H atom, is treated in Chapter 22. For

Physical Chemistry: How Chemistry Works, First Edition. Kurt W. Kolasinski.

Companion Website: www.wiley.com/go/kolasinski/physicalchemistry

the H atom problem, we will see how we can exploit similarities in the physics of models of rotational motion to solving the energy level structure of atoms.

21.1 Particle in a box

21.1.1 One-dimensional box

Consider a particle confined to a box where the potential energy is $V(x)$ inside the box and goes to infinity outside the box. The width of the box is given the value a. The particle cannot penetrate the walls of the box because its potential energy is infinite there. If the walls are not infinitely tall, some penetration of the walls will occur. This is the origin of tunneling, which we will investigate below. This *particle in a box* problem is also known as the *infinite square well.*

We begin by writing the Hamiltonian,

$$\hat{H} = -\frac{\hbar^2}{2m}\frac{d^2}{dx^2} + V(x) \tag{21.1}$$

where $V(x) = 0$ for $0 \le x \le a$ and $V(x) = \infty$ elsewhere. This potential confines the particle to be inside the box, which forces the wavefunction to go to zero at the walls and to vanish everywhere outside of the box. In the box the Schrödinger equation is

$$\hat{H}\psi = -\frac{\hbar^2}{2m}\frac{d^2\psi}{dx^2} = E\psi \tag{21.2}$$

To solve Eq. (21.2), we try inserting a general trial solution for $\psi(x)$,

$$\psi(x) = A\sin(kx) + B\cos(kx). \tag{21.3}$$

We simplify this solution further by considering the boundary conditions,

$$\psi(0) = \psi(a) = 0. \tag{21.4}$$

That the wavefunction goes to zero at the walls is clear in Fig. 21.1. Thus, at $x = 0$,

$$\psi(x = 0) = A\sin(0) + B\cos(0) = 0 \tag{21.5}$$

which requires that $B = 0$. At $x = a$,

$$\psi(x = a) = A\sin(ka) = 0, \tag{21.6}$$

this means that $ka = n\pi$ thus $k = n\pi/a$, where n is an integer.

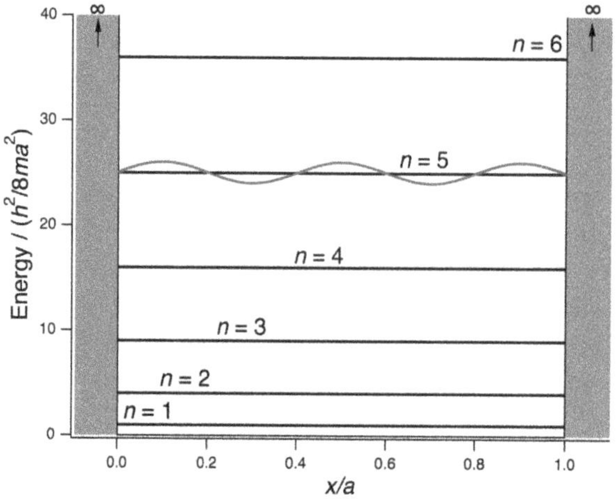

Figure 21.1 Potential energy diagram of the particle in a one-dimensional box with infinitely high walls and box length a. Classically all values of energy are allowed, but quantum mechanically only quantized energy levels are allowed. The energy of these levels increases with n^2, where n is a quantum number that defines the state. The first six allowed levels are included. The wavefunction of the $n = 5$ level is also shown. The wavefunction goes to zero at the walls of the box and is identically zero outside of the box.

We use this infinite set of wavefunctions to solve the Schrödinger equation, which reveals a formula for the energies of the levels associated with each wavefunction.

$$-\frac{\hbar^2}{2m}\frac{d^2}{dx^2}\left[A\sin\left(\frac{n\pi x}{a}\right)\right] = E_n A\sin\left(\frac{n\pi x}{a}\right) \tag{21.7}$$

This has the solution

$$E_n = \frac{n^2h^2}{8ma^2}, \quad n = 1, 2, 3 \cdots. \tag{21.8}$$

The allowed energies are not continuous. Only certain values are allowed. That is, now that the particle is confined to the box rather than translating freely, the energy is quantized.

To find the complete form of the wavefunction, we must normalize our trial solution. Normalization ensures that the probability of finding the particle in the box is exactly one.

$$1 = \int_0^a \psi^*\psi dx = \int_0^a A^2\sin^2\left(\frac{n\pi x}{a}\right)dx = \frac{1}{2}aA^2 \tag{21.9}$$

$$A = \sqrt{2/a} \tag{21.10}$$

The trial solution reduces to

$$\psi_n(x) = \sqrt{2/a}\sin(n\pi x/a) \tag{21.11}$$

and the complete set of normalized wavefunction is $\psi_n(x) = \sqrt{2/a}\sin(n\pi x/a)$ for $0 \le x \le a$ and $\psi_n(x) = 0$ otherwise. Note that this is a case where the derivative of the wavefunction is not continuous in two spots, namely, at $x = 0$ and $x = a$. Nonetheless, the wavefunction is continuous at these points and the wavefunction is a good wavefunction for describing the particle *in* the box. There is no wavefunction outside the box so that there is nothing to describe there.

Notice a few characteristics of this solution:

- Confinement along one coordinate in the problem leads to one quantum number (n) being required to define the energy.
- There are an infinite number of wavefunctions, the quantum number n ranges from 1 to ∞.
- Each wavefunction has a different energy and each energy corresponds to only one wavefunction. This means that the eigenstates are non-degenerate.
- The lowest energy state (the ground state) is not at $E = 0$; instead, it has a finite zero point energy (ZPE) associated with it.
- There is zero probability of finding the particle at the nodes.
- Higher energy corresponds to a greater number of nodes. This is true for any potential.
- The wavefunction changes sign at nodes.
- Wavefunctions are symmetrical about the middle of the box and alternate being even and odd with respect to the center of the well. This is true for any symmetric potential.
- The overlap between the different wavefunctions is zero; they are orthogonal.

Orthogonality means that the wavefunctions corresponding to different quantum numbers (i.e., n = 1, 2, 3...) do not mix with each other. Mathematically, *orthogonality* is expressed by the following equation:[1]

$$\int_0^a \psi_i^*(x)\,\psi_j(x)\,dx = 0 \quad \text{when} \quad i \neq j. \tag{21.12}$$

21.1.1.1 Example

Find the wavelength of light emitted when a 1×10^{-27} g particle in a 3-Å one-dimensional box goes from the $n = 2$ to the $n = 1$ level.

The wavelength λ can be found from the frequency ν. The quantity $h\nu$ is the energy of the emitted photon and equals the energy difference between the two levels involved in the transition

$$h\nu = E_{\text{upper}} - E_{\text{lower}} = \frac{2^2h^2}{8ma^2} - \frac{1^2h^2}{8ma^2} \Rightarrow \nu = \frac{3h}{8ma^2}$$

Then, from $\lambda = c/\nu$,

$$\lambda = \frac{8ma^2c}{3h} = 109\,\text{nm}.$$

This wavelength is in the UV.

21.1.1.2 Directed practice

Confirm that the wavefunction in Eq. (21.11) satisfies the Schrödinger equation inside of the box.

21.1.2 Three-dimensional box

Perhaps only a string theorist could consider a one-dimensional box a box. Boxes in the macroscopic world correspond to three-dimensional boxes. How does the problem change should it be extended to three dimensions? Let's first assume that our box has edges placed along the Cartesian coordinate axes x, y, and z with edge length L_x, L_y, and L_z. The origin is placed at one of the vertices of this box. The potential is zero inside the box and infinite outside of it, such that

$$\begin{aligned} \hat{V} &= 0 \quad \text{for } 0 \le x \le L_x, \quad 0 \le y \le L_y \quad \text{and} \quad 0 \le z \le L_z \\ \hat{V} &= \infty \quad \text{elsewhere.} \end{aligned} \tag{21.13}$$

Our particle has mass m. Its kinetic energy is equal to the total energy of the system; thus the Hamiltonian is

$$\hat{H} = \hat{T} + \hat{V} = -\frac{\hbar^2}{2m}\left(\frac{\partial^2}{\partial x^2} + \frac{\partial^2}{\partial y^2} + \frac{\partial^2}{\partial z^2}\right) = \frac{\hbar^2}{2m}\nabla^2 \tag{21.14}$$

and the Schrödinger equation is

$$-\frac{\hbar^2}{2m}\nabla^2\psi = E\psi \tag{21.15}$$

where ∇^2 is the Laplacian operator introduced in the last chapter. To solve this problem we can again introduce separation of variables. The Hamiltonian contains no cross-terms, that is, no terms containing x and y or any other combination of coordinates. Thus, it can be written as the sum of three terms, each involving only one coordinate,

$$\hat{H} = \hat{H}_x + \hat{H}_y + \hat{H}_z \tag{21.16}$$

where

$$\hat{H}_\xi = -\frac{\hbar^2}{2m}\frac{\partial^2}{\partial \xi^2} \tag{21.17}$$

with $\xi = x$, y or z as appropriate.

Each term in the Hamiltonian only acts on one coordinate. We found that when we applied separation of variables to the time-dependent Schrödinger equation, we obtained a product wavefunction with one term for each set of separable variables (space and time in that case). This suggests that our wavefunction for the three-dimensional box should also be a product wavefunction of the form

$$\psi(x, y, z) = \psi_x(x)\psi_y(y)\psi_z(z) \tag{21.18}$$

The solution of the Schrödinger equation for any one of the component terms in the Hamiltonian is of exactly the same form as the solution of the one-dimensional problem in the previous section. For example,

$$\hat{H}_x\psi_x = -\frac{\hbar^2}{2m}\frac{\partial^2\psi_x}{\partial x^2} = E_x\psi_x \tag{21.19}$$

Furthermore, since a component term only acts on one coordinate, the other coordinates can be treated as constants for that operator. For example,

$$\hat{H}_x\psi = -\frac{\hbar^2}{2m}\frac{\partial^2(\psi_x\psi_y\psi_z)}{\partial x^2} = -\frac{\hbar^2}{2m}\frac{\mathrm{d}^2\psi_x}{\mathrm{d}x^2}(\psi_y\psi_z) = E_x\psi_x(\psi_y\psi_z) \tag{21.20}$$

The energies and wavefunctions for the y- and z-components must have the same behavior. Substituting this into the Schrödinger equation, we find that the total energy must simply be the sum of the x-, y- and z-contributions

$$E = E_x + E_y + E_z. \tag{21.21}$$

21.1.2.1 Directed practice

Substitute into the Schrödinger equation and prove that the energy of the three-dimensional box is equal to the sum of the x, y and z energy components.

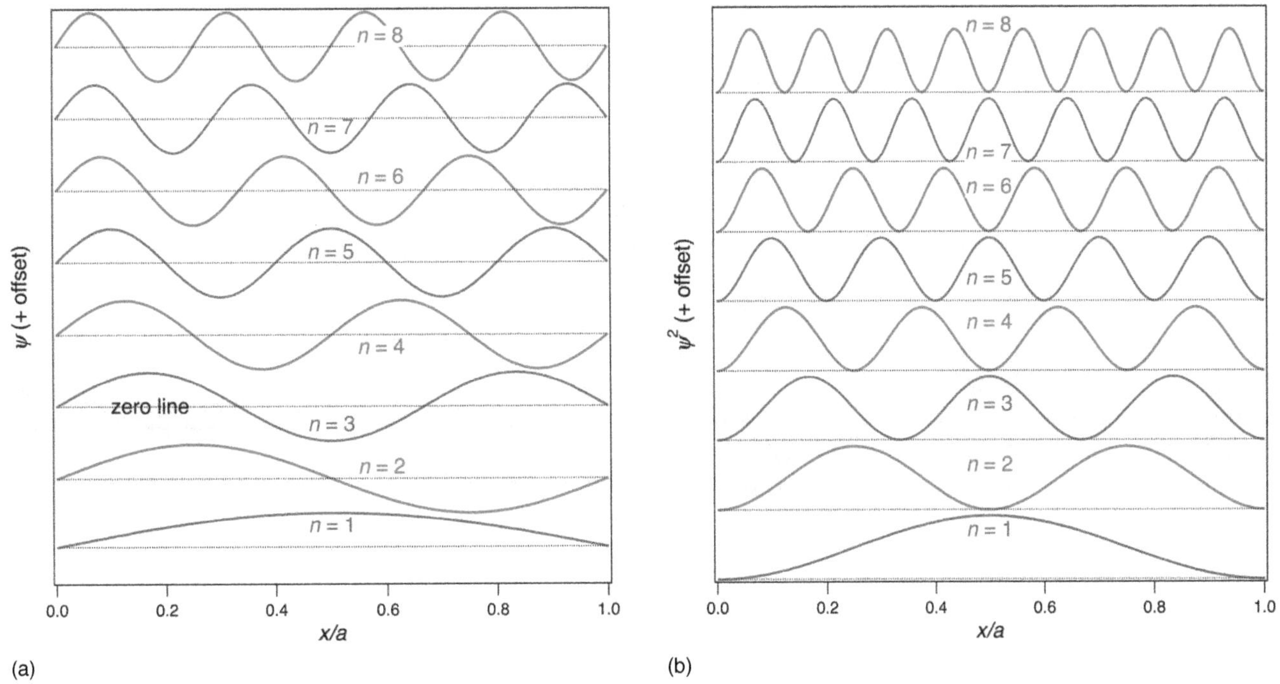

Figure 21.2 (a) Wavefunctions and (b) the square of the wavefunction for the particle in a box from $n = 1$ to $n = 8$. Note how the symmetry about the center of the box at $x/a = 0.5$ alternates from even to odd as n increases. The wavefunction changes sign about either side of a node. The number of extrema equals n and the number of nodes is $n - 1$.

If our particle were confined to a surface, the box would be two-dimensional. The separation of variables treatment used above would also apply. This argument can be generalized to an arbitrary number of dimensions. This also allows us to formulate a general principle

When the Hamiltonian can be written as the sum of terms, each of which depends only on a set of coordinates unique to that term, the corresponding wavefunction can be written as the product of component wavefunctions, each of which depends only on the corresponding set of coordinates.

Since each contribution in Eq.(21.21) is of the form $E_{\xi,n} = n_\xi^2 h^2/8mL_\xi^2$, the total energy of the three-dimensional particle in a box is

$$E_n = \frac{h^2}{8m}\left(\frac{n_x^2}{L_x^2} + \frac{n_y^2}{L_y^2} + \frac{n_z^2}{L_z^2}\right) \tag{21.22}$$

Three independent integers now specify each level. Each starts at $n_\xi = 1$; thus, there is now a zeropoint energy associated with each coordinate. Specification of three independent integers also introduces the possibility of degeneracy of levels, that is, more than one state existing at the same energy. If the three lengths are all different, this degeneracy would be accidental. If, on the other hand, the box is cubic and $L = L_x = L_y = L_z$, then degeneracy will occur whenever the sum of the squares of the n_ξ values is equal. Thus, the state defined by $(n_x, n_y, n_z) = (1, 2, 3)$ is degenerate to the other five states defined by the permutations of 1, 2, and 3. In addition $1^2 + 1^2 + 5^2 = 27$ just as $3^2 + 3^2 + 3^2 = 27$; therefore, these states as well as the states represented by permutations of 1, 1, and 5 are all degenerate.

21.1.3 Quantum confinement and the correspondence principle

A particle in a cubic box of edge length L has energy levels given by

$$E_n = \frac{h^2}{8mL^2}\left(n_x^2 + n_y^2 + n_z^2\right). \tag{21.23}$$

For spherical confinement instead of cubic in which the particles are confined to executing circular orbits on the surface of a sphere with radius r, the energy levels are given by

$$E_l = \frac{l\,(l+1)\,\hbar^2}{2mr^2}. \tag{21.24}$$

Thus, the spacing between successive energy levels is given by

$$E(l+1) - E(l) = \frac{(l+1)\hbar^2}{m}\frac{1}{r^2}. \quad (21.25)$$

You will see below that this corresponds to a model of rotational motion and the angular dependence of electrons trapped in the potential of a nucleus; hence, the switch to the quantum number l.

The form of the equations changes based on the geometry. If the box were rod-like or ellipsoidal, the equations would be slightly different. However, notice the similarities: (i) the energy level spacing is inversely proportional to length squared; (ii) it is also inversely proportional to the mass; and (iii) in the limit of large quantum numbers, the spacing between successive energy levels becomes negligible.

The emergence of significant quantum mechanical behavior with decreasing size is known as *quantum confinement*. More precisely, as the de Broglie wavelength of a particle approaches the size of the container formed by the potential that confines it, the more important quantum effects become. Quantum confinement is one of the fundamental phenomena underpinning nanoscience and nanotechnology.

In solid-state systems, an electron excited across a band gap to the conduction band can interact with the hole it left behind in the valence band to form a collective excitation (quasi-particle) known as an *exciton*. An exciton is a pseudo-hydrogenic system that has a characteristic exciton Bohr radius

$$a = 4\pi\varepsilon_0\hbar^2/\mu e^2 \quad (21.26)$$

where μ is the reduced mass of the exciton. For reasons related to the band structure of the material – that is, the potential that binds the electrons – the effective mass of the electron and the hole do not equal the mass of a free electron. The exciton radius of Si is 4.9 nm, that of GaAs is 14 nm, and that of InSb is 69 nm. When the size of a semiconductor nanoparticle approaches and then becomes less than the exciton radius, quantum confinement effects set in. As demonstrated by Leigh Canham,[2] when silicon nanoparticles become smaller than about 5 nm, the properties of the nanoscale Si particles change dramatically: the particles become visibly luminescent (as shown in Fig. 21.3) and the wavelength of the luminescence shifts to the blue (higher energy) as the size of the particles become smaller. Theoretical calculations made by Birgitta Whaley's group[3] show that the increase of the band gap resulting from the shift of the conduction band upward and the valence band downward can explain the shift of the photoluminescence with decreasing

Figure 21.3 Photoluminescence emitted from nanocrystalline porous silicon under excitation with ultraviolet light. The crystal in the upper right corner is bulk crystalline Si. It does not luminesce. The other crystals are covered with layers of nanocrystalline porous silicon produced under different conditions to achieve different crystallite sizes. The luminescence spectrum of the freshly produced nanocrystals shifts to the blue (from red to yellow to green) with smaller size.[4]

nanocrystal size. The size dependence is much like that expected from the application of Eq. (21.24). Other properties, such as the absorption spectrum, phonon spectrum (how the solid vibrates), thermal and electrical conductivity, and chemical reactivity also become size-dependent. Such quantum-confined nanoparticles are known as *quantum dots*.

A quantum dot is a system that is confined in three dimensions. Therefore, it has zero degrees of freedom and a quantum dot is a zero-dimensional (OD) system. Quantum dots have also been called artificial atoms because they have energy level structures that are analogous to those of atoms. They have discrete energy levels with sharp energies. A quantum wire is confined in two dimensions and represents a one-dimensional (1D) system. A quantum well is confined in one dimension and is a 2D system. One-, two-, and three-dimensional structures exhibit energy bands rather than discrete states.

The antithesis of quantum confinement is Bohr's correspondence principle.[5] When r is large enough the energy levels are spaced continuously. The system evolves toward classical behavior as the system leaves the nanoscale and enters the microscale. In the limit of large radii, large quantum numbers and large masses quantum mechanical behavior converges on classical behavior.

21.2 Quantum tunneling

In classical mechanics a particle with energy E encountering a barrier of height V_0 cannot surmount the barrier unless it has sufficient energy to do so, $E > V_0$. But the rules of classical mechanics do not apply to quantum particles. In quantum mechanics, subatomic (and sometimes larger) particles can overcome a barrier even if they do not have sufficient energy to overcome it. This process, shown schematically in Fig. 21.4 is known as quantum mechanical *tunneling*. Tunneling is involved in the explanation of the radioactivity of atomic nuclei by alpha particle decay. It can contribute to kinetic isotope effects, especially in proton-transfer reactions, including those catalyzed by enzymes. Tunneling is also the basis of the technique known as scanning tunneling microscopy (STM), which allows us to 'see' atoms. More accurately, it

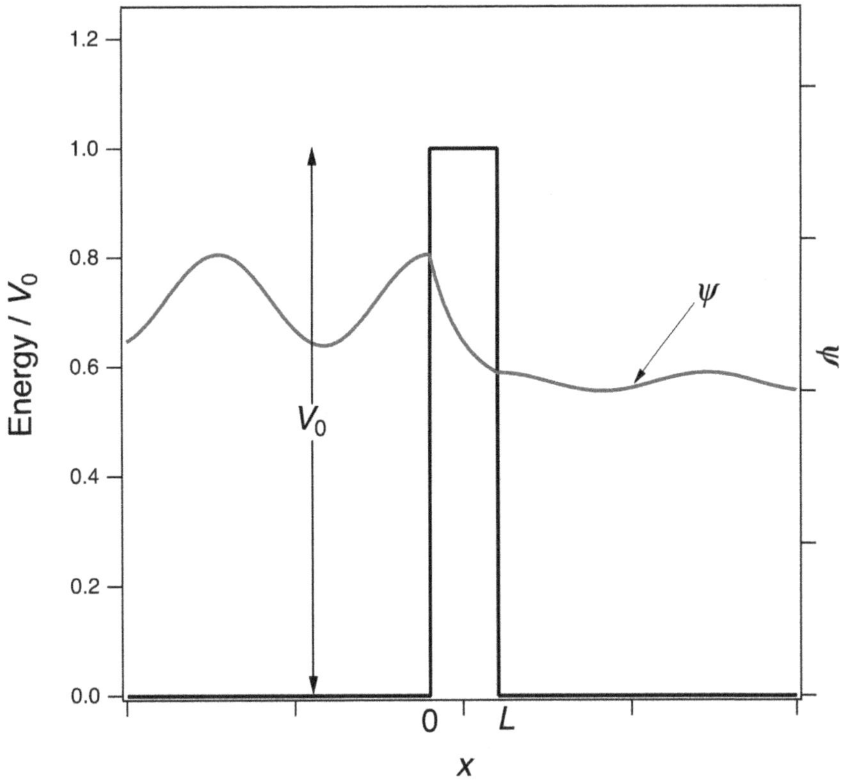

Figure 21.4 A particle approaching a barrier of height V_0 and width L from the left. The wavefunction decays exponentially inside the barrier and then re-emerges on the right side. Transmission through a barrier, even when the particle energy is less than the classical barrier height, is known as *tunneling*.

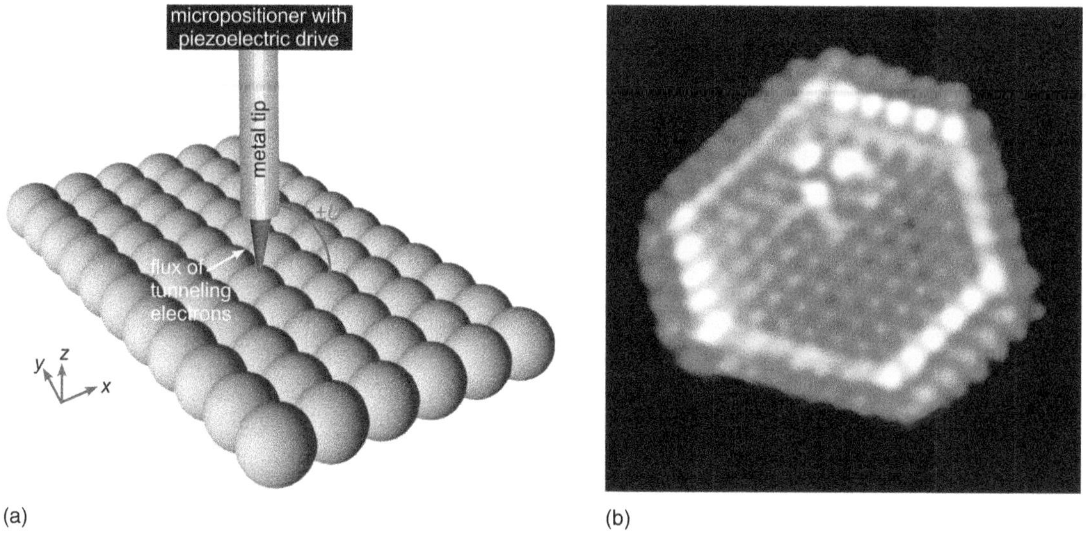

Figure 21.5 (a) Schematic representation of STM. A sharp metal tip is rastered across a surface in the x- and y-directions while held at potential U. Control of the potential allows electrons to tunnel either from the occupied electronic states of the tip into the unoccupied states of the surface, or vice versa. Holding this tunneling current constant by moving the tip up and down in the z direction while scanning in the xy plane, the contours of the surface are mapped with atomic resolution. (b). An STM image of a CoMoS nanocluster that acts as a desulfurization catalyst. Provided by Flemming Besenbacher. Reproduced with permission from Lauritsen, J.V., Vang, R.T., and Besenbacher, F. (2006) *Catal. Today*, **111**, 34; © 2006 Elsevier Science.

allows us to image the electronic states of atoms and molecules. Since the electronic states are often (though not always) strongly correlated with the positions of atoms, the contours of these images can reflect the positions of the atomic nuclei at the core of atoms, as shown in Fig. 21.5(b).

Figure 21.4 illustrates an important characteristic of the wavefunction:

The wavefunction and it first derivative must be continuous and finite.

This may seem like formal mathematical details. However, if the wavefunction changed discontinuously at the barrier wall, the first derivative would be infinite. From the Schrödinger equation we know that an infinite second derivative translates into an infinite energy. This cannot be, so the wavefunction approaching the barrier from the left must be equal to the wavefunction approaching the barrier from the right at the point where they meet. In the coordinate system of Fig. 21.4, this point is $x = 0$, thus

$$\psi_{\text{left}}(x = 0) = \psi_{\text{inside}}(x = 0) \tag{21.27}$$

Similarly,

$$\frac{\mathrm{d}}{\mathrm{d}x}\psi_{\text{left}}(x = 0) = \frac{\mathrm{d}}{\mathrm{d}x}\psi_{\text{inside}}(x = 0). \tag{21.28}$$

The two analogous boundary conditions apply on the right side of the barrier as the particle exits (if it can exit) the barrier at $x = L$.

When the free particle meets the barrier as it travels from left to right, a portion of the wave is reflected and a portion enters the barrier. The wavefunction on the left is

$$\psi_{\text{left}} = Ae^{ikx} + Be^{-ikx}. \tag{21.29}$$

As before, when we treated free particle motion in Section 20.6, the relationship between energy and wavevector is

$$k^2 = \frac{2mE}{\hbar^2}. \tag{21.30}$$

On the right side of the barrier, the particle only translates to the right and the wavefunction is

$$\psi_{\text{right}} = Fe^{ikx}. \tag{21.31}$$

The constant A is the incident amplitude, B is the reflected amplitude, and F is the transmitted amplitude. The *transmission coefficient* T is given by the ratio of probability flux

$$T = \frac{\hbar k|F|^2}{\hbar k|A|^2} = \frac{|F|^2}{|A|^2}. \tag{21.32}$$

These coefficients are determined by applying the boundary conditions as in Eqs (21.27) and (21.28).

Tunneling occurs because the wavefunction of the particle is able to penetrate the barrier into what is energetically a classically forbidden region. The transmission probability depends on the energy of the particle E compared to the barrier height V_0, the mass of the particle m, as well as the shape and thickness L of the barrier. Tunneling is more important for energies close to the top of the barrier, light particles (muons, electrons and oftentimes even protons) and thin smooth barriers. In particular, tunneling is important when the thickness of the barrier becomes comparable to the de Broglie wavelength of the particle.

Within the classically forbidden region inside the barrier, the wavefunction of the particle decays exponentially

$$\psi_{\text{inside}}(x) = Ae^{-\kappa x} \tag{21.33}$$

where the *decay constant* κ is

$$\kappa^2 = \frac{2m(V_0 - E)}{\hbar^2} \tag{21.34}$$

The transmission coefficient is determined by applying to continuity conditions from Eqs (21.27) and (21.28) at both $x = 0$ and $x = L$, from which we obtain

$$T = \frac{4E}{V_0}e^{-2\kappa L} = \frac{4E}{V_0}e^{-L/\Gamma} \tag{21.35}$$

where Γ is the *decay length*,

$$\Gamma = 1/2\kappa. \tag{21.36}$$

Typical values of κ lie in the range of about 0.8–1.2 Å^{-1} for many electron transfer events of interest in chemistry. Thus, tunneling usually is effective over a range of a few Å. One such example of this is outer-sphere electron transfer in electrochemistry. Electron transfer occurs when an electron hops from the donor to the acceptor, which are separated by a distance r. There is a Gibbs energy of activation associated with electron transfer $\Delta^{\ddagger}G^{\circ}_{m}$. The rate of electron transfer is proportional to the product of the Arrhenius activation term and the exponential distance dependence,

$$k_{\text{et}} \propto e^{-\beta r}e^{-\Delta^{\ddagger}G^{\circ}_{m}/RT}. \tag{21.37}$$

If a sharp metal tip is brought close to a surface, as shown in Fig. 21.5(a), a tunneling current will flow when the wavefunction of an electron in the tip overlaps spatially and energetically with the wavefunction of the surface. Charge transfer always occurs from occupied to unoccupied (or partially occupied) states, and the direction of flow (tip to surface or surface to tip) can be controlled by placing a voltage on the tip relative to the surface. The current that flows is highly sensitive to the distance between the two. It is proportional to the transmission coefficient

$$I \propto T \propto e^{-2\kappa L}. \tag{21.38}$$

This exponential dependence of the tunneling current on the distance L between the tip and the surface is the basis of scanning tunneling microscopy, for which its inventors Gerd Binnig and Heinrich Rohrer were awarded the Nobel Prize in Physics 1986.[6]

For $V_0 - E = 4.5$ eV, a change from $L = 0.2$ nm to $L = 0.3$ nm changes the current by a factor of 8.8. At $L = 0.2$ nm, changing from the mass of an electron to the mass of a proton decreases the current by 1.2×10^{-79}. It is easy to understand why tunneling probe microscopy is exceedingly sensitive to height variations and involves the tunneling of electrons rather than more massive particles.

21.3 Vibrational motion

One of the reasons that the harmonic oscillator is so versatile is that *in the vicinity of a local minimum almost any potential is approximately harmonic.* This is shown in Fig. 21.6 and can be explained in the following manner. Expand the potential $V(r)$ in a Taylor series about the local minimum located at r_e,

$$V(r) = V(r_e) + V'(r_e)(r - r_e) + \frac{1}{2}V''(r_e)(r - r_e)^2 + \cdots, \tag{21.39}$$

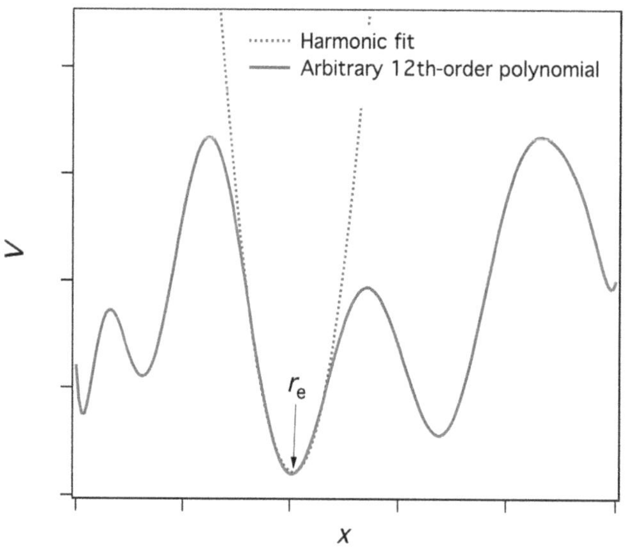

Figure 21.6 The harmonic approximation to a local minimum in an arbitrary potential constructed from a 12th-order polynomial.

where V' is the first derivative with respect to r and V'' the second. The constant is inconsequential because it does not change the force, which is given by the negative of the derivative of the potential

$$F = -\frac{\mathrm{d}V}{\mathrm{d}r}. \tag{21.40}$$

At the minimum, the first derivative is zero. Neglecting higher order terms, which should be small as long as the displacement from the minimum, $x = r - r_e$, is small, the potential is approximately,

$$V(r) \approx \frac{1}{2}V''(r_e)(r - r_e)^2 = \frac{1}{2}kx^2. \tag{21.41}$$

Taking the derivative we obtain,

$$F = -\frac{\mathrm{d}V}{\mathrm{d}r} = -kx, \tag{21.42}$$

which is Hooke's law for harmonic motion of a frictionless spring. The force constant k is then identified as

$$k = V''(r_e). \tag{21.43}$$

The force constant is related to the fundamental vibrational angular frequency of the oscillator by[7]

$$\omega_0 = (k/\mu)^{1/2} \tag{21.44}$$

where μ is the reduced mass

$$\mu = \frac{m_1 m_2}{m_1 + m_2} \tag{21.45}$$

of the two masses connected by the spring as shown in Fig. 21.7.

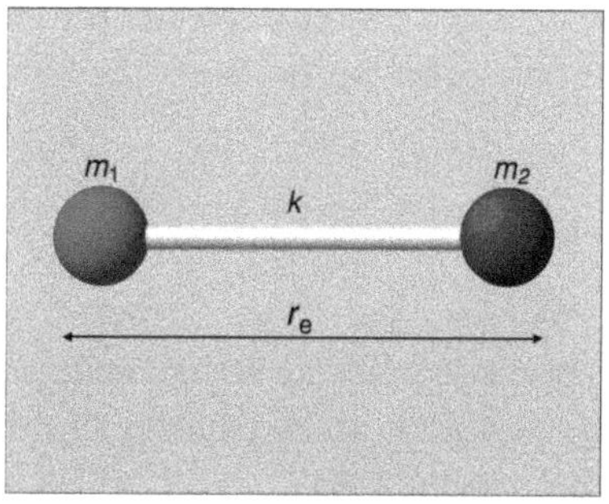

Figure 21.7 A harmonic oscillator with an equilibrium separation r_e between masses m_1 and m_2. The coordinate x along the spring axis measures the displacement from this equilibrium value. The masses are connected by a massless spring of force constant k.

The *harmonic oscillator* is the basic model for vibrational motion and is a good approximation to vibrational motion in small molecules for low levels of excitation. In this case we identify r_e as the equilibrium bond length and r is instantaneous length.

To work out the quantum mechanical solution to the harmonic oscillator, we first write the potential operator as

$$\hat{V} = \tfrac{1}{2} m\omega_0^2 x^2. \tag{21.46}$$

Writing the translational energy operator in terms of the reduced mass, the Schrödinger equation is then

$$-\frac{\hbar^2}{2\mu}\frac{\mathrm{d}^2\psi}{\mathrm{d}x^2} + \frac{1}{2}\mu\omega_0^2 x^2\psi = E\psi. \tag{21.47}$$

To aid in the solution of this equation we introduce

$$\xi \equiv \sqrt{\mu\omega_0/\hbar}\, x \tag{21.48}$$

$$K \equiv 2E/\hbar\omega_0. \tag{21.49}$$

Equation (21.47) then becomes

$$\frac{\mathrm{d}^2\psi}{\mathrm{d}x^2} = (\xi - K)\psi \tag{21.50}$$

the solution of which is known but the details are not particularly enlightening for our purposes. (See Griffths for details of the solution.) It is nonetheless exactly soluble in terms of a function of the form of a product of a normalization constant, a polynomial and a Gaussian,

$$\psi_v(x) = \left(\frac{\mu\omega_0}{\pi\hbar}\right)^{1/4} \frac{1}{\sqrt{2^v v!}} H_v(\xi)\, e^{-\xi^2/2}. \tag{21.51}$$

The functions $H_v(\xi)$ are the *Hermite polynomials*, the first few of which are detailed in Table 21.1. There are various methods of calculating the Hermite polynomials, including the *Rodrigues formula*

$$H_v(\xi) = (-1)^v e^{\xi^2}\left(\frac{\mathrm{d}}{\mathrm{d}\xi}\right)^v e^{-\xi^2} \tag{21.52}$$

and the recursion relationship

$$H_{v+1}(\xi) = 2\xi H_v(\xi) - 2vH_{v-1}(\xi). \tag{21.53}$$

The Hermite polynomials and therefore the wavefunctions are alternately even and odd functions. They change sign upon reflection through the origin. The quantum number v is equal to the number of nodes in the function. The harmonic oscillator wavefunctions, displayed in Fig. 21.7, constitute an orthonormal set of functions – that is, they are all normalized and orthogonal to one another.

The wavefunctions correspond to regularly spaced levels with energy

$$E_v = \left(v + \tfrac{1}{2}\right)\hbar\omega_0. \tag{21.54}$$

As before

$$\omega_0 = (k/\mu)^{1/2}. \tag{21.55}$$

Table 21.1 The Hermite polynomials, $H_v(\xi)$, where $\xi \equiv \sqrt{\mu\omega_0/\hbar}\, x$ and $\omega_0 = (k/\mu)^{1/2}$.

v	$H_v(\xi)$
0	1
1	2ξ
2	$4\xi^2 - 2$
3	$8\xi^3 - 12\xi$
4	$16\xi^4 - 48\xi^2 + 12$
5	$32\xi^5 - 160\xi^3 + 120\xi$
6	$64\xi^6 - 480\xi^4 + 720\xi^2 - 120$

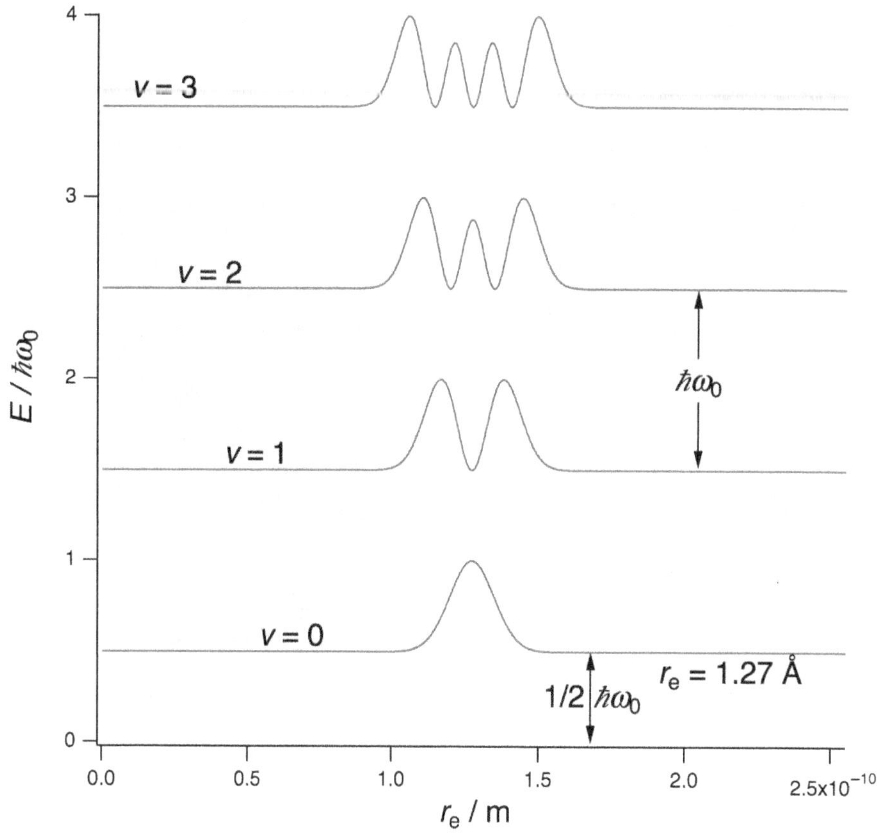

Figure 21.8 A harmonic oscillator with an equilibrium length r_e chosen to be 1.27 Å as in HCl. The potential extends to infinite energy. The first level with quantum number $v = 1$ does not lie at $E = 0$. Instead, it has a zero point energy of $\frac{1}{2}\hbar\omega_0$, where ω_0 is the fundamental angular frequency. The energy spacing between levels is $\hbar\omega_0$ for all levels. Note how the Gaussian term localizes the wavefunction to the center of the well, while the Hermite polynomial introduces an increasing number of nodes as v increases.

The numbering now begins with zero, $v = 0, 1, 2, \ldots$. The energy spacing is constant,

$$E_{v+1} - E_v = \hbar\omega_0. \tag{21.56}$$

Since the motion is confined, the energy levels are quantized. The lowest energy level does not have an energy of zero. Instead, the lowest energy level contains *zero point energy* of $\frac{1}{2}\hbar\omega_0$ and the atoms are always in motion, even at absolute zero. Zero point motion is required by the Heisenberg uncertainty principle since the vibration relates linear motion to position.

21.3.1 Directed practice

Inspecting the wavefunctions shown in Fig. 21.8, how does the number of nodes correlate with whether the wavefunction is even or odd with respect to reflection about the middle of the well?

21.4 Angular momentum

21.4.1 The operators

There is a common algebra that links all angular momenta, regardless of whether that angular momentum is generated by rotation, occupation of a specific orbital, spin (electron or nuclear) or the coupling of more than one of these types of angular momentum. Mastering this algebra and its consequences is essential to understanding spectroscopy, bonding and periodic behavior related to the filling of orbitals. Because of this we take a moment to look at the angular momentum operators in detail. A number of different types of angular momentum are encountered, as specified in Table 21.2. One thing you should notice is that J is particularly popular and its meaning is context-dependent; for instance, symbols may change their meaning based on whether an atomic, diatomic or polyatomic system is being considered. These context

Table 21.2 Angular momentum operators and associated quantum numbers for their magnitude and projection along a space-fixed z-axis. As a simplification the subscript in the z-axis quantum number is often dropped if there is no ambiguity.

		Quantum number symbol	
Angular momentum	**Operators**	**total**	**z-axis**
Electron orbital	$\hat{L}, \hat{L}_z$	L	M_L
Single electron orbital	$\hat{l}, \hat{l}_z$	l	m_l
Electron spin	$\hat{S}, \hat{S}_z$	S	M_S
Single electron spin	$\hat{s}, \hat{s}_z$	s	m_s
Rotation	$\hat{J}, \hat{J}_z$	J	M_J
Nuclear spin	$\hat{I}, \hat{I}_z$	I	M_I
Sum of $L + S$	$\hat{J}, \hat{J}_z$	J	M_J
Sum of $l + s$	$\hat{j}, \hat{j}_z$	j	m_j
Sum of $J + I$	$\hat{F}, \hat{F}_z$	F	M_F

dependencies complicate general definitions. Simply be aware that definitions depend on context, and specific notation should always be made clear. We will often deal with the angular momenta of single electrons that then couple to form resultant angular momenta for the system. A general convention is to use lower-case letters for a single-electron (or particle) and upper-case letters to refer to the coupled system.

A general unspecified angular momentum is usually denoted either L or J in classical mechanics. All of these angular momenta obey corresponding eigenvalue equations of the form $\hat{J}^2\psi = J(J+1)\hbar^2\psi$ and $\hat{J}_z\psi = M_J\hbar\psi$. For diatomics, IUPAC suggests using R for rotation, N for $R + L$ and J for $R + L + S$. This might be clearer but the weight of usage still favors J for rotational angular momentum, particularly since most molecules have $L = 0$ and $S = 0$ in their ground state; hence $J = R$. When molecules do not have L and S equal to zero care must be take to keep track of the total angular momentum.

Here, we will treat angular momentum in terms of three-dimensional motion of a particle orbiting a central point at a distance $\boldsymbol{r}$. If the linear momentum is $\boldsymbol{p}$, the angular momentum from classical mechanics is

$$\boldsymbol{L} = \boldsymbol{r} \times \boldsymbol{p}. \tag{21.57}$$

The relations of angular momentum to linear momentum and the coordinate system are specified in Fig. 21.9. Note that angular momentum is a vector quantity.

Equation (21.57) can be transformed into quantum mechanical operators by substituting the operators for $\boldsymbol{r}$ and $\boldsymbol{p}$. For position, the operator is simply multiplication by the coordinate. For linear momentum the operators are

$$\hat{p}_x = -i\hbar\frac{\partial}{\partial x} \tag{21.58}$$

$$\hat{p}_y = i\hbar\frac{\partial}{\partial y} \tag{21.59}$$

$$\hat{p}_z = i\hbar\frac{\partial}{\partial z}. \tag{21.60}$$

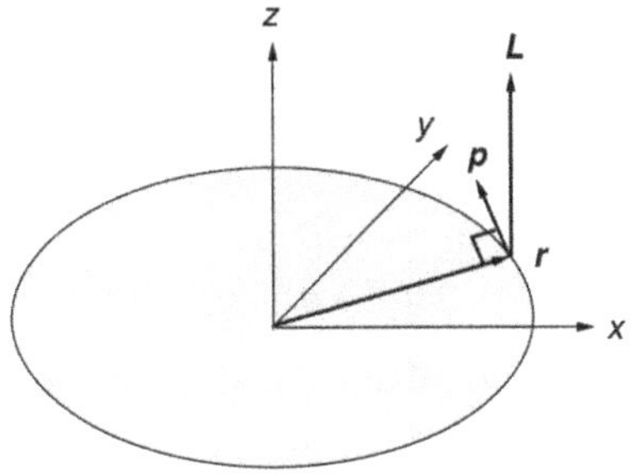

Figure 21.9 Coordinate system for the particle on a ring. A particle at position $\boldsymbol{r}$ and linear momentum $\boldsymbol{p}$ has an angular momentum given by the cross-product $\boldsymbol{L} = \boldsymbol{r} \times \boldsymbol{p}$. The linear momentum is tangential to the circle circumscribed by the position vector $\boldsymbol{r}$. The angular momentum is perpendicular to the plane containing $\boldsymbol{r}$ and $\boldsymbol{p}$.

We use these to generate the three operators for the components of angular momentum by substitution into the cross-product of Eq. (21.57),

$$\hat{L}_x = -i\hbar\left(y\frac{\partial}{\partial z} - z\frac{\partial}{\partial y}\right) \tag{21.61}$$

$$\hat{L}_y = -i\hbar\left(z\frac{\partial}{\partial x} - x\frac{\partial}{\partial z}\right) \tag{21.62}$$

$$\hat{L}_z = -i\hbar\left(x\frac{\partial}{\partial y} - y\frac{\partial}{\partial x}\right). \tag{21.63}$$

When these operators act on the wavefunctions of the system, they return the eigenvalues for the corresponding component of the total angular momentum. As mentioned at the bottom of Table 21.2, all of the angular momentum operators obey a similar set of eigenvalue equations. With an eye towards finding the solution for the H atom in which a single electron interacts with a nucleus, let us consider the operator for the orbital angular momentum of a single electron. When $\hat{l}_z$ operates on the wavefunction $\psi(\theta, \varphi)$, it returns the eigenvalue $m_l\hbar$. Or, in equation form the z-component of orbital angular momentum is

$$\hat{l}_z\psi = m_l\hbar\,\psi. \tag{21.64}$$

Similarly, the magnitude of the orbital angular momentum vector is

$$\hat{l}^2\psi = l(l+1)\hbar^2\psi. \tag{21.65}$$

Taking the commutator of any pair of the components, we arrive at the interesting and important result that no two components of angular momentum commute. For example,

$$\left[\hat{l}_x, \hat{l}_y\right] = \hat{l}_x\hat{l}_y - \hat{l}_y\hat{l}_x = i\hbar\hat{l}_z, \quad \left[\hat{l}_y, \hat{l}_z\right] = i\hbar\hat{l}_x, \quad \left[\hat{l}_z, \hat{l}_x\right] = i\hbar\hat{l}_y \tag{21.66}$$

A nonvanishing value of the commutator means that *no two components of the angular momentum can be simultaneously specified with absolute precision.* The component that is specified is by convention taken to be the z component.

Finding the wavefunctions of systems with central potentials such as those in atoms and molecules is much easier – well, less difficult – if the coordinates are expressed in spherical polar coordinates rather than linear coordinates. Making the transformation, the operator for the z-component of the angular momentum in spherical polar coordinates is

$$\hat{l}_z = i\hbar\frac{\partial}{\partial\phi}. \tag{21.67}$$

The operator for the magnitude of the angular momentum is found by using the definition

$$\hat{l}^2 = \hat{l}_x^2 + \hat{l}_y^2 + \hat{l}_z^2 \tag{21.68}$$

Upon transformation to spherical polar coordinates, we find

$$\hat{l}^2 = -\hbar^2\left[\frac{1}{\sin\theta}\frac{\partial}{\partial\theta}\left(\sin\theta\frac{\partial}{\partial\theta}\right) + \frac{1}{\sin^2\theta}\frac{\partial^2}{\partial\phi^2}\right] = -\hbar^2\Lambda^2 \tag{21.69}$$

You may recall that the term in brackets is the Legendrian operator from Eq. (20.102). The order of operation must be maintained in quantum mechanics. Thus, the first term in the square brackets requires first that the derivative with respect to θ is taken, then multiplication by $\sin\theta$, then another derivative, and finally division by $\sin\theta$. Unlike the commutator among the components, the commutator between the magnitude and a component does vanish, for example,

$$\left[\hat{l}^2, \hat{l}_z\right] = \hat{l}^2\hat{l}_z - \hat{l}_z\hat{l}^2 = 0. \tag{21.70}$$

This means that the magnitude and one component of angular momentum can be specified simultaneously. In the parlance of quantum mechanics, we say that the quantum numbers that determine the magnitude and the projection of the angular momentum vector are *good quantum numbers* (they are well-defined and defined simultaneously as eigenvalues of the wavefunction). Thus, for orbital angular momentum we can specify both l and m_l simultaneously; for rotation, we can specify both J and m_J; and for spin both s and m_s can be specified simultaneously.

21.4.2 Particle on a ring

To a first approximation, the rotational and vibrational motions of a molecule are independent, and it is instructive to consider the model of a *rigid rotor* to describe rotational motion. Consider first a particle rotating in two dimensions (the xy plane). This motion corresponds to the *particle on a ring* problem. The energy is all kinetic as the potential $V = 0$ everywhere. The energy is $E = p^2/2m$. The angular momentum j_z is directed perpendicular to the plane of rotation along the z axis,

$$j_z = \pm pr, \tag{21.71}$$

where r is the distance from the origin at which the particle is orbiting and opposite signs correspond to opposite senses of rotation. The *moment of inertia* is

$$I = mr^2. \tag{21.72}$$

Substituting, the energy is

$$E = j_z^2/2mr^2 = j_z^2/2I. \tag{21.73}$$

Up to this point our treatment has been completely classical. To obtain the quantum mechanical solution, we write the Hamiltonian as

$$\hat{H} = -\frac{\hbar^2}{2mr^2}\frac{\mathrm{d}^2}{\mathrm{d}\phi^2} = -\frac{\hbar^2}{2I}\frac{\mathrm{d}^2}{\mathrm{d}\phi^2} \tag{21.74}$$

and, by inspection, the Schrödinger equation,

$$-\frac{\hbar^2}{2I}\frac{\mathrm{d}^2\psi}{\mathrm{d}\phi^2} = E\psi, \tag{21.75}$$

has solutions

$$\psi_{m_j}(\phi) = Ne^{\mathrm{i}m_j\phi}. \tag{21.76}$$

21.4.2.1 Directed practice

Confirm that $\psi_{m_j}(\phi)$ satisfies the Schrödinger equation.

Wavefunctions must be single-valued. The azimuthal angle varies from 0 to 2π and the angles ϕ and $\phi + 2\pi$ represent the same point. Therefore, $\psi_{m_j}(\phi) = \psi_{m_j}(\phi + 2n\pi)$, which can only be satisfied if

$$m_j = 0, \pm 1, \pm 2, \ldots. \tag{21.77}$$

Since m_j can only assume specific integer values, the energy E is also quantized to certain values

$$E_{m_j} = m_j^2\,\hbar^2/2I, \tag{21.78}$$

which can be confirmed by direct substitution into the Schrödinger equation. Similarly the projection of the angular momentum is also quantized

$$\hat{j}_z\psi = m_j\hbar\psi. \tag{21.79}$$

Specifically for the particle on a ring, there are only two projections of rotational angular momentum for each state. This is a consequence of the confinement of rotational motion to a plane. The projection can either point up or down. The particle can either rotate clockwise or counter clockwise. The same quantum number is used to determine both the energy and the projection. This is unique to the particle on a ring – a consequence of its reduced dimensionality – and we will see that this same simplicity does not apply to the particle on a sphere problem.

Note the complete equivalence between this as a model for either the rotational angular momentum of a rigid rotor (a mass m rotating at a fixed distance from the origin) or the orbital angular momentum of an electron orbiting in a ring about a fixed nucleus at the origin.

Unlike the particle in a box system, the ground-state energy is $E_0 = 0$. There is no zero point energy. However, the ground-state wavefunction is $\psi_0(\phi) = Ne^0 = N$. It is simply a constant spread out over the entire ring. The position is indeterminate as required by the uncertainty principle.

21.4.3 Particle on a sphere

Now let us generalize the motion to three dimensions. We must account for rotational motion changing not only the azimuth ϕ but also the colatitude θ. r is again assumed to be fixed. The potential is zero everywhere and we need to use the Laplacian operator for kinetic energy in three dimensions (with r fixed we will be able to factor out this dependence). This is equivalent to the motion of a *particle on a sphere*.

Anticipating that we can separate the r dependence from the angular dependence of the problem, we write out trial wavefunction in variable separable form

$$\psi(r,\theta,\phi) = R(r)Y(\theta,\phi). \tag{21.80}$$

This is fully justified since the Laplacian in spherical polar coordinates from Eq. (20.100) involves a radial term that is separate from the angular terms. This allows us to write the Schrödinger equation

$$-\frac{\hbar^2}{2m}\nabla^2\psi + V\psi = E\psi \tag{21.81}$$

as

$$-\frac{\hbar^2}{2m}\left[\frac{1}{r^2}\frac{\partial}{\partial r}\left(r^2\frac{\partial\psi}{\partial r}\right) + \frac{1}{r^2\sin\theta}\frac{\partial}{\partial\theta}\left(\sin\theta\frac{\partial\psi}{\partial\theta}\right) + \frac{1}{r^2\sin^2\theta}\frac{\partial^2\psi}{\partial\phi^2}\right] + V\psi = E\psi. \tag{21.82}$$

Substituting the function RY for ψ, we obtain

$$-\frac{\hbar^2}{2m}\left[\frac{Y}{r^2}\frac{\partial}{\partial r}\left(r^2\frac{\partial R}{\partial r}\right) + \frac{R}{r^2\sin\theta}\frac{\partial}{\partial\theta}\left(\sin\theta\frac{\partial Y}{\partial\theta}\right) + \frac{R}{r^2\sin^2\theta}\frac{\partial^2 Y}{\partial\phi^2}\right] + VRY = ERY \tag{21.83}$$

Dividing by RY and multiplying by $-2mr^2/\hbar^2$, Eq. (21.83) breaks into two terms – one that depends only on r, and one that depends only on θ and ϕ,

$$\left[\frac{1}{R}\frac{\partial}{\partial r}\left(r^2\frac{\partial R}{\partial r}\right) - \frac{2mr^2}{\hbar^2}(V-E)\right] + \frac{1}{Y}\left[\frac{1}{\sin\theta}\frac{\partial}{\partial\theta}\left(\sin\theta\frac{\partial Y}{\partial\theta}\right) + \frac{1}{\sin^2\theta}\frac{\partial^2 Y}{\partial\phi^2}\right] = 0 \tag{21.84}$$

The only way that these two terms, which depend on different variables, can sum to the constant zero is if they are both equal to a constant,

$$\frac{1}{R}\frac{\partial}{\partial r}\left(r^2\frac{\partial R}{\partial r}\right) - \frac{2mr^2}{\hbar^2}(V-E) = l(l+1) \tag{21.85}$$

$$\frac{1}{Y}\left[\frac{1}{\sin\theta}\frac{\partial}{\partial\theta}\left(\sin\theta\frac{\partial Y}{\partial\theta}\right) + \frac{1}{\sin^2\theta}\frac{\partial^2 Y}{\partial\phi^2}\right] = -l(l+1). \tag{21.86}$$

We chose to write the constant in the suggestive form $l(l+1)$ with no loss of generality because for any constant l, $l(l+1)$ is also a constant. Shortly, we shall prove that l must also be an integer.

We will return to the radial term when we treat the H atom. For now, we can forget it and simply treat it as a constant. We concentrate on the solution of the *angular equation*. Multiplying both sides of Eq. (21.86) by $Y\sin^2\theta$ yields

$$\sin\theta\frac{\partial}{\partial\theta}\left(\sin\theta\frac{\partial Y}{\partial\theta}\right) + \frac{\partial^2 Y}{\partial\phi^2} = -l(l+1)\sin^2\theta Y. \tag{21.87}$$

This is a satisfying result because if you take a course in classical electrodynamics, you encounter this equation in the solution of the Laplace equation. Again, we see two terms on the LHS, one involving operation on θ and one only on ϕ. This suggests that we can use the same tricks: introduce separation of variables,

$$Y(\theta,\phi) = \Theta(\theta)\Phi(\phi) \tag{21.88}$$

then divide by $\Theta\Phi$ and collect terms to find that the sum of a θ-dependent term and a ϕ-dependent term is a constant,

$$\left\{\frac{1}{\Theta}\left[\sin\theta\frac{\partial}{\partial\theta}\left(\sin\theta\frac{\partial\Theta}{\partial\theta}\right) + l(l+1)\sin^2\theta\right]\right\} + \frac{1}{\Phi}\frac{\partial^2\Phi}{\partial\phi^2} = 0. \tag{21.89}$$

Again, this equation can only be satisfied if the two terms are both equal to a constant, which we now chose to call $m_l{}^2$,

$$\frac{1}{\Theta}\left[\sin\theta\frac{\partial}{\partial\theta}\left(\sin\theta\frac{\partial\Theta}{\partial\theta}\right) + l(l+1)\sin^2\theta\right] = m_l^2 \tag{21.90}$$

$$\frac{1}{\Phi}\frac{\partial^2\Phi}{\partial\phi^2} = -m_l^2. \tag{21.91}$$

Equation (21.91) is of the same form as Eq. (21.75). We have already solved that and, indeed, it represents the same part of the physics,

$$\Phi(\phi) = e^{\mathrm{i}m_l\phi}. \tag{21.92}$$

Again $m_l = 0, \pm 1, \pm 2, \ldots$. An upper limit exists that is constrained by the solution of the θ component.

The equation in θ is not so easy to solve. Fortunately, because it is a well-known equation from electrodynamics, the solution has been worked out and is given by

$$\Theta(\theta) = AP_l^{m_l}(\cos\theta) \tag{21.93}$$

where $P_l^{m_l}(\cos\theta)$ are the associated Legendre functions defined by

$$P_l^{m_l}(x) = (1 - x^2)^{|m_l|/2}(\mathrm{d}/\mathrm{d}x)^{|m_l|}P_l(x) \tag{21.94}$$

and $P_l(x)$ are the Legendre polynomials, defined by Rodrigues' formula

$$P_l(x) = \frac{1}{2^l l!}\left(\frac{\mathrm{d}}{\mathrm{d}x}\right)^l (x^2 - 1)^l. \tag{21.95}$$

It is the shape of the $P_l^{m_l}(\cos\theta)$ that are responsible for giving orbitals their characteristics shapes.

For Rodrigues' formula to make sense, l must be a non-negative integer

$$l = 0, 1, 2, \ldots \tag{21.96}$$

Moreover if $|m_l| > l$, then $P_l^{m_l} = 0$. Consequently, for any given l there are $(2l + 1)$ possible values of m_l,

$$m_l = -l, -l + 1, \ldots, 1, 0, 1, \ldots, l - 1, l. \tag{21.97}$$

Finally, the constant A is set by the normalization condition

$$\int_0^{2\pi}\int_0^{\pi} |Y|^2 \sin\theta \mathrm{d}\theta\mathrm{d}\phi = 1. \tag{21.98}$$

The normalized angular wavefunctions are known as the *spherical harmonics*

$$Y_{lm}(\theta,\phi) = \varepsilon\left(\frac{(2l+1)}{4\pi}\frac{(l-|m|)!}{(l+|m|)!}\right)^{1/2} e^{\mathrm{i}m\phi}P_l^m(\cos\theta) \tag{21.99}$$

where $\varepsilon = (-1)^m$ for $m \geq 0$ and $\varepsilon = 1$ for $m \leq 0$. It is conventional to drop the subscripted l from m_l in the notation of the spherical harmonics. The spherical harmonics form a set of orthonormal function, which means that they satisfy the condition

$$\int_0^{2\pi}\int_0^{\pi} Y_{lm}^* Y_{l'm'} \sin\theta\, \mathrm{d}\theta\, \mathrm{d}\phi = \delta_{ll'}\delta_{mm'}. \tag{21.100}$$

Quantization along the two angular coordinates θ and ϕ has thus led to the necessity to use two quantum numbers – l the azimuthal quantum number, and m_l the magnetic quantum number – in order to describe the states that arise. The first few members of the set of spherical harmonics are given in Table 21.3.

That the projection of angular momentum along the z axis is confined to certain values, as shown in Fig. 21.10, means that the orientation of a rotating body is quantized when the direction of the z-axis is imposed upon the body by, for

Table 21.3 The first few members of the set of spherical harmonics.

$Y_{lm}(\theta,\phi)$	
$Y_{00} = \left(\frac{1}{4\pi}\right)^{1/2}$	$Y_{2\pm2} = \mp\left(\frac{15}{32\pi}\right)^{1/2}\sin^2\theta e^{\pm2\mathrm{i}\phi}$
$Y_{10} = \left(\frac{3}{4\pi}\right)^{1/2}\cos\theta$	$Y_{30} = \left(\frac{7}{16\pi}\right)^{1/2}(5\cos^3\theta - 3\cos\theta)$
$Y_{1\pm1} = \mp\left(\frac{3}{8\pi}\right)^{1/2}\sin\theta e^{\pm\mathrm{i}\phi}$	$Y_{3\pm1} = \mp\left(\frac{21}{64\pi}\right)^{1/2}\sin\theta(5\cos^2\theta - 1)e^{\pm\mathrm{i}\phi}$
$Y_{20} = \left(\frac{5}{16\pi}\right)^{1/2}(3\cos^2\theta - 1)$	$Y_{3\pm2} = \left(\frac{105}{32\pi}\right)^{1/2}\sin^2\theta\cos\theta e^{\pm2\mathrm{i}\phi}$
$Y_{2\pm1} = \mp\left(\frac{15}{8\pi}\right)^{1/2}\sin\theta\cos\theta e^{\pm\mathrm{i}\phi}$	$Y_{3\pm3} = \mp\left(\frac{35}{64\pi}\right)^{1/2}\sin^3\theta e^{\pm3\mathrm{i}\phi}$

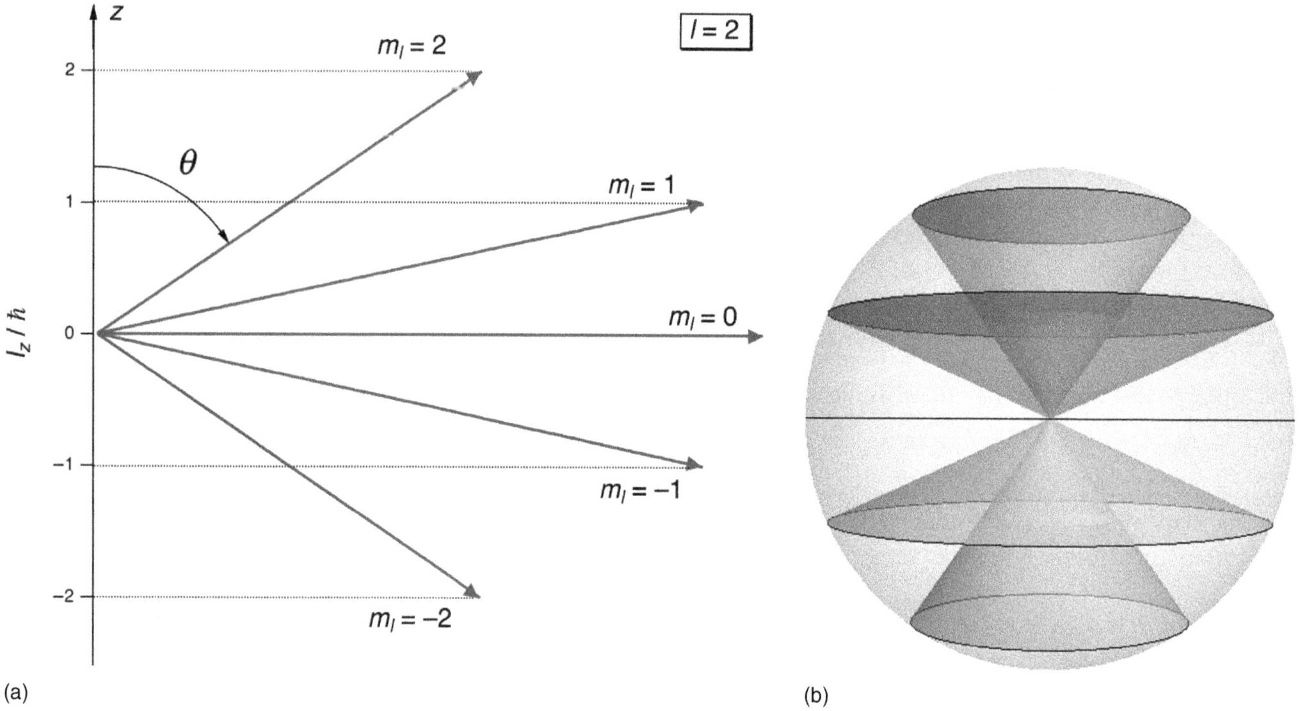

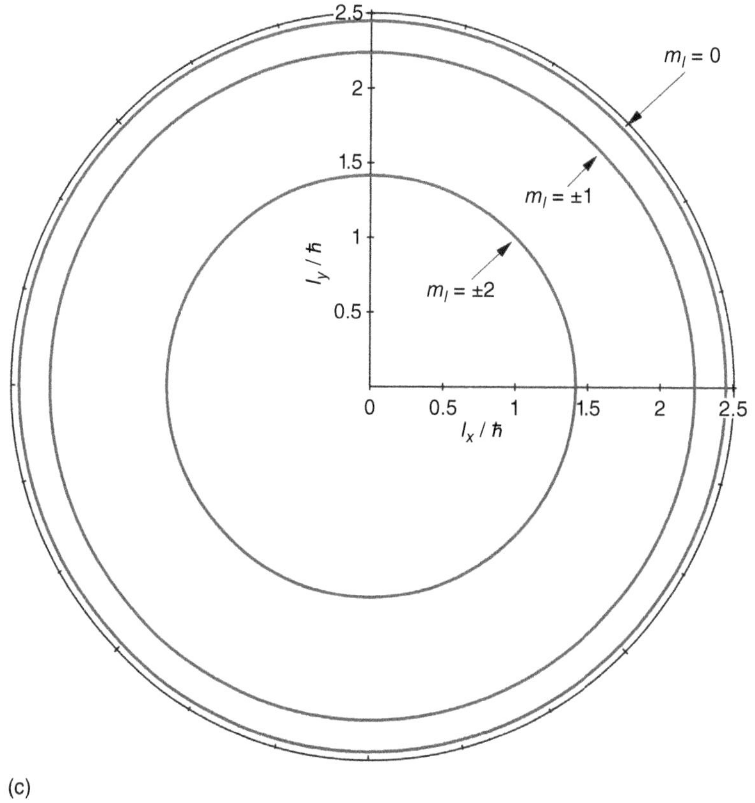

Figure 21.10 Solutions to three-dimensional rotation of a particle at fixed radius r allow for quantized angular momentum l and projection of this vector along the z-axis. (a) For the $l = 2$ solution, five projections are allowed corresponding to the values $m_l = 0, \pm 1, \pm 2$. (b) A depiction of the precession model viewed from the front, which emphasizes that while the projection along z is well-defined, the projections in the xy-plane are indeterminate. (c) A view of the precession model from above looking down into the xy-plane.

Table 21.4 Graphical representations of the absolute value of the spherical harmonics. Note that while the shapes are symmetric about $m = 0$, there are phase changes between the positive and negative values of m.

l	$m = -3$	$m = -2$	$m = -1$	$m = 0$	$m = 1$	$m = 2$	$m = 3$
0							
1							
2							
3							

instance, an external electric or magnetic field (*space quantization*).[8] The properties of all sorts of angular momentum (rotation, orbital, nuclear, spin) are similar and are ruled by the same types of algebra. As shown in Table 21.4, Y_{lm} functions have characteristic shapes. In the next chapter we will demonstrate how these functions are related to the shapes of the orbitals of the hydrogen atom.

Now that we have the wavefunctions, we can return to the Schrödinger equation to solve for the energies,

$$\frac{\hat{l}^2}{2mr^2} Y_{lm} = \frac{l\,(l+1)\,\hbar^2}{2mr^2} Y_{lm} = E\, Y_{lm}. \tag{21.101}$$

Thus, the allowed energies are

$$E_l = \frac{l\,(l+1)\,\hbar^2}{2mr^2} = \frac{l\,(l+1)\,\hbar^2}{2I}. \tag{21.102}$$

The energy does not depend on m_l, only l. For each value of l there are $2l + 1$ degenerate sublevels. These sublevels corresponds to the $2l + 1$ different directions in which the angular momentum vector can point and precess about the z-axis.

21.4.4 Spin

The intrinsic angular momentum of an electron is called *spin*. Predicted by Pauli,[9] it was confirmed experimentally by George Ulenbeck and Samuel Goudsmit[10] to have only two possible values, $+^1/_2$ and $-^1/_2$. As we shall see when we solve the Schrödinger equation for the H atom, quantization in a space of the three spatial degrees of freedom (r, θ, ϕ) leads to the necessity of three quantum numbers – n (principal), l (orbital angular momentum or azimuthal) and m_l (magnetic) – to define the wavefunction, and therefore, the energy of an orbital. Dirac developed a relativistic version of the Schrödinger equation that treats time in complete equivalence with the three spatial dimensions as demanded by relativity. This is known as the Dirac equation. Quantization in a four-dimensional spacetime demands a fourth quantum number, the spin quantum number s and its projection the spin magnetic quantum number m_s.

Spin is a purely quantum mechanical construct and classical analogies to it are fraught. For a classical body such as the Moon, we talk about the orbital angular momentum of the motion of its center of mass about the Earth. This is differentiated from its spin angular momentum by considering the spin as characteristic of motion about the Moon's center of mass. However, the Moon's spin angular moment results from the orbital angular momentum of all the rock and soil that are orbiting the Moon's center of mass. *In a classical body, both orbital and spin angular momenta are functions of r, θ, and ϕ. In quantum mechanics, orbital angular momentum is a function of r, θ, and ϕ; however, spin is not a function of r, θ, and ϕ. Spin is an intrinsic property of elementary particles.* Just as mass and charge are intrinsic properties of an elementary particle, so too is its spin, as shown in Table 21.5.

Spin follows the same set of commutation relations as orbital (or any other) angular momentum, namely,

$$\left[\hat{s}_x, \hat{s}_y\right] = i\hbar\hat{s}_z,\ \left[\hat{s}_y, \hat{s}_z\right] = i\hbar\hat{s}_x,\ \left[\hat{s}_z, \hat{s}_x\right] = i\hbar\hat{s}_y,\ \left[\hat{s}^2, \hat{s}_z\right] = 0 \tag{21.103}$$

Table 21.5 Intrinsic properties of elementary particles.

Name	Symbol	Spin	Charge/e	Mass/u
photon	γ	1	0	0
neutrino	ν_e	½	0	3.44×10^{-10}
electron	e^-	½	−1	5.49×10^{-4}
muon	$\mu^\pm$	½	±1	0.113
pion	$\pi^\pm$	0	±1	0.150
pion	π^0	0	0	0.145
proton	p	½	1	1.007
neutron	n	½	0	1.009
deuteron	d	1	1	2.014
helion	h	½	2	3.015
α-particle	α	0	2	4.002
Z-boson	Z^0	1	0	97.893
W-boson	$W^\pm$	1	±1	86.297
Higgs boson	H^0	0	0	134.29

Spin also satisfies analogous eigenvalue equations

$$\hat{s}^2|sm_s\rangle = s(s+1)\hbar^2|\,sm_s\rangle,\quad \hat{s}_z|\,sm_s\rangle = m_s\hbar|\,sm_s\rangle. \tag{21.104}$$

A distinguishing feature of spin is that it takes only one value specific to an elementary particle. A diatomic molecule can take on any integer value of the rotational quantum number J. Changing J does not change the identity of the molecule, only its energy (and maybe its bond length just a bit). An electron is particularly simple because with a value of $s = {}^1\!/_2$ there can only be two z-components $m_s = \pm{}^1\!/_2$. It is possible to change the projection of spin (the value of m_s) of the electron (a *spin flip*). This is the basis of electron spin resonance spectroscopy (ESR). However, it is not possible to change the spin s of the electron. To change the spin, you need to change the particle.

The wavefunctions of spin are not functions of (r,θ,ϕ); therefore, they are more appropriately called eigenvectors that can only be represented by symbols. Because they are so important a number of different symbols are used to denote the spin state of an electron. The $m_s = +{}^1\!/_2$ or spin up state is denoted by the ket $|\frac{1}{2}\frac{1}{2}\rangle$, and up arrow ↑, or the Greek letter α. The $m_s = -{}^1\!/_2$ or spin down state is denoted by the ket $|\frac{1}{2} - \frac{1}{2}\rangle$, a down arrow ↓, or the Greek letter β.

As we have already encountered in our study of statistical mechanics, spin also determines the intrinsic statistics of elementary particles as well as composite particles including nuclei. Spin ${}^1\!/_2$ particles (and all other particles with half-integer spin) are *fermions*. Looking at Table 21.5, we see that these include electrons, protons, and neutrons. Nuclei are *composite* particles that can also possess spin, such as ^{13}C. The spin states of nuclei provide the basis for their probing by nuclear magnetic resonance spectroscopy (NMR). That $s = {}^1\!/_2$ is essential for producing the structure of proton NMR spectra. No more than two fermions can occupy the same state, and if there are two they must have different values of m_s. This, of course, is the basis of the *Pauli exclusion principle*, and why no more than two electrons can occupy a given orbital. Since nuclei are composite particles, their spin is a property of their state rather than an intrinsic property of the nucleus. Therefore, excited states of a nucleus can have a different spin than the ground state.

Particles with integral spin are *bosons*. Arbitrarily, many bosons can occupy the same state. The photon has spin $s = 1$, which is important for explaining black-body radiation. The deuterium nucleus (the deuteron ^2H) also has $s = 1$. The helium nucleus changes its spin state in going from the helion ^3He with $s = {}^1\!/_2$ to the α-particle ^4He with $s = 0$.

21.4.5 Spin in a magnetic field

Spin produces a magnetic dipole. The magnetic dipole moment $\boldsymbol{\mu}$ is proportional to the spin $\boldsymbol{S}$ according to

$$\boldsymbol{\mu} = \gamma \boldsymbol{S}. \tag{21.105}$$

where γ is the *gyromagnetic ratio*. A magnetic field $\boldsymbol{B}$ exerts a torque $\boldsymbol{\mu}\times\boldsymbol{B}$ on a magnetic dipole. This torque makes it more favorable for the dipole to line up parallel to the magnetic field. The Hamiltonian of a spin at rest in a magnetic field is

$$\hat{H} = -\gamma \boldsymbol{B}\cdot\boldsymbol{S}. \tag{21.106}$$

Take a spin $^1/_2$ particle and place it in a uniform magnetic field. There are two orientations that the $s = {}^1/_2$ particle can assume – parallel and antiparallel with respect to the field – and these two orientations have different energies

$$E_+ = -\gamma B_0 \hbar/2 \tag{21.107}$$

and

$$E_- = +\gamma B_0 \hbar/2. \tag{21.108}$$

If we were to calculate the expectation value of the of $\boldsymbol{S}$ as a function of time, we would find the following results for the three components

$$\langle S_x \rangle = (\hbar/2)\sin\theta\cos\left(\gamma B_0 t\right) \tag{21.109}$$

$$\left\langle S_y \right\rangle = -(\hbar/2)\sin\theta\sin\left(\gamma B_0 t\right) \tag{21.110}$$

$$\left\langle S_z \right\rangle = (\hbar/2)\cos\theta. \tag{21.111}$$

The conclusion we must draw is that $\langle \boldsymbol{S} \rangle$ is inclined with respect to the z-axis by an angle θ and the spin precesses about the field direction with a constant frequency known as the *Larmor frequency*

$$\omega_\mathrm{L} = \gamma B_0. \tag{21.112}$$

If instead, the magnetic field is inhomogeneous, then a force

$$\boldsymbol{F} = \boldsymbol{\nabla}(\boldsymbol{\mu} \cdot \boldsymbol{B}) \tag{21.113}$$

is applied to the magnetic dipole. We can isolate the effect of this force on a spin alone by shooting a beam of heavy particles through a set of magnets as shown in Fig. 21.11. In this configuration there is no force applied along the y-axis. Larmor precession causes the x-component to oscillate rapidly and average to zero. This leaves only a force in the z-direction

$$F_z = \gamma \alpha S_z. \tag{21.114}$$

The beam experiences a deflection in the z-direction and the magnitude of the deflection is proportional to the z-component of spin. A particle with spin S has $2S + 1$ M_S components; therefore, the beam splits into $2S + 1$ separate beams. An $M_S = 0$ component experiences no deflection.

This is precisely the experiment performed by Walther Gerlach and Otto Stern.[11] Stern received the Nobel Prize in Physics 1943 for his contribution to the development of molecular beam methods and his discovery of the magnetic moment of the proton.[12] His work in many ways led physical chemistry on a 'stainless steel road' stretching up to discovery (see molecular beam apparatus in the photograph at the beginning of Chapter 17). The Stern–Gerlach experiment clearly demonstrates the phenomenon of space quantization – that is, not only the quantization of angular momentum but also its projection into space. It also provides a method of producing beams with a pure spin state.

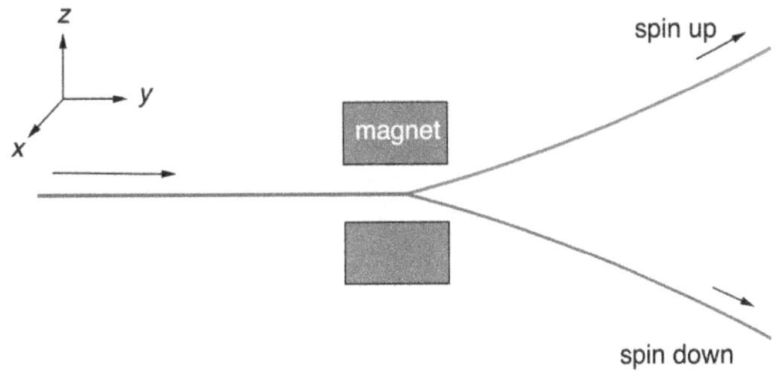

Figure 21.11 The Stern–Gerlach experiment. Passing a heavy atom with $s = {}^1/_2$ through an inhomogeneous magnetic field causes the two different M_S components to be split from one another.

Not only electrons but also nuclei can possess nonzero spin. The nuclear magnetic moment is given by

$$\boldsymbol{m}_{\rm N} = g_{\rm N}\mu_{\rm N}\boldsymbol{I}/\hbar \tag{21.115}$$

where $g_{\rm N}$ is the nuclear g factor, $\mu_{\rm N}$ the nuclear magneton, and $\boldsymbol{I}$ the *nuclear spin*. Each isotope has a characteristic value of I and if $I \neq 0$, then a characteristic value of $g_{\rm N}$ as well. Each value of I has $2I + 1$ values of M_I associated with it.

Nuclei in molecules act approximately as isolated systems interacting with a magnetic field. The Hamiltonian is

$$\hat{H} = -\boldsymbol{m}_{\rm N} \cdot \boldsymbol{B} = -(g_{\rm N}\mu_{\rm N}/\hbar)\boldsymbol{I} \cdot \boldsymbol{B} \tag{21.116}$$

which leads to an energy splitting based on the relative alignment of the nuclear spin with respect to the magnetic field

$$E_{M_I} = -g_{\rm N}\mu_{\rm N}M_I B_0. \tag{21.117}$$

This energy splitting provides the basis for *nuclear magnetic resonance* (NMR) spectroscopy. The selection rule for NMR is

$$\Delta M_I = \pm 1. \tag{21.118}$$

Therefore, the transition frequency is

$$\nu_{\rm NMR} = E/h = |g_{\rm N}|\mu_{\rm N}B_0/h \tag{21.119}$$

for a magnetic field with magnitude B_0 defining the z-axis. In principle, NMR can be performed by measuring the absorption of radiation of constant frequency when B_0 is scanned or by holding B_0 fixed and scanning ν. The quantity $\nu_{\rm NMR}/B_0$ is characteristic of a nucleus and is sensitive to the environment in which the nucleus is found.

The magnetic field interacts not only with the nucleus but also with the whole molecule. The interaction of the molecule with the external field generates a small local field about the nucleus. This local field depends on the specific response of the chemical environment – that is, it is sensitive to chemical bonding in the neighborhood of the nucleus. The local field, more often than not, opposes the external field. This *shielding* effect results in a reduced field at the nucleus. The local field B_i for nucleus i can then be written in terms of the shielding constant σ_i as

$$B_i = B_0(1 - \sigma_i). \tag{21.120}$$

The frequency of the NMR transition for nucleus i is then

$$\nu_i = |g_{\rm N}|\mu_{\rm N}B_i/h = |g_{\rm N}|\mu_{\rm N}B_0(1 - \sigma_i)/h = g_{\rm eff}\mu_{\rm N}B_0/h. \tag{21.121}$$

Each chemically distinct environment leads to peaks at distinct frequencies. These are usually displayed in terms of the chemical shift δ_i, which for proton NMR is

$$\delta_i = 10^6\frac{\nu_i - \nu_{\rm TMS}}{\nu_{\rm TMS}} = 10^6\frac{\sigma_{\rm TMS} - \sigma_i}{1 - \sigma_{\rm TMS}} \cong 10^6(\sigma_{\rm TMS} - \sigma_i) \tag{21.122}$$

where $\nu_{\rm TMS}$ is the frequency of the NMR transition associated with the protons in the standard tetramethylsilane, $Si(CH_3)_4$.

21.4.6 Directed practice

When placed in a magnetic field, how many orientations can the Z boson spin exhibit?

SUMMARY OF IMPORTANT EQUATIONS

$E_n = \frac{n^2h^2}{8ma^2}, \quad n = 1, 2, 3 \cdots$	Energy of particle of mass m in 1-D box of length a
$\psi_n(x) = \sqrt{2/a}\,\sin(n\pi x/a)$	Wavefunction of level n for particle in 1-D box
$T = \frac{4E}{V_0}e^{-2\kappa L} = \frac{4E}{V_0}e^{-L/\Gamma}$	Tunneling transmission coefficient T, particle energy E, barrier height V_0
$\kappa^2 = \frac{2m(V_0 - E)}{\hbar^2}$	Decay constant κ
$\Gamma = 1/2\kappa$	Decay length Γ
$F = -dV/dr$	Force is related to the gradient of the potential
$\omega_0 = (k/\mu)^{1/2}$	Fundamental angular frequency of harmonic oscillator ω_0, force constant k, reduced mass μ
$E_v = \left(v + \frac{1}{2}\right)\hbar\omega_0$	Energy levels of the harmonic oscillator
$\hat{l}_z\psi = m_l\hbar\psi$	Projection eigenvalue equation
$\hat{l}^2\psi = l(l+1)\hbar^2\psi$	Magnitude eigenvalue equation
$\left[\hat{l}_x, \hat{l}_y\right] = i\hbar\hat{l}_z, \quad \left[\hat{l}_y, \hat{l}_z\right] = i\hbar\hat{l}_x, \quad \left[\hat{l}_z, \hat{l}_x\right] = i\hbar\hat{l}_y$	Angular momentum commutation relations
$E_{m_j} = m_j^2\hbar^2/2I$	Energy levels of a particle on a ring
$I = mr^2$	Moment of inertia I of mass m rotating in a ring of radius r
$E_l = \frac{l(l+1)\hbar^2}{2mr^2}$	Energy levels of a particle on a sphere
$E_{M_I} = -g_N\mu_N M_I B_0$	Energy splitting of a nuclear spin level depends on nuclear g factor g_N, nuclear magneton μ_N, component of nuclear spin M_I and magnetic field strength B_0
$\Delta M_I = \pm 1$	Selection rule for NMR
$\delta_i \cong 10^6(\sigma_{TMS} - \sigma_i)$	Chemical shift δ_i is related to the shielding compared the standard TMS

Exercises

21.1 Use the conditions in Eqs (21.27)–(21.31) to show that the requirement of continuous wavefunctions and first derivatives is insufficient but that including continuous second derivatives is necessary to set the value of κ. Also show that $N = A$ because of these conditions.

21.2 Make the appropriate substitutions and derive the expression for the operator for the z-component of the angular momentum in spherical polar coordinates.

21.3 Discuss the origin of zero point energy.

21.4 Find the wavefunctions and an expression for the energies of a system for which the potential is $V(x) = -V_0$ for $-a \leq x \leq a$ and zero outside of this region. Solve the system in the limit that V_0 approaches infinity.

21.5 Prove that the particle in a box wavefunctions are orthogonal.

21.6 What are the most likely locations of a particle in a box of length L in the $n = 4$ state?

21.7 The wavefunction for a particle confined in a one-dimensional box of length L is

$$\psi_n(x) = \left(\frac{2}{L}\right)^{1/2} \sin\left(\frac{n\pi x}{L}\right).$$

Take $L = 3$ nm. Calculate the probability that a ground-state electron (i.e., an electron in the $n = 1$ state) is between $x = 1.95$ nm and 2.05 nm.

21.8 (a) Why is it necessary to apply a bias voltage between the tip and the surface in a scanning tunneling microscope? (b) In the mitochondrial electron-transport chain the redox centers need not come in contact because electron transfer occurs via tunneling between protein embedded redox groups that are separated by <14 Å. Calculate the decay length κ for electron transport in this chain based on the observation that the rate of electron transfer drops by a factor of 10 for each 1.7 Å increase in distance.

21.9 The work function of the (111) plane of Pt is 5.93 eV. Taking this as the barrier height that retains electrons in the bulk – that is, $V - E$ – calculate the decay constant and decay length.

21.10 Assume that the nanoparticles that make up porous silicon are spherical and the spacing of the electronic energy levels is given by

$$E(l) = \frac{l(l+1)\hbar^2}{2m_e}\frac{1}{r^2}.$$

Calculate the radius of the nanoparticles that would give rise to emission at 650 nm by calculating ΔE involved in the transition from the ground state to the first excited state. Note that $Z = 14$ for Si, that the electrons must obey the Pauli exclusion principle, and that l is determined by the electron count of a single atom. In other words, to determine l of the highest occupied state in ground-state configuration, use the fact that only two electrons can go in one energy level and that each level contains $2l + 1$ degenerate states. The first transition occurs between the highest occupied l level and the lowest unoccupied l level.

21.11 Derive a formula for the energy difference between successive levels for the particle in a spherical box problem.

21.12 Hexatriene, $C_6H_8^+$, has five π electrons with two each in the lowest three orbitals. Using the particle in the box model, assume that the carbon atom backbone of the molecule forms the box and that the molecule is linear with average bond length of 145 pm. What is the wavelength in nm of the light required to induce a transition from the ground state to the first excited state? What is the wavenumber in cm^{-1} of the photon that is emitted when an electron in the first excited state relaxes into the ground state?

21.13 Name three factors that favor tunneling.

21.14 Apply the boundary conditions and derive Eq. (21.35) for the dependence of tunneling current on conditions.

21.15 Evaluate the average kinetic and potential energies $\langle E_k \rangle$ and $\langle E_{pot} \rangle$ for the ground state ($n = 0$) of the harmonic oscillator by carrying out the appropriate integrations.

21.16 Assume that $^{12}C^{16}O$ acts like a harmonic oscillator and that the fundamental vibrational wavenumber is $\tilde{\nu}_0 =$ 2169.82 cm^{-1}. (a) Calculate the zero point energy and force constant of the molecule. (b) Calculate the wavelength in nm of a photon needed to excite from the $n = 1$ to $n = 2$ level.

21.17 Confirm that the wavefunction for the ground state of a one-dimensional linear harmonic oscillator, $\psi_0 = N_0 e^{-x^2/2\alpha^2}$, (a) is a solution of the Schrödinger equation; and (b) that its energy is $\frac{1}{2}\hbar\omega$.

21.18 Assume that $^1H^{35}Cl$ acts like a harmonic oscillator and that the fundamental vibrational wavenumber is $\tilde{\nu}_0 =$ 2989.74 cm^{-1}. (a) Calculate the zero point energy and force constant of the molecule. (b) Calculate the wavelength of a photon needed to excite from the $n = 1$ to $n = 3$ level.

21.19 Assume that the chlorine molecule can be treated as a harmonic oscillator with a force constant of $k = 329$ N m^{-1} and use the reduced mass

$$\mu = \frac{m_1 m_2}{m_1 + m_2}$$

for the mass of the oscillator. The mass of a ^{35}Cl atom is 34.9688 u and 36.9670 u for ^{37}Cl (a) Calculate the zero point energy of the $^{35}Cl_2$ molecule. (b) Calculate the energy in cm^{-1} of the $v = 0$ to $v = 1$ transition (also known as the fundamental transition) for $^{35}Cl_2$. (c) Calculate the wavenumber of the fundamental transition for $^{35}Cl–^{37}Cl$ isotopologue of the chlorine molecule.

21.20 Using the harmonic oscillator wavefunctions, assign each of the following integrals the appropriate value of 0, 1, finite or infinite. State with justification why each integral (bracket) has the value you assigned: $\langle 1 \mid 3 \rangle$, $\langle 1 \mid 1 \rangle$, $\langle 2 | x | 2 \rangle$, $\langle 2 | x^2 | 3 \rangle$, $\langle 2 | e^{-y^2} | 2 \rangle$.

21.21 **a** The wavefunction for the motion of a particle on a ring is

$$\psi = Ne^{\mathrm{i}m_l\phi}.$$

Determine the normalization constant N. That is, evaluate the integral

$$\int_0^{2\pi} \psi^*\psi \, \mathrm{d}\phi = 1.$$

b A point mass rotates in a circle with $l = 2$. Calculate the magnitude of its angular momentum and the possible projections of the angular momentum (in units of $\hbar$) onto an arbitrary axis (by convention we usually call this the z axis).

21.22 (a) Calculate the first five energy levels for a $^{12}C^{16}O$ molecule which has a bond length of 112.8 pm if it rotates freely in three dimensions. (b) Now calculate the first five levels if it is absorbed.

21.23 The $^{1}H^{35}Cl$ molecule has a bond length of 127.46 pm and is in the $J = 4$ rotational level. Calculate the magnitude of the rotational angular momentum, the possible projections of the angular momentum (in units of $\hbar$), the degeneracy of the level, the moment of inertia, and its rotational energy.

21.24 (a) Confirm that the wavefunction found in Eq. (21.76) yields the energy $E = m_j^2\hbar^2/2I$ when substituted into the Schrödinger equation for the particle on a ring. (b) Determine the normalization constant for this wavefunction.

21.25 For each model system below draw a diagram of the potential and energy level structure. Include the first three energy levels, their quantum numbers and appropriate coordinates in the diagram. Write equations to describe the potential and total energy in each case. (a) Particle in a 1-D box. (b) H atom. (c) Harmonic oscillator.

21.26 Calculate: (a) the mean potential energy and (b) the mean kinetic energy of an electron in the ground state of a hydrogenic atom. Recall that the mean value corresponds to an expectation value. The ground-state wavefunction is

$$\psi = \left(\frac{1}{\pi a_0^3}\right)^{1/2} e^{-r/a_0}.$$

$$\Lambda^2\psi = 0$$

Endnotes

1. Formally, the limits of the orthogonality integral should extend from $-\infty$ to ∞. However, since we know that the wavefunction goes to zero outside of the box, the only relevant portion of all space is 0 to a.
2. Canham, L.T. (1990) *Appl. Phys. Lett.*, **57**, 1046.
3. Hill, N.A. and Whaley, K.B. (1995) *Phys. Rev. Lett.*, **75**, 1130.
4. Wolkin, M.V., Jorne, J., Fauchet, P.M., Allan, G., and Delerue, C. (1999) *Phys. Rev. Lett.*, **82**, 197; Kolasinski, K.W., Aindow, M., Barnard, J.C., Ganguly, S., Koker, L., Wellner, A., Palmer, R.E., Field, C., Hamley, P., and Poliakoff, M. (2000) *J. Appl. Phys.*, **88**, 2472.
5. Bohr, N. (1913) *Philos. Mag.*, **26**, 1; Bohr, N. (1920) *Z. Phys.*, **2**, 423.
6. Binnig, G., Rohrer, H., Gerber, C., and Weibel, E. (1982) *Phys. Rev. Lett.*, **49**, 57.
7. Frequently, especially in the context of diatomic spectroscopy k and ω_0 will be denoted k_e and ω_e.
8. Space itself is not quantized; however, the *projection of angular momentum in space* is quantized and this is what we mean by space quantization.
9. Pauli, W. (1925) *Z. Phys.*, **31**, 765.
10. Ulenbeck, G.E. and Goudsmit, S. (1925) *Naturwissenschaften*, **13**, 953; Ulenbeck, G.E. and Goudsmit, S. (1926) *Nature (London)*, **117**, 264.
11. Gerlach, W. and Stern, O. (1922) *Z. Phys.*, **9**, 349; Gerlach, W. and Stern, O. (1922) *Z. Phys.*, **9**, 353.
12. Gerlach received 30 nominations for a Nobel Prize but never received one. Gilbert N. Lewis had 41, but Arnold Sommerfeld holds the record of 84 nominations without an award.

Further reading

Atkins, P.W. and Friedman, R.S. (2010) *Molecular Quantum Mechanics*, 5th edition. Oxford University Press, Oxford.

Griffiths, D.J. (2005) *Introduction to Quantum Mechanics*. Pearson Education, Upper Saddle River, NJ.

Metiu, H. (2006) *Physical Chemistry: Quantum Mechanics*. Taylor & Francis, New York.

Schatz, G.C. and Ratner, M.A. (2002) *Quantum Mechanics in Chemistry*. Dover Publications, New York.

Winn, J.S. (1995) *Physical Chemistry*. HarperCollins College Publishers, New York.

Zare, R.N. (1988) *Angular Momentum: Understanding Spatial Aspects in Chemistry and Physics*. John Wiley & Sons, New York.

CHAPTER 22

Atomic structure

> **PREVIEW OF IMPORTANT CONCEPTS**
>
> - The Schrödinger equation can be solved exactly for the simple hydrogen atom and hydrogenic ions.
> - For the H atom, solutions to the Schrödinger equation are composed of a radial part, on which the energy depends, and an angular part, which describe classes of shapes.
> - The classes of shapes define the s, p, d, f, … orbitals of chemical lore. These solutions give rise to electron shells, subshells, and the form of the Periodic Table.
> - The energy of the levels in hydrogenic atoms depends to a first approximation only on *n*, the principal quantum number.
> - In multi-electron atoms the dependence of screening on the azimuthal quantum number *l* lifts the degeneracy of orbitals with the same value of *n*.
> - High-resolution spectroscopy of the H atom has led to refinements of the Schrödinger equation, improvements to the Dirac equation, pushed the development of quantum electrodynamics, and is now providing a stringent test of the Standard Model of particle physics.
> - A total wavefunction must also take into account the spin of the electron and be antisymmetric with respect to the exchange of electrons.
> - The Pauli exclusion principle dictates that no two electrons in a system can have the same set of quantum numbers, and that no more than two electrons can occupy the same orbital.
> - Pauli repulsion means that electrons of the same spin naturally exhibit correlated motion that allows them to avoid one another.
> - The dependence of orbital energy and screening on *n* and *l*, together with Pauli repulsion, allow us to understand the energetic ordering of orbitals (Aufbau principle) as well as some exceptions to it.
> - The presence of electron correlation and exchange mean that the Schrödinger equation is not analytically soluble for multi-electron atoms, and we must find approximate solutions. Approximate orbitals can be constructed from linear combinations of basis functions.
> - Koopmans' theorem allows us to approximate the ionization energy as being equal to the negative of the orbital energy of the orbital from which the electron originated.

We have now been introduced to a number of empirical properties of atoms and molecules. We have seen the 'old' quantum theory, which was the first set of *ad hoc* explanations of quantum theory that bolted on quantum mechanical elements to classical physics because it had to, rather than because they were derived from first principles. One of the most successful and insightful models in 'old' quantum theory is the Bohr atom. Electron orbits are quantized because they have to be, and they do not fall into the nucleus because there must be a minimum value of the angular momentum. These orbits have associated with them a principal quantum number *n* because we need it. Then Pauli says there must also be a half-integer spin of the electron, which makes no sense, but fits the data and seems to make things work. Another extremely successful idea is that vibrations and photons are quantized with an energy directly proportional to their frequency.

The importance of the old quantum theory is that it allowed us to get used to the strange consequences that result from the mathematics of proper quantum mechanics. We have also seen quantum theory applied from first principles to model systems: free particle, particle in a box, harmonic oscillator, and particle on a sphere. It is time to bring the quantum mechanical machinery to bear on the atom and derive its structure from first principles. Thereby, we will discover what the quantum description of atoms, molecules and bonding really entails. We will also discover even more nuance and richness in the amazing world of light–matter interactions (spectroscopy, lasers and photochemistry) and the insight they reveal on the workings of the universe. This is our task over the next several chapters.

Physical Chemistry: How Chemistry Works, First Edition. Kurt W. Kolasinski.

Companion Website: www.wiley.com/go/kolasinski/physicalchemistry

22.1 The hydrogen atom

The H atom corresponds to a radial Coulomb field with a potential proportional to the inverse of distance,

$$V = -\frac{e^2}{4\pi\varepsilon_0}\frac{1}{r}. \tag{22.1}$$

Classically this is unstable. The lowest energy in the system corresponds to the electron spiraling into the nucleus where it crash-lands and stays. Bohr was able to rectify this situation by asserting that the angular momentum of the electron must be quantized and its minimum value is not zero. Doing so allowed him to derive an expression for the lines observed in the H atom emission (or absorption) spectrum. He surmised that the energy level structure of the H atom corresponds to levels with distinct energies. Transitions between levels correspond to quantum jumps that only occur at well-defined energies. The energy difference between levels is made up by the absorption or emission of a photon in order to conserve energy. Now, we seek a more fundamental *derivation* of the H atom structure that arises from the postulates of quantum mechanics. Such a derivation should contain within the mathematics additional predictions and deeper insight.

Within Bohr's model, the radius of the electron's orbit is also quantized. The motion of an electron in a Coulomb field constrained to concentric shells is analogous to the particle on a sphere model that we solved in the last chapter. Thus, much of the heavy lifting in terms of the mathematics has already been done. From this we understand the relevance of spherical harmonics to atomic orbitals. At least, the angular part has been solved. So far we have a collection of concentric spheres but we do not know how they relate to one another as a function of radius. That is, there is a radial dependence to the problem, which we need to understand.

22.1.1 Set up hamiltonian and solve Schrödinger equation

The procedure is to define a complete Hamiltonian, and then to solve the resulting Schrödinger equation for the infinite set of wavefunctions that describe the energy level structure. With the coordinate system defined as in Fig. 22.1 we can write the Hamiltonian in terms of the kinetic energy operator plus the electrostatic potential of the electron–proton attraction,

$$\hat{H} = \hat{T} + \hat{V} = -\frac{\hbar^2}{2\mu}\nabla^2 - \frac{e^2}{4\pi\varepsilon_0}\frac{1}{r} \tag{22.2}$$

where μ is the reduced mass

$$\mu = \frac{m_e m_p}{m_e + m_p}. \tag{22.3}$$

Spherical symmetry calls for the Laplacian to be written in spherical polar coordinates as done in Eq. (21.82), which we repeat here

$$-\frac{\hbar^2}{2\mu}\left[\frac{1}{r^2}\frac{\partial}{\partial r}\left(r^2\frac{\partial\psi}{\partial r}\right) + \frac{1}{r^2\sin\theta}\frac{\partial}{\partial\theta}\left(\sin\theta\frac{\partial\psi}{\partial\theta}\right) + \frac{1}{r^2\sin^2\theta}\frac{\partial^2\psi}{\partial\phi^2}\right] + V\psi = E\psi \tag{22.4}$$

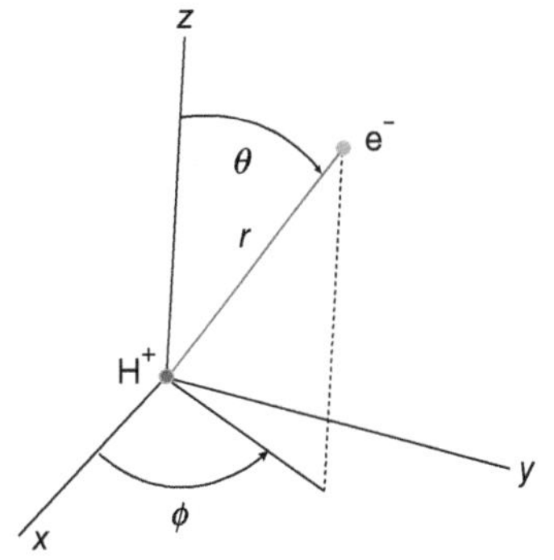

Figure 22.1 Coordinate system for the hydrogen atom in spherical polar coordinates.

Because of the analogy to the particle on a sphere model, we expect the angular part of the wavefunction to be solved by the spherical harmonics $Y_{lm}(\theta,\phi)$. Assume a separable solution composed of the product of radial and angular parts,

$$\psi(r,\theta,\phi) = R_{nl}(r)\, Y_{lm}(\theta,\phi). \tag{22.5}$$

We do not know a priori that the radial wavefunction $R_{nl}(r)$ will depend on n and l, but this will indeed be the case.

As we demonstrated in Chapter 21 in Eq. (21.86), the spherical harmonics $Y_{lm}(\theta,\phi)$ are eigenfunctions of the Legendrian operator (the angular part of the Laplacian operator) with eigenvalue $l(l+1)$, thus

$$\Lambda^2\, Y_{lm}(\theta,\phi) = -l(l+1)Y_{lm}(\theta,\phi). \tag{22.6}$$

Now we need to solve the *radial equation*, which from Eq. (21.85) we know to be

$$\frac{1}{R}\frac{\partial}{\partial r}\left(r^2\frac{\partial R}{\partial r}\right) - \frac{2\mu r^2}{\hbar^2}(V-E) = l(l+1) \tag{22.7}$$

While perhaps not immediately obvious unless you sit around looking at differential equations all the time, the solution to this second-order differential equation is made easier by the substitution

$$P(r) = rR(r), \tag{22.8}$$

which yields

$$-\frac{\hbar^2}{2\mu}\frac{\mathrm{d}^2P}{\mathrm{d}r^2} + \left[\frac{-e^2}{4\pi\varepsilon_0 r} + \frac{l(l+1)\hbar^2}{2\mu r^2}\right]P = EP. \tag{22.9}$$

Equation (22.9) is a one-dimensional Schrödinger equation in which the Coulomb potential has been replaced by an effective potential energy

$$V_{\text{eff}} = -\frac{e^2}{4\pi\varepsilon_0 r} + \frac{l(l+1)\hbar^2}{2\mu r^2}. \tag{22.10}$$

The first term is the Coulomb potential. The second is a repulsive centrifugal angular momentum dependent term. Substituting

$$\kappa = (-2\mu E)^{1/2}/\hbar \tag{22.11}$$

and dividing by E, which is negative for bound states, we can rewrite Eq. (22.9) as

$$\frac{1}{\kappa^2}\frac{\mathrm{d}^2P}{\mathrm{d}r^2} = \left[1 - \frac{\mu e^2}{2\pi\varepsilon_0\hbar^2\kappa}\frac{1}{\kappa r} + \frac{l(l+1)}{(\kappa r)^2}\right]P. \tag{22.12}$$

To make this even more compact, we introduce

$$\rho = \kappa r,\ \text{and}\ \rho_0 = \frac{\mu e^2}{2\pi\varepsilon_0\hbar^2\kappa}, \tag{22.13}$$

which allows us to write

$$\frac{\mathrm{d}^2P}{\mathrm{d}\rho^2} = \left[1 - \frac{\rho_0}{\rho} + \frac{l(l+1)}{\rho^2}\right]P \tag{22.14}$$

The form of the $R_{nl}(r)$ that provides the solution to Eq. (22.14) is given by

$$R_{nl}(r) = \frac{1}{r}\rho^{l+1}\, e^{-\rho}\, v(\rho). \tag{22.15}$$

The $v(\rho)$ term is related to the associated Laguerre polynomials, which are polynomials in ρ of degree $(n-l-1)$. For an excellent review of the method of solving for these functions, see the book by Griffiths in Further Reading. The form of $R_{nl}(r)$ is (exponential) × (polynomial). Examples of the $R_{nl}(r)$ are given in Table 22.1. The exponentially decaying term constrains the wavefunction to be localized in the region of space close to the nucleus. The polynomial of order $(n-l-1)$ introduces $(n-l-1)$ radial nodes into the wavefunction. A general rule will be that the greater the number of nodes, the higher the energy of the associated state. More nodes correlate with greater curvature in the wavefunction and, therefore, higher kinetic energy.

The radial wavefunction is found to depend on both the principal quantum number n,

$$n = 1, 2, 3, \ldots \tag{22.16}$$

Table 22.1 Hydrogenic radial wavefunctions derived from solution of the Schrödinger equation.

Orbital	n	l	$R_{n,l}$
1s	1	0	$R_{10} = 2a^{-3/2}\exp(-r/a)$
2s	2	0	$R_{20} = \frac{1}{\sqrt{2}}a^{-3/2}\left(1-\frac{r}{2a}\right)\exp(-r/2a)$
2p	2	1	$R_{21} = \frac{1}{\sqrt{24}}a^{-3/2}\left(\frac{r}{a}\right)\exp(-r/2a)$
3s	3	0	$R_{30} = \frac{2}{\sqrt{27}}a^{-3/2}\left(1-\frac{2r}{3a}+\frac{2}{27}\left(\frac{r}{a}\right)^2\right)\exp(-r/3a)$
3p	3	1	$R_{31} = \frac{8}{27\sqrt{6}}a^{-3/2}\left(1-\frac{r}{6a}\right)\left(\frac{r}{a}\right)\exp(-r/3a)$
3d	3	2	$R_{32} = \frac{4}{81\sqrt{30}}a^{-3/2}\left(\frac{r}{a}\right)^2\exp(-r/3a)$
4s	4	0	$R_{40} = \frac{1}{4}a^{-3/2}\left(1-\frac{3r}{4a}+\frac{1}{8}\left(\frac{r}{a}\right)^2-\frac{1}{192}\left(\frac{r}{a}\right)^3\right)\exp(-r/4a)$
4p	4	1	$R_{41} = \frac{\sqrt{5}}{16\sqrt{3}}a^{-3/2}\left(1-\frac{r}{4a}+\frac{1}{80}\left(\frac{r}{a}\right)^2\right)\left(\frac{r}{a}\right)\exp(-r/4a)$
4d	4	2	$R_{42} = \frac{1}{64\sqrt{5}}a^{-3/2}\left(1-\frac{r}{12a}\right)\left(\frac{r}{a}\right)^2\exp(-r/4a)$
4f	4	3	$R_{43} = \frac{1}{768\sqrt{35}}a^{-3/2}\left(\frac{r}{a}\right)^3\exp(-r/4a)$
			$a = \frac{4\pi\varepsilon_0\hbar^2}{\mu e^2}$ [footnote 1]

and the orbital angular momentum (azimuthal) quantum number l. The value of l is constrained by n such that

$$l = 0, 1, 2, \ldots, n-1. \tag{22.17}$$

For each l value there are $(2l+1)$ possible values of m. This means that the *degeneracy* of each level is

$$g(n) = \sum_{l=0}^{n-1}(2l+1) = n^2 \tag{22.18}$$

However, the energy of the H atom levels depends only on n according to

$$E_n = -\frac{\mu e^4}{32\pi^2\varepsilon_0^2\hbar^2}\left(\frac{1}{n^2}\right). \tag{22.19}$$

There is no l dependence to the energy of the H atom. Therefore, only R_{nl} is required to derive its energy level structure. If the reduced mass is calculated specifically for the H atom using $\mu = \mu_H$, then the value of the constants out front is equal to $R_H = 109\,677.581\ \text{cm}^{-1}$. If an infinite nuclear mass is used such that $\mu = m_e$, then we obtain $R_\infty = 109\,737.31534\ \text{cm}^{-1}$. The number of digits may seem extreme and, indeed it is. However, the *Rydberg constant* is arguably the most accurately measured constant in the pantheon of physical constants. It is known with a relative uncertainty of about 8×10^{-13} from the work of Theodor W. Hänsch and coworkers.[2] Hänsch was awarded the Nobel Prize in Physics 2005 for his work in high-resolution laser spectroscopy. At this level of accuracy, spectroscopy on the H atom and its exotic cousin muonic hydrogen (a muon bound by a proton) is able to probe the structure of the proton and provide a stringent test to the Standard Model of particle physics.

The H atom spatial wavefunction is represented by

$$\psi_{nlm}(r,\theta,\phi) = R_{nl}(r)Y_{lm}(\theta,\phi). \tag{22.20}$$

These wavefunctions form an orthonormal set that obeys the condition

$$\int_0^{2\pi} d\phi \int_0^{\pi} \sin\theta\, d\theta \int_0^{\infty} r^2\, \psi^*_{nlm}\, \psi_{n'l'm'}\, dr = \delta_{nn'}\delta_{ll'}\delta_{mm'}.$$
$$\text{or equivalently} \int_0^{\infty}\int_0^{\pi}\int_0^{2\pi} \psi^*_{nlm}\, \psi_{n'l'm'}\, r^2 \sin\theta\, dr\, d\theta\, d\phi = \delta_{nn'}\delta_{ll'}\delta_{mm'}. \tag{22.21}$$

In the next section we will see that this wavefunction is not yet complete because it does not include the electron spin.

22.2 How do you make it better? The Dirac equation

Great excitement accompanied the verification that the Schrödinger equation returned energies that correctly predicted the structure of the H atom spectrum. It won Erwin the Nobel Prize in Physics, which he shared with Dirac in 1933. However, as experimental resolution and sensitivity increased, systematic deviations from the predictions of the Schrödinger equation became apparent almost immediately. These deviations indicated that the gross structure of the H atom must be as predicted but that an underlying fine and even hyperfine structure needed further explanation.

To improve the predictions of the Schrödinger equation as solved above, we need to fix the Hamiltonian by making it consistent with relativity. Schrödinger had attempted this but abandoned the effort because of what appeared to be internal inconsistencies. Wolfgang Pauli delivered an essential clue about how to proceed when he proposed[3] the necessity for a new quantum number to describe the intrinsic magnetic moment of the electron. For historical reasons, he called this quantum number 'spin,' even though this is conceptually misleading. Importantly, he showed that the spin of an electron is half-integer, a concept that had previously been thought to be untenable. This insight allowed Paul Dirac to formulate the Dirac equation in 1930 – a fully relativistic form of the Schrödinger equation.[4] The Dirac equation combines quantum theory with Maxwell's theory of the electromagnetic field to produce quantum field theory, from which the dual nature (wave/particle as well as matter/antimatter) of subatomic particles can be deduced directly. The Dirac equation is much more complicated than the Schrödinger equation and we will not show it here. However, the resulting first-order corrections can be simply explained.

The Dirac equation shows us how to fix the Hamiltonian to account for relativistic effects. In so doing it will also show us a basic principle of how to treat systems quantum mechanically.

- Work out the Hamiltonian that contains the most important interactions (the minimum set of interactions that give the system its fundamental structure).
- Then add smaller perturbations to this minimal set of interactions by adding additional terms to the Hamiltonian.

To improve the accuracy of the Hamiltonian for the H atom this is done by adding three terms to the zeroth-order kinetic + potential energy term $\hat{H}_0$ derived from the Schrödinger equation,

$$\hat{H} = \hat{H}_0 + \hat{H}_{\text{m}} + \hat{H}_{\text{D}} + \hat{H}_{\text{so}}. \tag{22.22}$$

From the $\hat{H}_0$ term, the energy only depends on n. The three corrections are the mass-velocity term $\hat{H}_{\text{m}}$, which accounts for relativistic variation of the inertia of the electron with velocity. Written with the fine structure constant $\alpha = e^2/\hbar c$

$$\hat{H}_{\text{m}} = -(\alpha^2/4)(E - V)^2. \tag{22.23}$$

$E - V$ is the kinetic energy of the electron. Thus, this term is the relativistic correction in the form of $-E_{\text{k}}(E_{\text{k}}/2mc^2)$. The Darwin term is physically obscure, though some state that it corrects for the relativistic electric moment of the electron and nonlocalizability[5]

$$\hat{H}_{\text{D}} = -(\alpha^2/4)\left(\frac{\partial V}{\partial r}\right)\frac{\partial}{\partial r}. \tag{22.24}$$

The third term is the *spin–orbit term*,

$$\hat{H}_{\text{so}} = (\alpha^2/2)\,(1/r)\left(\frac{\partial V}{\partial r}\right)\vec{l}\bullet\vec{s}. \tag{22.25}$$

It represents the energy associated with the interaction of the intrinsic magnetic moment of the electron and the magnetic field generated by the orbital motion of the electron as it moves in the electric field of the nucleus. To simplify Eqs (22.24–22.25), we use atomic units in which distance is measured in terms of the Bohr radius a_0 and energy is in Rydberg units.

Quantization within a four-dimensional space-time leads to the natural introduction of a fourth quantum number, the spin s, just as quantization within a three-dimensional space leads to the need for the three quantum numbers n, l and m.[6] As expected, the energies of the states become dependent on the spin when a Hamiltonian that takes into account relativistic effects is solved in four-dimensional space-time. In fact, the spin–orbit term makes the energy dependent on the coupling of l and s, represented by $j = l + s$. The total energy of each level, now specified by the quantum numbers n, l and j is

$$E_{nlj}/\left(e^2/2a_0\right) = -\frac{Z^2}{n^2} + E_{\text{m}} + E_{\text{D}} + E_{\text{so}} = -\frac{Z^2}{n^2} - \frac{\alpha^2 Z^4}{4n^4}\left[\frac{4n}{j+\frac{1}{2}} - 3\right] \tag{22.26}$$

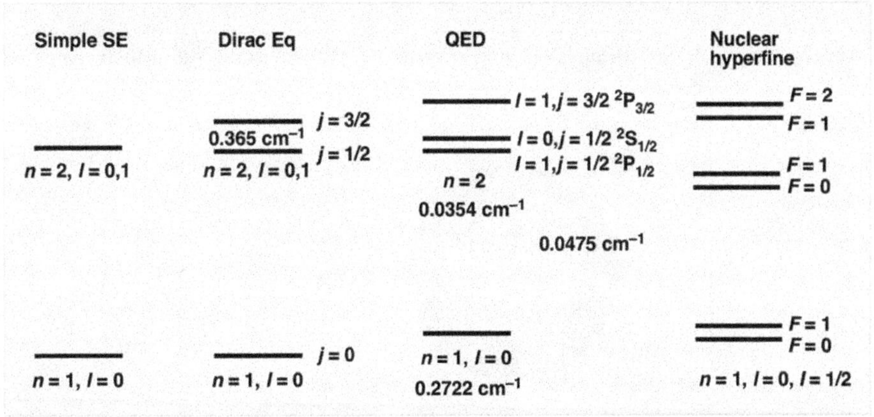

Figure 22.2 The progressive development of the H atom spectrum as higher levels of theory from the simple Schrödinger equation to the Dirac equation to quantum electrodynamics (QED) to the coupling of QED corrections with nuclear hyperfine splitting match higher levels of experimental resolution and sensitivity.

The term

$$\frac{e^2}{2a_0} = \frac{\hbar^2}{2m_e a_0^2} = 13.6058 \text{ eV} \tag{22.27}$$

is the atomic units unit of energy known as the *Rydberg*, Ry. The spin–orbit term increases in size with increasing atomic number Z. For heavier atoms, this term results in large shifts that will be dealt with in more detail in Chapter 23.

The simple Schrödinger equation leads to a single ground-state level with $n = 1$ and $l = 0$. There are two degenerate excited states, $n = 2$, $l = 0, 1$. The Dirac equation leads to a ground state with $n = 1$, $l = 0$, $j = {}^1\!/_2$. The two excited states with $n = 1$ and $l = 0, 1$ are now split into two different energies in which the $j = {}^1\!/_2$ state lies 0.365 cm^{-1} below the $j = 3/2$ state. These splitting are shown in Fig. 22.2.

Quantum electrodynamics and accounting for the finite size of the nucleus provides the next correction. The ground state $^2S_{1/2}$ shifts up by 0.2722 cm^{-1}. There are now three excited states as the $j = {}^1\!/_2$ state splits by 0.0354 cm^{-1} between the state derived from $l = 1$ (lower E, $^2P_{1/2}$) and the state from $l = 0$ ($^2S_{1/2}$). This is the *Lamb shift*, which is related to vacuum fluctuations (i.e., the fabric of the universe or the quantization of the electromagnetic field). These fluctuations cause deviations in electron trajectories from spherical symmetry. They are also responsible for spontaneous emission, that is, fluorescence. The two $j = {}^1\!/_2$ states both couple to the continuum of states resulting from vacuum fluctuations. The difference in coupling of theses states via a photon to the ground state leads to a slightly different lifetime for each state. Because there is an uncertainty relationship between lifetime and energy, the slight difference in lifetime leads to a slight shift in energy. The $l = 1$, $j = 3/2$ state ($^2P_{3/2}$) shifts upward.

The next correction is *nuclear hyperfine splitting*, which results from the interaction of the nuclear spin I with the electron. $F = 0$ and $F = 1$ states in ground state split by 0.0475 cm^{-1}.

The perturbations introduced as corrections to the Hamiltonian also slightly change the wavefunctions of the H atom. These changes are extremely small in comparison to the structure imposed by the zeroth-order Hamiltonian. We need not consider the changes in any of our discussions below. However, it will be worth keeping them in mind, as well as the ideas of perturbations and the emergence of more levels at high resolution when we explore atomic spectroscopy in the next chapter. While the splittings and corrections introduced are quite small, there are three very important concepts to take away from this section:

- A fully relativistic treatment of the H atom demands the inclusion of spin in the complete wavefunction, which makes the Pauli exclusion principle a natural consequence of the electron being a fermion.
- An extremely effective method of modeling the behavior of complex systems is to establish a Hamiltonian that contains the essence of the system, and then bolt on to this smaller perturbations that improve the description of the system.
- Radiative relaxation of an excited state (called in different contexts spontaneous emission, fluorescence or phosphorescence) is caused by the interaction of excited states with fluctuations in the vacuum field that arise from the quantization of the electromagnetic field that fills all space.

Table 22.2 A comparison of different terms of reference to solutions to the Schrödinger equation.

<table>
<tr><th>Energy levels – Place in energy space</th><th>Orbitals – Region of space</th><th>Wavefunctions – Probability amplitude</th><th>Dirac Notation</th></tr>
<tr><td>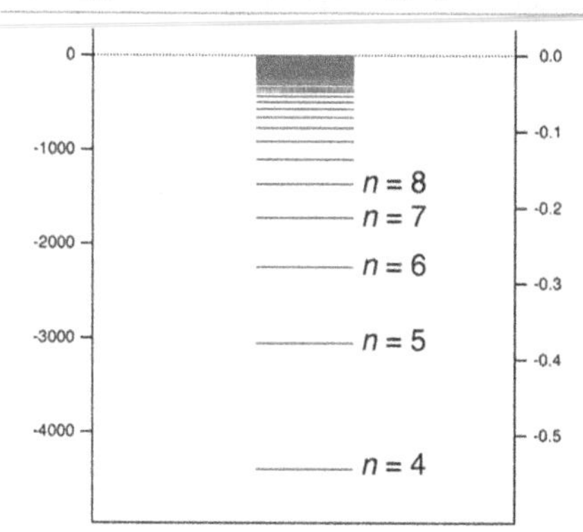
</td><td></td><td>$\psi_{100}(r,\theta,\phi) = 2\left(\frac{Z}{a_0}\right)^{3/2}\left(\frac{1}{4\pi}\right)^{1/2} e^{-\rho/2}$</td><td>$|100\rangle$</td></tr>
</table>

22.2.1 Directed practice

Why are the corrections imposed by the Dirac equation much larger for the hydrogenic Cs ion than for the H atom?

22.3 Atomic orbitals

The Dirac equation tells us that the solution to the Schrödinger equation is incomplete. The complete form of the wavefunction is a product not only of the radial and angular parts but also a spin eigenvector

$$\psi_{nlmsm_s}(r,\theta,\phi,s) = R_{nl}(r)\; Y_{lm}(\theta,\phi)\; \sigma_{m_s}(s)\,. \tag{22.28}$$

That spin needed to be included was already known from the work of Pauli. Indeed, his proposal of a half-integer quantum number was an essential clue that aided Dirac in formulating and solving his equation. However, the spin part of the overall wavefunction is distinct from the radial and angular parts in that it does not depend on spatial coordinates. It only depends on whether the spin of the electron is up ($m_s = {}^1\!/_2$) or down ($m_s = -{}^1\!/_2$). It is this portion of the total wavefunction that is responsible for only two electrons being able to occupy the same orbital.

22.3.1 Comparing representations

Solutions to the Schrödinger equation are given different names to emphasize different aspects depending on context, see Table 22.2. Energy levels, orbitals and wavefunctions are closely related. They emphasize different aspects of the same thing. Quantum numbers are used to describe all of these. The values of the quantum numbers determine the values of quantities such as energy, angular momentum, magnetic moment, and so forth.

For the H atom $E_n = -hcR_H/n^2$ and the explanation of the Rydberg formula,

$$h\nu = \Delta E = E(n_2) - E(n_1) = hcR_H\left(\frac{1}{n_1^2} - \frac{1}{n_2^2}\right), \tag{22.29}$$

is the same obvious explanation given by the Bohr model. The energy levels are quantized and the energy depends on n. There are an infinite number of bound states, $n = 1 \rightarrow \infty$. The energy of the $n = \infty$ level is 0. Transitions correspond to jumps between these quantized energy levels and are called *bound-bound transitions*. Bound-bound transitions correspond to sharp lines in spectra because they have well-defined energies. The photon that connects these levels through absorption or emission must match the energy difference between the states to ensure energy conservation. Small corrections to the Rydberg formula reflect corrections introduced by higher-level theories.

Above the ionization limit the electron is no longer bound to the nucleus. Because it is not confined to the region of the nucleus, the electron's energy is no longer quantized. Excitation above $n = \infty$ corresponds to ionization of the atom

Ionization $\quad H \rightarrow H^+ + e^-.$

Ionization is a *bound-free transition* that removes the electron from the atom into a continuum of states that can have any value of kinetic energy. The electron has been taken out of its spherical box. Such a transition is still subject to the limits of energy conservation, just as in the photoelectric effect. However, these transitions do not give rise to sharp lines; instead, bound-free transition gives rise to continuous spectra. They may still exhibit structure because the strength of the transition into the continuum will depend on the frequency of the light that excites it; but this structure resembles broad overlapping features rather than sharp peaks.

22.3.1.1 Example

Determine the *ionization energy* E_i (also known as the *ionization potential I*) of the first three levels of the H atom. *Koopmans' theorem* states that the ionization energy is equal to the negative of the orbital energy of the orbital from which the electron originated. For a one-electron system, this theorem is exact. We will discuss deviations from this in the next chapter. Numerical results are given in Table 22.3.

$$E_n = -\frac{hcR_\mathrm{H}}{n^2} \quad \therefore I_n = \Delta E(n_1, n_2 = \infty) = hcR_\mathrm{H}\left(\frac{1}{n_1^2} - \frac{1}{\infty}\right) = hcR_\mathrm{H}\left(\frac{1}{n_1^2}\right)$$

Table 22.3 The ionization energies of the first three H atom levels.

n	n^2	I/cm^{-1}	I/eV
1	1	109 679	13.599
2	4	27 420	3.400
3	9	12 187	1.511

22.3.2 Shells and subshells

The solutions of the Schrödinger equation impose a structure on the electronic states that is summarized in Fig. 22.3. This is, of course, the basis of the structure of the Periodic Table. Orbitals with the same nl are called *equivalent orbitals*, while electrons having the same nl are *equivalent electrons*. The value of n determines the *shell* to which an orbital belongs. For each value of nl there is a set of $2l + 1$ equivalent orbitals that forms a *subshell* that can hold a maximum of $2(2l + 1)$ electrons. The 'extra' factor of 2 arises from the Pauli exclusion principle, which dictates that one electron has $m_s = {}^1/_2$, the other $m_s = -{}^1/_2$, and that there are no more than two in an orbital because no two electrons can have the same set of quantum numbers.

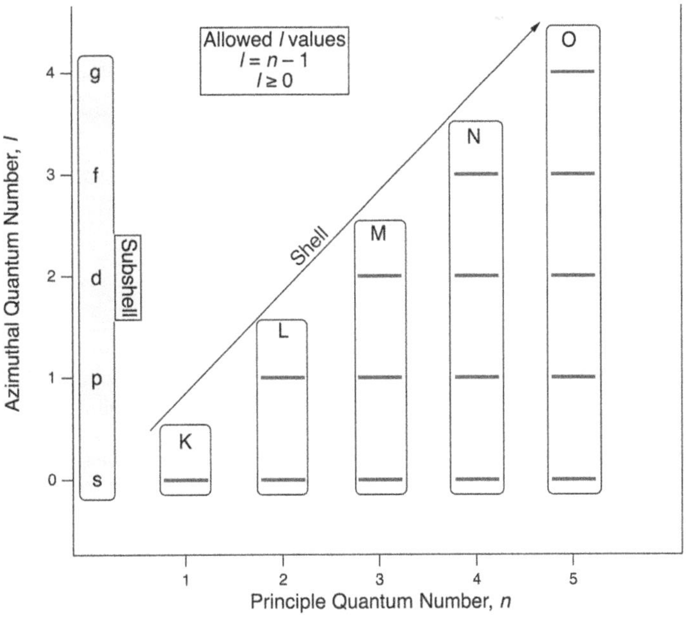

Figure 22.3 The structure of orbitals categorized by shell (determined by the value of n) and subshell (determined by the value of l).

Table 22.4 Allowed values of orbital angular momentum l, subshells and shells as function of the principle quantum number n.

n	l	Subshells	Shell
1	0	s	K
2	0, 1	s, p	L
3	0, 1, 2	s, p, d	M
4	0, 1, 2, 3	s, p, d, f	N
5	0, 1, 2, 3, 4	s, p, d, f, g	O
etc.			

A subshell with the maximum number of electrons is a closed subshell. For example, in the 2p subshell, there are three p orbitals – commonly referred to as p_x, p_y and p_z in chemistry – that can hold a maximum of six electrons to form the closed subshell denoted $2p^6$. Note that $2l + 1$ corresponds to the number of different m_l values for that value of l. There are $2l + 1$ orbitals for each value of l because there are $2l + 1$ allowed projections of l.

For a given value of n, the allowed values of l are $l = 0, 1, 2, \ldots, n - 1$; thus, there are n subshells with that value of n. Taken together, the subshells of a given n form a shell of the atom. Shells are referred to by upper-case letters starting with K for $n = 1$ and proceeding alphabetically beyond that. Half-filled and fully filled (closed) subshells are particularly important as points of stability and in angular momentum coupling. Tables 22.4 and 22.5 summarize the details described here.

22.3.3 Shapes of orbitals

Wavefunctions of atomic states describe probability amplitudes of finding the electron that occupies that state. Their absolute square is associated with orbitals, the regions of space in which electrons reside. Their energy and shape determine how they interact with one another and, as we shall see, these will also determine how the orbitals interact regarding chemical bonding. The shapes of s, p, and d orbitals are shown in Fig. 22.4. It is the number of nodal planes, equal to the value of l, that distinguishes these characteristic shapes. Any given orbital of the same l will have the same symmetry about the nucleus regardless of the value of n. However, as n increases both the size of the orbital and the number of nodes, equal to $n - l - 1$, increases. The remaining nodes are radial nodes.

Most of the charge density from the electron distribution is centered about the outermost antinode, and therefore, within a spherical shell of moderate thickness. As a function of n, the lower the value of n, the closer the electron density is centered about the nucleus. In other words, orbital size for a given orbital class increases with increasing n. For a given value of n, the maximum in charge density is further from the nucleus the smaller the value of l. However, the probability of an electron lying close to the nucleus is greatest for small l. Indeed, only for $l = 0$ (s electron) is there any probability of finding the electron at $r = 0$. This means that only s electrons exhibit appreciable interactions with the nucleus, which also means that they are the electrons most susceptible to isotopic shifts. In hydrogenic atoms, these two effects counterbalance each other and the orbital energy depends only on n not l.

This is not the case for multi-electron atoms. For multi-electron atoms, the radial distribution function is similar to but not identical to those shown for hydrogenic atoms in Fig. 22.4. It is still true that only an s electron has a nonvanishing probability at the nucleus. However, the trend in the peak of electron density with l is reversed. Higher l orbitals (for the

Table 22.5 Number of orbitals and designations of half-filled and closed subshells for the most commonly encounters subshells. The value of orbital angular momentum l is synonymous with a letter that designates the subshell.

Subshell	l	# of orbitals = $2l + 1$	Half-filled subshell	Closed subshell
s	0	1	s	s^2
p	1	3	p^3	p^6
d	2	5	d^5	d^{10}
f	3	7	f^7	f^{14}
g	4	9	g^9	g^{18}

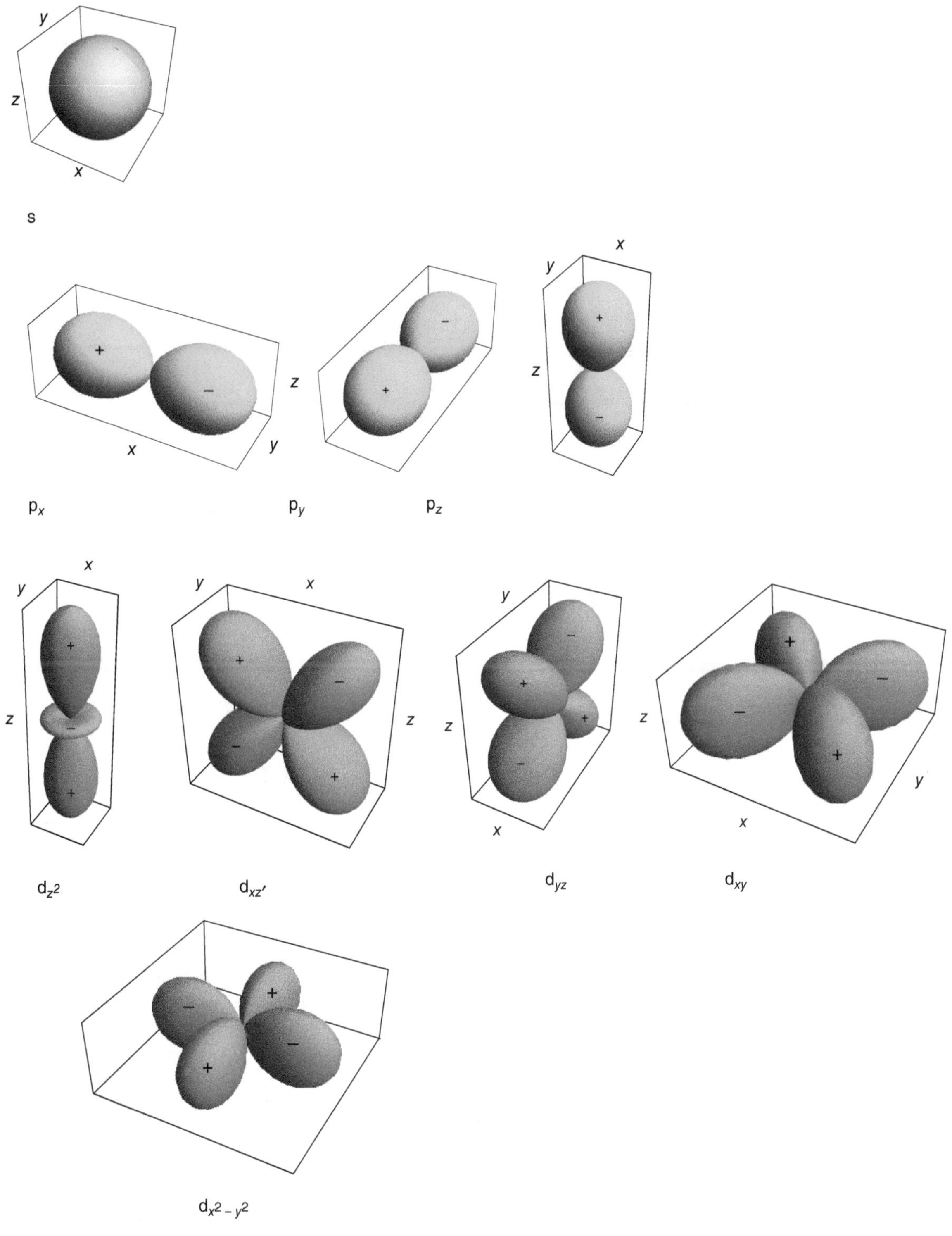

Figure 22.4 (a) An s orbital has $l = 0$ and $m_l = 0$. The origin lies at the center of the sphere. It has no nodal planes and $n - 1$ radial nodes. The angular shape shown here of an s orbital is determined by the absolute square of the corresponding spherical harmonic, $|Y_0^0(\theta, \phi)|^2$. Orbitals have infinite extent but are predominantly localized to a region of space near the origin. What is plotted here is a figure that contains most of the probability distribution. See Fig. 22.5 for a dimensionally correct depiction of the radial probability distribution of the H atom. (b) The p orbitals have $l = 1$ and $m_l = 0, +1$, and -1. They are dumbbell-shaped, with one nodal plane and $n - 2$ radial nodes. The angular shape of a p orbital shown here is determined by the absolute square of the corresponding spherical harmonic, $|Y_1^0(\theta, \phi)|^2$ for a p$_z$ orbitals but $\frac{1}{2}|Y_1^1(\theta, \phi) - Y_1^{-1}(\theta, \phi)|^2$ for p$_x$ and $-\frac{1}{2}|Y_1^1(\theta, \phi)+ Y_1^{-1}(\theta, \phi)|^2$ for p$_y$. The p$_x$ and p$_y$ orbitals have the same shape as p$_z$ but they are directed along the x and y axes, respectively. The phase of the wavefunction (arbitrarily shown as + and − in the figure) changes across a node. (c) The d orbitals have $l = 2$ and $m_l = 0, \pm 1$, and ± 2. They have two nodal planes and $n - 3$ radial nodes. (a) d$_{z2}$. (b) d$_{xz}$. (c) d$_{yz}$. (d) d$_{xy}$. (e) d$_{x2-y2}$ is similar to d$_{xy}$ but rotated by 90°. The phase of the wavefunction (arbitrarily shown as + and − in the figure) changes across a node.

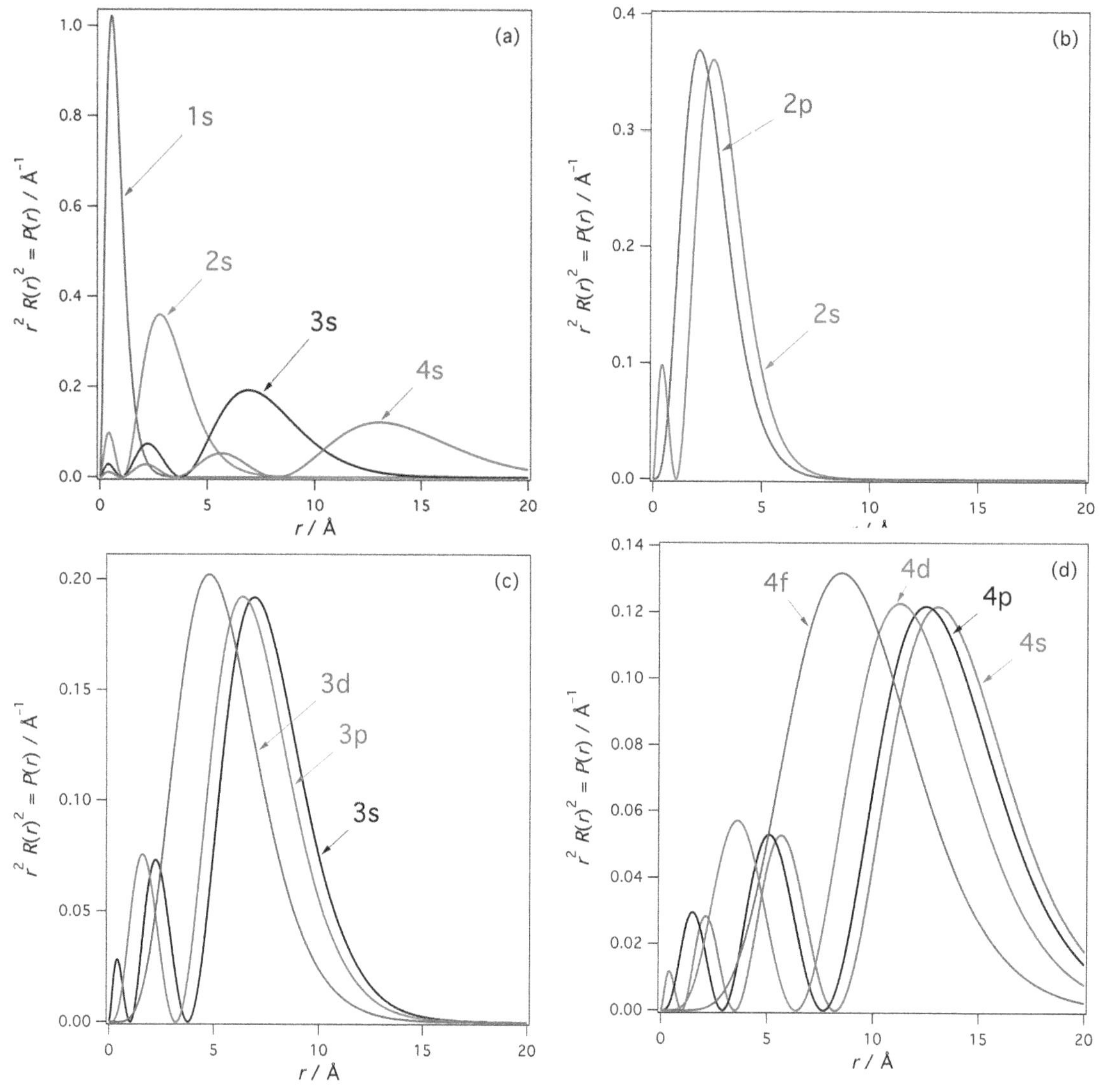

Figure 22.5 Radial distribution function $[P_{nl}(r)]$ for the lowest energy states of the H atom. (a) Comparing the first four of the s orbitals, we see how the maximum in the probability distribution shifts to longer radial distances with increases in the principal quantum number n. The number of radial nodes (points at which the probability distribution goes to zero) also increases as $n - 1$ because there are no angular nodes in the s function. (b) The L shell with $n = 2$ is composed of both an s and a p functions. The total number of nodes is $n - 1$ thus, the p function has an angular node rather than a radial node. In panels (c) and (d) these patterns continue. It can also be seen that the maximum occurs at smaller r as the azimuthal quantum number l increases, but that this is compensated by smaller maxima at even shorter r.

same value of n) have a peak in the electron density further from the nucleus than lower l orbitals. The result is that the degeneracy of the orbitals is lifted. The energy depends on both n and l and it increases with increases in both n and l.

22.3.4 Directed practice

Why do not most of the d orbitals in Fig. 22.4 look like the $l = 2$ solutions of the spherical harmonics in Table 21.3?

22.4 Many-electron atoms

22.4.1 Set up the hamiltonian

A multi-electron atom (or ion) is a system with N electrons about a nucleus with charge $+Z$, where Z is the atomic number. To obtain the energies of all allowed electronic states, we begin by figuring out the appropriate Hamiltonian within a coordinate system such as that shown in Fig. 22.6. The electrons each have kinetic and potential energies given by terms analogous to the one-electron terms derived previously in Eq. (22.2). To this we must add terms to account for

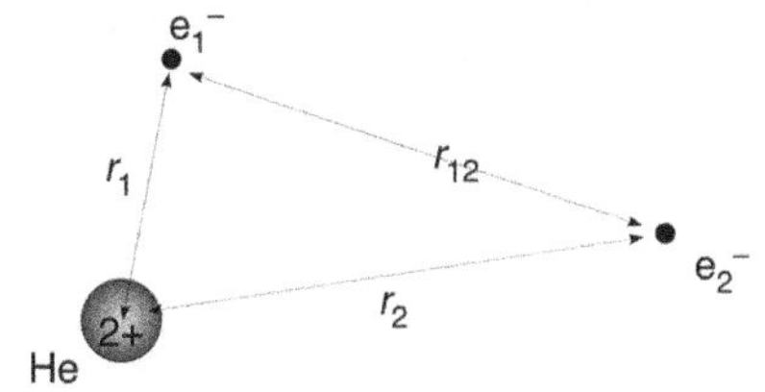

Figure 22.6 The coordinate system for a multi-electron atom. The specific example of the He atom, which has a nuclear charge of 2+ is shown. The distances from the nucleus to the electrons are r_1 and r_2. The inter-electron distance is r_{12}.

the electron–electron repulsions (like charged particles repel) and for the spin–orbit interaction (the interaction of spin and orbital angular momenta of each individual electron introduced by the Dirac equation).

$$\hat{H} = \hat{T} + \hat{V} + \hat{H}_{ee} + \hat{H}_{so} \tag{22.30}$$

$$\hat{T} + \hat{V} = -\frac{\hbar^2}{2\mu}\sum_i \nabla_i^2 - \frac{e^2}{4\pi\varepsilon_0}\sum_i \frac{Z}{r_i} \tag{22.31}$$

$$\hat{H}_{ee} = \frac{e^2}{4\pi\varepsilon_0}\sum_{i>j}\sum \frac{1}{r_{ij}} \tag{22.32}$$

$$\hat{H}_{so} = \sum_i \xi_i\left(r_i\right)\left(\boldsymbol{l}_i \cdot \boldsymbol{s}_i\right) \tag{22.33}$$

In the above, $r_i = |\boldsymbol{r}_i|$ is the position of the ith electron relative to the nucleus, $\boldsymbol{r}_{ij} = |\boldsymbol{r}_i - \boldsymbol{r}_j|$ is the distance between the ith and j^{th} electrons, and the summation over $i > j$ is over all pairs of electrons. Mass-velocity and Darwin terms could also be added to our Hamiltonian. We neglect them here for brevity and because the small corrections they represent only shift the absolute energies of groups of related energies levels. They do not introduce splittings and addition energy levels, as does the spin–orbit interaction. Neither the mass-velocity nor the Darwin term is as large as the electron–electron repulsion or spin–orbit term. The electron–electron repulsion and spin–orbit terms are essential to understanding the structure of atomic energy levels.

Next, we substitute the Hamiltonian into the Schrödinger equation,

$$\hat{H}\psi_k = E_k\psi_k, \tag{22.34}$$

and solve for the infinite series of wavefunctions ψ_k that satisfy this equation. These wavefunctions are functions of $4N$ variables: the three spatial dimensions and one spin coordinate for each electron. Further complication to the mathematics arises from the dreaded electron repulsion term. This term includes the relative distance between two electrons, which means that correlations in the motions of particles must be properly handled. The consequence is that the presence of *electron correlation* in any atom with more than one electron precludes finding exact, analytical solutions to the Schrödinger equation. Instead, approximations must be introduced.

22.4.2 The central field model

s, p, d, ... orbitals are, strictly speaking, only exact for the one electron case. For multi-electron atoms we make the approximation that the atoms are composed of a set of one-electron-like orbitals. The presence of electron–electron interactions means that the true orbitals are modified and slightly different than the H atom system; nonetheless, we can use the naming convention derived from the H atom case to identify the orbitals in multi-electron atoms. The repulsive interaction between electrons leads to electron correlation: the motions of electrons are not truly independent. Rather, they attempt to avoid each other to reduce repulsions.

The *central field model* of the atom is constructed as follows. Assume that each electron moves independently in an electric field generated by the nucleus and the $N - 1$ other electrons. The nucleus, because of it large mass relative to the electrons, is taken to be stationary. The $N - 1$ other electrons produce a time-averaged field that is spherically symmetric about the nucleus. If their motion were truly uncorrelated with the ith electron (the one that we are following), this assumption would hold exactly and it would generate a central field. The probability distribution of the ith electron moving in this central potential is described by a one-electron wavefunction (sometimes also called a spin–orbital)

$$\phi_i(r_i, \theta_i, \phi_i) = R_{n_i l_i}(r_i)\; Y_{l_i m_{l_i}}(\theta_i, \phi_i)\; \sigma_{m_{s_i}}(s_i). \tag{22.35}$$

This wavefunction is of the same form as the H atom wavefunctions with the addition of a factor to account for the spin of the electrons, $\sigma_{m_{s_i}}(s_i)$. We will discuss this factor further in the next section.

The angular momentum of electron i is a constant of motion. Thus, the wavefunction defined in Eq. (22.35) is an eigenfunction of the one-electron operators $\hat{l}_i^2$, $\hat{l}_{z_i}$, $\hat{s}_i^2$, and $\hat{s}_{z_i}$. The eigenvalues for these angular momentum operators are

$$\langle \phi_i | \hat{l}_i^2 | \phi_i \rangle = l_i(l_i + 1)\hbar^2 \quad (22.36)$$

$$\langle \phi_i | \hat{l}_{z_i} | \phi_i \rangle = m_{l_i}\hbar \quad (22.37)$$

$$\left\langle \phi_i \middle| \hat{s}_i^2 \middle| \phi_i \right\rangle = s_i(s_i + 1)\hbar^2 = 3/4\,\hbar^2 \quad (22.38)$$

$$\left\langle \phi_i \middle| \hat{s}_{z_i} \middle| \phi_i \right\rangle = m_{s_i}\,\hbar^2 = \pm 1/2\,\hbar. \quad (22.39)$$

The quantum numbers l and m_l can take on any allowed values but the only allowed values of the spin and projection of spin are $s = {}^1/_2$ and $m_s = \pm{}^1/_2$. These answers are exactly the same as for the H atom; thus, the angular part of the wavefunction has the same form.

The differences compared to the H atom wavefunctions are to be found in the radial factor $R_{nl}(r)$. These functions still need to satisfy a differential equation of the form of Eq. (22.4). In the multi-electron case, however, the potential energy $V(r)$ is no longer a simple Coulomb function proportional to $-Z/r$. An exact analytical solution of the differential equation is no longer possible. Instead, numerical methods must be employed to calculate $R_{nl}(r)$. We need not concern ourselves with these methods here. More on the exact manner of such calculations can be found in the references listed in Further Reading.

22.4.3 Pauli exclusion and Aufbau principles

In hydrogenic atoms and ions, all orbitals of a given n are degenerate, as shown in Fig. 22.7(a). We have stated above that, for multi-electrons atoms, the radial distribution function is similar to but not identical to those shown for hydrogenic atoms in Fig. 22.4. The consequences are particularly important when we consider the sequential filling (or depopulation by ionization) of orbitals. It is still true that only an s electron has a nonvanishing probability at the nucleus. Higher l orbitals (for the same value of n) have a peak in the electron density further from the nucleus than lower l orbitals. The result is that orbitals of the same n but different l are screened to different degrees by the electrons in other orbitals. This lifts the degeneracy of the orbitals, as shown in Fig. 22.7(b) and (c). The energies of orbitals in multi-electron atoms depend on both n and l, and increase with increases in both. The greater the penetration of the screening cloud of charge by a given orbital, the less effective the screening. Orbitals with the same value of n but lower l are less effectively screened. They experience more of the nuclear charge and their energy is lowered.

The *Aufbau principle* introduced by Bohr[7] explains the order of orbital filling on the basis of this screening effect. The lower the value of n, the lower the energy. The lower the value of l for a given n, the lower the energy. No more than two electrons can be placed in an orbital as a result of the exclusion principle. Fill from lowest energy to highest until all the required electrons have found a home. We need to add to this principle *Hund's rule* of maximum multiplicity: always choose the highest spin state – that is, maximize the number of unpaired electrons. We will discuss this in more detail below. Due to Pauli repulsion, electrons with the same spin automatically avoid each other and reduce their mutual repulsion. This leads to a reduction in the energy of higher spin configurations.

The Aufbau principle, coupled to the orbital occupancies sketched in Table 22.4, explains the regularities and pattern of the *periodic table*. Ground-state *electron configurations* are given in the periodic table found inside the back cover. Deviations from the electron configurations predicted by the Aufbau principle are indicated in bold. Some of these can be explained on the basis of the enhanced stability of fully occupied and half-occupied subshells. However, as shown in Fig. 22.7(c), the energetic difference between levels with different quantum numbers decreases with increasing n. Thus, the simple patterns found at low n exhibit increasing deviations higher in the periodic table. Further discussion of the energetics of electronic states will be found below after we discover how to write a proper wavefunction.

22.4.4 Directed practice

What is the atomic number of the first atom to exhibit an anomalous electron configuration?

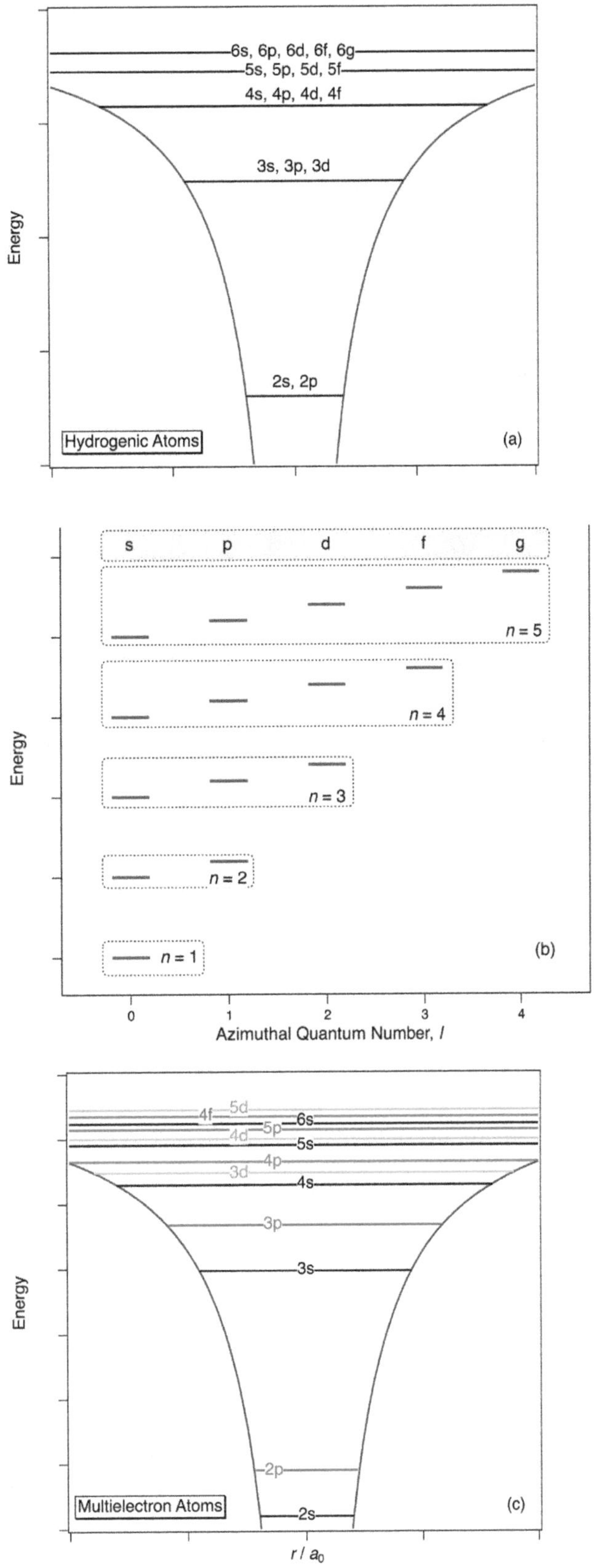

Figure 22.7 According to the Aufbau principle, orbitals are filled with no more than two electrons from lowest energy to highest until all available electrons are placed. (a) In hydrogenic atoms and ions, all levels of the same n are degenerate and the progression of orbital energies is quite regular. (b) The dependence of screening on l – resulting in the lowering of energy with lower l for the same value of n – lifts the degeneracy within a given value of n. Multi-electron interactions can lead to subtle shifts in orbital energy that are increasingly important as the spacing between levels decreases with increasing n. (c) The effects of screening, coupled with the decreasing separation between successively levels as n increases, lead to irregularities in orbital filling compared to the simple expectations in panel (b). The periodic table should be consulted for the filling order of orbitals.

22.5 Ground and excited states of He

22.5.1 Ground state

The ground state configuration of He is $1s^2$, that is, two electrons in the 1s orbital. The spins of the electrons are antiparallel, in accord with the Pauli exclusion principle. Thus, the total spin angular momentum is

$$S = |{}^1\!/\!_2 - {}^1\!/\!_2| = 0.$$

The *spin multiplicity* is the number of different spin states corresponding to a certain value of S and is given by

$$\text{spin multiplicity} = 2S + 1. \tag{22.40}$$

For the ground state $2S + 1 = 1$ and, as expected, there is only one way to combine two electrons in a 1s orbital and only one energy level associated with this combination.

22.5.2 Excited states

Now consider one electron in the 1*s* orbital and the second electron in the 2s orbital. The configuration is 1s 2s. The electron can either have antiparallel spins

$$S = |{}^1\!/\!_2 - {}^1\!/\!_2| = 0 \Rightarrow 2S + 1 = 1, \textbf{ singlet}$$

or parallel spins

$$S = |{}^1\!/\!_2 + {}^1\!/\!_2| = 1 \Rightarrow 2S + 1 = 3, \textbf{ triplet}.$$

The antiparallel spin configuration leads to only one state, which is therefore called a singlet state. The parallel spin configuration leads to three degenerate spin states with the same electronic configuration; this is the triplet state, which has a slightly lower energy.

Higher excited states are made by putting one or both electrons in higher energy orbitals. To describe these states we must understand how the orbital angular momenta of the two electrons couple (i.e., interact) and how spin and orbital angular momenta couple. We will investigate this topic in detail in Chapter 23.

22.5.3 He atom wavefunctions

We know that a He atom in its ground state has the electron configuration $1s^2$, but how do we write a proper wavefunction for it? Using the central field model, each electron occupies a one-electron orbital – in this case ϕ_1 and ϕ_2. The probability that an electron is at the point $\boldsymbol{r}_i$ is given by the square of the wavefunction, $|\phi_i(r_i)|^2$. If the motion of the electrons is uncorrelated, then the probability of finding electron 1 at point $\boldsymbol{r}_1$ and electron 2 at $\boldsymbol{r}_2$ is simply the product of the squares of the wavefunctions, $|\phi_1(\boldsymbol{r}_1)|^2|\phi_2(\boldsymbol{r}_2)|^2$. Consequently, we begin to write the spatial part of the wavefunction (excluding for the moment the spin part) as a simple product of orbitals,

$$\psi_{\text{spatial}}(\boldsymbol{r}_1, \boldsymbol{r}_2) = \phi_1(\boldsymbol{r}_1)\phi_2(\boldsymbol{r}_2). \tag{22.41}$$

Specifically for the ground state of the He atom, the subscript 1 is shorthand for $n_1 = 1,\ l_1 = 0,\ m_{l_1} = 0$. That is, orbital 1 is a 1s orbital. An analogous statement holds for the subscript 2. Thus, a more explicit description of the spatial part of the wavefunction is

$$\psi_{\text{spatial}}(\boldsymbol{r}_1, \boldsymbol{r}_2) = \phi_{100}(\boldsymbol{r}_1)\phi_{100}(\boldsymbol{r}_2). \tag{22.42}$$

There is a problem with this description. Electrons do not come with nametags, they are indistinguishable particles. Therefore, when two electrons are interchanged, the probability density has to remain the same,

$$|\psi_{\text{spatial}}(\boldsymbol{r}_1, \boldsymbol{r}_2)|^2 = |\psi_{\text{spatial}}(\boldsymbol{r}_2, \boldsymbol{r}_1)|^2. \tag{22.43}$$

Furthermore, the wavefunctions must satisfy this *antisymmetrization* postulate of quantum mechanics:

A good multi-electron wavefunction must be antisymmetric with respect to electron interchange.

This postulate demands that

$$\psi_{\text{total}}(\boldsymbol{r}_1, \boldsymbol{r}_2) = -\psi_{\text{total}}(\boldsymbol{r}_2, \boldsymbol{r}_1) \tag{22.44}$$

However, a wavefunction of the form given in Eq. (22.42) does not satisfy this postulate. There is also a second problem with the form of this wavefunction. The *Pauli exclusion principle* states that[8]

> *No two electrons can occupy the same spin orbital.*

More accurately, the exclusion principle states that

> *No antisymmetric wavefunction describing an N-electron atom can be made up of one-electron functions, two of which have the same set of one-electron quantum numbers.*

More generally yet, as we encountered in Chapter 6, the exclusion principle holds for all fermions while bosons have total wavefunctions that are symmetric with respect to particle exchange. Obviously, the wavefunction in Eq. (22.42) does not satisfy the exclusion principle because $n = 1$, $l = 0$, $m_l = 0$ for both electrons.

The solution to these problems – and the explanation of why the word 'total' appears in the subscript in Eq. (22.44) – requires recognition that the complete (total) wavefunction is composed of spatial and spin parts. Correctly handling these will allow us to properly satisfy both the Pauli exclusion principle and antisymmetrization of the wavefunction. We need to recognize that ψ_{spatial} describes the *orbital* containing the two electrons, but it does not completely describe the *electrons*. The complete wavefunction is a wavefunction that describes both electrons in the atom not just the region of space they occupy.

Previously we used $\sigma_{m_{s_i}}(s_i)$ to denote the spin part of the wavefunction. This notation is precise but cumbersome, especially since all electrons have $s = {}^1/_2$ and only two orientations, $m_s = \pm{}^1/_2$. To simplify and focus on the important aspects of the spin part of the wavefunction, we introduce α to represent the $m_s = +{}^1/_2$, spin-up state, and β to represent the $m_s = -{}^1/_2$, spin-down state. For example, if electron 1 is in the spin-up state, we write $\alpha(1)$.

Spin is an intrinsic quantum mechanical angular momentum. It is a vector quantity. As for all quantum mechanical angular momenta, there are restrictions on the orientation (i.e., the projection along the z-axis) of this vector. This restriction also leads to restrictions in how two spins couple to form a total spin angular momentum. We use a convention that the spin of a single particle corresponds to a lower-case s and the coupled spin of the system corresponds to upper-case S. Similarly for projection, m_s is a value pertaining to one electron, and M_S is a value for a system of coupled spins. There are only two ways to couple two $s = {}^1/_2$ particles. They can either be both in the same direction (parallel) or in opposing directions (antiparallel). However, we are coupling vector quantities, and the addition of the spin of electron 1, s_1 and that of electron 2, s_2, to form the total spin S is a vector addition,

$$S = s_1 + s_2. \tag{22.45}$$

In other words, $S = |\frac{1}{2} + \frac{1}{2}| = 1$ or $S = |\frac{1}{2} - \frac{1}{2}| = 0$. That may seem trivial (remember that thought when you try later to couple two or more angular momenta with values $> {}^1/_2$), but the vector addition means that there are four combinations arising from this coupling. There are two ways to conceptualize this. The first is that there must be four combinations because the $S = 1$ combination must have three projections $M_S = 0, \pm 1$ and the $S = 0$ combination only has one projection corresponding to $M_S = 0$. The geometrical way to visualize this is shown in Fig. 22.8. The electrons may either be both up ↑↑ or both down ↓↓, both of which obviously correspond to the $S = 1$ case. Or the electrons can be antiparallel; however, since they are indistinguishable, we cannot tell whether electron 1 is on the left or the right. We must make a state that includes both possibilities. Further, we can either add or subtract these possibilities, that is, ↑↓ + ↓↑ or ↑↓ − ↓↑ are both good combinations. More formally, the four allowed combinations of two electrons in the two spin states are

$$\alpha(1)\alpha(2), \beta(1)\beta(2), 2^{-1/2}[\alpha(1)\beta(2) + \beta(1)\alpha(2)], \tag{22.46}$$

$$2^{-1/2}[\alpha(1)\beta(2) - \beta(1)\alpha(2)]. \tag{22.47}$$

	S	**M_S**	**Exchange**
↑↑	1	1	symmetruc
↓↓	1	−1	symmetric
$\frac{1}{\sqrt{2}}$(↑↓ + ↓↑)	1	0	symmetric
$\frac{1}{\sqrt{2}}$(↑↓ − ↓↑)	0	0	antisymmetric

Figure 22.8 The M_S states associated with the coupling of s_1 and s_2 into S.

The factor of $2^{-1/2}$ is required for normalization. Note that the first three are symmetric with respect to electron exchange, but the fourth linear combination is antisymmetric,

$$2^{-1/2}[\alpha(2)\beta(1) - \beta(2)\alpha(1)] = -2^{-1/2}[\alpha(1)\beta(2) - \beta(1)\alpha(2)]. \tag{22.48}$$

In addition, the first three correspond to an $S = 1$ configuration with $M_S = 1, -1$ and 0, respectively. The fourth corresponds to an $S = 0$, $M_S = 0$ configuration.

Equations (22.46) and (22.47) also show us how we can make the overall wavefunction antisymmetric as required by the Pauli exclusion principle. The total wavefunction is constructed from the product of a spatial part and a spin part

$$\psi_{\text{total}} = \psi_{\text{spatial}}\,\psi_{\text{spin}} \tag{22.49}$$

If the spatial part is symmetric, as it is for the He atom ground state,

$$\phi_{100}(\boldsymbol{r}_1)\phi_{100}(\boldsymbol{r}_2) = \phi_{100}(\boldsymbol{r}_2)\phi_{100}(\boldsymbol{r}_1), \tag{22.50}$$

then the antisymmetric spin part from Eq. (22.47) must be chosen as the other part of the product to form the total wavefunction,

$$\psi_{\text{total}} = \phi_{100}(\boldsymbol{r}_1)\phi_{100}(\boldsymbol{r}_2)\, 2^{-1/2}[\alpha(1)\beta(2) - \beta(1)\alpha(2)]. \tag{22.51}$$

There is only one way to write this wavefunction so the ground state of the He atom corresponds to a singlet state.

The first excited state of He has an electron configuration of 1s 2s. The corresponding wavefunctions are formed from combinations of these component orbitals. The appropriate spin wavefunctions are then attached to form the overall antisymmetric wavefunctions.

Symmetric
$$2^{-1/2}[1s(1)2s(2) + 1s(1)2s(2)] \tag{22.52}$$

Antisymmetric
$$2^{-1/2}[1s(1)2s(2) - 1s(1)2s(2)] \tag{22.53}$$

22.5.3.1 Directed practice

Write out the complete wavefunctions for the excited state configuration 1s 2s of He. Hint: There will be a singlet combination and a triplet state.

Antisymmetrization of the wavefunction has one other unintended advantage. In a set of properly antisymmetrized wavefunctions, for any two electrons in an atom $\psi_{\text{total}} = 0$ whenever $\boldsymbol{r}_1 = \boldsymbol{r}_2$ and $s_1 = s_2$. In other words, any two electrons with the same spin naturally avoid each other (they never are at the same point $\boldsymbol{r}$ at the same time). This means that antisymmetrized wavefunctions implicitly include electron correlation for electrons with the same spin. This *Pauli repulsion* – the avoidance of two electrons simultaneously being in the same space based on their quantum numbers – is different than the electrostatic repulsion, which is the repelling of two like-charged particles. Therefore, the correlation exhibited by antisymmetrized wavefunctions is incomplete. It also does not apply to electrons of opposite spin.

Recall that spin arises from handling time on an equal footing with the other three dimensions of space–time as required by relativity. Therefore, we should perhaps not be surprised that Pauli repulsion occurs for quantum mechanical objects. Classical objects cannot occupy the same space simultaneously. Is this also true for quantum particles represented by wavefunctions? What Pauli repulsion tells us is that by the nature of their wavefunction, fermions (such as electrons) cannot occupy the same space simultaneously if they have the same spin, regardless of their other quantum numbers. Classical matter always has some number of electrons of the same spin, which implies that Pauli repulsion naturally means that classical objects cannot occupy the same space simultaneously. Pauli repulsion does not apply to bosons or spin-paired fermions. They are not excluded from the same point in space–time by the nature of their wavefunctions. However, to understand their overlap and collision properties at the required spatial and temporal resolution, we must descend into the nether regions of Bose–Einstein condensates and particle physics.

22.6 Slater–Condon theory for approximating atomic energy levels

With knowledge of how to write a proper wavefunction for a multi-electron atom, we can now approach the use of approximations to accurately calculate the energies of these wavefunctions. The approach used here was originally developed by Slater[9] and then extended by Condon and Shortley,[10] among others. The aim is to find the wavefunctions that solve the Schrödinger equation in Eq. (22.34). The approach used here is to expand the unknown wavefunctions ψ_k in terms of a set of known functions ϕ_b. The known functions are called *basis functions*. In principle, we can choose any sort of function to be contained in the set of functions ϕ_b, but in practice we should choose something that makes the calculation easier. One choice is to start with hydrogenic orbitals. This makes sense within the approximations used above; furthermore, we know the answers to the integrals for the hydrogen-like wavefunctions. It turns out that these wavefunctions are not particularly easy to integrate. Therefore, Gaussian functions, which are easier to integrate and still centered about the nucleus, are often chosen.

The wavefunctions of the system are constructed by taking linear combinations of the basis functions. In other words, we multiply each basis function by some coefficient c_b and add all of the terms together according to

$$\psi_k = \sum_b c_{k,b}\,\phi_b. \tag{22.54}$$

The basis functions are chosen to be members of a complete set of orthonormal functions for which

$$\langle \phi_b | \phi_{b'} \rangle = \delta_{bb'} \tag{22.55}$$

where $\delta_{bb'}$ is the Kronecker delta function, with $\delta_{bb'} = 1$ for $b = b'$ but $\delta_{bb'} = 0$ for $b \neq b'$. In principle, the set of basis functions is infinite and will, therefore, return the correct answer for a given Hamiltonian. In practice, however, we cannot work with an infinite set, and truncation of the set to some finite number of functions leads to an approximate answer. For a finite set of basis functions, the quality of the basis set is also important. The better the basis set, the fewer basis functions that are required to achieve a given level of accuracy in the calculation.

Assume that we truncate our set of basis functions so that there is a total of M basis functions ϕ_b that we will consider. Thus, $1 \leq b \leq M$ and we need to determine the M values of $c_{k,b}$ that are required by Eq. (22.54) to define each wavefunction ψ_k. If we were solving the first M levels of the H atom, then we would choose the ϕ_b to be the first M wavefunctions that we determined from solving the Schrödinger equation. In this case, $c_{1,1} = 1$ and $c_{1,b} = 0$ for all $b \neq 1$. In other words, the first wavefunction would simply be made up of 100% of the 1s wavefunction and no other wavefunctions would contribute to it.

If we move on to the He atom, we now have two electrons to consider. In the previous section we approached this problem by placing the two electrons each in a 1s orbital. We then constructed the total spatial part of the wavefunction from the product of these two 1s orbitals. Then, we had to multiply this by the appropriate spin wavefunction to obtain a total product wavefunction. Here, for electron one $c_{1,1}(1) = 1$ and $c_{1,b}(1) = 0$ for all $b \neq 1$ and for electron two $c_{1,1}(2) = 1$ and $c_{1,b}(2) = 0$ for all $b \neq 1$. Under the assumption that the interactions of the two electrons do not change the orbitals (the basis functions) from what they were like when we found them by solving the H atom case (apart from changing $Z = 1$ to $Z = 2$ for He), we then substitute the product wavefunction back into the Schrödinger equation, and will obtain the correct answer to the Hamiltonian we have chosen.

The problem is that electrons do interact. Therefore, the character of the orbitals in He is slightly different than in the H atom. Electrons in s orbitals still have some value of n, $l = 0$, $m_l = 0$ and $s = {}^1\!/_2$; and p electrons still have some value of n, $l = 1$, $m_l = 0$ or ± 1 and $s = {}^1\!/_2$; but the charge distribution is slightly different than compared to hydrogenic orbitals. In the H atom, levels with the same value of l were degenerate for any given value of n. In a multi-electron atom, for any value of n screening leads to an increase of the energy of orbitals with increasing l. That is, s electrons lie lower in energy than p electrons, which are lower than d, and so forth.

The way we can account for these changes is to allow other orbitals, which correspond to different electron distributions, to mix with the first orbital that we chose. The amount they add to the mix is determined by the coefficients $c_{k,b}$. The problem now is to determine the set of coefficients that we need to describe each orbital. Then, we construct a product wavefunction using the rules described in the previous section to describe the wavefunction of the atom.

Substitution of the wavefunctions obtained in Eq. (22.54) into the Schrödinger equation yields

$$\hat{H} \sum_{b'=1}^{M} c_{k,b'}\,\phi_{b'} = E_k \sum_{b'=1}^{M} c_{k,b'}\,\phi_{b'}. \tag{22.56}$$

The Hamiltonian operator has to act on each individual basis function, so we move it inside the summation. Now, we multiply from the left with the complex conjugate of one of the basis functions $\phi_b{}^*$,

$$\phi_b^* \sum_{b'=1}^{M} \hat{H}\, c_{k,b'}\, \phi_{b'} = \phi_b^* E_k \sum_{b'=1}^{M} c_{k,b'}\, \phi_{b'}. \quad (22.57)$$

We integrate both sides of Eq. (22.57) over all $3N$ space coordinates (and sum over both possible directions of each of the N spins). We can write this in Dirac notation as

$$\sum_{b'=1}^{M} \langle \phi_b | \hat{H}\, | c_{k,b'}\, \phi_{b'} \rangle = E_k \sum_{b'=1}^{M} \langle\, \phi_b | c_{k,b'}\, \phi_{b'} \rangle \quad (22.58)$$

Note that each value of E_k is a constant so it could be pulled out of the integral. Similarly, each value of $c_{k,b'}$ is a constant, thus,

$$\sum_{b'=1}^{M} c_{k,b'} \langle \phi_b | \hat{H}\; | \phi_{b'} \rangle = E_k \sum_{b'=1}^{M} c_{k,b'} \langle\, \phi_b |\, \phi_{b'} \rangle. \quad (22.59)$$

From Eq. (22.55) we know that $\langle \phi_b | \phi_{b'} \rangle = 0$ for all cases except when $b = b'$. We introduce the notation

$$H_{bb'} \equiv \langle \phi_b | \hat{H}\; | \phi_{b'} \rangle, \quad (22.60)$$

which is called the *matrix element* of the Hamiltonian operator between the basis functions ϕ_b and $\phi_{b'}$. It is called the matrix element because we can use these values to define the matrix $\boldsymbol{H}$,

$$\boldsymbol{H} = \begin{pmatrix} H_{11} & H_{12} & \cdots & H_{1M} \\ H_{21} & H_{22} & \cdots & \vdots \\ \vdots & \vdots & \ddots & \vdots \\ H_{M1} & \cdots & \cdots & H_{MM} \end{pmatrix}. \quad (22.61)$$

This is a *Hermitian matrix* – that is, its elements are real numbers, $H_{bb'} = H^*_{bb'}$, and the matrix is symmetric such that $H_{b'b} = H_{bb'}$.

Equation (22.59) can then be written as

$$\sum_{b'=1}^{M} c_{k,b'}\, H_{bb'} = E_k c_{k,b}. \quad (22.62)$$

Equation (22.62) represents a set of M simultaneous linear homogeneous equations in M unknowns $c_{k,b'}$. This set of equations has a nontrivial solution (in other words at least one $c_{k,b'} \neq 0$) only if the determinant of the matrix with elements defined by $H_{bb'} - E_k\, \delta_{bb'}$ is zero,

$$|\boldsymbol{H} - E_k \boldsymbol{I}| = 0, \quad (22.63)$$

where $\boldsymbol{I}$ is the unit matrix – a matrix with the value 1 along the diagonal but elsewhere zero.

If M is no more than 2 or 3, this system of simultaneous equations can be solved by hand by finding the roots of the second- or third-order polynomial that results when the determinant is expanded. However, if high accuracy is required M will be on the order of 10^2, 10^3 or even much more. In this case, the only practical method available is numerical diagonalization of the $\boldsymbol{H}$ matrix. This is one reason why very efficient algorithms have been developed for diagonalization of matrices. Accordingly, we rewrite the Schrödinger equation as a matrix equation,

$$\boldsymbol{H}\boldsymbol{C}_k = E_k\, \boldsymbol{C}_k, \quad (22.64)$$

where $\boldsymbol{C}_k$ is a column vector containing all of the coefficients that define the wavefunction ψ_k,

$$\boldsymbol{C}_k = \begin{pmatrix} c_{k,1} \\ c_{k,2} \\ \vdots \\ c_{k,M} \end{pmatrix}. \quad (22.65)$$

A further restriction on the values of $c_{k,b'}$ results from the requirement that each wavefunction ϕ_k must be normalized. Thus,

$$\langle \psi_k | \psi_k \rangle = \left\langle \sum_b c_{k,b}\, \psi_b \middle| \sum_{b'} c_{k,b'}\, \psi_{b'} \right\rangle \tag{22.66}$$

$$= \sum_b \sum_{b'} (c_{k,b})^* c_{k,b'} \langle \psi_{b'} | \psi_b \rangle \tag{22.67}$$

$$= \sum_b |c_{k,b}|^2 = 1. \tag{22.68}$$

The Schrödinger equation has been transformed by the techniques of linear algebra into the matrix equation found in Eq. (22.64). Formally then, instead of solving second-order differential equations to find wavefunctions and the associated energies, the problem is now to find the M eigenvalues E_k of the matrix $\boldsymbol{H}$ as well as the corresponding eigenvectors $\boldsymbol{C}_k$, that represent the coefficients of the basis functions. This is achieved by calculating numerical values of the matrix elements $H_{bb'}$ (this is a computationally intensive step). With the $\boldsymbol{H}$ matrix defined, we have to find the matrix $\boldsymbol{T}$ that diagonalizes it.

The kth diagonal element of the diagonalized Hamiltonian matrix is the eigenvalue E_k,

$$\boldsymbol{T}^{-1}\boldsymbol{H}\boldsymbol{T} = E_k\, \delta_{kb}, \tag{22.69}$$

and the kth column of the matrix $\boldsymbol{T}$ represents the corresponding eigenvector $\boldsymbol{C}_k$. This can be proved by multiplying Eq. (22.69) from the left by $\boldsymbol{T}$ to obtain

$$\boldsymbol{H}\boldsymbol{T} = \boldsymbol{T}(E_k\, \delta_{kb}). \tag{22.70}$$

The kth column of the matrix $\boldsymbol{T}(E_k\, \delta_{kb})$ is equal to E_k times the kth column of $\boldsymbol{T}$, which is identical to Eq. (22.64).

The difficult step is computing the matrix elements of $\boldsymbol{H}$. For this, we need to define the Hamiltonian and choose the form and number of basis functions. Once we accomplish this, we 'simply' turn over the $\boldsymbol{H}$ matrix to a computer algorithm, which then solves for the $\boldsymbol{T}$ matrix that diagonalized $\boldsymbol{H}$ and gives us the corresponding eigenvectors.

22.6.1 Directed practice

Consider three different orbitals constructed from linear combinations as follows, $c_1 = 1$ $c_2 = 0$, $c_3 = 0$; $c_1 = c_2 = 1/\sqrt{2}$, $c_3 = 0$; and $c_1 = \sqrt{7/8}$, $c_2 = c_3 = {}^1\!/_4$. What do these tell us about the linear combination wavefunctions compared to the basis set function?

22.6.2 Directed practice

Show that if the basis set functions in the previous example are normalized, the linear combinations are also normalized.

22.7 Electron configurations

From the quantum mechanical treatment above we know the nature of the orbitals formed in atoms, and how their energy changes. The energy increases in substantial steps as the principle quantum number n increases, but the size of the step gets progressively smaller as the levels approach the ionization limit. The energy also increases (for a given value of n) with increasing orbital angular momentum l. This increment is generally much smaller than the increment related to n. In a sense we should think of all orbitals existing at all times. They are solutions of the Schrödinger equation (or if we look at higher resolution the Dirac equation) and all are available to the system. Some are occupied and some are unoccupied, depending on the nuclear charge, the charge state of the atom/ion, and its degree of excitation. An empty orbital or vacancy (also called a 'hole') in an orbital can sometimes be just as important as the presence of an electron.

The lowest energy state of a system is the ground state. The arrangement of electrons in orbitals is called an *electron configuration*. The number of electrons in an orbital is the occupation number, w_j. The electron configuration that corresponds to the ground state of the system is the ground-state configuration. All higher-energy configurations are excited configurations. An electron configuration of N electrons distributed over q orbitals is denoted

$$n_1 l_1^{w_1}\; n_2 l_2^{w_2} \cdots n_q l_q^{w_q} \tag{22.71}$$

where

$$\sum_{j=1}^{q} w_j - N. \tag{22.72}$$

The principal quantum number is used explicitly, but the orbital angular momentum is represented by the letter associated with the orbital. *To a first approximation*, the ground-state electron configuration can be read from the Periodic Table. Hence, the Ne atom has the configuration $1s^2\ 2s^2\ 2p^6$. Shells completely filled up to a rare gas atom are often abbreviated by placing the symbol of the corresponding rare gas in square brackets to avoid repetition. In this notation, the Na atom has the ground-state configuration [Ne] 3s. Orbitals with occupation $w = 0$ are usually omitted unless you are trying to emphasize the point. Occupations of $w = 1$ are indicated simply by the orbital description, for example 3s in the example above. Ground-state electron configurations are found in the periodic table inside the back cover of the book. Exceptions to regular ordering of electron configurations are shown in bold-face type.

An electron configuration written as shown in Eq. (22.71) is a semiclassical shorthand notation that effectively handles the book-keeping of electrons in atoms and ions. It comports with our one-electron, central-field approximation of electron wavefunctions. It tells us the correct number of electrons. It tells us about the orbital angular momentum of these electrons as a system and the number of principal quantum numbers over which they are distributed. Quantum mechanically, an electron configuration is not a strictly defined idea.

The one-electron, central-field approximation of electron wavefunctions is not an independent electron model. It explicitly includes electron–electron interactions through both electrostatic and Pauli repulsion. However, we cannot identify one electron in an atom and say it is a 1s electron and another electron in the same atom as a 2p electron. The electrons are indistinguishable and the wavefunctions are antisymmetrized, which allows for (+) and (−) combinations of spatial and spin components. A better way to conceptualize an electron configuration is to consider it as the basis set of orbital functions that need to be combined (properly antisymmetrized and in accord with the Pauli exclusion principle) to describe the atom. This will be particularly important in the next chapter when we deal with atomic spectroscopy. We will see that one configuration can have many energy levels associated with it. Furthermore, when one electron is excited or ionized, it directly affects all the other electrons in the system.

Keeping in mind that the electrons interact as a system also allows us to understand the exceptions to filling of orbitals predicted by position in the periodic table (see inside back cover). Pauli repulsion allows electrons to correlate their motion to avoid one another. This means that electrons in equivalent subshells naturally assume the same spin state because correlation allows them to lower their energy by minimizing electrostatic repulsion. This also means that half-filled subshells are unusually stable: a half-filled subshell of electrons with the same spin exhibits maximum correlation. Fully filled subshells with their spherically symmetric charge distributions (as will be shown in Eq. (23.84)) are also particularly favorable. Combining these two tendencies, we can explain a number of the anomalies in electron configurations. For instance, Cr and Mo both assume s d^5 configurations in their ground state. Cu, Ag and Au all assume s d^{10} configurations. But Ni is $4s^2\ 3d^8$, Pd is $4d^{10}$ and Pt 6s $5d^9$. Buyer beware of simple rules to explain the most intricate details of electron–electron interactions! It should be noted that exceptions to periodic behavior in electron configurations are less pronounced for +1 cations, and almost completely absent for more positively charged cations as a result of the increasing effective charge allowing the n dependence of binding energy to overcome the more subtle effects of the l dependence.

The concept and notation of electron configurations helps to organize the information contained in them by making a classical mental picture of boxes on shelves that can only hold two electrons, each shelf at the energy of that level, with one arrow up for one electron and one arrow down for its spin-paired partner. Remember the first rule of classical analogies to quantum phenomena: *we will make classical analogies but use them no further than they are valid.* The classical picture allows us to conceptualize the initial state of the system before, for example, photoionization. Then, a photon is absorbed, grabs one of the electrons and ejects it from the orbital it occupies. This is essentially how we treated photoionization previously in this chapter. A simple application of the conservation of energy allowed us to infer the initial energy of the electron, which to our first approximation is the orbital energy of the orbital it originally occupied; this is known as *Koopmans' theorem*. However, the reality is more complex than that, as we shall see in the next chapter. The photon shakes the whole tree of electrons. Often, one pops out just as we would expect from our first-order approximation. On other occasions, 'extra' peaks appear in our spectra, and these can only be explained by couplings between the electrons.

Within these constraints, the ground state and low-lying excited states of atoms and atomic ions are characterized well by a single electron configuration. This is not necessarily so of highly excited states. For large values of n, the spacing between each shell becomes comparable to, or even smaller than, the difference due to different values of l. Highly excited states may need a basis set of multiple configurations to properly account for their energy. This effect is called *configuration mixing*. One way to think of this is that there is a set of energy levels corresponding to each single configuration. When these levels are very close in energy (if they are at the same energy we would say that they are degenerate), they perturb each other. The perturbation is a result of the mixing of configurations, which shifts the energy of the perturbed levels compared to the energies predicted for the single configuration levels. Configuration interaction leads to configuration–interaction *perturbations*. Fortunately, selection rules forbid the interaction of many levels with each other and perturbations are not as common as would be predicted by simple chance.

22.7.1 Rydberg series

A particularly interesting set of highly excited configurations is a *Rydberg series* of levels. The Rydberg series of the H atom occurs naturally – it is truly a one-electron system within a central potential. In a multi-electron system, an effectively one-electron system can be made if the one electron is excited to a high n-state. Such a state exists primarily outside of the 'core' electrons with lower n. Consider a system such as the Na atom with a ground-state configuration of $1s^2\ 2s^2\ 2p^6\ 3s$ or [Ne] 3s, the set of states [Ne] 3s, [Ne] 4s, [Ne] 5s, [Ne] 6s, [Ne] 7s … all differing by 1 in the principal quantum number form a Rydberg series. This series converges on the behavior of a one-electron system. The energy in the series follows a modified Rydberg formula

$$E_{nl}/\text{Ry} = -\frac{Z_c^2}{(n^*)^2} = -\frac{Z_c^2}{(n-\delta)^2} \tag{22.73}$$

where Z_c is the net charge on the nucleus plus the $N-1$ electrons

$$Z_c = Z - N + 1, \tag{22.74}$$

n^* is the effective principal quantum number and the quantum defect is defined by

$$\delta = n - n^*. \tag{22.75}$$

A more insightful way to formulate this is in terms of the effective nuclear charge

$$Z^* = Z - S \equiv Z_c + p, \tag{22.76}$$

that is, the nuclear charge minus the effective screening charge of the $N-1$ electrons. Thus the energy in Rydberg units is

$$E_{nl}/\text{Ry} = -\frac{(Z^*)^2}{n^2} = -\frac{(Z-S)^2}{n^2} = -\frac{(Z_c+p)^2}{n^2}. \tag{22.77}$$

The screening defect S or penetration p

$$p = Z - Z_c - S = (N-1) - S \tag{22.78}$$

is the amount by which the $N-1$ core electrons deviate from perfect screening of the nuclear charge. The penetration depends on n and l; tending to zero as n and l increase. In other words, for increasingly large values of n, the Rydberg electron is progressively less likely to be found in the vicinity of the nucleus and the screening of the nuclear charge by the core electrons becomes more complete.

SUMMARY OF IMPORTANT EQUATIONS

$V = -\dfrac{e^2}{4\pi\varepsilon_0}\dfrac{Z}{r}$	Coulomb potential: elementary charge e, nuclear charge Z, vacuum permittivity ε_0, radial distance r
$\mu = \dfrac{m_1 m_2}{m_1 + m_2}$	Reduced mass of two particles with masses m_1 and m_2
$\psi_{nlm\,sm_s}(r,\theta,\phi,s) = R_{nl}(r)\,Y_{lm}(\theta,\phi)\,\sigma_{m_s}(s)$	Total atomic wavefunction composed of radial, angular (spherical harmonic Y_{lm}) and spin parts
$E_n = -\dfrac{\mu e^4}{32\pi^2\varepsilon_0^2\hbar^2}\left(\dfrac{Z}{n^2}\right)$	Energy levels of hydrogenic system, principal quantum number n
$I_n = -E_n$	Koopmans' theorem. The ionization energy of the nth orbital is equal to the negative of the orbital energy
$\dfrac{e^2}{2a_0} = \dfrac{\hbar^2}{2m_e a_0^2} = 13.6058\text{ eV}$	Rydberg atomic energy unit
$a_0 = \dfrac{4\pi\varepsilon_0\hbar^2}{m_e e^2} = 5.29177 \times 10^{-11}\text{ m}$	Bohr radius
$E_{nl}/\text{Ry} = -\dfrac{Z_c^2}{(n-\delta)^2}$	Energy of a Rydberg series
$E_{nl}/\text{Ry} = -\dfrac{(Z-S)^2}{n^2} = -\dfrac{(Z_c+p)^2}{n^2}$	
$\delta = n - n^*$	Quantum defect
$p = Z - Z_c - S = (N-1) - S$	Screening defect S or penetration p

Exercises

22.1 State the Pauli exclusion principle including its implications for quantum numbers and orbital occupancy.

22.2 Why is the s, p, d, f... nomenclature of the H atom orbitals also valid for multi-electron orbitals?

22.3 What is the orbital angular momentum of an electron in the orbitals (i) 1s, (ii) 3s, (iii) 3d? Give the numbers of angular and radial nodes in each case.

22.4 (a) Write out the electron configuration of the ground state of a As atom. (b) For As, write out the electron configurations of two excited states, which are made by promoting only one electron. The two excited configurations should conform to what are likely to be the first two excited states.

22.5 (a) Describe what screening is. (b) Explain why shielding is more effective by electrons in a shell of lower principal quantum number than by electrons having the same principal quantum number.

22.6 Why is a d orbital less stable than an s orbital with the same value of n?

22.7 Electrons are indistinguishable and the total wavefunction, which is written as the product of a spatial part with a spin part, must be antisymmetric with respect to particle exchange. Writing a two-electron wavefunction as a linear combination properly handles the indistinguishability.

a Complete the linear combination wavefunction below so that it is symmetric re exchange.

$$\psi_{\text{spatial}}(1,2) = A(1)B(2) \quad A(2)B(1)$$

b Complete the linear combination wavefunction below so that it is antisymmetric re exchange.

$$\psi_{\text{spatial}}(1,2) = A(1)B(2) \quad A(2)B(1)$$

c The spin part of a two-electron wavefunction can be written in four different ways:

Symmetric $\{\alpha(1)\alpha(2), \beta(1)\beta(2), 2^{-1/2}[\alpha(1)\beta(2) + \beta(1)\alpha(2)]\}$
Antisymmetric $\{2^{-1/2}[\alpha(1)\beta(2) - \beta(1)\alpha(2)]\}$

Write out the four proper combinations of the spatial and spin parts to form total wavefunctions.

d Let $\hat{S}^2$ represent the operator for the square of the total spin angular momentum with eigenvalue $s(s+1)\hbar^2$. $\hat{S}_z$ is the operator for its projection on the z axis with eigenvalue $m_s\hbar$. Determine the eigenvalues for $\hat{S}^2$ and $\hat{S}_z$ for the four wavefunction from part (c).

22.8 Given that 1s and 2s represent normalized spatial wavefunctions and that α and β represent normalized spin wavefunctions, explain why the following are either good or bad wavefunctions for a two electron system.

$$\psi = 1s(1)1s(2)2^{-1/2}[\alpha(1)\beta(2) - \beta(1)\alpha(2)]$$

$$\psi = 1s(1)1s(2)\alpha(1)\alpha(2)$$

$$\psi = 1s(1)1s(2)2^{-1/2}[\alpha(1)\beta(2) + \beta(1)\alpha(2)]$$

$$\psi = 2^{-1/2}[1s(1)1s(2) + 1s(1)1s(2)]2^{-1/2}[\alpha(1)\beta(2) - \beta(1)\alpha(2)]$$

$$\psi = 2^{-1/2}[1s(1)2s(2) - 1s(1)2s(2)]2^{-1/2}[\alpha(1)\beta(2) - \beta(1)\alpha(2)]$$

$$\psi = 2^{-1/2}[1s(1)2s(2) + 1s(1)2s(2)]\alpha(1)\beta(2)$$

$$\psi = 2^{-1/2}[1s(1)2s(2) + 1s(1)2s(2)]\alpha(1)\alpha(2)$$

$$\psi = 2^{-1/2}[1s(1)2s(2) + 1s(1)2s(2)]2^{-1/2}[\alpha(1)\beta(2) - \beta(1)\alpha(2)]$$

22.9 Calculate $\langle r \rangle$ and the most probably value of r for the H atom in its ground state. Explain why they differ with a drawing.

22.10 Calculate (a) the mean potential energy and (b) the mean kinetic energy of an electron in the ground state of a hydrogenic atom. Recall that the mean value corresponds to an expectation value. The ground-state wavefunction is

$$\psi = \left(\frac{1}{\pi a_0^3}\right)^{1/2} e^{-r/a_0}.$$

22.11 Discuss the steps involved in the calculation of the energy of a system by using the variation principle.

22.12 Take as a trial wavefunction for the ground state of the H atom

$$\psi = e^{-kr}.$$

Use the variation principle to find the optimum value of k. Note that the wavefunction is not normalized as written and that the only part of the Laplacian that need be considered is the part that involves radial derivatives.

22.13 Draw a diagram of the energy level structure of H atom. Include in your diagram the first four energy levels (the lowest four in terms of energy), including the principal quantum number, values of the energy in eV and the zero of energy. Calculate the wavenumber in cm^{-1} of two highest energy transitions between these levels.

22.14 The operator for the projection of the angular momentum is $\hat{l}_z = \frac{\hbar}{i}\frac{d}{d\varphi}$. The wavefunctions of one of the $2d$ orbitals ψ_{21-1} and a $1s$ orbital ψ_{100} are given by

$$\psi_{21-1} = \left(8\sqrt{\pi}\right)^{-1}\left(\frac{1}{a_0}\right)^{3/2}\left(\frac{r}{a_0}\right)e^{-r/2a_0}\sin\theta\, e^{-i\phi} \quad \text{and} \quad \psi_{100} = \left(\sqrt{\pi}\right)^{-1}(1/a_0)^{3/2}e^{-r/a_0}$$

Show that these wavefunctions are eigenfunctions of the operator $\hat{l}_z$. Calculate the eigenvalue. Set up the definite integral in spherical polar coordinates that shows that these wavefunctions are orthogonal. Simplify them sufficiently and explain why the integral vanishes. You do not need to evaluate the integral analytically, use reasoning.

22.15 Describe and explain why the 3s and 3d orbitals are degenerate in the H atom but not in the He atom.

22.16 In three short answers state what you can conclude from the following three integrals.

$$\int \psi_j\,\psi_k\,d\tau = 0$$

$$\int \psi_j\,\psi_k\,d\tau = 0.42$$

$$\int \psi_j\,\psi_k\,d\tau = 1$$

Endnotes

1. a is almost the Bohr radius a_0. In a the reduced mass is used. In a_0 the mass of the electron is used. $a = 0.529465$ pm. $a_0 = 0.529177$ pm. In many situations this is immaterial; however, the spectroscopy of the H atom has been performed to amazingly high resolution that easily measures the difference between these two values.
2. Matveev, A., *et al.* (2013) *Phys. Rev. Lett.*, **110**, 230801; Antognini, A., *et al.* (2013) *Science*, **339**, 417; Pohl, R., *et al.* (2010) *Nature (London)*, **466**, 213.
3. Pauli, W. (1925) *Z. Phys.*, **31**, 765.
4. Dirac, P.A.M. (1958) *The Principles of Quantum Mechanics*, 4th edition. Clarendon Press, Oxford.
5. Cowan, R.D. (1981) *The Theory of Atomic Structure and Spectra*. University of California Press, Berkeley.
6. Space-time is continuous not quantized. Quantum theory is constructed within continuous four-dimensional space-time.
7. Bohr, N. (1922) *Z. Phys.*, **9**, 1.
8. Pauli, W. (1925) *Z. Phys.*, **31**, 765.
9. Slater, J.C. (1929) *Phys. Rev.*, **34**, 1293.
10. Condon, E.U. and Shortley, G.H. (1935) *The Theory of Atomic Spectra*. Cambridge University Press, Cambridge.

Further reading

Cowan. R.D. (1981) *The Theory of Atomic Structure and Spectra*. University of California Press, Berkeley.
Eyring, H., Walter, J., and Kimball, G.E. (2015) *Quantum Chemistry*. Scholar's Choice, Rochester, NY.
Metiu, H. (2006) *Physical Chemistry: Quantum Mechanics*. Taylor & Francis, New York.

CHAPTER 23

Introduction to spectroscopy and atomic spectroscopy

PREVIEW OF IMPORTANT CONCEPTS

- Each state has an energy and any two states can combine to give rise to a spectral line in either absorption or emission. The energy difference is equal to the photon energy, which is the same for absorption and emission.
- Once all instrumental factors are eliminated, spectral lines have characteristic lineshapes. The shape and width of the line are determined by physical processes such as the natural lifetime of the state, collisions, and the Doppler effect.
- The strength of a transition depends on the magnitude of the transition moment. The strongest transitions are usually electric-dipole-induced.
- By symmetry some transition moments vanish.
- Selection rules determine which transitions are allowed (large transition moment) and which are forbidden (small transition moment). Selection rules arise from symmetry principles and the conservation of angular momentum.
- There are three types of radiative processes: absorption; spontaneous emission; and stimulated emission.
- Absorption and stimulated emission are both initiated by the presence of a photon that is resonant with the transition.
- Spontaneous emission corresponds to fluorescence and is stimulated by fluctuations in the vacuum field.
- The rate of stimulated emission increases as the cube of the transition frequency. Therefore, fluorescence lifetimes are very long for transitions in the infrared but very short in the ultraviolet.
- The total amount of light absorbed by a sample depends on the wavelength of the light, the absorption coefficient of the sample, as well as the sample density (concentration) and the pathlength.
- In hydrogenic ions, the energy of a level is determined by the principal quantum number n alone. Orbitals with different values of the azimuthal quantum number l but the same n are degenerate.
- In multi-electron atoms, screening lifts the degeneracy of levels with the same n but different l.
- Further splitting of atomic levels occurs because of spin–orbit coupling, which increases with increasing atomic number Z.
- The spin and orbital angular momenta of all electrons in unfilled shells couple to form the total angular momentum. The manner in which they couple depends on Z and determines the type of energy levels that form for a given electron configuration.
- Photoelectron spectroscopy (photoemission) can be used to determine the binding energy of electrons in atoms. The binding energy of atomic core levels only experiences small chemical shifts based on the phase and bonding of the atom.
- To a first approximation the binding energy is the negative of the orbital energy.
- An Auger transition relies on coupling between electrons. An Auger electron is an electron that is excited by the relaxation of another electron.

Spectroscopy is the study of transitions between energy levels. Commonly, these transitions involve the interaction of photons with atoms or molecules, as we have discussed, for example, in the spectroscopy of the H atom. However, because of wave-particle duality there is also spectroscopy performed with the excitation provided by electrons, which can act as waves to engender many of the same transitions as photons but without the restrictions on changing electron spin. This was the basis of the experiments by Franck and Hertz in which light emission from Hg atoms was excited by electron irradiation. Spectroscopy can also be performed between bound and unbound levels; this is the case in photoelectron spectroscopy. Spectroscopy is the most important method of investigating the world of atoms and molecules, and its importance has increased since the invention of the laser. The applications of lasers in atomic, molecular and chemical physics continue to develop and refine our understanding of chemical processes and the interactions between atoms and molecules.

Physical Chemistry: How Chemistry Works, First Edition. Kurt W. Kolasinski.

Companion Website: www.wiley.com/go/kolasinski/physicalchemistry

You are already aware of the fundamentals of spectroscopy. You are already familiar with the role of conservation of energy. Conservation of energy and angular momentum set the rules for spectroscopy. In this chapter we will delve into the ramification of angular momentum in more detail.

23.1 Fundamentals of spectroscopy

23.1.1 Terminology

The first thing to remember about spectroscopy is that it involves matter–wave interactions. Thus, the *Planck equation*

$$E = h\nu = \hbar\omega = hc\tilde{\nu} = hc/\lambda \tag{23.1}$$

and the fundamental relationships of the speed of light c, frequency ν, wavenumber $\tilde{\nu}$, wavelength λ and angular frequency ω

$$c = \lambda\nu, \quad \tilde{\nu} = 1/\lambda, \quad \omega = 2\pi\nu \tag{23.2}$$

have to kept in mind. Recall the discussion in Chapter 19 that the frequency of light ν is independent of the medium. However, in accord with Eq. (19.3) the wavelength of light in the medium λ must be reduced from its value *in vacuo* λ_0 such that $\lambda = \lambda_0/n$. Correspondingly, the wavenumber is increases by a factor of n. Thus, the relations in Eq. (23.1) hold strictly only in vacuum. Much of the remainder of spectroscopy is respecting the implications of conservation of energy and angular momentum.

How much structure we see in spectroscopy depends on how closely we look – that is, on the resolution and the energy scale at which we probe the molecule. We have already encountered this in the spectroscopy of the electronic states of the H atom. As we are beginning with atomic spectroscopy, we need only concern ourselves with the electronic degree of freedom. Most transitions involving the electronic ground state and the first few excited states in atoms occur either in the visible or ultraviolet portions of the electromagnetic spectrum. The density of electronic states increases dramatically as the atom approaches its first ionization limit. In other words, energy level spacings decrease with increasing principal quantum number n. The energies of transitions between these high-lying energy levels can be found at lower energies in the infrared, or below.

As Z increases, the binding of the K shell electrons rapidly increases. In multi-electron atoms we distinguish between core electrons and valence electrons. Core electrons are little influenced by the chemical environment and are not directly involved in bonding. They lie at very low energies and their ionization energies correspond to the vacuum ultraviolet or X-ray regions. Valence electrons are found in the highest energy normally occupied electronic states. Their ionization energies are lower than those of core electrons. Valence electrons participate in bonding. Because they are strongly linked to bonding, the energetic position of valence electron states shift sensitively in response to the chemical environment around an atom.

The shifts in energy of electronic levels is detected by performing spectroscopy of various sorts. There is a great deal of spectroscopic terminology that you need to familiarize yourself with. A number of conventions surround the use of the words 'red' and 'blue' and the use of energy scales. It is conventional in many settings to place the origin of wavelength on the left. But the wavelength is *inversely* proportional to photon energy, which means that the origin of photon energy, frequency and wavenumber is often placed on the right. The red side of the visible spectrum is the long-wavelength/low-energy side. The blue end of the visible spectrum is the short-wavelength/high-energy side. Spectroscopists will often speak of the 'blue end' or 'red end' of a spectrum or of a 'blue shift' or 'red shift' of peaks. This terminology is used even when referring to other parts of the electromagnetic spectrum. A *blue shift* is a shift to higher energy, while a *red shift* is a shift to lower energy. Thus, we say that the a peak in the X-ray photoemission spectrum of the Ti $2P_{3/2}$ level is blue shifted from 453.8 eV in Ti metal to 458.5 eV in TiO_2, which means that the peak lies at higher energy in TiO_2 than in Ti.

When a photon is absorbed, the atom makes a transition from one energy state to an excited state (a state of higher energy). When a photon is emitted, the atom makes a transition from an excited state to a state of lower energy. The lowest energy state in the atom is the ground state.

There are many types of transitions, excitations, and excitation mechanisms. Here, we deal only with atoms, and the only transitions we are interested in involve moving electrons between different energy levels. Other transitions involving rotation, vibration, nuclear spin, and so forth will be discussed in subsequent chapters.

The *Ritz combination principle*[1] states that each atomic state has an energy, and any two states can combine to give rise to a spectral line in either absorption or emission. The energy difference is equal to the photon energy. While all combinations are in principle possible, selection rules mean that some are forbidden and have very low intensities. The strength of a transition is a measure of how likely the transition is to occur. An 'allowed' transition is a strong absorber or emitter of photons. A 'forbidden' transition is a weak absorber or emitter. There are different degrees of how forbidden a transition is. Lasers, because of their high intensity, allow us to probe very weak transitions that were once thought to be totally forbidden. Alternatively, we might try to use very long pathlengths and the interaction of light with as many atoms as possible to observe a strongly forbidden transition. Selection rules provide indicators of whether a transition is allowed or forbidden. Later, we will develop specific examples of selection rules and determine why they arise. As we shall see, they are related to the conservation of angular momentum and symmetry. The quantitation of how big peaks are gives us information both on how strong a transition is (through the selection rules) and how many atoms are in a particular state. Therefore, spectroscopy can also be used to determine how atoms are distributed among the available quantum states.

Spectroscopy generally involves some type of energy resolution (energy dispersion). Either the excitation source or the emitted radiation (and sometimes both) will be used in an energy-resolved fashion. In other words, we need to figure out where the peaks are in energy space. Where peaks are tells us something about the properties of the atom and its energy levels.

The spectrum emitted by a neutral atom of an element is called the *first spectrum* of that element. It is denoted with the Roman numeral I. The singly ionized ion of that element emits the *second spectrum* of the element, and it is denoted with Roman numeral II. This pattern is repeated for all possible ions of the element. Therefore, we label the spectra of Na, Na^+ and Na^{2+} with the notation Na I, Na II and Na III, respectively.

23.1.2 Lineshape and linewidth

Absorption and emission spectra are composed of peaks in regions where bound–bound transitions are observed and continuous features when ionization, a bound-free transition, occurs. As shown in Fig. 23.1, peaks contain four characteristics that need to be specified to define them. First is the intensity. Intensity is related to the peak height; however, the most accurate value of intensity is related to the integrated area of a peak. Second is the position of the peak center, which in Fig. 23.1 is specified in terms of the wavenumber $\tilde{\nu}_0$. Third is the linewidth of the peak. Fourth is the peak shape. Three characteristic shapes are shown in Fig. 23.1 – Lorentzian, Gaussian, and Fano. A fourth lineshape, a Voigt profile, is a convolution of Lorentzian and Gaussian lineshapes.

A *Lorentzian* peak profile centered at a wavenumber $\tilde{\nu}_0$ is described by the equation

$$L(\tilde{\nu}) = \frac{A}{(\tilde{\nu} - \tilde{\nu}_0)^2 + \gamma^2}. \tag{23.3}$$

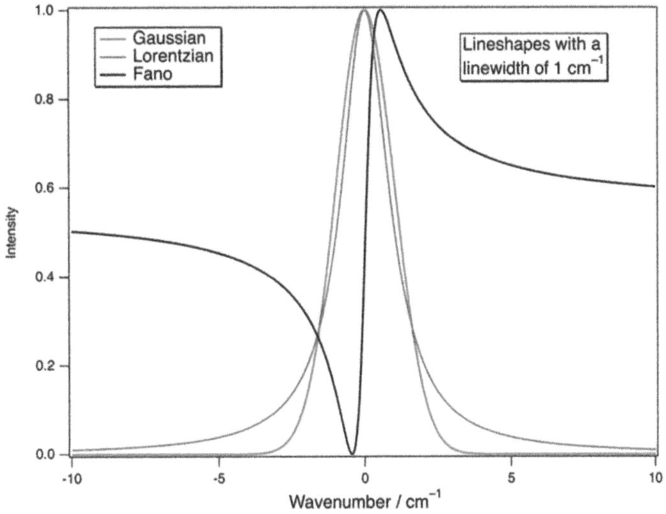

Figure 23.1 A comparison of Gaussian, Lorentzian and Fano profiles for peaks centered at $\tilde{\nu}_0 = 0$ with linewidth parameters $\gamma = \sigma = \Gamma = 1$ cm^{-1}. In each case the peaks have been normalized to their maximum intensity. The Fano parameter is taken as $q = 0.9$.

A *Gaussian* peak profile similarly centered is described by the equation

$$G(\tilde{\nu}) = A\exp\left(-\left(\frac{\tilde{\nu}-\tilde{\nu}_0}{\sigma}\right)^2\right). \tag{23.4}$$

A *Voigt* peak profile is a convolution of Lorentzian and Gaussian profiles.

$$V(\tilde{\nu}) = \int_{-\infty}^{\infty} G(\tilde{\nu}')L(\tilde{\nu}-\tilde{\nu}')dv'. \tag{23.5}$$

It lies somewhere in between the two, depending of the widths of the two components. A Fano profile is not as commonly observed; it depends on the interference between two competing pathways, usually a continuum background and a resonant component. The equation of a *Fano* profile is

$$F(\tilde{\nu}) = \frac{(0.5q\Gamma + \tilde{\nu}-\tilde{\nu}_0)^2}{(0.5\Gamma)^2 + (\tilde{\nu}-\tilde{\nu}_0)^2}, \tag{23.6}$$

where q is the Fano parameter, a ratio of the resonant event to the continuum process. The peak width is related to the parameter γ, σ, or Γ, appropriately. Linewidth is often referred to in terms of either the full-width at half-maximum (FWHM) or half-width at half-maximum (HWHM). For a Gaussian peak, the FWHM $= 2\sqrt{\ln 2}\sigma$ and for a Lorentzian, FWHM $= 2\gamma$.

The lineshape of a peak and its width are determined by a number of competing physical processes. If we observe the light emitted from an isolated stationary atom, the spectral line emitted gets narrower and narrower as we improve our instrumental resolution. At some point, the linewidth reaches a minimum value and gets no narrower, regardless of how much further we improve our instrument. This limit is known as the *natural linewidth* of the transition. Light is emitted when the excited state relaxes to a lower energy state. This means that the excited state has a characteristic finite lifetime. From the energy–lifetime uncertainty relationship (see Section 20.10.3) we know that this translates into a finite uncertainty in the energy of the corresponding energy level. In neutral atoms, the lifetime is usually not less than about 10 ns but sometimes is as small as 1 ns. Measured in units of wavenumber, this natural linewidth is given by the uncertainty relationship as

$$\frac{\Delta E}{hc} = \frac{1}{2\pi c\,\Delta t} \leq \frac{1}{6(3\times 10^{10}\ \text{cm s}^{-1})\,(10^{-9}\ \text{s})} \approx 5\times 10^{-3}\ \text{cm}^{-1} \tag{23.7}$$

Because all atoms in a particular excited state experience the same broadening from the effect of excited state lifetime, this is an example of *homogeneous broadening*. Homogeneous broadening leads to a Lorentzian lineshape. The Lorentzian linewidth parameter γ is related to the excited state lifetime Δt by

$$\gamma = (2\pi c\,\Delta t)^{-1}. \tag{23.8}$$

The *lifetime* is often represented by the symbol τ.

The value in Eq. (23.7) represents an extremely narrow linewidth. Furthermore, most excited states have longer lifetimes than this lower limit, which means that their natural linewidth is even narrower. Under most circumstances, and unless unusual experimental measures are taken, observed spectral lines are much broader than this limit.

A normal sample is a collection of atoms rather than an isolated atom. These atoms are bathed in the light of emission of other atoms or else the light from our source that is causing absorption. Stimulated emission and absorption couple the atoms to the radiation field. The atoms are also colliding with one another. Stimulated emission reduces the effective lifetime of the excited state. Similarly, collisions reduce this lifetime further. Both of these mechanisms lead to further sources of homogeneous broadening that increase the linewidth by one, two, or even more orders of magnitude depending on the strength of the radiation field and the pressure of the gas.

The atoms in the gas phase are not at rest; rather, they are traveling with speeds governed by the Maxwell distribution. Any atom moving with a component of velocity parallel to the observer's line of sight $v_{||}$ is subject to a *Doppler shift* of its emitted radiation (or equivalently of the wavenumber of the light it absorbs). The apparent wavenumber of the observed radiation $\tilde{\nu}$ is Doppler-shifted from the actual wavenumber in the atomic reference frame $\tilde{\nu}_0$ according to

$$\tilde{\nu} - \tilde{\nu}_0 = \tilde{\nu}_0(v_{||}/c). \tag{23.9}$$

For a Maxwell distribution of velocities, the resulting normalized lineshape is that of a Gaussian function given by

$$I(\tilde{\nu}) = \left(\frac{\ln 2}{\pi\Gamma^2}\right)^{1/2} \exp\left(-\ln 2\left(\frac{\tilde{\nu}-\tilde{\nu}_0}{\Gamma}\right)^2\right). \tag{23.10}$$

The half-width at half-maximum (HWHM) Γ depends on the atomic mass and absolute temperature according to

$$\Gamma = (2\ln 2k_B T/mc^2)^{1/2}\tilde{\nu}_0. \tag{23.11}$$

For the Kr spectral line at 605.78 nm, which is used as the international length standard with a source temperature of 4000 K, this leads to a full-width at half-maximum (FWHM) of 0.0817 cm^{-1}. This is at least an order of magnitude larger than the natural linewidth as expected. Since they depend only on the parallel component of velocity, Doppler widths can be substantially reduced by using molecular beams with observation at a right-angle to the beam axis. Molecular beams downstream from the point of expansion represent a collision-free zone, which also greatly reduces the collisional broadening.

Further broadening or even splitting of spectral lines can be induced by the presence of an electric or magnetic field. The electric field effect is known as the *Stark effect*,[2] and that of the magnetic field is the *Zeeman effect*.[3] Stray electric fields can lead to broadening and, in some cases, shifts of the center wavenumber of peaks due to the Stark effect. This is usually only noticeable in high-resolution work unless an electrical discharge is used to generate the spectrum. The Zeeman effect is usually not important in laboratory spectra unless a strong magnetic field is intentionally applied. However, the spectra observed from some stars show clearly resolvable Zeeman splittings. A further magnetic interaction is that of the electrons with the nuclear spin. This leads to *hyperfine splitting* of atomic lines which, as the name implies, can only be observed at ultrahigh-resolution for light elements. Hyperfine splitting becomes increasingly important as the atomic number increases and can lead to observable line splitting in, for example, actinides. In general, we will not consider any of these field effects and hyperfine splittings unless stated explicitly.

23.1.2.1 Directed practice

Calculate the Doppler linewidth of the Kr transition at 605.78 nm at 300 K.

[Answer: 0.0224 cm^{-1}]

23.1.3 Types of spectroscopy

In emission spectroscopy we excite the system with light, electrons, collisions, heat, electrical discharge, explosion, and so forth, and measure the intensity and/or energy of the photons or electrons that are given off in a transition corresponding to the down arrow in Fig. 23.2. Fluorescence, phosphorescence, photoluminescence, and photoemission are all types of emission spectroscopy. From radioactive nuclei there is also the emission of subatomic particles and gamma radiation.

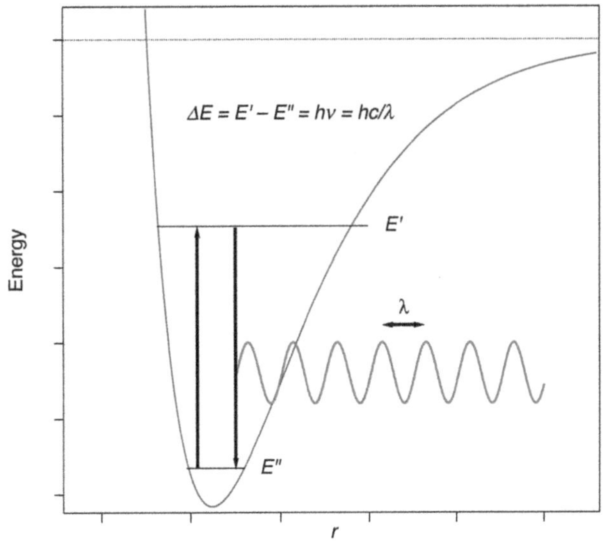

Figure 23.2 Transitions between upper (energy E') and lower (energy E'') states are connected by photons. The photon energy is the same regardless of the direction of the transitions; thus, the peaks in emission and absorption spectra occur at the same energy/frequency/wavelength.

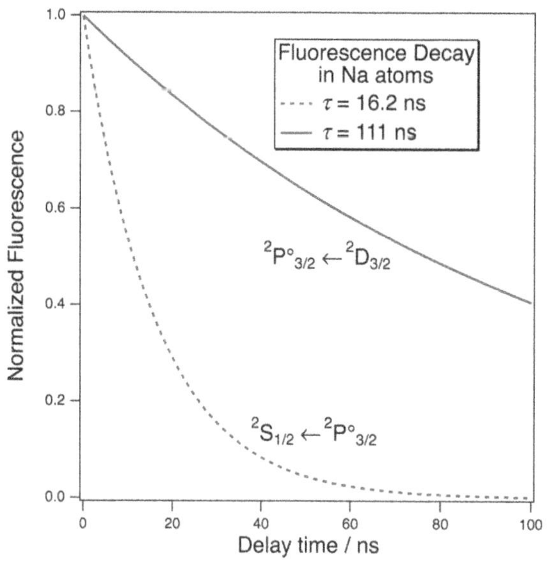

Figure 23.3 Simulated time-resolved fluorescence spectra. The influences of short (16.2 ns) and long (111 ns) lifetimes for allowed transitions in the Na atom. The short lifetime corresponds to a transition known as the D-line of Na.

In absorption spectroscopy we measure how much radiation is absorbed by a sample in a transition corresponding to the up arrow in Fig. 23.2 as a function of the energy of the radiation. Infrared (IR) and ultraviolet/visible (UV/vis) spectroscopy are usually performed as absorption spectroscopy. Nuclear magnetic resonance (NMR) and electron spin resonance (ESR) are peculiar types of absorption spectroscopy performed in a strong magnetic field.

Fluorescence spectroscopy involves both photons in and photons out. The conventional emission mode is to excite with one wavelength of light and then to disperse the spectrum of emitted light (dispersed fluorescence). The fluorescence will change depending on the wavelength of the excitation light. Fluorescence spectroscopy performed in excitation mode corresponds to detecting at a particular wavelength while the excitation wavelength is scanned. Synchronous mode fluorescence corresponds to scanning both the excitation and emission wavelengths simultaneously with a constant separation between the two being maintained.

In time-resolved spectroscopy we measure absorption or emission as a function of time. This is a direct probe of dynamics: how a system evolves in time. This is how we can probe the natural lifetime of a particular atomic state, as shown in Fig. 23.3. Probing the lifetime of a state is greatly simplified if the duration of the excitation is much shorter than that the lifetime of the state, so as to avoid interferences from scattered light. The simultaneous absorption and emission of light, and possible saturation of the transition, must be considered if the signal level during the excitation pulse is to be analyzed. Saturation occurs when stimulated emission begins to compete with spontaneous emission.

A particularly refined type of spectroscopy is known as *two-dimensional (2D) spectroscopy*, which is becoming of increasing importance as we seek more detailed information on the dynamical behavior of systems. Two-dimensional spectroscopy involves following how a spectrum evolves as a function of two variables. For instance, we might detect emission spectra as a function of time or as a function of excitation wavelength. By its nature, 2D spectroscopy requires pulsed excitation sources and following signals as a function of the delays and frequencies between these pulses. In nuclear magnetic resonance (NMR) spectroscopy there are a number of 2D methods including correlation spectroscopy (COSY) and nuclear Overhauser effect spectroscopy (NOESY). COSY identifies which spins are coupled to one another by applying a single radiofrequency (RF) pulse, waiting for a specific length of time before applying a second RF pulse before making a measurement with a third pulse.

23.2 Time-dependent perturbation theory and spectral transitions

We need to introduce a time-dependent formalism to describe transitions between states. We write the wavefunction of a system as the product of a time-independent wavefunction and a time-dependent term

$$\Psi_i^0(\tau, t) = \psi_i(\tau)e^{-iE_i t/\hbar}. \tag{23.12}$$

The spatial coordinates are represented by τ and time by t.

Assume that we can solve this for all allowed values of E_i. The associated wavefunctions are called the eigenstates of the system. The superscript 0 denotes that the state is in the initial unperturbed state i before we apply the electromagnetic perturbation (the light) that is going to cause the transition to the final state.

Now apply an oscillating electromagnetic field (light). Either the electric field or magnetic field couples to the molecule to cause a transition. Molecules having oscillating dipole moments emit light. Any and all electric and magnetic moments (dipole, quadrupole, hexapole, etc.) can, in principle, couple to an electromagnetic wave. Electric moments couple to the electric field and magnetic moments couple to the magnetic field. Not all molecules have all types of moments.

The *electric dipole moment* μ_x results from the separation of electric charge δq by a distance x (along the x axis) and is given by

$$\mu_x = x\,\delta q. \tag{23.13}$$

The electric dipole moment is the most commonly encountered moment, and is most commonly responsible for atomic absorption and emission, pure rotational transitions, vibrational and electronic spectroscopy. Raman spectroscopy depends on polarizability. Spins generate magnetic moments and are responsible for ESR and NMR spectroscopies.

Consider the coupling of the electric moments to be a small perturbation. First, we write the time-dependent Schrödinger equation

$$\hat{H}(t) = \Psi = \mathrm{i}\hbar\frac{\partial \Psi}{\partial t}. \tag{23.14}$$

This describes how a system evolves in time. Now we write

$$\hat{H}(t) = \hat{H} + \hat{H}'(t). \tag{23.15}$$

Thus

$$[\hat{H} + \hat{H}'(t)]\Psi = \mathrm{i}\hbar\frac{\partial \Psi}{\partial t}. \tag{23.16}$$

The time-dependent Hamiltonian is composed of the usual time-independent term and a time-dependent perturbation, a first-order perturbation that is sinusoidally oscillating in time

$$\hat{H}'(t) = -\mu_x E_x^{\circ} \cos \omega t \tag{23.17}$$

This represents the coupling of the electric dipole moment of the atom (or molecule) with the electric component of the incident wave (light or other particle).

The radiation mixes the unperturbed stationary states (the eigenstates) and forms a superposition of them

$$\Psi(\tau, t) = \sum_i c_i(t)\Psi_i^0(\tau, t). \tag{23.18}$$

In the absence of the radiation, the perturbation vanishes and the eigenstates exist as stationary states ψ_i. They do not evolve in time. When the perturbation (radiation) is turned on, the states are mixed to form Ψ and the system can make a transition from ψ_i to ψ_k. The time-dependent behavior of ψ_k is governed by the coefficients $c_i(t)$ that define the linear combination of stationary states that describe ψ_k. The probability of finding the system in state ψ_k is

$$P_k(t) = c_i^*(t)c_i(t) = |c_i(t)|^2. \tag{23.19}$$

The *transition rate* from initial state i to final state k, W_{ki}, is the rate of change of the probability,

$$W_{ki} = \mathrm{d}P_k/\mathrm{d}t \tag{23.20}$$

In order for a transition to occur two conditions must be met. The first condition corresponds to the conservation of energy, the second relates to the conservation of angular momentum. The *Bohr condition* states that the photon has to have the same energy as the difference in energy between the initial state ψ_i and the final state ψ_k

$$E_k - E_i = h\nu_{ki}. \tag{23.21}$$

If an electron (or other particle) causes the transition, then the loss of energy by the electron equals the energy gained by the molecule. At very high energies – that is, in the gamma range – the photon can also exchange substantial momentum and this must be taken into account. This is the basis of the Compton effect, but we will neglect this here as we are more concerned with the interactions of lower-energy photons (soft X-ray and below) for which the linear momentum exchange is negligible.

23.2.1 Einstein transition probabilities

The second condition for a transition to occur is that the matrix element of the *transition dipole moment*

$$\mu_{ki} = \langle \psi_k | \hat{\mu} | \psi_i \rangle = \int \psi_k^* \, \hat{\mu} \, \psi_i \, \mathrm{d}\tau \tag{23.22}$$

that couples the initial and final states under the perturbation of electromagnetic field cannot be equal to zero because the transition rate is given by

$$W_{ki} = \frac{1}{6\varepsilon_0 \hbar^2 c} |\langle \psi_k | \hat{\mu} | \psi_i \rangle|^2 I(\nu_{ki}) = B_{ki}\, \rho(\nu_{ki}) \tag{23.23}$$

where $I(\nu_{ki})$ is the intensity of the radiation at the frequency of the transition and B_{ki} is the *Einstein B coefficient*[4]

$$B_{ki} = \frac{|\langle \psi_k | \hat{\mu} | \psi_i \rangle|^2}{6\varepsilon_0 \hbar^2} = \frac{|\mu_{ki}|^2}{6\varepsilon_0 \hbar^2} \tag{23.24}$$

The intensity of radiation is proportional to the square of the electric field strength. The *energy density of radiation* $\rho(\nu_{ki})$ at the frequency of the transition ν_{ki} is equal to the energy per unit volume per unit frequency. It is related to the intensity by

$$\rho(\nu_{ki}) = I(\nu_{ki})/c \tag{23.25}$$

Equation (23.23) is a special case of *Fermi's golden rule*:

> *The transition rate is proportional to the product of the absolute square of the matrix element of the perturbing potential and the strength of the perturbation at the transition frequency. More generally, the transition rate is equal to the product of the absolute square of the matrix element of the perturbing potential and the density of states at the transition frequency.*

The absolute square of the matrix element is symmetrical, that is, $|\langle \psi_k | \hat{\mu}_x | \psi_i \rangle|^2 = |\langle \psi_i | \hat{\mu}_x | \psi_k \rangle|^2$ because the operator is Hermitian. Therefore, there is only one value of the Einstein B coefficient for a given set of nondegenerate levels, irrespective of the direction of the transition, that is, $B_{ki} = B_{ik}$.

If there are N_i particles in state i at $t = 0$, the rate of change in population in state i is equal in magnitude but of opposite sign to the rate of photon absorption as given by

$$\frac{\mathrm{d}N_i^{\mathrm{abs}}}{\mathrm{d}t} = -N_i\, B_{ki}\, \rho(\nu_{ki}). \tag{23.26}$$

The absorption of photons with an energy equal to the transition energy is how radiation stimulates a transition from the initial to the final state.

The transition from upper state back to the lower state can occur in two ways as shown in Fig. 23.4. *Spontaneous emission* is governed by the *Einstein A coefficient* (Einstein coefficient of spontaneous emission)

$$\frac{\mathrm{d}N_k^{\mathrm{spon}}}{\mathrm{d}t} = -N_k\, A_{ik}. \tag{23.27}$$

This rate does not depend on the energy density or frequency of the radiation. The rate of *stimulated emission* does and is given by

$$\frac{\mathrm{d}N_k^{\mathrm{stim}}}{\mathrm{d}t} = -N_k\, B_{ik}\, \rho(\nu_{ki}). \tag{23.28}$$

At steady state, there is no net absorption: the rate of absorption is balanced by the sum of stimulated and spontaneous emission

$$N_i B_{ki} \rho(\nu_{ki}) = N_k [A_{ik} + B_{ik} \rho(\nu_{ki})]. \tag{23.29}$$

The relative rate of spontaneous emission to stimulated emission is related to the ratio of A and B coefficients and can be derived from statistical mechanics (the Boltzmann distribution) and the Planck radiation law (used to describe

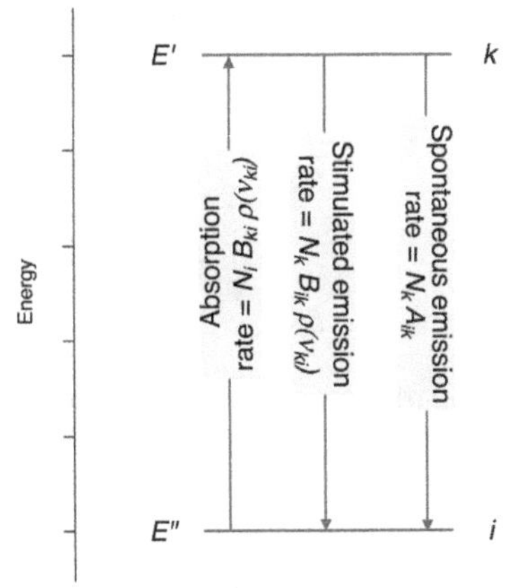

Figure 23.4 Radiative transitions between a lower state i at energy E'' and an upper state k at energy E' occur at the same frequency ν_{ki} in absorption and emission. The rate of photon absorption depends on the number of molecules in the lower state N_i, the Einstein B coefficient B_{ki} and the energy density of the radiation at the transition frequency $\rho(\nu_{ki})$. The rate of stimulated emission depends on the upper state population N_k as well as B_{ik} and $\rho(\nu_{ki})$. The rate of spontaneous emission is independent of the energy density and depends only on N_k and the Einstein A coefficient A_{ik}.

black-body radiation). We assume that the molecules and the radiation field are in thermal equilibrium at temperature T. Thus, the ratio of populations in the upper state k and the lower state i is given by the Boltzmann ratio

$$\frac{N_k}{N_i} = \frac{g_k}{g_i} \exp\left(-\frac{E_k - E_i}{k_B T}\right) = \frac{g_k}{g_i} \exp\left(-\frac{h\nu_{ki}}{k_B T}\right) \tag{23.30}$$

where g_k and g_i are the degeneracies of states k and i, respectively. Thus, the expression for N_k/N_i from Eq. (23.30) can be set equal to the same ratio derived from Eq. (23.29). Solving these for $\rho(\nu_{ki})$ and setting them equal to the radiation energy density per unit frequency given in the Planck radiation law

$$\rho(\nu_{ki}) = \frac{8\pi h\nu_{ki}^3}{c^3} \frac{1}{\exp(h\nu_{ki}/k_B T) - 1} \tag{23.31}$$

we see that the following relations must hold

$$g_i B_{ki} = g_k B_{ik} \tag{23.32}$$

$$\frac{A_{ik}}{B_{ik}} = \frac{8\pi h}{c^3} \nu_{ki}^3. \tag{23.33}$$

23.2.1.1 Directed practice

Perform the algebra to prove that Eqs (23.30) and (23.31) can be derived from Eqs (23.27–23.29).

Thus, *the rate of spontaneous emission increases with the cube of the frequency*. This has important implications for the production of lasers. It becomes progressively harder to maintain a population inversion as the frequency of the transition becomes larger. Hence, an IR laser is easier to make than a visible laser, and the much sought-after X-ray laser is a very difficult proposition indeed. The A and B coefficients depend only on the states i and k. They do not depend on thermal equilibrium, the temperature, or any other molecular property.

23.3 The Beer–Lambert law

23.3.1 Transmittance and absorbance

Quantitative spectroscopy or spectrophotometry measures the amount of light absorbed or transmitted. Johann Lambert figured out that the amount of light absorbed is independent of the light intensity (true for conventional sources but not high-powered lasers) and proportional to the pathlength through the sample. Thus, each successive layer of a sample dx decreases the intensity by an equal fraction –dI/I, where I is the radiant intensity. Therefore,

$$-\mathrm{d}I/I = b\mathrm{d}x. \tag{23.34}$$

We now integrate this equation using the following boundary conditions. For a cell of pathlength l, the initial intensity at $l = 0$ is I_0. Hence,

$$\int \frac{dI}{I} = -b\int_0^l dx \tag{23.35}$$

and

$$I = I_0 e^{-bl}. \tag{23.36}$$

It is conventional to use common (base 10) logarithms and to introduce the *absorbance A* and the *transmittance T* according to

$$T = I/I_0 \tag{23.37}$$

$$\lg(I_0/I) = \lg(1/T) = \frac{bl}{2.303} = A. \tag{23.38}$$

August Beer found that there is also a linear relationship between concentration and absorbance. Combining these two results, we derive the *Beer–Lambert law*

$$A = \varepsilon cl, \tag{23.39}$$

where ε is the *absorption coefficient*. A is dimensionless. The units of ε will depend on the units of c and l. By convention for solutions, c is given in molarity and l is in cm, so the molar absorption coefficient has units of $\mathrm{l\ mol^{-1}\ cm^{-1}}$. Note that ε is a frequency-dependent quantity $\varepsilon(\nu)$; indeed, it is the frequency dependence of $\varepsilon(\nu)$ that describes the shape of the absorption peak.

Instead of the decadic absorbance (the one used above related to the use of the base 10 logarithm), the naperian absorbance A_e (related to the use of the natural logarithm) is also encountered. It is defined in terms of the absorption cross-section σ according to

$$I/I_0 = \exp(-\sigma l\rho) \tag{23.40}$$

and

$$A_e = \ln(I_0/I) = \sigma l\rho, \tag{23.41}$$

where l is again the pathlength and ρ is the density of the absorbing species. Again, remember that the absorption cross-section $\sigma(\nu)$ is a frequency-dependent quantity that describes the shape of the absorption spectrum.

23.3.1.1 Directed practice

Show that the molar absorption coefficient and the absorption cross-section are related by

$$\sigma = (c\ln 10/\rho)\varepsilon. \tag{23.42}$$

23.3.2 Relating absorption coefficient, Einstein coefficients and transition dipole moment

Accelerated charged particles emit radiation. If an electron is accelerated at a rate $\ddot{r}$, it can be shown (see for example Heitler in Further Reading) that the classical rate of energy loss by irradiation is

$$\dot{E} = -\frac{e^2}{6\pi\varepsilon_0 c^3}|\ddot{r}|^2. \tag{23.43}$$

Assume that the electron experiences simply harmonic motion as if it were a mass held between two springs. If the electron oscillates parallel to the z-axis with amplitude r_0, then

$$r = r_0 \cos\omega t \tag{23.44}$$

$$\ddot{r} = -\omega^2 r_0 \cos\omega t = -\omega^2 r. \tag{23.45}$$

Substituting this into Eq. (23.43) and averaging over one cycle of the harmonic motion, we obtain

$$\dot{E}_{av} = -\frac{e^2\omega^4}{6\pi\varepsilon_0 c^3}|r|^2_{av}. \tag{23.46}$$

The same result is found for the analogous harmonic circular motion in the xy-plane. However, the energy of a simple harmonic oscillator is also proportional to $|r|^2_{\mathrm{av}}$. Thus, the average energy loss is proportional to the negative of the energy of the oscillator. This implies that the energy decays exponentially in time,

$$E = E_0 e^{-at} \tag{23.47}$$

where E_0 is the energy originally available and a is the fractional decay rate. The classical weighted average loss rate over the entire duration of the radiative decay is

$$\langle \dot{E}_{\mathrm{av}} \rangle = \frac{\int_0^\infty \dot{E}_{\mathrm{av}} E \mathrm{d}t}{\int_0^\infty E \mathrm{d}t} = -\frac{a}{2} E_0. \tag{23.48}$$

The corresponding decay rate is

$$a = -\frac{2\langle \dot{E}_{\mathrm{av}} \rangle}{E_0} = \frac{e^2 \omega^4}{3\pi\varepsilon_0 c^3 E_0} |r|^2_{\mathrm{av}} \tag{23.49}$$

where $|r|^2_{\mathrm{av}}$ is averaged over the entire duration of radiative decay.

Quantum mechanically, the weighted average value of $\boldsymbol{r}$ in the state described by the wavefunction Ψ is given by the expectation value

$$\langle \boldsymbol{r} \rangle = \langle \Psi | \boldsymbol{r} | \Psi \rangle \tag{23.50}$$

Here, $\boldsymbol{r}$ has been generalized to a vector. As a transition is being made, the wavefunction is a time-dependent wavefunction

$$\Psi = \psi e^{-\mathrm{i}Et/\hbar}. \tag{23.51}$$

Substituting into Eq. (23.50) we obtain the expectation value of $\boldsymbol{r}$ during irradiation

$$\langle \boldsymbol{r} \rangle = \langle \Psi_k | r | \Psi_i \rangle = \langle \psi_k | \boldsymbol{r} | \psi_i \rangle e^{-\mathrm{i}(E_k - E_i)t/\hbar} \tag{23.52}$$

The analog of the classical angular frequency is

$$\omega_{ki} = 2\pi\nu_{ki} = |E_k - E_i|/\hbar. \tag{23.53}$$

Now interpret that the value $|\boldsymbol{r}|^2_{\mathrm{av}}$ averaged over the period of irradiation is given quantum mechanically by

$$|r|^2_{\mathrm{av}} \equiv |\langle \boldsymbol{r}_{ki} \rangle|^2 = |\Psi_k | \boldsymbol{r}_{ki} | \Psi_i|^2 = |\langle \psi_k | \boldsymbol{r}_{ki} | \psi_i \rangle|^2, \tag{23.54}$$

from which the decay rate is

$$a_{ki} = -\frac{e^2 \omega_{ki}^4}{3\pi\varepsilon_0 c^3 h\nu_{ki}} |\langle \psi_k | \boldsymbol{r}_{ki} | \psi_i \rangle|^2 = -\frac{16\pi^3 e^2 \nu_{ki}^3}{3\varepsilon_0 c^3 h} |\langle \psi_k | \boldsymbol{r}_{ki} | \psi_i \rangle|^2. \tag{23.55}$$

From Eq. (23.55), we see that the transition dipole moment is equal to

$$\boldsymbol{\mu}_{ki} = \langle \psi_k | -e\boldsymbol{r}_{ki} | \psi_i \rangle, \tag{23.56}$$

from which we can make the identification that the decay rate is, in fact, the Einstein A coefficient for the transition

$$a_{ki} = \frac{16\pi^3 \nu_{ki}^3}{3c^3 \varepsilon_0 h} |\mu_{ki}|^2 = A_{ik}. \tag{23.57}$$

Whereas, Eq. (23.55) was ambiguous as to whether the frequency dependence of the ratio of A_{ik} to B_{ik} was due to A_{ik}, Eq. (23.57) makes it clear that it is the rate of spontaneous emission that increases with increasing transition frequency. Also from Eq. (23.55), the Einstein B coefficient is

$$B_{ik} = \frac{2\pi^2}{3\varepsilon_0 h^2} |\mu_{ki}|^2, \tag{23.58}$$

which is, of course, equivalent to Eq. (23.24).

We now relate the Einstein B coefficient to the absorption cross-section. The differential form of Eq. (23.40) relates the decrease in light intensity to the absorption cross-section and pathlength

$$-\mathrm{d}I(\nu)\,\mathrm{d}\nu = I(\nu)\,\mathrm{d}\nu\,\sigma N_i\,\mathrm{d}x, \tag{23.59}$$

where $dI(\nu)\,d\nu$ is the intensity of light in the range from ν to $\nu + d\nu$. Assuming that there is little excited state population, that is $N_k \ll N_i$, then the decrease in number density of absorbers in the ground state $-dN_i$ is equal to the decrease in the number density of photons $-dN_\gamma$. The intensity divided by the speed of light is equal to the energy density, and the energy density divided by the photon energy is equal to the number density of photons; hence,

$$\frac{I(\nu)}{ch\nu} = \frac{\rho(\nu)}{h\nu} = N_\gamma \tag{23.60}$$

and

$$-\frac{dI(\nu)d\nu}{ch\nu} = -dN_i \tag{23.61}$$

The change in initial state population is related to B_{ki} and the energy density of the light by

$$-dN_i = N_i B_{ki} \rho(\nu) dt. \tag{23.62}$$

Equating the last two expressions for $-dN_i$ and substituting from Eq. (23.59), we find

$$\frac{I(\nu)\,d\nu\,\sigma N_i\,dx}{ch\nu} = N_i B_{ki} \rho(\nu) dt \tag{23.63}$$

$$\frac{\sigma d\nu}{h\nu}\frac{dx}{dt} = B_{ki}\frac{c\rho(\nu)}{I(\nu)}. \tag{23.64}$$

The derivative $dx/dt = c$ and from the definition in Eq. (23.25), $c\rho(\nu)/I(\nu) = 1$. Therefore,

$$B_{ki} = \frac{c\sigma d\nu}{h\nu}. \tag{23.65}$$

Equation (23.65) can be interpreted in two ways. Integration over frequency yields the value of the Einstein B coefficient for the entire absorption band. Alternatively, taking $d\nu$ to be the bandwidth of the light centered at the frequency ν (assumed to be much narrower than the absorption peak), then we can calculate B_{ki} at this particular value of ν.

23.4 Electronic spectra of atoms

23.4.1 Transition dipole moment

The previous section demonstrates that the *transition dipole moment* is a particularly important parameter. Transitions can take place by various mechanisms: electric dipole, magnetic dipole, electric quadrupole, and so on. The probability of a transition taking place by these different mechanisms is very different, with probabilities for these mechanisms being in the approximate ratio $1{:}10^{-5}{:}10^{-7}$. However, due to symmetry constraints on the states involved, certain transitions are forbidden between certain states. This is the basis of selection rules.

Whether a transition is allowed or forbidden, and the degree to which it is allowed or forbidden, is determined by the magnitude of the transition matrix element, μ_{fi}

$$\mu_{fi} = \int \psi_f^* \hat{\mu}_{tr} \psi_i \, d\tau. \tag{23.66}$$

where $\hat{\mu}_{tr}$ is an operator specific to the type of transition (electric dipole, etc.) that connects the initial and final state wavefunctions, ψ_i and ψ_f.

A forbidden transition might be experimentally observed, especially when high-power lasers are used to interrogate the transition. However, the transition will have a low intensity compared to an allowed transition. The selection rules for the different mechanisms are different. For the particle in a 1D box the selection rule for electric dipole transitions is Δn = an odd integer. To understand the genesis of selection rules, we will first have to understand more about electronic structure and the coupling of different sorts of angular momenta.

23.4.2 Hydrogenic atoms

The electronic absorption and emission spectra of the H atom and hydrogenic ions are singularly simple, especially if we limit ourselves to a description in terms of the Schrödinger equation and disregard the small corrections introduced by the Dirac equation. Hydrogenic ions are ions that have only one electron. The wavenumber of the lines in the spectra of H and hydrogenic ions with nuclear charge Z can be described by the *Rydberg formula*

$$\tilde{\nu} = ZR_X\left(\frac{1}{n_1^2} - \frac{1}{n_2^2}\right). \tag{23.67}$$

where

$$R_X = \frac{\mu e^4}{8\varepsilon_0^2 h^3} \tag{23.68}$$

$R_H = 109\,677.581$ cm^{-1} is the *Rydberg constant* for H atoms, and each hydrogenic atom has its own characteristic value of R because of slight shifts in the reduced mass,

$$\mu = \frac{m_e m_{nucleus}}{m_e + m_{nucleus}}. \tag{23.69}$$

As the mass of the nucleus increases, the Rydberg constant converges on the value for an infinitely massive nucleus, $R_\infty = 109\,737$ cm^{-1}.

The form of the Rydberg equation suggests that it results from the difference of two terms

$$T_n = ZR_X/n^2 \tag{23.70}$$

such that the wavenumber of the emitted light is

$$\tilde{\nu} = T_1 - T_2 = ZR_X\left(\frac{1}{n_1^2} - \frac{1}{n_2^2}\right). \tag{23.71}$$

What does this tell us about the structure of atoms? We know that photons have an energy of $h\nu$. Therefore, by conservation of energy, the energy difference of two electronic states must be equal to the energy of the emitted photon

$$E_{photon} = \Delta E_{atomic\ levels} = h\nu. \tag{23.72}$$

This is the Bohr frequency condition.

Therefore, atoms have electronic states with well-defined (quantized) energies and the terms that we defined above are, in fact, the energies of the electronic states. That is, we can write

$$E_n = -hcR_X\frac{Z}{n^2}. \tag{23.73}$$

23.4.3 Multi-electron atoms

Figure 23.5 is a *Grotrian diagram* for atomic Na. The objective of the next few sections is to explain and understand the principles behind this diagram and what it means in terms of atomic structure. The diagram contains numerous energy levels that are identified by term symbols. Lines denote allowed transitions and connect the states involved in these transitions. Note that not all states are connected to all other states because of selection rules.

23.5 Spin–orbit coupling

The Dirac equation is a relativistic form of the Schrödinger equation that allows for quantization within four dimensions of space–time in which time and the three spatial dimensionals are treated equivalently. A natural result of this treatment is that the electron possesses an intrinsic angular momentum that we call 'spin.' The spin of an electron is $s = \frac{1}{2}$, which means that it can only be oriented in two direction, denoted by $m_s = \pm\frac{1}{2}$. Use of the term spin is in some ways unfortunate. The term was used in an attempt to draw parallels with a classical analogy of a particle with mass m_e and charge $-e$ spinning about its axis. However, a half-integer quantum number means that this 'spinning wave' has some peculiar properties. For instance, when an electron spins around by 2π radians, it does not return to the same configuration as would a classical body. Instead, it has changed sign and must spin by 4π radians to return to its original orientation. Besides, how does a wave spin? Electron spin is more accurately described as the intrinsic magnetic moment of the electron.

Because the electron has a nonzero value of spin, it has associated with it an intrinsic magnetic field. An electron occupying an orbital that possesses a nonvanishing angular momentum, that is an orbital with $l > 0$, also has a magnetic

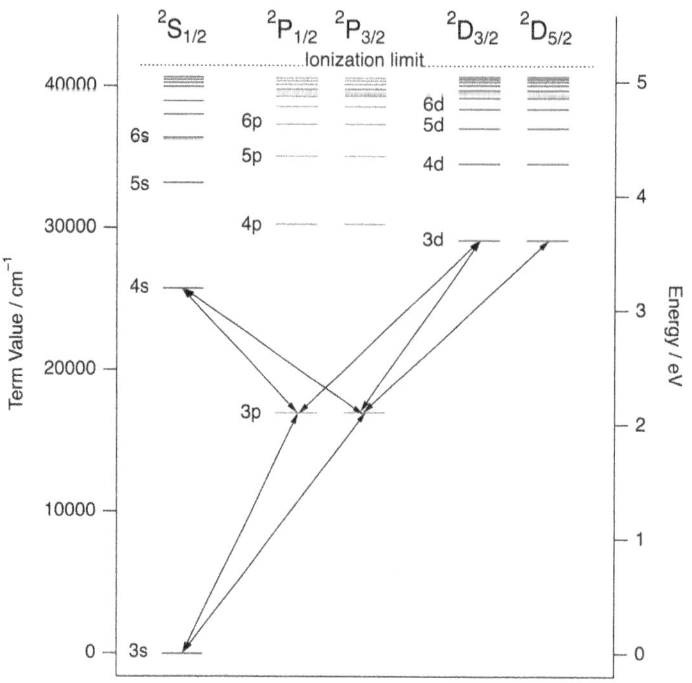

Figure 23.5 The Grotrian diagram of Na I depicts the energy states of the neutral atom and lines indicate allowed transitions. There is no restriction on Δn; however, for simplicity only $\Delta n = \pm 1$ is shown for a few selected transitions.

field associated with its orbit. These two magnetic fields interact. *Spin–orbit coupling* is the interaction of the electron's intrinsic magnetic moment with the orbital magnetic moment. We have seen in Section 22.2 that this interaction gives rise to an additional term in the Hamiltonian that describes the interaction of an electron with the nucleus. Not only does spin–orbit coupling lead to a shift in energy levels, it also leads to splitting of energy levels. Here, we look at this interaction in more detail in order to understand quantitatively the effect of spin–orbit coupling on spectra.

The spin–orbit interaction energy naturally results from a relativistic calculation. However, the correct energy can also be obtained nonrelativistcally if one computes the electrostatic potential energy of the nuclear charge q_N in the electric field produced by the moving electron magnetic moment μ. The spin–orbit energy is obtained by imagining the nuclear charge is brought from infinity to the position of the nucleus along a straight-line path directed towards the electron at r and through the electric field generated by the electron. This energy is given by the integral of the dot product of the force $\boldsymbol{F}$ with the position vector $\boldsymbol{r}$,

$$E_{\text{so}} = -\int_{\infty}^{r} \boldsymbol{F} \cdot \mathrm{d}\boldsymbol{r} = -(q_{\text{N}}/2r^3 c)(\boldsymbol{r} \times \boldsymbol{v}) \cdot \boldsymbol{\mu} \tag{23.74}$$

However

$$(\boldsymbol{r} \times \boldsymbol{v}) = \boldsymbol{l}\hbar/m_{\text{e}} \tag{23.75}$$

and

$$\boldsymbol{\mu} = -e\hbar\boldsymbol{s}/m_{\text{e}}c. \tag{23.76}$$

Since the potential energy of the electron in the electrostatic field of the nucleus is

$$V(r) = -q_{\text{N}}e/r, \tag{23.77}$$

the spin–orbit energy is

$$E_{\text{so}} = \frac{\hbar^2}{2m_{\text{e}}^2 c^2}\left(\frac{1}{r}\frac{\mathrm{d}V}{\mathrm{d}r}\right)(\boldsymbol{l} \cdot \boldsymbol{s}). \tag{23.78}$$

This energy correction depends on a dot product of the angular momentum vectors $\boldsymbol{l}$ and $\boldsymbol{s}$. Therefore, the strength of the coupling depends on the *orientation* of the two angular momenta with respect to one another. An antiparallel alignment

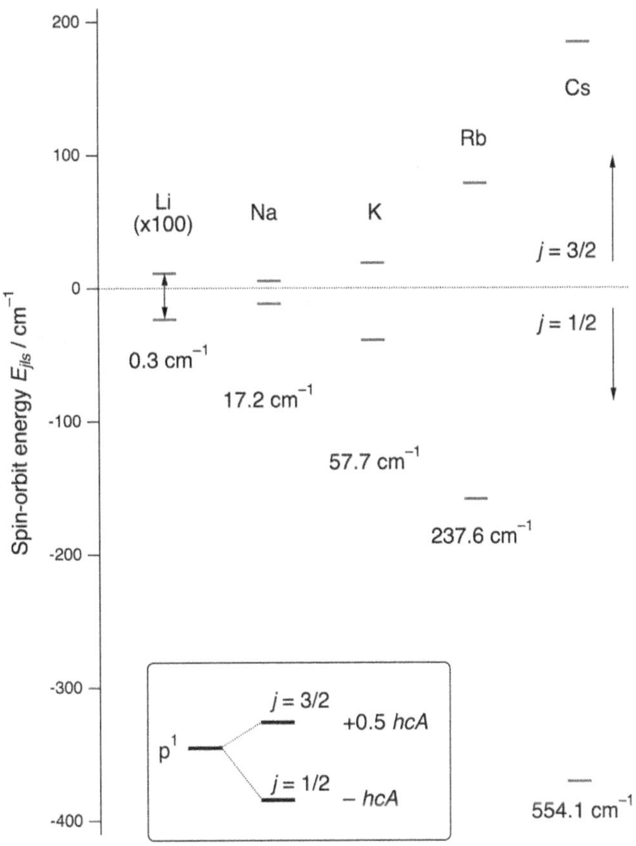

Figure 23.6 Spin–orbit coupling splits a p^1 electron configuration into two levels distinguished by their values of j, the total angular momentum of the electron. The spin–orbit coupling constant increases with atomic number, causing the spin–orbit energy of the $j = 3/2$ level to increase and the $j = \frac{1}{2}$ level to decrease to a greater extent as shown for the alkali metals Li, Na, K, Rb, and Cs.

is favorable (lower energy), that is, ↑↓ is a favorable alignment of magnetic dipoles. A parallel alignment is unfavorable, that is, ↑↑ is an unfavorable alignment of magnetic dipoles

The energy of the spin–orbit interaction is not determined solely by either the angular orbital moment or the spin of an electron. Instead, it is determined by how they couple to form the total angular momentum $\boldsymbol{j}$, which is the vector sum of these two quantities. This energy E_{jls}, is given by

$$E_{jls} = {}^1/_2 hcA\{j(j+1) - l(l+1) - s(s+1)\}, \tag{23.79}$$

where A is the spin–orbit coupling constant. The value of A can be derived explicitly for hydrogenic ions for which the potential is known exactly. It is difficult to derive from theory in a multi-electron atom as it depends on both the orbital and the atom. The strength of spin–orbit coupling as measured by the magnitude of A for a given type of orbital increases with atomic number roughly as Z^4. For the lowest energy p orbital above the ground state of alkali metals, the spin–orbit coupling constant increases from 0.23 cm^{-1} for Li(2p) to 11.5, 38.5, 158, and 370 cm^{-1} for Na(3p), K(4p), Rb(5p) and Cs(6p), respectively. The resulting spin–orbit energy of the split levels is shown in Fig. 23.6. As can be seen by the values for Na I in Table 23.1, the spin–orbit splitting decreases for any one orbital type as the principal quantum number n

Table 23.1 The term values and spin–orbit splittings in cm^{-1} of the electronic states of the neutral Na atom (Na I) with $3 \le n \le 6$ and $0 \le l \le 2$. Values taken from the NIST Atomic Spectroscopy Database.

n	$^2S_{1/2}$	$^2P_{1/2}$	$^2P_{3/2}$	Splitting	$^2D_{3/2}$	$^2D_{5/2}$	Splitting
3	0.000	16956.170	16973.366	17.196	29172.887	29172.837	−0.050
4	25739.999	30266.990	30272.580	5.590	34548.764	34548.729	−0.035
5	33200.673	35040.380	35042.850	2.470	37036.772	37036.752	−0.020
6	36372.618	37296.320	37297.610	1.290	38387.268	38387.255	−0.013

increases. Obviously, we need to learn how to calculate the total angular momentum, not only of one electron but for any electron configuration.

23.5.1 Total angular momentum

The total angular momentum in a one-electron system is denoted j with projection m_j. It is calculated from the vector sum of l and s,

$$\boldsymbol{j} = \boldsymbol{l} + \boldsymbol{s} \tag{23.80}$$

Note that $\boldsymbol{j}$, $\boldsymbol{l}$ and $\boldsymbol{s}$ are *vectors* and must be added *vectorially*. The magnitude of j is a positive number, whereas m_j can be positive or negative (as was also the case for the pairs l and m_l; s and m_s). Thus, for an electron $j = |l + {}^1/_2|$ (when the l and s are in the same direction) or $j = |l - {}^1/_2|$ (when they point in opposite directions). The value of j labels different levels of a term. The two levels of the same term associated with different values of j have slightly different energies. This leads to *fine structure* in atomic spectra caused by the splitting of energy levels.

For example, an s electron has $j = 1/2$ and only one level associated with it. A p electron has $j = 1/2$ or $3/2$ and two spin–orbit fine-structure levels associated with it. This effect is shown for the alkali metals in Fig. 23.6. Alkali metals have a closed shell core surrounded by one electron; thus, they act effectively as a one-electron system if only the outermost electron is excited. Consider the first excited p orbital above the ground state. There is a splitting of 17 cm^{-1} between the two j levels of the upper state for Na. The level with $j = 3/2$ is pushed to higher energy by the spin–orbit interaction. In the ground state, derived from an s orbital, there is only one level. Transitions from the excited state, which is split by spin–orbit coupling into two fine-structure levels, to the ground state, which is not split, have almost the same energy. When observed at sufficient resolution, the emission lines for all of the alkali metals will appear as doublets. These are easily resolved for Cs but require high resolution for Li.

23.6 Russell–Saunders (*LS*) coupling

In the Hamiltonian for multi-electron atoms,

$$\hat{H} = \hat{T} + \hat{V} + \hat{H}_{ee} + \hat{H}_{so}, \tag{23.81}$$

the only term that depends on the spin quantum number is the spin–orbit coupling term $\hat{H}_{so}$, which couples $\boldsymbol{l}$ and $\boldsymbol{s}$. This term is negligible for light atoms (low Z). The three remaining terms describe the kinetic energy, potential energy and the electron–electron repulsions. A Hamiltonian with only these three terms commutes with the operators for total orbital, total spin and total angular momentum, $\hat{L}^2$, $\hat{S}^2$ and $\hat{J}^2$. In other words, L, S and J are all good quantum numbers and the associated angular momenta are constants of motion. The total angular momentum J is equal to the *vector* sum of L and S,[5]

$$\boldsymbol{J} = \boldsymbol{L} + \boldsymbol{S} \tag{23.82}$$

For low-Z atoms, when the spin–orbit interaction is negligible, we have what we call *LS* coupling. The orbital angular momenta of individual electrons in any open subshells couple to form values of L for the atom. Separately, the spins of individual electrons couple to form values of S for the atom. This coupling to form the vectors $\boldsymbol{L}$ and $\boldsymbol{S}$ is shown graphically for the case of two p electrons in Fig. 23.7.

$\boldsymbol{L}$ and $\boldsymbol{S}$ then couple – that is, we perform the vector sum in Eq. (23.82) to form the resulting $\boldsymbol{J}$ values for the atom. This is known as the *Russell–Saunders* or *LS coupling scheme.* It is valid when spin–orbit coupling is weaker than electron–electron repulsions. That is, the Russell–Saunders coupling scheme is valid as long as the spin–orbit interaction represents a small perturbation that splits levels of J by an amount that is generally much smaller than the splittings between levels of different L.

The various processes contributing to the energy level structure are depicted in Fig. 23.8. We begin with the energy of the spherically averaged electron distribution about the atom. Next, we include splitting according to the L value that results from the electron–electron Coulomb interaction. This is known as the direct interaction. The Pauli exclusion principle introduces the exchange interaction that splits levels according to their value of S. Spin–orbit interactions, presumed to be weak when *LS* coupling is applicable, lead to a much smaller J-dependent splitting. Finally, the *Zeeman*

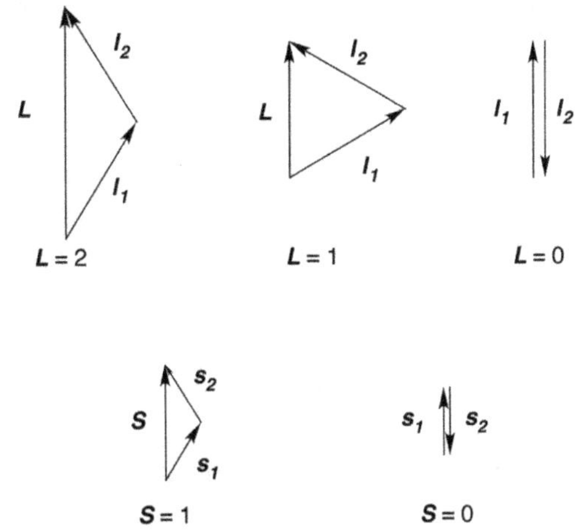

Figure 23.7 Coupling of orbital and spin angular momentum vectors l_1 and s_1 with l_2 and s_2 from two p electrons (p^2 configuration). Each electron has $l = 1$ and $s = 1/2$, such that the magnitude of the components are $|l_1| = |l_2| = [l(l+1)]^{1/2}\hbar = \sqrt{2}\hbar$ and $|s_1| = |s_2| = [s(s+1)]^{1/2}\hbar = (\sqrt{3}/2)\hbar$. The diagrams show the only ways of coupling the vectors, such that they satisfy $|L| = [L(L+1)]^{1/2}\hbar$ and $|S| = [S(S+1)]^{1/2}\hbar$.

effect leads to splitting according to M values (the projection of J on the z-axis) that is proportional to an applied magnetic field applied along the z-axis.

The Zeeman effect splits each line associated with a particular J value into $2J + 1$ components. Counting these lines provides a direct experimental method of determining J. As long as the Zeeman splitting is small compared to the splittings associated with the stronger interactions, the center of gravity of each group of split levels is the same as the position of the unsplit parent level. Similarly, the center of gravity of the spin–orbit split levels is same as energy of the levels resulting from the exchange interaction.

23.6.1 Total orbital angular momentum

The first step in LS coupling is to add the orbital angular momenta of individual electrons vectorially to create the allowed values of total orbital angular momentum L of the atom. The allowed values are given by the *Clebsch–Gordon series*. For

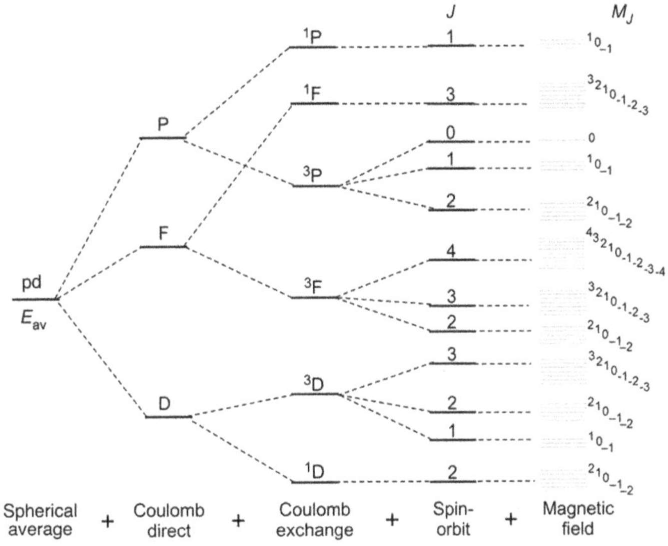

Figure 23.8 The energy of a pd configuration in the LS coupling scheme is shown. It starts with the spherically averaged energy derived from the central field approximation. Electron–electron repulsions first act directly on the product wavefunctions to split the energy into three levels. They then act on the determinant wavefunctions to split the terms according to spin multiplicity. The spin–orbit interaction is weak relative to the e$^-$–e$^-$ repulsion, and the splittings of the states associated with individual J values within each term are much smaller than shown. The splitting of individual J states into M_J sublevels by a magnetic field is further exaggerated so as to make it visible on this scale.

Table 23.2 The correspondence between the value of L and the letter used to represent it in term symbols.

$L =$	0	1	2	3	4	5	6	...
Λ	S	P	D	F	G	H	I	...

the coupling of two values l_1 and l_2, this series looks like

$$L = l_1 + l_2, l_1 + l_2 - 1, \ldots, |l_1 - l_2|. \tag{23.83}$$

That is, all *positive* values from $l_1 + l_2$ to $|l_1 - l_2|$ differ by one. Each one of these values of L corresponds to a *term symbol*. Rather than specifying the magnitude of L directly, the letter in the term symbol corresponds to the magnitude of L, as specified in Table 23.2.

If the atom has more than two electrons we need to sum the contributions of all the electrons. This looks like a rather onerous task if we must make such a sum for, say, the 92 electrons of U. Fortunately, *Unsöld's theorem*[6] tells us that the total probability density of electrons of both spins at the point (r, θ, ϕ) for a closed subshell nl is given by

$$|\psi(r, \theta, \phi)|^2 = \frac{2l+1}{4\pi^2 r^2}|P_{nl}(r)|^2. \tag{23.84}$$

This probability density does not depend on the angular variables, which means that it is spherically symmetric. Consequently, the total orbital angular momentum of a closed subshell must vanish – that is, $L = 0$ for all subshells with a full complement of electrons. Similarly, all of the electrons are paired in a closed subshell, which means that the spin angular momentum vanishes, $S = 0$. The total angular momentum is the sum of these two vectors; hence, it too must vanish, $J = 0$ for all closed subshells. Therefore, any closed subshell (and, by extension, any closed shell) contributes nothing to the orbital, spin or total angular momentum of a system. Only partially filled subshells need be considered when calculating the angular momenta of a system. All closed subshells and shells are represented by a $^1S_0$ term symbol.

For nonequivalent electrons – that is, electrons that differ either in the value of n or l – all values derived from the Clebsch–Gordon series correspond to different terms symbols and to different allowed levels. For equivalent electrons – that is, electrons with the same value of nl – the Pauli exclusion principle does what the name says and excludes those terms which correspond to two electrons possessing the same set of quantum numbers. A procedure for determining the allowed values was delineated first by Slater[7] and later using group theory by Curl and Fitzpatrick.[8] The results, which can also be derived by hand by filling in boxes with arrows to represent the electrons and painstakingly ensuring adherence to the exclusion principle, are given in Table 23.3. Should you need values of nf^w configurations (including

Table 23.3 Allowed terms for nl^w configurations of equivalent electrons as derived by Curl and Fitzpatrick. A number appearing below a term indicates that more than one term of the given type occurs. For instance, for nd^3 two different 2D terms arise.

nl^w	Allowed terms						
s	^{2}S						
p or p^5		^{2}P					
p^2 or p^4	^{1}S	^{3}P	^{1}D				
p^3	^{4}S	^{2}P	^{2}D				
d or d^9			^{2}D				
d^2 or d^8	^{1}S	^{3}P	^{1}D	^{3}F	^{1}G		
d^3 or d^7		^{2}P	^{2}D	^{2}F	^{2}G	^{2}H	
			2				
		^{4}P		^{4}F			
d^4 or d^6	^{1}S		^{1}D	^{1}F	^{1}G		^{1}I
		^{3}P	^{3}D	^{3}F	^{3}G	^{3}H	
			^{5}D				
d^5	^{2}S	^{2}P	^{2}D	^{2}F	^{2}G	^{2}H	^{2}I
			3	2	2		
		^{4}P	^{4}D	^{4}F	^{4}G		
	^{6}S						

the 119 terms associated with the nf^7 configuration!), these can be found in Curl and Fitzpatrick and higher yet can be found elsewhere.[9]

23.6.1.1 Example

Find the terms that can arise from the configuration d^2.

The two values of orbital angular momenta are $l_1 = 2$ and $l_2 = 2$. The maximum value of L is $2 + 2 = 4$. The minimum value is $2 - 2 = 0$. Therefore, the allowed values of L span from 4 to 0 in unit steps,

$$L = 4, 3, 2, 1, 0.$$

The corresponding terms are G, F, D, P and S. Next treat the spin. For two electrons, this is always the same

$$S = s_1 + s_2 = 1, \text{ spin multiplicity } 2S + 1 = 3, \text{ triplet.}$$
$$S = s_1 - s_2 = 0, \text{ spin multiplicity } 2S + 1 = 1, \text{ singlet.}$$

Therefore, one term of each type (singlet and triplet) is possible for each letter designation. If the two electrons have different values of n, all terms are allowed. If the value of n is the same for both, the Pauli exclusion principle (see Table 23.3) allows only the ^{1}S, ^{3}P, ^{1}D, ^{3}F and ^{1}G terms.

23.6.1.2 Example

Find the terms arising from the configuration p^3.

With more than two electrons, we need to pick any two, couple those to find an intermediate value, and then couple the third electron to the intermediate value to obtain the final result (repeating this procedure until all electrons have been coupled). Couple the first two electrons.

$$l_1 = 1, l_2 = 1$$
$$L' = 2, 1, 0$$

Now couple the third electron to *each* of these intermediate results, again creating total L values that range from the maximum $(L' + l_3)$ to the minimum $|L' - l_3|$ in unit steps,

$$l_3 = 1, L' = 2$$
$$L = 3, 2, 1$$
$$l_3 = 1, L' = 1$$
$$L = 2, 1, 0$$
$$l_3 = 1, L' = 0$$
$$L = 1$$

The overall result is

$$L = 3, 2, 2, 1, 1, 1, 0.$$

yielding one F, two D, three P, and one S terms.

23.6.2 Spin multiplicity

The spin multiplicity is the value of $2S + 1$. The possible values of S are similarly given by a Clebsch–Gordon series.

$$S = s_1 + s_2, s_1 + s_2 - 1, \ldots, |s_1 - s_2|. \tag{23.85}$$

Again, recall that $S = 0$ for a closed shell and that these electrons can be ignored for total angular momentum calculations.

23.6.3 Total angular momentum

You know the drill. J is formed by combining each value of L with each value of S. The allowed values are given by the Clebsch–Gordon series, which again ranges from the maximum value $(L + S)$ to the minimum value $|L - S|$ in unit steps,

$$J = L + S, L + S - 1, \ldots, |L - S|. \tag{23.86}$$

23.6.4 Russell–Saunders coupling term and level symbols

Term and level symbols are used to describe the energy states of atoms. While lower-case letters denote orbitals, upper-case letters denote overall states in atoms. The *term symbol* informs us of the values of L and S while the level symbol contains three pieces of information:

- The letter (S, P, D, F, etc.) specifies the total orbital angular momentum quantum number ($L = 0, 1, 2, 3$, etc.) as in Table 23.2.
- The left superscript denotes the spin multiplicity, $2S + 1$.
- The right subscript indicates the total angular momentum, J.

Consequently, a general level symbol can be described by

$$^{2S+1}\Lambda_J. \tag{23.87}$$

Designating just the values of L and S specifies an *LS term*. The values of L, S and J specify a *level*. The values of L, S, J and M specify a *state*. The value of $2S + 1$ is the multiplicity of the term. The number of levels in the term is given by $2S + 1$ unless $L < S$, and then the number of levels in the term is $2L + 1$. One additional designation is added to some term symbols. The ground state of the B atom and each atom in the corresponding column of the periodic chart is $^2\mathrm{P}^\circ_{1/2}$. The superscript ° indicates that arithmetic sum of l for all the electrons is an odd number. If the sum is even, no indication is made.

23.6.4.1 Examples

Determine the complete level symbol for the configurations s^1 and p^2.

The configuration s^1 has $l = 0$ and $s = \frac{1}{2}$. Hence, $J = 0 + \frac{1}{2} = \frac{1}{2}$ and the only level symbol possible is $^2\mathrm{S}_{1/2}$.

The configuration p^2 has

$$l_1 = 1, l_2 = 1$$
$$L = 2, 1, 0$$
$$s_1 = 1/2, s_2 = 1/2$$
$$S = 1, 0$$

From the allowed L values there are D, P, and S terms. Now, combine each L value with each S value to determine all allowed level symbols.

$L = 2,\ S = 1$ $J = 3,\ 2,\ 1$	$L = 2,\ S = 0$ $J = 2$
$L = 1,\ S = 1$ $J = 2,\ 1,\ 0$	$L = 1,\ S = 0$ $J = 1$
$L = 0,\ S = 1$ $J = 1$	$L = 0,\ S = 0$ $J = 0$

The p^2 configuration has thus generated six terms (^{1}D, ^{1}S, ^{3}P, ^{3}D, ^{3}S, and ^{1}P). The degeneracy of a term is given by $(2S + 1)(2L + 1)$, and therefore, these six terms correspond to $5 + 1 + 9 + 15 + 3 + 3 = 36$ quantum states. If the two p electrons have different values of n – that is, if the electrons are in two different p orbitals – all of these combinations are allowed. If, however, the two p electrons have the same values of the principal quantum number n (they are in the same orbital), then the number of terms and states is limited by the Pauli exclusion principle, which states that no two electrons can have the same set of quantum numbers. The allowed terms can be found in Table 23.3.

23.6.5 Hund's rules

Friedrich Hund,[10] on the basis of empirical evidence, developed a series of rules to establish the ordering of energy levels that result from LS coupling.

1. The lowest energy term is that which has the greatest spin multiplicity (greatest value of S).
2. Among all of the lowest terms (one for each value of L), the term with the greatest value of L has the lowest energy.

A further rule of energy ordering was added later.

3 For any given $^{2S+1}\Lambda$ term, the order of the J-levels associated with it are given by the following:
   i The levels increase in energy from lowest J value to highest J value if the subshell is less than half full. The J levels are said to form a *normal multiplet*.
   ii If the subshell is more than half full, the highest J value corresponds to the lowest energy state. The J levels are said to form an *inverted multiplet*.

Based on available data at the time, Hund believed these rules to be general, but they are not. They can be confidently applied only to equivalent electrons from one open subshell in a configuration l^w or one open subshell coupling to an s electron l^w s. In other words, they apply to configurations of the type p^2 or sp configurations, but not mixed ones such as pd.

When LS coupling applies, the splitting between J levels will be much smaller than the splitting between terms. For example, the C atom has a $2p^2$ configuration. The splitting between the $^3P_0$ and $^3P_2$ terms is 45.3 cm^{-1}, whereas the splitting between $^3P_0$ and $^1D_2$ is 10 194 cm^{-1}.

23.7 *jj*-coupling

When spin–orbit coupling is strong (high-Z atoms), the spin–orbit term in the Hamiltonian is no longer small compared to the electron–electron interactions. In this case we must use *jj*-coupling as illustrated in Fig. 23.9. In the limit that the spin–orbit splitting is large compared to the Coulomb terms, pure *jj*-coupling has been achieved. In *jj*-coupling, l and s for each electron are coupled to give j for that electron. Then all of the j values are coupled to give the final value of the total angular momentum J. Unless the atom is very heavy ($Z > 80$), l^w configurations – that is, when all of the electrons are in a single sub-shell – are found to follow LS coupling. However, when the configuration involves two or more sub-shells, coupling closer to that of pure *jj*-coupling is expected at somewhat lower Z.

The terms derived according to Russell–Saunders coupling are often used to label groups of states, even when *jj*-coupling is present. Then, when needed, the total angular momentum for each state is specified. Labeling according to *jj*-coupling follows the notation

$$(j_1, j_2)_J \quad \text{(two electrons)}, \tag{23.88}$$

$$(j_1, j_2, j_3)_J \quad \text{(three electrons)}, \tag{23.89}$$

and so forth. The Pauli exclusion principle still applies in the presence of *jj*-coupling. No two electrons may have the same set of one-electron quantum numbers $n_i l_i j_i m_{j,i}$.

Other coupling schemes (pair coupling and intermediate coupling) are possible. These are most likely to be found for highly excited states or high orbital angular momentum orbitals (f, g, ...). They will not be discussed further here.

23.7.1 Example of *jj*-coupling

Find the values of J that result from *jj*-coupling in the configuration s^1p^1.

In *jj*-coupling, the terms are generated by finding the maximum J from $(j_1 + j_2)$, the minimum J from $|j_1 - j_2|$ and all values in between in unit steps.

$$j_1 = 0 + {}^1/_2 = {}^1/_2, j_2 = 1 + {}^1/_2 = 3/2 \quad \text{or} \quad 1 - {}^1/_2 = {}^1/_2$$
$$j_1 = {}^1/_2 \quad \text{and} \quad j_2 = 3/2 \Rightarrow J = 2, 1$$

Thus, $\left(\frac{1}{2}, \frac{3}{2}\right)_2$ and $\left(\frac{1}{2}, \frac{3}{2}\right)_1$ terms are generated as well as

$$j_1 = {}^1/_2 \quad \text{and} \quad j_2 = 1/2 \Rightarrow J = 1, 0$$

$\left(\frac{1}{2}, \frac{1}{2}\right)_1$ and $\left(\frac{1}{2}, \frac{1}{2}\right)_0$ terms.

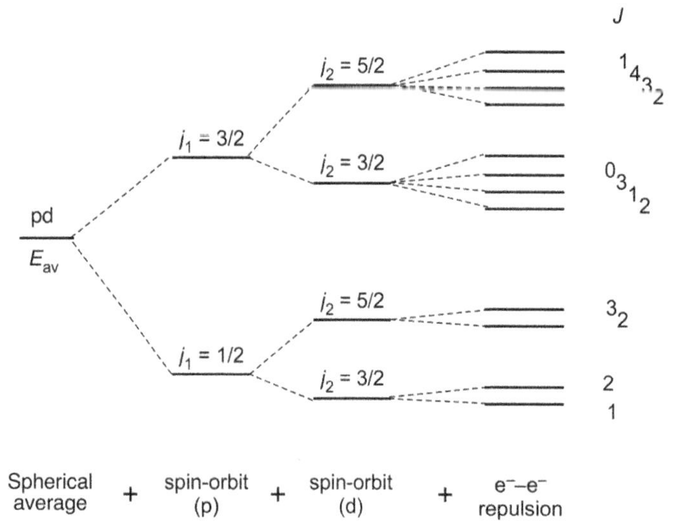

Figure 23.9 In *jj*-coupling strong spin–orbit interactions first result in four states, two associated with the p electron and two with the d electron. These four pairs of (j_1, j_2) values then couple under the influence of the weaker electron–electron Coulomb repulsion to form the states associated with much smaller splittings according to the final set of $(j_1, j_2)_J$ values.

23.8 Selection rules for atomic spectroscopy

23.8.1 Hydrogenic atoms

Not all transitions are allowed. Whether a transition can occur depends on the properties of the wavefunctions of the states involved, the type of transition, and certain physical principles such as the conservation of energy and angular momentum. Energy conservation is ensured by following the Bohr frequency condition. As long as the photon energy matches the energy difference between states, states of any n can be connected. There is no selection rule on n – that is, Δn can take on any value.

Angular momentum is particularly important for determining selection rules. The photon has an intrinsic spin angular momentum of $s = 1$. In an electronic transition, the change in angular momentum of the electron must compensate for the angular momentum associated with the photon. Absorption or emission of a photon by a hydrogenic atom requires the electron to change its orbital angular momentum by ± 1. The associated selection rule is

$$\Delta l = \pm 1. \tag{23.90}$$

It follows that the change in the magnetic quantum number is also constrained, such that a further selection rule also applies

$$\Delta m_l = 0, \pm 1. \tag{23.91}$$

23.8.1.1 Example

Consider an electron in the 4p orbital of the Nb^{40+} ion (a hydrogenic ion). The absorption of one photon can lead to the occupation of which orbitals?

Initially, the electron has $l = 1$ and $m_l = 0$ or ± 1. The $\Delta l = \pm 1$ selection rule constrains the final orbital angular momentum to be $1 \pm 1 = 0$ or 2. The $\Delta m_l = 0, \pm 1$ selection rule constrains m_l to values of $(0 \text{ or } \pm 1) \pm 1 = 0, \pm 1$, or ± 2. There is no restriction on n, therefore, the electron can enter into any m_l sublevel of any ns ($l = 0$, $m_l = 0$) or nd ($l = 2$, $m_l = 0, \pm 1$, or ± 2) orbital. Transitions to p and f orbitals are forbidden.

23.8.2 Multi-electron atoms

Selection rules arise from the need to conserve angular momentum (total and orbital). In addition, a photon cannot directly change the spin of an atom in the absence of spin–orbit coupling. Combining these requirements, we can derive the following selections rules for multi-electron atoms:

$$\Delta S = 0, \Delta L = 0, \pm 1 \text{ but } L = 0 \not\leftrightarrow L = 0,\ \Delta l = \pm 1, \Delta J = 0, \pm 1 \text{ but } J = 0 \not\leftrightarrow J = 0 \tag{23.92}$$

In LS coupling, it is usually true – particularly for large values of L – that the strongest lines of a multiplet are the ones for which $\Delta J = \Delta L$. These lines are denoted the principal lines of the multiplet. The principal line of maximum J is always the strongest line. The strength of the other lines in the multiplet decreases monotonically with J.

In pure LS coupling (when spin–orbit coupling is weak), $\Delta S = 0$ is a strict selection rule. However, if a degree of intermediate coupling mixes basis sets of different values of S into one or more of the states involved in the transition, violations of this selection rule are found. The strength of the transition depends on the extent of mixing. It can also happen that one state belongs to nearly pure LS coupling while the other is better represented by almost pure jj-coupling. In this case, multiple violations of selection rules may be observed and the only true selection rules would involve quantum numbers that are common to both coupling schemes. Other violations of selection rules are engendered by configuration mixing in one (or both) of the states involved in the transition.

23.9 Photoelectron spectroscopy

The photoelectric effect was explained by Einstein when he recognized that conservation of energy could be coupled with the Planck equation. This enabled him to state that if the minimum binding energy of an electron in a metal is the work function Φ, then the maximum kinetic energy of electrons emitted by photon absorption must be the photon energy minus the work function. This analysis can be generalized to help explain photoionization of atoms in any phase.

Multiphoton photoelectron spectroscopy is discussed in Chapter 27. Here, we focus on conventional one-photon photoemission. As depicted in Fig. 23.10, an electron in an atom is bound in a potential well. Photoelectron spectroscopy is performed with photons of sufficient energy to overcome the electron's binding energy. Such photons are generally in the UV or X-ray regions of the electromagnetic spectrum. The spectrometer that collects the electrons usually also measures their velocity. In other words, it detects the electrons in an energy-resolved manner. The temporal structure of the electromagnetic waves used to excite photoemission can also be exploited to measure the electrons in a time-resolved manner. In some cases this is used to measure the electron kinetic energy. In much more refined experiments (as discussed in Chapter 27) this temporal structure is used to investigate the time-dependence of the dynamics of photoemission.

Ultraviolet photons probe valence electrons, which are particularly important for and influenced by chemical bonding. Valence electrons are strongly influenced by the chemical environment in which they are found. UV photoemission is essential for probing the molecular orbitals of molecules, complexes, and adsorbed species.

X-rays probe core electrons, which are much less influenced by chemical bonding. They are most strongly influenced by the nucleus to which they are bound. Core levels are only weakly affected by changes in oxidation states, and are insensitive to the finer details of chemical bonding. Because they are so low in energy and localized about the nucleus, to a first approximation we can treat core-level photoelectron spectroscopy as a form of atomic spectroscopy, even if the

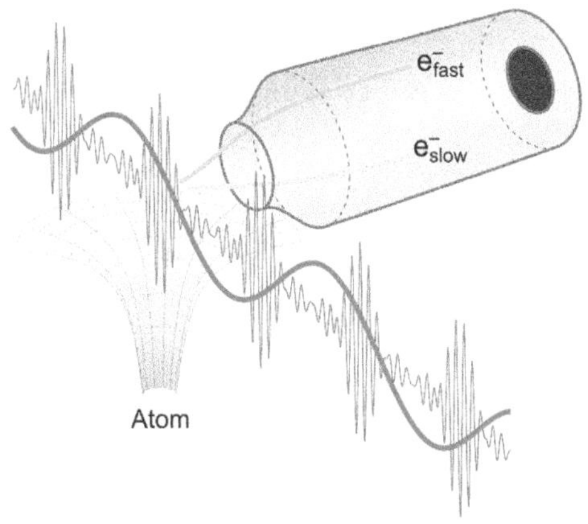

Figure 23.10 Schematic representation of photoelectron spectroscopy with pulsed light sources. The electron kinetic energy is measured in an electron spectrometer (in this case by measuring how long it takes to travel from the atom to the detector; a fast electron arrives earlier). If two pulses of light irradiate the sample with a controllable delay (such as a conventional laser field shown in red and high harmonics of it shown in blue; see Chapter 27), then not only the kinetic energy but also the time-resolved nature of the ionization process from a metal, molecule or atom can be investigated. Figure provided by Anne L'Huillier and Erik Månsson. Reproduced with permission from Månsson, E.P., Guénot, D., Arnold, C.L., Kroon, D., Kasper, S., Dahlström, J.M., Lindroth, E., Kheifets, A.S., L'Huillier, A., Sorensen, S.L., and Gisselbrecht, M. (2014) *Nat. Phys.*, **10**, 207. © 2014, Macmillan Publishers Ltd.

atoms are in the solid phase. Hence, as we are currently investigating atomic spectroscopy we concentrate on core-level spectroscopy for the remainder of this chapter.

Here, we consider two types of electron spectroscopy in which excitation leads to the ejection of an electron from an atom. The energy and possibly angular distribution of the ejected electrons are analyzed using photoelectron spectroscopy. When X-rays are used, this technique is termed *X-ray photoelectron spectroscopy* (XPS) or, especially in the older and analytical literature, electron spectroscopy for chemical analysis (ESCA). XPS involves the kinetic energy analysis of electrons directly emitted from a sample. Usually, this has required the use of ultra-high vacuum so that the electrons reach the detector without scattering. However, an exciting recent advance is the ability to perform XPS at ambient pressures,[11] which facilitates, for example, the *in-situ* observation of chemical reactions and catalysis under 'real world' conditions.

Auger electron spectroscopy (AES) involves excitation with either X-rays or, more usually, with electrons of similar energy. The ejection of an Auger electron is an indirect process that leads to the emission of a secondary electron after excitation. We will describe this process in detail below.

23.9.1 X-ray photoelectron spectroscopy (XPS)

Deep core electrons have binding energies corresponding to the energies of photons that lie in the X-ray region. When an atom absorbs a photon with an energy in excess of the binding energy of an electron, a photoelectron is emitted, regardless of the phase in which the atom is present. The kinetic energy of the photoelectron is related to the energy of the photon by the *Einstein equation*,

$$E_\mathrm{B} = h\nu - E_\mathrm{k}. \tag{23.93}$$

This is a generalization of the equation involving the work function. In its place is the binding energy of the electron E_B, $h\nu$ is the photon energy, and E_k the kinetic energy of the photoelectron. The binding energy is referenced to the vacuum level. That is, we set the zero of our energy scale to be the point at which the electron is far removed and not interacting with the atom from which it came.

Deep core electrons do not participate in bonding, and their energies are characteristic of the atom from which they originate. Therefore, XPS is particularly useful for elemental analysis. Not only can XPS identify the composition of a sample but also it can be used to determine the composition *quantitatively*. The modern application of XPS owes much of its development to Siegbahn and coworkers in Uppsala, Sweden. Kai Siegbahn, whose father Manne became a Nobel laureate in physics in 1924 for his discoveries and research in the field of X-ray spectroscopy, was awarded the Nobel Prize in Physics 1981 for his contributions to the development of XPS/ESCA.

When the energies of core levels are investigated in detail it is found that small, but easily detected, shifts occur. These *chemical shifts* depend on the bonding environment around the atom, in particular, on the oxidation state of the atom.

The *binding energy* of an electron is equal to the energy difference between the final and initial states of the atom. That is, the binding energy of an electron in the kth state is equal to the difference in energies between the atom with N electrons $E_i(N)$ and the ion with $N-1$ electrons $E_f(N-1)$

$$E_\mathrm{B}(k) = E_f(N-1) - E_i(N). \tag{23.94}$$

If there were no rearrangement of all other electrons in the system (the spectator electrons), the binding energy would exactly equal the negative of the orbital energy of the initial state of the electron, $-\varepsilon_k$. This approximation is known as *Koopmans' theorem*:

$$E_\mathrm{B}(k) = -\varepsilon_k. \tag{23.95}$$

Koopmans' theorem is based on an effective one-electron approximation in which the initial state wavefunction is described by a product of the single-particle wavefunction of the electron to be removed, φ_j, and the $(N-1)$ electron wavefunction of the resulting ion, $\Psi_j(N-1)$,

$$\psi_i(N) = \varphi_j \Psi_j(N-1). \tag{23.96}$$

Within this picture the final state wavefunction of the ion is simply

$$\psi_f(N-1) = \Psi_j(N-1). \tag{23.97}$$

The energies required for Eq. (23.94) are

$$E_i(N) = \langle \psi_i(N) | \hat{H} | \psi_i(N) \rangle \tag{23.98}$$

and

$$E_f(N-1) = \langle \Psi_j(N-1) | \hat{H} | \Psi_j(N-1) \rangle \tag{23.99}$$

Substitution of Eqs (23.98) and (23.99) into Eq. (23.94) demonstrates that only one peak is expected in the photoelectron spectrum since both $E_i(N)$ and $E_f(N-1)$ are single-valued.

The electrons are not frozen. The final state arising from the removal of one electron corresponds to an ionic state in which a hole exists in place of the ejected photoelectron. This does not correspond to the ground state of the ion. The remaining electrons relax and, thereby, lower the energy of the final state. This type of relaxation occurs regardless of the phase (gas, liquid, or solid) in which the atom exists. If the atom is located in a solid, then in addition to relaxation within the atom, there can be extra-atomic relaxation. In other words, not only the electrons in the ionized atom can relax but also the electrons on neighboring atoms can relax in response to the ionization event.

These relaxation phenomena lead us to reconsider our description of the final state wavefunction, which we now write in terms of the eigenstates of the ion,

$$\psi_f(N-1) = u_k \Phi_{jl}(N-1) \tag{23.100}$$

where u_k is the wavefunction of the excited electron with momentum k and Φ_{jl} are the wavefunctions of the l ionic states that have a hole in the jth orbital. The relaxation of the ion core in the final state means that the wavefunction $\Psi_j(N-1)$ is not an eigenstate of the ion. Therefore, there is not a unique $\Psi_j(N-1)$ that corresponds to $\Phi_{jl}(N-1)$. Instead, the $\Psi_j(N-1)$ must be projected onto the true eigenstates of the ion. This results in the formation of one or more possible final state wavefunctions such that the final state energy required for Eq. (23.94) is

$$E_{fl}(N-1) = \langle \Phi_{jl}(N-1) | \hat{H} | \Phi_{jl}(N-1) \rangle, \tag{23.101}$$

which has between 1 and l solutions. This means that between 1 and l peaks appear in the spectrum.

The energy of an electronic state is determined not only by the electron configuration but also by angular momentum coupling. In the LS coupling scheme, the total angular momentum is given by the vector sum $J = L + S$. Any state with an orbital angular momentum $L > 0$ and one unpaired electron is split into two states, a doublet, corresponding to $J = L \pm {}^1/_2$. More complex multiplet splittings in the initial state arise from states with higher total spins. In X-ray spectroscopy, a nomenclature based on the shell is commonly used. The shell derives its name from n according to K, L, M, N, ... for $n = 1, 2, 3, 4, \ldots$. A subscript further designates the subshell. The numbering starts at 1 for the lowest (L, J) state and continues in unit steps up to the highest (L, J) state. Thus, the shell corresponding to $L = 2$ has levels L_1, L_2 and L_3 corresponding to 2s, $2p_{1/2}$, and $2p_{3/2}$.

Both, the initial state and final state effects influence the binding energy. Initial state effects are caused by chemical bonding, which influences the electronic configuration in and around the atom. Thus, the energetic shift caused by initial state effects is known as a *chemical shift*, ΔE_b. To a first approximation, all core levels in an atom shift to the same extent. For most samples, it is a good approximation to assume that the chemical shift is completely due to initial state effects

The chemical shift depends on the oxidation state of the atom. Generally, as the oxidation state becomes more positive the binding energy increases. The greater the electron-withdrawing power of the substituents bound to an atom, the higher the binding energy. This can be understood on the basis of simple electrostatics. The first ionization energy of an atom is always lower than the second ionization energy. Similarly, the higher the effective positive charge on the atom, the higher the binding energy of the photoelectron. Returning to the example of Ti used at the beginning of this chapter, in Ti metal the Ti atoms have an oxidation state of zero. In TiO_2, they have an oxidation state of 4+. Therefore, the binding energy of the $2P_{3/2}$ state shifts from 453.8 eV in Ti metal to 458.5 eV in TiO_2. The chemical shift is small (<1%) compared to the binding energy, but characteristic and distinct enough to be able to distinguish the two oxidation states.

Most of the atomic relaxation results from rearrangement of outer shell electrons. Inner shell electrons of higher binding energy make a small contribution. The nature of a material's conductivity determines the nature of extra-atomic relaxation. In a conducting material such as a metal, valence electrons are free to move from one atom to a neighbor to screen the hole created by photoionization. In an insulator, the electrons do not possess such mobility. Hence, the magnitude of the extra-atomic relaxation in metals (as much as 5–10 eV) is greater than that of insulators.

Several other final state effects result in what are known as *satellite features*. Satellite features arise from multiplet splitting, shake-up events and vibrational fine structure. Multiplet splittings result from spin–spin interactions when unpaired electrons are present in the outer shells of the atom. The unpaired electron remaining in the ionized orbital interacts with any other unpaired electrons in the atom/molecule. The energy of the states formed, just as for any other case of angular momentum coupling, depends on whether the spins are aligned parallel or anti-parallel. In a *shake-up event*, the outgoing photoelectron excites a valence electron to a previously unoccupied state. This unoccupied state may be either a discrete electronic state, such as a normally unoccupied antibonding orbital or, especially in metals, is better thought of as electron–hole pair formation. By energy conservation, the photoelectron must give up some of its kinetic energy in order to excite the shake-up transition; hence, shake-up features always lie on the high-binding energy (lower kinetic energy) side of a direct photoemission transition. Occasionally, the valence electron is excited above the vacuum level. Such a double ionization even is known as *shake-off*. Shake-off events generally do not exhibit distinct peaks; instead, they lead to peak broadening. Shake-up transitions involve excitation to discrete states and do exhibit distinct peaks.

Given sufficient instrumental resolution, the width of photoemission peaks is determined to a large extent by the lifetime of the core hole. Intrinsic lifetime broadening is an example of homogeneous broadening and has a Lorentzian lineshape. Through the Heisenberg uncertainty relation, the *intrinsic peak width*, Γ, is inversely related the core hole *lifetime*, τ, by

$$\Gamma = h/\tau, \tag{23.102}$$

where h is the Planck constant. The lifetime generally decreases the deeper the core hole because core holes are filled by higher-lying electrons, and the deeper the core hole, the more de-excitation channels there are that can fill it. Analogously, for a given energy level, the lifetime decreases (the linewidth increases) as the atomic number increases.

23.9.1.1 Directed practice

State some experimentally observable manifestations of breakdowns in Koopmans' theorem.

23.9.2 Auger electron spectroscopy (AES)

In XPS, the electron that is detected and energy-analyzed is the electron that is excited. This electron is called a *primary electron*. In Auger electron spectroscopy (AES) a secondary electron is detected. A *secondary electron* is an electron that was excited by a primary electron. The initial excitation in an Auger transition can be excited by photons, electrons or even by ion bombardment, and is depicted in detail in Fig. 23.11. The phenomenon was first described by Pierre Auger in 1925.[12] Conventionally, electrons with kinetic energies in the 3–5 keV range bombard the sample. Auger transitions are also simultaneously observed during XPS measurements. Like XPS, AES can be used for quantitative elemental analysis.

Auger spectroscopy requires coupling between electrons. As shown in Fig. 23.11, after ionization of a core level a higher-lying (core or valence) electron fills the resulting hole in a radiationless transition (a transition that does not emit

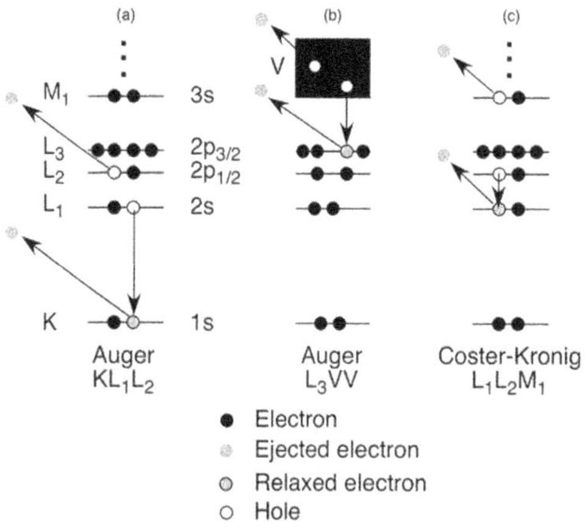

Figure 23.11 A detailed depiction of Auger transitions involving: (a) three core levels; (b) two core levels and the valence band; and (c) a Coster–Kronig transition in which the initial hole is filled from the same shell. Reproduced with permission from Kolasinski, K.W. (2012) *Surface Science: Foundations of Catalysis and Nanoscience*, 3rd edition. John Wiley & Sons, Chichester.

a photon). This process leaves the atom in an excited state. The excited state energy is removed by the ejection of a second electron. An inspection of Fig. 23.11 makes clear that a convenient method of labeling Auger transitions is to use the X-ray notation of the levels involved. Thus, the transition in Fig. 23.11(a) is a KL_1L_2 transition, and that in Fig. 23.11(b) is a L_3VV transition in which the V stands for a valence band electron. In some cases, more than one final state is possible and the final state is added to the notation to distinguish between these possibilities – for example, $KL_1L_3(^3P_1)$ as opposed to $KL_1L_3(^3P_2)$. *Coster–Kronig transitions,* such as the example of an L_1L_2M transition in Fig. 23.11(c), involve excitation and a first relaxation step from the same shell. These transitions are usually particularly strong.

Auger spectroscopy can be used to detect virtually any element, apart from H and He. With increasing atomic number (Z), however, the probability of radiative relaxation of the core hole (i.e., X-ray fluorescence) increases. The incident electron energy (the primary electron energy E_p) is often only 3 keV. Thus, the Auger transitions of interest generally have energies below 2 keV, which means that the primary excitation shell shifts upward as Z increases, for example K for lithium, L for sodium, M for potassium, and N for ytterbium.

The energy of an Auger transition is difficult to calculate precisely as many-electron effects and final-state energies have to be considered. Fortunately, for analytical purposes the exact energy and lineshape of an Auger transition need only be considered in the highest resolution specialist applications. The low resolution required for quantitative analysis also means that the small energy differences between final-state multiplets can be neglected. Auger transitions from different elements tend to be fairly widely spaced in energy and have been tabulated in a comprehensive reference work.[13]

The observed energy of an Auger transition is best approximated with reference to Fig. 23.12. Note that the energy scale is referenced to the Fermi level. The primary electron e_p^- must have sufficient energy to ionize the core level of energy E_W. An electron initially at energy E_X fills the initial hole, and the energy liberated in this process is transferred to an electron at E_Y, which is then ejected into the vacuum. This electron must overcome the work function of the sample, Φ_e, to be released into the vacuum; hence, the kinetic energy is reduced by this amount. However, the electron is detected by an analyzer with a work function Φ_{sp}. The vacuum level is constant and electrical contact between the sample and the analyzer leads to an alignment of the sample and analyzer Fermi levels. Thus, it is actually Φ_{sp} that must be subtracted from the energy of the electron to arrive at the final kinetic energy

$$E_{WXY} = E_K - E_L - E_V - \Phi_{sp}. \tag{23.103}$$

The spectrometer work function is only 4–5 eV. Many Auger transitions of interest are well above 100 eV. Thus, the correction for the spectrometer work function will in many practical applications be negligible. If the sample is an insulator, the Fermi levels do not automatically align. Sample charging can easily occur and special care must be taken in

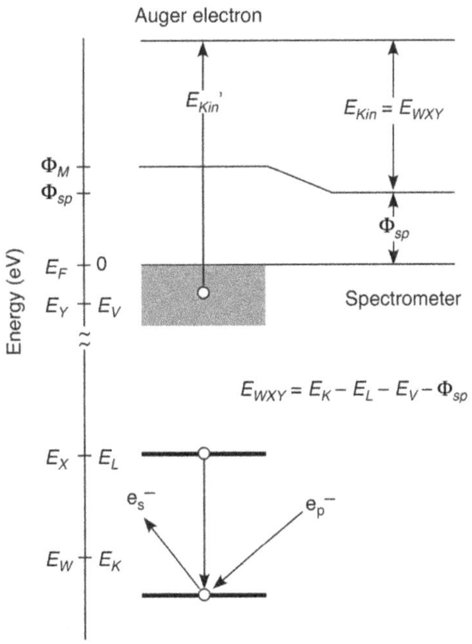

Figure 23.12 Auger transition, including the influence of the spectrometer work function Φ_{sp}. Reproduced with permission from Kolasinski, K.W. (2012) *Surface Science: Foundations of Catalysis and Nanoscience*, 3rd edition. John Wiley & Sons, Chichester.

interpreting the spectra. As shown by Eq. (23.103), the energy of the electrons created through the Auger process, the Auger electrons, is not dependent on the primary electron energy.

SUMMARY OF IMPORTANT EQUATIONS

$E = h\nu = hc\tilde{\nu} = hc/\lambda = \hbar\omega$	Planck equation
$c = \lambda\nu,\ \tilde{\nu} = 1/\lambda,\ \omega = 2\pi\nu$	c = speed of light, λ = wavelength, ν = frequency, $\tilde{\nu}$ = wavenumber, ω = angular frequency
$L(\tilde{\nu}) = \dfrac{A}{(\tilde{\nu}-\tilde{\nu}_0)^2+\gamma^2}$	Lorentzian lineshape with FWHM = 2γ
$G(\tilde{\nu}) = A\exp\left(-\left(\dfrac{\tilde{\nu}-\tilde{\nu}_0}{\sigma}\right)^2\right)$	Gaussian lineshape with FWHM = $2\sqrt{\ln 2}\sigma$
$\gamma = (2\pi c\tau)^{-1}$	Homogeneous broadening, Lorentzian linewidth parameter γ, excited state lifetime τ
$E_k - E_i = h\nu_{ki}$	Bohr frequency condition. Energy difference between states matches photon energy.
$\Gamma = (2\ln 2k_BT/mc^2)^{1/2}\tilde{\nu}_0$	HWHM of a Doppler broadened peak Γ for atom of mass m and central wavenumber $\tilde{\nu}_0$
$\mu_{ki} = \langle\psi_k\|\hat{\mu}\|\psi_i\rangle = \int \psi_k^*\hat{\mu}\psi_i\mathrm{d}\tau$	Transition dipole moment μ_{ki}. Electric dipole moment operator $\hat{\mu}$.
$W_{ki} = \dfrac{1}{6\varepsilon_0\hbar^2 c}\langle\psi_k\|\hat{\mu}\|\psi_i\rangle^2 I(\nu_{ki}) = B_{ki}\rho(\nu_{ki})$	Fermi's golden rule. Transition rate for absorption of light from state $\psi_i \rightarrow \psi_k$, W_{ki}^{abs}.
$\rho(\nu_{ki}) = I(\nu_{ki})/c$	Energy density of radiation at the frequency of the transition $\rho(\nu_{ki})$, Intensity of radiation $I(\nu_{ki})$.
$B_{ki} = \dfrac{\|\langle\psi_k\|\hat{\mu}\|\psi_i\rangle\|^2}{6\varepsilon_0\hbar^2} = \dfrac{\|\mu_{ki}\|^2}{6\varepsilon_0\hbar^2}$	Einstein B coefficient
$\dfrac{A_{ik}}{B_{ik}} = \dfrac{8\pi h}{c^3}\nu_{ki}^3$	Ratio of Einstein A coefficient to Einstein B coefficient
$A_{ik} = \dfrac{16\pi^3\nu_{ki}^3}{3c^3\varepsilon_0 h}\|\mu_{ki}\|^2$	Einstein A coefficient.
$A = \varepsilon c l$	Beer–Lambert law. A = absorbance, ε = molar absorptivity (absorption coefficient, c = concentration, l = pathlength
$I/I_0 = \exp(-\sigma l\rho)$	Transmitted intensity I, incident intensity I_0, absorption cross-section σ, density ρ
$E_{lsj} = \frac{1}{2}hcA\{j(j+1) - l(l+1) - s(s+1)\}$	Spin–orbit coupling energy, A = spin–orbit coupling constant; total, orbital and spin angular momentum quantum numbers j, l and s
$E_B = h\nu - E_k$	Einstein equation. Binding energy E_B, kinetic energy E_k after photoemission

Exercises

23.1 A Nd:YAG laser used for laser ablation produces 2 W (time averaged) at 532 nm. The repetition rate is 20 pulses per second. The pulsewidth is 6 ns. (a) What is the pulse energy in mJ. (b) What is the peak power in W of a single pulse?

23.2 The absorption coefficient of Si at 632 nm (the wavelength of a HeNe laser) is 12 250 cm^{-1}. Neglecting reflection, how far does the light penetrate before its intensity is reduced by 33%?

23.3 (a) Calculate the photon energy in eV and for each of the following types of radiation: (i) microwaves at 2450 MHz. (ii) infrared light at 2100 cm^{-1}. (iii) ultraviolet light at 192.6 nm. (iv) x-rays at 2.30 nm. (b) For (i) and (iv) calculate the ratio of Einstein A to Einstein B coefficient. What is the implication for the rate of spontaneous emission?

23.4 Show that the full-width at half-maximum of a Gaussian lineshape is $\sqrt{\ln 16}\sigma$ and that of a Lorentzian lineshape is 2γ.

23.5 Calculate the value of the Rydberg constant for muonic hydrogen, a form of hydrogen in which the electron is replaced by a muon.

23.6 (a) The energy levels for hydrogenic ions (one-electron systems) are given by

$$E_n = -Z^2e^2/8\pi\varepsilon_0 a_0 n^2$$

with $n = 1, 2, 3, 4, \ldots$. Calculate the ionization energies of H^+, He^+, Li^{2+}, Ca^{19+} and Rn^{85+} in their ground states in units of electron-volts (eV). (b) Core electrons shield valence electrons so that they experience an effective nuclear charge Z_{eff} rather than the full nuclear charge. Given that the first ionization energies of He and Li are 24.587 and 5.392 eV, use the formula derived in part (a) to estimate the effective nuclear charge experience by the first 1s electron in He and the 2s electron in Li.

23.7 Draw a diagram of the energy level structure of H atom. Include in your diagram the first four energy levels (the lowest four in terms of energy), including the principal quantum number, values of the energy in eV, and the zero of energy. Calculate the wavenumber in cm^{-1} of two lowest energy transitions between these levels.

23.8 (a) What is the origin of the chemical shift in X-ray photoelectron spectroscopy (XPS)? (b) Explain the effect of the chemical shifts for Fe(0), Fe(II) and Fe(III).

23.9 Write out the electron configuration of vanadium. Determine all numerically possible term symbols based on this configuration. What is the ground-state term symbol of the vanadium atom? When the X-ray photoelectron spectra of several vanadium compounds were acquired, the following binding energies were measured for the V($2p$) core electrons: 512.4, 515.7, 515.9 and 516.9 eV. Appropriately assign these binding energies to V, VO_2, V_2O_3 and V_2O_5. Explain your reasoning.

23.10 Using Russell–Saunders coupling, calculate the allowed values of L, S and J for the following three configurations and write out all allowed term symbols:

$1s^2\ 2s^2\ 2p^6\ 3s^2\ 3p^6\ 4s^2\ 3d^{10}$

$1s^2\ 2s^2\ 2p^6\ 3s^2\ 3p^6\ 4s^0\ 3d^2$

$1s^2\ 2s^2\ 2p^6\ 3s^2\ 3p^6\ 4s^1\ 3d^1\ 4p^1$

23.11 Write out all allowed term symbols that arise from the d^1f^1 configuration. Considering only the two lowest orbital angular momentum terms found, indicate all allowed transitions between these two levels.

23.12 Write out all allowed terms (just the letters) that arise from the d^8 configuration. Considering only the two highest orbital angular momentum terms found, determine all allowed terms symbols. Put the resulting states in energetic order according to Hund's rules – that is, draw an energy level diagram. Indicate all allowed transitions from the ground state.

23.13 The intensities of spectroscopic transitions between vibrational states of a molecule are proportional to the square of the integral

$$\int_{-\infty}^{\infty} \psi_{v'}\, x\, \psi_v\, d\tau.$$

Use the relations between the Hermite polynomials given in Table 21.1 to show that the only permitted transitions are those for which $v' = v \pm 1$ and evaluate the integral in these cases.

23.14 Calculate the transition dipole moment in Debye (D) $\mu_z^{mn} = \int \psi_m^*(\tau)\mu_z\psi_n(\tau)d\tau$, where $\mu_z = -er\cos\theta$ for a transition from the 2s level to the 3d, $m_l = 0$ level in H. Is this transition is allowed? The integration is over r, θ, and ϕ.

23.15 Calculate the transition dipole moment in Debye (D)

$$\mu_z^{mn} = \int \psi_m^*(\tau)\mu_z\psi_n(\tau)d\tau$$

where $\mu_z = -er\cos\theta$ for a transition from the 1s level to the $2p_z$ level in H. Show that this transition is allowed. The integration is over r, θ, and ϕ. Use

$$\psi_{210}(r,\theta,\phi) = \frac{1}{\sqrt{32\pi}}\left(\frac{1}{a_0}\right)^{3/2}\frac{r}{a_0}e^{-r/2a_0}\cos\theta$$

for the $2p_z$ wavefunction.

23.16 Using the selection rules $\Delta S = 0, \Delta L = 0, \pm 1, \Delta J = 0, \pm 1$ but not $J = 0$ to $J = 0$, state whether each transition is allowed or forbidden and why.

Transition	Allowed?	Reason
${}^4F_{7/2} \rightarrow {}^2F_{5/2}$		
${}^4F_{7/2} \rightarrow {}^2D_{5/2}$		
${}^4F_{7/2} \rightarrow {}^4D_{1/2}$		
${}^4F_{7/2} \rightarrow {}^4D_{5/2}$		
${}^4F_{7/2} \rightarrow {}^4P_{5/2}$		

23.17 The transition Al[Ne]$(3s^2)(3p^1) \rightarrow$ Al[Ne]$(3s^2)(4s^1)$ has two lines given by $\tilde{\nu} = 25\,354.8$ cm^{-1} and $\tilde{\nu} = 25\,342.7$ cm^{-1}. The transition Al[Ne]$(3s^2)(3p^1) \rightarrow$ Al[Ne]$(3s^2)(4d^1)$ has three lines given by $\tilde{\nu} = 32\,444.8$ cm^{-1}, $\tilde{\nu} = 32\,334.0$ cm^{-1} and $\tilde{\nu} = 32\,332.7$ cm^{-1}. Explain all lines and sketch an energy level diagram of the states involved. Recall that only $\Delta J = 0, \pm 1$ is allowed.

23.18 The principal line in the emission spectrum of Na is yellow. At high resolution the line is found to be a doublet with wavelengths of 588.9950 and 589.5924 nm. (a) Explain the source of this doublet and calculate the energy differences in cm^{-1} that make this phenomenon possible. (b) In K, a similar structure is found but with an energy difference of 223.0 cm^{-1}. Explain the difference.

23.19 The oxygen atom exhibits the structure given below. The only three terms that arise are ${}^1S$, ${}^3P$, and ${}^1D$. Three transitions are observed with the wavelengths shown in the figure below. Assign complete term symbols (including J value) to each level. Placing the ground state at zero, calculate the energy in cm^{-1} above the ground state for the remaining states. State with justification whether the transitions are weak or strong. Why are there only five levels?

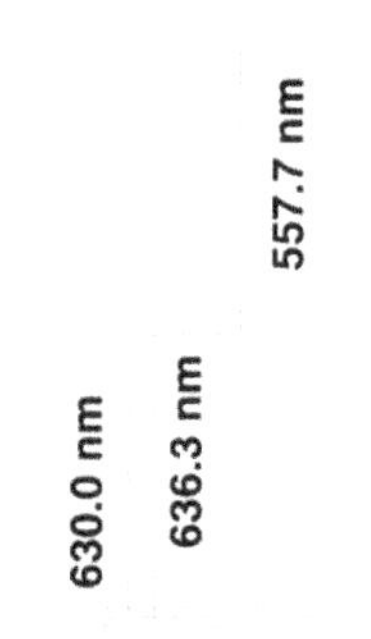

23.20 The three lowest energy electron configurations of the B atom are $1s^2\,2s^2\,2p^1$, $1s^2\,2s^1\,2p^2$, and $1s^2\,2s^2\,3s^1$. ${}^4S$, ${}^4D$ and ${}^2P$ terms are not allowed to result from the second configuration. Determine all of the allowed term symbols for these three configurations, including their J values. Then draw an energy level diagram with the resulting states in the proper order.

23.21 The low-resolution, one-photon absorption spectrum of the ground state of the Si^+ ion has peaks at the wavelengths given in the table below. The lowest energy peak corresponds to the excitation of an s electron. The next four peaks involve excitation of a p electron to higher orbitals. Placing the ground state at $E = 0$ eV, complete the table below. In the Selection Rules column, refer to the values of ΔS and Δl that make the transition allowed. State the reasons for the energetic ordering of the final configurations. You do not need to determine the full term symbol.

λ/nm	Initial e$^-$ config	Final e$^-$ config	E/eV	Selection Rules
181.6				
153.3				
126.5				
102.3				
99.3				

23.22 Fill out the following table; answer whether each transition is allowed or forbidden for a one-photon electric dipole transition, and why. Using the given wavenumbers for each absorption transition as a measure of energy (in other words, don't convert to J or some other unit) draw an energy level diagram roughly to scale that correctly lists the named states from lowest to highest energy.

Transition	Wavenumber/cm^{-1}	Allowed?	Reason
$^1F_0 \rightarrow {}^1D_1$	101,250		
$^3F_1 \rightarrow {}^1F_0$	3,450		
$^1S_0 \rightarrow {}^1F_0$	68,250		

23.23 The $^4F_{3/2} \rightarrow {}^4I_{3/2}$ transition of Nd^{3+}:YAG has a radiative lifetime of 1.22 ms. The population lifetime (determined by the sum of radiative and nonradiative processes) is 230 μs. Determine the natural linewidth of the transition.

23.24 For spin multiplicity of four, what are the allowed values of M_S?

23.25 Discuss why Russell–Saunders coupling is not valid at high Z.

23.26 Write the following complete set of wavefunctions in matrix notation:

$$\psi_1 = a\varphi_1 + b\varphi_2 + c\varphi_3, \quad \psi_2 = d\varphi_1 + e\varphi_2 + f\varphi_3, \quad \psi_3 = g\varphi_1 + h\varphi_2 + j\varphi_3$$

Endnotes

1. Ritz, W. (1908) *Phys. Z.*, **9**, 521.
2. Stark, J. (1919) *Sitzber. preuss. Akad. Wiss.* Berlin, p. 932.
3. Zeeman, P. (1897) *Philos. Mag.*, **43**, 226; Zeeman, P. (1897) *Philos. Mag.*, **44**, 55, 255.
4. Einstein, A. (1917) *Phys. Z.*, **18**, 121.
5. Recall the convention that lower-case j, l and s correspond to values for single electrons, and upper-case J, L and S correspond to total values.
6. Unsöld, A. (1927) *Ann. Phys.*, **387**, 355.
7. Slater, J.C. (1960) *Quantum Theory of Atomic Structure*. McGraw-Hill, New York.
8. Curl, R.F. and Kilpatrick, J.E. (1960) *Am. J. Phys.*, **28**, 357.
9. Gilleron, F. and Pain, J.-C. (2009) *High-Energy Density Physics*, **5**, 320.
10. Hund, F. (1927) *Linienspektren und periodisches System der Elemente*. Julius Springer, Berlin.
11. Salmeron, M. and Schlögl, R. (2008) *Surf. Sci. Rep.*, **63**, 169.
12. Auger, P. (1925) *J. Phys. Radium*, **6**, 205.
13. Childs, K.D., Carlson, B.A., LaVanier, L.A., Moulder, J.F., Paul, D.F., Stickle, W.F., and Watson, D.G. (1995) *Handbook of Auger Electron Spectroscopy: A Book of Reference Data for Identification and Interpretation in Auger Electron Spectroscopy*, 3rd edition. Physical Electronics, Inc., Eden Prairier, MN.

Further reading

Atkins, P.W. and Friedman, R.S. (2010) *Molecular Quantum Mechanics*, 5th edition. Oxford University Press, Oxford.

Cowan, R.D. (1981) *The Theory of Atomic Structure and Spectra*. University of California Press, Berkeley.

Ertl, G. and Küppers, J. (1985) *Low Energy Electrons and Surface Chemistry*, 2nd edition. VCH, Weinheim.

Heitler, W. (2010) *The Quantum Theory of Radiation*, 3rd edition. Dover Publications, New York.

Kolasinski, K.W. (2012) *Surface Science: Foundations of Catalysis and Nanoscience*, 3rd edition. John Wiley & Sons, Chichester.

NIST Atomic Spectroscopy Databases: http://www.nist.gov/pml/data/atomspec.cfm

NIST Atomic Spectra Database Levels Form: http://www.physics.nist.gov/PhysRefData/ASD/levels_form.html

Steinfeld, J.I. (2012) *Molecules and Radiation: An Introduction to Modern Molecular Spectroscopy*, 2nd edition. Dover Publications, New York.

CHAPTER 24

Molecular bonding and structure

PREVIEW OF IMPORTANT CONCEPTS

- The Hamiltonian of a molecule is obtained by adding the potential energy to the kinetic energy. Contributions are made by both the electrons and the nuclei. The kinetic energy sums an electronic term T_e and a nuclear term T_n. The potential energy sums the repulsive electron–electron interactions V_{ee} and internuclear interactions V_{nn}, as well as the attractive electron–nuclei term V_{en}.
- The Born–Oppenheimer approximation assumes that the nuclei are fixed and allows us to separate nuclear motion from the motion of electrons.
- Any wavefunction that can be created by a product of terms corresponds to an energy that is the sum of the energies for each of those terms.
- A good first approximation is that rotational, vibrational and electronic wavefunctions are separable, and the energy of a molecule is the sum of rotational plus vibrational plus electronic energies.
- A proper wavefunction is composed of a spatial part times a spin part. The overall wavefunction must be antisymmetric with respect to exchange of any two electrons. No more than two electrons can occupy an orbital and if two electrons occupy an orbital, their spins must be paired (antiparallel).
- Nuclei brought together in the arrangement of a molecule create a molecular potential that gives rise to molecular orbitals.
- A convenient way to conceptualize the formation of chemical bonds and generate the vocabulary to describe them is given by the valence bond model.
- Molecular orbitals come in three flavors: bonding; antibonding; and nonbonding.
- Bonding molecular orbitals enhance density in the internuclear region.
- Antibonding molecular orbitals result from destructive interference in the internuclear region and have a node perpendicular to the bond axis.
- σ bonds are axially symmetric and exhibit no nodes.
- π bonds are antisymmetric about a nodal plane containing the bond axis.
- Once a set of molecular orbitals is formed, they are filled using the Aufbau principle from the lowest energy up.
- A chemical bond is formed by maximizing electron–nucleus attractions and minimizing electron–electron and internuclear repulsions while simultaneously minimizing the electron kinetic energy.
- A qualitative approximation of bond strength is given by the bond order, which is calculated by subtracting the number of electron in antibonding orbitals from the number in bonding orbitals, and then multiplying by one-half.
- The sharing of valence electrons between nuclei is equal in nonpolar bonds and unequal in polar bonds.
- The variational principle states that, for a given Hamiltonian, the lowest energy is the energy given by the correct wavefunction. All other wavefunctions result in a higher energy.
- Molecular orbitals are not essential for the calculation and description of chemical bonding. The electron density and the shape of the density contours can instead be used. This is the basis of density functional theory (DFT).

Bonding and the changes in bonding that occur in molecular encounters are obviously at the heart of chemistry. We have covered a lot of ground in chemistry without every really defining chemical bonds more precisely than involving either the sharing of electrons (covalent bonding) or the binding of ions by electrostatic interactions (ionic bonding). In Chapter 3, we investigated all kinds of weaker intermolecular interactions, for example, hydrogen bonding and multipole interactions. We have discussed electrostatic interactions, which hold together ionic solids and are essential for understanding electrolyte solutions, in Chapter 14. While we have a good idea that covalent bonding – the most prevalent form of bonding in molecules and which is also important in many solids – involves the sharing of electrons, we have yet

Physical Chemistry: How Chemistry Works, First Edition. Kurt W. Kolasinski.

Companion Website: www.wiley.com/go/kolasinski/physicalchemistry

to really understand it or treat it fundamentally. This is because we need to understand quantum mechanics first before we can really approach covalent bonding.

There are several flavors of *covalent bonds*. We will start out discussing *two-center bonding*. A single bond between two H atoms in H_2, of course, is the simplest case and the place to start. Two-center bonding is also extremely common – at least, many bonds are predominantly two-center. However, there are many counter examples. You have encountered delocalized bonding in conjugated organic molecules such as benzene. You may have encountered *three-center bonding* in boron hydrides and some compounds involving halogens, such as HF_2^-. Metallic bonding is also highly delocalized. In two-center bonding we generally think of two atoms coming together, each contributing an electron, then sharing these electrons to form a bond. If the electrons are shared equally, the bond is nonpolar. If the sharing is unequal, the bond is polar. A *dative bond* is a type of covalent bond formed when both electrons are contributed by one atom. This concept is common in coordination chemistry. Once we have a clearer picture of what molecular orbitals are and discover how molecules minimize their energies by filling them and adjusting their structure, we will see that these semantic distinctions are sometimes just that. However, as we have learned from our introduction to quantum mechanics, part of our learning how to understand the interactions of subatomic particles, atoms and molecules is learning how to graft classical terms and analogies onto intrinsically quantum phenomena.

24.1 Born–Oppenheimer approximation

To write the Hamiltonian of a molecule we add the potential energy to the kinetic energy. Contributions to the sum are made by both the electrons and the nuclei. The kinetic energy is a simple sum of an electronic term T_e and a nuclear term T_n. The potential energy has contributions from the repulsive electron–electron interactions V_{ee} and internuclear interactions V_{nn}, as well as the attractive electron–nuclei term V_{en}. Thus, the total Hamiltonian is

$$\hat{H} = \hat{T}_e + \hat{T}_n + \hat{V}_{ee} + \hat{V}_{nn} + \hat{V}_{en}. \tag{24.1}$$

Nuclei are much heavier and faster than electrons. Therefore, we can treat the nuclei as fixed and separate nuclear motion from electronic motion. This means that the electrons react so quickly to changes in nuclear positions, that they are always able to relax to their lowest energy configuration for any arrangement of the nuclei. This approximation was proposed by Max Born and J. Robert Oppenheimer.[1] It is a good approximation for ground-state molecules and most excited states. Some excited states of polyatomic molecules and the electronic states of cations are where violations are most likely to be found. The *Born–Oppenheimer approximation* allows us to construct molecular potential energy curves and potential energy surfaces (PESs) using the following procedure.

For fixed nuclei, the nuclear kinetic energy term is zero and the internuclear potential is constant. There exists a set of electronic wavefunctions that satisfy the electronic Schrödinger equation

$$\hat{H}_e \psi_e = E_e \psi_e \tag{24.2}$$

where

$$\hat{H}_e = \hat{T}_e + \hat{V}_{ee} + \hat{V}_{en}. \tag{24.3}$$

The electronic wavefunctions ψ_e depend on the nuclear coordinates. However, the Born–Oppenheimer approximation allows us to use these coordinates as a parameter. The nuclear coordinates are set to a given value and the Schrödinger equation is solved in the electronic coordinates to arrive at a value of the total energy. The procedure is repeated until we have calculated values for the energy for every value of the nuclear coordinates. For a diatomic molecule such as SrH – a random example depicted in Fig. 24.1 – this corresponds to calculating the energy as a function of the internuclear distance between the Sr atom and the H atom. The results of such a calculation are shown in Fig. 24.1. The ground state, denoted with X, is bound and has a well depth of D_e. A bound electronically excited state, denoted C in this example, is also shown.

As far as the motion of the nuclei is concerned, the rapid response of the electrons means that the electronic energy E_e can be treated as part of the potential field in which the nuclei move. The nuclear Hamiltonian is then

$$\hat{H}_n = \hat{T}_n + \hat{V}_{nn} + E_e \tag{24.4}$$

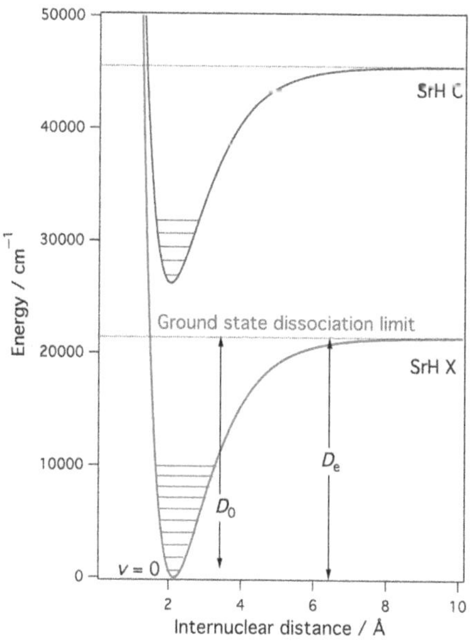

Figure 24.1 Born–Oppenheimer potential energy curves for the ground state of SrH, also called the X state, and one particular excited state designated as the C state. Total energy is plotted along the y-axis and the internuclear distance between the Sr and H nuclei is plotted along the x-axis. The horizontal lines represent different vibrational levels within each electronic state. The lowest energy vibrational state is labeled with the vibrational quantum number $v = 0$. D_0 is the dissociation energy from the $v = 0$ level. D_e is the dissociation energy measured from the minimum of the potential energy curve.

and the Schrödinger equation for nuclear motion,

$$\hat{H}_n \psi_n = E_n \psi_n, \tag{24.5}$$

can be solved independently for the nuclear wavefunctions ψ_n and the nuclear energy E_n. The total wavefunction can be factored and is given by the product of the electronic and nuclear wavefunctions,

$$\psi(\tau_n, \tau_e) = \psi_e(\tau_e, \tau_n)\psi_n(\tau_n), \tag{24.6}$$

where τ_e and τ_n are the electronic and nuclear coordinates, respectively. The energy of such a factored wavefunction is given by the sum of electronic and nuclear contribution,

$$E = E_e + E_n. \tag{24.7}$$

To a first approximation, the vibrational and rotation degrees of freedom are independent. Therefore, the nuclear wavefunction can be further factored into vibrational and rotational components, and the energy is given by the sum of vibrational and rotational energies. In summary,

$$\psi_n = \psi_v \psi_r \tag{24.8}$$

$$E_n = E_v + E_r \tag{24.9}$$

$$\psi = \psi_e \psi_v \psi_r \tag{24.10}$$

$$E = E_e + E_v + E_r. \tag{24.11}$$

We will find that to, a first approximation, the vibrational contribution can be treated as a harmonic oscillator near the ground state and then as a Morse oscillator closer to the dissociation limit. The rotational contribution is that of a rigid rotor with centrifugal distortion added at high levels of excitation. We will return to these components in detail when we discuss molecular spectroscopy in the next chapter. For now, we concentrate on the electronic term and how it is related to bonding.

24.2 Valence bond theory

How is a chemical bond formed? The first successful quantum mechanical theory of bond formation was the *valence bond theory*, in which electrons occupying atomic orbitals in the constituent atoms are brought together and combined either by adding or subtracting their wavefunctions. In constructing the wavefunctions that describe the combination of two orbitals we observe several things:

- A bond is formed by combining atomic orbitals from the two atoms that are bonding.
- Hybridization is introduced to explain the bonding of polyatomic molecules.
- The orbitals form a bonding combination by enhancing electron density between the nuclei caused by favorable orbital overlap (constructive interference between constituent orbitals).
- The orbitals form an antibonding combination when electron density is insufficiently built up between the nuclei (destructive interference between constituent orbitals).
- The complete wavefunction is composed of a spatial part multiplied by a spin part.
- Electrons must be paired so that they obey the Pauli exclusion principle – no two electrons have the same quantum numbers. The total wavefunction must be antisymmetric with respect to particle exchange.
- Good wavefunctions form an orthonormal set.
- σ bonds are axially symmetric and exhibit no nodes.
- π bonds are antisymmetric about a nodal plane containing the bond axis.
- Antibonding orbitals (either σ or π) have a nodal plane perpendicular to the internuclear axis along the bond.

24.2.1 Homonuclear diatomic molecules

We start by trying to form an H_2 molecule from two H atoms – H_A and H_B – each with one electron in a 1s orbital. We will attempt to construct a wavefunction that describes the system in terms of the two wavefunctions that describe the atoms. The system is the sum of two electrons and two protons placed in the same region of space. Nuclei of like atoms are indistinguishable, just as are all electrons. When we bring two atoms together, we cannot distinguish whether it is electron 1 or electron 2 that occupies the orbital on atom A, and vice versa for atom B. We account for this in the system wavefunction by allowing for both possibilities in the spatial part of the wavefunction

$$\psi = A(1)B(2) \pm A(2)B(1) \tag{24.12}$$

where A is a 1s orbital on atom H_A and B is a 1s orbital on atom H_B. The (+) combination corresponds to constructive interference between the atomic wavefunctions. We will see that this forms the bonding combination. The (–) combination corresponds to the antibonding combination formed by destructive interference of the wavefunctions in the internuclear region.

To complete the wavefunction we now multiply the spatial part by the spin component. We have already encountered the four combinations of two spin states when we dealt with making proper wavefunctions for multi-electron atoms. The same combination rules are applicable to molecules,

Symmetric $$\sigma_+ = \{\alpha(1)\alpha(2),\quad \beta(1)\beta(2),\quad 2^{-1/2}[\alpha(1)\beta(2) + \beta(1)\alpha(2)]\} \tag{24.13}$$

Antisymmetric $$\sigma_- = \{2^{-1/2}[\alpha(1)\beta(2) - \beta(1)\alpha(2)]\}. \tag{24.14}$$

To couple the proper spin component with the proper spatial component, we consider the symmetry upon electron exchange

$$\psi_+(1,2) = A(1)B(2) + A(2)B(1) \tag{24.15}$$

$$\psi_+(2,1) = A(2)B(1) + A(1)B(2) \tag{24.16}$$

$$\psi_+(1,2) = \psi_+(2,1). \tag{24.17}$$

Therefore, $\psi_+(1,2)$ is symmetric with respect to electron exchange. For the minus combination,

$$\psi_-(1,2) = A(1)B(2) - A(2)B(1) \tag{24.18}$$

$$\psi_-(2,1) = A(2)B(1) - A(1)B(2) \tag{24.19}$$

$$\psi_-(1,2) = -\psi_-(2,1) \tag{24.20}$$

Therefore $\psi_-(1,2)$ is antisymmetric with respect to electron exchange.

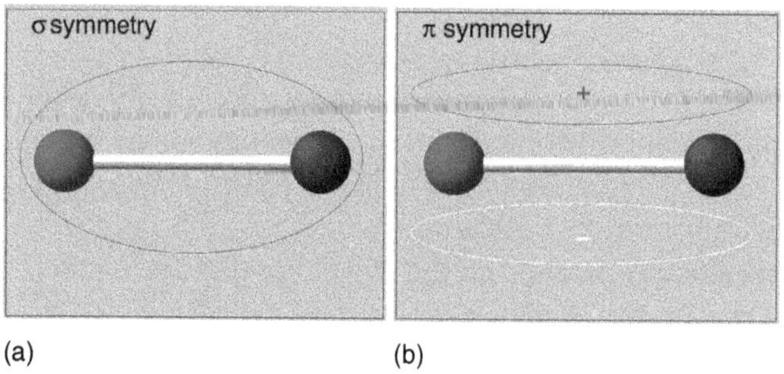

Figure 24.2 (a) An example of a bonding orbital that has σ symmetry. (b) An example of a bonding orbital that has π symmetry.

The *total wavefunction* is formed from the product of the spatial and spin parts and must be *antisymmetric*. Therefore, the proper full wavefunctions are

$$\Psi_{\text{bonding}}(1,2) = \psi_{+}(1,2)\sigma_{-} \tag{24.21}$$

$$\Psi_{\text{antibonding}}(1,2) = \psi_{-}(1,2)\sigma_{+} \tag{24.22}$$

There is only one way to write the bonding wavefunction. It corresponds to a singlet state. There are three ways to write the antibonding wavefunction. These correspond to three degenerate states for this one-electron configuration, a triplet state. Degenerate states have the same energy.

Bonds and orbitals are classified based on the symmetry of the spatial distribution of electrons in the bond or orbital. A σ bond has cylindrical symmetry about the internuclear axis and no nodal plane along this axis, as shown in Fig. 24.2(a). A π bond has a nodal plane containing the internuclear axis, as in Fig. 24.2(b).

The presence of nodal planes is not the only way to characterize orbitals. Three symmetry operations are considered when naming orbitals:

1 Reflection in a plane placed at the midpoint between nuclei and perpendicular to the internuclear axis.
2 Reflection in a plane in which the internuclear axis lies.
3 Inversion through the midpoint on the internuclear axis.

A diatomic molecule also contains an axis of rotational symmetry along the internuclear axis (it is a C_∞ axis, more on that in the next chapter). This is not considered in the symmetry designation of orbitals. However, it does define the vertical plane. The vertical plane is defined as the plane containing the highest-fold rotational axis, which in this case is the internuclear axis. Reflection is denoted by the operator $\hat{\sigma}$ (sorry, some letters are too popular). Thus, reflection in the vertical plane (the one containing the internuclear axis) is $\hat{\sigma}_v$. This is the operation defined in item (2). Reflection in a plane perpendicular to the internuclear axis is denoted reflection through a horizontal plane with operator $\hat{\sigma}_h$. Inversion is denoted by the operator $\hat{\imath}$. All three symmetry operations are denoted in Figure 24.3.

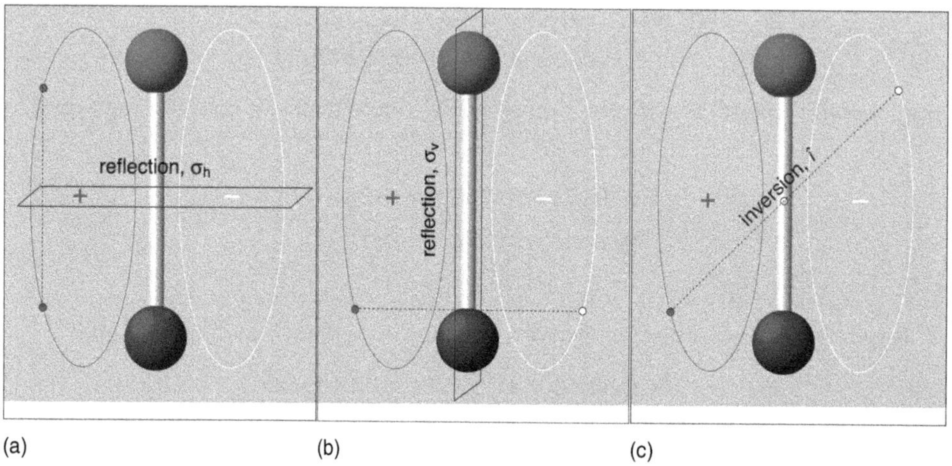

Figure 24.3 Symmetry elements commonly encountered in diatomic molecules. (a) Reflection through a horizontal plane denoted by the operator $\hat{\sigma}_h$. (b) Reflection through a vertical plane denoted by the operator $\hat{\sigma}_v$. (c) Inversion through a center of inversion denoted by the operator $\hat{\imath}$.

When a symmetry operation acts on an object, the object must be transformed into a configuration that is indistinguishable from the original configuration apart from its sign. For example, the symbol ↔ is symmetric upon rotation by π radians about an axis perpendicular to the page (a C_2 rotation). This rotation is a symmetry operation of the symbol. Rotation by only $\pi/2$ leads to ↕. This is distinguishable from the original and does not correspond to a symmetry operation of the object. If an object does not exhibit a certain symmetry element it is *asymmetric* with respect to that symmetry element. If the object is exactly the same after application of the symmetry operation, the object is *symmetric* with respect to the symmetry element. For example, $\overset{+\,+}{\leftrightarrow}$ is transformed to $\overset{+\,+}{\leftrightarrow}$ by a C_2 rotation, the plus signs indicating that the two ends of the arrow both have the same sign. An object that is exactly the same apart from a change in sign after the application of a symmetry operation is said to *antisymmetric*. For example, $\overset{+\,-}{\leftrightarrow}$ is transformed into $\overset{-\,+}{\leftrightarrow}$ by a C_2 rotation. This object – with opposite signs at the two ends – is antisymmetric with respect to the symmetry operation.

Let us apply this to diatomic molecules. We are concerned with the symmetry of the wavefunctions formed by the bonding and antibonding combinations. The three cases described above are represented by the relations

$$\text{asymmetric} \qquad \hat{\Omega}\psi \neq \psi \tag{24.23}$$

$$\text{symmetric} \qquad \hat{\Omega}\psi = \psi \tag{24.24}$$

$$\text{antisymmetric} \qquad \hat{\Omega}\psi = -\psi. \tag{24.25}$$

The last two equations represent eigenvalue equations. The symmetry operators have eigenvalues of either 1 or −1. In Eq. (24.23) the wavefunction does not correspond to an eigenfunction of the symmetry operator. The symmetry with respect to inversion determines whether a wavefunction is gerade (German for even) or ungerade (odd). Gerade symmetry corresponds to $\hat{\imath}\psi = \psi$ (no change in sign upon inversion) and is denoted g. Ungerade symmetry corresponds $\hat{\imath}\psi = -\psi$ (change in sign upon inversion) and is denoted u.

Antibonding orbitals contain a nodal plane perpendicular to the internuclear axis. This corresponds to a $\hat{\sigma}_h$ operation. Since the wavefunction changes sign on either side of a nodal plane, we see that $\hat{\sigma}_h\,\psi = -\psi$ corresponds to an antibonding orbital and $\hat{\sigma}_h\,\psi = \psi$ corresponds to a bonding orbital.

The final element in the designation of orbitals does not involve symmetry elements; rather, it is based upon the electron angular momentum. Just as we had s, p, d, f, ... atomic orbitals, we designate the orbitals in molecules with (switching to the Greek alphabet) σ, π, δ, ϕ, ... In atoms, the letter designation was determined by orbital angular momentum quantum number l. In molecules, the letter designation is determined by the projection of the angular momentum along the internuclear axis, which also takes on integer values 0, 1, 2, 3, ... in units of $\hbar$. In your garden-variety organic molecules, only σ and π orbitals are encountered. Involving higher-Z atoms can lead to the formation of δ or even ϕ orbitals, though higher projections of angular momentum become progressively less likely.

24.2.2 Polyatomic molecules

For polyatomic molecules in the valence bond description, we think of two or more orbitals from one of the atoms undergoing the process of hybridization or promotion to form a set of hybrid orbitals. This is how we form the familiar sp, sp^2 and sp^3 hybrids of the C atom. A variety of hybridization schemes have been rationalized. These are illustrated in Table 24.1.

Hybridization does not happen and valence bond theory is rather inaccurate. A carbon atom simply does not have a conversation with itself, think that it wants to bond with four H atoms, spontaneously form four sp^3 orbitals, and then have these hybrid orbitals form bonds with the individual H atoms along the way to making CH_4. Nonetheless, both the concept of hybridization and the valence bond approach are highly successful and useful schemes. How can that be? Remember what we said during the construction of Old Quantum Theory, we will use classical analogies and push them as far as we can. Hybridization accepts a quantum idea (atomic orbitals) and tries to push it into an explanation of molecules. We need to be able to collect our thoughts and conceptions of what a covalent bond is. With our vocabulary drawn from valence bond theory and hybridization we can make a first-order approximation toward a more well-defined concept of what a covalent bond is. This puts us in a position to explore how we can improve upon these theoretical underpinnings.

Table 24.1 Hybridization schemes constructed from s, p and d orbitals.

Component orbitals	Geometry	Coordination number
sp, pd	linear	2
sd, p^2 d^2	angular	
sp^2, p^2d	trigonal planar	3
spd	unsymmetrical planar	
p^3, pd^2	trigonal pyramidal	
sp^3, sd^3	tetrahedral	4
spd^2, p^3d^3	irregular tetrahedral	
p^2d^2, sp^2d	square planar	
sp^3d, spd^3	trigonal bipyramidal	5
sp^2d^2, sd^4, pd^4, p^3d^2	tetragonal pyramidal	
p^2d^3	pentagonal planar	
sp^3d^2	octahedral	6
spd^4, pd^5	trigonal prismatic	
p^3d^3	trigonal antiprismatic	
sp^3d^4	dodecahedron	8

Source: Eyring, H., Walter, J., and Kimball, G.E. (1944) *Quantum Chemistry*. John Wiley & Sons.

24.3 Molecular orbital theory

What does an atomic orbital have to do with a molecule or a molecular bond? An *atomic orbital* is related to the wavefunction that describes the result of an *atomic potential*. A molecule has a potential that is constructed from the sum of the interactions of all the nuclei and all the electrons. For the valence electrons, this potential does not look anything like a central-field-atomic potential. Instead of producing atomic orbital states, *a molecular potential forms molecular orbitals*. Rather than confining bonding to a conception of hybrid orbitals centered on atoms that combine only by addition or subtraction between two centers, imagine now that we allow the electrons to occupy orbitals that can spread throughout the molecule and occupy the space that allows them to lower their energy as fully as possible. The resulting molecular orbitals will often be localized between atoms approximating a two-center bond. They are then associated with one particular bond. However, we do not a priori constrain them to do so.

To make a bond with molecular orbital theory, we generalize the procedure of valence bond theory by recognizing that it is the molecular potential that determines the nature of molecular orbitals in the valence region.

- A bond is formed by combining two or more basis functions.
- The basis functions can correspond to atomic orbitals, but they do not have to.
- The resulting orbitals are not constrained to be between only two atoms but can extend over the whole molecule.
- We define the Hamiltonian and solve for the best linear combination wavefunctions that describe the system, that is, give the lowest energy.
- The wavefunctions will be antisymmetric with respect to electron exchange, orthonormal, and obey the Pauli exclusion principle.
- Bonding, antibonding and nonbonding combinations are possible.

The strategy is pretty clear. Define the Hamiltonian then attempt to find solutions in terms of well-defined basis functions. Early work concentrated on using atomic orbitals as the basis set. This led to the idea of finding solutions in terms of linear combinations of atomic orbitals (LCAO), or *LCAO-MO theory*. Such orbitals are also known as *Slater-type orbitals* (STO), and are intuitive and familiar but difficult computationally. The functions in atomic orbitals are difficult to integrate, and we need to integrate to calculate expectation values. Therefore, an alternative is to abandon atomic orbitals and instead use a functional form that is easy to integrate and still localized about the atoms. For this, *Gaussian-type orbitals* (GTO) or variations on Gaussian functions are chosen. These do not correspond to atomic orbitals but they are much easier to implement computationally.

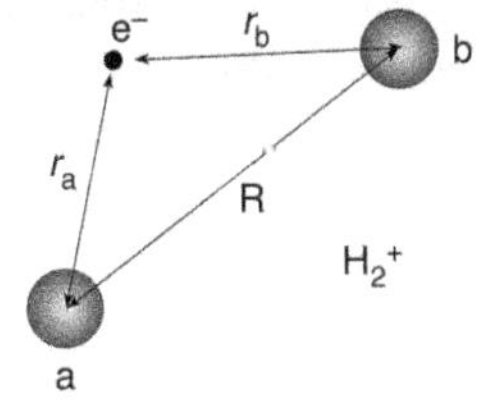

Figure 24.4 The H_2^+ molecular ion.

We will now look in depth at two model systems to establish the foundations of molecular orbital theory. The two models are the H_2^+ molecular ion, which can be solved exactly, and the H_2 molecule. The coordinate system for H_2^+ is shown in Fig. 24.4.

24.4 The hydrogen molecular ion H_2^+

The strategy is to set up the Hamiltonian including the all interactions with the aid of the Born–Oppenheimer approximation (electrons move, nuclei do not). As shown in Fig. 24.4, one electron is separated by distance r_a from nucleus a and r_b from nucleus b. The internuclear separation is R. The Hamiltonian includes the kinetic energy of the one electron and the potential between all three charged particles,

$$\hat{H} = -\frac{\hbar^2}{2m_e}\nabla^2 + \hat{V}, \tag{24.26}$$

$$\hat{V} = -\frac{e^2}{4\pi\varepsilon_0}\left(\frac{1}{r_a} + \frac{1}{r_b} - \frac{1}{R}\right). \tag{24.27}$$

Keeping track of these various constants can become cumbersome, which has motivated the use of *atomic units* (au) when performing quantum chemical calculations. Atomic units form a natural system of units when dimensions approach the atomic scale. These units are defined in Table 24.2. The atomic units for mass, charge and angular momentum are fairly obvious. The length scale is set by the *Bohr radius* a_0, which is the diameter of the $n = 1$ state within the Bohr model of the atom. The energy scale is set by the *Hartree* E_h, which is the potential energy associated with the $n = 1$ state of the H atom. The unit of time t_0 is equal to how long it would take an electron with the mean velocity of a 1s electron, which is the atomic unit of speed v_0, to travel the length of one Bohr radius.

Rewriting the Hamiltonian in terms of atomic units of Hartree for the energy, the Bohr radius for length, and the charge and mass of the electron, we obtain

$$\text{Atomic units} \qquad \hat{H} = -\frac{1}{2}\nabla^2 - \frac{1}{r_a} - \frac{1}{r_b} + \frac{1}{R}. \tag{24.28}$$

Table 24.2 Selected atomic units and their SI equivalents.

Quantity	Atomic unit	Symbol	Value in SI units
Mass	electron mass	m_e	9.109×10^{-31} kg
Charge	elementary charge	e	1.602×10^{-19} C
Angular momentum (action)	Reduced Planck constant	$\hbar = h/2\pi$	1.055×10^{-34} J s
Length	Bohr radius	$a_0 = \dfrac{4\pi\varepsilon_0\hbar^2}{m_e e^2}$	5.292×10^{-11} m
Energy	Hartree energy	$E_h = \hbar^2/m_e a_0^2$	4.360×10^{-18} J = 27.21 eV
Time		$t_0 = \hbar/E_h$	2.419×10^{-17} s
Speed		$v_0 = a_0 E_h/\hbar$	2.188×10^{6} m s^{-1}
Speed of light		c/au of speed $= 1/\alpha$	137.036
Fine-structure constant		$\alpha = a_0 E_h/c\hbar$	
Force		E_h/a_0	8.239×10^{-8} N
Linear momentum		$\hbar/a_0$	1.992×10^{-24} N s

The Schrödinger equation for H_2^+ can be solved analytically (exactly) by introducing confocal elliptical coordinates. Unfortunately, this solution cannot be extended to any other system with more than one electron because of electron correlation. Therefore, let us explore approximate answers to the $H_2{}^+$ problem that will be applicable and illustrative of what we can do for any diatomic molecule.

24.4.1 Directed practice

Write out the Hamiltonian using a form similar to Eq. (24.28), that is, using atomic units, for $Li_2{}^{5+}$.

24.4.2 Molecular orbitals from linear combinations of atomic orbitals (LCAO-MO)

The idea here is that we are going to make an orbital for the system (a molecular orbital, MO), out of component orbitals that each atom brings to the system (the atomic orbitals). We will add (or subtract) the atomic orbitals and multiply them by factors (for normalization), but we will not raise them to powers. Therefore, the combinations of atomic orbitals are linear combinations. For the H_2^+ system we start off with the simplest approximation: The ground state is composed of a linear combination of a 1s orbital associated with the nucleus a, which we call A for short, and a 1s orbital from the nucleus b, which we call B for short.

There are two possible linear combinations (this should look pretty familiar from valence bond theory)

$$\psi_+ = N_+(A + B) \tag{24.29}$$

$$\psi_- = N_-(A - B). \tag{24.30}$$

Normalization requires

$$1 = \int |\psi_+|^2 \, d\tau = N_+^2 \int (A^2 + B^2 + 2AB) \, d\tau. \tag{24.31}$$

Each constituent orbital is normalized

$$S_{AA} = \int A^2 \, d\tau = 1 \tag{24.32}$$

$$S_{BB} = \int B^2 \, d\tau = 1 \tag{24.33}$$

But the *overlap integral* or *overlap density* will depend on the specific orbital combination and we call its value S

$$\int AB \, d\tau = S. \tag{24.34}$$

The value of the overlap integral S depends on the internuclear separation R. Note that in Eqs. (24.31) and (24.34) we have made use of the identity that the overlap integral is symmetrical with respect to whether the integral is taken over the product AB or BA, that is, $S_{AB} = S_{BA} = S$. The normalization constant is then found to be

$$N_+ = [2(1 + S)]^{-1/2}. \tag{24.35}$$

Similarly

$$N_- = [2(1 - S)]^{-1/2}. \tag{24.36}$$

24.4.2.1 Directed practice

Evaluate the integrals in Eq. (24.31) to confirm the values of the normalization constant.

By looking at the squares of the wavefunctions, remembering that it is proportional to the probability distribution for the electron and noting that S is a positive number, we can interpret the meaning of the (+) and (−) combinations

Bonding LCAO-MO $$\psi_+^2 = N^2(A^2 + B^2 + 2AB) \tag{24.37}$$

Antibonding LCAO-MO $$\psi_-^2 = N^2(A^2 + B^2 - 2AB). \tag{24.38}$$

The (+) combination enhances the electron density between the two nuclei and the (−) combination decreases the electron density in the internuclear region. Both combinations are cylindrically symmetric about the internuclear axis. As in valence bond theory, we identify this as *σ symmetry*. However, they are differentiated by their behavior under inversion. The (+) combination has gerade symmetry, denoted g, and the (−) combination has ungerade symmetry, denoted u. We will discover shortly that the bonding orbital is lower in energy than the antibonding orbital so the (+)

combination is a σ_g MO and the (−) combination, which is the next lowest orbital of σ symmetry, is a σ_u^* MO, where the * denotes antibonding character. Since both orbitals are derived from a 1s configuration, we further label them as the $1\sigma_g$ and the $1\sigma_u^*$. This naming convention is used for the molecular orbitals generated by homonuclear diatomics, for which the MOs are generated by degenerate orbitals of the same n value. Again, just as in valence bond theory, orbitals that are antisymmetric with respect to a plane of symmetry that includes the internuclear axis have *π symmetry*.

To develop our discussion beyond the phenomenological and the qualitative, we need to solve for the energies of these wavefunctions. We want to develop a method we can use more generally than for just the H_2^+ molecular ion. For this we use the variational theorem introduced in Chapter 20 and the matrix notation introduced in Chapter 22. First, we write the wavefunction as a linear combination of the 1s wavefunction centered on nucleus A and one centered on B as

$$\psi(r_a, r_b, R) = c_A\, A + c_B\, B. \tag{24.39}$$

The coefficients c_A and c_B will be determined in a *variational calculation*. The two 1s basis functions depend only on the electronic coordinates. They possess no dependence on the internuclear separation R. This R-dependence is to be found in the coefficients c_A and c_B.

With the Hamiltonian in Eq. (24.28) and the wavefunctions from Eq. (24.39), we calculate the expectation value of the total energy as a function of the internuclear distance R,

$$\langle E(R)\rangle = \frac{\int \psi^*(r_a, r_b, R)\, \hat{H}\, \psi(r_a, r_b, R)\, d\tau}{\int \psi^*(r_a, r_b, R)\, \psi(r_a, r_b, R)\, d\tau}. \tag{24.40}$$

The variational principle can be stated as

> *For a given Hamiltonian, the lowest energy is the energy given by the correct wavefunction. All other wavefunctions result in a higher energy.*

Therefore, the values of c_A and c_B that minimize the energy at a given value of R from Eq. (24.40) represent the best possible wavefunction for the system in terms of the two basis functions that we have chosen.

We perform this minimization by solving two coupled equations for c_A and c_B. In solving this system of equations we will need the overlap integrals defined above, and the Hamiltonian integrals defined by

$$H_{jk} = \int \psi_j^*\, \hat{H}\, \psi_k\, d\tau, \quad j, k = A \text{ or } B. \tag{24.41}$$

The solution is obtained by finding the roots of the secular determinant

$$\begin{vmatrix} H_{AA} - S_{AA}\langle E\rangle & H_{AB} - S_{AB}\langle E\rangle \\ H_{BA} - S_{BA}\langle E\rangle & H_{BB} - S_{BB}\langle E\rangle \end{vmatrix} = 0 \tag{24.42}$$

As we have already determined above, $S_{AA} = S_{BB} = 1$. Integration of the cross-terms of the overlap integral yield

$$S_{AB} = S_{BA} = S = e^{-R}(1 + R + R^2/3) \tag{24.43}$$

The overlap integrals behave intuitively at their limits. As $R \to \infty$, there is no overlap $S \to 0$. Complete overlap occurs as $R \to 0$, at which point $S \to 1$.

The Hamiltonian integrals are also symmetrical

$$H_{AA} = H_{BB} = H = -1/2 + 1/R + J \tag{24.44}$$

$$H_{AB} = H_{BA} = H' = -S/2 + S/R + K. \tag{24.45}$$

In atomic units, the −1/2 term is the energy of the H atom 1s orbital, $1/R$ is the nuclear repulsion term, and J and K are variants on the Coulomb and exchange terms that we encountered in the solution of the multi-electron atom problem. The J integral represents the electrostatic energy of a nucleus (a proton in this case) at distance R with the charge density of $|1s|^2$,

$$J = \int A \frac{1}{R} A\, d\tau = e^{-2R}(1 + 1/R) - 1/R. \tag{24.46}$$

Recall that A is just the 1s orbital located on atom A. Thus, The J integral is a measure of the Coulombic interaction between a nucleus and the electron density centered on the *other* atom.

The K integral is a two-center integral. Just like H', it involves integration with respect to two nuclear positions

$$K = -\int A\frac{1}{r_a}B\,d\tau = -e^{-R}(1+R), \tag{24.47}$$

The K integral is a measure of the Coulombic interaction between a nucleus and the charge density that builds up in the internuclear region (the region of overlap of A and B). Just like S, J and K vanish at large R.

We expand the secular determinant in terms of S, H, H', and $\langle E\rangle$. Solving for the expectation value of the total energy as a function of internuclear distance, we find two solutions

$$\langle E(R)\rangle = \frac{H \pm H'}{1 \pm S}. \tag{24.48}$$

The coefficients that lead to the these results are

$$c_A = \pm c_B. \tag{24.49}$$

When substituted back into Eq. (24.39), we obtain two wavefunctions

$$\psi_{1\sigma_g} = \frac{1s_A + 1s_B}{[2(1+S)]^{1/2}} \tag{24.50}$$

$$\psi_{1\sigma_u^*} = \frac{1s_A - 1s_B}{[2(1-S)]^{1/2}}. \tag{24.51}$$

Both have σ symmetry (they are cylindrically symmetric along the internuclear axis) but they are differentiated by their behavior under inversion. The (+) combination has gerade symmetry and the (−) combination has ungerade symmetry.

Expanding Eq. (24.48) using the Coulomb and resonance integrals, the two energies associated with these wavefunctions are represented by the sum of three terms

$$\left\langle E_{1\sigma_g}\right\rangle = E_{H1\,s} + \frac{1}{R} + \frac{J+K}{1+S} \tag{24.52}$$

and

$$\left\langle E_{1\sigma_u^*}\right\rangle = E_{H1s} + \frac{1}{R} + \frac{J-K}{1-S}. \tag{24.53}$$

The first term is simply the energy of the H atom in the 1s state. This ensures the proper asymptotic behavior as both tend to this value as $R \to \infty$. The behavior of the $1/R$, J, K and S are depicted in Fig. 24.5(a). The $1/R$ term represents the internuclear repulsion. This term dominates at very small internuclear distances and leads to the repulsive wall shown in Fig. 24.5(b) at small R. The third term represents the chemical interaction.

If there were no overlaps, both $S = 0$ and $K = 0$. The bonding and antibonding combinations have the same energy, which is true at large separation. In other words, the bonding and antibonding orbitals have the same energy asymptote at large separation. As R decreases, the state described by $\psi_{1\sigma_g}$ goes down in energy because increased electron density in the internuclear region screens the nuclear charges from one another while enhancing the electron–nucleus attraction. This wavefunction corresponds to the bonding state. The state described by $\psi_{1\sigma_u*}$ is pushed up in energy because of increased repulsion between the nuclei not being offset by attractive electron–nucleus interactions as the electrons are pushed away from the internuclear region. Near the equilibrium bond distance (the minimum in the $1\sigma_g$ potential energy curve), the $1\sigma_u$ state is pushed higher in energy than the $1\sigma_g$ state is pushed down.

For the H_2^+ molecular ion, we took two atomic orbitals and found that we could form two LCAO-MOs from them – two orbitals in, two out. As with any orbitals, only two electrons are allowed in molecular orbitals and they fill in accord with the Aufbau principle: fill from the bottom up, pairing spins as you go. If degenerate orbitals are available, maximize the spin and only pair when necessary. The $1\sigma_u$ orbital is not only ungerade but also antibonding (∗), thus a more complete notation is $1\sigma_u^*$. There is only one electron in the H_2^+ system and it will occupy the lowest energy orbital. Therefore, the molecular orbital configuration for H_2^+ is labeled $1\sigma_g$.

24.5 Solving the H_2 Schrödinger equation

Moving on to the simplest neutral molecule H_2, what has changed and what stays the same compared to H_2^+? We know that the system is bound and that the same mechanisms are driving the formation of the bond: (i) minimization of internuclear repulsion; (ii) maximization of electron–nucleus attraction; and (iii) minimization of the total energy, which also includes minimizing the contribution from the electron kinetic energy. In discussing model quantum systems

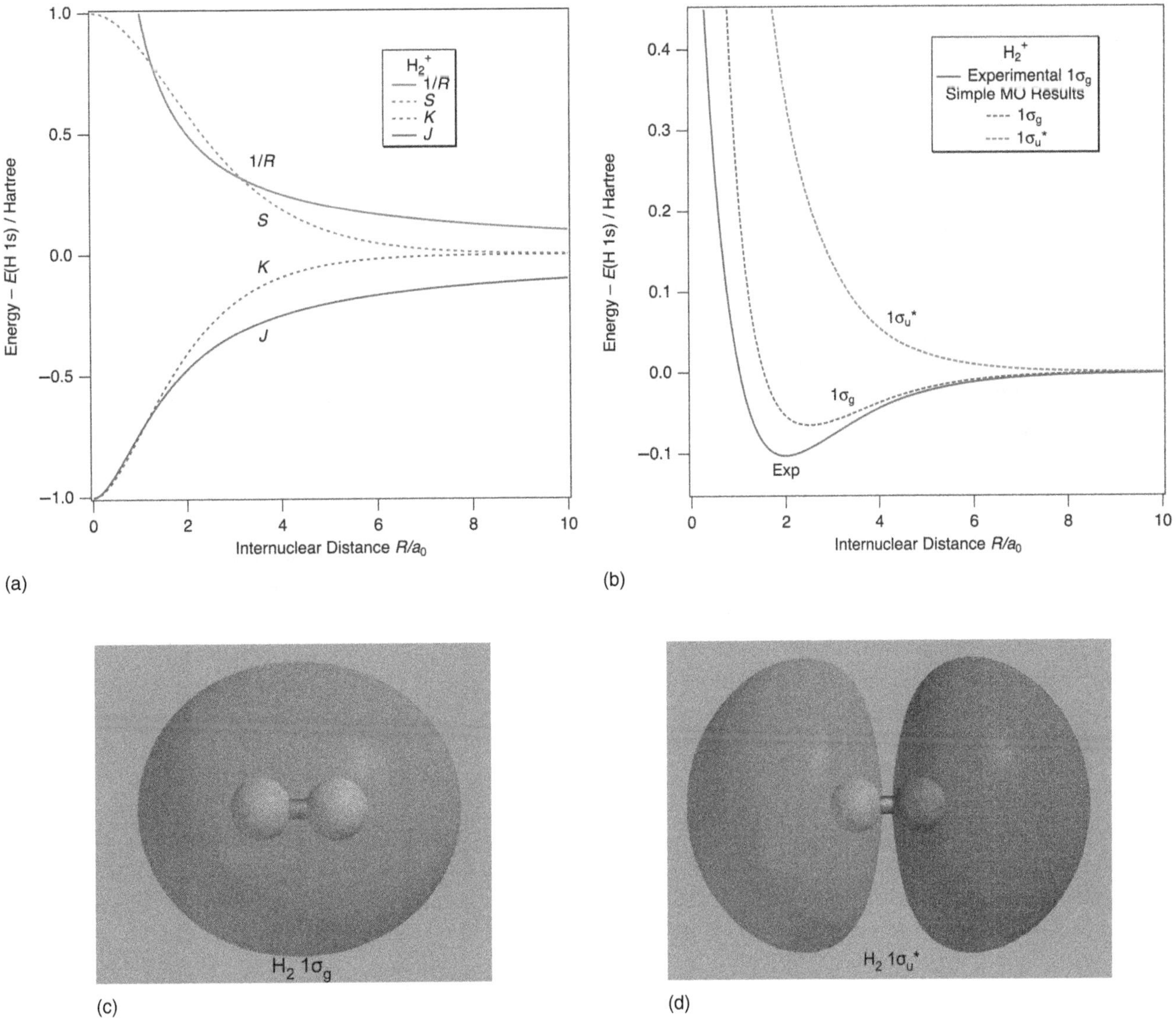

Figure 24.5 (a) The behavior of nuclear repulsion $1/R$, the Coulomb integral J, the exchange integral K and orbital overlap S as a function of internuclear distance R for the H_2^+ molecular ion. (b) The results of Eqs (24.52) and (24.53) for the energies of the H_2^+ ground and first excited electronic states are plotted as a function of internuclear distance in atomic units (dashed curves). These are compared to the experimentally determined ground-state potential energy curve (solid curve) obtained by inserting the parameters from the NIST WebBook into a Morse potential. Note how the variational principle is followed in that the energy of the curve calculated for an approximate wavefunction is always greater than or equal to the energy of the more accurate experimentally determined curve. (c) Electron density associated with the H_2 $1\sigma_g$ molecular orbital. (d) Electron density associated with the H_2 $1\sigma_u^*$ molecular orbital.

we found that wavefunctions with high curvature and more nodes are associated with higher kinetic energy. This is still the case when it comes to molecular orbitals. What has changed by the inclusion of two electrons is the introduction of electron correlation as well as the necessity for antisymmetrization of the total wavefunction. Now that we have two electrons, we must ensure that electron spins are paired if two electrons occupy the same orbital, that no more than two electrons can fill an orbital, and that the overall wavefunction (spatial part times spin part) is antisymmetric with respect to exchange of electrons.

We start by writing the Hamiltonian. We now have to keep track of two electrons and their interaction. Anticipating that we will eventually move on to systems with even more than two electrons and two nuclei, the most general form of the Hamiltonian for n electrons and N nuclei with charge Z and mass m is

$$\hat{H} = -\frac{1}{2}\sum_{i=1}^{n}\nabla_i^2 - \frac{1}{2}\sum_{j=1}^{N}\frac{m_e}{m_j}\nabla_j^2 - \sum_{i=1}^{n}\sum_{j=1}^{N}\frac{Z_j}{r_{ij}} + \frac{1}{2}\sum_{i=1}^{n}\sum_{\substack{i'=1\\ i'\neq i}}^{n}\frac{1}{r_{ii'}} + \frac{1}{2}\sum_{j=1}^{N}\sum_{\substack{j'=1\\ j'\neq j}}^{N}\frac{Z_j Z_{j'}}{R_{jj'}} \qquad (24.54)$$

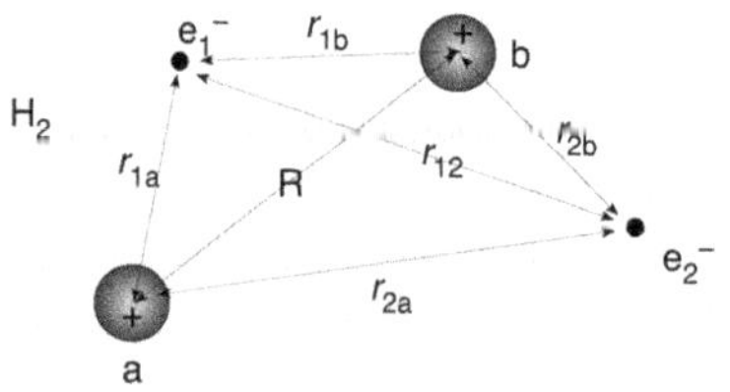

Figure 24.6 The H_2 molecule. Nucleus a is separated from nucleus b by distance R and electron 1 is separated from electron 2 by distance r_{12}. Other separations are denoted as indicated.

The first term is the kinetic energy of the electrons; electrons are indexed with the letter i. The second term is the nuclear kinetic energy with nuclei indexed by j. Within the Born–Oppenheimer approximation we set this term equal to zero, which is the equivalent to setting $m_j = \infty$ and $m_e/m_j = 0$. The third term sums over all electrons and all nuclei. It represents the attractive potential energy of the interaction of the ith electron with the jth nucleus. The change of sign indicates that the fourth term represents a repulsive interaction – in this case the electron–electron repulsion. The factor of $\frac{1}{2}$ ensures that double counting does not occur and the $i' \neq i$ notation ensures that particles do not repel themselves. Similarly, the fifth term represents the repulsive potential energy of internuclear interactions. Note that the form of the Hamiltonian does not assign an electron to any particular atom. This must be so because electrons are indistinguishable.

As mentioned above, the first step of the Born–Oppenheimer approximation is to set the second term (the nuclear kinetic energy) to zero. The second step is to treat internuclear distance as a parameter. We set it to a certain value and solve the Schrödinger equation. Then we repeat this procedure for every value required. Within the Born–Oppenheimer approximation (which we will always use so I will not change the notation of the Hamiltonian operator to indicate this) and with reference to Fig. 24.6, we write the Hamiltonian operator for the H_2 molecule as

$$\hat{H} = -\frac{1}{2}\nabla_1^2 - \frac{1}{2}\nabla_2^2 - \frac{1}{r_{1a}} - \frac{1}{r_{1b}} - \frac{1}{r_{2a}} - \frac{1}{r_{2b}} + \frac{1}{r_{12}} + \frac{1}{R}. \tag{24.55}$$

The simplest valence bond approach to constructing wavefunctions is to use the idea that we know where the electrons came from and where they end up to form a bond. Two 1 s orbitals contain the electrons when they are far apart, and we can either add these two orbitals together constructively or deconstructively. However, we must also account for electron indistinguishability and *antisymmetrization* just as we did for the multi-electron atom. This leads us to two wavefunctions

$$\psi_{VB}^{+} = \frac{1}{[2(1+S)]^{1/2}}[A(1)B(2) + A(2)B(1)] \tag{24.56}$$

$$\psi_{VB}^{-} = \frac{-1}{[2(1+S)(1-S)]^{1/2}}[A(1)B(2) - A(2)B(1)] \tag{24.57}$$

The total wavefunction is then formed by multiplying each spatial part above by the appropriate spin wavefunction. The overall wavefunction must be antisymmetric with respect to electron exchange. The (+) combination is symmetric and must be multiplied by the one and only antisymmetric spin wavefunction. This corresponds to a singlet state. The (−) combination is antisymmetric and can be multiplied by any one of the three symmetric spin wavefunctions. The (−) combination describes a triplet state. The energies of these two combinations are given by summation of terms involving the Coulomb, exchange and overlap integrals just as we derived before for the $H_2{}^+$ molecular ion. The shapes of the curves look very much the same as in Fig. 24.5(b). ψ_{VB}^{-} describes a state that is purely repulsive, while ψ_{VB}^{+} describes a state with a well with a minimum at $R_e = 0.84$ Å and a dissociation energy of $D_e = 2.65$ eV. These values can be compared to the experimental values of $R_e = 0.741$ Å and a dissociation energy of $D_e = 4.75$ eV. The agreement is encouraging; we are on the right path, but far from chemically accurate. Chemically accurate is a loosely defined term, which means that a value is sufficiently accurate to provide quantitative predictions about chemistry. 'Chemically accurate' is often taken to mean that a theoretical value is <1 kcal mol^{-1} (<4 kJ mol^{-1}) away from the true value.

24.5.1 Directed practice

Write out the four total wavefunctions associated with ψ_{VB}^{+} and ψ_{VB}^{-} by multiplying each by the appropriate spin part.

What is wrong and how can we improve it? This simple valence band model is highly instructive. It gets the symmetry of the orbitals correct. The two molecular orbitals obtained have the same symmetry as the two discussed for $H_2{}^+$. The ground state is $1\sigma_g$ and the first excited state is $1\sigma_u^*$. The shapes of the potential energy curves are qualitatively correct and the energies are not far off. However, the electron configurations we have chosen are highly restrictive. We have chosen a form of the wavefunctions that is solely based on a covalent solution. We have only considered the combination of a 1s + 1s configuration to the exclusion of ionic contributions from $1s^2 + 1s^0$ – that is, the combination of $H^- + H^+$. This is not to say that the bond in H_2 has polar character. On the contrary, the H–H bond must by symmetry be completely nonpolar (at least on a time-averaged basis as described by the stationary states that are solutions of the time-independent Schrödinger equation). Rather, combination of wavefunctions associated with $H^- + H^+$ and $H^+ + H^-$ lead to nonpolar states derived from ionic wavefunctions. Making new linear combination wavefunctions from covalent and ionic electron configurations leads to what is called a *configuration interaction* (CI) molecular orbital treatment.

Simply adding in ionic contributions but still only using 1s basis functions in our basis set, an improved bonding ground state $1\sigma_g$ wavefunction is calculated. This improved state has a well with a minimum at $R_e = 0.880$ Å and a dissociation energy of $D_e = 3.23$ eV. This is an improvement but still some way off of the accepted values. The change is small because the ionic contribution to bonding is only a few percent of the whole. Next, we could add in more basis functions, the 2s and 2p orbitals, for instance. Then we could add yet more atomic orbitals and all possible covalent and ionic configurations associated with them. Eventually, we have to look at our Hamiltonian and see whether new terms are needed, similar to what we found was the case with the H atom and the improvements that resulted from the Dirac equation.

The drive to improve the theoretical description of the H_2 bond and its dissociation energy has a fascinating history. For a time, the theoretical value obtained from the calculations of Włodzimierz Kołos and Lutosław Wolniewicz was more accurately known than the experimental value, which forced Herzberg and others to re-evaluate experiments and improve their accuracy.[2] The best theoretical calculations completely discard the use of basis functions that resemble atomic orbitals. Instead, they use as many as 1800 optimized Gaussian basis functions. The expansion of the basis set and taking account of configuration interaction improve the accuracy sufficiently that the zeroth-order Hamiltonian from Eq. (24.55) needs be corrected. These corrections include adiabatic and nonadiabatic corrections to the Born–Oppenheimer potential as well as relativistic effects, and finally corrections related to quantum electrodynamics (QED) to arrive at $D_e = 36\,118.0695(10)\ \text{cm}^{-1} = 4.478\,069\,53 \pm 0.000\,000\,12$ eV.[3] This value is essentially in perfect agreement with the best experimental value.[4]

In summary then, what all goes in to making a chemical bond? Putting aside now any small corrections from QED and such, the major contributions to the chemical bond are to be found in the Hamiltonian and the total wavefunction. There are kinetic energy terms, a potential energy term determined by electrostatic interactions, and there is the spin part of the wavefunction. In a covalent bond formed by placing two electrons in a bonding molecular orbital spins must be paired – that is, one up and one down. Through electrostatic interactions, the most stable state minimizes the repulsions of like-charged particles and maximizes the attractions of oppositely charged species. The charge density distribution is given by the square of the wavefunction, and we have seen how constructive interference that places electron density between nuclei leads to lower energy electrostatic interactions than the removal of charge from the internuclear region. Then, in the mathematical treatment of the Hamiltonian and construction of the wavefunction we need to properly account for quantum mechanical effects such as exchange and correlation.

Finally, there is the contribution of the electron kinetic energy (nuclear kinetic energy is absent within the Born–Oppenheimer approximation and its effects represent one of those small nonadiabatic corrections that needs to be made to the Hamiltonian to get things absolutely right). The kinetic energy is related to the shape of the wavefunction. As stated above, more curvature and more nodes correlate with higher kinetic energy. We can discuss the balance between kinetic and potential energy by discussing the implications of the *molecular virial theorem*.

The total energy of the molecule as a function of the internuclear distance $\langle E(R)\rangle$ is given by the sum of the average total electron kinetic energy $\langle T(R)\rangle$ and the average total electrostatic potential energy $\langle V(R)\rangle$,

$$\langle E(R)\rangle = \langle T(R)\rangle + \langle V(R)\rangle. \tag{24.58}$$

The molecular virial theorem states that any such $\langle E(R)\rangle$ curve can be separated into kinetic plus potential energy contributions according to

$$\langle T(R)\rangle = -\langle E(R)\rangle - R\frac{\mathrm{d}\langle E(R)\rangle}{\mathrm{d}R} \tag{24.59}$$

and

$$\langle V(R)\rangle = 2\langle E(R)\rangle + R\frac{\mathrm{d}\langle E(R)\rangle}{\mathrm{d}R}. \tag{24.60}$$

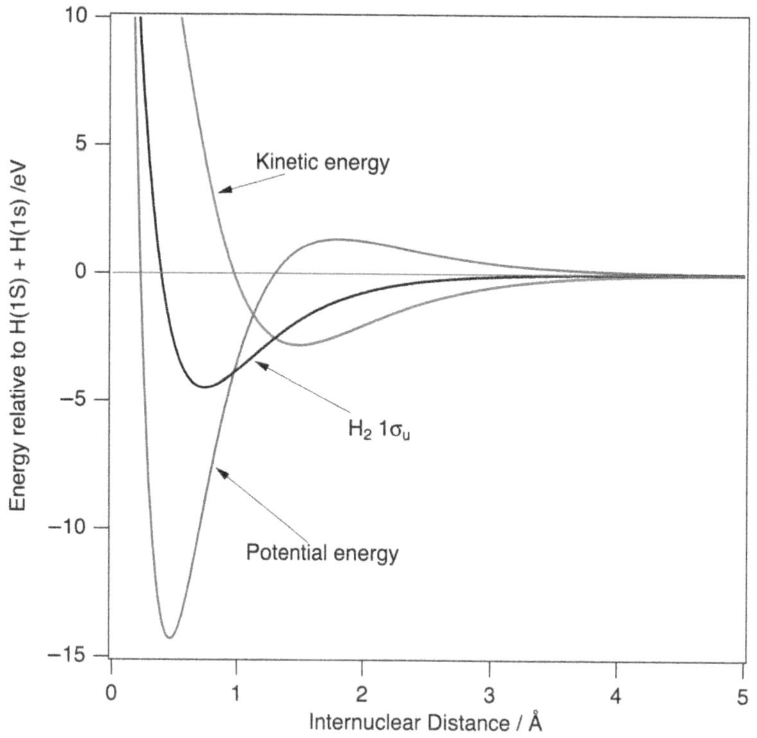

Figure 24.7 The sum of electron kinetic and electrostatic potential energy add to give the total energy of H_2 molecule in its ground electronic state. The energies are shifted such that zero corresponds to their values in the separated atoms limit (large R).

The results of the virial calculation for H_2 in its ground electronic state are shown in Fig. 24.7. The energy asymptote has been chosen to correspond to zero at large separation; in other words, all of the energy curves depicted converge to their values for the sum of two independent H atoms in the 1s ground state. This is done to emphasize how these values change under the influence of chemical bonding interactions. Both the kinetic and potential energies exhibit minima. However, these two minima are displaced from one another. Neither corresponds to the minimum in the total energy curve.

The kinetic energy drops at long range as the H 1s wavefunctions begin to overlap. This is the result of *electron delocalization*. As the wavefunctions begin to overlap, the electrons are able to extend their range over the entire molecular system, which lowers their kinetic energy. This is a general result of quantum confinement that we have observed before. Increasing the size of the box in which electrons are confined by a potential decreases their kinetic energy. However, as the nuclei move closer to one another the region of space over which the electrons are confined becomes smaller, the kinetic energy passes through a minimum at 1.49 Å and begins to rise until it eventually is greater than in the two isolated atoms. Remember that the kinetic energy is always positive, and that the kinetic energy relative to the sum of kinetic energy in two isolated atoms is plotted in Fig. 24.7.

The electrostatic energy produces a well that is much deeper than the bond dissociation energy with a minimum at a distance (0.46 Å) shorter than the equilibrium bond distance (0.74 Å). This minimum is not approached monotonically as the atoms approach each other. Instead, the potential energy first passes through a maximum at 1.80 Å. The rearrangements in the electron clouds are not sufficient at long distance to make the attractive forces outweigh the repulsive forces. At smaller separations, the electron density enhancement between the nuclei eventually causes the electron–nucleus attractions to become the dominant interaction. At even shorter distance the internuclear repulsion takes over and eventually turns the potential repulsive.

The effect of kinetic energy is not only to increase the bond length and decrease the binding energy of the H_2 molecule. The effect of kinetic energy also avoids the presence of an activation barrier for the formation of an H_2 molecule from two H atoms of opposite spin. This is because in the large separation region, the kinetic energy decreases faster than the rise in potential energy. As a consequence, the total energy drops monotonically at the atoms with opposite spin approach each other.

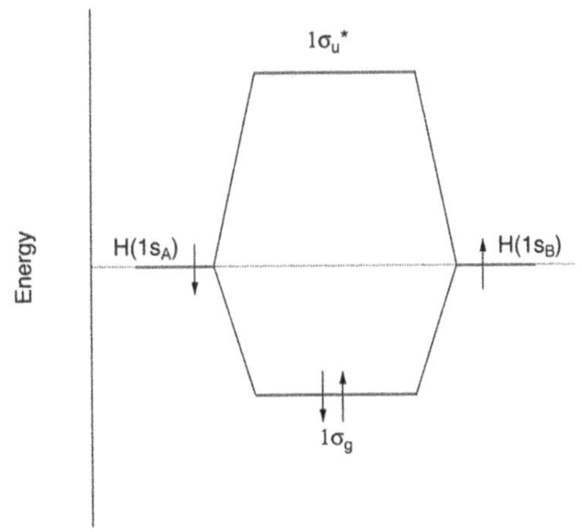

Figure 24.8 A molecular orbital diagram for the H_2 molecule. The two arrows indicate the occupation of the bonding orbital by two spin-paired electrons.

24.6 Homonuclear diatomic molecules

Having generated two molecular orbitals we can use these to interpret the bonding in H_2 and He_2. For these molecules we simply use the orbitals generated from the examples above and fill them according to the Aufbau process. For H_2 we begin with two electrons, one from each 1s orbital, and obtain $1\sigma_g^2$, as shown in Fig. 24.8. For He_2 we begin with four electrons from combining two $1s^2$ atoms. The result is that, after filling the bonding molecular orbital, the remaining two electrons must be placed in the antibonding molecular orbital. The configuration is $1\sigma_g^2\, 1\sigma_u^{*2}$.

We know empirically that H_2 forms a covalent bond and He_2 does not. The bond dissociation energy of H_2 is 4.478 eV and is a single bond. He_2 is the most fragile ground state bound species. Its binding energy of only 0.1 μeV and long bond distance of 5.2 nm are due to exceedingly weak van der Waals interactions rather than a covalent bond.[5] To explain this trend we introduce the concept of *bond order*. Bond order p_{AB} between the atoms A and B is calculated by adding up half the number of electrons in bonding orbitals minus half the number of electrons in antibonding orbitals,

$$p_{AB} = \frac{1}{2}(w - w^*) \tag{24.61}$$

where w is the occupancy of bonding orbitals and w^* is the occupancy of antibonding orbitals. For a given pair of atoms, we expect stronger and shorter bonds with higher fundamental vibrational wavenumber with increasing bond order.

For poly-electron systems (more than two electrons in the constituent atoms), the molecular orbital structure is obviously more complicated. Our first approximation is that we do not need to consider *core electrons* when it comes to bonding. Core electrons are the electrons in closed shells with the lowest values of the principal quantum number n. The core electrons are tightly localized about the nucleus they are associated with. They are little influenced by the chemical environment. Slight chemical shifts in their binding energies accompany changes in the oxidation state of the atom. Only the *valence electrons* need to be considered with reference to bonding.

In principle, all orbitals from the constituent atoms can contribute to each MO. There are simplifying rules that help us sort out which orbitals will make the greatest contributions to a given molecular orbital. These rules also apply to the interactions of the electronic states associated with orbital configurations.

- The closer two orbitals are in energy, the more likely they are to interact and contribute.
- Orbitals that are far apart in energy have little interaction and make small contributions.
- Only orbitals of an appropriate symmetry can interact (σ with σ, π with π but not mixed).
- The spatial part of the wavefunction will contain contributions from two or more AOs (assuming that we are using atomic orbitals as our basis set).

The interaction between charged particles is dominated by the electrostatic interaction. The greater the charge, the greater the interaction. Also, the closer two charged particles, the greater the interaction. This is also true for electrons in orbitals. The greater the nuclear charge, the greater the attraction of the electron (the lower the energy of the orbital). If there are other electrons in the way of the nucleus (in between and electron and the nucleus), the effective nuclear

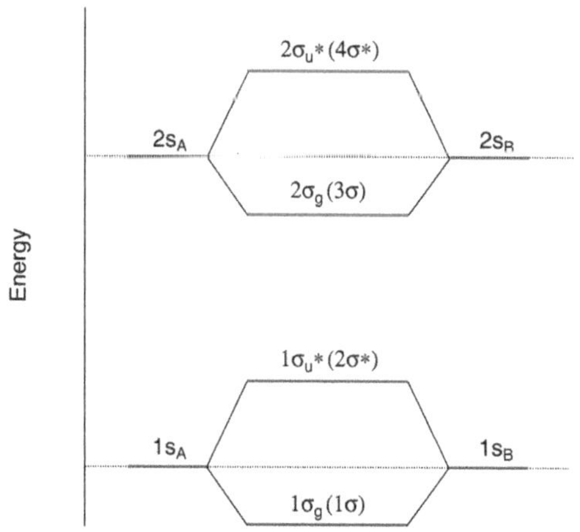

Figure 24.9 In a homonuclear diatomic molecule, each pair of s orbitals leads to a pair of σ bonding and antibonding molecular orbitals with progressively higher energy. In homonuclear diatomics, the number in the label is the same for a bonding-antibonding pair made from a pair of atomic orbitals. In heteronuclear diatomics the labeling is consecutive (as shown in parenthesis) and often the contributing atomic orbitals will not have the same value of n.

charge is less. This process is known as *screening* and we have discussed it at length in the Chapter 22. It explains the ordering of atomic orbitals for a given n (s orbitals are lower in energy than p, which are lower than d, etc.) and it also plays a role in determining the relative energies of molecular orbitals.

In a poly-electron molecule we have multiple filled atomic orbitals that lead to multiple molecular orbitals. Here, we are obviously thinking in a valence bond picture, which we will use as a conceptual crutch. Recall that when we were constructing electron configurations of atoms, we stated that we can think of all orbitals being present at all times, regardless of whether they are filled or not. The same is true in molecules. We have seen this above with H_2^+ and H_2, an unfilled $1\sigma_u^*$ orbital was created. This antibonding orbital turns out to be essential for understanding the dissociative adsorption of H_2 on a metal surface or the interaction of two H atoms with parallel spins as they approach each other. The filling of normally unoccupied orbitals corresponds to the creation of electronically excited states. Therefore, they are also of importance to understand molecular spectroscopy.

The formation of two σ molecular orbitals from two s atomic orbitals as we observed for H_2^+ and H_2 is the same for all values of n. Progressively higher energy results from progressively higher values of n. At least at low to moderate values of n, the energy spacing between successive states with increasing n is such that a pair of σ orbitals formed from a pair s orbitals are well separated from the pair formed by the $n + 1$ pair of atomic s orbitals. This is illustrated in Fig. 24.9.

The combination of p orbitals is more complex. In the atom, these three levels are degenerate. However, there is a fundamental difference between the p_z orbital, which points along the internuclear axis, and the p_x and p_y orbitals, which are perpendicular to the internuclear axis. The difference can be seen in the symmetry that results from their combinations. The interaction of two p_z orbitals with one another leads to two molecular orbitals, one bonding and the other antibonding, with σ symmetry. They are cylindrically symmetric with no node along the internuclear axis. To emphasize their parentage, such orbitals are sometimes labeled as pσ orbitals. The p_x and p_y orbitals, on the other hand, interact with their opposite number on another nucleus with π symmetry to produce pπ orbitals. Each produces a bonding and an antibonding combination, and because their interactions are symmetry equivalent, the resulting bonding orbitals are degenerate. The two resultant antibonding orbitals are also degenerate but lie at higher energy compared to the bonding combination.

The combination of two energetically well-isolated sets of p orbitals is shown in Fig. 24.10. The bonding π_u combination is the first (lowest energy) π orbital formed and receives the designation $1\pi_u$. The higher energy antibonding combination has gerade symmetry and is denoted the $1\pi_g^*$. Note the difference in the correlation of u and g with bonding and antibonding character compared to σ orbitals: π bonding orbitals all have u inversion symmetry; σ bonding orbitals all have g inversion symmetry. The lowest energy σ molecular orbital formed from the p atomic orbitals has been labeled the $3\sigma_g$ because there are two sets of σ orbitals formed from the 1s and 2s orbitals at lower energy. This brings us to a very important point: How do the orbitals formed in Fig. 24.10 (and yet higher energy molecular orbitals) align

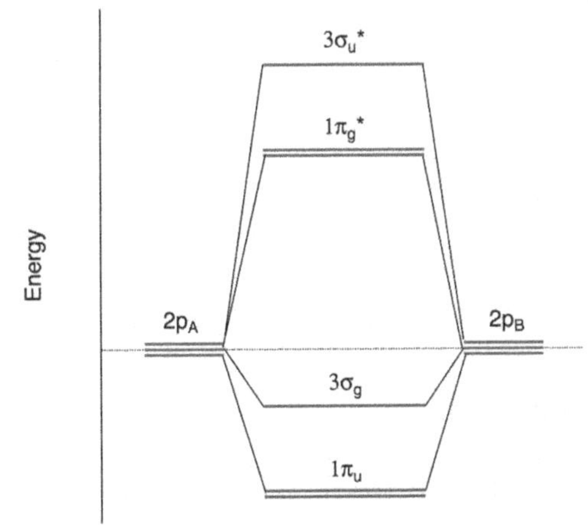

Figure 24.10 The combination of two isolated sets of atomic p orbitals to form molecular orbitals in a homonuclear diatomic.

energetically with the orbitals shown in Fig. 24.9 (and yet higher energy sσ orbitals)? The answer is that it depends on Z, n and screening. However, just as for atoms, we can use photoelectron spectroscopy to determine the energy of molecular orbitals. Again, our first-order approximation is Koopmans' theorem: The binding energy of an electron is the negative of the orbital energy from which it came.

Recall that screening changes the relative energetic positioning of s, p, d, ... atomic orbitals. As n increases from 3 to 4, this effect is strong enough that the 4s orbitals drop below the energy of the 3d orbitals. A similar mechanism is at work with molecules. All orbitals shift to lower energy as the nuclear charge Z increases. However, less well-screened orbitals shift more rapidly than more well-screened orbitals. The first observation of this effect is evident with the behavior of the $1\pi_u$ and the $3\sigma_g$ orbitals. σ molecular orbitals are less effectively screened than π molecular orbitals, just as s atomic orbitals are less well-screened than p atomic orbitals. This means that as the nuclear charge Z increases, σ orbitals will shift to lower energy more rapidly than π orbitals. At the beginning of the second row of the periodic chart, the $1\pi_u$ is lower in energy than the $3\sigma_g$ orbital. However, the $3\sigma_g$ orbital decreases in energy more rapidly with increasing Z than does the $1\pi_u$ orbital. The switch in ordering occurs between N_2 and O_2. The results are shown in Table 24.3, along with the correlations between electron configuration, bond order, bond length and bond energy.

We have already encountered configuration interaction both in the construction of atomic orbitals of poly-electron atoms and in the covalent/ionic contributions to the bonding in H_2. It should, therefore, come as no surprise that the parentage of molecular orbitals can be affected by configuration interaction. The simple 'pure' additions of atomic orbitals shown in Figs 24.9 and 24.10 are ideal cases not always observed, and particularly are not observed at higher values of n when the spacing between orbitals of successive values of n becomes small compared to the spacings between levels

Table 24.3 Ground-state electron configurations, equilibrium bond length R_e, dissociation energy and bond order p for selected homonuclear diatomics. Source for bond lengths and dissociation enthalpies 96th edition of the *CRC Handbook* except for H_2 and He_2, which are from references found in the text. R_e and $\tilde{\nu}_0$ for Ne_2 taken from Herzberg and Huber.

Diatomic	Configuration	p	R_e/Å	D_e/eV	$\tilde{\nu}_0$/cm^{-1}
H_2	$1\sigma_g^2$	1	0.74144	4.478 069 53	4401.21
He_2	$1\sigma_g^2 1\sigma_u^{*2}$	0	52	1×10^{-7}	
Li_2	$[He_2]2\sigma_g^2$	1	2.6733	1.050	351.41
Be_2	$[He_2]2\sigma_g^2 2\sigma_u^{*2}$	0		0.57	
B_2	$[He_2]2\sigma_g^2 2\sigma_u^{*2} 1\pi_u^2$	1	1.590	2.97	1051.3
C_2	$[He_2]2\sigma_g^2 2\sigma_u^{*2} 1\pi_u^4$	2	1.24244	6.2322	1855.01
N_2	$[He_2]2\sigma_g^2 2\sigma_u^{*2} 1\pi_u^4 3\sigma_g^2$	3	1.09769	9.7544	2358.56
O_2	$[He_2]2\sigma_g^2 2\sigma_u^{*2} 3\sigma_g^2 1\pi_u^4 1\pi_g^{*2}$	2	1.20752	5.12761	1580.19
F_2	$[He_2]2\sigma_g^2 2\sigma_u^{*2} 3\sigma_g^2 1\pi_u^4 1\pi_g^{*4}$	1	1.41264	1.60596	916.93
Ne_2	$[He_2]2\sigma_g^2 2\sigma_u^{*2} 3\sigma_g^2 1\pi_u^4 1\pi_g^{*4} 3\sigma_u^{*2}$	0	3.1	0.003644	13.7

of different orbital angular momentum quantum number l. The bulleted list that appears above will be used to help us determine which atomic orbitals are most likely to make contributions. This is particularly important in heteronuclear molecules because the constituent atomic orbitals will no longer be degenerate or even of the same value of n.

The electron configurations in Table 24.3 are quite useful for making predictions about related species concerning the bond order, bond length, dissociation energy and vibrational wavenumbers. A greater difference between the number of electrons in bonding orbitals compared to the number in antibonding orbitals leads to the following correlations: (i) greater bond order; (ii) shorter equilibrium bond distance; (iii) larger bond dissociation energy; and (iv) higher vibrational wavenumber.

Excited states are made by moving an electron (or more than one) from one molecular orbital to another. Multiple electronic states can be associated with one electronic configuration. This is the result of angular momentum coupling, as we will discuss in the next chapter. Moving an electron from one bonding orbital to another does not change the bond order. Such an excitation will have subtle effects on the bond length, dissociation energy and vibrational wavenumber. Excited states generally have higher polarizability than the ground state since their formation involves moving an electron to a larger radius orbital. Being further from the nuclei, such an electron is also less tightly bound, so we would expect the chemical bond to be weakened somewhat as well. Moving an electron from a bonding to an antibonding orbital will decrease the bond order with concomitant reduction in the dissociation energy and vibrational wavenumber and an increase in the bond length. Excitation of an antibonding electron to a bonding orbital will have just the opposite effect. The electron configuration also determines whether a species is a radical containing unpaired electrons. For example, the O_2 molecule in its ground state has two electrons occupying two different degenerate $1\pi_g^*$ orbitals. It is, therefore, a diradical with two unpaired electrons.

Anions and cations are formed by adding or subtracting electrons from the ground-state electron configuration. Thus, Be_2 is very weakly bound because of its $2\sigma_g^2 2\sigma_u^{*2}$ configuration and zero bond order. But $Be_2{}^-$ with ground-state configuration $2\sigma_g^2 2\sigma_u^{*2} 1\pi_u$ and $Be_2{}^+$ with a $2\sigma_g^2 2\sigma_u^*$ configuration both have a bond order of 0.5. This means that they are more strongly bound than the Be_2 molecule.

It should be noted that some orbitals that are nominally labeled as bonding or antibonding actually have little effect on the bond length or dissociation energy. This means that the orbital is more accurately described as nonbonding. Electrons lone pairs form nonbonding molecular orbitals.

24.6.1 Directed practice

Discuss whether all $2\sigma_g$ orbital have the same energy.

24.7 Heteronuclear diatomic molecules

When two atoms of the same element form a bond, electrons are shared equally to create a nonpolar covalent bond. The atomic orbitals that join together to form molecular orbitals initially have the same energy and contribute equally to both the bonding and antibonding molecular orbitals that result, as shown in Figs 24.9 and 24.10.

This is not the case for a diatomic molecule formed from two dissimilar elements. When two atoms of different elements form a bond, they share the electrons approximately according to their electronegativities. What this means in terms of molecular orbitals is that the orbitals formed in heteronuclear molecules are not symmetric (as they were for homonuclear diatomics). The molecular orbitals are not composed of 50/50 contributions from each atom and are not evenly distributed in space between the atoms.

An extremely polar covalent bond is made between H atoms, with an electronegativity of $\chi = 2.20$, and F with $\chi = 3.98$. The large electronegativity difference translates into much more of the charge density associated with bond formation residing near the F atom, which is partially negative, than near the H atom, which is partially positive. This is also reflected in the form of the molecular orbitals, as depicted in Fig. 24.11. The bond between H and F (again with our first-approximation being in a valence bond model) is made by combining the H 1s orbital with the F $2p_z$ orbital. If the bond is polarized toward the F atom, then the occupied bonding orbital responsible for the bond must have character that resembles the F $2p_z$ much more so than the H 1s. This also implies that the molecular orbital shifts little in energy from the energy of its F $2p_z$ parent. This is confirmed in the energy diagram found in Fig. 24.11. Note that the molecular orbitals in HF, as for all heteronuclear diatomics, lack inversion symmetry so there is no subscript u or g in the molecular

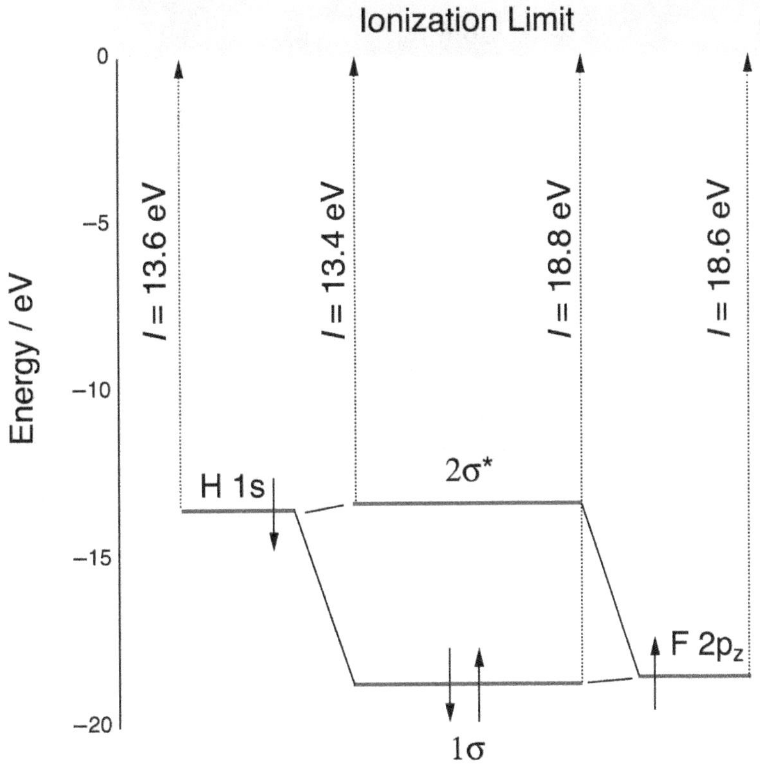

Figure 24.11 A molecular orbital diagram of the polar diatomic molecule HF. Ionization energies from Atkins, P. and de Paula, J. (2006) *Physical Chemistry*, 8th edition. W.H. Freeman and Company, New York.

orbital designation. Note also that the molecular orbitals of heteronuclear diatomics are numbered consecutively within a symmetry class.

What is the contribution of each atomic orbital to the HF bond? The way to determine this is to perform a variational calculation. We construct an LCAO-MO such as

$$\psi = c_A A + c_B B, \tag{24.62}$$

where A is the H 1s wavefunction and B is the F $2p_z$ wavefunction. Recall that the coefficients represent how large of a contribution the orbital A or B makes to the overall wavefunction. In a polar bond, the coefficients are different.

$$|c_A|^2 = |c_B|^2 \tag{24.63}$$

In a nonpolar bond, the coefficients are different and the greater the difference (allowed range is 0–1), the greater the polarity. For the simple basis set chosen here, the coefficients would be $c_A = 0.19$ and $c_B = 0.98$.

The CO molecule is an example of a heteronuclear diatomic. It demonstrates that the simple valence bond picture is good as a rough sketch of molecular bonding, but may be quite misleading in detail. The only complete description of the bonding in a molecule is the one made with knowledge of the shape of the wavefunctions. The 1s electrons are core electrons tightly held and localized about either the C or the O atom. They do not participate significantly in bonding. The O atom has a higher electronegativity ($\chi = 3.44$) than C ($\chi = 2.55$). Therefore, we *expect* the molecule to be polar with more electron density on the O atom. The 2s electrons form bonding and antibonding combinations that are fully occupied. The 2p orbitals combine to form the expected bonding and antibonding pσ orbitals as well as two degenerate pπ and two degenerate pπ^* orbitals. There are six 2p electrons (four from O and two from C) to be apportioned into these orbitals, which allows for the complete filling of the three bonding molecular orbitals. This results in a triple bond. Now let's see where our expectations go wrong.

CO has a very large *dynamic* dipole moment, which makes it a strong absorber of IR radiation, but a surprisingly small *static* dipole moment. Indeed, the dipole points in the 'wrong' direction compared to the expectation. There is slightly more electron density on the C atom because of its molecular orbital structure, contrary to what electronegativities would

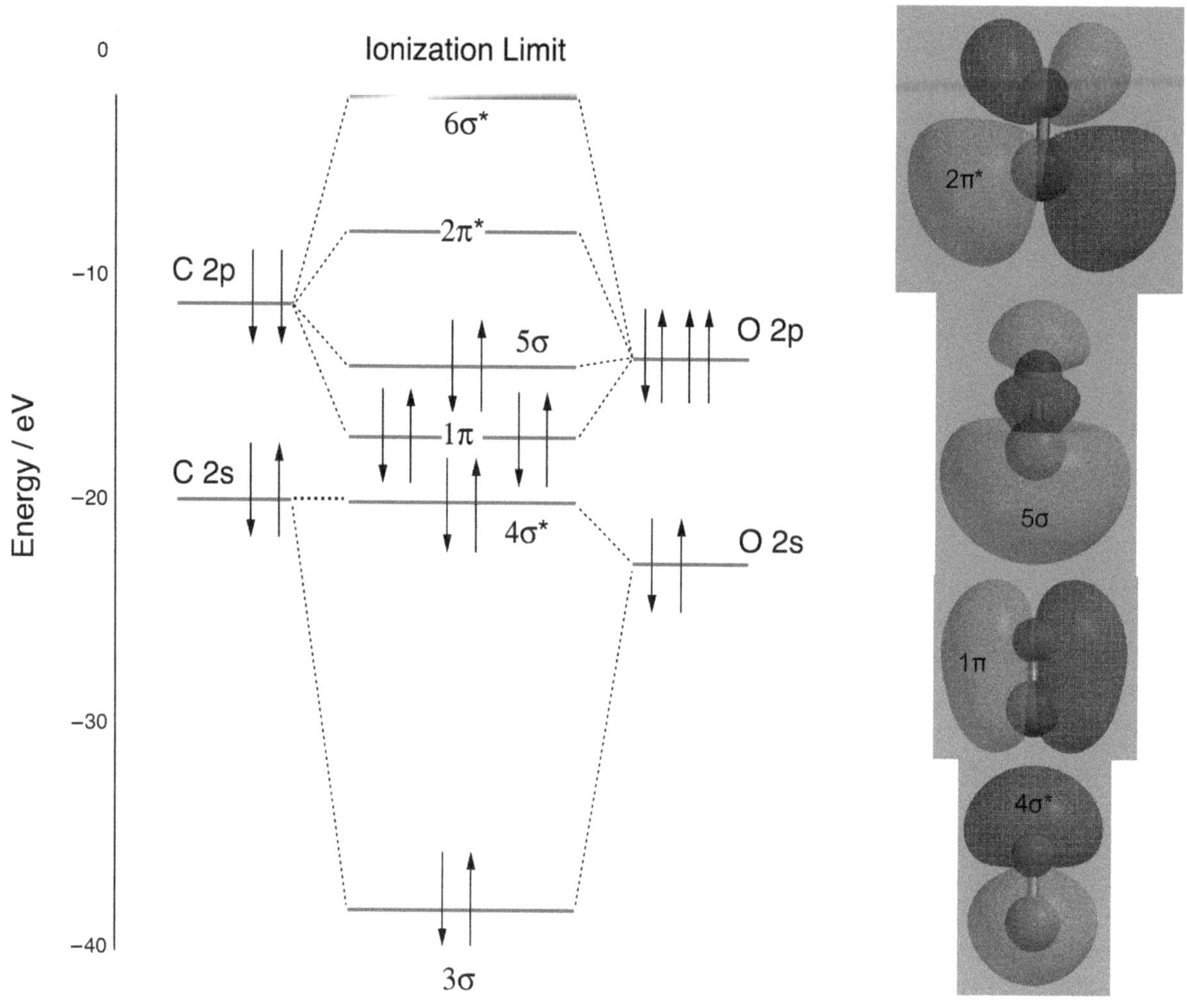

Figure 24.12 The molecular orbital diagram for CO. The orbital energies are taken (or estimated) from Winn, J.S. (1995) *Physical Chemistry*. HarperCollins College Publishers, New York, the NIST XPS database, and the *CRC Handbook*. Molecular orbital shapes calculated using the Møller–Plesset scheme with a 6-31G* basis set within the Spartan quantum chemical package.

predict. With reference to Fig. 24.12, this is because the 5σ orbital is more heavily localized about the C atom than the O atom. This cannot be predicted on the basis of a valence bond picture. It is a result that is consistent with experiment and *ab initio* calculations of the wavefunctions. The molecule is considered to have a triple bond. Two of these bonds result from the fully filled 1π orbitals. But we have just stated that the 5σ looks much like a lone pair on the C side of the molecule, which makes it nonbonding. Similarly, the 3σ looks very much like a lone pair on the O side of the molecule. This leaves the $4\sigma^*$ orbital, which we expected to be antibonding but which must be lending some of the bonding character in this specific example, perhaps, as can be seen in Fig. 24.12, because the nodal plane perpendicular to the internuclear axis (a hallmark of the antibonding designation) is not in the center of the C–O bond but displaced toward the O atom.

The ideas used here to explain polar covalent bonding can be extended to ionic bonding in molecules as well. After all, the case of HF discussed above is quite close to the limit of ionic bonding. For ionic systems, configuration interaction is usually required to obtain an accurate description of the bonding. The configurations to be considered may include not only covalent contributions but also multiple ionic configurations.

24.7.1 Directed practice

The 5σ orbital of CO contains nodes along the internuclear axis. How can it still be a bonding orbital?

24.8 The variational principle in molecular orbital calculations

The *variational principle* has been mentioned in several places in this book. However, it is of sufficient importance that we should take a closer look at it. We state it here again

> *For a given Hamiltonian, the lowest energy is the energy given by the correct wavefunction. All other wavefunctions result in a higher energy.*

A proof of the variational principle was given in Chapter 20. The variational principle allows us to answer the following fundamental questions: What wavefunctions describe the molecular orbitals responsible for molecular bonding? How do we know which orbitals contribute? And what are the coefficients of the components in the linear combinations that define the molecular orbitals? Furthermore, when do we know we have the correct answer? In principle, we already know the answer. If we have the correct Hamiltonian then once we find the correct wavefunctions we will get the right energies. The variation principle is a powerful tool in that it gives us a way to get the right answer (or at least our best approximation to it) and know if we are approaching it in the right way.

It is a difficult task to find the correct wavefunction. What the variation principle does is give us an orderly way to go about it. The method goes something like the following:

- Measure one (or more) easily accessible energy values experimentally.
- Write out a complete (or complete as feasible) Hamiltonian for the system.
- Choose a wavefunction composed of any arbitrary mathematical function (preferably one that looks something like the orbital in question and one that it easy to handle mathematically). This is the trial wavefunction. The set of functions (basis functions) that we use to mix by way of linear combination into the trial wavefunction is the basis set.
- To find the energy associated with this combination of Hamiltonian and wavefunction use the expectation value equation,

$$E = \frac{\langle\psi|\hat{H}|\psi\rangle}{\langle\psi|\psi\rangle}. \tag{24.64}$$

- Make a change to the wavefunction by adjusting the coefficients of the basis functions.
- Recalculate the energy.
- Compare the second energy to the first energy. If the second is better – that is, closer to the correct answer – discard the first wavefunction.
- Repeat the procedure until the minimum energy has been found. The minimum energy corresponds to the best answer for that Hamiltonian and that basis set.
- The coefficients that define the linear combination that gives the minimum energy define the best possible wavefunction for the given Hamiltonian and basis set.

When the correct answer is obtained (within your tolerance for error), you have the correct wavefunction. You can now use this wavefunction to calculate other properties of the system with confidence.

The size of the mathematical problem that we have to solve will depend on the size of the basis set we chose. The bigger the basis set, the greater the chance that we will eventually get the right wavefunction, the better the answer will be. However, the more functions, the harder is the mathematical equation that we have to solve.

Imagine for the moment that we make the simple approximation that our system is described by

$$\psi = aA + bB. \tag{24.65}$$

satisfying

$$a^2 + b^2 = 1. \tag{24.66}$$

Thus, a and b are related by

$$a = \sqrt{1 - b^2}. \tag{24.67}$$

In principle, we could just calculate E as a function of b and find the minimum by trial and error. This is illustrated in Fig. 24.13. The variational principle says that the lowest energy is the best. So, we calculate the energy for a given set of a and b and then repeat the calculation for a different set of a and b. If the energy is better (i.e., lower) we keep the second set and throw out the first. Now, repeat with a third set and repeat the comparison until the minimum is found.

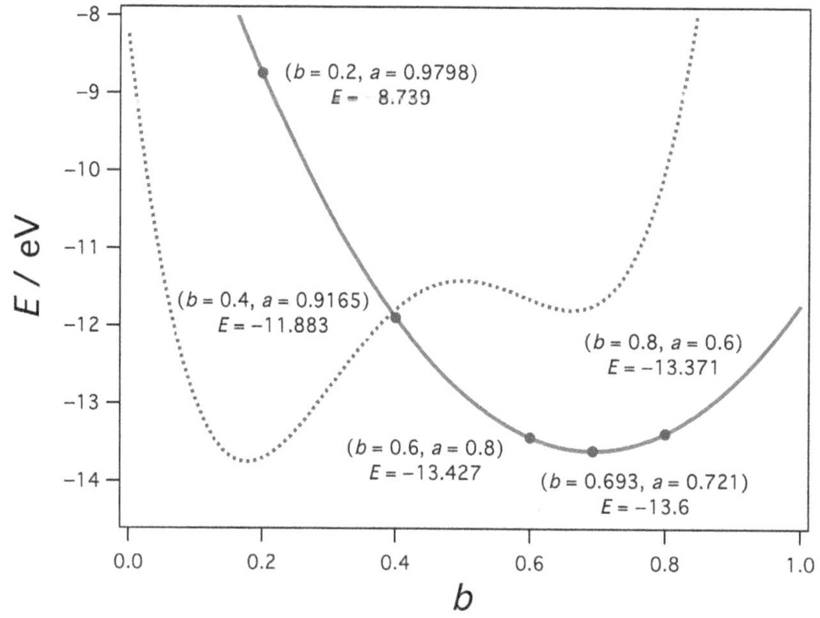

Figure 24.13 Arbitrary second-order and third-order polynomials are drawn to illustrate the concept of searching for global minima while avoiding local minima.

This type of search routine works very well for a second-order polynomial. For higher-order polynomials, we have to be aware of the problem illustrated by the dotted curve in Fig. 24.13. We need to find a global minimum not a local minimum. In other words, we have to make sure that the coefficients we determine do not give us a solution that is 'stuck' in the higher energy well on the right rather than the lower energy global minimum on the left. This is a nontrivial problem.

In Section 24.4, we have already encountered a more systematic method to determining the coefficients. This method is to use the eigenvalue equation in Eq. (24.64) to determine the form of the energy as a function of the mixing coefficients a and b. Writing this in its most general form, we start with N basis functions ϕ_i and form the linear combinations ψ_k with the arbitrary coefficients c_{ik}

$$\psi_k = \sum_i^N c_{ik}\,\phi_i. \tag{24.68}$$

Then we determine the partial derivatives of energy with respect to these coefficients and set them equal to zero,

$$\frac{\partial E}{\partial c_{ik}} = 0. \tag{24.69}$$

This must be done for all the coefficients c_{ik} that define the linear combinations of the N basis function that we have chosen. This leads to a set of N equation that must be satisfied simultaneously for Eq. (24.69) to be true,

$$\sum_i^N c_{ik}(H_{ki} - ES_{ki}) = 0. \tag{24.70}$$

Again, for each i and k. The H_{ki} and S_{ki} are the Hamiltonian and overlap integrals defined in Section 24.4. The H_{ki} with $k \neq i$ are also sometimes called *resonance integrals*, while when $k = i$ they are also called *Coulomb integrals*.

This set of N equations with N unknowns (the coefficients c_{ik}) is solved using techniques from linear algebra. This system has a nontrivial answer (the trivial answer is that $c_{ik} = 0$ for all ik) if, and only if, the *secular equation* is equal to zero. The secular equation is the determinant formed from the coefficients $(H_{ki} - ES_{ki})$ according to

$$\begin{vmatrix} H_{11} - ES_{11} & H_{12} - ES_{12} & \cdots & H_{1N} - ES_{1N} \\ H_{21} - ES_{21} & H_{22} - ES_{22} & \cdots & H_{2N} - ES_{2N} \\ \vdots & \vdots & \ddots & \vdots \\ H_{N1} - ES_{N1} & H_{N2} - ES_{N2} & \cdots & H_{NN} - ES_{NN} \end{vmatrix} = 0 \tag{24.71}$$

In general, the secular equation has N roots – that is, N values of the energy E_j. Each value of the energy E_j leads to a different set of optimum coefficients c_{ij} The coefficients are found by substituting the energies E_j, found by solving the secular determinant, back into Eq. (24.70). These coefficients then define the optimum wavefunctions ψ_j within the chosen basis set,

$$\psi_j = \sum_{i=1}^{N} c_{ij}\phi_i. \tag{24.72}$$

In a one-configuration system, the lowest energy corresponds to the ground state and the higher energy solutions correspond to excited states. The variational principle also holds for each excited state: the calculated energy of a trial wavefunction is always greater than or equal to the exact value for the chosen Hamiltonian.

Summarizing, the method of determining the best molecular wavefunctions for our system are the following.

1 Select a set of N basis functions.
2 Determine the N^2 values of both H_{ki} and S_{ki}.
3 Generate the secular determinant and determine the N roots that correspond to the energies E_j.
4 For each value of E_j, solve the set of linear equations defined by Eq. (24.70), and determine the coefficients c_{ij} that define the corresponding molecular orbital.

The problem involving two basis functions can be solved analytically by hand. Imagine if we were to try this for CO_2. To account for all of the electrons we would have to have 6 + 8 + 8 = 22 spin orbitals (for the 22 electrons in the system). If we treat the 1s electrons as core electrons and assume that they are not involved in bonding, then we obtain a minimum of 16 spin orbitals that have to go into our basis set. We then need to solve a 16 × 16 matrix, and that does not yet even account for the influence of excited states or configuration interaction. Methods of approximation (e.g., neglecting overlap, that is $S = 0$), some practical rules (e.g., orbitals far from each other in energy do not interact) and using optimized orbitals are important to simplify the problem. Even with these, modern computations sometimes require the diagonalizing of matrices that have dimensions of one thousand or more.

24.9 Polyatomic molecules: The Hückel approximation

Computational power continues to grow. Because of this ever larger basis sets and numbers of configurations can be handled in full-blown quantum chemical calculations in a reasonable amount of time. Nonetheless, the highest accuracy calculations can still only be performed on relatively small molecules. To obtain a better understanding of how bonding works in larger molecules it is instructive to investigate approximate models. One such approximate model is the *Hückel approximation*, which was proposed by Erich Hückel in 1931[6] in an attempt initially to understand the bonding in benzene and, subsequently, other extended conjugated systems.

Molecular orbitals extend over the entire molecule. This is particularly true for planar conjugated systems of organic molecules that we treat here. Our goal is to describe the π-bonding network in planar hydrocarbons, such as that shown in Fig. 24.14. We write our molecular orbitals ϕ_i in terms of basis functions χ_r with coefficients c_{ri}

$$\phi_i = \sum_{r} c_{ri}\chi_r. \tag{24.73}$$

Having established our basis set, we now need to accomplish steps 1–4 above. In Hückel theory, this is accomplished with the aid of these five conventions:

a The basis set is formed exclusively from parallel C 2p orbitals. One orbital is added for each C atom in the system.

Figure 24.14 The simplest π-system is that of ethene. When the 2p orbitals are properly aligned, as they are on the left, they overlap and their energy is equal to E_π. A rotation by 90° shown on the right eliminates the π-bonding interaction; the energy is then just the sum of two orbital energies $2E_p$.

b The overlap integrals have values defined by a delta function,

$$S_{ir} = \delta_{ir} \tag{24.74}$$

In other words, the overlap of any C 2p orbital with itself is equal to 1 because it is normalized but we neglect any overlap between any two different p orbitals.

c The matrix elements H_{rr} (the coulomb integrals) are defined by

$$H_{rr} = \int \chi_r^* \hat{H} \chi_r \, d\tau = \alpha_r \tag{24.75}$$

The value of α_r is a negative number. With reference to Fig. 24.14, the value of the Coulomb integral is given by the 2p orbital energy

$$\alpha_r = E_p. \tag{24.76}$$

d The matrix elements H_{rs} (the resonance integrals) are defined by

$$H_{rs} = \int \chi_r^* \hat{H} \chi_s \, d\tau = \beta_{rs} \tag{24.77}$$

As shown in Fig. 24.14, the 90° rotation is assumed to completely remove the π-bonding interaction between neighboring 2p orbitals. The energy cost of such a rotation is given by $\Delta E = 2E_p - E_\pi$. The negative of ΔE is the stabilization energy for the π-bond. It is distributed symmetrically over the two p orbitals. This defines the value of the resonance integral, that is,

$$\beta_{rs} = -\frac{1}{2}\Delta E. \tag{24.78}$$

β_{rs} is a negative number.

e Matrix elements H_{rs} only have a nonvanishing value for nearest neighbors. For more distant neighbors $H_{rs} = 0$.

Note that the Hückel model does not start with a Hamiltonian. Instead of being an *ab initio* theory – a theory that starts from first principles with the only input being the number of atoms and electrons, their charges and masses – it is a semi-empirical theory. It is a theory that uses a portion of first principles theory and a portion of experimentally determined numbers to approximate the values that the theory cannot calculate on its own. In this case, the empirical parts are the value of the matrix elements H_{rr}, H_{rs} and S_{ir}. Instead of being evaluated from a Hamiltonian and wavefunctions, these matrix elements are determined by experimental values or assumptions.

24.9.1 Applying Hückel molecular orbital theory to allyl C_3H_3

Allyl, shown in Fig. 24.15, has a π-system composed of three 2p orbitals, one on each C atom. The wavefunctions of the type defined by Eq. (24.73) can all be formulated as matrix equations. The row matrix (vector) of linear combination wavefunctions can be represented by the matrix multiplication of a basis set vector of dimension N with an $N \times N$ matrix of coefficients,

$$(\phi_1 \quad \phi_2 \quad \phi_3) = (\chi_1 \quad \chi_2 \quad \chi_3)\begin{pmatrix} c_{11} & c_{12} & c_{13} \\ c_{21} & c_{22} & c_{23} \\ c_{31} & c_{32} & c_{33} \end{pmatrix}. \tag{24.79}$$

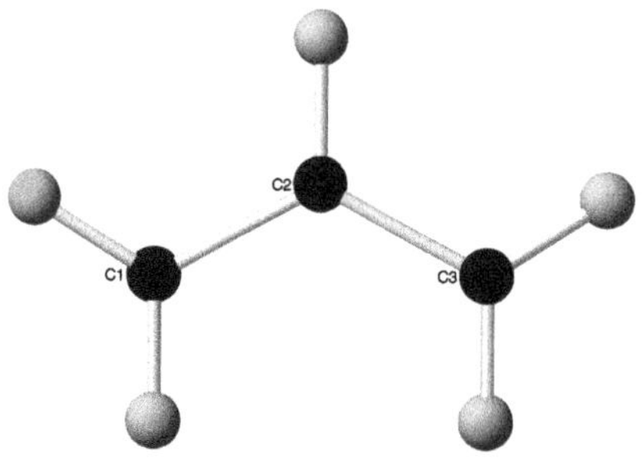

Figure 24.15 The allyl radical C_3H_3 is planar. The basis set for the Hückel MO calculation is composed of three 2p orbitals, one from each C atom labeled C1, C2, and C3 as shown.

In other words, each wavefunction can be represented by a column vector that contains the coefficients for each of the N basis functions

$$\mathbf{c}_r = \begin{pmatrix} c_{1r} \\ c_{2r} \\ c_{3r} \end{pmatrix} \tag{24.80}$$

and the complete set of wavefunctions is represented by $N \times N$ matrix constructed by putting all of the columns together

$$\mathbf{C} = (\mathbf{c}_1 \quad \mathbf{c}_2 \quad \mathbf{c}_3) = \begin{pmatrix} c_{11} & c_{12} & c_{13} \\ c_{21} & c_{22} & c_{23} \\ c_{31} & c_{32} & c_{33} \end{pmatrix}. \tag{24.81}$$

Similarly, we can construct a matrix of all of the terms (matrix elements) involving the Hamiltonian operator (**H**), a matrix of all of the overlap terms (**S**) and a matrix of the energies corresponding to each wavefunction (**E**). Within the assumption of the Hückel model these become

$$\mathbf{H} = \begin{pmatrix} H_{11} & H_{12} & H_{13} \\ H_{21} & H_{22} & H_{23} \\ H_{31} & H_{32} & H_{33} \end{pmatrix} = \begin{pmatrix} \alpha_r & \beta & 0 \\ \beta & \alpha_r & \beta \\ 0 & \beta & \alpha_r \end{pmatrix} \tag{24.82}$$

$$\mathbf{S} = \begin{pmatrix} S_{11} & S_{12} & S_{13} \\ S_{21} & S_{22} & S_{23} \\ S_{31} & S_{32} & S_{33} \end{pmatrix} = \begin{pmatrix} 1 & 0 & 0 \\ 0 & 1 & 0 \\ 0 & 0 & 1 \end{pmatrix} \tag{24.83}$$

$$\mathbf{E} = \begin{pmatrix} E_1 & 0 & 0 \\ 0 & E_2 & 0 \\ 0 & 0 & E_3 \end{pmatrix}. \tag{24.84}$$

The secular equations correspond to the matrix problem

$$\mathbf{HC} = \mathbf{SCE}. \tag{24.85}$$

In the Hückel approximation $H_{AA} = H_{BB} = \alpha$, $H_{AB} = H_{BA} = \beta$ for nearest neighbors only. Since we neglect overlap, $\mathbf{S} = 1$. Thus,

$$\mathbf{HC} = \mathbf{CE}, \tag{24.86}$$

which has as a solution

$$\mathbf{C}^{-1}\mathbf{HC} = \mathbf{E}. \tag{24.87}$$

To find the energies, we need to find the inverse matrix $\mathbf{C}^{-1}$. We find the eigenvalues of **H** by finding a matrix transformation that diagonalizes **H**. In general this is difficult. With the approximation of the Hückel model for allyl it means that we have to solve the secular determinant

$$\begin{vmatrix} \alpha - E & \beta & 0 \\ \beta & \alpha - E & \beta \\ 0 & \beta & \alpha - E \end{vmatrix} = 0. \tag{24.88}$$

We simplify by introducing

$$x = (E - \alpha)/\beta, \tag{24.89}$$

which transforms the secular determinant into

$$\begin{vmatrix} x & 1 & 0 \\ 1 & x & 1 \\ 0 & 1 & x \end{vmatrix} = 0. \tag{24.90}$$

Expanding the determinant in Eq. (24.90), we obtain the third-order equation

$$x^3 - 2x = 0 \tag{24.91}$$

which has solutions $x = -\sqrt{2}, 0, \sqrt{2}$. This corresponds to energies

$$E = \alpha + \sqrt{2}\beta,\ \alpha,\ \alpha - \sqrt{2}\beta. \tag{24.92}$$

Since both α and β are negative, the ground-state energy is $E_1 = \alpha + \sqrt{2}\beta$. The other two energies correspond to two excited states.

The wavefunctions are determined by solving the matrix equation Eq. (24.86) or equivalently, the three equations defined by Eq. (24.70) An infinite number of combinations satisfies these conditions. However, there is only one unique set of coefficients for each energy that also ensures that the wavefunctions are normalized. Combining these calculations, we obtain the coefficient matrix,

$$\mathbf{C} = \begin{pmatrix} \frac{1}{2} & \frac{\sqrt{2}}{2} & \frac{1}{2} \\ \frac{\sqrt{2}}{2} & 0 & -\frac{\sqrt{2}}{2} \\ \frac{1}{2} & -\frac{\sqrt{2}}{2} & \frac{1}{2} \end{pmatrix}. \tag{24.93}$$

Or explicitly in terms of the basis functions, the ground-state wavefunction is

$$\phi_1 = \tfrac{1}{2}\chi_1 + \tfrac{\sqrt{2}}{2}\chi_2 + \tfrac{1}{2}\chi_3, \tag{24.94}$$

where χ_1 is a 2p orbital on atom 1, and so forth.

We started with three basis orbitals. Three atomic orbitals in, three molecular orbitals out. Call it conservation of orbitals (or basis functions). Originally, the three orbitals had the same energy (α). They correspond to a set of degenerate orbitals. Their combined energies for occupation by three electrons within a single-electron approximation is 3α. By single-electron, we mean not that the electrons do not interact (their interaction is somewhat accounted for in the semi-empirical constant β). Rather, we mean that the total energy of the system is equal to the sum of the orbital energies of the three electrons. As shown in Fig. 24.16, after the formation of the π-bonding system, the total energy of the electrons is $3\alpha + 2\sqrt{2}\beta$. The molecular orbitals have different energies than the p orbitals had. The system is able to stabilize itself by adding the three electrons 'below the center of gravity' of the original orbitals. The center of gravity is the average orbital energy of the three 2p orbitals. Since the orbitals were degenerate, their center of gravity is the same as the orbital energy of one of the orbitals.

The initial energy of the three 2p electrons was 3α. The final energy is $3\alpha + 2\sqrt{2}\beta$. Looking at the structure of the allyl radical, it is held together by one double bond and one single bond. The single bond and one of the bonds in the double bond are sigma bonds. They are not the bonds involved in the Hückel model. The Hückel model is describing the bonds made by the three 2p orbitals. Two of these electrons made the 'second' bond, the π-bond, in the double bond, while the

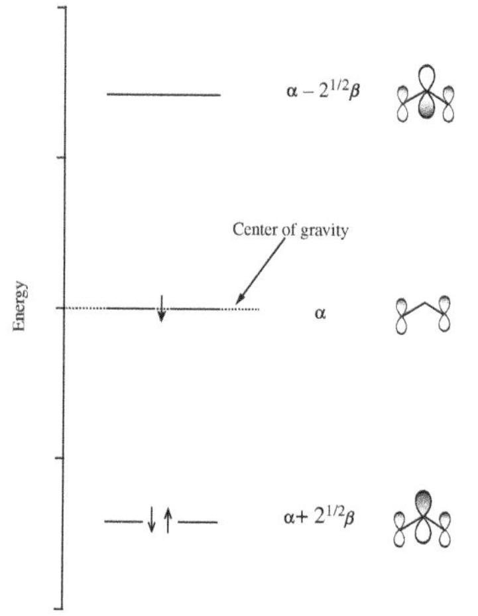

Figure 24.16 A diagram of the energy of the Hückel molecular orbitals generated for allyl. The formation of orbitals in the π system leads to molecular orbitals with energies shifted from the energies of the constituent atomic orbitals. If the occupancy of molecular orbitals below the center of gravity of the atomic orbitals energy is greater than that of molecular orbitals above the center of gravity, the system is stabilized by molecular orbital formation.

third resides in the orbital described by c_2 (the second column in the **C** matrix). Since this orbital has the same energy as a 2p orbital, it is a nonbonding molecular orbital with energy α. If the π-bond were localized such that it is formed just between two of the orbitals, its energy would be $2\alpha + 2\beta$ and the total energy of the system would be $3\alpha + 2\beta$. Instead, the energy of the system is lower than this by 0.828β. This extra stabilization caused by the structure of the molecular orbitals is the *delocalization energy* of the π-system. In a valence bond picture this is called the *resonance energy*.

24.10 Density functional theory (DFT)

In progressing from the valence bond model to the molecular orbital model, we found that we could greatly simplify the mathematics by dropping the notion of atomic orbitals and instead using easy to integrate functions in our basis set. Okay, 'simplify' and 'easy to integrate' are relative terms, but certainly this change in outlook leads to much faster computations. Having abandoned the idea that we need to keep track of atomic orbitals, we can now make an even more radical change in strategy.

What is it about the wavefunction that we really need to know? We have mentioned several times that the kinetic and potential energies are related to the shapes of the wavefunctions or, even more precisely, to the shapes of the wavefunctions squared. What we really need to know is the shape and density of the charge distribution about the nuclei. Is it possible then to abandon orbitals and wavefunctions all together and to perform calculations involving the curvature and density of the electron distribution? Might this be a way to perform calculations on extremely large systems where keeping track of hundreds or thousands of atoms, each requiring hundreds of basis functions, becomes impossible even with the largest supercomputers? This is the realm of *density functional theory* (DFT).

The heart of DFT is the *Hohenberg–Kohn existence theorem.*[7]

> *For molecules with a non-degenerate ground state, the ground-state molecular energy, wavefunction and all other molecular electronic properties are uniquely determined by the ground state electron probability density $\rho_0(x,y,z) = \rho(\mathbf{r})$.*

Another way to state this is that the density determines the external potential, which itself determines the Hamiltonian. The Hamiltonian then determines the wavefunction. The density is a function of three variables instead of $3N$ spatial $+N$ spin variables for an N electron system in a molecular orbital-based scheme. This is a promising start to what can be a more efficient algorithm for the calculation of molecular energies than molecular orbital calculations.

We begin with the proposition that there exists a functional of the electron density $E[\rho(\mathbf{r})]$ such that energy E_0 can be calculated from the electron probability density

$$E_0 = E[\rho(\mathbf{r})]. \tag{24.95}$$

A function whose argument is a function is a functional. A functional $F[f]$ is a rule that associates a number with each function f. As an example, consider the variational integral

$$W[\phi] = \frac{\langle\phi|\hat{H}|\phi\rangle}{\langle\phi|\phi\rangle}. \tag{24.96}$$

This is a functional of the function ϕ and gives one number for each ϕ.

None of the above leads to an advance compared to molecular orbital methods. The problem is that the Hamiltonian still needs to be solved and the electron–electron interaction term (electron correlation) makes this difficult. The real advance comes from what is known as the Kohn–Sham method.[8] Start with a hypothetical system of *non-interacting* electrons. Kohn–Sham orbitals are orbitals that describe a noninteracting electron gas. These 'easily' found orbitals are used to calculate the energy, electron density, and so forth. The ideal noninteracting electrons have the same density distribution as the real electrons in the system of interest.

The energy is then given by a functional which is the sum of three easily soluble terms and two correction functionals.

$$E_0 = E[\rho(\mathbf{r})] = T_{ni}[\rho(\mathbf{r})] + V_{ne}[\rho(\mathbf{r})] + V_{ee}[\rho(\mathbf{r})] + \Delta T[\rho(\mathbf{r})] + \Delta V_{ee}[\rho(\mathbf{r})] \tag{24.97}$$

1. T_{ni} = kinetic energy of the noninteracting electrons.
2. V_{ne} = the nuclear–electron interaction.
3. V_{ee} = classical electron–electron repulsion.

4 ΔT = the correction to the kinetic energy deriving from the interacting nature of the electrons.
5 ΔV_{ee} = all non-classical corrections to the electron–electron repulsion energy.

The two correction terms are often lumped into one exchange–correlation functional to calculate E_{xc} the exchange–correlation energy. The pursuit of the exchange–correlation functional is an active area of research. This term also includes corrections for the classical self-interaction energy and for the difference in kinetic energy between the hypothetical noninteracting electron gas and the real system.

The Kohn–Sham method determines the energy through a self-consistent field procedure with the form:

i Guess ρ.
ii Calculate energy.
iii Iterate to self-consistency.

This is similar to how variational calculations are performed, but it has some significant differences. DFT is exact in principle, but the method of solution is approximate. We know that a functional for E_{xc} must exist, but we do not know what its form is. DFT optimizes the electron density and does not need to know what the wavefunctions are.

One way to improve the exchange–correlation functional is to make it depend not only on the local value of the density, but also on the extent to which the density is locally changing. This is known as the *generalized gradient approximation* (GGA) and is essential for accurate calculation on molecules.

DFT has a number of advantages and disadvantages compared to molecular orbital calculations. A big advantage is that the number of Coulomb integrals scales as N^3 instead of N^4, where N is the number of basis functions. DFT is generally the lower cost option on big systems. DFT has problems with van der Waals forces and hydrogen bonding. Excited states are more difficult in DFT and require time-dependent calculations. The best form of the E_{xs} and GGA functionals are unknown and may depend on the system being considered. There also is a lack of rigorous variational principle because an approximate form of the exchange–correlation functional must be used.

SUMMARY OF IMPORTANT EQUATIONS

$$\hat{H} = -\frac{1}{2}\sum_{i=1}^{n}\nabla_i^2 - \frac{1}{2}\sum_{j=1}^{N}\frac{m_e}{m_j}\nabla_j^2 - \sum_{i=1}^{n}\sum_{j=1}^{N}\frac{Z_j}{r_{ij}} + \frac{1}{2}\sum_{i=1}^{n}\sum_{\substack{i'=1\\ i'\neq i}}^{n}\frac{1}{r_{ii'}} + \frac{1}{2}\sum_{j=1}^{N}\sum_{\substack{j'=1\\ j'\neq j}}^{N}\frac{1}{R_{jj'}}$$

The molecular Hamiltonian is formed from the sum of electron kinetic energy, nuclear kinetic energy, the attractive potential energy of the interaction of the electrons and nuclei, the electron–electron repulsion, and the repulsive potential energy of internuclear interactions.

$\langle E(R)\rangle = \langle T(R)\rangle + \langle V(R)\rangle$ — The total energy is the sum of kinetic and potential energy

$\langle T(R)\rangle = -\langle E(R)\rangle - R\frac{d\langle E(R)\rangle}{dR}$ — Molecular virial theorem relates kinetic and potential energy to total energy and its derivative with respect to internuclear distance R

$\langle V(R)\rangle = 2\langle E(R)\rangle + R\frac{d\langle E(R)\rangle}{dR}$

$p_{AB} = \frac{1}{2}(w - w^*)$ — Bond order between atoms A and B is half the difference of occupation of bonding and antibonding molecular orbitals.

Exercises

24.1 How does valence bond theory describe chemical bond formation?

24.2 How does molecular orbital theory describe bonding?

24.3 A chemical bond is formed by doing what with respect to the interactions (forces) between particles (electrons and nuclei) in the system and the kinetic energy of the electrons?

24.4 Drawn below are Born–Oppenheimer potential energy curves for three species: AlO, SiO and Na_2. All three curves approach the zero of energy asymptotically at infinite separation. State your reasoning using molecular orbital occupations and bond orders to assign which species corresponds to each potential energy curve. Label the curves to indicate the species and how to determine the values of R_e and D_e. Determine the numerical values of R_e and D_e for each species.

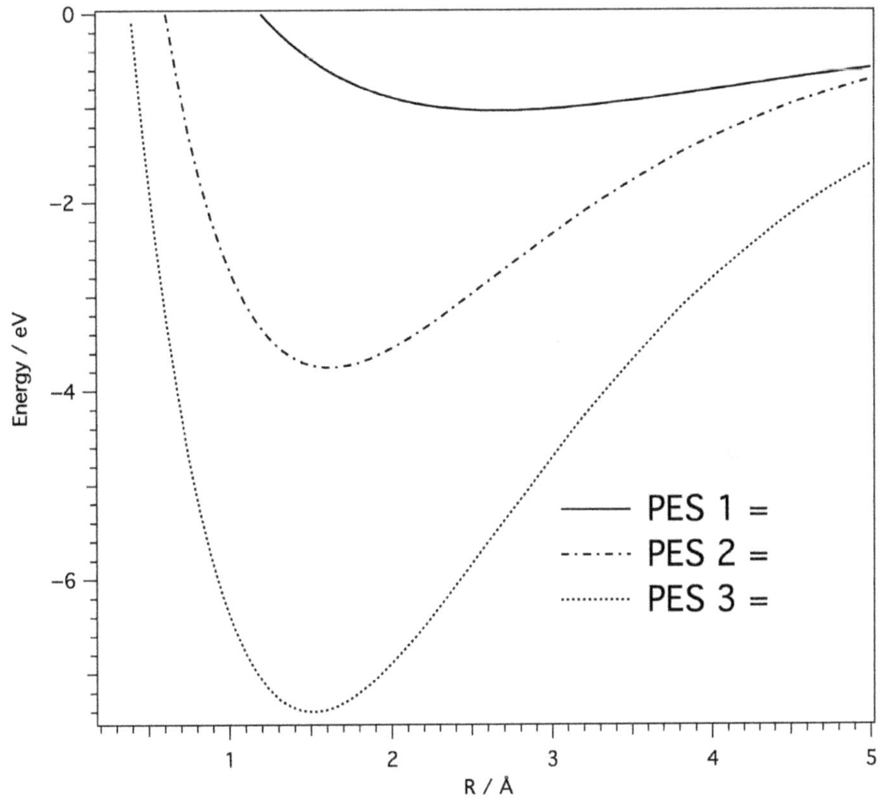

24.5 Drawn below are Born–Oppenheimer potential energy curves for three species: B_2, BN and BO. All three curves approach the zero of energy asymptotically at infinite separation. State your reasoning using molecular orbital occupations and bond orders to assign which species corresponds to each potential energy curve. Label the curves to indicate the species and how to determine the values of R_e and D_e. Enter the numerical values of R_e and D_e for each species in the table below.

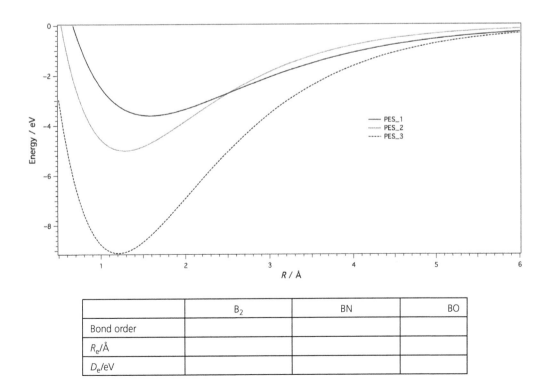

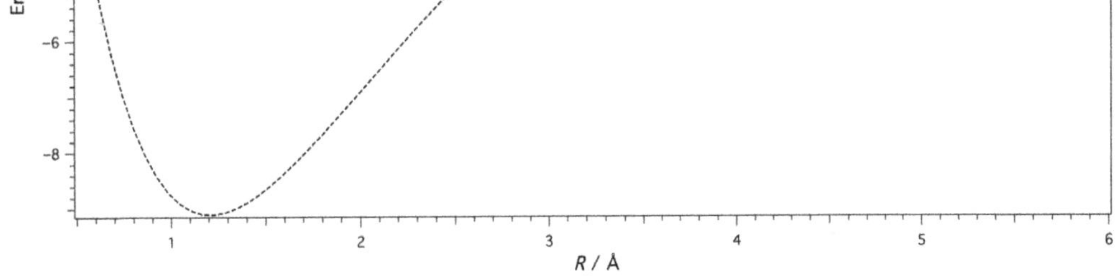

	B_2	BN	BO
Bond order			
R_e/Å			
D_e/eV			

24.6 Consider the diatomic species OF, OF^+ and OF^-. Write out the electron configurations for each species. Calculate the bond order. Draw Born–Oppenheimer potential energy curves for each species. Draw all three species on the same set of axes and draw them such that they all approach the zero of energy asymptotically at infinite separation. On the basis of the molecular orbital occupation, assign the following data to the appropriate species.

R_e/nm	1.204	1.332	1.492
$\tilde{\nu}_{01}$/cm^{-1}	863	1393	1719
D_e/eV	1.41	2.28	2.81

24.7 What spatial characteristic differentiates a bonding MO from an antibonding MO?

24.8 Distinguish between semi-empirical, *ab initio*, and density functional theory methods of electronic structure determination.

24.9 What is the Born–Oppenheimer approximation and what does it allow us to do?

24.10 With the aid of a diagram to define all charges and distances, write out the Hamiltonian for the HHe molecule.

24.11 Give the ground-state electron configurations (i.e., the fillings of the molecular orbitals) and bond orders of Li_2, B_2, Be_2, C_2^+, CO, NO, N_2, and CN^-.

24.12 (a) Which has the greater bond dissociation energy, B_2 or CO? Why? (b) Which has the shorter bond length NO or CO based on the electronic configuration of the ground state? Why? (c) What is the correlation between bond order and vibrational frequency? (d) What do the ground states of NO and O_2 have in common, which differentiates them from, for instance, N_2?

24.13 What determines the bond order of a particular molecular orbital configuration?

24.14 Sketch drawings of orbitals with σ_g, σ_u, π_g and π_u symmetry, including reasoning for why they have the assigned symmetry. Indicate which orbitals are bonding and which are antibonding, and indicate why they are either antibonding or bonding.

24.15 Discuss the steps involved in the calculation of the energy of a system by using the variation principle.

24.16 Take as a trial wavefunction for the ground state of the H atom

$$\psi = e^{-kr}.$$

Use the variation principle to find the optimum value of k. Note that the wavefunction is not normalized as written and that the only part of the Laplacian that need be considered is the part that involves radial derivatives.

24.17 A normalized wavefunction is constructed from an orthonormal basis set such that

$$\psi = c_A A + c_B B + c_C C.$$

When a variational treatment is used to calculate the energy, the energy is found to depend on the coefficients according to

$$\frac{\partial E}{\partial c_A} = c_A - \frac{1}{\sqrt{3}}, \quad \frac{\partial E}{\partial c_B} = c_B^2 - \frac{1}{5}.$$

(i) State how the variational theorem allows you to calculate the correct energy for the system described by ψ.
(ii) Determine the three coefficients and the correct energy of the system.

24.18 Write the following complete set of wavefunctions in matrix notation:

$$\psi_1 = a\varphi_1 + b\varphi_2 + c\varphi_3, \quad \psi_2 = d\varphi_1 + e\varphi_2 + f\varphi_3, \quad \psi_3 = g\varphi_1 + h\varphi_2 + j\varphi_3$$

24.19 Koopmans' theorem allows us to state that to a first approximation, the binding energy of an electron is equal to the negative of the orbital energy of the orbital from which the photoelectron originated. When ethanol (CH_3CH_2OH) is irradiated with X-rays from Al K_α radiation (with an energy of 1487.0 eV), photoelectrons are emitted in three peaks with kinetic energies of 954.2, 1200.5 and 1202.0 eV. Two peaks are associated with C 1s levels, and one peak is associated with the O 1s level. Calculate the electron binding energies associated with the three peaks. Draw an energy level diagram. Label it with the vacuum energy E_{vac}, the three binding energies and the kinetic energies of the photoelectrons. Identify which levels correspond to which elements and explain why on the basis of screening, there are two C 1s peaks.

$E_b = E_{photon} - E_k$

E_k/eV	E_b/eV	Assignment

Endnotes

1. Born, M. and Oppenheimer, J.R. (1927) *Ann. Phys.*, **84**, 457.
2. Kołos, W. and Wolniewicz, L. (1968) *J. Chem. Phys.*, **49**, 404; Wolniewicz, L. (1995) *J. Chem. Phys.*, **103**, 1792.
3. Piszczatowski, K., Łach, G., Przybytek, M., Komasa, J., Pachucki, K., and Jeziorski, B. (2009) *J. Chem. Theory Comp.*, **5**, 3039.
4. Liu, J., Salumbides, E.J., Hollenstein, U., Koelemeij, J.C.J., Eikema, K.S.E., Ubachs, W., and Merkt, F. (2009) *J. Chem. Phys.*, **130**, 174306.
5. Zhao, B.S., Meijer, G., and Schöllkopf, W. (2011) *Science*, **331**, 892.
6. Hückel, E. (1931) *Z. Phys.*, **70**, 204.
7. Hohenberg, P. and Kohn, W. (1964) *Phys. Rev.*, **136**, B864.
8. Kohn, W. and Sham, L.J. (1965) *Phys. Rev.*, **140**, A1133. Walter Kohn was awarded the Nobel Prize in Chemistry 1998 for his contributions to density functional theory.

Further reading

Atkins, P.W. and Friedman, R.S. (2010) *Molecular Quantum Mechanics*, 5th edition. Oxford University Press, Oxford.

Cramer, C.J. (2002) *Essentials of Computational Chemistry*. John Wiley & Sons, Chichester, UK.

Cook, D.B. (2012) *Quantum Chemistry: A Unified Approach*. Imperial College Press, London.

Eyring, H., Walter, J., and Kimball, G.E. (2015) *Quantum Chemistry*. Scholar's Choice, Rochester, NY.

Metiu, H. (2006) *Physical Chemistry: Quantum Mechanics*. Taylor & Francis, New York.

NIST Computational Chemistry Comparison and Benchmark Database; http://cccbdb.nist.gov/Summary.asp

Schatz, G.C. and Ratner, M.A. (2002) *Quantum Mechanics in Chemistry*. Dover Publications, New York.

Winn, J.S. (1995) *Physical Chemistry*. HarperCollins College Publishers, New York.

Zewail, A.H. (1992) *The Chemical Bond – Structure and Dynamics*. Academic Press, London.

CHAPTER 25

Molecular spectroscopy and excited-state dynamics: Diatomics

PREVIEW OF IMPORTANT CONCEPTS

- Molecules undergo transitions that correspond to changing the rotational state; rotational and vibrational state; or rotational, vibrational and electronic state.
- Each of these transitions typically occurs at characteristic energies that are successively higher.
- Each type of transition has its own set of selection rules, which result from angular momentum conservation and molecular symmetry.
- To exhibit a pure rotational transition, the molecule must have a permanent dipole moment.
- $\Delta J = \pm 1$ is observed for rotational transitions of most diatomic molecules (those with Σ ground states).
- All rigid rotors have the same pattern of energy levels but different values of the rotational constant lead to spacings that are characteristic of a particular pair of isotopes in a particular molecule.
- Centrifugal distortion leads to shifts in the rotational levels of real molecules.
- Vibrations of bonds are close to harmonic near the bottom of the well, but anharmonicity becomes increasingly important with increasing excitation.
- Harmonic oscillator energy levels are evenly spaced.
- An anharmonic oscillator is represented by a Morse potential. The levels of a Morse oscillator are progressively more closely spaced toward the top of the well.
- There must be a change of dipole moment for a vibrational transition to be allowed.
- $\Delta v = 1$ is usually the most intense vibrational transition, with less intense overtones ($\Delta v > 1$) sometimes observed.
- A Raman transition occurs via a virtual state – a state of the molecule that only occurs in the presence of photons.
- There must be a change in the polarizability of the molecule for a Raman transition to occur.
- Electronic transitions are vertical. An electronic transition can go from the ground state to any vibrational state in the excited state; however, it is the state with the greatest vibrational overlap with the ground state that has the highest intensity. This is the Franck–Condon principle.
- The electronic states of molecules are represented by state designations of the form ${}^{2S+1}\Lambda^{\pm}_{u/g}$, which specify the spin multiplicity, orbital angular momentum, reflection symmetry (parity) and inversion symmetry.
- Dissociation energies, dissociation limits, transition energies and trends in the vibrational and rotational state characteristics can be read off of plots of potential energy curves.
- Excited states can relax radiatively (fluorescence) and nonradiatively. Radiative lifetimes are usually on the order of 100 ns for allowed transitions, but can be substantially longer for forbidden transitions.
- Intramolecular nonradiative processes include internal conversion, intersystem crossing and photochemistry.
- Intermolecular nonradiative processes involve collisional energy transfer.
- Quantum yield is used to quantify the branching probability between different radiative and nonradiative processes.
- Electron–molecule collisions can lead to many of the same excitations as photon–molecule collisions, but with differences caused by the mass and spin of the electron.

In our treatment of atomic spectroscopy we have already introduced the fundamentals of all spectroscopy: spectra are made up of lines from bound–bound transitions and continuous regions from bound–free transitions. Lines have widths determined by lifetimes, collisions and any other relaxation or broadening mechanisms. Conservation of energy leads

Physical Chemistry: How Chemistry Works, First Edition. Kurt W. Kolasinski.

Companion Website: www.wiley.com/go/kolasinski/physicalchemistry

to the resonance condition in which the photon energy in either absorption or emission must exactly match the energy difference between the states involved in the transition. Conservation of angular momentum enforces selection rules on the states that can be connected by different types of transitions.

In atomic spectroscopy all transitions were between different electronic states. In molecules there are additional degrees of freedom: rotations and vibrations. There is also the possibility for photodissociation – the breaking of bonds caused by photon absorption. Molecules also exhibit characteristic symmetries. Molecular symmetry is important in determining new sets of selection rules. These are the details that we will investigate in more depth in this chapter. Here, we will concentrate on diatomic molecules. Brief reference is made to the spectroscopy of polyatomic molecules, which are discussed in more detail in Chapter 26.

25.1 Introduction to molecular spectroscopy

In the study of statistical mechanics we found that the energy of a molecule is distributed between translational energy ε_t and internal energy ε_i. There are three components to the internal energy: rotational energy ε_r, vibrational energy ε_v and electronic energy ε_e. To a first approximation we can consider these four contributions independent of each other; hence

$$\varepsilon = \varepsilon_t + \varepsilon_{int} = \varepsilon_t + \varepsilon_r + \varepsilon_v + \varepsilon_e. \tag{25.1}$$

Here, we neglect the nuclear degrees of freedom. We will add these as and when we need them, but we will neglect them as a first approximation. The independence of the four contributions allows us to factor the partition function Z for the molecule. We have already seen that the effect of the translation degree of freedom is observed in the linewidth of a transition through the Doppler effect. Otherwise, we can neglect translations as we consider the spectroscopy of internal degrees of freedom. In our introduction to quantum theory we also found that spin is another degree of freedom. Spin, which is not related to spatial coordinates, is quite important in spectroscopy. We will return to its effects later, but for now we concentrate on the influence of the spatial degrees of freedom on rotational, vibrational and electronic excitations.

The ability to factor the partition function extends to the wavefunctions. To a first approximation, the internal degrees of freedom do not interact and we can factor the wavefunction into separate rotational, vibrational and electronic contributions,

$$\psi(\boldsymbol{r}, \boldsymbol{R}) = \psi_r(\boldsymbol{R})\psi_v(\boldsymbol{R})\psi_e(\boldsymbol{r}, \boldsymbol{R}). \tag{25.2}$$

This means we can separate the discussion of rotational, vibrational and electronic spectroscopy, at least initially. The approximation that the electrons will always be able to find the lowest energy configuration as the nuclear coordinates change, for example as a result of vibration, is known as the *Born–Oppenheimer approximation.*[1] We used it in the last chapter to construct potential energy curves for each electronic state. Its impact resonates throughout spectroscopy as well. Importantly, the Born–Oppenheimer approximation allows us to treat the nuclear coordinates $\boldsymbol{R}$ as a parameter that can be separated from the electronic coordinates $\boldsymbol{r}$. The electrons are much lighter and faster than the nuclei. They respond almost instantaneously to changes in the nuclear coordinates.

Each electronic state has an associated potential energy curve, which if the state is bound usually has a shape similar to that shown in Fig. 25.1. Each bound electronic state contains several vibrational levels with energy spacings typically on the order of several tens of cm^{-1} up to 4000 cm^{-1} (in the infrared). On top of the vibrational levels (at an even finer energy scale) are superimposed the rotational levels, with energies in the microwave range. A dissociative state, also called a repulsive state, has no potential well and does not support bound vibrational levels such as those shown in Fig. 25.1. A repulsive potential decreases monotonically with increasing internuclear separation.

The nucleus has a spin. When the value of nuclear spin is nonzero, further energy level splittings arise. This also leads to the spectroscopy of nuclear magnetic resonance (NMR) in the radio wave region. The nucleus has a shell structure much like that of the electrons. Excited nuclear states are usually associated with gamma rays (high-energy excitations).

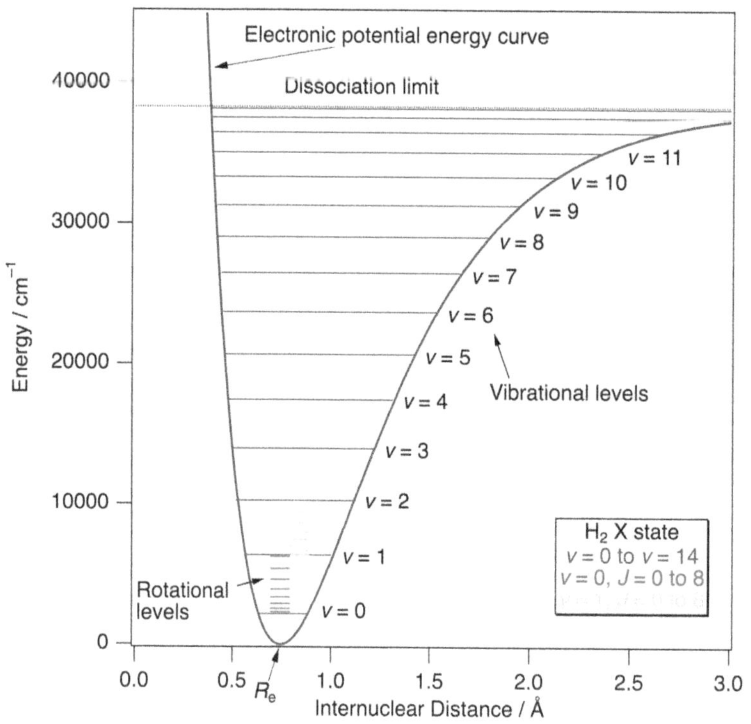

Figure 25.1 A potential energy curve representative of the ground electronic state of H_2. Each electronic state has such a curve associated with it. The vibrational levels of a Morse oscillator designated by the vibrational quantum number v converge on the dissociation limit, which is positioned at the energetic origin in this diagram. The minimum potential occurs at R_e. At an even finer level of energy resolution rotational states are associated with each vibrational level.

25.2 Pure rotational spectra of molecules

25.2.1 Diatomic molecules

For a rotational transition to occur by the interaction of light with a molecule, there must be an oscillating dipole moment in the molecule. This can only happen if the molecule has a permanent dipole. Only molecules with a permanent dipole moment exhibit pure rotational transitions. Therefore, homonuclear diatomic molecules do not have pure rotational spectra.

The pre-condition for a pure rotational transition is that the molecule must have a permanent dipole moment.

The first-order approximation for the rotational motion of a diatomic molecule is that of a *rigid rotor*: a massless rod of length R rotating about its center of mass, as shown in Fig. 25.2. The two atoms with masses m_1 and m_2 reside at

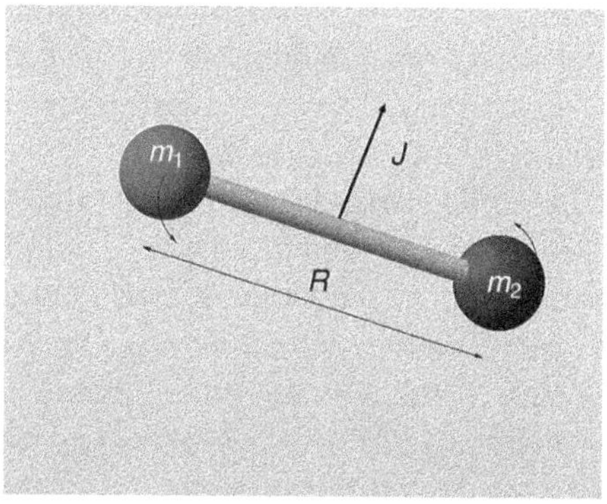

Figure 25.2 In the rigid rotor model of a diatomic molecule, two masses m_1 and m_2 are held together by a massless rod. The length of the rod is fixed such that the distance between the centers of mass of the two atoms is R. The rotational axis J is perpendicular to the internuclear axis.

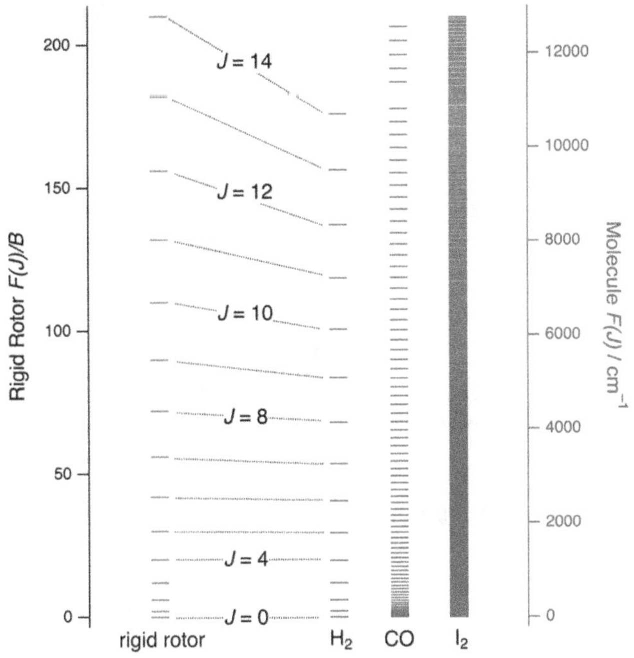

Figure 25.3 Rotational levels of a rigid rotor are spaced in a pattern. Different molecules have characteristic values of the rotational constant B, which leads to decidedly different spacings between rotational levels. H_2 with $B = 60.9\ \text{cm}^{-1}$ has very few levels, while the levels of I_2 with $B = 0.037\ \text{cm}^{-1}$ look like a quasi-continuum at this resolution. Centrifugal distortion shifts the energy of the H_2 states compared to the prediction of the rigid rotor model. The shift increases with increasing rotational quantum number J.

either end of the rod. The rotational dynamics of this body are the same as the particle on a ring system that we solved in Chapter 21. Thus, the *rotational energy* of a diatomic molecule as modeled by a rigid rotor is given by

$$\varepsilon_{\text{r}} = E_J = \frac{\hbar^2}{2I} J(J+1) = \frac{\hbar^2}{2\mu R^2} J(J+1) = hcBJ(J+1) \tag{25.3}$$

where

$$I = \mu R^2 \tag{25.4}$$

is the *moment of inertia*, μ is the reduced mass

$$\mu = \frac{m_1 m_2}{m_1 + m_2}, \tag{25.5}$$

J the rotational quantum number and B the *rotational constant*,

$$B = \frac{\hbar}{4\pi\mu c R^2}. \tag{25.6}$$

The associated energy level structure is given in Fig. 25.3. When expressed in wavenumbers, the energy of a rotational level is known as the *rotational term* $F(J)$

$$F(J) = E_J / hc = BJ(J+1). \tag{25.7}$$

The magnitude of the *rotational angular momentum* is given by the expectation value of the rotational angular momentum operator $\langle \hat{J}^2 \rangle^{1/2}$

$$\langle \hat{J}^2 \rangle^{1/2} = [J(J+1)]^{1/2}\,\hbar. \tag{25.8}$$

To each rotational quantum number J, there are $2J + 1$ degenerate rotational states that are designated by the M_J quantum number. J is a positive integer but M_J ranges in units steps from $-J$ to $+J$. Here, we are dealing only with rotation. Because it is unambiguously associated just with J, we will refer to M_J as M to simplify the notation. These degenerate states arise from space quantization. The projection of the rotational angular momentum is determined by the quantum number M,

$$\langle \hat{J}_z \rangle = M\hbar. \tag{25.9}$$

The degeneracy of these levels can be lifted by an electric field, which is known as the *Stark effect*. In other words, when a rotating molecule is placed in an electric field, the field direction defines the orientation of the z-axis in the laboratory frame. The $2J + 1$ different projections along the z axis of the rotational angular momentum of the states with different values of M cause the states to have different energies.

Each rotational quantum state can be identified by the JM quantum numbers, and to determine the selection rules we have to evaluate the matrix element of the transition dipole moment operator (as with atomic spectroscopy, we are assuming all transitions to be electric dipole transition unless we specify otherwise)

$$M_r = \langle J'M' | \hat{\mu} | J''M'' \rangle. \tag{25.10}$$

Spectroscopic notation by convention denotes the upper state with a single prime and the lower state with a double prime. In other words, we have to evaluate the integral

$$M_r = \int \psi^*_{J'M'} \, \hat{\mu}_R \, \psi_{J''M''} \, d\Omega, \tag{25.11}$$

where $d\Omega = \sin\theta \, d\theta \, d\phi$. The wavefunction for rotation is a spherical harmonic $Y_{JM}(\theta, \phi)$. It does not depend on the radial coordinate. $\hat{\mu}_R$ is the dipole moment vector. The R indicates that it is directed along the bond. To determine which transitions are allowed, we argue just as we did for atoms. The angular momentum of the photon is one, and angular momentum must be conserved during absorption or emission of the photon. Thus, rotational angular momentum must change by $\Delta J = \pm 1$ while $\Delta M = 0, \pm 1$ is possible. Most molecules have a ground electronic state that does not possess an electronic angular momentum. In Table 25.1 it can be seen that an exception (the one with a Π state instead of a Σ state) is NO. Molecules such as NO that do not have a Σ ground state exhibit $\Delta J = 0$ pure rotational transitions.

The energy of a transition between rotational levels differing by one unit of J is given by

$$\Delta E = E_{J+1} - E_J = 2(J+1)\frac{h}{8\pi^2 I}. \tag{25.12}$$

Table 25.1 Ground-state spectroscopic constants for selected diatomic molecules taken from the 96th edition of the *CRC Handbook of Chemistry and Physics*. All values are in units of cm^{-1} except the centrifugal distortion constant D, which is in units of 10^{-6} cm^{-1}, and r_e in units of Å. The traditional notation from Herzberg and Huber is to denote the fundamental wavenumber ω_e, the harmonic bond length r_e, and the centrifugal distortion constant D_e. The vibration–rotation interaction constant is α_e and x_e is the anharmonicity constant

Molecule	Ground state	$\tilde{\nu}_e = \omega_e$ / cm^{-1}	$\omega_e x_e$ / cm^{-1}	B_e / cm^{-1}	α_e / cm^{-1}	D / 10^{-6} cm^{-1}	$R_e = r_e$ Å
$^1H_2$	$^1\Sigma_g^+$	4401.21	121.34	60.853	3.062	47100	0.74144
$^2H_2$	$^1\Sigma_g^+$	3115.50	61.82	30.444	1.0786	11410	0.74152
$^1H^{19}F$	$^1\Sigma^+$	4138.39	89.94	20.953712	0.7933704	2150	0.91685
$^1H^{35}Cl$	$^1\Sigma^+$	2990.92	52.80	10.5933002	0.3069985	531.94	1.27456
$^1H^{127}I$	$^1\Sigma^+$	2309.01	39.64	6.4263650	0.1689	206.9	1.60916
$^{19}F_2$	$^1\Sigma_g^+$	916.93	11.32	0.889294	0.0125952	3.3	1.41264
$^{35}Cl_2$	$^1\Sigma_g^+$	559.75	2.69	0.24415	0.00152	0.186	1.9872
$^{79}Br_2$	$^1\Sigma_g^+$	325.32	1.08	0.082107	0.0003187	0.02092	2.2811
$^{127}I_2$	$^1\Sigma_g^+$	214.50	0.61	0.03737	0.000114	0.0043	2.666
$^{12}C^{16}O$	$^1\Sigma^+$	2169.81	13.29	1.931280985	0.01750439	6.1216	1.12832
$^{14}N_2$	$^1\Sigma_g^+$	2358.56	14.32	1.998236	0.017310	5.737	1.09769
$^{14}N^{16}O$	$^2\Pi_{1/2}$	1904.20	14.07	1.67195	0.0171	0.5	1.15077
$^{16}O_2$	$^3\Sigma_g^-$	1580.19	11.98	1.445622	0.015933	4.839	1.20752
$^{23}Na_2$	$^1\Sigma_g^+$	159.09	0.71	0.15473537	0.0086375	0.58	3.07858
$^{138}Ba^{127}I$	$^2\Sigma^+$	152.14	0.27	0.02680587	0.00006634	0.00333	3.08476

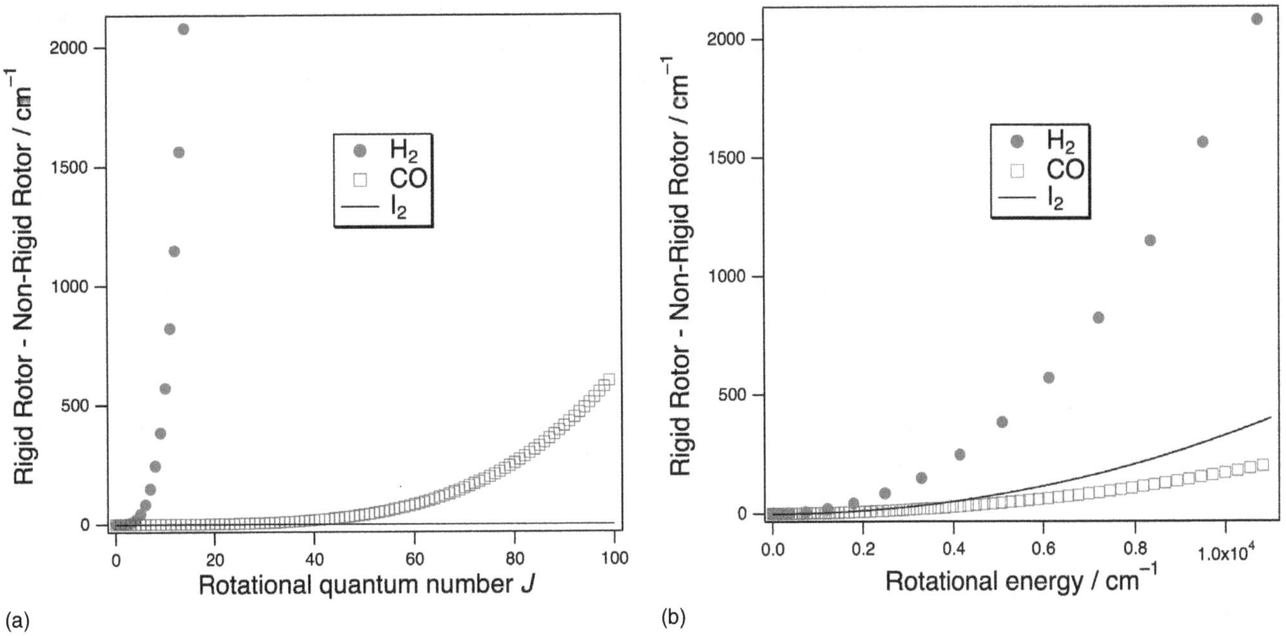

Figure 25.4 Centrifugal distortion leads to increasingly large deviations from the predictions of the rigid rotor energy levels as shown here for H_2, CO and I_2. (a) The deviation is plotted versus the rotational quantum number. (b) The deviation is plotted against rotational energy. It is clear that H_2 is much less rigid even when compared at equal levels of rotational energy.

The wavenumber of the transition between these two states is given by the difference in the rotational terms

$$\tilde{\nu}_J = F(J+1) - F(J) = \Delta E/hc = 2(J+1)B. \tag{25.13}$$

Therefore, *the pure rotational spectrum is composed of equally spaced lines with a spacing of 2B between them.*

The above equation is strictly true for a rigid rotor – that is, one in which the bond length R does not change. If a diatomic molecule were a simple harmonic oscillator, each vibrational level would have exactly the same bond length. A diatomic molecule is not a perfect harmonic oscillator, and anharmonicity leads to a lengthening of bond length with increasing vibrational quantum number v. Therefore, each vibrational level has its own value of rotational constant, which is denoted B_v because each vibrational level has its own value of equilibrium bond length R_v. Rotational excitation also stretches the bond. This lengthening of the bond as the rotational quantum number increases is known as *centrifugal distortion*. Accounting for the vibrational state dependence of the rotational constant and centrifugal distortion leads to the addition of a term that is second order in $J(J+1)$ multiplied by the centrifugal distortion constant D,

$$F(J) = B_v J(J+1) - D[J(J+1)]^2 \tag{25.14}$$

The effects of centrifugal distortion are shown in Fig. 25.4 for three molecules – H_2, CO and I_2 – with much different rotational constants. Some molecules have been characterized well enough that a vibrational-state-specific value $D_v = D_e +$ corrections is used in place of a constant value, but we will not use D_e in this context so as to avoid confusion with the dissociation energy measured from the bottom of the well. The value of D can be approximated from the harmonic oscillator in terms of the ground vibrational state values of rotational constant and fundamental wavenumber as

$$D = 4B_0^3/\tilde{\nu}_e. \tag{25.15}$$

A third-order correction to the rotational term that is multiplied by a constant H_v is known for some molecules. The *vibration–rotation interaction constant* α_e is used to calculate the value of vibrational-level-specific value B_v from

$$B_v = B_e - \alpha_e\left(v + \tfrac{1}{2}\right) \tag{25.16}$$

The constant B_e is the rotational constant of a hypothetical harmonic state that exists at the minimum of the potential energy curve. However, as we have seen, the Heisenberg uncertainty principle leads to the presence of a zeropoint energy. Thus, B_0 is ever so slightly shifted from B_e.

Selected values of the spectroscopic constants of diatomics are given in Table 25.1. Note that because of the reduced mass dependence in Eq. (25.6), the rotational constant is extremely sensitive to the mass combination. Therefore, pure rotational spectroscopy is quite sensitive to isotopic composition. Pure rotational spectra form uniquely characteristic and easily interpretable fingerprints of molecules. These spectra are the primary feature by which molecules are identified in astrochemical studies.[2]

Inspection of the values in Table 25.1 reveals some expected trends. The behavior of H_2 is exceptional because of its low mass and small size. There is a change in the fourth decimal place in the bond length of the H_2 molecule upon isotopic substitution of ^2H for ^1H, but there is a sizeable shift in all other constants. Hydrides have very high rotational constants. Heavy–heavy diatomics have very low rotational constants. Bond lengths increase with increasing atomic number, as shown dramatically by the hydrogen halides and halogens. As reflected in the ratios $\omega_e x_e/\omega_e$ and D/B_e, hydrides are also considerably more anharmonic and strongly affected by centrifugal distortion than are the other diatomics.

The intensity of rotational absorption peaks varies with the population in the initial rotational state of the transition. If the sample is described by a temperature T, then the population follows the Boltzmann distribution. The relative number density of molecules in state J compared to the number density in the ground state is

$$N_J/N_0 = (2J+1)\,e^{-E_J/k_BT}. \tag{25.17}$$

Nascent molecules produced in a chemical reaction might not be fully equilibrated. This means that the rotational state distribution might not be Boltzmann, or that even though the population follows a Boltzmann distribution the corresponding temperature need not equal the temperature of the container, or be the same as the vibrational or translational temperature. The degeneracy term $(2J+1)$ increases with J but the exponential term decreases with J since the rotation energy is increasing. This means that the population, and therefore line intensities, go through a maximum at a value of J_{max} that is related to the temperature and rotational constant by

$$J_{max} = \left(\frac{k_B T}{2hcB}\right)^{1/2} - \frac{1}{2}. \tag{25.18}$$

25.2.1.1 Directed practice

Why does centrifugal distortion lead to closer spacing of rotational states?

25.2.2 Polyatomic molecules

Linear molecules can be treated just like diatomic molecules because they also only have one rotational axis. Symmetric linear polyatomics such as O=C=O do not exhibit a pure rotational spectrum because they do not possess a permanent dipole moment. Asymmetric polyatomics such as O=C=S, which have a permanent dipole moment, exhibit a pure rotational spectrum. Because of centrifugal distortion, tetrahedral molecules such as CH_4 and SiH_4 exhibit similar pure rotational spectra.

Nonlinear molecules can rotate about three principal axes of inertia, some of which may or may not be equal. There are three types of polyatomic rotors, as defined in Table 25.2.

The rotational structure of polyatomic molecules is complex. Whereas the spherical top requires two quantum numbers to describe it (J and M, in analogy to the diatomic case), other tops require three quantum numbers to describe them, namely J, M and K. The last is associated with the projection of J onto the symmetry axis of the molecule. To a first approximation we can think of the rotational spectrum of a polyatomic as the sum of J stacks and K stacks, independent ladders of the rotational levels for each vibrational state. However, as the number of atoms increases and the level of excitation increases, the density of states rapidly increases. This leads to significant coupling between different rotational levels and eventually different vibrational levels as well. This is discussed further in Chapter 26.

Table 25.2 Classification of rotors based on the magnitudes of their moments of inertia I_A, I_b, and I_c.

Type of rotor/molecule	Moments of inertia
Spherical top	$I_a = I_b = I_c$
Symmetric top	$I_a = I_b \neq I_c$ or $I_a \neq I_b = I_c$
Asymmetric top	$I_a \neq I_b \neq I_c$

25.3 Rovibrational spectra of molecules

25.3.1 Diatomic molecules

Because vibrational excitations (at low levels of excitation) tend to have much larger energy spacings than rotations, we cannot independently excite a vibration without considering the change of rotational state at the same time. To a first approximation we take rotations and vibrations to be independent. That means that we can simply add rotational energies to vibrational energies to determine the energies of transitions involving the excitation of both. At high values of the rotational quantum number J and the vibrational quantum number v, we will find considerable vibration–rotation coupling. Modeling molecular vibrations as simple harmonic oscillators will allow us to establish the basic features of rovibrational transitions, which we can then modify to include deviations from ideal behavior.

As shown in Chapter 21, when modeled as a *harmonic oscillator*, the vibrational energy of a diatomic molecule is given by

$$E_v = \left(v + \tfrac{1}{2}\right) hc\tilde{\nu}_e, \quad v = 0, 1, 2... \tag{25.19}$$

with the fundamental wavenumber given by

$$\omega_e = \tilde{\nu}_e = \frac{1}{2\pi c}\sqrt{k/\mu}, \tag{25.20}$$

We will use wavenumber $\tilde{\nu}$ to avoid confusion between the frequency ν (Greek nu) and the vibrational quantum number v (italicized v). In much of the vibrational spectroscopy literature, in which the nomenclature of Herzberg is used, ω is used as the symbol for wavenumber. This is reflected in Table 25.1 for the value ω_e because this is the IUPAC recommended notation used in data tables. Here, we will use ω to represent the angular frequency with the only exception being ω_e and $\omega_e x_e$. Thus, $\tilde{\nu}_e = \omega_e$.

When performed at low resolution (or in the liquid or solid phase), a broad peak corresponding to the vibrational energy level change is observed. When performed at higher resolution, a fine structure appears that corresponds to rotational level structure. The internal energy of a molecule is the sum of the rotational and vibrational energies

$$E_{vJ} = \left(v + \tfrac{1}{2}\right) hc\tilde{\nu}_e + hcB_v J\left(J+1\right). \tag{25.21}$$

The *vibrational term* $G(v)$ in wavenumbers is then

$$G\left(v\right) = \left(v + \tfrac{1}{2}\right) \tilde{\nu}_e. \tag{25.22}$$

When molecules make transitions from one vibrational state to another, the change in rotational state must also be taken into account. The pre-condition for a vibrational transition is that there must be *a change in the dipole moment* during the transition. For a harmonic oscillator, the selection rules for the quantum numbers are

$$\Delta v = \pm 1 \quad \text{and} \quad \Delta J = \pm 1. \tag{25.23}$$

As always, the energy of a transition is given by the difference in energy of the states involved

$$\Delta E = E_{v'J'} - E_{v''J''}. \tag{25.24}$$

Alternatively, the wavenumber of the transition is

$$\tilde{\nu} = G\left(v'\right) + F\left(J'\right) - G\left(v''\right) - F\left(J''\right). \tag{25.25}$$

The prime refers to the upper level; the double prime refers to the lower level.

As the resolution is increased, the lines will continue to get narrower until some limit is reached. The linewidth of a rovibrational peak is determined, just as for atomic spectroscopy (as discussed in Section 23.1.2) by several factors that include the Doppler effect and the lifetime of the state. Relaxation of the excited state can result from collisions with other molecules (including a solvent) or the walls of the container. Assuming that only *lifetime broadening* is significant, the lifetime τ (in ps) of a state is related to the linewidth in cm^{-1} by

$$\delta\tilde{\upsilon} = \frac{5.31\,\text{cm}^{-1}}{\tau} \tag{25.26}$$

Table 25.3 Vibrational lifetimes and predicted natural linewidths for CO in various environments.

	τ/ps	$\delta\tilde{\nu}$/cm^{-1}
CO(g)	33×10^9	1.6×10^{-10}
CO(a)/NaCl	4.3×10^9	1.2×10^{-9}
CO(a)/Si	2300	2.3×10^{-3}
CO(a)/Pt	2.2	2.41

The wavenumbers of rovibrational transitions are sensitive reporters of the local, dynamic environment of the molecule. They provide information on vibrational energy exchange and solvent restructuring as the system approaches equilibrium after reaction or excitation by collisions or irradiation.

25.3.1.1 Example

The vibrational lifetime of the CO stretch vibration is 33 ms in the gas phase,[3] 4.3 ms when physisorbed on NaCl,[4] 2.3 ns when chemisorbed on Si(100),[5] and 2.2 ps when chemisorbed on Pt(111).[6] Calculate the natural linewidths expected in each case if the only mechanism of line broadening is lifetime broadening due to population relaxation.

Substituting into Eq. (25.26) we obtain the results found in Table 25.3. The linewidth increases from values that are difficult, if not impossible, to measure directly due to experimental constraints for the gas phase and physisorbed CO, to values that are easily accessible in the chemisorbed phase. Note how the linewidth scales with the strength of interactions of CO with its environment. Isolated in the gas phase, CO only interacts with fluctuations in the vacuum field. CO interacts very weakly with the NaCl surface in the physisorbed phase, with neither its vibrational frequency nor lifetime changed by a large amount. However, it interacts much more strongly with Si and Pt when chemisorbed, which leads to both shifts in the vibrational frequency and reduction of the vibrational lifetime. The difference between Si and Pt can be traced back to their electronic structures. Si is a semiconductor with a band gap. Therefore, the primary means of vibrational relaxation is through the interaction of various vibrational modes. Pt is a metal. The absence of a band gap means that excitations of the electrons near the Fermi energy (so-called electron–hole pair formation) are also possible and quite efficient at inducing vibrational relaxation.

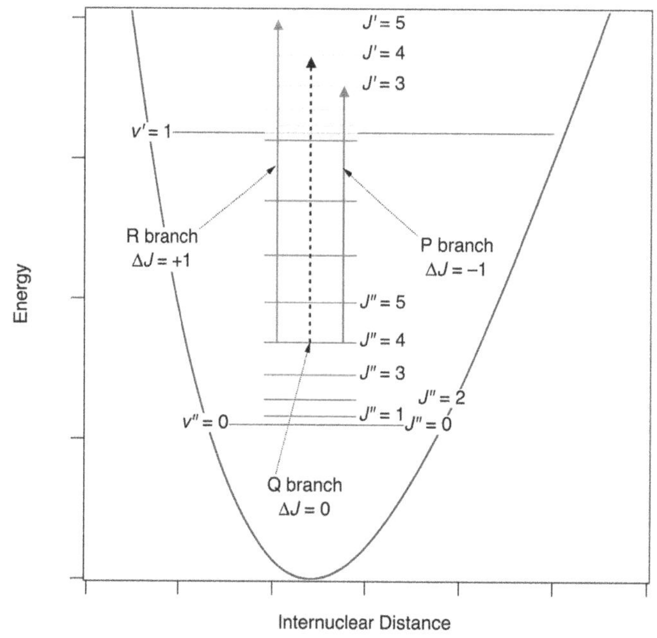

Figure 25.5 The three possible rotational branches arise from the restrictions on simultaneously combining the rotational and vibrational selection rules. The longer arrow of the R branch corresponds to higher wavenumber for the transition. The R branch falls on the high-energy side of the Q branch. The P branch falls on the low-energy side of the Q branch.

25.3.2 P, Q, and R branches

The combination of rotational selection rules with vibrational selection rules leads to a spectrum that can be broken up into three regions known as the *rotational branches*. As illustrated in Fig. 25.5, the R branch corresponds to the $\Delta J = +1$ transitions, the P branch to $\Delta J = -1$ transitions. The Q branch corresponds to $\Delta J = 0$ transitions, which are forbidden for most diatomics, those with Σ ground states, but not necessarily for polyatomic molecules. The lettering scheme for rotational branches is easily remembered as follows. The Q branch is thought of as the origin (*Quelle*) of the band (also called the band center) that is associated with a particular v'–v'' transition. Then assign letters in alphabetical order beginning with $\Delta J = -1$ (P), $\Delta J = 0$ (Q), then $\Delta J = +1$ (R). Should $\Delta J = -2$ be possible (as in Raman spectroscopy or two-photon transitions), then the lettering would go O, P, Q, R, S for the branches spanning from $\Delta J = -2$ to $\Delta J = +2$.

Under the assumption that B is independent of v and J, that is for a harmonic oscillator-rigid rotor, the wavenumber of a rovibrational transition is

$$\tilde{\nu} = \tilde{\nu}_e + B\left[J'\left(J'+1\right) - J''\left(J''+1\right)\right] \tag{25.27}$$

For the R branch $J' = J'' + 1$ and the allowed wavenumbers are

$$\tilde{\nu}_R = R\left(J\right) = \tilde{\nu}_e + 2\left(J''+1\right)B = \tilde{\nu}_e + 2J'B \tag{25.28}$$

For the P branch $J' = J'' - 1$ and the allowed wavenumbers are

$$\tilde{\nu}_P = P\left(J\right) = \tilde{\nu}_e - 2\left(J'+1\right)B = \tilde{\nu}_e - 2J''B \tag{25.29}$$

When labeling the transitions with the notation $R(J)$ and $P(J)$ it is understood that J refers to the lower state value J''.

25.3.3 Anharmonicity and the morse oscillator

The vibrations of real molecules are close to harmonic near the ground state, but exhibit progressively larger deviations for higher vibrational levels, as shown in Figure 25.1. The harmonic oscillator has the obviously unphysical characteristic that it does not dissociate – a poor representation of a molecule if we are looking for a model that allows for chemistry to occur. A Morse oscillator more accurately describes the potential energy curve, in particular because it has a finite binding energy D_e measured from the bottom of the well. The *Morse potential* is defined by

$$V\left(R\right) = D_e\left(1 - e^{-\alpha\left(R-R_e\right)}\right)^2, \tag{25.30}$$

$$\alpha = \sqrt{k/2D_e}, \tag{25.31}$$

$$k = \left.\frac{\partial^2 V}{\partial x^2}\right|_{R_e}, \tag{25.32}$$

where D_e is the classical *dissociation energy* (as opposed to D_v, the dissociation energy of the $(v, J = 0)$ state). D_e differs from D_0 by the *zero point energy*, which is half the fundamental wavenumber

$$D_0 = D_e - \tfrac{1}{2}\omega_e. \tag{25.33}$$

The *force constant k*, which is defined as the curvature of the potential in the vicinity of the minimum at an internuclear separation R_e, and α can be treated as fitting parameters to get the shape of the potential correct. To calculate representative Morse potentials requires the constants D_e, α, k and the equilibrium bond distance. Selected values of D_e, α and k are found in Table 25.4. The equilibrium bond distance is found in Table 25.1.

Anharmonicity leads to a decreasing spacing between levels as v increases. We account for this to a first approximation by adding a correction factor to the vibrational term

$$G\left(v\right) = \omega_e\left[\left(v+\tfrac{1}{2}\right) - x_e\left(v+\tfrac{1}{2}\right)^2\right] = \left[\omega_e - \omega_e x_e\left(v+\tfrac{1}{2}\right)\right]\left(v+\tfrac{1}{2}\right). \tag{25.34}$$

where x_e is the *anharmonicity constant*. It is typically quite small; for the molecules in Table 25.1 it ranges from 0.002 to 0.03. It is usually reported as the product $\omega_e x_e$.

The selection rule is also modified. To see how, let us derive the selection rule explicitly. We need to evaluate the transition dipole moment connecting the lower state with quantum numbers v'', J'' and M'' with the upper state that has v', J' and M'. In Dirac notation this is

$$M_{rv} = \left\langle v'J'M'\right|\hat{\mu}\left|v''J''M''\right\rangle = \left\langle J'M'\right|\hat{\mu}\left|J''M''\right\rangle\left\langle v'\right|\hat{\mu}\left|v''\right\rangle. \tag{25.35}$$

Table 25.4 Ground-state constants for selected diatomic molecules taken from the 96th edition of the *CRC Handbook of Chemistry and Physics*. The value of the Morse parameter α is calculated from Eq. (25.31) and classical dissociation energy D_e from Eq. (25.33). The value of the force constant k for BaI is calculated from ω_e using Eq. (25.20).

Molecule	D_e / eV	D_0 / kJ mol^{-1}	α / Å^{-1}	k / N m^{-1}
$^1H_2$	4.7508	435.78	1.943	575
$^2H_2$	4.7508	439.75	1.943	575
$^1H^{19}F$	6.1223	569.68	2.219	966
$^1H^{35}Cl$	4.6176	431.36	1.867	516
$^1H^{127}I$	3.1958	298.26	1.751	314
$^{19}F_2$	1.6628	158.67	2.970	470
$^{35}Cl_2$	2.5131	242.85	2.003	323
$^{79}Br_2$	1.9908	193.86	1.964	246
$^{127}I_2$	1.5527	152.25	1.859	172
$^{12}C^{16}O$	11.2545	1076.63	2.297	1902
$^{14}N_2$	9.9006	944.87	2.690	2295
$^{14}N^{16}O$	6.6149	630.57	2.743	1595
$^{16}O_2$	5.2256	498.46	2.651	1177
$^{23}Na_2$	0.7466	74.8	0.843	17
$^{138}Ba^{127}I$	3.2978	321.	0.925	90.5

Because of the separability of the wavefunctions, we can factor the first bracket into the product of two brackets. The first bracket on the RHS is precisely the same integral that we solved in Eq. (25.10). It leads to precisely the same selection rules as it did for pure rotational spectroscopy, $\Delta J = \pm 1$, $\Delta M = 0, \pm 1$. This selection rule is strict for diatomics. For polyatomics we have to add the caveat that some vibrational modes carry angular momentum, which makes the Q branch ($\Delta J = 0$) allowed when these vibrations are excited (as is also the case for molecules that do not have a Σ ground state). The selection rules for rovibrational transitions are then the result of the product of selection rules for rotations and vibrations

$$M_{rv} = M_r \langle v' | \hat{\mu} | v'' \rangle = M_r M_v, \tag{25.36}$$

which means that we now only have to solve the vibrational part.

To calculate the vibrational transition dipole moment, we expand the dipole moment function about the equilibrium internuclear separation R_e in the form of a Taylor series,

$$\mu(R) = \mu(R_e) + \mu' [R - R_e] + \dots, \tag{25.37}$$

where $\mu\prime = (d\mu/dR)_{R_e}$. Substituting this expansion into the integral for M_v, we obtain,

$$M_v = \int \psi_{v'}^* \{\mu(R_e) + \mu' [R - R_e] + \dots\} \psi_{v''} \, d\boldsymbol{R} \tag{25.38}$$

$$= \mu(R_e) \int \psi_{v'}^* \psi_{v''} \, dR + \mu' \int \psi_{v'}^* [R - R_e] \psi_{v''} \, dR + \int \psi_{v'}^* \{\dots\} \psi_{v''} \, dR \tag{25.39}$$

The first integral is zero by orthogonality. The second integral is zero if the derivative of the dipole moment as a function of the internuclear distance R vanishes. In other words, *if the vibration does not lead to a change in dipole moment, the transition is not allowed*. The second term also vanishes if the integrand is zero because of symmetry. For the harmonic oscillator, this term only has a finite value (and all higher-order terms vanish) if $\Delta v = \pm 1$. The strength of the vibrational transition is directly proportional to the size of the 'dynamic dipole,' that is, the magnitude of the change in dipole moment that occurs during the transition.

Anharmonicity leads to small changes in the behavior predicted for a harmonic oscillator. These deviations result from finite values in the terms in Eq. (25.39) that are supposed to vanish. As these deviations are reflected in progressively higher-order correction terms, they become progressively smaller. This means that for an anharmonic oscillator, the $\Delta v = \pm 1$ is by far the strongest transition. The $\Delta v = \pm 2$ transition is weaker by one or two orders of magnitude, depending on the extent of anharmonicity. Each progressively larger change in v is again weaker by one or two orders of magnitude.

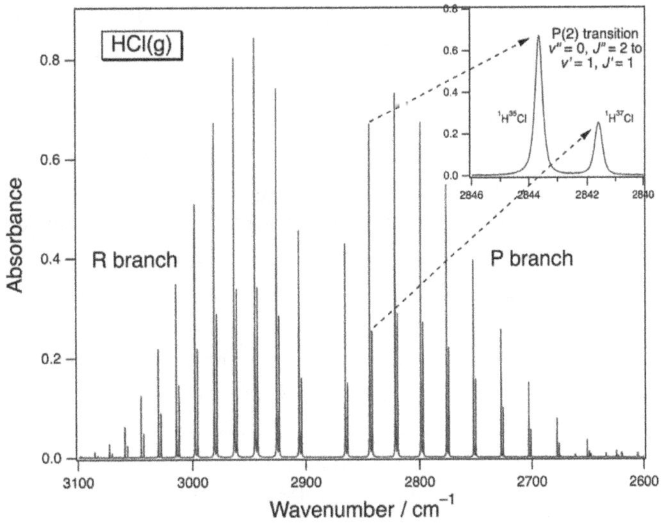

Figure 25.6 The rovibrational spectrum of HCl for the $v'' = 0$ to $v' = 1$ fundamental transition. Each line appears as a doublet (two closely spaced lines, see inset) because of the presence of two Cl isotopes. The molecules in the naturally occurring sample exist as two different isotopologues $^1H^{35}Cl$ and $^1H^{37}Cl$. Some of the small peaks near 2600 cm^{-1} are due to impurities of HBr in the sample.

The transition beginning in $v'' = 0$ and ending in $v' = 1$ band is known as the fundamental band. In fact, all transitions with $\Delta v = 1$ are known as fundamental transitions. The $\Delta v > 1$ transitions are overtones; $\Delta v = +2$ is the first overtone, and so forth. A transition that starts from a level above $v = 0$, such as $v = 1$ to $v = 2$, is known as a 'hot band.' The anharmonicity is also related to the dissociation energy by

$$D_e \approx \frac{\omega_e}{4x_e} = \frac{\omega_e^2}{4x_e\omega_e}. \quad (25.40)$$

The Boltzmann population of a rovibrational level depends on the energy and temperature as before with the inclusion of the zero-point energy,

$$\frac{N_{vJ}}{N_{00}} = \frac{g_{vJ}}{g_{00}} e^{-(E_{vJ}-E_{00})/k_B T}. \quad (25.41)$$

Within any given vibrational band, the intensity of rovibrational transitions follows the Boltzmann factor, just as for pure rotational transitions. This variation in intensity is present in Fig. 25.6. When comparing different vibrational bands, for example comparing the band originating in $v = 0$ and ending in $v = 1$ with the $v = 0$ to $v = 2$ band, the intensity is a product of the line strength factor and the Boltzmann factor.

The vibrational term value $G(v)$ is determined by one quantum number and two molecular constants (the fundamental wavenumber and the anharmonicity constant). These two constants cannot be determined by one transition; however, they can be determined by measuring any two transitions. The wavenumber of a transition is given by the difference of $G(v)$ values.

$$\Delta G(v) = G(v_f) - G(v_i) = G(v') - G(v''). \quad (25.42)$$

The wavenumber difference for a transition between any two states differing by $\Delta v = 1$ is given by

$$\Delta G_{v+1/2} = G(v+1) - G(v) = (1 - 2x_e)\,\omega_e - 2vx_e\omega_e. \quad (25.43)$$

Thus, a plot of $\Delta G_{v+1/2}$ versus the vibrational quantum number v should have a slope of $-2x_e\omega_e$ and an intercept of $(1 - 2x_e)\,\omega_e$.

25.3.3.1 Directed practice

Why does anharmonicity lead to decreased spacing between vibrational levels?

25.3.4 Vibrations of polyatomic molecules

As we learned in Chapter 6 on statistical mechanics, an atom has three translational degrees of freedom. N atoms have $3N$ translational degrees of freedom. When these atoms are bound into a molecule, the number of degrees of freedom must be conserved. The molecule has three translational degrees of freedom, leaving $3N - 3$ for rotations and vibrations. A linear molecule (whether diatomic or polyatomic) only has two rotational degrees of freedom. Thus, a linear molecule has $3N - 5$ vibrational modes. A nonlinear molecule has three rotations and $3N - 6$ vibrations.

Vibrational modes (also called the normal modes) are named simply by placing a number after ν (Greek nu). Sometimes, vibrations are further classified as being one of four characteristic types of vibrations. Stretches change the bond length and are also denoted with ν. In-plane motions that change bond angles but not bond lengths are bends. These are denoted with δ and are sometimes further subdivided between rocks, twists, and wags. Out-of-plane bends are assigned the symbol γ. Torsions (τ) change the angle between two planes containing the atoms involved in the motion.

Each vibrational mode has a characteristic frequency and each mode has rotational branches associated with it. The rovibrational spectra of polyatomic molecules quickly become very complicated (the spectra are highly congested) as the number of atoms increases (or when the quantum numbers become large). For molecules with around 10 or more atoms, the room-temperature spectra are impossible to resolve. The rovibrational states form a continuum and only broad bands rather than sharp lines are found.

A mode is infrared-active (it appears in normal absorption and emission spectra) only if the mode creates an oscillating dipole moment. This means that the mode must change the dipole moment of the molecule, or else it is inactive. The dipole moment only has to change along any one of the symmetry axes of the molecule and this condition can be expressed in terms of the symmetry of the molecule:

> *There will be a rovibrational spectrum corresponding to a normal mode of vibration only if that mode belongs to the same symmetry species as one or more of the three coordinates x, y and z.*

The symmetry of polyatomic molecules is discussed further in Chapter 26.

25.4 Raman spectroscopy

Spectroscopy can be performed on scattered light, not just absorbed or emitted light. Scattering occurs when light collides with an atom or molecule and changes its direction. Until now we have only considered the role of *real states* – all those states that the isolated molecule can occupy. However, *virtual states* – states that only exist when the molecule is in the presence of an optical field – may also play a role, particularly in laser spectroscopy for which the optical field strength can be considerable. Virtual states play a role in a range of nonlinear spectroscopies. Linear spectroscopy is spectroscopy in which the response of the system depends linearly on the intensity of the light. Up to this point all of the absorption and emission of light that we have studied has involved linear spectroscopy. In nonlinear spectroscopy, the response depends on some higher power (square, cube, etc.) of the light intensity. Frequently, the number of photons involved will equal the order in this intensity dependence. Examples include two-photon absorption and the Raman effect. This effect is named after Chandrasekhara V. Raman, who in 1928 published[7] the first observation of the effect, for which he was awarded the Nobel Prize in Physics 1930.

In the *Raman effect*, the incident radiation is not resonant with the any transition in the molecule. The incident light can, however, polarize the electron distribution in the molecule. The size of the induced dipole is proportional to the electric field strength in the light and the polarizability of the molecule. At low field strength, the interaction is generally elastic and not of much consequence. The probability of any energy exchange is quite low. However, as the field strength increases, the polarization in the molecule increases and the chance of energy exchange with the field increases nonlinearly. This is the basis of a number of nonlinear and multiphoton phenomena.

If one looks carefully at the light coming out of the system (particularly in a direction not along the incident beam), there are peaks on either the red (low-energy) side or blue (high-energy) side of the incident frequency, as shown in Fig. 25.8. The low-energy lines are the Stokes lines. The high energy lines are the anti-Stokes lines. These lines are the Raman lines and they correspond in energy to molecular transitions. There can be electronic, vibrational and rotational Raman transitions. The utility of Raman spectroscopy was greatly enhanced by the invention of the laser for several reasons. One reason is that a laser naturally forms a well-collimated beam. The detector for Raman scattering is placed at 90° with respect to the propagation direction of this beam, as shown in Fig. 25.9, where the contrast between incident and scattered light is best. Lasers can have an extremely well-defined wavelength, and this facilitates the implementation of wavelength-selective detection with filters and gratings that allows one to detect light very close to the wavelength of the laser. Finally, lasers can be operated at extremely high intensity compared to conventional light sources.

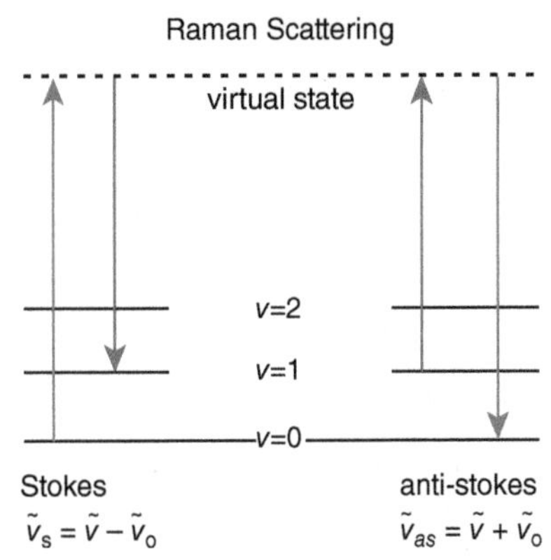

Figure 25.7 Raman scattering involves two photons. The molecule is excited by the first photon to a virtual state (dashed line). A second photon is scattered from the molecule. The difference in energy between the two photons is made up by the molecule. Stokes scattering leaves the molecule excited and produces a photon at lower wavenumber. Anti-Stokes scattering relaxes an excited state of the molecule and produces a photon at higher wavenumber.

The intensity of scattered light varies inversely with the wavelength of incident light raised to the fourth power,

$$I_s = \lambda^{-4}. \tag{25.44}$$

This was shown first by Rayleigh, who studied elastically scattered light; hence the elastic peak is called the Rayleigh peak. The same is true of Raman scattering. Because of this strong wavelength dependence, blue light scatters and more effectively generates Raman scattering than does red light. Rayleigh scattering gives a cloudless sky its blue appearance, as red light from the sun tends to pass through the atmosphere more easily while blue light is scattered to Earth.

The Raman effect is essentially a two-photon event mediated by a virtual state, as shown in Fig. 25.7. Because it contains both 'up' and 'down' transitions, the pure rotational or rovibrational transitions associated with it (at least in diatomics) correspond to $\Delta J = 0, \pm 2$, that is, to Q, S, and O branches. The pre-condition for Raman spectroscopy requires a *change in the polarizability of the molecule*. This changes as a quadratic function of the symmetry axes. Thus, vibrations that transform as x^2, y^2, z^2, xy, xz and yz are Raman-active. Some modes are both Raman- and infrared-active.

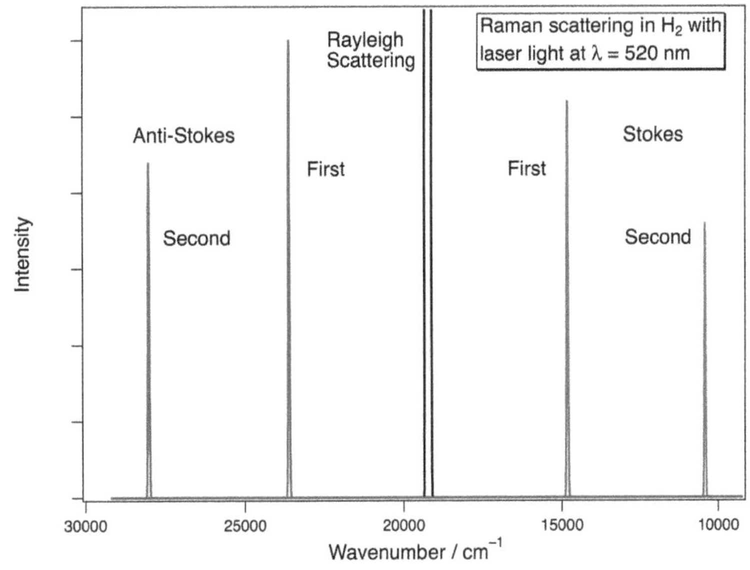

Figure 25.8 Vibrational Raman scattering in H_2 leads to large shifts from the central wavenumber of the light used to induce the scattering. Light scattered with no change in wavenumber appears in the Rayleigh scattering line. In the spectrum above, the intensity of the Rayleigh scattering is off-scale compared to the Raman lines. Stokes shifting produces peaks on the red (lower wavenumber) and anti-Stokes shifting produces peaks on the blue (high wavenumber) side of the central line.

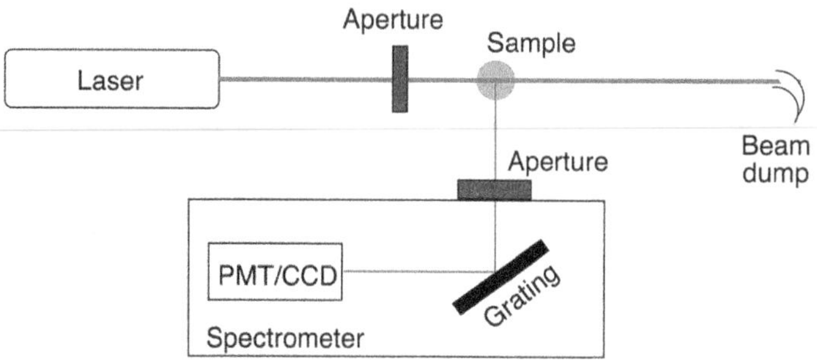

Figure 25.9 Raman scattering from the sample is detected at right-angles to the direction of propagation of the laser beam used to excite it. Apertures, an effective beam dump and others optics (lenses, polarizers, filters) help to suppress scattered light. A grating in front of the photomultiplier tube (PMT) or charge-coupled device (CCD) allow one to obtain the spectrum of scattered light.

For pure rotational Raman from a diatomic excited by a laser at wavenumber $\tilde{\nu}_L$, lines at wavenumber $\tilde{\nu}$ appear on either side of the Rayleigh peak with a wavenumber shift

$$\Delta\tilde{\nu} = \tilde{\nu} - \tilde{\nu}_L \tag{25.45}$$

given by

$$|\Delta\tilde{\nu}| = F(J+2) - F(J). \tag{25.46}$$

Neglecting centrifugal distortion and substituting from Eq. (25.7), the separation of a peak due to Raman scattering away from the Rayleigh scattered light – that is, the *Raman shift* – is

$$|\Delta\tilde{\nu}| = 4B_0 J + 6B_0. \tag{25.47}$$

25.4.1 Directed practice

Confirm the result for the Raman shift of pure rotational lines in Eq. (25.47).

Vibrational spectroscopy utilizing the Raman effect is one of the most widespread applications of Raman spectroscopy. In part this is because the selection rules are different than those of infrared absorption spectroscopy, which allows for complementary analysis. Some modes that are not IR-active are Raman-active, and vice versa. Vibrational Raman spectroscopy is also allowed on homonuclear diatomics. Whereas conventional vibrational spectroscopy by absorption or emission requires a change in dipole moment, Raman spectroscopy requires a change in polarizability with displacement along the vibrational coordinate. Analogous to Eq. (25.37), we expand along $x = R - R_e$,

$$\boldsymbol{\alpha} = \boldsymbol{\alpha}(R_e) + \left(\frac{d\boldsymbol{\alpha}}{dx}\right)_{R_e} x + \ldots \tag{25.48}$$

The vibrational Raman transition dipole moment then becomes,

$$\boldsymbol{M}_v = \left(\frac{d\boldsymbol{\alpha}}{dx}\right)_{R_e} \int \psi_{v'}^{*}\, x\, \psi_{v''}\, dx + \ldots. \tag{25.49}$$

The first and strongest term is nonzero only if

$$\Delta v = \pm 1. \tag{25.50}$$

Anharmonicity again allows the observation of much weaker overtones as in conventional vibrational absorption.

25.4.2 Directed practice

Why can vibrational Raman spectroscopy be performed on homonuclear diatomics whereas infrared absorption spectroscopy cannot?

25.5 Electronic spectra of molecules

25.5.1 Born–Oppenheimer meets Franck–Condon

The separability of the wavefunction

$$\psi(\mathbf{r},\mathbf{R}) = \psi_{\mathrm{r}}(\mathbf{R})\,\psi_{\mathrm{v}}(\mathbf{R})\,\psi_{\mathrm{e}}(\mathbf{r},\mathbf{R}) \tag{25.51}$$

has important implications for electronic spectroscopy. In particular, the separation of the electronic coordinates from the nuclear coordinates has a profound effect on the transition dipole moment. This separation arises because electrons respond almost instantaneously to changes in the nuclear coordinates. This is the basis of the Born–Oppenheimer approximation. It means that an electronic transition occurs in the following order. An electron changes its wavefunction from that of one state to another with the nuclei frozen in position. The nuclei then react and relax to new equilibrium positions that correspond to the electronically excited state.

Neglecting for the moment the rotational part (we will factor it in later), the transition dipole moment is calculated from

$$\boldsymbol{M}_{\mathrm{ev}} = \langle \psi'(\mathbf{r},\mathbf{R})\,|\hat{\boldsymbol{\mu}}|\,\psi''(\mathbf{r},\mathbf{R})\rangle = \iint \psi_{\mathrm{e}}'^{*}\psi_{\mathrm{v}}'^{*}\,\hat{\boldsymbol{\mu}}\,\psi_{\mathrm{e}}''\,\psi_{\mathrm{v}}''\,\mathrm{d}\mathbf{r}\,\mathrm{d}\mathbf{R}. \tag{25.52}$$

First, we integrate over the electronic coordinates $\mathbf{r}$ to obtain

$$\boldsymbol{M}_{\mathrm{ev}} = \int \psi_{\mathrm{v}}'^{*}\boldsymbol{M}_{\mathrm{e}}\,\psi_{\mathrm{v}}''\,\mathrm{d}\mathbf{R} \tag{25.53}$$

where $\boldsymbol{M}_{\mathrm{e}}$ is the electronic transition moment

$$\boldsymbol{M}_{\mathrm{e}} = \int \psi_{\mathrm{e}}'^{*}\,\hat{\boldsymbol{\mu}}\,\psi_{\mathrm{e}}''\,\mathrm{d}\mathbf{r} = -e\int \psi_{\mathrm{e}}'^{*}\mathbf{r}\,\psi_{\mathrm{e}}''\,\mathrm{d}\mathbf{r} \tag{25.54}$$

We can do this because the vibrational part of the wavefunction is independent of the electronic coordinates $\mathbf{r}$. Thus, it factors out of the integral involving the nuclear coordinates. The Born–Oppenheimer approximation allows us to consider the nuclear coordinates as fixed with respect to electronic motion. Thus, $\boldsymbol{M}_{\mathrm{e}}$ is a constant that can be pulled out of the integral over the nuclear coordinates $\mathbf{R}$, to yield

$$\boldsymbol{M}_{\mathrm{ev}} = \boldsymbol{M}_{\mathrm{e}}\int \psi_{\mathrm{v}}'^{*}\psi_{\mathrm{v}}''\,\mathrm{d}\mathbf{R} = \boldsymbol{M}_{\mathrm{e}}\,\boldsymbol{M}_{\mathrm{v}} \tag{25.55}$$

The transition dipole moment is composed of two terms. The first, $\boldsymbol{M}_{\mathrm{e}}$, the electronic transition dipole moment, describes the strength of how well the two electronic states are coupled to one another by the radiation. The second, $\boldsymbol{M}_{\mathrm{v}} = \langle \psi_{\mathrm{v}}' | \psi_{\mathrm{v}}'' \rangle$, describes the vibrational overlap of the two electronic states. By vibrational overlap we mean how closely the vibrational levels in the two electronic states resemble each other; specifically, it compares whether the two vibrational state wavefunctions have significant amplitude at the internuclear separation where the transition occurs.

Vibrational overlap is the basis of the *Franck–Condon principle*:

> *Electronic transitions are vertical. In principle, an electronic transition can go from the ground state to any vibrational state in the excited state; however, it is the state with the greatest vibrational overlap with the ground state that has the highest intensity.*

The Franck–Condon principle holds for transitions with energies well into the extreme UV and even soft X-ray regions. By a *vertical transition*, we mean that the linear momentum transfer from the photon is negligible and does not change the positions of the nuclei during the transition. Thus, transitions between Born–Oppenheimer potential energy curves for the electronic states can be drawn as vertical lines.

A strong transition is said to have a good *Franck–Condon factor* as opposed to a weaker transition with a bad Franck–Condon factor. The Franck–Condon factor is quantified by the square of the vibrational overlap integral

$$\text{FC factor} \equiv |\boldsymbol{M}_{\mathrm{v}}|^{2} = \left|\langle \psi_{\mathrm{v}}' | \psi_{\mathrm{v}}'' \rangle\right|^{2}. \tag{25.56}$$

If the value of the equilibrium bond distance R_{e} is the same in the lower and upper states, the Franck–Condon factor is greatest for the $\Delta v = 0$ transition but then drops off to higher values, exactly as we would expect for an anharmonic oscillator. When the R_{e} values in the two states are displaced, there will be a maximum Franck–Condon factor at some value of $\Delta v \neq 0$, the larger the displacement, the greater the value of the difference. The intensity of transitions on either side of this maximum decays to lower intensity.

25.5.2 Rovibrational structure of electronic transitions

We have now established that a transition between two electronic states can be accompanied by a range of changes in the vibrational quantum number. The strength of the electronic transition depends on the strength of the coupling between the electronic states multiplied by the Franck–Condon factor. Next, we must account for the rotational state structure. It is accounted for in exactly the same way as for a rovibrational transition. Energy and angular momentum must be conserved. Therefore, for each vibrational band within each electronic transition, there will be rotational branch structure. For a one-photon transition, we will always observe P and R branches. If there is a change in the orbital angular momentum, then a Q branch can also be observed. By far the most common electronic transitions in diatomic molecules are Σ–Σ and Σ–Π transitions. For a Σ–Σ transition, there can be no change in orbital angular momentum, therefore electronic transitions must obey $\Delta J = \pm 1$. For a Σ–Π transition, the selection rule is broadened to $\Delta J = 0, \pm 1$.

Energy conservation demands that the photon energy is resonant with the difference in energy between the upper and lower states. The total term value in cm^{-1} T for a given rovibrational state in a particular electronic state is now the sum of the electronic term T_e, the vibrational terms $G(v)$ and the rotational term $F(J)$,

$$T = T_e + G(v) + F(J). \tag{25.57}$$

The ground electronic state is always called the X state. The electronic term value of the X state is used as the origin, that is $T_e(\text{X}) = 0$. Electronic term values are measured from the position of the minimum of the potential energy curve. The wavenumber of an electronic transition between a rovibrational level in the lower electronic state characterized by (e", v", J") and an upper level corresponding to (e′, v', J') is

$$\tilde{v} = T\left(\text{e}', v', J'\right) - T\left(\text{e}'', v'', J''\right). \tag{25.58}$$

Just as for the ground electronic state, the rovibrational states of electronically excited states are well-characterized as being anharmonic oscillators and centrifugally distorted rotors.

25.5.3 State designations for linear molecules

We previously encountered *term symbols* for atoms. The total orbital angular momentum L corresponds to a particular upper-case Roman letter, and the spin multiplicity $(2S + 1)$ is denoted in a leading superscript. Similar information is contained in a state designation for a linear molecule.

State designations for linear molecules (diatomics and polyatomics) are determined by the value of Λ, the projection of electron orbital angular momentum along the internuclear axis. The projection must be used because its total magnitude is undefined – that is, L is a bad quantum number. Λ is a positive number and its value is represented by an upper-case Greek letter according to the naming scheme presented in Table 25.5. All states with $\Lambda > 0$ are doubly degenerate. This can be thought of as Λ, when it has a non-zero value, being equal regardless of whether it points to the left or right along the internuclear axis.

The total spin S is a good quantum number in a molecule as is its projection along the internuclear axis Σ.

$$\Sigma = S, S-1, \cdots - S. \tag{25.59}$$

The spin multiplicity $(2S + 1)$ is added to the Greek letter denoting the term as a superscript on the left, much like it was for atoms.

Table 25.5 The letter portion of term symbols is related to the magnitude of orbital angular momentum in the state. Atoms and linear molecules follow analogous schemes for relating the letter to magnitude. Roman letters are used for atoms. Greek letters are used for molecules.

Atoms	Term symbol	Linear molecules	State designation
$L = 0$	S	$\Lambda = 0$	Σ
$L = 1$	P	$\Lambda = 1$	Π
$L = 2$	D	$\Lambda = 2$	Δ
$L = 3$	F	$\Lambda = 3$	Φ

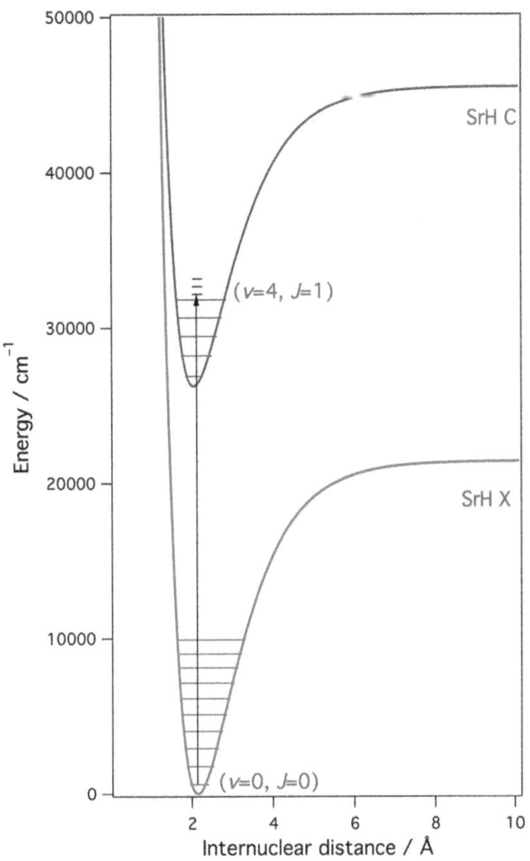

Figure 25.10 An electronic transition in a diatomic molecules occurs between specific rovibrational states in the lower and upper states. The absorption transition shown here is from the ($v = 0$, $J = 0$) rovibrational level of the ground electronic state of SrH (the X state) to the ($v = 4$, $J = 1$) rovibrational level of the electronically excited C state. The C state is unusual in that it is an excited state with a shorter bond length than the ground state.

The component of total angular momentum along the internuclear axis takes on values $\Omega\hbar$ where the quantum number Ω is given by

$$\Omega = |\Lambda + \Sigma| \,. \tag{25.60}$$

If we wish to differentiate the various components of a spin multiplet, the value of $|\Lambda + \Sigma|$ is used as a right subscript. For example, a state with $\Lambda = 1$ and $S = 1$ has $\Sigma = 1, 0, -1$ and the three components of the $^3\Pi$ term correspond to $\Omega = 2, 1, 0$. The resulting state designations are $^3\Pi_2$, $^3\Pi_1$ and $^3\Pi_0$. Spin–orbit coupling shifts these components by an amount

$$\Delta E = A\Lambda\Sigma, \tag{25.61}$$

where A is the spin–*orbit coupling constant.*

In addition to the quantum numbers Λ, S and Ω, we need to consider either one (for heteronuclear diatomics) or two (for homonuclear diatomics) symmetry properties to define the total state designation. The symmetry operations are shown in Fig. 25.11. All of these properties together will also be used to define the selection rules.

If the molecule has a center of inversion, then we need to add an indication of whether the state is even (gerade) or odd (ungerade) with respect to inversion. This letter is placed as a following subscript. If the molecule does not contain a center of inversion, no letter is indicated. Only homonuclear diatomics and symmetric linear polyatomics such as CO_2 and C_2H_2 exhibit this type of symmetry.

The second symmetry property applies to all diatomics, but only for Σ states (because they have $\Lambda = 0$). We can also consider the symmetry of the molecule upon reflection through any plane containing the internuclear axis; this symmetry is called *parity*. If there is no change in sign upon reflection, the state is designated +, but if the sign changes the state is designated –. If the molecule lacks a plane of symmetry in such a plane, no sign is indicated.

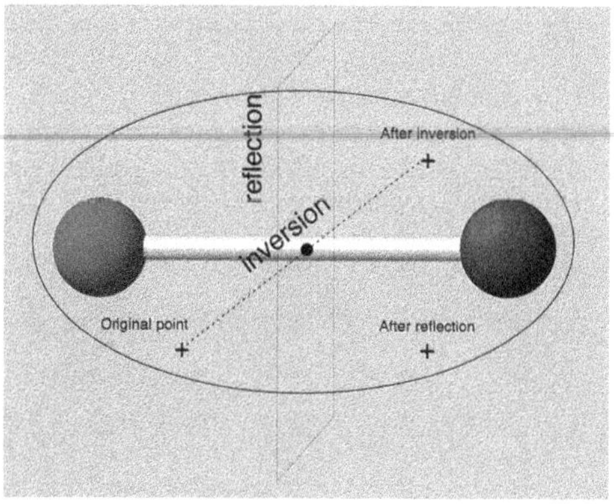

Figure 25.11 The transformations of molecular electronic states under inversion and reflection are used to define the gerade/ungerade and +/– (parity) character in molecular state designations.

Hence, for the H_2 molecule in its ground state, $\Sigma = 0$, $\Lambda = 0$, the inversion symmetry is gerade and the parity is positive. Therefore, the ground electronic state of the H_2 molecule is denoted $H_2X\ ^1\Sigma_g^+$. The ground state is always denoted the X state. Excited states with the same spin multiplicity as the ground state are denoted by upper-case letters in alphabetical order, A, B, C, and so on. Excited states with a different spin multiplicity are designated by a lower-case letter in alphabetical order, a, b, c, and so on. Several exceptions to the naming rules occur. The B state of H_2 is so high in energy that investigators originally felt there must be an A state lurking below, but none exists. Thus, the first excited state of H_2 is the B $^1\Sigma_u^+$. The states of N_2 are a mess with the order X $^1\Sigma_g^+$, A $^3\Sigma_u^+$, B $^3\Pi_g$, W $^3\Delta_u$, B′ $^3\Sigma_u^-$, and a′ $^1\Sigma_u^-$. Similarly, misinterpretation led to the nonsystematic naming of the first few states of I_2 because usage of nonsystematic names was already too common to allow for changing the names.

25.5.4 Selection rules

The selection rules for molecular transitions are analogous to those of atoms: energy and angular momentum must still be conserved. To this we add additional symmetry constraints.

- The spin selection rule is the same as for atoms: the spin is not allowed to change, $\Delta S = 0$ (or $\Delta\Sigma = 0$).
- For atoms $\Delta L = 0, \pm 1$. For molecules $\Delta\Lambda = 0, \pm 1$.
- $\Delta\Omega = 0, \pm 1$ for transitions between multiplet components.
- For Σ–Σ transitions, the parity must not change

$$+ \leftrightarrow +$$
$$- \leftrightarrow -$$
$$+ \not\leftrightarrow -$$

- The inversion symmetry must change

$$u \leftrightarrow g$$
$$u \not\leftrightarrow u$$
$$g \not\leftrightarrow g$$

25.5.4.1 Example

The O_2 molecule is particularly important because of its abundance in the atmosphere. It is quite interesting spectroscopically because is a diradical and has a number of low-lying electronic states. The properties of the first five electronic states of O_2 are depicted in Table 25.6 and Fig. 25.12.

The triplet to singlet transitions are forbidden. This means that their intensity is quite low. The b←X transition occurs in the red to IR (peak at 760 nm); this is known as the *sunset band*. It is triply forbidden by the spin change and g ← g and + ← –. The a←X transition is further in the IR at about 1300 nm and is triply forbidden because in addition to the spin and g←g, $\Delta\Lambda = 2$. The Schumann–Runge band is a strong absorption band that corresponds to the B←X transition at around 200 nm. Absorption in this band leads to some structured absorption but also photodissociation of O_2 because of the large Franck–Condon shift in electronic transitions. A large shift in bond distance also accompanies the X state to

Table 25.6 The spectroscopic constants of the first five states of O_2. Values obtained from Herzberg and Huber as now collected within the NIST Webbook http://webbook.nist.gov/.

State	T_e / cm^{-1}	ω_e / cm^{-1}	$x_e\omega_e$ / cm^{-1}	B_0 / cm^{-1}	r_e / Å
$B^3\Sigma_u^-$	49802.1	700.36	8.002	0.819	1.60
$A^3\Sigma_u^+$	36096	819	22.5	1.05	1.42
$b^1\Sigma_g^+$	13195.222	1432.687	13.9500	1.40041	1.22675
$a^1\Delta_g$	7918.1	1509.3	12.9	1.4264	1.2155
$X^3\Sigma_g^-$	0	1580.361	12.0730	1.44566	1.20739

B state transition. Its Franck–Condon factor thus favors a large change Δv in the vibrational quantum number. The A←X transition is forbidden on the basis of parity.

On examining the O_2 potential energy curves displayed in Fig. 25.12 we see a wealth of information. The first four states are all approaching the same asymptote. This is because all four states correlate with the same dissociation limit, which corresponds to the O atom in its ground state,

$$O_2(\Lambda) \rightarrow O(^3P) + O(^3P).$$

The B state correlates with dissociation into one ground-state O atom and one excited into the ^{1}D state

$$O_2(B) \rightarrow O(^3P) + O(^1D).$$

Because $O(^1D_2)$ lies 15 867 cm^{-1} above $O(^3P_0)$, the dissociation limit of the B state lies 1.97 eV above that of the X state.

The dissociation energy D_e correlates with the equilibrium bond length. On the scale of Fig. 25.12 it is difficult to pick up the slight shifts in the a and b states compared to the X state. However, the much more weakly bound A and B states exhibit quite large shifts in the potential well minimum. O_2 is unusual in having two low-lying excited states nested within the X state potential. In this respect, the A and B state potentials are more typical in that excited states generally either correspond to: (i) a significantly more weakly bound molecule (or even a repulsive state) that correlates to the same dissociation limit; or (ii) a molecule with a different dissociation limit. In exceptional cases such as the SrH C state depicted in Fig. 25.10, an excited state can have a greater D_e value and shorter r_e than in the ground state.

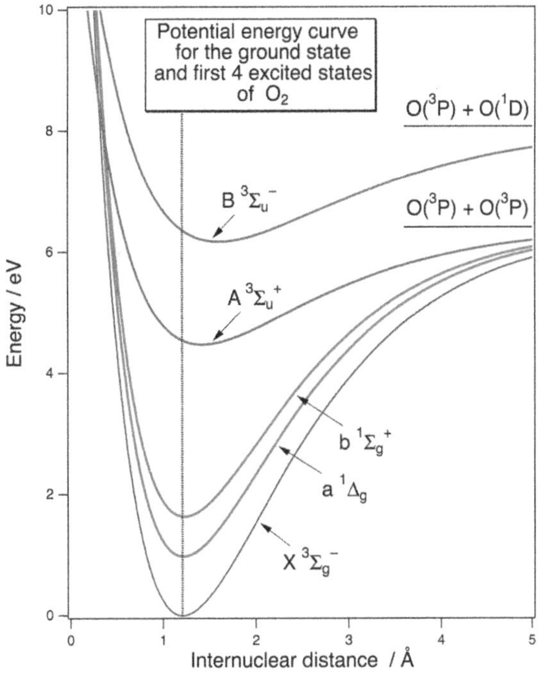

Figure 25.12 Potential energy curves for the ground electronic state and first four electronically excited states of O_2. Oxygen is unusual in that its ground state is a diradical and the ground state is a triplet. It is also unusual in that low-lying electronic states are nested within the ground state.

From Eq. (25.32) we see that the force constant of a bond is related to the curvature of the potential energy well near its minimum. From Eq. (25.20) a larger force constant leads to a larger spacing between vibrational levels. Therefore, a narrow potential such as that of the X state has a larger ω_e value than a broad potential such as that of the B state. The curvature of the potential usually correlates with depth of the well – a larger well depth leads to a greater curvature and greater fundamental vibrational frequency. These trends are confirmed in Table 25.6. Furthermore, since the rotational constant is inversely proportional to r^2, shifts to longer r_e in the excited states also result in smaller rotational constants and closer spacings between rotational levels.

25.6 Excited-state population dynamics

When an excited state is formed by the absorption of a photon, a chemical reaction or collision with a molecule or electron, it will eventually relax to a lower energy state. It can do so by either radiative or nonradiative processes. In all cases energy and momentum must be conserved. Here, we investigate the processes that change the population of excited states.

25.6.1 Radiative decay (photon emission)

In Chapter 23 we learned that the rate of spontaneous emission increases as the cube of the frequency of the transition. In terms of the energy, this can be expressed as

$$\frac{A_{ik}}{B_{ik}} = \frac{8\pi}{h^2c^3}\Delta E_{ki}^3. \tag{25.62}$$

The rate of spontaneous emission of a photon from an excited states increases with the cube of the energy difference between the states involved in the transition. This has a profound effect on the dynamics of excited states. It means that making a laser operate between rotational levels is so easy that it even happens in Nature.[8] On the other hand, achieving laser output in the X-ray region is exceedingly difficult.

Figure 25.13 demonstrates another implication of Eq. (25.62). Once an excited state has been populated, it is possible for it to relax by the spontaneous emission of a photon. Spontaneous emission is caused by the interaction of a molecule with the quantized electromagnetic field that pervades all space. Fluctuations in this vacuum field couple to the excited state and cause it to be unstable with respect to radiative relaxation. In principle, the excited state in Fig. 25.13 could relax via: (a) a rotational; (b) a rovibrational; or (c) an electronic transition. Because of the cubic dependence in Eq. (25.62), the radiative lifetime for an allowed pure rotational transition is extremely long, that of an allowed rovibrational transition is long, and that of an electronic transition is significantly shorter. A normal allowed electronic transition in a diatomic has a radiative lifetime on the order of 100 ns. Therefore, radiative relaxation of a rovibrational level in an excited electronic state occurs exclusively through an electronic transition.

25.6.2 Nonradiative relaxation

Conservation of energy applies to the transition of a molecule from one state to another. If the initial and final states of a molecule have different energies, the difference has to be made up by the emission or absorption of a photon (a *radiative transition*) or by exchange of energy with a collision partner (a *nonradiative transition*). In the absence of collisions this means that a molecule can only change its rovibrational state (or electronic state) if the initial and final states are degenerate. If allowed by symmetry, degenerate states perturb each other, which leads to a quantum mechanical mixing of their wavefunctions. Accidental degeneracies occur frequently if the density of states is high. For diatomic molecules (and small polyatomics in low-energy states for which the density of states is not too high) such accidental degeneracies are infrequent. This means that there is almost never a nonradiative transition between rovibrational states or electronic states. For larger polyatomics (as discussed in the next chapter) such degeneracies and mixing are common. They play a key role in intramolecular energy redistribution.

A rare counterexample is CN. When observing low-pressure gas-phase fluorescence from the CN A $^2\Pi$ ($v' = 7$) vibrational level to the ground X $^2\Sigma^+$ state, perturbations are found as reflected in the spectral linewidth $\delta\tilde{\nu}$ and fluorescence quantum yield ϕ_f. The *fluorescence quantum yield* of a particular state is defined as the ratio of the number fluorescence events to the total number of excitations to that state,

$$\phi_f = \frac{n_{\text{events}}}{n_{\text{excitations}}} \tag{25.63}$$

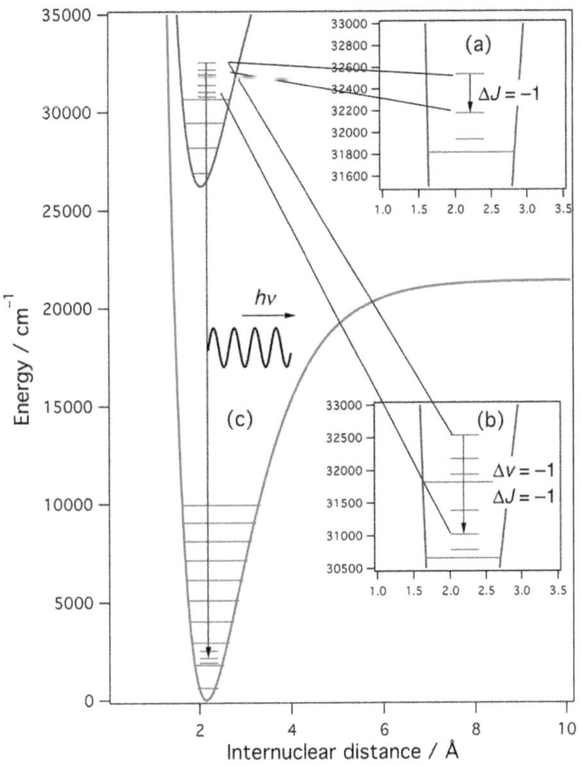

Figure 25.13 Once a rovibrational state of an excited electronic state is populated it can relax radiatively by (a) changing rotational state, (b) changing both rotational and vibrational states, or (c) by changing the electronic state as well as the rovibrational state. Because of the larger transition energy, the electronic transition is much faster than the other two alternatives.

The fluorescence quantum yield is unity if all excitations lead to fluorescence. The quantum yield is less than one if other relaxation channels besides fluorescence are available to the molecule.

For rotational levels in CN A $^2\Pi$ ($v' = 7$) with $20 \le J \le 50$, there is a broadening of the lines and a decrease in ϕ_f. The rotational and vibrational constants of the A and X states lead to an accidental degeneracy for rovibrational levels between A $^2\Pi$ ($v' = 7$) and X $^2\Sigma^+$ ($v'' = 11$) for $20 \le J \le 50$. All other rovibrational levels in ($v' = 7$) exhibit narrow lines with near unit fluorescence quantum yield. Those rovibrational levels in the A state that are degenerate with rovibrational levels in the X state form perturbed states. Once populated, they relax as mixed states, which can transfer population from the A state to the X state. Therefore, their lifetime decreases. This increases the linewidth of the transition in accord with the energy–lifetime uncertainty relationship. It also decreases the quantum yield.

The type of nonradiative process described here, where a molecule makes an intramolecular transfer from one electronic state to another electronic state of the *same spin multiplicity*, is called an *internal conversion*. An intramolecular transfer from one electronic state to another electronic state of a *different spin multiplicity*, is called an *intersystem crossing*. For highly excited small polyatomics and large polyatomics even in the first excited state, we shall find that nonradiative transfer is quite common (this subject will be discussed in Chapter 26).

In diatomics, nonradiative transitions almost always are associated with collisions. Whereas radiative transitions become more probable the greater the energy difference between the initial and final states, collisional energy transfer is more likely the closer the transition in the collision partners is to degenerate and the smaller the energy transfer. Translational states are continuous (at least as long as we do not put our molecules in a box of molecular dimensions). Therefore, it is very easy for molecules to exchange translational energy. Virtually every collision leads to changes in translational energy. Rotations and vibrations are quantized and the spacings between them are generally significant for diatomics. The most probable collisions to exchange energy are like with like. A CO ($v = 0$, $J = 15$) colliding with a CO($v = 0$, $J = 14$) is extremely likely to exchange rotational energy. That is, the first molecule experiences $\Delta\vartheta = -1$ and the second $\Delta\vartheta = +1$ in the collision. There is no restriction on the translational energy exchange. It is less probable for these two collision partner to undergo a $\Delta J = \pm 14$ exchange. Because of the small energy difference between rotational levels, rotation-changing collisions are very frequent. For example, virtually every collision of N_2 with CN in either the

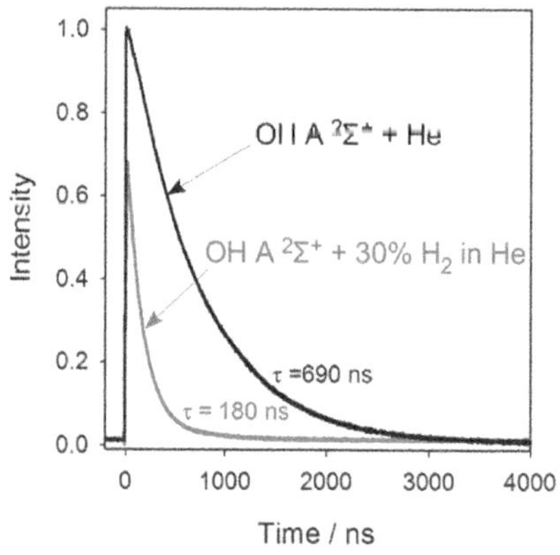

Figure 25.14 Fluorescence decays for OH A $^2\Sigma^+$ ($v' = 0$, $N'' = 0$) in a pure He carrier gas (black curve) and 30% H_2 in an He gas mixture (red curve) under similar experimental conditions. The decreased fluorescence lifetime and integrated intensity with H_2 are attributed to enhanced quenching compared to He, which is sufficiently ineffective that the observed fluorescence lifetime is approximately equal to the radiative lifetime. Provided by Marsha Lester and adapted with permission from Lehman, J.H. and Lester, M.I. (2014) *Annu. Rev. Phys. Chem.*, **65**, 537. © Annual Reviews.

A $^2\Pi$ or B $^2\Sigma$ state changes the rotational state of the CN molecule.[9] Also, particularly in heavy-atom molecules, it is easy to exchange energy between rotations and translations.

Vibrations have much larger spacings. Except for accidental degeneracies and like-with-like transitions, it is difficult to relax vibrational energy efficiently with collisions. This effect is particularly pronounced in supersonic molecular beams. Whereas it is trivial to cool translations, and rotations are efficiently cooled, the vibrations of diatomics and small polyatomics generally experience little to no cooling in a supersonic jet.

An exception is H_2 because of its unusually large rotational constant and vibrational energy spacing. The same can be said of all hydrides. Another unusual relaxation pathway that is particularly effective for C—H bonds is what is known as a Fermi resonance. By coincidence, the vibrational frequency of a C—H stretch is quite close to twice the frequency of a bend. This means that there is a coincidental degeneracy between one quantum of stretch and two quanta of bend, which facilitates population transfer from stretch to bend modes.

Collisional quenching of electronic states is not an issue at low pressure if there are allowed radiative transitions connecting the excited electronic state to other states because of the short radiative lifetime of electronic states. If transitions to all lower states are forbidden, then collisional deactivation is much more efficient. Figure 25.14 shows an example from Marsha Lester's group of how a molecule – H_2 in this case – and its better matched energy level structure is more efficient at quenching the fluorescence of OH A $^2\Sigma^+$ than is He, which has no rovibrational structure and extremely widely spaced electronic levels (~20 eV from ground to first excited state). Of course, H_2 also has another channel open to it that is not available to He, in that H_2 can react chemically with OH. Only about 15% of the quenching by H_2 in collisions with OH A $^2\Sigma^+$ is unreactive. The majority of the excited state population is lost to reactive channels.

25.6.3 Photodissociation and pre-dissociation

Photon absorption can lead to bond breaking, which is known as *photodissociation*. The minimum photon energy required for photodissociation is equal to the bond dissociation energy minus any internal energy content of the rovibrational state photodissociated. Usually, much greater photon energy is required to induce bond breakage, and this is easily seen in Fig. 25.15. Photodissociation is usually mediated by excitation to an electronic state. Direct photodissociation can occur by one of two routes: (i) as shown in Fig. 25.15, excitation to a bound excited state above its dissociation limit; or (ii) by direct excitation to a repulsive state.

A third path for photodissociation is *pre-dissociation*. Pre-dissociation involves excitation to a bound state that lies below the dissociation limit. If a repulsive curve intersects the bound state, then the following scenario can arise. After photoexcitation, the bound electronically excited molecule begins to vibrate. When it reaches its outer turning point, if the energy of the bound excited state equals the energy of the unbound repulsive state, there is a certain probability that the molecule will make a transition from the bound state to the unbound state. Thereafter, the molecule propagates on

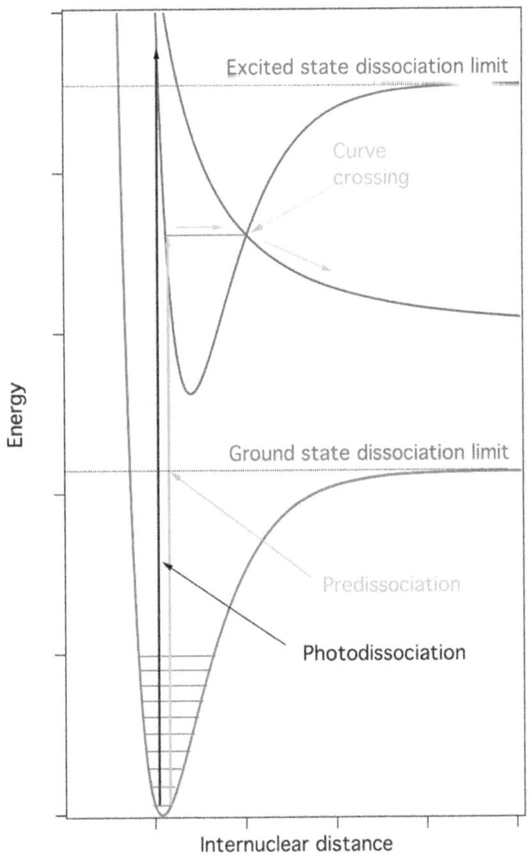

Figure 25.15 Direct photodissociation occurs when excitation to a bound state occurs above its dissociation limit or when excitation is made to a repulsive potential. In either case the excited state wavepacket simply propagates until the bond is broken. In pre-dissociation, the excited state wavepacket first propagates on a bound state potential before encountering a curve crossing where some fraction of the population will transfer to the dissociative potential.

the repulsive potential energy curve, which means that the two atoms involved in the bond are accelerated away from each other. The bond falls apart.

A useful way to think of excited state dynamics is in terms of wavepackets. A wavepacket, as we saw when we solved the free particle problem, is a propagating state. Excitation corresponds to moving a wavepacket from the ground state to the excited state. The wavepacket then propagates on the excited state potential energy curve. In radiative decay, the wavepacket propagates in the bound potential of the excited state, bouncing back and forth with a period equal to the vibrational period of the excited state bond between the inner turning point and the outer turning point. Eventually, the wavepacket drops back down to the ground state and a photon is emitted simultaneously.

In the case of pre-dissociation, or for the CN example noted above, the wavepacket propagates toward the curve crossing. When it encounters a curve crossing at the turning point, some portion of the wavepacket moves onto the other potential energy curve. The amplitude squared of the wavepacket that moves onto the other potential divided by the amplitude squared of the wavepacket that was on the original potential prior to curve crossing corresponds to the probability of the curve crossing per encounter. If the radiative lifetime of the original state is long enough, the wavepacket can move back to the inner turning point, then propagate back toward the curve crossing. With each encounter of the curve crossing, some fraction of the wavepacket will hop onto the dissociative potential. We will see in Chapter 27 that such dynamics has been observed in real time on the femtosecond timescale using laser techniques.

25.6.3.1 Directed practice

Do we expect to see sharp lines in the absorption spectrum when photodissociation occurs?

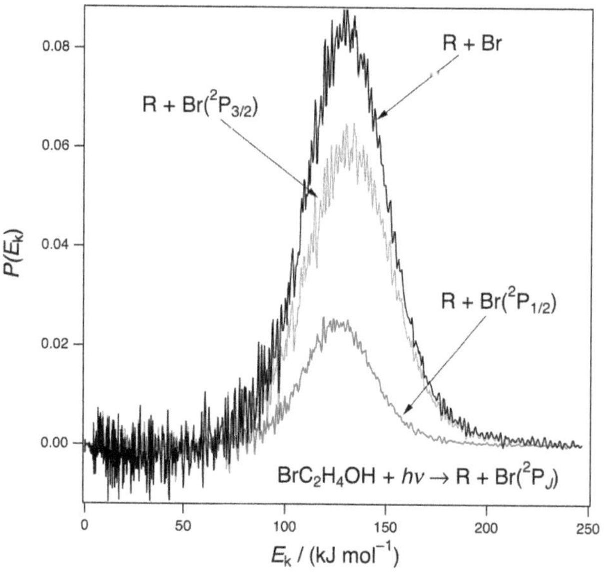

Figure 25.16 Kinetic energy distributions for the two spin–orbit states of Br produced in the photodissociation of BrC_2H_4OH at 193.3 nm. These non-normalized distributions have been adapted from data provided by Matt Brynteson and Laurie Butler. Reproduced with permission from Brynteson, M.D., Womack, C.C., Booth, R.S., Lee, S.H., Lin, J.J., and Butler L.J. (2014) *J. Phys. Chem. A*, **118**, 3211. © 2014, American Chemical Society.

25.6.4 Quenching and quantum yield

The relative probability that a particular event occurs after photoexcitation is called the *quantum yield*. Above, we defined it specifically for fluorescence. For each and every process we can define the quantum yield of that process in terms of the number of events or the rate of events compared to the number or rate of excitations as

$$\phi = \frac{n_{\text{events}}}{n_{\text{excitations}}} = \frac{n_{\text{product}}}{n_{\text{photons absorbed}}} = \frac{r_{\text{process}}}{r_{\text{photon absorption}}}. \tag{25.64}$$

Photodissociation resulting in a halogen atom as a product exhibits two channels – one channel that creates the halogen atom in the ground $^2P_{3/2}$ state and a second that creates the excited $^2P_{1/2}$ state. The spin–orbit splitting of the halogens is 404 cm^{-1}, 881 cm^{-1} and 3685 cm^{-1} for F, Cl, and Br, respectively, and roughly 5240 cm^{-1} for I. An example if the photodissociation at 193.3 nm of BrC_2H_4OH,

$$BrC_2H_4OH + h\nu \rightarrow C_2H_4OH + Br(^2P_J).$$

The kinetic energy distributions of the Br atoms produced by photodissociation have been measured by Laurie Butler and coworkers, and are reproduced in Fig. 25.16. Integration of the peaks leads to a quantity that is proportional to the number of atoms produced in each channel. By taking the ratio of the area of one peak to the total area we can then determine a relative quantum yield of roughly 0.75 for the channel that produces $Br(^2P_{3/2})$, and 0.25 for the $Br(^2P_{1/2})$ channel. Note that the kinetic energy distributions are quite close to Gaussian in shape. This is *not* a Maxwell–Boltzmann distribution – that is, the distributions are not determined by collisional equilibration. The distributions are nascent distributions determined by the dynamics of the photodissociation process.

Summation of the quantum yields of all processes that involve a particular excited state must equal 1 to account for the fates of all the molecules excited to that state by absorbed photons. Thus, for a system that can fluoresce (f), undergo internal conversion (ic) or intersystem crossing (isc), react photochemically (pc) and quench due to collisions with a third body (q), we have

$$\sum_k \phi_k = 1 = \phi_f + \phi_{ic} + \phi_{isc} + \phi_r + \phi_q. \tag{25.65}$$

Imagine now that we have a system for which excitation from the ground state S_0 to an electronic excited state S_1 is described by a rate constant k_a. Fluorescence (with rate constant k_f) competes with quenching by collisions with some species Q. The rate expressions for the process are

$$S_0 \xrightarrow{k_a} S_1 \tag{25.66}$$

$$S_1 \xrightarrow{k_f} S_0 + h\nu \tag{25.67}$$

$$S_1 + Q \xrightarrow{k_q} S_0 + Q. \tag{25.68}$$

The quenching rate is

$$R_q = k_q\left[S_1\right][Q] \tag{25.69}$$

Under constant illumination, S_1 can be thought of as an intermediate and the steady-state approximation allows us to write

$$\frac{d\left[S_1\right]}{dt} = 0 = k_a\left[S_0\right] - k_f\left[S_1\right] - k_q\left[S_1\right][Q]. \tag{25.70}$$

The fluorescence intensity is given by

$$I_f = k_f\left[S_1\right] = k_a\left[S_0\right]k_f\tau_f \tag{25.71}$$

where the *fluorescence lifetime* τ_f is defined by

$$\frac{1}{\tau_f} = k_f + k_q[Q]. \tag{25.72}$$

The ratio of the fluorescence intensity in the absence of quencher (I_0) to the fluorescence in the presence of quencher I_f depends on the concentration of the quencher according to

$$\frac{I_0}{I_f} = 1 + \frac{k_q}{k_f}[Q] = 1 + K_{SV}[Q]. \tag{25.73}$$

A plot of this ratio versus [Q] has a slope of k_q/k_f and is known as a *Stern–Volmer plot*. The ratio is called the *Stern–Volmer constant* K_{SV}. The rate constant for fluorescence k_f is a property of the molecule alone – that is, it is a characteristic of the molecule determined by the Einstein A coefficient for that particular transition. The value of k_f is related to the natural fluorescence lifetime τ_f^0 by

$$\frac{1}{\tau_f^0} = k_f = A_{10}. \tag{25.74}$$

Note the difference between the observed fluorescence lifetime τ_f, which depends on the conditions according to Eq. (25.72), and the natural fluorescence lifetime τ_f^0, which depends only on the specific molecular transition. On the other hand, the quenching rate constant k_q depends on the identities of both the molecule and its collision partner.

The collision rate changes slowly with temperature. On the other hand, internal conversion and intersystem crossings can be strongly influenced by vibrational excitation of the excited state, which often depends strongly on temperature. Under such conditions, thermal quenching of fluorescence greatly reduces fluorescence as the temperature increases.

If the molecule is excited with a short pulse (much shorter than the lifetime of the excited state), the rate equation analysis shows that the fluorescence intensity decays exponentially in time because

$$\left[S_1\right] = \left[S_1\right]_0 e^{-t/\tau_f} \tag{25.75}$$

and according to Eq. (25.71) the fluorescence intensity depends directly on $[S_1]$. Equation (25.72) shows that the inverse of the fluorescence lifetime is directly proportional to the concentration of quencher. Therefore, if the fluorescence lifetime can be measured as a function of quencher concentration, then a plot of $1/\tau_f$ versus [Q] will lead to values for k_f from the intercept and k_q from the slope.

25.6.4.1 Directed practice

Do we expect the Stern–Volmer constant to have only one value for each molecule?

25.7 Electron collisions with molecules

Electron collisions with molecules are important in plasmas. Electrons can also intentionally be scattered from molecules to perform spectroscopy. Because of wave-particle duality, electron–molecule collisions share some similarities to photon–molecule collisions. Because they are massive, the particle nature is much more prominent. Electrons are also involved in bonding, thus they can interact with molecules in ways that are unique to electrons.

Electrons can interact with molecules as electromagnetic waves. When scattered from molecules, they can induce exactly the same sort of vertical Franck–Condon transitions as photons; this is called *dipole scattering*. No linear momentum is exchanged in such a scattering event so the electrons scatter into the same direction as the incident electrons. Energy and angular momentum must be conserved. Importantly, electrons carry a spin $s = \frac{1}{2}$. This spin can couple directly with the spin of the molecule. Therefore, unlike photons, electrons are quite capable of changing the spin state of a molecule: $\Delta S = 0$ is *not* a selection rule for election–molecule collisions. Both spin-preserving (e.g., singlet–singlet) and spin-changing (e.g., singlet–triplet) transitions are possible.

Electrons can act like massive particles. The mass of the electron makes it much easier for the electron to exchange linear momentum with molecules; this process is known as *impact scattering*. Nuclear motion accompanies the transition, making nonvertical transitions possible. The conservation of momentum is reflected in the angular distribution of the scattered electrons, which now differs from that of the incident beam. That is, electrons that undergo impact scattering are scattered off of the incident beam axis, which allows facile separation of signals arising from dipole scattering and impact scattering.

Electrons can also act as electrons. Electrons are able to bind to molecules; this is known as *resonance scattering*. The electron becomes trapped in a bound (or quasi-bound) electronic state. The trapping state can be a real excited state of the isolated anion. Such a state is called a *Feshbach resonance*. Alternatively, a centrifugal barrier, which arises from the angular momentum associated with the electron–molecule scattering event, can trap the electron; this is known as a *shape resonance*. A shape resonance is very short-lived state (on the order of a few femtoseconds). However, even a Feshbach resonance generally has a lifetime of only 10^{-10} to 10^{-15} s. Resonance scattering is sensitively dependent on the incident electron energy – that is, resonances in the excitation probability are observed as a function of the incident electron energy E_0; this is also called the *primary energy*. Resonances have characteristic angular dependencies for both their excitation and decay. The angular distribution of the electrons scattered through a resonance provides information about the symmetry of the resonance, for example, whether a σ or π symmetry state is involved.

The conservation of energy connects the transition made in the molecule to the kinetic energy of the electron. Electron spectroscopy using excitation with electron beams is commonly performed in electron energy loss mode; hence the name *electron energy loss spectroscopy* (EELS). The kinetic energy of scattered electrons E is less than the incident kinetic energy E_0 by exactly the same amount as the energy of the molecular transition that is excited $h\nu$,

$$E = E_0 - h\nu. \tag{25.76}$$

Rotational, vibrational and electronic transitions can all be excited and studied by EELS. In practice, it is difficult to obtain sufficiently high energy resolution to study rotational excitations. Most commonly vibrational and electronic excitation is studied. Vibrational spectroscopy analogous to IR absorption spectroscopy is quite frequently used in the study of surfaces and adsorbed species. Electronic EELS is increasingly being used in electron microscopy to perform elemental analysis as a type of core level spectroscopy. It can also be used to probe valence electronic states.

Once an excited state is formed, its dynamics is independent of the method of excitation. Therefore, electron-induced excitations are subject to exactly the same dynamics as those discussed for photon-induced excitations.

25.7.1 Directed practice

When electrons with a kinetic energy of 1 eV are incident on an adsorbed CO layer they excite vibrations at 2050 cm^{-1}. Calculate the final kinetic energy of the electrons.

[Answer: 0.746 eV]

SUMMARY OF IMPORTANT EQUATIONS

$F(J) = B_v J(J+1) - D_v [J(J+1)]^2$	Rotational term $F(J)$ with centrifugal distortion
$B_v = \dfrac{\hbar}{4\pi\mu c r_{e,v}^2}$	Rotational constant, $r_{e,v}$ = equilibrium bond length of vibrational level v
$B_v = B_e - \alpha_e\left(v + \frac{1}{2}\right)$	B_e = harmonic rotational constant, vibration-rotation interaction constant α_e
$D_v = 4B_0^3/\tilde{v}_e$	Centrifugal distortion constant
$I = \mu r^2$	Moment of inertia
$\mu = \dfrac{m_1 m_2}{m_1 + m_2}$	Reduced mass
$\langle J^2 \rangle^{1/2} = [J(J+1)]^{1/2}\hbar$	Magnitude of rotational angular momentum
$\langle J_z \rangle = M\hbar$	Projection of rotational angular momentum
$N_J/N_0 = (2J+1)\,e^{-E_J/k_B T}$	Boltzmann factor for rotational state with energy E_J
$D_0 = D_e - \frac{1}{2}\omega_e$	Dissociation energy of $v = 0$, D_0, dissociation energy from potential minimum, D_e, harmonic fundamental wavenumber, ω_e
$D_e \approx \dfrac{\omega_e^2}{4x_e\omega_e}$	
$G(v) = \left[\omega_e - \omega_e x_e\left(v + \frac{1}{2}\right)\right]\left(v + \frac{1}{2}\right)$	Vibrational term, $G(v)$, anharmonicity constant x_e
$\omega_e = \tilde{v}_e = \dfrac{1}{2\pi c}\sqrt{k\mu}$	Fundamental wavenumber is related to the force constant k and reduced mass μ
$\delta\tilde{v} = \dfrac{5.31\,\mathrm{cm}^{-1}}{\tau}$	Linewidth in cm^{-1} of a lifetime-broadened state with lifetime τ in ps

Exercises

25.1 For transitions with the wavelength given below, calculate the wavenumber in cm^{-1}, the energy in eV and the frequency in Hz of the corresponding photons. State which part of the electromagnetic spectrum each transition corresponds to and what type of molecular excitation is most likely responsible for each transition.
(a) 1.54 Å (b) 2.27 μm (c) 2.59 mm (d) 193 nm

25.2 The wavefunction for the motion of a particle on a ring is

$$\psi = N e^{\mathrm{i} m_l \phi}.$$

Determine the normalization constant N. That is, evaluate the integral

$$\int_0^{2\pi} \psi^* \psi \, \mathrm{d}\phi = 1.$$

25.3 A point mass rotates in a circle with $l = 2$. Calculate the magnitude of its angular momentum and the possible projections of the angular momentum (in units of $\hbar$) onto an arbitrary axis (by convention we usually call this the z axis).

25.4 (a) Calculate the first five energy levels for a $^{12}C^{16}O$ molecule which has a bond length of 112.8 pm if it rotates freely in three dimensions. (b) Now calculate the first five levels if it is adsorbed onto a surface and forced to rotate in two dimensions

25.5 Calculate the transition dipole moment in Debye (D) $\mu_z^{J'J''} = \int \psi^*_{J'M'}(\tau)\,\mu_z \psi_{J''M''}(\tau)\,\mathrm{d}\tau$, where $\mu_z = -er\cos\theta$ for a transition from the $J'' = 0$, $M'' = 0$ level to the $J' = 2$, $M' = 2$ level of a rigid rotor. Is this transition allowed? The integration is over r, θ, and ϕ.

25.6 The first three transitions in the pure rotational spectrum of $^{12}C^{16}O$ occur at 3.8626, 7.7252 and 11.5878 cm^{-1}. Calculate the rotational constant and bond length of $^{12}C^{16}O$ as well as the wavenumbers of the first three pure rotational transitions of $^{12}C^{18}O$.

25.7 Which of the following molecules exhibit a pure rotational microwave absorption spectrum: H_2, HCl, CH_4, CH_3Cl and CH_2Cl_2? State on what basis you make your decision.

25.8 Given that the spacing of lines in the microwave spectrum of $^{27}Al^{1}H$ is constant at 12.604 cm^{-1}, calculate the moment of inertia and bond length of the molecule.

25.9 Given that the rotational constant of $^{1}H^{35}Cl$ is 10.5909 cm^{-1} and the bond length is 1.27460 Å, calculate where in cm^{-1} the first two pure rotational transitions are and where you expect the first two transitions for $^{2}H^{35}Cl$ to occur.

25.10 For HD $\tilde{\nu}_0 = 3632.1$ cm^{-1}. Calculate the relative population of $v = 1$ to $v = 0$ at $T = 500$ K.

25.11 For the ground electronic state of H_2, $B(v = 0) = 59.3392$ cm^{-1} and $B(v = 5) = 44.9580$ cm^{-1}, calculate the change in the bond length between $v = 0$ and $v = 5$.

25.12 If $H_2(v = 1)$ lies 4161.14 cm^{-1} above $H_2(v = 0)$ and $H_2(v = 2)$ lies 8087.11 cm^{-1} above $H_2(v = 0)$, calculate the term values $G(v)$ for H_2 in $v = 0$, 1 and 2 as well as $\tilde{\nu}_0$, x_e and the zero point energy.

25.13 State the Heisenberg uncertainty relation with respect to energy and lifetime, its implication, and how it is reflected in spectra.

25.14 (a) Estimate the lifetime of a state that gives rise to a line of width (i) 0.03 cm^{-1}, (ii) 2.0 cm^{-1}. (b) A molecule in a liquid undergoes about 1.0×10^{13} collisions per second. Suppose that: (i) every collision is effective in deactivating the molecule vibrationally; and (ii) that one collision in 100 is effective. Calculate the width in cm^{-1} of vibrational transitions in the molecule.

25.15 Evaluate the average kinetic and potential energies $\langle E_K \rangle$ and $\langle E_{pot} \rangle$ for the ground state ($n = 0$) of the harmonic oscillator by carrying out the appropriate integrations.

25.16 Confirm that the wavefunction for the ground state of a one-dimensional linear harmonic oscillator, $\psi_0 = N_0 e^{-x^2/2\alpha^2}$, (a) is a solution of the Schrödinger equation and (b) that its energy is $\frac{1}{2}\hbar\omega$.

25.17 Assume that the chlorine molecule can be treated as a harmonic oscillator with a force constant of $k = 329$ N m^{-1} and use the reduced mass

$$\mu = \frac{m_1 m_2}{m_1 + m_2}$$

for the mass of the oscillator. The mass of a ^{35}Cl atom is 34.9688 u and 36.9670 u for ^{37}Cl. (a) Calculate the zero point energy of the $^{35}Cl_2$ molecule. (b) Calculate the energy in cm^{-1} of the $v = 0$ to $v = 1$ transition (also known as the fundamental transition) for $^{35}Cl_2$. (c) Calculate the wavenumber of the fundamental transition for ^{35}Cl—^{37}Cl isotopologue of the chlorine molecule.

25.18 Assume that $^{1}H^{35}Cl$ acts like a harmonic oscillator and that the fundamental vibrational wavenumber is $\tilde{\nu}_0 = 2989.74$ cm^{-1}. (a) Calculate the zero point energy and force constant of the molecule. (b) Calculate the wavelength of a photon needed to excite from the $v = 1$ to $v = 3$ level.

25.19 The first five vibrational energy levels of HCl are at 1175.49, 3505.18, 5806.50, 8079.44 and 10324.01 cm^{-1}. Calculate the dissociation energy of the molecule in cm^{-1} and eV.

25.20 Assuming that it acts as a rigid rotor, harmonic oscillator and using the molecular constants for $^{14}N^{16}O$ given below, calculate the wavenumbers of the first transitions in both the P and the R branch of the fundamental vibrational transition, the first overtone and the first hot band. $B = 1.7046$ cm^{-1}, $\tilde{\nu}_0 = 1904.03$ cm^{-1}

25.21 (a) How can you observe vibrational transitions in Raman spectroscopy using visible laser light even though the photon energy of the laser is much greater than the vibrational energy spacing? (b) If a pulsed dye laser operating at 525 nm is used for excitation of H_2, calculate the positions of the first three Stokes and the first three anti-Stokes lines related to vibrational transitions. Use $\tilde{\nu}_e = 4395.2$ cm^{-1} and $\tilde{\nu}_e x_e = 117.99$ cm^{-1}.

25.22 For a system that responds at frequency ω, the intensity ratio of anti-Stokes to Stokes Raman scattering is given by

$$I_{\text{anti-Stokes}}/I_{\text{Stokes}} = e^{-\hbar\omega/k_B T}.$$

Explain why the intensity of Stokes and anti-Stokes transitions does not have the same temperature dependence and therefore exhibits the observed exponential relationship.

25.23 Show that the pure rotational Raman shift in the presence of centrifugal distortion is given by

$$|\Delta\tilde{\nu}| = (4B_0 - 6D_0)\left(J + \tfrac{3}{2}\right) - 8D_0\left(J + \tfrac{3}{2}\right)^3$$

25.24 State and justify the Franck–Condon principle and its consequences.

25.25 The potential energy curves for the first five electronic states of O_2 are listed in Fig. 25.12. The spectroscopic constants are given in Table 25.6. Detail each selection rule and describe the relative strengths of the four electronic transitions listed below. Justify your answer.

Transition	$\Delta\Lambda$	Allowed	ΔS	Allowed	Parity	Allowed	Inversion	Allowed	Strength
B–X									
B–a									
B–b									
B–A									

NA = not applicable; ex = extremely; v = very.

25.26 The potential energy curves for the first five electronic states of O_2 are listed in Fig. 25.12. The rotational constants of the X, a and b states are all close to each other, but the A and B states have significantly different rotational constants. How can you tell this by looking at the potential energy curves?

25.27 The spectroscopic constants of the first five electronic states of O_2 are given in Table 25.6. Calculate the dissociation energies D_e and D_0 in eV of each the five electronic states listed in the table. Why do these values differ?

State	D_e / eV	D_0 / eV
$B^3\Sigma_u^-$		
$A^3\Sigma_u^+$		
$b^1\Sigma_g^+$		
$a^1\Delta_g$		
$X^3\Sigma_g^-$		

25.28 The spectroscopic constants of the first five electronic states of O_2 are given in Table 25.6. Calculate the term value in cm^{-1} of the $v = 0$, $J = 5$ level in each of the five electronic states.

25.29 Compare the vibrationally but not rotationally resolved fluorescence spectrum of the $X(v'') \leftarrow a(v' = 0)$ and $X(v'') \leftarrow B(v' = 0)$ bands of O_2. Calculate the wavenumber of the first three transitions for both bands (ignoring the rotational energy). How do you expect the relative intensities of the vibrational transitions to differ for the two bands?

25.30 The Einstein A coefficient of a typical electric dipole allowed transition is 10^7 to 10^8 s^{-1}. For O_2 b → X, the value is $A = 0.14$ s^{-1}. Calculate the lifetime of the O_2 b state as compared to a typical state. Discuss the factors that determine the magnitude of the A coefficient.

25.31 Discuss the processes that occur in the X ← B band of O_2 as the photon wavenumber is tuned from 50 000 to 60 000 cm^{-1}.

25.32 What do we mean by a vertical transition, and why are they usually observed?

25.33 The $^{138}Ba^{19}F$ molecule has the molecular constants given in the table below. All are in cm^{-1} except r_e, which is given in Å.

State	T_e	$\tilde{\nu}_0$	$\tilde{\nu}_0 x_e$	B	r_e
B $^2\Sigma^+$	14062.5	424.4	1.88	0.2071	2.208
X $^2\Sigma^+$	0	468.9	1.79	0.2158	2.162

(a) Draw and label appropriately the X and B state potential energy curves (assume different dissociation limits). Indicate in this diagram the T_e values, the zero point energies, the D_0 value of the ground state, the D_e value of the excited state and the r_e values for both states. Only labels not numerical values are required. (b) Calculate the term energies in cm^{-1} of the X $v = 0$ $J = 2$, 3 and 4 states and the B $v = 2$ $J = 2$, 3 and 4 states. (c) Draw a diagram with the energy levels involved in the P, Q and R branch transitions that begin in the X $v = 0$ $J = 3$ state and end in the appropriate J state of the B $v = 2$ level. Draw arrows to indicate these three transitions. (d) Calculate the wave numbers of the transitions referenced in part (c). (e) Are these three transitions allowed? Justify your answer. (f) Do you expect the R branch transition from $J = 3$ to be stronger for B $(v = 2) \leftarrow X$ $(v = 0)$ or for B $(v = 0) \leftarrow X$ $(v = 0)$? Justify your answer.

25.34 For CN, the first excited state is A $^2\Pi$ and the ground state is X $^2\Sigma^+$. The $(n' = 7)$ level lies below the dissociation limit of the A state. The A state and the X state both have the same dissociation limit of C(^{2}P) + N(^{4}S). There is no other intersecting state that need be considered. Draw a potential energy diagram of both states on the same set of axes and state why transitions between the two states are allowed.

25.35 From integration of the peaks in Fig. 25.16 the total area is 4.39, the Br($^2P_{3/2}$) peak is 3.28, and the Br($^2P_{1/2}$) peak is 1.11 (arbitrary units). (a) Determine the relative quantum yields of the two channels. (b) Fitting the peaks to Gaussian profiles, the kinetic energy of the maximum of the distribution is 126.6 kJ mol^{-1} for Br($^2P_{1/2}$) and 131.5 kJ mol^{-1} for Br($^2P_{3/2}$). Give a reasonable explanation for this shift.

25.36 The nanoputians, distraught over their thrashing during the Ashes, are back in the laboratory attempting to make a better cricket ball. Inspired by sausage-making, they have decided to investigate what is known as endohedral fullerenes. These are carbon cages such as C_{60} or larger, that have stuffed inside of them a molecule or atom. They have isolated two different compounds $Ce_2@C_{80}$ and $Sc_3N@C_{80}$. (a) Describe a spectroscopic method to tell the difference between these two compounds (you can consider the trapped species in the center of the cage to be a free molecule. (b) The nanoputians have photon sources that emit light near 1 nm, 100 nm and 10 000 nm. Calculate the photon energy in eV for each of these wavelengths and state to which part of the electromagnetic spectrum each photon corresponds. Describe the type of information that can be obtained by performing spectroscopy in these three wavelength regimes.

25.37 Two stainless steel tubes are closed off with fused silica windows that are transparent almost everywhere from the infrared to the near UV. One tube is filled with H_2. One is filled with HCl. Describe and explain spectroscopic methods to identify both gases.

Endnotes

1. Born, M. and Oppenheimer, J.R. (1927) *Ann. Phys.*, **84**, 457.
2. Tielens, A.G.G.M. (2013) The molecular universe. *Rev. Mod. Phys.*, **85**, 1021.
3. Millikan, R.C. (1963) *J. Chem. Phys.*, **38**, 2855.
4. Chang, H.C. and Ewing, G.E. (1990) *Phys. Rev. Lett.*, **65**, 2125.
5. Lass, K., Han, X., and Hasselbrink, E. (2005) *J. Chem. Phys.*, **123**, 051102.
6. Beckerle, J.D., Casassa, M.P., Cavanagh, R.R., Heilweil, E.J., and Stephenson, J.C. (1990) *Phys. Rev. Lett.*, **64**, 2090.
7. Raman, C.V. (1928) *Indian J. Phys.*, **2**, 387.
8. 'Easy' is in the eye of the beholder. Maser action, the microwave equivalent of the laser, has been observed to occur in interstellar clouds. Nonetheless, the first time you demonstrate this effect in the lab, you still get a well-deserved Nobel prize for it.
9. Pratt, D.W. and Broida, H.P. (1969) *J. Chem. Phys.*, **50**, 2181.

Further reading

Demtröder, W. (2014) *Laser Spectroscopy 1. Basic Principles*, 5th edition. Springer-Verlag, Berlin.

Demtröder, W. (2015) *Laser Spectroscopy 2: Experimental Techniques*, 5th edition. Springer-Verlag, Berlin.

Heitler W., *The Quantum Theory of Radiation*, 3rd ed. (Dover Publications, New York, 2010).

Herzberg, G. (1950) *Molecular Spectra and Molecular Structure. I. Diatomic Molecules*, 2nd edition. Van Nostrand, New York.

Herzberg, G. (1945) *Molecular Spectra and Molecular Structure. II. Infrared and Raman Spectra of Polyatomic Molecules*. Van Nostrand, New York.

Herzberg, G. (1967) *Molecular Spectra and Molecular Structure. III. Electronic Spectra and Electronic Structure of Polyatomic Molecules*. Van Nostrand, New York.

Hollas, J.M. (2004) *Modern Spectroscopy*, 4th edition. John Wiley & Sons, New York.

NIST Diatomic Spectral Database; http://www.nist.gov/pml/data/msd-di/index.cfm

Schinke, R. (1993) *Photodissociation Dynamics*. Cambridge University Press, Cambridge.

Steinfeld, J.I. (2012) *Molecules and Radiation: An Introduction to Modern Molecular Spectroscopy*, 2nd edition. Dover Publications, New York.

Zare, R.N. (1988) *Angular Momentum: Understanding Spatial Aspects in Chemistry and Physics*. John Wiley & Sons, New York.

CHAPTER 26

Polyatomic molecules and group theory

PREVIEW OF IMPORTANT CONCEPTS

- Molecules can be classified by the set of symmetry operations that are preserved when they operate on the molecule. These classifications are called point groups.
- Symmetry operations include rotation, reflection through a plane, inversion and rotation–reflection.
- The number of degrees of freedom of a molecule include translations, vibrations, rotations, as well as its electronic and spin states.
- Information on the set of operations contained within a point group is collected in a character table.
- The character tables tell us how rotations, vibrations, translations, wavefunctions and polarizability transform under the various symmetry operations. All of these have a symmetry species to which they belong.
- The symmetry properties of the initial and final state wavefunctions and the transition operator that couples them determine the selection rules for polyatomic molecules.
- $\Delta S = 0$ is a good selection rule for electric dipole transitions in polyatomics in the absence of strong spin–orbit coupling (i.e., in the absence of high-Z elements).
- An electronic transition in which no vibration is excited is allowed if the cross-product of the irreducible representations of the initial state, the transition dipole and the final state contains a totally symmetric irreducible representation.
- If a vibration is excited in the electronic transition, then its irreducible representation must also be included in the cross-product. The complete cross-product must contain a totally symmetric irreducible representation for the transition to be allowed.
- A vibrational mode is active for electric dipole transitions in the infrared (i.e., IR-active) if the cross-product of the irreducible representations of the initial state, the transition dipole and the final state contains a totally symmetric irreducible representation.
- A mode is Raman-active if the cross-product of the irreducible representations of the initial state, the polarizability tensor and the final state contains a totally symmetric irreducible representation.
- A molecule has a permanent electric dipole moment if any of the components of translational symmetry of the molecule's point group is totally symmetric.
- A linear molecule composed of N atoms has $3N - 5$ vibrational modes, and nonlinear molecules have $3N - 6$ modes.
- The rovibrational spectroscopy of linear polyatomics shares much in common with diatomics, with P, R and in some cases Q rotational branches accompanying the vibrational transition.
- Coupling between vibrational modes is quite common in polyatomics.
- The rotational fine structure in rovibrational and electronic transitions depends on the type of rotor to which the molecule corresponds.
- State designations for linear molecules follow rules much like those for diatomics. State designations for other polyatomics are based on the irreducible representation of the electronic state wavefunction.
- Radiative lifetimes for allowed transitions (fluorescence) in polyatomics range from roughly 0.1 to 100 ns. Phosphorescence corresponds to forbidden transitions with significantly longer lifetimes.
- Intramolecular vibrational redistribution (IVR) becomes increasingly fast and more likely for bigger molecules. This means that a local vibrational excitation is quickly delocalized across the molecule.
- Quenching of electronic state excitation can occur both by intramolecular and intermolecular (collisionally induced) mechanisms.
- The density of states in large polyatomics is so large that the electronic and vibrational structure become band-like. This makes intermolecular vibrational energy relaxation extremely efficient.
- Internal conversion corresponds to intramolecular nonradiative conversion of electronic energy to vibrational energy.
- Intersystem crossing corresponds to an intramolecular nonradiative transition from a state of one spin multiplicity to another.
- Fermi resonance refers to the interaction of an overtone of one vibration and the fundamental of another vibration. Fermi resonance leads not only to population transfers; it also causes shifts in vibrational frequencies.

Physical Chemistry: How Chemistry Works, First Edition. Kurt W. Kolasinski.

Companion Website: www.wiley.com/go/kolasinski/physicalchemistry

Concentrating first on atoms and diatomic molecules allowed us to focus on the fundamental physics of poly-electron and molecular systems. Increasing the number of atoms in a molecule introduces not only increased complexity but also new phenomena. Two of the most important concepts that need to be developed more fully are those of *delocalization* and *symmetry*. In diatomic molecules, bonding is necessarily two-centered. In molecules, this need not be the case as we have seen in extended π-bonding systems. Just as for clusters, which were discussed in Chapter 5, coupled systems develop band structure in both electronic and vibrational states as the number of atoms increases. The development of band structure strongly influences energy relaxation dynamics, charge transport and the absorption and emission spectra of molecules.

The symmetry of diatomic molecules is also quite straightforward. There are heteronuclear and homonuclear diatomics. The former require a rotation of 2π radians to transform into a symmetry equivalent configuration. The latter require only a rotation of π radians. The introduction of more atoms into molecules introduces much more intricate symmetries. We have already seen that molecular symmetry is important in determining selection rules for spectroscopic transitions, and we need to develop that in more detail for polyatomics. The symmetry of molecules is important for determining numerous properties such as dipole moments and optical activity, in addition to their spectroscopic properties and selection rules.

26.1 Absorption and emission by polyatomics

Recall, as shown in Fig. 26.1, that two molecular energy levels are connected radiatively in three ways: (i) absorption; (ii) stimulated emission; and (iii) spontaneous emission. Stimulated emission, which is essential for the operation of lasers, will be discussed in Chapter 27. Here, we concentrate on absorption and spontaneous emission. Spontaneous emission is also called *fluorescence* or *phosphorescence*, and will be discussed later in more detail.

Just as for atoms and diatomics, the transition rates between levels are determined by the magnitude of Einstein coefficients and transition dipole moment matrix elements. As discussed in more detail in Chapter 23, we assume that the transitions observed are electric dipole transitions. Thus, the *transition rate* for absorption is given by

$$W_{ki}^{\mathrm{abs}} = \frac{1}{6\varepsilon_0 \hbar^2 c} |\langle \psi_k | \hat{\mu} | \psi_i \rangle|^2 I(\nu_{ki}) = B_{ki}\, \rho(\nu_{ki}) \tag{26.1}$$

where $I(\nu_{ki})$ is the intensity of the radiation at the frequency of the transition and B_{ki} is the Einstein B coefficient,[1]

$$B_{ki} = \frac{|\langle \psi_k | \hat{\mu} | \psi_i \rangle|^2}{6\varepsilon_0 \hbar^2} = \frac{|\mu_{ki}|^2}{6\varepsilon_0 \hbar^2}. \tag{26.2}$$

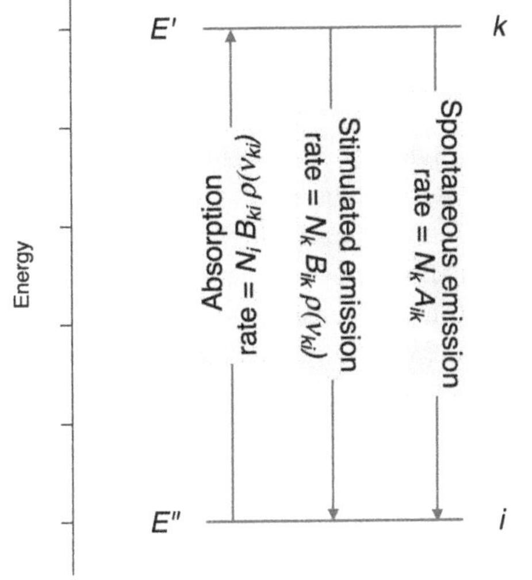

Figure 26.1 Radiative transitions between levels *i* and *k* can be stimulated by a radiation field (absorption and stimulated emission) or occur spontaneously by relaxation from the excited state *k* to the lower state *i* (spontaneous emission). Each process has a characteristic rate expression to describe the transfer of population from one state to the other.

The intensity of radiation is proportional to the square of the electric field strength. The *energy density of radiation* $\rho(\nu_{ki})$ is related to the intensity by

$$\rho(\nu_{ki}) = I(\nu_{ki})/c. \tag{26.3}$$

Equation (26.1) is a special case of *Fermi's golden rule*. It tells us the rate of change of the probability of being in one state or another for a molecule starting in state i and transitioning to state k. For the rate of change of the population of state i, we need to multiply by the number of molecules in state i.

The transition from upper state back to the lower state can occur in two ways. Spontaneous emission is governed by the Einstein A coefficient

$$W_{ik}^{\text{spon}} = A_{ik}, \tag{26.4}$$

from which we make the identification that the probability decay rate is, in fact, the Einstein A coefficient for the transition

$$A_{ik} = \frac{16\pi^3 \nu_{ki}^3}{3c\varepsilon_0 h} |\mu_{ki}|^2. \tag{26.5}$$

A_{ik} has units of s^{-1}. Its meaning is two-fold. *The Einstein A coefficient represents the rate of probability decay. It is also the rate constant for population decay of the upper state.* The kinetics of population decay follows a first-order rate law whenever spontaneous emission is the only means of relaxation of the excited state.

The ratio of the Einstein A and B coefficients is

$$\frac{A_{ik}}{B_{ik}} = \frac{8\pi h}{c^3} \nu_{ki}^3. \tag{26.6}$$

For two nondegenerate states i and k, $B_{ik} = B_{ki}$. The Einstein B coefficient is determined by the magnitude of the transition dipole moment matrix element according to

$$B_{ik} = \frac{2\pi^2}{3\varepsilon_0 h^2} |\mu_{ki}|^2. \tag{26.7}$$

As discussed in Chapter 23, each transition has an intrinsic linewidth (also called natural linewidth), which for an isolated molecule in the gas phase in the absence of all other line broadening mechanisms is a Lorentzian lineshape. A Lorentzian peak profile centered at a wavenumber $\tilde{\nu}_0$ is described by the equation

$$L(\tilde{\nu}) = \frac{A}{(\tilde{\nu} - \tilde{\nu}_0)^2 + \gamma^2}. \tag{26.8}$$

Because all atoms in a particular excited state experience the same broadening from the effect of excited state lifetime, this is an example of *homogeneous broadening*. The Lorentzian linewidth parameter γ is related to the excited state lifetime τ by

$$\gamma = (2\pi c\tau)^{-1} \tag{26.9}$$

and the full-width at half-maximum, FWHM, of a spectral line by

$$\text{FWHM} = 2\gamma. \tag{26.10}$$

The natural linewidth of a transition involving nondissociating excited electronic states is much less than 1 cm^{-1}, usually <0.01 cm^{-1}. Because of the ν_{ki}^3 dependence of the Einstein A coefficient, radiative lifetimes increase and linewidths decrease even further for transitions involving only rovibrational and rotational states. The absorption and emission of photons from two levels are connected by the Einstein coefficients. Just as an emission transition has a natural linewidth, so too does an absorption transition. However, as a consequence of Eq. (26.1) the linewidth of an absorption feature is a convolution of the natural linewidth and the linewidth of the excitation source. This is stated mathematically as

$$W_{ki} \propto |\mu_{ki}|^2 \rho(\nu) L(\nu). \tag{26.11}$$

$L(\nu)$ is the linewidth of the transition and $\rho(\nu)$ is the energy density profile of the excitation source – that is, the energy density across the bandwidth of the laser. To calculate the exact absorption rate we need to take account of the vector nature of the transition dipole moment and the electric field. More specifically, we need to consider the polarization of the excitation source and how it lines up with the transition dipole vector. This need not concern us here. Conventional light sources suffer from a direct relationship between decreased linewidth and decreasing intensity. On the other hand, lasers can exhibit extraordinarily high intensity in extremely narrow bandwidth.

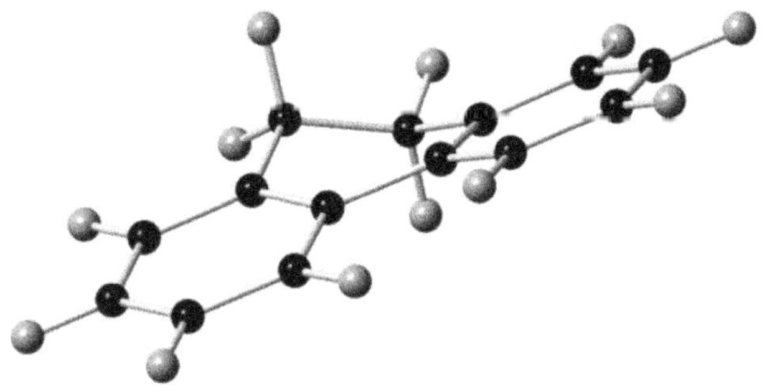

Figure 26.2 The structure of 9,10-dihydrophenanthrene (DHPH).

Analogously, conventional emission spectroscopy – in which a broadband source (or other means of excitation such as electrical discharge) excites a sample and then a spectrophotometer disperses and detects the emitted light – only obtains high resolution at the expense of signal intensity. With the use of laser excitation, we can perform high-resolution spectroscopy based on the absorption of photons with a photon source that is intense but also has a narrow bandwidth. As discussed further in Chapter 27, these techniques include laser-induced fluorescence (LIF), fluorescence excitation spectroscopy (FES) and resonance-enhanced multiphoton ionization (REMPI). Here, we focus on an example of FES. The principle behind FES is that a laser with a bandwidth smaller than the spacing between spectral lines (indeed, the laser may have a bandwidth even narrower than the natural linewidth) excites a molecule quantum-state specifically. Fluorescence usually involves excitation to an excited electronic state. The structure of the excited electronic state can be probed by state-specific excitation, followed by measuring the intensity of fluorescence as a function of the excitation wavelength. As the excitation light is tuned, different transitions are probed. Fitting the spectrum acquired in this manner to a Hamiltonian that includes sufficient interactions allows us to determine intimates details of the structure of the ground and excited states.

Consider the example of 9,10-dihydrophenanthrene (DHPH) with molecular formula $C_{14}H_{12}$. The structure of DHPH is shown in Fig. 26.2. An extreme example of the ultrahigh resolution afforded by laser spectroscopy is displayed in Fig. 26.3. Because this molecule contains 26 atoms, it has an incredibly high density of rovibrational states. When observed in the gas phase at room temperature, its absorption spectrum consists of only broad bands, as is typical of sufficiently large polyatomics (this point is discussed in more detail in subsequent sections). When the molecule is expanded in a supersonic molecular beam, its rotational temperature is cooled to roughly 5 K, which means that very few states are populated initially. A laser is used to excite the molecule from the ground electronic state to the first singlet electronically excited state. This is denoted the $S_0 \rightarrow S_1$ transition. Figure 26.3(a) shows that transitions from the ground electronic state to a number of different vibrational states in the S_1 state are resolved with the use of a tunable pulsed dye laser that engenders excitation in the UV with a spectral resolution of $\sim$0.6 cm^{-1}. This narrow (but easily obtainable) bandwidth is sufficient to resolve vibrational transitions, but not transitions involving individual rotational levels. To do so requires a more elaborate continuous wave dye laser with an effective spectral resolution of $\sim$1 MHz, which corresponds to $\sim 3 \times 10^{-5}$ cm^{-1} in this spectral range. Lasers and their characteristics are discussed in detail in Chapter 27.

The integral that determines the value of the transition dipole moment (and therefore also the values of the Einstein coefficients) is dependent on the *changes in electron distribution* that occur when the molecule absorbs light. It can be evaluated explicitly only in the simplest cases (for atoms, diatomics and some small polyatomics). In order to interpret complex spectra, such as that for DHPH and other polyatomics, we need to develop means of evaluating the transition dipole moment in general and approximately. Fortunately, the symmetry of the wavefunctions and the transition operator determine whether or not the integral has a finite value, or if it vanishes. This motivates us to investigate the symmetry of molecules systematically (as shown in the next few sections). We will also apply the Born–Oppenheimer approximation to factor the Hamiltonian. This allows us to treat each degree of freedom independently in the first approximation. The Franck–Condon factor will again be introduced, as it was for diatomics.

26.2 Electronic and vibronic selection rules

To understand spectra such as those in Fig. 26.3 in detail, we need to develop our understanding of selection rules and rovibrational structure beyond what we know already for atoms and diatomics. Rovibrational structure is discussed in Section 26.7. Here, we focus on selection rules arising from the electronic and vibrational degrees of freedom.

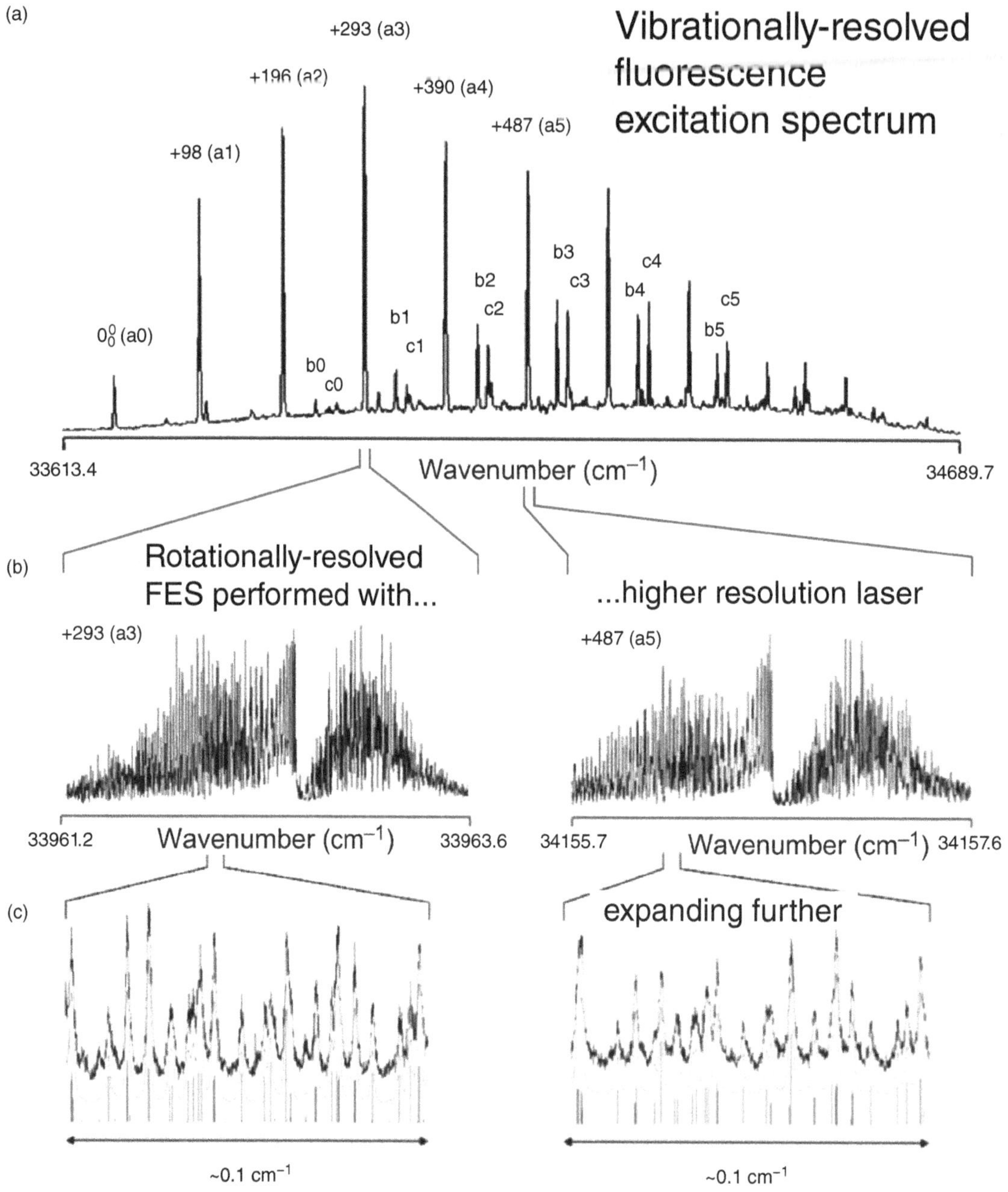

Figure 26.3 Spectroscopic investigation of 9,10-dihydrophenanthrene (DHPH) that has been cooled rotationally to ~5 K. (a) A conventional high-resolution laser fluorescence excitation scheme resolves various vibrational transitions that accompany excitation from the ground electronic state to the first excited singlet state. (b, c) Ultrahigh-resolution fluorescence excitation spectroscopy resolves individual rotational transitions even in this molecule with 26 atoms ($C_{14}H_{12}$). Figure provided by David Pratt. Reproduced with permission from Alvarez-Valtierra, L. and Pratt, D.W. (2007) *J. Chem. Phys.*, **126**, 224308. © 2007 AIP Publishing LLC.

Selection rules for atomic transitions are particularly simple. Atoms are characterized by a set of good quantum numbers that are sufficient to express all of the selection rules in terms of these. For diatomic and linear polyatomic molecules, the selection rules are framed in terms of the good quantum numbers. However, it is necessary to include one or two other symmetry properties to fully establish their selection rules. For homonuclear diatomics and symmetrical linear molecules, we need to add an indication of whether a state is even (gerade) or odd (ungerade) with respect to inversion. Inversion symmetry must change during a transition (g to u or u to g). For all diatomic and linear molecules in Σ states (for which the projection of electron orbital angular momentum along the internuclear axis is zero, that is, $\Lambda = 0$), we must also consider the symmetry of the molecule upon reflection through any plane containing the internuclear axis. This symmetry is called *parity*. If there is no change in sign upon reflection, the state is designated +, but if the sign changes the state is designated –. If the molecule lacks symmetry in such a plane, no sign is indicated. For Σ–Σ transitions, the parity must not change.

In the absence of strong spin–orbit coupling, spin changing transitions are forbidden in atoms and diatomics. This is the only quantum number-based selection rule that applies to polyatomics as well. In the absence of high-Z atoms the total electron spin quantum number S should not change,

$$\Delta S = 0. \tag{26.12}$$

Just as for diatomics, the Born–Oppenheimer approximation is our starting point for calculations involving the molecular wavefunction. We use this and the Franck–Condon principle to make our first approximations of how to treat the spectroscopy of polyatomics. We begin by factoring of the wavefunction,

$$\psi(r, \boldsymbol{R}) = \psi_{\mathrm{r}}(\boldsymbol{R})\psi_{\mathrm{v}}(\boldsymbol{R})\psi_{\mathrm{e}}(r, \boldsymbol{R}). \tag{26.13}$$

The rotational part is relatively simple. It leads to a transition dipole moment component, denoted $\boldsymbol{M}_{\mathrm{r}}$, the absolute square of which is also called the *Hönl–London factor*. As discussed in more detail below, it is always zero unless $\Delta J = 0, \pm 1$. This leads to the familiar P, Q and R branches.

The transition dipole moment is calculated for the electronic and vibronic parts from

$$\boldsymbol{M}_{\mathrm{ev}} = \langle \psi'(\boldsymbol{r}, \boldsymbol{R})|\hat{\boldsymbol{\mu}}|\psi''(\boldsymbol{r}, \boldsymbol{R})\rangle = \int\!\!\int \psi_{\mathrm{e}}'^{*}\psi_{\mathrm{v}}'^{*}\,\hat{\boldsymbol{\mu}}\,\psi_{\mathrm{e}}''\,\psi_{\mathrm{v}}''\,\mathrm{d}\boldsymbol{r}\,\mathrm{d}\boldsymbol{R}. \tag{26.14}$$

Vibronic is used to differentiate the change in vibrational state that occurs during an electronic transition, as is the case here, from changes in vibrational state that occur in the absence of electronic excitation. First, we integrate over the electronic coordinates $\boldsymbol{r}$ to obtain

$$\boldsymbol{M}_{\mathrm{ev}} = \int \psi_{\mathrm{v}}'^{*}\boldsymbol{M}_{\mathrm{e}}\,\psi_{\mathrm{v}}''\,\mathrm{d}\boldsymbol{R} \tag{26.15}$$

where $\boldsymbol{M}_{\mathrm{e}}$ is the electronic transition moment

$$\boldsymbol{M}_{\mathrm{e}} = \int \psi_{\mathrm{e}}'^{*}\,\hat{\boldsymbol{\mu}}\,\psi_{\mathrm{e}}''\,\mathrm{d}\boldsymbol{r} = -e\int \psi_{\mathrm{e}}'^{*}\,\boldsymbol{r}\,\psi_{\mathrm{e}}''\,\mathrm{d}\boldsymbol{r} \tag{26.16}$$

The vibronic part of the wavefunction is independent of the electronic coordinates $\boldsymbol{r}$. Thus, it factors out of the integral involving the nuclear coordinates. The Born–Oppenheimer approximation allows us to consider the nuclear coordinates as fixed with respect to electronic motion. Thus, $\boldsymbol{M}_{\mathrm{e}}$ is a constant that can be pulled out of the integral over the nuclear coordinates $\boldsymbol{R}$, to yield

$$\boldsymbol{M}_{\mathrm{ev}} = \boldsymbol{M}_{\mathrm{e}}\int \psi_{\mathrm{v}}'^{*}\psi_{\mathrm{v}}''\,\mathrm{d}\boldsymbol{R} = \boldsymbol{M}_{\mathrm{e}}\,\boldsymbol{M}_{\mathrm{v}} \tag{26.17}$$

The transition dipole moment is composed of two terms. The first term $\boldsymbol{M}_{\mathrm{e}}$, the electronic transition dipole moment, describes how well the two electronic states are coupled to one another by the radiation. The second term $\boldsymbol{M}_{\mathrm{v}} = \langle \psi_{\mathrm{v}}'|\psi_{\mathrm{v}}''\rangle$ describes the *vibrational overlap* of the two electronic states. Vibrational overlap compares whether the two vibronic wavefunctions have significant amplitude at the internuclear separation where the transition occurs, and is the basis of the *Franck–Condon principle*:

> *Electronic transitions are vertical. It is the excited state with the greatest vibrational overlap with the initial state that has the highest intensity.*

The Franck–Condon principle holds for transitions with energies well into the extreme UV and even soft X-ray region. In a *vertical transition* the linear momentum transfer from the photon is negligible and does not change the positions of the nuclei during the transition. Thus, transitions between Born–Oppenheimer potential energy curves for the electronic states can be drawn as vertical lines.

A strong transition is said to have a good *Franck–Condon factor* as opposed to a weaker transition with a bad Franck–Condon factor. The Franck–Condon factor is quantified by the square of the vibrational overlap integral

$$\text{FC factor} \equiv |\boldsymbol{M}_{\mathrm{v}}|^2 = |\langle \psi_{\mathrm{v}}'|\psi_{\mathrm{v}}''\rangle|^2. \tag{26.18}$$

If the value of the equilibrium bond distance R_{e} is the same in the lower and upper states, the Franck–Condon factor is greatest for the $\Delta v = 0$ transition and then drops off to higher values, mimicking the behavior of an anharmonic oscillator.

When the R_e values in the two states are displaced, there will be a maximum Franck–Condon factor at some value of $\Delta v \neq 0$, the larger the displacement, the greater the value of the difference. The intensity of transitions on either side of this maximum decays to lower intensity.

In Section 26.1, we found that transition rates are proportional to the absolute square of the transition dipole moment, $|\mu_{kl}|^2$. The value of the transition dipole moment is proportional to the product of the electronic transition moment and the Franck–Condon factor,

$$|\mu_{ki}|^2 \propto |\boldsymbol{M}_e|^2 |\boldsymbol{M}_v|^2. \tag{26.19}$$

The electronic and vibronic selection rules for polyatomics are based on finding those circumstances under which the product in Eq. (26.19) has a finite value as opposed to zero. The integrals that define $\boldsymbol{M}_e$ and $\boldsymbol{M}_v$ are integrals over symmetrical limits. Therefore, they only have finite values if the integrand is totally symmetric (in the cases of nondegenerate states) or contains a symmetric species (in the case of degenerate states). Whether or not the integral vanishes can be determined by looking at the symmetry representations, Γ, of the initial and final wavefunctions and the transition dipole operator that connects them.

The electronic transition moment is written in Dirac notation as

$$\boldsymbol{M}_e = \langle \psi_e'^* | \hat{\boldsymbol{\mu}} | \psi_e'' \rangle. \tag{26.20}$$

This integral only has a finite value if the cross-product of the symmetry representations of the final wavefunction with the operator and the initial wavefunction is totally symmetric

$$\Gamma(\psi_e') \times \Gamma(\hat{\boldsymbol{\mu}}) \times \Gamma(\psi_e'') = \text{A}. \tag{26.21}$$

The designation A indicates the totally symmetric representation. Equation (26.21) holds for nondegenerate states. For degenerate states, the cross-product produces more than one representation. For a transition involving a degenerate state to be allowed, the cross-product must 'contain' a totally symmetric representation – that is, at least one of the representations generated must be totally symmetric,

$$\Gamma(\psi_e') \times \Gamma(\hat{\boldsymbol{\mu}}) \times \Gamma(\psi_e'') \supset \text{A}. \tag{26.22}$$

The electric dipole transition operator is a vector with components along the Cartesian axes x, y, and z. Therefore, the symmetry of each of these components is the same as a translation along the appropriate axis. We say that the components transform as the translations along these axes, which are labeled T_x, T_y and T_z, respectively. Labeling the components of the electronic transition moment $M_{e,x}$, $M_{e,y}$ and $M_{e,z}$, the absolute square of the total electronic transition moment is equal to the sum of the squares of the components

$$|\boldsymbol{M}_e|^2 = |M_{e,x}|^2 + |M_{e,y}|^2 + |M_{e,z}|^2, \tag{26.23}$$

where the absolute squares of the components are defined analogously to Eq. (26.20) with the total operator replaced by the component operator. Because Eq. (26.23) contains a sum rather than a product, *an electronic transition is allowed if any one of the components is non-zero*. Written in terms of symmetry representations, this is

$$\Gamma(\psi_e') \times \Gamma(T_x) \times \Gamma(\psi_e'') = \text{A}, \tag{26.24}$$
$$\Gamma(\psi_e') \times \Gamma(T_y) \times \Gamma(\psi_e'') = \text{A}, \tag{26.25}$$
$$\Gamma(\psi_e') \times \Gamma(T_z) \times \Gamma(\psi_e'') = \text{A}, \tag{26.26}$$

(for degenerate states replace = with $\supset$). What these symmetry representations are and how we determine them are discussed in Section 26.3.

The product of two symmetry representations is totally symmetric when those two species are the same. Therefore, Eqs (26.24–26.25) are only satisfied when the cross-product of the representations of the wavefunctions is the same as the representations of one of the components of translation. Formally, we write this condition as

$$\Gamma(\psi_e') \times \Gamma(\psi_e'') = \Gamma(T_x) \quad \text{and/or} \quad \Gamma(T_y) \quad \text{and/or} \quad \Gamma(T_z). \tag{26.27}$$

The cross-product of the symmetry species of the initial and final wavefunctions must transform as one of the translational axes. Just as for diatomics, most polyatomics have closed shells and a Σ ground state, which is totally symmetric. Because the product of any representation with A is equal to the same representation, the selection rule is expressed in the most common circumstance as

$$\Gamma(\psi'_e) \times A = \Gamma(\psi'_e) = \Gamma(T_x) \quad \text{and/or} \quad \Gamma(T_y) \quad \text{and/or} \quad \Gamma(T_z). \tag{26.28}$$

In other words, *for any system with a totally symmetric ground state, the electronic transition is allowed as long as the excited state wavefunction transforms as one of the translational axes.*

If vibrations are excited in either the initial state or the final state (or both), then we need to look at the symmetry of the full integral in Eq. (26.14) to determine its value. In Dirac notation this is

$$\boldsymbol{M}_{ev} = \langle \psi'^*_e \psi'^*_v | \hat{\mu} | \psi''_e \psi''_v \rangle = \langle \psi'_{ev} | \hat{\mu} | \psi''_{ev} \rangle. \tag{26.29}$$

Using the same arguments as above, in the most common case of a Σ ground state, the selection rule is found from

$$\Gamma(\psi'_{ev}) \times \Gamma(\psi''_{ev}) = \Gamma(T_x) \quad \text{and/or} \quad \Gamma(T_y) \quad \text{and/or} \quad \Gamma(T_z). \tag{26.30}$$

The symmetry of the electronic vibronic state is given by the product

$$\Gamma(\psi_{ev}) = \Gamma(\psi_e) \times \Gamma(\psi_v). \tag{26.31}$$

Consequently, the full selection rule is that the cross-product of the composite wavefunctions must transform as at least one of the components of translation along *x*, *y* or *z*,

$$\Gamma(\psi'_e) \times \Gamma(\psi'_v) \times \Gamma(\psi''_e) \times \Gamma(\psi''_v) = \Gamma(T_x) \quad \text{and/or} \quad \Gamma(T_y) \quad \text{and/or} \quad \Gamma(T_z). \tag{26.32}$$

If the same vibration is excited in the initial and final states, $\Gamma(\psi'_v) = \Gamma(\psi''_v)$. The selection rule is then the same as the electronic selection rule. If no vibration is excited (i.e., in the vibrational ground state) the representation of that vibrational component is totally symmetric.

What are these symmetry representations and how do we determine them? For atoms this is extremely simple. All atoms belong to point group K_h. Linear molecules (diatomic and polyatomics) belong to either $D_{\infty h}$ or $C_{\infty v}$. Polyatomics in general can belong to a wide range of point groups. In the next section we investigate how to determine which symmetry elements pertain to a given molecule, and to which point group that molecule belongs.

26.3 Molecular symmetry

Figures 26.4 and 26.5 display a number of different molecules with characteristic shapes. The shapes demonstrate the types of transformations – called *symmetry elements* – that we need to define to treat molecular symmetries. Terms for these symmetry operations and their classification are given in the figure captions, and they are defined in the subsequent sections. A linear molecule such as CO_2 has symmetry much like a homonuclear diatomic molecule. A linear molecule such as HCN has symmetry more akin to a heteronuclear diatomic. Categorizing and classifying molecules by symmetry facilitates the understanding of a number of molecular properties, such as any involving dipole moments, and molecular spectroscopy.

From geometrical studies it has been determined that five types of symmetry elements must be defined to describe the symmetry of molecules. These symmetry elements are listed, along with their systematic notation in terms of Schönflies symbols in Table 26.1.

26.3.1 Rotation – *n*-fold axis of symmetry, C_n

A molecule possesses an *n*-fold axis of symmetry if you can stick a fork into it, rotate it clockwise by $2\pi/n$ radians, and return the molecule to an orientation that looks indistinguishable from the initial geometry. The fork is called a *rotational axis* or, more accurately, an *n*-fold axis of symmetry, and is shown in Fig. 26.6. For instance, the CO_2 molecule has a C_2 axis perpendicular to the internuclear axis passing through the center of the C atom. Rotation by π radians (180°) leaves the central C atom where it was (atoms by themselves are spherically symmetric) and moves one O atom to where the other one was initially. Since the two O atoms are indistinguishable (assuming they correspond to the same isotope of O), the molecule has been transformed from an initial configuration into a symmetry equivalent configuration. As also shown in Fig. 26.6(b), placing an axis parallel to and centered upon the molecular axis, the molecule can be rotated by

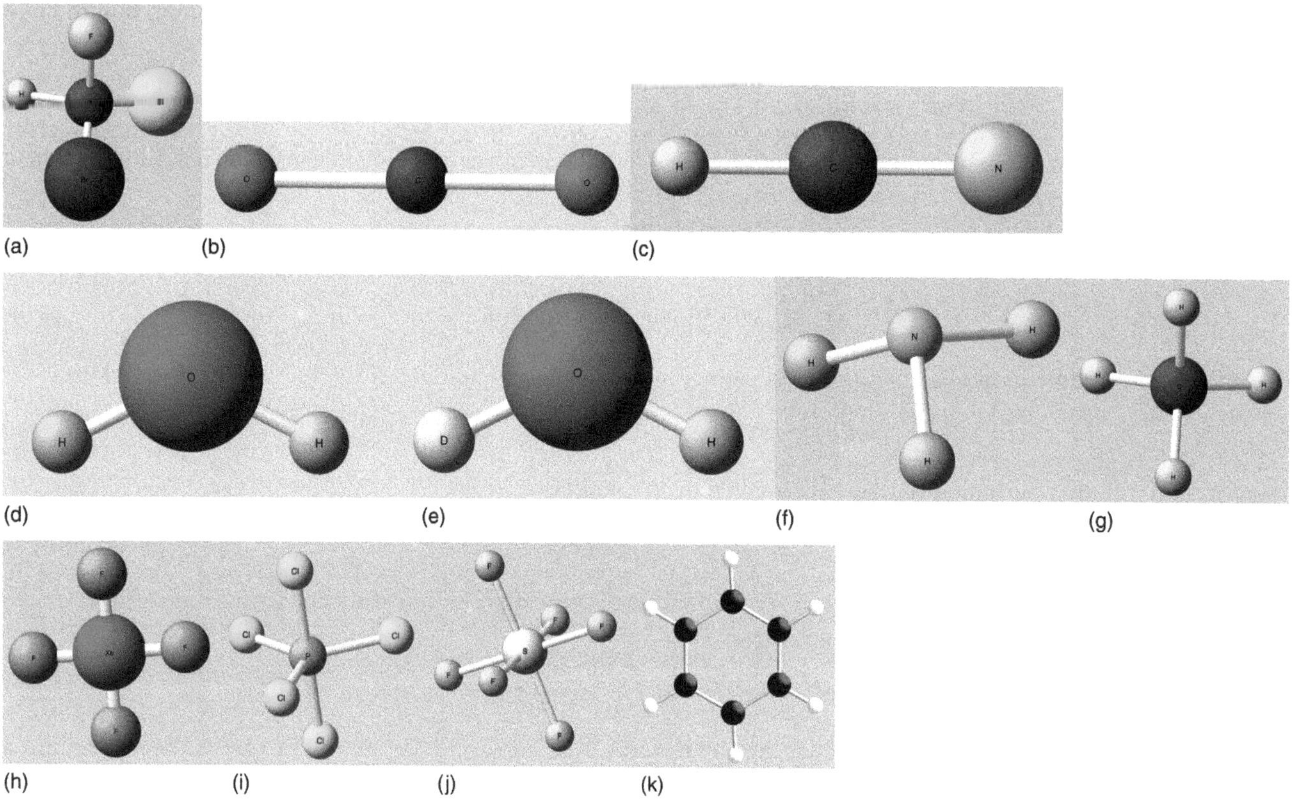

Figure 26.4 Molecules representative of a number of different symmetry classes. (a) CHFClBr, C_1 with only a C_1 axis of rotation, which is the same as doing nothing (identity symmetry element). (b) CO_2, $D_{\infty h}$ with a C_∞ axis along the internuclear axis, C_2 axis perpendicular to it, and vertical planes of reflection. (c) HCN, $C_{\infty v}$ with C_∞ and vertical planes. (d) H_2O, C_{2v} with a C_2 axis and a vertical plane of reflection. (e) HDO, C_s with only a plane of symmetry (f) NH_3, trigonal pyramid, C_{3v} with a C_3 axis and vertical planes. (g) CH_4, tetrahedral, T_d. (h) XeF_4, square planar, D_{4h} with C_4 and C_2 axes, vertical, horizontal, dihedral planes. (i) PCl_5, trigonal bipyramidal, D_{3h} with C_3, C_2 and S_3 axes, vertical and horizontal planes. (j) SF_6, octahedral, O_h. (k) Benzene, D_{6h}, C_6 axis, vertical and horizontal planes and much more.

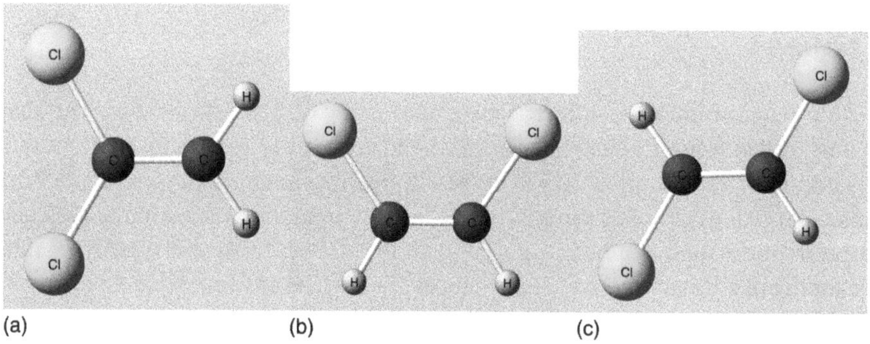

Figure 26.5 Isomerization and its effect on symmetry. Dichloroethene, $C_2H_2Cl_2$, exists in three forms with different symmetries. (a) 1,1-Dichloroethene, C_{2v}. (b) *cis*-1,2-dichloroethene, C_{2v} and (c)*trans*-1,2-dichloroethene, C_{2h}. All three forms are planar (contain a horizontal plane of reflection) with a C_2 axis (though along three different axes for the different molecules) but the *trans* (or *E*)isomer does not contain a vertical plane.

Table 26.1 Symmetry operators in molecule-fixed coordinates, known as Schönflies symbols.

Operation	Symbol
identity	E
rotation by $2\pi/n$	C_n
reflection	σ, σ_v, σ_d, σ_{dh}
inversion	i
rotation-reflection	S_n (= $C_n\sigma_h$)

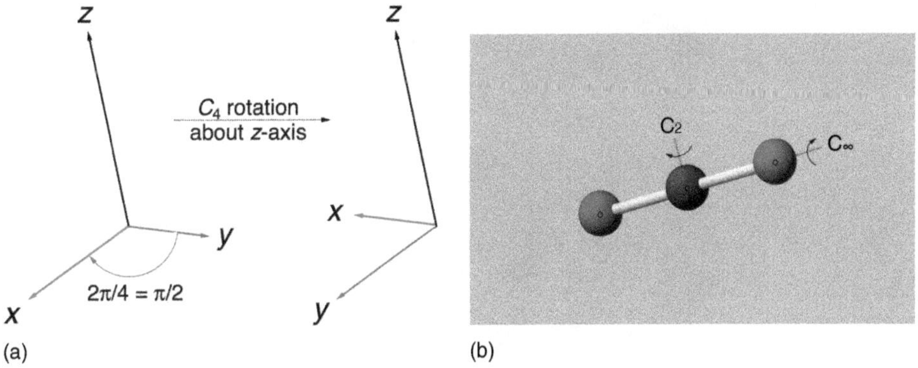

Figure 26.6 Clockwise rotation about a C_n axis looking down the z-axis transforms x and y by rotating them through $2\pi/n$ radians within the xy-plane. (a) A C_4 rotation is shown. (b) CO_2 possesses a C_2 axis perpendicular to the internuclear axis and a C_∞ axis along it.

an arbitrary amount and it still looks the same. Such an axis is known as a C_∞ axis. The value of n change in integer steps from $n = 1, 2, 3, \dots \infty$.

Corresponding to every *symmetry element* (a property of the object) is a *symmetry operation* (the thing that is done to the object) with the same symbol. Thus, we speak of a CO_2 molecule possessing a C_2 axis and state that when transformed under a C_2 symmetry operation by rotation of π radians about the C_2 axis (or more simply, under C_2), the CO_2 molecule is symmetric. The highest-fold axis of a molecule defines the vertical axis, labeled as the z-axis. If a C_n operation is performed twice, we denote this by C_n^2 and so forth for other multiples. C_n corresponds to a clockwise rotation by $2\pi/n$ radians, whereas C_n^{-1} corresponds to a counterclockwise rotation by $-2\pi/n$ radians,

Some molecules contain C_2 axes at equal angles to each other and perpendicular to the main axis. This is true, for example of the benzene molecule. This combination is common enough that axes of this type are called *dihedral axes*.

Rotation about an axis transforms a point with initial coordinates (x, y, z) to the point (x', y', z'). The new coordinates are obtained from a linear combination of the original coordinates according to the set of equations

$$x' = c_{11}x + c_{12}y + c_{13}z \tag{26.33}$$

$$y' = c_{21}x + c_{22}y + c_{23}z \tag{26.34}$$

$$z' = c_{31}x + c_{32}y + c_{33}z. \tag{26.35}$$

This set of equations can be represented by a matrix equation in which the points are given as column vectors and the coefficients form a (3×3) matrix,

$$\begin{pmatrix} x' \\ y' \\ z' \end{pmatrix} = \begin{pmatrix} c_{11} & c_{12} & c_{13} \\ c_{21} & c_{22} & c_{23} \\ c_{31} & c_{32} & c_{33} \end{pmatrix} \begin{pmatrix} x \\ y \\ z \end{pmatrix} \tag{26.36}$$

A C_2 rotation about the z-axis leaves z unchanged but rotates x to $-x$ and y to $-y$. The transformation matrix for this C_2 operation along z is thus,

$$\begin{pmatrix} -1 & 0 & 0 \\ 0 & -1 & 0 \\ 0 & 0 & 1 \end{pmatrix}. \tag{26.37}$$

Inspection of Figure 26.6(a) reveals that we can represent a rotation about z by an angle $2\pi/n$ by the matrix

$$(C_n)_z = \begin{pmatrix} \cos 2\pi/n & \sin 2\pi/n & 0 \\ -\sin 2\pi/n & \cos 2\pi/n & 0 \\ 0 & 0 & 1 \end{pmatrix}. \tag{26.38}$$

If we were to change the rotation to be about the x-axis, the zeros and ones would move to keep x unchanged and the block of sine and cosine terms would move down an over to transform y and z appropriately.

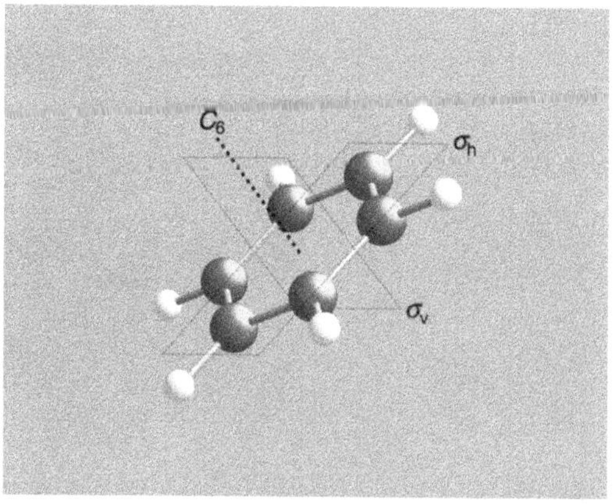

Figure 26.7 The highest-fold axis, a C_6 axis for this example of benzene, defines the z-axis and the vertical plane. In benzene the vertical plane is a plane of symmetry, denoted σ_v. The horizontal plane in benzene is also a plane of symmetry, denoted σ_h.

26.3.1.1 Directed practice

The methane molecule CH_4 contain a C_3 axis directed along a C–H bond. Looking down this axis (in other words making this axis the z-axis in a molecule-fixed frame) you will see the three other H atoms, which sit at the vertices of an equilateral triangle with the z-axis passing through its center. Starting at the top of the triangle and proceeding clockwise we label the H atoms H_1, H_2 and H_3. Write out the matrix equation that corresponds to performing the C_3 operation.

26.3.2 Reflection – plane of symmetry, σ

If reflection of all the nuclei in a molecule through a plane to an equal distance on the opposite side of the molecule produces a configuration indistinguishable from the initial configuration, the molecule is said to possess a *plane of symmetry*, σ; this is also called a *mirror plane*. A molecule may possess more than one plane of symmetry, just as it can have multiple axes of symmetry. A plane that contains the z-axis, which is defined to contain the highest-fold axis, is called a *vertical plane of symmetry*, σ_v. The H_2O molecule is planar with a C_2 axis through the central O atom. It also contains two vertical planes of symmetry containing this axis. They are named, according to convention, $\sigma_v(xz)$ and $\sigma_v(yz)$. The yz-plane is the plane of the molecule. Any planar molecule must contain at least one plane of symmetry. Any molecule with a plane of symmetry cannot be chiral.

The benzene molecule shown in Fig. 26.7 is planar with a C_6 axis perpendicular to this plane. The plane of the molecule is a horizontal plane, σ_h. However, if we look at the vertical planes of symmetry of the molecule – the ones in which the C_6 axis lies – we see that there are two ways that we can draw them. The most obvious is that we can draw a vertical axis through the center of opposing C–H bonds. We label these (there are three of them) the σ_v axes. However, we can also draw three vertical axes through the middle of opposing C–C bonds in the benzene ring. To distinguish these three axes, we label them σ_d, for dihedral plane of symmetry. Dihedral planes bisect the angles between dihedral axes.

Reflection in the xy-plane leaves the coordinates of x and y unchanged but transforms z to $-z$, $(x, y, z) \rightarrow (x, y, -z)$. Similarly, reflection in the yz plane leaves y and z unchanged while changing x to $-x$. The transformation matrix for the operation σ_{xy} is thus,

$$\sigma_{xy} = \begin{pmatrix} 1 & 0 & 0 \\ 0 & 1 & 0 \\ 0 & 0 & -1 \end{pmatrix} \qquad (26.39)$$

with appropriately similar matrices for σ_{yz} and σ_{xz}.

26.3.2.1 Directed practice

Describe why H_2O has two vertical planes of symmetry but HOD only has one.

26.3.3 Center of inversion, i

Inversion symmetry, i, involves reflection of atoms through the center of the molecule to a point an equal distance on the opposite side of the molecule, $(x, y, z) \rightarrow (-x, -y, -z)$. The transformation matrix is thus,

$$\boldsymbol{i} = \begin{pmatrix} -1 & 0 & 0 \\ 0 & -1 & 0 \\ 0 & 0 & -1 \end{pmatrix} \tag{26.40}$$

Any molecule possessing inversion symmetry cannot be chiral.

26.3.3.1 Directed practice

Benzene C_6H_6 contains a center of inversion. Draw isotopologues of benzene formed by the substitution of either one D or two D atoms for H atoms that either do or do not have a center of inversion.

26.3.4 Rotation–reflection axis of symmetry, S_n

A distinct symmetry operation can be generated by performing first a rotation of $2\pi/n$ radians, followed by reflection through a plane. The plane is perpendicular to the axis passing through the center of the molecule. The combined symmetry operation is known as n-fold rotation–reflection symmetry, denoted by S_n,

$$S_n = \sigma_\mathrm{h} \times C_n. \tag{26.41}$$

This is also known as an *improper rotation*. The order of operation is important for matrix multiplication. In other words, matrices do not in general commute and $M \times N \neq N \times M$. Therefore, to obtain the transformation matrix to perform an S_n operation about the z-axis, we must form the cross-product

$$(S_n)_z = \sigma_{xy} \times (C_n)_z \tag{26.42}$$

$$(\boldsymbol{S}_n)_z = \begin{pmatrix} \cos 2\pi/n & \sin 2\pi/n & 0 \\ -\sin 2\pi/n & \cos 2\pi/n & 0 \\ 0 & 0 & -1 \end{pmatrix} \tag{26.43}$$

Any molecule possessing an S_n axis cannot be chiral. In this context it is to be noted that $S_1 \equiv \sigma$ and $S_2 \equiv i$. Benzene contains both S_3 and S_6 rotation–reflection axes.

26.3.5 Identity element of symmetry, E

The symmetry operation E corresponds to leaving the molecule unchanged. It is sometimes also denoted I in order to avoid confusion between E (for the German *Einheit* formatted in italics) the identity symmetry element, and E (in Roman type), which is the name of a doubly degenerate irreducible representation. This may seem trivial but it is required by the mathematics of group theory as we shall see below. Performing any n-fold rotation n times corresponds to rotation by 2π and reproduces the initial configuration, for example,

$$C_2 \times C_2 = C_2^2 = E. \tag{26.44}$$

All molecules possess this element of symmetry. Equations (26.41) and (26.44) also demonstrate that we can generate a full range of other symmetry operations by performing cross-products of symmetry elements. The transformation matrix for the identity operation, also called the unit matrix,

$$\boldsymbol{E} = \begin{pmatrix} 1 & 0 & 0 \\ 0 & 1 & 0 \\ 0 & 0 & 1 \end{pmatrix} \tag{26.45}$$

is unusual in that it commutes with any other matrix.

26.4 Point groups

Any given molecule possesses a set of symmetry elements. These symmetry elements are not unique to the molecule; instead, they are unique to the shape of the molecule and the juxtapositions of similar and dissimilar atoms within the

Table 26.2 Point groups and the symmetry elements that define them. All point groups contain the identity operation *E*. Other symmetry elements generated by the defining elements may also be present in any given point group. Note that the names of point groups are in roman type while the symmetry elements are in italics because they represent mathematical operators

Point group	Symmetry elements	Examples
C_i	i	*trans*-$(CHFCl)_2$
C_n	C_n	C_1: CHFClBr, substituted alkanes and alkenes; C_2: H_2O_2
S_n	S_n	Rare apart from S_2, which is the same as C_i.
C_{nv}	C_n, $n\ \sigma$ planes	$C_{1v} \equiv C_s$: planar, no rotations, e.g., CH_2CHCl; C_{2v}: H_2O ; C_{3v} NH_3; $C_{\infty v}$: all linear molecules without σ_h such as HCN and heteronuclear diatomics
D_n	C_n, $n\ C_2$ axes	uncommon; D_3: tetramethylbutane
C_{nh}	C_n, σ_h, i (if n even)	C_{2h}: *trans*-1,2-dichloroethene
D_{nd}	C_n, S_{2n}, $n\ C_2$ axes $\perp$ to C_2, $n\ \sigma_d$ planes, i (if n odd)	D_{2d}: allene, staggered C_2H_6, otherwise rare
D_{nh}	C_n, $n\ C_2$ axes $\perp$ to C_n, σ_h, n other σ planes, i (if n even)	D_{2h}: C_2H_4; D_{3h}: BF_3; D_{4h}: $PtCl_4{}^{2-}$; D_{6h}: C_6H_6; $D_{\infty h}$: symmetric linear molecules such as C_2H_2 and homonuclear diatomics
$\mathrm{T_d}$	four C_3, three C_2 and six σ_d	all tetrahedral molecules such as CH_4
$\mathrm{O_h}$	three C_4, four C_3, six C_2, three σ_h, six σ_d, i	all octahedral molecules such as SF_6, cubane
$\mathrm{K_h}$	infinite C_∞, i	any sphere including atoms

molecule. Water has the symmetry elements I, C_2, $\sigma_v(xz)$ and $\sigma_v(yz)$. This is true for all bent triatomic molecules of the formula AB_2. Similarly, both C_6H_6 and C_6F_6 have the symmetry elements I, C_6, C_3, C_2, i, S_3, S_6, σ_h, σ_d and σ_v. What these sets have in common is that when all of the operations in a set are performed on the molecule, there is a point (more accurately at least one point) in the molecule that does not move. In the case of H_2O, every point along the C_2 axis is unaffected by the operations in the set. In the benzene molecule, the center of inversion does not move under any operation. For this reason, we call the set of operations a *point group*.

The members of a point group follow a number of rules. The product of any two operations is defined and is also a member of the group. The inverse of each operation is in the group. The identity operator is a member of every group. All operations of the group are associative,

$$LMN = (LM)N = L(MN). \tag{26.46}$$

The C_n group contains just E and C_n operations. Since n repetitions of C_n bring the object back to its original configuration – that is, $C_n^n = E$ – there are a total of n operations in the group. We say the *order* of the group is n. Molecules with this point group are chiral.

Adding a horizontal plane of symmetry σ_h to a group containing C_n generates various S_n operations, including i if n is even; this creates the C_{nh} group. The order of a C_{nh} group is $2n$.

If a mirror plane contains the rotational axis, the group is called C_{nv}. This vertical mirror plane σ_v is repeated n times. The order of a C_{nv} group is $2n$.

Adding a C_2 axis perpendicular to a C_n axis generates one of the dihedral groups. D_n arises from adding $n\ C_2$ axes to the C_n group. This group has order $2n$ and molecules with this symmetry are optically active.

Adding a horizontal plane of symmetry σ_h to D_n generates the D_{nh} group with $4n$ symmetry elements. When n is even, the group also contains an inversion center i.

Adding a vertical plane that bisects adjacent C_2 axes generates the D_{nd} group. The vertical plane is called a dihedral plane σ_d. There will be n of them and $4n$ elements in total. An inversion center is found when n is odd.

S_n groups contain an S_n axis with n even and no mirror planes. If n is odd, $S_n^{2n} = E$ and the group is identical to the C_{nh} group. The group is then named as a C_{nh} group. Similarly, if the group contains mirror planes, then it is named as a D_{nd} group.

$\mathrm{C}_{\infty v}$ and $\mathrm{D}_{\infty h}$ are known as the linear groups. Linear molecules symmetrical about their center such as N_2, CO_2 and HCCH exhibit $\mathrm{D}_{\infty h}$ symmetry, while linear molecules that are not symmetrical about their center such as HCl or HCN exhibit $\mathrm{C}_{\infty v}$ symmetry.

Table 26.3 Character table for the T_d point group.

T_d	E	$8C_3$	$3C_2$	$6S_4$	$6\sigma_d$		
A_1	1	1	1	1	1		$\alpha_{xx}+\alpha_{yy}+\alpha_{zz}$
A_2	1	1	1	−1	−1		
E	2	−1	2	0	0		$(\alpha_{xx}+\alpha_{yy}-2\alpha_{zz}, \alpha_{xx}-\alpha_{yy})$
$T_1 \equiv F_1$	3	0	−1	1	−1	(R_x, R_y, R_z)	
$T_2 \equiv F_2$	3	0	−1	−1	1	(T_x, T_y, T_z)	$(\alpha_{xy}, \alpha_{xz}, \alpha_{yz})$

26.5 Character tables

Information on the set of operations contained within a point group is collected in a *character table*. Up to this point we have concerned ourselves only with the symmetries of the nuclear coordinates. The symmetry properties of the wavefunctions that describe the electronic and vibrational degrees of freedom are also of interest. These wavefunctions need not obey all of the symmetry operations in the symmetry group of the molecule. Their behavior under one or more operations will be of importance, and that behavior is captured in the character table of the point group that describes the molecule. Character tables for five important and common point groups, C_{2v}, $C_{\infty v}$, D_{6h}, T_d and O_h, can be found in Appendix 7. C_{2v} is an example of a nondegenerate point group. All of the other groups are degenerate point groups. $C_{\infty v}$ is an example of an infinite group. The T_d and C_{3v} character tables (see Tables 26.3 and 26.4) are repeated here to aid our description of the contents.

When a wavefunction does not exhibit a certain symmetry operation, we say that it does not preserve that element of symmetry. When a wavefunction preserves an element of symmetry, one of two outcomes is observed. A wavefunction that is *symmetric* with respect to the symmetry element is equal to itself under the symmetry operation, for example for the operation O_s.

$$\psi_k \xrightarrow{O_s} \psi_k. \tag{26.47}$$

A wavefunction is *antisymmetric* if under the operation, the sign of the wavefunction changes,

$$\psi_k \xrightarrow{O_s} -\psi_k. \tag{26.48}$$

The values +1 and −1 in front of the wavefunctions in Eqs (26.47) and (26.48) are known as the *character* of the wavefunction with respect to the symmetry operation O_s. The character with respect to the identity operator E is equal to the degeneracy of the wavefunction. Hence, we see that this operation is nontrivial in any degenerate point group. It can be seen in Tables 26.3 and 26.4 that the different rows of characters are grouped together and named with a letter. This letter is called the *Mulliken label*, in accord with the conventions for the labeling of polyatomic molecules and their spectra.[2] Each row corresponds to an *irreducible representation*. A_1 is the name of the totally symmetric irreducible representation. Each character table begins with this irreducible representation. The letters A and B are used to denote nondegenerate (also called singly degenerate) irreproducible representations, while E and T are used to denote doubly and triply degenerate representations, respectively. Higher degeneracies follow the order G, H,...for quadruply, quintuply, and so forth degenerate representations.

A further note on the naming convention in the first column is required. If C_n is the primary axis of symmetry, electronic wavefunctions of *polyatomics* that are unchanged under the operator C_n are given species label A. If operation by C_n changes the sign of the wavefunction, it is labeled B. If under C_n the wavefunction is multiplied by $\exp(\pm 2\pi i/n)$, it is given the species label E. Wavefunctions that are unchanged or change sign under inversion i are labeled with the subscript g for gerade or u for ungerade symmetry, respectively. Wavefunctions that are unchanged (changed) under

Table 26.4 Character table for the C_{3v} point group.

C_{4v}	E	$2C_3$	$3\sigma_v$		
A_1	1	1	1	T_z	$\alpha_{xx}+\alpha_{yy}, \alpha_{zz}$
A_2	1	1	−1	R_z	
E	2	−1	0	$(T_x, T_y)(R_x, R_y)$	$(\alpha_{xx}-\alpha_{yy}, \alpha_{zz})(\alpha_{xz}, \alpha_{yz})$

Table 26.5 Correspondence of irreducible representation names to the state designations of diatomic molecules.

$C_{\infty v}$	$D_{\infty v}$	
$A_1 \equiv \Sigma^+$	$A_{1g} \equiv \Sigma_g^+$	$A_{2u} \equiv \Sigma_u^+$
$A_2 \equiv \Sigma^-$	$A_{2g} \equiv \Sigma_g^-$	$A_{1u} \equiv \Sigma_u^-$
$E_1 \equiv \Pi$	$E_{1g} \equiv \Pi_g$	$E_{1g} \equiv \Pi_u$
$E_2 \equiv \Delta$	$E_{2g} \equiv \Delta_g$	$E_{2u} \equiv \Delta_u$
$E_3 \equiv \Phi$	$E_{3g} \equiv \Phi_g$	$E_{3u} \equiv \Phi_u$
⋮	⋮	⋮

σ_h have species labels with a prime ("double prime"). Vibrational wavefunctions (and the symmetry of the corresponding vibrational modes) are labeled with lower-case rather than upper-case letters to distinguish them from electronic wavefunctions.

For *diatomic* molecules, the naming convention of electronic states was already well-established before Mulliken labels had been introduced. Diatomics follow a convention derived from atomic spectroscopy, in which the value of the total orbital angular momentum Λ determines the symmetry label of the state in the form of an upper-case Greek letter. The correspondence between Mulliken labels and the symmetry labels of diatomic molecules is given in Table 26.5. This table also serves to reinforce the relationship of symmetry properties to the magnitude of angular momentum.

The character of nondegenerate irreducible representations is easily obtained by inspection. The value is always ± 1. This is not so for degenerate irreducible representations. Under a symmetry operation, a degenerate mode is transformed into a linear combination of the degenerate coordinates. Thus, for a doubly degenerate irreducible representation under O_s, a normal coordinate Q with components Q_a and Q_b is transformed into Q' according to

$$Q_a \xrightarrow{O_s} Q_a' = q_{aa}Q_a + q_{ab}Q_b \tag{26.49}$$

$$Q_b \xrightarrow{O_s} Q_b' = q_{ba}Q_a + q_{bb}Q_b. \tag{26.50}$$

This system of linear equations lends itself to more compact representation in the form of a matrix equation,

$$\begin{pmatrix} Q_a' \\ Q_b' \end{pmatrix} = \begin{pmatrix} q_{aa} & q_{ab} \\ q_{ba} & q_{bb} \end{pmatrix} \begin{pmatrix} Q_a \\ Q_b \end{pmatrix} \tag{26.51}$$

The sum $q_{aa} + q_{bb}$ is the trace of the (2×2) matrix. It is also equal to the character of the irreducible representation under the symmetry operation O_s. A doubly degenerate irreducible representation is associated with a (2×2) matrix, a triply degenerate with a (3×3) matrix, and so on.

Consistent with the statement above, the trace of the identity operation is equal to the degeneracy of the representation. For example,

$$\begin{pmatrix} Q_a' \\ Q_b' \end{pmatrix} = \begin{pmatrix} 1 & 0 \\ 0 & 1 \end{pmatrix} \begin{pmatrix} Q_a \\ Q_b \end{pmatrix}. \tag{26.52}$$

Now with specific reference to the E irreducible representation in the C_{3v} character table, we can think in terms of the NH_3 molecule, which has this symmetry. Rotation about the C_3 axis leaves the N atom stationary but rotates an H atom from one vertex of a triangle to the next, a rotation of $2\pi/3$ radians. The ν_3 vibrational mode of NH_3 is an e species. It thus has two degenerate components ν_{3a} and ν_{3b} that transform under C_3 as

$$\begin{pmatrix} \nu_{3a}' \\ \nu_{3b}' \end{pmatrix} = \begin{pmatrix} \cos 2\pi/3 & \sin 2\pi/3 \\ -\sin 2\pi/3 & \cos 2\pi/3 \end{pmatrix} \begin{pmatrix} \nu_{3a} \\ \nu_{3b} \end{pmatrix} \tag{26.53}$$

$$\begin{pmatrix} \nu_{3a}' \\ \nu_{3b}' \end{pmatrix} = \begin{pmatrix} -\frac{1}{2} & \sqrt{\frac{3}{2}} \\ -\sqrt{\frac{3}{2}} & -\frac{1}{2} \end{pmatrix} \begin{pmatrix} \nu_{3a} \\ \nu_{3b} \end{pmatrix} \tag{26.54}$$

The trace of this matrix is −1, which is the character found in Table 26.4 for E under C_3.

The final element of the reflection about a σ_v plane passing through both the N and one of the H atoms, while bisecting the angle between the other two H atoms. This operation leaves ν_{3a} unchanged and changes the sign of ν_{3b},

$$\begin{pmatrix} \nu'_{3a} \\ \nu'_{3b} \end{pmatrix} = \begin{pmatrix} 1 & 0 \\ 0 & -1 \end{pmatrix} \begin{pmatrix} \nu_{3a} \\ \nu_{3b} \end{pmatrix}. \tag{26.55}$$

The trace of this matrix is 0 as is the character of E under σ_v.

The procedure of finding the irreducible representations and characters in the general case is not so easily explained. This may be found in the book by Cotton in Further Reading. Vibrational mode symmetries can be found for all symmetry classes in Herzberg's *Infrared and Raman Spectra*.

The symmetries of rotations and translations along the three Cartesian axes are indicated by (R_x, R_y, R_z) and (T_x, T_y, T_z), respectively. For the T_d point group all three components of rotation transform as a triply-degenerate T_1 irreducible representation. In the C_{3v} point group, the z-component of rotation transform as A_2, while the x- and y-components transform as an E irreducible representation. The translations transform as A_1 (z-component) and E (x- and y-components).

When we determine which vibrational modes are active for detection by IR absorption spectroscopy, we need to know the irreducible representations of the translations. A mode is IR-active if the cross-product of the irreducible representations of the initial state, the transition dipole and the final state contains a totally symmetric irreducible representation. The components of the transition dipole transform as the translations. Thus, the cross-product $\Gamma(\psi_{v'}) \times \Gamma(T_{x,y,\text{ or } z}) \times \Gamma(\psi_{v'})$ is equal to the totally symmetric irreducible representation A, the mode is active, and it is polarized along that component of translation. For a degenerate mode, the cross-product must contain the totally symmetric irreducible representation.

Note that the ground state in most molecules is totally symmetric. Any irreducible representation multiplied by the totally symmetric species is equal to itself. Any irreproducible representation multiplied by itself is equal to the totally symmetric irreducible representation (or contains it for degenerate modes). Therefore, any mode from a totally symmetric ground state is IR-active if the final state has the same symmetry as one of the components of translation.

The final column of the character table contains the elements of the symmetric polarizability tensor α. These elements are required for determining which vibrational modes are active in Raman spectroscopy. They are used in place of the components of translation in the cross-product. Again, if the cross-product is equal to the totally symmetric irreducible representation (or contains it for degenerate modes), the mode is Raman-active. By similar reasoning to the above, Raman-active modes are those that transform as one of the components of the polarizability tensor.

As an example consider CH_4. A nonlinear polyatomic molecule, CH_4 has $3N - 6 = 9$ vibrational modes. These modes are described in Table 26.6. There are only four modes listed; however, the total number of vibrational modes is obtained by summing the degeneracies of each mode, which sums to the required nine modes. The x-, y- and z-components of translation transform as F_2. Therefore, only an f_2 vibrational mode is IR-active. The components of the polarizability tensor transform as A_1, E and F_2; therefore, a_1, e and f_2 vibrations are Raman-active. Note that upper-case letters are used in the character tables but lower-case letters are used in the designation of vibrational modes. The entries in Table 26.6 confirm our expectations on which modes are IR- and Raman-active.

Now consider the symmetry of a vibration in which one quantum of the ν_1 deformation and one quantum of the ν_2 antisymmetric stretch or two quanta of the ν_2 mode are excited. The simultaneous excitation of two different modes results in a *combination mode*. The simultaneous excitation of two quanta of the same mode results in an *overtone*. This is where the distinction between a reducible and irreducible representation is made. The symmetry of such a combined state is given by the cross-product of the irreducible representations.

Table 26.6 The normal vibrational modes of CH_4. ω_e is the fundamental wavenumber of the mode, g is the degeneracy. IR or Raman in the last column indicates whether the mode is active in IR or Raman spectroscopy.

Mode	Description	Irreducible representation	ω_e/cm^{-1}	g	Active
ν_1	symmetric stretch	a_1	2917	1	Raman
ν_2	deformation	e	1534	2	Raman
ν_3	antisymmetric stretch	f_2	3019	3	IR, Raman
ν_4	deformation	f_2	1306	3	IR, Raman

We begin with the $(\nu_1\nu_2)$ combination mode and its representation $\Gamma(\nu_1\nu_2)$. The cross-product is given by

$$\Gamma(\nu_1\nu_2) = a_1 \times e = (1\ 1\ 1\ 1\ 1) \times (2\ -1\ 2\ 0\ 0) \quad (26.56)$$

$$\Gamma(\nu_1\nu_2) = (2\ -1\ 2\ 0\ 0) = e \quad (26.57)$$

This is a general result. The cross-product of any irreducible representation with a totally symmetric representation is equal to itself. The combination mode $(\nu_1\nu_2)$ corresponds to an e irreducible representation.

Now we turn to the cross-product of two degenerate modes. This brings with it two complications: (i) the cross-product generates a reducible representation; and (ii) if the same mode is excited twice, the result must conform to the Pauli exclusion principle. First, we generate the representation of the cross-product,

$$\Gamma(2\nu_2) = e \times e = (2\ -1\ 2\ 0\ 0) \times (2\ -1\ 2\ 0\ 0) \quad (26.58)$$

$$\Gamma(2\nu_2) = (4\ 1\ 4\ 0\ 0). \quad (26.59)$$

Comparing this result to the character table in Table 26.3, we see that no row contains these values. Instead, the cross-product is composed of a unique sum of irreducible representations. Because of this, the cross-product is called a *reducible representation*. Every cross-product of an irreproducible representation with itself generates the totally symmetric irreducible representation. Subtracting this from the result in Eq. (26.59) and inspecting Table 26.4, we see that the unique set of irreducible representations must correspond to

$$e \times e = a_1 + a_2 + e. \quad (26.60)$$

If the two excitations were from different modes, the resulting combination modes would correspond to a_1, a_2 and e symmetry, as indicated by Eq. (26.60). However, if the two excitations conform to an overtone of a single mode, then the Pauli exclusion principle must be taken into consideration. To differentiate these two cases, we write the cross-product as

$$(e)^2 = a_1 + e. \quad (26.61)$$

The a_2 symmetry mode violates the exclusion principle and is dropped. The remaining two terms form the symmetric part of the cross-product, so-called because they are symmetric with respect to particle exchange. A complete table for all degenerate point groups containing the symmetry species of vibrational combination states can be found in the classic tome of Herzberg[3] and the more recent book by Hollas (see Further Reading). Usually, the excluded mode that forms the antisymmetric part of the cross-product is an a mode, if possible, not the a_1 mode.

26.6 Dipole moments

The *electric dipole moment* $\boldsymbol{\mu}$ of a set of discrete charges Q_i separated by distances $\boldsymbol{r}_i$ is given by

$$\boldsymbol{\mu} = \sum_i Q_i \boldsymbol{r}_i. \quad (26.62)$$

For charge distributed with a charge density ρ, this becomes

$$\boldsymbol{\mu} = \int \rho \boldsymbol{r}\, \mathrm{d}\tau. \quad (26.63)$$

The electric dipole is a vector. When a dipole is composed of two point charges Q and $-Q$ separated by a distance r, the direction of the dipole vector is, according to the IUPAC recommendation displayed in Fig. 26.8, from the negative to the

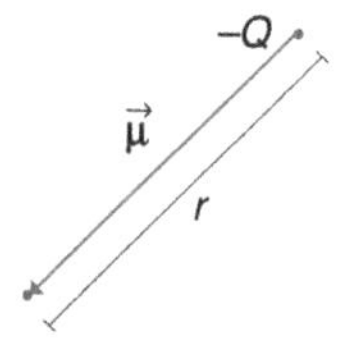

Figure 26.8 IUPAC convention on the direction of the electric dipole vector $\boldsymbol{\mu}$. Point charges $+Q$ and $-Q$ are separated by a distance r. The electric dipole vector points from negative charge to positive charge.

positive charge. The SI unit for dipole moment is C m; however, its magnitude lends itself to a nonstandard unit known as the Debye:

$$1 \quad D = 3.335\,64 \times 10^{-30}\ \text{C m}.$$

A molecule that contains an asymmetric charge distribution will exhibit a dipole moment. Each bond that contains two dissimilar atoms can potentially exhibit a *bond dipole moment.* The greater the electronegativity difference between the two atoms, the greater the bond dipole moment. Values of the electronegativity can be found in Appendix 8. The dipole moment of a molecule is obtained by the vector sum of all the bond dipole moments plus the component derived from electrons occupying lone pairs that are not involved in bonding. More exactly, the dipole moment is given by integration of Eq. (26.63) over the extent of the molecule.

For molecules such as HCl, H_2O, NaCl and NH_3 with simple structures and clear differences in the electronegativity between atoms, the presence and direction of a dipole is clear. The Cl atom pulls electron density from the H atoms, as do both O and N atoms in H_2O and NH_3, respectively. Even more pronounced is the charge separation in NaCl. In each case the electronegative atom has a higher charge density near it than it would if it existed as an isolated atom, which leads to an asymmetric charge distribution.

However, when the dipole moment is small, as in the cases of NF_3, CO and toluene, the direction of the dipole moment may not be obvious. This is most notoriously so for CO, in which the dipole points in the 'wrong' direction – that is, counter to expectations based on electronegativities, because the C end of the molecule is more negative. Oddly, there is no experimental method of determining the direction of the dipole moment. It is possible to measure its magnitude, but its direction is only given by high-level quantum mechanical calculations of the charge density in the appropriate molecule orbitals. Selected values of the dipole moment are given in Table 26.7.

Quantum mechanically, it is quite simple to see when a charge distribution has the appropriate symmetry to allow a dipole to form. The permanent dipole moment of a molecule in the state defined by the wavefunction ψ is given by the integral

$$\langle \mu \rangle = \int_{-\infty}^{\infty} \psi^* \, \hat{\mu} \, \psi \, d\tau, \tag{26.64}$$

where $\hat{\mu}$ is the dipole moment operator given by Eq. (26.62). If $\hat{\mu}$ is a symmetric function along any direction in the molecule, the integral will have a nonvanishing value and the dipole moment will have a component in that direction. The components of the dipole moment vector transform exactly like the components of the translational vector (x,y,z). That is, their representations are the same,

$$\Gamma(\mu_x) = \Gamma(T_x) \tag{26.65}$$

$$\Gamma(\mu_y) = \Gamma(T_y) \tag{26.66}$$

$$\Gamma(\mu_z) = \Gamma(T_z). \tag{26.67}$$

Inspection of all possible character tables representative of molecules shows that only a limited set of symmetry types lend themselves to permanent dipole moments, as shown in Table 26.8. This also makes it possible to formulate a rule on whether or not a molecule will possess an electric dipole moment based on symmetry:

A molecule has a permanent electric dipole moment if any of the components of translational symmetry of the molecule's point group is totally symmetric.

Table 26.7 Dipole moments of selected molecules. Values taken from the 96th edition of the *CRC Handbook of Chemistry and Physics.*

Molecule	μ/C m	μ/D
HCl	3.60×10^{-30}	1.08
H_2O	6.24×10^{-30}	1.87
NaCl	30.0×10^{-30}	9.00
NH_3	4.90×10^{-30}	1.47
NF_3	0.784×10^{-30}	0.235
CO	0.377×10^{-30}	0.110
$C_6H_5CH_3$	1.25×10^{-30}	0.375

Table 26.8 Symmetry classes that do not prohibit the establishment of a permanent electric dipole moment.

Character table	Totally symmetric component
C_1	μ_x, μ_y, μ_z
C_s	$\mu_x, \mu_y,$
C_n	μ_z
C_{nv}	μ_z

26.6.1 Directed practice

On the basis of point group symmetry, can NH_3, C_2F_4 and SF_6 exhibit electric dipole moments?

[Answer: C_{3v} yes; D_{2h} no; O_h no]

26.7 Rovibrational spectroscopy of polyatomic molecules

26.7.1 Vibrational structure of polyatomics

We delved into the spectroscopy of diatomic molecules in some detail. This sets a good basis for understanding quantum mechanics and light–matter interactions. The spectroscopy of diatomic molecules is also quite general. Intentionally, we cover here the spectroscopy of polyatomic molecules in much less detail. The reason for this is that the wide variety of molecular symmetries makes the spectroscopy quite specific to certain types of molecules. Polyatomic spectroscopy simply is a much more specialist occupation and here, rather than trying to convey a wide variety of cases, we are going to try to concentrate on a few central aspects. Much greater detail can be found in the references listed in Further Reading at the end of the chapter.

As shown in Figs 26.9 and 26.10, the rovibrational spectroscopy of linear molecules shares much in common with the spectroscopy of diatomics. A linear molecule has $3N - 5$ vibrational modes and nonlinear molecules have $3N - 6$

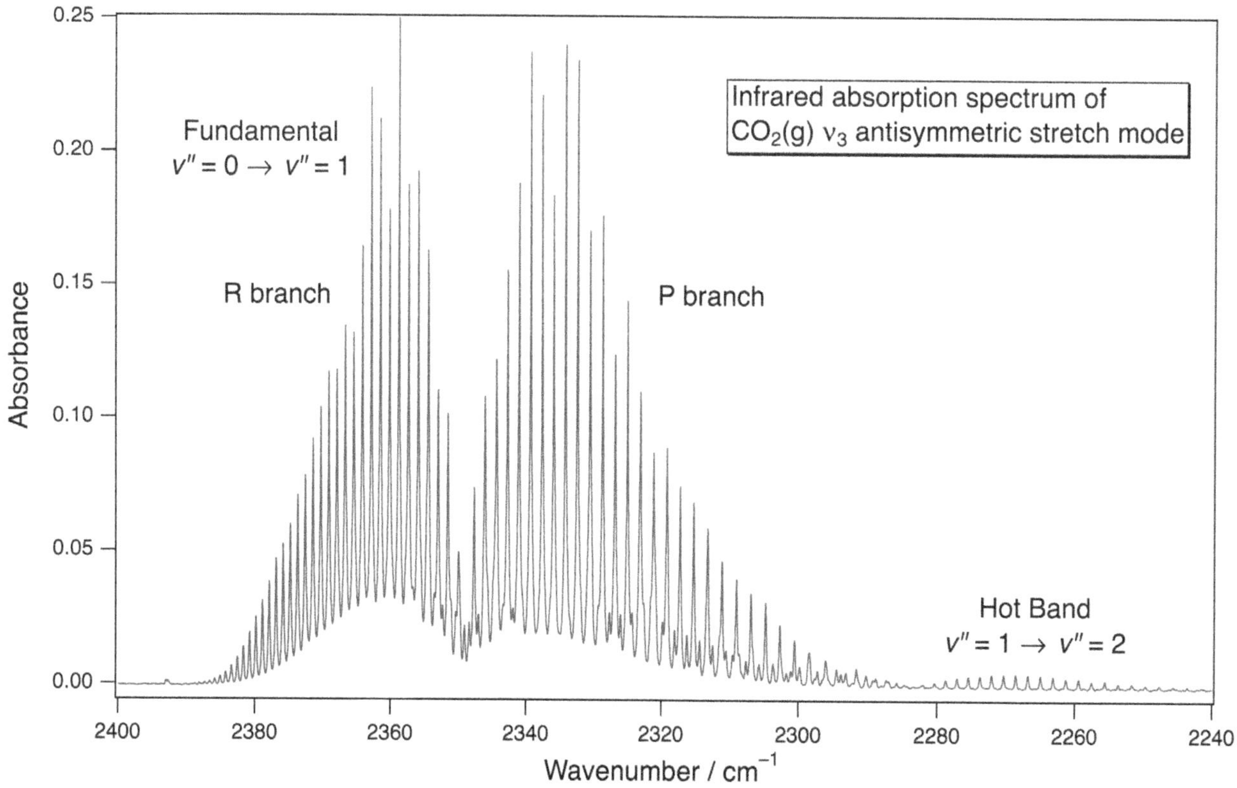

Figure 26.9 Gas-phase infrared absorption spectrum of CO_2 in the region where its antisymmetric stretch mode absorbs. Spectrum recorded at 0.25 cm^{-1} resolution.

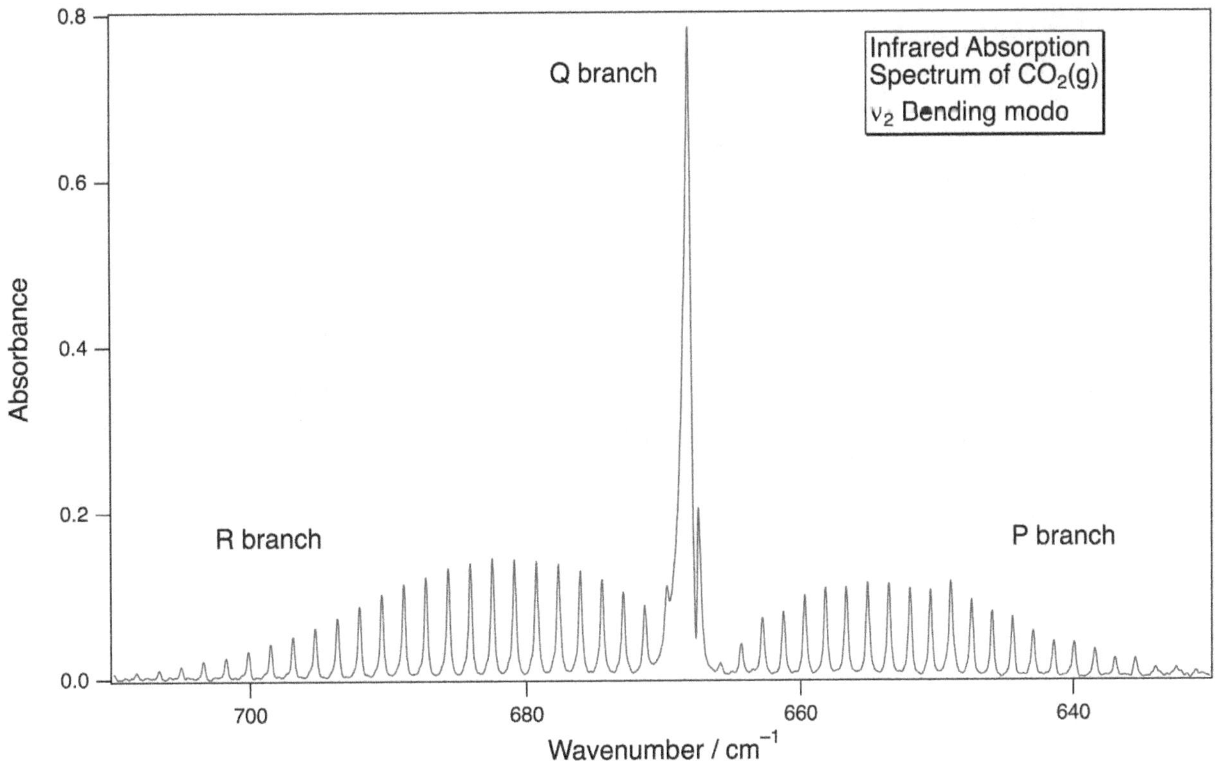

Figure 26.10 Gas-phase infrared absorption spectrum of CO_2 in the region where its bending mode absorbs recorded at 0.25 cm^{-1} resolution.

modes. The difference is that linear molecules rotate in two perpendicular planes – that is, they have two rotational axes. Nonlinear molecules rotate about three rotational axes. Thus, CO_2 has four vibrational modes while H_2O has three vibrational modes. Both exhibit symmetric and antisymmetric stretch modes. The difference is in the bend mode. The bend mode of CO_2 is doubly degenerate. Place a linear CO_2 molecule on the screen of your computer and it can either bend in the plane of the screen or in and out of the plane. These normal modes of CO_2 and H_2O are illustrated in Fig. 26.11. A vibration is a vibration and a translation is a translation. The center of mass of the molecule must not move during excitation of a normal vibrational mode. Thus, for instance, in the symmetric stretch of H_2O, the O atom has to move down a little to compensate for the upward motion of the two H atoms to keep the center of mass from translating. This also emphasizes that *normal modes* are vibrational modes of the whole molecule rather than local modes or vibrations of just one bond.

A mode is infrared-active (it appears in electric dipole absorption and emission spectra) only if the mode creates an oscillating dipole moment. This means that the mode must change the dipole moment of the molecule or else it is inactive. The dipole moment only has to change along any one of the symmetry axes of the molecule and this condition can be expressed in terms of the symmetry of the molecule:

There will be a rovibrational spectrum corresponding to a normal mode of vibration only if that mode belongs to the same symmetry species as one or more of the three coordinates x, y and z.

For a diatomic or small polyatomic, it is often easy to determine whether a change in dipole moment accompanies excitation of a normal mode. For larger molecules this is not the case. However, as shown in the previous section, we can restate this selection rule in terms of the irreducible representation of the normal mode symmetry:

A mode is infrared-active if the cross-product of the irreducible representations of the initial state, the transition dipole and the final state equals a totally symmetric irreducible representation, $\Gamma(\psi_{v'}) = \Gamma(T_{x,y,\text{ or } z}) \times \Gamma(\psi_{v'}) = \mathrm{A}$. *The mode is polarized along the Cartesian component that satisfies this relation. For a degenerate mode to be active, the cross-product must contain the totally symmetric irreducible representation.*

Returning to the specific cases of carbon dioxide and water, the fundamental wavenumbers of the CO_2 vibrations are 1333 cm^{-1} (ν_1 symmetric stretch), 667 cm^{-1} (ν_2 bend) and 2349 cm^{-1} (ν_3 antisymmetric stretch). For H_2O, the same modes have fundamental wavenumbers of 3657, 1595 and 3756 cm^{-1}, respectively. Vibrational modes are named simply

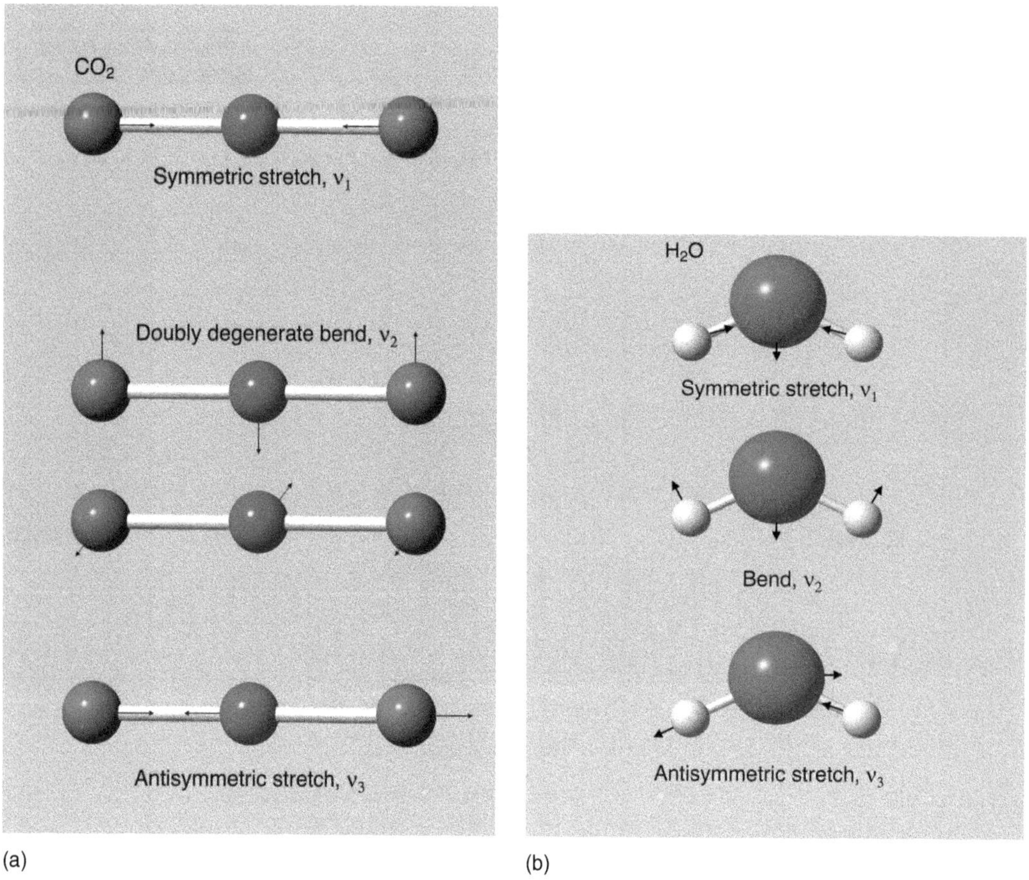

Figure 26.11 The normal modes of (a) CO_2 and (b) H_2O.

by placing a number after ν (Greek nu), as has been done above. Sometimes, vibrations are further classified as being one of four characteristic types of vibrations. Stretches change the bond length and are also denoted with ν. In-plane motions that change bond angles but not bond lengths are bends. These are denoted with δ and are sometimes further subdivided between rocks, twists, and wags. Out-of-plane bends are assigned the symbol γ. Torsions (τ) change the angle between two planes containing the atoms involved in the motion.

At low levels of excitation, and particularly if several vibrational modes have much different fundamental wavenumbers, each one of these vibrations can be treated independently and each will be involved in its own set of rovibrational transitions. At high levels of excitation, this approximation will certainly break down. The heavy overlap of levels and their interactions lead to perturbations, even for molecules as small as triatomics. *Perturbations* are interactions between levels that shift their energies and also change the intensities of spectral lines.

Unfortunately, our first-order approximation breaks down even at low levels of excitation. For example, when we look at the absorption spectrum of CO_2, we find a band centered at 2349 cm^{-1} that is associated with the excitation of the ν_3 antisymmetric stretch mode. This is precisely as we expect, and the band looks very much like that of a diatomic molecule. There is a P branch at lower wavenumber and an R branch at higher wavenumber. The Q branch is absent. Again, the branches are related to the value of $\Delta J = J' - J''$, where J' is the rotational quantum number in the upper state vibrational level and J'' is the rotational quantum number in the lower vibrational level. The P branch corresponds to $\Delta J = -1$, R to $\Delta J = +1$ and Q to $\Delta J = 0$.

However, this region of the spectrum also illustrates a *hot band*. A hot band is a set of transitions that originates from a thermally excited vibrational level above the ground state. If the vibration were completely harmonic, the hot band would fall right on top of the fundamental transition. This is obviously not the case for the antisymmetric stretch. It is clearly anharmonic with the origin of the first hot band (the position of the nonexistent Q branch) shifted by almost -70 cm^{-1} from the origin of the fundamental. The shift to smaller wavenumber is consistent with the closer spacing of anharmonic levels as the vibrational quantum number increases.

On the other hand, the IR absorption spectrum of CO_2 in the region of the ν_2 bend exhibits not only the P and R branches but also the Q branch. This is because the symmetry of the bend is different than that of the stretch, which makes the $\Delta J = 0$ transitions allowed. In linear polyatomic molecules a doubly degenerate bending mode has an effective rotational angular momentum l_2 associated with it. We call this angular momentum arising from vibration l_2 because it is the value of l associated with the ν_2 mode. This angular momentum makes the Q branch allowed and can lead to a splitting of lines when the degeneracy is broken by interactions with other levels. Notice how all of the intensity of the Q branch piles up into one large unresolved peak, whereas the P and R branches are easily resolved in individual lines at the 0.25 cm^{-1} resolution used. The congestion in the Q branch is caused by the nearly identical values of the rotational constant in both the ground state and the vibrationally excited state. Much higher resolution is required to clearly resolve the Q branch.

It is much more difficult to observe the ν_1 symmetric stretch because it is inactive in IR absorption. A normal mode must cause a change in dipole moment for it to be IR-active. The larger the dynamic dipole, the greater the strength of the IR absorption. The ν_3 antisymmetric stretch of CO_2 has a huge dynamic dipole moment and is an incredibly strong IR absorber. The ν_2 bend is also allowed but not as strong. The symmetric stretch can be observed in Raman spectroscopy, as can the first overtone of the bend. To keep these levels straight we introduce the notation $(\nu_1, \nu_2^{l_2}, \nu_3)$, where we indicate the value of how many quanta of each vibrational mode is excited. If $l_2 = 0$, it is often left out. The ground state is (0,0,0). Exciting one quantum of the ν_1 symmetric stretch corresponds to (1,0,0), and the first overtone of the bend corresponds to the transition $(0,0,0) \to (0,2^0,0)$ because the first overtone of the bend excited two quanta of the ν_2 mode with $l_2 = 0$.

We expect to find $(0,0,0) \to (1,0,0)$ at 1333 cm^{-1} and $(0,0,0) \to (0,2^0,0)$ at 2(667) = 1334 cm^{-1}. Instead, we find the excitation of the symmetric stretch at 1388.3 cm^{-1} and the first overtone of the bend at 1285.5 cm^{-1}. The reason is that these two modes are nearly degenerate and share the same symmetry. This allows them to interact in what is known as a *Fermi resonance*. This is an example of a spectral perturbation.

This discussion shows us that a harmonic approximation to the vibrational term value for polyatomics is a poor approximation. Even a Morse oscillator will be a poor approximation because it alone would not allow for couplings between modes. Therefore, we need to use a more complex formula (related to a more complex Hamiltonian) for the vibrational term of a polyatomic molecule,

$$G(\nu_1, \nu_2, \nu_3, \ldots) = \sum_i \tilde{\nu}_i \left(\nu_i + \frac{d_i}{2}\right) + \sum_i \sum_{j \le i} x_{ij} \left(\nu_i + \frac{d_i}{2}\right)\left(\nu_j + \frac{d_j}{2}\right) + \sum_i \sum_{j \le i} g_{ij} l_i l_j. \tag{26.68}$$

The first term is the energy of the harmonic normal mode i excited with ν_i quanta and characterized by degeneracy d_i and fundamental wavenumber $\tilde{\nu}_i$. The second term expresses the anharmonic couplings and the coupling between modes. The magnitude of x_{ij} describes the strength of anharmonic coupling when $i = j$, and mode-to-mode coupling when $i \neq j$. The final term quantifies the energy shifts associated with the vibrational angular momentum couplings with g_{ij} a constant and l_i and l_j the values of angular momentum for the levels i and j.

26.7.1.1 Directed practice

How many translational, rotational and vibrational modes does NH_3 have?

[Answer: 3, 3, and 6]

26.7.2 Rotational structure of polyatomics

The selection rule for pure rotational transitions is:

A molecule must have a permanent dipole moment to exhibit a pure rotational spectrum.

Because of centrifugal distortion, tetrahedral molecules such as CH_4 and SiH_4 exhibit weak pure rotational spectra. In the absence of centrifugal distortion, the results in Table 26.8 tell us that only molecules with C_1, C_s, C_n and $C_{n\nu}$ symmetry can have permanent electric dipole moments. Therefore, only molecules with these symmetries exhibit pure rotational spectra. Symmetric linear polyatomics such as O=C=O do not exhibit a pure rotational spectrum, but asymmetric polyatomics such as O=C=S, which have a permanent dipole moment, exhibit a pure rotational spectrum.

Table 26.9 Classification of rotors based on the magnitudes of their moments of inertia I_A, I_b, and I_c.

Type of rotor/molecule	Moments of inertia
Linear	$I_a = 0, I_b = I_c$
Spherical top	$I_a = I_b = I_c$
Symmetric top	$I_a = I_b > I_c$ (oblate)
	$I_a < I_b = I_c$ (prolate)
Asymmetric top	$I_a < I_b < I_c$

Nonlinear molecules can rotate about three principal axes of inertia, some of which may or may not be equal. There are three types of nonlinear polyatomic rotor, as defined in Table 26.9. By convention, the labels of the three axes are set by

$$I_c \geq I_b \geq I_a \tag{26.69}$$

Each *moment of inertia* has associated with it a rotational constant, $A = \hbar^2/2hcI_a$, $B = \hbar^2/2hcI_b$, $C = \hbar^2/2hcI_c$. Whereas for a diatomic the moment of inertia is simply $I = \mu R^2$, for polyatomics the more general expression

$$I = \sum_k m_k R_k^2, \tag{26.70}$$

where m_k is the mass of the kth atoms located a distance R_k from the center of mass, must be used. For a symmetrical linear molecule, such as Y–X–Y, this is simply $2m_Y R_Y^2$, where R_Y is the X–Y bond length.

The rotational energy expression depends on the type of rotor. A *linear molecule* has a single rotational constant B and a single rotational quantum number J, and the rotational energy expression is of familiar form

$$F(J) = BJ(J+1). \tag{26.71}$$

For a *spherical top* all three rotational constants are equal, $A = B = C$ and again we find that

$$F(J) = BJ(J+1), \tag{26.72}$$

For symmetric tops two quantum numbers are required: J and the component of the rotational angular momentum along the principal symmetry axis of the molecule K. The eigenvalue of the projection is $K\hbar$, as is to be expected, and the value of K is bounded by

$$-J \leq K \leq J. \tag{26.73}$$

The rotational term of an *oblate top* is

$$F(J,K) = BJ(J+1) + (C-B)K^2 \tag{26.74}$$

and that of a *prolate top* is

$$F(J,K) = BJ(J+1) + (A-B)K^2. \tag{26.75}$$

There is no simple expression for the rotational energy of an asymmetric top. Obviously, all of the above expressions are for *rigid rotors* and exclude the effects of centrifugal distortion. As shown by the fit to the rotationally-resolved spectrum of DHPH at the bottom of Fig. 26.3(c), a rigid rotor model with appropriate rotational constants can be implemented to accurately reproduce the spectrum of complex polyatomics molecules at low levels of excitation.

26.7.2.1 Directed practice

In an asymmetric top how are the rotational constants related to one another?

26.8 Excited-state dynamics

To discuss electronic states we first have to discuss the naming of electronic states. State designations for linear molecules (diatomics and polyatomics) are determined by the value of Λ, the projection of electron orbital angular momentum

Table 26.10 The correspondence between the projection of electron orbital angular momentum along the internuclear axis and state designation.

Linear molecules	Term symbol
$\Lambda = 0$	Σ
$\Lambda = 1$	Π
$\Lambda = 2$	Δ
$\Lambda = 3$	Φ

along the internuclear axis. The projection must be used because total magnitude is undefined, that is, L is a bad quantum number. Λ is a positive number and its value is represented by an upper-case Greek letter according to the naming scheme presented in Table 26.10. All states with $\Lambda > 0$ are doubly degenerate. This can be thought of as Λ, when it has a non-zero value, being equal regardless of whether it points to the left or right along the internuclear axis.

The total spin S is a good quantum number in a molecule as is its projection along the internuclear axis Σ.

$$\Sigma = S, S-1, \cdots - S. \tag{26.76}$$

The *spin multiplicity* $(2S+1)$ is added to the Greek letter designation as a superscript on the left, much like it was for atoms.

The component of *total angular momentum* along the internuclear axis takes on values $\Omega\hbar$ where the quantum number Ω is given by

$$\Omega = |\Lambda+\Sigma|. \tag{26.77}$$

If we wish to differentiate the different components of a spin multiplet, the value of $\Lambda + \Sigma$ is used as a right subscript. For example, a state with $\Lambda = 1$ and $S = 1$ has $\Sigma = 1, 0, -1$ and the three components of the $^3\Pi$ correspond to $\Omega = 2, 1, 0$. The resulting state designations are $^3\Pi_2$, $^3\Pi_1$ and $^3\Pi_0$. Spin–orbit coupling shifts these components by an amount

$$\Delta E = A\Lambda\Sigma, \tag{26.78}$$

where A is the spin–orbit coupling constant.

In addition to the quantum numbers Λ, S and Ω, we need to consider either one (for asymmetric linear molecules) or two (for symmetric linear molecules) symmetry properties to define the total state designation. All of these properties together will also be used to define the selection rules.

If the molecule has a center of inversion, then we need to add an indication of whether the state corresponding to the term symbol is even (gerade) or odd (ungerade) with respect to inversion. This letter is placed as a following subscript. If the molecule does not contain a center of inversion, no letter is indicated. Only symmetric linear polyatomics such as CO_2 and C_2H_2 exhibit this type of symmetry.

The second symmetry property applies to all linear molecules but only for Σ states (because they have $\Lambda = 0$). We can also consider the symmetry of the molecule upon reflection through any plane containing the internuclear axis. This symmetry is called *parity*. If there is no change in sign upon reflection, the state is designated +, but if the sign changes the state is designated –. If the molecule lacks a plane of symmetry in such a plane, no sign is indicated.

As an example, the ground state of CO_2 has $S = 0$, $\Lambda = 0$, even parity and gerade inversion symmetry. The ground state is thus a singlet sigma g plus state with term symbol $^1\Sigma_g^+$. HCN has the same set of quantum numbers and symmetry with the exception that inversion symmetry no longer applies. Its ground state is thus $^1\Sigma^+$.

For nonlinear molecules Λ, Ω and Σ are no longer good quantum numbers. Spin multiplicity is an acceptable quantum number designation, though not strictly good. Nonetheless, it is good enough to be used as a superscript just as for linear molecules. Parity (+ and –) no longer applies, but inversion symmetry can be specified. Since Λ is not a good quantum number, the letter to designate the state is instead taken from the irreducible representation to which the molecular orbital that describes the excited state corresponds. Thus, electronic states in polyatomics are given designations such at $^1A_{1g}$ for a single, gerade excited state with A_1 symmetry representation. Most polyatomic molecules have a singlet ground state. More simply and generally, a series of singlet states is denoted by S_0 (ground state), S_1 (first excited state), and so forth and a series triplet states as T_1, T_2, and so on.

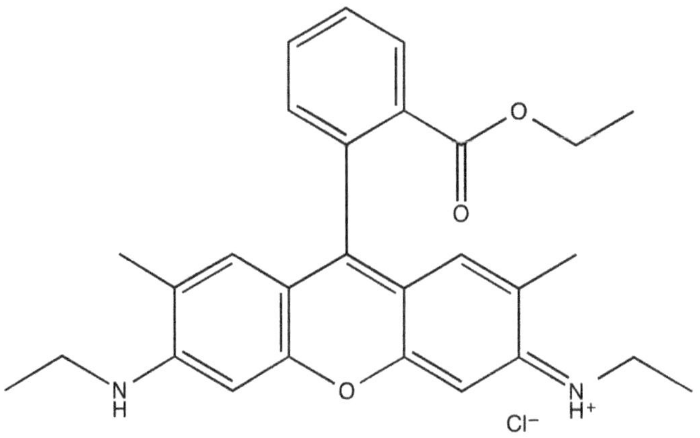

Figure 26.12 A schematic diagram of the structure of the laser dye rhodamine 6G.

Polyatomic molecules with more than a few atoms have such a high density of states that the simple picture of unperturbed rovibrational states needs to be evaluated critically. The Born–Oppenheimer approximation is still valid and the Franck–Condon principle still holds for electronic transitions. However, the rovibrational states within a given electronic state can exhibit strong coupling. If the molecule is buffeted by collisions with solvent molecules, the rovibrational levels overlap to form bands. This has implications for the excitation and relaxation dynamics of excited states. Also, whereas the internuclear separation along the bond is clearly the coordinate associated with the z axis for a diatomic molecule, the internuclear coordinate is a convolution over a number of motions in a polyatomic molecule.

26.8.1 Radiative transitions

Rhodamine 6G (R6G) is a venerable molecule in the lore of spectroscopy and lasers. Its chloride salt has the chemical formula $C_{28}H_{31}N_2O_3Cl$ and its structure is shown in Fig. 26.12. R6G absorbs and fluoresces strongly in the visible, as shown in Fig. 26.13. An important coincidence is that its absorption maximum lies close to 532 nm, the wavelength of a frequency-doubled Nd^{3+}:YAG laser. Because the molecule absorbs so strongly at 532 nm, is soluble in a number of common solvents (including water and ethanol), fluoresces with high efficiency, has low photochemical activity, and is not frighteningly toxic, R6G was tailor-made for laser spectroscopy studies in solution and for the creation of one of the first tunable lasers. Related rhodamine and fluorescein dyes (among others) are widely used in biological staining and for microscopy applications.

The structure of R6G exhibits a number of characteristic features. Organic molecules that absorb visible light strongly often have an extended π-bonding system, either in the form of fused rings (as for R6G) or in long chains with conjugated double and single bonds. Strong fluorophores tend not to contain high-Z atoms that can engage in spin–orbit coupling. They also often exhibit a good deal of structural rigidity such as that provided by fused aromatic rings. Xanthene – the heterocyclic group at the heart of R6G – forms the basis for a large class of dyes, which include a number of rhodamine and fluorescein derivatives. Reducing the number of heterocyclic rings to two creates a class of dyes known as coumarins, which absorb and fluoresce to the blue of the xanthene dyes. Adding an additional ring (and a N in the same ring as the O-containing ring) yields a class of oxazone dyes such as Nile blue, that absorb and fluoresce to the red of the rhodamines. The shifting of the absorption and fluorescence spectra with the size of the conjugated 'box' in which the radiative processes occurs conforms with our expectations from quantum confinement. A larger box has closer spacing between the ground and excited state and redder light corresponds to lower-energy photons.

Now we would like to explain the excited state dynamics of polyatomic molecules in more detail. The R6G cation has 64 atoms and therefore $3(64) - 6 = 186$ vibrational modes. Because of its large mass, its moment of inertia is quite small. Therefore, the rovibrational structure of large molecules such as R6G creates a continuum of levels within each electronic state. This is, in part, why the absorption and fluorescence spectra in Fig. 26.13 exhibit little structure, especially when compared to the infrared absorption spectra of CO_2 shown in Figs 26.9 and 26.10. Excitation from the ground state to the first singlet state (the first allowed transition) simply produces a broad peak. If we were able to put the molecule in a molecular beam, cool it rotationally, and excite it right near the threshold for absorption, we would see well-resolved lines in both absorption and fluorescence. This has been done, for example for 9,10-dihydrophenanthrene (DHPH), as shown in Fig. 26.3. This is a very special case but it does shed light on what is occurring, and we will now discuss this in some detail.

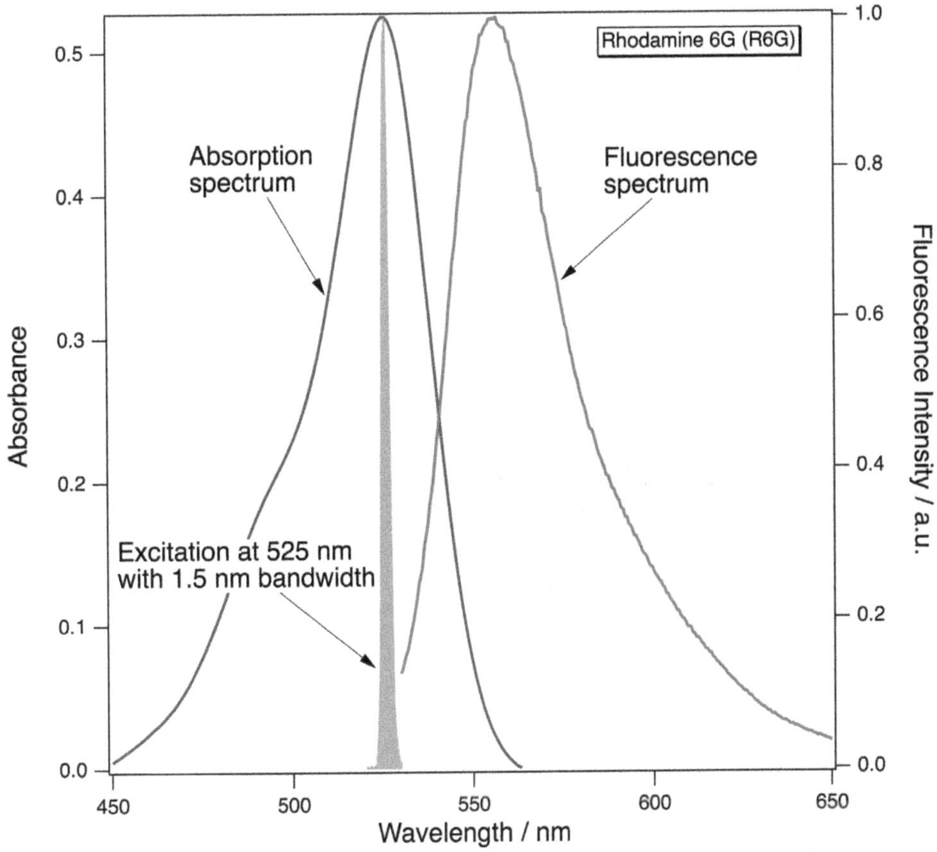

Figure 26.13 The absorption spectrum of an aqueous solution of R6G in the visible region peaks at 525 nm. When the solution is excited at this wavelength, R6G fluoresces in a structureless feature that peaks at 557 nm.

Another even more important reason for the lack of distinct structure in the fluorescence spectrum of Fig. 26.13 is that the R6G is in an aqueous solution. The molecule is experiencing rapid collisions with the solvent. These collisions facilitate intra- and intermolecular energy transfer. Inspection of the absorption and fluorescence spectra in Fig. 26.13 reveals that the fluorescence spectrum is red-shifted compared to the absorption spectrum. Fluorescence is shifted to lower-energy photons compared to absorption. A *red shift* is also known as a *Stokes shift*. The red shift caused by the presence of a solvent (or other change in the environment such as substitution) is also called a *bathochromic shift*. The opposite – a *blue shift* – is called an *anti-Stokes* or *hypsochromic shift*. The fluorescence spectrum is Stokes-shifted compared to the excitation wavelength, 532 nm in this case. Part of the shift is related to the Franck–Condon overlap between the excited state and the ground state. To understand this fully we need to better understand the competition between radiative and nonradiative processes in the excited state.

The fate of an excited state in a large molecule shares much in common with small molecules. An isolated molecule can relax radiatively, that is, by emitting a photon. Allowed transitions are associated with the short radiative lifetimes of *fluorescence* (0.1–100 ns depending on molecular structure; see Table 26.11). Long radiative lifetimes are associated with forbidden transitions and the phenomenon of *phosphorescence*. The most commonly encountered forbidden transition in organic molecules is the spin-forbidden transition between triplet and singlet states. The ground state of most molecules is a singlet electronic state. Above this lie numerous singlet and triplet excited states.

A typical set of excited states encountered in a normal organic molecule (as opposed to, for instance, an organometallic molecule containing a transition metal in which we would have to consider excitations of the organic framework, the transition metal that is being coordinated and transitions between these two systems) is shown in Fig. 26.14(a). Photon absorption leads to electronic excitation by a vertical transition. As long as the temperature is not ultralow, we can consider the electronic states to contain a continuum of rovibrational states, especially if the absorbing molecule is found in solution. The continua of states lead to broad absorption features. Depending on the structure of the molecule there may be some spectral features associated with the excitation of specific vibrations.

Table 26.11 Fluorescence lifetime of different classes of fluorescent molecules commonly used in lifetime imaging. Data from Berezin, M.Y. and Achilefu, S. (2010) *Chem. Rev.*, **110**, 2641.

Fluorophore class	Fluorescence lifetime range
endogenous fluorophores	0.1–0.7 ns
organic dyes	0.1–20 ns (up to 90 ns in pyrenes)
fluorescent proteins	0.1–4 ns
quantum dots	10–30 ns (direct gap), 10 μs–1 ms (Si, indirect gap)
organometallic complexes	10–700 ns
lanthanides	μs (Yb, Nd) to ms (Eu, Tb)
fullerenes	<1.2 ns
Single-wall nanotubes	10–100 ns, up to 10 μs for short nanotubes

To alleviate the ambiguity of what the internuclear coordinate is, the information contained in Fig. 26.14(a) is often summarized in the form of a *Jablonski diagram*,[4] such as that in Fig. 26.14(b). We introduce the notation S_0 for the singlet ground state, S_1 for the first excited singlet state, T_1 for the first excited triplet state, and so forth. The horizontal lines indicate the vibrational levels. Energy is plotted along the *y*-axis, but there is no *x*-axis. IC and ISC are nonradiative processes, as explained below.

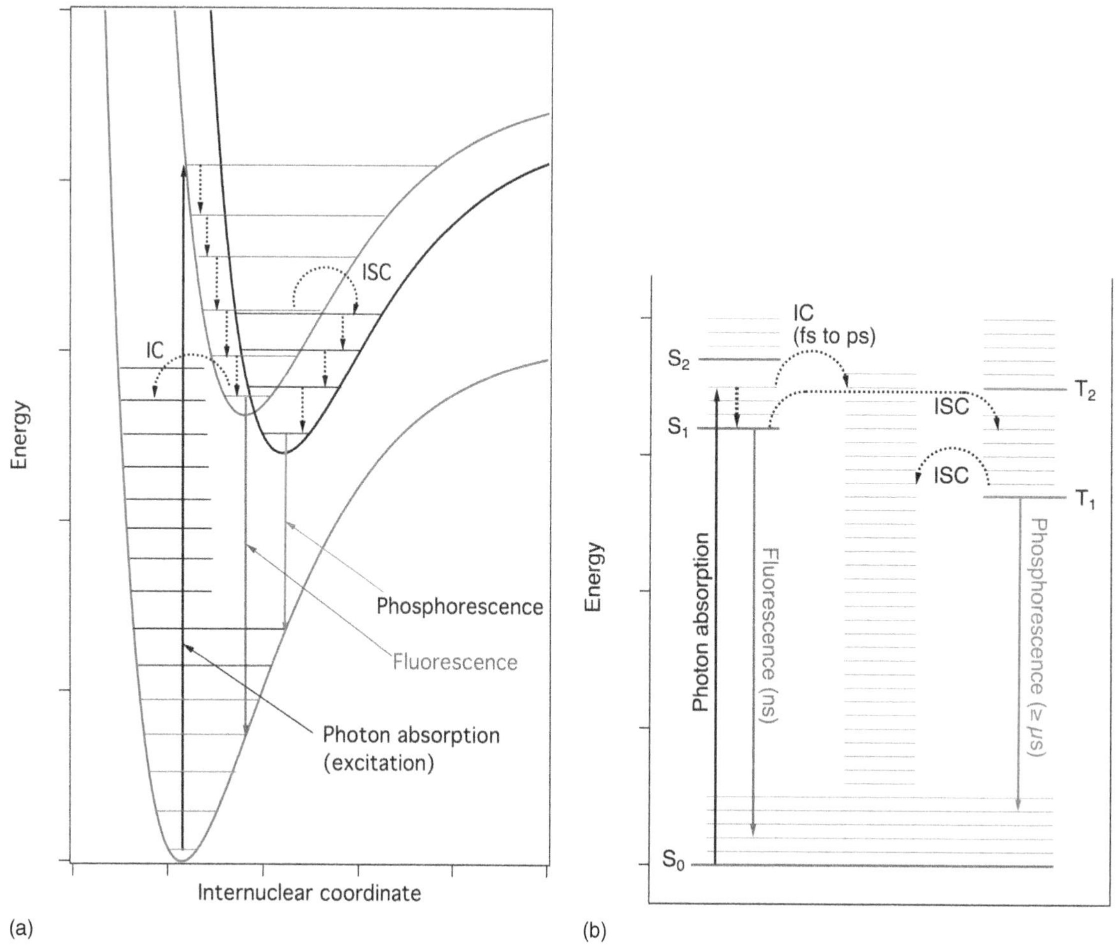

Figure 26.14 (a) A potential energy diagram of radiative (solid lines) and nonradiative (dashed lines) relaxation in a polyatomic molecule. IC = internal conversion; ISC = intersystem crossing. The red curve is the singlet ground state, the blue curve is the first excited singlet state, and the black curve is the first triplet excited state. (b) A Jablonski diagram conveying much of the same the same information. Phosphorescence corresponds to radiative relaxation between states with different spin multiplicities. Fluorescence corresponds to radiative relaxation between states with the same spin multiplicity. Typical lifetimes are given. Collisions and temperature influence the rate of energy redistribution and relaxation.

26.8.2 Nonradiative transitions

All radiative and nonradiative transitions must conserve energy. An isolated molecule can only make an intramolecular *nonradiative transition* if two states of the molecule happen to be degenerate. In a small molecule such accidental degeneracy is relatively rare. Large molecules can have enormous densities of states and accidental degeneracies, and the accompanying mixing of states are the rule rather than the exception. Degeneracies of this sort facilitate *intramolecular vibrational redistribution* (IVR) – the exchange of population between different vibrational modes within a molecule. IVR is quite common in large molecules but uncommon in small molecules. Of course, if a molecule is excited above the dissociation limit of a state, or if it is excited to a repulsive state directly, photodissociation can occur. As with small molecules, pre-dissociation is possible if a curve crossing from a bound state to a repulsive state occurs.

Collisions are required to engender vibrational energy relaxation in diatomic molecules. However, in larger molecules the coupling of different vibrational modes by perturbations that are usually small enough to be neglected in the Hamiltonians of small molecules becomes appreciable. These perturbations include the effects of anharmonicity and vibration–rotation (Coriolis) coupling. They facilitate the redistribution of energy between vibrational states within the molecule, which is the process of IVR. The rate of IVR rapidly increases with increasing density of vibrational states. A high density of states occurs naturally for large polyatomic molecules, especially in the presence of heavy atoms, which leads to low-frequency vibrations. However, even in small molecules at excitation levels approaching the dissociation limit, the density of states becomes adequate for efficient IVR.

The rate of IVR depends on the density of states and the strength of the interactions that couple the different vibrational modes. The smaller the spacing between vibrational levels, the easier it is to transfer energy between them. The spacing between levels is inversely proportional to the density of states. Let V be the interaction potential that represents the sum of the effect of anharmonicity and vibration–rotation interactions. The strength of the coupling is then given by the matrix element

$$V_{ik} = \langle \psi_i | \hat{V} | \psi_k \rangle. \tag{26.79}$$

V_{ik} has the units of energy. When the magnitude of V_{ik} becomes comparable to or larger than the spacing between the levels ψ_i and ψ_k, IVR becomes rapid. At this point, the excitation of one vibrational mode rapidly leads to the population of several neighboring vibrational modes. In other words, it becomes impossible to excite a single localized mode because that excitation rapidly exchanges throughout the molecule.

Three types of radiationless relaxation processes are depicted in Fig. 26.14: vibrational relaxation; internal conversion (IC); and intersystem crossing (ISC). *Vibrational relaxation* is the loss of vibrational energy, equivalently the moving of population from higher vibrational levels to lower levels. Just as in small molecules, the dependence of the Einstein A coefficient of spontaneous emission on photon energy demands that the radiative lifetime of a vibrational transition is much longer than an electronic transition. Therefore, vibrational relaxation is almost exclusively nonradiative. IVR can lead to the moving of a population between degenerate states. This is a redistribution of vibrational population and energy between modes rather than relaxation. Redistribution is an extremely rapid process that can occur on the subpicosecond timescale. In the presence of collisions with solvent molecules (or in the gas phase with a third body) vibrational energy transfer and relaxation are greatly facilitated because small differences in the energies of the initial and final states can be transferred to the kinetic energy of the collisions partners. In the presence of collisions we need to differentiate between intramolecular relaxation and intermolecular energy transfer. Intermolecular energy transfer can occur between any two degrees of freedom. As always, energy transfer is easiest when it is degenerate or involves small energy differences. However, when energy levels match (their degeneracy is exact), even electronic excitations can be transferred in collisions.

Resonant intermolecular molecular energy transfer is known as *Förster resonance energy transfer* (FRET). In FRET the acceptor level must lie at or slightly below the energy of the donor level, with the difference in energy being transferred to other degrees of freedom. FRET is a form of energy transfer usually involving excited electronic states. It involves only the transfer of the excitation. If an electron is transferred and this also leads to a transfer of the excitation energy, the event is known as *Dexter electron transfer*. FRET does not require a collision to occur but it is strongly distance-dependent. This means that FRET can be used as a probe of distances and intermolecular interactions when combined with fluorescence imaging and fluorescence lifetime imaging.

Vibrational energy exchange in the form of IVR is related to the phenomenon of *Fermi resonance* insofar as they both occur because of the coupling of modes within a molecule. Fermi resonance refers to the interaction of an overtone of one type of vibration and the fundamental of another vibration. An example is found in the first overtone of the ν_2 bend in CO_2, which couples to the ν_1 symmetric stretch. The fundamental wavenumber of the latter (1333 cm^{-1}) is almost exactly twice the fundamental wavenumber of the former (667 cm^{-1}). Fermi resonance leads not only to population transfers; it also causes slight shifts in the vibrational frequency and to intensity sharing between modes.

An *intersystem crossing* (ISC) is a radiationless transition between electronic states with different spin multiplicities. An ISC from a singlet to triplet excited state generally precedes phosphorescence, since direct photoexcitation of the triplet state from the singlet ground state is a low probability excitation. When electrons are used to irradiate the molecule (as can be done in the gas phase but not in aqueous solution), spin flip transitions occur readily. Strong spin–orbit coupling also enhances the likelihood of spin flip transitions. Molecules involving only H, C, N and O are not very good at inducing spin flip transitions. Heavier atoms such as Br, I or higher transition metals, lanthanides and actinides are much more likely to cause spin flip transitions because of their much greater spin–orbit coupling. Collisions with solvent molecules containing heavy atoms are effective at inducing intersystem crossings. Substitution of Br or I into a strong fluorophore can transform a fluorescent molecule into a strong phosphor because of the high efficiency of singlet to triplet conversion. Such is the case for rose Bengal, which is a tetraiodo-substituted fluorescein derivative.

Internal conversion (IC), sometimes also called internal quenching, is the conversion of electronic energy into vibrational energy in the absence of a collision. This is rather efficient in large molecules that exhibit excited singlet states that are nested in high vibrational states of the ground electronic state. Such a case is illustrated in Fig. 26.14. Degeneracy between the vibrational levels in the two different electronic states is essential for rapid internal conversion. Collisional *quenching* or external conversion is the loss of electronic energy induced by a collision with another molecule or surface. The energy is lost to translational and internal energy (or possibly electronic excitation) of the collision partner. Both of these processes quench radiative transitions. They provide pathways that compete with photon emission for relaxation of excited electronic states.

26.8.3 Quenching

We have mentioned quenching several times. Quantitative treatment of quenching is quite similar to that of diatomic molecules. For polyatomics, however, we are much more likely to encounter internal conversion and intersystem crossing. The *quantum yield* ϕ is defined as before

$$\phi = \frac{n_{\text{events}}}{n_{\text{excitations}}} = \frac{n_{\text{product}}}{n_{\text{photons absorbed}}} = \frac{r_{\text{process}}}{r_{\text{photon absorption}}}. \tag{26.80}$$

Summation of quantum yields over all radiative processes – in this case fluorescence ϕ_f – and nonradiative processes, ϕ_{nr}, involving the excited state must equal one. Thus, for an electronic state that can fluoresce, undergo internal conversion or intersystem crossing, react photochemically and be quenched by a third body, we have

$$\sum_k \phi_k = \phi_f + \phi_{nr} = \phi_f + \phi_{ic} + \phi_{isc} + \phi_{pc} + \phi_q = 1 \tag{26.81}$$

Excitation from the ground state S_0 to an electronic excited state S_1 is described by a rate constant k_a. Fluorescence (with rate constant k_f) competes with internal conversion (k_{ic}, transfer from S_1 to a vibrationally excited state in S_0) and intersystem crossing (k_{isc}, transfer from S_1 to an excited triplet state T_1). Phosphorescence with a rate constant k_p is radiative decay of T_1 to S_0. It occurs after an intersystem crossing. Phosphorescence does not affect the population of S_1. If there is no quenching of T_1 – either by collisions or by intersystem crossing back to S_0 – then the ϕ_{isc} would equal the fraction of initial excitations that lead to phosphorescence. However, this is a special case that needs to be investigated on a case-by-case basis.

As long as a molecule does not exhibit more than one conformation in the excited state, its radiative decay is a simple first-order process governed by a single lifetime. Multiple conformations lead to multiple lifetimes. We will assume a single conformation in the discussion below. In the absence of quenching, the *natural fluorescence lifetime* is equal to the inverse of the fluorescence rate constant

$$\tau_f^0 = 1/k_f. \tag{26.82}$$

The natural lifetime is related to the Einstein A and B coefficients by

$$\frac{1}{\tau_f^0} = A_{S_1S_0} = \frac{8\pi h}{\lambda^3} B_{S_1S_0}, \tag{26.83}$$

where h is the Planck constant and λ the wavelength of the transition from S_1 to S_0. An equivalent relationship holds for phosphorescence between T_1 and S_0,

$$1/\tau_p^0 = A_{T_1S_0} = k_p.$$

The excited state can also be quenched by collisions with some species Q. In the presence of quenching by internal conversion, intersystem crossing and collisions, the *fluorescence lifetime* τ_f is given by

$$\frac{1}{\tau_f} = k_f + k_{nr} = k_f + k_{ic} + k_{isc} + k_q[Q] = \frac{1}{\tau_f^0} + k_{ic} + k_{isc} + k_q[Q]. \tag{26.84}$$

The rate constant k_{nr} represents the rate constant for all non-radiative processes. If the quencher Q is the solvent, k_{nr} is a pseudo-first-order rate constant.

As before, a *Stern–Volmer plot* can be used to characterize quenching. This is a plot of the ratio of fluorescence intensity in the absence of quencher (I_0) to fluorescence in the presence of quencher as a function of [Q]. The slope of a Stern–Volmer plot is used to determine the Stern–Volmer constant K_{SV} and the ratio k_q/k_f,

$$\frac{I_0}{I_f} = 1 + \frac{k_q}{k_f}[Q] = 1 + K_{SV}[Q]. \tag{26.85}$$

Such a plot for the quenching of Nile blue fluorescence in aqueous solution by the presence of $CaCl_2$ is shown in Fig. 26.15. The Ca^{2+} ion is primarily responsible for quenching as the concentration increases. The alkali earth ions with 2+ charge are generally more effective quenchers of Nile blue than the corresponding alkali metal ions with 1+ charge; this suggests that the charge state of the quencher has an important influence on the de-excitation process.

The collision rate changes slowly with temperature. On the other hand, internal conversion and intersystem crossings can be strongly influenced by vibrational excitation of the excited state. Vibrational populations are strongly influenced by temperature. Under such conditions, thermal quenching of fluorescence strongly reduces fluorescence as the temperature increases, as is shown for Nile blue in the pseudo-Arrhenius plot shown in Fig. 26.16. Internal rotations are also known to influence fluorescence lifetimes and quenching. Because of this, viscous solvents which slow internal rotations are often observed to increase fluorescence lifetimes.

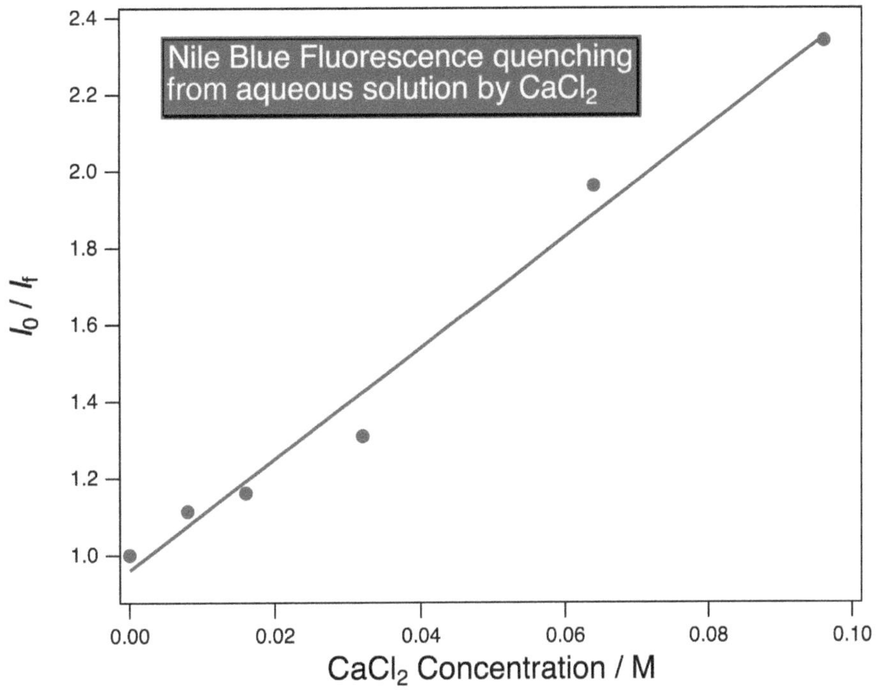

Figure 26.15 Nile blue dissolved in water fluoresces efficiently around $\lambda = 640$ nm. Addition of $CaCl_2$ to the solution leads to collisional deactivation of the excited state with a concentration dependence consistent with the Stern–Volmer analysis, as indicated by the straight-line fit.

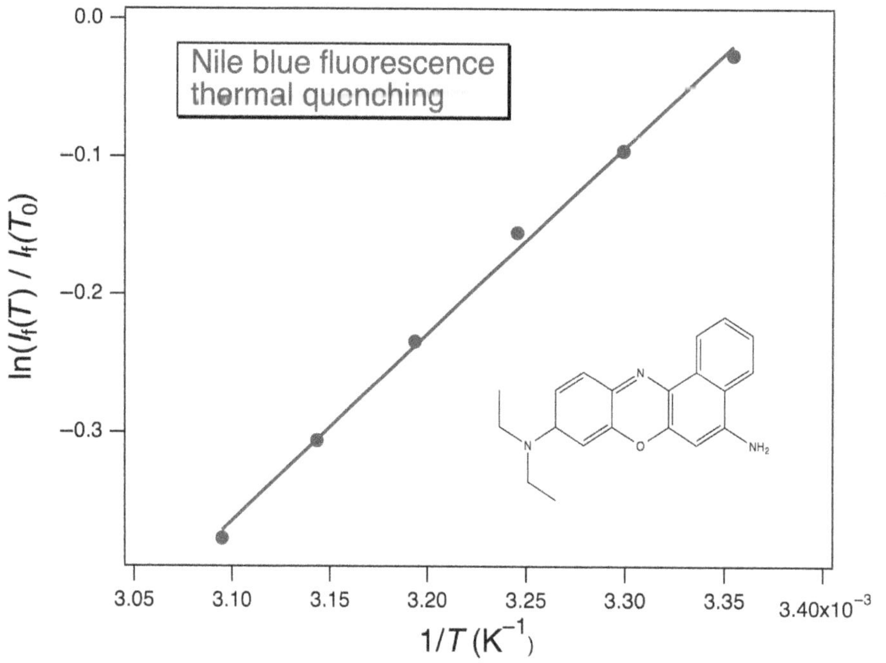

Figure 26.16 Thermal quenching of Nile blue in aqueous solution over the temperature range 293–323 K. Assuming that the slope is proportional to an activation energy for quenching, a value of 860 cm^{-1} is found, consistent (but not conclusive) with a role for vibrational energy transfer in the quenching process.

SUMMARY OF IMPORTANT EQUATIONS

$L(\tilde{\nu}) = \dfrac{A}{(\tilde{\nu}-\tilde{\nu}_0)^2+\gamma^2}$	Lorentzian lineshape with FWHM = 2γ
$G(\tilde{\nu}) = A\exp\left(-\left(\dfrac{\tilde{\nu}-\tilde{\nu}_0}{\sigma}\right)^2\right)$	Gaussian lineshape with FWHM = $2\sqrt{\ln 2}\,\sigma$
$\gamma = (2\pi c\tau)^{-1}$	Homogeneous broadening, Lorentzian linewidth parameter γ, excited state lifetime τ
$E_k - E_i = h\nu_{ki}$	Bohr frequency condition.
$W_{ki}^{abs} = B_{ki}\,\rho(\nu_{ki})$	Fermi's golden rule. Transition rate for absorption of light from state $\psi_i \rightarrow \psi_k$, W_{ki}^{abs}.
$B_{ki} = \dfrac{\lvert\mu_{ki}\rvert^2}{6\varepsilon_0\hbar^2}$	Einstein B coefficient B_{ki}.
$\rho(\nu_{ki}) = I(\nu_{ki})/c$	Energy density of radiation at the frequency of the transition $\rho(\nu_{ki})$, Intensity of radiation $I(\nu_{ki})$. Speed of light c.
$\mu_{ki} = \langle\psi_k\lvert\hat{\mu}\rvert\psi_i\rangle = \int \psi_k^*\,\hat{\mu}\,\psi_i\,d\tau.$	Transition dipole moment μ_{ki}. Electric dipole moment operator $\hat{\mu}$.
$\dfrac{A_{ik}}{B_{ik}} = \dfrac{8\pi h}{c^3}\nu_{ki}^3$	Ratio of Einstein A coefficient to Einstein B coefficient
$A_{ik} = \dfrac{16\pi^3\nu_{ki}^3}{3c\varepsilon_0 h}\lvert\mu_{ki}\rvert^2$	Einstein A coefficient.
FC factor $\equiv \lvert \boldsymbol{M}_v\rvert^2 = \lvert\langle\psi_v'\lvert\psi_v''\rangle\rvert^2$	Franck–Condon factor is given by the absolute square of the vibrational overlap.
$(\boldsymbol{C}_n)_z = \begin{pmatrix}\cos 2\pi/n & \sin 2\pi/n & 0\\ -\sin 2\pi/n & \cos 2\pi/n & 0\\ 0 & 0 & 1\end{pmatrix}$	Rotation matrix for a C_n rotation about the z-axis.
$\sigma_{xy} = \begin{pmatrix}1 & 0 & 0\\ 0 & 1 & 0\\ 0 & 0 & -1\end{pmatrix}$	Reflection matrix in the xy-plane.

$i = \begin{pmatrix} -1 & 0 & 0 \\ 0 & -1 & 0 \\ 0 & 0 & -1 \end{pmatrix}$	Inversion matrix.
$(S_n)_z = \begin{pmatrix} \cos 2\pi/n & \sin 2\pi/n & 0 \\ -\sin 2\pi/n & \cos 2\pi/n & 0 \\ 0 & 0 & -1 \end{pmatrix}$	Rotation–reflection matrix about the z-axis.
$\boldsymbol{\mu} = \int \rho \mathbf{r}\, d\tau$	Electrical dipole moment $\boldsymbol{\mu}$ of charge density distribution ρ.
$G(v_1, v_2, v_3, \ldots) = \sum_i \tilde{\nu}_i \left(v_i + \frac{d_i}{2}\right) + \sum_i \sum_{j \le i} x_{ij}\left(v_i + \frac{d_i}{2}\right)\left(v_j + \frac{d_j}{2}\right) + \sum_i \sum_{j \le i} g_{ij} l_i l_j$	Vibrational term, excited with v_i quanta, degeneracy d_i, fundamental wavenumber $\hat{\nu}_i$, anharmonic coupling constant x_{ij}, vibrational angular momentum couplings with g_{ij}, l_i and l_j the values of angular momentum for the levels i and j
$F(J) = BJ(J+1)$	Rotational term for linear molecules and spherical tops.
$F(J,K) = BJ(J+1) + (C-B)K^2$	Oblate top
$F(J,K) = BJ(J+1) + (A-B)K^2$	Prolate top
$\phi = \dfrac{n_{\text{events}}}{n_{\text{excitations}}} = \dfrac{n_{\text{product}}}{n_{\text{photons absorbed}}} = \dfrac{r_{\text{process}}}{r_{\text{photon absorption}}}$	Quantum yield of an event can be calculated in terms of the number of events or rates of events.
$\dfrac{1}{\tau_f^0} = k_f = A_{S_1S_0} = \dfrac{8\pi h}{\lambda^3} B_{S_1S_0}$	Natural lifetime τ_f^0. Fluorescence rate constant k_f. Einstein A and B coefficients for the transition $S_1 \rightarrow S_0$
$\dfrac{1}{\tau_f} = \dfrac{1}{\tau_f^0} + k_{ic} + k_{isc} + k_q[Q]$	Fluorescence lifetime τ_f in the presence of internal conversion, intersystem crossing and collisional quenching.
$\dfrac{I_0}{I_f} = 1 + \dfrac{k_q}{k_f}[Q] = 1 + K_{SV}[Q]$	Stern–Volmer equation for the dependence of fluorescence intensity I_f on quencher concentration [Q].

Exercises

26.1 Calculate the cross-products $A_2 \times B_1$, $A_2 \times B_2$ and $B_2 \times B_2$ in the C_{2v} point group.

26.2 Using transformation matrices, show that the S_1 operation is identical to reflection through a plane and that the S_2 operation is identical to inversion.

26.3 Write the transformation matrix for reflection through a plane defined by the z-axis and a line at 45° to the x- and y-axes.

26.4 (a) Estimate the lifetime of a state that gives rise to a line of width (i) 0.03 cm^{-1}, (ii) 2.0 cm^{-1}. (b) A molecule in a liquid undergoes about 1.0×10^{13} collisions per second. Suppose that: (i) every collision is effective in deactivating the molecule vibrationally; and (ii) that one collision in 100 is effective. Calculate the width in cm^{-1} of vibrational transitions in the molecule.

26.5 When gas-phase H_2O molecules are placed in an intense laser field, the lifetime of an excited electronic state changes from 10 ns to 1 ns due to ionization. Describe quantitatively what happens to the width of a line in the spectrum of this state as the lifetime changes.

26.6 Consider a polyatomic molecule in solution. Why is the fluorescence spectrum Stokes-shifted from the excitation wavelength? How does this make extremely low detection sensitivities possible?

26.7 Assuming that quenching is negligible and that a polyatomic molecule contains a strong fluorophore and no heavy atoms, calculate the fluorescence quantum yield ϕ_f if $\tau_f = 20$ ns and $k_{ic} = 7.5 \times 10^6$ s^{-1}.

26.8 For phenanthrene, the measured lifetime of the triplet state is $\tau_p = 3.3$ s. The measured lifetime of the singlet state is $\tau_f = 34$ ns. The fluorescence quantum yield is $\phi_f = 0.12$ and the phosphorescence quantum yield is $\phi_p = 0.13$. Assume that no collisional quenching and no internal conversion from the excited singlet state S_1 occurs.

Determine the fluorescence rate constant k_f and the rate constant for intersystem crossing from the excited singlet state to the triplet state k^S_{isc}. Also determine the phosphorescence rate constant k_p, and the rate constant for intersystem crossing back to the ground electronic state S_0 from the triplet state k^T_{isc}.

26.9 Hexatriene, $C_6H_8^+$, has five π electrons, with two each in the lowest three orbitals. Using the particle in the box model. Assume that the carbon atom backbone of the molecule forms the box and that molecule is linear with average bond length of 145 pm. Calculate the wavelength in nm of the light required to induce a transition from the ground state to the first excited state. Calculate the wavenumber in cm^{-1} of the photon that is emitted when an electron in the first excited state relaxes to the ground state.

26.10 Define internal conversion and intersystem crossing.

26.11 The fluorescence quantum yield of 1 μM rhodamine B in ethanol is 0.9. The rates of triplet state formation and internal conversion are zero. Calculate the quenching quantum yield.

26.12 Compare and contrast fluorescence and phosphorescence.

26.13 A large conjugated organic molecule containing C, O, H atoms and one I atom, exhibits a spectrum that looks like this:

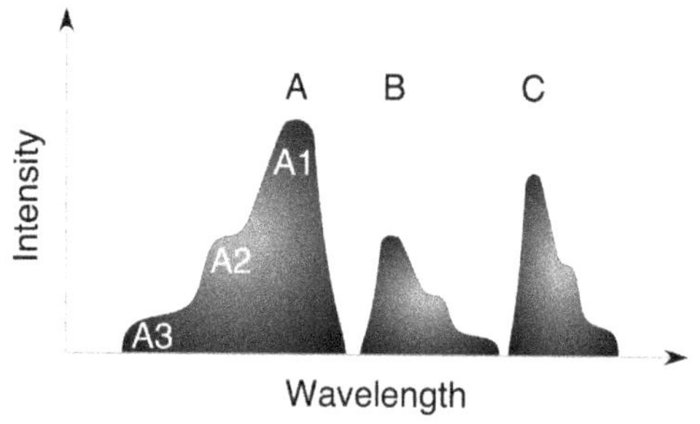

Band A is the absorption spectrum. Bands B and C are observed in emission. Band B exhibits a lifetime of 125 ns. Band C exhibits a lifetime of 1.6 μs. Answer all of the following:

(a) Draw a Jablonski diagram for this system. Identify each of the three states involved. (b) Give a likely explanation of the structure (A1, A2, A3) in band A and state what states are involved in the transition. (c) Explain why there are two emission bands, why they have different lifetimes, and why band C is at lower energy than band B and both are lower in energy than A. Identify the states involved and the type of radiative transition. (d) If the I atom were replaced by a H atom, how would the spectrum change? (e) A time-resolved experiment is conducted. An ultrashort pulsed laser is used to excite the molecule at the energy of the transition to the A2 feature. Describe the radiative and nonradiative processes that occur after excitation to A2. It is probably best to redraw the Jablonski diagram and use it in your explanation.

26.14 The fluorescence intensity of NO_2 exhibits a linear Stern–Volmer plot in the presence of a quencher Q. The Stern–Volmer equation is $I_0/I_f = 1 + (k_q/k_f)[Q]$. Explain why the slope of the Stern–Volmer plot is much steeper when the quencher is CO_2 as compared to Xe.

26.15 Why do all photosynthetic systems exhibit irreversible charge separation significantly faster than 1 ns?

26.16 Transitions between electronic states with different spin multiplicities are unlikely to occur in molecules that contain light atoms. Explain.

26.17 State three mechanisms for photodissociation.

26.18 When excited at 260 nm, $CH_3I(g)$ photodissociates in two channels:

i $CH_3I(g) + h\nu \rightarrow CH_3(g) + I(g)$

ii $CH_3I(g) + h\nu \rightarrow CH_3(g) + I^*(g)$.

Channel (i) has a quantum yield of 0.69. Channel (ii) has a quantum yield of 0.31. The fluorescence quantum yield of I* is 1. If 6.23×10^8 photons s^{-1} are emitted by I*, how many 260 nm photons are absorbed, what is the absorbed power, and what are the rates of $CH_3I(g)$ photodissociation and I(g) formation?

26.19 When acetone absorbs light at 313 nm it photodissociates with a quantum yield of 0.170. If a radiant power of 20.0 mW is absorbed how long will it take to photodissociate 1% of the acetone in a 100 ml container held at 2.00 kPa and 300 K?

26.20 Rhodamine 6G (R6G) has a fluorescence quantum yield from its first excited state of $\phi_f = 1.00$ when dissolved in ethanol, and $\phi_f = 0.92$ in water. In both cases no phosphorescence is observed. The measured fluorescence lifetime is 4.08 ns in water when excited at 532 nm and fluorescence intensity peaks at 555 nm. The natural fluorescence lifetime, intersystem crossing and internal conversion are properties of a molecule, not of the solvent. (a) State with justification what the quantum yields and rate constants for internal conversion and intersystem crossing are for R6G in both ethanol and water. (b) Calculate the rate constants for fluorescence and nonradiative relaxation. Calculate the natural fluorescence lifetime. What nonradiative process is responsible for the change in fluorescence quantum yield between the two solvents?

26.21 The data in Fig. 26.16 appear to show slight curvature and departure from linearity. Assuming that thermal fluorescence quenching is still primary mechanism for quenching, posit a plausible explanation for the curvature.

Endnotes

1. Einstein, A. (1917) *Phys. Z.*, **18**, 121.
2. Mulliken, R.S. (1955) *J. Chem. Phys.*, **23**, 1997.
3. Herzberg, G. (1945) *Molecular Spectra and Molecular Structure. II. Infrared and Raman Spectra of Polyatomic Molecules*. Van Nostrand, New York.
4. Jabłoński, A. (1933) *Nature (London)*, **131**, 839.

Further reading

Atkins, P.W. and Friedman, R.S. (2010) *Molecular Quantum Mechanics*, 5th edition. Oxford University Press, Oxford.

Cook, D.B. (2012) *Quantum Chemistry: A Unified Approach*. Imperial College Press, London.

Cotton, F.A. (1971) *Chemical Applications of Group Theory*, 2nd edition. John Wiley & Sons, New York.

Eyring, H., Walter, J., and Kimball, G.E. (2015) *Quantum Chemistry*. Scholar's Choice, Rochester, NY.

Herzberg, G. (1945) *Molecular Spectra and Molecular Structure. II. Infrared and Raman Spectra of Polyatomic Molecules*. Van Nostrand, New York.

Herzberg, G. (1967) *Molecular Spectra and Molecular Structure. III. Electronic Spectra and Electronic Structure of Polyatomic Molecules*. Van Nostrand, New York.

Hollas, J.M. (1972) *Symmetry in Molecules*. Chapman & Hall, London.

Hollas, J.M. (2004) *Modern Spectroscopy*, 4th edition. John Wiley & Sons, New York.

Lakowicz, J.R. (2006) *Principles of Fluorescence Spectroscopy*, 3rd edition. Springer, New York.

NIST list of molecules by point group. http://cccbdb.nist.gov/pglist.asp.

NIST Computational Chemistry Comparison and Benchmark Database; http://cccbdb.nist.gov/Summary.asp.

Pratt, D.W. (1998) High-resolution spectroscopy in the gas phase: Even large molecules have well-defined shapes. *Annu. Rev. Phys. Chem.*, **49**, 481.

Schäfer, F.P. (1990) *Dye Lasers* in *Topics in Applied Physics* (ed. H.K.V. Lotsch), Vol. 1, Springer-Verlag, Berlin.

Steinfeld, J.I. (2012) *Molecules and Radiation: An Introduction to Modern Molecular Spectroscopy*, 2nd edition. Dover Publications, New York.

Zewail, A.H. (1992) *The Chemical Bond – Structure and Dynamics*. Academic Press, London.

CHAPTER 27

Light–matter interactions: Lasers, laser spectroscopy, and photodynamics

PREVIEW OF IMPORTANT CONCEPTS

- Lasing occurs between levels with a population inversion when stimulated emission triggers the coherent generation of photons.
- Lasing occurs when gain exceeds loss in the optical cavity.
- Lasers emit light that is monochromatic, directional, bright, intense, and coherent.
- Lasers can operate in either a continuous-wave or pulsed mode.
- A laser is constructed from a three-level or four-level system. A two-level system cannot maintain a population inversion.
- Construction of a laser is increasingly difficult with increasing photon energy.
- The high intensity of lasers makes nonlinear spectroscopic techniques feasible.
- In nonlinear spectroscopy, the signal has a power law dependence on the laser intensity with an exponent greater than one.
- Second harmonic generation (SHG) and sum frequency generation (SFG) occur when two photons are added to create a new photon with an energy equal to the sum of the two photon energies.
- Two-photon transitions have selection rules that can be thought of as the product of one-photon selection rules.
- In laser-induced fluorescence (LIF), a laser is used to excite a molecule (usually quantum state specifically) and then the molecule is detected by collection of fluorescence.
- Lasers can be used to stimulate emission of photons, and this makes it possible to pump population into specific quantum states.
- Stimulated emission also makes super-resolution fluorescence microscopy possible. Super-resolution microscopy has resolution below the diffraction limit.
- Resonance-enhanced multiphoton ionization (REMPI) utilizes the absorption of more than one photon to state-specifically detect molecules.
- Both LIF and REMPI are capable of single molecule detection.
- Multi-pass absorption cells facilitate extremely sensitive absorption spectroscopy.
- Femtosecond pulsed lasers enable the observation of bond formation and bond breaking in real time.
- Attosecond pulses make possible the observation of electronic transitions in real time.
- Ultrafast pulses can only be made by sacrificing energy resolution.
- High harmonic generation (HHG) is an example of high-field physics – the physics of atoms and molecules in optical fields that have a strength comparable to the nucleus–electron Coulomb field.
- In photosynthesis, excitonic effects are used to efficiently remove an electron excited by photon absorption from the light-harvesting complex to the reaction center.
- Excitons are delocalized states that involve the coupling of electronic excitation to vibrational states in complex structures.
- Vision and color perception rely on the absorption of light in retinoid chromophores, the absorption spectra of which are highly sensitive to the protein environment in which they are located.
- Any physical or chemical process that changes the distribution of visible light wavelengths that enter the eye affects the color of an object.

Of all the phenomena studied by physical chemists, perhaps none has been more centrally important than light–matter interactions. There are many techniques by which physical chemists have interrogated matter. Nothing has taught us more and has more to teach us than light–matter interactions in the form of spectroscopy and photochemistry. It is

Physical Chemistry: How Chemistry Works, First Edition. Kurt W. Kolasinski.

Companion Website: www.wiley.com/go/kolasinski/physicalchemistry

the discovery of lasers and the application of lasers that has opened up the world of how atoms and molecules interact more than any other technique. And it is the application of new laser techniques, including microspectroscopy and spectromicroscopy, that has more to tell us about intermolecular and interatomic interactions than any other technique. Of course, lasers are not the only word in spectroscopy and microscopy. Nuclear magnetic resonance (NMR) forms the foundation of structural analysis along with diffraction techniques. These techniques are also being pushed ever more into the time-resolved domain and dynamical studies. Electron microscopy is becoming more relevant to chemistry, and mass spectrometry is essential in the chemist's arsenal. This chapter, however, will focus on lasers and their interactions with the frontiers of chemistry, physics, and biology.

27.1 Lasers

The introduction of lasers is one of several groundbreaking technologies that has transformed physical chemistry. The others are molecular beams, routine attainment of (ultra)high vacuum and high-speed electronics (along with the accompanying detectors and computers). The last two have also facilitated advances in electron microscopy, mass spectrometry and NMR. Lasers are transformative to research because they combine a number of exceptional traits:

- *Monochromaticity.* Most lasers operate at a wavelength with a well-defined central value and a small bandwidth about this central value.
- *Directionality.* Lasers naturally form beams with low divergence. This means that they are easily directed with conventional optics such as mirrors and can be tightly focused with lenses. Just try this with a light bulb.
- *Pulsed versus continuous-wave operation.* Lasers can be operated in a continuous-wave (CW) or pulsed mode, based on their construction without the need for mechanical modulation. CW operation is required for the highest monochromaticity. Pulse widths of lasers can range from microseconds to femtoseconds. Lasers can also be used to generate attosecond pulses.
- *Brightness and intensity.* Combining the first two properties, lasers deliver extremely high values of photons per unit bandwidth per unit area. When crammed into extremely short pulses this corresponds to extremely high energy density (J m^{-2}), power density (J s^{-1} m^{-2}) and electric field strengths (even exceeding the electric field that binds electrons to nuclei).
- *Coherence.* The photons in a laser beam are in phase. This happens both transversely and longitudinally (axially). That is, there is a well-defined phase relationship across the diameter of the beam as well as along the propagation direction. Another way to say this is that the coherence is both temporal and spatial.

Infrared and visible lasers were inspired by the work of Arthur Schawlow and Charles Townes.[1] The first laser was a ruby laser with millisecond pulses. This was demonstrated in 1960 by Theodore Maiman at the Hughes Research Laboratories. In the press conference announcing the discovery, Maiman said that the laser was 'a solution seeking a problem.'[2] It would not take long for chemists and physicists – who are never short of problems – to start matchmaking between lasers and problems. The first CW laser followed the next year. This was the HeNe laser.[3] Such a laser was used in the first laser-induced fluorescence experiments,[4] which was one of its first applications to chemistry. The dye laser invented by Mary Spaeth at the Hughes Aircraft Company in 1966 was followed independently and with improvements by Peter P. Sorokin at IBM and Fritz P. Schäfer at the Max-Planck-Institut für biophysikalische Chemie. The dye laser offered tunability, which was essential for its applications to chemical physics. The age of lasers was now set to take off.

The operation of a laser depends on the recognition that there are three ways to link two levels radiatively. The absorption of light is stimulated by the presence of on-resonance photons. The emission of light is either spontaneous, initiated by fluctuations in the vacuum field that fills the universe, or stimulated, initiated by the presence of on-resonance photons that we add directly to the system. The *net rate of stimulated emission* depends on the population difference between the upper and lower states.

$$R_{\text{stim em}} = (N_k - N_i)B_{ki}\rho(\omega_{ki}) \tag{27.1}$$

Normally (in thermal equilibrium) more population is in the lower energy state. When more population is in an excited state compared to a lower state, we call this a *population inversion*. A population inversion is one basic requirement for making a laser (unless something exotic is going on). Consider a two-level system with total population N and the population of the two individual states given by N_i and N_k,

$$N = N_i + N_k. \tag{27.2}$$

The population difference is

$$\Delta N = N_i - N_k. \tag{27.3}$$

Population of the two levels is changed by the absorption and emission of photons. The rate of change of the population difference is

$$\frac{d\Delta N}{dt} = \frac{dN_i}{dt} - \frac{dN_k}{dt} = -2B_{ki}\rho(\omega_{ki})\Delta N - A_{ik}(\Delta N - N) \tag{27.4}$$

However, since $\Delta N - N = -2N_k$ we re-write this as

$$\frac{d\Delta N}{dt} = -2\Delta N\, B_{ki}\rho(\omega_{ki}) - (\Delta N - N)\, A_{ik} \tag{27.5}$$

At steady state $d\Delta N/dt = 0$ and

$$\Delta N = \frac{N}{2(B_{ki}/A_{ik})\rho(\omega_{ki}) + 1}. \tag{27.6}$$

In the limit of a large driving field (strong irradiation), this is driven to saturation at $\Delta N = 0$. *At saturation there is no net emission of photons.* A two-level system can never lase as it kills its own inversion. A three- or four-level system must be used so that a population inversion can be sustained.

This idea of *saturation* has important implications beyond the construction of lasers. It means that once we have constructed a laser, we cannot use a photon field (no matter how intense) to move all the population from the ground state to the excited state of a two-level system. Once 50% of the population has been moved to the excited state, the probability that the laser beam excites more molecules to the excited state is the same as the probability that the laser beam stimulates the emission of photons and moves population back to the ground state. There is only so hard we can pump a transition before we start to pump population down as well as up. At this point applying more intensity will not change the net rate of absorption (which is zero); however, it may change the rate of other processes that previously were low-probability events (such as two-photon absorption).

Let us also plant a fertile idea that is rooted in stimulated emission and saturation. We see from this discussion that is possible not only to move population upwards by stimulated absorption, but also to move population downward by stimulated emission. Furthermore, if we introduce two different lasers with two different colors (which is therefore called a two-color experiment), the stimulated processes move population between two different sets of states.

The production of lasing action depends on pumping a population inversion in a three- or four-level system. The ruby laser is a three-level system. The four-level system of the much more important and popular Nd:YAG laser (often simply called a YAG, which is used to make the green beam in Fig. 27.1(a)) is illustrated in Fig. 27.1(b). A population inversion is required so that stimulated emission (gain) occurs at a higher rate than absorption (loss).

Pumping requires an energy source that can be directed into the optical medium. Just about anything can be used for this. The best source is a laser – which is, of course, difficult if you are trying to invent the first laser! Excitation can be provided by a discharge, particularly for gas lasers, or simply as current for semiconductor/diode lasers. Flashlamps for pulsed operation or arc lamps for CW operation can be used for optical excitation. A select few chemical reactions are able to produce a population inversion directly that can be used as the optical medium. A nuclear blast provides copious energy and has been used for the pumping of high-energy (X-ray) transitions. But nuclear bombs tend not to be very useful in a laboratory setting. The power to drive a laser scales as the fifth power of the frequency of the laser, $P_{\text{pump}} \propto \nu^5$. This effectively prevents conventional laser schemes from extending far beyond the UV.

The strategy for obtaining a population inversion is illustrated in Fig. 27.1(b). A pump source excites the Nd^{3+} ion. A flashlamp is a simple but inefficient means of pumping, as most of the light is not resonant with the pump band. Extremely efficient pumping with much less heat generation is performed by a diode laser tuned to 810 nm to overlap the pump band resonantly. Nd^{3+} is an ion with electron configuration $4f^3$. It is doped into crystalline $Y_3Al_5O_{12}$ (yttrium aluminum garnet or YAG) at a level of 0.5–2.0 wt%. Its electronic levels are sharp and do not interact strongly with the lattice. It is not tunable, but it can be operated either in a pulsed or CW mode. The excitation is strong and moves a lot of population to the pump levels. A fast transfer of population occurs from this set of levels to the upper state of the transition that will be involved in lasing. This $^4F_{3/2}$ state has a forbidden transition with the $^4I_{11/2}$ state. Importantly,

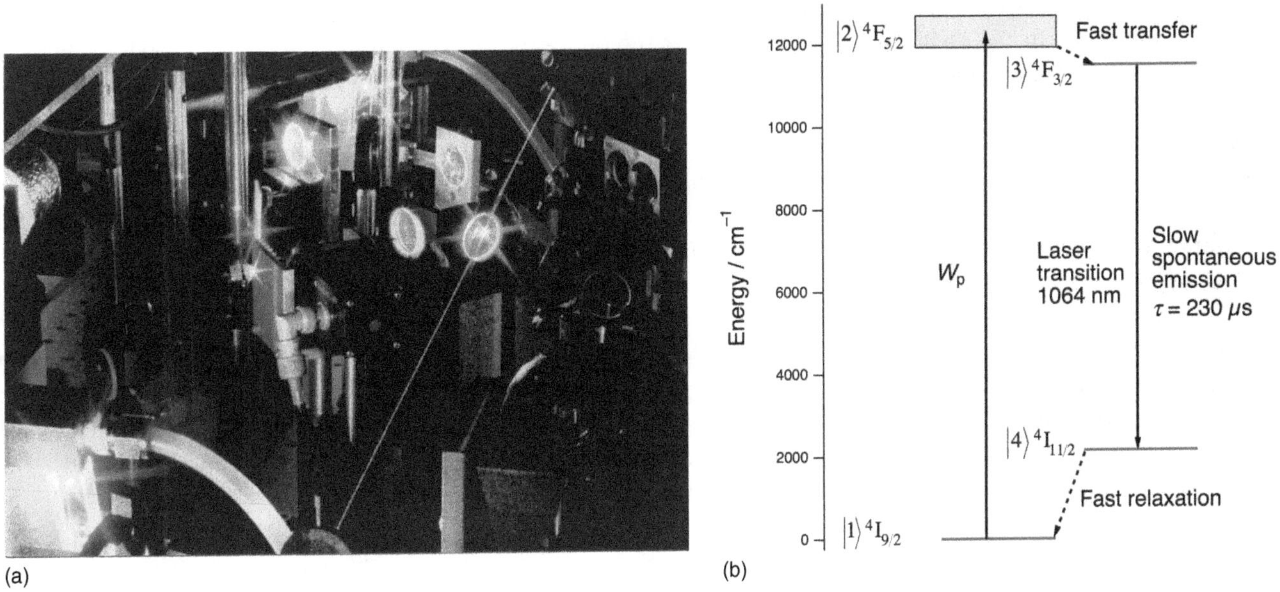

Figure 27.1 (a) A Nd^{3+}:YAG-pumped femtosecond dye laser. (b) The Nd^{3+}:YAG system is an example of a four-level laser system. When sufficient pump power W_p is applied a population inversion between levels $|3\rangle$ and $|4\rangle$ is established and lasing can occur. Fast nonradiative processes (dashed lines) move population into the upper state and remove it from the lower state to maintain the population inversion.

transition to the ground state is even more forbidden; this means that population can build up in the $^4F_{3/2}$ state rather than being lost to spontaneous emission.

Eventually, spontaneous emission lets loose a photon on the $^4F_{3/2} \rightarrow {}^4I_{11/2}$ transition. This photon encounters a Nd^{3+} ion that has been excited to the $^4F_{3/2}$ state. The photon is resonant with the transition to the $^4I_{11/2}$ state, and therefore it stimulates the emission of a second photon. Not only does it stimulate emission of a photon with the same wavelength, but the emitted photon has the same phase and the same direction of propagation as the first photon. These two coherent photons propagate and stimulate more emission. As long as there is more population in the upper state than the lower state, the number of new photons created by stimulated emission will outnumber the number of photons consumed by absorption. The result is a monochromatic, coherent, directional beam of light. The final requirement of the four-level system is that the lower level of the lasing transition is strongly coupled to the ground state so that the population inversion is easier to maintain.

The most efficient way to build a laser is to start with a laser. The photograph in Fig. 27.1 (a) is of a Nd^{3+}:YAG-pumped femtosecond pulsed dye laser. The YAG laser pumps both the oscillator and amplifier of the dye laser. The oscillator and the amplifier are two different optical elements that are important for the construction of lasers. Any region of space in which a population inversion exists can act as an *amplifier*. An amplifier is usually used in a single-pass or few-pass mode. The medium is pumped to obtain a population inversion. Light that enters the amplifier interacts with the population inversion to stimulate emission. The output intensity is amplified by some factor determined by the extent of the population inversion and the path length. Some lasers operate in this manner; for example, excimer lasers and CO_2 lasers are sometimes constructed of only an amplifier. An amplifier does not lead to particularly good beam quality in terms of divergence and coherence, especially if they rely on a single pass through the medium.

An *oscillator* is required to get the best beam quality. An oscillator, as the name implies, allows the light to circulate many times back and forth between highly reflective mirrors. This creates a beam with superior qualities of monochromaticity, directionality, brightness and coherence. In an oscillator, only those modes that fit exactly inside the cavity will be amplified. Only light for which an integer number of half-wavelengths n matches the cavity length d fulfills the resonance condition,

$$d = n\lambda/2. \tag{27.7}$$

Most lasers are run in multimode operation. The highest degree of monochromaticity requires single-mode operation, which will also generally improve the coherence and the uniformity of the intensity profile of the laser. It also comes with the cost of considerably reduced intensity.

If the oscillator mirrors are highly reflective, very little of the laser light escapes from the oscillator. While this provides for many round-trips in the oscillator, it also means that the optical power circulating in the optical cavity is possibly orders of magnitude higher than what is escaping out of the end mirror (also called an *output coupler*). Such an oscillator is said to have a high quality factor or high Q factor. If we could suddenly switch one mirror from begin 100% reflective to 0% reflective – that is, if we could switch the oscillator from being a high-quality optical cavity to a low-quality optical cavity – then we could let a pulse of laser light out of the oscillator. This is known as *Q switching*, and can be used to generate nanosecond pulses with repetition rates from single pulses to the kHz range. Often, this is done with polarization tricks and acousto-optic or electro-optic modulators. For instance, the mirror might be 100% reflective for vertical polarization of light but 0% reflective for horizontal polarization. By using the modulator to switch back and forth between the two linear polarizations, pulses can be coupled out of the oscillator as a regular pulse train.

An example of a laser that can be run in single-mode operation is the Ti:sapphire laser, which crucially is tunable in the range 670–1100 nm. Crucial because this extremely high-gain, broad bandwidth facilitates the production of pulses as short at 5 fs. Sapphire is crystalline Al_2O_3. This is doped with around 0.1 wt% Ti_2O_3. Oscillator and amplifier crystals are doped to different densities. The Ti^{3+} ion has the electron configuration $3s^2\ 3p^6\ 3d^1$. The 3d orbitals are crystal field split into a triply degenerate lower-energy T_2 orbital and an upper doubly degenerate E orbital. The ground state is $^2T_2$ with a ^2E excited state. These levels are separated by $\approx$19 000 cm^{-1}.

Figure 27.2(b) demonstrates how, even though only two electronic states are involved in the laser transition, the system still effectively acts as a four-level system. The upper state is split by the Jahn–Teller effect. The upper band can be pumped in the blue, but it is more efficient to pump the lower energy component in the green. Only the lower band is shown in Fig. 27.2(b). The Jahn–Teller distortion in the ^2E state leads to intrinsic coupling of the state to the lattice and a large shift in the Franck–Condon region for absorption from the excited state potential minimum. Relaxation in the excited state is an ultrafast process that occurs much more rapidly than radiative relaxation to the ground state. Population builds up at the minimum of the excited state, which then relaxes radiatively at wavelengths greatly red-shifted compared to

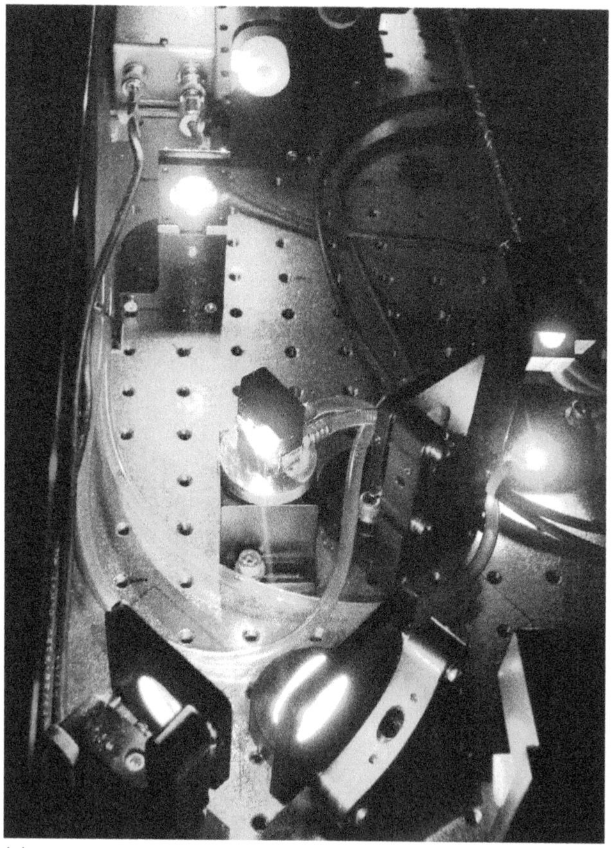

(a)

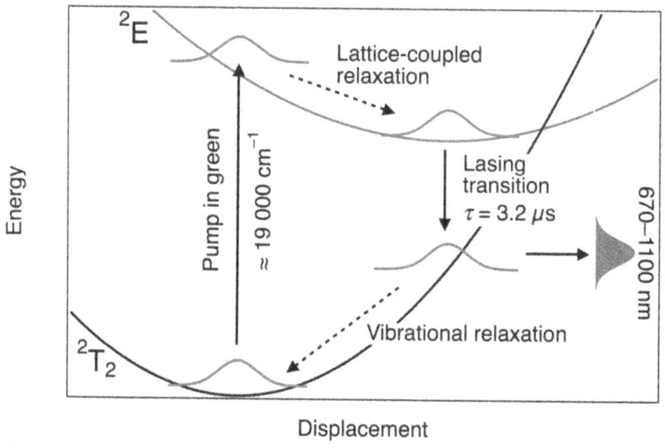

(b)

Figure 27.2 (a) A diode-laser-pumped regeneratively amplified Ti:sapphire laser system. The green light is from the pump laser. Two Ti-doped sapphire crystals fluoresce red and act as two stages of amplification. The oscillator is a separate laser in this system. (b) The level scheme of the Ti:sapphire laser.

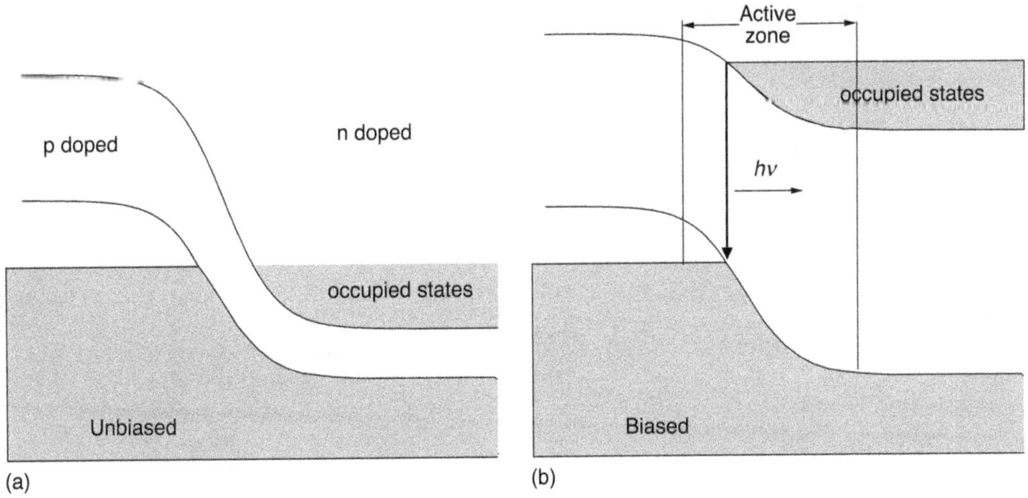

Figure 27.3 A semiconductor diode laser. (a) In the unbiased state there is no overlap between electron and hole state so there is no radiative recombination. (b) When biased, hole states in the valence band of the p-doped region become available for recombination with electrons from the n-doped region.

the pump. A broad Franck–Condon region for this transition leads to emission over a broad wavelength range. Efficient vibrational relaxation in the ground state removes population from the lower level of the lasing transition and returns it to the ground state. Because of the essential role of lattice coupling and vibrational relaxation, the Ti:sapphire laser is sometimes called a *vibronic laser*.

The Ti:sapphire laser can operate in either CW or pulsed mode. Mode locking is essential for operation in the femtosecond regime. Although normally it is not a good idea to hit your laser, it was discovered accidentally that vibration of the sapphire rod, which can be caused by striking the table on which the laser sits, can force the Ti:sapphire system to self-mode lock. Piezoelectric modulators are used to perform this task controllably. A broad range of laser systems are described in Table 27.1.

27.1.1 Directed practice

For a Nd^{3+}:YAG laser operating on its fundamental wavelength in a 10 cm cavity, calculate the wavelength spacing between two successive modes.

[Answer: 5.66 pm]

27.2 Harmonic generation (SHG and SFG)

An incident electromagnetic wave induces a polarization in the medium it strikes given by

$$\boldsymbol{P} = \varepsilon_0[\chi^{(1)}\boldsymbol{E} + \chi^{(2)}\boldsymbol{E}^2 + \chi^{(3)}\boldsymbol{E}^3 + \ldots] = \varepsilon_0 \sum_{k=1}^{n} \chi^{(k)}\boldsymbol{E}^k \qquad (27.8)$$

where ε_0 is the vacuum permittivity, $\chi^{(k)}$ is the kth order *nonlinear susceptibility* of the medium, and $\boldsymbol{E}$ is the electric field vector. Any phenomenon that depends on one of the $k \geq 2$ terms is called a *nonlinear effect* because it depends on the field strength raised to a power larger than one. In the multipole expansion of the nonlinear susceptibility $\chi^{(k)}$ each term of successively higher order is much smaller than the previous term. On symmetry grounds some terms vanish. For instance, in any centrosymmetric medium (a medium that contains an inversion center such as an elemental crystal or isotropic material) all even order terms are zero.

The linear term relates to the oscillating electric field with amplitude A and angular frequency ω,

$$E = A \sin \omega t. \qquad (27.9)$$

The second-order term oscillates according to

$$E^2 = A^2(\sin \omega t)^2 = \tfrac{1}{2}A^2(1 - \cos 2\omega t). \qquad (27.10)$$

Table 27.1 A survey of laser systems and their characteristics.

Laser material	λ/nm	$h\nu$/eV	Characteristics
Solid state			
Semiconductor laser diode	IR–visible ~0.4–20 μm		Usually CW but can be pulsed; wavelength depends on material; GaN for short λ, AlGaAs 630–900 nm, InGaAsP 1000–2100 nm, used in telecommunications, optical discs
Nd^{3+}:YAG (1st harmonic)	1064	1.16	CW or pulsed, ~10 ns pulses, most common, 150 ps versions (and shorter) available, 10–50 Hz rep rate, 1 J to many J pulse energies. Nd^{3+} can also be put in other crystalline media such as YLF (1047 and 1053 nm) or YVO_4 (1064 nm)
Nd^{3+}:YAG (2nd harmonic)	532	2.33	
Nd^{3+}:YAG (3rd harmonic)	355	3.49	
Nd^{3+}:YAG (4th harmonic)	266	4.66	
Nd^{3+}:glass	1062 or 1054	2.33	~10 ps, can be used to make terawatt systems for inertial confinement fusion studies
Ruby (Cr:Al_2O_3 in sapphire)	694	1.79	~10 ns pulses when Q switched, first laser system
Ti:sapphire	700–1000	1.77–1.24	fs to CW; from low repetition rate to 82 MHz
alexandrite (Cr^{3+} doped $BeAl_2O_4$)	700–820	1.77–1.51	tattoo removal
Liquid			
dye laser	300–1000	4.13–1.24	rep rate and pulse length depend on pump laser; fs, ps, ns up to CW
Gas			
CO_2	10600 (10.6 μm)	0.12	Long (many μs), irregular pulses, CW or pulsed at high rep rates, line tunable, few W to >1 kW
Kr ion	647	1.92	CW, line tunable, 0.1–100 W
HeNe	632.8	1.96	CW, 0.5–35 W
	543.5	2.28	
Ar ion	514.5	2.41	CW, line tunable
	488	2.54	
HeCd	441.6, 325	2.81,3.82	CW, 1–100 mW
ArF excimer	193	6.42	~20 ns, 1– >1000 Hz, several W to over 1 kW, 100 mJ to >1 J
KrF excimer	248	5.00	30–34 ns
XeCl excimer	308	4.02	22–29 ns
XeF excimer	351	3.53	12 ns
F_2	154	8.05	1– several kHz rep rate, 1–20 W, 10–50 mJ pulse energies, 10 ns
N_2	337	3.68	1–3.5 ns, 0.1–1+ mJ, 1–20 Hz rep rate
Exotic			
Free electron laser	1 nm–THz		Wide variety of characteristics depending of machine characteristics
X-ray lasers			Lasing medium is a highly ionized plasma producing light at one wavelength
High harmonic generation	UV–X-ray		High-powered laser pumps a gas to produce light at wavelengths that are harmonics of the pump wavelength. Capable of producing attosecond pulses.

Thus, radiation scattered as a result of the $\chi^{(2)}$ term oscillates at twice the incident frequency. It has been *frequency-doubled*. Equivalently, this *second harmonic* signal is generated at half the wavelength of the incident light. The intensity in the second harmonic signal increases with the square of the incident intensity. Effectively, the induced polarization field takes two photons from the incident field, adds them together and emits them as one photon of twice the energy.

Second harmonic generation (SHG) does not happen in a gas or a glass. There are a number of crystalline materials that are particularly effective at second harmonic generation that find widespread use in nonlinear optics. These include potassium dihydrogen phosphate (KDP), β-barium borate (β-Ba_2O_4) and lithium niobate ($LiNbO_4$). These crystals are often inserted into laser beams to generate light at frequencies that cannot be generated directly by the oscillator or amplifier. SHG is also possible at the interface of two materials. In this way, SHG can be used to probe molecules adsorbed at the interface. If a second laser is incident on the polarization induced by the first laser, the $\chi^{(2)}$ term of the material will, with a probability equal to the product of the intensity of the two lasers, generate a response at the *sum frequency*. In other words, the $\chi^{(2)}$ term can add two photons of different frequencies to form one photon with an energy equal to the sum of the two photon energies. This process is called *sum frequency generation* (SFG). Both, SHG and SFG are much more likely to occur if one of the frequencies involved in the process is resonant with a molecular process. For example, a common application of SFG is to perform time-resolved vibrational spectroscopy.

Since its introduction by Ron Shen,[5] a very common combination is to choose an infrared frequency, which can be tuned through vibrational resonances of an adsorbate, and to mix it with a visible or near-IR photon of fixed frequency to yield a visible photon. The fixed-frequency pulse is called the up-conversion pulse. In time-resolved studies, the tunable pulse is used to pump population into an excited state, and is called the *pump pulse*. The fixed-frequency pulse (probe pulse) then upconverts the polarization field at the frequency of the first pulse into the photon that is detected. The signal intensity in such a *pump–probe experiment* is determined not only by the intensities of the pump and probe pulses, but also by the lifetime of the excited state and the delay between the two pulses, which allows for direct determination of the lifetime of the excited state. The delay between the two pulses, as shown in Fig. 27.4(a) can be controlled by changing the position of a mirror, which introduces a controlled path length difference between the pump and probe beams.

27.2.1 Directed practice

Calculate the wavelength of the third harmonic of a Ti:sapphire laser operating at 800 nm. Calculate the sum-frequency photon created by mixing a 532 nm photon with a photon resonant with a vibration at 2100 cm^{-1}.

[Answers: 267 nm, 479 nm]

27.3 Multiphoton absorption spectroscopy

SHG, SFG and the Raman effect are all *nonlinear spectroscopies*. The signal level in nonlinear spectroscopies usually changes as I^n, where I is the laser intensity and n is an integer greater than one that often equals the number of photons in the process. There is a much broader branch of spectroscopy involving transitions mediated by more than one photon. The theory of multiphoton transitions in atoms was founded by Maria Goeppert-Meyer in her doctoral thesis. This was before the invention of the laser. Therefore, experimental verification of any of this theory appeared extremely unlikely at the time she submitted her thesis. Faced with this, she changed the focus of her research. And a good choice it was, because she was to become the first woman since Marie Curie in 1903 to be awarded the Nobel Prize in Physics. Goeppert-Meyer received hers in 1963 for her discoveries pertaining to nuclear shell structure.

The understanding of multiphoton transitions is greatly simplified with the following conceptual framework. Solutions of the Schrödinger equation for a molecule reveal the wavefunctions that describe the states of the molecule – the molecule in the absence of a photon field. When applied at a high enough intensity, states of the molecule + field system emerge. We call these molecule + field states *virtual states*. Another way to put it is that virtual states are states of the molecule that only exist in the presence of the photon field. Virtual states are available for transitions just as are real states. One interesting property of virtual states is that they draw their existence and properties from the real states of the molecule. In particular, they draw the largest contributions from the states nearest to them. The strength of transitions through virtual states differs from one state to the next, just as it does for transitions between real states. The more character a virtual state can draw from real states, the stronger transitions through it will be.

The *selection rules* for two-photon absorption are the same as for vibrational Raman transitions. The cross-product of the symmetry species that characterize the two electronic states and the two-photon tensor $\boldsymbol{S}$ must contain a totally symmetric character,

$$\Gamma(\psi_e') \times \Gamma(S_{ij}) \times \Gamma(\psi_e'') \tag{27.11}$$

The two-photon tensor $\boldsymbol{S}$ transforms as does the polarizability tensor $\boldsymbol{\alpha}$, that is,

$$\Gamma(S_{ij}) = \Gamma(\alpha_{ij}). \tag{27.12}$$

This is particularly important to recognize for determining two-photon transition selection rules in polyatomics, and whether the transitions are vibronically allowed. For multiphoton transitions of nth order, one can also apply a 'chain rule' to determine if multiphoton transitions are allowed in which the one-photon selection rules are applied for each

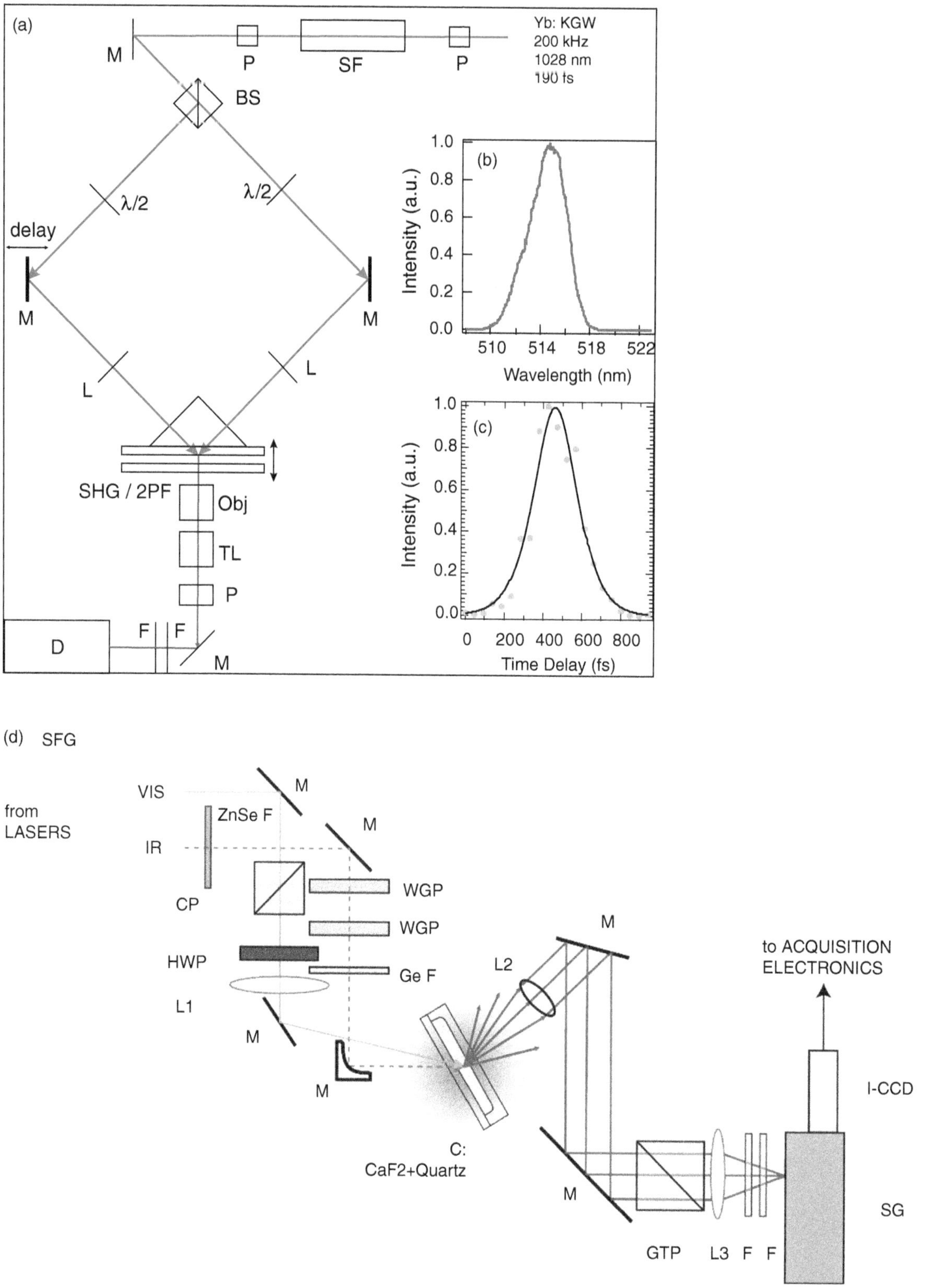

Figure 27.4 (a) Second harmonic generation (SHG) is illustrated. Two coherent beams, created by splitting the beam generated by one short-pulsed laser, are focused on the same spot in a sample. (b) The spectrum of the SHG beam, centered at 514 nm for 1028 nm incident wavelength, reflects the bandwidth of the incident light. (c) By varying the spatial and temporal overlap of the beams the signal is optimized. For nonresonant SHG, the cross-correlation signal in panel (c) is a measure of the pulse width. For resonant SHG through a real state with a finite lifetime, the cross-correlation signal is now a convolution of the pulse width and the intermediate state lifetime. (d) In sum frequency generation (SFG) two different wavelengths, visible (vis) and infrared (IR) in this example, are mixed on the sample. Panels (a–c): Images provided by Sylvie Roke and reproduced with permission from Macias-Romero, C., Didier, M.E., Jourdain, P., Marquet, P., Magistretti, P., Tarun, O.B., Zubkovs, V., Radenovic, A., and Roke, S. (2014) *Opt. Express*, **22**, 31102; © Optical Society of America. Panel (d): Roke, S. and Gonella, G. (2012) *Annu. Rev. Phys. Chem.*, **63**, 353. © 2012, Annual Reviews.

'link' in the multiphoton process. This is particularly simple for diatomics, which have a simple set of selection rules for one-photon transitions.

- First, spin conservation. The spin selection rule is the same for any n^{th} order transition: the spin quantum number does not change, either $\Delta S = 0$ or $\Delta\Sigma = 0$ as appropriate.
- Next, orbital angular momentum conservation. $\Delta\Lambda = 0, \pm 1$ for each photon involved in the multiphoton process. Thus, for a two-photon process, $\Delta\Lambda = 0, \pm 1, \pm 2$ result from all combinations of $\Delta\Lambda = 0, \pm 1$ taken twice. In a three-photon process, $\Delta\Lambda = 0, \pm 1, \pm 2, \pm 3$ are allowed.
- Next, parity conservation. In a one-photon Σ–Σ transition, the parity must not change. Since one photon cannot change parity, so too n photons cannot change parity. The parity does not change regardless of the number of photons involved (+ remains + or − remains −).
- Finally, consideration of inversion symmetry. Inversion symmetry must change in any one- or odd-photon transition (g or u must change). Inversion symmetry must not change in any even-photon transition (g or u must stay the same).

27.3.1 Laser-induced fluorescence (LIF)

In multiphoton spectroscopy we have to keep track of whether we are adding photons independently or coherently, as well as the bandwidth of the light source. Conventional fluorescence involves two independent photons. The first photon excites a molecule to an electronic excited state. A second photon is then emitted independently by spontaneous emission, as shown in Fig. 27.5(a). The fluorescence can be detected in one of two modes: (i) we can detect all of the light without dispersion of the wavelength (total fluorescence); or (ii) we can measure the spectrum of the emitted light by dispersing it with, for example, a grating (dispersed fluorescence). What we usually call a *fluorescence spectrum* is more precisely called a *dispersed fluorescence spectrum* with constant excitation wavelength. In this case, the spectrum is a plot of fluorescence intensity as a function of the fluorescence photon wavelength, as shown in Fig. 26.13 for rhodamine 6G (R6G). It probes how the excited state relaxes to a lower state.

We can also acquire an excitation spectrum. In one type of excitation spectrum, we disperse the fluorescence and detect the light emitted at one wavelength while changing the excitation wavelength. Alternatively, we collect the undispersed light while scanning the excitation wavelength. In either case, the spectrum obtained is a plot of fluorescence intensity versus excitation wavelength, and the spectrum resembles an absorption spectrum.

Conventional fluorescence spectroscopy utilizes a broadband source to excite a sample. A spectrometer disperses the light onto a detector. This technique obtains high resolution at the expense of signal intensity.[6] With the use of narrow

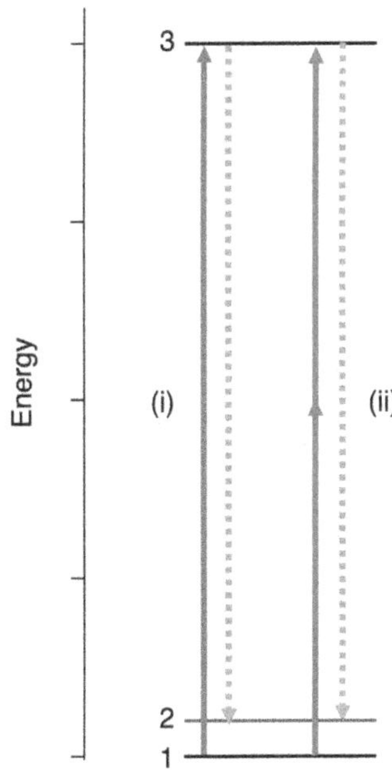

Figure 27.5 Laser-induced fluorescence (LIF). A laser is used to excite the system from state 1 to state 3 by either one-photon absorption, as in process (i), or two-photon absorption, as in process (ii). Subsequently, spontaneous emission (fluorescence) is detected between the excited state 3 and lower level 2.

bandwidth and intense laser excitation, we can perform high-resolution spectroscopy based on the absorption of photons that is also highly sensitive. A laser with a bandwidth smaller than the spacing between spectral lines excites the molecule quantum-state specifically. The excited state population is directly proportional to the initial state population. The fluorescence signal is then proportional to the initial state population. In this manner a spectrum of the initial state molecule is acquired by detecting either the dispersed or undispersed fluorescence emitted when the laser is tuned across the frequency range in which the initial states absorb light. If the electronic transition is excited to saturation, we move the maximum amount of population into the excited state. At this point it can be shown that the ratio of the intensity of two transitions in the same rotational band and the same band progression (i.e., Δv constant) I_i/I_j is directly proportional to the population ratio of the two initial states n_i/n_j,

$$\frac{n_i}{n_j} = \frac{I_i}{I_j}. \tag{27.13}$$

In the absence of saturation, the line strength of the transition (its Einstein B coefficient) and the laser power (in terms of the energy density) must also be taken into account according to

$$\frac{n_i}{n_j} = \frac{I_i B_j \rho_j}{I_j B_i \rho_i}. \tag{27.14}$$

This technique is known as *laser-induced fluorescence*. It is a sensitive method for the measurement of quantum state distributions, and was introduced by Richard N. Zare.[7] First performed with CW lasers such as the HeNe and Ar ion laser, its applications skyrocketed with the introduction of tunable lasers such as the dye laser. When applied to atoms it can resolve spin–orbit splittings between different J states in a given electronic level. After excitation, the undispersed fluorescence is collected to obtain the highest detection efficiency. High detection efficiency is achieved because we know when to look for a signal (after the laser pulse has excited the sample), and because we detect against zero background. A background of nothing but electronic noise is achieved because the time delay between excitation and detection allows us to wait for scattered laser light to be absorbed, and because the fluorescence is red-shifted compared to the excitation light to further enhance its separation. Single-molecule detection has been possible with LIF.

LIF is normally used to measure the population in the *initial state* after it has been populated by scattering, a chemical reaction, or expansion in a molecular beam. This is particularly true when undispersed fluorescence is acquired as a function of the excitation laser wavelength. Dispersing the fluorescence allows us to probe the transitions from a quantum state selected by the excitation laser to a lower electronic level. Thereby, we can probe the relaxation dynamics of the upper state. The fluorescence can either be collected by lenses and mirrors onto a photomultiplier, or else it can be imaged to obtain more detailed information about spatial distributions. The use of pulsed lasers also allows us to introduce delays and time resolution into our data analysis. Great sensitivity and the ability to produce detectable signal in the few nanoseconds of a pulsed laser also facilitates the detection of transient species, such as reactive intermediates produced in chemical reactions, even combustion. LIF is capable of detecting nascent reaction products – that is, reaction products with quantum state distributions as produced by the reaction and before they have collided with any other species. The polarization of the incident light and/or the fluorescence can also be analyzed. This allows us to determine whether the sample is aligned or oriented following a reaction[8] or scattering.[9] An aligned sample is one that rotates in a preferred plane. An oriented sample is one that not only rotates in a preferred plane but also has a preferred sense of rotation – that is, clockwise at opposed to counterclockwise.

An LIF signal can be collected from essentially any volume into which a laser can be focused. Thus, it can be used not only to detect molecules in a gas cell but also inside of a modified diesel engine to study combustion reactions, or in a molecular beam to measure the rotational temperature of expanding molecules. By translating the focal volume, spectroscopic information can be obtained with spatial resolution. As we will discuss below, by using more than one laser beam we can even beat the diffraction limit in spatial resolution. LIF also works in liquids. One area that has been particularly fruitful in this respect is the detection of molecules after they have been separated by capillary electrophoresis (CE). Indeed, LIF detection of CE effluent was used to readout the millions of bases in DNA required to sequence the human genome for the first time.

Laser excitation to induce fluorescence is not limited to one-photon processes. Coherent absorption of two or more photons can also be used to excite fluorescence. This is exploited in two-photon fluorescence microscopy, most commonly with a femtosecond-pulsed Ti:sapphire laser as the excitation source.

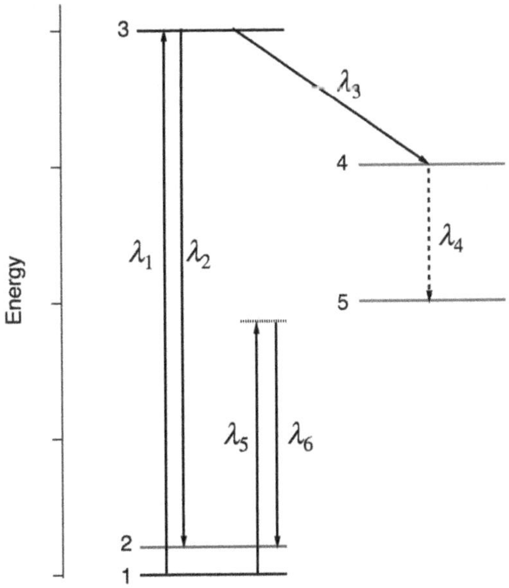

Figure 27.6 Stimulated emission pumping (SEP) of state 4 occurs by populating state 3 with the absorption of photon λ_1, then stimulating the transition $3 \rightarrow 4$ by resonant irradiation at wavelength λ_3. Population of state 4 can be followed by, for example, detecting its fluorescence at λ_4. Stimulated Raman scattering (SRS) can also be used to pump levels to quantum state specifically. Regardless of whether pumping is performed resonantly, for example using photons with wavelengths λ_1 and λ_2, or nonresonantly through a virtual state, using photons with wavelengths λ_5 and λ_6, the difference in photon energy must be resonant with the difference in energy between the initial state and the state pumped.

27.3.2 Stimulated emission pumping

Laser-induced fluorescence is most commonly implemented as an incoherent two-photon technique in which the first photon comes from a laser and the second photon is emitted from the molecule spontaneously and randomly based on the lifetime of the excited state that was created by absorption of the laser photon. If we introduce a second laser that is resonant with a transition from the excited state created by the absorption of the first photon to a lower energy state, then we do not need to wait for spontaneous emission. Instead, we force the system to relax. Moreover, we can determine the state to which the photon relaxes by tuning the wavelength of the photon to be resonant with a particular transition. This process of pumping population into an excited state with one laser, then dumping population into another (but lower energy) excited state with a second laser, is known as *stimulated emission pumping* (SEP) and is illustrated in Fig. 27.6. SEP has three important applications: (i) it selectively moves population to a specific quantum state; (ii) it facilitates pump-dump-and-probe detection; and (iii) it selectively turns off fluorescence in the region of space depleted of excited state population by the dump laser. The excited state does not have to be a real state. If the excited state is a virtual state instead of a real state, the technique is known as *stimulated Raman scattering* (SRS). In either case (SEP or SRS), the difference in energy between the pump and dump photons matches the energy difference between the initial and final molecular states. SRS has the disadvantage that practical schemes can only change the vibrational quantum number by 0 or ± 1. SEP can, in principle, be used to change the vibrational quantum number by an arbitrary amount, limited only by Franck–Condon factors to determine the efficiency. However, appropriately tunable lasers must be found in either case, which can be problematic depending on the molecule under consideration.

The first application of SEP is the preparation of a specific quantum state. Selective population of a quantum state allows us to study the behavior of a reactive system in a quantum-state-specific manner. For instance, suppose you would like to study how the rotational and vibrational states of a molecule change when, for instance H_2 or NO is scattered from a metal or insulator surface,[10] or to test whether the Born–Oppenheimer approximation is appropriate for describing these interactions and reactions on surfaces.[11] Then either SRS or SEP can be used to transfer population to specific quantum states in a molecular beam. The molecular beam is aimed at a surface, and the quantum state distribution before and after interaction with the surface can be measured using resonance-enhanced multiphoton ionization (REMPI; see below).

The second application is a pump, dump, and probe experiment. This is a three-color experiment in that three different photons are used. One photon prepares an excited state (pump), while the second photon moves population to a different quantum state by stimulated emission pumping (dump). The third or probe photon ionizes the molecule (or induces LIF) so that the ions can be detected and quantified. The use of three separate photons allows us to potentially introduce two variable delay times. Suppose, for instance, you want to study the dependence of the ionization rate of I_2 in a strong laser field as a function of the internuclear distance R.[12] You could pump the molecule to an excited state by a Franck–Condon transition. Since you know that the molecule is excited onto the repulsive wall of the excited state potential, you know that the molecule's bond will stretch after excitation. By waiting a known amount of time after the pump pulse, you can control how much the internuclear separation increases before you dump the molecule back to the ground electronic state. Subsequently, the third photon probes the ground state by ionization and the ionization rate can be measured as a function of R. Similarly, the reactivity of a prepared state can be measured by delaying the time between the pump, dump and probe pulses.

The third application is detailed in the next section.

27.3.3 Superresolution fluorescence microscopy

In Chapter 19 we mentioned that the ultimate resolution of a microscope $\Delta_{\min}$ is given by the *Abbe limit*, and is roughly $\Delta_{\min} \approx \lambda/2$. A more refined derivation[13] of this limit demonstrates that for a point source propagating along the z-axis, the resolution deviates slightly from this formula. More importantly, the optimum resolution is significantly worse in the z-coordinate than in the lateral xy-plane. In the xy-plane the ultimate lateral resolution is given by

$$(\Delta_{\min})_{x,y} = \frac{\lambda}{2n \sin \alpha} \tag{27.15}$$

while the axial resolution in the z-direction is

$$(\Delta_{\min})_z = \frac{\lambda}{2n(\sin \alpha)^2}, \tag{27.16}$$

where λ is the wavelength of light, n is the refractive index of the medium, and α is half the aperture angle under which the point source is observed. In practice, the value in the xy plane is still close to the $\lambda/2$ limit, but that in the z-direction is significantly worse. Conventional fluorescence microscopy is limited by this ultimate resolution, but usually does not reach it because of practical factors such as noise, scattered light, unfocussed background, and so forth. The resolution can be improved slightly by placing an additional aperture before the detector (confocal microscopy) or using two-photon excitation (two-photon fluorescence microscopy), but neither of these techniques fundamentally breaks the Abbe limit.

The first observation of a single fluorophore in a dense medium was made by in 1989.[14] Even considering that this was performed at 4 K, this was a breakthrough, in particular because Moerner and Kador measured the absorption of pentacene rather than its fluorescence. Another milestone was single-molecule detection using *near-field scanning optical microscopy* (NSOM).[15] However, both of these exploited techniques that restrained them from wide adoption. That more conventional microscopy techniques could be more routinely adapted to super-resolution (i.e., beyond the Abbe limit) microspectroscopy and the study of dynamics was presaged in the work of Zare and coworkers,[16] who observed the diffusion of a single molecule using confocal microscopy.

The widespread adoption of super-resolution fluorescence needed to await a brilliant idea by Stefan W. Hell[17] for one implementation, and the combination of a brilliant discovery by William E. Moerner[18] with a brilliant idea from Eric Betzig.[19] For their efforts they were awarded the Nobel Prize in Chemistry 2014.

Hell's seminal discovery was to recognize that as long we use one or two photons to simply turn on fluorescence, the diffraction limit was going to limit the ultimate resolution of microscopy and spatially-resolved spectroscopy. However, as we have discussed above with regard to stimulated emission pumping, photons can also be used to *turn off* fluorescence. The trick, which leads to the technique known as *stimulated emission depletion* (STED) microscopy, is the following. A well-focused laser beam is used to excite fluorescence. This low-power beam leads to diffraction-limited resolution, as demanded by the Abbe limit. A second laser with a more intense and redder beam is then used to illuminate the same region; this is known as the STED beam. If this beam had a Gaussian intensity profile, as does the first beam, it would simply turn off all of the fluorescence by stimulated emission back to the ground electronic state. This stimulated emission is red-shifted from the spontaneous emission (fluorescence) and is, therefore, easily separated from it. So far, we have created a more difficult way to obtain no better resolution. The trick is to use a STED beam that has a 'donut' intensity profile – an intensity profile that has a hole in the middle. Such a laser beam does nothing to the fluorescence intensity from the center of the beam, but it totally eliminates fluorescence from the intense parts of the beam. The resulting ultimate lateral resolution is determined by the intensity of the STED beam I_0 compared to the intensity required to saturate the stimulated emission I_{sat} by the function

$$(\Delta_{\min})_{x,y} = \frac{\lambda}{2n\sin\alpha\sqrt{1+I_0/I_{sat}}}. \tag{27.17}$$

Equation (27.17) shows that the resolution becomes infinitely good, $(\Delta_{\min})_{x,y} \to 0$, in the limit of high intensity. Of course, blowing up the sample will limit the resolution in practice. The axial resolution is also significantly enhanced.

Many fluorophores, including quantum dots, exhibit blinking. Under continuous excitation and relaxation they repeatedly emit photons by fluorescence, and the fluorescence intensity blinks on and off for many cycles until suddenly relaxing into a dark state. The dark state is an excited state of the system that is not connected radiatively to the fluorescent state by the photon that connects the ground state to the fluorescent state. Moerner and coworkers discovered blinking when green fluorescent protein (GFP) was irradiated at 488 nm. Just as importantly, they found that irradiation with 405 nm light was able to reactivate GFP. Therefore, GFP represented a system that could be switched between one state that was fluorescent and one that was not. Furthermore, the dark state could be switched back on at will.

Betzig's method, called *photoactivated localization microscopy* (PALM), of obtaining super-resolution goes something like this. If we know that each emitter is a single fluorophor – that is, a point source – then we can improve the resolution by counting the number of photons N emitted from each point source. The image of each point source will exhibit a Gaussian photon spatial distribution with a width determined by the Abbe diffraction limit Δ_{Abbe}. However, since we know that the source is a point source, the center of the photon number distribution – which is the actual position of the source – can be determined with an accuracy given by

$$(\Delta_{\min})_{x,y} = \frac{\Delta_{\text{Abbe}}}{\sqrt{N}} = \frac{1}{\sqrt{N}}\frac{\lambda}{2n\sin\alpha}. \tag{27.18}$$

The resolution is, in principle, limited only by the time it takes to count enough photons to achieve the desired resolution. However, this is of little utility unless the sample has extremely sparse density, which ensures that each fluorophore is isolated. What we need is a way to take a dense sample, turn on a very small fraction of the fluorophores, image them, turn them off, and then repeat the cycle until the whole of the sample has been imaged. This is precisely what the regenerative blinking state allows us to do.

27.3.3.1 Directed practice

Green fluorescent protein emits at 504 nm. Calculate the number of photons that have to observed to obtain 1 nm resolution assuming that the aperture angle is 70° and you are operating in water with refractive index 1.333.

[Answer: 4.05×10^4]

27.3.4 Resonance-enhanced multiphoton ionization (REMPI)

Visible photons are extremely easy to detect if they strike a photomultiplier tube. Even single photons can be detected. However, there are geometric limits on the fraction of photons that can be collected when they are created within the focal volume of a laser. Ions do not suffer from this limitation. Because of their charge, essentially all ions created in the focal volume of a laser can be collected by the proper application of focusing and acceleration voltages in front of a multichannel plate detector.

If we keep turning up the laser intensity in a laser-induced fluorescence experiment, we will often find that the fluorescence signal will decrease due to the formation of ions. The process that is occurring is shown in Fig. 27.7. In general, multiphoton ionization (MPI) can occur in three distinct ways: (i) nonresonant absorption of n photons; (ii) resonant absorption of n photons of one color followed by ionization caused by the absorption of m photons of the same color; and (iii) resonant absorption of n photons of one color followed by ionization caused by the absorption of m' photons of a different color.[20] In 1977, it was demonstrated that REMPI can be more sensitive than LIF.[21] The requirement of extracting ions from the focal volume demands that REMPI can only be performed in the gas phase.

A quantum-state resolved spectrum of the initial state population is obtained in REMPI by measuring the ion yield as the laser wavelength is scanned. Whenever the resonance condition is met, an ion signal is recorded and the number of ions collected is proportional to the population in the initial state. The formation of ions immediately brings forth the possibility of adding two additional aspects to REMPI detection. Passing the accelerated ions through a field-free region allows them to be separated by mass by means of time-of-flight mass spectrometry (TOF-MS). This can be used

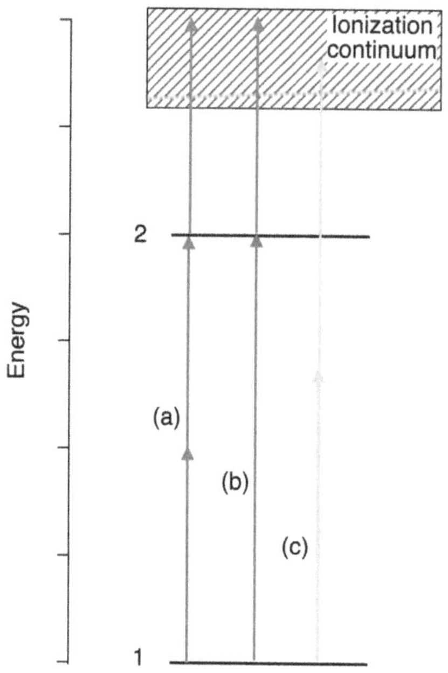

Figure 27.7 (a) Two photons are resonantly absorbed by the intermediate state, after which a third photon ionizes the molecule in (2 + 1) resonance-enhanced multiphoton ionization (REMPI) using one color (same wavelength) for all three photons. (b) (1 + 1′) REMPI, where the prime indicates that the second photon has a different wavelength than the first photon. (c) Nonresonant two-photon multiphoton ionization (MPI).

to enhance the signal-to-noise ratio of the detected signal by using gated detection over just the mass of interest and blocking out other signals such as those due to scattered light or electronic noise. If the ionization process is nonspecific (as is the case for nonresonant MPI, or if a number of different molecules with overlapping absorption bands are present) or dissociative, TOF-MS detection can be used to separate and identify the ion products.

A second aspect is that not only mass resolution but also velocity and angular resolution can be added to the detection scheme. Ion imaging combined with quantum-state-resolved ionization then provides a nearly complete analysis of the energy and momentum of molecules and reactive systems to provide an exceedingly detailed look into molecular reaction dynamics.[22]

27.3.5 Multiphoton photoelectron spectroscopy (MPPE)

Instead of detecting the ions, the sign of the focusing and acceleration voltages can be switched and the electrons produced through multiphoton ionization can be detected instead of the ions. In this way, electron spectroscopy can be performed on either gas-phase samples or on solids placed in a vacuum chamber. Time-resolved multiphoton photoelectron spectroscopy is essential for probing the dynamics of electrons. Recently, time resolution has been pushed below the femtosecond range and into the attosecond regime with the implementation of high harmonic generation. We discuss this in greater detail below. In this way, it has been possible to show that delays occur between photon absorption and electron emission, and that these delays depend on the electronic state from which the electron is derived. To obtain information about electron momentum in addition to its velocity, the angular distribution of electrons must be measured.

27.4 Cavity ring-down spectroscopy

In absorption spectroscopy, we usually operate under conditions in which the Beer–Lambert law holds. As long as the laser intensity is sufficiently low not to induce saturation of the transition or nonlinear absorption, and the bandwidth of the laser is small compared to the width of the absorption feature, the intensity of the transmitted light I is related to the intensity of incident light I_0 by

$$I/I_0 = \exp(-\sigma l_s \rho), \tag{27.19}$$

where σ is the absorption cross-section, l_s the path length, and ρ the number density of absorbers. For a constant absorption cross-section, increased absorption (decreased transmitted intensity) can be obtained by either increasing the path

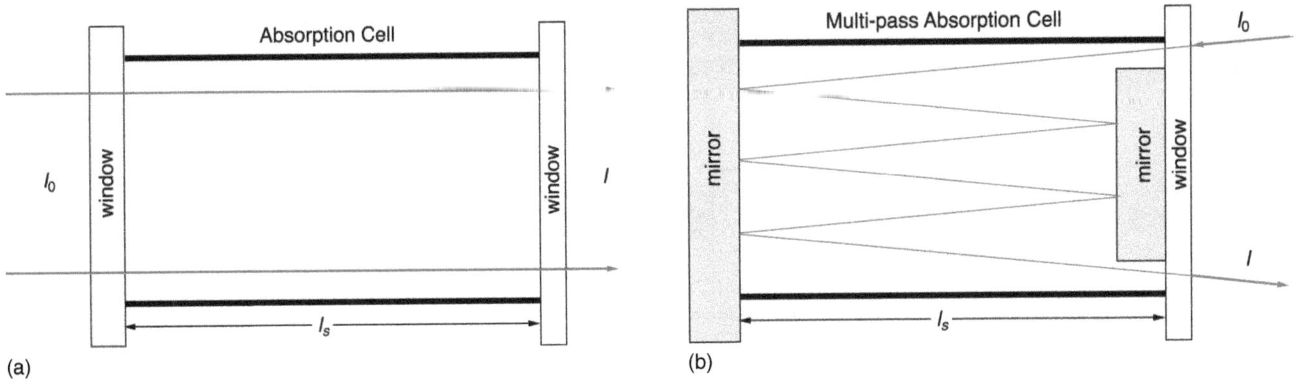

Figure 27.8 Absorption spectroscopy revolves around the comparison of the incident intensity I_0 to the transmitted intensity I after light from a source of known wavelength passes through a cell. The decrease in signal is dependent on the concentration of absorbers as well as the path length of the light through the cell. (a) A single-pass absorption cell of path length l_s. (b) A multi-pass cell of path length $2nl_s$, where n is the number of round-trips while transiting the cell.

length or the number density. Assuming that our sample is in the gas phase, we increase the number density by increasing the pressure. However, we have seen in Chapter 23 that if we increase the pressure too much, pressure broadening will increase the width of the absorption peaks. This may be detrimental to our experiment if it requires high resolution (non-overlapping peaks). There may be other limits on the pressure that can be obtained, for instance, if we are dealing with a sample that has a low vapor pressure, decomposes at high temperature, or that is produced by expansion in a supersonic molecular beam.

The alternative for obtaining more absorption is to increase the path length. The Earth's atmosphere provides a very long path length for the absorption of solar radiation. Thus, the extremely weak emission of the O_2 $b^1\,\Sigma_g^+ \rightarrow X^3\,\Sigma_g^-$ transition can be observed with the naked eye as the red glow of a glorious sunset – hence the name sunset band. By placing the gas in a cell we can control the pressure and temperature of the gas as well as l_s, as shown in Fig. 27.8(a). Furthermore, by sealing the cell and connecting it to a vacuum pump as well as a gas source, we can not only control p and T but also remove impurities from the atmosphere. Physical limits such as the size of a laboratory limit the length of the absorption cell. However, the path length through the sample can be increased if, instead of using simple windows, we change these windows to some combination of mirrors and lenses, as shown in Fig. 27.8(b). In this way, we can construct a multi-pass cell, in which the light bounces back and forth a controlled number of times. This increases the sampling length and the absorption by the sample. Depending on the combination of lenses and mirrors, multi-pass cells are known by several different names, including Pfund, White, and Herriott cells.

The longest possible path length is obtained if we have two perfectly flat mirrors with perfect reflectivities $R = 1$, and we were able to place a beam of collimated light along the optical axis of these mirrors. Unfortunately, if the reflectivity is perfect then no light can enter or exit the cavity. Instead, we take extremely good mirrors with $R > 0.99$, perhaps even $R = 0.9999$. For practical reasons we replace the flat mirrors with two matched concave mirrors, placed symmetrically about their focal point as shown in Fig. 27.9. We use a laser beam, conditioned by spatial filtering, to produce a beam with a Gaussian intensity profile, also known as a TEM_{00} beam. This geometry was first demonstrated in 1988.[23] Performing spectroscopy with such a cell is known as *cavity ring-down spectroscopy* (CRDS). This arrangement allows measurements over path lengths exceeding thousands of times the sample length, in a way that is immune to variations in the light source intensity. This combination makes it possible to measure absorption features as small as 10^{-7} of the incident intensity.

A theory of CRDS that works out details of spectroscopy and cell design was formulated by Zalicki and Zare.[24] It is important to avoid the effects of interference fringes and to ensure that the bandwidth of the laser is narrower than the linewidth of the absorption features. Under these conditions and others that ensure that the Beer–Lambert law governs absorption, the following derivation is valid. Our signal as a function of time $S(t)$ is measured (assuming that a narrow bandwidth pulsed laser is being used) by detecting the intensity of light transmitted through a cell with path length l_s. The intensity of signal $S(t)$ is given by the product of the intensity of light $I(t)$ at the transmitting cavity mirror times the mirror transmissivity T

$$S(t) = T\,I(t). \tag{27.20}$$

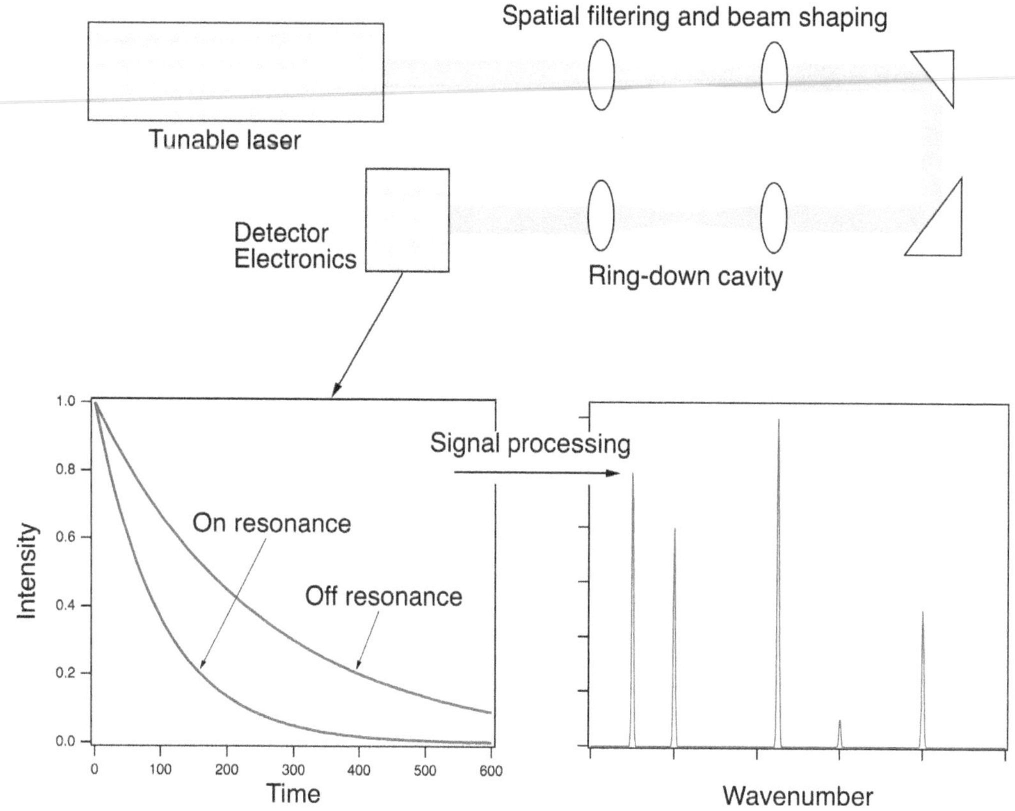

Figure 27.9 Schematic diagram for the experimental realization of a cavity ring-down spectroscopy (CRDS) experiment.

For a short input pulse, the intensity pattern of the cavity output consists of a train of pulses of decreasing intensity. If the input pulse length is longer than the cavity round-trip length, these pulses overlap and give a continuous decay waveform.

Consider the intensity decay of light in an evacuated cavity under the condition where no beats between cavity modes are observed. During a round-trip period t_r (i.e., the time it takes for light to complete a full round trip in the cavity), the intensity $I(t)$ decreases by the square of mirror's reflectivity R

$$I(t + t_r) = R^2 I(t) \tag{27.21}$$

If the output signal $S(t + nt_r)$ is monitored after n round trips starting from some time t, the signal measured has the form

$$S(t + nt_r) = R^{2n} S(t) = S(t)e^{2n \ln R}. \tag{27.22}$$

For R close to unity, $\ln R = -(1 - R)$ and Eq. (27.22) may be approximated by

$$S(t + nt_r) = S(t)e^{-2n(1-R)}. \tag{27.23}$$

We define

$$L_0 = 2(1 - R) \tag{27.24}$$

to be the round-trip loss coefficient for the empty cavity. Then,

$$S(t + nt_r) = S(t)e^{-nL_0}. \tag{27.25}$$

The exponential decay with cavity transit in Eq. (27.25) resembles the Beer–Lambert law. Note, however, that the decay depends on the number of passes and the reflectivity but not the physical length of the cavity. We can determine the value of L_0 and the mirror's reflectivity R by fitting the experimentally observed ring-down waveform to Eq. (27.25) and using for n the ratio of the observation time $[(t + nt_r) - t]$ to the round-trip period t_r.

When the cavity is filled with an absorbing medium, an extra loss is introduced on each round trip. There is no restriction on the phase of the absorbing medium (solid, liquid or gas). The sample need not even fill the cavity, as is the case if the cavity is constructed inside of a vacuum chamber and a molecular beam passes through it. As long as Beer's law behavior is valid, the loss in the presence of a sample is given by the sample absorbance $2\alpha l_s$ where α is the frequency-dependent absorption coefficient of the sample having a length l_s inside the cavity. Hence, the round-trip loss coefficient for the cavity with sample becomes

$$L = 2[(1 - R) + \alpha l_s] \tag{27.26}$$

An absorption spectrum is a plot of the absorption coefficient α as a function of wavelength (or frequency or wavenumber or energy as appropriate). The CRDS absorption spectrum is derived from the difference of the cavity loss coefficients

$$\alpha l_s = \tfrac{1}{2}(L - L_0). \tag{27.27}$$

Expressing these two loss coefficients by means of the ring-down time

$$\tau = \frac{t_r}{L} = \frac{t_r}{2[(1 - R) + \alpha l_s]} \tag{27.28}$$

and

$$\tau_0 = \frac{t_r}{L_0} = \frac{t_r}{2(1 - R)} \tag{27.29}$$

we obtain

$$\alpha l_s = \tfrac{1}{2} L_0(\Delta\tau/\tau) = (1 - R)(\Delta\tau/\tau), \tag{27.30}$$

with $\Delta\tau = \tau - \tau_0$. Consequently, the minimum absorbance measured with the CRDS apparatus is limited by the mirror reflectivity R and the minimum change in the time τ that can be detected. For mirror reflectivities as high as $R = 99.99\%$ and for the reported accuracy in the ring-down time determination of $(\Delta\tau/\tau)_{\text{min}} = (2 - 5) \times 10^{-3}$ the minimum measured absorbance $(\alpha l_s)_{\text{min}}$ is on the order of a few parts in 10^7. An absorption spectrum is obtained by tuning the laser to a particular wavelength at which no absorption is observed to obtain L_0 and τ_0. Then, the laser is tuned point-by-point across the whole of the absorption spectrum, and L and τ are recorded at each point. Plugging the data into Eq. (27.30) and plotting α versus λ yields the absorption spectrum.

27.5 Femtochemistry

27.5.1 Following molecular dynamics in real time

Manfred Eigen, Ronald G.W. Norrish and George Porter were awarded the Nobel Prize in Chemistry 1967 "...for their studies of extremely fast chemical reactions, effected by disturbing the equilibrium by means of very short pulses of energy." Their work was largely performed before the invention of the laser, and because of this it was limited to the millisecond to microsecond timescale. This was the beginning of fast chemistry – microchemistry if you will. The natural timescale for motion that leads to reaction is the period of a vibration. Molecular vibrational frequencies range roughly from that of I_2 to H_2, that is, wavenumbers from 215 cm^{-1} to 4400 cm^{-1}. This corresponds to periods of 1.5×10^{-13} s to 7.6×10^{-15} s – in other words, 150 fs down to 7.6 fs. Another way to look at this is that the time that it takes a molecule with a mean velocity of a room-temperature gas ($\approx$500 m s^{-1}) to move the length of a bond (say 1 Å) is 0.2 ps. Clearly, we need to produce laser pulses in the range of a few femtoseconds to a few picoseconds if we are to follow atomic and molecular motions directly. Of course, what we gain in temporal resolution we will lose in energy resolution. The minimum pulse width of a laser τ_p is inversely proportional to the coherent bandwidth ($\Delta\omega$ in angular frequency or $\Delta\lambda$ in wavelength) of a laser operating at wavelength λ according to

$$\tau_p > \frac{C}{\Delta\omega} \approx \frac{C}{2\pi}\frac{\lambda^2}{c\,\Delta\lambda}, \tag{27.31}$$

where $C = 4\ln 2$ for a Gaussian pulse and c is the speed of light. The minimum value of the pulsewidth is known as the *transform-limited pulsewidth*. To obtain such high temporal resolution demands a corresponding loss in energy resolution.

Observation of the transition state – direct observation of the making and breaking of bonds – is a 'Holy Grail' of physical chemistry.[25] But as lasers with pulse widths of a few hundred femtoseconds and less became available in the laboratory we obtained the ability to do just this. For his relentless pursuit of technological and experimental advances that made this possible, Ahmed Zewail was awarded the Noble Prize in Chemistry in 1999. Now, we will look in some

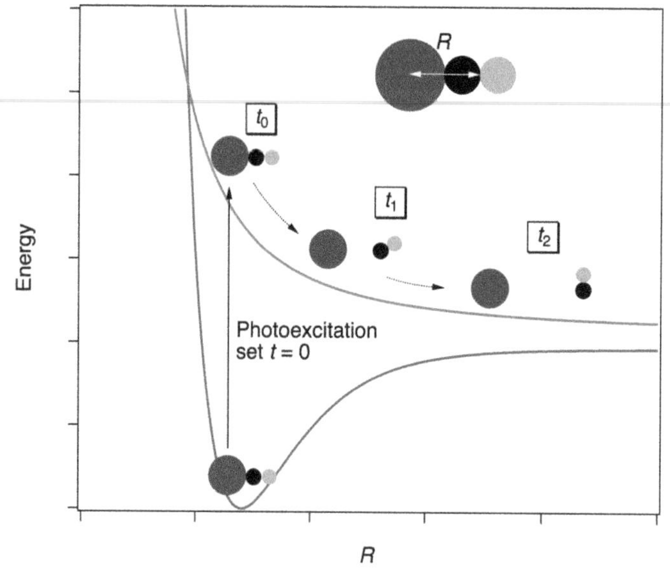

Figure 27.10 Photodissociation of ICN can be followed in real time by initiating breaking of the I–CN bond with 100 fs pulses of 388 nm light. The dynamics proceeds on a repulsive excited state. It takes about 200 fs for the bond to break, as can be observed by detecting LIF from the product CN. Adapted from Rosker, M.J., Dantus, M., and Zewail, A.H. (1988) *Science*, **241**, 1200.

detail at two results from his laboratory that not only demonstrate the power of *femtochemistry* (the study of chemistry on the timescale of chemical change) but also illustrate many of the concepts that we have used to understand reactivity and light–matter interactions.

Let us begin with a simple molecule, the triatomic ICN. This molecule is nice because we can perform spectroscopy on its fragments and because it has a very accessible simple reaction: the photochemical breaking of one bond (the I–CN bond) by direct excitation to a repulsive state, as illustrated in Fig. 27.10. Whenever we perform time-resolved experiments, we need a clock. Not only do we need a clock, we need a clock that we can set to zero and that has sufficient resolution. For the observation of the breaking of a bond, photochemistry is well suited because we know precisely when $t = 0$ is. The zero of time is set by the absorption of a photon that puts the molecule on the repulsive potential. Rosker, Dantus and Zewail[26] used a 60 fs pulse at $\lambda = 306$ nm to excite the ICN to a dissociative state.

Having set $t = 0$, we now need to read the clock. We read the clock with laser-induced fluorescence. The excitation of CN with 100 fs pulses of tunable light near 388 nm induces fluorescence that can be used to detect the presence of CN. CN is obviously only formed once as the ICN falls apart. In order to get the (femto)second hand of our clock to move, we must controllably introduce a delay between the pump and the probe pulses. This is done using a delay line similar to the one shown in Fig. 27.4(a). An actuator with submicrometer resolution must be used because a delay time Δt is related by the speed of light c to a change in optical path length Δl by

$$\Delta t = \Delta l / c, \tag{27.32}$$

which means that a 100 fs delay corresponds to an optical path length change of only 30 μm. What they found is that there is a delay between the excitation of ICN to the dissociative electronic state and the formation of free CN. The time it takes to break the I–CN bond is 205 ± 30 fs.

Not content with watching bonds break, Zewail's group went on to perform spectroscopy on the transition state while also watching the bond break.[27] The example discussed here is the pre-dissociation of NaI and NaBr. Pre-dissociation requires an avoided curve crossing of the type shown in Fig. 27.11(a). We discuss the dynamics of the excited state in terms of a wavepacket. Excitation was performed with a variety of wavelengths either in the range 275–364 nm, for which the pulse width was 150–200 fs, or around 388 nm with a pulse width of 100 fs. The probe wavelength was either at 598 nm with a pulse width of ~100 fs or at 620 nm with a pulse width of 60 fs. At 598 nm, the laser is resonant with atomic Na. At 620 nm, the laser excites a transition state species – that is, a Na atom that is perturbed by its interaction with the halogen.

Excitation places a wavepacket on the excited state potential curve. A wavepacket, as we have discussed before, is not a stationary state of the system. As shown in Fig. 27.11(a), it propagates on the excited state potential energy curve

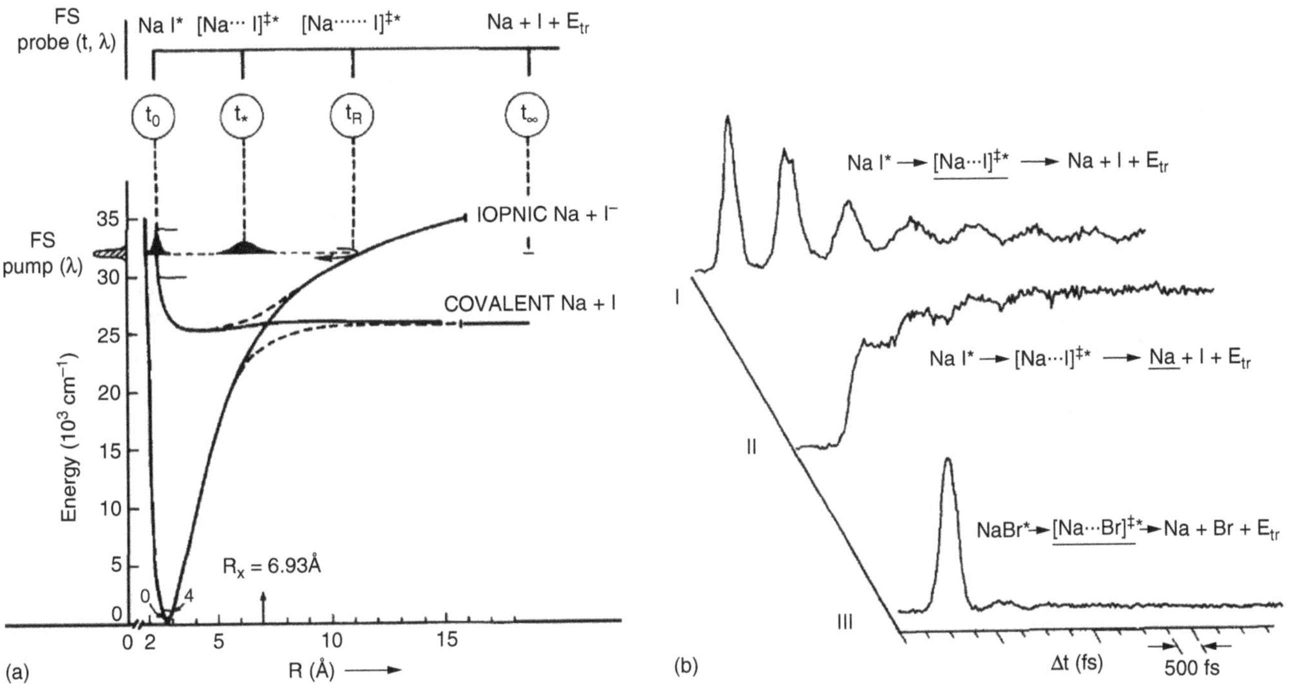

Figure 27.11 (a) Photoexcitation of NaI or NaBr leads to the possibility for pre-dissociation when the excited state species stretches to reach the avoided crossing of the ionic and covalent potential energy curves. (b) Detection of transition state species $[Na\cdots I]^{\ddagger *}$ or $[Na\cdots Br]^{\ddagger *}$ or the dissociation fragment Na allows the photodissociation dynamics to be followed in real time. Reproduced with permission from Rose, T.S., Rosker, M.J., and Zewail, A.H. (1988) *J. Chem. Phys.*, **88**, 6672. © 1988 AIP Publishing LLC.

beginning in the Franck–Condon region near the repulsive wall. Wavepacket propagation corresponds to vibration of the Na–X bond. As the Na–X bond expands, the wavepacket nears the avoided crossing of the ground state covalent and ionic potential energy curves. When a wavepacket encounters such a curve crossing, there is a certain probability that the wavepacket will hop from one potential energy curve to the other, and if that happens the Na–X bond will break. If hopping does not occur, the vibration will return the wavepacket to the repulsive wall and the cycle will start again. We predict that we should observe product Na atoms forming after a short delay, and then puffs of Na atoms will be formed periodically each time the molecule stretches in the excited state to reach the classical outer turning point.

The reaction involves photoexcitation to produce an excited state species, NaX*. This species stretches to form a transition state species $[Na\cdots X]^{\ddagger *}$ with a stretched bond indicated by the dots. As the bond stretches, the interaction between the Na and the halogen atom also changes. The bound species experiences bonding that is some combination of covalent and ionic interactions. The bonding interaction, as we have learned, perturbs the electron distribution around the Na atom. In the bound molecule it clearly is perturbed from the free atom. The dissociated species has a broken bond and the Na atom is unperturbed at this point with a spectrum that corresponds to the free atom. In between – near the transition state – the Na atom is perturbed by its interaction with the halogen, and if we performed spectroscopy on it we would expect the peak of its fluorescence signal to be shifted from the unperturbed atom.

Looking at the data in Fig. 27.11(b) we find that our expectations are met. The free Na signal recorded by measuring the LIF that is obtained by resonant excitation of Na atoms builds up puff by puff each time the bond expands. Diffusion of the Na out of the laser focal volume is slow on the timescale of the experiment, so the signal does not drop in between puffs. The transition state species $[Na\cdots X]^{\ddagger *}$ is formed by vibrational motion in the excited state. This vibrational period is such that it can be observed on the timescale of the experiment. LIF from this species is excited by 620 nm light (light that is off-resonance for free Na but on resonance for the transition state species of a particular bond length). A peak is noted in the transition state species LIF each time the vibration brings the internuclear distance to the correct value to make the species resonant with the incident light. The difference between the $[Na\cdots I]^{\ddagger *}$ and the $[Na\cdots Br]^{\ddagger *}$ signals is due to a subtle difference in the potential energy curves. The curve crossing is extremely efficient for NaBr. Almost all of the molecules pre-dissociate on the first attempt and only a few are left for the second transit. NaI, on the other hand, exhibits a curve-crossing probability of only about 0.1. This means that many more round-trips are possible and more bumps are observed in the spectrum.

27.5.2 Coherent control

Up to this point we have not made full use of the coherent nature of laser light. Coherence is the well-defined phase relationship between photons in a beam of laser light. Coherence is essential for the observation of various interference, diffraction, and multiphoton phenomena. In the context of femtochemistry, we can add another dimension to coherent phenomena: the control of wavepackets and their propagation. In our discussion of photochemistry and photodissociation, we have been primarily concerned with the energy of photons and, on occasion, their time-dependent behavior. We have used photons as a means to transfer energy into atomic and molecule systems to change how they emit light (fluorescence and phosphorescence), to change their charge state (photoionization), to break bonds (photodissociation), and to change how they react compared to a system in which only thermal energy is present (photochemistry). In Chapter 18, for example, we discussed how the mechanism of the reaction of HI + HI can be changed from a thermal reaction that occurs in the dark to a chain reaction that is induced by light with wavelength shorter than 400 nm. This is something of a brute force approach to photon stimulated dynamics. We dump energy into a system by photon absorption, and the energy is allowed to redistribute according to IVR if necessary, the lifetime of electronic states, and the timescale of collisional (or Förster) energy transfer dynamics. The experimenter initiates the chemistry but allows the system to evolve at its own pace and as is most probable for that given excitation.

Now let us consider more active experimental input in which not only the photon energy and the timing of pulses but also the phase of the electric field within the pulses are controlled. The theoretical basis of these ideas was framed nicely by David Tannor, Ronnie Kosloff and Stuart Rice.[28] It is built on the concepts described in Section 27.3.2, that we can not only use a laser field to pump population into an excited state but also use laser fields to dump excited state population into lower levels.

Consider a multiphoton process in which two or more photons resonantly excite a molecule to an excited electronic state. Excitation of the molecular system can be thought of in terms of placing a wavepacket onto the excited state potential energy surface. *Propagation* of the wavepacket *on the excited-state potential* energy surface can be used to *assist chemistry on the ground-state potential* energy surface. By controlling the delay between a pair of ultrashort laser pulses, it is possible to control the propagation time on the excited-state potential energy surface. Further control can be obtained by controlling the phase structure of the laser pulses. This is known as *pulse shaping*.[29] Different propagation times and pulse shaping can be used to push the system down different reaction paths. Such techniques have been used to change branching ratios in radiative decay, ionization, and chemical reactions.[30]

27.6 Beyond perturbation theory limit: High harmonic generation

Up until this point, all of the above light–matter interactions we have discussed have been in the linear or perturbative regimes. This means that the field strength of the electromagnetic field is small compared to the field in the atoms and molecules that it interacts with. This allows us to use an approach based on Eq. (27.8) to describe the phenomena that we have encountered. At the lowest laser intensities we are in the linear regime, and the Beer–Lambert law holds. The presence of the laser field does not change the characteristics of the molecule. But, increase the number of photons by a factor of two and (all other things being equal) the number of photons absorbed increases by a factor of two. For example, in photoemission one photon in produces one electron out (as long as the photon has an energy greater than the ionization potential/work function). These are the characteristics of the linear regime.

As we increase the laser intensity, the laser field begins to leave an imprint on the molecular system. The laser polarizes the molecule. The laser can create virtual states – molecule/field states that only exist while the laser interacts with the molecule. It causes the higher-order terms in the expansion found in Eq. (27.8) to become actively involved in describing the system; this is the nonlinear regime. As long as the process is not saturated, increasing the power by a factor of three leads to a factor of nine increase in the amount of excited-state population formed by two-photon absorption. The higher the power, the more terms are required to describe the response. These are the hallmarks of the perturbative regime.

If we keep increasing the laser power, eventually the laser field becomes appreciable (say 1% or more) in comparison to the field that is holding the electrons close to the nucleus. We leave the perturbative regime and *high-field physics* emerges. With new physics comes new phenomena. Here, we will discuss a few aspects of high-field physics including high harmonic generation and the generation of attosecond pulses.

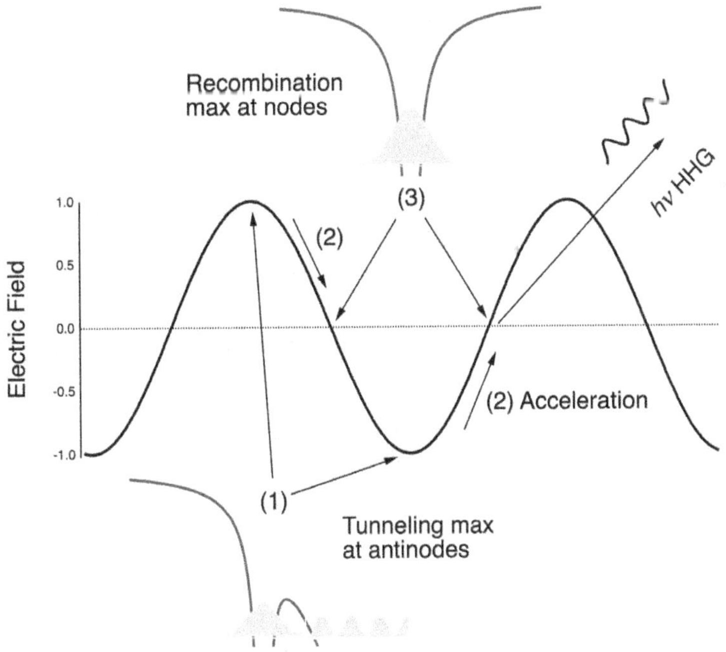

Figure 27.12 A schematic depiction of the three-step model that describes high harmonic generation (HHG). The process occurs twice each optical cycle. Step (1): Near the maximum in electric field strength the electron tunnels out of the atom. Step (2): The free electron is accelerated by the optical field. Step (3): Near the node in the optical field strength radiative recombination occurs emitting the harmonic photons. Adapted from Dahlström, J.M., L'Huillier, A., and Maquet, A. (2012) *J. Phys. B: At. Mol. Opt. Phys.*, **45**, 183001.

The process of *high harmonic generation* (HHG), which was discovered in the late 1980s,[31] is displayed in Fig. 27.12. A semiclassical model is most often used to describe HHG,[32] called the 'three-step' model. As the laser fluence approaches and exceeds $I_L \approx 10^{17}$ W m^{-2}, the electric field of the laser's electromagnetic wave approaches the strength of the Coulomb potential that holds the electrons in an atom or molecule. This regime is routinely accessible with commercially available Ti:sapphire lasers when they are focused. An electric field of this magnitude increasingly suppresses the potential barrier during the first part of the optical pulse (the portion of the wave with increasing amplitude).

The three steps of the model are: (1) tunneling; (2) acceleration; and (3) radiative recombination. Step (1) is that the electron tunnels out of the atom, leaving the atom when the laser field is near its peak value and the suppression of the potential barrier is maximal. The electrons that will contribute to HHG have little or no kinetic energy and are still in the vicinity of the nucleus after tunneling. In step (2), the electron, which is essentially in free space, accelerates in the laser potential. While this is happening, the laser field is dropping back to zero as the electric field amplitude approaches the node in the optical cycle. When the node is reached, the field is zero and the barrier has returned to its full height. In step (3) the electron recombines with the nucleus, but to do so it must lose the energy it acquired while being accelerated by the laser. The acceleration is quantized because energy can only be taken out of the optical field in units of the photon energy of the fundamental laser radiation. When the electron recombines with the ion core, the energy gained in this round-trip process is lost as radiation. The maximum in photon emission occurs near the node in the optical cycle. For a spherically symmetric medium such as an atom, symmetry constraints (i.e., parity conservation) demand that only odd harmonics are generated. Most of the electrons simply detach from the ion core, resulting in photoionization. However, under optimal conditions about 1 in 10^6 electrons will generate HHG photons. From this three-step model, it is predicted that photon emission should exhibit two maxima per optical cycle, one at each node.

The highest harmonic order N_{max} and photon energy $N_{max}\, h\nu_L$ that can be reached in a nonphase-matched geometry can be predicted by solving the Schrödinger equation from which the three-step model can be derived.[33] From this, we obtain the *cut-off law*

$$N_{max} h\nu_L = 1.3 I_p + 3.2 U_p \tag{27.33}$$

where $h\nu_L = 1.55$ eV is the photon energy of the laser fundamental if a Ti:sapphire laser operating at 800 nm is used, and I_p is the ionization energy of the gas. When performed in a waveguide to limit plasma-induced laser beam defocusing, this cut-off can be substantially extended. U_p is the *ponderomotive energy*, which denotes the energy that the free electron can acquire in the laser field after the tunneling process, and is given by

$$U_p = \frac{e^2 E_L^2}{16\pi^2 m_e \nu_L^2} = \frac{e^2 \lambda_L^2 I_L}{8\pi^2 \varepsilon_0 c^3 m_e}. \tag{27.34}$$

In Eq. (27.34) E_L is the electric field amplitude and ν_L is the frequency of the laser light.

Equations (27.33) and (27.34) show that the cut-off energy increases linearly with the laser intensity I_L, but quadratically with the laser wavelength λ_L. The number of photons in the harmonics scales with the polarizability of the gas used as the generation medium. Of the noble gases, Xe exhibits the highest conversion efficiency but the lowest energy cut-off because it has the highest polarizability and lowest ionization energy. Using He as the plasma gas leads to lower conversion efficiencies but a significantly higher energy cut-off. The flatness of the plateau depends on experimental parameters such as the gas pressure and the focusing geometry.

27.7 Attosecond physics

The range of a few hundred femtoseconds to a few picoseconds is the time scale of atomic motions related to reactivity. The atomic unit of time is the time it would take for an electron with the mean kinetic energy of an electron in a H 1s orbit to travel the length of the Bohr radius. This duration is $t_0 = 24$ as ('as' is the abbreviation for attosecond). Thus, the attosecond time scale is the natural time scale of electronic motions. Just as it took the introduction of picosecond and femtosecond pulses to probe atomic rearrangements directly, it has taken the introduction of attosecond pulses to probe electronic rearrangements directly.

High harmonic generation facilitates the production of photons that span from the vacuum ultraviolet to the X-ray region. This corresponds to making harmonics that range from just a few (say the 9th or 11th) to well over the 291st. The most common laser system for producing the fundamental radiation is a mode-locked Ti:sapphire laser delivering several mJ pulses into ~100 fs, or shorter pulses operating at a fundamental wavelength of ~800 nm. It is best not to focus too tightly to achieve the optimum fluence of $\sim 10^{18}$ W m^{-2} by using higher pulse energy and shorter pulses. Fluences much above 10^{19} W m^{-2} are counterproductive because ionization becomes too facile and coherence is not easily maintained over a wide range of photon energies in the highly ionized plasma. An extremely interesting phenomenon is that the intensity of the harmonics can be made to form a plateau over an extended range of harmonics.

Now consider this. The light generated in a plateau extending from, say, the 21st to the 101st harmonic is generated in a coherent fashion. The harmonics have specific phase relationships and their energy range (assuming a fundamental wavelength of 800 nm) is 124 eV. If we have a pulse with an energy bandwidth of 124 eV, then the energy-lifetime uncertainty relation $\Delta E \Delta t \geq \hbar/2$, tells us that we should be able to make a pulse with a width of less than 5×10^{-18} s = 5 as. This is a very rough estimate. Indeed, it does not take into account, for example, that the emission of high harmonics only occurs during the least intense parts of the optical cycle. A more accurate estimate of the coherent bandwidth is 15 eV, which still corresponds to a 100 as pulse width. Therefore, high harmonic generation provides the first controllable pathway to the generation of attosecond pulses and has ushered in a new era of attosecond physics.[34]

27.7.1 Directed practice

For a Ti:sapphire laser operating at 800 nm, calculate the energies in eV of the 21st and 101st harmonics and the corresponding transform-limited pulse width.

[Answers: 32.5 eV, 156.5 eV, 3.1 as]

The attosecond time scale brings us face to face with some philosophical questions, if not barriers. Pauli showed that time is not a directly observable quantity. There is not an operator for time. We encountered this before when we discussed the energy–time uncertainty relation. We got around any problem by stating that what we really meant was lifetime or the time for the system to make a significant change. We do not know absolute time, but we can measure a duration.

We spent a great deal of effort trying to convince ourselves that electrons are waves. They do not orbit in Bohr obits with a well-defined trajectory. However, if we were to measure the time it took for an electron to move the length of one Bohr orbit, we would know that it was there at one time and here at a later time. And that is okay. We just do not know what orbital it is in since we do not know its energy exactly.

What if we try to measure the amount of time it takes to photoionize an atom? We have always assumed that the absorption of light is instantaneous. On the time scale of nuclear motions, the assumption of not only instantaneous absorption but also instantaneous electron transition from one orbital to another led us to the Born–Oppenheimer approximation. Now, in a certain sense we have come full circle. At the beginning of our journey (or descent depending on your perspective and my ability to excite your imagination!!) into the quantum world we found that one of the primary objections to Bohr's atom and the foundations of 'old' quantum theory was the utter disgust voiced in opposition to the idea of quantum jumps. At the precipice of attosecond physics, we are now in the position to ask, 'How fast is instantaneous?'

We are not yet in a position to answer this unambiguously. However, experiments are beginning to address questions about the differences in ionization times between two different states in the same ion. Argon has a $3s^2\ 3p^6$ configuration and it takes longer to remove a 3s electron than it does to remove a 3p electron. A delay of 110 as was measured when the atoms were ionized with a train of attosecond pulses with a photon energy of ~35 eV.[35] A delay of approximate 20 as was measured between 2s and 2p electrons removed from Ne atoms by 100 eV photons in single attosecond pulses.[36] In both cases, the s electron of the same value of n is photoionized more slowly. This comports with a simple idea that the s electron is initially closer to the nucleus in a less diffuse orbital. Therefore, the Coulomb field from the nucleus is less well screened for the s electron than it is for the p electron. However, multi-electron interactions also play a role in determining the absolute time scale for ionization.

27.8 Photosynthesis

One of the most important light–matter interactions is photosynthesis. Photosynthesis is a photostimulated process that involves not only photon absorption, energy transfer between electronic states, and nonradiative relaxation, but also charge transfer. Charge transfer then promotes a string of chemical reactions that are essential for plant life. The net reaction of photosynthesis in plants is

$$6CO_2(g) + 6H_2O(l) \rightarrow C_6H_{12}O_6(s) + 6O_2(g) \tag{27.35}$$

This reaction is rather endergonic with $\Delta G° = 2870\ \text{kJ mol}^{-1}$ (29.8 eV), which is an energy deficit much greater than the energy of a single visible photon. Photosynthesis, therefore, cannot be a simple photochemical process like the ones we have encountered before in which the absorption of one photon triggers a chain of events that leads to a complete reaction cycle. We will follow the path of how a single photon initiates electron transfer. However, several iterations of the cycle and several photons are required to complete all of the steps required for glucose synthesis.

The photosynthetic chain begins when chlorophyll absorbs light. It is relatively easy to figure out what occurs on a 'long' timescale. Around 100 μs after photon absorption, an electron lands in a B-branch quinone. This electron is used to split water

$$2H_2O \rightarrow O_2 + 4\,H^+ + 4\,e^-. \tag{27.36}$$

Several electrons and H^+ ions are transported in a series of sequential reactions, resulting in the production of carbohydrates and O_2, which are used to drive ATP synthesis. Ultrafast dynamics is essential for the efficient use of absorbed photons. Now what we want to look at the first 100 ps or so after photon absorption in order to discover why photosynthetic systems are so efficient.

Because the reaction in Eq. (27.35) is so far uphill, thermal excitation will not drive it. Photoexcitation can be used to provide the 'missing' electrode potential that is required to push an endergonic redox reaction to products if an excited electron can be transferred to a reaction center before it loses the energy gained by photon absorption. However, microscopic reversibility also pertains to photoexcitation. A strong absorber is a strong emitter. So how do we efficiently turn photoexcited electrons into redox available electrons before they simply relax by fluorescence or internal conversion to the ground state?

To answer these questions and obtain some insight into the dynamics of complex systems, we turn to the photosynthesis system of green sulfur bacteria.[37] This is biochemistry, so we need lots of acronyms: RC = reaction center; PPC = pigment–protein complexes; LHC = light-harvesting complex; EET = excitation energy transfer from antenna to RC; FMO = Fenna–Matthews–Olson protein. The FMO protein was the first chlorophyll protein structure to be determined using atomic resolution by X-ray crystallography. Such accurate structural information is crucial in determining the dynamics of this photosynthetic system.

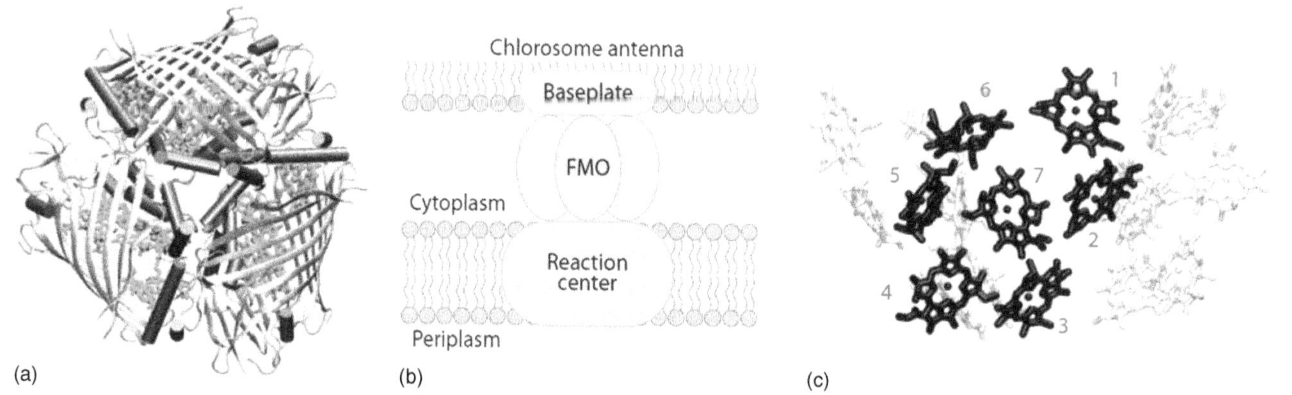

Figure 27.13 (a) Top-view of the Fenna–Matthews–Olson (FMO) protein trimer from green sulfur bacterium *Prosthecochloris aestuarii*. The protein is depicted in yellow, and the bacteriochlorophyll (BChl) molecules are in green. (b) The FMO protein is located between the light-harvesting antenna (chlorosome) and the reaction center, with the C_3 symmetry axis of the trimer perpendicular to the membrane plane of the baseplate. (c) Side view of the BChl arrangement in the FMO trimer. Seven BChl a molecules belonging to one of the monomeric subunits are highlighted in black. Reproduced with permission from Cheng, Y.C. and Fleming, G.R. (2009) *Annu. Rev. Phys. Chem.*, **60**, 241. © 2009, Annual Reviews.

Essential components of the photosynthetic system in green sulfur bacteria are shown in Fig. 27.13. Chlorosome antennas act as the LHC; these are composed of chlorophylls and carotenoids, as well as bilins in marine systems. Examples of each structure class are shown in Fig. 27.14. Chlorophylls absorb in the range 400–475 nm (3.10–2.61 eV). Accessory pigments such as carotenoids, phycoerythins (absorption maxima at 498 and 565 nm, 2.49 and 2.19 eV, respectively), and phycocyanins (absorb in the red/orange around 620 nm, 2.00 eV, and fluoresce at 650 nm) fill out the absorption spectrum across the visible. Seven bacteriochlorophylls are arranged around each FMO protein within ~1 nm (9–12 Å) of each other.

The chain of events that occur in this photosystem are shown schematically in Fig. 27.15. Photosynthesis begins with photon absorption. Based on the wavelength of the light absorbed, different electronic excited states in the different

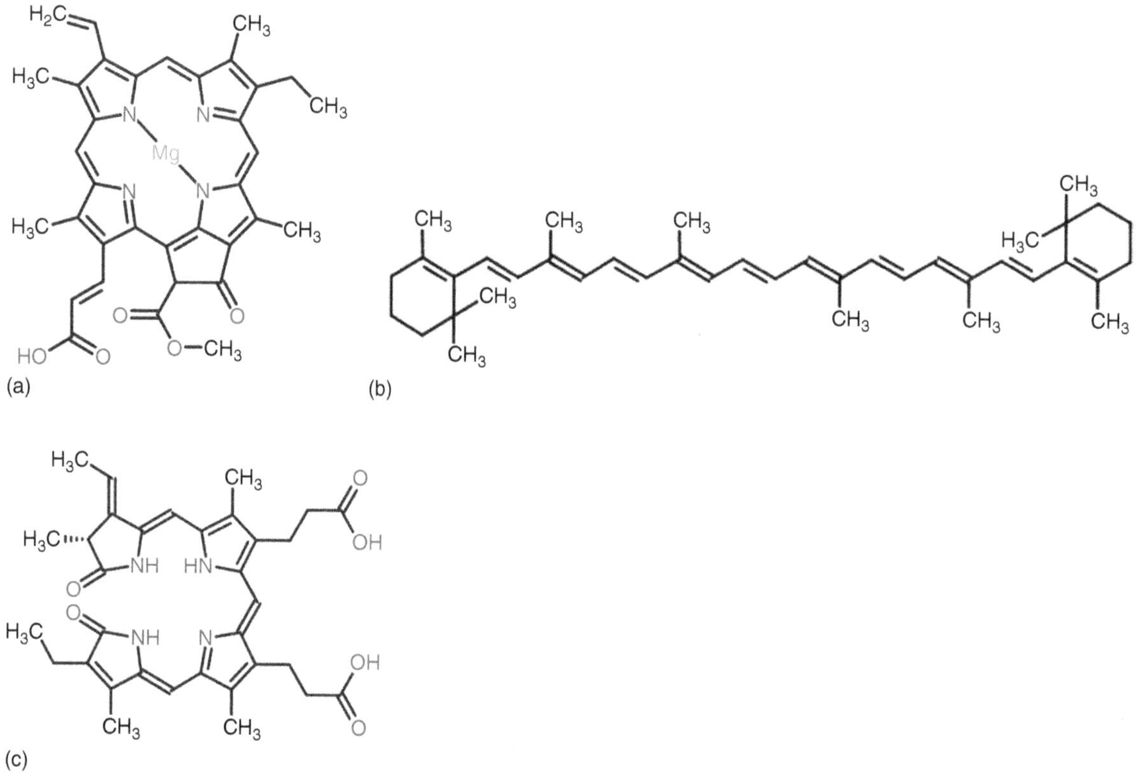

Figure 27.14 Structures of three typical molecules that make up light-harvesting complexes (LHC) in photosynthetic systems. (a) Chlorophyll c1. (b) β-carotene. (c) Phycocyanobilin. Structures taken from Chem Spider.

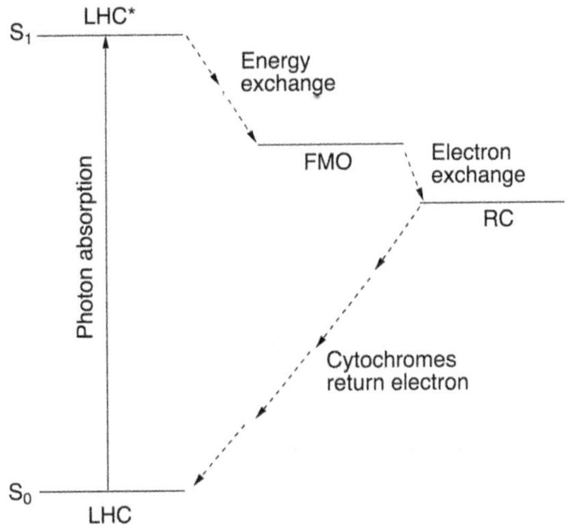

Figure 27.15 The photoexcitation within the light-harvesting complex (LHC), energy transfer to the Fenna–Matthews–Olson (FMO) protein, electron transfer to the reaction center (RC), and return pathways are represented schematically for the green sulfur bacteria photosynthesis system.

chromophores that make up the light-harvesting complex are excited. While we use a Jablonski diagram to describe excitation in Fig. 27.15, we should not confuse the chlorosome antenna with small molecules. The LHC is a highly complex, extended system of coupled molecular species. Indeed, it has been optimized by evolution to perform a rather extraordinary task with extreme efficiency. Light to charge conversion efficiency is about 95%. Instead of a localized picture of one electron excited by a Franck–Condon transition within an isolated molecule, the antenna should be thought of as something more like a solid-state system. The electronic excitation is in the form of an exciton, a collective electronic excitation between chlorophylls (or other chromophores).

Multiple ultrafast steps involving small energy exchange are essential to preclude radiative and nonradiative transfer back to the ground state. These steps make electron transfer essentially irreversible. The electron is passed along the chlorosome antenna until it reaches the FMO. These steps take 60–70 ps in total. Exchange from the FMO to the reaction center RC occurs in 30–40 ps. Thus, it takes around 100 ps for an electron from the chlorophyll to land in the RC. Within the reaction center the excited electron is exchanged to complete the photo-oxidation reaction. A series of cytochrome complexes then engage in incremental energy relaxation as they return the electron back to the ground state of the LHC. The incremental nature of energy loss is important. Cell function could be disrupted if too much heat were generated too quickly in any relaxation step.

Let us look at the excitonic state in more detail. If $|n\rangle$ is a local excitation, then the exciton wavefunction can be written as a superposition of these local excitations,

$$|\psi_\alpha\rangle = \sum_{n=1}^{N} \phi_n^\alpha |n\rangle. \tag{27.37}$$

Therefore, the transition dipole is also delocalized over the full range of local excitations,

$$\vec{\mu}_\alpha = \sum_{n=1}^{N} \phi_n^\alpha \vec{\mu}_n. \tag{27.38}$$

Excitons couple to phonons, and this coupling aids the directed redistribution of energy. Redistribution among the coupled modes delocalizes the initial photo-excitation in less than 100 fs. As a collective excitation, the wavefunction of the excited state is delocalized over multiple molecular entities within the antenna. This allows the excitation to sample a number of energy transfer pathways simultaneously and pick out the most favorable. This is essential because the system needs to get the excited electron away from the hole that it left behind to avoid radiative recombination (fluorescence). The system also needs to avoid internal conversion back to the ground state, which would mean that the photon energy is simply lost as heat.

The coupling between different regions of the antenna differs. This leads to a range of different relaxation times, which may help in making energy transfer unidirectional toward the exit site. Ultimately, the system must transfer an excited electron to the FMO and then the RC. Thus, after using delocalization to avoid radiative recombination and internal conversion, somehow the system must relocalize the excitation and deposit an electron in the FMO.

27.9 Color and vision

On comparing Fig. A5.1 (in Appendix 5) and Fig. 19.2, we see that we give the visible region an outsized emphasis if we compare its small slice of wavelength space to the complete electromagnetic spectrum. Because we can see it, it simply is more interesting and makes a more immediate impact on our perception. It also is an extremely important region of the electromagnetic spectrum because solar radiation peaks in the middle of the visible region. The eye is also a wonderful spectrometer that can be used to analyze materials. White and black are not really colors. White is the presence of all colors (or at least enough to saturate our optical detectors) and black is the absence of color. When we see a white powder, we immediately know that we are looking at an insulator (or at least a semiconductor with a band gap wider than the blue end of the visible spectrum). When we see a black powder, we immediately know that we are looking at a metal or other materials without a band gap (or at least a direct band gap much smaller than the red end of the visible spectrum). A white object must backscatter and not absorb light. A black object must forward-scatter and absorb light.

There are 15 distinct sources of color production in matter. Several derive from geometrical and physical optics: diffraction (gratings); dispersive refraction (rainbows); interference (films); and scattering (sky). Colors generated from regular structures that arise from diffraction, interference and scattering are known as *structural colors*. Many natural organisms have evolved to optimize their visual appearance using layers of modulated refractive index. They can have not only very vivid colors but also, if they are able to change the optical path length (the product of refractive index and length) between structural elements, they can change their color in response to external stimuli. *Opalescence* is produced by scattering from structures with a size comparable to the wavelength of light. This is known as *Tyndall scattering*, and is wavelength-dependent; consequently, opalescent objects can appear to have different colors in reflection and transmission. Scattering from ordered structures, as well as interference and diffraction, are also angle of incidence-dependent, which can lead to changes in color when an object is viewed from different perspectives. A photonic glass contains scatterers with local order (a narrow distribution of near-neighbor distances) but not long-range order. This leads to structural color that is independent of the angle of observation. Blue and green bird feathers often exhibit this phenomenon.

There are chemical means of color production, including incandescence (flames), chemiluminescence (luminol reaction), and gas excitations (vapor lamps). Vibrational overtone excitation (e.g., iodine and the blue tint in glacial ice) can lead to visible colors in a few examples. However, much more commonly electronic excitations are responsible for color generation. This is certainly true of organic compounds (e.g., dyes, chlorophyll). Charge transfer (redox) reactions change the oxidation state of ions (e.g., Prussian Blue). Electronic transitions can also be tied to ligand field effects in transition metal complexes (e.g., chrome green) and transition metal impurities in crystals (e.g., Cr in ruby, Ti and Fe in blue sapphire, see Fig. 27.16).

Transitions between energy bands lead to four distinct processes: color in metals (e.g., copper and gold); in pure semiconductors (galena, PbS in an isometric hexoctahedral phase or iron pyrites, FeS_2 also known as 'fool's gold' shown in Fig. 27.16(d)); in doped semiconductors (light-emitting diodes; LEDs); and in color centers (amethyst, Fig. 27.16(a)). Bulk metals have contiguous conduction and valence bands. The large number of electrons in these bands acts much like a free electron gas. These electrons interact strongly with visible light, the penetration depth of which is generally only a few tens of nanometers. The energy of the photon is absorbed in the free electron gas. Most of the energy is reradiated (reflected) along the specular direction, leading to the familiar mirror finish of polished metals. Since most metals reflect light across the visible spectrum evenly, they appear silvery (white). In a few cases, the reflectivity spectrum is not flat across the visible. This occurs when light is absorbed by the creation of a collective oscillation of electrons known as a *plasmon*. Plasmons lie at the heart of a number of nanoscale phenomena and are an area of active research.[38] Copper and gold have anomalous reflectivity spectra compared to other metals with an enhancement in the red and reduction in the blue, which leads to their color. The absorption spectrum of plasmons is sensitive to the size of metal nanoparticles. This is responsible for a shift in the color of, for instance, gold nanoparticles (i.e., colloidal gold). Unknowingly, medieval craftsman stumbled upon this effect and used it to create stained glass windows of various colors.

27.9.1 Transition metal impurities and charge transfer

Insulators and semiconductors have a gap between their valence band (normally occupied states) and conduction band (normally unoccupied states); this is known as the *band gap*. The band gap is large for insulators and small or moderate for semiconductors. Absorption occurs when a photon excites an electron across the gap. The only way for a photon to

Figure 27.16 Images of (a) amethyst, (b) sapphire, (c) ruby, and (d) iron pyrites. The color of amethyst is related to color centers, while that of sapphire and ruby originates in the levels of transition metal impurities (Ti and Fe in sapphire, Cr in ruby). The band gap of the semiconductor iron pyrites is just right to make it look golden and lustrous; hence, the name 'fool's gold.' Jewels provided for photography by Ivan Kaplan and Teresa Martin. Image of iron pyrites provided by Astro Gallery of Gems. Reproduced with permission.

be absorbed at an energy smaller than the band gap energy is if a (chemical or structural) defect creates an electronic state in the gap.

Most oxides are either insulators or semiconductors at room temperature. Impurities can introduce *gap states* – states associated with the impurity that are located in the band gap. For instance, addition of a small amount of substitutional Cr^{3+} into crystalline Al_2O_3 (corundum) leads to the formation of ruby. The deep red color results from the gap states introduced by Cr^{3+} ions, the d levels of which are split and broadened by ligand field interactions with the rest of the lattice.

If the Al^{3+} is substituted with a few hundredths of a percent of Fe^{2+} and Ti^{4+}, blue sapphire results. The blue color, corresponding to a photon energy of ~2.2 eV results from the charge transfer transition

$$Fe^{2+} + Ti^{4+} \rightarrow Fe^{3+} + Ti^{3+}.$$

27.9.2 Color centers

Defects in alkali halide crystals lead to coloration: LiF turns pink; NaCl, yellow; and KCl, blue-violet. The most common type of defect is the F center (*Farbenzentrum*). An F center is made by removing an anion from the lattice, which would leave the crystal positively charged. Instead, the anion is replaced by an electron to maintain neutrality. The trapped electron has associated with it a set of electronic states much like those of a H atom. The energy of the transition from the ground state to the first excited state is

$$E_a = \frac{0.257\ \text{eV}}{d} \tag{27.39}$$

where d is the anion–cation distance.

Other color centers include: M centers (two adjacent F centers); F′ centers (an F center with two electrons); V_K centers (two adjacent F centers with only one electron); and an electron color center (an electron that is trapped interstitially). The formation of color centers is often associated with radiation damage caused by high-energy irradiation.

27.9.3 Color perception

How does the (normal) eye function as a spectrometer? There are two morphologically distinct photoreceptor cells in the human retina – the *rods* and *cones*. The photoreceptor cells express 30–60 kDa receptor proteins that are known as *opsins*. Cones are further differentiated based upon the type of opsins that they express. S cones respond to short-wavelength light because they express blue opsin that has an absorption peak in the range 420–440 nm. M cones express green opsin and respond to medium wavelengths peaking in the range of 530–540 nm. Long-wavelength response is provided by the L cones, which express red opsin and have maximum absorption in the range 560–580 nm. The absorption spectra of these opsins are broad and overlap; hence, one wavelength of light may be able to excite more than one receptor. Color perception is derived from which receptors are stimulated and how strongly they are stimulated. Exciting all three receptors gives the impression of white light. However, extremely strong green light can also look effectively white, caused by saturation of the M cone response and excitation bleeding over into the tails of its neighboring cones.

Cones are better at handling high-brightness conditions, while rods are more sensitive. Appropriately, rods are capable of single photon detection but saturate at low levels of excitation, while cones have poor response at low photon intensity but saturate at much higher intensities. Vision is initiated by the absorption of light in photoreceptors. The 11-*cis*-retinylidene chromophore is extremely photosensitive. The long conjugated chain provides a highly polarizable π system of electrons, which you will recall from our discussion of the particle in a box problem and R6G, has an appropriate length to absorb in the visible part of the spectrum. The high polarizability also means that its absorption spectrum is extremely sensitive to the chemical environment around it. Hence, the manner in which it nestles into the protein network of the opsins shifts its optical behavior and allows us to perceive color. The photoisomerization of 11-*cis*-retinal to an all-*trans* form (see Fig. 27.17) is the first step in the sensing of light by the eye.

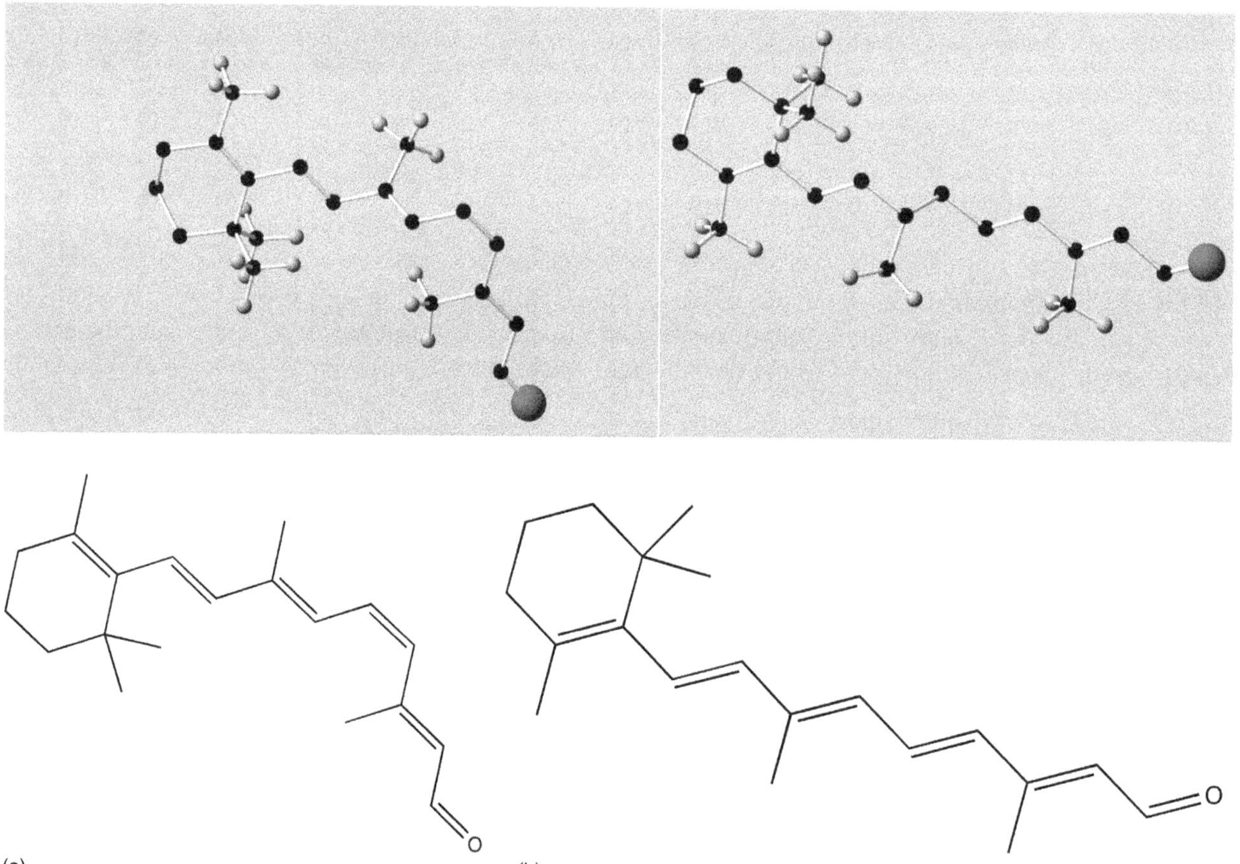

(a) (b)

Figure 27.17 (a) 11-*cis*-retinal. (b) 11-*trans*-retinal.

SUMMARY OF IMPORTANT EQUATIONS

$R_{\text{stim em}} = (N_k - N_i)B_{ki}\rho(\omega_{ki})$	Net rate of stimulated emission depends on the population difference, the Einstein B coefficient and the optical power density at the frequency of the transition
$d = n\lambda/2$	A half-integer number of optical cycles must fit into an optical cavity if round-trip constructive interference is to occur.
$(\Delta_{\min})_{x,y} = \dfrac{\lambda}{2n\sin\alpha}$	Abbe (diffraction) limit of lateral resolution.
$(\Delta_{\min})_z = \dfrac{\lambda}{2n(\sin\alpha)^2}$	Abbe limit of axial resolution.
$I/I_0 = \exp(-\sigma l_s \rho)$	Transmitted intensity decreases exponential with sample path length l_s, absorption cross section σ, density ρ.
$\tau_p \approx \dfrac{4\ln 2}{2\pi}\dfrac{\lambda^2}{c\Delta\lambda}$	Transform-limited pulse width τ_p of a laser with wavelength bandwidth $\Delta\lambda$.

Exercises

27.1 A Ti:sapphire laser operates at a wavelength of 800 nm with a pulse width of 120 fs and a pulse energy of 1.00 mJ. (a) Calculate the frequency in Hz as well as the energy per photon in eV and J and the energy of a mole of photons in kJ mol^{-1}. (b) What is the power associated with one pulse? (c) How many photons are in one pulse?

27.2 (a) What is the photon energy in eV and for each of the following types of radiation? (b) For (i) and (iv) calculate the ratio of Einstein A to Einstein B coefficient. What is the implication for the rate of spontaneous emission? (c) Discuss briefly the effect of each type of radiation on a hockey puck. (i) Microwaves at 2450 MHz. (ii) Infrared light at 2100 cm^{-1}. (iii) Ultraviolet light at 192.6 nm. (iv) X-rays at 2.30 nm.

27.3 The Nd^{3+}:YAG laser operates as a four-level system. The lasing transition occurs from the $^4F_{3/2}$ level at 11 502 cm^{-1} to a $^4I_{11/2}$ level at 2146 cm^{-1}. The wall plug efficiency (laser power out compared to electrical power in) is typically 1% for a flashlamp pumped system. Calculate the wavelength of the fundamental transition of the laser. Draw an energy level diagram and explain why the efficiency is so low and what the requirements are for each of the levels in the laser system with reference to kinetics.

27.4 A Nd:YAG laser operating at 532 nm is used to pump the Ti-sapphire laser that emits at 800 nm. The lasing transition is between a ^2E upper state and a $^2T_2$ lower state. (a) Express the pump and lasing photon energies in cm^{-1} and calculate what the maximum possible efficiency of the laser is if every pump photon leads to a lasing photon. (b) Where does the excess energy go? (c) Take the pulsewidth of 20 fs as the lifetime of the upper state, and calculate the linewidth of the laser $\delta\tilde{\nu}$ in cm^{-1}. How much broader is this than the linewidth of a stabilized version of the CW laser, which has a linewidth in frequency units of $\delta\nu = 15$ kHZ? Compare by calculating the ratio of $\delta\tilde{\nu}/\tilde{\nu}$ to $\delta\nu/\nu$. What type of spectroscopy is each laser most suited for?

27.5 When observing low-pressure gas-phase fluorescence from the CN A $^2\Pi$ ($v' = 7$) vibrational level to the ground X $^2\Sigma^+$ state, perturbations are found as reflected in the spectral linewidth $\delta\nu$ and fluorescence quantum yield ϕ_f. For rotational levels with $20 \le J \le 50$, there is a broadening of the lines and a decrease in ϕ_f. The rotational and vibrational constants of the two states lead to an accidental degeneracy of rovibrational levels between A $^2\Pi$ ($v' = 7$) and X $^2\Sigma^+$ ($v'' = 11$) for $20 \le J \le 50$. All other rovibrational levels in ($v' = 7$) exhibit narrow lines with near unit fluorescence quantum yield. Explain the cause of the perturbation, the nature of the nonradiative process, and the reasons for the broadening and decrease in fluorescence quantum yields.

27.6 Light at 306 nm excites ICN into an excited repulsive state ICN*. This time scale of photodissociation is measured in a pump-probe experiment in which laser-induced fluorescence (LIF) is used either to detect the CN fragment or the ICN* excited state. [Zewail, A.H. (2000) *J. Phys. Chem. A.*, **104**, 5660] (a) Explain why the ICN* curves exhibit a peak with a width of ~200 fs whereas the CN curve simply rises on the time scale of the experiment (900 fs). (b) How far does a mirror have to be moved in order to introduce a delay of 15 fs?

27.7 The radiative lifetime in nanoseconds of a classical electron oscillator in air is approximately

$$\tau_{\text{rad}} = 45\lambda_{ki}^2$$

where λ_{ki} is the transition wavelength in μm. The radiative decay rate is the inverse of the radiative lifetime and is exactly equal to the Einstein A coefficient for the transition. The HeNe laser has a transition at 632.8 nm. Calculate the minimum resonant energy density $\rho(\omega_{ki})$ required to make the system lase. Assume that lasing occurs if the rate of stimulated emission exceeds the rate of spontaneous emission.

27.8 For the Nd^{3+}:YAG laser (λ_{ki} = 1064 nm) and the Ar fluoride excimer laser (λ_{ki} = 193 nm), will the pump rate have to be higher or lower than that of the HeNe laser (λ_{ki} = 632.8 nm)? Explain your answer in terms of the properties of the states involved. All three are four-level laser systems.

27.9 Explain why it is much easier to make a laser that operates in the infrared than in the X-ray region of the electromagnetic spectrum.

27.10 Why can a two-level system never be made to lase?

27.11 Describe what is meant by a virtual state and a real state of an atom or molecule.

27.12 In a REMPI experiment, gas-phase H_2O molecules are ionized through a (2 + 1) scheme. As the laser field is intensified, the lifetime of the excited electronic state of the neutral molecule changes from 10 ns to 1 ns due to ionization. Describe quantitatively what happens to the width of a line in the spectrum of this state as the lifetime changes. Describe the trade-off between detection sensitivity and resolution that results.

27.13 Light to charge transfer efficiency in photosynthesis is 0.95 efficient. Assume that radiative decay via fluorescence is responsible for the loss, and that the natural lifetime of fluorescence is that of a typical allowed transition, say, 10 ns. Calculate the fluorescence quantum yield, the rate constant for fluorescence, the observed fluorescence lifetime, and the rate constant for the nonradiative pathway that leads to charge transfer.

27.14 (a) Compare the inverse of the actual time it takes to move an electron from LHC* to FMO in green sulfur bacteria to the value of k_{nr} you derived in exercise 27.13. On the basis of this comparison make an argument as to whether the 5% loss of efficiency in photosynthesis is due to radiative recombination or other losses along the charge transfer chain. (b) What would the observed fluorescence lifetime be if the inverse of the LHC* to FMO time scale is used as k_{nr}?

27.15 Why do all photosynthetic systems exhibit irreversible charge separation significantly faster than 1 ns?

27.16 The shimmer of butterfly wings is made by a combination of pigments and schemochromes. Schemochromes are composed of structures with critical dimensions that are close in size to the wavelength of light. Discuss the different roles of pigments and schemochromes in producing the color of butterfly wings.

Endnotes

1. Schawlow, A.L. and Townes, C.H. (1958) *Phys. Rev.*, **112**, 1940.
2. Bloembergen, N. (1999) *Rev. Mod. Phys.*, **71**, S283.
3. Javan, A., Bennett, W.R., and Herriott, D.R. (1961) *Phys. Rev. Lett.*, **63**, 106.
4. Tango, W.J., Link, J.K., and Zare, R.N. (1968) *J. Chem. Phys.*, **49**, 4264.
5. Zhu, X.D., Suhr, H., and Shen, Y.R. (1987) *Phys. Rev. B*, **35**, 3047.
6. Hence proving the adage that spectroscopy if a trade-off between resolution, signal-to-noise ratio and patience.
7. Tango, W.J., Link, J.K., and Zare, R.N. (1968) *J. Chem. Phys.*, **49**, 4264.
8. Greene, C.H. and Zare, R.N. (1982) Photofragment alignment and orientation. *Annu. Rev. Phys. Chem.*, **33**, 119.
9. Luntz, A.C., Kleyn, A.W., and Auerbach, D.J. (1982) Observation of rotational polarization produced in molecule-surface collisions. *Phys. Rev. B*, **25**, 4273.
10. Sitz, G.O. (2002) *Rep. Prog. Phys.*, **65**, 1165.
11. Golibrzuch, K., Bartels, N., Auerbach, D.J., and Wodtke, A.M. (2015) *Annu. Rev. Phys. Chem.*, **66**, 399.
12. Chen, H., Tagliamonti, V., and Gibson, G.N. (2011) *Phys. Rev. Lett.*, **109**, 193002.
13. Born, M. and Wolf, E. (1999) *Principles of Optics: Electromagnetic Theory of Propagation, Interference and Diffraction of Light*, 7th edition. Cambridge University Press, Cambridge.

14. Moerner, W.E. and Kador, L. (1989) *Phys. Rev. Lett.*, **62**, 2535.
15. Betzig, E. and Chichester, R.J. (1993) *Science*, **262**, 1422.
16. Nie, S., Chiu, D.T., and Zare, R.N. (1994) *Science*, **266**, 1018.
17. Hell, S.W. and Wichman, J. (1994) *Opt. Lett.*, **19**, 780; Hell, S.W. and Kroug, M. (1995) *Appl. Phys. A*, **60**, 495
18. Dickson, R.M., Cubitt, A.B., Tsien, R.Y., and Moerner, W.E. (1997) *Nature (London)*, **388**, 355.
19. Betzig, E., Patterson, G.H., Sougrat, R., Lindwasser, O.W., Olenych, S., Bonifacino, J.S., Davidson, M.W., Lippincott-Schwartz, J., and Hess, H.F. (2006) *Science*, **313**, 1642.
20. Ashfold, M.N.R. and Howe, J.D. (1994) *Annu. Rev. Phys. Chem.*, **45**, 57.
21. Feldman, D.L., Lengel, R.K., and Zare, R.N. (1977) *Chem. Phys. Lett.*, **52**, 413.
22. Ashfold, M.N., Nahler, N.H., Orr-Ewing, A.J., Vieuxmaire, O.P., Toomes, R.L., Kitsopoulos, T.N., Garcia, I.A., Chestakov, D.A., Wu, S.M., and Parker, D.H. (2006) *Phys. Chem. Chem. Phys.*, **8**, 26.
23. O'Keefe, A. and Deacon, D.A.G. (1988) *Rev. Sci. Instrum.*, **59**, 2544.
24. Zalicki, P. and Zare, R.N. (1995) *J. Chem. Phys.*, **102**, 2708.
25. Polanyi, J.C. and Zewail, A.H. (1995) Direct Observation of the Transition State. *Acc. Chem. Res.*, **28**, 119.
26. Rosker, M.J., Dantus, M., and Zewail, A.H. (1988) *Science*, **241**, 1200.
27. Rose, T.S., Rosker, M.J., and Zewail, A.H. (1988) *J. Chem. Phys.*, **88**, 6672; Rose, T.S., Rosker, M.J., and Zewail, A.H. (1989) *J. Chem. Phys.*, **91**, 7415.
28. Kosloff, R., Rice, S.A., Gaspard, P., Tersigni, S., and Tannor, D.J. (1989) *Chem. Phys.*, **139**, 201; Tannor, D.J., Kosloff, R., and Rice, S.A. (1986) *J. Chem. Phys.*, **85**, 5805.
29. Warren, W.S. (1988) *Science*, **242**, 878.
30. Brinks, D., Hildner, R., van Dijk, E.M., Stefani, F.D., Nieder, J.B., Hernando, J., and van Hulst, N.F. (2014) *Chem. Soc. Rev.*, **43**, 2476; Ohmori, K. (2009) *Annu. Rev. Phys. Chem.*, **60**, 487; Shapiro, M. and Brumer, P. (2003) *Rep. Prog. Phys.*, **66**, 859; Assion, A., Baumert, T., Bergt, M., Brixner, T., Kiefer, B., Seyfried, V., Strehle, M., and Gerber, G. (1998) *Science*, **282**, 919.
31. Ferray, M., L'Huillier, A., Li, X.F., Lompre, L.A., Mainfray, G., and Manus, C. (1988) *J. Phys. B: At., Mol. Opt. Phys.*, **21**, L31; McPherson, A., Gibson, G., Jara, H., Johann, U., Luk, T.S., McIntyre, I.A., Boyer, K., and Rhodes, C.K. (1987) *J. Opt. Soc. Am. B: Opt. Phys.*, **4**, 595.
32. Corkum, P.B. (1993) *Phys. Rev. Lett.*, **71**, 1994.
33. Lewenstein, M., Balcou, P., Ivanov, M.Y., L'Huillier, A., and Corkum, P.B. (1994) *Phys. Rev. A*, **49**, 2117.
34. Krausz, F. and Ivanov, M. (2009) *Rev. Mod. Phys.*, **81**, 163; Kling, M.F. and Vrakking, M.J.J. (2008) *Annu. Rev. Phys. Chem.*, **59**, 463; Sansone, G., Benedetti, E., Calegari, F., Vozzi, C., Avaldi, L., Flammini, R., Poletto, L., Villoresi, P., Altucci, C., Velotta, R., Stagira, S., De Silvestri, S., and Nisoli, M. (2006) *Science*, **314**, 443; Baker, S., Robinson, J.S., Haworth, C.A., Teng, H., Smith, R.A., Chirila, C.C., Lein, M., Tisch, J.W.G., and Marangos, J.P. (2006) *Science*, **312**, 424; Kapteyn, H.C., Murnane, M.M., and Christov, I.P. (2005) *Phys. Today*, **58**, 39; Antoine, P., L'Huillier, A., and Lewenstein, M. (1996) *Phys. Rev. Lett.*, **77**, 1234; Corkum, P.B., Burnett, N.H., and Ivanov, M.Y. (1994) *Opt. Lett.*, **19**, 1870.
35. Klünder, K., Dahlström, J.M., Gisselbrecht, M., Fordell, T., Swoboda, M., Guénot, D., Johnsson, P., Caillat, J., Mauritsson, J., Maquet, A., Taïeb, R., and L'Huillier, A. (2011) *Phys. Rev. Lett.*, **106**, 143002.
36. Schultze, M., Fiess, M., Karpowicz, N., Gagnon, J., Korbman, M., Hofstetter, M., Neppl, S., Cavalieri, A.L., Komninos, Y., Mercouris, T., Nicolaides, C.A., Pazourek, R., Nagele, S., Feist, J., Burgdorfer, J., Azzeer, A.M., Ernstorfer, R., Kienberger, R., Kleineberg, U., Goulielmakis, E., Krausz, F., and Yakovlev, V.S. (2010) *Science*, **328**, 1658.
37. Schlau-Cohen, G.S., Ishizaki, A., and Fleming, G.R. (2011) *Chem. Phys.*, **386**, 1; Cheng, Y.C. and Fleming, G.R. (2009) *Annu. Rev. Phys. Chem.*, **60**, 241.
38. Wang, Y., Plummer, E.W., and Kempa, K. (2011) *Adv. Phys.*, **60**, 799.

Further reading

Altkorn, R. and Zare, R.N. (1984) Effects of saturation on laser-induced fluorescence measurements of population and polarization. *Annu. Rev. Phys. Chem.*, **35**, 265.

Berezin, M.Y. and Achilefum S. (2010) Fluorescence Lifetime Measurements and Biological Imaging. *Chem. Rev.*, **110**, 2641.

Dahlström, J.M., L'Huillier, A., and Maquet, A. (2012) Introduction to attosecond delays in photoionization. *J. Phys. B: At. Mol. Opt. Phys.*, **45**, 183001.

Demtröder, W. (2014) *Laser Spectroscopy 1: Basic Principles*, 5th edition. Springer-Verlag, Berlin.

Demtröder, W. (2015) *Laser Spectroscopy 2: Experimental Techniques*, 5th edition. Springer-Verlag, Berlin.

Hollas, J.M. (2004) *Modern Spectroscopy*, 4th edition. John Wiley & Sons, New York.

Houston, P.L. (2001) *Chemical Kinetics and Reaction Dynamics*. Dover Publications, Mineola, NY.

Kling, M.F. and Vrakking, M.J.J. (2008) Attosecond electron dynamics. *Annu. Rev. Phys. Chem.*, **59**, 463.

Krausz, F. and Ivanov, M. (2009) Attosecond physics. *Rev. Mod. Phys.*, **81**, 163.

Lakowicz, J.R. (2006) *Principles of Fluorescence Spectroscopy*, 3rd edition. Springer, New York.

Schinke, R. (1993) *Photodissociation Dynamics*. Cambridge University Press, Cambridge.

Siegman, A.E. (1986) *Lasers*. University Science Books, Mill Valley, CA.

Young, M. (2000) *Optics and Lasers: Including Fibers and Optical Waveguides*, 5th edition. Springer-Verlag, Berlin.

Zewail, A.H. (1992) *The Chemical Bond – Structure and Dynamics*. Academic Press, London.

APPENDIX 1

Basic calculus and trigonometry

Characteristics of derivatives

1 The derivative d*y*/d*x* has units given by the (*y*-axis units) / (*x*-axis units).

2 The reciprocal relationships

$$\frac{dy}{dx} = \left(\frac{dx}{dy}\right)^{-1} = \frac{1}{dx/dy} \qquad \left(\frac{dy}{dx}\right)_z \left(\frac{dx}{dz}\right)_y \left(\frac{dz}{dy}\right)_x = -1.$$

3 If u is a function of y and y is a function of x, then

$$\frac{d}{dx}(u) = \frac{du}{dx} = \frac{d}{dy}(u)\frac{d}{dx}(y) = \frac{du}{dy}\frac{dy}{dx}.$$

4 Further differentiation is possible. The second derivative with respect to the same variable is

$$\frac{d}{dx}\left(\frac{dy}{dx}\right) = \frac{d^2y}{dx^2}.$$

The second derivative may be taken with respect to a different variable:

$$\frac{d}{dx}\left(\frac{du}{dx}\right) = \frac{d^2u}{dy\,dx}$$

5 The derivative of a product of functions is

$$\frac{d}{dx}(uv) = u\frac{dv}{dx} + v\frac{du}{dx}.$$

6 Frequently encountered derivatives in which a is a constant and u is a function of x.

$$\frac{d}{dx}(x^n) = nx^{n-1} \qquad \frac{d}{dx}(a) = 0$$

$$\frac{d}{dx}(e^{ax}) = ae^{ax} \qquad \frac{d}{dx}(e^u) = e^u\frac{d}{dx}(u)$$

$$\frac{d}{dx}(\ln x) = \frac{1}{x} \qquad \frac{d}{dx}(au) = a\frac{d}{dx}(u)$$

7 Partial derivatives

If u is a function of two variables, i.e., $u = f(x,y)$, we can use the basic expressions by holding x constant and taking the derivative with respect to y. The result is the partial derivative of u with respect to y with x constant and is written

$$\left(\frac{\partial u}{\partial y}\right)_x.$$

The total differential of a function $u = f(x,y)$ is the sum of its partial derivatives, each multiplied by the infinitesimal change dx or dy:

$$du = \left(\frac{\partial u}{\partial x}\right)_y dx + \left(\frac{\partial u}{\partial y}\right)_x dy$$

Physical Chemistry: How Chemistry Works, First Edition. Kurt W. Kolasinski.

Companion Website: www.wiley.com/go/kolasinski/physicalchemistry

Integrals

The indefinite integral always includes a constant of integration. The definite integral requires limits of integration. The definite integral represents the area bordered by the graph of a function, the horizontal axis, and vertical lines that intersect the graph at two selected endpoints (limits of integration). Some basic forms of the indefinite integrals are listed below.

$$\int ax^n\,dx = a\int x^n\,dx = \frac{ax^{n+1}}{n+1} + c$$

$$\int \frac{a}{x}\,dx = a\ln x + c$$

$$\int (u+v)\,dx = \int u\,dx + \int v\,dx + c$$

$$\int u^n\,dx = \frac{u^{n+1}}{n+1} + c \quad n \neq -1$$

$$\int x^m e^{ax}\,dx = \frac{x^m e^{ax}}{a} - \frac{m}{a}\int x^{m-1} e^{ax}\,dx + c$$

$$\mathrm{erf}(x) = 1 - \mathrm{erfc}(x) = \frac{2}{\sqrt{\pi}}\int_0^x e^{-u^2}\,du$$

$$\int u^{-1}\,dx = \int \frac{1}{u}\,dx = \int d\ln u = \ln u + c$$

$$\int e^u\,dx = e^u + c$$

$$\int e^{-x}\,dx = -e^{-x} + c$$

$$\int_0^\infty x^{2n} e^{-ax^2}\,dx = \frac{(2n-1)!}{2^{n+1}a^n}\left(\frac{\pi}{a}\right)^{1/2}$$

$$\int_0^\infty e^{-ax^2}\,dx = \left(\frac{\pi}{4a}\right)^{1/2}$$

A volume integral in 3D, linear coordinates $\int_{-\infty}^{\infty}\int_{-\infty}^{\infty}\int_{-\infty}^{\infty} f(x,y,z)\,dx\,dy\,dz$

A volume integral in spherical polar coordinates $\int_0^{\infty}\int_0^{\pi}\int_0^{2\pi} f(r,\theta,\phi)\,r^2\sin\theta\,dr\,d\theta\,d\phi$

A volume integral of a variable separable function $f(r,\theta,\phi) = f(r)f(\theta)f(\phi)$ can be separated into the product of three integrals

$$\int_0^{\infty}\int_0^{\pi}\int_0^{2\pi} N f(r)f(\theta)f(\phi)\,r^2\sin\theta\,dr\,d\theta\,d\phi = N\int_0^{\infty} r^2 f(r)\,dr\int_0^{\pi} f(\theta)\sin\theta\,d\theta\int_0^{2\pi} f(\phi)\,d\phi$$

Mathematical relations

$$\ln x + \ln y = \ln xy$$

$$\ln x - \ln y = \ln\left(\frac{x}{y}\right)$$

$$a\ln x = \ln x^a$$

$$e^x e^y e^y = e^{x+y+z}$$

$$e^x/e^y = e^{x-y}$$

$$(e^x)^a = e^{ax}$$

$$\ln x = 2.303\,\lg x$$

$$\ln(n!) = n\ln n - n$$

The quadratic formula:

$$x = \frac{-b \pm \sqrt{b^2 - 4ac}}{2a}$$

Lorentzian profile:

$$L(\tilde{\nu}) = \frac{A}{(\tilde{\nu}-\tilde{\nu}_0)^2 + \gamma^2}$$

Gaussian profile:

$$G(\tilde{\nu}) = A\exp\left(-\left(\frac{\tilde{\nu}-\tilde{\nu}_0}{\sigma}\right)^2\right)$$

$$e^x = 1 + \frac{x}{1!} + \frac{x^2}{2!} + \frac{x^3}{3!} + \ldots \quad (x^2 < \infty)$$

$$e^{-x} = 1 - \frac{x}{1!} + \frac{x^2}{2!} - \frac{x^3}{3!} + \frac{x^4}{4!} - \ldots \quad (x^2 < \infty)$$

$$\ln(1+x) = x - \frac{x^2}{2} + \frac{x^3}{3} - \frac{x^4}{4} + \ldots \quad -1 < x \leq 1$$

$$\frac{1}{1-x} = 1 + x + x^2 + x^3 + \ldots \quad |x| < 1$$

$$\sin\theta = \theta - \frac{\theta^3}{3!} + \frac{\theta^5}{5!}\ldots$$

$$\cos\theta = 1 - \frac{\theta^2}{2!} + \frac{\theta^4}{4!} - \frac{\theta^6}{6!}$$

A complex number has a real and an imaginary part,

$\cos x = \frac{1}{2}\left(e^{ix} + e^{-ix}\right)$, $e^{\pm ix} = \cos x \pm i\sin x$

where $i^2 = -1$

Trigonometry

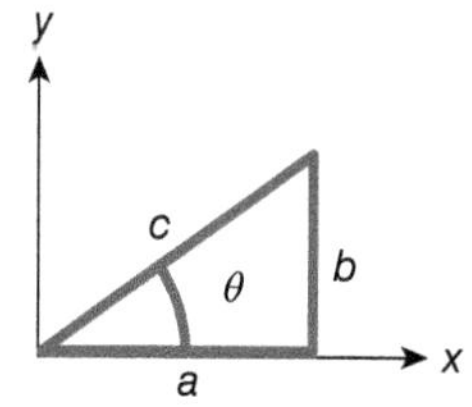

$\sin^2\theta + \cos^2\theta = 1$
$\cos 2\theta = \cos^2\theta - \sin^2\theta$
$\sin 2\theta = 2\sin\theta\ \cos\theta$
$a^2 + b^2 = c^2$
$\cos\theta = a/c$
$\sin\theta = b/c$
$\tan\theta = \frac{\sin\theta}{\cos\theta} = \frac{b}{a}$

Further reading

Bronshtein, I.N. and Semendyayev, K.A. (1985) *Handbook of Mathematics,* 3rd ed. Verlag Harri Deutsch, Thun.
Dwight, H.B. (1961) *Tables of Integrals and Other Mathematical Data,* 4th ed. Macmillan, New York.
Ladd, M. (1998) *Symmetry and Group Theory in Chemistry.* Horwood Publishing, Chichester.
McQuarrie, D.A. (2008) *Mathematics for Physical Chemistry: Opening Doors.* University Science Books, Mill Valley, CA.

APPENDIX 2

The method of undetermined multipliers

Here, we apply the method of undetermined multipliers to derive an expression for the most probable value of the number of microstates Ω. At large values of the total number of particles N, only the greatest term $t_{\max}$ makes a contribution to Ω, which allows us to write

$$\Omega = t_{\max} = \frac{N!}{n_1!n_2!\dots n_j!}. \tag{A2.1}$$

We need to find the values of the populations that maximize Ω subject to the constraints on N and the internal energy U

$$N = \sum_j n_j \tag{A2.2}$$

and

$$U = \sum_j n_j \, \varepsilon_j \tag{A2.3}$$

involving the population of the jth state n_j and its energy ε_j.

To simplify, we work with $\ln \Omega$ rather than Ω directly. Thus, Eq. (A2.1) becomes

$$\ln \Omega = \ln N! - \left(\ln n_1! + \ln n_2! + \cdots + \ln n_j!\right) = \ln N! - \sum_j \ln n_j! \tag{A2.4}$$

Applying Stirling's theorem to the summation yields

$$\ln \Omega = \ln N! - \sum_j \left(n_j \ln n_j - n_j\right). \tag{A2.5}$$

To calculate the maximum in $\ln \Omega$, we first use partial derivatives to express how $\ln \Omega$ responds to infinitesimal changes in the populations δn_1, δn_2, ... This is given by

$$\delta \ln \Omega = \frac{\partial \ln \Omega}{\partial n_1} \delta n_1 + \frac{\partial \ln \Omega}{\partial n_2} \delta n_2 + \cdots. \tag{A2.6}$$

At the maximum in $\ln \Omega$, $\delta \ln \Omega = 0$ for all small changes in δn_j, thus,

$$\frac{\partial \ln \Omega}{\partial n_1} \delta n_1 + \frac{\partial \ln \Omega}{\partial n_2} \delta n_2 + \cdots = \sum_j \frac{\partial \ln \Omega}{\partial n_j} = 0. \tag{A2.7}$$

The constraints that both N and U are constant mean that

$$\sum_j \delta n_j = 0 \tag{A2.8}$$

and

$$\sum_j \varepsilon_j \, \delta n_j = 0. \tag{A2.9}$$

Multiplication of Eq. (A2.8) by the undetermined multiplier α and Eq. (A2.9) by β also leads to equations that are equal to zero as long as α and β are finite constants. Thus, we can add these to Eq. (A2.7) to obtain

$$\sum_j \frac{\partial \ln \Omega}{\partial n_j} \delta n_j + \sum_j \alpha \, \delta n_j + \sum_j \beta \, \varepsilon_j \, \delta n_j = \sum_j \left[\frac{\partial \ln \Omega}{\partial n_j} + \alpha + \beta \varepsilon_j\right] \delta n_j = 0. \tag{A2.10}$$

Physical Chemistry: How Chemistry Works, First Edition. Kurt W. Kolasinski.

Companion Website: www.wiley.com/go/kolasinski/physicalchemistry

Equations (A2.8) and (A2.9) allow us to select two n_j values and to set these values in terms of the other n_j – 2 independent values in the sums. The two we choose are δn_1 and δn_2. We then fix the values α and β to satisfy the following two conditions

$$\frac{\partial \ln \Omega}{\partial n_1} + \alpha + \beta\varepsilon_1 = 0 \tag{A2.11}$$

and

$$\frac{\partial \ln \Omega}{\partial n_2} + \alpha + \beta\varepsilon_2 = 0. \tag{A2.12}$$

The sum in Eq. (A2.10) then becomes

$$\sum_{j>2} \left[\frac{\partial \ln \Omega}{\partial n_j} + \alpha + \beta\varepsilon_j\right] \delta n_j = 0. \tag{A2.13}$$

However, since all of the δn_1 in Eq. (A2.13) are independent, the only way to satisfy Eq. (A2.13) is

$$\frac{\partial \ln \Omega}{\partial n_j} + \alpha + \beta\varepsilon_j = 0 \tag{A2.14}$$

for all $j > 2$. With the combination of Eqs (A2.11) and (A2.12) with Eq. (A2.14), we see that the proper choice of α and β means that Eq. (A2.14) actually holds for all values of j.

Returning to Eq. (A2.5), differentiation gives

$$\frac{\partial \ln \Omega}{\partial n_j} = -\frac{\partial}{\partial n_j}\left(n_j \ln n_j - n_j\right) = -\ln n_j. \tag{A2.15}$$

Substituting this result into Eq. (A2.14) yields

$$-\ln n_j + \alpha + \beta\varepsilon_j = 0. \tag{A2.16}$$

Hence

$$n_j = e^{\alpha} e^{\beta\varepsilon_j}, \tag{A2.17}$$

which is the Boltzmann distribution, the n_j being the values of the populations of the j states that maximize the number of microstates Ω.

APPENDIX 3

Stirling's theorem

We wish to evaluate $N!$

$$N! = 1 \times 2 \times 3 \times \rightleftharpoons \times N \tag{A3.1}$$

in the limit of large N. Thus,

$$\ln N! = \ln 1 + \ln 2 + \ln 3 + \cdots + \ln N = \sum_{x=1}^{N} \ln x. \tag{A3.2}$$

Take a rectangle of unit width as shown in Fig. A3.1. Place one rectangle centered on each integer along the x axis. Then draw the height of each rectangle such that the top of the rectangle intersects the $\ln x$ versus x curve at the center of the rectangle. The sum of the areas of the rectangles approximate the area under the $\ln x$ curve. This approximation becomes more accurate the greater the value of x. In other words, we can approximate the value of $\ln N!$ by integrating the function $\ln x$ from 1 to N.

$$\ln N! \simeq \int_1^N \ln x \, \mathrm{d}x \tag{A3.3}$$

$$= [x \ln x - x]_1^N \tag{A3.4}$$

$$= N \ln N - N \text{ for } N \gg 1 \tag{A3.5}$$

Stirling actually approximated the value more accurately as

$$\ln N! = N \ln N - N + \tfrac{1}{2} \ln 2\pi N. \tag{A3.6}$$

However, the last term is negligible for the extremely large values of N associated with macroscopic samples.

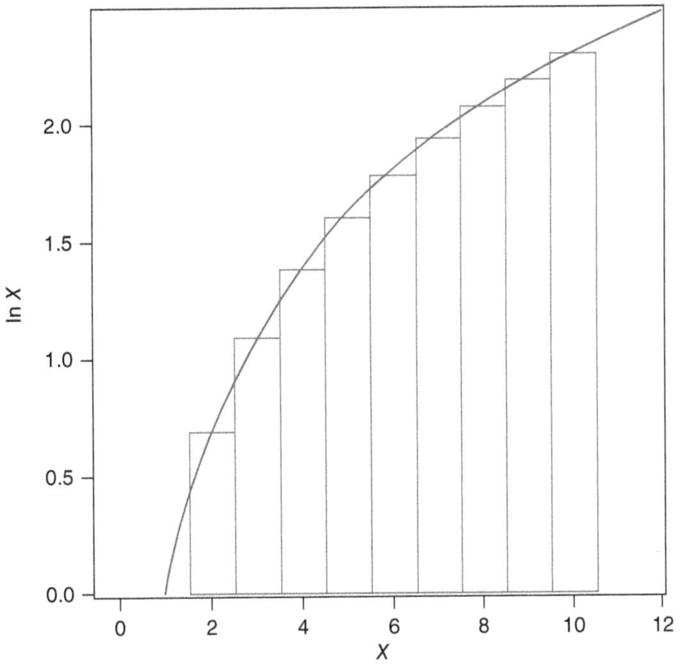

Figure A3.1 A histogram of $\ln x$ for integral values of x along with a curve of $\ln x$ generated from continuous values of x. This illustrates how $\ln N!$ converges on Stirling's approximation for large N.

Physical Chemistry: How Chemistry Works, First Edition. Kurt W. Kolasinski.

Companion Website: www.wiley.com/go/kolasinski/physicalchemistry

APPENDIX 4

Density of states of a particle in a box

A particle confined to a cubic box of side length a is described quantum mechanically by the wavefunction

$$\psi(x, y, z) = A \sin(n_x \pi x/a) \sin(n_y \pi y/a) \sin(n_z \pi z/a). \tag{A4.1}$$

The three quantum numbers n_x, n_y and n_z take on positive integer values beginning at 1. These are stationary wave solutions. They require that: (i) the wavefunctions vanish at the walls; and (ii) an integer number of half-wavelengths fit within the box along the three axes. Thus, the wavelengths along the x, y and z axes must satisfy

$$\lambda_x = 2a/n_x, \quad \lambda_y = 2a/n_y \quad \text{and} \quad \lambda_z = 2a/n_z. \tag{A4.2}$$

Alternatively, the wavefunction can be written in terms of the wavevector $\boldsymbol{k}$, which has components defined by

$$k_x = n_x \pi/a, \quad k_y = n_y \pi/a \quad \text{and} \quad k_z = n_z \pi/a, \tag{A4.3}$$

as

$$\psi(x, y, z) = A \sin(k_x x) \sin(k_y y) \sin(k_z z). \tag{A4.4}$$

We can think of the allowed states in real space – describing the states by their spatial extent in x, y, and z. Another representation is in n-space, in which a plot of the various values of (n_x, n_y, n_z) form a simple cubic lattice representing the various energy levels in terms of their quantum numbers. A cross-section of this lattice is shown in Fig. A4.1. A third representation is in k-space – in which a point of coordinates (k_x, k_y, k_z) represents the k-state of the wavefunction.

The linear momentum p is given by the de Broglie relationship

$$p = h/\lambda. \tag{A4.5}$$

Therefore, the components of the linear momentum are

$$p_x = n_x h/2a, \quad p_y = n_y h/2a \quad \text{and} \quad p_z = n_z h/2a. \tag{A4.6}$$

The resultant linear momentum of the jth state is then

$$p_j^2 = p_x^2 + p_y^2 + p_z^2 = \left(n_x^2 + n_y^2 + n_z^2\right) h^2/4a^2 = n_j^2 h^2/4a^2 \tag{A4.7}$$

and its energy is

$$\varepsilon_j = p_j^2/2m = \left(n_x^2 + n_y^2 + n_z^2\right) h^2/8ma^2 = n_j^2 h^2/8ma^2. \tag{A4.8}$$

Note that there are possibly several combinations of n_x, n_y and n_z that correspond to the same value of n_j^2. That is, many of the states have a degeneracy $g_j > 1$. Since the volume is $V = a^3$, Eq. (A4.8) can be rewritten

$$\varepsilon_j = n_j^2 h^2/8mV^{2/3}. \tag{A4.9}$$

Written in this form, the equation for the energy is independent of the shape of the box. Instead, the energy depends on the quantum number n_j and the volume V. If there are multiple particles in the box, the energy levels available to these particles are still described by Eq. (A4.9). How many particles are allowed in any given level will depend on whether the particles are bosons or fermions.

Now, we calculate the density of states directly, which is the number of states per unit energy. This is done with the aid of the three-dimensional plot that is behind Fig. (A4.1). Solving Eq. (A4.8) for the quantum number, we find

$$n^2 = 8mV^{2/3}\varepsilon/h^2 = R^2. \tag{A4.10}$$

Physical Chemistry: How Chemistry Works, First Edition. Kurt W. Kolasinski.

Companion Website: www.wiley.com/go/kolasinski/physicalchemistry

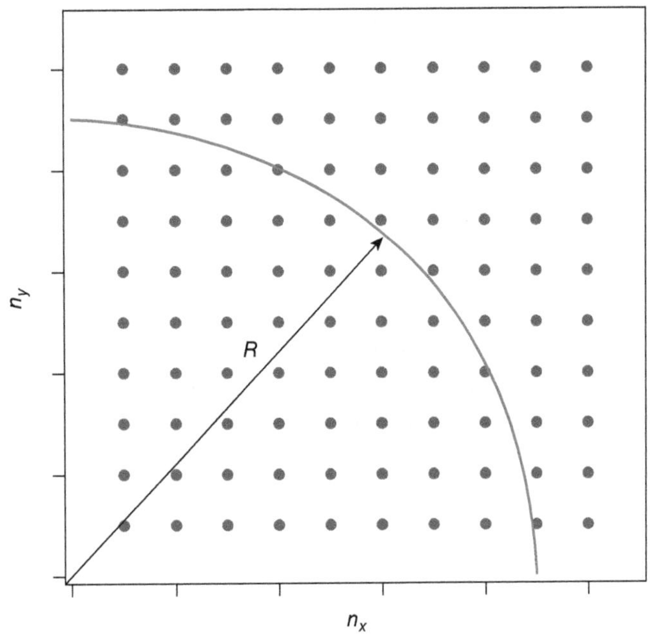

Figure A4.1 A representation of the states of the particle in a box in n-space. A cross-section through the positive octant of radius R in the $n_x n_y$ plane is shown.

The n-space radius R is introduced. This allows us to calculate the number of states contained within an n-space volume, as shown below. Note that we have dropped the subscript j to simplify the notation. Looking back at Fig. A4.1, we see that if we set R to some arbitrary value, a number of lattice points are contained within the region defined by the arc of this radius. The quantum numbers must be positive. Therefore, points corresponding to energy levels are only found in the positive octant of the coordinate system defined by n_x, n_y and n_z. The number of points contained within this octant of a sphere of radius R is equal to the number of energy levels with an energy less than ε as defined in Eq. (A4.10). The number of energy levels with an energy less than ε is, therefore,

$$N = \tfrac{1}{8}\left(4\pi R^3/3\right) = \tfrac{4\pi}{3} V \left(2m\varepsilon/h^2\right)^{3/2}. \qquad \text{(A4.11)}$$

From which the number of states contained within a range of energies dε is

$$\mathrm{d}N = \frac{2\pi V}{h^3}(2m)^{3/2}\varepsilon^{1/2}\,\mathrm{d}\varepsilon. \qquad \text{(A4.12)}$$

The function appearing before dε is known as the *density of states* in energy $\rho(\varepsilon)$. It tells us the number of states in the range between ε and $\varepsilon + \mathrm{d}\varepsilon$

$$\rho(\varepsilon)\,\mathrm{d}\varepsilon = \left(2\pi V/h^3\right)(2m)^{3/2}\varepsilon^{1/2}\,\mathrm{d}\varepsilon. \qquad \text{(A4.13)}$$

As written, this version of the density of states assumes only one particle per energy level. Should the quantum statistics of the particle allow two particles per state, as is the case for spin $\frac{1}{2}$ electrons, then Eq. (A4.13) needs to be multiplied by the appropriate factor of two.

For determining the spectral distribution of black-body radiation, we will actually need the density of states in k space. To do so we need to convert between ε and k using the relation

$$\varepsilon = k^2h^2/8m\pi^2, \qquad \text{(A4.14)}$$

which means that $\mathrm{d}\varepsilon = \left(2kh^2/8m\pi^2\right)\mathrm{d}k$. Upon substitution for ε and dε into Eq. (A4.13), we obtain

$$\rho(k)\,\mathrm{d}k = \left(V/2\pi^2\right)k^2\,\mathrm{d}k. \qquad \text{(A4.15)}$$

APPENDIX 5

Black-body radiation: Treating radiation as a photon gas

Photons have a spin of $s = 1$. Thus, they are bosons and follow Bose–Einstein statistics. However, they are massless bosons and because of relativistic effects, they only display two polarization states $m_s = \pm 1$. A black body contains a cavity with volume V held at temperature T. At equilibrium, electromagnetic radiation with a continuous spectrum is emitted from the walls of the cavity. Planck was the first to apply statistical mechanics to derive an expression for the spectral distribution of the light emitted from a black body. Photons are emitted and reabsorbed from the walls of the cavity. Therefore, their number is not constant, which means that the multiplier α in the Bose–Einstein distribution law is identically zero. The Bose–Einstein distribution function can be written as a continuous distribution in energy as

$$f(\varepsilon) = \frac{1}{\exp(-\beta\varepsilon) - 1}. \tag{A5.1}$$

The energy of a photon can be expressed in terms of its frequency ν,

$$\varepsilon = h\nu. \tag{A5.2}$$

Therefore, the number of photons per energy state can also be expressed in terms of ν,

$$f(\nu) = \frac{1}{\exp^{-\beta h\nu} - 1}. \tag{A5.3}$$

As shown in Appendix 3, the particle in a box model can be used to derive the density of states in k; however, we must account for the two polarization states of the photon by multiplying our previous result by a factor of 2,

$$\rho(k)\,\mathrm{d}k = \left(V/\pi^2\right) k^2\,\mathrm{d}k. \tag{A5.4}$$

This is transformed to frequency space using

$$k = 2\pi\nu/c \tag{A5.5}$$

and, thus, $\mathrm{d}k = (2\pi/c)\,\mathrm{d}\nu$, which leads to

$$\rho(\nu)\,\mathrm{d}\nu = \left(8\pi V/c^3\right)\nu^2\mathrm{d}\nu. \tag{A5.6}$$

The photon energy $h\nu$ multiplied by the number of photons $N(\nu)$ in the range ν to $\nu + \mathrm{d}\nu$ equals the radiant energy in this frequency range $u(\nu)\,\mathrm{d}\nu$,

$$u(\nu)\,\mathrm{d}\nu = h\nu\,N(\nu)\,\mathrm{d}\nu. \tag{A5.7}$$

However, according to Bose–Einstein statistics, the number of photons at a given frequency must also equal the distribution function $f(\nu)$ times the density of states in ν,

$$N(\nu)\,\mathrm{d}\nu = f(\nu)\,\rho(\nu)\,\mathrm{d}\nu. \tag{A5.8}$$

Consequently

$$u(\nu)\,\mathrm{d}\nu = h\nu f(\nu)\,\rho(\nu)\,\mathrm{d}\nu \tag{A5.9}$$

$$u(\nu)\,\mathrm{d}\nu = \frac{8\pi V h\nu^3}{c^3}\frac{1}{\exp\left(h\nu/k_\mathrm{B}T\right) - 1}\,\mathrm{d}\nu. \tag{A5.10}$$

Physical Chemistry: How Chemistry Works, First Edition. Kurt W. Kolasinski.

Companion Website: www.wiley.com/go/kolasinski/physicalchemistry

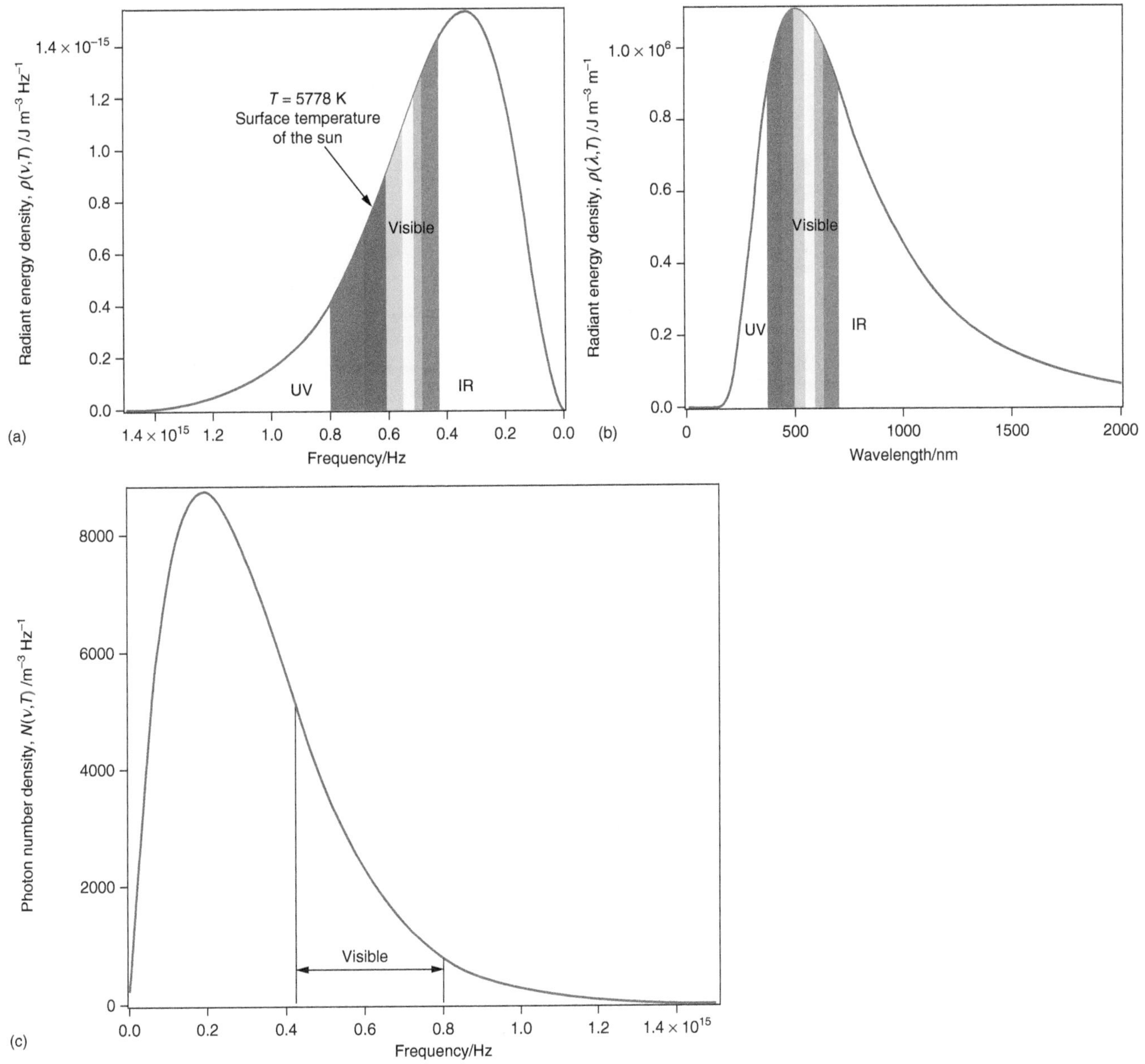

Figure A5.1 (a) Radiant energy density per unit frequency bandwidth. (b) Radiant energy density per unit wavelength bandwidth. (c) Number density of photons per unit frequency bandwidth. Data are plotted for a black body radiating at the surface temperature of the Sun. Note how three curves peak at different wavelengths: 882 nm, 500 nm and 1563 nm, respectively.

This is the *Planck radiation formula* for the radiant energy within a constant temperature cavity. To express this formula in wavelength, we use $c = \nu\lambda$ and $|\mathrm{d}\nu| = \left(c/\lambda^2\right)|\mathrm{d}\lambda|$ to arrive at

$$u(\lambda)\,\mathrm{d}\lambda = \frac{8\pi Vhc}{\lambda^5\left[\exp\left(hc/\lambda k_\mathrm{B}T\right)-1\right]}\,\mathrm{d}\lambda. \tag{A5.11}$$

The radiant energy density per unit frequency bandwidth in J m^{-3} Hz^{-1}, denoted $\rho(\nu)$, is obtained by dividing $u(\nu)$ by V. The radiant energy density per unit wavelength bandwidth in J m^{-3} m^{-1}, denoted $\rho(\lambda)$ is obtained by dividing $u(\lambda)$ by V. These radiant energy densities as well as the number density of photons per unit frequency bandwidth $N(\nu)$ are plotted in Fig. A5.1.

APPENDIX 6

Definitions of symbols used in quantum mechanics and quantum chemistry

Name	Symbol	Definition	Note
linear momentum operator	$\hat{p}$	$\hat{p} = -i\hbar\nabla$	[1]
z-component	$\hat{p}_z$	$\hat{p}_z = -i\hbar(\partial/\partial z)$	
angular momentum operator	$\hat{L}^2$	$\hat{L}^2 = -\hbar^2\left[\frac{1}{\sin\theta}\frac{\partial}{\partial\theta}\left(\sin\theta\frac{\partial}{\partial\theta}\right)+\frac{1}{\sin^2\theta}\frac{\partial^2}{\partial\phi^2}\right]$	
z-component	$\hat{L}_z$	$\hat{L}_z = -i\hbar\left(x\frac{\partial}{\partial y}-y\frac{\partial}{\partial x}\right) = -i\hbar\left(\frac{\partial}{\partial\varphi}\right)$	
kinetic energy operator	$\hat{T}$	$\hat{T} = -(\hbar^2/2m)\nabla^2$	
potential energy operator	$\hat{V}$		[2]
Hamiltonian operator	$\hat{H}$	$\hat{H} = \hat{T} + \hat{V}$	
wavefunction, state function time-dependent	Ψ	$\Psi(\tau, t) = \psi(\tau)\exp(-iEt/\hbar)$	
wavefunction, state function time-independent	$\psi,\ \phi$	$\hat{H}\psi = E\psi$	
hydrogen-like wavefunction	$\psi_{nlm}(r, \theta, \phi)$	$\psi_{nlm} = R_{nl}(r)\,Y_{lm}(\theta, \phi)$	
spherical harmonic function	$Y_{lm}(\theta, \phi)$	$Y_{lm} = N_{l\lvert m\rvert}P_l^{\lvert m\rvert}\cos\theta\ e^{im\phi}$	[3]
probability density	P	$P = \psi^*\psi = \lvert\psi\rvert^2$	[4]
normalization		$1 = \langle\psi\vert\,\psi\rangle = \int_{-\infty}^{\infty}\psi^*\psi\ d\tau$	
orthogonality		$0 = \langle\psi_n\vert\,\psi_m\rangle = \int_{-\infty}^{\infty}\psi_n^*\psi_m\ d\tau$	$n \neq m$
electron charge density	ρ	$\rho = -eP$	
integration element	$d\tau$	$d\tau = dx\,dy\,dz$	[5]
matrix element of operator $\hat{A}$	$A_{ij},\ \langle i\vert\hat{A}\,\vert j\rangle$	$A_{ij} = \int\psi_i^*\hat{A}\psi_j\ d\tau$	
expectation value of operator $\hat{A}$	$\langle A\rangle$	$\langle A\rangle = \int\psi_i^*\hat{A}\psi_j\ d\tau$	[6]
commutator of $\hat{A}$ and $\hat{B}$	$[\hat{A}, \hat{B}]$	$[\hat{A}, \hat{B}] = \hat{A}\hat{B} - \hat{B}\hat{A}$	
uncertainty principle for operators $\hat{A}$ and $\hat{B}$		$(\Delta A)^2(\Delta B)^2 \geq \frac{1}{4}\langle[\hat{A}, \hat{B}]\rangle$	[7]
spin wavefunctions	$\alpha,\ \beta$		[8]
Bohr radius	a_0	$a_0 = 4\pi\varepsilon_0\hbar^2/m_e e^2$	
Hartree energy	E_h	$E_h = \hbar^2/m_e a_0^2$	
Rydberg constant	R_∞	$R_\infty = E_h/2hc$	
fine-structure constant	α	$\alpha = e^2/4\pi\varepsilon_0\hbar c$	
de Broglie wavelength	λ	$\lambda = h/p$	

[1] The hat (circumflex) distinguishes an operator from an algebraic quantity.
[2] The potential energy operator depends on the interactions present in the system.
[3] $P_l^{\lvert m\rvert}$ denotes the associated Legendre function of degree l and order $\lvert m\rvert$. $N_{l\lvert m\rvert}$ is the normalization factor.
[4] Integration over a region of space is required to obtain the probability within that region.
[5] The integration element varies with the dimensionality of the system and the coordinate system of choice.
[6] Here it is assumed that the wavefunction is normalized.
[7] For any pair of operators whose commutator forms an operator of the form $[\hat{A}, \hat{B}] = -i\hbar$, where $\Delta A = \left[\langle A^2\rangle - \langle A\rangle^2\right]^{1/2}$ and similarly for ΔB.
[8] The spin wavefunctions of a single electron are defined by the matrix elements of the z component of the spin angular momentum $\hat{s}_z$, by the relations $\langle\alpha\vert\,\hat{s}_z\,\vert\alpha\rangle = +\frac{1}{2}\hbar$, $\langle\beta\vert\,\hat{s}_z\,\vert\beta\rangle = -\frac{1}{2}\hbar$ and $\langle\beta\vert\,\hat{s}_z\,\vert\alpha\rangle = \langle\alpha\vert\,\hat{s}_z\,\vert\beta\rangle = 0$.

Physical Chemistry: How Chemistry Works, First Edition. Kurt W. Kolasinski.

Companion Website: www.wiley.com/go/kolasinski/physicalchemistry

The hydrogen atom wavefunction is formed from the product of radial and angular parts,

$$\psi(r,\theta,\phi) = R_{nl}(r)\, Y_{lm}(\theta,\phi).$$

Table A6.1 Hydrogenic radial wavefunctions derived from solution of the Schrödinger equation.

Orbital	n	l	$R_{n,l}$
1s	1	0	$R_{10} = 2a^{-3/2}\exp(-r/a)$
2s	2	0	$R_{20} = \frac{1}{\sqrt{2}}a^{-3/2}\left(1-\frac{r}{2a}\right)\exp(-r/2a)$
2p	2	1	$R_{21} = \frac{1}{\sqrt{24}}a^{-3/2}\left(\frac{r}{a}\right)\exp(-r/2a)$
3s	3	0	$R_{30} = \frac{2}{\sqrt{27}}a^{-3/2}\left(1-\frac{2r}{3a}+\frac{2}{27}\left(\frac{r}{a}\right)^2\right)\exp(-r/3a)$
3p	3	1	$R_{31} = \frac{8}{27\sqrt{6}}a^{-3/2}\left(1-\frac{r}{6a}\right)\left(\frac{r}{a}\right)\exp(-r/3a)$
3d	3	2	$R_{32} = \frac{4}{81\sqrt{30}}a^{-3/2}\left(\frac{r}{a}\right)^2\exp(-r/3a)$
4s	4	0	$R_{40} = \frac{1}{4}a^{-3/2}\left(1-\frac{3r}{4a}+\frac{1}{8}\left(\frac{r}{a}\right)^2-\frac{1}{192}\left(\frac{r}{a}\right)^3\right)\exp(-r/4a)$
4p	4	1	$R_{41} = \frac{\sqrt{5}}{16\sqrt{3}}a^{-3/2}\left(1-\frac{r}{4a}+\frac{1}{80}\left(\frac{r}{a}\right)^2\right)\left(\frac{r}{a}\right)\exp(-r/4a)$
4d	4	2	$R_{42} = \frac{1}{64\sqrt{5}}a^{-3/2}\left(1-\frac{r}{12a}\right)\left(\frac{r}{a}\right)^2\exp(-r/4a)$
4f	4	3	$R_{43} = \frac{1}{768\sqrt{35}}a^{-3/2}\left(\frac{r}{a}\right)^3\exp(-r/4a)$
			$a = \frac{4\pi\varepsilon_0\hbar^2}{\mu e^2}$ [1]

[1] a is almost the Bohr radius a_0. In a the reduced mass is used. In a_0 the mass of the electron is used. $a = 0.529465$ pm. $a_0 = 0.529177$ pm.

Table A6.2 The first few members of the set of spherical harmonics.

$Y_{lm}(\theta,\phi)$	
$Y_{00} = \left(\frac{1}{4\pi}\right)^{1/2}$	$Y_{2\pm2} = \mp\left(\frac{15}{32\pi}\right)^{1/2}\sin^2\theta\, e^{\pm 2i\phi}$
$Y_{10} = \left(\frac{3}{4\pi}\right)^{1/2}\cos\theta$	$Y_{30} = \left(\frac{7}{16\pi}\right)^{1/2}(5\cos^3\theta - 3\cos\theta)$
$Y_{1\pm1} = \mp\left(\frac{3}{8\pi}\right)^{1/2}\sin\theta e^{\pm i\phi}$	$Y_{3\pm1} = \mp\left(\frac{21}{64\pi}\right)^{1/2}\sin\theta\,(5\cos^2\theta - 1)\, e^{\pm i\phi}$
$Y_{20} = \left(\frac{5}{16\pi}\right)^{1/2}(3\cos^2\theta - 1)$	$Y_{3\pm2} = \left(\frac{105}{32\pi}\right)^{1/2}\sin^2\theta\cos\theta\, e^{\pm 2i\phi}$
$Y_{2\pm1} = \mp\left(\frac{15}{8\pi}\right)^{1/2}\sin\theta\cos\theta e^{\pm i\phi}$	$Y_{3\pm3} = \mp\left(\frac{35}{64\pi}\right)^{1/2}\sin^3\theta\, e^{\pm 3i\phi}$

APPENDIX 7

Character tables

Table A7.1 Character table for the C_{2v} point group.

C_{2v}	E	C_2	$\sigma_v(xz)$	σ_v		
A_1	1	1	1	1	T_z	α_{xx}, α_{yy}, α_{zz}
A_2	1	1	−1	−1	R_z	α_{xy}
B_1	1	−1	1	−1	T_x, R_y	α_{xz}
B_2	1	−1	−1	1	T_y, R_x	α_{yz}

Table A7.2 Character table for the C_{3v} point group.

C_{3v}	E	$2C_3$	$3\sigma_v$		
A_1	1	1	1	T_z	$\alpha_{xx} + \alpha_{yy}$, α_{zz}
A_2	1	1	−1	R_z	
E	2	−1	0	T_x, T_y, R_x, R_y	$\alpha_{xx}-\alpha_{yy}$, α_{xy}, α_{xz}, α_{yz}

Table A7.3 Character table for the $C_{\infty v}$ point group.

$C_{\infty v}$	E	$2C_\infty^\phi$	···	$3\sigma_v$		
$A_1 \equiv \Sigma^+$	1	1	···	1	T_z	$\alpha_{xx} + \alpha_{yy}$, α_{zz}
$A_2 \equiv \Sigma^-$	1	1	···	−1	R_z	
$E_1 \equiv \Pi$	2	$2\cos\phi$	···	0	T_x, T_y, R_x, R_y	α_{xz}, α_{yz}
$E_2 \equiv \Delta$	2	$2\cos 2\phi$	···	0		$\alpha_{xx}-\alpha_{yy}$, α_{xy}
$E_3 \equiv \Phi$	2	$2\cos 3\phi$	···	0		
⋮	⋮	⋮	···	⋮		

Table A7.4 Character table for the $D_{\infty h}$ point group.

$D_{\infty h}$	E	$2C_\infty^\phi$	···	$\infty\sigma_v$	i	$2S_\infty^\phi$	···	$3\sigma_v$		
$A_{1g} \equiv \Sigma_g^+$	1	1	···	1	1	1	···	1		$\alpha_{xx}+\alpha_{yy}$, α_{zz}
$A_{2g} \equiv \Sigma_g^-$	1	1	···	−1	1	1	···	−1	R_z	
$E_{1g} \equiv \Pi_g$	2	$2\cos\phi$	···	0	2	$-2\cos\phi$	···	0	R_x, R_y	α_{xz}, α_{yz}
$E_{2g} \equiv \Delta_g$	2	$2\cos 2\phi$	···	0	2	$2\cos 2\phi$	···	0		$\alpha_{xx}-\alpha_{yy}$, α_{xy}
$E_{3g} \equiv \Phi_g$	2	$2\cos 3\phi$	···	0	2	$-2\cos 3\phi$	···	0		
⋮	⋮	⋮	···	⋮	⋮			⋮		
$A_{1u} \equiv \Sigma_u^+$	1	1	···	1	−1	−1	···	−1	T_z	
$A_{2u} \equiv \Sigma_u^-$	1	1	···	−1	−1	−1	···	1		
$E_{1u} \equiv \Pi_u$	2	$2\cos\phi$	···	0	−2	$2\cos\phi$	···	0	T_x, T_y	
$E_{2u} \equiv \Delta_u$	2	$2\cos 2\phi$	···	0	−2	$-2\cos 2\phi$	···	0		
$E_{3u} \equiv \Phi_u$	2	$2\cos 3\phi$	···	0	−2	$2\cos 3\phi$	···	0		
⋮	⋮	⋮	···	⋮	⋮			⋮		

Physical Chemistry: How Chemistry Works, First Edition. Kurt W. Kolasinski.

Companion Website: www.wiley.com/go/kolasinski/physicalchemistry

Table A7.5 Character table for the T_d point group.

T_d	E	$8C_3$	$3C_2$	$6S_4$	$6\sigma_d$		
A_1	1	1	1	1	1		$\alpha_{xx}+\alpha_{yy}+\alpha_{zz}$
A_2	1	1	1	−1	−1		
E	2	−1	2	0	0		$\alpha_{xx}+\alpha_{yy}-2\alpha_{zz}$, $\alpha_{xx}-\alpha_{yy}$
$T_1 \equiv F_1$	3	0	−1	1	−1	R_x, R_y, R_z	
$T_2 \equiv F_2$	3	0	−1	−1	1	T_x, T_y, T_z	α_{xy}, α_{xz}, α_{yz}

Table A7.6 Character table for the O_h point group.

O_h	E	$8C_3$	$6C_2$	$6C_4$	$3C_2' = 3C_4^2$	i	$6S_4$	$8S_6$	$3\sigma_h$	$6\sigma_d$		
A_{1g}	1	1	1	1	1	1	1	1	1	1		$\alpha_{xx}+\alpha_{yy}+\alpha_{zz}$
A_{2g}	1	1	−1	−1	1	1	−1	1	1	−1		$\alpha_{xx}+\alpha_{yy}-2\alpha_{zz}$, $\alpha_{xx}-\alpha_{yy}$
E_g	2	−1	0	0	2	2	0	−1	2	0		
$T_{1g} \equiv F_{1g}$	3	0	−1	1	−1	3	1	0	−1	−1	R_x, R_y, R_z	
$T_{2g} \equiv F_{2g}$	3	0	1	−1	−1	3	−1	0	−1	1		α_{xy}, α_{xz}, α_{yz}
A_{1u}	1	1	1	1	1	−1	−1	−1	−1	−1		
A_{2u}	1	1	−1	−1	1	−1	1	−1	−1	1		
E_u	2	−1	0	0	2	−2	0	1	−2	0		
$T_{1u} \equiv F_{1u}$	3	0	−1	1	−1	−3	−1	0	1	1	T_x, T_y, T_z	
$T_{2u} \equiv F_{2u}$	3	0	1	−1	−1	−3	1	0	1	−1		

APPENDIX 8

Periodic behavior

Physical Chemistry: How Chemistry Works, First Edition. Kurt W. Kolasinski.

Companion Website: www.wiley.com/go/kolasinski/physicalchemistry

Periodic Properties of the elements

Group

nl^w = ground state configuration
Term = ground state term symbol
E_i = first ionization energy in eV

symbol
nl^w
Term
E_i

1 I	2 II	3 IIIB	4 IVB	5 VB	6 VIB	7 VIIB	8 VIIIB	9 VIIIB	10 VIIIB	11 IB	12 IIB	13 III	14 IV	15 V	16 VI	17 VII	18 VIII
H 1s $^2S_{1/2}$ 13.598																	He $1s^2$ $^1S_0$ 24.587
Li 2s $^2S_{1/2}$ 5.3917	Be $2s^2$ $^1S_0$ 9.3227											B 2p $^2P^o_{1/2}$ 8.2980	C $2p^2$ $^3P_0$ 11.260	N $2p^3$ $^4S^o_{3/2}$ 14.534	O $2p^4$ $^3P_2$ 13.618	F $2p^5$ $^2P^o_{3/2}$ 17.423	Ne $2p^6$ $^1S_0$ 21.565
Na 3s $^2S_{1/2}$ 5.1391	Mg $3s^2$ $^1S_0$ 7.6462											Al 3p $^2P^o_{1/2}$ 5.9858	Si $3p^2$ $^3P_0$ 8.1517	P $3p^3$ $^4S^o_{3/2}$ 10.487	S $3p^4$ $^3P_2$ 10.360	Cl $3p^5$ $^2P^o_{3/2}$ 12.968	Ar $3p^6$ $^1S_0$ 15.760
K 4s $^2S_{1/2}$ 4.3407	Ca $4s^2$ $^1S_0$ 6.1132	Sc $4s^23d$ $^2D_{3/2}$ 6.5615	Ti $4s^23d^2$ $^3F_2$ 6.8281	V $4s^23d^3$ $^4F_{3/2}$ 6.7462	Cr $\mathbf{4s3d^5}$ $^7S_3$ 6.7665	Mn $4s^23d^5$ $^6S_{5/2}$ 7.4340	Fe $4s^23d^6$ $^5D_4$ 7.9024	Co $4s^23d^7$ $^4F_{9/2}$ 7.8810	Ni $4s^23d^8$ $^3F_4$ 7.6399	Cu $\mathbf{4s3d^{10}}$ $^2S_{1/2}$ 7.7264	Zn $4s^23d^{10}$ $^1S_0$ 9.3942	Ga 4p $^2P^o_{1/2}$ 5.9993	Ge $4p^2$ $^3P_0$ 7.8994	As $4p^3$ $^4S^o_{3/2}$ 9.7886	Se $4p^4$ $^3P_2$ 9.7524	Br $4p^5$ $^2P^o_{3/2}$ 11.814	Kr $4p^6$ $^1S_0$ 14.000
Rb 5s $^2S_{1/2}$ 4.1771	Sr $5s^2$ $^1S_0$ 5.6949	Y $5s^24d$ $^2D_{3/2}$ 6.2173	Zr $5s^24d^2$ $^3F_2$ 6.6339	Nb $\mathbf{5s4d^4}$ $^6D_{1/2}$ 6.7589	Mo $\mathbf{5s4d^5}$ $^7S_3$ 7.0924	Tc $5s^24d^5$ $^6S_{5/2}$ 7.28	Ru $\mathbf{5s4d^7}$ $^5F_5$ 7.3605	Rh $\mathbf{5s4d^8}$ $^4F_{9/2}$ 7.4589	Pd $\mathbf{4d^{10}}$ $^1S_0$ 8.3369	Ag $\mathbf{5s4d^{10}}$ $^2S_{1/2}$ 7.5762	Cd $5s^24d^{10}$ $^1S_0$ 8.9938	In 5p $^2P^o_{1/2}$ 5.7864	Sn $5p^2$ $^3P_0$ 7.3439	Sb $5p^3$ $^4S^o_{3/2}$ 8.6084	Te $5p^4$ $^3P_2$ 9.0096	I $5p^5$ $^2P^o_{3/2}$ 10.451	Xe $5p^6$ $^1S_0$ 12.130
Cs 6s $^2S_{1/2}$ 3.8939	Ba $6s^2$ $^1S_0$ 5.2117	*	Hf $6s^25d^2$ $^3F_2$ 6.8251	Ta $6s^25d^3$ $^4F_{3/2}$ 7.5496	W $6s^25d^4$ $^5D_0$ 7.8640	Re $6s^25d^5$ $^6S_{5/2}$ 7.8335	Os $6s^25d^6$ $^5D_4$ 8.4382	Ir $6s^25d^7$ $^4F_{9/2}$ 8.9670	Pt $\mathbf{6s5d^9}$ $^3D_3$ 8.9588	Au $\mathbf{6s5d^{10}}$ $^2S_{1/2}$ 9.2255	Hg $6s^25d^{10}$ $^1S_0$ 10.438	Tl 6p $^2P^o_{1/2}$ 6.1082	Pb $6p^2$ $^3P_0$ 7.4167	Bi $6p^3$ $^4S^o_{3/2}$ 7.2855	Po $6p^4$ $^3P_2$ 8.414	At $6p^5$ $^2P^o_{3/2}$	Rn $6p^6$ $^1S_0$ 10.749
Fr 7s $^2S_{1/2}$ 4.0727	Ra $7s^2$ $^1S_0$ 5.2784	#	Rf 6.0														

* Lanthanide series	La $\mathbf{5d}$ $^2D_{3/2}$ 5.5769	Ce $\mathbf{5d4f}$ $^1G^o_4$ 5.5387	Pr $4f^3$ $^4I^o_{9/2}$ 5.473	Nd $4f^4$ $^5I_4$ 5.5250	Pm $4f^5$ $^6H^o_{5/2}$ 5.582	Sm $4f^6$ $^7F_0$ 5.6437	Eu $4f^7$ $^8S^o_{7/2}$ 5.6704	Gd $\mathbf{5d4f^7}$ $^9D^o_2$ 6.1498	Tb $4f^9$ $^6H^o_{15/2}$ 5.8638	Dy $4f^{10}$ $^5I_8$ 5.9389	Ho $4f^{11}$ $^4I^o_{15/2}$ 6.0215	Er $4f^{12}$ $^3H_6$ 6.1077	Tm $4f^{13}$ $^2F^o_{7/2}$ 6.1843	Yb $4f^{14}$ $^1S_0$ 6.2542	Lu $\mathbf{5d4f^{14}}$ $^2D_{3/2}$ 5.4259
# Actinide series	Ac $\mathbf{6d}$ $^2D_{3/2}$ 5.3807	Th $\mathbf{6d^2}$ $^3F_2$ 6.3067	Pa $\mathbf{6d5f^2}$ $(4,3/2)_{11/2}$ 6.2657	U $\mathbf{6d5f^3}$ $(9/2,3/2)_6$ 6.2657	Np $\mathbf{6d5f^4}$ $(4,3/2)_{11/2}$ 6.2657	Pu $5f^6$ $^7F_0$ 6.0260	Am $5f^7$ $^8S^o_{7/2}$ 5.9738	Cm $\mathbf{6d5f^7}$ $^9D^o_2$ 5.9914	Bk $5f^9$ $^6H^o_{15/2}$ 6.1979	Cf $5f^{10}$ $^5I_8$ 6.2817	Es $5f^{11}$ $^4I^o_{15/2}$ 6.3067	Fm $5f^{12}$ $^3H_6$ 6.50	Md $5f^{13}$ $^2F^o_{7/2}$ 6.58	No $5f^{14}$ $^1S_0$ 6.65	Lr 4.9

Periodic Properties of the elements

Group

Key: symbol / fcc_i / R_{cov} / χ — crystal structure; R_{cov} = covalent radius; χ = electronegativity

1 I	2 II	3 IIIB	4 IVB	5 VB	6 VIB	7 VIIB	8 VIIIB	9 VIIIB	10 VIIIB	11 IB	12 IIB	13 III	14 IV	15 V	16 VI	17 VII	18 VIII
H hex 0.32 2.20																	He hcp 0.37 –
Li bcc 1.30 0.98	Be hcp 0.99 1.57											B rhom 0.84 2.04	C hex 0.75 2.55	N hex 0.71 3.04	O cub 0.64 3.44	F cub 0.60 3.98	Ne fcc 0.62 –
Na bcc 1.60 0.93	Mg hcp 1.40 1.31											Al fcc 1.24 1.61	Si cub 1.14 1.90	P orth 1.09 2.19	S orth 1.04 2.58	Cl orth 1.00 3.16[5]	Ar fcc 1.01 –
K bcc 2.00 0.82	Ca fcc 1.74 1.00	Sc hcp 1.59 1.36	Ti hcp 1.48 1.54	V bcc 1.44 1.63	Cr bcc 1.30 1.66	Mn bcc 1.29 1.55	Fe bcc 1.24 1.83	Co hcp 1.18 1.88	Ni fcc 1.17 1.91	Cu fcc 1.22 1.90	Zn hcp 1.20 1.65	Ga orth 1.23 1.81	Ge cub 1.20 2.01	As rhom 1.20 2.18	Se hex 1.18 2.55	Br orth 1.17 2.96	Kr fcc 1.16 –
Rb bcc 2.15 0.82	Sr fcc 1.90 0.95	Y hcp 1.76 1.22	Zr hcp 1.64 1.33	Nb bcc 1.56 1.6	Mo bcc 1.46 2.16	Tc hcp 1.38 2.10	Ru hcp 1.36 2.2	Rh fcc 1.34 2.28	Pd fcc 1.30 2.20	Ag fcc 1.36 1.93	Cd hcp 1.40 1.69	In tetra 1.42 1.78	Sn tetra 1.40 1.96	Sb rhom 1.40 2.05	Te hex 1.37 2.1	I orth 1.36 2.66	Xe fcc 1.36 2.60
Cs bcc 2.38 0.79	Ba bcc 2.06 0.89	57–71 *	Hf hcp 1.64 1.3	Ta bcc 1.58 1.5	W bcc 1.50 1.7	Re hcp 1.41 1.9	Os hcp 1.36 2.2	Ir fcc 1.32 2.2	Pt fcc 1.30 2.2	Au fcc 1.30 2.4	Hg rhom 1.32 1.9	Tl hcp 1.44 1.8	Pb fcc 1.45 1.8	Bi rhom 1.50 1.9	Po cub 1.42 2.0	At fcc 1.48 2.2	Rn fcc 1.46 –
Fr bcc 2.42 0.7	Ra bcc 2.11 0.9	89– 103 #	Rf hcp 1.57	Db bcc 1.49	Sg bcc 1.43	Bh hcp 1.41	Hs hcp 1.34	Mt fcc 1.29	Ds bcc 1.28	Rg bcc 1.21	Cn hcp 1.22	Nh 1.36	Fl 1.43	Mc 1.62	Lv 1.75	Ts 1.65	Og 1.57

* Lanthanide series	La dhcp 1.94 1.10	Ce dhcp 1.84 1.12	Pr dhcp 1.90 1.13	Nd dhcp 1.88 1.14	Pm dhcp 1.86 –	Sm rhom 1.85 1.17	Eu bcc 1.83 –	Gd hcp 1.82 1.20	Tb bcp 1.81 –	Dy bcp 1.80 1.22	Ho hcp 1.79 1.23	Er hcp 1.77 1.24	Tm hcp 1.77 1.25	Yb fcc 1.78 –	Lu hcp 1.74 1.0
# Actinide series	Ac fcc 2.01 1.1	Th fcc 1.90 1.3	Pa tetra 1.84 1.5	U orth 1.83 1.7	Np orth 1.80 1.3	Pu mon 1.80 1.3	Am dhcp 1.73	Cm dhcp 1.68	Bk dhcp 1.68	Cf dhcp 1.68	Es fcc 1.65	Fm 1.67	Md 1.73	No 1.76	Lr hcp 1.61

bcc: body-centered cubic; fcc: face-centered; hcp hexagonal close-packed; dhcp double hexagonal close-packed; orth: orthorhombic; tetra tetragonal; rhom: rhombohedral; cub: cubic; hex hexagonal; mon: monoclinic

APPENDIX 9

Thermodynamic parameters

Heat capacity coefficients for the Shomate equation from the NIST WebBook.

	A	*B*	*C*	*D*	*E*	*T*/*K*
Al(s)	28.1	−5.41	8.56	3.43	−0.277	298–933
γ-Al_2O_3(s)	108.7	37.2	−14.2	2.19	−3.21	298–2327
Cl_2(g)	33.1	12.2	−12.1	4.39	−0.159	298–1000
Fe(s)	24.0	8.37				298–1809
Fe_2O_3(s)	93.4	108.4	−50.9	25.6	−1.61	298–950
H_2(g)	33.1	−11.4	11.4	−2.77	−0.159	298–1000
HCl(g)	32.1	−13.5	19.9	−6.85	−0.050	298–1200
HF(g)	30.1	−3.25	2.87	0.458	−0.025	298–1000
H_2O(g)	30.1	6.83	6.79	−2.53	0.082	500–1700
H_2SO_4(g)	47.3	190	−148	43.9	−0.740	298–1200
N_2(g)	29.0	1.85	−9.65	16.6		100–500
NH_3(g)	20.0	49.8	−15.4	1.92	0.189	298–1400
NaCl(s)	50.7	6.67	−2.52	10.2	−0.201	298–1073
NaOH(s)	419.5	−1718	2954	−1597	−6.05	298–572
O_2(g)	31.3	−20.2	57.9	−36.5	−0.007	100–700
SO_2(g)	21.4	74.4	−57.8	16.4	0.087	298–1200
Si(s)	22.8	3.90	−0.083	0.042	−0.354	298–1685
SiO_2(s)	72.8	1.29	−0.004	0.001	−4.14	298–1996
V_2O_5(s)	161.9	33.3	−12.9	5.13	−3.58	298–943
CO(g)	25.6	6.10	4.05	−2.67	0.131	298–1300
CO_2(g)	25.0	55.2	−33.7	7.95	−0.137	298–1200
$CO(NH_2)_2$(s)	94.0					303–413
CH_4(g)	−0.703	108.5	−42.5	5.86	0.679	298–1300
C_2H_6(g)	−0.303	194.6	−86.1	14.2	0.175	298–2000
C_2H_4(g)	−6.39	184.4	−113.0	28.5	0.316	298–1200
$C_2H_4Cl_2$(l)	129.4					298
C_3H_6(g)	0.853	244.6	−126.5	25.5	0.095	298–1500
C_6H_6(l)	135.7					298
$C_6H_5C_2H_3$(l)	183.2					298
$C_6H_5C_2H_5$(l)	185.6					298

Physical Chemistry: How Chemistry Works, First Edition. Kurt W. Kolasinski.

Companion Website: www.wiley.com/go/kolasinski/physicalchemistry

Values for $\Delta_f H_m{}^{\circ}$, $S_m{}^{\circ}$ and $C_{p,m}$ are for 298.15 K. Phase transition enthalpies are taken at the phase transition temperature. Values taken from the 96th edition of the *CRC Handbook of Chemistry and Physics*.

	$\Delta_f H_m{}^{\circ}$ / kJ mol^{-1}	$S_m{}^{\circ}$ / J K^{-1} mol^{-1}	$C_{p,m}$ / J K^{-1} mol^{-1}	T_f / K	$\Delta_{fus} H_m{}^{\circ}$ / kJ mol^{-1}	T_b / K	$\Delta_{vap} H_m{}^{\circ}$ / kJ mol^{-1}
Al(s)	0	28.3	24.2	933.47	10.71	2792.15	294
Al_2O_3(s)	−1675.7	50.9	79	2327	111.1	3250.15	
Cl_2(g)	0	223.1	33.9	171.7	6.40	239.11	20.41
Fe(s)	0	27.3	25.1	1811.05	13.81	3134.15	
Fe_2O_3(s)	−824.2	87.4	103.9	1812	89		
H_2(g)	0	130.7	28.8	13.56	0.12	20.38	0.9
HCl(g)	−92.31	186.9	29.14	159.01	2.00	188.15	16.15
HF	−273.3	173.8	29.14	189.58	4.58	292.68	
H_2O(l)	−285.8	70.0	75.3	273.15	6.01	373.15	40.65
H_2O(g)	−241.8	188.8	33.6				
H_2SO_4	−814	156.9	138.9	283.46	10.71	610.15	
N_2(g)	0	191.6	29.1	63.15	0.71	77.35	5.57
NH_3(g)	−45.9	192.8	35.1	195.42	5.66	239.82	23.33
NaCl(s)	−411.2	72.1	50.5	1074	28.16	1738	204.5
NaOH(s)	−425.8	64.4	59.5	596	6.60	1661.15	175
O_2(g)	0	205.15	29.39	54.36	0.44	90.17	6.82
SO_2(g)	−296.8	248.2	39.9	200.75		263.1	24.94
Si(s)	0	18.8	20	1687.15	50.21	3538.2	359
SiO_2(s)	−910.7	41.5	44.4	1995	9.6	3223.15	
V_2O_5(s)	−1550.6	131	127.7	954	64	2023.15	
CO(g)	110.5	197.7	29.1	68.13	0.833	81.65	6.04
CO_2(g)	−393.5	213.8	37.1	216.59*	9.02*	194.7*	sublimes
$CO(NH_2)_2$	−333.1	104.26	92.79	406.5	13.9	dec	
CH_4(g)	−74.6	186.3	35.7	90.69	0.94	111.66	8.19
C_2H_6(g)	−84.0	229.2	52.5	90.36	2.72	184.5	14.69
C_2H_4(g)	52.4	219.32	42.9	104.00	3.35	169.38	13.53
$C_2H_4Cl_2$(g)	−126.4	308.4	78.7	176.25	7.87	330.45	28.85
C_3H_6(g)	20.0		64.3	87.91	3.003	225.46	18.42
C_6H_6(l)	49.1	173.4	136	278.68	9.87	353.24	30.72
$C_6H_5C_2H_3$(l)	103.8	240.5	182	242.5	10.9	418.15	38.7
$C_6H_5C_2H_5$(l)	−12.3	255.01	183.2	178.19	9.18	409.31	35.57

Index

Physical Chemistry: How Chemistry Works, First Edition. Kurt W. Kolasinski.

Companion Website: www.wiley.com/go/kolasinski/physicalchemistry

	Energy	Pressure
1 eV	1.602176×10^{-19} J	1 atm = 101325 Pa = 1013.25 mbar = 760 torr
1 eV/*hc*	8065.53 cm^{-1}	1 bar = 10^5 Pa
1 meV/*hc*	8.06553 cm^{-1}	1 torr = 1.3332 mbar = 133.32 Pa
1 eV/particle	96.485 kJ mol^{-1}	
1 kcal mol^{-1}	4.184 kJ mol^{-1}	

Thermal Energy	4 K	77 K	298 K	1000 K
RT/kJ mol^{-1}	0.03254	0.6264	2.424	8.134
$k_B T$/meV	0.3447	6.635	25.68	86.17

Quantity	Symbol	Value		Units
Boltzmann constant	k_B	1.3806504	10^{-23}	J K^{-1}
		8.61741	10^{-5}	eV K^{-1}
Planck constant	h	6.62606896	10^{-34}	J s
	$\hbar = h/2\pi$	1.05457163	10^{-34}	J s
Avogadro constant	N_A	6.02214179	10^{23}	mol^{-1}
Bohr radius	a_0	5.29177209	10^{-11}	m
Rydberg constant	R_∞	1.09737316	10^5	cm^{-1}
Speed of light	c	2.99792458	10^8	m s^{-1}
Elementary charge	e	1.60217649	10^{-19}	C
Faraday constant	$F = N_A e$	9.64853399	10^4	C mol^{-1}
Gas constant	$R = N_A k_B$	8.314472		J K^{-1} mol^{-1}
		8.20574	10^{-2}	L atm K^{-1} mol^{-1}
Vacuum permittivity	ε_0	8.85418782	10^{-12}	J^{-1} C^2 m^{-1}
Atomic mass unit	u	1.66054	10^{-27}	kg
Electron mass	m_e	9.10939	10^{-31}	kg
Proton mass	m_p	1.67262	10^{-27}	kg
Neutron mass	m_n	1.67493	10^{-27}	kg
Standard acceleration	g	9.80665		m s^{-2}
Debye-Hückel Constant	B	0.5092	(for $\lg\gamma_\pm$ in water at 25°C)	

Greek alphabets

A, α alpha	B, β beta	Γ, γ gamma	Δ, δ delta	E, ε epsilon	Z, ζ zeta
H, η eta	Θ, θ theta	I, ι iota	K, κ kappa	Λ, λ lambda	M, μ mu
N, ν nu	Ξ, ξ xi	Π, π pi	P, ρ rho	Σ, σ sigma	T, τ tau
Y, υ upsilon	Φ, ϕ phi	X, χ chi	Ψ, ψ psi	Ω, ω omega	

Periodic table of the elements

nl^w = ground state configuration

Key: Z
symbol
g mol^{-1}
nl^w

Group

1 I	2 II	3 IIIB	4 IVB	5 VB	6 VIB	7 VIIB	8 VIIIB	9 VIIIB	10 VIIIB	11 IB	12 IIB	13 III	14 IV	15 V	16 VI	17 VII	18 VIII
1 H 1.0079 1s																	2 He 4.0026 $1s^2$
3 Li 6.94 2s	4 Be 9.0122 $2s^2$											5 B 10.81 2p	6 C 12.011 $2p^2$	7 N 14.007 $2p^3$	8 O 15.999 $2p^4$	9 F 18.998 $2p^5$	10 Ne 20.180 $2p^6$
11 Na 22.990 3s	12 Mg 24.305 $3s^2$											13 Al 26.982 3p	14 Si 28.085 $3p^2$	15 P 30.974 $3p^3$	16 S 32.06 $3p^4$	17 Cl 35.45 $3p^5$	18 Ar 39.948 $3p^6$
19 K 39.098 4s	20 Ca 40.078 $4s^2$	21 Sc 44.956 $4s^23d$	22 Ti 47.867 $4s^23d^2$	23 V 50.942 $4s^23d^3$	24 Cr 51.996 $\mathbf{4s3d^5}$	25 Mn 54.938 $4s^23d^5$	26 Fe 55.845 $4s^23d^6$	27 Co 58.933 $4s^23d^7$	28 Ni 58.693 $4s^23d^8$	29 Cu 63.546 $\mathbf{4s3d^{10}}$	30 Zn 65.38 $4s^23d^{10}$	31 Ga 69.723 4p	32 Ge 72.63 $4p^2$	33 As 74.922 $4p^3$	34 Se 78.96 $4p^4$	35 Br 79.904 $4p^5$	36 Kr 83.798 $4p^6$
37 Rb 85.468 5s	38 Sr 87.62 $5s^2$	39 Y 88.906 $5s^24d$	40 Zr 91.224 $5s^24d^2$	41 Nb 92.906 $\mathbf{5s4d^4}$	42 Mo 95.96 $\mathbf{5s4d^5}$	43 Tc (98) $5s^24d^5$	44 Ru 101.07 $\mathbf{5s4d^7}$	45 Rh 102.91 $\mathbf{5s4d^8}$	46 Pd 106.42 $\mathbf{4d^{10}}$	47 Ag 107.87 $\mathbf{5s4d^{10}}$	48 Cd 112.41 $5s^24d^{10}$	49 In 114.82 5p	50 Sn 118.71 $5p^2$	51 Sb 121.76 $5p^3$	52 Te 127.60 $5p^4$	53 I 126.90 $5p^5$	54 Xe 131.29 $5p^6$
55 Cs 132.91 6s	56 Ba 137.33 $6s^2$	57–71*	72 Hf 178.49 $6s^25d^2$	73 Ta 180.95 $6s^25d^3$	74 W 183.84 $6s^25d^4$	75 Re 186.21 $6s^25d^5$	76 Os 190.23 $6s^25d^6$	77 Ir 192.22 $6s^25d^7$	78 Pt 195.08 $\mathbf{6s5d^9}$	79 Au 196.97 $\mathbf{6s5d^{10}}$	80 Hg 200.59 $6s^25d^{10}$	81 Tl 204.38 6p	82 Pb 207.2 $6p^2$	83 Bi 208.98 $6p^3$	84 Po (209) $6p^4$	85 At (2?0) $6p^5$	86 Rn (222) $6p^6$
87 Fr (223) 7s	88 Ra (226) $7s^2$	89–103#	104 Rf (265)	105 Db (268)	106 Sg (271)	107 Bh (270)	108 Hs (277)	109 Mt (276)	110 Ds (281)	111 Rg (280)	112 Cn (285)	113 Nh (284)	114 Fl (289)	115 Mc (288)	116 Lv (293)	117 Ts (294)	118 Og (294)

*Lanthanoids	57 La 138.91 **5d**	58 Ce 140.12 **5d4f**	59 Pr 140.12 $4f^3$	60 Nd 144.24 $4f^4$	61 Pm (145) $4f^5$	62 Sm 150.36 $4f^6$	63 Eu 151.96 $4f^7$	64 Gd 157.25 $\mathbf{5d4f^7}$	65 Tb 158.93 $4f^9$	66 Dy 162.50 $4f^{10}$	67 Ho 164.93 $4f^{11}$	68 Er 167.26 $4f^{12}$	69 Tm 168.9 $4f^{13}$	70 Yb 173.05 $4f^{14}$	71 Lu 174.97 $\mathbf{5d4f^{14}}$
#Actininoids	89 Ac (227) **6d**	90 Th 232.04 $\mathbf{6d^2}$	91 Pa 231.04 $\mathbf{6d5f^2}$	92 U 238.03 $\mathbf{6d5f^3}$	93 Np (237) $\mathbf{6d5f^4}$	94 Pu (244) $5f^6$	95 Am (243) $5f^7$	96 Cm (247) $\mathbf{6d5f^7}$	97 Bk (247) $5f^9$	98 Cf (251) $5f^{10}$	99 Es (252) $5f^{11}$	100 Fm (257) $5f^{12}$	101 Md (258) $5f^{13}$	102 No (259) $5f^{14}$	103 Lr (262)

Deviations from expected electron configurations are indicated in **bold**.